IDENTIFICATION AND
SYSTEM PARAMETER
ESTIMATION

Identification and System Parameter Estimation

Part 2

Proceedings of the 3rd IFAC Symposium,
The Hague/Delft, The Netherlands,
12–15 June 1973.

A symposium organized by

the Royal Institution of Engineers in The Netherlands (KIvI);
Division for Automatic Control.

and sponsored by

the International Federation of Automatic Control (IFAC);
IFAC–Theory Committee
IFAC–Technical Committee on Applications

edited by

P. EYKHOFF

University of Technology
Eindhoven, The Netherlands

N·H
P&C

1973

NORTH-HOLLAND PUBLISHING COMPANY – AMSTERDAM · LONDON
AMERICAN ELSEVIER PUBLISHING COMPANY, INC. – NEW YORK

Library of Congress Catalog Number: 73-77709
ISBN North-Holland 0 7204 20830
ISBN American Elsevier 0 444 10516 6

PUBLISHERS:
NORTH-HOLLAND PUBLISHING COMPANY – AMSTERDAM
NORTH-HOLLAND PUBLISHING COMPANY, LTD. – LONDON

SOLE DISTRIBUTORS FOR THE U.S.A. AND CANADA
AMERICAN ELSEVIER PUBLISHING COMPANY, INC.
52 VANDERBILT AVENUE
NEW YORK, N.Y. 10017

These proceedings of the 3rd IFAC Symposium on Identification and System Parameter Estimation were reproduced by means of the photo-offset process, using the manuscripts supplied by the authors of the different papers. This way of producing allowed preprints to be published before the opening of the Symposium. The manuscripts of the papers had been typed with different typewriters/types. The layout, the figures as well as tables of some papers did not agree completely with the standard requirements; consequently the reproduction does not display complete uniformity. To ensure a rapid publication this discrepancy could not be changed; the English could not be checked either. Therefore, the readers are asked to excuse kindly the deficiencies of this publication due to the above-mentioned reasons.

The Editor.

EDITORIAL

"Identification is coming of age ..."

After the fine and successful IFAC symposia:

- "Identification in Automatic Control Systems", Prague, 1967
- "Identification and Process Parameter Estimation", Prague, 1970

the third symposium in this sequence clearly indicates the truth of the device given above. This may be judged from these proceedings ...

The number of applications is clearly growing. The same holds for the variety of applications. A number of techniques have resulted in "standard" procedures and routines. Progress is being made with fundamental problems like order-determination. Comparison of different methods is feasible ...

Contributions to the symposium that are given in these proceedings are:

code	type		number of contributions
S	Survey papers		5
C	Case studies		6
P	Practically	oriented technical papers	53
T	Theoretically	oriented technical papers	86
R	Round Tables		
E	Exercises/test cases		

Note that the code has a mnemonic character. This holds also for the technical papers:

code	category	number of contributions
Practically oriented papers		
PB	Biological objects	12
PC	Chemical processes	9
PE	Economic and ecologic systems	5
PP	Power systems	8
PT	Transportation systems	7
PV	Various processes	12
Theoretically oriented papers		
TA	Adaptive control	9
TC	Correlation-like techniques	9
TD	Distributed-parameter systems	6
TM	Model adjustment techniques	18
TN	Nonlinear systems	11
TO	Observability, identifiability, state estimation	11
TS	Problems on structures	15
TT	Testsignals	7

These categories are used just as a convenient means for ordering the papers according to some point of view. Not too much emphasis nor an expression of appreciation should be attached to this ordering.

The large number and the high quality of the majority of the submitted abstracts required some special measures for coping with available number of sessions and the available printing volume. The Committees involved considered that probably the scientific/engineering community would be served best if a wide variety of interesting papers were published. The limits imposed by the size of the papers can easily be circumvented by personal contacts at and contacts by mail after the Symposium.

It is a pleasant duty to express the gratitude

- to the members of the International Program/Paper Selection Committee for their contribution to the paper selection procedures;
- to Dr. Ruppel, former Honorary Secretary of IFAC, for his expert advice;
- to the members of the Local Organizing Committee for their time, ideas and effort put into the many tasks connected with organizing a Symposium;
- to the Congress Office of the Royal Institution of Engineers in the Netherlands (KIvI) for its fine assistance in technical matters.

It proved possible to include in these proceedings some special material (address of welcome; opening address) as well as some reports on important discussions during the Symposium (cf. "Industry-university confrontation on process identification", p. 1077 ff.; "Results for test-case A", p. 1103 ff). To those who assisted in such a fine way to improve the value of these proceedings I want to extend my sincere thanks.

These volumes present the results of work done in countries all over the world. They indicate the willingness of engineers/scientists to cooperate on an international scale; a cooperation with human well-being in mind.

The Editor

CONTENTS

part 1

---

a) pag. 1107; b) pag. 1123; c) pag. 1131. part 2 (late arrivals)

CONTENTS

PB-7 Estimation of cardiovascular parameters. 231
R. Aaslid (Norway)

PB-8 Comparison of identification methods in the cardiovascular system. 235
B. Szücs, E. Monos and J. Szutrély (Hungary)

PB-9 Parameter estimation for medical diagnosis. 239
A. Cohen (Israel)

PB-10 The estimation of the "state" of the heart. 243
J.L. Talmon and J.H. van Bemmel (the Netherlands)

PB-11 Some aspects of human operator identification in real time. 247
S.S. Stanković and N.G.M. Kouwenberg (Yugoslavia, the Netherlands)

PB-12 Application of random search techniques and stochastic approximation in human operator modelling. 251
G. Johannsen (FRG)

PC Chemical processes

PC-1 Estimation of the diffusion coefficient of a fluidised bed catalytic reactor using non linear filtering techniques. 255
J. Aguilar Martin, G. Alengrin and C. Hernandez (France)

PC-2 Variations of process parameters; a pilot study of adaptive control. 265
L.G.M. van Aarle (the Netherlands)

PC-3 Choice of the weight coefficient of the running regression analysis with exponential weighting function. 273
K. Velev (Bulgaria)

PC-4 Evaluation of five parameter estimation algorithms for a nonlinear reactor model via Monte Carlo simulation. 277
H.D. Hensley and D.M. Himmelblau (USA)

PC-5 Application of stochastic approximation to the on-line estimation of the parameters of noisy processes. 281
F.G. Abbosh (UK)

PC-6 Identification of a bauxite digestion process by Fourier analysis. 285
J.P. Riffaud and P. Magistry (Canada)

PC-7 Model structure in system identification for fermentation process design. 289
T. Takamatsu, I. Hashimoto, S. Shioya and A. Manago (Japan)

PC-8 Identification of process models for plastics extrusion systems. 299
J. Parnaby, A.K. Kochhar, K.C. Chow and B. Wood (UK)

PC-9 Estimation of spatially distributed decay parameters in a chemical reactor. 307
C. McGreavy and A. Vago (UK, N. Ireland)

PE Economic and ecologic systems

PE-1 Parameter estimation of mass transfer in compartmented aquatic ecosystems. 317
F. Argentesi, G. Di Cola and N. Verheyden (Italy)

PE-2 Optimal estimation of polluted stream variables. 327
H.N. Koivo and A.J. Koivo (Canada)

PE-3 Some methodological problems encountered in applying correlation techniques to Forrester-Koenig models. 331
R.A. Brown (USA)

PE-4 Demand and cost forecasting for postal system optimization. 339
J.T. Wagner (USA)

PE-5 On the identification of arrivals in certain economic systems. 347
A.A. Voronov, V.V. Skaletsky and Y.V. Chistyakov (USSR)

PP Power systems

PP-1 Identification and control of a pilot-scale boiling rig. 355
D.W. Clarke, D.A.J. Dyer, R. Hastings-James, R.P. Ashton and J.B. Emery (UK)

PP-2 Estimation of power generator dynamics from normal operating data. 367
S. Lindahl and L. Ljung (Sweden)

PP-3 Modeling and identification of nuclear power reactor dynamics from multivariable experiments. 375
G. Olsson (Sweden)

PP-4 On-line identification of continuous systems from sampled data. 385
N.K. Sinha (Canada)

PP-5 Estimation of fault locations on power lines. 393
J. Kohlas (Switzerland)

PP-6 Real-time nuclear power plant parameter identification with a process computer. 403
H. Roggenbauer (Norway)

CONTENTS

d) pag. 1147; e) pag. 1151. part 2. (late arrivals)

CONTENTS

part 2

T THEORETICALLY - ORIENTED PAPERS

TA Adaptive Control

TA-1 Control of multivariable systems with unknown but constant parameters. 535
V. Peterka and K.J. Åström (Czechoslovakia, Sweden)

TA-2 Digital adaptive controllers using second order models with transport lag. 545
S. Joshi and H. Kaufman (USA)

TA-3 A deterministic type controller for a class of uncertain plants with partial identification. 555
P.N. Nikiforuk, M.M. Gupta, K. Kanai and L. Wagner (Canada)

TA-4 Plant-adaptive pulse-frequency modulated control systems. 567
M.B. Broughton (Canada)

TA-5 Open-loop-feedback-optimal adaptive stochastic control of linear systems. 571
R. Ku and M. Athans (USA)

TA-6 Adaptive identifier transfer functions. 579
V.M. Chadeev (USSR)

TA-7 Adaptive control and identification by a model. 583
H. Nitsche (FRG)

TA-8 Plants identification by the statistical regularization method. 587
A.I. Beliaevsky, R.A. Poluektov and G.N. Solopchenko (USSR)

TA-9 Adaptive estimation and control of linear systems. f)
O.J.M. Smith (USA)

TC Correlation - like techniques

TC-1 The identification of linear multivariable systems from frequency response data. 591
M. Marchand (FRG)

TC-2 Some properties of second order correlation functions in relation to biological rhythm research. 599
M. ten Hoopen (the Netherlands)

TC-3 Analysis of a spike generator. 607
H.R. de Jongh (the Netherlands)

TC-4 Data-compression for adaptive model identification of nonlinear systems. 613
K.B. Norkin (USSR)

TC-5 Dispersional identification. 619
A.L. Bunich and N.S. Rajbman (USSR)

TC-6 Identification of heteroscedastic plants. 623
F.A. Ovsepian and N.A. Vardanian (USSR)

TC-7 Recursive estimation of the states and their higher order moments in nonlinear systems. 627
N. Ramani and R. Balasubramanian (Canada)

TC-8 On information measure in identification and parameter estimation. 631
P.A. Devijver (Belgium)

TC-10 Identification of parametric models using correlation analysis. g)
D.C. Williams and G.G. Delaney (UK)

TD Distributed - parameter systems

TD-1 Different methods for estimation of thermal diffusivity of a heat diffusion process. 639
B. Leden, M.H. Hamza and M.A. Sheirah (Sweden, Canada, Egypt)

TD-2 Identification of distributed parameters. 649
G. Chavent (France)

TD-3 Identification of nonlinear distributed systems by using an inverse describing function method. 661
G. Jumarie (France)

TD-4 Combined state and parameter estimation for distributed-parameter systems using discrete observations. 671
H. Sherry and D.W.C. Shen (USA)

TD-5 Identification and stochastic control of non-dynamic systems. 679
A.V. Balakrishnan (USA)

TD-6 Identification of distributed parameter systems using the epsilon method: a new algorithm. 687
G. Di Pillo, L. Grippo and M. Lucertini (Italy)

---

f) pag. 1155; g) pag. 1165 (late arrivals).

CONTENTS

TM Model adjustment techniques

TM-1 Convergence properties of the generalized least squares identification method. 691
T. Söderström (Sweden)

TM-2 Two stage least squares estimators and their recursive approximations. 701
R.N. Pandya and B. Pagurek (India, Canada)

TM-3 On the estimation of the transfer function parameters of process- and noise dynamics using a single-stage estimator. 711
J.L. Talmon and A.J.W. van den Boom (the Netherlands)

TM-4 The delta-LMS adaptive algorithm. 721
J.E. Timm (USA)

TM-5 On the identification of linear discrete time system models using the instrumental variable method. 729
B. Finigan and I.H. Rowe (Canada)

TM-6 Maximum likelihood identification of Åström model by quasilinearization. 737
B.P. Furht (Yugoslavia)

TM-7 The design of model reference parameter estimation systems using hyperstability theories. 741
Chang-Chieh Hang (UK)

TM-8 Convergence properties in linear parameter tracking systems. 745
S.J. Merhav and E. Gabay (Israel)

TM-9 A method for discrete system identification. 751
R. Clarke and J.W. Green (UK)

TM-10 System identification using stochastic approximation. 755
N.K. Sinha (Canada)

TM-11 Equivalent filter identification from autocovariance estimates. 759
K. Gerdin (Sweden)

TM-12 Determination of changes in the properties of random processes by the Bayes method. 763
L. Telksnys (USSR)

TM-13 Analogue model identification of system dynamic error transfer functions. 767
D.R. Towill (UK)

TM-14 Identification of very noisy dynamic processes using models with few parameters. 771
R. Isermann (FRG)

TM-15 Simultaneous identification of the dynamic system and restoration of an input signal. 775
C. Paulauskas (USSR)

TM-16 An identification technique for multi-input multi-output systems. 779
A.Y. Bilal and M.F. Hassan (Egypt)

TM-17 Numerical algorithms for the identification of system, via splines. 783
A. Caprihan (Brasil)

TM-18 Application of cubic splines to system identification. 787
M. Shridhar and N.A. Balatoni (Canada)

TN Nonlinear systems

TN-1 Some estimation methods for nonlinear discrete time identification. 793
Cs. Bányász, R. Haber and L. Keviczky (Hungary)

TN-2 Identification of power blocks as non-linear dynamic objects. 803
V. Kaminskas and A. Nemura (USSR)

TN-3 The identification of processes with direction-dependent dynamic responses. 809
K.R. Godfrey and P.A.N. Briggs (UK)

TN-4 Least-squares estimation of restricted parameters in physical and chemical models. 821
H. van Houwelingen, J. Petiet and C. Schweigman (the Netherlands)

TN-5 On-line estimation of non-linear process parameters. 827
A.McN. Jackson and D.W.T. Rippin (UK, Switzerland)

TN-6 Frequency domain identification of nonlinear systems. 831
A.B. Gardiner (UK)

TN-7 Application of three-level-pseudorandom-signals for parameter estimation of nonlinear systems. 835
R. Krempl (FRG)

TN-8 Identification of nonlinear systems through the utilisation of discontinuous orthogonal filters in applying a multilevel pseudorandom signal. 839
L. Šutek and M. Varga (Czechoslovakia)

TN-9 An improved performance criterion for pseudorandom sequences in the measurement of 2-nd order Volterra kernels by crosscorrelation. 843
F.L. Kadri and J.D. Lamb (UK)

TN-10 Identification by pseudosensitivity functions and quasilinearization. 847
M.R. Mataušek and M.D. Milovanović (Yugoslavia)

TN-11 Identification of multivalued parameters in a class of noisy dynamic processes. 851
R.E. Klein (USA)

CONTENTS

TO Observability, identifiability, state estimation

TO-1 Limit identifiability of control systems. 857
J. Štěpán (Czechoslovakia)

TO-2 On the identifiability of linear dynamical systems. 867
K. Glover and J.C. Willems (USA, the Netherlands)

TO-3 Identifiability of linear dynamical systems. 871
H.E. Berntsen and J.G. Balchen (Norway)

TO-4 Test signals and nonlinear observability. 875
K.S.P. Kumar (USA)

TO-5 Treatment of time varying stochastic bias in recursive filtering. 879
M.M. Shah (Kenya)

TO-6 Nonlinear filtering error-analysis algorithm. 893
A. Vaněček (Czechoslovakia)

TO-7 A sequential method for the estimation of parameters in nonlinear models of multivariable systems. 897
M. Mendes and C. De Polignac (France)

TO-8 Design of state estimator for systems originally described by input-output relation. 901
Y. Funahashi and K. Nakamura (Japan)

TO-9 The non-linear dynamic objects parameters estimation based on the sensitivity-algorithm. 905
A.I. Rouban (USSR)

TO-10 A sequential method for system identification in hierarchical structure. 909
N.J. Smith and A.P. Sage (USA)

TO-11 Identification and system parameter estimation. 913
Huynh Huu Thanh (France)

TS Problems on structures

TS-1 Application of different statistical tests for the determination of the most accurate order of the model in parameter estimation. 917
H. Unbehauen und B. Göhring (FRG)

TS-2 The determination of the order of process- and noise dynamics. 929
A.J.W. van den Boom and A.W.M. van den Enden (the Netherlands)

TS-3 An application of realization theory to identification of multivariable process. 939
K. Furuta (Japan)

TS-4 Canonical structures in the identification of multivariable systems. 943
R.P. Guidorzi (Italy)

TS-5 Minimal realization and approximation of linear systems from normal operating records. 953
A. Barraud and P. de Larminat (France)

TS-6 Multidimensional industrial plant identification. 957
A.A. Dorofeyuk, A.D. Kasavin and I.Sh. Torgovitsky (USSR)

TS-7 Optimal model building. 965
J. Gordesch and P.P. Sint (Austria)

TS-8 Stochastic identification of digital feedback loops using a table approach. 969
N.A. Lindberger (USA)

TS-9 An automatic model adjustment technique. 973
L.C.W. Dixon (UK)

TS-10 Identification of complex systems in terms of reduced models. 977
F. Donati and E. Canuto (Italy)

TS-11 Modeling a particular class of multiple-input/multiple-output black boxes with stochastic integral equations and identifying the required parameters. 983
E.D. Eyman, T.H. Kerr and N.K. Loh (USA)

TS-12 Estimation of the order of nonlinear systems. 987
C.M. Woodside (Canada)

TS-13 Errors in orthogonal expansions and relevance to modelling. 991
J. Hiller (Australia)

TS-14 Estimation of state variable models using input-output data. 995
S.P. Bingulac and M. Djorović (Yugoslavia)

TS-16 Reduction of multivariable systems. h)
M. Labarrere, J.P. Krief and B. Gimonet (France)

---

h) pag. 1169 (late arrival)

CONTENTS

CONTROL OF MULTIVARIABLE SYSTEMS WITH UNKNOWN BUT CONSTANT PARAMETERS

V. Peterka

Institute of Information Theory
and Automation
Czechoslovak Academy of Sciences
Prague, Czechoslovakia

K.J.Åström

Division of Automatic Control
Lund Institute of Technology
Lund, Sweden

The problem of designing selfadjusting regulators is approached from the viewpoint of control of systems with constant but unknown parameters. An algorithm is proposed which can be applied to multivariable systems including those with unknown time delay and/or nonminimum phase. The main feature of the approach is the application of orthogonal transformations in order to reduce the observed input-output data.The use of orthogonal transformations makes it possible to obtain an appropriate information state of relatively low dimension. This information state asymptotically gives all second order statistical characteristics which are required for the solution of the control problem when formulated as control of system with uncertain parameters.

1. INTRODUCTION

The difficulty in obtaining appropriate mathematical models of industrial processes is one of the major obstacles when attempting to apply the highly developed modern control theory to practical industrial control problems. The control designer has to determine both the structure and the parameters of the process and the characteristics of the disturbances. The derivation of a reliable mathematical model from basic natural laws is possible only for relatively simple and well understood systems. Examples are found e.g. in the aerospace field. In most other cases a model calculated from natural laws and construction data is only a crude approximation and the control system designed from the model must often be carefully adjusted on-line in order to achieve a reasonable performance of the closed loop system. The identification of the process from experimental data measured in open loop under the influence of natural disturbances does not always give reliable results. Moreover there are examples when the optimal controller gives a closed loop system which is very near to the stability boundary. In such cases even small model inaccuracies could have a significant influence. Hence, a combination of system identification and synthesis does not always solve the control problem. Regulators designed in this way must in fact often be adjusted on-line to achieve good performance.

The significant experience that is today available about PID regulators clearly indicate that good performance can often be obtained by a simple controller provided that it is properly adjusted. Experience has, however, also indicated that it is not easy to keep PID regulators in industrial environments properly adjusted.

There is consequently a significant incentive in finding selfadjusting regulators. The problem of designing self-adjusting regulators can be formulated mathematically as control of systems with constant but unknown parameters. Several approaches to this problem are known [1],[2],[3] . In particular it has been demonstrated that self-adjusting digital regulators can be obtained for single-input single--output minimum phase systems. See [2], [4] . An interesting feature of these regulators is that they will adjust to minimum variance regulators for large classes of systems. However, some apriori information such as minimum--phase property and time delay of the system are still required.

This paper proposes self-adjusting digital regulators for multivariable systems including those with unknown time delay and nonminimum phase. The main feature of the approach is the application of orthogonal transformations in order to reduce the observed input-output data. The use of orthogonal transformations makes it possible to obtain an information state of reasonable low dimension. For a system with 3 inputs, 3 outputs governed by a vector-difference equation with 3 delayed terms the chosen information state requires storage of 294 numbers while the conditional distribution obtained using a Bayesian approach [5] under gaussian assumptions requires the storage of 2039 numbers. The self--adjusting control algorithm proposed in this paper do not rely on gaussian assumptions.

The paper is organised as follows. The problem formulation is covered in section 2 which contains a discussion of the mathematical model for the system, the selection of criterion and a discussion of the information state which defines the admissible controls. The proposed algorithm is given in Section 3. It is shown that orthogonal transformations of the input-output data gives a field of reduced data that can be used as an information state. It is shown that the unknown parameters can be expressed in terms of this information state and certain random variables whose second order statistical properties are asymptotically given by the information state. The control problem then reduces to the problem of control of a linear system with uncertain parameters which can be solved by Dynamic Programming. The properties of the algorithm are illustrated by simulated example in Section 4.

2. PROBLEM FORMULATION

The problem of controlling a system with constant but unknown parameters will now be discussed. The model of the system is first given as a linear vector difference equation with uncorrelated disturbances. The selection of the criterion is then given and the section ends with a discussion of the chosen information state which defines the admissible controls.

System Model

Consider a multivariable system described by the vector difference equation

$$y(t) = \sum_{i=1}^{n} A_i y(t-i) + \sum_{i=0}^{n} B_i u(t-i) + e(t) \qquad (2.1)$$

where the control vector u has dimension μ the output y and the disturbance e have dimensions ν. It is assumed that $\{e(t)\}$ is a sequence of uncorrelated random vectors with a common covariance R. The elements of the matrices $A_1, A_2, \dots, A_n, B_0, B_1, \dots, B_n$ and R are assumed to be constant but unknown. It is assumed that n is known or that an upper bound can be given.

The model (2.1) can be given a state representation in many different ways. A state vector z can e.g. be introduced in the following way.

$$z(t) = \begin{bmatrix} u(t-n) \\ y(t-n) \\ u(t-n+1) \\ \vdots \\ u(t-1) \\ y(t-1) \end{bmatrix} \qquad (2.2)$$

The model can then be represented as follows.

$$z(t+1) = \begin{bmatrix} 0 & I \\ 0 & 0 \\ \multicolumn{2}{c}{A^T} \end{bmatrix} z(t) + \begin{bmatrix} 0 \\ I_\mu \\ B^T \end{bmatrix} u(t) + \begin{bmatrix} 0 \\ 0 \\ I_\nu \end{bmatrix} e(t) \qquad (2.3)$$

$$y(t) = [0 \mid 0 \mid I_\nu]\, z(t+1)$$

where $B^T = B_0$

$$A^T = [B_n, A_n, B_{n-1}, A_{n-1}, \dots, B_1, A_1] \qquad (2.4)$$

The state vector z in (2.3) is of dimension $n(\nu+\mu)$. It is of course possible to find other state representations of lower dimension. This will, however, require that the coefficients of the model (2.3) are known. Notice that the state vector (2.2) does not depend on the parameters of the model. It will also be seen that the dimension of the chosen state space is actually minimal for the overall closed loop including the constant controller to which the selfadjusting procedure converges.

Choice of Criterion

As was mentioned in the introduction the purpose of this paper is to develop a self-adjusting regulator which in the limit as $t \to \infty$ will give a controller that is optimal in some sense. If the parameters were known the criterion could e.g. be chosen as

$$V_1 = \mathcal{E}[y^T(s) Q_1 y(s) + u^T(s) Q_2 u(s)] \quad s = 1, \dots, N \qquad (2.5)$$

or

$$V_2 = \mathcal{E} \sum_{s=1}^{N} [y^T(s) Q_1 y(s) + u^T(s) Q_2 u(s)] \qquad (2.6)$$

For both criteria it is straightforward to find the optimal controls using well known methods [6],[7] . It is also known that the strategy which minimizes V_2 will converge to the optimal strategy for V_1 as $N \to \infty$.

When looking for selfadjusting regulators we could thus attempt to find strategies which will converge to the optimal strategies either for V_1 or for V_2. A straightforward way of approaching the problem of selfadjusting regulators would thus be to find the controls which minimize V_1 or V_2 for systems with constant but unknown parameters. When this is attempted it appears that the criteria will no longer give the same strategies [1]. In fact it turns out that:

1. The selfadjusting control which minimizes V_1 does not converge when the system is nonminimum phase.
2. The control strategy which minimizes V_1 for systems with constant but unknown parameters can be determined. It turns out however that the computation of the control strategy will require computing facilities far exceeding what is today available even for off time computations. The strategy which minimizes V_2 is also a dual control strategy in Feldbaum's sense [8].

Hence if we are looking for a practical selfadjusting procedure which can be implemented on a process control computer none of the above approaches is satisfactory. Instead the strategy will be chosen in the following way. At each time $t+1$ the data observed up to time t is used to substitute the lack of knowledge of the system parameters. The future control actions $u(t+1), u(t+2), \dots, u(t+n)$ are then determined in such a way that the criterion

$$V_3 = \mathcal{E} \sum_{i=1}^{N} [y^T(t+i) Q_1 y(t+i) + u^T(t+i) Q_2 u(t+i)] \tag{2.1}$$

is minimal when the system model used in each step is based on the data observed up to time t. The control is then applied and the whole procedure is then repeated when the next output is obtained. The control strategy obtained in this way is not a dual control in Feldbaums sense. This procedure apparently requires N iterations of the Riccati equation in each step. This computation is significantly smaller than the computation required to determine the full dual control strategy. It is also possible to reduce the number of iterations of the Riccati equation.

The Information State

To arrive at a control problem it is necessary to specify the information available for the selection of the control variable at time t. This problem was already briefly mentioned above but we will now discuss it further. It is assumed that at time t the state $z(t)$ is known to the controller. This means that the input-output pairs $\{u(s), y(s), s = t-n, \dots, t-1\}$. are known. If the system parameters were known it is easily shown that nothing would be gained by knowing additional input-output pairs. When the system parameters are unknown it is, however, advantageous to know other input-output pairs i.e. $\{u(s), y(s), s=1, \dots, t-n-1\}$. The set of all past input output pairs will, however, grow as time increases. From a practical point of view it is also unreasonable to store all past inputs and outputs. The problem is then if it is possible to reduce the amount of data stored without increasing the loss function. The word information state will be used to denote the stored data.

If the unknown parameters at the system were assumed to be random variables and if the criterion were to minimize the mathematical expectation of (2.5) it can be shown that it is sufficient to store the conditional probability distribution of the parameters given all observed inputs and outputs. Choosing the information state as a probability distribution means in general that it is infinite dimensional. In the particular case of the model (2.1) or (2.3) under the additional assumption that the disturbances and the initial state of (2.3) are gaussian and the covariance matrix R is known it is known that the conditional distribution of the parameters $A_1, A_2, \dots, A_n, B_0, B_1, \dots, B_m$ are gaussian. See e.g. [5] [9]. The information state can thus be characterized by the conditional meanvalues and the conditional covariances and it is thus finite dimensional. The updating of the information state is then given by equations very similar to the equations for the Kalman-Bucy filter. See e.g. [5] and [9]. To avoid assuming that the disturbances $\{e(t)\}$ are gaussian and that their covariance R is known we will choose a certain finite dimensional information state obtained by orthogonal transformations of the input output pairs. The control problem will then be solved for this postulated information state.

3. PROPOSED ALGORITHM

In this section the proposed algorithm is given. The orthogonal transformations are first discussed and the chosen information state is then defined precisely. It is then shown that the unknown parameters can be expressed in terms of the information state and certain random variables whose first and second moment can also be expressed asymptotically in terms of the chosen information state. Using the information state the unknown parameters can also be eliminated from the system model thereby giving a linear system with uncertain coefficients. The quadratic control problem for such a system is then given thereby completing the solution.

Orthogonal Transformations

The equation describing the model (2.1) can be written in the following way

$$[z^T(t)\; u^T(t)\; y^T(t)] \begin{bmatrix} -A \\ -B \\ I_\nu \end{bmatrix} = e^T(t) \tag{3.1}$$

Assuming that the set of all input-output pairs obtained up to time t are known, then the following equations can be written.

$$\begin{bmatrix} z^T(n) & u^T(n) & y^T(n) \\ z^T(n+1) & u^T(n+1) & y^T(n+1) \\ \vdots & & \\ z^T(t) & u^T(t) & y^T(t) \end{bmatrix} \begin{bmatrix} -A \\ -B \\ I_\nu \end{bmatrix} = \begin{bmatrix} e^T(n) \\ e^T(n+1) \\ \vdots \\ e^T(t) \end{bmatrix} \tag{3.2}$$

Using orthogonal transformations these equations can be reduced to triangular form. See [10],[11]. The following equations are then obtained

$$\begin{bmatrix} D_{11}(t) & D_{12}(t) & D_{13}(t) \\ 0 & D_{22}(t) & D_{23}(t) \\ 0 & 0 & D_{33}(t) \\ 0 & 0 & 0 \end{bmatrix} \begin{bmatrix} -A \\ -B \\ I_\nu \end{bmatrix} = \begin{bmatrix} E_1(t) \\ E_2(t) \\ E_3(t) \\ 0 \end{bmatrix} \tag{3.3}$$

Notice that D_{11}, D_{22} and D_{33} are upper triangular matrices. The orthogonal transformations can be performed in real-time using a sequence of plane rotations as was described in [12]. The information state is now selected as

$$J(t) = [D_{11}, D_{12}, D_{13}, D_{22}, D_{23}, D_{33}] \tag{3.4}$$

Since the vectors z, u and y are of dimensions $n(\mu+\nu)$, μ and ν, respectively, we find that the information state has dimension

$$N = \frac{1}{2}[(n+1)(\nu+\mu)+1](n+1)(\mu+\nu) \tag{3.5}$$

The number of unknown parameters in the model (2.1) are

$$N_p = n\nu^2 + (n+1)\nu\mu + \frac{1}{2}\nu(\nu+\mu) \tag{3.6}$$

It is thus a considerable saving in using the information state (3.4) in comparison with the information state consisting of the meanvalues and the covariances of all the parameters. It follows from (3.3) that the unknown parameters A and B can be expressed in terms of the information state and the stochastic variables E_1 and E_2 as follows.

$$A = D_{11}^{-1}(D_{13} - D_{12}B - E_1) = D_{11}^{-1}[D_{13} - D_{22}^{-1}(D_{23} - E_2) - E_1] \tag{3.7}$$

$$B = D_{22}^{-1}(D_{23} - E_2) \tag{3.8}$$

Notice that the actual computation of A and B is simple since the matrices D_{11} and D_{22} are triangular and nonsingular if a sufficient number of independent rows in (3.2) have been transformed. The matrices E_1, E_2 and E_3 are related to the vectors $\{e(s), s = n,\ldots t\}$ in the following way

$$\begin{bmatrix} E_1 \\ E_2 \\ E_3 \\ 0 \\ \vdots \\ 0 \end{bmatrix} = T \begin{bmatrix} e^T(n) \\ e^T(n+1) \\ \vdots \\ e^T(t) \end{bmatrix} \tag{3.9}$$

where the vectors $e(i)$ have the property

$$\mathcal{E}[e(i)e^T(j)] = \delta_{ij} R \tag{3.10}$$

and where T is an orthogonal matrix. Since the elements of T depend on the $\{e(n), \cdots, e(t)\}$ it is not straightforward to determine the statistical properties of E_1, E_2 and E_3. It can however be shown that the following properties hold asymptotically

$$\begin{aligned} &\mathcal{E} E_1 \sim 0 \\ &\mathcal{E} E_2 \sim 0 \\ &\mathcal{E} E_3 \sim 0 \\ &\mathcal{E}[E_1 Q E_1^T] \sim I_\nu \operatorname{tr}(QR) \\ &\mathcal{E}[E_2 Q E_2^T] \sim I_\mu \operatorname{tr}(QR) \\ &\mathcal{E}[E_1 E_2^T] \sim 0 \\ &[E_3^T E_3] \sim (t - n(\mu+\nu) - \mu) R \end{aligned} \tag{3.11}$$

where Q is an arbitrary matrix having dimensions compatible with the formulas. Since $E_{33} = D_{33}$ and $\frac{1}{t-n(\nu+\mu)-\mu} E^T E$ converges to R we thus find that the statistical properties of E_1 and E_2 are known asymptotically.

Also notice that since the matrices D_{11} and D_{22} grow linearly with t the terms $D_{11}^{-1} E_1$ and $D_{22}^{-1} E_2$ will actually converge to zero as t increases.

Control of Systems with Uncertain Parameters

The system to be controlled (2.1) can be described by

$$y(t) = B^T u(t) + A^T z(t) + e(t) \tag{3.12}$$

where the unknown parameters A and B can be expressed in terms of the information state $J(t-1)$ and the stochastic matrices E_1, E_2 and E_3 through (3.7) and (3.8). To find the control strategy which minimizes the criterion (2.6) it is than necessary to solve a quadratic control problem for a system with uncertain parameters. This can be done in a straightforward way using dynamic programming. The solution is given by the linear feedback

$$u(t) = -L(t) z(t) \tag{3.13}$$

where

$$L(k) = [\Gamma^T(t) S(k) \Phi(t) - R_B(t,k)]^T [\Gamma^T(t) S(k) \Gamma(t) + R_B(t,k)]^{-1} \tag{3.14}$$

$$\begin{cases} S(k-1) = \Phi^T(t) S(k) \Phi(t) - L(k)[\Gamma^T(t) S(k) \Phi(t) - R_B(t,k)] + \\ \qquad + Q + R_A(t,k) \\ S(t+N) = Q \qquad\qquad , \ k = t+N, \dots, t+1 \end{cases} \tag{3.15}$$

$$\Phi(t) = \begin{bmatrix} 0 & I \\ 0 & 0 \\ \multicolumn{2}{c}{\hat{A}^T(t)} \end{bmatrix}, \quad \Gamma(t) = \begin{bmatrix} 0 \\ I \\ \hat{B}^T(t) \end{bmatrix}, \quad Q = \begin{bmatrix} 0 & 0 & 0 \\ 0 & Q_2 & 0 \\ 0 & 0 & Q_1 \end{bmatrix} \tag{3.16}$$

$$\hat{B}(t) = D_{22}^{-1}(t) D_{23}(t) \tag{3.17}$$

$$\hat{A}(t) = D_{11}^{-1}(t) [D_{13}(t) - D_{12}(t) \hat{B}(t)] \tag{3.18}$$

$$R_B(t,k) = D_{22}^{-1}(t) (D_{22}^{-1}(t))^T \gamma(t,k) \tag{3.19}$$

$$R_A(t,k) = D_{11}^{-1}(t) [D_{12}(t) R_B(t) D_{12}^T(t) + I \cdot \gamma(t,k)] (D_{11}^{-1}(t))^T \tag{3.20}$$

$$\gamma(t,k) = \frac{1}{t - n\nu - (n+1)\mu} \operatorname{tr}(S_3(k) D_{33}^T(t) D_{33}(t)) \tag{3.21}$$

$$S(k) = \begin{bmatrix} S_1(k) & S_{12}(k) & S_{13}(k) \\ S_{21}(k) & S_2(k) & S_{23}(k) \\ S_{13}(k) & S_{23}(k) & S_3(k) \end{bmatrix} \tag{3.22}$$

Notice that the results only hold asymptotically because the formulas (3.11) for the moments of E_1, E_2 and E_3 are only valid asymptotically.

The Algorithm

Summarizing we now obtain the following algorithm for the control of a system with unknown parameters.

1. At each step t determine the information state given by (3.3) and (3.4) for $t-1$.
2. Knowing $J(t-1)$ the parameters of the Riccati equation (3.15) are known and the Riccati equation can then be iterated $N-1$ steps in order to obtain $L(t)$ which then gives the control to be applied at time t.

Notice that the algorithm can be simplified by reducing the number of iterations of the Riccati equation. This will, of course, correspond to a different criterion.

4. EXAMPLE

It would be of considerable interest to analyse the properties of the closed loop systems obtained when the self-adjusting algorithm is applied to systems of various types. Since the terms $D_1^{-1} E_1$ and $D_{22}^{-1} E_2$ will converge to zero the estimates of A and B given by (3.17) and (3.18) will probably converge to the true parameter values if the system to be controlled is actually governed by a model (2.1) with the correct structure. Under reasonable subsidiary conditions the selfadjusting algorithm will then probably converge to the optimal control that can be computed if the parameters were known. Neither a proof nor the precise conditions are yet available. If the order has to be reasonably small the model (2.1) is fairly restrictive due to the fact that the stochastic variables $\{e(t)\}$ are assumed uncorrelated. It is an interesting and still open problem to find out what will happen if the selfadjusting algorithm is applied to such a system. The complete analysis of the properties of the selfadjusting algorithm thus is largely lacking. It is, however, the authors belief that the assumptions are not as restrictive as they might first appear and algorithm can give a solution to many industrial control problems. Several simulated examples support this viewpoint. One of them is given in the following.

Head Box of a Paper Machine

We will discuss the application of the selfadjusting algorithm to control of a headbox of a paper machine.A schematic diagram of a headbox is shown in Fig.1.

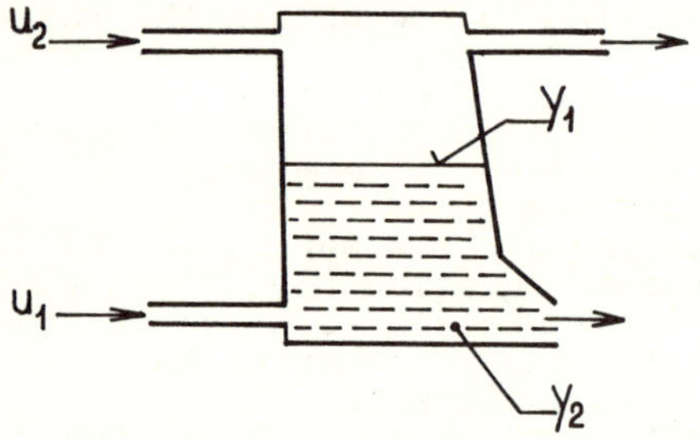

Fig.1 Schematic diagram of the headbox of a paper machine.The control variables are stockflow u_1 and airflow u_2 into the headbox. The outputs are stock level y_1 and total pressure y_2

If the sensor and actuator dynamics are neglected the headbox can be described as a second order dynamical system.Linearizing the equations around a given operating point the following model is obtained.

$$\frac{d}{dt}\begin{bmatrix}x_1\\x_2\end{bmatrix} = \begin{bmatrix}a_{11} & a_{12}\\a_{21} & a_{22}\end{bmatrix}\begin{bmatrix}x_1\\x_2\end{bmatrix} + \begin{bmatrix}b_{11} & 0\\b_{21} & b_{22}\end{bmatrix}\begin{bmatrix}u_1\\u_2\end{bmatrix} \quad (4.1)$$

where $x_1 = h - h_0$ deviation in stock level from steady state value [m]

$x_2 = (p-p_0)/\varrho_2 g$ normalized deviation in airpad pressure from steady state value

u_1 deviation in stockflow from steady state value

u_2 deviation in airflow from steady state value

$y_1 = x_1$ stock level

$y_2 = x_1 + x_2$ total pressure

The parameters $a_{11}, a_{12}, a_{21}, a_{22}, b_{11}, b_{21}$ and b_{22} are easily determined from construction data [13] . Typical values are given below

$a_{11}=-0.01$ $a_{12}=-0.01$

$a_{21}=-0.21$ $a_{22}=-0.25$

$b_{11}= 0.10$ $b_{21}= 2.10$

$b_{22}= 1.30$

A block diagram of the system is shown in Fig. 2. The stepresponses of the system to 10 % changes in the control variables from their nominal values are given in Fig. 3 and Fig. 4. The eigenvalues of the system matrix are both real and correspond to time constants of 4 s and 600 s which is clearly reflected in the figures.

Now assume that the control signals are kept constant over sampling intervals of constant lenght. The system can then be sampled and the following relation is obtained between the values of the inputs and the outputs at the sampling intervals.

$$y(t) = A_1' y(t-1) + B_1' u'(t-1) \quad (4.2)$$

If a sampling interval of one second is chosen we get the following numerical values

$$A_1' = \begin{bmatrix}0.9998 & -0.008805\\0.03522 & 0.7709\end{bmatrix}, \quad B_1' = \begin{bmatrix}0.08989 & -0.005971\\1.9390 & 1.1447\end{bmatrix}$$

To complete the description of the system it is also necessary to give a description of the disturbance. It is simply assumed that the disturbance can be represented as an autoregression in the stockflow in the sampled model i.e.

$$u_1(t) = u_1'(t) + V(t)$$

where $v(t+1) = c\, v(t) + e(t)$

and $\{e(t)\}$ is white noise with zero mean and variance R. After simple calculations we then find that the system can be represented as follows

$$y(t) = A_1 y(t-1) + A_2 y(t-2) + B_1 u(t) + B_2 u(t-2) + e(t) \tag{4.3}$$

If it is assumed that c = 0.95 the following numerical values are obtained

$$A_1 = \begin{bmatrix} -1.9489 & -0.008805 \\ 0.003522 & 1.7209 \end{bmatrix}, \quad A_2 = \begin{bmatrix} -0,9498 & 0,008365 \\ -0,03346 & -0,7323 \end{bmatrix}$$

$$B_1 = \begin{bmatrix} 0.0898 & -0.005971 \\ 1.9390 & 1.1446 \end{bmatrix}, \quad B_2 = \begin{bmatrix} -0.08539 & 0.005673 \\ -1.8420 & -1.0874 \end{bmatrix}$$

$$R = \begin{bmatrix} 7.9 \times 10^{-8} & 1.7 \times 10^{-6} \\ 1.7 \times 10^{-6} & 3.69 \times 10^{-5} \end{bmatrix}$$

In Fig. 5 we show a simulation of the system obtained when no control is applied. It is clear from this Figure that the simple closed headbox can not be used without control. Fig. 6 shows the results obtained when the headbox is controlled with a selfadjusting regulator. The selfadjusting regulator was designed for a loss function with the parameters

$$Q_u = \begin{pmatrix} 1 & 0 \\ 0 & 1 \end{pmatrix}, \quad Q_y = \begin{pmatrix} 10^3 & 0 \\ 0 & 10^4 \end{pmatrix}$$

A first order regulator was used.

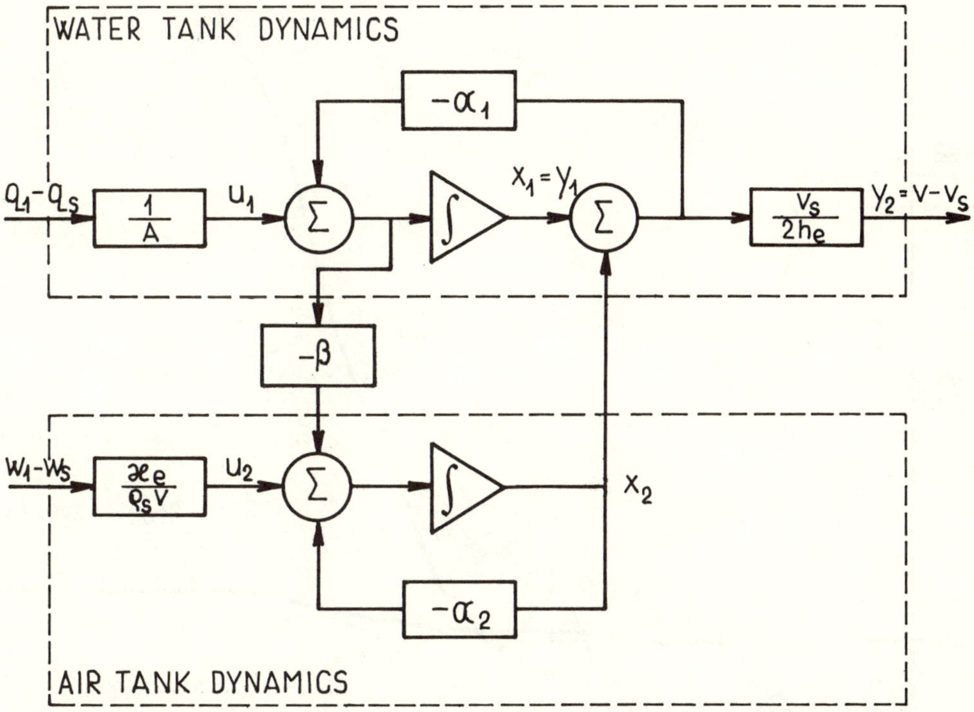

Fig. 2 Block diagram of the headbox of a paper machine. The system is of second order and the state variables are chosen as the stock level x_1 and the pressure in the air cushion x_2

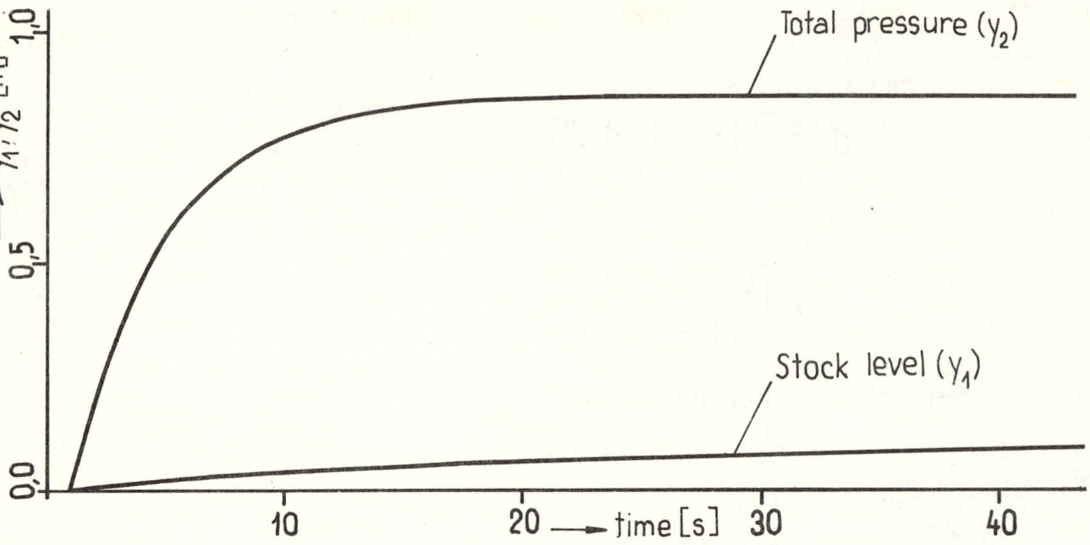

Fig. 3 Responses of stock level and total pressure to a stepchanges in the stockflow of 0.1 m^3/s corresponding to 10 % of the steady state value

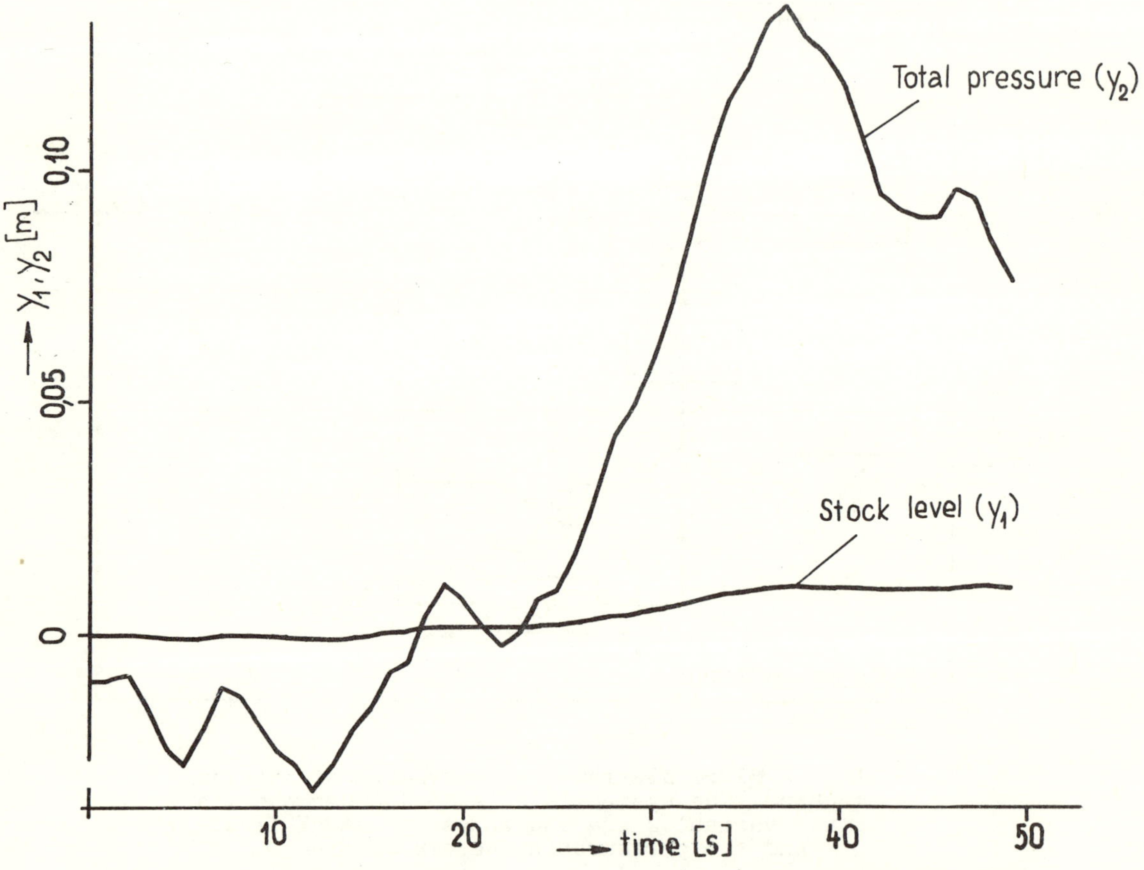

Fig. 5 Simulated responses of the system (4.3) when the controls are zero and the disturbance $\{e(t)\}$ is a sequence of uncorrelated random variables

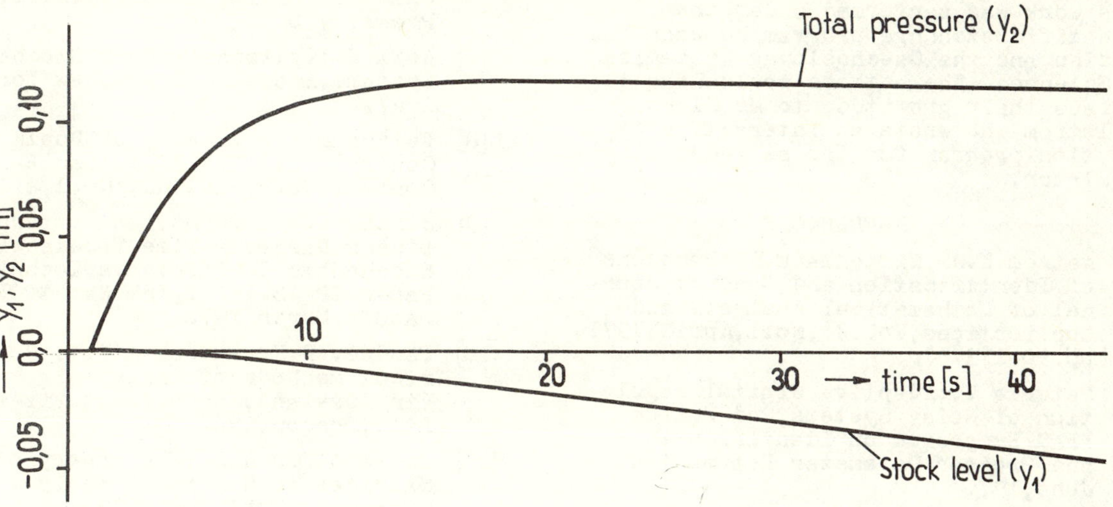

Fig. 4 Responses in stock level and total pressure to a stepchange in the airflow of 0.0245 corresponding to 10 % of the steady state value

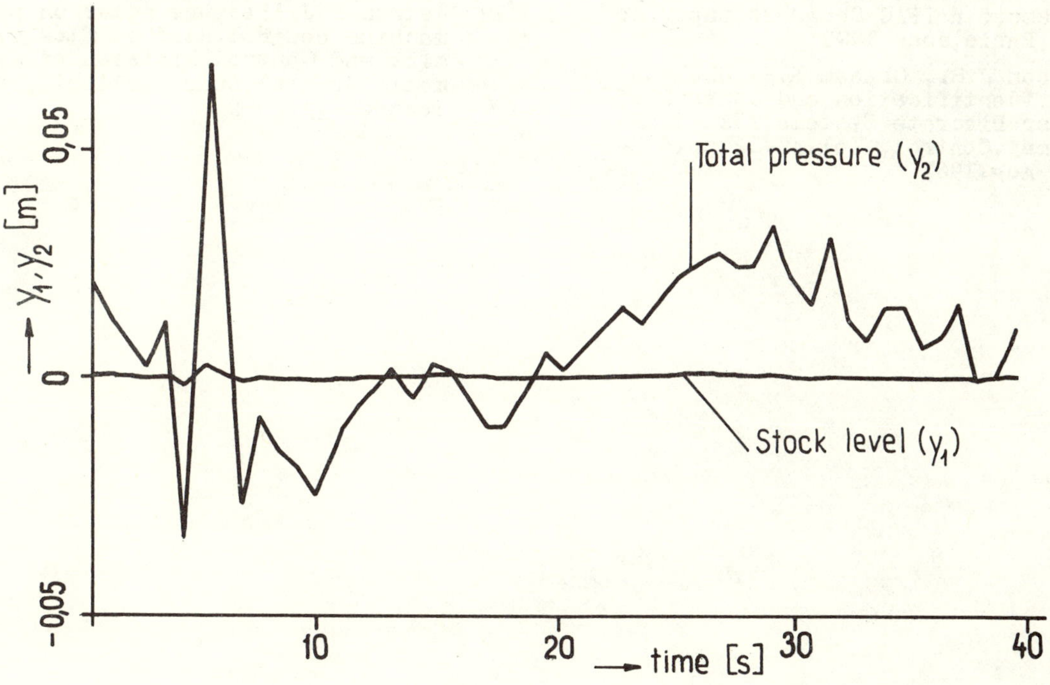

Fig. 6 Simulation of headbox control using the selfadjusting regulator. The disturbance $\{e(t)\}$ in (4.3) is a sequence of uncorrelated random variables

5. ACKNOWLEDGEMENTS

This work was performed under the scientific exchange program between the Swedish and the Czechoslovak Academies of Sciences. The authors would like to express their gratitude to Mr Claes Källström who wrote an interactive simulation program for the selfadjusting regulator.

REFERENCES

[1] Åström K.J.,Wittenmark B.:Problems of Identification and Control,Journal of Mathematical Analysis and Applications,Vol.34,No.1,April,1971, pp.90-113

[2] Peterka V.:Adaptive digital regulation of Noisy Systems,2nd Prague IFAC Symposium on Identification and Process Parameter Estimation, June,1970

[3] Tse E., Athans M.:Adaptive Stochastic Control for a Class of Linear Systems, IEEE Trans.Automat.Control, Vol. AC-17,No.1,February 1972

[4] Åström K.J.,Wittenmark B.:On the Control of Constant but Unknown Systems,5th IFAC Congress,paper No. 37.5,Paris,June 1972

[5] Farison J.B., Graham R.E.,Shelton R.C.:Identification and Control of Linear Discrete Systems,IEEE Trans. Automat.Control, Vol.AC-12,pp.438-442, Aug.1967

[6] Åström K.J.:Introduction to Stochastic Control Theory,Academic Press,1970

[7] Aoki M.:Optimization of Stochastic Systems,Academic Press,New York 1967

[8] Feldbaum A.A.:Theory of Dual Control,I,II,III,IV.Automn Remote Control No.9,11.1961; No.1,2.1962

[9] Bohlin T.: Information Pattern for Linear Discrete-Time Models with Stochastic Coefficients,Technical Paper TP 18.192.1,IBM Nordic Laboratory,March 1970

[10] Faddeev D.K.,Faddeeva V.N.:Numerical Methods of Linear Algebra (in Russian),Gosud.Izdat.Fiz-Mat. Lit.,Moscow,1960

[11] Householder A.S.:The Theory of Matrices in Numerical Analysis, Blaisdell Publ.Comp.,1964

[12] Peterka V., Šmuk K.:On-Line Estimation of Dynamic Model Parameters from Input-Output Data,4th IFAC Congress, paper 26.1, Warsaw , June 1969

[13] Åström K.J.:Lecture notes on paper machine control.Headbox flow dynamics and Control,Division of Automatic Control,Lund Institute of Technology,1972

DIGITAL ADAPTIVE CONTROLLERS USING SECOND ORDER MODELS WITH TRANSPORT LAG*

S. Joshi and H. Kaufman**
Rensselaer Polytechnic Institute
Troy, New York
United States of America

Design of a discrete optimal regulator requires the a priori knowledge of a mathematical model for the system of interest. Because a second order model with transport lag is very amenable to control computations and because this type of model has been used previously to represent certain high order single input-single output processes, an adaptive controller was designed based upon adjustment of controls computed for such a model. An extended Kalman filter was utilized for tracking the model parameters which were subsequently used to update a set of optimal control gains. Favorable results were obtained in applying this procedure to the control of several examples including a ninth order nonlinear process.

1. INTRODUCTION

Development of a mathematical model for a physical system is often the first step undertaken in the design of a controller. A mathematical model is usually obtained after a detailed study and a thorough understanding of the underlying physical phenomena, and in many practical cases, turns out to be of high order, nonlinear and/or stochastic in nature.

Modern control systems often include a digital computer in the loop for processing the output measurements of the system or the plant, and synthesizing the optimal control law. If the original mathematical model of the plant is complex, or of high order, the requirements on the size and the speed of this computer can be very demanding. Consequently attempts are often made to obtain a low order model which represents the plant to some accuracy.

In particular it has been found that high order overdamped systems as often encountered in chemical process control can be represented to a fair accuracy by a second order model containing transport lag (see Coughanour, 1965; Cox, 1966; Gallier, 1969). The simple philosophy behind this structure, shown in Fig. 1, is that a portion of the phase lag in the system due to the large numbers of poles can be lumped into a single puretime delay. It should be noted that this time delay gives an extra degree of freedom without increasing the order of the model.

Consequently, computation and synthesis of the optimal control law for this model is a relatively simple task. Furthermore, if such a model can be determined and updated on line as the process evolves, an adaptive controller can be easily synthesized and interfaced between the plant output and the control input.

Development of such an adaptive controller requires that consideration be given to the:
Offline determination of nominal model parameters,
Online estimation of the model parameters,
Selection of a performance index and
Design of adaptation logic.

To this effect, section 2 considers both online and offline methods for estimating the model parameters including the time delay from measurements of the plant output and knowledge of the control input. In particular, the application of a modified extended Kalman filter was found to be quite useful.

Section 3 next discusses two types of adaptive controllers based on a quadratic performance index and a summed absolute error performance index.

Finally, Section 4 presents five numerical examples, including a ninth order nonlinear nuclear reactor model. Results of these experiments lead to the conclusion (summarized in Section 5) that adaptive controllers based upon a second order model with a time lag can give very satisfactory results.

2. DETERMINATION OF MODEL PARAMETERS

Although the proposed adaptive control scheme requires on-line estimation of the model parameter, it is also desirable to be able to judiciously select an initial set of model parameters based upon previously recorded operating data. To this effect it is proposed that either the method of quasilinearization or a graphical analysis (as described below) be used to determine initial model parameters and that an extended Kalman filter be used for on-line updating.

2.1 Model Structure

The model under consideration, shown in Figure 1

*The work described in this paper was supported by Project Themis, Contract #DAAB07-69-C-0365. The United States Government reserves the right to separately reproduce and distribute published materials which result from research under this contract.

** Prof. Systems Engineering Division

is described by a synchronous sampler and zero order hold in series with the continuous transfer function

$$G(s) = \frac{Ke^{-hs}}{(s+p_1)(s+p_2)} \tag{1}$$

Thus the complete transfer function for the model is

$$G(s) = \left(\frac{Ke^{-hs}}{(s+p_1)(s+p_2)}\right)\left(\frac{1-e^{-sT}}{s}\right) \tag{2}$$

Impulse invariant discretization of this transfer function using the Jordan canonical form shown in Fig. 2, results in the equations:

$$x_{k+1} = A\, x_k + b\, u_{k-m} \tag{3a}$$

$$y_k = c^T x_k \tag{3b}$$

where x_k is the 2-dimensional state vector defined in Fig. 2, u_k is the scalar input, y_k is the scalar output, and under the assumption that h is an integral multiple of the sample period T, m=h/T. A, b and c are given by

$$A = \begin{bmatrix} e^{-p_1 T} & 0 \\ 0 & e^{-p_2 T} \end{bmatrix} \tag{4a}$$

$$b = \frac{K}{p_2 - p_1} \begin{bmatrix} \dfrac{1-e^{-p_1 T}}{p_1} \\ \dfrac{1-e^{-p_2 T}}{p_2} \end{bmatrix} \tag{4b}$$

$$c^T = [1, \ -1] \tag{4c}$$

Using techniques to be described in the sequel, the parameters p_1, p_2, K, and m are to be determined.

Alternately, the continuous phase variable canonical form, with states y and $\dot{y}$ as shown in Fig.3, could have been used and subsequently discretized. In fact, this form has been found most useful if complex or repeated poles are anticipated (Joshi, 1973). Furthermore, for real distinct poles, the Jordan form was observed experimentally to be more satisfactory for identification because of its having fewer terms which are nonlinear functions of the parameters.

2.2 Offline Determination of Model Parameters

Two methods are proposed for determing initial model parameters namely:

(1) A graphical analysis of the system's step response valid for overdamped system. This can be easily extended to the underdamped case. Noise is not analytically considered.

(2) Quasilinearization, valid for any input (other than a step) that sufficiently excites the dominant modes. A summed squared error index is used for noise smoothing.

2.2.1 Graphical Analysis

Coughanour, 1965 discusses a method for determining from an experimentally recorded step response of an overdamped system, the parameter h, T_1 and T_2 which appear in the transfer function

$$\frac{e^{-hs}}{(T_1S+1)(T_2S+1)}$$

Only trivial modifications are needed in order to extend these procedures to estimating the gain K as well.

Basically the method requires that the response's inflection point be located and that the slope and ordinate at this point be recorded. Using these two quantities in the appropriate formulae, the parameters can be easily determined.

It should be noted that this method is valid only for overdamped systems and does not take into account the effects of measurement noise. Clearly, if a step response is available, this method can quickly yield a reasonable set of nominal parameters.

2.2.2 Quasilinearization

In order to identify the model parameter using quasilinearization, it is first necessary to note that the true process output y_k^P can be regarded as the model output y_k plus a systematic noise term n_k ; that is to say,

$$y_k^P = c^T x_k + n_k \tag{5}$$

where x_k is the two dimensional model state vector (from either the Jordan or phase canonical forms). In this form, the systematic noise term n_k can also be made to contain contributions from actual noise inherent in measuring the plant output y_k^P; this measurement noise, however, should, in general, be much less than the systematic modelling errors. Estimation of the model parameters is to be performed using a given set of output samples ($y_0^P, \ldots, y_N^P$) recorded offline in response to a known signal u_k.

Because of equation 5 it is now necessary to estimate the model state vector x_k as well as the parameters p_1, p_2, K, and m. (Larson, 1969, Jazwinski, 1970). These parameters can be regarded as elements of an additional state vector $q = (p_1, p_2, K, h)^T$ satisfying the equation:

$$q_{k+1} = q_k \tag{6}$$

Thus the complete model can be defined by the augmented system of nonlinear state equations

$$\begin{pmatrix} x_{k+1} \\ q_{k+1} \end{pmatrix} = \begin{pmatrix} f(x_k, q_k, u_{k-m}) \\ q_k \end{pmatrix} \tag{7}$$

whereas before h = mT.

Estimates $\hat{x}_k$ and $\hat{q}_k$, for x_k and q_k respectively are now determined offline in order to minimize

$$I = \sum_{i=1}^{N} (y_k^P - c^T\hat{x}_k)^2 \quad (8)$$

where

$$\hat{x}_{k+1} = f(\hat{x}_k, \hat{q}_k, u_{k-\hat{m}_k}) \quad (9a)$$

$$\hat{q}_{k+1} = \hat{q}_k \quad (9b)$$

Basically the quasilinearization procedure begins with a set of estimated initial conditions $(\hat{x}_0^o, \hat{q}_0^o)$ which in turn define the solution to Eqn. 9. Linearization of Eqn. 7 about this solution leads to a set of linear difference equations, the solution $(\hat{x}_k, \hat{q}_k)$ of which is readily available and linear with respect to the initial values. Estimates of these initial values $\hat{x}_0^1$ and $\hat{q}_0^1$ are determined by minimizing the index I of equation 8. This process is repeated until convergence results. (Joshi, 1973, Sage, 1968)

Because the linearization requires the computation of the derivative $\frac{\partial u(kT-h)}{\partial h} = -u\,(kT-h)$, a step input will not yield information sufficient for determining the delay h by the above procedures.

2.3 Online Estimation, The Kalman Filter

Having obtained an initial set of model parameters it is necessary to be able to update these parameters during plant operation in order to account for true process variation and for the dependence of the parameter value upon the type of input signal caused by the nonlinear nature of the augmented system (7).

Thus, at any sample time N, it is desireable to be able to compute the parameter estimates based on the control and output measurement record (i.e., $y_0^P, y_1^P, \;\; , y_N^P$) received to date.

One procedure which has sucessfully been used previously by (Kaufman, 1969) for online state and parameter extinction is the linearized (or extended) Kalman Filter. The equations for this filter are developed using the same performance index, process equations, and measurement equations that define the quasilinearization technique, i.e., Eqs. 8, 9 and 5 respectively. It can be shown that at a given sample time N the algorithm computes the state and parameter estimates $\hat{x}_N$ and $\hat{q}_N$ using the predicter corrector formulae given by Kaufman (1969) and Jazwinski, (1970):

$$x_N = x_N^P + G_N^1(y_N^P - c^T x_N^P) \quad (10a)$$

$$\hat{q}_N = q_N^P + G_N^2(y_N^P - c^T x_N^P) \quad (10b)$$

where x_N^P and q_N^P are the predicted estimates computed from Eqs. 9 as:

$$x_N^P = f(\hat{x}_{N-1}, \hat{q}_{N-1}, u_{N-1-\hat{m}_{N-1}}) \quad (11a)$$

$$q_N^P = \hat{q}_{N-1} \quad (11b)$$

and G_N^1 and G_N^2 are the corrector gains (computed on-line) that multiply the residual or difference between the received measurement at time N, y_N^P and the predicted measurement, $c^T x_N^P$. Computation of gains G_N^1 and G_N^2 requires the apriori selection of:

- $\hat{x}_0, \hat{q}_0$: The initial state and parameter estimates
- P_0^x, P_0^q: The initial covariance of the errors in the initial estimates $\hat{x}_0$ and $\hat{q}_0$ respectively.
- R : The covariance of the measurement noise.

As with quasilinearization, a series of successive linearizations about some priming trajectory is required in order to compute these gains. In particular, it has been shown by Kaufman (1970) that with the proper initialization procedures and selection of the appropriate priming trajectory, the linearized Kalman filter used offline can yield the same estimates as quasilinearization.

3. ADAPTIVE CONTROL PROCEDURES

In order to evaluate the usefulness of the second order model as the basis for controller design, two types of performance indices were considered; namely, a sum squared error index, and a summed absolute error index. For both the performance indices the model control was computed as a linear function of the state vector. This enabled a choice of two configurations, namely open-loop and closed-loop control of the plant. In the open-loop configuration shown in Fig. 4, the plant control input is synthesized using the model control gains and the model state variables while in the closed-loop configuration (Fig. 5) the plant control input is synthesized using the model control gains and the plant output.

For the sum squared performance index, as will be shown in section 3.1, the optimal input is of the form

$$u_k^* = K_k^{*T} x_k \quad (12)$$

Thus if the closed-loop configuration is to be used, the components in the two-dimensional model state vector x_k must be in correspondence with the true process output. This is true for the phase canonical form; however, for the Jordan form given in Eqs. 3, 4, it can be shown that

$$\begin{pmatrix} x_1 \\ x_2 \end{pmatrix}_k = (V) \begin{pmatrix} y \\ \dot{y} \end{pmatrix}_k \quad (13a)$$

where

$$V = \frac{1}{p_2 - p_1} \begin{pmatrix} p_2 & 1 \\ p_1 & 1 \end{pmatrix} \quad (13b)$$

Consequently, to apply the optimal gain K_k^* to the true process, the plant output vector

$$\begin{pmatrix} \hat{y}(k) \\ \dot{\hat{y}}(k) \end{pmatrix}$$

must be premultiplied by V in order to establish a compatible state vector. Thus the appropriate expression becomes

$$U_k^* = K_k^{*T} \, V \begin{pmatrix} \hat{y}(k) \\ \dot{\hat{y}}(k) \end{pmatrix}$$

where K_k^* is the optimal gain for the model

V is the above transformation matrix

$\hat{y}(k)$ is the true process output

$\dot{\hat{y}}(k)$ is the derivative of the process output (approximated in practice using first order differences)

3.1 Sum Squared Error Performance Index

3.1.1 The Control Structure

The problem under consideration is to minimize

$$J_1 = \sum_{k=0}^{\infty} (x_k^T \, Q \, x_k + u_k^2) \qquad (14)$$

subject to the dynamic constraint

$$x_{k+1} = A \, x_k + b \, u_{k-m} \qquad (15)$$

where $Q = cc^T \, . \, q$, $q>0$, a scalar.

An analysis very similar to that performed by Sage (1968) when m=0 yields the optimal control law:

$$u_k^* = K_k^{*T} x_k \qquad (16)$$

where

$$K_k^{*T} = \begin{cases} -b^T \, P \, T^{\,k+1} A^{m-k} & 0 \le k \le m-1 \\ -b^T \, P \, T^{\,m+1} & k \ge m \end{cases} \qquad (17)$$

where

$$T = (I + bb^T P)^{-1} A \qquad (18)$$

and P is the solution of

$$P = Q + A^T P[I+bb^T P]^{-1} A \qquad (19)$$

and T, P and K_k^* are 2x2, 2x2 and 2x1 dimensional matrices respectively.

Whereas the above control is designed under the assumption that the desired output is zero and that $u_k=0 \quad -m \le k \le -1$ a more general problem (often encountered in practice) would be to minimize

$$J = \sum_{k=0}^{\infty} (x_k - x_D)^T Q (x_k - x_D) + (u_k - u_D)^2 \qquad (20)$$

for $u_k = u^o \quad -m \le k \le -1$

where

Q is the same as above

and x_o and u_D are the respective desired steady state trajectory and control satisfying

$$x_D = A \, x_D + b \, u_D \qquad (21a)$$

$$y_D = c^T \, x_D \qquad (21b)$$

For this case, Joshi (1973) has in a similar fashion shown that

$$u_k = u_D - b^T P T^{k+1} [A^{m-k}(x_k - x_D) - \sum_{i=0}^{m-k-1} A^i b(u^o - u_D)] \qquad 0 \le k \le m-1 \qquad (22a)$$

and

$$u_k = u_D - b^T P T^{m+1} (x_k - x_D) \qquad k \ge m \qquad (22b)$$

where P and T are as defined in Eqs. 18 and 19.

3.1.2 Adaptation Logic For Summed Squared Error Index

In the adaptive control loop, the optimal gain vector K_k^* is updated using the latest estimate of A, b and m. This requires an updating scheme for the Riccati matrix P, which is a solution of the nonlinear Eq. (19). Since P must be updated from its old value in an on-line fashion, a first order approximation was used for δP, the change in P, due to the changes δA and δb in A and b. This of course assumes that the per sample changes in P, A, and b are sufficiently small. By using the matrix approximation

$$(P+\delta P)^{-1} \simeq P^{-1} - P^{-1} \, \delta P \, P^{-1} \qquad (23)$$

and retaining only the first order terms, the following equation is obtained after some algebraic manipulation:

$$\delta P - B^T \, P^{-1} \, \delta P \, P^{-1} B = B^T \delta A + \delta A^T B - B^T (b \delta b^T + \delta b \, b^T) B \qquad (24)$$

where

$$B = (P^{-1} + bb^T)^{-1} A$$

Since (24) is a linear equation in δP which is a 2x2 symmetric matrix, the necessary correction to P can easily be mechanized as the solution to three simultaneous linear equations.

It should be noted, however, that equation (24) is based on a first order approximation and is valid only for small changes in the model parameters. If the model parameters vary rapidly, equation (24) may not give a good approximation to the change in P. Under these circumstances, it is advisable to compute P more accurately every N samples using the latest model parameters. This may very well be performed by backwards differencing of standard Riccati equation:

$$P_k = Q + A^T P_{k+1} (I + bb^T P_{k+1})^{-1} A \qquad (25)$$

until convergence occurs. Such a procedure would be initialized with P_{k+1} set to the value of P, corresponding to the most recent parameter estimates.

3.2 Summed Absolute Error Index

As an alternative to the previous controllers designed using linear-quadratic optimal control theory, the direct digital control structure developed by Gallier and Otto (1969) was also considered. The procedure to be used for designing such a controller is valid provided that the process can be described by the transfer function

$$G(s) = \frac{Ke^{-hs}}{(s+p_1)(s+p_2)} = \frac{y(s)}{u(s)} \qquad (26)$$

where p_1 and p_2 are real.

3.2.1 Controller Structure

For a process described by the transfer function (26) a feedback control of the form

$$u_k^* = K_p[y_k + \frac{T}{\tau_I} \sum_{j=0}^{\infty} y_{k-j}] \qquad (27)$$

is to be found in order to minimize

$$J = \sum_{k=0}^{\infty} |y_k{}^P| \qquad (28)$$

where

K_p = Proportional gain

τ_I = Discrete equivalent of reset time

and T = sampling period

Appropriate value of K_p and τ_I can be easily obtained using the graphical analysis developed by Gallier and Otto (1969). This procedure makes use of two distinct graphs - the first being a plot of $K_o = K_pP_1P_2/K$ versus normalized dead time $T_D = h/(h+P_1^{-1}+P_2^{-1})$ for various values of P_2/P_1, and the second plot of $T_I = \tau_I/(h+P_1^{-1}+P_2^{-1})$ versus T_D, again for various values of P_2/P_1. Thus computation of τ_I and K_p can be performed using the following steps:

1. Record K, P_1, P_2, h
2. Compute P_2/P_1 and T_D
3. Record from the appropriate plots K_o and T_I
4. Compute $K_p = K_oK/P_1P_2$ and $\tau_I = T_I(h+P_1^{-1}+P_2^{-1})$

3.2.2 Adaptation Logic

In the adaptive control scheme, the gains K_p and τ_I must be selected on-line using the most recent model parameters. To implement this procedure, the graphs discussed in the previous section (3.2.1) were stored in the computer as polynomial fits and used in conjunction with the on-line parameter estimation scheme described in section 2.3. At each stage, the updated optimal gains were computed by substituting the updated model parameters into these stored polynomials.

4. RESULTS

In order to demonstrate the application of the proposed adaptive control logic, five numerical examples were considered. Four of these examples were linear and included an unstable process and a process with a free integrator in its continuous transfer function. The fifth example considered the control of a nuclear power reactor simulated as a ninth order nonlinear system. Effect of measurement noise was studied using this last example.

4.1 Third Order Process

The third order plant selected for consideration was defined by the transfer function

$$G(s) = \frac{1}{(s+1)(s+2)(s+4)}\left(\frac{1-e^{-sT}}{s}\right)$$

with T=0.05 seconds and y_0=1.0. The corresponding discretized state space equations are:

$$\begin{pmatrix} x_1 \\ x_2 \\ x_3 \end{pmatrix}_{k+1} = \begin{pmatrix} e^{-T} & 0 & 0 \\ 0 & e^{-2T} & 0 \\ 0 & 0 & e^{-4T} \end{pmatrix} \begin{pmatrix} x_1 \\ x_2 \\ x_3 \end{pmatrix}_k + \begin{pmatrix} 1 - e^{-T} \\ (1 - e^{-2T})/2 \\ (1 - e^{-4T})/4 \end{pmatrix} uk$$

and the output is

$$y_k = \frac{1}{3} x_1(k) - \frac{1}{2} x_2(k) + \frac{1}{6} x_3(k).$$

Using a second order model, controllers were determined to minimize

$$J_1 = \sum_{k=0}^{\infty} 100\, y_k^2 + u_k^2$$

and

$$J_2 = \sum_{k=0}^{\infty} |y_k|$$

Initial model parameters were obtained by the graphical analysis of the unit step response of the plant, as described in Section 2.2.1. The plant was controlled using this model for both the above indices J_1 and J_2. The following configurations were used:

Open-loop, adaptive and nonadaptive control.
Closed-loop, adaptive and nonadaptive control.

The model parameters and the control law were updated at every sampling instant.

As shown in Fig. 6 for the sum squared error index, the adaptive control showed a marked improvement over the nonadaptive control only for the open loop configuration. In contrast, it was observed that in using the summed absolute error index J_2, both adaptive and nonadaptive control gave good results starting from the graphically obtained parameters. Thus, the model parameters were set somewhat away from these values and the above configurations were again tested. In this case, as shown in Fig. 7, the adaptive control was shown to be significantly better than the nonadaptive control, for both the closed loop and open loop configurations.

Finally, in order to investigate the extent to which the presence of the time delay in the model affects the performance, three runs were made for the index J_1:

a. With time delay=0, other model parameters arbitrarily chosen, and the plant controlled nonadaptively.
b. Under the same conditions as (a), the plant controlled adaptively, but the time delay in the model kept constant at zero.
c. Under the same conditions as in (a), the plant controlled adaptively, but the model allowed to adapt the delay as the system evolved.

As shown in Fig. 8, the adaptive control was found to be much better than the nonadaptive control, and the presence of delay resulted in a significant improvement in the performance.

A typical pattern of model parameter variation is shown in Fig. 9 (recorded for the open loop implementation of the controller for the summed absolute error index). This pattern of convergence to a final set of values was observed in the adaptive control of all systems tested.

4.2 Fifth Order Process

To further test the proposed adaptive control algorithm, a fifth order process described by the transfer function

$$G_p(s) = \frac{120}{(s+1)(s+2)(s+3)(s+4)(s+5)}$$

was considered. This is the same system controlled by Chidambara and Schainker, 1971, using a third order model developed by aggregation.

Using a sample time of T=0.05 seconds and a unit ramp input, quasilinearization gave as the transfer function for the initial model

$$G_m(s) = \frac{1.77\, e^{-.465S}}{(s+1.061)(s+1.86)}$$

For the summed squared error index

$$J = \sum_{k=0}^{\infty} (y_k^2+u_k^2)$$

and $y_0=1.0$
the following controls were computed:
- .True optimal controller
- .Open and closed loop non-adaptive controls using the above model $G_m(s)$
- .Open and closed loop adaptive controls initialized with the model $G_m(s)$

Trajectories resulting from the suboptimal controllers were all very close to the optimal and essentially indistinguishable as shown in Fig.10. Slight improvements(given in Fig. 10) in the performance index were noted when the adaptation logic was used.

4.3 Fourth Order Process With a Free Integrator

The applicability of the proposed control configuration to a type one process was tested using the transfer function

$$\frac{Y(s)}{U(s)}=G(s)= \frac{0.482}{s(\frac{s}{3.5}+1)(\frac{s}{9.55}+1)(\frac{s}{19.2}+1)}$$

This transfer function was used previously by Semmelhack, 1963, to model a gun mount system; the output y(t) is the azimuth in degrees, and u(t) is the command input in volts. Of interest was the minimization of

$$J_1 = \sum_{k=0}^{\infty} (y_k-y_D)^2\, q + u_k^2$$

with

$$y_0 = 0$$

and

$$y_D=40, \text{ the desired azimuth in degrees.}$$

The weight q was found experimentally as 121 in order to limit the maximum control effort to ±450 volts. The initial model parameters were chosen arbitrarily, except for the fact that a free integrator was included in the continuous transfer function of the model in order to take into account the presence of a free integrator in the plant, that is, one of the model poles was restricted to be zero. The initial model parameters were: $p_1=0$, $p_2=8$, $K/(p_2-p_1)=0.482$(same gain as process) and m=0. Using a sample period of 0.05 seconds, the following controls were computed:
- . True optimal control of the plant.
- . Open-loop, nonadaptive and adaptive control of the plant.
- . Closed-loop, nonadaptive and adaptive control of the plant.

In all the adaptive runs, the model and the control law were updated at each sampling instant. As shown in Figure 11, the open loop nonadaptive response was somewhat close to optimal and improved even further when adaptation was used. The closed-loop nonadaptive response was observed to be oscillatory, with a peak overshoot of about 37%. This was improved significantly using adaptation. The minimum degradation in performance was found to be less than 1.5% and occurred in the open loop adaptive case.

4.4 Control of An Unstable Plant

In order to investigate the stabilizing properties of the adaptive scheme, the third order unstable transfer function

$$\frac{Y(s)}{U(s)} = G(s)= \frac{1}{(s-1)(s+1)(s+2)} \quad \text{was considered.}$$

Using a sampling period of 0.05 seconds, controls were determined to minimize:

$$J_1 = \sum_{k=0}^{\infty} (100\, y_k^2 + u_k^2)$$

for

$$y_0 = 1.0$$

Initial model parameters were $p_1= -1$, $p_2=1$, k=1 and mT=0.

Unstable oscillations were observed in the responses resulting from both the adaptive and nonadaptive open-loop configurations and the nonadaptive closed loop configurations. The closed-loop adaptive controller gave a response characterized by decaying oscillations; this was significantly more severe than the true optimal response (Joshi, 1973). No conclusions could be made from these results regarding stabilizing properties.

4.5 Control of a Nuclear Power Reactor

The reactor dynamics were those used by Bereznai and Sinha, 1972, who used a controller based upon a second order model with no time delay.

The plant under consideration, shown in Fig. 12, is described by the following ninth order system of differential equations:

$$\frac{dn}{dt} = \frac{-\beta}{\ell}n + \sum_{i=1}^{6} \lambda_i c_i + \frac{n}{\ell}[\delta k_T - nT_c + t\cdot F(\nu)]$$

$$\frac{dc_i}{dt} = \frac{\beta_i}{\ell}n - \lambda_i c_i, \quad i=1,2,\ldots,6$$

$$\frac{d\delta k_T}{dt} = -\frac{1}{T_t}\delta k_T + \frac{Tc}{T_t}\cdot\nu$$

$$\frac{d\nu}{dt} = -\frac{1}{T_m}\nu + G(u-n)$$

where n is the neutron power level as a fraction of the full power shown as actual power in Fig. 12, c_i is the concentration of the ith precursor, δk_T is the reactivity contribution due to the temperature change and ν is the effective motor drive signal. β_i, λ_i, ℓ are the usual constants in nuclear physics, and T_c, T_t, T_m and G are the temperature coefficient of reactivity, the associated time constant, the motor time constant and the amplifier gain respectively. Values for these parameters are available in the thesis by Joshi, 1973. $F(\nu)$ is the nonlinear absorber rod characteristic, given by

$$F(\nu) = .02 \quad |\nu| \leq 15$$
$$= 0.3 \quad |\nu| \geq 15$$

For purposes of application, it was necessary to use G=9.45, a value 100 times that used by Bereznai and Sinha, in order to obtain convergence of the numerical solutions.

The temperature coefficient of reactivity T_c greatly influenced the stability of the reactor and can vary over a wide range. Results shown are for T_c=.0454; similar results for T_c=-0.005 can be found in Joshi's thesis, 1973.

Using phase variable canonical form in order to account for the possibility of complex poles and the graphical analysis of section 2.21, the initial model parameters were determined to be:

$$A_0^0 = K = 5.48$$
$$A_1^0 = p_1 p_2 = 8.14$$

and $h = 0.05$

For a sample period of 0.05 seconds controls, u_k were determined in order to improve the response to a change in the neutron power level set point demand n_D. Fig. 13 shows that without compensations that is $u_k \equiv n_D$, the response to a demand from n_D = .9 to n_D=1.0 takes more than 10 seconds to reach steady state.

In order to improve upon this response, controls u_k were determined in order to minimize:

$$J = \sum_{k=0}^{\infty} [q\cdot(n_k - 1.0)^2 + (u_k - 1.0)^2]$$

The weight q in the performance index, was varied from the value of q=1 to $q=10^4$. It was found that the open loop nonadaptive and adaptive responses improved successively until q reached a value of 100. Above this value, the plant response was no longer close to the model response since the latter converged to the desired steady state very rapidly. The adaptive open loop response was very satisfactory for q=100, and the convergence to the desired steady-state occurred in less than 5 seconds as shown in Fig. 14. The initial covariance matrix (discussed in section 2.3) for the augmented state vector used for the adaptive run was: P_0 = diag. (1000, 1000, 5, 10, 5, 0.1). In an attempt to improve the response further, P_0 was increased to: P_0=Diag. (1000, 1000, 50, 100, 50, 1) and the open-loop adaptive run was again made. The response showed a significant improvement, and the convergence to the desired steady-state was obtained within 4 seconds. Further increase in P_0 caused deterioration in the performance.

However, in the closed-loop configuration, the adaptive response was satisfactory for q=1, but both adaptive and nonadaptive responses showed stable oscillations about the new steady state for higher values of q. The magnitude of the oscillations increased with q. This was apparently because of the high initial rise rate of the plant response, and the fact that the plant output and output rate were used for the synthesis of the control law. The initial covariance matrix P_0 used was the same as in the open-loop case.

The reactor output measurement, which is the neutron power level, is likely to be contaminated by noise, small in comparison to the modelling error. However, to test the effect of this inherent measurement disturbance, the neutron power level signal n_k was contaminated with Gaussian white noise and the reactor was controlled adaptively for the case T_c=0.0454. The utilized disturbance had standard deviations of (i) 0.005, which is 5% of the step change in the load demand; (ii) 0.01, which is 10%; and (iii) 0.02 which is 20%. For open-loop control, these disturbances did not deteriorate the performance anymore than 5%.

For closed-loop control, the corresponding responses for the 5% and 10% noise disturbances were acceptable in that less than a 5% performance degradation resulted, however, the response for 20% noise exhibited oscillations and took excessive time to reach the new steady state.

As an alternative to the closed-loop control in presence of large noise, the control law was synthesized using the model optimal control gains, and the Kalman estimates of the output and output rate. The latter are available as a by-product of the parameter estimation scheme. This scheme resulted in a substantial improvement in the response, with no oscillations, a faster convergence to the steady-state, and performance degradation of 4.5% with respect to the zero noise case.

5. CONCLUSIONS

An adaptive control scheme has been developed which can be applied to a wide class of single input - single output plants. The algorithm can be used with two types of performance functions- quadratic function, and summed absolute error function. The schemes depend on a linear second order model with a time-delay in the input. The methods for determining such a model from finite length input-output operating data of the plant were discussed, and an optimal control law was developed for this type of model. Methods were developed for online updating of the model and the control law. Both open-loop and closed-loop configurations of the control law were considered. Several numerical examples were given in order to demonstrate the adaptive schemes. In particular, an operating nuclear power reactor was considered and it was shown that this complex ninth order, nonlinear, time varying system with multiple nonlinearities could be controlled very satisfactorily using these adaptive schemes.

Since the model used for approximating the plants is of order two, it will in general, be a poor representation of a plant which has two or more free integrators. Also, as pointed out in Section 3, in its present form, the adaptive scheme based on the summed absolute error performance function is not applicable to cases where the model is unstable, or has complex characteristic roots. The scheme based on quadratic performance functions is applicable to all cases if the model state equations corresponding to phase-variable canonical form are used. It appears that the adaptive scheme based on quadratic performance functions should be recommended because of its advantages - applicability to cases where the initial model is unstable or has complex characteristic roots, and the possibility of adjusting the relative weighting (q) on the error and the control effort. In addition, the response seems to be superior to that obtained using summed absolute error performance function at least in the case of the example discussed in Section 4.

REFERENCES

Bereznai, G. T. and Sinha, N., (1972), "Application of Optimum Low-Order Models To The Adaptive Control of Nuclear Reactors", Fifth IFAC World Congress, Paris, France.

Chidambara, M. R. and Schainker, R. B., (1971), "Lower-order Generalized Model and Suboptimal Control", IEEE Transactions on Automatic Control, Vol. AC-16, No. 2, p 175.

Coughanour, D. R., and Koppel, L. B. (1965), "Process Systems Analysis and Control", McGraw-Hill, Vol. 97, p 225.

Cox, J. B., Hellums, L. J., and Williams, T. J. (1966), "Algorithms for Direct Digital Control of Chemical Processes", IFAC, Session 4B, paper 43A.

Gallier, P.W., and Otto, R. E., (1969), "Self-tuning Computer Adapts DDC Algorithms", Progress in Direct Digital Control, ISA 1969, edited by T. J. Williams and F. M. Ryan, p 235.

Jazwinski, A. H. (1970), "Stochastic Processes and Filtering Theory", Academic Press, New York.

Joshi, S. (1973), "Digital Adaptive Control of High Order Plants Using Second Order Time Delay Models", Ph.D. Thesis, Systems Engineering Div., Rensselaer Polytechnic Institute, Troy, N.Y.

Kaufman, H. (1969), "Aircraft Parameter Identification Using Kalman Filtering", Proceedings of 1969 National Electronics Conference, Chicago, Dec. 1969, p 85.

Kaufman, H. (1970), "Kalman Filtering and Quasilinearization; A Comparative Discussion of Two Procedures for Parameter Estimation", 1970 International Conference on Information Theory, Noordwijk, The Netherlands.

Larson, D. (1969), "Quasilinearization Techniques" Proceedings of the 1969 National Electronics Conference, Chicago, Ill., p 95.

Sage, A. P. (1968), "Optimum Systems Control", Prentice-Hall, Inc.

Semmelhack, H. P. and Martins, H. R., (1963), "Optimal Control of Sampled Data Systems with Deterministic Inputs", Cornell Aeronautical Laboratory Report No. IH-1250-P-10.

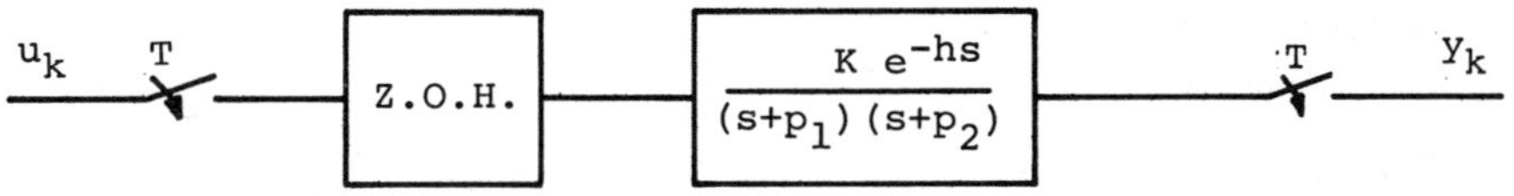

Fig. 1 - Model of Process to be Controlled

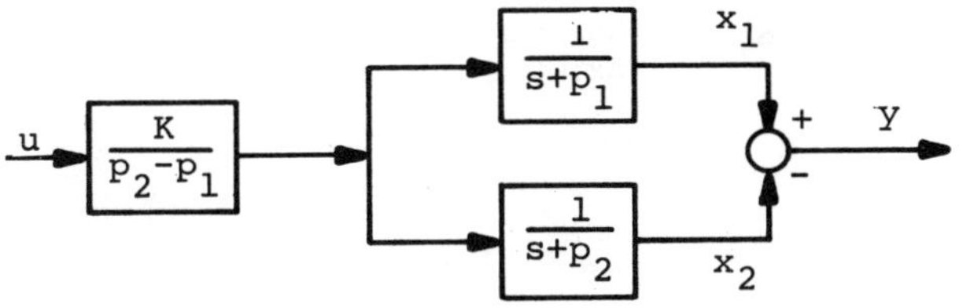

Fig. 2 - Jordan Canonical Form

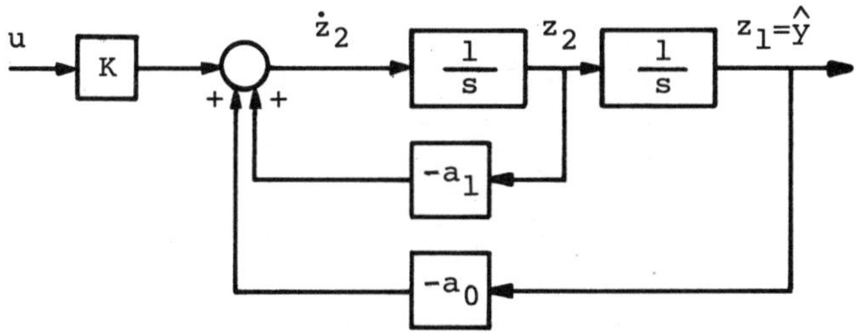

Fig. 3 - Phase-Variable Canonical Form

DEFINITION OF STATE VARIABLE

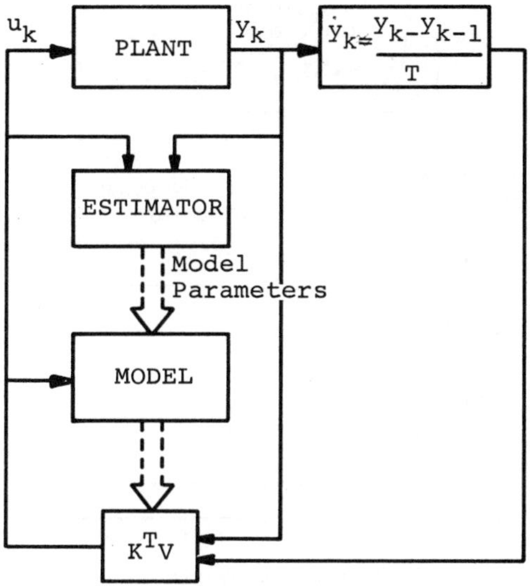

Fig. 5 - Closed-loop Control

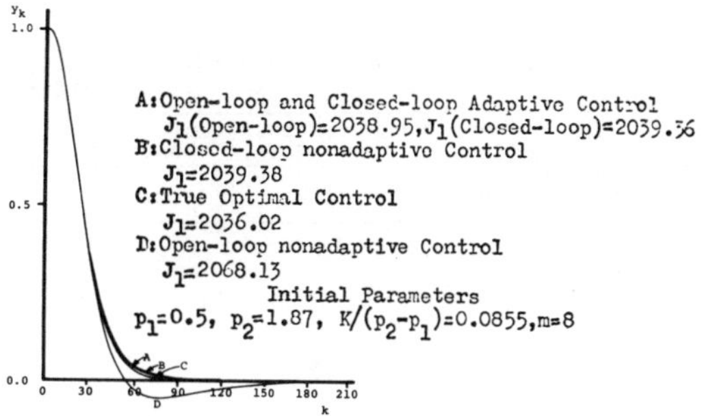

Fig. 6 Third Order Response for Sum Square Error Index

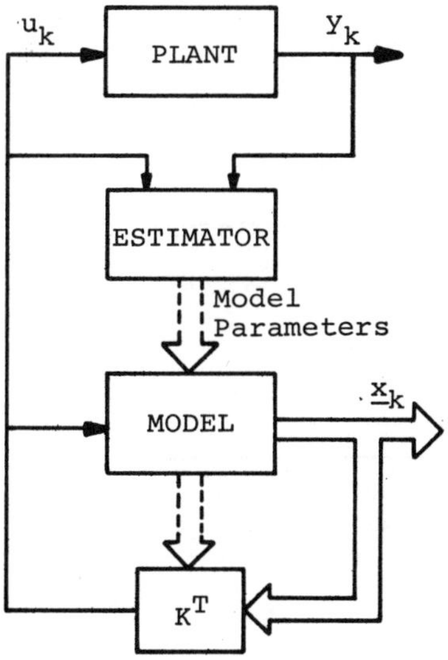

Fig. 4 - Open-loop Control

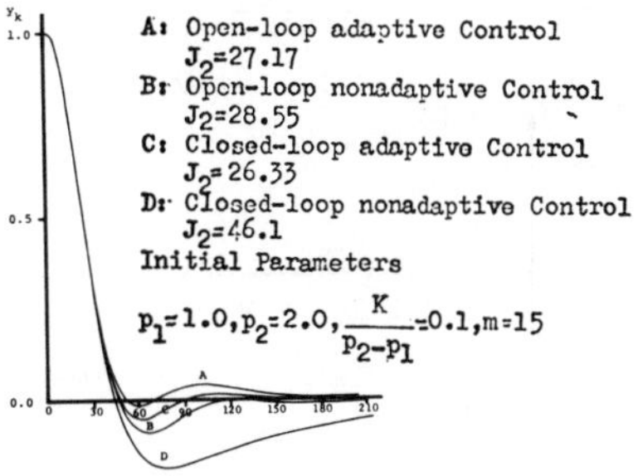

Fig. 7. Third Order Plant Response for summed absolute error index

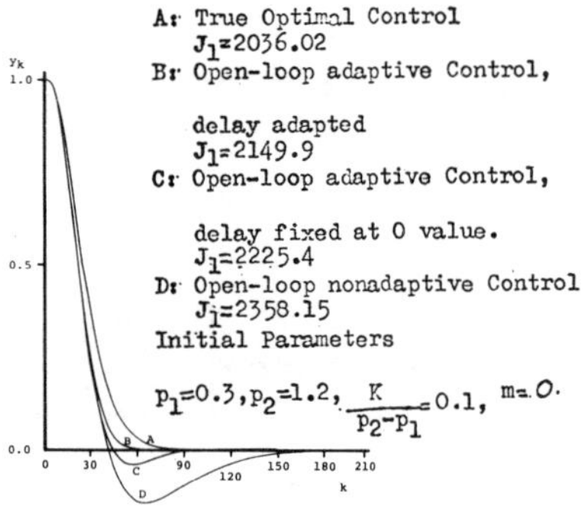

Fig. 8 Third Order Plant Response with and without delay

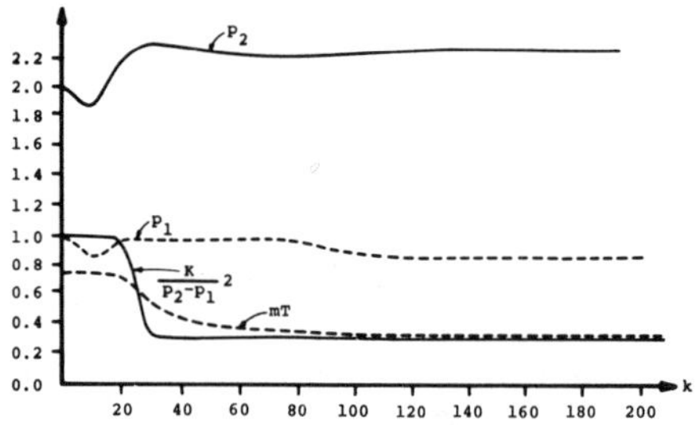

Fig. 9. Typical Parameter Variation for the Third-Order Plant Open loop Adaptive Case for Summed Absolute Error Index

Initial Model Parameters: $p_1=1.061, p_2=1.86$
$K/(p_2-p_1)=2.212, mT=0.465$

	Open-loop Nonadaptive	Open-loop Adaptive	Closed-loop Nonadaptive	Closed-loop Adaptive
Performance Function	33.78	33.43	35.02	34.25

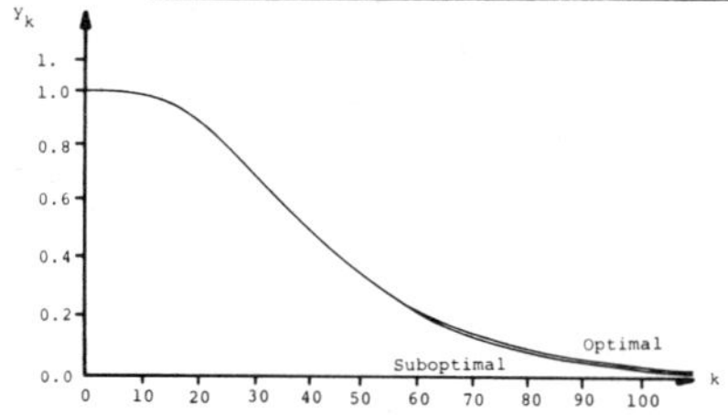

Fig. 10. Responses of the Fifth-Order Plant

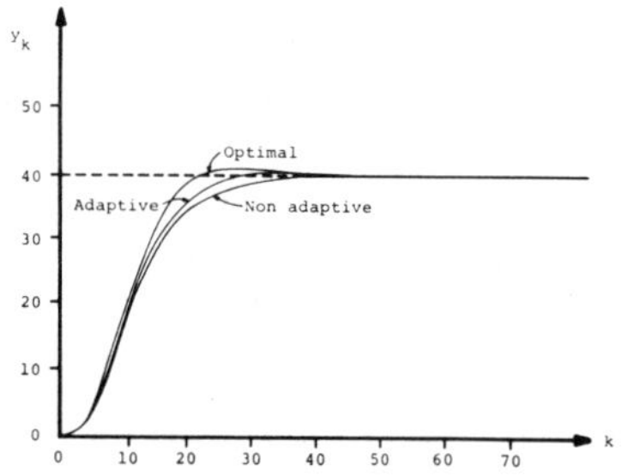

Fig. 11. Open-loop Control of 4th order type system

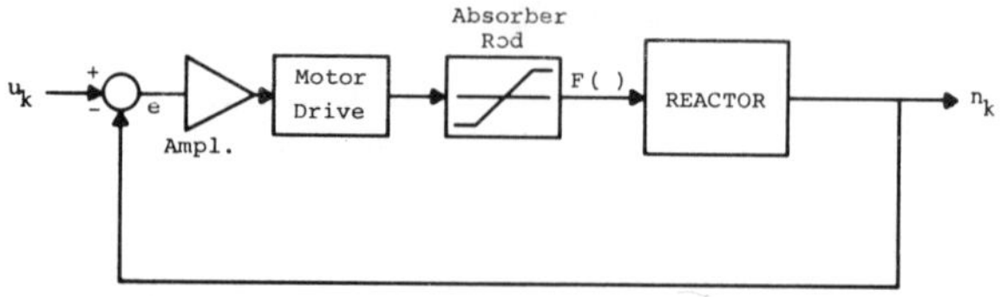

Fig. 12. Nuclear Reactor and the Existing Control System

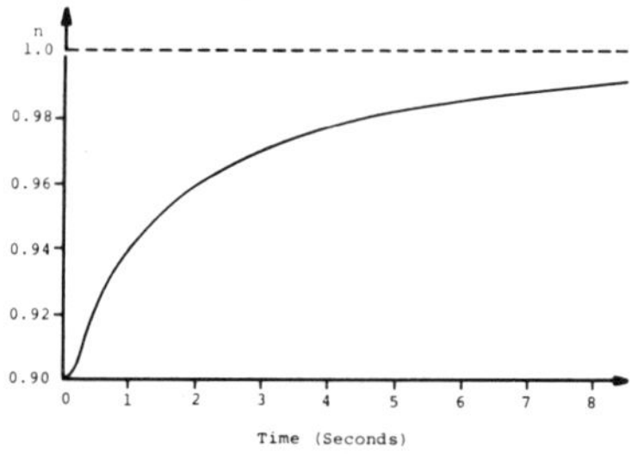

Fig. 13. Step Response of the Reactor for $T_c=0.0454$

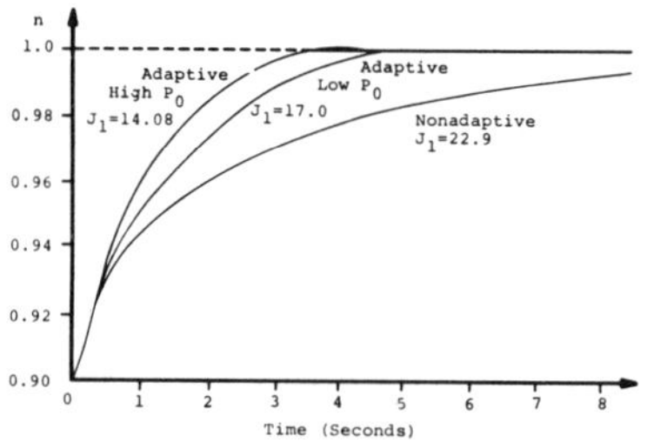

Fig. 14. Open-loop Control for $T_c=0.0454$

A DETERMINISTIC TYPE CONTROLLER FOR A CLASS OF UNCERTAIN PLANTS WITH PARTIAL IDENTIFICATION

P.N. Nikiforuk, M.M. Gupta, K. Kanai and L. Wagner
Systems and Adaptive Control Research Group
Department of Mechanical Engineering
University of Saskatchewan,
Saskatoon, Sask., Canada

Summary

This paper presents a deterministic design approach for controlling a class of unknown nonlinear multivariable plants opertaing in an uncertain environment. The approach uses a "two-level" control technique, with a feedforward controller on its first level, and a conditional feedback controller on its second-level. The procedure employs the Liapunov type signal synthesis procedure. By introducing a characteristic vector, which is characteristic of the unknown plant and its environment, the design of the controller bypasses the rigid requirements of plant identification as needed in the implementation of a controller using conventional techniques.

The applicability of this control approach is demonstrated by means of hybrid computer simulation studies which were carried out for an exothermic chemical reactor.

1. Introduction

In recent years much interest has arisen in the application of the Liapunov signal synthesis approach to the solution of large scale control problems [Grayson, 1965; Lindorff, 1967, 1972; Landau, 1972; Monopoli, 1966]. A large body of literature in this area is devoted to the "almost known situation", where information concerning the dynamics of the plant, the statistical properties of the uncertainties in the plant dynamics and measurements of the state variables [Grayson 1965; Landau, 1972; IEEE AC - 1971] is available. Most of these design aspects are not , however, applicable to many practical situations although, theoretically, they are expected to give superior performance, simply because most of this information is available fuzzily.

From a practical viewpoint, it is desirable that a simple design procedure exists for the design of feedback control systems from available information concerning the plant dynamics, its uncertainties and measurements, and which (i) guarantees a bounded input-bounded output stability of the system, or stability in the Lagrange sense of the overall controlled system; and (ii) yields the output response characteristics as desired by the designer.

A design technique employing the Liapunov signal synthesis procedure that is applicable to single-input single-output systems and which considers some of the above aspects has been reported previously[Nikiforuk,1969;Choe, 1971]. In it a class of unknown plants are made to track a stationary linear model. Recently, the authors have reported [Nikiforuk, 1973a] a design procedure applicable to a class of linear multivariable unknown plants in an uncertain environment. This synthesis procedure has employed a two-level controller technique [Preusche, 1972], with a feedforward controller on its first-level, and a conditional feeback controller on its second-level [Nikiforuk, 1973a]. The practical application of this synthesis procedure has been reported in a recent paper [Nikiforuk, 1973b].

The problem of concern in this paper is that of designing a two-level controller for a class of unknown, nonlinear, multi-input multi-output plants operating in an uncertain environment. This procedure employs the Liapunov synthesis approach via the introduction of a vector, called "the characteristic vector", which is the characteristics of the unknown plant and its environment. The system is expressed in the state-space form, with an unknown functional relationship. However, the design problem is treated from an input-output viewpoint. This design procedure, as will be shown, does not require an explicit knowledge of the plant or its uncertain ties, and the design approach may be viewed as a deterministic one. This bypasses some of the rigid restrictions of the stochastic control theory and the existing adaptive techniques.

The general control problem is formulated in Section 2. Section 3 describes the procedure for designing the two-level controller. In Section 4 some practical aspects of controller

---

This research is supported by the National Research Council of Canada, Grants A-5625 and A-1080, and the Defence Research Board of Canada. Grants 4003-02 and 9781-04.

implementation are discussed. Hybrid simulation results for an exothermic chemical reactor are presented in Section 5, and conclusions are given in Section 6.

2. Problem Formulation

The problem of concern is that of a general multivariable nonlinear plant described by

$$\dot{x}(t) = f_x\{x(t),\ w(t),\ u(t),\ t\},\ x(t)\varepsilon\ R^n \quad (1)$$

and with measurements,

$$y(t) = f_y\{x(t),\ u(t),\ t\},\ y(t)\varepsilon R^m \quad (2)$$

The plant is operating in an uncertain environment where $w(t)\varepsilon\ R^w$ represents the measurable and unmeasurable environmental (additive type) and the behavioral (multiplicative type) bounded disturbances on the plant. In (1) and (2), the functions $f_x\{\cdot\}$ and $f_y\{\cdot\}$ are, in general, unknown nonlinear and timevarying vector valued functions of their arguments.

The control problem to be solved for this unknown nonlinear plant is that of designing a controller of the form

$$u(t) = f_u\{y(t),\ y_m(t),\lambda(t),\ \gamma(t),t\},\ u(t)\varepsilon R^u \quad (3)$$

where $u(t)\varepsilon\ \mathbf{\Omega}$ is a bounded allowable set of controls, $\gamma(t)\ \varepsilon R^\gamma$ is a vector of reference inputs, and $\lambda(t)\varepsilon R^m$ is the characteristic vector to be defined later.

On the basis of the available output measurements $u(t)$ is such that it guarantees at least the Lagrange stability of (1) and which forces the measurable plant outputs $y(t)$ to track the desired response characteristics $y_m(t)\varepsilon R^m$ as closely as desired.

In general, existing type of controllers may be classified into two categories:
(i) one-level controllers, or controllers with one-degree of freedom as used in the stochastic optimal control theory, (ii) two-level controllers, or controllers with two-degrees of freedom as used in the conventional adaptive type controllers.

There are many deficiencies associated with the 1st type of controllers for, in general, they are unable to simultaneously cope with the following two problems [Horowitz, 1963],
(i) attaining the desired response characteristics as specified by the designer , and
(ii) achieving sufficient feedback to counteract inaccuracies in the mathematical modelling and environmental uncertainties.
To simultaneously attain these two control objectives, a controller with at least two-levels of control is necessary, [Horowitz, 1963; Preusche 1972].

The control structure proposed in this paper has a two-level control structure, with a feedforward path on its first-level, and a conditional feedback path on its second-level. The first-level controller is designed so as to satisfy a given set of dynamic response characteristics for an assumed characteristic model of the plant and its environmental uncertainties. The second-level controller is designed to cope with the unknown characteristics or uncertainties of the plant and its environment. This work does not use conventional adaptive techniques* in the synthesis procedure, rather it employs a conditional feedback technique to cope with the unknown factors that are associated with the plant and its environment.

3. Synthesis of Two-Level Controller

A general form of the control law is given in (3). The equivalent control law in the two-level structural form may be written as a linear combination of the feedforward control $u_m(t)\varepsilon R^{um}$ on its first-level, and the conditional feedback control $u_c(t)\varepsilon R^m$ on its second-level,

$$u(t) = K_m u_m(t) + K_c u_c(t),\ u(t)\varepsilon R^m \quad (4)$$

where K_m and K_c are linear gain matrices in the controller, of compatitable dimensions which are to be determined. The design procedure for the feedforward and conditional feedback control laws will now be illustrated,

(a) First-Level Feedforward Control Law, $u_m(t)\varepsilon R^{um}$

For the design of the first-level feedforward controller a characteristic model representing certain opertaing charateristics of the unknown plant and its uncertainties is desired. The selection of the mathematical model is mainly at the discretion of the designer and may be chosen so as to represent certain input-output operating characteristics of the plant. Further, for the purpose of achieving certain design objectives for the controlled plant the characteristic model may be selected to be a linear time-invariant model with no uncertainties. In other words, the selection of the characteristic model will depend upon the control objectives and the simplicity of the design. For the purpose of this work, let a time-invariant characteristic model of the plant, which is controllable and observable, be given by

---

* The phrase "adaptive technique" has been used in the literature in the sense of identification and feedback on the basis of this identification. This is achieved by introducing a measurable characteristic vector which represents the characteristic of the unknown plant and its environment.

$$\dot{x}_m(t) = A_m x_m(t) + B_m u_m(t),\ x_m(t)\ \varepsilon R^m \tag{5}$$

where m, the order of the characteristic model, $\leq$ n, the order of the plant. The control law,

$$u_m(t) = u_m\{x_m(t), \gamma(t), t\} \tag{6}$$

may be designed using conventional (classical or optimal) control theory so as to satisfy the desired output response characteristics, $y_m(t)\varepsilon R^m$

$$y_m(t) = c_m x_m(t),\quad y_m(t)\varepsilon R^m \tag{7}$$

where each element of the vector $y_m(t)$ has the same physical meaning as the corresponding elements of the plant output vector y(t). In the present work, it is assumed that the order of the model is equal to the order of the vector $y_m(t)$, the desired response characteristics, however, using the concept of "pseudoinverse", one may proceed with the design without this assumption.

In general, $u_m(t)$ is a linear control law of the form

$$u_{\gamma m}(t) = F_m x_m(t) + F_\gamma \gamma(t) \tag{8}$$

where $\gamma(t)\varepsilon R^\gamma$ is the vector of reference inputs, and F_m and F_γ are constant linear gain matrices of compatible dimensions. Since the model chosen is asymptotically stable, it implies that

$$Re(\sigma_i)<0,\ i = 1,2,\ .\ .\ .\ m$$

where σ_i are the characteristic roots of the matrix $(A_m + B_m F_m)$. It will be shown that the control $u_m(t)$ given in (8) is used to control the plant in a feedforward mode.

(b) Second-Level Conditional Feedback Control Law, $u_c(t)\varepsilon R^m$

The configuration of the second-level conditional feedback controller will depend upon the essential properties of the plant, its uncertainties and the assumed characteristic model. This feedback controller is viewed as a conditional controller in the sense that it is operative only an error exists between the plant output and the output of the assumed characterisitc model. Thus, if the unknown plant and the model are dynamically equivalent with the same initial conditions, $u_c(t) = 0$ and $u(t) = K_m u_m(t)$; thus, the model control $u_m(t)$ will be able to control the plant in the feedforward mode. This dynamic equivalence of the plant and that of the model may be asserted by defining the plant tracking error as follows. With the plant output vector $y(t)\varepsilon R^m$ as given in (2), and the model output vector $y_m(t)\varepsilon R^m$ as defined by (7), the plant tracking error is

$$e(t) = y_m(t) - y(t),\ e(t)\varepsilon R^m \tag{9}$$

The derivation of the conditional control structure is given below. As will be seen, it depends upon the following assumptions for the plant characteristics:

Assumption 1, Let the essential dynamic behavior of the plant defined in (1) and (2) be such that the plant is controllable and observable, and that the input and output properties of the plant be given by the following vector differential equation

$$\dot{y}(t) = g\{y(t),\ u(t), w(t), t\}+D_o u(t),\ y(t)\varepsilon R^m \tag{10}$$

where the exact form of $g\{\cdot\}$ is not necessarily known, $w\varepsilon W$, a bounded set of disturbances, are the plant uncertainties, and D_o is a known matrix. The choice of D_0 is not very critical except that the rank of D_0 should be the same as the order of the control vector u(t).

The overall control objective for the unknown plant defined in (10) may now be formally stated as follows:

"Given the plant input-output characteristics as defined in (10), with unknown functions $g\{\cdot\}$, and the characteristic model defined in (5), synthesize a control law $u(t)\varepsilon U$ of the form given in (3) so as to force the plant tracking error e(t) to lie within some bounded region $||e(t)||\varepsilon \mathbf{E}$ in the neighborhood of the origin in the m-dimensional error space". In the limit, by making a judicious choice for the controller parameters, e(t) may be forced asymptotically to the origin making the set $\mathbf{E}\to 0$.

For the purpose of synthesizing the controller, the error differential equation is written as,

$$\dot{e}(t) = \dot{y}_m(t) - \dot{y}(t) \tag{11}$$

Substitution for $\dot{y}_m(t)$ and $\dot{y}(t)$ from (5)(7) and (10) yield

$$\dot{e}(t) = C_m A_m C_m^{-1} y_m(t) + C_m B_m u_m(t) - g\{y(t), w(t), u(t), t\} - D_0 u(t)$$

or eqivalently, the output error equation may be written as

$$\dot{e}(t) = A_e e(t) + D_m u_m(t) + A_e y(t) - g\{y, w, u, t\} - D_o u(t) \tag{12}$$

where $A_e = C_m A_m C_m^{-1}$, $D_m = C_m B_m$.

Equation (12) embodies all the desired response characteristics plus the known and unknown characteristics of the plant. The overall control objectives for (1) may now be directed at (12) in order to design u(t) so as to force the e(t) trajectory into the proximity of the origin. Since (12) possesses the inherent unknown characteristics of the plant, there is no control theory which can be applied to (12) to make the equation stable. To circumvent this difficulty, a vector function $\lambda(t)$, called the characteristic vector, which is characteristic to the unknown plant and its environment, is introduced. Defining $\lambda(t)$ as

$$\lambda(t) = A_e y(t) - g\{y,w,u,t\},\ \lambda(t)\varepsilon R^m \quad (13)$$

(12) may be written as

$$\dot{e}(t) = A_e e(t) + D_m u_m(t) + \lambda(t) - D_o u(t) \quad (14)$$

This output error differential equation is a function of the known parameters in A_e, D_m and D_o, the measurable variables $e(t)$, $u_m(t)$, $u(t)$, and an unknown vaiable $\lambda(t)$. If $\lambda(t)$, which is an explicit function of the error, and contains all of the information which pertains to the uncertainties of the plant as defined in (13), is generated by some indirect means, (14) can be viewed as a deterministic equation. Thus, employing the relation of (14), $\lambda(t)$ may be generated as,

$$\lambda(t) = \dot{e}(t) - A_e e(t) - D_m u_m(t) + D_0 u(t) \quad (15)$$

In the implementation of the controller, this characteristic vector equation will be used in the generation of $\lambda(t)$.

Controller Synthesis

In order to accomplish the overall stability of the output differential equation (14), the Liapunov synthesis procedure is employed. For this purpose a positive definite scalar function $V(e,t)$ of the tracking error is selected. By selecting a suitable control $u(t)$, the time derivative $\dot{V}(e,t)$ along the error trajectory may be made negative definite outside some regions $||e|| < \mathbf{E}$ assuring the Lagrange stability of (14).

To illustrate the design procedure, define a scalar functional* $V(t) = V(e,t)$ as

$$V(e,t) = e'(t) P e(t) \quad (16)$$

where P is a real positive definite symmetric matrix. Taking the time derivative of $V(e,t)$, and the substitution for $\dot{e}(t)$ from (14) yields

$$\dot{V}(e,t) = \dot{e}'(t) P e(t) + e'(t) P \dot{e}(t) = e'(t)[A_e' P + P A_e] e(t) + 2M(t)$$

where $M(t) = M(u(t), u_m(t), \lambda(t), t)$ is a scalar functional defined as (17)

$$M(t) = e'(t) P[D_m u_m(t) + \lambda(t) - D_0 u(t)] \quad (18)$$

The Lagrange stability outside certain bounded region $\mathbf{E}$ is guaranteed if $V(t)$ is a Liapunov function in that region. The condition for $V(t)$ to be a Liapunov function is assured by the following theorem.

Theorem 1

Consider the output error differential equation (14) with forcing function $\{\lambda(t) + D_m u_m(t) - D_0 u(t)\}$ where $\lambda(t)$ is the characteristic vector defined in (15). For this equation, let $V(e,t)$, defined in (16), be a scalar functional with its time derivative given by

$$\dot{V}(e,t) = -e'(t) Q e(t) + 2M(e,t) \quad (19a)$$

where for a stable A_e, $Q = (A_e' P + P A_e)$ and P are positive definite matrices and $M(e,t)$, is defined in (18).

Let $u(t) = u\{e(t), u_m(t), \lambda(t), t\}$ be the control policy given * by

$$u(t) = K_m u_m(t) + K_c u_c(t) \quad (19b)$$

with $K_m = D_0^{-1} D_m$, $K_c = D_0^{-1}$, and $u_m(t)$ as defined in (8) and

$$u_c(t) = G_1 \lambda(t) \psi\{G_2 \sigma(t)\} \quad (19c)$$

where $G_1 \geq 1$ and $G_2 > 0$ are scalar gain constants,

$$\sigma(t) \triangleq e'(t) p \lambda(t),\ \sigma(t)\varepsilon R^1 \quad (19d)$$

and $\psi\{\cdot\}$ is a scalar nonlinear functional having the following properties:

(i) $\psi\{0\}=0$

(ii) $\dfrac{\psi(\sigma_2) - \psi(\sigma_1)}{\sigma_2 - \sigma_1} \geq 0$ for all real σ_1 and σ_2, $\sigma_1 \neq \sigma_2$

(iii) the functional $\psi\{\cdot\}$ is bounded by a constant $1 \leq \beta \leq \beta_0 < \infty$, that is, there exists a positive constant $\beta \geq 1$ such that $|\psi\{\sigma\}| \leq \beta$ for all real σ.

Then. for a stable A_e, $V(e,t)$ is a Liapunov function whenever $||e(t)|| > \rho$, ρ is a small real constant, yielding Lagrange stability of the output tracking error equation (14) outside the hypersphere of radius ρ.

[Proof]: Using (18) and (19), $M(t) = M(\sigma,t)$ can be written as

$$M(\sigma,t) = \sigma(t)[1 - G_1 \psi\{G_2 \sigma(t)\}] \quad (20)$$

where $M(\sigma,t)$ is a continuous mapping from TxR^m into R^1 for all $t\ \varepsilon [t_0, \infty)$.

Let $s = T \times R^q$, and define s_-, s_{+A} and s_{+B} to be mutually disjoint sub-sets of the set S such that

$$S = S_- \cup S_{+A} \cup S_{+B} \quad (21a)$$

where

$$S_- = \{(t,e); \sigma(t) \leq 0,\ (t,e)\varepsilon S\} \quad (21b)$$

$$S_{+A} = \{(t,e); \sigma(t) > 0,\ 0 < G_1 \psi\{G_2 \sigma(t)\} < 1,\ (t,e)\varepsilon S\} \quad (21c)$$

$$S_{+B} = \{(t,e); \sigma(t) > 0,\ 1 < G_1 \psi\{G_2 \sigma(t)\} \leq G_1 \beta < \infty, (t,e)\varepsilon S\} \quad (21d)$$

$G_1, \beta \geq 1$

---

* To illustrate the basic synthesis procedure, a simplest form of V-functional is selected. To improve the transient response of (14), however, a judicious choice of V is required, and is a subject for further investigations.

* If D_0 is a non-square matrix having full rank, then one can extend this approach using the concept of the pseudo inverse.

If $(t,e) \varepsilon S_- \cup S_{+B}$, and noting that $G_1 \geq 1$, $G_2 > 0$, and $\mathrm{sign}[\Psi\{G_2\sigma(t)\}] = \mathrm{sign}[\sigma(t)]$, it can be seen from (20) that

$$M(t) \leq 0$$

Thus

$$\dot{V}(e,t) \leq e'(t)Qe(t) < 0, \quad (t,e) \varepsilon S_- \cup S_{+B}, \quad (22)$$

Therefore, in the region $(t,e) \varepsilon S_- \cup S_{+B}$, $V(e,t)$ is a Liapunov function yielding asymptotic stability with $e \to 0$ as $t \to \infty$.

If $(t,e) \varepsilon S_{+A}$, then since $\sigma(t) > 0$

$$M(\sigma,t) = \sigma(t)[1-G_1\Psi\{G_2\sigma(t)\}] > 0$$

This yields

$$\dot{V}(e,t) = -e'(t)Qe(t) + 2M(\sigma,t)$$

$$\leq -C_0\|e\|^2 + 2\sigma(t) \quad (23)$$

where $C_0\|e\|^2 = \inf[e'(t)Qe(t)]$, $(t,e) \varepsilon S_{+A}$, which implies that

$$\dot{V}(e,t) < 0 \text{ whenever } C_0\|e\|^2 > 2\sigma, \forall (e,t) \varepsilon S_{+A} \quad (24)$$

with maximum bounds on $\|e\|$ given by

$$\|e\|^2_{max} = \frac{2\sigma_{max}}{C_0}, \quad 0 < G_1\Psi\{G_2\sigma(t)\} < 1$$

$$= \frac{2}{G_2C_0}\Psi^{-1}(1/G_1) < \rho^2 \quad (25)$$

Thus the relations (22), (24) and (25) prove the theorem.

The following corollary establishes the maximum bounds on the tracking error.

Corollary 1

Let Ω be a subset of S defined as

$$\Omega = \{(t,e), \|e\|^2 < \frac{2}{G_2C_0}\Psi^{-1}\left(\frac{1}{G_1}\right) = \rho^2, (t,e) \varepsilon S\} \quad (26)$$

Then $\dot{V}(e,t) \leq 0$ on $\bar{\Omega}$, the complement of Ω relative to the set S.

[Proof]: This proof follows directly from the relations (24) and (25).

Comments:

(i) It is seen from (25) and (26) that the region $\mathbf{E}$ can be made arbitrarily small by choosing the nonlinearity $\Psi\{\cdot\}$ suitably and by making the gain constants G_1 and G_2 sufficiently large. In the limit, if $G_2 \to \infty$ (a bang-bang controller), $\|e\|_{max} \to 0$ implying that ρ, $\mathbf{E} \to 0$ yielding an asymptotic stability of the error differential equation (14)

(ii) As seen from the relations (15) and (19), and Figure 1, $\lambda(t) = \lambda(u,e,t)$ and $u(t) = u(e,\lambda,t)$, the controller itself has a feedforward path. This requires the boundedness of $u(t)$ and $\lambda(t)$, which is difficult to prove theoretically, however, it has been shown by means of simulation results of a large class of stable and unstable plants (both linear and nonlinear) that $\lambda(t)$ and $u(t)$ remain bounded.

4. Controller Implementation

For the unknown plant defined in (1), with available measurements $y(t)$ given in (2), the structural form of the two-level controller is given by (4) and (19c). Theorem 1 suggests an admissible control law $u(t) = u(u_m,u_c,t)$ of the form given in (19), this control law guarantees the Lagrange stability, and forces the plant to track the desired response characteristics $y_m(t)$ with some bounded error in the region Π.

The control law is

$$u(t) = D_0^{-1}D_m u_m(t) + D_0^{-1}G_1\lambda(t)\Psi\{G_2 e'(t)P\lambda(t)\} \quad (27)$$

where $G_1 \geq 1$, $G_2 > 0$ are scalar gain constants, and $\lambda(t)$, the characteristic vector, is generated using the relation

$$\lambda(t) = \dot{e}(t) - A_e e(t) + u_c(t) \quad (28)$$

where $u_c(t) = G_1\lambda(t)\Psi\{G_2 e'(t)P\lambda(t)\}$, the conditional control. The structural configuration of the controller is shown in Figure 1, with controller implementation in Figure 2.

Note that in the implementation of the controller an explicit knowledge of the process dynamics is not required except for a judicious choice of D_0, the nominal values of the plant input matrix. As discussed previously, however, the parameters of D_0 may be selected arbitrarily subject to the following constraint

$$\mathrm{Rank}[D_0] = m, \text{ the order of } u(t) \quad (29)$$

Theorem 1 specifies certain conditions on the nonlinear function $\Psi(\sigma) = \Psi\{G_2 e'(t)P\lambda(t)\}$ in the implementation of the conditional feedback control law $u_c(t)$. This function may be realized in various ways, and Table 1 summarizes some possible types of controllers.

Although extensive simulation studies have shown the boundedness of $\lambda(t)$ and $u(t)$, there may be a few extreme cases, for example in the case of noisy measurements, when the variables may exhibit unbounded behavior. Under these situations one may use a saturation type nonlinearity, with acceptable maximum bounds, at the input to the process, without effecting the results of Theorem 1. Some of the practical aspects of the controller implementation may be found in Nikiforuk [1969] and Choe[1971].

5. Hybrid Simulation Results - Control of an Exothermic Chemical Reactor

In order to demonstrate some of the important aspects and useful applications of the two-level controller developed in this paper, some hybrid simulation results obtained for a stirred tank exothermic chemical reactor [Aris, 1958, Lindorff, 1967] will now be presented.

Plant Description

For this chemical reactor, the process may be described as a continuous first-order exothermic chemical reaction. Referring to Figure 3, the output variables are defined as the concentration $x_1(t)$ of a particular component C of the outflow, and the temperature $x_2(t)$ of the tank. The control inputs are the inflow concentration $u_1(t)$ of the component C, and the flow rate of the coolant $u_2(t)$. Based upon the assumption of ideal mixing, the concentration $x_1(t)$ and the temperature $x_2(t)$ are taken to be uniform throughout the tank.

The equations characterizing the process are derived in [Aris, 1958] from heat and mass relationships. Assuming constant volume (inflow = outflow), the heat-mass balance equations in normalized form are

$$\dot{x}_1(t) = -[1+g(x_2,t)]x_1(t) + u_1(t)$$

$$\dot{x}_2(t) = g(x_2,t)x_1(t) - [x_2(t)-x_0+w_1(t)][1+u_2(t)+w_2(t)] \quad (30)$$

where x_0 = constant temperature of inflow coolant,

$g(x_2,t)$ = reactor-rate term which is a nonlinear function of the temperature $x_2(t)$,

$w_1(t)$ = temperature disturbance at the inflow, and

$w_2(t)$ = input disturbance at the coolant rate flow.

Using the numerical values cited in Aris [1958], it will be assumed that

$$x_0 = 1.75, \text{ and } g(x_2) = \exp(25-50/x_2(t))$$

The process equation (30) has a strong nonlinear intercoupling by virture of the reaction term $g(x_2)$. Furthermore, due to its exothermic nature, the uncontrolled process is inherently unstable. For example, for the nominal values of $u_1 = u_2=1$,

the uncontrolled process can be shown to have an unstable equilibrium at $x_1 = 0.5$ and $x_2 =2$.

The output variables for this plant are defined as $y(t) = [\mathbf{I}]\, x(t),\ y(t)\varepsilon R^2$ (31)

The desired input-output characteristics of the process are assumed to be represented by the following uncoupled second-order characteristic model,

$$\dot{x}_m(t) = A_m x_m(t) + B_m u_m(t), u_m,\ x_m \varepsilon R^2 \quad (32)$$

where

$$A_m = \begin{bmatrix} -0.5 & 0 \\ 0 & -0.5 \end{bmatrix}, \qquad B_m = \begin{bmatrix} 1 & 0 \\ 0 & 1 \end{bmatrix}$$

with $u_m(t) = [1\ 1]'\ \gamma(t),\ \ \gamma(t)\varepsilon R^1,\ u_m(t)\varepsilon R^2$

and $y_m(t)=C_m x_m(t),\ C_m = [\mathbf{I}],\ y_m \varepsilon R^2$ (33)

Controller Design

Following the design procedure given in Sections 3 and 4, the important parameters of the controller employed in its implementation are given below:

$$Q = P = \begin{bmatrix} 1 & 0 \\ 0 & 1 \end{bmatrix}, \qquad D_0 = [\mathbf{I}],$$

(a) The first-level controller, $u_m(t)$:

$u_m(t) = [1\ 1]'\gamma(t)$,

$\gamma(t)$ is the reference input, $\gamma(t)\varepsilon R^1$

(b) The second-level controller, $u_c(t)$:

$$A_e = C_m A_m C_m^{-1} = A_m,\ D_m = C_m B_m = [\mathbf{I}],$$

$$\lambda(t) = \dot{e}(t) - A_m e(t) + u_c(t)$$

$$u_c(t) = G_1 \lambda(t)\psi\{G_2 e(t) P\lambda(t)\},\ G_1 \geq 1,\ G_2 > 0$$

Note that the design of the first-level and second-level controllers is independent of the parameters of the chemical reactor to be controlled.

This problem was studied on a hybrid computer, with chemical reactor simulated on the analog part, and with on-line controller implemented on the digital part of the computer. Some typical results which were obtained from these studies will now be described.

The performance of the controlled process was studied using a step reference input. Figure 4 shows the results obtained employing the bang-bang nonlinearity in the implementation of the second-level controller, and subjected to an input disturbance. It may be noted that the process output $y_1(t)$ and $y_2(t)$ follow the desired output variables with very small tracking errors $e_1(t)$ and $e_2(t)$, and that the control variables are bounded.

Figure 5 shows the performance of the process with a saturation type nonlinearity in the implementation of the second-level controller. It may be noted that the tracking performance characteristics of the process are improved by increasing the gain constants G_1 and G_2, and by choosing the parameters of the matrix P. In Figure 5(a), with small G_1 and G_2, the tracking performance is very poor, but it was improved considerably by increasing the values of G_1 and G_2 as shown in Figure 5(b).

These simulation results demonstrate the control capabilities of the two-level controller in controlling an unknown nonlinear plant in the presence of input disturbances.

6. Conclusions

A deterministic type two-level controller consisting of a feedforward controller and a conditional feedback controller has been described for controlling a class of unknown nonlinear multivariable plants in an uncertain environment. The implementation of the controller is relatively simple, and does not require an explicit knowledge of the plant dynamics or its uncertainties. The example that is given in this paper is very important in the chemical process industry, the conclusions drawn from the simulation studies of this inherently unstable and interacting process demonstrate that the control approach may prove to be effective in controlling large scale industrial processes.

The design procedure outlined in this paper assumes noise free measurements. Noisy measurements, if they exist, may create some problems in the generation of the characteristic vector and hence in the implementation of the controller. This may cause some stability problem in the tracking.

The above conclusions are mainly drawn from intuitive theoretical developments and extensive hybrid simulation studies of various examples, however, more decisive conclusions are subject to further investigations on a physical process. Some investigations in this direction including the theoretical studies on the judicious selection of Liapunov function in the design of the controller, and the effect of measurement noise and controller imperfection on control performance are underway.

References

(1) Aris, R., and Amundson, N., (1958), "On Analysis of Chemical Reactor Stability and Control, I and II," Chem. Eng. Sci., Vol. 7, No. 3, PP. 121-147.

(2) Choe, H.H., and Nikiforuk, P.N., (1971), "Inherently Stable Feedback Control of a Class of Unknown Plants," Automatica, Vol. 7, pp. 607-625.

(3) IEEE Trans. on Automatic Control, Special Issue, "On the Linear-Quadratic-Gausian Estimation and Control Problem" (1971), Vol. AC-16, No.6.

(4) Grayson, L.P., (1965), "The Status of Synthesis Using Liapunov's Method", Automatica, Vol. 3, PP. 91-121.

(5) Horowitz, I.M., (1963), "Synthesis of Feedback System", Academic Press, New York.

(6) Landau, I.D., (1972), "Model Reference Adaptive Systems-A Survey -what is possible and why?", Trans. of the ASME, June, PP. 119-133.

(7) Lindorff, D.P., (1967), "Control of Nonlinear Multivariable Systems", IEEE Trans. on Automatic Control, Vol. AC-12, No. 5, PP. 506-515.

(8) Lindorff, D.P., and Carroll, R.L., (1972), "Survey of Adaptive Control Using Liapunov Design", TR-72-7, the University of Conneticut.

(9) Monopoli, R.V., (1966), "Engineering Aspects of Control System Design via the Direct Method of Liapunov", NASA Contractor Rept., CR-564, Dec.

(10) Nikiforuk, P.N., Gupta, M.M. and Choe, H.H. (1969), "Control of Unknown Plants in Reduced State Space", IEEE Trans. on Automatic Control, Vol. AC-14, Oct. PP. 489-496.

(11) Nikiforuk, P.N., Gupta, M.M., Kanai, K. and Wagner, L., (1973a), "Synthesis of Two-level Controller for a Class of Linear Plants in an Unknown Environment" To be presented at the 3rd IFAC Symposium on Sensitivity, Adaptivity and Optimality, Italy.

(12) Nikifouk, P.N., Gupta, M.M. and Kanai, K. (1973b), "On-line Two-Level Gust Alleviation Control System for Aircraft in an Unknown Environment", To be presented at the 5th IFAC Symposium on Automatic Control in Space, Italy.

(13) Preusche, G. (1972), Two-level Model Following Control System and its Application to the Power Control of a Steam-cooled Faster Reactor, Automatica, Vol. 8, No. 2, PP. 143-151.

TABLE 2.1

REALIZATION OF NONLINEAR FUNCTION $\psi\{\cdot\}$

$$u_c(t) = G_1\lambda(t)\psi\{G_2\sigma(t)\}, \sigma(t) = e^T(t)P\lambda(t) \in R^1$$

$$G_1 \geq 1, \quad G_2 > 0$$

Types of Controller		Comments on Stability in (t-e) Space
(1) Bang-Bang $\psi\{G_2\sigma(t)\} = \text{Sign}\{\sigma(t)\}$	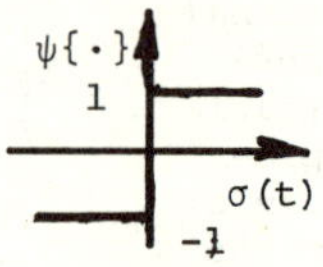	Asymptotic Stability is guaranteed **E** =0
(2) Saturation $\psi\{G_2\sigma(t)\} = \text{Sat}\{G_2\sigma(t)\}$	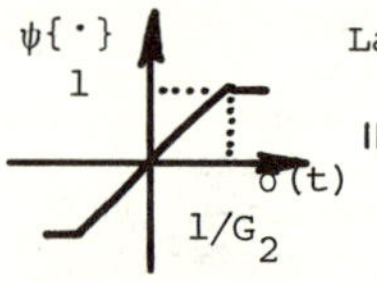	Lagrange stability (with bounded error) $\|e\|^2_{max} = \frac{1}{2C_0G_1G_2}$
(3) Hyperbolic Tangent $\psi\{G_2\sigma(t)\} = \tanh\{G_2\sigma(t)\}$	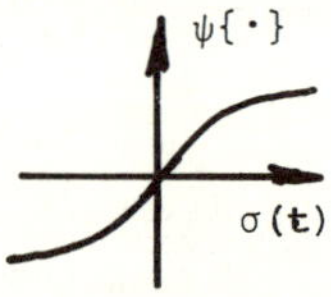	Lagrange stability (with bounded error) $\|e\|^2_{max} = \frac{2}{C_0G_2}\tanh^{-1}\left(\frac{1}{G_1}\right)$
(4) Arctangent $\psi\{G_2\sigma(t)\} = \tan^{-1}\{G_2\sigma(t)\}$	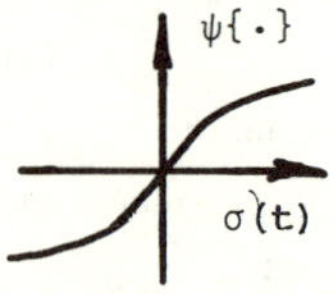	Lagrange stability (with bounded error) $\|e\|^2_{max} = \frac{2}{C_0G_2}\tan\left(\frac{1}{G_1}\right)$

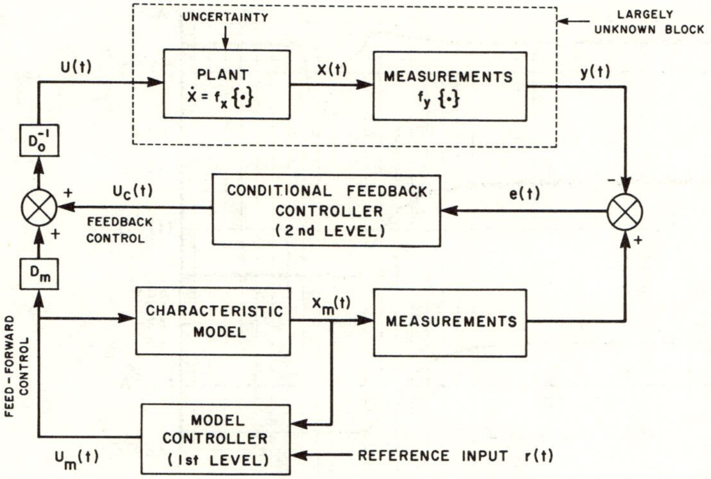

Figure 1 TWO - LEVEL CONTROLLER FOR A CLASS OF UNKNOWN PLANTS

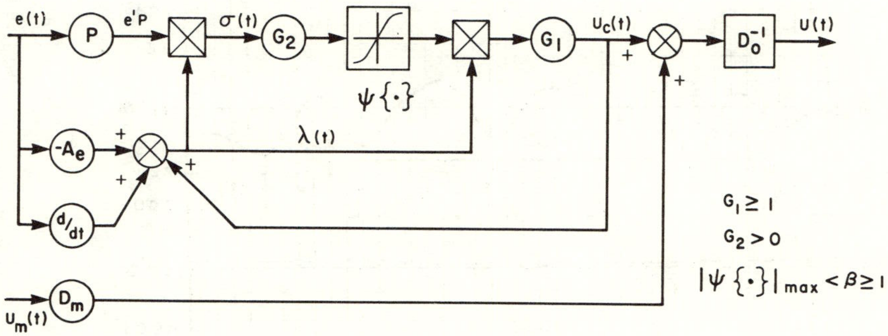

Figure 2 CONTROLLER IMPLEMENTATION

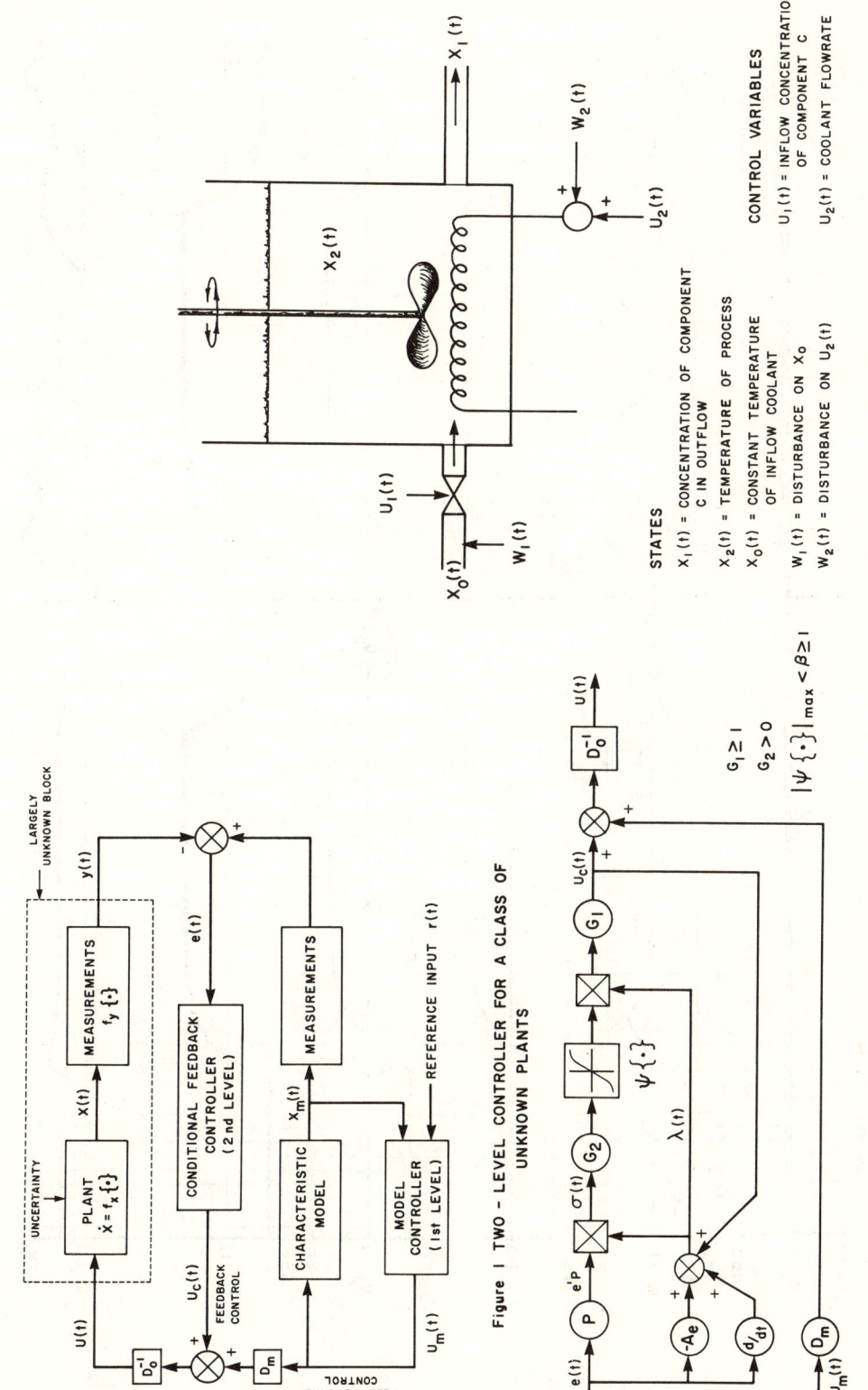

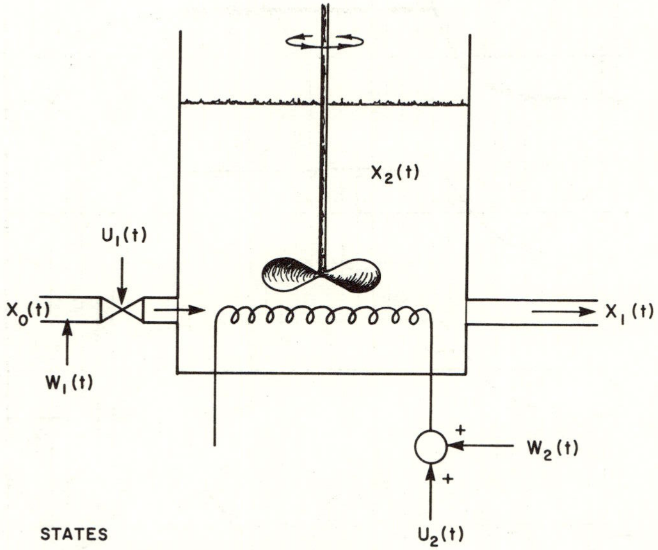

STATES

$X_1(t)$ = CONCENTRATION OF COMPONENT C IN OUTFLOW

$X_2(t)$ = TEMPERATURE OF PROCESS

$X_0(t)$ = CONSTANT TEMPERATURE OF INFLOW COOLANT

$W_1(t)$ = DISTURBANCE ON X_0

$W_2(t)$ = DISTURBANCE ON $U_2(t)$

CONTROL VARIABLES

$U_1(t)$ = INFLOW CONCENTRATION OF COMPONENT C

$U_2(t)$ = COOLANT FLOWRATE

Figure 3 EXOTHERMIC CHEMICAL REACTOR

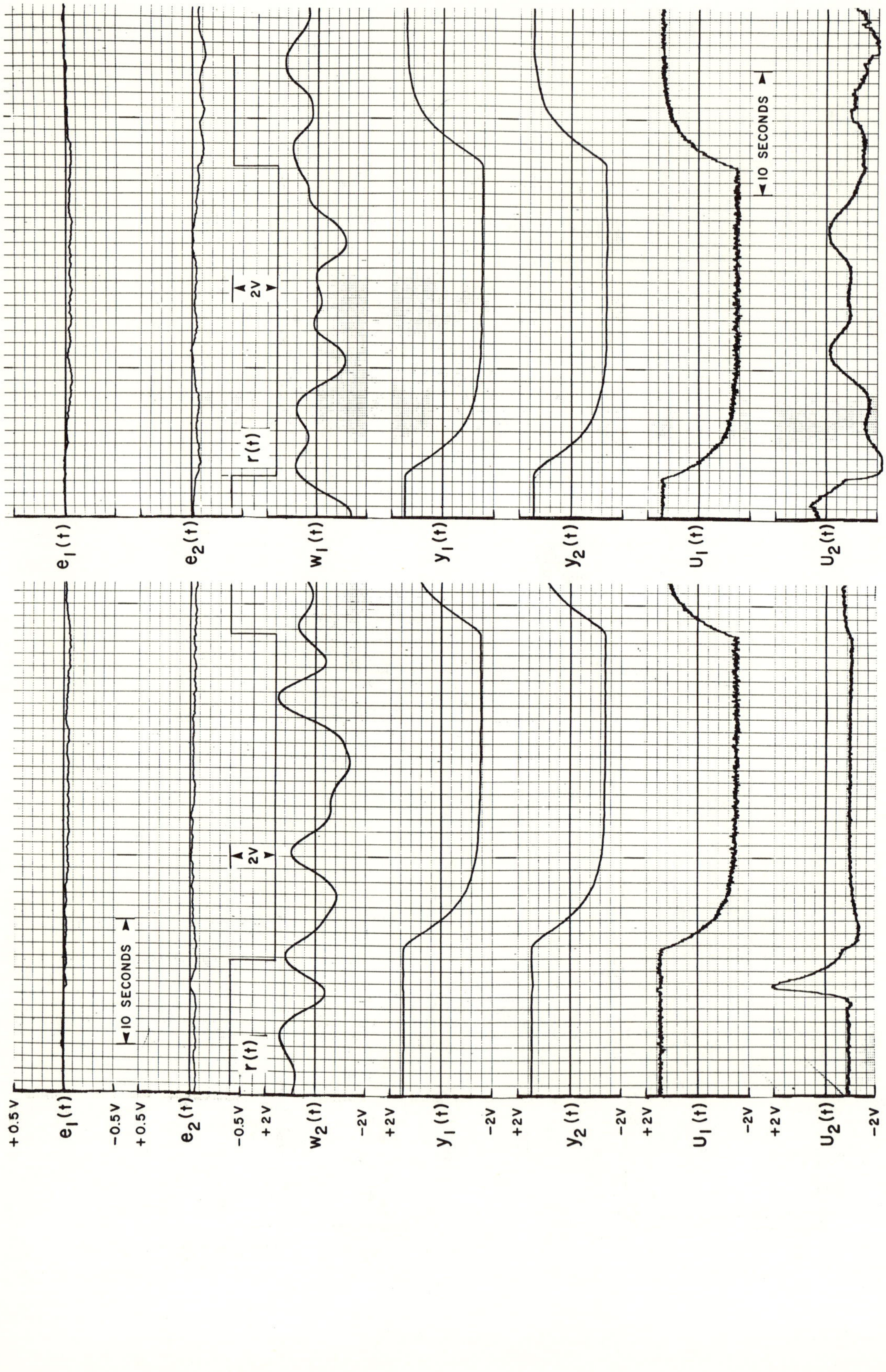

(a) $w_1(t) = 0$, $w_2(t)$ = GAUSSIAN DISTURBANCE

(b) $w_1(t)$ = GAUSSIAN DISTURBANCE, $w_2(t) = 0$

Figure 4 BANG - BANG CONTROLLER PERFORMANCE WITH STEP INPUT

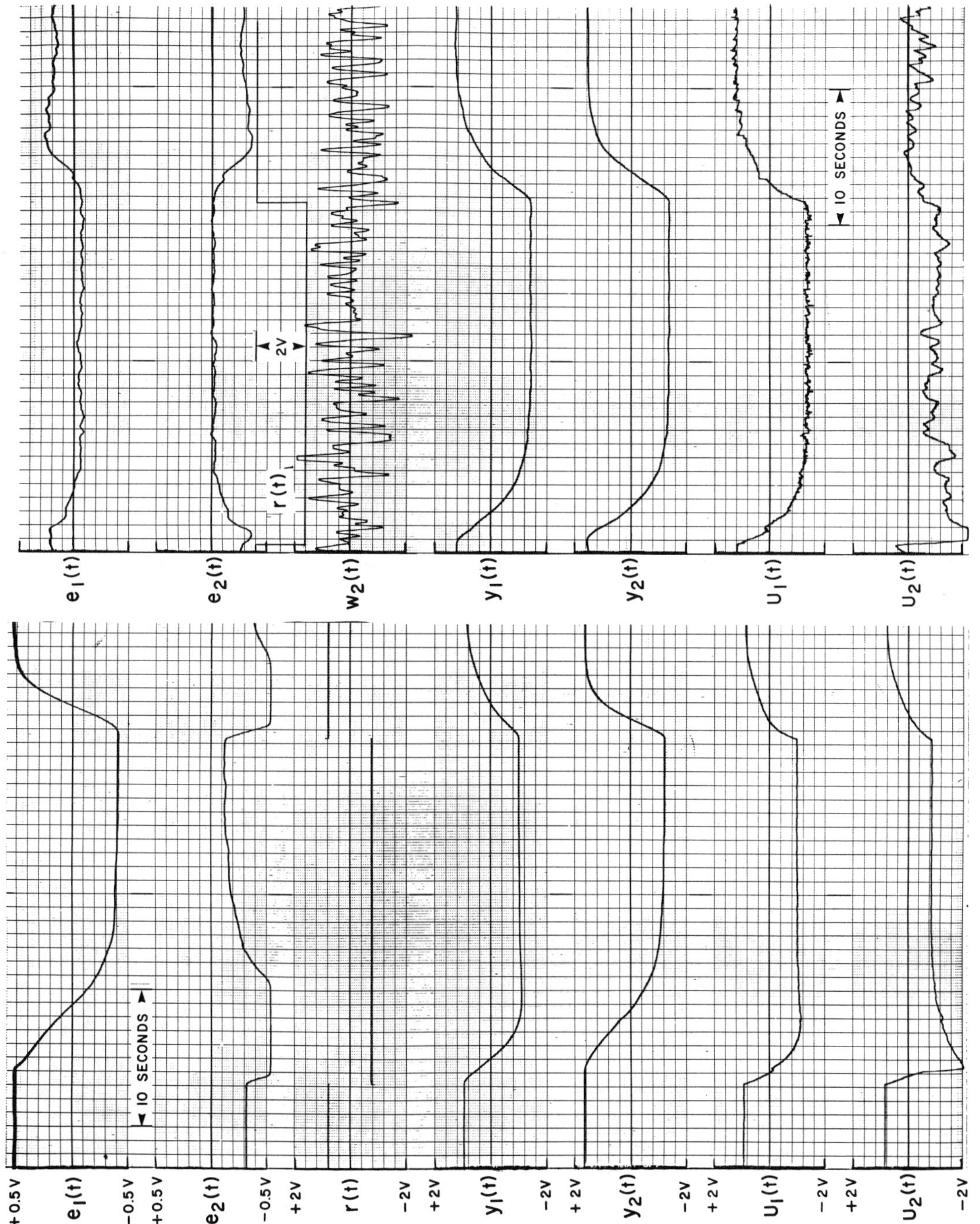

(a) NO DISTURBANCE, $G_1 = 1$, $G_2 = 4$, $P = [I]$

(b) $W_1(t) = 0$, $W_2(t)$ = GAUSSIAN DISTURBANCE
$G = 5$, $G = 8$, $P = \begin{bmatrix} 2 & 1 \\ 1 & 2 \end{bmatrix}$

Figure 5 SATURATION TYPE CONTROLLER PERFORMANCE WITH STEP INPUT

PLANT-ADAPTIVE PULSE-FREQUENCY MODULATED CONTROL SYSTEMS

M. Blythe Broughton
Royal Military College of Canada
Kingston, Ontario, Canada

By adding an asynchronous discrete-time identifier, estimates of the slowly-varying plant parameters in a single-input, single-output feedback system become available for adaptive adjustment of the pulse-frequency modulator/controller pulse-emission threshold. The inherent tendency of some pulse-frequency modulators to operate in a (non-periodic) limit-cycling mode for small system errors is exploited to facilitate parameter estimation without introducing additional "probe" signals.

1. INTRODUCTION

The study of pulse-frequency modulated (pfm) systems is motivated by similarities which signals in pfm systems bear to neural signals in biological systems [Meyer, 1961; Pavlidis, 1964]. The application of pulse-frequency modulators to the control of satellite attitude has also been investigated [Farrenkopf, et al, 1963; Clark, 1969].

Pulse-frequency modulated control systems tend to exhibit non-periodic limit-cycle (limit annulus) behaviour [Pavlidis and Jury, 1965]. In many pfm systems this is a disadvantage, since control effort (fuel) appears to be wasted when system error is small. However, when plant parameters vary with time, limit-cycling in the pfm system facilitates estimation of plant parameters by forcing the plant state to be nonzero, thereby satisfying a necessary condition for parameter identification. Because it is driven by an impulse train, all of the plant's characteristic modes are excited by each pulse. Hence, we are able to estimate slowly-changing plant parameters directly from the impulse response. By encouraging limit-cycling, perhaps by the addition of positive feedback within the modulator, when the system error is small, parameter estimation and tracking takes place under most circumstances [Broughton, 1971]. These estimates may be used for adaptive control of the modulator/controller pulse-emission threshold in accord with a selected performance criterion.

Figure 1 is a block diagram of the adaptive pfm system under consideration. The integral pulse-frequency (ipf) modulator [Li, 1964] emits a fixed-area impulse $\delta(t-t_i)$ whenever the integral of its input reaches either of two zero-symmetric thresholds $\pm q(t,\underline{\alpha},\underline{\beta})$; the impulse polarity ϵ_i is that of the threshold at the instant of its emission t_i, $i=1,2,\ldots$, and the integrator within the modulator is reset to zero in zero time, ie, $p(t_i^+) = 0$. Hence, the plant is driven by the impulse train

$$u(t) = \sum_i \epsilon_i \delta(t-t_i), \tag{1}$$

where t_i and ϵ_i are defined by

$$\left|p(t_i)\right| = \left|\int_{t_{i-1}}^{t_i} e(t)\,dt\right| = q(t_i); \epsilon_i = \mathrm{Sgn}\, p(t_i^-), \tag{2}$$

and $e(t)=r(t)-y(t)$ is the system error.

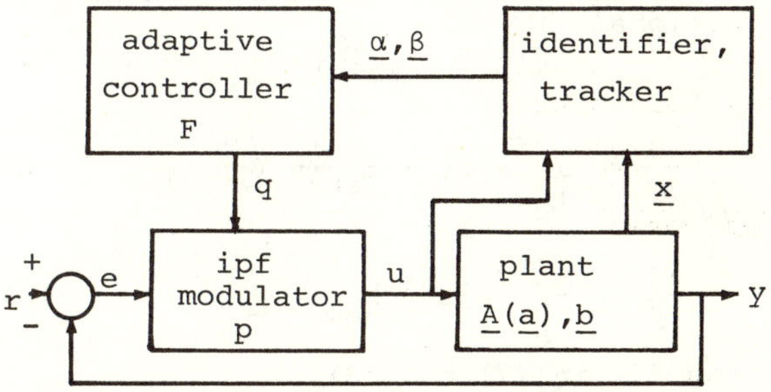

Fig.1 Plant-Adaptive Integral Pulse-Frequency (IPF) Modulated Feedback Control System.

The single-input, single-output linear plant is assumed to be described by

$$\dot{\underline{x}} = \underline{A}(\underline{a}(t)\underline{x} + \underline{b}(t)u(t), \quad \underline{x}(0) = x_o, \tag{3}$$

$$y = x_1,$$

where $u(t)$ is given by (1) and

$$\underline{A} = \begin{bmatrix} -a_1(t) & 1 & 0 & \ldots & 0 \\ -a_2(t) & 0 & 1 & \ldots & 0 \\ \vdots & \vdots & \vdots & & \vdots \\ -a_n(t) & 0 & 0 & \ldots & 0 \end{bmatrix}, \quad \underline{a} = \begin{bmatrix} a_1 \\ a_2 \\ \vdots \\ a_n \end{bmatrix}, \quad \underline{b} = \begin{bmatrix} b_1 \\ b_2 \\ \vdots \\ b_n \end{bmatrix}. \tag{4}$$

By changing the impulse emission threshold $q(t,\underline{\alpha},\underline{\beta})$ with changes in the discrete-time estimates $\underline{\alpha},\underline{\beta}$ of the plant parameters $\underline{a},\underline{b}$, adaptive controlof the pfm system is exercised, thereby keeping a selected performance function

$$F(q,\underline{a},\underline{b}) = 0 \tag{5}$$

insensitive to changes in plant parameters $\underline{a}$ and $\underline{b}$.

When the error signal is large compared with the modulator pulse-emission threshold, the ipf modulator may be replaced by its linear equivalent gain K_e, given by

$$K_e = q^{-1} \tag{6}$$

For illustrative purposes, we require the closed-loop damping of a linear second-order system, which is the large-signal equivalent of the given pfm system, to be insensitive to plant parameter changes.

2. PLANT PARAMETER IDENTIFICATION

Based on Bellman's method of differential approximation [Bellman, et al, 1963], we define the parameter estimation error $\underline{E}_{ab}$ by

$$\underline{E}_{ab} = \int_{t_o}^{t_f} \underline{\varepsilon}_{ab} dt , \tag{7}$$

where

$$\underline{\varepsilon}_{ab} = \dot{\underline{x}} - \underline{A}(\underline{a}(t)\underline{x} - \underline{b}(t)u , \tag{8}$$

and $t_f - t_o$ is an identification interval in which values of $\underline{a}(t)$ and $\underline{b}(t)$ may be considered constant. The "correct" values of $\underline{a}$ and $\underline{b}$ are those which minimize the (nonconvex) identifier performance functional

$$J = \frac{1}{2} \underline{E}'_{ab} \underline{E}_{ab} , \tag{9}$$

ie, for which

$$\frac{\partial J}{\partial \underline{a}} = \int_{t_o}^{t_f} x_1 dt \cdot E_{ab} = \underline{0} \tag{10}$$

and

$$\frac{\partial J}{\partial \underline{b}} = -\int_{t_o}^{t_f} u dt \cdot \underline{E}_{ab} = \underline{0} , \tag{11}$$

using (4) in (7) and (8).

The use of a "square-mean" identifier performance functional in (9) eliminates the need for time-function multiplication in the identifier unit. Equations (10) and (11) indicate that $\underline{a}$-identification and $\underline{b}$-identification may take place if

$\int_{t_o}^{t_f} x_1 dt \neq 0$ and $\int_{t_o}^{t_f} u dt \neq 0$ respectively; $\underline{a}$ and $\underline{b}$ must satisfy

$$\underline{E}_{ab} = \underline{0}. \tag{12}$$

Owing to the impulse nature of u(t), simplification of $\underline{a}$-identifier mechanization results if the interval $t_f - t_o$ does not include modulator output impulses. Then, for a_k constant in $t_f - t_o$, k=1,2,...n, and using (4),(7) and (8) in (12),

$$a_k = \frac{x_k(t_o) - x_k(t_f) + \int_{t_o}^{t_f} x_{k+1} dt}{\int_{t_o}^{t_f} x_1 dt} , t_{i-1} \leq t_o < t_f < t_i. \tag{13}$$

$\underline{b}$-parameters may be estimated at impulse times only; discrete changes in plant state are proportional to $\underline{b}(t_i)$, ie, from (12), (7) and (4),

$$b(t_i) = [\underline{x}(t_i^+) - \underline{x}(t_i^-)] \epsilon_i , t_{i-1} < t_o < t_i < t_f < t_{i+1}. \tag{14}$$

Implementation of (13) requires re-initialization of all identifier integrators when a modulator impulse occurs. Inaccuracies at the beginning of the identification interval, as well as the need for dividing one time function by another, may be removed by sampling and storing the numerator on the right side of (13) whenever

$$\left| z_o \right| = \left| \int_{t_{j-1}}^{t_j} x_1 dt \right| = h , z_o(t_j^+) = 0 , t_{i-1} \leq t_{j-1} < t_j < t_i , \tag{15}$$

where h is a fixed sampling threshold, $h < q_{min}$. $\underline{a}$-identification now occurs at times t_j which depend on the system state; in this respect the identifier is compatible with the modulator. From (13) and (14), we write discrete-time parameter tracking equations (16) and (17):

$$\alpha_k(t) = a_k(t_j^+) = [x_k(t_{j-1}) - x_k(t_j) + \int_{t_{j-1}}^{t_j} x_{k+1} dt] \times$$

$$\times \text{Sgn } z_o \ (t_j^-)/h , \tag{16}$$

$$\underline{\beta}(t) = \underline{b}(t_i). \tag{17}$$

Estimates $\underline{\alpha}$ and $\underline{\beta}$ are now available to replace $\underline{a}$ and $\underline{b}$ in the implementation of the adaptive control law (5). Since output $y = x_1$, in view of (15), we expect $\underline{a}$-identification to take place more often then the output is large; $\underline{b}$-identification occurs more often when the system error e is large.

3. ADAPTIVE CONTROL EXAMPLE

The feedback system of Fig. 1 containing the plant described by

$$\dot{x}_1 = -a_1(t)x_1 + x_2 + b_1(t)u, \quad x_1(0) = 0; \tag{18}$$

$$\dot{x}_2 = -a_2(t)x_1 + b_2(t)u, \quad x_2(0) = 0, \tag{19}$$

is driven by an ipf modulator in response to unit reference input. Fluctuations of a_1, a_2, b_1, b_2 with time are shown in Figs. 2 and 3. We wish to maintain the closed-loop damping $\zeta(q,\underline{a},\underline{b})$ of the equivalent large-signal linear system constant at $\zeta_o = 0.5$; ie,

$$\zeta(q,\underline{a},\underline{b}) - \zeta_o = 0. \tag{20}$$

Replacing the modulator by its linear equivalent gain (6), we obtain the second-order characteristic equation of this linearized system:

$$s^2 + (a_1 + b_1/q)s + a_2 + b_2/q = 0, \quad s \equiv \frac{d}{dt}, \tag{21}$$

wherein the damping ζ and natural frequency ω_n are defined by

$$2\zeta\omega_n = a_1 + b_1/q, \tag{22}$$

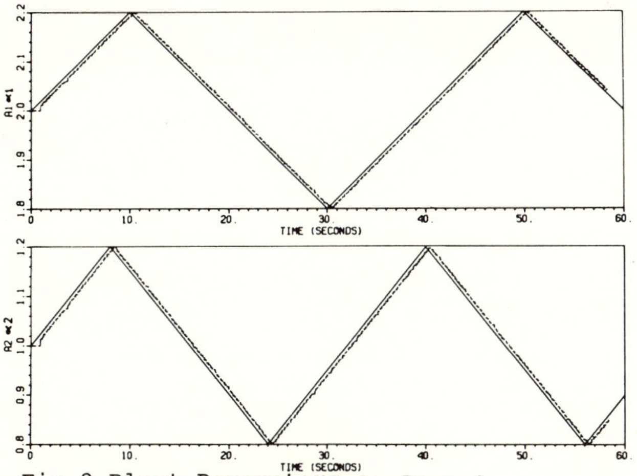

Fig.2 Plant Parameters a_1, a_2 ——; Estimates α_1, α_2 - - -.

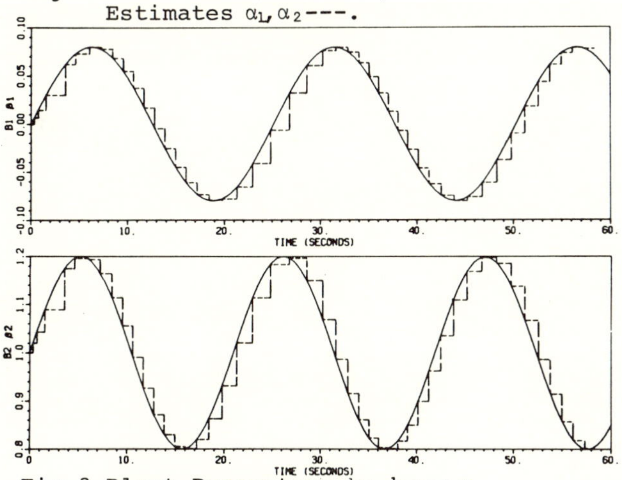

Fig.3 Plant Parameters b_1, b_2 - - -; Estimates β_1, β_2 - - -.

$$\omega_n^2 = a_2 + b_2/q. \qquad (23)$$

In view of adaptive control law (20), taking linear changes in the parameters about reference values $a_{10}=2$, $a_{20}=1$, $b_{10}=0$, $b_{20}=1$, ($q_o = 1/3$), on solution of (22) and (23), the adaptive control equation becomes

$$q-\frac{1}{3} = -\frac{4}{9}(a_1-2.)+\frac{1}{9}(a_2-1.)+\frac{4}{3}(b_1-0.)+\frac{1}{3}(b_2-1.) \qquad (24)$$

Whence, using discrete-time estimates α_1, $\alpha_2, \beta_1, \beta_2$ of a_1, a_2, b_1, b_2, we obtain the adaptive control equation actually implemented:

$$q = \frac{7}{9} - \frac{4}{9}\alpha_1+\frac{1}{9}\alpha_2-\frac{4}{3}\beta_1+\frac{1}{3}\beta_2 \quad . \qquad (24)$$

Estimates $\alpha_1, \alpha_2, \beta_1, \beta_2$ are shown in Figs. 2 and 3. Figure 4 displays modulator internal state p(t) and pulse-emission threshold q(t); the latter modulates the amplitude of the former. Figure 5 shows the plant state.

The adaptive system with unit reference input was found to reduce the maximum fluctuation $\Delta\zeta$ in closed-loop damping to about 0.034 from 0.21 for the unadapted system with the same reference input.

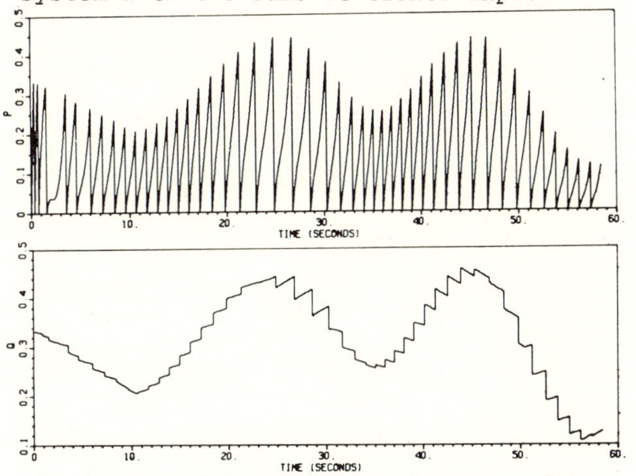

Fig.4 IPF modulator internal state p(t), pulse-emission threshold q(t).

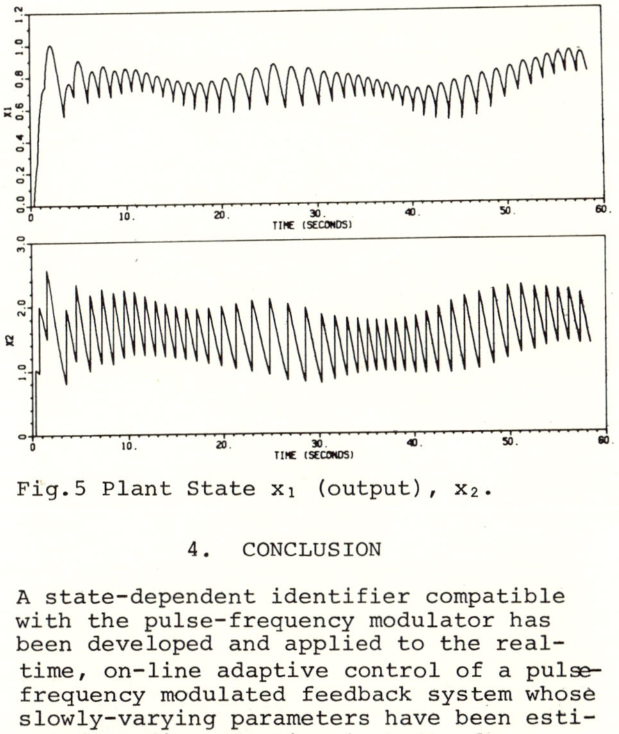

Fig.5 Plant State x_1 (output), x_2.

4. CONCLUSION

A state-dependent identifier compatible with the pulse-frequency modulator has been developed and applied to the real-time, on-line adaptive control of a pulse-frequency modulated feedback system whose slowly-varying parameters have been estimated at discrete time instants from plant impulse response with the aid of a dual-function controller.

The research reported here was supported in part by the Defence Research Board of Canada.

5. REFERENCES

1. Bellman, Kalaba, Sridar (1963): Adapt-

ive control via quasilinearization and differential approximation. Rand Corp. Memorandum RM-3928-PR.

2. Broughton (1971): Plant-Adaptive Pulse-Frequency Modulated Feedback Control Systems. Ph.D. Thesis, Queen's University, Kingston, Canada.

3. Clark (1969): Analysis of Oscillations in Pulse Frequency Modulated Control Systems. Ph.D. Thesis, Stanford University, California, USA.

4. Farrenkopf, Sabroff, Wheeler (1963): Integral pulse-frequency on-off attitude control. AIAA Guidance and Control Conference, Cambridge, Massachusetts, USA. Preprint 63-328.

5. Li, Jones (1964): Integral pulse frequency modulated control systems. Proc. Second IFAC Congress; Butterworths, London, 186.

6. Meyer (1961): Pulse Frequency Modulation and Its Effect in Feedback Systems. Ph.D. Thesis, Northwestern University, Illinois, USA.

7. Pavlidis (1964): Analysis and Synthesis of Pulse-Frequency Feedback Systems. Ph.D. Thesis, University of California, Berkeley, USA.

8. Pavlidis, Jury (1965): Analysis of a new class of pulse frequency modulated feedback systems. IEEE Trans AC-10, 35.

OPEN-LOOP-FEEDBACK-OPTIMAL ADAPTIVE STOCHASTIC CONTROL OF LINEAR SYSTEMS*

Richard Ku
Intermetrics, Inc.
Cambridge, Mass. 02138

Michael Athans
Dept. of Electrical Engineering
Massachusetts Institute of Technology
Cambridge, Mass. 02139

This paper considers the suboptimal stochastic control of linear discrete-time dynamical systems with unknown or stochastically varying parameters. The suboptimal scheme is based upon the use of the open-loop-feedback-optimal (O.L.F.O.) method. The state and parameter estimates are generated by an extended Kalman filter algorithm. Numerical results for first-order systems are presented.

1. INTRODUCTION

This paper considers the stochastic control of linear discrete-time systems with unknown or stochastically varying parameters. The design of the stochastic control system is based upon

(a) a suboptimal state estimation and parameter estimation scheme, which utilizes the so-called extended Kalman filter JAZWINSKI (1970)

(b) a suboptimal stochastic control technique, which is commonly referred to as the O.L.F.O. method, and which minimizes a particular quadratic cost criterion DREYFUS (1965).

The resultant structure of the adaptive control system in real time contains

(a) A <u>learning</u> subsystem (the extended Kalman filter) which generates estimates of the plant system state variables and unknown parameters based upon past noisy measurements;

(b) A <u>control</u> subsystem which consists of two parts (i) an adaptive time-varying gain that multiplies the instantaneous state estimate, and (ii) a control correction term; both the adaptive gain and control correction term depend upon the current and future estimates of the unknown parameters <u>as well as</u> their covariance matrices.

This research represents a significant extension of the philosophy and results reported by TSE (1970) TSE and ATHANS (1970),(1972); in that research the only unknown parameters were those of the gain vector which multiplied the control. In this paper the natural system dynamics are also not known with certainty. In addition, this approach seems to bear close resemblance to a recent contribution by E. CHASE-MCRAE (1972) in the stochastic economic literature; in that paper, however, the measurements of the state variables were assumed exact.

Unfortunately, the state-of-the-art of knowledge in the area of adaptive systems and the complex equations that characterize them preclude an analytical type of comparison. Hence, one can only rely upon simulation results. It is for this reason that in Section 4 we present our experience in using these methods to one of the simplest but nontrivial examples that one can think of, namely the control of a first-order system with two unknown but constant parameters; and present the numerical comparisons under three modes of control.

2. PROBLEM STATEMENT

Suppose we have the following discrete-time n-dimensional single-input linear stochastic dynamical system

S1:

<u>Plant:</u> $\underline{x}(k+1)=\underline{A}(k)\underline{x}(k)+\underline{b}(k)u(k)+\underline{\xi}(k)$ k=0,1..., N-1 (2.1)

<u>Measurement:</u> $\underline{z}(k)=\underline{C}\ \underline{x}(k)+\underline{\theta}(k)$ k=0,1,...,N (2.2)

where we assume that $\underline{x}(k)$, $\underline{b}(k)$ and $\underline{\xi}(k) \in R^n$, $\underline{z}(k)$ and $\underline{\theta}(k) \in R^m$, $\underline{C}$ is a known constant mxn matrix, and u(k) is an unconstrained <u>scalar</u> input.* Furthermore, we assume that the actual (but unknown) nxn plant companion matrix $\underline{A}(k)$ has the following canonical representation or is convertible to this equivalent system by a linear nonsingular transformation.

---

* This research was conducted at the Decision and Control Sciences group, of the M.I.T. Electronic Systems Laboratory with support extended by the U.S. Air Force Office of Scientific Research under grants AFOSR-70-1941 and 72-2273 and by NASA Ames Research Center under grant NGL-22-009-124.

* The extension to the vector case is conceptually straightforward; but the equations are extremely complex.

$$\underline{A}(k) = \begin{bmatrix} \underline{0} & \vdots & \underline{I}_{n-1} \\ \hline \leftarrow & \underline{a}'(k) & \rightarrow \end{bmatrix}, \quad \underline{a}(k) = \begin{bmatrix} a_1(k) \\ a_2(k) \\ \cdot \\ \cdot \\ \cdot \\ a_n(k) \end{bmatrix} \in R^n \quad (2.3)$$

The unknown plant parameter vector $\underline{a}(k)$ and gain vector $\underline{b}(k)$ are assumed to satisfy the difference equations**

$$\underline{a}(k+1) = \underline{a}(k) + \underline{\delta}(k) \quad (2.4)$$

$$\underline{b}(k+1) = \underline{b}(k) + \underline{\gamma}(k) \quad (2.5)$$

respectively, where $\underline{\delta}(k)$ and $\underline{\gamma}(k) \in R^n$. We shall assume also that the triplet $[\underline{A}(k),\underline{b}(k),\underline{C}]$ of Eqs. 2.1 and 2.2 represents a completely controllable and completely observable system.

We shall assume that the random sequences $\{\underline{\xi}(\cdot), \underline{\theta}(\cdot), \underline{\delta}(\cdot), \underline{\gamma}(\cdot)\}$ are white Gaussian, zero-mean, and uncorrelated with each other and with the Gaussian random vectors $\underline{x}(0)$, $\underline{a}(0)$, $\underline{b}(0)$. The statistical laws of the underlying random vectors are assumed known for all k. The symbol $N[\underline{m},\underline{\Sigma}]$ denotes normal distribution with mean equal to $\underline{m}$ and finite covariance matrix equal to $\underline{\Sigma}$. Thus, let

$$\underline{x}(0) \sim N[\underline{x}_o,\underline{\Sigma}_{xo}];\underline{a}(0)\sim N[\underline{a}_o,\underline{\Sigma}_{ao}];\underline{b}(0)\ N[\underline{b}_o,\underline{\Sigma}_{bo}] \quad (2.6)$$

$$\underline{\xi}(k)\sim N[\underline{0},\underline{\Xi}(k)];\ \underline{\theta}(k)\sim N[\underline{0},\underline{\Theta}(k)];$$

$$\underline{\delta}(k)\sim N[\underline{0},\underline{\Delta}(k)];\ \underline{\gamma}(k)\sim N[\underline{0},\underline{\Gamma}(k)] \quad (2.7)$$

where all covariance matrices are symmetric and at least positive semidefinite, except that $\underline{\Theta}(k)$ is positive definite.

We shall assume that the true cost functional is quadratic

$$\bar{J}(u) \triangleq \tfrac{1}{2} E\,\{\underline{x}'(N)\underline{Q}(N)\underline{x}(N) + \sum_{k=0}^{N-1} \underline{x}'(k)\underline{Q}(k)\underline{x}(k)+r(k)u^2(k)\} \quad (2.8)$$

with N fixed and finite. The expectation is taken over all the underlying random processes, $\underline{a}(0)$, $\underline{b}(0)$, $\underline{x}(0)$, $\underline{\xi}(\cdot)$, $\underline{\theta}(\cdot)$, $\underline{\delta}(\cdot)$, and $\underline{\gamma}(\cdot)$. We assume that $\underline{Q}(k) = \underline{Q}'(k) \geq \underline{0}$ and $r(k) > 0$. Thus given the noise-corrupted unknown linear dynamic system S1 the objective is to find the admissible control sequence $U(0,N-1) = \{u(j)\}_{j=0}^{N-1}$ which performs best "on the average" such that it minimizes the performance measure $\bar{J}(u)$ of Eq. 2.8 subject to the dynamic constraints Eqs. 2.1 and 2.2.

** One can readily extend the theory to the case where $\underline{a}(k)$ and $\underline{b}(k)$ satisfy more complex equations, e.g., $\underline{a}(k+1) = \underline{M}(k)\underline{a}(k) + \underline{\delta}(k)$, $\underline{b}(k+1) = \underline{L}(k)\underline{b}(k) + \underline{\gamma}(k)$, where $\underline{M}(k)$ and $\underline{L}(k)$ are known.

To complete the problem statement we must now define what we mean by admissible control sequence. For a stochastic optimal control problem it is very important to specify precisely that data or information pattern which is available for determining the control input. Depending on what measurements the computation of the control sequence U(0,N-1) is based on exactly, different formulations of the optimal stochastic control problem are possible. In the general case, one cannot obtain explicit solutions to the optimal adaptive control problem, so that one has to resort to a suboptimal controller. We shall, therefore, restrict our attention to the open-loop-feedback-optimal (O.L.F.O.) controller. (See KU (1972) AOKI (1967) BAR-SHALOM (1969) DREYFUS (1965) for explanations).

3. OPEN-LOOP-FEEDBACK-OPTIMAL CONTROL

Let the present time be indexed by k. Denote the accumulative observation statistic, that is, the (random) output observations or the available information at time k by

$$Z_k \triangleq \{\underline{z}(0),\ \underline{z}(1),\ldots,\underline{z}(k)\}$$

Let us also assume that the optimal control sequence

$$U^*(0,k-1) \triangleq \{u^*(j)\}_{j=0}^{k-1}$$

has been applied to the system and is part of the information structure. Let us then define for $j \geq k$ for the augmented state system the conditional expectations

$$\hat{\underline{x}}(j|k) \triangleq E\{\underline{x}(j)|Z_k\};\hat{\underline{a}}(j|k) \triangleq E\{\underline{a}(j)|Z_k\};$$

$$\hat{\underline{b}}(j|k) \triangleq \{\underline{b}(j)|Z_k\} \quad (3.1)$$

the error vectors

$$\underline{e}_x(j|k) \triangleq \hat{\underline{x}}(j|k)-\underline{x}(j);\ \underline{e}_a(j|k) \triangleq \hat{\underline{a}}(j|k) -\underline{a}(j);\ \underline{e}_b(j|k) \triangleq \hat{\underline{b}}(j|k)-\underline{b}(j), \quad (3.2)$$

and the conditional error covariance matrix

$$\underline{\Sigma}(j|k) \triangleq E\{\begin{bmatrix} \underline{e}_x(j|k) \\ \cdots\cdots \\ \underline{e}_a(j|k) \\ \cdots\cdots \\ \underline{e}_b(j|k) \end{bmatrix} [\underline{e}_x'(j|k) \vdots\ \underline{e}_a'(j|k) \vdots \underline{e}_b'(j|k)]|Z_k\} \quad (3.3)$$

The suboptimal closed-loop control policy we are solving consists of replacing the closed-loop controls in Eq. 2.8 with open-loop controls for $j \geq k$. It does not take into account the knowledge that future measurements will be made.

Using Eqs. 3.1 to 3.3 we then obtain the conditional cost

$$\bar{J}_k = 1/2\ \hat{\underline{x}}'(N|k)\underline{Q}(N)\hat{\underline{x}}(N|k) + 1/2\ \mathrm{tr}[\underline{Q}(N)\underline{\Sigma}_{xx}(N|k)]$$

$$+ 1/2 \sum_{j=k}^{N-1} \{\hat{\underline{x}}'(j|k)\underline{Q}(j)\hat{\underline{x}}(j|k) + \mathrm{tr}[\underline{Q}(j)\underline{\Sigma}_{xx}(j|k)]$$

$$+ r\ (j)u^2(j)\} \qquad k = 0,1\ldots,N-1 \qquad (3.4)$$

To complete the formulation of the deterministic open-loop control problem, we will need in Eq. 3.4 the deterministic dynamical equations satisfied by $\hat{\underline{x}}(j|k)$ and $\underline{\Sigma}_{xx}(j|k)$ $(j \geq k)$, the respective estimate and covariance matrix of $\underline{x}(k)$ conditioned upon all past measurements Z_k and the past controls of $U^*(0,k-1)$. We shall, however, use only the approximate† expressions for $\hat{\underline{x}}(j|k)$ and $\underline{\Sigma}_{xx}(j|k)$, to define a completely deterministic optimal control problem for the kth step, whose solution would yield the optimal open-loop controls for $k \leq j \leq N-1$.

The aim is to find a deterministic control sequence $\{u(k),\ldots,u(N-1)\}$ such that it minimizes in (N-k) steps the average value of the cost-to-go

$$\bar{J}_k = 1/2\{\hat{\underline{x}}'(N|k)\underline{Q}(N)\hat{\underline{x}}(N|k) + \mathrm{tr}[\tilde{\underline{Q}}(N)\underline{\Sigma}(N|k)]$$

$$+ \sum_{j=k}^{N-1} \hat{\underline{x}}'(j|k)\underline{Q}(j)\hat{\underline{x}}(j|k) + \mathrm{tr}[\tilde{\underline{Q}}(j)\underline{\Sigma}(j|k)]$$

$$+ r(j)u^2(j)\} \quad k = 0,1,\ldots,N-1 \qquad (3.5)$$

where

$$\tilde{\underline{Q}}(j) = \begin{bmatrix} \underline{Q}(j) & \underline{0} & \underline{0} \\ \underline{0} & \underline{0}_n & \underline{0} \\ \underline{0} & \underline{0} & \underline{0}_n \end{bmatrix} \qquad (3.6)$$

and $\tilde{\underline{Q}}(N)$, $\tilde{\underline{Q}}(j) \geq \underline{0}$ and $r(j) > 0$. We shall also assume that $\underline{Q}(j)$ and $\underline{Q}(N)$ are not both the zero matrix. The terminal "states" $\hat{\underline{x}}(N|k)$ and $\underline{\Sigma}(N|k)$ are not specified.

We shall now state the solution to the deterministic optimal control problem formulated above. It can be shown that the O.L.F.O. control $\{u^*(j)\}_{j=k}^{N-1}$ exists and is unique. The detailed derivation via dynamic programming is given in KU (1972) and is not presented here (it is straightforward but very lengthy). The symbol $u^o(j|k)$ denotes the optimal open-loop control conditioned on the observations up to and including time k, and is given for $j \geq k$ below. (It should be stressed that $\hat{\underline{x}}(j|k)$ and $\underline{\Sigma}(j|k)$ are not the exact conditional means and error covariance matrix, but those generated by the extended Kalman predictor.)

$$u^o(j|k) = -\{[\tilde{r}(j|k)+\tilde{\underline{b}}'(j|k)\tilde{\underline{K}}(j+1|k)\tilde{\underline{b}}(j|k)]^{-1}\tilde{\underline{b}}'(j|k)\tilde{\underline{K}}(j+1|k)\underline{\Phi}(j|k) + \tilde{r}^{-1}(j|k)\underline{d}'(j+1|k)\} \begin{bmatrix} \hat{\underline{x}}(j|k) \\ \underline{\sigma}(j|k) \\ \underline{\rho}(j|k) \end{bmatrix} \qquad (3.7)$$

where the $n(2n+1) \times n(2n+1)$ symmetric matrix $\tilde{\underline{K}}(j+1|k)$ is the unique solution of **a certain** recursive backward difference equation **for** $k+1 \leq j \leq N-1$

$$\tilde{\underline{K}}(j|k) = \underline{\Phi}'(j|k)[\tilde{\underline{K}}(j+1|k)-\tilde{\underline{K}}(j+1|k)\tilde{\underline{b}}(j|k)(\tilde{r}(j|k) + \tilde{\underline{b}}'(j|k)\tilde{\underline{K}}(j+1|k)\tilde{\underline{b}}(j|k))^{-1}\tilde{\underline{b}}'(j|k)\tilde{\underline{K}}(j+1|k)]\underline{\Phi}(j|k) + \underline{V}(j|k) \qquad (3.8)$$

satisfying the boundary condition

$$\underline{K}(N\ k) = \begin{bmatrix} \underline{Q}(N) & \underline{0} & \underline{0} \\ \underline{0} & \underline{0}_{n^2} & \underline{0} \\ \underline{0} & \underline{0} & \underline{0}_{n^2} \end{bmatrix} \qquad (3.9)$$

where the parameters $\tilde{r}(j|k)$, $\tilde{\underline{b}}(j|k)$, $\underline{\Phi}(j|k)$, $\underline{d}(j+1|k)$, and the modified weighting $\underline{V}(j|k)$ are defined in KU (1972).

The open-loop feedback optimal control actually applied at time k is given by

$$u^*(k) = u^o(k|k) \qquad (3.10)$$

To find the open-loop feedback optimal control sequence, we have to solve the open-loop control problem for $k = 0,1,\ldots,N-1$. It can be shown that we can write the open-loop feedback optimal control as

$$u^*(k) = \underline{\phi}'(k)\hat{\underline{x}}(k|k) + u_c(k) \qquad k = 0,1,\ldots,N-1 \qquad (3.11)$$

where we define the $1 \times n$ (row) vector

$$\underline{\phi}'(k) \triangleq -\{[\tilde{r}(k|k)+\tilde{\underline{b}}'(k|k)\tilde{\underline{K}}(k+1|k)\tilde{\underline{b}}(k|k)]^{-1}\tilde{\underline{b}}'(k|k)\tilde{\underline{K}}$$

† Because the filter that generates the true conditional means and covariances is an infinite dimensional one; we shall rather employ the extended Kalman filter and accept its estimates as being the true conditional means.

$$(k+1|k)\underline{\Phi}(k|k) + \tilde{r}^{-1}(k|k)\underline{d}'(k+1|k)\} \begin{bmatrix} \underline{I}_n \\ \underline{0}_{n^2} \\ \underline{0}_{n^2} \end{bmatrix} \quad (3.12)$$

to be the optimal open-loop feedback adaptive gain. We shall call the scalar

$$u_c(k) \triangleq$$

$$-\{[\tilde{r}(k|k)+\tilde{\underline{b}}'(k|k)\tilde{\underline{K}}(k+1|k)\tilde{\underline{b}}(k|k)]^{-1}\tilde{\underline{b}}'(k|k)\tilde{\underline{K}}$$

$$(k+1|k)\underline{\Phi}(k|k) + r^{-1}(k|k)\underline{d}'(k+1|k)\}$$

$$\begin{bmatrix} \underline{0} & \underline{0} & \underline{0} \\ \underline{0} & \underline{I}_{n^2} & \underline{0} \\ \underline{0} & \underline{0} & \underline{I}_{n^2} \end{bmatrix} \begin{bmatrix} \hat{\underline{x}}(k|k) \\ \underline{\sigma}(k|k) \\ \underline{\rho}(k|k) \end{bmatrix} \quad (3.13)$$

the adaptive control correction term.

We note that the adaptive control gain $\underline{\phi}'(k)$, Eq. 3.12 is, by definition, independent of the current estimate of the state vector $\hat{\underline{x}}(k|k)$. It depends on $\hat{\underline{b}}(k|k)$ and $\hat{\underline{b}}(j|k)$; $\hat{\underline{a}}(k|k)$ and $\hat{\underline{a}}(j|k)$; $\underline{\Sigma}_{bb}(k|k)$ and $\underline{\Sigma}_{bb}(j|k)$; $\underline{\Sigma}_{ab}(k|k)$ and $\underline{\Sigma}_{ab}(j|k)$; and $\underline{\Sigma}_{aa}(k|k)$ and $\underline{\Sigma}_{aa}(j|k)$ all evaluated along the open-loop feedback optimal augmented trajectory for $k \leq j \leq N-1$.

The control correction term $u_c(k)$, Eq. 3.13 is independent of $\hat{\underline{x}}(k|k)$. It depends on $\hat{\underline{b}}(k|k)$ and $\hat{\underline{b}}(j|k)$; $\hat{\underline{a}}(k|k)$ and $\hat{\underline{a}}(j|k)$; $\underline{\Sigma}_{xb}(j|k)$; and $\underline{\Sigma}_{xb}(j|k)$ $\underline{\Sigma}_{xa}(k|k)$ and $\underline{\Sigma}_{xa}(j|k)$; $\underline{\Sigma}_{bb}(k|k)$ and $\underline{\Sigma}_{bb}(j|k)$; and $\underline{\Sigma}_{ab}(k|k)$ and $\underline{\Sigma}_{ab}(j|k)$ all evaluated along the open-loop feedback optimal control augmented trajectory for $k < j < N-1$. If the cross error covariances $\underline{\Sigma}_{xb}(k|k)$, $\underline{\Sigma}_{xa}(k|k)$, and $\underline{\Sigma}_{ab}(k|k)$ are zero, then the adaptive control correction term $u_c(k) = 0$ in Eq. 3.13.

In Eq. 3.12 we have the explicit variation of the adaptive gain as a function of the future expected uncertainty of the parameters. In Eq. 3.13 the O.L.F.O. control correction term is affected by the estimation accuracy of the $\underline{a}$ and $\underline{b}$ vector through $\underline{\Sigma}_{xa}(\cdot|k)$, $\underline{\Sigma}_{xb}(\cdot|k)$, $\underline{\Sigma}_{ab}(\cdot|k)$, and $\underline{\Sigma}_{bb}(\cdot|k)$.

Asymptotic Behavior

In this section we study the asymptotic properties of the overall system by considering the behavior of $\underline{\Sigma}_{bb}(k|k)$ and $\underline{\Sigma}_{aa}(k|k)$ as $k \to \infty$.

Lemma 3.1: Let $\underline{\delta}(k) = \underline{0}$, that is, there is no stochastic variation in parameter vector $\underline{a}(k)$, and hence the $\underline{a}(k)$ vector is constant. Then, given any nonzero control sequence, the error covariance $\underline{\Sigma}_{aa}$ is monotonically decreasing.

$$\underline{\Sigma}_{aa}(k+1|k+1) \leq \underline{\Sigma}_{aa}(k|k) \quad (3.14)$$

Proof:

$$\underline{\Sigma}_{aa}(k+1|k+1) = \underline{\Sigma}_{aa}(k|k) - [\underline{0} \vdots \underline{I}_n \vdots \underline{0}]\underline{G}(k+1)$$

$$[\tilde{\underline{C}}\,\underline{\Sigma}(k+1|k)\tilde{\underline{C}}' + \underline{\Theta}(k+1)\underline{G}'](k+1) \begin{bmatrix} \underline{0} \\ \cdots \\ \underline{I}_n \\ \cdots \\ \underline{0} \end{bmatrix} \quad (3.15)$$

where $\underline{G}(k+1)$ is the filter gain matrix. The Lemma then follows immediately. Intuitively, since

$$\underline{a}(k+1) = \underline{a}(k)$$

the uncertainty in $\underline{a}(k)$ cannot grow.

As a result of the Lemma, there exists then a $\underline{\Sigma}_{aa}$ such that

$$\lim_{k\to\infty} \underline{\Sigma}_{aa}(k|k) = \underline{\Sigma}_{aa} \quad (3.16)$$

Lemma 3.2: For $\underline{b}(k+1) = \underline{b}(k)$, given any control sequence we have

$$\underline{\Sigma}_{bb}(k+1|k+1) \leq \underline{\Sigma}_{bb}(k|k) \quad (3.17)$$

Proof: The proof is similar to that for Lemma 3.1. There exists then a $\underline{\Sigma}_{bb}$ such that

$$\lim_{k\to\infty} \underline{\Sigma}_{bb}(k|k) = \underline{\Sigma}_{bb} \quad (3.18)$$

In analogy with the deterministic case, we can thus say that the parameters $\underline{a}$ and $\underline{b}$ are observable since the variance of the estimation error of $\underline{a}$ and $\underline{b}$ can be decreased by operation on $\underline{z}$.

It can be shown that if $\underline{\delta}(k) = \underline{0}$, $\underline{\gamma}(k) = \underline{0}$, and the system completely observable, then for any bounded but nonzero control $u(k)$, $k = 0,1\ldots$ TSE(1970) TSE and ATHANS (1970)

$$\lim_{k\to\infty} \underline{\Sigma}_{aa}(k|k) = \underline{0}; \quad \lim_{k\to\infty} \underline{\Sigma}_{bb}(k|k) = \underline{0} \quad (3.19)$$

Since $\underline{\Sigma}(k|k) \geq \underline{0}$, this result implies that

$$\lim_{k\to\infty} \underline{\Sigma}_{xb}(k|k) \to \underline{0}; \quad \lim_{k\to\infty} \underline{\Sigma}_{xa}(k|k) \to \underline{0}; \text{ and}$$

$$\lim_{k\to\infty} \underline{\Sigma}_{ab}(k|k) \to \underline{0}\,. \quad (3.20)$$

Hence, we can design a reasonable controller for a completely observable system with unknown parameters

$$\underline{a}(k+1) = \underline{a}(k); \; \underline{b}(k+1) = \underline{b}(k)$$

using an ad hoc control law $\xi_k(\hat{\underline{x}}(k|k), \hat{\underline{a}}(k|k), \hat{\underline{b}}(k|k), \underline{\Sigma}_{aa}(k|k), \underline{\Sigma}_{bb}(k|k))$ for $k \geq 0$ given by

(1) $\xi_k(\underline{x},\underline{a},\underline{b},\underline{\Sigma}_{aa},\underline{\Sigma}_{bb}): R^n x R^n x R^n x M_{nn} x M_{nn} \to R$

$\underline{x} \varepsilon R^n$, $\underline{a} \varepsilon R^n$, $\underline{b} \varepsilon R^n$, $\underline{\Sigma}_{aa} \varepsilon M_{nn}$, $\underline{\Sigma}_{bb} \varepsilon M_{nn}$

(2) $\xi_k(\underline{x},\underline{a},\underline{b},\underline{\Sigma}_{aa},\underline{\Sigma}_{bb}) \neq 0$, $\underline{\Sigma}_{bb} \neq \underline{0}$, $\underline{\Sigma}_{aa} \neq \underline{0}$, $\underline{x} \neq \underline{0}$

(3) $\xi_k(\underline{x},\underline{a},\underline{b},\underline{0},\underline{0}) = -(r(k)+\underline{b}'\underline{K}(k+1)\underline{b})^{-1}\underline{b}'\underline{K}(k+1)$

$\underline{A}\ \underline{x}$

Condition 2 satisfies Eq. 3.19 and condition 3 implies that the ad hoc control will converge to the optimal control when a and b become known. Hence, the ad hoc control scheme for system with unknown parameters can provide reasonable control system response.

4. SIMULATION RESULTS

The main purpose of the simulation studies is to provide the quantitative measures on the convergence rate of both the O.L.F.O. control and the enforced separation schemes to the truly optimal stochastic control when the parameters are known. The simulation studies will compare the performance measure of (1) the truly optimal stochastic control system when the full dynamics are known, (it will give the unachievable lower bound $\bar{J}$), (2) the O.L.F.O. adaptive control system, and (3) the (ad hoc) enforced separation scheme. The extended Kalman filter is used for real-time identification of parameters.

We shall consider specifically the first-order linear dynamical system described by the stochastic difference equation

$$x(k+1) = ax(k) + bu(k) + \xi(k) \qquad (4.1)$$

with noisy measurements given by

$$z(k) = cx(k) + \theta(k) \qquad (4.2)$$

We assume that the unknown parameters a and b are constant.

The initial values x(0), a(0), and b(0) are assumed to be Gaussian random variables with known a priori statistics $x(0) \sim N[x_o, \Sigma_{xo}]$, $a(0) \sim N[a_o, \Sigma_{ao}]$, $b(0) \sim N[b_o, \Sigma_{bo}]$ where Σ_{xo}, Σ_{ao}, and Σ_{bo} are positive semidefinite. The scalar zero-mean white Gaussian driving noise sequence has the known covariance $\Xi(k)$, and the scalar zero-mean white Gaussian observation noise has the known covariance $\theta(k)$, where $\Xi(\cdot) \geq 0$ and $\theta(\cdot) > 0$. The random variables $\{x(0), \underline{a}(0), b(0), \xi(\cdot), \theta(\cdot)\}$ are mutually independent.

The objective of the problem is to find an optimal control sequence such that the expected cost functional

$$\bar{J} = E\{\frac{1}{2} q(N)x^2(N) + \frac{1}{2}\sum_{k=0}^{N-1} q(k)x^2(k) + r(k)u^2(k)\} \qquad (4.3)$$

is minimized based on some information set. The difficulty lies in that both the pole and zero of the system are unknown. The system is initially at rest.

Table I contains a summary of the results for different stable and unstable plants, different initial values of parameter uncertainty, different values of plant and measurement covariance matrices, and different weights in the cost functional. The results are summarized by presenting the average numerical values of the cost, computed on the basis of 20 Monte Carlo runs for the two suboptimal adaptive control schemes; the cost when the parameters are known exactly are given for comparison.

The simulation results of U1 using the crude Monte Carlo method are shown in Figs. 1-4. We assume that the a priori distribution of a(0) is given by

$$a(0) \sim N[1.2,\ 0.0049]$$

Thus, the system is assumed to be strongly unstable. From the plot of state trajectories in Fig. 1, we see that the open-loop feedback optimal trajectory has an overshoot at k=1 due to the large control u(0). The large control magnitude is used for identification and control purposes. The average of the parameter estimates $\hat{a}(k|k)$ and $\hat{b}(k|k)$ are shown in Figs.2-3. They approach nearly the initial mean values. The identification of b was better using the O.L.F.O. control than the enforced separation scheme. In the plot of the feedback gains, Fig. 4, we have the experimental result that the open-loop feedback initial adaptive gains are non-zero and, surprisingly, large compared to the truly optimal feedback gain. The open-loop feedback optimal control sequence proved to be more "aggressive" on the average than the enforced separation design. Both the open-loop feedback optimal and the enforced separation control sequence were able to stabilize the system, although the identification was not exact. Not shown here is the O.L.F.O. correction term which does go to zero as $k \to N$ in all the simulation runs.

Large controls help in the identification of the unknown parameters. We note that the larger the control $u^*(k)$, the faster $\underline{\Sigma}_{aa}(k|k)$ and $\underline{\Sigma}_{bb}(k|k)$ decrease. The estimates $\underline{\hat{a}}(k|k)$ and $\underline{\hat{b}}(k|k)$ of the parameter vectors $\underline{a}(k)$ and $\underline{b}(k)$ themselves will depend on the particular control law, since the recursive filter contains $u^*(k)$ as a parameter. This is verified by the simulation results. Conversely, the goodness of the estimates $\underline{\hat{x}}(k|k)$, $\underline{\hat{a}}(k|k)$, and $\underline{\hat{b}}(k|k)$ will affect the control law actually used.

For both the unstable and stable systems we remark that exact identification of $\underline{a}(k)$ and $\underline{b}(k)$ is not necessary from the control viewpoint. This was shown experimentally to be more so in the case of stable systems. Simulation results showed that the open-loop feedback optimal control systems can work well even if the parameter estimates are bad. The use of feedback also reduces the effect of parameter variations or the system's sensitivity to parameter inaccuracy. We recall that our objective functional rewards the system for good control performance, but not for good estimation of parameters.

Simulation results via Monte Carlo method compare the average cost incurred from combined identification and control, using, first, the open-loop feedback technique in Section 3, and, second, the enforced separation scheme. The experimental results seem to indicate that in the stable systems, the open-loop feedback optimal method on the average incurred a smaller performance index that the enforced separation scheme. For unstable systems, the results seem to indicate that the enforced separation scheme will incur small average cost in doing the job than the open-loop feedback optimal control design. This partial ordering seems to originate in the large control magnitude that the open-loop feedback optimal techniques uses to probe the parameters and stabilize the system in the beginning.

5. CONCLUSION

We have considered in detail both analytically and experimentally the problem of controlling a discrete stochastic linear system on the basis of noisy measurements, using the quadratic cost criterion. In addition, the system dynamics is imperfectly known, and the uncertain parameters may be varying in a random manner. The analytical results showed that the O.L.F.O. adaptive control gains are "modulated" by the current and future uncertainty of the parameter estimation. A second suboptimal feedback technique was used to obtain the enforced separation scheme controller in the form of a cascade of the extended Kalman filter with a deterministic actuator. This arbitrary use of the Separation Theorem led to a feedback controller design that performed on the average a little inferior than the O.L.F.O. control for stable systems and much superior than the O.L.F.O. control for unstable systems. The performance comparison on a first order system was based on the evaluation of the original cost functional provided by the Monte Carlo method for lack of analytical tools.

REFERENCES

(1) Tse, E., "On the Optimal Control of Linear Systems with Incomplete Information", M.I.T. Electronic Systems Laboratory report ESL-R-412, January 1970.

(2) Tse, E., Athans, M., "Adaptive Stochastic Control for Linear Systems", Pt. I and Pt. II in Proceedings 1970 IEEE Conference on Decision and Control, Austin, Texas 1970.

(3) Tse, E., and Athans, M., "Adaptive Control for a Class of Linear Systems", IEE Trans. Auto. Control, pp. 38, 1972.

(4) Chase-McRae, E., "Linear Decisions with Experimentation", presented at the National Bureau of Economic Research on Stochastic Control and Economic Systems, Princeton, New Jersey, May, 1972.

(5) Lainiotis, D., "Optimal Adaptive Control: A Nonlinear Separation Theorem", International J. Control, 15, No. 5, pp. 877, 1972.

(6) Saridis, G.N., and Lobbia, R.N., "Parameter Identification and Control of Linear Discrete Time Systems", IEEE Trans. on Auto, Control, 17, pp. 52, 1972.

(7) Ku, R., "Adaptive Control of Stochastic Linear Systems with Unknown Parameters", M.I.T. Electronic Systems Laboratory report, ESL-R-477, May, 1972.

(8) Aoki, M., Optimization of Stochastic Systems, New York, Academic Press, 1967.

(9) Bar-Shalom, Y., and Sivan, R., "The Optimal Control of Discrete Time Linear Systems with Random Parameters", IEEE Trans. on Auto. Control, 14, pp. 3, 1969.

(10) Dreyfus, S.E., Dynamic Programming and the Calculus of Variations, New York, Academic Press, 1965.

(11) Ku, R., and Athans, M., "On the Adaptive Control of Linear Systems Using the Open-Loop-Feedback Optimal Approach", M.I.T. Electronic Systems Laboratory, paper ESL-P-484, June, 1972.

(12) Gunckel, T.L., and Franklin, G.F., "A General Solution for Linear Sampled-Data Control", Transactions ASME, Journal of Basic Engineering, 85 D, pp. 197, 1963.

(13) Joseph, P.D., and Tou, J., "On Linear Control Theory", AIEE Transactions (Applications an Industry), Pt. II, 80, p. 193, 1961.

(14) Levis, A., "On the Optimal Sampled-Data Control of Linear Processes", M.I.T., Ph.D. Thesis in M.E., 1968.

(15) Jazwinski, A.H., Stochastic Process and Filtering Theory, New York, Academic Press, 1970.

(16) Lainiotis, D.G., Upadhyay, T.N.,and Deshpande, J.G., "Optimal Adaptive Control of Linear Systems", Proceedings 1971 IEEE Conference on Decision and Control, Miami Beach, 1971.

(17) Stein, G., and Saridis, G.N., "A Parameter Adaptive Control Technique", Automatica, 5, No. 6, pp. 731, 1969.

(18) Tse, E., Meier, L., and Bar-Shalom, Y., "Dual Control of Stochastic Nonlinear Systems", Proceedings 1971 IEEE Conference on Decision and Control, Miami Beach, 1971.

Table I

Summary of the Monte Carlo Simulation (Sampling)

Sample size = 20, $c = 1$, $b_o = 1$, $x_o = 0$,

$\Sigma_{xo} = 3$, $\Xi = 0.004$, $r = 1$, $N = 30$

Simulation	a_o	Σ_{ao}	Σ_{bo}	Θ	q	Cost when Parameters are Known	Cost of O.L.F.O.	Cost of Enforced Separation
U1	1.2	0.0049	0.25	1.0	10	109.93	301.93	148.27
U2	1.2	0.0009	0.25	1.0	10	109.93	307.33	146.52
U3	1.2	0.0009	0.25	1.0	2	25.22	84.81	45.97
U4	1.2	0.0049	0.25	1.0	2	25.22	85.26	45.95
U5	1.2	0.0009	0.25	4.0	10	333.55	802.48	420.75
U6	1.2	0.0009	0.25	9.0	10	676.11	1631.40	862.33
U7	1.2	0.0049	0.0	1.0	10	109.93	126.05	119.37
U8	1.2	0.0	0.25	1.0	10	109.93	155.09	144.18
S1	0.8	0.0049	0.25	1.0	10	30.55	34.99	35.65
S2	0.8	0.0049	0.25	4.0	10	37.21	42.41	43.97

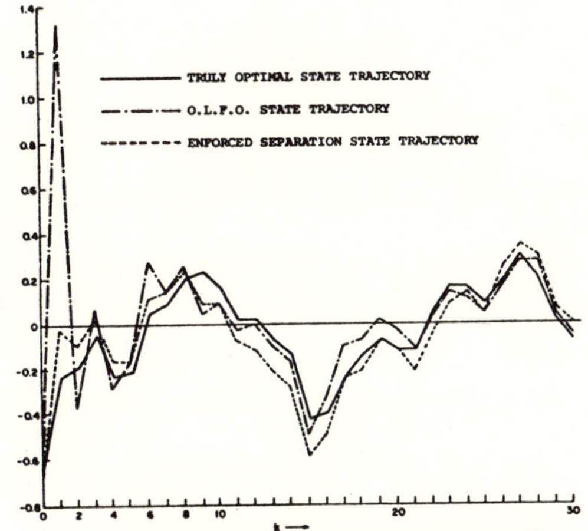

Fig. 1 Comparison of the average response of the unstable systems U1 when the parameters a(k) and b(k) are known (optimal stochastic control) and when the parameters are unknown (O.L.F.O. method and enforced separation scheme). The sample noise sequence was the same. Sample size = 20.

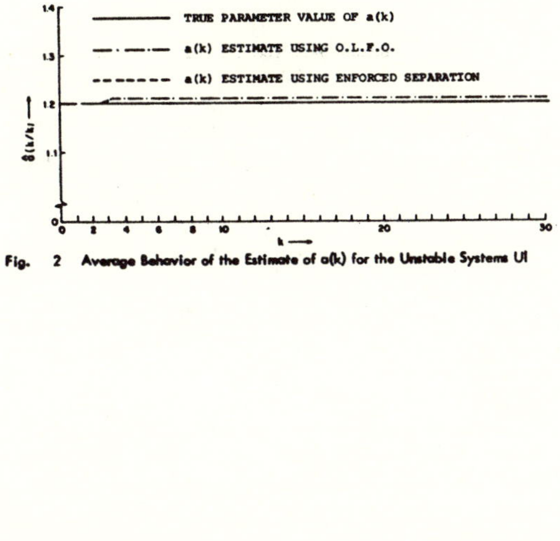

Fig. 2 Average Behavior of the Estimate of a(k) for the Unstable Systems U1

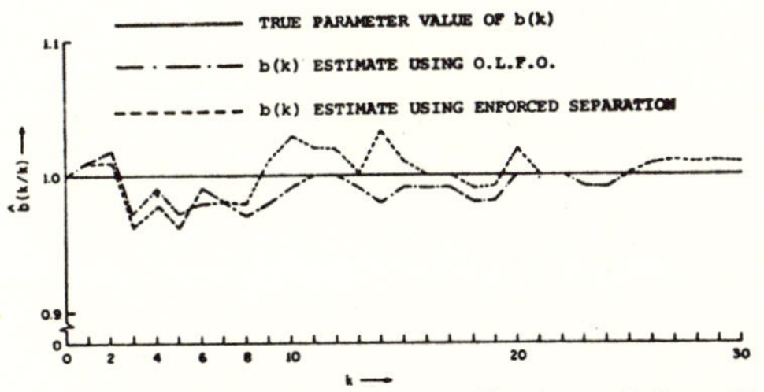

Fig. 3 Average Behavior of the Estimate of b(k) for the Unstable Systems U1

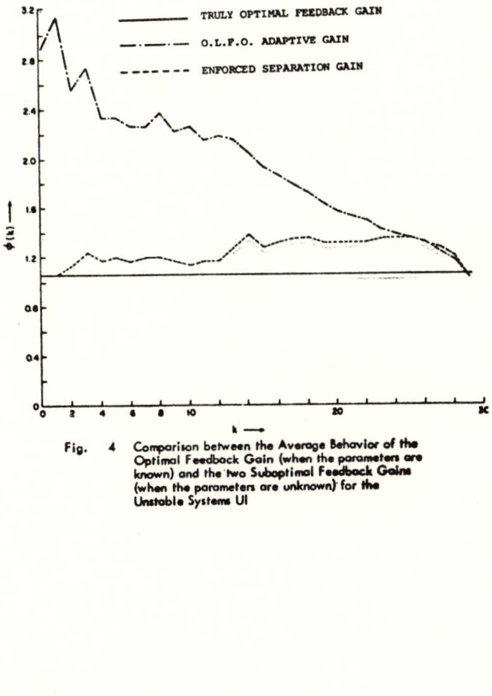

Fig. 4 Comparison between the Average Behavior of the Optimal Feedback Gain (when the parameters are known) and the two Suboptimal Feedback Gains (when the parameters are unknown) for the Unstable Systems U1

ADAPTIVE IDENTIFIER TRANSFER FUNCTIONS

V.M Chadeev
Institute of Control Sciences
Moscow, USSR

Facing a decision concerning the use of a control system in a technological plant the manager would first wish to have an estimate of the expected economic return. For stationary plant control systems such an estimate can be obtained if a plant mathematical model and the statistical characteristics of input signals and disturbances are available. For nonstationary plants, which industrial plants usually are, an estimate of the control system efficiency would require also the knowledge of the statistical characteristic of the plant internal parameters.

For industrial systems the most reasonable system is, in our view, a system with an identifier (Rajbman, Chadeev, 1971). Such a system can be made insensitive to plant parameter variations as well as to disturbances. A circuit use in stabilization system of a linear dynamic nonstationary plant is shown in Fig. 1

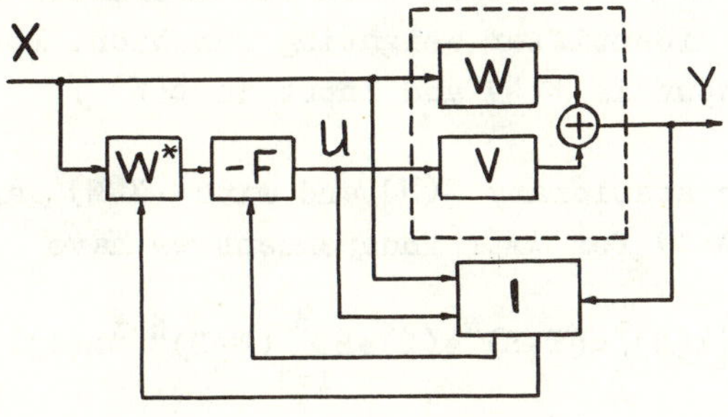

where X the vector of input disturbances; Y the plant output; U control V the control channel transfer function: I _ the identifier; W^* an estimate of the disturbance channel transfer function; F an estimate of the inverse transfer function of the control channel.

To calculate the effect of stabilization in such a system, we should know the statistic characteristics of the transfer functions W and V . Let us assume that V is known for certain and the problem consists in determining the statistical characteristics of the transfer function W parameter (or, equivalently, in estimating the current accuracy of determining W by the identifier).

Let us consider a circuit for determination of statistical characteristics for a linear multidimensional statistical plant parameters

$$Y(N)=h'(N)X(N)$$

where the random variation of the parameters X is a stationary process.

An identifier can be regarded as a parametric filter which converts plant nonmeasurable parameters h into their estimates k(N) . The general approach to obtaining the plant statistical characteristics is as follows. On the knowledge of samples of the input, $X(N)$ and output, $Y(N)$ a certain identification algorithm represents the estimation of the plant parameters. The resultant process $k(N)$ is in a way a representation of variations in the plant parameters $h(N)$.

Determining the process statistical characteristics $k(N)$ and knowing the "transfer function" of identifier I, certain conclusions concerning the process characteristics $h(N)$ can be made. Note that since the "transfer function" depends on random values of input signals, $h(N)$ cannot usually be restored unambiguously in the presence of $k(N)$.

Let it be emphasized that the identifier response depends on the probabilistic distribution of the input signal $X(N)$ although variations of the plant parameters h and input signals X may be independent. A wider spectrum of the input signals X leads, as a rule, to increased response of the identifier; conversely, when in the case of slow variation of X the estimates $k(N)$ follow the plant parameter slower.

Obtain the adaptive identifier transfer function. For (Rajbman, Chadeev, 1966)

$$k(N)=k(N-1) + \frac{Y(N)-Y^*(N)}{X'(N)\cdot X(N)}X(N) \tag{1}$$

where $X(N)$ the plant input vector; $Y(N)$ the output, $(Y(N)=h'(N)\cdot X(N))$; $h(N)$ the plant parameter estimate vector; $Y^*(N)=k'(N)\cdot X(N)$ the plant output estimate.

Introduce the notation

$$A(N)=X(N)X'(N)(X'(N)X(N))^{-1} \tag{2}$$

Then the algorithm can be rearranged as

$$k(N)=(E-A(N))k(N-1)+A(N)h(N). \tag{3}$$

The solution of equation (3) is

$$k(N)=\prod_{i=1}^{N}(E-A(i))k(0)+A(N)h(N)+ \\ +\sum_{s=1}^{N-1} A(s)\prod_{i=1}^{N}(E-A(i))h(s). \tag{4}$$

At arbitrary inputs the relation of $k(N)$ and $h(N)$ is generally random. Of the greatest practical interest is the case where $X(N)$ is a stationary process. Then for the conditional mathematical expectation $M[k(N)/h(N), h(N-1),\ldots,h(0),k(0)]$ we can write

$$M[k(N)]=M\left[\prod_{i=1}^{N}(E-A(i))\right]k(0)+M[A(N)]h(N)+ \\ +\sum_{s=1}^{N-1} M\left[A(s)\prod_{i=s+1}^{N}(E-A(i))\right]h(s). \tag{5}$$

For stationary processes all mathematical expectations in the right-hand part of the equality can be computed on the knowledge of experimental data. The term $M[A\prod(E-A)]$ represents the ordinates of the identifier weighting function. Its output is $k(N)$ and input is $h(N)$.

For stationary $X(N)$ and with $X(N)$ and $X(N-i)$ for $i\neq 0$ independent we have

$$M[k(N)]=(E-B)^N k(0)+B\sum_{i=1}^{N}(E-B)^{N-i}h(i) \tag{6}$$

where $B=M\left[\frac{XX'}{X'X}\right]$. This solution corresponds to the difference equation

$$M[k(N)]=(E-B)k(N-1)+Bh(N). \qquad (7)$$

Obtain estimates for the maximal and minimal response of the identifier.

The maximal response will be obtained when the plant receives series of orthogonal input vectors. In this case the identification equation for each channel will be of the form

$$k_j(N)=h_j(N-\bar{N}), \qquad (j=1,2,\ldots,n) \qquad (8)$$

where $\bar{N}$ is modulo n subtraction.

The plot for a three-dimensional plant is shown in Fig. 2.

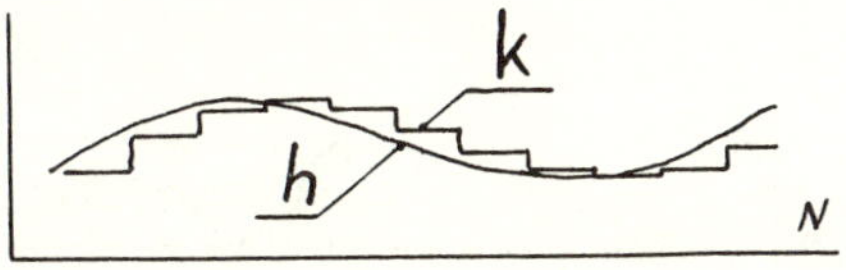

Parameter estimates are true ones with a shift averaging $(N+1)/2$ cycles.

The minimal speed represents a the constant input vector $X(N)$. In this case the difference equation (3) is solved simpler

$$k(N)=Ak(0)+(E-A)h(N). \qquad (9)$$

The maximal and minimal responses are obtained at deterministic input signals. In practie the disturvbances are random more often than not; therefore the formula (5) should be used directly in order to compute the actual weighting function.

To compute the current error of plant parameter fillow up would also be of interest. This error

$$M[(h(N)-k(N-1))'(h(N)-k(N-1))]$$

can be computed by using the formula (5) where $h(N)$ is a stationary random process with a known correlation function. The resultant formulae are too unwieldy to be cited here.

As an example let us compute the resifual follow up error when inputs are statistically independent and $M[X]=0$ $M[X(N)X'(N)]=E\sigma^2/n$ then equation (6) can be rearranged as

$$M[k(N)]=(1-1/n)^N k(0)+(1/n)\sum_{i=1}^{N}(1-1/n)^{N-i}h(i)$$

Thence follows that at independent inputs the identifier is on the average an inertial element whose response depends on the plant dimensionality alone. The higher the dimensionality the worse is the response.

We cannot go into detailed discussion of the effect of moises and measurement errors on the identifier response but would note that as a rule higher noise level reduces response.

An adaptive identifier was used, in particular, to predict the wall thickness of a tube leaving a mill on the knowledge of the slab wall thickness and diameter. An adaptive algorithm was used.

Fig. 3 protrays the plots of inputs, and Fig. 4, of outputs. Figs 5, 6, and 7 show graphs of plant parameter current estimates obtained by an adaptive identifier. Estimates obtained by the least squares method are shown as dotted lines. Adaptive identification permitted increasing the coefficient correlating the actual and predicted wall thickness from 0.45 for the least square method to 0.6 which is equivalent to reducing the residual dispersion by 20%. The economic efficiency of adaptation can be easily computed on the knowledge of these data.

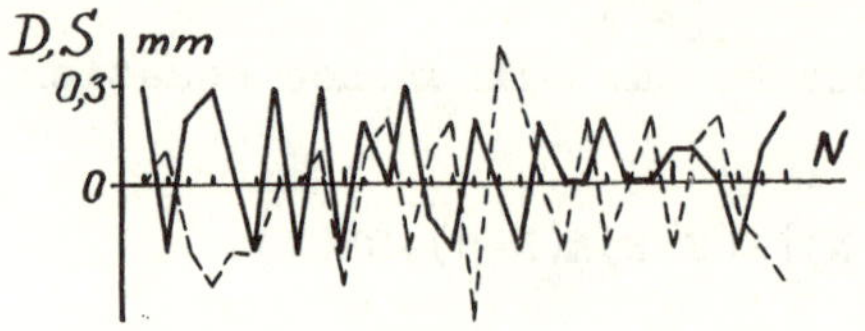

Fig. 3.

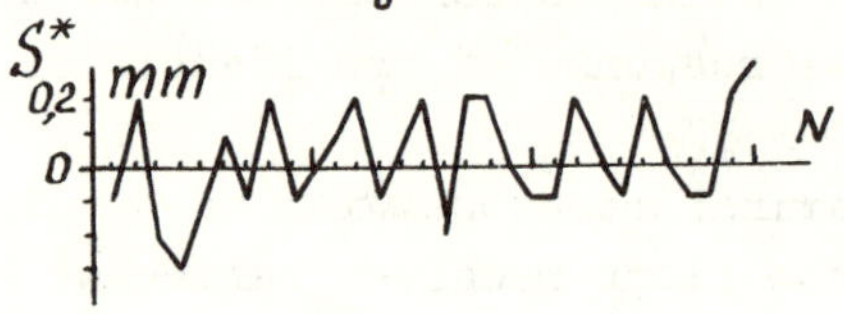

Fig. 4.

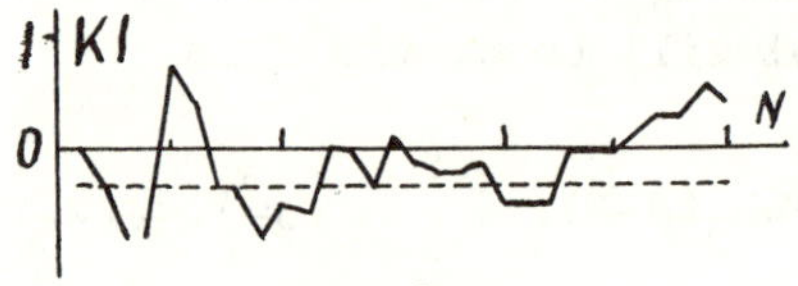

Fig. 5.

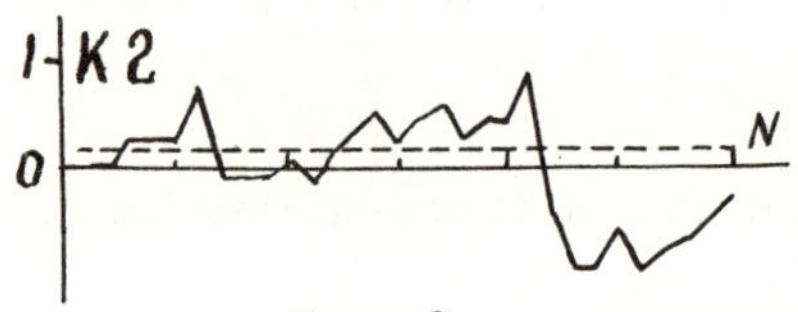

Fig. 6.

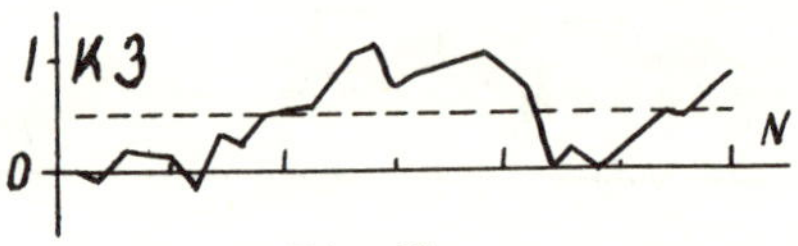

Fig. 7.

REFERENCES

(1) Rajbman N.S., Chadeev V.M Tube rolling mill adaptive control system.

3rd IFAC/IFIP Conference on Digital Computer Application to Process Control, Helsinki, 1971.

(2) Rajbman N.S., Chadeev V.M. Adaptivnye modeli v sistemakh upravleniya. "Sovetskoe radio". M. 1966.

ADAPTIVE CONTROL AND IDENTIFICATION BY A MODEL

H. Nitsche
SIEMENS AG
Karlsruhe, FRG

A method for parameter identification of linear or linearizable control systems is presented. The method is distinguished by simplicity and a relatively short measure-time. After this the results of analog simulation are shown. Finally you will find the analysis of the efficiency of the method with a reference to special applications in adaptive control systems.

1. INTRODUCTION

If wide parameter changes make worse the dynamic behaviour of automatic control systems or even result in instability, an appreciable improvement can be obtained by adaptive control. In such systems the controller adapts itself automatically to changed operating conditions.

A basic problem in adaptive systems is the parameter identification which must meet three essential requirements:

1. The system shall as far as possible not be stimulated by test signals.
2. Disturbances which are not accessible shall not invalidate the measurement inadmissibly.
3. The measurement must be carried out with sufficient speed and in an as simple as possible way.

A well known method uses a model which may be simulated on an analog computer. When now both inputs of controlled system and model will be excited by the same stochastic signal, it can be ascertained by comparison of both outputs, whether the parameters of the model are correctly adjusted.

The follow-up of one single model parameter can be carried out by a superimposed control loop. If, however, several parameters are to be measured, no comparable simple method has been known until now /1/. For such cases the general procedure has been to form a quality criterion and to balance it to a maximum by varying the model parameters /2/. As, however, the search process connected with that takes much time, such methods could not succeed in practice, all the more since they are mostly associated with a considerable expenditure. That is why below a new method of identification shall be presented for the measurement of the parameters of linear or linearizable systems.

2. WAY OF SOLUTION

Starting from the simple method for the measurement of one single parameter, it shall be demonstrated how a most simple solution can be found even at the identification of several parameters. It is not any restriction of universality, if at first only systems with pure low-pass character are examined:

$$F_s(s) = \frac{1}{\alpha_0 + \alpha_1 s + \dots + \alpha_n s^n} \qquad (1a)$$

The model may have the same structure:

$$F_s'(s) = \frac{1}{\alpha_0' + \alpha_1' s + \dots + \alpha_n' s^n} \qquad (1b)$$

To identify n+1 model parameters, n+1 equations would be needed. When transferred to measuring technique this means that n+1 separate measurements have to be made.

These measurements can be carried out in the frequency range which means that always special spectral contents have to be picked out of the power density spectrum of the output signals. If these are further processed as is routine with the method for identification of only one parameter, finally n+1 different signals are obtained which then can be used immediately for influencing always one specified model parameter (Fig. 1). The input variable y may show a sufficient wide spectrum. With the help of selecting filters of the centre frequencies ω_ν then special parts can be picked out of the signals x and x' and be compared in an interpretation unit. The result is the signal $P_v(t)$ whose mean value influences directly the parameter α_v' of the model.

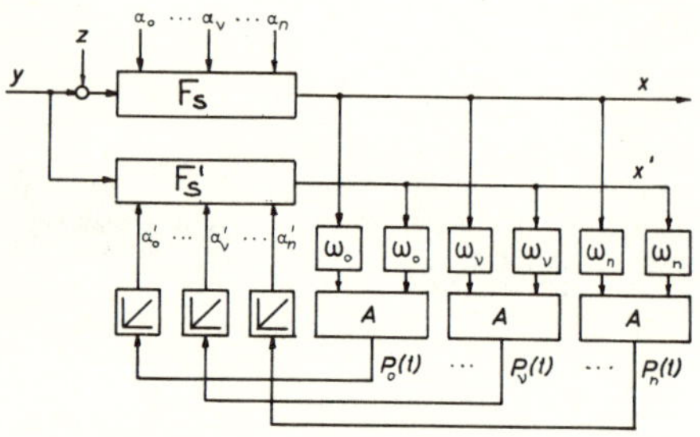

Fig. 1 Principle of model comparison in the frequency range.

That such a solution is absolutely reasonable shows a simple example of 1st order and the parameters α_o and α_1. It is evident that signals of the frequency $\omega_o = 0$ 1/s provide directly a measure for the parameter α_o, whilst signals of a higher frequency ω_1 can provide a measure for the parameter α_1.

As regards the interpretation units, known solutions can be reverted to. In this context, however, two solutions have been especially suitable since they are leading to linear system equations (Fig. 2). The signals x and x' are compared by subtraction. To eliminate the disturbance z on the average, the output will be correlated with the undisturbed signal x'. The second solution shown is equivalent to the first one. Since, however, the multiplier is replaced by a switch, this solution is more economic.

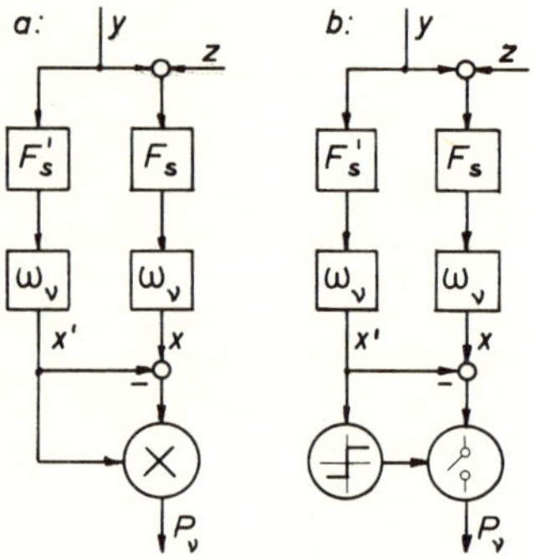

Fig. 2 Two equivalent solutions for the interpretation unit.

3. SYSTEM EQUATIONS FOR AN EXAMPLE

The derivation of the state equations becomes particularly simple when starting from the following assumptions:

1. The selective filters may permit passing of practically one single frequency only.
2. The adjustment of the model parameters must not be too rapid.

Under these assumptions fundamental points of view of the method shall be demonstrated by an example. Thereby especially the question shall be of interest, how the filters resp. their centre frequencies ω_v must be selected so that the system is stable. In order that conditions remain easy to survey, controlled system and model shall have the following simple transfer functions:

$$F_S(s) = \frac{1}{\alpha_o+\alpha_1 s} \qquad F_S'(s) = \frac{1}{\alpha_o'+\alpha_1' s} \tag{2}$$

For stability the deviations from the operating point are of interest only:

$$\alpha_o' = \alpha_o + \delta_o \qquad \alpha_1' = \alpha_1 + \delta_1 \tag{3}$$

When considering the characteristics of the interpretation unit, the following non-linear state equations will be arrived at after an intermediate calculation /3/:

$$\begin{aligned} \frac{d\delta_o}{dt} &= -K_o C_o (\alpha_o \delta_o + \omega_o^2 \alpha_1 \delta_1) \\ \frac{d\delta_1}{dt} &= -K_1 C_1 (\alpha_o \delta_o + \omega_1^2 \alpha_1 \delta_1) \end{aligned} \tag{4}$$

The factors K_o and K_1 are proportional to the root out of the power density spectrum of the input signal at the spots ω_o and ω_1. C_o and C_1 are functions of the model parameters:

$$\begin{aligned} C_o &= |F_S'(j\omega_o)|^2 |F_S(j\omega_o)|^2 \\ C_1 &= |F_S'(j\omega_1)|^2 |F_S(j\omega_1)|^2 \end{aligned} \tag{5}$$

The variation of the factors K_o and K_1 must not have any influence on the stability. If now C_o and C_1 are varying subsequent on the non-linear relation (5), this has practically the same effect. Therefore the influences of K_v and C_v can be combined to a parameter D_v and calculation can be continued as for a linear system which would simplify the examination of the stability essentially. Only the variation of D_v must be observed. Equation (4) then gets following form:

$$\begin{bmatrix} \delta_o \\ \delta_1 \end{bmatrix} = \underbrace{\begin{bmatrix} -D_o\alpha_o & -D_o\omega_o^2\alpha_1 \\ -D_1\alpha_o & -D_1\omega_1^2\alpha_1 \end{bmatrix}}_{A} \begin{bmatrix} \delta_o \\ \delta_1 \end{bmatrix} \qquad (6)$$

Matrix A alone is decisive for the stability, and so finally, after a short intermediate calculation, the following simple condition for dimensioning the filters is obtained:

$$\omega_o < \omega_1 \qquad (7)$$

Therefore, as expected, parameter α_o' is influenced by the lower frequency contents of the output signals and parameter α_1', however, by the higher ones.

4. RESULTS OF SIMULATION

Example (2) has been simulated on an analog computer. The result is plotted in the form of phase trajectories in the two-dimensional space of state (Fig. 3).

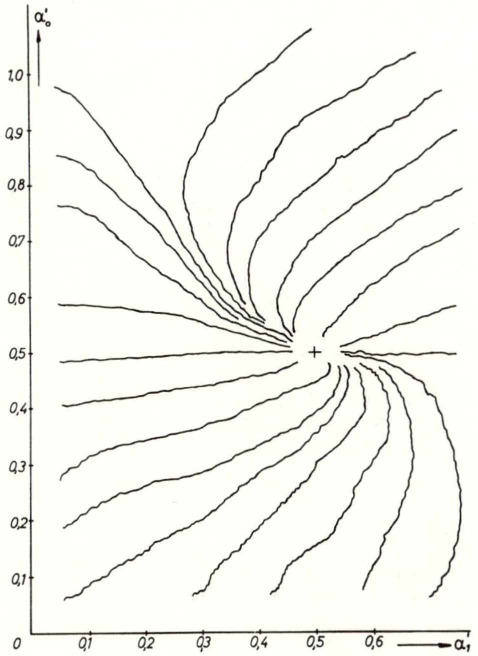

Fig. 3 Field of phase trajectories with white noise as disturbance. The operating point is $\alpha_o = \alpha_o' = 0{,}5$ and $\alpha_1 = \alpha_1' = 0{,}5$.

The coordinates are the model parameters α_o' and α_1'. The irregular course of the trajectories is due to a stochastic disturbance. They all aim at the operating point, the system therefore is stable.

Thereupon a system of the 2nd order with the three parameters α_o, α_1 and α_2 has been simulated. Here the state space is three-dimensional. Fig. 4 shows how, starting from zero, parameter α_2' adapts itself to various operating points. The behaviour of the other two parameters is quite similar. As regards the duration of the adaption, it can be roughly estimated that this process must at least by factor 10 be slower than the controlled system itself.

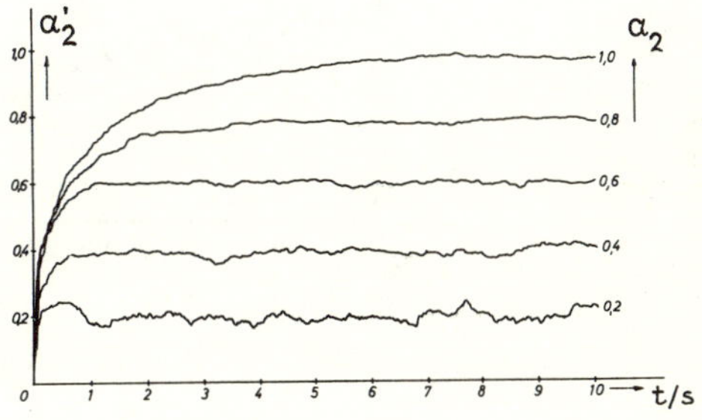

Fig. 4 Adaptation of α_2' with the initial conditions $\alpha_o' = \alpha_1' = \alpha_2' = 0$.

When selecting $\alpha_o = \alpha_o' =$ const., this model parameter can be fixedly adjusted, and only just α_1' and α_2' need be determined automatically. Thus a two-dimensional space of state will be reached (Fig. 5). The reason for the smooth course of the trajectories lies in the absence of a disturbance.

5. GENERAL SYSTEM EQUATIONS

It is of particular interest whether the described method is of general validity or is limited to special cases. Indeed, matrix A shows a regular design. As can be proved, this is also the case with more complicated examples. But then it is more practical to write equations (1) in a complex form:

$$F_s(j\omega_\nu) = \frac{1}{Re_\nu + jIm_\nu}$$

$$F_s'(j\omega_\nu) = \frac{1}{Re_\nu' + jIm_\nu'} \qquad (8)$$

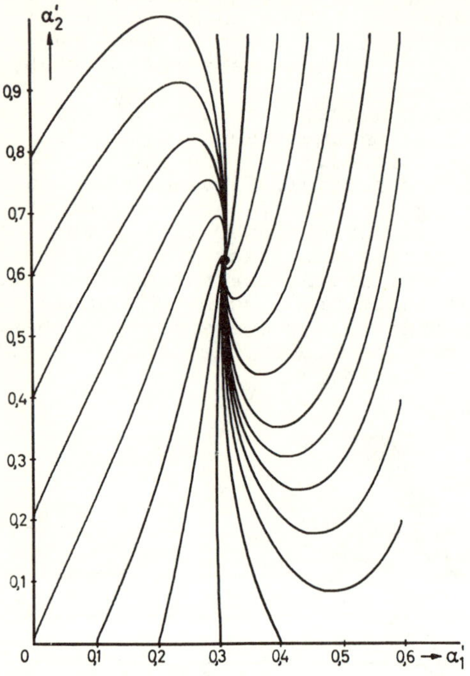

Fig. 5 Idealized field of phase trajectories.

The result is following matrix:

$$A = \begin{bmatrix} -D_oRe_o & -D_o\omega_oIm_o & +D_o\omega_o^2Re_o & \cdot\cdot \\ -D_1Re_1 & -D_1\omega_1Im_1 & +D_1\omega_1^2Re_1 & \cdot\cdot \\ \cdot\cdot\cdot & \cdot\cdot\cdot & \cdot\cdot\cdot & \\ -D_nRe_n & -D_n\omega_nIm_n & +D_n\omega_n^2Re_n & \cdot\cdot \end{bmatrix} \quad (9)$$

This matrix of order n can be examined in general. With a not very simple argumentation /3/ the confirmation will finally be reached that the method is valid for low-pass systems of any order.

6. SOME SPECIAL PROBLEMS

A highly important problem is the approximation of systems of higher order by a simpler model. Theoretical investigations and practical experiments have shown that the appearing models are always quite reasonable. This will mean that the model shows for instance a transient function which is throughout very similar to that of the system /3/.

The method is not applicable in any case, but always for identification of parameters in the numerator polynomial only or the denominator polynomial only.

Finally a reference shall once more be made to the opening of this report where adaptive control has been mentioned. Thereafter indeed the identification stood in the foreground. It shall, however, still be pointed out that the described method without additional effort can be thus modified that it performs all the problems of adaptive control simultaneously /3/. A description of the connecting problems, however, shall be renounced at this point.

REFERENCES

/1/H.Buchta, G.Dörfel, P.Rieger: Modellverfahren bei der Kennwertermittlung mit stochastischen Signalen. Messen-Steuern-Regeln 1970,H.3/4.

/2/K.J.Aström, P.Eykhoff: System Identification - A Survey. IFAC-Symposium Prag 1970.

/3/H.Nitsche: Adaption und Identifikation von Regelkreisen im Frequenzbereich. Stuttgarter Dissertation 1971.

PLANTS IDENTIFICATION BY THE STATISTICAL REGULARIZATION METHOD

A.I.Beliaevsky, R.A.Poluektov, G.N.Solopchenko

Agrophysical Institute

Leningrad, USSR

To identify linear plants an iterative calculating procedure is suggested which is based on the use of a regularizing sequence of the estimators of the plant operator and a statistic stopping rule. The notion of a point and uniform statistical equivalence of models is introduced so as to solve the identification problem. The convergence of the method is investigated and some examples of its realization are considered.

I. INTRODUCTION

Let us consider the problem of estimating an impulse response function on the basis of measurements at discrete moments of the output resulting from a specified input signal. The input-output relation of the system has the form

$$\int_0^{t_i} u(t_i-v)h(v)dv = y(t_i)+n(t_i), \quad i = \overline{1,l} \qquad (I.I)$$

where $u(t)$ is a specified input signal, $h(v)$ is an unknown impulse response function, $y(t_i)$ is an actual value of the output signal at times t_i, $t_i \in [0,T]$ $n(t_i)$ is random error of a measurement, $[0,T]$ is an observation interval. Assume that (i) input signal is generated by a special device the error of which is negligible, (ii)set T_l of values t_i is fixed and depends on measuring equipment, (iii) measurement errors $n(t_i)$ form a random sequence with a zero mean value and a covariance matrix W having the elements

$$W_{ij} = E(n(t_i)\cdot n(t_j)),$$

where $E(\cdot)$ is expectation.
The problem of estimating $h(v)$ from (I.I) is illegitimate. Therefore, in order to solve it a special method of regularization is to be developed. This paper is devoted to the description of one of the possible methods for constructing a stable solution of equation (I.I). This method of regularization differs from those already reported in that the statistical nature of the measurement error is allowed for and the choice of the regularization parameter cannot be arbitrary.

2. STATEMENT OF THE IDENTIFICATION PROBLEM

Let us state the identification problem following Åström (I97I), a few remarks would be in order, however. One of the reasons why the problem under study is illegimate is the infinite dimensionality of the description of non-parametric model that, generally speaking, results in the non-uniqueness of the solution of (I.I) even in the absence of the measurement error. In this situation, as was indicated by Stratonovich (I970), one should look for an estimate of the desired function in some finite-dimensional (l-dimensional) subspace of a initial Banach space. Let us denote this space by H_l and $f(v)=\{f_k(v),\ k=\overline{0,l-1}\}$ be the basis in H_l. Then the function $h(v)$ can be found in the form

$$h(v)=\sum_{k=0}^{l-1} c_k f_k(v) = (c,f(v)), \qquad (2.I)$$

where $c = \{c_k,\ k = \overline{0,l-1}\}$ is an unknown parameters vector. This kind of representation permits using instead of (I.I) a system of linear algebraic equations:

$$\sum_{k=0}^{l-1} c_k \int_0^{t_i} u(t_i-v)\, f_k(v)dv = y(t_i)+n(t_i), \quad i = \overline{1,l}, \qquad (2.2)$$

or, in the matrix form,

$$Fc = y+n. \qquad (2.3)$$

If the basic functions are chosen so that the l x l-matrix F is non-singular, there is the only solution of equation

$$c = F^{-1}(y+n). \qquad (2.4)$$

In practical situations, however, the matrix F appears to be ill-conditionned, which makes this solution sensitive to errors n. Therefore, regularizing algorithms should be used for solving (2.2). It should be mentioned that the sensitivity of the solution of (2.4) to disturbances can be estimated with the help of a conditionality number of the matrix F (Collatz (1964)):

$$k(F) = \sqrt{\frac{\lambda_{max}(F^T F)}{\lambda_{min}(F^T F)}}, \qquad (2.5)$$

where T is the symbol of matrix transponizing.

In order to obtain an error-immune solution we define a parametric family of matrices $\{F_a, a > 0\}$, possessing the property:

j. if $a_1 > a_2 > 0$, $k(F_{a_1}) < k(F_{a_2})$,

jj. at $a \to 0$, $F_a \to F$ (at least in every point) and at a=0 $F_o = F$.

Let us call the parameter a a generalized parameter of regularization. This sequence of matrices is associated with a sequence of c_n^W, of parameter vector estimators, the weighting function $h_a^W(v)$ and output signals y_a^W obtained by substituting c_a^W in (2.3).

The values of c_a^W, h_a^W and y_a^W depend, of course, on the errors intensity w=SpW. Let us a system model be any its approximate mathematical description with a known or estimated accuracy of approximation. Let us assume now a class of models, a class of input signals and define the meaning of model equivalence.

1. A class of models, M, is formed by a set of estimators, h_a^W or c_a^W, with w=const, a=var. In general, these estimators are statistical.

11. A class of input signals, U, is formed by a set of physically realizable signals completed with a set of limit points.

111. Assume that the models h and h_a^W are statistically equivalent in a point if at a fixed input signal $u(t) \in U$, the distance between the vectors of discrete values of the output of these models, y and y_a^W, is statistically negligible, i.e. if in checking a statistical hypothesis H_o: $y = y_a^W$ against an alternative H_I: $y \neq y_a^W$ with a prescribed level of probability, P, the zero hypothesis is not rejected. The notion of point statistical equivalence is, obvious, transformed into the notion of uniform statistical equivalence over the whole set U or over its subsets.

The definitions introduced 1.-111. completely the specific nature of the identification method in accordance with Zadeh's formulation cited by Åström (1971). The method which is based on notions 1. and 111. we call "the method of statistical regularization" (Solopchenko (1970), Poluektov (1971)).

3. IMPLEMENTATION

The specific form of the statistical regularization algorithm depends upon experimental conditions under which data for solving the identification problem are obtained. Let us consider first the case when conditions i.-iii. take place.

3.I. Let us use the matrix equation (2.3).

Let $f_k(v)$ be the basic functions in (2.I) which are determined over $[0,T]$. Form a parametric family of models $\{h_r(v), r=\overline{1,l}\}$ in the following way

$$h_r(v) = \sum_{k=0}^{r-1} c_k f_k(v), \quad r=\overline{1,l}. \qquad (3.I)$$

If $a = \frac{l-r}{lr}$, then from (3.I) and (2.2) we deduce a family of matrices $\{F_a, a > 0\}$. Property j. of this family follows from Shturm's separation theorem (Bellman (1960)); property jj. is evident.

The sequence of estimators c_a^W can be obtained by the least squares method:

$$c_a^W = (F_a^T W^{-1} F_a)^{-1} F_a^T W^{-1}(y+n). \qquad (2.3)$$

The sequence y_a^W is obtained by substitution

$$y_a^W = F c_a^W. \qquad (3.3)$$

Decreasing a monotonically by successive increasing r by a unit we shall test the hypothesis H_o: $y_a^W = y$ against H_I: $y_a^W \neq y$ with the help of the criterion χ^2 with (l-r) degrees of freedom and a confidence level P (Rao (1965)). If at some $a=a_o$ the hypothesis H_o has not been rejected, estimate $c_{a_o}^W$ and an associated model $h_{a_o}^W(v)$ is assumed to be as the solution to the problem.

3.2. Eorm a family $\{F_a, a>0\}$ in the following way:

$$F_a = (F^T)^{-1}(F^TF+aI), \qquad (3.4)$$

where I is a unit matrix, $a>0$.
The properties j., jj. of the received family are proved easily.
Sequences $\{c_a^w\}$, $\{h_a^w(v)\}$, $\{y_a^w\}$ are defined by the formulas:

$$c_a^w=(F^TF+aI)^{-1}F^T(y+n), \qquad (3.5)$$

$$h_a^w(v)=(c_a^w,f(v)), \qquad (3.6)$$

$$y_a^w = Fc_a^w. \qquad (3.7)$$

As the initial value of the parameter a one can take $a_o=0,1\max_i f^2_{ii}$ where f^2_{ii} is a diagonal element of the matrix F^TF. A successive decrease in the parameter a is conveniently performed using the rule:

$$a_{s+1} = \frac{a_s}{\sqrt{2\frac{\chi_s^2}{G}}}, \qquad (3.8)$$

where χ_s^2 is the value of statistics of the χ^2-criterion at the s-th iteration, G is the critical value of the criterion which is chosen from tables. The stopping rule of this procedure is similar to the rule (3.I), where for the statistics of criterion χ^2 one should take 1 degrees of freedom.
Here, it is possible to use the functions which are piecewise constant over $(t_i, t_{i+1}]$ or functions resulting in a piecewise linear approximation of h(v) to act as the functions $f_k(v)$.
Note that the estimators (3.5) are no longer statistical.
Of course, the above examples cover far from all implementations. Consider briefly the questions connected with removing constraints placed upon i-iii.
If the covariance matrix W is unknown, it is necessary to perform $N>1$ experiments and, as a result, are obtained N-samples of the vectors y+n. Then an estimator S of the matrix W is calculated which is then used in evaluating c_a^w and in testing the hypothesis H_o. In this case the hypothesis should be tested by using Hotelling's criterion T^2 (Rao (I965)).
If continuous form of the output signal can be obtained and the set T be changed, a convenient method of forming a family $\{F_a, a>0\}$ can be such in which the time between readings successively decreases and, as a result of it, the dimensionality of the matrix F increases. Here, the value T/l is used as the parameter a.
In the case when one cannot neglect the measurement error of the input signal in each value of parameter a, these errors should be reduced to the output with the help of (I.I) by replacing the variable $t_i-v=z$. If the measurement errors of the input and output signals are independent, then as the part of the matrix W in evaluating vector c and testing hypothesis H_o can be played by a sum of the covariance matrix of the output signal measurement error with the covariance matrix of the measurement errors of the input signal referred to the output.

4. CONVERGENCE

Convergence of statistical regularization algorithms should be studied in two stages:
- estimation of probability $P(H_o)$ that the hypothesis H_o will not be rejected at fixed w,u(t) and an accuracy q of calculations;
- establishment of the fact of the convergence of $h^w_{a_o}(v)$ to h(v) as $w\to 0$ and fixed u(t) and a_o.

The methods for studying the convergence are illustrated with two examples concerning the algorithms 3.I and 3.2 in the assumption that the W-matrix has the form $\sigma^2 I$. This assumption does not restrict the generality of the results.

4.I. The upper bound of the statistics of the criterion χ^2 for the algorithm 3.I can be made:

$$\frac{1}{\sigma^2}(y+n-Fc_a^w)^T(y+n-Fc_a^w) \leq \frac{3}{2}\Big[\chi^2_{l-r} + \frac{1}{4}\Big(\frac{\lambda_1}{\lambda_r}\Big)^2\frac{q}{\lambda_r}\chi^2_g\Big], \qquad (4.I)$$

where χ_f^2 is a random quantity which is subjected to the χ^2-distribution with f degrees of freedom, $\lambda_1=\lambda_{max}(F_a^TF_a)$, $\lambda_r=\lambda_{min}(F_a^TF_a)$, q=rd, d is a round-off error in calculations

$$g=ent\Big[\frac{y^Ty}{\sigma^2}+1\Big],\ \sigma^2=\frac{w}{l},\ r=\frac{l}{al-1}.$$

The probability that the value of statistics is below the critical value G is estimated by the inequality:

$$P(H_0) \geq P\{[\chi^2_{\ell-r} + \frac{1}{4}\left(\frac{\lambda_1}{\lambda_r}\right)^2 \frac{q}{\lambda_r}\chi^2_g] < < \frac{2}{3} G\sigma^2\}. \quad (4.2)$$

From (4.2) it is seen that if the round-off error d is equal to zero such a (or r) will be found for any $\sigma^2 \neq 0$ that $P(H_0)=1$. The situation $d\neq 0$ is usually occured. Therefore some cases are possible when as $a\to 0$ ($r\to 1$) and w is small, the value of statistics of χ^2, once the minimum is reached, will not in an acceptance region of the zero hypothesis. This implies that a high accuracy of measurement does not correspond to the accuracy of calculations which should be improved.

The value of r which is recommended as a tentative value for the algorithm 3.I minimizes the expression

$$\chi^2_{\ell-r} + \frac{1}{4}\left(\frac{\lambda_1}{\lambda_r}\right)^2 \frac{q}{\lambda_r}\chi^2_g.$$

4.2. The statistics of the criterion χ^2 for the algorithm 3.2 can have an upper bound:

$$\frac{1}{\sigma^2}(y+n-Fc^w_a)^T(y+n-Fc^w_a) \leq$$
$$\leq y^Ty\left[\left(\frac{a}{\lambda_\ell+a}\right)^2 + \frac{1}{4}\left(\frac{\lambda_1}{\lambda_\ell+a}\right)^2 \frac{q}{\lambda_\ell+a}\right]\cdot\frac{1}{\sigma^2}, \quad (4.3)$$

where $\lambda_1=\lambda_{min}(F^TF)$; the other designations are similar to those in (4.I).

The probability that the value of statistics is less than a critical value can be estimated by the inequality

$$P(H_0) \geq P\left\{\chi^2_g\left[\left(\frac{a}{\lambda_\ell+a}\right)^2 + \frac{1}{4}\left(\frac{\lambda_1}{\lambda_\ell+a}\right)^2 \frac{q}{\lambda_\ell+a}\right] < < \frac{2}{3}G\sigma^2\right\}. \quad (4.4)$$

Comparing (4.3) with (4.I) and (4.4) with (4.2), it is seen that arguments concerning (4.I) and (4.2) are applicable to (4.3) and (4.4) almost without changes. Inequality (4.4) gives an opportunity to estimate roughly the initial value of parameter a:

$$a = \frac{\lambda_\ell}{2}\left(\sqrt{1+\frac{3}{2}\left(\frac{\lambda_1}{\lambda_\ell}\right)^2\frac{q}{\lambda_\ell}} - 1\right). \quad (4.5)$$

Convergence $h^w_a(v) \to h(v)$ as $w\to 0$ for (4.I), (4.2) follows from a continuity and a good conditionality of the mappings F^{-1}_a.

5. SOME OTHER APPLICATIONS

The method of statistical ragularization can be successively used in all cases when it is necessary to estimate an unknown function from an operator equation with a quite continuous or ill-conditionned operator. The examples of such problems may be:

- estimation of a density function from the convolution of two distributions,
- estimation of input signal by a known output and impulse response function - Brigham (I968) calls this problem inverse filtering, and Poluektov calls it dynamic error correction;
- numerical inversion of the Fourier transform.

REFERENCES

I. Åström K.-J., Eykhoff P. (I97I). System identification - a survey. Automatica, 7, No 2, p.I23.

2. Bellman R.(I960), Introduction to matrix analysis. McGraw-Hill Book Company, Inc.

3. Brigham E.Oraw, Smith Harold W., Bostick Francis X. (I968), Duesterhoeft William C. An iterative technique for determination inverse filters. IEEE Trans.Geosci.Electron.6; No 2, p.86.

4. Collatz L. (I964), Funktionalanalysis und numerische Mathematik. Springer-Verlag.

5. Poluektov R.A., Solopchenko G.N. (I97I). Metody correktsii dinamicheskih pogreshnostey. Avtometriia, No5, str.3.

6. Rao C.R. (I965). Linear statistical inference and its applications. John Wiley & Sons.

7. Solopchenko G.N. (I970). Metod statisticheskoi regularizatsii. Trudy VNII electroismeritelnyh priborov, No5, str.43.

8. Stratonovich R.L. (I970). Optimalnoe rasshirenie functsionalnogo prostranstva v algoritmah vosstanovleniia plotnosti i functsii raspredeleniia. Isv.AN SSSR, "Technitscheskaia kibernetika", No2, str.57.

THE IDENTIFICATION OF LINEAR MULTIVARIABLE SYSTEMS FROM FREQUENCY RESPONSE DATA

Martin Marchand

Institut für Flugmechanik
Deutsche Forschungs- und Versuchsanstalt für Luft- und Raumfahrt e.V. (DFVLR)
Braunschweig, Germany

Two methods for the determination of parameters of linear multivariable systems from frequency response data are presented. The first method yields an analytical expression for the frequency response curves in the form of rational functions of $j\omega$, from which then further parameters, such as the location of poles and zeros of transfer functions, can be determined. The second method yields the coefficients of the differential equations. The lecture contains some examples of application in the field of flight mechanics.

1. INTRODUCTION

The present paper deals with the determination of the parameters of linear multivariable systems from measured values of the input and output signals. It is assumed that the physical system to be investigated can be described by a system of linear differential equations with constant coefficients. Two evaluation procedures are described which work in the frequency domain and lead to different kinds of characteristic values. Procedure A provides the coefficients of the transfer functions and the frequency responses respectively, procedure B provides the coefficients of the system of differential equations (see fig. 1).

Procedures for the least-square fitting of a single frequency response have already been published by several authors, for instance [1, 2]. Section 4 shows how the several frequency response curves of a multivariable system can be fitted simultaneously by means of procedure A. Compensation of the frequency response errors is made according to the maximum-likelyhood method.

The equation coefficients can be obtained by inversion of the equation system [3, 4]. The procedures so far used, however, do not provide good results, if the input and output signals are very noisy. On the contrary, when statistically calculating frequency response values, relatively strong noise can still often be suppressed so far that the inversion in the frequency domain still provides good results according to the procedure B described in section 5. In this procedure compensation of the equation errors is also made by using the maximum-likelyhood method.

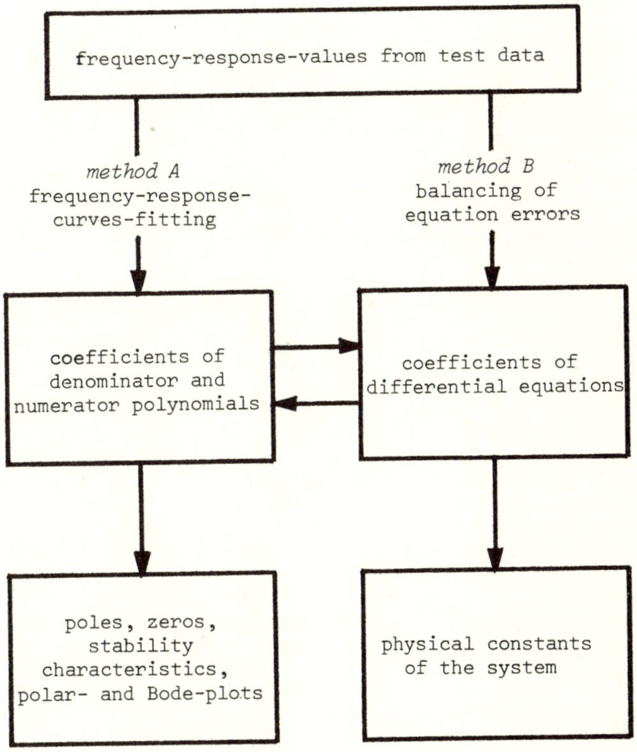

Fig. 1. Schematic diagram

2. CALCULATION OF FREQUENCY RESPONSE VALUES

At the beginning of evaluation, frequency response values are obtained from the test data by using the statistical method. The most important stations of this calculation are (details see, for instance, [5, 6, 7]):

time histories, e.g.

$$x(t),\ y(t) \tag{1}$$

correlation function

$$R_{xy}(\tau) = \frac{1}{n} \sum_{t=t_1}^{t_n} x(t)\ y(t+\tau) \tag{2}$$

application of lag-window, for instance, "Hanning window"

$$W(\tau) = \frac{1}{2}\left(1 + \cos \pi \frac{\tau}{\tau_{max}}\right) \tag{3}$$

$$\tilde{R}_{xy}(\tau) = R_{xy}(\tau) \cdot W(\tau)$$

spectral density function

$$S_{xy}(\omega) = \sum_{-\tau_m}^{\tau_m} \tilde{R}_{xy}(\tau)\ e^{-j\omega\tau} \tag{4}$$

frequency response

$$F^*(\omega) = \frac{S_{xy}(\omega)}{S_{xx}(\omega)} \tag{5}$$

The asterisk shall distinguish the frequency response determined by measurement from the corresponding true function.

The advantages of this statistical method shall be illustrated here only as far as they are of importance for the further evaluation of the frequency response values. There are two items:

i) For each frequency, the whole measuring time can be used for the evaluation. By this, disturbances and errors in measurement which are statistically independent of the input signals are largely eliminated when calculating the correlation functions with eq. (2).

ii) The number of data to be evaluated is already essentially reduced after the calculation of the correlation functions. For the frequency responses, a further reduction can be obtained by arranging the frequencies ω equidistantly on the logarithmic scale, so that

$$\frac{\omega_{k+1}}{\omega_k} = \frac{\omega_k + \Delta\omega}{\omega_k} = \text{const.} \tag{6}$$

Then, each frequency response can be well described with relatively few (20 - 50) points, irrespectively of its position in the frequency domain. In order to fully utilize the information lying between the frequencies - in spite of the reduced number of values - the spectral windows which are the Fourier-transformed lag-windows of eq. (3), must have a width comparable to the distance of frequencies $\Delta\omega$. As the width of the main maximum of the spectral filter is $4\pi/\tau_{max}$, for instance

$$\frac{4\pi}{\tau_{max}} = 2\Delta\omega \tag{7}$$

can be equated. Therefore, with increasing frequency, $\Delta\omega$ becomes larger (eq. (6)) and the lag window more narrow (eq. (7)).

3. THE FREQUENCY RESPONSE ERRORS

The values calculated according to eq. (5) are estimations of the frequency response which are subject to statistical errors. Causes for these errors are:

1. inadequacies of measurement
2. not measurable noise
3. finite measuring time.

Statements on the kind and magnitude of these errors are made in [2, 6, 8]. The results described in these papers largely agree in spite of differing starting points, so that a limitation on the statements made in [6] does not restrict the general validity of the following investigations. The results important for the following sections are summarized here. On the assumption, that the measuring time is sufficiently long, the linearized theory yields:

The real and imaginary parts of the frequency response errors

$$\Delta F = \Delta F_{Re} + j\ \Delta F_{Im} = F^* - F \tag{8}$$

are normally distributed and independent of each other, and it applies

$$E\{\Delta F_{Re}(\omega_k)\ \Delta F_{Im}(\omega_h)\} = 0 \quad \text{for all } k, h$$

$$E\{\Delta F_{Re}(\omega_k)\ \Delta F_{Re}(\omega_h)\} = E\{\Delta F_{Im}(\omega_k)\ \Delta F_{Im}(\omega_h)\} = \begin{cases} \sigma^2 & \text{for } k=h \\ 0 & \text{for } k\neq h \end{cases} \tag{9}$$

where $E\{...\}$ is the statistical expectation value of $\{...\}$.

Under the same conditions of evaluation, σ^2 is proportional to $|F|^2$, so that the relative variance

$$\sigma_{rel}^2 = \frac{\sigma^2}{|F|^2} \tag{10}$$

is constant for each frequency response. To be exact,

$$\sigma^2_{rel} = K_1 \, K_2 \, K_3 \qquad (11)$$

$$\text{with } K_1 = \frac{\text{spectrum of noise}}{\text{spectrum of signal}} ,$$

$$K_2 = \frac{\text{duration of test}}{\tau_{max}} ,$$

$$K_3 = \begin{cases} 1 & \text{without lag window} \\ 0.38 & \text{in the case of the "Hanning window", eq. (3).} \end{cases}$$

In the following sections, calculations will be performed with constant σ^2_{rel}. This is meaningful, if one i) has no statements on the spectrum of noise, ii) operates with constant τ_{max}. The general applicability of the evaluation methods is not restricted by this. If the spectrum of noise shall be taken into consideration or if τ_{max} is variable according to eq. (7), a frequency dependent weighting function can be used additionally.

4. COMPENSATION OF FREQUENCY RESPONSE ERRORS

The frequency response values determined from measured data shall be approximated by rational functions of $j\omega$ (procedure A of fig. 1). We equate

$$F_{ik} = F_i(\omega_k) = \frac{Z_{ik}}{N_k} \qquad (12)$$

with Z_i and N being polynomials in $j\omega$ and the index i denoting one of the frequency responses of the multivariable system. All F_i have the same denominator polynomial N. It is equated

$$Z_{ik} = \sum_{\nu} a_{\nu i} \, (j\omega_k)^{\nu}$$

$$N_k = \sum_{\nu} b_{\nu} \, (j\omega_k)^{\nu} \qquad (13)$$

To be determined are the coefficients $a_{\nu i}$, b_{ν} with the exception of a freely choosable value. For instance, it can be equated:

$$b_o = 1 \quad . \qquad (14)$$

For compensating the frequency response errors, the simple least square criterion

$$\sum_i \sum_k |\Delta F_{ik}|^2 \rightarrow \min. \qquad (15)$$

could be used. This, however, has two disadvantages: i) it leads to a nonlinear system of equations, ii) it suppresses too much frequency response values having a small magnitude. More meaningful from the physical point of view is the application of the maximum-likelyhood method, using the results represented in section 3. When doing so, the coefficients of the frequency responses are determined in such a way, that the occurrence of the set of the resulting errors possesses the maximum probability. The probability density function for the occurrence of a single error in measurement, for instance $\Delta F_{Re,ik}$, is

$$\frac{1}{\sqrt{2\pi}\,\sigma_{ik}} \exp\{-\frac{\Delta F^2_{Re,ik}}{2\sigma^2_{ik}}\}$$

Because of the statistical independency of the frequency response errors (eq. (9)), we obtain the joint probability density function for the occurrence of a pair of values $\Delta F_{Re,ik}$, $\Delta F_{Im,ik}$ by multiplication

$$\frac{1}{2\pi\,\sigma^2_{ik}} \exp\{-\frac{|\Delta F_{ik}|^2}{2\sigma^2_{ik}}\} \qquad (16)$$

and for the occurrence of all frequency-response errors

$$\frac{1}{\prod_i \prod_k (2\pi\,\sigma^2_{ik})} \exp\{-\sum_i \sum_k \frac{|\Delta F_{ik}|^2}{2\sigma^2_{ik}}\} , \qquad (17)$$

index i denoting the frequency responses and index k denoting the frequencies. Maximization of (17) is equivalent with the minimization of the sum S

$$S = \sum_i \sum_k \frac{|\Delta F_{ik}|^2}{\sigma^2_{ik}} \rightarrow \min. \qquad (18)$$

According to eq. (10) and the motivation given in section 3, we can equate

$$\sigma^2_{ik} = \sigma^2_{rel,i} \, |F_{ik}|^2 \approx \sigma^2_{rel,i} \, |F^*_{ik}|^2 \qquad (19)$$

in which, for $\sigma^2_{rel,i}$, only one value for each frequency response is to be mentioned and a factor common to all can be arbitrarily established, since it plays no part in (18).

From (18, 19) and with (12) it follows

$$S = \sum_i \sum_k \frac{|Z_{ik} - F^*_{ik} \, N_k|^2}{\sigma^2_{rel,i} \, |Z_{ik}|^2} \qquad (20)$$

If the values of Z_{ik} in the denominator were known, then S were a quadratic function of the unknown coefficients $a_{\nu i}$ and b_{ν}. The equations

$$\frac{\partial S}{\partial a_{\nu i}} = 0 \quad ; \quad \frac{\partial S}{\partial b_{\nu}} = 0 \qquad (21)$$

then would yield a linear system of equations for the $a_{\nu i}$ and b_ν. One of the variables can be arbitrarily established, e.g. $b_o = 1$. Now, in an iterative process the sum

$$\tilde{S} = \sum_i \sum_k \frac{|Z_{ik} - F^*_{ik} N_k|^2}{\sigma^2_{rel,i} \, |\tilde{Z}_{ik}|^2} \tag{22}$$

can be minimized instead of the sum S. When doing so, $|\tilde{Z}_{ik}|$ is inserted from the preceding iteration step. In the course of the calculation, $\tilde{Z}_{ik} \to Z_{ik}$ and thus $\tilde{S} \to S$, so that we obtain the desired results. Table 1 shows the system of equations to be solved for $b_o = 1$.

For the first iteration, all $\tilde{Z}_{ik}$ must be advanced, a common factor being without importance. As the dependency on the frequency ω is not yet known and $F_{ik} \to Z_{ik}$ for $\omega \to 0$, if b_o can be equated to 1, we can, for instance, use the values of $|F|$ determined at the smallest frequency, thus equating for all ω_k

$$|\tilde{Z}_{ik}| = |F^*_{i1}| \tag{23}$$

It has, however, been warned already in [2] to apply this procedure to insert $\tilde{Z}_{ik}$ independently of the frequency. When approximating a single frequency response, already the first approximation of the polynomial coefficients can be so bad that no good values are obtained at all. This danger is still greater when analysing multivariable systems, as the very numerators constitute the distinctions between the individual frequency responses. But obviously there does not yet exist a safe method to evade these difficulties automatically. It is, therefore, absolutely necessary to check the agreement between the measured values and the fitting curves. In the case of poor agreement, the calculation should be repeated. When doing so, the points which are very badly approximated after the first iteration are given additional weighting factors. If good agreement has been attained in this way, further iterations can be done without the weighting factors.

Example:
Fig. 2 shows an example for the simultaneous fitting of three frequency response curves of the longitudinal motion of an aircraft. It represents the evaluation of a flight of a DC-8 lasting 15 min, which was simulated at the digital

$\frac{\partial \tilde{S}}{\partial a_{\nu i}} = 0$	ν even	$\sum'_{\mu} S_{1i}(\nu,\mu)a_{\mu i} - \sum'_{\mu>0} S_{2i}(\nu,\mu)b_\mu - j\sum''_{\mu} S_{3i}(\nu,\mu)b_\mu = S_{2i}(\nu,0)$
one equation for each pair ν,i	ν odd	$-j\sum''_{\mu} S_{1i}(\nu,\mu)a_{\mu i} - \sum'_{\mu>0} S_{3i}(\nu,\mu)b_\mu + j\sum''_{\mu} S_{2i}(\nu,\mu)b_\mu = S_{3i}(\nu,0)$
$\frac{\partial \tilde{S}}{\partial b_\nu} = 0$	ν even, $\nu > 0$	$\sum_i\sum'_{\mu} S_{2i}(\nu,\mu)a_{\mu i} - j\sum_i\sum''_{\mu} S_{3i}(\nu,\mu)a_{\mu i} - \sum'_{\mu>0} S_4(\nu,\mu)b_\mu = S_4(\nu,0)$
one equation for each $\nu>0$	ν odd	$-\sum_i\sum'_{\mu} S_{3i}(\nu,\mu)a_{\mu i} - j\sum_i\sum''_{\mu} S_{2i}(\nu,\mu)a_{\mu i} + j\sum''_{\mu} S_4(\nu,\mu)b_\mu = 0$

Comments: $j = \sqrt{-1}$; $\sum'_{\mu}$ = sum for even values of μ ; $\sum''_{\mu}$ = sum for odd values of μ

$$S_{1i}(\nu,\mu) = \sum_k \frac{j^\mu \, \omega_k^{\nu+\mu}}{\sigma^2_{rel,i} \, |\tilde{Z}_{ik}|^2} \; ; \; S_{2i}(\nu,\mu) = \sum_k \frac{j^\mu \, \omega_k^{\nu+\mu} \, \mathrm{Re}\, F^*_{ik}}{\sigma^2_{rel,i} \, |\tilde{Z}_{ik}|^2} \; ; \; S_{3i}(\nu,\mu) = \sum_k \frac{j^\mu \, \omega_K^{\nu+\mu} \, \mathrm{Im}\, F^*_{ik}}{\sigma^2_{rel,i} \, |\tilde{Z}_{ik}|^2}$$

$$S_{4i}(\nu,\mu) = \sum_k \frac{j^\mu \, \omega_k^{\nu+\mu} \, |F^*_{ik}|^2}{\sigma^2_{rel,i} \, |\tilde{Z}_{ik}|^2} \; ; \; S_4(\nu,\mu) = \sum_i S_{4i}(\nu,\mu)$$

Table 1. Equations for the coefficients of frequency-response-functions of multivariable systems

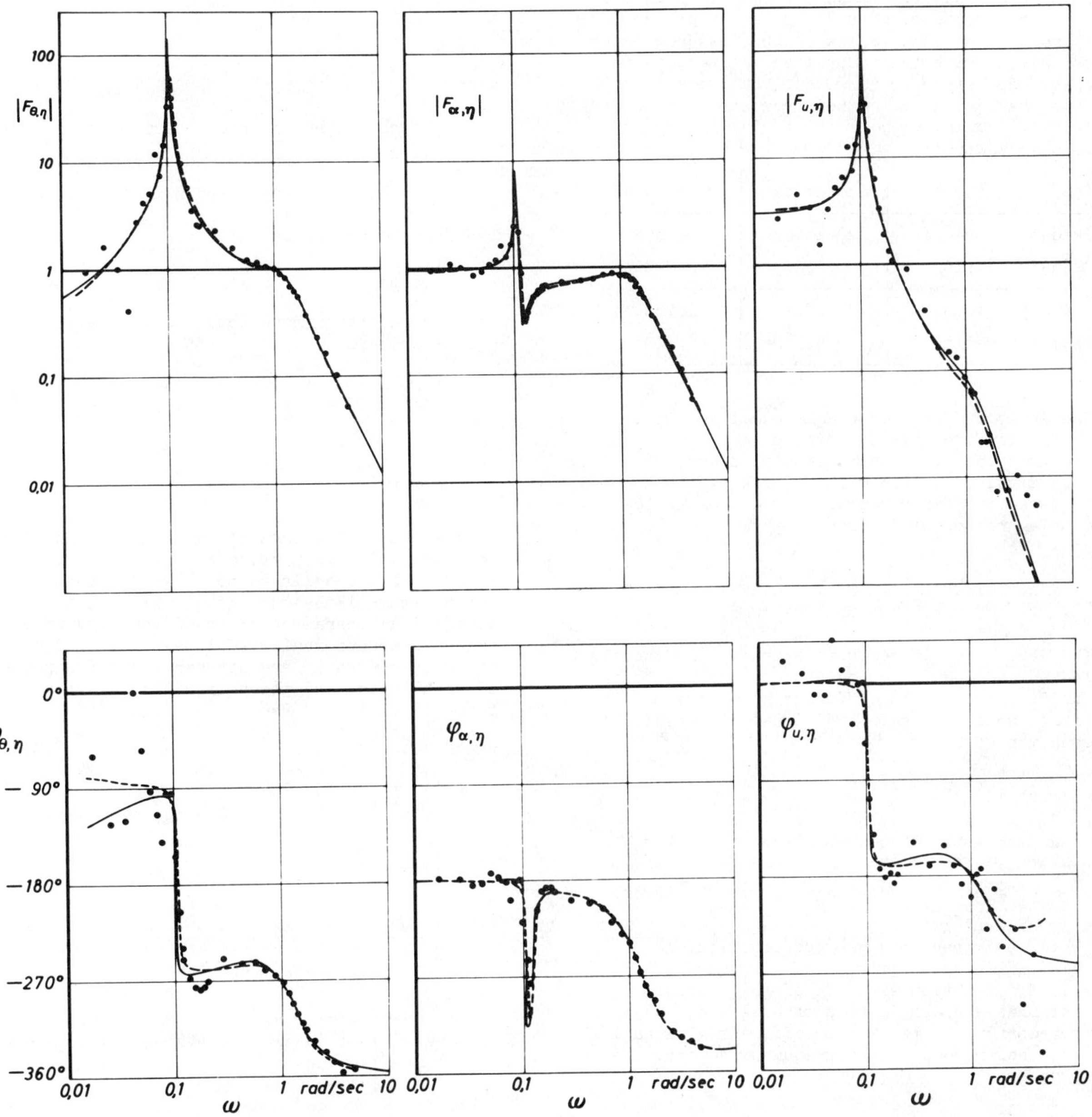

Fig. 2. Maximum-likelyhood curve fitting of three frequency responses (simulated longitudinal motion of a DC-8 aircraft). Bode-plots of the frequency responses of pitch angle Θ, angle of attack α, and velocity u with respect to elevator deflection η.

● test results
--- fitting curves
—— theoretical curves

computer. The motion of the aircraft was excited by a statistical elevator signal and disturbed by statistical gusts. Apart from some points at very low frequencies, good agreement of the approximating functions with the measured points as well as with the theoretical curves of the frequency responses is found. The two pairs of poles have been determined from the coefficients of the denominator polynomial. The following values resulted:

	theoretical	test result
undamped frequency	0.1046	0.1063
damping ratio	0.0136	0.0093
undamped frequency	1.276	1.289
damping ratio	0.487	0.500

The analysis provided nearly exact results, except for the damping ratio of the lower frequency pole. It must be pointed out that it is very difficult, also with other known procedures, to determine experimentally so extremely low damping ratios of disturbed systems.

5. DETERMINATION OF THE COEFFICIENTS OF THE DIFFERENTIAL EQUATIONS

The procedure denoted with B in fig. 1 directly leads from the frequency response values to the coefficients of the differential equations. At first, we consider an individual differential equation

$$\sum_i x_i(t)\, p_i = z(t) \tag{24}$$

with x_i = input and output signals and their derivations

p_i = unknown coefficients of the differential equation

z = known component of equation.

By multiplication with one of the control variables and application of equations (1) - (5) to each term of (24), the differential equation is transformed into the frequency domain. Then it reads

$$\sum_i F_i^*(\omega)\, p_i = r(\omega) \tag{25}$$

Here it needs not be assumed that the estimated values F_i^* obtained with the equations (1) to (5) are identical with the true frequency response values. With absolutely faultless measurement they satisfy equation (25) exactly, even if, due to the limited measuring time, the estimated value does not agree with the true frequency response values.

Thus, we obtain one complex or two real equations for each frequency ω_k, which form the system of equations

$$\underline{\underline{M}}\ \underline{p} = \underline{r} \tag{26}$$

where

$\underline{\underline{M}}$ = matrix known from measurement

$\underline{r}$ = vector known from measurement

$\underline{p}$ = vector of the unknown coefficients.

The equation error is

$$\underline{\varepsilon} = \underline{\underline{M}}\ \underline{p} - \underline{r} \tag{27}$$

For compensating these equation errors, the least square criterion

$$\sum_\nu \varepsilon_\nu^2 = \underline{\varepsilon}^T\, \underline{\varepsilon} \rightarrow \min. \tag{28}$$

leads to the solution

$$\underline{p} = (\underline{\underline{M}}^T\, \underline{\underline{M}})^{-1}\, \underline{\underline{M}}^T\, \underline{r} \tag{29}$$

This solution is, for the same reasons as already indicated with respect to eq. (15), not ideal. Also in this case it is more expedient to apply the maximum-likelyhood method and to utilize the statistical characteristics of the frequency response errors indicated in section 3. The equation errors ε_ν are assumed to be independent of each other, and the probability density function of the ε_ν is

$$\frac{1}{\prod_\nu (2\pi\, \sigma_{\varepsilon\nu}^2)^{1/2}} \exp\left(- \sum_\nu \frac{\varepsilon_\nu^2}{2\sigma_{\varepsilon\nu}^2}\right) \tag{30}$$

The maximization of (30) leads to

$$\sum_\nu \frac{\varepsilon_\nu^2}{\sigma_{\varepsilon\nu}^2} = \underline{\varepsilon}^T\, \underline{\underline{G}}\, \underline{\underline{G}}\, \underline{\varepsilon} \rightarrow \min. \tag{31}$$

with

$$G_{\nu\mu} = \begin{cases} \sigma_{\varepsilon\nu}^{-1} & \text{for } \nu = \mu \\ 0 & \text{for } \nu \neq \mu \end{cases} \tag{32}$$

and from (27)

$$\sigma_{\varepsilon\nu}^2 = E\{ \sum_i \underline{M}_{\nu i}\, \underline{p}_i - \underline{r}_\nu \}^2 \tag{33}$$

We obtain the unknown coefficients from (31) and (27):

$$\underline{p} = (\underline{\underline{M}}^T\, \underline{\underline{G}}\, \underline{\underline{G}}\, \underline{\underline{M}})^{-1}\, \underline{\underline{M}}^T\, \underline{\underline{G}}\, \underline{\underline{G}}\, \underline{r} \tag{34}$$

We also obtain this solution, if we multiply the system of equations (26) by the matrix $\underline{\underline{G}}$ and then compensate the equation errors according to the least square criterion, as the equations (29) and (34) are then identical.

A linearized theory for small errors yields, as estimation for the covariance matrix,

$$E\{\underline{\Delta p}\ \underline{\Delta p}^T\} = (\underline{\underline{M}}^T\ \underline{\underline{G}}\ \underline{\underline{G}}\ \underline{\underline{M}})^{-1} \quad {}^{*}) \qquad (35)$$

The diagonal elements of this matrix (35) are the variances of the sought solutions, σ^2_{pi}, the other elements are a measure for the statistical interdependencies of the coefficients.

In order to calculate $\underline{p}$ according to (34), $\underline{\underline{G}}$ must be known. On the other hand, $\underline{p}$ is necessary for the calculation of $\underline{\underline{G}}$ according to (32) and (33). Therefore, we must calculate iteratively and use $\underline{p}$ from the preceding approximation.

At the first iteration, all p_i are equated to 1. The procedure converges very rapidly so that mostly 4 iterations are sufficient.

Examples:
The procedure described herein was used for the determination of the characteristic values of aircrafts. These values (e.g. the so-called derivatives) appear as coefficients in the differential equations of longitudinal motion. The evaluation of the simulation test (DC 8) already mentioned in section 4 yielded the following results:

equation	derivative	theoretical value	test result
longitudinal	C_{Xu}	0.092	0.082
force equation	$C_{W\alpha}$	0.685	0.72
vertical force	$C_{A\alpha}$	4.84	4.85
equation	$C_{A\eta}$	0.277	0.35 +)
pitch moment	C_{mu}	-0.0204	-0.071 +)
equation	$C_{m\dot{\alpha}} + C_{mq}$	-8.01	-8.57
	$C_{m\alpha}$	-0.96	-0.93
	$C_{m\eta}$	-0.85	-0.88

*) Derivation of eq. (35). We equate:

$\underline{p}_o$ = true solution ; $\underline{p} = \underline{p}_o + \underline{\Delta p}$;

$\underline{\underline{M}} = \underline{\underline{M}}_o + \Delta\underline{\underline{M}}$; $\quad \underline{r} = \underline{r}_o + \underline{\Delta r}$;

$\underline{\underline{Q}} = \underline{\underline{Q}}_o + \Delta\underline{\underline{Q}}$; $\quad \underline{\underline{Q}}_o = \underline{\underline{M}}_o^T\ \underline{\underline{G}}\ \underline{\underline{G}}\ \underline{\underline{M}}_o$

Then $p = \underline{\underline{Q}}^{-1}\ \underline{\underline{M}}^T\ \underline{\underline{G}}\ \underline{\underline{G}}\ \underline{r}$.

The linearization leads to

$\underline{\Delta p} = \underline{\underline{Q}}_o^{-1}\ \underline{\underline{M}}_o^T\ \underline{\underline{G}}\ \underline{\underline{G}}\ (\underline{\Delta r} - \Delta\underline{\underline{M}}\ \underline{p}_o) = -\ \underline{\underline{Q}}_o^{-1}\ \underline{\underline{M}}_o^{-1}\ \underline{\underline{G}}\ \underline{\underline{G}}\ \underline{\varepsilon}$,

$E\{\underline{\Delta p}\ \underline{\Delta p}^T\} = \underline{\underline{Q}}_o\ \underline{\underline{M}}_o^T\ \underline{\underline{G}}\ \underline{\underline{G}}\ E\{\underline{\varepsilon}\ \underline{\varepsilon}^T\}\ \underline{\underline{G}}\ \underline{\underline{G}}\ \underline{\underline{M}}_o\ \underline{\underline{Q}}_o^{-1}$

From (32) it follows $E\{\underline{\varepsilon}\ \underline{\varepsilon}^T\} = (\underline{\underline{G}}\ \underline{\underline{G}})^{-1}$ and with $\underline{\underline{Q}}_o = \underline{\underline{M}}_o^T\ \underline{\underline{G}}\ \underline{\underline{G}}\ \underline{\underline{M}}_o$ finally

$E\{\underline{\Delta p}\ \underline{\Delta p}^T\} = \underline{\underline{Q}}_o^{-1} = (\underline{\underline{M}}_o^T\ \underline{\underline{G}}\ \underline{\underline{G}}\ \underline{\underline{M}}_o)^{-1}$

The cross-marked values have only little influence on the motion of the aircraft and can therefore be determined only badly or not at all.

For practically testing this procedure, a flight test was performed with a Do 27 aircraft. The elevator was activated pseudostatistically for 20 min. Since no exact values for a comparison are available, the responses of the model obtained by the flight test were compared with the test data. Fig. 3 shows that the found coefficients provide good correspondence.

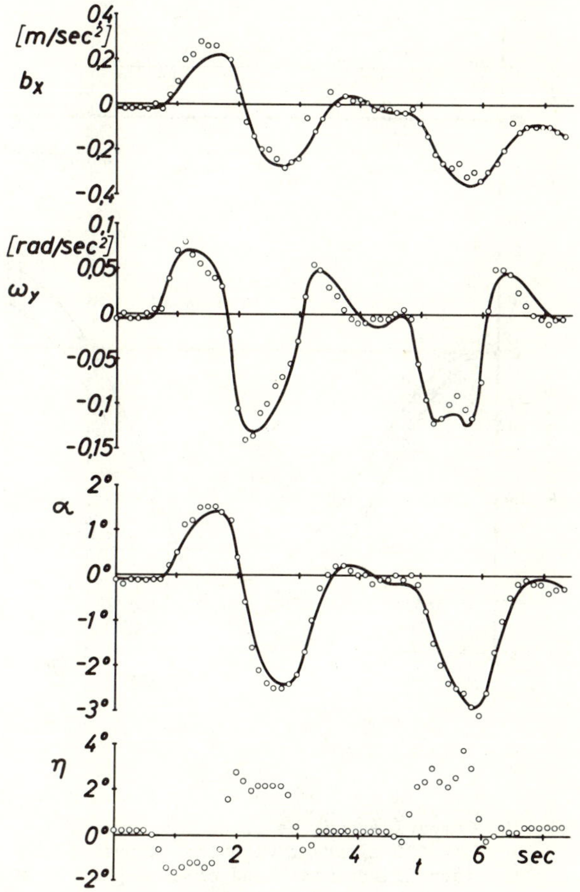

Fig. 3. Comparison of flight test data with the output signals of the identified model (airplane Do 27)

o flight test data
—— model

b_x = signal of accelerometer in longitudinal direction
ω_y = pitch rate
α = angle of incidence
η = elevator angle

The method is also applicable for the evaluation of short-time tests with deterministic inputs. As an example, fig. 4 shows the pitch responses of a HFB 320 Hansa aircraft and of the identified model. For determining the parameters, the results of 4 flight tests lasting each 12 s were averaged. The model thus obtained yields a better correspondence with test data than the model determined by wind tunnel measurements, the response of which is also plotted in fig. 4.

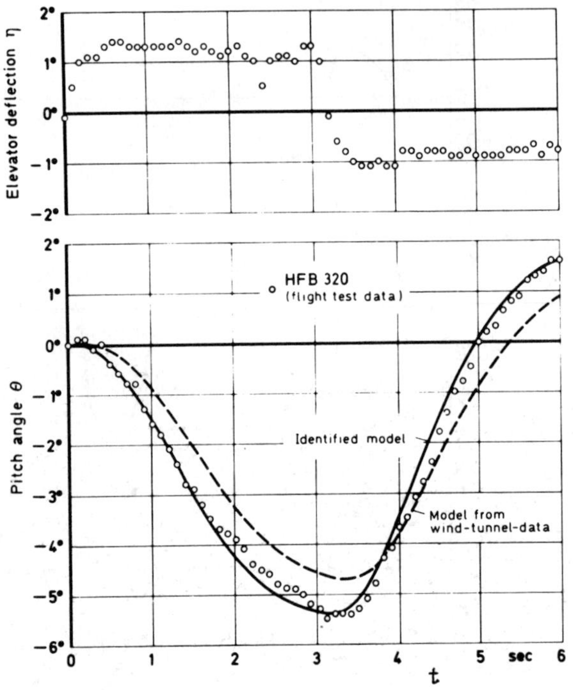

Fig. 4. Comparison of the pitch response of a HFB 320 aircraft with the output of the identified model and of an a-priori model known from wind tunnel tests

o flight test data
——— identified model
– – – a-priori model

6. REFERENCES

[1] LEVY, E.C. Complex-Curve-Fitting. IRE-Trans. Autom. Control 4 (1959) May, p.37-43.

[2] STROBEL, H. Systemanalyse mit determinierten Testsignalen (System Analysis with Deterministic Test Signals). Berlin: VEB Verlag Technik, 1968.

[3] KLEIN, V. Application of Several Methods Based on Least-Squares Principle for Determining Longitudinal Aerodynamic Derivatives of Aeroplane from Flight Test Data (Summary Report). Report of the ARTI (Aeronautical Research and Test Institute, Prag), Z-12, May 1969.

[4] FRIEDRICH, H. Ermittlung der Stabilitätsderivative aus den Flugversuchsergebnissen mit Hilfe der Regressionsanalyse (Determination of Stability Derivatives from Flight Test Data Using Regression Analysis). Dornier-Bericht 68/7, Friedrichshafen, 1968.

[5] BLACKMAN, B.R. The Measurement of Power Spectra. New York: Dover Publications, Inc. 1959.

[6] MARCHAND, M. Über den statistischen Fehler von Frequenzgangmessungen bei digitaler Auswertung (On the Statistical Error of Frequency Response Measurements in the Case of Digital Evaluation). Deutsche Luft- und Raumfahrt, Forschungsbericht 69-28, 1969.

[7] GILOI, W. Simulation und Analyse stochastischer Vorgänge (Simulation and Analysis of Stochastic Processes). München, Wien: Oldenbourg, 1967.

[8] RAKE, H. Identifizierung linearer Systeme mit statistischen Methoden (Identification of Linear Systems Using Statistical Methods). Habilitationsschrift, Technical University Aachen, 1969.

SOME PROPERTIES OF SECOND ORDER CORRELATION FUNCTIONS IN RELATION TO BIOLOGICAL RHYTHM RESEARCH

M. ten Hoopen

Institute of Medical Physics TNO
Utrecht, The Netherlands

Some properties of the second order correlation function $R(\tau_1,\tau_2)$ have been studied with the help of a model that is believed to mimic certain physiological time series containing more than one periodicity. The model consists of the superposition of two series of oscillatory wave packets, occurring randomly and more or less in synchrony, while the degree of coupling is variable. In contrast to the first order correlation function $R(\tau)$, which is hardly able to differentiate between loose and tight coupling of the two components, $R(\tau_1,\tau_2)$ is able to do so, at least within a limited range of parameter values. It is suggested that one does not necessarily need to compute $R(\tau_1,\tau_2)$ in the same detail for both τ_1 and τ_2. It may suffice, in the first instance, to calculate the said quantity for only one, or a few values of τ_2, for example for $\tau_2 = 0$; thus restricting oneself to $R(\tau_1,0)$.

INTRODUCTION

Correlation techniques are frequently used to detect periodicities of a signal in a noisy background, in the search for dependencies within one time function or between time functions, and as an intermediary step in the computation of the power spectrum. Among others, biomedical research has much benefited from this method of analysis. However, besides restrictions of practical nature, the procedure also has fundamental limitations in the sense that the information it provides by no means gives a unique description of the phenomenon under consideration. So, a continuously changing Markow process, the highly discontinuous bivalued random telegraph signal and a sequence of Poisson triggered exponentially shaped wave forms all possess an identical (exponentially decaying) autocorrelation function. In part, differences can be measured by means of the amplitude distribution or its moments. Still more information is obtained by evaluating higher order correlation functions, which is particularly desirable if conditions of stationarity, normal distribution of amplitudes or mutual independence of different frequency components are violated or suspected to be violated.

In the present note, merits and draw-backs of the second order correlation function are investigated when applied to a signal containing two or more rhythms appearing simultaneously and superimposed on the record. This situation is encountered in a variety of processes of physiological origin. A few recent references have been listed in the DISCUSSION. If the presence of more than one frequency component has been ascertained by inspection of the autocorrelation function, the power spectrum, or by other means, the question often arises whether or not these rhythms are mutually independent, or in part so. With respect to this issue, the conventional first order correlation function hereafter designated as $R(\tau)$ is of little help in most instances while the second order version, shortened to $R(\tau_1,\tau_2)$ may give valuable indications.

To illustrate the thesis that items which are not reflected in $R(\tau)$ but may be revealed by $R(\tau_1,\tau_2)$ consider the following situation:
Each event of a stationary stochastic univariate point process $f_1(t) \equiv \Sigma\, \delta(t-t_i)$ is associated with a wave packet $h_1(\tau)$ which might be regarded as the unit-impulse response of a linear system, so that one has the function $f_I(t) \equiv \Sigma\, h\,(t-t_i)$. Each event is preceeded or followed in a random fashion, according to a distribution function $\psi(\tau)$ with standard deviation σ, by one other event. The thus formed second point process is denoted by $f_2(t) \equiv \Sigma\, \delta(t-t_j)$. The latter series of events is transformed by a unit-response function $h_2(\tau)$ into $f_{II}(t) \equiv \Sigma\, h_2(t-t_j)$. To simplify matters, and without much detracting from the essence, the first sequence $f_1(t)$ is taken as a Poisson process and the mean of $\psi(\tau)$ is set equal to zero.
Generalizations can be performed without many fundamental difficulties, if so wished, but numerical evaluation then becomes disproportionately more tedious.

Properties of $f(t) = f_I(t) + f_{II}(t)$ are investigated by means of $R(\tau)$ — the average product of $f(t)$ and $f(t+\tau)$ —, and of $R(\tau_1,\tau_2)$ — the average product of $f(t)$, $f(t+\tau_1)$ and $f(t+\tau_2)$ — for different degrees of coupling, dependent on the choice of $\psi(\tau)$. If $\psi(\tau)$ represents a highly narrow distribution, $f_I(t)$ and $f_{II}(t)$ are tightly coupled; if a very broad distribution is taken, the two said functions are nearly independent of each other.
Dumermuth et al. (1971) have performed simulations on a model based on an analogous concept which is comprised of combinations of nine components $h(\tau)$. These were either all strictly synchronized

($\sigma = 0$ in our nomenclature), or divided into two groups which were totally independent ($\sigma = \infty$). There were no situations in between these two extreme conditions.

THEORY

The first order (auto-) correlation function is defined as

$$R(\tau) = \lim_{T\to\infty} \int_{-T}^{T} f(t)f(t+\tau)d\tau/(2T) \ .$$

It contains four terms from combinations of $f_I(t)$, $f_{II}(t)$, $f_I(t+\tau)$ and $f_{II}(t+\tau)$. After substituting

$$f_I(t) = \int_{-\infty}^{\infty} h_1(\nu)f_1(t-\nu)d\nu$$

and likewise for $f_{II}(t)$, the result is expressed in the properties of the point processes $f_1(t)$, $f_2(t)$ and their mutual inter-dependency by means of the auto- and crosscovariances $R_{11}(\tau)$, $R_{22}(\tau)$, $R_{12}(\tau)$ and $R_{21}(\tau)$. The auto-covariance of $f_1(t)$ is defined as

$$R_{11}(\tau) = \lim_{T\to\infty} \int_{-T}^{T} f_1(t)f_1(t+\tau)dt/(2T) =$$

$$= k_1\left[\delta(\tau) + \sum_{n=1}^{\infty} p_n(\tau)\right] \quad \text{for } \tau \geqslant 0,$$

where $\delta(\tau)$ denotes the Dirac function, k_1 the mean number of events per unit of time in $f_1(t)$, and $p_n(\tau)$ the probability density function of the sum of n consecutive internal durations between the events. Differently stated, $R_{11}(\tau)/k_1$ equals the probability that an event is present in a time interval $(\tau, +d\tau)$, given an event at $\tau = 0$. For a Poisson process, $\Sigma\, p_n(\tau) = k_1$ thus:

$R_{11}(\tau) = k_1[\delta(\tau) + k_2]$. Similarly,

$R_{22}(\tau) = k_2[\delta(\tau) + k_2]$,

$R_{12}(\tau) = k_1[\psi(\tau) + k_2]$ and

$R_{21}(\tau) = k_2[\psi(\tau) + k_1]$, where k_2 denotes the mean number of events per unit of time in $f_2(t)$. By way of example and in preparation for the slightly more complicated derivation of $R(\tau_1,\tau_2)$, consider that term of $R(\tau)$ which arises from the product of $f_I(t)$ and $f_{II}(t+\tau)$ and which is written as

$$[I.II] = \lim_{T\to\infty} \int_{-T}^{T} dt/(2T) \int_{-\infty}^{\infty} h_1(\nu)f_1(t-\nu)d\nu \cdot \int_{-\infty}^{\infty} h_2(\nu_1)f_2(t+\tau-\nu_1)d\nu_1 .$$

Upon changing the order of integration and substituting $t-\nu \to t'$ one has

$$[I.II] = \int_{-\infty}^{\infty} h_1(\nu)d\nu \int_{-\infty}^{\infty} h_2(\nu_1)d\nu_1 \times$$

$$\times \lim_{T\to\infty} \int_{-T}^{T} f_1(t')f_2(t'+\nu+\tau-\nu_1)dt'/(2T) \ .$$

The last factor amounts to $R_{11}(\nu+\tau-\nu_1)$. With $\nu_1 \to t_1+\nu$ and inserting the expression for $R_{11}(\tau)$ stated before, one has

$$[I.II] = \int h_1(\nu)h_2(t_1+\nu)d\nu \times k_1 \int \psi(\tau-t_1)dt_1 +$$

$$+ k_1 \int h_1(\nu)d\nu \times k_2 \int h_2(t+\nu_1)d\nu_1 \ .$$

For briefness' sake, integration bounds are omitted. If the mean of $f(t)$ is thought subtracted, the last term cancels in the overall outcome for $R(\tau)$. If one sets

$h_2'(\nu+\tau) = \int h_2(t_1+\nu).\psi(\tau_1-t_1)dt_1$, it follows

$[I.II] = k_1 \int h_1(\nu)h_2'(\nu+\tau)d\nu$. Similarly,

$[II.I] = k_2 \int h_2(\nu)h_1'(\nu+\tau)d\nu$,

$[I.I\ ] = k_1 \int h_1(\nu)h_1(\nu+\tau)d\nu$ and

$[II.II] = k_2 \int h_2(\nu)h_2(\nu+\tau)d\nu$.

The sum of the latter two terms is shortened to $T(\tau)$, the sum of the former two terms to $S(\tau)$, or $R(\tau) = S(\tau) + T(\tau)$. If $f_1(t)$ and $f_2(t)$ are independent of each other: $R(\tau) = T(\tau)$.

In the numerical computations it has been assumed that an event of $f_1(t)$, occurring at t_i, is associated with a wave form

$$h_1(t-t_i) = A_1\exp\{-(t-t_i)^2/(2a_1^2)\}\cos\{b_1(t-t_i)\}$$

for all t. If the wave form is to be generated via a physical filter one would have to exclude $t < t_i$. Events of $f_2(t)$, occurring at t_j, are transformed analogously. Furthermore,

$\psi(\tau) = \exp\{-\sigma^2/(2\sigma^2)\}/(\sigma\sqrt{2\pi})$. The particular form chosen for $\psi(\tau)$, $h_1(\tau)$ and $h_2(\tau)$, in combination with the assumption that these functions are defined for all τ, reduces the computational effort greatly. In fact, $h_1'(\tau)$ and $h_2'(\tau)$ are of the same shape as $h_1(\tau)$ and $h_2(\tau)$ but the parameters A, a and b differ by a factor that depends on the value of σ. Subsequently it has been taken $k_1 = k_2 = k$ and $a_1 = a_2 = a = 1.2$ unit of time.

The second order correlation function is defined as

$$R(\tau_1,\tau_2) = \lim_{T\to\infty} \int_{-T}^{T} f(t)f(t+\tau_1)f(t+\tau_2)dt/(2T) \ .$$

It contains eight terms. One of these, resulting from $f_I(t)$, $f_I(t+\tau_1)$ and $f_{II}(t+\tau_2)$, is written as

$$[\mathrm{I.I.II}] = \lim_{T\to\infty} \int_{-T}^{T} dt/(2T) \int h_1(\nu) f_1(t-\nu)d\nu \, .$$

$$\cdot \int h_1(\nu_1) f_1(t+\tau_1-\nu_1)d\nu_1 \int h_2(\nu_2) f_2(t+\tau_2-\nu_2)d\nu_2$$

After changing the order of integration one has

$$[\mathrm{I.I.II}] = \int h_1(\nu)d\nu \int h_1(\nu_1)d\nu_1 \int h_2(\nu_2)d\nu_2 \times$$

$$\times \lim_{T\to\infty} \int_{-T}^{T} f_1(t-\nu) f_1(t+\tau_1-\nu_1) f_2(t+\tau_2-\nu_2)dt/(2T)$$

The last factor is set equal to $R_{112}(\nu+\tau_1-\nu_1, \nu+\tau_2-\nu_2)$. $R_{112}(\tau_1,\tau_2)d\tau_1 d\tau_2/k_1$ denotes the probability of an event of $f_1(t)$ in $(\tau_1,\tau_1+d\tau_1)$ and an event of $f_2(t)$ in $(\tau_2,\tau_2+d\tau_2)$, given an event of $f_1(t)$ at $\tau = 0$. For $f_1(t)$ obeying an Poisson process, and $f_2(t)$ depending on $f_1(t)$ as defined above, it holds that, for $\tau_1,\tau_2 > 0$ and $k_1 = k_2 = k$,

$$R_{112}(\tau_1,\tau_2) = k\Big[\delta(\tau_1)\psi(\tau_2) + k\delta(\tau_1) + k\psi(\tau_2) + + k\psi|\tau_1-\tau_2| + k^2\Big]$$

If the mean of f(t) is subtracted, only the first term contributes to $R(\tau_1,\tau_2)$. Therefore, when changing the variables under the integral sign and substituting $\nu_1 \to t_1+\nu$, $\nu_2 \to t_2+\nu$, one has

$$[\mathrm{I.I.II}] = \int h_1(\nu) h_1(t_1+\nu) h_2(t_2+\nu) \times$$

$$\times k_1 \iint \delta(\tau_1-t_1)\psi(\tau_2-t_2)dt_1 dt_2$$

$$= k\int h_1(\nu) h_1(\nu+\tau_1)d\nu \int h_2(t_2+\nu)\psi(\tau_2-t_2)dt_2$$

$$= k\int h_1(\nu) h_1(\nu+\tau_1)\, h_2'(\nu+\tau_2)d\nu$$

By the same reasoning one obtains for the other contribution to $R(\tau_1,\tau_2)$:

$$[\mathrm{I.II.II}] = k\int h_1(\nu) h_2'(\nu+\tau_1) h_2'(\nu+\tau_2)d\nu \, ,$$

$$[\mathrm{I.II.I}\;] = k\int h_1(\nu) h_2'(\nu+\tau_1) h_1(\nu+\tau_2)d\nu \, ,$$

$$[\mathrm{II.I.II}] = k\int h_2(\nu) h_1'(\nu+\tau_1) h_2(\nu+\tau_2)d\nu \, ,$$

$$[\mathrm{II.II.I}] = k\int h_2(\nu) h_2(\nu+\tau_1) h_1'(\nu+\tau_2)d\nu \, ,$$

$$[\mathrm{II.I.I}\;] = k\int h_2(\nu) h_1'(\nu+\tau_1) h_1'(\nu+\tau_1)d\nu \, ,$$

$$[\mathrm{I.I.I}\;] = k\int h_1(\nu) h_1(\nu+\tau_1) h_1(\nu+\tau_2)d\nu \quad \text{and}$$

$$[\mathrm{II.II.II}] = k\int h_2(\nu) h_2(\nu+\tau_1) h_2(\nu+\tau_2)d\nu \, .$$

The sum of the latter two terms is denoted by $T(\tau_1,\tau_2)$, the sum of the remaining six terms by $S(\tau_1,\tau_2)$. Thus, $R(\tau_1,\tau_2) = S(\tau_1,\tau_2) + T(\tau_1,\tau_2)$. If $f_1(t)$ and $f_2(t)$ are independent of each other $R(\tau_1,\tau_2) = T(\tau_1,\tau_2)$.

RESULTS

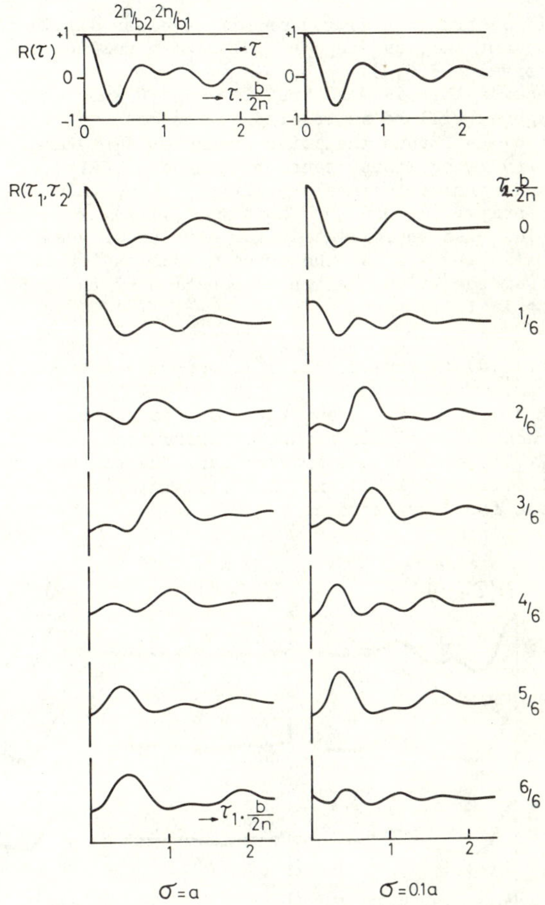

Fig.1. First order and second order correlation functions, $R(\tau)$ and $R(\tau_1,\tau_2)$ for two superimposed rhythms with frequencies $b_1 = b$ and $b_2 = 3/2$ b.
Left-hand side for loose coupling ($\sigma = a$); right-hand side for tight coupling(σ=0.1a).

The curves in Fig. 1 refer to $R(\tau)$ and $R(\tau_1,\tau_2)$ for $A_1 = A_2 = A$; $b_1 = b$, $b_2 = 3/2$ b, b = 4 per unit of time. The left side holds for $\sigma/a = 1$ and the right side for $\sigma/a = 0.1$. $R(\tau)$ is expressed in units of 1.77 akA^2 and 1.78 akA^2, respectively; $R(\tau_1,\tau_2)$ in units of 0.02 akA^3 and 0.31 akA^3, respectively. $R(\tau_1,\tau_2)$ is presented as a function of τ_1 for several values of τ_2, increasing from $\tau_2 = 0$ in steps of $2\pi/(6b)$ downwards to $6 \times 2\pi/(6b)$.
Comparing left and right, it is seen that $R(\tau)$ shows hardly any difference, even when plotted on a semi-logarithmic scale or when transformed into

the power spectrum. In either case, $R(\tau)$ is almost entirely dominated by the terms (I.I) and (II.II) and the curves are, therefore, composed of the sum of two decaying cosine functions with periods $2\pi/b_1$ and $2\pi/b_2$.

Bij contrast, $R(\tau_1,\tau_2)$ permits a clear distinction not only by shape but also with respect to the value $R(0,0)$.
Because this is also true for $\tau_2 = 0$ subsequent figures will be restricted to the case $\tau_2 = 0$.
If $\sigma \to \infty$, thus the point processes $f_1(t)$ and $f_2(t)$ being independent of each other, $R(\tau_1,\tau_2)$ is determined entirely by $T(\tau_1,\tau_2)$ and this quantity, in turn, by the term (I.I.I). This holds also approximately for a value of σ as small as $\sigma = a$ for the present combination of parameter values. After elaboration of the formulae it follows that

$$R(\tau_1,0) :: \exp\left[-\tau_1^2/(3a^2)\right]\left[2\cos(\tfrac{4}{6}b\tau_1)+\cos(\tfrac{4}{3}b\tau_1)\right]$$

If $\sigma \to 0$, or $f_1(t)$ and $f_2(t)$ are completely in synchrony, one finds that contributions to $R(\tau_1,\tau_2)$ are for the greater part due to the terms (I.I.II), (I.II.I) and (II.I.I). This is valid, too, for σ as large as $\sigma = 0.1$ a. One finds that

$$R(\tau_1,0) :: \exp\left[-\tau_1^2/(3a^2)\right]\left[2\cos(\tfrac{5}{6}b\tau_1)+\cos(\tfrac{5}{3}b\tau_1)\right]$$

The periods of the oscillations in $R(\tau_1,\tau_2)$ for $\sigma = 0$ and $\sigma = a_1$ differ by 25%.

More generally, if $R(\tau)$ demonstrates two oscillations with frequencies b_1 and b_2 ($> b_1$), both decaying to about the same degree as a function of τ ($a_1 = a_2$), and if the magnitude of the oscillation belonging to b_2 is not much larger than that of b_1 ($A_2 \leq A_1$), $R(\tau_1,0) = R(0,\tau_2) = R(\tau_1,\tau_1)$ will show two oscillations with frequencies $\frac{4}{3}b_1$ and $\frac{2}{3}b_1$ if the components are weakly synchronized, and frequencies $\frac{2}{3}(b_1+b_2)$ and $\frac{1}{3}(b_1+b_2)$ if the components are strongly synchronized ($\sigma/a \ll 1$).
In both cases, the lower frequency component is two times larger in magnitude than the higher frequency component, whereas the decay time of the oscillations is slightly smaller than that in $R(\tau)$. This rule of thumb may, if so wished, be expressed more quantitatively in a nomogram with parameters a_1/a_2, A_1/A_2 and σ/a_1.

Fig. 2 shows in more detail how the gradual change in inter-dependency between $f_1(t)$ and $f_2(t)$ becomes transparant in $R(\tau_1,0)$ by letting σ/a decrease in steps from $\sigma = a$ to $\sigma = 0.1$ a,

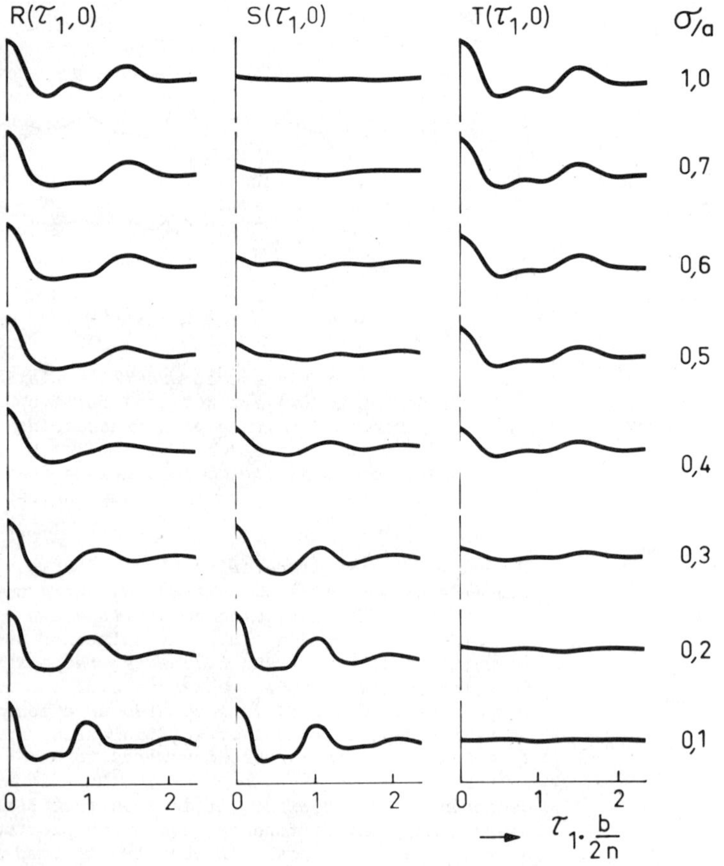

Fig. 2.
Function R(1,0) for two superimposed rhythms with frequencies $b_1 = b$ and $b_2 = 2b$ for different degrees of coupling σ/a.

and how the difference in shape of the curves is caused by a change in magnitude of $S(\tau_1,0)$ relative to $T(\tau_1,0)$. The curves are computed for $b_1 = b$, $b_2 = 2b$, $b = 4$ and $A_1 = A_2 = A$. The ordinate of $R(\tau_1,0)$ is expressed un units of c aKA^3 where c increases downwards from 0.02 to 0.8. In all cases, $R(\tau)$ again changes only minutely. It must be admitted that, leaving all other parameter values unaltered, the more the ratio of b_1 and b_2 approaches unity, the smaller the difference in $R(\tau_1,\tau_2)$with regard to the degree of inter-dependence. In fact, if $b_1 = b_2$ curves are identical in shape.

Another point concerns the influence of A_1 and A_2 which have been kept equal so far. The condition A_1 smaller than A_2 will be illustrated with the help of Fig. 3 which shows from left to right $R(\tau_1,0)$ for $\sigma = a$, $\sigma = 0.4$ a, $\sigma = 0.1$ a, as well as for $R(\tau)$ for $\sigma = 0.1$ a, the latter once more hardly affected by the value of σ. The full-drawn curves to positive values of A_1, the dashed curves to negative values of A_1, with $A_2 = A$; $|A_1|/A$ decreases downwards from 1 to 0.5, 0.4, 0.3, 0.2 and 0.1. Furthermore $b_1 = b$, $b_2 = \frac{7}{4}b$ and $b = 4$. In connection with Fig. 1 it has been pointed out that for $\sigma = a$, or loose coupling between two contributing elements, $R(\tau_1,\tau_2)$ is primarily determined by the term (I.I.I), incorporating the factor A_1^3 and for $\sigma = 0.1$ a, or tight coupling, by three terms all with the factor $A_1^2A_2$. It follows that if A_1 is diminished moderately, say by a factor of two, such a change will not much influence the shape of $R(\tau_1,\tau_2)$. Reversely, for $R(\tau)$, governed mainly by the terms (I.I) and (II.II) with factors A_1^2 and A_2^2, respectively, a reduction of A_1 or A_2 by one half will be good distinguishable; see upper two curves of Fig. 3.
By the same reasoning, if A_1 gets still smaller, $R(\tau)$ is dominated by A_2 and $h_2(\tau)$, and its shape does not change much any more. Consequently, the presence of a frequency with period $2\pi/b_1$ is difficult to detect. The value of A_1 being

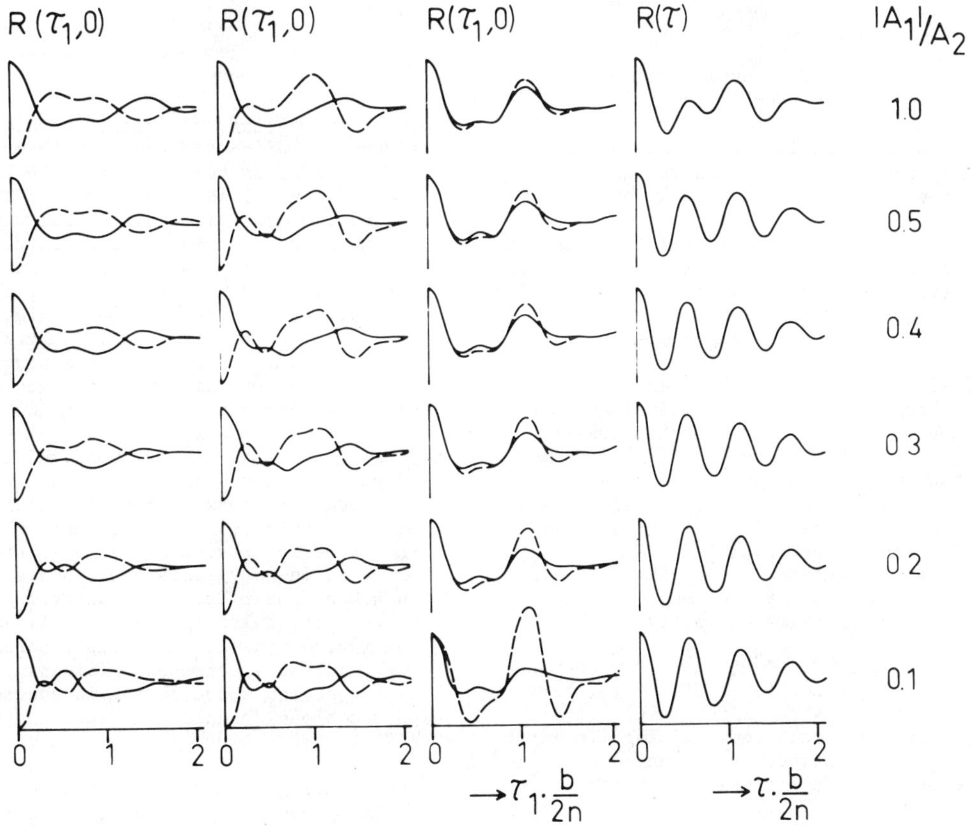

Fig. 3. Functions $R(\tau_1,0)$ for $\sigma/a = 1$, 0.4, 0.1 (from left to right) and $R(\tau)$ for $\sigma/a = 0.1$ for two superimposed rhythms with frequencies $b_1 = b$ and $b_2 = \frac{7}{4}$ b for different values of $|A_1|/A_2$. Full-drawn curves A_1 positive, dashed curves A_1 negative.

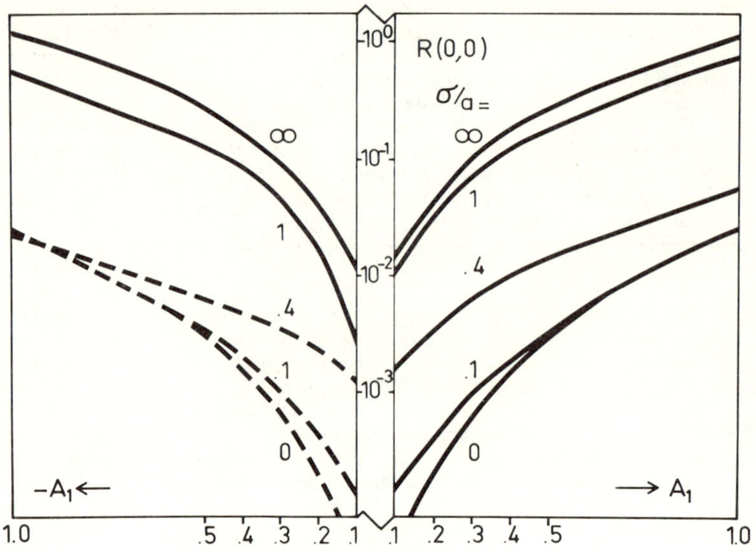

Fig. 4.
Quantity R(0,0) as a function of $|A_1|$ for different values of σ/a; $b_1 = b$, $b_2 = \frac{7}{4}$ b.

reduced by a factor of five or so, implies with regard to $R(\tau_1,\tau_2)$that other terms than those listed above come into play and that the component $h_2(\tau)$ will exert its influence in the final outcome. Symptoms are revealed in the lower most two curves for $\sigma = a$.

One shortcoming of $R(\tau)$ is that from its time course one cannot distinguish $f_I(t) + f_{II}(t)$ from $f_I(t) - f_{II}(t)$ if the contributions occur independent of each other, or for $\sigma \to \infty$ in our nomenclature. For the particular choice of $h_1(\tau)$ and $h_2(\tau)$ this property exists to a considerable extent also in case of inter-dependence, including $\sigma = 0$. Here, $R(\tau_1,\tau_2)$ is definitely of advantage, as demonstrated by the dashed curves which apply to negative values of A_1. Only for $\sigma = 0.1$ a and larger absolute values of A_1 the sign is difficult to recover when comparing pairs of curves. As a side-issue it is remarked that, unlike $R(\tau)$ which possesses at least for a stationary process a maximum for $\tau = 0$, $R(\tau_1,\tau_2)$ is not necessarily maximal and may even be negative. Both features are demonstrated in Fig. 3.

Values of R(0,0) and R(0,0) , in as far as negative values are concerned (dashed curves), have been pictured in Fig. 4 for a few values of σ/a on a semi-logarithmic scale emphasizing its large range when A_1 varies between 0.1 A and A with $A_2 = A$. The ordinate is expressed in units of akA^3.

DISCUSSION

In the foregoing, the situation has been considered in which two randomly occurring wave forms appear in summated form. It is believed that this model is related to problems of interaction and control of biological rhythms. The following recent reports in this field — by no means a complete list — tend to support the supposition: a) the respiration-linked variations in heart rate in man and some animals (Womack, 1971), b) two distinct periodicities in the spontaneous fluctuation of the perfusion flow in the isolated rat kidney (Basar and Weiss, 1970), c) the superposition of higher and lower frequencies in depolarization of membrane potentials in frog skeletal fibres (Sinz and Isenberg, 1972), d) periodic components in steady-state discharges of cuneate neurons in the cat (Amassian and Giblin, 1968), e) the simultaneous occurrence of the circadian rhythm and an acquired meal rhythm observed in body temperature and sodium excretion in humans under isolated conditions and meals served at regular intervals (Sollberger, 1971), and f) the presence of various harmonic components in the electro-encephalogram, just as the alpha, mu and theta rhythm in adults and non-phasic sleep spindles in newborns (c.f. Rémond, 1973).

When dealing with records of time series of composite phenomena, several techniques are available to establish the presence of more than one prefered periodicity and the value of the periods or frequencies. The next question often posed is whether or not the sequential ordering of these components is related. If the amplitude distribution and other characteristics of the signal are such that the elements cannot be separated from each other, or not easily so, and if one is not willing or able to resort to more advanced pattern recognition machinery, difficulties arise in order to find the degree of coupling. In this respect, the second order correlation function has been shown to at least provide additional information and, to some extent, to be able to differentiate between alternatives.

The model under consideration — consisting of the sum of two series of oscillatory wave packets and occurring more or less in synchrony while the degree of coupling can be manipulated with in a well-organized way — is probably relevant to the above-mentioned phenomena of bio-periodicity. At present, it is intended primarily as a tool and a test-object for an orienting study of properties of the second order correlation function. The continuous functions have been thought to be generated with the help of point processes, in part because of its relevance to certain biological systems, and in part for reasons of mathematical convenience. The properties of a point process derived from another point process through translation have been reported in some detail elsewhere (Ten Hoopen and Reuver, 1967).

In view of the preliminary nature, and to avoid unnecessary computations at this stage, several simplifications have been introduced which do not, however, detract from the essential features. The results are restricted to a Poisson distribution of the elementary events, each generating an harmonically varying wave form which increases and decreases according to a Gaussian function. The resulting first order auto-correlation resembles those encountered in biomedical research. The auto-correlation function of experimental data is also frequently described by an oscillatory function that decays consistent with a negative exponential. Therefore, one may alternatively use $h(\tau) = \exp(-\alpha\tau)\cos(b\tau)$ for $\tau > 0$. To simplify matters further, the delay between the events of the two point processes has been assumed to be normally distributed with mean of zero. Other functions, including a non-zero delay, may be used at the expense of more effort in computation. Again it is conjectured that no essentially new viewpoints are to be expected.

One concludes from Fig. 1 that in practise it will be possible to distinguish between loose coupling ($\sigma > a$) and tight coupling ($\sigma \ll a$) on the basis of a comparison of $R(\tau_1,\tau_2)$, whereas this is not so for $R(\tau)$. This distinction is feasible if: a) the wave packets have about the same amplitudes A_1 and A_2, b) if they occur with about the same mean frequency k_1 and k_2, and c) if the periods of oscillation $2\pi/b_1$ and $2\pi/b_2$ differ by a factor not much less than two. Upon elaboration of the formulae, however, one finds that the range in degree of synchronization that can be detected by this measure is small. Fig. 2 illustrates that $R(\tau_1,\tau_2)$ differs significantly for values of $\sigma/a = 0.3$, 0.4 and 0.5. For smaller or larger ratios one can only conclude that the two constituting processes possess either weak or strong interdependency. Moreover, it must be kept in mind dat advantages of $R(\tau_1,\tau_2)$ over $R(\tau)$ are valid for a limited range of values of the other parameters involved. Above all, a fairly strong oscillatory character of $h_1(\tau)$ and $h_2(\tau)$ is required. For the same reasons, a proviso with respect to the amplitudes of the wave packets is due here. From Fig. 3 it follows that, although $R(\tau_1,\tau_2)$ has clearly dissimilar shapes for $\sigma = 0.1$ a and σ = a at any particular ratio of A_1 and A_2, for either $\sigma = 0.1$ a or σ = a differences in A_1/A_2 are, occasionally, better revealed through $R(\tau)$.

No attempt has been made to perform some kind of normalization with respect to $R(\tau_1,\tau_2)$, except that the mean of $f(t) = f_I(t) + f_{II}(t)$ has been subtracted. The term correlation function used so far, instead of covariance function, might have suggested so. To compensate for this in some degree, for the curves of Fig. 3 the quantity $R(0,0)$ has been included in Fig. 4. A means of normalization is suggested by the consideration that the second order auto-covariance of $f(t)$ may be described also as the first order cross-covariance of $f(t)$ and $f(t).f(t+\tau_2)$. Accordingly, $R(\tau_1,\tau_2)$ might be estimated relative to the first order auto-covariances of the two said functions.

Because in the area of application we have in mind analysis is predominantly performed in the time domain, at any rate until a short while ago, the counterpart of the second order correlation function in the frequency domain — the bispectrum — and its normalized measure — the bicoherence — has not been included. However, a large limitation of the correlation function lies in its inability to distinguish more than two or three frequencies in a mixed signal and these frequencies must be well separated if the researcher is to see them at all as remarked and demonstrated by Hord et al. (1965). In these cases, an investigation with the help of the bispectrum is more suitable.

If $R(\tau_1,\tau_2)$is plotted as a continuous function of τ_1, and for discrete values of τ_2, it is found that there exists a noticeable difference in outcome when comparing tight and loos coupling between the occurrence of the two basic components. This is in contrast to $R(\tau)$ whose shape is practically in variant to this property. Because the discrimination is about equally good for a large range of τ_2, one is inclined to restrict the analysis to a few values of τ_2, perhaps to those corresponding to the periods observed in $R(\tau)$, or even to only one value, e.g. to $\tau_2 = 0$. The latter operation effectively amounts to a cross-correlation between $f(t)$ and $f^2(t)$. In addition to the facilities for the ordinary cross-correlation between different time series signals, one has to square the signal. If the said reduction in data-compilation would hold generally, the simplified procedure removes a main objection against the second order correlation function; that is,it requires so much more effort to obtain it. It involves a quadratic increase in the number of computations in comparison with the first order correlation function. With regard to the frequency domain, Dummermuth et al. (1971) state that the computer time for the estimation of bispectra is about twenty times longer than for power spectra.

A natural extension of this line of thought suggests a compression of the third order correlation function $R(\tau_1,\tau_2,\tau_3)$ in a similar way. Taking $\tau_1 = 0$ and $\tau_2 = \tau_3 = \tau$, one would have to calculate just the first-order correlation function of $f^2(t)$. Again, in case of three or more superimposed rhythms, it may be profitable to apply the Fourier transformation.

REFERENCES

AMASSIAN, V.E. and GIBLIN, D.R. (1968). Periodic components in steady-state discharges of cuneate post-synaptic neurones. J. Physiol. 194, 36P.

BASAR, E. and WEISS, C. (1970). Time series analysis of spontaneous fluctuations of the flow in the perfused rat kidney. Pflüchers Arch. 319, 205.

DUMERMUTH, G., HUBER, P.J., KLEINER, B. and GASSER, Th. (1971). Analysis of the interrelationships between frequency bands of the EEG bij means of the bispectrum. A preliminary study. Electroenceph. clin. Neurophysiol. 31, 137.

HORD, D.J., JOHNSON, L.C., LUBIN, A. and AUSTIN, M.T. (1965). Resolution and stability in the autospectra of EEG. Electroenceph. clin. Neurophysiol. 19, 305.

REMOND, A., Ed. Methods of electrophysiological exploration, Paris, 1973. In Press.

SINZ, R. and ISENBERG, G. (1972). Minutenrhythmische Spontandepolarisation des Ruhe-Membranpotentials von Skeletmuskeln. Acta biol. med. Germ. 29, 247.

SOLLBERGER, A. (1971). Biological rhythms and their control in neurobehavioral perspective. p. 101-163. In: Neuroscience Research, Vol.4, S. Ehrenpreis and O.C. Solnitzky, Eds. Acad. Press, New York.

TEN HOOPEN, M. and REUVER, H.A. (1967). Analysis of sequences of events with random displacements applied to biological systems. Math. Biosc. 1, 599.

WOMACK, B.F. (1971). The analysis of respiratory sinus arrhythmia using spectral analysis and digital filtering. IEEE Trans. Bio-Medical Engng BME-18, 399.

ANALYSIS OF A SPIKE GENERATOR

Herman Robert de Jongh

Department of Otolaryngologie
Wilhelmina Hospital, Amsterdam
The Netherlands

This research was subsidized by the Neth. Organization for the advancement of pure research (ZWO)

Summary:
The model to be discussed is one which transforms an analog signal into a series of events (point process). It consists of a linear filter followed by a triggerable spike generator and finally a jitter source (fig. 2). For the analysis of this system we are free to chose a suitable input signal $\underline{x}(t)$ but have only access to the output signal $\underline{z}(t)$. A theoretical method is given which enables one, under rather wide assumptions, to obtain estimates of the three relevant parameters:
1) the impulse response of the linear part,
2) the threshold value of the trigger, and
3) the properties of the jitter generator.
In practice difficulties are encountered because sampling of the signals is required for the analyzing procedure via a digital computer. In part I the theoretical approach is described. Part II deals with practical considerations and an experimental result is given, obtained from a study on the mechanism of analysis of acoustic signals by the ear.

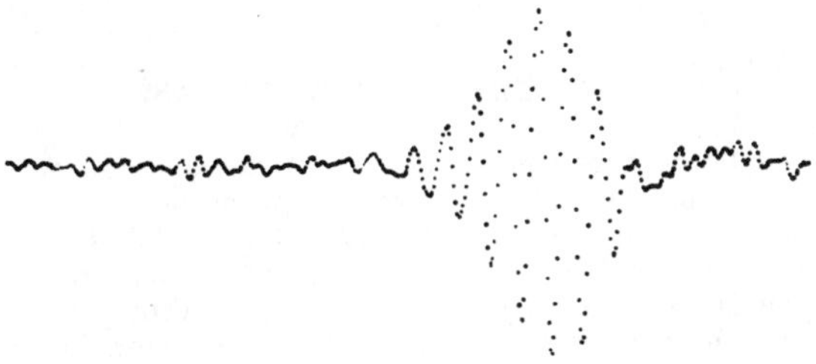

Fig. 1

INTRODUCTION

The model of fig. 2 originates from the field of the peripheral hearing theory (Weiss, 1964; de Boer 1967 and 1969). It shows how the acoustic stimulus is transformed into a series of uniform pulses as they can be observed in a single fiber of the auditory nerve.

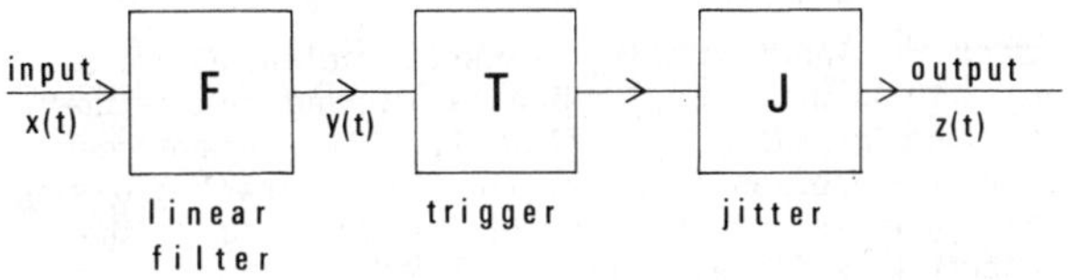

Fig. 2

The linear circuit F stands for the frequency selectivity performed in the inner ear (cochlea). It is known to be a sharply tuned band-pass filter. The trigger mechanism is an idealization of the spike-generating action of the receptor cell: a spike is generated on a moment t when for $\underline{y}(t)$ we have the condition:

$$\underline{y}(t)=\underline{b} \text{ and } \underline{y}'(t) \geqslant 0$$

thus on upward going $\underline{b}$-level crossings. The jitter source is included to make this hypothetical trigger a more realistic one: the observable spike moments θ_n, n=1, 2, ... are derived from the trigger moments t_n, n=1, 2, ... via:

$$\theta_n = t_n + \nu_n$$

where the jitter amounts ν_n are independent random numbers obeying a probability distribution $\underline{p}_J(\nu)$. The problem is to obtain estimates of the filter characteristic, the trigger level and the jitter properties. This is done by stimulating the system (actually a physiological experimental preparation) with a suitable signal and processing the resulting pulse train; of course we are not able to measure somewhere in between input and output.

List of used symbols:

$\underline{x}(t)$:input signal; white Gaussian noise.
$\underline{y}(t)$:the filtered signal; input to the trigger.
$\underline{z}(t)$:output signal; point process.
$\underline{h}(\tau)$:impulse response of the linear part F
$\underline{b}$:trigger threshold level relative to the s.d. of $\underline{y}(t)$.
θ_n :observed spike moments.
$\underline{t}_n$:trigger moments before jitter (not observable)
$\underline{p}_J()$:probability distr. characterizing the jitter source.
' :differentiation and normalisation, for instance:

$$\underline{h}'(\tau) = c.\frac{d\underline{h}(\tau)}{d\tau} \quad \text{with} \int_0^\infty \underline{h}'^2(\tau)d\tau = \int_0^\infty h^2(\tau)d\tau$$

$\underline{h}*_J(\tau)$: result of cross correlating $\underline{x}(t)$ and $\underline{z}(t)$.

1. THEORETICAL APPROACH

That the choice of Gaussian noise is a natural one in view of problems involving linear filtering is well-known. Let us now consider the model of fig. 2 as a known network and analyze it. The output signal is considered as the sum of Dirac δ-functions:

$$\underline{z}(t) = \sum_n \delta(t - \theta_n)$$

where the moments θ_n are the observed spike moments (which constitute a "disturbed" version of the exact trigger moments t_n). Let us first try to get a general insight in the problem via a heuristic approach. When for a moment we discard the trigger + jitter and thus reduce the problem to determining the impulse response of a linear filter, we have the following well-known method to our disposal:
Stimulate with white Gaussian noise, and determine the crosscorrelation function between this input and the resulting output signal; the result is the impulse response (J.S. Bendat, 1958) but time-reversed:

$$\varphi_{xy}(\tau) = \underline{h}(-\tau)$$

What happens now when we take the cascade of the filter and a trigger? Knowing that considerable mutilation of $\underline{x}(t)$ or $\underline{y}(t)$ or even both may take place before the correlation disappears, we may start with the ad hoc assumption that also in this case the crosscorrelation function between input and output of the system is proportional to the impulse response of the linear part. The mutilation concerns here $\underline{y}(t)$ and consists of replacing it by a series of spikes. We assume that $\varphi_{xz}(\tau) = c.\ \varphi_{xy}(\tau)$. If that assumption holds true we are able to extract $h(-\tau)$ from a measurement on input and final output of the system of fig. 2. Then we can mimic the action of the filter F by a simulated filter F* with this estimated $h^*(-\tau)$ as its impulse response. This situation is shown by fig. 3. Now comparison of the original set of spike moments $\{\theta_n\}_{n=1}^{N}$ with the corresponding values of the simulated signal $\underline{y}^*(t)$ should mark the latter as positive going $\underline{b}$ crossings, because under the assumption we should have reconstructed the trigger initiating signal $\underline{y}(t)$! In this case we might be able to estimate $\underline{b}$ directly. However when doing this in an actual simulation, we get quite a different result: the original spike moments do not coincide with $\underline{b}$ level crossings of the simulated signal $\underline{y}^*(t)$, but quite accurately with tops or relative maxima of it! This holds for all $\underline{b}$ levels that can be introduced into the model. Thus we must conclude that the crosscorrelation function in this case is not exactly equal to $\underline{h}(-\tau)$, though it is closely related to it. The result for the crosscorrelation function (called $\underline{h}^*(\tau)$ because of its correspondence to $\underline{h}(\tau)$) is (de Boer & Kuyper 1967):

$$\underline{h}^*(\tau) = \underline{b}.\ \underline{h}(\tau) + \sqrt{\frac{2}{\pi}}\,\underline{h}'(\tau)$$

When we take $\underline{b}$ equal to zero, thus generating spikes on positive-going zero crossings, we get as the crosscorrelation function:

$$\varphi_{xy}(\tau) = \sqrt{\frac{2}{\pi}}\,\underline{h}'(-\tau),$$

thus the simulated signal $\underline{y}^*(t)$ equals $\sqrt{\frac{2}{\pi}}\underline{y}'(t)$, which, in the narrow-band case, has indeed relative maxima on the moments θ_n at which $\underline{y}(\theta_n) = 0$. When $\underline{b}$ is not equal to zero the situation is more complicated. With a rather tedious derivation it can be shown that the spike moments lie on or very near to the tops of $y^*(t)$.
Our problem of determining $\underline{b}$ is not affected very much by this complication as we will see in the next section. A second type of perturbation is the inclusion of time jitter. This also will be discussed in the sequel.

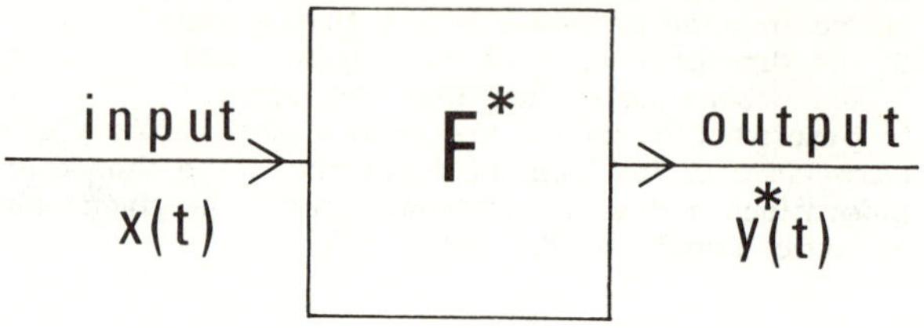

Fig. 3

1-A. THE JITTERLESS CASE

In the undisturbed case (no jitter) three steps suffice for the determination of the parameters.
First: cross correlation of $\underline{x}(t)$ and $\underline{z}(t)$ yields an approximation $\underline{h}^*(\tau)$ of $\underline{h}(\tau)$, the impulse response of the linear filter $\underline{F}$. With the aid of this $\underline{h}^*(\tau)$ we are able to generate a signal $\underline{y}^*(t)$ which in its turn approximates $\underline{y}(t)$, the input to the trigger.
Second: the simulated signal $\underline{y}^*(t)$ has characteristic properties on the spike moments θ_n (here $\theta_n = t_n$) which make the determination of $\underline{b}$ possible.
Third: when $\underline{b}$ is known, the solution of a simple differential equation suffices to find $\underline{h}(\tau)$.
These three steps will be clarified in A1, A2 and A3 respectively. The procedure will be illustrated with results obtained from a hardware realization of the model of fig. 2.

A1.
It is a well-known fact that the crosscorrelation function $\varphi_{xy}(\tau)$ of in- and output of a linear filter reads

$$\varphi_{xy}(\tau) = \underline{h}(-\tau),$$

when the input is white Gaussian noise. Quite a lot of nonlinear distorsion is allowed before the correlation disappears (Nuttal, 1958). Even when only the moments of upward going $\underline{b}$-level crossings are processed in the correlation procedure, the result is still very close to $\underline{h}(\tau)$:

$$\underline{h}^*(\tau) = \varphi_{xz}(-\tau) = \underline{b}.\ \underline{h}(\tau) + \sqrt{\frac{2}{\pi}}\,\underline{h}'(\tau)$$

(see de Boer & Kuyper 1967, though a much shorter derivation is possible).

Now we notice that crosscorrelation of $\underline{x}(t)$ and $\underline{z}(t)$ is equivalent to an averaging operation on the input signal:

$$\overline{\underline{x}(t-\tau)\,\underline{z}(t)} = \frac{1}{T}\int_0^T \underline{x}(t-\tau)\sum^{N}\delta(t-\theta_n)dt$$

$$= \frac{1}{N}\sum^{N}\underline{x}(\theta_n-\tau)$$

thus we need only to average the pieces of $\underline{x}(t)$ just before the spikes in order to obtain the desired crosscorrelation function.

The fact that $\underline{h}^*(\tau)$ is a linear combination of $\underline{h}(\tau)$ and $\underline{h}'(\tau)$ is illustrated in fig. 4. The upper two figures are $\underline{h}(\tau)$ and $\underline{h}'(\tau)$ the impulse response and its (normalized) derivative of the used filter (See part II for further details).

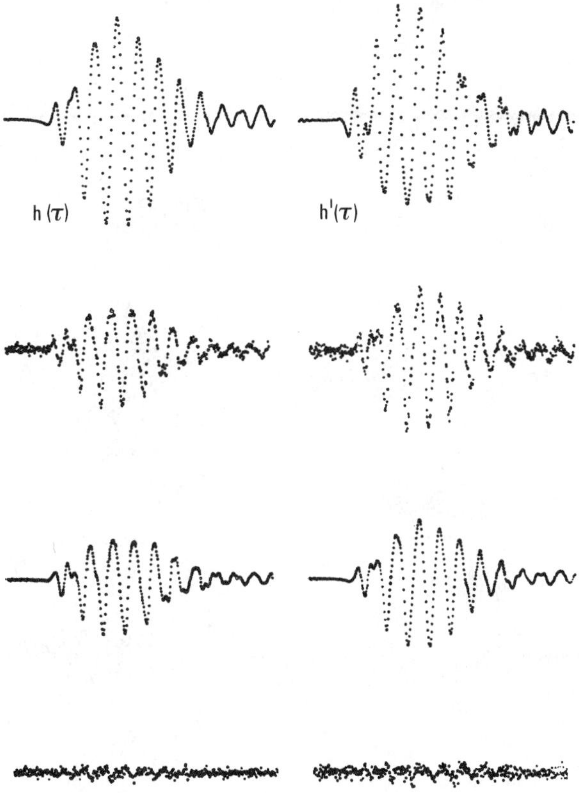

Fig. 4

In the second row we see two examples of $\underline{h}^*(\tau)$ functions: the left came out in a set up with trigger level $\underline{b}$ set to zero, the right one with $\underline{b}$ equal to 1,1. The third row displays the corresponding combinations of $\underline{h}(\tau)$ and $\underline{h}'(\tau)$ namely

$$\sqrt{\frac{2}{\pi}}\cdot\underline{h}'(\tau) \quad \text{and}$$

$1.1.\underline{h}(\tau) + \sqrt{\frac{2}{\pi}}\,\underline{h}'(\tau)$ respectively.
The lowermost traces constitute the differences between the second and third.

The function $\underline{h}^*(\tau)$ can be considered as the impulse response of a filter $\underline{F}^*$. The original $\underline{x}(t)$ signal can be filtered with $\underline{F}^*$ yielding a signal $\underline{y}^*(t)$ which approximates $\underline{y}(t)$ and evidently has the same relation to $\underline{y}(t)$ and $\underline{y}'(t)$ as $\underline{h}^*(\tau)$ has to $\underline{h}(\tau)$ and $\underline{h}'(\tau)$:

$$\underline{y}^*(t) = \underline{b}.\ \underline{y}(t) + \sqrt{\frac{2}{\pi}}\ \underline{y}'(t)$$

A2.

We have to look at $\underline{y}^*(t)$ more carefully to discover where the $\underline{b}$ value is hidden. $\underline{y}(t)$ has on the spike moments θ_n the characteristic property that $\underline{y}(\theta_n) = \underline{b}$. Also $\underline{y}^*(t)$ does have on these moments some special property though more intricate.
First we note that:

$$\underline{y}^*(\theta_n) = \underline{b}.\underline{y}(\theta_n) + \sqrt{\frac{2}{\pi}}\,\underline{y}'(\theta_n) = \underline{b}^2 + \sqrt{\frac{2}{\pi}}\,\underline{y}'(\theta_n)$$

Furthermore, we easily see that $\underline{y}(t)$ and $\underline{y}'(t)$ are statistically uncorrelated:

$$\overline{\underline{y}(t).\underline{y}'(t)} = \frac{d}{d\tau}\,\overline{\underline{y}(t).\underline{y}(t+\tau)}\Big|_{\tau=0} = \rho'_{yy}(\tau)\Big|_{\tau=0} = 0$$

Since $\underline{y}(t)$ and $\underline{y}'(t)$ are Gaussian variables they are also uncorrelated.
Thus the $\underline{y}'(\theta_n)$ values are distributed according to the distribution:

$$p(\underline{y}')d\underline{y}' = \begin{cases} \sqrt{\frac{2}{\pi}}\, e^{-\underline{y}'^2/2}\, d\underline{y}' & \underline{y}' \geq 0 \\ 0 & \underline{y}' < 0 \end{cases}$$

(remember the condition $\underline{y}'(\theta_n) \geq 0$

When we construct a histogram of $\underline{y}^*(t)$ values on spike moments θ_n, we will arrive at a figure like fig. 5:

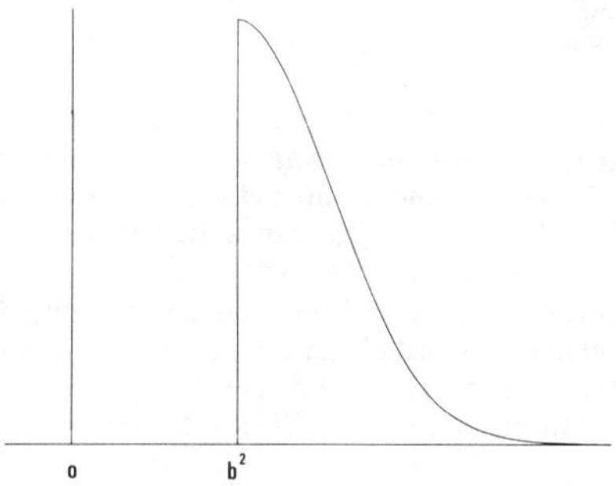

Fig. 5

all $\underline{y}^*(\theta_n)$ values are larger than $\underline{b}^2$ and distributed as one half of a Gaussian distribution.

The left boundary thus gives (theoretically) directly the value $\underline{b}^2$.

A3.
The only thing left to solve is the differential equation

$$\underline{h}^*(\tau) = \underline{b}.\underline{h}(\tau) + \sqrt{\frac{2}{\pi}}\,\underline{h}'(\tau)$$

which is readily integrated with the aid of the Fourier transform $\mathcal{F}$:

$$\underline{h}(\tau) = \mathcal{F}^{-1}\left[\frac{\mathcal{F}\,[h^*]\,(i\omega)}{b + \sqrt{\frac{2}{\pi}}.\, i\omega}\right]$$

This step is self-explanatory.

B. INTRODUCING JITTER

What effect does time jitter have on the crosscorrelation function? As a first step to the answer we ask ourselves what the result would be when all spikes would be shifted an amount ν in time:

$$\theta_n = t_n + \nu$$

then:

$$\underline{h}^*_J(\tau) = \overline{\underline{x}(\theta_n - \tau)} = \overline{\underline{x}(\theta_n + \nu - \tau)} = \underline{h}^*(\tau - \nu)$$

thus the result $\underline{h}^*_J(\tau)$ is just a shifted version of $\underline{h}^*(\tau)$. The derivation when the spikes are shifted by independent stochastic amounts ν_n, having the probability density function $\underline{p}_J(\nu)$ is straightforward:

$$\underline{h}^*_J(\tau) = \int_{-\infty}^{\infty} \underline{h}^*(\tau - \nu)\, \underline{p}_J(\nu)\, d\nu$$

In the following we take for $p_J(\nu)$ the Gaussian distribution with zero mean and s.d. σ_J.
In fig.'s 6 and 7 the effect of jitter on the crosscorrelation function is shown. They are obtained, as before, from a hardware model. Fig. 6a displays a $\underline{h}^*(\tau)$, obtained without jitter. The filter in question was narrow-band filter (1/3 octave filter) with a central frequency of 317 Hz.
In fig. 6b we see the corresponding $\underline{h}^*_J(\tau)$ where σ_J was .25 msec., while in fig. 6c σ_J was 1 msec.
The curves in fig. 7 are obtained via the known $\underline{b}$ and $\underline{h}(\tau)$:
The upper curve is the theoretical jitterless

$$\underline{h}^*(\tau) = \underline{b}.\underline{h}(\tau) + \sqrt{\frac{2}{\pi}}\,\underline{h}'(\tau)$$

The lower ones (7b and 7c) are the same function but convolved with the Gaussian distribution with a σ of .25 and 1 msec. inserted, respectively.
Fig.'s 8 and 9 show the same but with another $\underline{b}$ value in the experimental set up.
We conclude that, provided the mean amplitude of the time jitter is much smaller than the distance of successive peaks in the correlation function, the jitter leaves us with a sufficiently accurate estimate of $\underline{h}^*(\tau)$.
It is, of course, more difficult to read $\underline{b}$ from a histogram that is perturbed by the jitter. But under the same proviso no spike moment will deviate from its normal position, a top of y*(t), too much.
Hence all spike moments can be shifted to coincide with the nearest-by top and the effect of jitter has largely been eliminated.

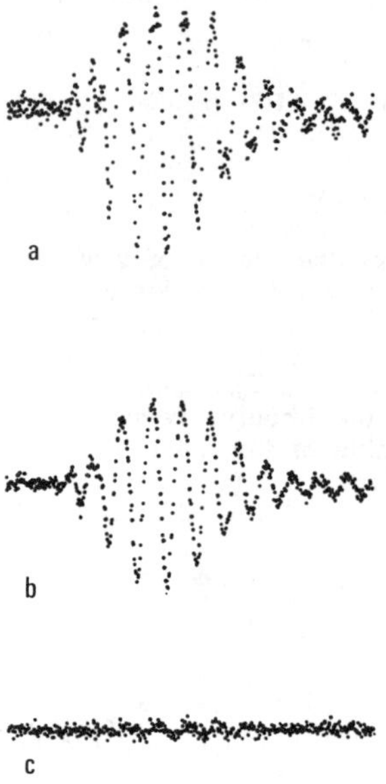

Fig. 6

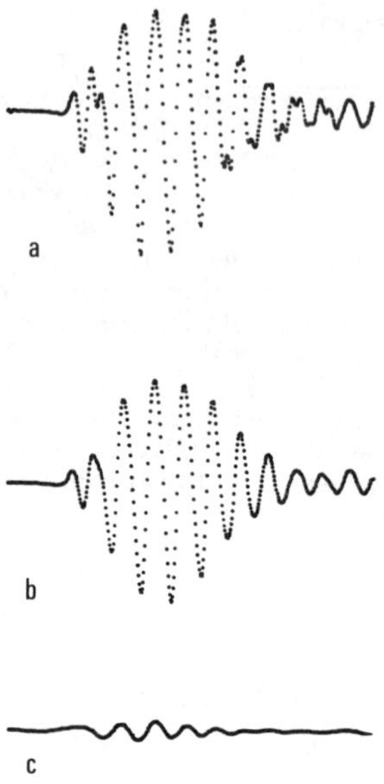

Fig. 7

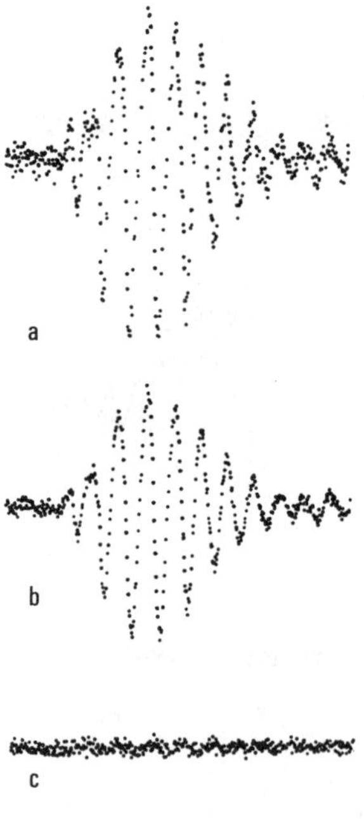

Fig. 8

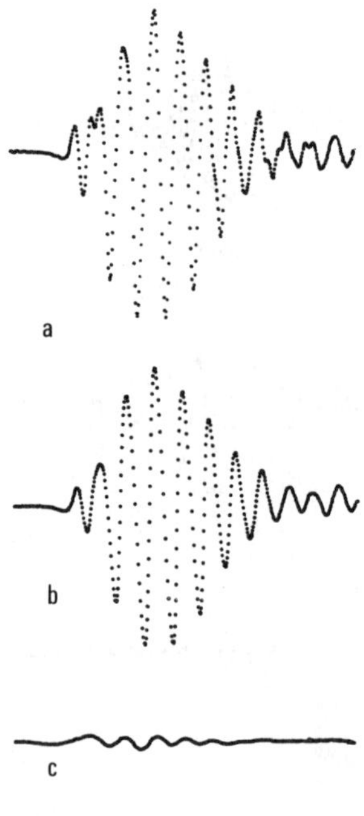

Fig. 9

II. PRACTICAL CONSIDERATIONS.

The analysis requires three main steps: obtaining the crosscorrelation function $\underline{h}^*(\tau)$, generating a filtered version $\underline{y}^*(t)$ of the original x(t), and finally, producing the histogram of $\underline{y}^*(t)$ values on the maxima closest to the spike moments. Practical problems will be met, some of them have to do with the digital computer processing of the signals, others are just practical problems. Let us first discuss the obtaining of $\underline{h}^*(\tau)$:

The stimulus is read into a digital computer via a digital to analog converter (on-line or after recording it together with the resulting spike train on magnetic tape), with a uniform sampling rate, the sample period being Δt. The analog signal $\underline{x}(t)$ is reduced to the values $x(n.\Delta t)$ n = 0, 1, ...

Each moment a spike occurs, an interrupt is generated which initiates the averaging procedure. The moment of the first sample of $\underline{x}(t)$ after the interrupt is considered to be the spike moment. This moment is at most Δt seconds later than the interrupt or exact spike moment. Evidently the delay is a stochastic variable uniformly distributed on $(0, \Delta t)$. In the averaging procedure this is equivalent to a time jitter, and the first inaccuracy we meet in the analysis is that the in this way experimentally found $\underline{h}^*(\tau)$ is the convolution of the exact one with the above mentioned square distribution. It is thus important to take care that Δt is much smaller than the inverse of the central frequency of the filter F. The averaging action just consists of adding the N most recent samples $\underline{x}(\theta_n)$, $\underline{x}(\theta_n - \Delta t)$, ..., $\underline{x}(\theta_n - (N-1).\Delta 1)$ to the result in an "averaging buffer". The signal to noise ratio is proportional to the square root of the total number of averagings. The experimentally obtained results in fig. 4 required 40.000 spikes. Even when we "clean" the noisy correlate by a windowing procedure, the result will always be contaminated by an additive noise. This will of course also hold for the signal $\underline{y}^*(t)$ which must be generated via the obtained $\underline{h}^*(\tau)$. This second inaccuracy may be more disturbing than the first, especially when only few spikes can be processed. To see what effect these disturbing factors do have on the most important figure, namely the histogram of $\underline{y}^*(t)$ values on the spike moments (or the derived maximal value moments), we simulated two situations with known $\underline{h}(\tau)$ and $\underline{b}$. The same computer programs and electronic equipment where used here as for the analyses of the experimentally obtained data, thus arriving at comparable results.

In fig.'s 10 and 11 we see the histograms of $\underline{y}^*(t)$ values on derived spike moments in two situations: The top one originates from an experimental set up with $\underline{b}$ = 1.1, the lower one had the same filter but $\underline{b}$ = 2.2. A Gaussian distribution function is depicted simultaniously, being the amplitude distribution of the $\underline{y}^*(t)$ signal (normalized to have a s.d. of 1). Since the s.d. of $\underline{y}^*(t)$ is a factor:

$$\sqrt{b^2 + \frac{2}{\pi}}$$

larger than that of $\underline{y}(t)$ and $\underline{b}$ is defined relative to the s.d. of $\underline{y}(t)$, a threshold $\underline{b}$ will have the numerical value of:

$$\frac{b}{\sqrt{b^2 + \frac{2}{\pi}}}$$

on the scale in the figure. Thus a value of $\underline{b}$ = 1.1 corresponds with 0,9 in the figure and b = 2.2 with 1,96. A comparison of the theoretical curve (fig. 5) and the experimentally obtained ones shows the slight distorsion of the latter. However the estimation of the pertinent $\underline{b}$-values by means of the position of the left steep slope of the curves is quite accurate.

Let us remember how we intended to get information about the jitter properties: it was claimed that the nearest-by maximum to a jittered spike moment is a good estimate of the real trigger moment, so that the following 2 conclusions are justified:

1) A histogram of the $\underline{y}^*(t)$ values on these maxima will be a good approximation to the desired but inaccessable histogram of $\underline{y}^*(t)$ values on the trigger moments.
2) A histogram of the time shifts necessary to bring the observed spike moments to the nearest by maxima gives information about the jitter.

A justification of this claim is given in fig. 12.

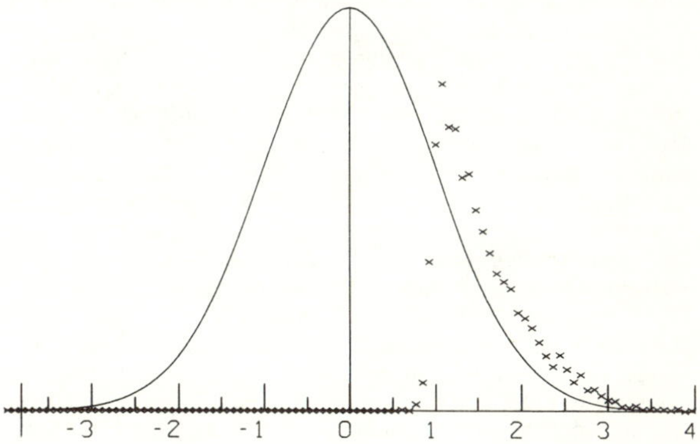

Fig. 10

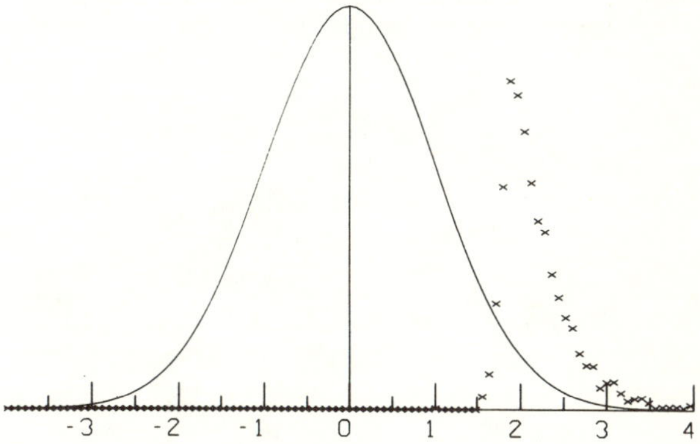

Fig. 11

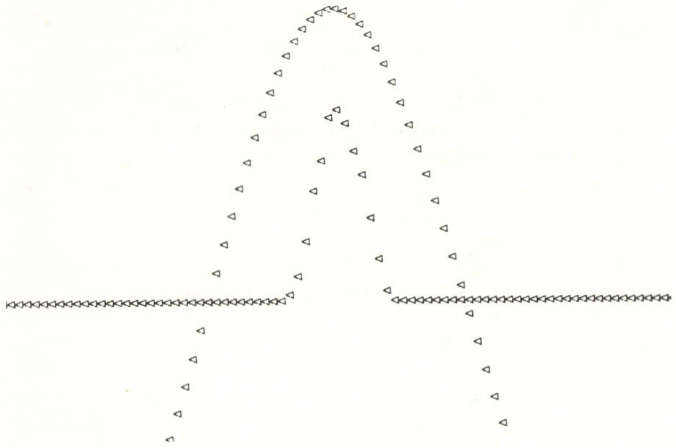

Fig. 12

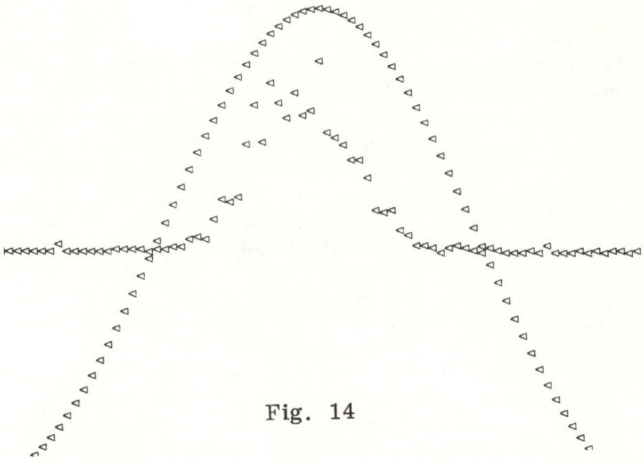

Fig. 14

The upper trace displays $\overline{\underline{y}^*(\theta_n + \tau)}$, thus value of $\underline{y}^*(t)$ around the spike moments in the (jitterless) simulation with $\underline{b}$ = 2.2.*) The point τ = 0 coincides with the maximum thus visualizing the fact that in the mean spikes lay on tops of $\underline{y}^*(t)$. In order to get an idea of the accuracy with which the spikes coincide with maxima (in our experimental set up) we displayed in the lower trace of the same figure the histogram of the distances (in sample points) from spike moment to nearest top. The correspondence is quite good, but what is more important, the $\underline{y}^*(t)$ values do not vary much in this interval around the top! Thus, a histogram of $\underline{y}^*(t)$ values taken at spike moments is nearly equal to one taken at the nearest-by maxima.

To conclude we give one experimental result: the Gaussian noise signal x(t) was input to a cat's ear. The action potentials recorded from a single fibre in the eighth nerve constituted the $\underline{z}(t)$ spike signal. The analysis was done: $\underline{h}^*(\tau)$ was acquired, $\underline{y}^*(t)$ generated and the histogram of $\underline{y}^*(t)$ values on corrected spike moments (maxima of $\underline{y}^*(t)$) calculated. The result is depicted in fig. 13. It is rather encouraging, showing a threshold between 1 and 2. The jitter histogram is depicted in the next figure (fig. 14).

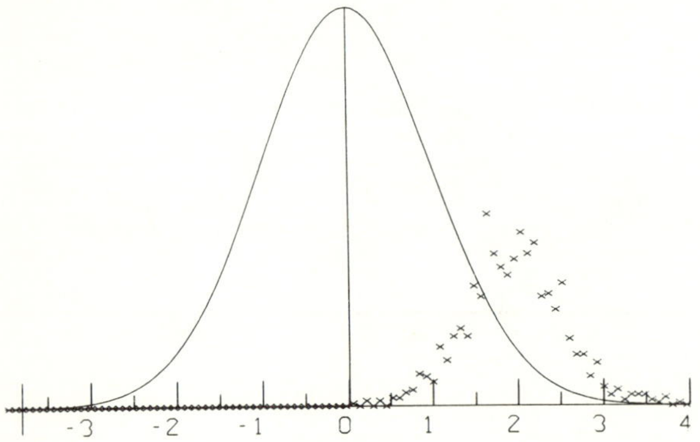

Fig. 13

It evidently is much broader th n the expected one in the jitterless case (fig. 12). The physiological implications of these findings are discussed in a forthcoming paper.

) Note that this is an estimate of the autocorrelation function of y(t). (Cf. de Boer and de Jongh, 1971).

REFERENCES

1) J.S. Bendat, "Principles and Applications of Random Noise Theory." New York: Wiley, 1958, p. 70.

2) E. de Boer, "Correlation studies applied to the frequency resolution of the cochlea." The Journal of Auditory Research p. 209-217, 1967.

3) E. de Boer, "Encoding of frequency information in the discharge pattern of auditoy nerve fibers." International Audiology, Vol. VIII, No. 4, 1969, p. 547-556.

4) E. de Boer, H.R. de Jongh, "Computer simulation of cochlear filtering." Seventh international congress on acoustics, Budapest 1971.

5) E. de Boer, P. Kuyper. "Triggered Correlation. IEEE Transactions on bio-medical engineering." Vol. BME-15, Number 3, July 1968. p. 169-179.

6) A.H. Nuttal, "Invariance of correlation functions functions under nonlinear transformations," M.I.T. Research Lab. of Electronics, Cambridge, Mass., Quart. Prog. Rept., p. 69, April 1957.

7) Th. Weiss, "A model for firing patterns of auditory nerve fibers", M.I.T. Research Lab. of Electronics, Cambridge, Mass., Tech. Rept. 418, 1964; and "A model of the peripheral auditory system", Kybernetik, vol. 3, p. 153-175, 1966.

DATA-COMPRESSION FOR ADAPTIVE MODEL IDENTIFICATION OF NONLINEAR SYSTEMS

K.B. Norkin
Institute of Control Sciences
Moscow, USSR

The problem of nonlinear system identification is considered. An information "compressing" method earlier is suggested that applied only to systems described by linear differential equations. An example is given where this method is applied to a problem with a non-linearity of the type sign $\dot{x}$. The scope of this method is discussed.

1. INTRODUCTION

The paper will be concerned with the identification problem stated as follows: For the equation:

$$x(t)=A\{x(t),u(t),\mu\}$$
$$0 \le t \le T$$
$$x(0)=x^0 \qquad (1)$$

find μ_e which leads to

$$Q(\mu_e) = \min_{\mu} Q(\mu) \qquad (2)$$

if A - a known operator
$u(t)$ - a known function (input variable)
$\mu=\{\mu_1,\mu_2,\dots,\mu_n\}$, t.i
$\mu \in R^n$ - an unknown parameters
$x(t)$ the solution of equation (1)
$Q(\mu)=\|x^*(t)-x(t)\|$,
$x^*(t)$ - a known function (output variable).

The constraints assumed are:

1. A is nonlinear, but of a particular form outlined below.
2. $x^*(t), u(t)$ are "very long" functions, such that are too expensive or impossible, to store in fast computer memory.
3. The only practical way of solving the problem (2) is to iterate with μ^i using the gradient-like approach and $Q(\mu)$ as a performance index.

Such restrictions occur, for example, in identification of mechanical system (tractor suspendor), using real disturbances as $u(t)$ and real shifts as $x^*(t)$.

The system was described by the equations:

$$\mu_0\ddot{x}+\mu_1\dot{x}+\mu_2 x+\mu_3 \operatorname{sign}\dot{x}=y(t)$$
$$x(0)=x^0, \quad \dot{x}(0)=\dot{x}^0 \qquad (3)$$

Disturbances were, naturally, such that

$$|x(t)| \le C < \infty, \qquad (4)$$

where $C, x^0, \dot{x}^0$ are some known constants.

It was certainly impossible to connect the tractor with a computer, and we need to use a magnetograph to memorize $x^*(t)$ and $u(t)$. We received records but to make sure that in the experiment was representative we hand to take very long sample. Simple calculation convinced us the in impossibility of using these records directly for identification by the iterative method. That is why the data had to be compressed.

Our experience in identification has convinced us that such a situation may occur very often and special ways of identification are worth elaborating in these cases.

We will explain for the sake of brevity and clarity the method with a specific example. The scope of this method will be discussed separately.

2. METHOD

In systems, described by linear differential equations, an initial information can be compressed by introduction of the following convolutions:

$$K_{yy}(\tau) = \lim_{T\to\infty} \frac{1}{2T}\int_{-T}^{T} y(t)\cdot y(t-\tau)\,dt \tag{5}$$

$$Z_{xy}(\tau) = \lim_{T\to\infty} \frac{1}{2T}\int_{-T}^{T} x(t)\cdot y(t-\tau)\,dt \tag{6}$$

Using the fact that $x(t)$, $y(t)$ are constrained and the obvious relation :

$$\frac{d^i Z_{xy}(\tau)}{d\tau^i} = \lim_{T\to\infty} \frac{1}{2T}\int_{-T}^{T} x^{(i)}(t)y(t-\tau)\,dt \tag{7}$$

it is easy to see that a relation between $K_{yy}(\tau)$ and $Z_{xy}(\tau)$ analogous to the initial differential equation should hold.

Therefore in the identification problem one may use the signal $K_{yy}(\tau)$ instead of $u(t)$ and $Z_{xy}(\tau)$ in place of $x^*(t)$.

Obviously the replacement of the functions $x(t)$ and $y(t)$ to be stored by $Z_{xy}(\tau)$ and $K_{yy}(\tau)$ permits essential decrease of the amount of data for memorization and storage.

Conditions are also known in which $K_{yy}(\tau)$ exists and is "representative".

A similar approach may be used in development of the identification method for some non-linear differential equations.

Multiplying both parts of equation (3) by $y(t-\tau)/2T$ and integrating in the limits $[-T,T]$ we obtain

$$\mu_0 I_0(\tau) + \mu_1 I_1(\tau) + \mu_2 I_2(\tau) + \mu_3 I_3(\tau) = I(\tau) \tag{8}$$

where

$$I_0(\tau) = \frac{1}{2T}\int_{-T}^{T} \ddot{x}(t)y(t-\tau)\,dt \tag{8.1}$$

$$I_1(\tau) = \frac{1}{2T}\int_{-T}^{T} \dot{x}(t)y(t-\tau)\,dt \tag{8.2}$$

$$I_2(\tau) = \frac{1}{2T}\int_{-T}^{T} x(t)y(t-\tau)\,dt \tag{8.3}$$

$$I_3(\tau) = \frac{1}{2T}\int_{-T}^{T} \operatorname{sign}\dot{x}(t)y(t-\tau)\,dt \tag{8.4}$$

$$I(\tau) = \frac{1}{2T}\int_{-T}^{T} y(t)\,y(t-\tau)\,dt \tag{8.5}$$

Obviously, if $|\tau| \leq \tau^*$ then we must observe $y(t)$ during $|t| \leq T+\tau^*$, and all others during $|t| \leq T$.

Equation (8) may be regarded as the equation of the plant model and the identification problem may be formulated accordingly.

It will be explained below what functions is ought to be considered as given disturbances and what are to be considered as the solution of equation (8).

As a rule the functions (8.1) - (8.5) are considerably "smoother" than initial functions and, besides decrease comporatively fast with increasing $|\tau|$ therefore their storage and repeated reproduction are more economical than for the initial functions. Memory reduction makes even expensive fast memory reasonable. Note that calculation of these functions nevertheless requires repeated reproduction of $x^*(t)$ and $y(t)$. But usually the calculation of (8.1) - (8.5) requires tens of references to $x(t)$ and $y(t)$ there are also devices such as some types of correlators, permiting this operator with a single reference to $x^*(t)$ and $y(t)$, and the solution of the identification problem requires several hundreds and more iterations and the gain is about an order higher. Note that a quite analogous situation may occur for linear systems as well. The examples are known, where data-compression gives considerable gain for linear systems.

Let us see how functions, represented in (8) can be obtained.

Note that:

$$I_1(\tau) = \frac{x(t)\,y(t-\tau)}{2T}\Bigg|_{-T}^{T} + \frac{1}{2T}\int_{-T}^{T} x(t)\cdot \dot{y}(t-\tau)\,dt = \frac{\varphi_1(\tau,T)}{2T} + \frac{dI_2(\tau)}{d\tau}; \tag{9}$$

$$I_0(\tau) = \frac{\dot{x}(t)\cdot y(t-\tau) - x(t)\dot{y}(t-\tau)}{2T}\Bigg|_{-T}^{T} + \frac{1}{2T}\int_{-T}^{T} x(t)\cdot \ddot{y}(t-\tau)\,dt = \frac{\varphi_2(\tau,T)}{2T} + \frac{d^2 I_2(\tau)}{d\tau^2} \tag{10}$$

Consequently (8) is transformated into:

$$\mu_0 \ddot{I}_2(\tau) + \mu_1 \dot{I}_2(\tau) + \mu_2 I_2(\tau) + \mu_3 I_3(\tau) + \mu_1 \frac{\varphi_1(\tau,T)}{2T} + \mu_0 \frac{\varphi_2(\tau,T)}{2T} = I(\tau) \tag{11}$$

The functions $I(\tau)$ and $I_3(\tau)$ represented in (11) can be received by processing the experimental data and are considered as the disturbances $u(t)$. The function $I_2^*(\tau)$ is also obtained from experimental data and considered as the solution $x^*(t)$. The functions $\varphi_1(\tau,T)$ and $\varphi_2(\tau,T)$ can also be obtained in principle, from experimental data. However, since to obtain $\varphi_2(\tau,T)$ the differentiation should first be performed, the acquisition and processing of experimental data may be arranged so that these functions may be neglected. In particular this is possible if T is large enough (since $\varphi_1(\tau,T)$, $\varphi_2(\tau,T)$ are constrained functions), or if $y(t)$ is such that at $|t \pm T| \le \tau^*$ $y(t)=\dot{y}(t)\equiv 0$, or if $y(t)$ is such that at $t = \pm T$.

$$x(t) = \dot{x}(t) = 0.$$

Denoting in analogy with the auto- and cross-correlation functions $I_2(\tau) = Z_{xy}(\tau)$ and $I(\tau) = K_{xy}(\tau)$ then if one of the above conditions holds we will have

$$\mu_0 \ddot{Z}_{xy}(\tau) + \mu_1 \dot{Z}_{xy}(\tau) + \mu_2 Z_{xy}(\tau) + \mu_3 I_3(\tau) = K_{yy}(\tau) \tag{12}$$

The solution of equation (12) was obtained in a hybrid computer"KAC-2" (Norkin, 1972) by the "open-close" method (Fig. 1) (Norkin, 1969). The "open" method a single extremum found, and then the coefficients of the differential equation were defined more precisely by the "closed"method. The input actions were records of the functions $K_{yy}(\tau)$ and $I_3(\tau)$ in the block-diagram used for finding the unknown parameters of equation under study. The model output $\ddot{Z}(t)$ was compared with a sample record $\ddot{Z}_{xy}(\tau)$ in the unit where the criterion was calculated. The control unit generated control actions $\alpha, \beta, \gamma, \delta$ for of coefficients in order to minimize the criterion. The switch S implemented the "open-close" method. The convergence of the experimental and theoretical curves is shown in Fig. 2b.

3.THE DISCUSSION

Consequently the method proposed permits identification of some non-linear differential equations and reciving full conformity of a theoretical model and an actual plant. Considerable saving of computer time is achieved by preliminary processing of the data. The process of solution was accelerated roughly ten times in the above example. It may be noted that the method is not a completely "close" one in the sense of (Norkin, 1968). It would be "close if $I_3(\tau)$ and $Z_{xy}(\tau)$ could be analytically related. If "opening" for a given coordinate is undesirable, the mathematical description by "close" method can be updated by using samples. Because this usually requires few references to the records of samples, the computer time will still be saved in most cases.

In conclusion let us define the scope of the method. Althouth the functions obtained are in some sense near to correlation functions, the analogy with the correlation functions is purely superficial. The transformation (8) is possible for a very large class of functions, even for nonstationary ones.

The signals to be processed are not required in fact to have the properties ensuring the existence of correlation functions. It would be sufficient for the transformation (8) to have a sense and "smooth" the initial functions; the functions resulting from (8) should be obtained either by processing the experimental data,or by the analytical relations of the type (9), (10). The effect of measuring disturbances is reduced by the transformation (8), but the disturbances will still have the effect as was described in particularly in (Norkin, 1968) and finally, a few words about representability. Some types of correlators allow calculating functions of the type (8.1) — (8.5) using a single sample of the initial functions (other types perform this operation for different values of τ in different parts of current samples). This device allows obtaining the functions (8) with T enough large for the initial functions to be representative. Then the functions (8) will be most the "smoothest" and decrease the fastest on $|\tau|$. "The represantability" of the functions (8) will also be ensured. If the possibilities of processing samples with a large T are limited, then "the represantability" of the functions (8) should be checked on the knowledge of differences in (12) and, it necessary, to increase , or to choise more suitable samples.

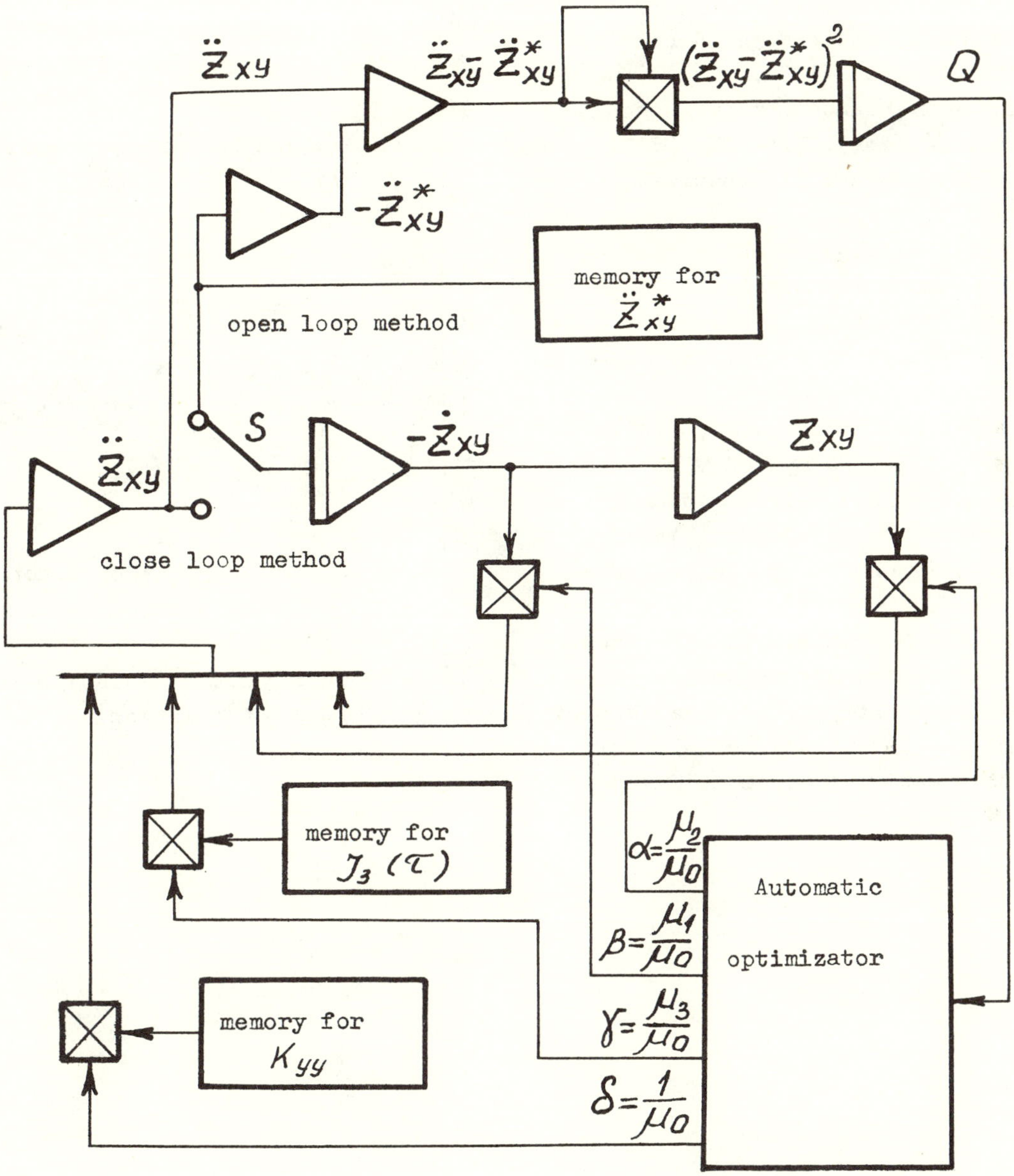

memory for
open loop method
S
close loop method
memory for
memory for
Automatic
optimizator
Q

Fig. 1

It is a pleasure for me to thank
V.M. Prut, Ye.N. Faleeva, N.L. Logunova.

REFERENCES

(1) Rajbman N.S., Shto takoye identifikatsiya M., "Nauka", 1970.

(2) Osovskiy L.M., Samonastraivayushchiesya modeli (obzor). Izv. AN. SSSR, OTN "Tekhnicheskaya kibernetika", No.1, 1963.

(3) Norkin K.B., Poiskovye metody nastroyki upravlyaemykh modyeley v zadachakh opryedyelyeniya paramyetrov ob"yektov, "Avtomatika i tyelyemyekhanika", No.11, 1968.

(4) Norkin K.B., Spiridonov V.D., Isslyedovaniye poiskovykh myetodov nastroyki upravlyayemykh modyeley v zadachakh opryedyelyeniya paramyetrov lineynykh ob"yektov, Avtomatica i tyelyemyekhanika", No.1, 1969.

(5) Norkin K.B., Shubin A.B., Razrabotka komplyeksa apparatury avtomatichyeskogo sintyeza. Sb. st. pod ryedaktsiey Tsypkina Ya.Z. "Myetody optimizatsii avtomatichyeskikh sistyem", "Enyergiya", 1972.

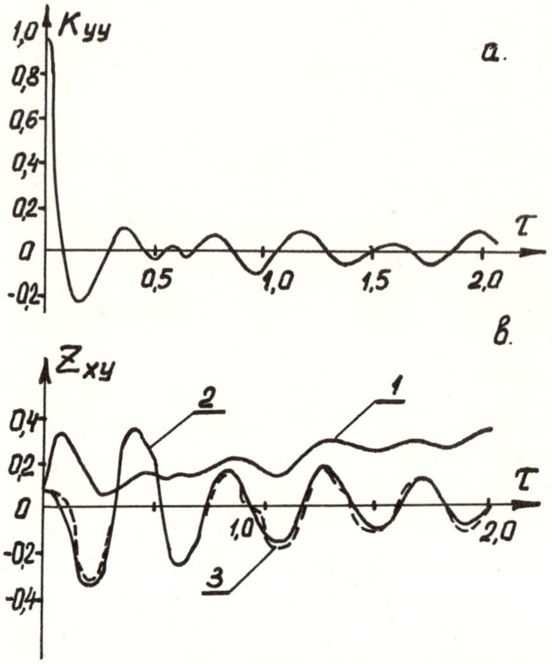

The graph of correlation function (a) and cross-correlation functions (b):

1) at the beginning of solution
2) at the end of solution
3) experimental

Fig. 2

DISPERSIONAL IDENTIFICATION

A.L. Bunich, N.S. Rajbman
Institute of Control Sciences
Moscow, USSR

Methods of the random process correlation theory prove insufficient and an approach and method are needed that would describe the structure of a complex nonlinear plant in sufficient detail. The concepts of accompanying noise and useful component are introduced and a uniform method is developed that uses the first-and second-order moment conditional characteristics to tackle a number of identification, nonlinear filtration and control problems. These concepts permit uniform interpretation of some earlier results which can be regarded as particular cases of the more general approach.

1. INTRODUCTION

Linear plants are normally identified by using random function correlation methods (Pugachev V.S., 1962, Rajbman N.S., 1970) which, however, prove inefficient for nonlinear plants and so new approaches are needed that would yield the plant structure and parameters in sufficient detail. One possible approach is to use random function dispersion methods whereby first-and second-order conditional moment characteristics are used (Bunich A.L., 1972; Rajbman N.S., 1966, 1970). This paper will cite certain new results obtained by this method involving the concepts of accompanying noise and useful component. A number of identification adaptation, nonlinear filtration and control performance estimation problems can now be solved by the same technique. Certain known results can be interpreted from the same point of view.

2. NONLINEAR PLANT IDENTIFICATION

Let $x_s, s=1,2,\dots,n$ denote input actions and y, the response of a multidimensional static plant. Assume that from a total of $x_s, s=1,2,\dots,n$ inputs only $x_\tau, \tau\in T$ are directly observable. Assume also that optimal r.m.s. estimates of inputs in terms of the observables $x_\tau, \tau\in T$ are known

$$\hat{x}_s = M(x_s / x_\tau, \tau\in T) \qquad (2.1)$$

Determine the optimal mean square estimate $\hat{y}$ of the plant response in a class of observable input functions

$$\hat{y} = M(y / x_\tau, \tau\in T) \qquad (2.2)$$

where $M\{\cdot/\cdot\}$ is conditional mathematical exprectation.

The random quantity $\hat{y} = M(y / x_\tau, \tau\in T)$ will be termed useful component of the response to observable inputs $x_\tau, \tau\in T$ the random quantity $\zeta = y - M(y / x_\tau, \tau\in T)$ will be referred to as accompanying noise of the response to these inputs.

A plant is weakly stochastic if the accompanying noise and input signals are non-correlated.

In identification of nonlinear plants the class of linear models is too narrow;

therefore let us introduce the concept of a plant linear on the average, or a plant whose useful component is

$$\hat{y}=\sum_{s=1}^{n} h_s \hat{x}_s \tag{2.3}$$

where h_s-expansion coefficients.

As a result, the equation of a linear on the average plant can be given in the form

$$\hat{y}=\sum_{s=1}^{n} h_s x_s+\eta$$

where the additive noise η satisfied the condition $M(\eta/x_\tau, \tau\in T)=0$.

A linear plant is evidently linear on the average.

A static plant model optimal in terms of the least r.m.s. error in the class of linear on the average models is determined from the mean square optimality criterion

$$\Delta(h)=D(y-\sum_{s=1}^{n} h_s \hat{x}_s)=min \tag{2.4}$$

The unknown expansion coefficients h_s are then determined from the dispersion equation

$$\sum_{s=1}^{n} \theta_x(\sigma,s)h_s=\theta_{yx}(\sigma),\ \sigma=1,2,\dots,n,$$

where $\theta_x(\sigma,s)$ - covariant of x_σ and x_s and $\theta_{yx}(\sigma)$- coveriant of y and $\hat{x}_\sigma$. The functions $\theta_x(\sigma,s)$ and $\theta_{yx}(\sigma)$ are referred to as the cross-dispersion functions of the response and input signals respectively (Bunich A.L., 1972).

The identification r.m.s. error $\Delta(h)$ of a weakly stochastic plant can be shown to be below the statistic linearization error in the least squares method.

Example. Statistical linearization can be regarded as a particular case for ourmethod for plants where all inputs are observable.

The results can be extended to multidimensional dynamic plants (Bunich A.L., 1972), In particular, the definition of a plant, linear on the average, with an allowance for the requirement of feasibility can be represented in the following form: $M(Y_t/X_{t-\tau},\tau\in T)=\int_0^\infty H(t,s)M(X_{t-s}/X_{t-\tau},\tau\in T)ds$

For a dynamic plant linear on the average we have $Y_t=\int_0^\infty H(t,s)X_{t-s}\,ds+\eta_t$, where Y_t - vector of responses, X_t -vector of input actions, $H(t,s)$ -matrix weighting function and T, as above, is a certain finite subset of the positive half-axis.

The definition of the useful component and the accompanying noise are extended to the class of dynamic plants without changes as is the definition of a weakly stochastic plant.

3. ADAPTIVE APPROACH

The quadratic functional (2.4) can be minimized by the algorithm

$$h_s^{t+1}=h_s^t+\gamma_{t+1}\Big[y^{t+1}-\sum_{\sigma=1}^{n} h_\sigma^t\varphi_\sigma(x_\tau^{t+1},\tau\in T)\Big]\varphi_s(x_\tau^{t+1},\tau\in T) \tag{3.1}$$

where $\varphi_s(x_\tau^{t+1},\tau\in T)=M(x_{t-s}/x_\tau^{t+1},\tau\in T)$, $\gamma_{t+1}=\dfrac{1}{\sum_{\sigma=1}^{n}\varphi_\sigma^2(x_\tau^{t+1},\tau\in T)}$

and $x_\tau^{t+1}, \tau\in T$ and y^{t+1} are current values of observed inputs and the plant response at the time t+1. Consequently, for plants linear on the average the hypothesis of representability (Ayzerman M.A. 1970) can be justified and the selection of baxic functions can be related to the input action structure. In the particular case of a plant with independent inputs the algorithm (3.1) coincides with the Kaczmarz algorithm
Tsypkin Ya.Z., 1968).

The adaptive algorithm (3.1) can be use for training a plant model in normal operation. This algorithm can also be applied to the problem of nonlinear forecast of a random process that can be interpreted as adaptive identification of the extrapolator.

4. CONTROL PERFORMANCE

Control performance can be characterized in terms of entropy of the accompanying noise at specified values of observable inputs as $H_\zeta(x_\tau, \tau \in T)$. For a very wide class of conditional distribution densities the following formula is true

$$H_\zeta(x_\tau, \tau\in T) = \log\sqrt{\frac{D(\zeta/x_\tau, \tau\in T)}{D\zeta}} + const$$

Consequently, the static plant control problem is reduced to, firstly, determination of the conditional dispersion of the accompanying noise in learning and, secondly, to minimization of conditional dispersion on the knowledge of observed inputs $x_\tau, \tau\in T$. The latter problem can be handled by the method of stochastic approximation. May it be noted that extended Dvoretsky procedure of stochastic approximation

$$Z_{n+1} = T_n(Z_1, \dots, Z_n) + \zeta_n$$

can be interpreted as decomposition of the iteration Z_{n+1} into a useful component $T_n(Z_1,\dots,Z_n)$ and accompanying noise ζ_n relative to preceding iterations $Z_1, Z_2, \dots, Z_n$.

This method permits minimizing the effect of observable inputs on the plant response and obtaining the information characteristic of control performance.

5. NONLINEAR FILTRATION

Let X_t be the observed random process and it is required to obtain the estimate y_t of the random process X_t which is stochastically related to y_t. Decomposing the stochastic differential of the processes y_t and $\hat{X}_t$ into the useful component and accompynying noise we have

$$dy_t = M(dy_t|\hat{x}_{t-\tau}, \tau\in T) + dW_t \qquad (5.1)$$

$$d\hat{x}_t = M(d\hat{x}_t|\hat{x}_{t-\tau}, \tau\in T) + d\zeta_t \qquad (5.2)$$

The r.m.s. limit $\lim_{h\to 0} M\left(\frac{y_{t+h}-y_t}{h}/\hat{x}_{t-\tau}, \tau\in T\right)$ is referred to (if it exists) as conditional derivative of the process y_t relative to the sections $\hat{x}_{t-\tau}, \tau\in T$ and is denoted as $M\left(\frac{dy_t}{dt}/\hat{x}_{t-\tau}, \tau\in T\right)$.

In a similar way the conditional derivative $M\left(\frac{d\hat{x}_t}{dt}/\hat{x}_{t-\tau}, \tau\in T\right)$ of the process $\hat{x}_t$ with respect to the sections $\hat{x}_{t-\tau}, \tau\in T$ is determined.

Assume that W_t is a Wiener process and at fixed values of the accompanying noises of stochastic differentials are linearly related as

$$d\zeta_t = K(t, \hat{x}_{t-\tau}, \tau\in T)\, dW_t \qquad (5.3)$$

where $K(t, \hat{x}_{t-\tau}, \tau\in T)$ is a matrix function of t and the sections $\hat{x}_{t-\tau}, \tau\in T$.

The conditional derivative of a random process can also exist where the process is not differentiable in terms of the r.m.s. error. By using the above definition, (5.1) and (5.2) can be stringently interpreted

$$y_t - y_{t_0} = \int_{t_0}^{t} M\left(\frac{dy_s}{ds}|\hat{x}_{t-\tau}, \tau\in T\right)ds + W_t - W_{t_0}$$

$$; \hat{x}_t - \hat{x}_{t_0} = \int_{t_0}^{t} M\left(\frac{d\hat{x}_s}{ds}|x_{t-\tau}, \tau\in T\right) + \zeta_t + \zeta_{t_0}$$

Substituting (5.3) into (5.2) and making an allowance for (5.2) we will have the following stochastic differential equation with an argument delayed with respect to $\hat{x}_t$.

$$d\hat{x}_t = M(d\hat{x}_t|\hat{x}_{t-\tau}, \tau\in T) + K(t\, \hat{x}_{t-\tau}, \tau\in T)[dy_t - M(dy_t|\hat{x}_{t-\tau}, \tau\in T)] \qquad (5.4)$$

Example. Let the joint distribution density for the processes $y_t, \hat{x}_t$ be Gaussian and the filter matrix gain $K(t, \hat{x}_{t-\tau}, \tau\in T)$ be independent of the sections $\hat{x}_{t-\tau}, \tau\in T$. For simplicity assume also that the processes y_t and $\hat{x}_t$ are centered and the set T consists of one point, $T=\{0\}$. Then the conditional derivatives of these processes will be linear functions

$$M\left(\frac{dy_t}{dt}/\hat{x}_t\right) = B(t)\,\hat{x}_t,\; M\left(\frac{d\hat{x}_t}{dt}/\hat{x}_t\right) = A(t)\,\hat{x}_t \qquad (5.5)$$

where matrices $A(t), B(t)$ are expresses in terms of the covariant matrices for y_t and $\hat{x}_t$ and their derivatives. Substituting (5.5) into (5.5) gives the Kalman-Bucy filter equation.

$$d\hat{x}_t = A(t)\hat{x}_t dt + K(t)[dy_t - B(t)\hat{x}_t dt]$$

6. STATISTICAL CHARACTERISTICS OF NONLINEAR PLANTS

Let us determine statistical characteristics of nonlinear plants in terms of a useful component and accompanying noise. Then in the dispersion decomposition function (Pugachev V.S., 1962).

$$DM(y_t|x_{t-\tau}, \tau\in T) + MD(y_t|x_{t-\tau}, \tau\in T) = Dy_t \quad (6.1)$$

the first addend in the left-hand part is dispersion of the useful component of the response y_t with respect to the sections $\hat{x}_{t-\tau}, \tau\in T$ and is termed the cross-dispersion function of the sections y_t and $x_{t-\tau}, \tau\in T$ while the second addend is dispersion of the accompanying noise of the response y_t relative to the sections $\hat{x}_{t-\tau}, \tau\in T$ of the input signal and is referred to as the plant stochasticity measure. The auto-dispersion function of random process, can be determined in a similar way.

The function $\eta_{yx} = \sqrt{\dfrac{DM(y_t|x_{t-\tau}, \tau\in T)}{Dy_t}}$ is known as normalized cross-dispersion function of the sections y_t and $x_{t-\tau}, \tau\in T$. From (6.1) follows that $0 \le \eta_{yx}(t) \le 1$. For an inertialess plant i.e. when $T=\{0\}$ one can obtain a more accurate lower estimate: for . This estimate is where $r_{yx}(t)$ is the correlation coefficient of $y_t x_t$. It is an equality if and only if the useful component of the response is a linear function of x_t. This estimate permits determining the nonlinearity degree of an inertialess plant (Pugachev V.S., 1962).

REFERENCES

(1) Ayserman M.A., Braverman E.M., Rosonoer L.I. Metod potentsial'nykh funktsiy v teorii obucheniya mashin. Izd-vo "Nauka", 1970.

(2) Bunich A.L., Rajbman N.S. A dispersion equation of nonlinear plant identification. Problems of control and information Theory, V.1 (1). Publishing House of the Hungarian Academy of Science, 1972.

(3) Pugachev V.S. Teoriya sluchaynykh funktsiy i ee primenenie k zadacham avtomaticheskogo upravleniya. Fismatgiz, 1962.

(4) Rajbman N.S., Chadeev V.M. Adaptivnye modeli v sistemakh upravlenia. Izd-vo "Sovetskoe radio", 1966.

(5) Rajbman N.S. Chto takoe identificatsiya. Izd-vo "Nauka", 1970.

(6) Tsypkin Ya.Z. Adaptatsiya i obuchenie v avtomaticheskikh sistemakh. Izd-vo "Nauka", 1968.

IDENTIFICATION OF HETEROSCEDASTIC PLANTS

F.A. Ovsepian, N.A. Vardanian
Institute of Control Sciences
Moscow, USSR

Two new measures of statistical dependence are introduced, their properties and applicability to verification of the hypothesis that the conditional dispersion of the plant output variable versus input variable is constant are studied. A criterion for estimation of the form of the conditional dispersion function on experimental evidence is suggested. Simulation results are given for the class of plants under study.

1. INTRODUCTION

At present plants are frequently identified by correlation-regression analysis methods. These methods and, above all, the least squares method, are theoretically legitimate provided that certain conditions are observed. The latest studies, e.g. (Oyvazian 1), reveal that the most important requirement is homoscedasticity of the conditional dispersion of the random quantity y (the plant output) at a fixed value of the random quantity X (the plant input)

$$D(y/x) = const \tag{1.1}$$

Estimates of regression coefficients by the least squares method preserve all favorable properties of statistical estimates only if Eq. (1.1) holds or the function $D(y/x)$ is somehow known (Hald 2). If the conditional dispersion is heterscedastic and the form of $D(y/x)$ is unknown, there are no satisfactory methods to handle the problem of estimating the regression parameters. Consequently, a method for estimating the form of the function $D(y/x)$ would be useful statistically. Also a criterion for checking Eq. (1.1) would also prove useful because the criterion reported in the literature has numerous disadvantages that reduce its practical applicability.

Such criteria are essential for solution of a number of problems in control of heterscedastic plants. Thus in solving the optimization problem for such plants one cannot confine to search for a condition optimal on the average because this condition may prove unfeasible where the conditional dispersion varies. Consequently, even without a correlation between y and X , or at $M(y/x) =$ = const, the optimization problem remains legitimate. The search for an optimum reduces in this case to finding an operating condition with the least dispersion.

2. MEASURES OF RELATION AND THEIR PROPERTIES

The problem of estimating the form the

function $D(y/x)$ will be solved by collation of two regression functions $M(y^2/x)$ and $M(y/x)$ starting from the well-known relation

$$M(y^2/x) = M^2(y/x) + D(y/x) \tag{2.1}$$

Let us consider the following two definitions:

Definition 1.

The quadratic correlation coefficient $\omega(y,x)$ of the random quantity y with the random quantity X is

$$\omega(y,x) = \sqrt[4]{\frac{D[M(y^2/x)] - A}{Dy^2 - A}}, \tag{2.2}$$

where $A = 2\,cov[M^2(y/x), D(y/x)]$.

Let us cite the properties of the quadratic correlation coefficient $\omega(y,x)$ proved (Ovsepian in 3).

Property 1.

$$0 \le \omega(y,x) \le 1$$

Property 2.

The condition $\omega(y,x) = 1$ is necessary and sufficient for the relation of y versus X have the form $y = \pm g(x)$, where $g(x)$ is any single-valued function.

Property 3.

The condition $\omega(y,x) = 0$ is necessary and sufficient for $M(y/x) = const$, and $D(y/x) = const$.

Analysis of the above properties shows that the quadratic correlation coefficient $\omega(y,x)$ characterizes that part of the measure of relation of y^2 versus X which depends on $M(y/x)$ and $D(y/x)$.

Definition 2.

The correlativity coefficient of the random quality y and the random quality X is

$$v(y,x) = \sqrt[4]{\frac{D[M^2(y/x)]}{Dy^2 - A}}, \tag{2.4}$$

where A is defined over (2.3).

Let us give the properties of the correlativity coefficient $v(y,x)$ [3].

Property 1.

$$0 \le v(y,x) \le 1$$

Property 2.

The condition $v(y,x) = 1$ is necessary and sufficient for a single-valued functional dependence of y on X .

Property 3.

The condition $v(y,x) = 0$ is necessary and sufficient for $M(y/x) = const$ to be constant.

Analysis of properties of $v(y,x)$ reveals that the correlativity coefficient characterizes that part of the measure of y versus X which depends on $M(y/x)$ alone.

Property 4.

$$v(y,x) \le \omega(y,x), \tag{2.5}$$

with the equality taking place only at $D(y/x)$ = const.

3. USING RELATION MEASURES IN SOLUTION OF IDENTIFICATION PROBLEMS

The relation (2.5) may be used as a criterion for checking the homoscedasticity of the conditional dispersion in the following way.

The experimental findings are employed to calculate the estimates of $\tilde{\omega}(y,x)$ and $\omega(y,x)$ by (2.3) and (2.4) respectively. On the knowledge of sampled distributions of the quadratic correlation coefficient and the correlativity coefficient at a specified confidence probability P_{gob} the confidence intervals

$$\omega_1 \leq \tilde{\omega}(y,x) \leq \omega_2, \qquad (3.1)$$

$$v_1 \leq \tilde{v}(y,x) \leq v_2. \qquad (3.2)$$

are found.

The criterion for verification of the hypothesis (1,1) may be formulated in two ways

$$\tilde{v}(y,x) \in [\omega_1, \omega_2], \qquad (3.3)$$

$$\tilde{\omega}(y,x) \in [v_1, v_2]. \qquad (3.4)$$

It has been mentioned above that the corcelation coefficient $v(y,x)$ (2.4) characterizes the relation of y versus $M(y/x)$ and is kind of an analogue of a the correlation ratio well-known in the leterature. Therefore the correlation coefficient $v(y,x)$ may be used as a measure of the model $M^2(y/x)$ adequacy to plant.

The quadratic correlation coefficient $\omega(y,x)$ (2.2) characterizing the dependence of y on $M(y/x)$ and $D(y/x)$ may also be used as a measure of the model $M(y^2/x)$ adequacy to the plant. The algorithm of their usage is as follows.

For some reasons a type of the $M(y/x)$ and $M(y^2/x)$ functions is chosen. Using the experimental data the unknown coefficients of these regressions are defined. The experimental evidence find the functions $M(y^2/x)$ and $M(y^2/x)$ are used to find the values of the quadratic correlation coefficient $\omega_P(y,x)$ and of the correlation coefficient $v_P(y,x)$ which are compared with the confidence intervals $\tilde{\omega}(y,x)$ and $\tilde{v}(y,x)$ according (3.1) and (3.2) respectively:

$$\omega_P(y,x) \in [\omega_1, \omega_2],$$

$$v_P(y,x) \in [v_1, v_2]. \qquad (3.5)$$

If (3.5) is not satisfied then the above procedure is repeated starting from the of the type of $M(y/x)$ and $M(y^2/x)$ functions and continues till (3.5) is true. If (3.5) holds the type selection hypothesis for $M(y^2/x)$ and $M(y/x)$ functions does not contradict the experimental data when P_{gob} is given. The $D(y/x)$ function type is defined by the relation (2.1).

The degree of heteroscedasticity of the plant may be evaluated by the formula

$$\mathcal{L}(y,x) = \sqrt[4]{\frac{\omega^4(y,x) - v^4(y,x)}{\omega^4(y,x)}} \qquad (3.6)$$

It is obvious that $\mathcal{L}(y,x)$ varies from 0 to 1; $\mathcal{L}(y,x) = 0$ only when $D(y/x) = const$ while $\mathcal{L}(y,x) = 1$ only when $M(y/x) = const$ and $D(y/x) = var$.

These results are easily extendible to the case when x is a vector.

4. SIMULATION RESULTS

Let us consider a system of two plants connected in cascade. The first one will be a plant whose input X and output Z are normally distributed with the parameters $MX=MZ=0$, $DX=DZ=1$, ρ - being the correlation coefficient. The second plant will be standard nonlinear converter.

I $y=az^n$

II $y=\begin{cases} a & z \geq b, \\ 0 & z \leq b. \end{cases}$

Let us define the values of the quadratic correlation coefficient and of the correlation coefficient for each of the plants and for the system as a whole.

For the first plant we have

$$\omega(z,x)=\vartheta(z,x)=|\rho|. \quad (4.1)$$

It should be noted that the ration (4.1) characterizes one more feature of the dependence measures (2.2) and (2.4).

For the second object irrespective of the type of the converter we have

$$\omega(y,z)=\vartheta(y,x)$$

The analysis shows that the values of the quadratic correlation coefficient $\omega(y,x)$ and of a correlation coefficient $\vartheta(y,x)$ for a system depend only on the parameters n and b of the nonlinear converter, when these increase the heteroscedasticity degree (3.6) increases for nearly all ρ.

The simulation results show that if due regard is given to the heteroscedasticity of a system, then the adequacy of the model to the original plant increases by 15 per cent on the average.

REFERENCES

(1) Ayvazyan S.A., Rozanov U.A. Nekotorye zamechaniya k asimptoticheski effektivnym lineym otsenkam regressii. Tr. Matem. in-to im V.A. Steklova, t. 71, str. 3. Nauka, 1964.

(2) Hald A. Statistical Theory with Engineering Applications, page 472. Foreign literature, 1956.

(3) Oysepyan F.A. Nekotorye kriterii klassifikafsii neleneynostey modeli ob'ektov uprovleniya. Trudy II Vsesoyuznogo soveshchaniya po informasionnym metodam v teorii upravleniya, izmereniy i kontrolya. Vladivostok, 1972.

RECURSIVE ESTIMATION OF THE STATES AND THEIR HIGHER ORDER MOMENTS IN NONLINEAR SYSTEMS

N. Ramani
R. Balasubramanian

Department of Electrical Engineering
University of New Brunswick
Fredericton, N. B., Canada

The conditional probability density of the states given the observations is first written as a truncated Gram-Charlier series and secondly as the sum of a finite number of Gaussian-like densities. The orthogonal polynomials associated with each of these densities as kernels are derived, and the nonlinear functions are expanded in these polynomials. Retaining terms upto second order in this expansion, recursive algorithms are derived for the state estimates and their second, third and fourth order moments.

1. INTRODUCTION

The problem of determining the minimum mean squared unbiased estimates of the states of a system from noisy measurement data is the subject of this paper. When the system and observation equations are linear and the noise sequences are white Gaussian, the solution is well known and is generally referred to as the Kalman filter [Kalman, 1960]. This idea can easily be extended to the estimation of the states of nonlinear systems using linear perturbation theory and the resulting estimator is called the extended Kalman filter [Sage, 1968]. However, this has been found to be inadequate in many instances and a variety of complex algorithms have been suggested in the literature. Almost all of these algorithms assume that the conditional probability density of the states given the observation is Gaussian. However, moments of order higher than two associated with this density are not always negligible. Here, two algorithms are derived using approximations for this conditional density which take into account, these higher order moments.

Attempts to model a biological system using other methods resulted in negative variances for the parameter estimates. The problem involved modeling the generation of myoelectric signals in terms of the force produced by the muscle and the length of the muscle. The force produced, the length of the muscle and the myoelectric signal were all measured in situ in the case of a sheep and the data used to estimate the parameters associated with an assumed model. The algorithms of this paper resulted in better estimates provided that the initial values of the parameters were chosen correctly.

2. STATEMENT OF THE PROBLEM

It is assumed that the system is characterized by the nonlinear difference equation

$$x(k+1) = f[x(k)] + u(k) \qquad (1)$$
$$k = 1,2....$$

and the observations are given by

$$z(k+1) = h[x(k)] + v(k) \qquad (2)$$
$$k = 1,2....$$

where $x(k)$, $u(k)$ are n-vectors and $z(k)$, $v(k)$ are m-vectors. $U(k) = \{u(1), u)2) \,..\, u(k)\}$ and $V(k) = \{v(1) \,...\, v(k)\}$ are assumed to be sequences of white Gaussian vectors with zero means and

$$E[u(i)u^T(j)] = Q(i)\delta_{ij}$$
$$E[v(i)v^T(j)] = R(i)\delta_{ij}$$
$$E[u(i)v^T(j)] = 0$$

The probability density of the initial state $x(1)$ is known and is not necessarily Gaussian. Also $x(1)$ is independent of the noise sequences. The estimate of $x(k)$, given the observation sequences $Z(k) = \{z(1), \ldots z(k)\}$ is denoted by $\hat{x}(k/k)$ and can be obtained for any optimality criterion if $p[x(k)/Z(k)]$ is known. Since one is interested in a recursive solution, the problem reduces to one of evaluating $p[x(k+1)/Z(k+1)]$ from $p[x(k)/Z(k)]$ and $z(k+1)$ using equations (1) and (2). If $\hat{x}(k/k)$ is to be the minimum variance estimate, i.e., it is to minimize $E[\tilde{x}^T(k/k)\ \tilde{x}(k/k)]$ where $\tilde{x}(k/k) = x(k) - \hat{x}(k/k)$, it is given by the first moment associated with $p[x(k)/Z(k)]$.

3. METHOD OF SOLUTION

One can find $p[x(k+1)/Z(k+1)]$ from $p[x(k)/Z(k)]$ and $z(k+1)$ using the equations

$$p[x(k+1)/Z(k)] = \int p[x(k+1)/x(k)]p[x(k)/Z(k)]\, dx(k)$$

$$p[x(k+1)/Z(k+1)] = \frac{p[x(k+1)/Z(k)]\ p[z(k+1)/x(k+1)]}{p[z(k+1)/Z(k)]}$$

In order to obtain convenient recursive algorithms, one needs two approximations:

i) an approximation for $p[x(k)/Z(k)]$
ii) approximations for f( ) and h( )

Two approximations of $p[x(k)/Z(k)]$ are considered in the following two sections and $f(\ )$ and $h(\ )$ are approximated using orthogonal polynomials.

For convenience, it will first be assumed that $x(k)$ and $z(k)$ are scalars.

4. GRAM-CHARLIER EXPANSION

This method is based on the fact that any probability density with mean 'a' and standard deviation 'σ' can be written as [Sorenson, 1968]

$$p(x) = \frac{1}{\sqrt{2\pi}\,\sigma} \exp\left[\frac{(x-a)^2}{2\sigma^2}\right]$$

$$\left[1 + c_3 H_3\left(\frac{x-a}{\sigma}\right) + c_4 H_4\left(\frac{x-a}{\sigma}\right) + \ldots\ldots\right] \quad (3)$$

where $H_n(\ )$ = Hermite polynomial of order n

$$c_3 = -\frac{\mu_3}{3!\,\sigma^3}$$

$$c_4 = \frac{\mu_4 - 3\sigma^4}{4!\,\sigma^4}$$

μ_n = central moment of nth order

Let $p[x(k)/Z(k)]$ have

a mean	$\hat{x}(k/k) = E[x(k)/Z(k)]$
a covariance	$P(k/k) = E[\tilde{x}^2(k/k)/Z(k)]$
a third moment	$S(k/k) = E[\tilde{x}^3(k/k)/Z(k)]$
a fourth moment	$T(k/k) = E[\tilde{x}^4(k/k)/Z(k)]$

where $\tilde{x}(k/k) = x(k) - \hat{x}(k/k)$

$f[x(k)]$ is now approximated as

$$f[x(k)] \simeq a(k) + A(k)\tilde{x}(k/k) + 1/2B(k)\tilde{x}^2(k/k) \quad (4)$$

where $a(k)$, $A(k)$ and $B(k)$ are chosen to minimize

$$E[\{f[x(k)] - a(k) - A(k)\tilde{x}(k/k) - 1/2B(k)\tilde{x}^2(k/k)\}^2]$$

It can be easily shown that

$$a(k) = [T(k/k)P(k/k)a_o + P^2(k/k)a_2 - S^2(k/k)a_o - P(k/k)S(k/k)a_1]/b_o \quad (4a)$$

$$A(k) = [T(k/k)a_1 - S(k/k)a_2 - P^2(k/k)a_1 + P(k/k)S(k/k)a_o]/b_o \quad (4b)$$

$$B(k) = 2[T(k/k)a_2 - S(k/k)a_1 - T(k/k)a_o]/b_o \quad (4c)$$

$$b_o = T(k/k)P(k/k) - S^2(k/k) - P^3(k/k) \quad (4d)$$

$$a_o = E[f[x(k)]/Z(k)] \quad (4e)$$

$$a_1 = E[\tilde{x}(k/k)f[x(k)]/Z(k)] \quad (4f)$$

$$a_2 = E[\tilde{x}^2(k/k)f[x(k)]/Z(k)] \quad (4g)$$

From equation (4), one gets

$$\hat{x}(k+1/k) = E[x(k+1)/Z_k] = a(k) + 1/2\,B(k)P(k/k) \quad (5a)$$

which on using equations (1) and (4), gives

$$\tilde{x}(k+1/k) = A(k)\tilde{x}(k/k) + 1/2\,B(k)[\tilde{x}^2(k/k) - P(k/k)] + u(k)$$

Thus

$$P(k+1/k) = A^2(k)P(k/k) + 1/4\,B^2(k)[T(k/k) - P^2(k/k)] + A(k)B(k)S(k/k) + Q(k) \quad (5b)$$

$$S(k+1/k) = A^3(k)S(k/k) + 1/8\,B^3(k)[17\,P^3(k/k) - 3\,T(k/k)\,P(k/k)] + \frac{3}{2}A^2(k)B(k)[T(k/k) - P^2(k/k)] - \frac{3}{2}A(k)B^2(k)S(k/k)P(k/k) \quad (5c)$$

$$\begin{aligned} T(k+1/k) = {} & A^4(k)T(k/k) + \frac{1}{16}B^4(k)[42\,P^4(k/k) + 6\,T(k/k)P^2(k/k)] \\ & + 3Q^2(k) + \frac{3}{2}A(k)B^3(k)S(k/k)P^2(k/k) \\ & - 2A^3(k)B(k)S(k/k)P(k/k) \\ & + 6A(k)B(k)S(k/k)Q(k) + 6A(k)Q(k)P(k/k) \\ & + 3A^2(k)B^2(k)[8\,P^3(k/k) - T(k/k)P(k/k)] \\ & + \frac{3}{2}B^2(k)Q(k)[T(k/k) - P^2(k/k)] \end{aligned} \quad (5d)$$

Here it is assumed that

$$E[\tilde{x}^5(k/k)] = E[\tilde{x}^7(k/k)] = 0$$

$$E[\tilde{x}^6(k/k)] = 15\,P^3(k/k),$$

$$E[\tilde{x}^8(k/k)] = 105\,P^4(k/k)$$

Using equations (5a) to (5d) which specify the first four moments of $p[x(k+1)/Z(k)]$, one can write this density in the form of equation (3) assuming that $c_n = 0$ for $n > 5$. This can be rewritten as

$$p[x(k+1)/Z(k)] = C_o \exp\left[\frac{\tilde{x}^2(k+1/k)}{2\,P(k+1/k)}\right] [g_o + g_1\tilde{x}(k+1/k) + \ldots + g_4\tilde{x}^4(k+1/k)] \quad (6)$$

where g's are known in terms of the moments of $p[x(k+1)/Z_k]$ and C_o ensures that

$$\int p[x(k+1)/Z(k)] = 1$$

Now $h[x(k+1)]$ is approximated as

$$h[x(k+1)] \simeq g(k+1) + C(k+1)\tilde{x}(k+1/k) + 1/2\, D(k+1)\tilde{x}^2(k+1/k) \quad (7)$$

where the constants $g(k+1)$, $D(k+1)$ are found as before and can be written in terms of the moments of $p[x(k+1)/Z(k)]$ similar to equations (5a) to (5g).

One can now write using equation (7)

$$p[z(k+1)/x(k+1)] = C_1 \exp[\frac{\{z(k+1) - h[x(k+1)]\}^2}{2\, R(k+1)}$$

$$c_1 \exp[- \frac{\{z(k+1) - g(k+1) - C(k+1)\tilde{x}(k+1/k)\}^2}{2\, R(k+1)}]$$

$$[1 + \ell_2 \tilde{x}^2(k+1/k) + \ldots + \ell_4 \tilde{x}^4(k+1/k)] \quad (8)$$

where exp ( ) is approximated using power series and only terms up to the fourth power in $\tilde{x}(k+1/k)$ are retained.

One also has

$$p[z(k+1)/Z(k)] = C_2, \text{ independent of } x(k+1) \quad (9)$$

Substituting equations (7), (8) and (9) in the equation

$$p[x(k+1)/Z(k+1)] = \frac{p[z(k+1)/x(k+1)]}{p[z(k+1)/Z(k)]}\, p[x(k+1)/Z(k)]$$

one gets

$$p[x(k+1)/Z(k+1)] = C \exp[- \frac{\tilde{e}^2(k+1)}{2\, F(k+1)}] [b_o + b_1 \tilde{e}(k+1) + \ldots + b_4 \tilde{e}^4(k+1)] \quad (10)$$

where b's are known in terms of g's, ℓ's, $F(k+1)$ etc and

$$F(k+1) = P(k+1/k)C(k+1)[P(k+1/k)C^2(k+1) + R(k+1)]^{-1}$$

$$\tilde{e}(k+1) = \tilde{x}(k+1/k) - F(k+1)C(k+1) [z(k+1) - g(k+1)]/R(k+1)$$

From equation (10), one can easily obtain the moments of $p[x(k+1)/Z(k+1)]$ and it can be shown that

$$C = [b_o + b_2F(k+1) + 3b_4F^2(k+1)]^{-1}$$

$$d(k+1) = C[b_1F(k+1) + 3b_3F^2(k+1)]$$

$$\hat{x}(k+1/k+1) = \hat{x}(k+1/k) + F(k+1)C(k+1) [z(k+1) - g(k+1)]/R(k+1) + d(k+1) \quad (11a)$$

$$P(k+1/k+1) = F(k+1) + C[2\, b_2\, F^2(k+1) + 12\, b_4\, F^3(k+1)] - d^2(k+1) \quad (11b)$$

$$S(k+1/k+1) = 2\, d^3(k+1) + 6\, CF^2(k+1) [b_3F(k+1) - b_2d(k+1) - 6b_4F(k+1) d(k+1)] \quad (11c)$$

$$T(k+1/k+1) = 3F^2(k+1) - 6F(k+1) d^2(k+1) - 3\, d^4(k+1) + 12CF^2(k+1) [b_2d^2(k+1) + b_2F(k+1) + 8\, b_4F^2(k+1) - 2\, b_3F(k+1)d(k+1) + 6\, b_4\, F(k+1)d^2(k+1)] \quad (11d)$$

Thus the recursive estimation problem is complete, since the moments of $p[x(1)/Z(0)]$ are assumed to be known to start with.

5. A SECOND APPROXIMATION FOR $p[x(k+1)/Z(k)]$

Consider a random variable x with mean $\bar{x}$, variance P, a third moment S, and a fourth moment T. Write

$$p(x) = \sum_1^p c_m \frac{1}{\sqrt{2\pi\, P_m}} \exp[- 1/2 \frac{(x - x_m)^2}{P_m}] \quad (12)$$

The constants c_m, x_m and P_m are chosen to satisfy the equations

$$1 = \Sigma\, c_m \quad ; \quad \bar{x} = \Sigma\, c_m x_m$$

$$P = \Sigma\, c_m [P_m + (x_m - \bar{x})^2]$$

$$S = \Sigma\, c_m [3P_m (x_m - \bar{x}) + (x_m - \bar{x})^3]$$

$$T = \Sigma\, c_m [3P_m^2 + 6P_m (x_m - \bar{x})^2 + (x_m - \bar{x})^4] \quad (13)$$

Thus 3xp constants are to be chosen to satisfy 5 equations. With p = 2 (and considering only three moments) one has

$$p(x) = 1/2 \frac{1}{\sqrt{2\pi\, P_1}} \exp[- \frac{(x - x_1)^2}{2P_1}] + 1/2 \frac{1}{\sqrt{2\pi\, P_2}} \exp[- \frac{(x - x_2)^2}{2P_2}]$$

where c_1 and c_2 have been set equal to 1/2 and the first equation of (10) is satisfied. Choosing

$$x_1 = \bar{x} + a \qquad P_1 = P$$

$$x_2 = \bar{x} - a \qquad P_2 = P - 2a^2$$

the second and third equations of (10) are satisfied. The fourth equation of (10) gives $S = 3a^3$, which determines the value of a. Since P_2 has to be positive, $2a^2 < P$ and thus this form can accommodate all distributions for which

$S^2 < 9P^3/8$.

To obtain the estimation algorithms, one approximates $f[x(k)]$, as before, using two terms in an orthogonal expansion, as in equation (4). The moments of $p[x(k+1)/Z(k)]$ are again given by equations (5), though the constants $a(k)$, $B(k)$ etc. are different now. $p[x(k+1)/Z(k)]$ is now written in the form

$$p[x(k+1)/Z(k)] = \Sigma\, c_m \frac{1}{\sqrt{2\pi\, P_m}} \times \exp[-1/2\{x(k+1) - \hat{x}(k+1/k),m\}^2 / P(k+1/k),m] \qquad (14)$$

where the constants c_m, $\hat{x}(k+1/k),m$ and $P(k+1/k),m$ are determined from the moments of $p[x(k+1)/Z(k)]$ as described in the previous paragraphs.

$h[x(k+1)]$ is now approximated as in equation (7) and one can write

$$p[z(k+1)/x(k+1)] \simeq c_1 \exp[-1/2\{\tilde{z}(k+1) - g(k+1) - C(k+1)\tilde{x}(k+1/k)\}^2/R(k+1)\} \times [g_o + g_1 x(k+1) + \ldots + g_4 x^4(k+1)] \qquad (15)$$

where $g_o, \ldots g_4$ are known in terms of $D(k+1)$, $C(k+1)$, the moments of $p[x(k+1)/Z(k)]$ etc. Proceeding as before, one can obtain

$$p[x(k+1)/Z(k+1)] = [c_m G_m] \times [g_o + \ldots + g_4 x^4(k+1)] \qquad (16)$$

where

$$G_m = \frac{1}{\sqrt{2\pi\, P(k+1/k+1),m}} \exp[-\frac{x(k+1) - x(k+1/k+1),m}{P(k+1/k+1),m}]$$

$$\hat{x}(k+1/k+1),m = \hat{x}(k+1/k),m + G(k+1),m\, [z(k+1) - g(k+1) - c(k+1)\tilde{x}(k+1/k)$$

$$\tilde{x}(k+1/k),m = \hat{x}(k+1/k),m - \hat{x}(k+1/k)$$

$$G(k+1),m = P(k+1/k),m\, C(k+1)\,[P^2(k+1/k),m\; C(k+1) + R(k+1)]^{-1}$$

$$P(k+1/k+1),m = P(k+1/k),m - G(k+1),m\; C(k+1)\, P(k+1/k),m$$

It is straightforward now, to obtain the moments of $p[x(k+1)/Z(k+1)]$ from equation (16) and since $\hat{x}(1/0)$ and $P(1/0)$ are known, the recursive estimation algorithm is complete.

Extension of the above results to the case when $x(k)$, $z(k)$ are vectors is straightforward although this introduces some notational problems. The algorithms can be simplified considerably if it is assumed that all moments of order greater than two are such that all cross moments are negligible. For example, the third moment S is given by $S = E[\tilde{x}_1^3 \ldots \tilde{x}_n^3]^T$. Simulation studies show that this assumption is quite reasonable.

6. AN EXAMPLE

A scalar example is considered

$$x(k+1) = x(k) - 0.1\, x^3(k) + u(k)$$

$$x(1) = 2.0\ ;\ Q(k) = 0.01 \text{ for all } k$$

$$z(k) = x^3(k) + v(k)$$

$$\hat{x}(1/0) = 2.0\ ;\ R(k) = 0.01 \text{ for all } k$$

The above example was simulated on a digital computer. Random number sequences $U(k)$ and $V(k)$ were generated on the computer and $\hat{x}(k/k)$ was determined for each k. The simulation was repeated 25 times using different random numbers each time and the square root of the average of $[x(k) - \hat{x}(k/k)]^2$, denoted by M_k, is plotted in fig. 1 for the algorithms of this paper and for the extended Kalman filter.

7. CONCLUSIONS

Tests on the algorithms discussed above show that in some cases, considerable improvement can be obtained using higher order moments. However, in general, these more complicated algorithms tend to make the estimates very sensitive to $\hat{x}(1/0)$ and the actual noise in the measurement and of course increase the computation time. Consequently, it seems reasonable to conclude that, not unlike many other nonlinear estimation schemes, the above schemes will be suitable only for some specific problems.

REFERENCES

1. Kalman, R. E., (1960), "A new approach to linear filtering and prediction problems", J. Basic Eng., Trans. ASME, 82D, 35.
2. Sage, A. P., and Melsa, J. L., (1971), "Estimation theory with applications to communications and control", New York: McGraw Hill.
3. Sorenson, H. W., (1966), "A nonlinear perturbation theory for estimation and control of time-discrete stochastic systems", Ph.D. dissertation, Dept. of Elec. Engg., UCLA.

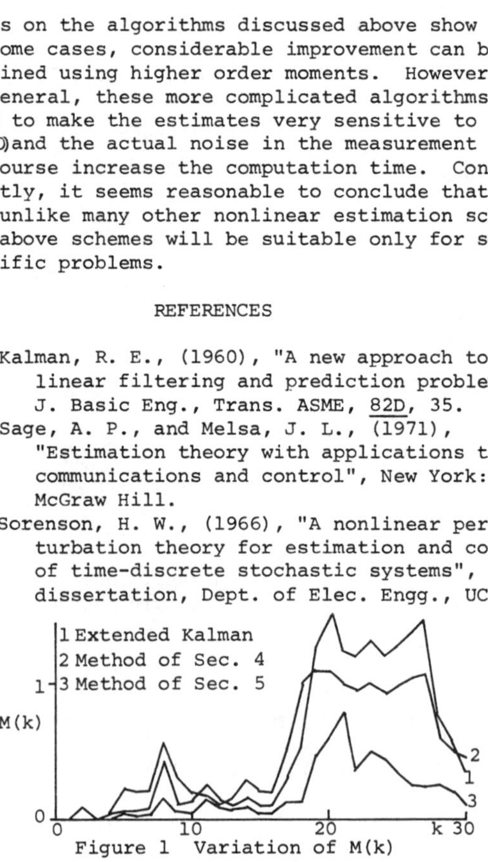

Figure 1 Variation of M(k)

ON INFORMATION MEASURE IN IDENTIFICATION AND PARAMETER ESTIMATION

Pierre A. Devijver
MBLE Research Laboratory
Brussels BELGIUM

O. ABSTRACT

A new definition for a measure of the amount of information associated with probability distributions is proposed. This measure, called Bayesian distance, appears to be particularly appropriate in statistical decision theory for it is tightly connected to the important concept of error probability of decision making associated with the Bayes' rule. In this respect the Bayesian distance is shown to be more efficient than the logarithmic Equivocation.

The properties of the Bayesian distance are investigated in the context of the finite-parameter estimation problem. Some of the potentialities it offers in identification problems are also suggested.

I. INTRODUCTION

The information theoretic concept of average conditional Entropy (Equivocation) has proved very useful in several problems areas related to identification and parameter estimation. Its usefulness for identification purposes - as defined in Eykhoff (1963) - is exemplified by the work reported in Rajbman (1971), in which the Equivocation serves to characterize the isomorphism of a mathematical model to the actual plant. In estimation problems, it has been used as a measure of the missing information about a parameter after observing a sample : Renyi (1967). Besides, the Equivocation may also be interpreted as a measure of insufficiency of a statistic with respect to another one, Kovalevsky (1968), a notion which plays a very important role in estimation and control problems.

All these works are largely inspired by the remarkable success of information theory, when applied to communication problems, and by the fact that all these techniques call for application of methods belonging to the statistical decision theory.

As is pointed out in Lindley (1956), Shannon's theory relies upon two major ideas. The first is that the information concept is of a statistical nature and the second is that there exists an essentially unique definition for a function of the statistical distribution which measures the amount of information. The first of these ideas is clearly not questionable. However, it is important to observe that the justification of the second idea cannot be found within the relationships evolving entirely from the framework of the definition of the measure. At the contrary, it must be emphasized that the definition of Shannon's measure of the amount of information is mainly justified by the fact that it provides quantitative and mathematically sound answers to communication problems. In other words, the mathematical model corresponding to the theoretical framework of information theory has proved quite adequate to describe the physical world the theory was interested in. Furthermore, the model exhibits enough generality - or flexibility - so as to be of major interest in a wide variety of scientific disciplines.

Today, there is at no doubt, a better understanding not only of the specific complexities of the various techniques in which information theoretic methods were initially introduced, but also of the nature of the interdependence between these techniques. Moreover, it has now become evident (Arimoto (1971)) that most of the attracting properties of the logarithmic information measure - imposed by the axiomatic approach of the theory - are due to the concavity of the logarithmic function. One should therefore not wonder that, in recent years, attempts to widening the theoretical framework of the definition of an information measure have aroused interest. Several of these are presently scattered in the literature (see, for example, the references in Arimoto (1971)). Thus, it becomes a natural question to ask whether other measures than the logarithmic Equivocation might prove more adequate or performing in statistical decision theory from the point of view of its specific problems.

It is the purpose of this paper to discuss the properties of such a measure which we have proposed recently, (Devijver (1972 a and b)) and to point out some of its possible applications.

The presentation of our subject matter may be conceived in either of two, not necessarily exclusive ways :

a) First, postulate a set of desirable properties of the measure and then derive its corresponding functional form.

b) Assume a known functional form for the measure associated with a probability distribution and then, justify its usefulness for the problem under consideration.

Our approach is basically of the second type. For the sake of completeness, the set of axioms, upon which the theory may be established, is given in Appendix I. In Section II, the theory is developed in the context of the finite-parameter estimation problem : one of the simplest but most important problem in statistical decision theory. In Section III, the usefulness of the measure is briefly suggested in the framework of the identification problem.

II. INFORMATION MEASURE IN FINITE-PARAMETER ESTIMATION PROBLEMS.

II.1. The Bayesian distance.

In this section we discuss a measure of the average information which is given by the observation x of a random variable X, concerning another random variable Θ. The following developments are restricted to the case where Θ may have a finite (or denumerable) number m of values :

$$\Theta = \{\theta_1, \theta_2, \ldots, \theta_m\},\ 2 \leq m \leq \infty .$$

As is usual in estimation problems, we express our prior knowledge about Θ through and an *a priori* probability distribution :

$$P(\theta = \theta_i) = p_i \ ; \ i=1, \ldots, m \ ; \qquad (1)$$

$$p_i \geq 0 \ , \text{ for all } i \ ; \ \textstyle\sum_{i=1}^m p_i = 1.$$

In Eq. (1), θ denotes any arbitrary value taken by the random variable Θ. We further assume that the distribution of X depends on Θ and let $p(x|\theta_i)$ be the conditional probability density function of X, given that $\theta = \theta_i$, $i=1, \ldots, m$. The finite-parameter estimation problem is that of deciding which value of θ is associated with the occurrence of the observation x.
Let $\delta(\theta|x)$ be any arbitrary decision rule :

$$\delta(\theta|x) = \theta_i \text{ if one decides } \theta = \theta_i \text{ after observing } x \ ;$$

and let $Pe(\delta)$ be the error probability of decision making associated with the decision rule $\delta(\theta|x)$. As is well known, the decision rule which minimizes the error probability of decision making is the Bayes' rule $\delta_B(\theta|x)$ which is defined by :

$$\delta_B(\theta|x) = \theta_i \text{ if } p(\theta_i|x) = \max_j \left[p(\theta_j|x)\right]. \qquad (2)$$

The error probability of Bayes' rule $Pe(\delta_B)$ plays a very important role in statistical decision theory. It will be a concept of major concern in all that follows. $Pe(\delta_B)$ is given by :

$$Pe(\delta_B) = 1 - E_X\left[\max_j p(\theta_j|x)\right]. \qquad (3)$$

In Eq. (3) E_X denotes expectation with respect to the unconditional or mixture probability density of X :

$$p(x) = \textstyle\sum_{i=1}^m p_i \, p(x|\theta_i) \ . \qquad (4)$$

Let us now define a new functional associated with the statistical distributions of X and Θ, and which we have called the Bayesian distance $B(\Theta|X)$, (Devijver (1972.a))* :

$$B(\Theta|X) = E_X\left[\textstyle\sum_{i=1}^m p^2(\theta_i|x)\right] \ . \qquad (5)$$

We have demonstrated that the Bayesian distance which characterizes the distributions of X and Θ and the corresponding error probability of Bayes' decision rule are two intimately connected concepts. Their relationship is expressed in the following theorem :

Theorem 1. *For any arbitrary statistical distributions of the random variables X and Θ, the following inequalities always hold :*

$$\frac{1}{2}\left[1-B(\Theta|X)\right] \leq 1 - \left[B(\Theta|X)\right]^{1/2} \leq \qquad (6)$$

$$\leq \frac{m-1}{m}\left[1 - \left[\frac{mB(\Theta|X)-1}{m-1}\right]^{1/2}\right] \leq Pe(\delta_B) \leq 1-B(\Theta|X).$$

The proof of this basic theorem may be found in Devijver (1972.a). A direct and important consequence of this theorem is given by the following corollary :

Corollary 1. *The error probability associated with the Bayesian decision rule tends to zero if and only if the Bayesian distance tends to unity.*

Proof. From noting the inequalities

$$0 \leq Pe(\delta_B) \leq 1 - \frac{1}{m} \qquad (7)$$

and the inequalities (6), it is a simple matter to set bounds on the Bayesian distance :

* The reasons for calling $B(\Theta|X)$ the *Bayesian distance* are specific to the field in which the concept was defined, namely the Pattern Recognition theory. Those reasons are indicated in Devijver (1972.a).

$$\frac{1}{m} \leqslant B(\Theta|X) \leqslant 1 . \qquad (8)$$

Now, let us rewrite the loosest bounds on $Pe(\delta_B)$ given in the inequalities (6) :

$$\frac{1}{2}[1-B(\Theta|X)] \leqslant Pe(\delta_B) \leqslant 1-B(\Theta|X) . \qquad (9)$$

It is obvious that the "if" statement of the corollary is a direct consequence of the upper inequality of (9) whereas the "only if" statement results from the lower inequality.

Q.E.D.

The above corollary lends itself to a straightforward information theoretic interpretation: When the error probability of decision making associated with Bayes' rule tends to zero, it may be said that the observation of X tends to *give us full information about* Θ. On the other hand, as we have indicated above,

$$Pe(\delta_B) \leqslant Pe(\delta), \quad \text{for any } \delta(\theta|x) , \qquad (10)$$

or clearly, there does not exist any decision rule for which the error probability is less than $Pe(\delta_B)$. Thus if $B(\Theta|X)<1$, one also has $Pe(\delta)>0$ for any $\delta(\Theta|X)$ and, in this case, the observation of X cannot give full information on Θ. Following the viewpoint expressed in Renyi (1967), these considerations lead us to propose the quantity $[1-B(\Theta|X)]$ as a measure of the *missing information on* Θ *given by the observation of* X. In what follows this latter quantity will be referred to as the D-measure*

$$D(\Theta|X) = 1 - B(\Theta|X) . \qquad (11)$$

It is well known that the information provided by an experiment - the observation of X - cannot be made independent, in general, of our prior knowledge - the *a priori* probability distribution of Θ. It is a simple matter to show that our approach through the Bayesian distance does not fail to the rule. However, on the basis of what we have obtained thus far this question can only be treated in an indirect manner. In order to obtain an expression for the amount of missing information corresponding to our *a priori* knowledge, it suffices to consider the case where Θ and X are two statistically independent random variables. Indeed, this is essentially the situation in which the observation of X does not add anything to our prior knowledge. Clearly we must have :

$$p(\theta_i|x) = p_i \ , \ i=1,\dots,m \ ; \text{ for all } x \ ; \qquad (12)$$

so that by Eqs (5) and (12) the Bayesian distance corresponding to the *a priori* probability of Θ, $B(\Theta)$ is simply given by :

$$B(\Theta) = \sum_{i=1}^{m} p_i^2 \qquad (13)$$

and the measure of missing information, prior to any experimentation is :

$$D(\Theta) = 1 - \sum_{i=1}^{m} p_i^2 \ . \qquad (14)$$

In order to be intuitively consistent, our approach must satisfy the condition expressed in the following theorem.

Theorem 2. *For any arbitrary distributions of* X *and* Θ, *the following inequality always holds :*

$$B(\Theta|X) \geqslant B(\Theta) \quad ,$$

with equality if and only if X *and* Θ *are statistically independent.*

Clearly, this theorem means that the observation of X can only reduce the amount of missing information about Θ.

Proof. First observe that the expectation and the summation operators may be interchanged in Eq.(5), for the expectation operation is taken over the entire observation space Ω_X of the random variable X.
Thus we also have :

$$B(\Theta|X) = \sum_{i=1}^{m} E_X[p(\theta_i|\,x)]^2 \ . \qquad (15)$$

Eq. (15) may be transformed into :

$$B(\Theta|X) = \sum_{i=1}^{m} p_i^2 \, E_X\Big[\frac{p(x|\theta_i)}{p(x)}\Big]^2 \ . \qquad (16)$$

The term $E_X\Big[\frac{p(x|\theta_i)}{p(x)}\Big]^2$ is of the general functional form

$$\int f(z)\, g(y(z))\, dz \ ,$$

* The author is indebted to Prof. C. H. Chen for calling his attention towards the similarity which exists between the D-measure and the "quadratic entropy" which appears in Vajda (1968).

where $f(z) = p(x)$, with $\int_{\Omega_Z} f(z)dz = 1$;

$$y(z) = \frac{p(x|\theta_i)}{p(x)} \quad ;$$

and $g(y) = y^2$.

But since $g(y)$ is a strictly downward convex function, direct application of Jensen's inequality for integral :

$$\int f(z)\, g(y(z))\, dz \geq g\left(\int f(z)\, y(z)\, dz\right)$$

yields the following inequality :

$$E_X\left[\frac{p(x|\theta_i)}{p(x)}\right]^2 \geq \left[E_X \frac{p(x|\theta_i)}{p(x)}\right]^2 = 1, \; i=1,\ldots,m. \quad (17)$$

Making use of this latter inequality into Eq.(16), we obtain :

$$B(\Theta|X) \geq \sum_{i=1}^{m} p_i^2 = B(\Theta) \qquad (18)$$

which is the desired result. It must be noticed that the equality sign in (17) - and thus in (18) - is obtained if and only if

$$\frac{p(x|\theta_i)}{p(x)} = \text{Const} \;, \quad \text{for any } i=1, \ldots, m \;; \quad \text{for all } x \in \Omega_X \;.$$

This condition obviously requires statistical independence of X and Θ.
This completes the proof of Theorem 2.

The above arguments lead us to an evaluation of the decrease of missing information on Θ which results from the observation of X. This quantity is given by :

$$B(\Theta|X) - B(\Theta) \quad ,$$

or, in virtue of Eq.(11) :

$$D(\Theta) - D(\Theta|X) \; .$$

II.2. Bayesian distance and Equivocation.

It is enlightening to compare the functional expressions of the D-measure and of the equivalent concept in the framework of the classical information theory, namely the logarithmic Equivocation $H(\Theta|X)$ which is defined by :

$$H(\Theta|X) = E_X\left[\sum_{i=1}^{m} - p(\theta_i|x) \log p(\theta_i|x)\right]. \quad (19)$$

In Eq. (19), as well as in all that follows , logarithms are taken to base 2.

Combining Eqs (11) and (5) we have :

$$D(\Theta|X) = 1 - E_X\left[\sum_{i=1}^{m} p^2(\theta_i|X)\right].$$

The right side term may be transformed as follows :

$$D(\Theta|X) = E_X\left[\sum_{i=1}^{m} p(\theta_i|X)\; [1-p(\theta_i|X)]\right]. \quad (20)$$

From the comparison of Eqs (19) and (20), it is clear that the expression of $D(\Theta|X)$ may be obtained from that of $H(\Theta|X)$ by simply replacing the function $- \log p(\theta_i|X)$ by $[1-p(\theta_i|X)]$.

It is worth noting that this latter function satisfies the conditions required by the definition of a generalized entropy function, (Arimoto (1971)) : it is continuous and continuously decreasing for $0<p(\theta_i|x)\leq 1$, it is concave (yet not in the strict sense) and it is equal to zero for $p(\theta_i|x) = 1$.

It has now become evident that, in reality, the two functionals are of much the same nature. However, in the context of the finite-parameter estimation problem, or more generally, in statistical decision theory the Bayesian distance is a markedly more powerful tool than the logarithmic Equivocation. This statement is supported by the following considerations.

In a statistical sense, the most efficient measure of missing information is the error probability of decision making when optimal use is made of the available information about probabilistic situation, that is when Bayes' rule is applied. However it is well known that $Pe(\delta_B)$ is most often an extremely complex concept to deal with. It turns out that the main reason why the Equivocation has been used in several fields related to statistical decision theory is that, like the Bayesian distance, it permits to set bounds on $Pe(\delta_B)$. It is evident from works like those of Lindley (1956), Renyi (1967) and Arimoto (1971) that using the Equivocation is but an alternative to direct work with $Pe(\delta_B)$. It is the purpose of the remainder of this section to show that the Bayesian distance does better characterize $Pe(\delta_B)$ than does the Equivocation. (For the clarity of the presentation, $Pe(\delta_B)$ is from now on simply written as Pe, it being understood that it refers to the error probability of the Bayesian decision rule).

In terms of Equivocation, the bounds on Pe are given by

$$2Pe \leq H(\Theta|X) \leq H(Pe,\ 1-Pe)+Pe\ \log(m-1)\ . \quad (21)$$

The upper inequality of (21) is the well-known Fano bound.

We define

$$B(Pe,\ 1-Pe) = Pe^2 + (1-Pe)^2, \quad (22)$$

and correspondingly

$$D(Pe,\ 1-Pe) = 2Pe(1-Pe)\ . \quad (23)$$

Then, it follows from the inequalities (6), Eqs.(11) and (23) that

$$Pe \leq D(\Theta|X) \leq 2D(Pe,\ 1-Pe)+2Pe^2\ \frac{m-2}{m-1}\ . \quad (24)$$

The comparison of the bounds given by the inequalities (21) and (24) is the subject matter of the following two theorems.

Theorem 3. *For any arbitrary distributions of* X *and* Θ, *the following inequality always holds :*

$$2D(\Theta|X) \leq H(\Theta|X)\ ,$$

with equality if and only if

a) m=2 *(binary hypothesis problem)* ,

and b) $p(\theta_i|x)=0,.5,1$; *for all* $x \in \Omega_X$,

$i=1,2.$

The rigorous proof of this theorem is long and tedious and is therefore not given here. The interested reader may satisfy himself with a simpler but somewhat less rigorous method of proof which is given in Appendix 2.
From the upper inequality of (6), the lower one of (21) and (25) we obtain :

$$Pe \leq D(\Theta|X) \leq \frac{1}{2}\ H(\Theta|X)\ . \quad (26)$$

This results indicates that the D-measure, or Bayesian distance provides a tighter upper bound on the error probability of the Bayesian decision rule than does the Equivocation. In the following theorem, the tightness of the lower bounds on Pe is discussed in terms of upper bounds on $D(\Theta|X)$ and $H(\Theta|X)$ expressed in functions of Pe as in the inequalities (21) and (24).

Theorem 4. *For any arbitrary distributions of* X *and* Θ, *the following inequality always holds :*

$$2D(Pe,1-Pe) + 2Pe^2\ \frac{m-2}{m-1} \leq H(Pe,1-Pe)+Pe\ \log(m-1), \quad (27)$$

with equality if and only if :

Pe = 0 , *for any* m=2,3, ...

or

Pe =.5 , *and* m=2 .

Proof. As a direct consequence of theorem 3 we have

$$2D(Pe,\ 1-Pe) \leq H(Pe,\ 1-Pe)\ , \quad (28)$$

with equality iff Pe=0, .5, 1.
It is also clear that

$$2Pe^2\ \frac{m-2}{m-1} \leq Pe\ \log(m-1)\ , \quad (29)$$

with equality if Pe=0 or if m=2.
Summing the inequalities (28) and (29) side by side gives (27)

Q.E.D.

It is worth mentioning that when m becomes arbitrary large, the left side term in inequality (29) tends to $2Pe^2$ whereas the right side term tends logarithmically to infinity (providing that Pe>0). Thus, for large m, the upper bound on $D(\Theta|X)$ is much smaller than the upper bound on $H(\Theta|X)$. Consequently the lower bound on Pe given by the Bayesian distance or D-measure is much tighter than the lower bound given by the Equivocation.

Since both the upper and lower bounds on the error probability of the Bayesian decision rule are tighter when expressed in terms of the Bayesian distance or D-measure, this latter concept seems to be a fair candidate for a measure of missing information in statistical decision theory and related fields.

Some reader may wonder why the presentation has been made, thus far, in terms of *missing* information. The essential reason is that, as suggested above, we adopted the viewpoint expressed in Lindley (1955) : In statistical decision theory, complete information is available when the probability distribution is concentrated on a single value of θ. Moreover the amount of available information decreases when the distribution of θ "spreads" over several possible values of the parameter. It must be observed that this interpretation is precisely the reverse of that encountered in the classical information theory as applied to communication engineering problems.
It results from the above considerations that $[1-B(\Theta|X)]$ appears to satisfy the most important intuitive requirements for a useful measure of missing statistical information. Moreover, it shares with $Pe(\delta)$ the property to range between 0 and 1, one feature which is especially appealing. This latter point clearly suggests that the Bayesian distance - the com-

plement to unity of the amount of missing information - be taken as a measure of the amount of available statistical information which is associated with a set of probability distributions.

Up to now, the distributions of the random variable X and Θ were assumed to be known and quite arbitrary. It must be emphasized that this is generally not the case, and that all information about the distributions is contained within sets of samples Θ and x. It is therefore appropriate to mention that the Bayesian distance has already been shown to play a major role in connection with practical and extensively developed methods for solving the finite-parameter estimation problem. One such application related to the use of the least-mean-square criterion is described in Devijver (1972.c).

III. INFORMATION MEASURE IN IDENTIFICATION PROBLEMS.

As is clearly pointed out in Eyckhoff (1963), the identification problem, considered as the determination of the topology of a process, is generally much more difficult to deal with than the associated process parameter estimation problem. The essential reason lies in the difficulty to develop an efficient measure of the isomorphy which exists between the actual process and the assumed mathematical model. This question lends itself quite naturally to an information theoretic formulation for the ultimate aim is to optimize our knowledge about the process under consideration. Furthermore, the identification problem raises most important questions which can be best answered in terms of information measures : The selection of most informative process parameters is such a question which may turn out to be of major economic interest.

In this section we intend to briefly suggest that the Bayesian distance may be a useful concept in this important problem area. The framework of our approach is basically inspired by the work reported in Rajbman (1971).
We slightly modify the conventional notations which we have been using so far, and we consider a stationary plant with the random variable X at the input and the random variable Y at the output. X and Y may be regarded either as scalars or as vectors whose components are measurements of the plant input and output parameters respectively.

It results from the theory developed in Section II, that B(Y) reflects our *a priori* information about the output variable which is not due to the observation of the input variable. On the other hand, B(Y|X) measures the information available about the process when both input and output variables are under control. One may thus define a ratio R(X), $0 \leq R(X) \leq 1$ which characterizes our degree of knowledge about the plant under consideration :

$$R(X) = \frac{B(Y|X) - B(Y)}{1-B(Y)}$$

For $R(X)=0$, (and $B(Y)<1$), the input variable X does not provide any information on the output variable Y and the plant is completely undertermined. At the contrary, for $R(X)=1$, the input variable gives us full information about Y and the plant is completely determined.

In view of the remarkable properties of the Bayesian distance, the ratio R(X) may be regarded as an improved version of the "Relative Informational Measure of Certainty" (RIMC) defined in Rajbman (1971).

A straightforward extension of the above considerations concerns the evaluation of the informative content of input process parameters. This is a subtle and interesting question, unfortunately lack of space does not allow us to discuss it in detail. Let us however mention that, with an obvious use of subscripts, the following inequality always holds :

$$B(Y|x_1,\ldots, x_n) \leq B(Y|x_1,\ldots, x_n, x_{n+1})$$

with equality if and only if Y and x_{n+1} are statistically independent. This statement is a direct consequence of Theorem 2. There follows that

$$B(Y|x_1,\ldots, x_n, x_{n+1})-B(Y|x_1,\ldots, x_n)$$

may be viewed as the contribution of the parameter x_{n+1} to the total information provided by the set $(x_1,\ldots, x_{n+1})$ of process input parameters.

Detailed mathematical expressions for the measures introduced in this section and algorithms for sequential selection of informative parameters may be derived on the basis of the theory of Section II.1.

IV. CONCLUSION.

In this paper, a new functional has been proposed as a measure of the amount of information associated with probability distributions. This new concept the Bayesian distance, has been shown to satisfy the intuitive requirements for such a measure in the context of statistical decision theory, and, in particular, in finite-parameter estimation problems. We have also shown that a tight relationship exists between the Bayesian distance and the important concept of error probability of decision making of the Bayesian decision rule. We have proved that the Bayesian distance does better characterize this error probability than does the information theoretic concept of logarithmic Equivocation.

These properties of the Bayesian distance have enabled us to propose a method for characterizing our "degree of knowledge" about a plant under control. This degree of knowledge may be interpreted as a measure of the isomorphism of the mathematical model to the actual plant. Finally, a method has been suggested for evaluating the informative content of process parameters.

APPENDIX I.

Axiomatic definition of the D-measure.

Consider any finite probability scheme $(p_1, \ldots, p_n)$ with

$$p_i \geq 0 \, , \; \sum_{i=1}^{n} p_i = 1, \quad n<\infty . \qquad (I.1.)$$

Let $D(p_1,\ldots,p_n)$ be a function defined for any integer n and for all p_k satisfying the conditions (I.1)
Let it further be required that

a) $D(p_1,\ldots,p_n)$ be continuous in p_k for all $0 \leq p_k \leq 1$; $k=1, \ldots, n$;

b) $D(p_k, 1-p_k) = D(1-p_k, p_k)$, $k=1,2,\ldots,n$;

c) maximum of $D(p_1,p_2,\ldots,p_n) = D(\frac{1}{n}, \frac{1}{n}, \ldots, \frac{1}{n})$;

d) $D(p_1,p_2,\ldots, p_{n-1}, q_1,q_2) =$

$$D(p_1,p_2,\ldots,p_n)+p_n^2 \, D(\frac{q_1}{p_n} , \frac{q_2}{p_n}) \qquad (I.2)$$

where $p_n = q_1+q_2$.

Then it may be shown that $D(p_1,\ldots,p_n)$ is uniquely defined by

$$D(p_1, p_2,\ldots,p_n) = \lambda \sum_{i=1}^{n} p_i(1-p_i) \qquad (I.3)$$

where λ is a positive multiplying constant.

Eq. (I.3) with $\lambda=1$ corresponds with Eq.(20) in Sec. II.2.

APPENDIX II.

Proof of theorem 3.

In this appendix we present a simple method of proof of the equivalent inequality :

$$2D(p_1, p_2,\ldots,p_n) \leq H(p_1,p_2,\ldots,p_n). \qquad (II.1)$$

Let us first consider the case n=2.
We have

$$2D(p_1,p_2) = 4p_1(1-p_1)$$

and

$$H(p_1,p_2) = -p_1 \log p_1 - (1-p_1) \log(1-p_1)$$

Inspection of these two functions for $0 \leq p_1 \leq 1$ reveals that

$$2D(p_1, p_2) \leq H(p_1, p_2) \qquad (II.2)$$

with equality if $p_1 = 0, .5, 1$.

Now the theorem may be easily established by induction :

Assuming the inequality (II.1) holds for any given n, we show that it also holds for n+1. We have by the grouping axiom of the classical information theory :

$$H(p_1,p_2,\ldots,p_{n-1}, q_1,q_2) =$$

$$H(p_1,p_2,\ldots, p_n) + p_n \, H(\frac{q_1}{p_n} , \frac{q_2}{p_n}), \qquad (II.3)$$

where as in Eq. (I.2) $p_n = q_1+q_2$.

We have $\quad 2D(p_1,p_2,\ldots,p_n) \leq H(p_1,p_2,\ldots,p_n) \quad (II.4)$

by assumption, and

$$2p_n^2 \, D(\frac{q_1}{p_n} , \frac{q_2}{p_n}) \leq p_n \, H(\frac{q_1}{p_n} , \frac{q_2}{p_n}) \qquad (II.5)$$

as a consequence of inequality (II.2.). Then summing the inequalities (II.4) and (II.5) side by side gives the desired result. Since the inequality (II.1) holds for n=2 it also holds for any finite n.

REFERENCES.

S. Arimoto ; (1971) ; Information - Theoretical Considerations on Estimation Problems ; Information and Control, 19, 181.

C. H. Chen ; (1973) ; personal communication.

P. A. Devijver ; (1972.a) ; On a New Class of Bounds on Bayes Risk in Multihypothesis Pattern Recognition ; submitted for publication in IEEE Trans. on Computers.

P. A. Devijver ; (1972.b) ; The Bayesian Distance. A New Concept in Statistical Decision Theory ; Proc. 1972 IEEE Conf. Decision and Control, 543.

P. A. Devijver ; (1972.c) ; On an Asymptotic Property of the Least-Mean-Square Error Design Criterion in Pattern Recognition ; to appear in Philips Research Reports.

P. Eyckhoff ; (1963) ; Some Fundamental Aspects of Process Parameter Estimation ; IEEE, Trans. Automatic Control, AC-8, 347.

V.A. Kovalevsky; (1968) ; The Problem of Character Recognition from the Point of View of Mathematical Statistics ; in Character Readers and Pattern Recognition ; V.A. Kovalevsky Edt ; Spartan book, N. Y., 3.

D. V. Lindley ; (1956) ; On a Measure of the Information Provided by an Experiment ; Ann. Math. Stat., 27, 986.

N. S. Rajbman, M. I. Shpunt, F. A. Ovsepian , I. S. Durgarian ; (1971) ; Informational Measure of Certainty and Its Use in the Identification of Plants ; J.O.T.A. ; 8 ; (3) , 212.

A. Renyi ; (1967) ; On Some Basic Problems of Statistics from the Point of View of Information Theory ; Proc. 5th. Berkeley Symp. Math. Stat. and Prob. ; Univ. California Press; 531.

I. Vajda ; (1968) ; Bounds of the Minimal Error Probability on Checking a Finite or Countable Number of Hypotheses ; Problemij peredachi informatsii ; 4, (1), 6.

DIFFERENT METHODS FOR ESTIMATION OF THERMAL DIFFUSIVITY OF A HEAT DIFFUSION PROCESS†

B. Leden
Division of Automatic Control
Lund Institute of Technology
Lund, Sweden

M. H. Hamza
The University of Calgary
Calgary, Alberta, Canada

M. A. Sheirah
Ain-Shams University
Cairo, Egypt

Different methods for determining parametric models of a one dimensional heat diffusion process are presented. Process inputs are the end temperatures of a long copper rod. Process outputs are the temperatures in seven equidistant points on the rod.

The purpose of the study is to compare three methods for estimation of the thermal diffusivity of the rod. The methods are a periodic temperature method, an on-line least squares method and a maximum likelihood method. The accuracy, the amount of computation, the storage capacity and in general the advantages and limitations of the methods are discussed. The study is entirely based on experimental results.

1. INTRODUCTION

Several mathematical and experimental methods have been developed to determine thermal properties of solid materials during the past hundred years. The methods can be separated into stationary and non-stationary techniques. The stationary methods require measurements of heat fluxes and temperature gradients while the non-stationary methods require measurements of lengths and time intervals. In principle, lengths and time intervals can be measured more accurately than heat fluxes and temperature gradients.

Since parameters like thermal diffusivity strongly influence the dynamic properties of heat conduction, it is natural to use dynamic measurements and system identification techniques. In this paper it is attempted to use such tools to determine the thermal diffusivity. The results are based on experiments made on an experimental process. Three different methods are compared.

The first method is a straightforward application of frequency response techniques proposed by Sidles and Danielson in 1954. This method has gained widespread acceptance as an accurate method for determining the thermal diffusivity.

The second method is an application of on-line least squares. This method has the advantages of being applicable to arbitrary distributed parameter systems and of requiring a small amount of computation and storage.

The third method is an application of the maximum likelihood method. This is shown to be the most accurate method. The accuracy is limited by errors in the determination of sensor position. If those errors can be accurately controlled this method makes it possible to determine the thermal diffusivity to within 0.01%.

2. EXPERIMENTAL PROCESS AND MEASUREMENTS

2.1 Conduction of Heat in a Long Rod

Consider a long rod surrounded by a gas and subjected to small temperature variations around a constant temperature. The conduction of heat in the rod is given by

$$\kappa\frac{\partial^2\theta}{\partial z^2} = \frac{\partial\theta}{\partial t} + \mu\theta \tag{2.1}$$

where

$$\theta = \theta(z,t) \tag{2.2}$$

is the temperature, κ the thermal diffusivity, z the position and t the time. The coefficient of surface-heat loss is μ which takes into account any heat loss by radiation, conduction or convection.

In a case where the temperature difference between the rod and the surrounding medium is proportional to the temperature of the rod, with a constant of proportionality equal to a, we have

$$\mu = a\frac{2H}{\rho c r} \quad . \tag{2.3}$$

†This work has partially been supported by the Swedish Board for Technical Development under Contract 72-202/U 137, also by the National Research Council of Canada under Grant A-5102.

In Eq. (2.3) H is the coefficient of heat transfer, ρ the density, c the specific heat and r the radius of the rod. For a sample of pure copper [8] gives

$$\kappa = 1.185 \text{ cm}^2/\text{sec} \quad . \tag{2.4}$$

Putting

$a = 0.15$

$r = 0.70$ cm

$$H = 5 \cdot 10^{-4} \text{ W/cm}^2 \text{ °C} \tag{2.5}$$

Eq. (2.3), (2.5) and [5] give for a sample of pure copper

$$\mu = 0.31 \cdot 10^{-4} \text{ 1/sec} \quad . \tag{2.6}$$

The studied material is commercial copper.

2.2 Block Diagram

The experimental process is schematized in Fig. 2.1.

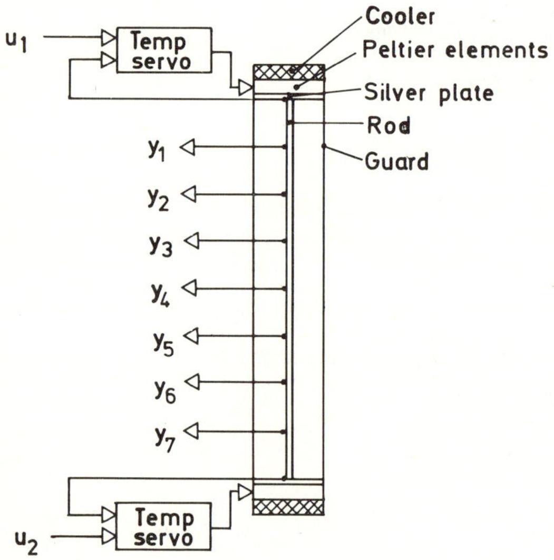

Fig. 2.1 - A schematic diagram of the experimental process

The rod is 1.40 cm in diameter and 45.0 cm long. These dimensions adequately fulfill the requirements for a thin one dimensional rod. The sensor separation is

$$\ell = 5.625 \pm 0.005 \text{ cm} \quad . \tag{2.7}$$

This distance also separates the heat source from the nearest sensor, i.e. any irregularity in the cross-sectional distribution of a heat pulse is smoothed out before it reaches the first sensor. The rod is placed into a guard cylinder, fabricated of the same material as the rod and attached to the guard by a silver plate. The end temperatures of the rod are controlled by inputs u_1, u_2 using Peltier elements. Transducers convert the temperatures in seven equidistant points on the rod to outputs $y_1, y_2, \ldots, y_7$. The temperature T (°C) corresponds to the input/output voltage v (V), where

$$v = T - 25 \quad . \tag{2.8}$$

The temperature range of operation is 20°C - 30°C.

2.3 Electronics

The temperature transducer consists of a one thermistor resistive Wheatstone bridge and a differential amplifier detecting the unbalance of the bridge. The resistance versus temperature characteristic of a thermistor is nonlinear. Leden (1970) has shown that by a proper choice of the Thevenin resistance with respect to the thermistor terminals the transducer can be designed to yield an almost linear voltage versus temperature characteristic in a fairly large temperature range. The linearity error of the transducer is 0.01°C in the temperature range 20°C - 30°C. The 6 hours drift and the short time stability (1 min) of the transducers are 0.001°C and 0.0002°C respectively.

The temperature servo consists of a cascade connection of a PID-regulator and a power amplifier. The PID-regulator is made up of an integrating compensation and a lead compensation. The integrating compensation improves the transient behaviour of the servo. The regulator contains 2 nonlinear compensations, viz. a voltage dependent gain factor and a limiter. The gain factor of the regulator is reduced during a heat cycle of the rod. The bounds of the limiter are adjusted to give the same Peltier cooling and heating effect.

The power amplifier supplies a maximum power of 80 W at an output of 10 A. The Peltier element is made of a semiconductor material. The current through the element pumps heat from one side of the element to the other. The direction of flow of heat is altered by a reversed current. The power transferred through the element is for small input currents proportional to the magnitude of the current. The maximum cooling effect obtained at a current of 10 A and zero temperature difference across the element is 23 W. Each servo contains 2 Peltier elements which gives a maximum cooling effect of 46 W.

The solution time of the servo (5% of final value) is 5 sec. The 6 hours drift and the short time stability (1 min) of the servo are 0.001°C and 0.0002°C, respectively. A detailed description of the diffusion process is given by Leden (1970).

2.4 Measurements

2.4.1 Periodic Temperature Method

A process computer PDP-15 was employed to carry out the experiments. The sampling rate was 20 msec. The time elapsed between the readings of two arbitrary analogue input signals of the interface of the computer was 20 μsec. The resolution of the A/D- and D/A-converters of the interface was 3 decimal digits.

2.4.2 On-line Least Squares Method and Maximum Likelihood Method

The measurements of the input and output variables of the process were performed with a data logger. A start command connected the input signals of the logger, through a multiplexer, to the voltmeter of the logger at each sampling event. The start command was synchronized to the shifts of the input signal. Aitken's scheme for Lagrange interpolation was employed to synchronize the readings of the different channels within the same sampling period to the shifts of the input signal. The long term accuracy of the voltmeter was 0.01% of full scale and 0.02% of reading.

The initial termperature profiles of the rod were stationary but nonzero. The input u_2 was kept constant at 25°C during the experiments. The signals u_1, u_2, y_1, y_2,...,y_7 were recorded.

The series analysed using the on-line least squares method was S1. The sampling period of the series was 10 sec. The input signal u_1 consisted of two consecutive steps of magnitude 1.8°C and of a mean value corresponding to 25°C. Stationary conditions were achieved when the second step was applied to the input. The length of the record was 320.

The series used for maximum likelihood identification was S2. The input signal u_1 was a PRBS sequence of maximum period. The minimum pulse length was 60 sec and the period of the sequence 255. The amplitude of the sequence was 1.8°C. The sequence had a mean value corresponding to 25°C. The length of the record was 920.

3. PERIODIC TEMPERATURE METHOD

3.1 Outline of the Method

In the periodic temperature method the energy supplied to the same rod is modulated with a fixed period. The thermal diffusivity is determined from measurements of the amplitude and the phase relationships in the rod. The solution of Eq. (2.1) subjected to the boundary conditions

$$\theta(0,t) = A_0 + A_1 \sin(\omega t+\phi) \tag{3.1}$$

$$\theta(\infty,t) = 0$$

is

$$\phi(z,t) = A_0 e^{-\alpha_0 z} + A_1 e^{-\alpha z} \sin(\omega t-\beta z+\phi) \tag{3.2}$$

where

$$\alpha_0 = \sqrt{\mu/k}$$

$$\alpha = \{\tfrac{1}{2}\kappa^{-1}(\sqrt{\mu^2+\omega^2}-\mu)\}^{\frac{1}{2}}$$

$$\beta = \{\tfrac{1}{2}\kappa^{-1}(\sqrt{\mu^2+\omega^2}-\mu)\}^{\frac{1}{2}} \tag{3.3}$$

see Sidles (1954), Sidles (1969).

Equation (3.2) yields that the temperature oscillation produced at $z = 0$ will propagate along the rod with a velocity

$$v = \omega/\beta \tag{3.4}$$

and have an amplitude decrement between $z = z_1$ and $z = z_2$ of

$$q = e^{\alpha \ell} \tag{3.5}$$

where

$$\ell = z_2 - z_1 \quad . \tag{3.6}$$

Sidles (1954) has shown that

$$\alpha\beta = \frac{\omega}{2\kappa} = \frac{\omega \ln q}{\ell v} \quad . \tag{3.7}$$

Equation (3.7) implies

$$\kappa = \frac{\ell^2}{2T_\ell \ln q} \tag{3.8}$$

where T_ℓ is the time required for a temperature wave to move from $z = z_1$ to $z = z_2$.

3.2 Experimental Approach

The block diagram of an experimental arrangement suggested for determining the amplitude decrement q and the time T_ℓ appears in Fig. 3.1.

The synchronized sine and cosine signals of the signal generator are generated in software by a process computer. The sine signal is converted to analogue form and this signal serves as input signal to the process. The outputs y_1, y_2 are converted to digital form and the multiplication and the integration are performed in software by the computer.

If we neglect the disturbances operating on the system we have

$$y_k = y_{0_k} \sin(\omega t+\varphi_k) \qquad k = 1,2 \quad . \tag{3.9}$$

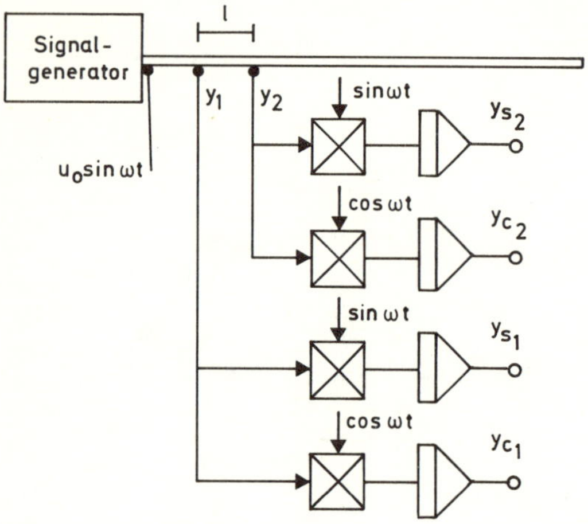

Fig. 3.1 - Block diagram of an experimental arrangement for determining the amplitude decrement q and the time T_ℓ.

The outputs of the integrators are then given by

$$y_{s_k} = \int_0^T y_k(t) \sin\omega t \, dt = \tfrac{1}{2}T \, y_{0_k} \cos\varphi_k$$

$$y_{c_k} = \int_0^T y_k(t) \cos\omega t \, dt = \tfrac{1}{2}T \, y_{0_k} \sin\varphi_k$$

$$k = 1,2 \qquad (3.10)$$

if ωT is an integer multiple of π.

Equation (3.10) yields

$$y_{o_k} = 2\sqrt{\frac{1}{T_p}(y_{s_k}^2 + y_{c_k}^2)}$$

$$\varphi_k = \operatorname{arctg}(y_{c_k}/y_{s_k}) \qquad k = 1,2 \qquad (3.11)$$

if ωT is an integer multiple of π.

Denoting the time period of the input signal T_p, the amplitude decrement q and the time T_ℓ are thus given by

$$q = \sqrt{(y_{s_1}^2 + y_{c_1}^2)/(y_{s_2}^2 + y_{c_2}^2)}$$

$$T_\ell = \frac{1}{2\pi} \operatorname{arctg} \{(y_{c_1}/y_{s_1} - y_{c_2}/y_{s_2})/(1 + y_{c_1}/y_{s_1} \cdot y_{c_2}/y_{s_2})\} T_p \qquad (3.12)$$

provided that ωT is an integer multiple of π.

This method effectively eliminates noise of the signals y_1, y_2. Constant measurement errors will not influence the results, if ωT is an integer multiple of 2π. Further the time T_ℓ is not affected by static nonlinearities of the transducers of the interface, if ωT is an integer multiple of 2π. The influence of a static nonlinearity on the amplitude decrement q can be determined by connecting the analogue output to the analogue inputs and making one measurement.

On a process computer PDP-15, with floating point arithmetic in hardware, a storage capacity of 1.8 k words is required to run the program. In those 1.8 k words, system subroutines are included but the executive and the programs for data acquisition are excluded.

3.3 Results

A sinusoidal input signal of amplitude 2.3°C and period 50 sec was applied to the left end point of the rod and the system was allowed to reach equilibrium. The input signal closely met the requirements for a semi-infinite rod, i.e. the amplitude of the temperature oscillations produced at the right end point of the rod was 0.0001°C.

The thermal diffusivity was determined from the outputs y_1, y_2. The amplitude decrement q and the time T_ℓ obtained from the measurements were

$$q = 3.704$$

$$T_\ell = 10.413 \quad . \qquad (3.13)$$

The values of q and T_ℓ depend slightly on the choice of the time period T_p. The errors in the temperature transducer characteristics (2.8) were 0.1% and the errors in ℓ^2 were 0.2%, according to Eq. (2.7). Equations (3.8), (3.13) thus yield

$$\kappa = 1.160 \pm 0.003 \qquad (3.14)$$

where the errors in the time period T_p were neglectable.

4. ON-LINE LEAST SQUARES METHOD

4.1 Theory

The model of the diffusion process is taken to be (2.1). It is required to determine an estimate for κ and μ. Let

$$e_{i,j}(\kappa,\mu) = \left[\kappa \frac{\partial^2\theta}{\partial x^2} + \mu\theta - \frac{\partial\theta}{\partial t}\right]_{i,j} \qquad (4.1)$$

where i and j indicate discretization in time and space, respectively.

κ and μ will be selected such that

$$E(K-k) = \sum_{i=K-k-N}^{K-k} \left[\sum_{j=1}^{J} e_{i,j}^2(\kappa,\mu) \right] W_{K-k-i} \qquad (4.2)$$

is minimized. W_{K-k-i} is a weighting function, N is the number of past samples considered and k is an integer which depends upon the finite difference formulae used to approximate the partial derivatives. In general, determining the minimum of E(K-k) is a nonlinear programming problem. Since the model is linear in its parameters and since the parameters are constant, satisfactory results can be obtained if W_{K-k-i} is taken to be constant, Hamza (1971). For convenience, W_{K-k-i} is selected equal to unity. Under these conditions a closed form expression for the unknown parameters κ and μ can be derived. Taking the partial derivatives of (4.2) with respect to κ and μ, we obtain

$$\frac{\partial E(K-k)}{\partial \kappa} = 2 \sum_{i=K-k-N}^{K-k} \sum_{j=1}^{J} e_{i,j}(\kappa,\mu) \frac{\partial}{\partial \kappa} e_{i,j}(\kappa,\mu) = 0 \qquad (4.3)$$

and

$$\frac{\partial E(K-k)}{\partial \mu} = 2 \sum_{i=K-k-N}^{K-k} \sum_{j=1}^{J} e_{i,j}(\kappa,\mu) \frac{\partial}{\partial \mu} e_{i,j}(\kappa,\mu) = 0. \qquad (4.4)$$

Substituting (4.1) into (4.3) and (4.4) and taking the partial derivatives yields

$$\begin{bmatrix} \phi_{\theta_{xx},\theta_{xx}}(K-k) & \phi_{\theta,\theta_{xx}}(K-k) \\ \phi_{\theta_{xx},\theta}(K-k) & \phi_{\theta,\theta}(K-k) \end{bmatrix} \begin{bmatrix} \kappa \\ \mu \end{bmatrix} = \begin{bmatrix} \phi_{\theta_t,\theta_{xx}}(K-k) \\ \phi_{\theta_t,\theta}(K-k) \end{bmatrix} \qquad (4.5)$$

where

$$\phi_{a,b}(K-k) = \sum_{i=K-k-N}^{K-k} \sum_{j=1}^{J} a_{i,j}\, b_{i,j} \quad . \qquad (4.6)$$

$\phi_{a,b}(K-k)$ can be evaluated efficiently using the recurrence relation, Hamza (1971),

$$\phi_{a,b}(K+1-k) = \phi_{a,b}(K-k) - \sum_{j=1}^{J} a_{K-k-N,j}\, b_{K-k-N,j} + \sum_{j=1}^{J} a_{K+1-k,j}\, b_{K+1-k,j} \quad . \qquad (4.7)$$

ϕ_{xx} and θ_t denote $\frac{\partial^2\theta}{\partial x^2}$ and $\frac{\partial\theta}{\partial t}$ respectively. For approximating the partial derivatives, (4.8) and (4.9),

$$\left.\frac{\partial^2\theta}{\partial x^2}\right|_{i,j} \simeq \frac{\theta_{i,j+1} - 2\theta_{i,j} + \theta_{i,j-1}}{(\Delta x)^2} \qquad (4.8)$$

$$\left.\frac{\partial\theta}{\partial t}\right|_{i,j} \simeq \frac{\theta_{i+1,j} - \theta_{i-1,j}}{2\Delta t} \qquad (4.9)$$

which together have an error of order $\Delta(\cdot)^2$, are used.

4.2 Results

The signals y_2 to y_4 of S1 were used for the identification. The input was a step applied at time t = 5T, where T = 10 sec. y_3 is shown in Fig. 4.1. u_1, u_2, y_1 to y_7 were given to four significant figures.

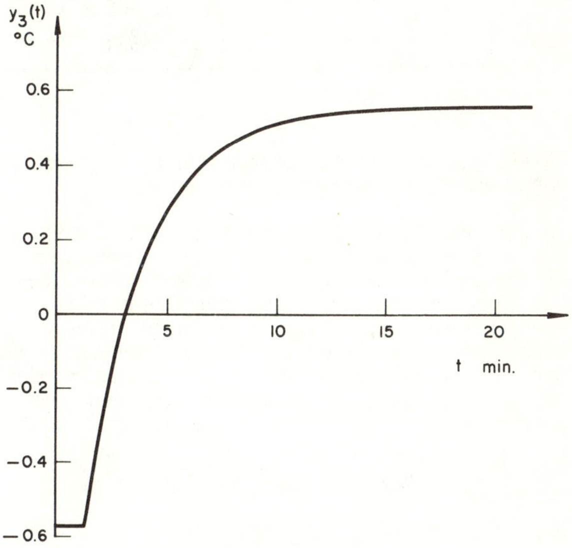

Fig. 4.1 - Output temperature $y_3(t)$

In computing the percentage error in κ, the exact value of κ was taken to be 1.16. The results are given in Table 4.1 where $k = \frac{t}{T}$.

The time taken to compute κ for a record length N equal to 150 samples, was 0.7 sec. The computer used was the CDC 6400 of The University of Calgary. Recalling that the sampling period T was 10 sec for this experiment, it becomes evident that even using a very long record length the computation can still be done on-line. The percentage error in μ seems large when compared with the results in (2.6) or with those obtained using the maximum likelihood method, section 5.3. This, however, does not have an appreciable effect on the system response, since the contribution of $\mu\theta$ in (2.1) is very small compared to that of $\kappa \frac{\partial^2\theta}{\partial x^2}$. The computation of κ was repeated using as a model

$$\kappa\frac{\partial^2\theta}{\partial x^2} = \frac{\partial\theta}{\partial t} \tag{4.10}$$

and the error in κ was less than 2% for $k \geq 12$. The storage capacity used to perform the identification and obtain the results in Table 4.1 was 0.029 k words.

Table 4.1

k	Percentage error in κ	μ
10	8.00	0.5159×10^{-5}
20	2.36	0.6859×10^{-5}
50	1.77	0.5768×10^{-5}
100	0.99	0.4468×10^{-5}
150	1.04	0.4534×10^{-5}

This suggests that the method can readily be implemented on a minicomputer.

In the heat conduction experiment considered, the parameters to be identified were constant. If the parameters varied with time, as would be the case for example during the flight of a space vehicle especially at re-entry, then a closed form expression for the unknown parameters cannot be obtained. In such a case, the on-line mathematical programming approach presented by Hamza (1971) can be used.

5. MAXIMUM LIKELIHOOD METHOD

5.1 Diffusion Process Model

The model (2.1) of the diffusion process is transformed into a finite dimensional system of ordinary differential equations by approximating the partial derivative in the z-direction by finite difference formulae. Two different sets of formulae are used

$$\left.\frac{\partial^2\theta}{\partial z^2}\right|_{i,j} = \frac{1}{h_z^2}(\theta_{i-1,j} - 2\theta_{i,j} + \theta_{i+1,j}) \quad i = 1,n-1$$

$$\left.\frac{\partial^2\theta}{\partial z^2}\right|_{i,j} = \frac{1}{12h_z^2}(-\theta_{i-2,j} + 16\theta_{i-1,j} - 30\theta_{i,j} + \tag{5.1}$$

$$+ 16\theta_{i+1,j} - \theta_{i+2,j}) \quad i = 2,3,\ldots,n-2$$

which have an error of order h_z^2 and h_z^4 respectively. The rod of length L is divided into n equal intervals h_z, where $h_z = L/n$. The temperature $\theta_{i,j}$ approximates the temperature in the ith point on the rod at time j.

Introduce the state vector

$$x = \begin{bmatrix} \theta_{1,j} \\ \theta_{2,j} \\ \cdot \\ \cdot \\ \cdot \\ \theta_{n-1,j} \end{bmatrix} \tag{5.2}$$

and the input vector v

$$v = \begin{bmatrix} \theta_{0,j} \\ \theta_{n,j} \end{bmatrix} \tag{5.3}$$

Then

$$\dot{x} = \begin{bmatrix} -a & b & 0 & 0 & & & & 0 \\ d-c & d-e & \ddots & & & & & \\ -e & \ddots & \ddots & \ddots & & & & \\ 0 & \ddots & \ddots & \ddots & \ddots & & & \\ & \ddots & \ddots & \ddots & \ddots & \ddots & & 0 \\ & & \ddots & \ddots & \ddots & \ddots & \ddots & -e \\ & 0 & & & -e & d-c & & d \\ & & & & & 0 & 0 & b-a \end{bmatrix} x + \begin{bmatrix} b & 0 \\ -e & 0 \\ 0 & 0 \\ \cdot & \\ \cdot & \\ \cdot & \\ 0 & 0 \\ 0 & -e \\ 0 & b \end{bmatrix} v \triangleq Ax + Bv \tag{5.4}$$

where

$$a = 2\kappa/h_z^2 + \mu$$
$$b = \kappa/h_z^2$$
$$c = 30\kappa/(12h_z^2) + \mu$$
$$d = 16\kappa/(12h_z^2)$$
$$e = \kappa/(12h_z^2) \tag{5.5}$$

approximates the model (2.1). In fact it can be shown that the difference between the step responses of the models (2.1) and (5.4) in the midpoint of the rod is less than 0.0001°C for all time t, when a unit step is applied to the left end point and the right end point is kept constant.

The discrete model of the system (5.4) becomes

$$x(t+1) = \Phi x(t) + \Gamma v(t) \tag{5.6}$$

where

$$\Phi = e^{A}$$

$$\Gamma = \int_0^1 e^{As}ds\, B \quad . \qquad (5.7)$$

The sampling period is for convenience chosen to be unity. The responses of the models (5.4), (5.6) are the same to a piecewise constant input signal. The input vector v considered is not piecewise constant. By approximating the input vector, within each sampling period, with a finite polynomial in t a piecewise constant input vector can be obtained. Lagrange polynomials of order 3 defined by

$$P_i(k-1) = \theta_{i,k-1}$$

$$P_i(k) = \theta_{i,k}$$

$$P_i(k+1) = \theta_{i,k+1}$$

$$P_i(k+2) = \theta_{i,k+2} \qquad i = 0,n \qquad (5.8)$$

in the interval $k < t \leq k + 1$, are employed to perform the approximation.

The model used for maximum likelihood identification now becomes

$$\dot{\tilde{x}} = \left[\begin{array}{ccc|c|ccc} 0&0&0 & & & & \\ 1&0&0 & 0 & & 0 & \\ 0&1&0 & & & & \\ \hline 0&0&b & & 0&0&0 \\ 0&0&-e & & 0&0&0 \\ 0&0&0 & & 0&0&0 \\ &\cdot& & & &\cdot& \\ &\cdot& & A & &\cdot& \\ &\cdot& & & &\cdot& \\ 0&0&0 & & 0&0&0 \\ 0&0&0 & & -e&0&0 \\ 0&0&0 & & b&0&0 \\ \hline & & & & 0&1&0 \\ &0& & 0 & 0&0&1 \\ & & & & 0&0&0 \end{array}\right] \tilde{x} + \left[\begin{array}{cc} 1&0 \\ 0&0 \\ 0&0 \\ \hline 0&0 \\ 0&0 \\ 0&0 \\ &\cdot \\ &\cdot \\ &\cdot \\ 0&0 \\ 0&0 \\ 0&0 \\ \hline 0&0 \\ 0&0 \\ 0&1 \end{array}\right] \tilde{v} = \tilde{A}\tilde{x} + \tilde{B}\tilde{v} \qquad (5.9)$$

where the matrix A and the parameters b, e are defined by Eq. (5.4) and (5.5) respectively. The null matrix is denoted by 0. The model (5.6) is employed to generate a temperature trajectory, for a given boundary condition v. The matrices Φ, Γ of the model (5.6) are defined by Eq. (5.7), where $A = \tilde{A}$ and $B = \tilde{B}$. The components $\tilde{x}_1$, $\tilde{x}_2$, $\tilde{x}_3$, $\tilde{x}_{19}$, $\tilde{x}_{20}$, $\tilde{x}_{21}$, of the state vector $\tilde{x}$ and the input vector $\tilde{v}$ should be updated according to Eq. (5.8) at each sampling event. The order of the model (5.9) is 21.

5.2 Estimation Technique

The unknown parameter vector α is identified from the single output stochastic model

$$x(t+1) = \Phi x(t) + \Gamma v(t) + K\varepsilon(t)$$

$$y(t) = Cx(t) + Dv(t) + \varepsilon(t) \qquad (5.10)$$

where y is the output variable and ε is a sequence of independent gaussian random variables. The principle of maximum likelihood is used to estimate the unknown parameters. Åström (1970) has shown that maximization of the likelihood function is equivalent to minimization of the loss function.

$$V(\alpha) = \tfrac{1}{2}\sum_{t=1}^{N} \varepsilon(t)^2 \qquad (5.11)$$

where N is the length of the record. An estimate of the variance of the residuals is given by

$$\lambda^2 = \frac{2}{N}\min_{\alpha} V(\alpha) \quad . \qquad (5.12)$$

Under the restrictive assumption that the data were actually generated by a model (5.10), where ε is a sequence of independent gaussian random variables, it is possible to pose and answer several statistical problems. With mild additional assumptions Åström (1965) has shown that the likelihood estimates are consistent, asymptotically efficient and normal $(\alpha, \lambda^2 V_{\alpha\alpha}^{-1})$.

The matrix of second order partial derivatives $V_{\alpha\alpha}$ is calculated using finite difference formulae of order h^4. The sets

$$\left.\frac{\partial^2 V}{\partial \alpha_j^2}\right|_{.,i,j,k,.} = \frac{1}{12h^2}\,(-V_{.,i,j-2,k,.} + 16V_{.,i,j-1,k,.} - 30V_{.,i,j,k,.} + 16V_{i,j+1,k,.} - V_{.,i,j+2,k,.})$$

$$\left.\frac{\partial^2 V}{\partial \alpha_j \partial \alpha_k}\right|_{.,i,j,k,\ell,.} = \frac{1}{48hk}\,(-V_{.,i,j-2,k-2,\ell,.} + V_{.,i,j-2,k+2,\ell,.} - 16V_{.,i,j-1,k+1,\ell,.} + 16V_{.,i,j-1,k-1,\ell,.} + 16V_{.,i,j+1,k+1,\ell,.} - 16V_{.,i,j+1,k-1,\ell,.} + V_{.,i,j+2,k-2,\ell,.} - V_{.,i,j+2,k+2,\ell,.}) \qquad (5.13)$$

are used to estimate the diagonal and the non-diagonal elements of the matrix $V_{\alpha\alpha}$.

On a Univac 1108 computer the storage capacity required for the programs and the data are 5.6 k words and 9.9 k words respectively. The system subroutines require extra 4.4 k words. The time required to compute the optimum parameter vector α is 6 min for a reasonable initial guess. The algorithm used to minimize the loss function is based on [8].

5.3 Results

The parameters to be identified were the thermal diffusivity κ, the coefficient of surface-heat loss μ, the deviation from unity slope γ and the zero adjustment error δ of the midpoint transducer. The signals y_2, y_4, y_6, of series S2 were used for identification. The temperatures y_2, y_6, determined the boundary conditions for the heat equation. The choice of boundary conditions eliminated end effects of the rod and influences of the servodynamics on the results of identification. The output variable y was y_4, which gave

$$C = (0\ 0\ 0\ 0\ \cdots\ 0\ 1{+}\gamma\ 0\ \cdots\ 0\ 0\ 0\ 0)$$

$$D = \begin{bmatrix} 0 & 0 \\ 0 & 0 \end{bmatrix} \tag{5.14}$$

where C and D were defined by Eq. (5.10). The matrix K of Eq. (5.10) was put equal to zero, i.e. the model was assumed to contain no state variable noise.

To save computer time the computations were based on a record of length 205. The first part of the series S2 was not used, which eliminated the transient caused by the initial nonzero steady-state profile of the rod. The chosen record contained 18 bad readings. The readings were substituted using Lagrange interpolations.

The results of the identification were

$$\kappa = 1.1589 \pm 0.0001$$

$$\mu = 0.356\cdot10^{-4} \pm 0.005\cdot10^{-4}$$

$$\gamma = 0.6\cdot10^{-4} \pm 0.1\cdot10^{-4}$$

$$\delta = 0.392\cdot10^{-2} \pm 0.001\cdot10^{-2} \tag{5.15}$$

The estimated values of the standard deviations should be corrected for errors in the position of the sensors and errors in the sampling period. Errors in the position of the sensors influenced κ quadratic, according to Eq. (2.1). Remembering that the sensor separation was 2ℓ, Eq. (2.7) yields

$$\kappa = 1.159 \pm 0.001 \tag{5.16}$$

where the errors in the sampling period were neglectable. The estimated value of κ and μ were consistent with Eq. (2.6), (3.14). Notice the small value of γ.

The estimated value of the standard deviation of the residuals was

$$\lambda = 1.419\cdot10^{-4} \tag{5.17}$$

This standard deviation is extremely small, compare Fig. (5.2).

The statistical properties of the residuals ε were examined. The sample covariance function r_ε defined by

$$r_\varepsilon(\tau) = \frac{1}{N-\tau}\sum_{t=1}^{N-\tau} \varepsilon(t)\varepsilon(t+\tau) \quad , \quad \tau \geq 0 \tag{5.18}$$

is shown in Fig. (5.1).

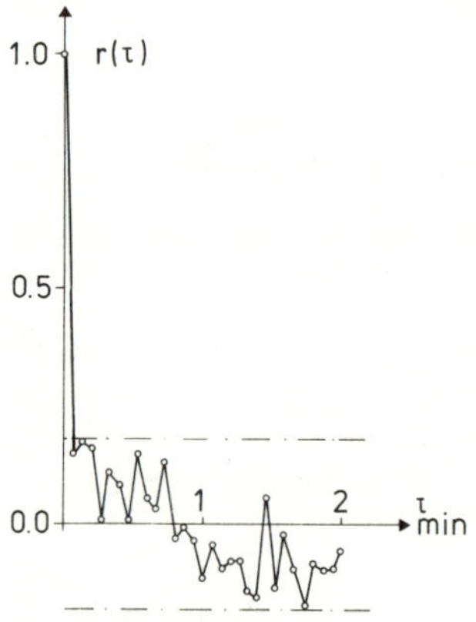

Fig. 5.1 - Normalized sample covariance function $r(\tau)$ of the residuals of the identified model. The dashed lines give the 99% confidence interval for $r(\tau)$, $\tau \neq 0$.

The residuals were tested for normality using a chi-square goodness-of-fit test. The test quantity χ was

$$\chi^2 = 33.6 \quad . \tag{5.19}$$

The number of degrees of freedom was 21. Provided that χ^2 is less than 38.9 the hypothesis that the residuals are normally distributed is accepted at a risk level of 1%. The assumptions made on the residuals were fulfilled at a risk level of 1%.

In Fig. (5.2) we show:

-- the boundary temperatures y_2, y_6

-- the output temperature y_4
-- the model output y_{m_4}

$$\dot{\tilde{x}} = A\tilde{x} + Bv$$

$$y_{m_4} = C\tilde{x} + Dv$$

-- the model error e_4

$$e_4 = y_4 - y_{m_4}$$

for the identified model.

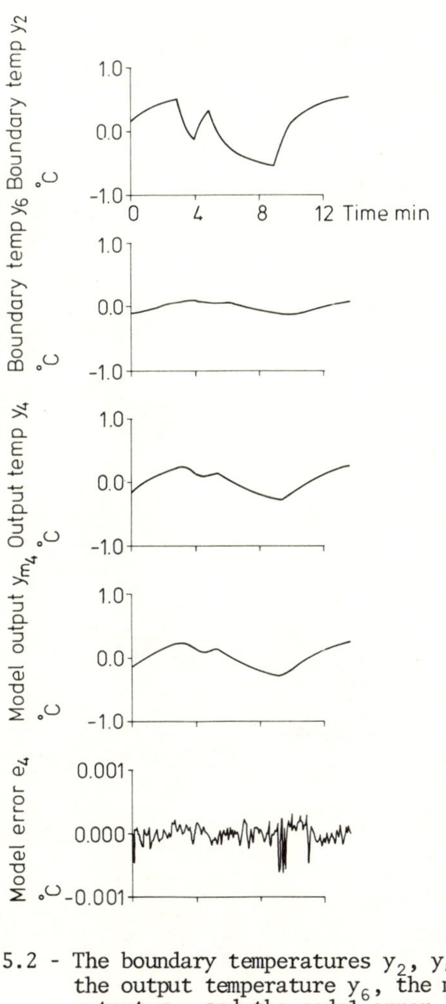

Fig. 5.2 - The boundary temperatures y_2, y_4, the output temperature y_6, the model output y_{m_4} and the model error e_4 of the identified model.

6. CONCLUSIONS

Three different methods for determining the thermal diffusivity of copper at room temperature are compared. The advantages and limitations of the methods are given below.

6.1 Periodic Temperature Method

This method applies to a semi-infinite rod. Errors in the characteristics of the temperature transducers influence the thermal diffusivity. This method requires a relative small sensor separation. The thermal diffusivity can be calculated on-line using a process computer. The storage capacity required is 1.8 k words. Errors in determining the thermal diffusivity are of the order of 0.3%.

6.2 On-line Least Squares Method

This method applies to an arbitrary model of a distributed parameter system. Errors in determining the thermal diffusivity are of the order of 1%. The storage capacity required is 0.029 k words and the computing time is less than one second. The least squares method is simpler, requires much less storage and computing time than the maximum likelihood method, however, it is less accurate. The accuracy achieved, however, could be considered satisfactory for many applications. Although not demonstrated in this paper, if the data are very noisy, the maximum likelihood method gives much better results, since the least squares method yields biased estimates.

6.3 Maximum Likelihood Method

A high speed computer must be used to calculate the parameters which maximize the likelihood function. The computation must be done off-line and the storage capacity required is 20 k words. This method allows the validity of the heat equation model to be checked. Errors in the characteristics of the temperature transducers can be estimated and do not influence the thermal diffusivity. The accuracy of this method is limited by errors in the sensor position. If those errors can be accurately controlled, this method makes it possible to determine the thermal diffusivity to within 0.01%. In the present experimental process errors in determining the thermal diffusivity are of the order of 0.1%.

7. ACKNOWLEDGEMENT

The authors wish to express their gratitude to Dr. S. Lindahl for several valuable discussions of the problem.

8. REFERENCES

1. Åström K.J. (1970): Introduction to Stochastic Control Theory. Academic Press.

2. Åström K.J., Bohlin T. (1965): Numerical Identification of Linear Dynamic Systems

from Operating Records. Proc. of IFAC Symposium on The Theory of Self-Adaptive Control Systems, Teddington, England.

3. Åström K.J., Eykhoff P. (1971): System Identification - A Survey. Automatica 1, 123.

4. Hamza M.H., Sheirah M.A. (1971): On-line Identification of Distributed Parameter Systems. Proc. IFAC Symposium on The Control of Distributed Parameter Systems, Banff, Canada.

5. Handbook of Thermophysical Properties of Solid Materials. Pergamon Press, 1961.

6. Leden B. (1970): Linear Temperature Scales from One Thermistor Reciprocal Networks. Report 7009, Lund Institute of Technology, Division of Automatic Control.

7. Leden B. (1970): The Design of a One Dimensional Heat Diffusion Process. Report 7010, Lund Institute of Technology, Division of Automatic Control.

8. Powell M.J.D. (1964): An Efficient Method of Finding the Minimum of a Function of Several Variables without Calculating Derivatives. Comput.J.

9. Sidles P.H., Danielson G.C. (1954): Thermal Diffusivity of Metals at High Temperatures. J. Appl. Phys., 25, 58.

10. Sidles P.H., Danielson G.C. (1969): Thermal Diffusivity and Other Non-steady-state Methods. Thermal Conductivity, 2, 147.

IDENTIFICATION OF DISTRIBUTED PARAMETERS

G. CHAVENT

Institut de Recherche d'Informatique et d'Automatique
I.R.I.A. - LABORIA
BP N° 5 - 78150 LE CHESNAY FRANCE

We present here a method specially designed for the off-line estimation of distributed parameters in distributed systems, and based upon the use of modern control theory in function spaces. We minimize a non quadratic cost function by a standard gradient method. The special features of this way of working are :

- no assumption about the algebraic form of the unknown parameter function is made in calculating the gradient with respect to that function,
- if one wants to try different algebraic given forms (with only a few unknown coefficients) this can be done very easily by inserting only a few cards in the programm,
- the computation of the gradient requires only two resolution of P.D.E., even if the unknown parameter is a function,
- the method works even if there are only a few points of measurements in space and/or time.

Applications are given to the estimation of functions of the space variables (in water and petroleum fields), and to the identification of a function of the solution of the system.

I. INTRODUCTION

The problem of parameter estimations can be considered as a control problem, the control being the unknown parameters, the cost index being a quadratic form on the difference between the measures and the computed outputs. So it was natural to transpose to the estimation of distributed parameters in distributed systems the techniques of functional analysis developped by J.L. LIONS ([1] [2]) for the control of distributed systems. The main difference is that, in our case, the functionnal to minimize is not quadratic, so that a lot of theoretical results cannot be transposed. But this approach gives us a systematic way of computing the gradient of that functionnal, which is efficient.
In § II, some functionnal analysis notations are summarized ; in § III, the theory is sketched, and in § IV, three applications are given.

II . NOTATIONS

Let X and Y be Banach spaces ; then :
- X' and Y' denote the dual spaces of X and Y
- $\mathcal{L}(X,Y)$ denotes the space of all continuous linear applications from X into Y.
- $\| \ \|_X$, $\| \ \|_Y$ denote the norms in spaces X and Y.
- if $A \in \mathcal{L}(X;Y)$, then $A^* \in \mathcal{L}(Y';X')$ denotes the transposed of A.
- $(\ , \ )_{X'X}$ denotes a duality between X' and X
- $(\ , \ )_X$ denotes, when X is an Hilbert space, the scalar product in X.

Let Ω be an open set of $\mathbb{R}^n$; then :
- $\mathcal{C}(\Omega)$ = space of continuous real functions on Ω
- $L^\infty(\Omega)$= space a.e. bounded real function on
- $L^2(\Omega)$= space of Lebesgue square integrable functions on Ω.
- $H^1(\Omega) = \{u \in L^2(\Omega) \mid \frac{\partial u}{\partial x_i} \in L^2(\Omega) \ i=1,2,\dots n\}$
- $H^1_o(\Omega) = \{u \in H^1(\Omega) \mid$ "u=0 on the boundary Γ of Ω"$\}$.

Throughout the paper, derivable will always mean Frechet-derivable.

III. THEORY

We suppose that the equations governing the system to be identified are completely known, up to a set of (distributed or discrete) parameters.
Our state-equation will be :

$$\Psi (y,a) = f . \tag{1}$$

where (Y, $\mathcal{a}$ and F being Banach spaces)
- $y \in Y$ is the state of the system,
- $a \in \mathcal{a}$ is the set of unknown parameters
- $f \in F$ is the second number of the state equation
- Ψ is some known application from $Y \times \mathcal{a}$ into F.

We will then suppose that there exists an open subset $\mathcal{a}_c$ of $\mathcal{a}$ such that :

$$\left.\begin{array}{l}\text{for every } a \in \mathcal{a}_c \text{, the state-equation}\\ \text{(1) has one and only one solution } y \in Y;\\ \text{we denote this solution by } y(a).\end{array}\right\} \tag{2}$$

Remark 1 : Time is apparently lacking in equation (1). But Y can be a set of functions of time (see Applications II and III), so that dynamic systems can still be modelled by (1). ■
In order to identify the parameter a , we must have some information about the system (1). We consider an Hilbert space $\mathcal{H}$, which will be the space of the

observations, and an application $\mathcal{C}$ from Y into $\mathcal{H}$. Though the theory can be developed for $\mathcal{C}$ nonlinear, but continuously differentiable, we will here suppose for simplicity that :

$$\mathcal{C} \in \mathcal{L}(Y ; \mathcal{H}) \qquad (3)$$

If we denote by $a_o \in \mathcal{A}_c$ the true, but unknown, parameter value, the observation consists in some given element :

> $z \in \mathcal{H}$, which is supposed to be a "measure" of $\mathcal{C} y(a_o)$ (4)

It is then possible to associate to every parameter $a \in \mathcal{A}_c$ some cost index $J(a)$ by setting :

$$J(a) = \| \mathcal{C} y(a) - z \|^2_{\mathcal{H}} \quad \forall\ a \in \mathcal{A}_c .$$

Remark 2 : If we consider the observation as the output of the system, this cost function is simply the quadratic cost function on the output of the system. ■

We then formulate the Identification problem as :

$$\left. \begin{array}{l} \text{Find } \hat{a} \in \mathcal{A}_{ad} \text{ such that :} \\ J(\hat{a}) \leqslant J(a) \quad \forall\ a \in \mathcal{A}_{ad} \end{array} \right\} \qquad (6)$$

where :

> $\mathcal{A}_{ad}$, the set of all admissible parameters, is a given closed, generally bounded, subset of $\mathcal{A}_c$. (7)

At least three questions arise now :

i) Under what conditions does the identification problem (6) have at least one solution ?

ii) Under what conditions does the identification problem (6) have at most one solution ?

iii) How to really minimize J ?

Question i) : would be easy to answer to, if we suppose J to be quadratic with respect to a, but this is not the case, even in the simplest applications ([1]). It is still clear that, if the "measure" appearing in (4) is without error, at least a_o minimize J. But, in the realistic case where J is not quadratic and where z is different from $\mathcal{C} y(a_o)$, the only known general result concerning question i) is the (trivial) following one :

Proposition 1 : If :

- Ψ is continuously differentiable from $Y \times \mathcal{A}_c$ into F (8)
- for every $(y,a) \in Y \times \mathcal{A}_c$, $\frac{\partial \Psi}{\partial y}(y,a)$ is an invertible linear applica-application of Y on F (9)
- $\mathcal{A}$ is finite-dimensional (10)

then there exists at least one $\hat{a} \in \mathcal{A}_{ad}$ solution of (5).

Proof : (2)(3)(8)(9) and the implicit functions theorem imply that J is continuous on $\mathcal{A}_c$. (7) and (10) imply that $\mathcal{A}_{ad}$ is compact. ■

If $\mathcal{A}$ is infinite dimensional (which is the case when the unknown parameters are functions), there is no general result more. But in that case, the existence of a minimum has mainly a theoretical interest because when numerically dealing with functions, they are always dsicretized and Prop. 1 applies to the discretized problem.

Question ii) has more practical implications : if there is an unique solution $\hat{a}$ to problem (5), we can hope that, under reasonable other assumptions, $\hat{a}$ will be "close" to a_o if z is "close" to $\mathcal{C}$ $y(a_o)$. Unfortunately, there is no general result of uniqueness of the solution of (5). Even in each particular application, it is very difficult to prove such a result (for instance, partial uniqueness results for the following applications exist only for Application I). When theoretical uniqueness results are lacking, one can minimize J using different values of a to have an idea of the degree of confidence we can have in the physical meaning of $\hat{a}$.

Concerning Question iii), we fortunately have more results, which enable us, in practical applications, to compute an $\hat{a}$ even if we do not have results of existence and/or uniqueness.

In order to minimize J, we will use a gradient method. Our choice was the steepest-descent method motivated by the fact that the function to minimize was not quadratic, even near the minimum (in some of our applications ([2]), a numerical comparison with the Fletcher-Powell algorithm gave equality between the two methods). But if needed, every gradient method can be employed.
On the other hand, we must choose a good way of computing the gradient of J with respect to a because when a is a function, it can be discretized in many points, so that the gradient of J may have many components (a few hundred in some applications), and the computation of that gradient by a non-adapted method would take a prohibitive time.

The theoretical answer to that problem is given by :

Proposition 2 : Under the assumptions (8) and (9), the function J is continuously derivable from $\mathcal{A}_c$ into $\mathbb{R}$. Its derivative $J'(a) \in \mathcal{A}'$ is given by :

$$J'(a) = \frac{\partial \Psi}{\partial a}(y(a),a).\rho \quad \forall\ a \in \mathcal{A}_c \qquad (11)$$

where :

([1]) think of the system $\frac{dy}{dt} + ay = 0$, $y(o)=1$, $t \in [0,1]$, the observation being $\mathcal{C}y = y(1)$!

([2]) See [5].

- $y(a) \in Y$ is the state of the system, solution of (1) with the value a of the unknown parameters,
- $\rho \in F'$ is the adjoint state, solution of:

$$\frac{\partial \Psi^*}{\partial y}(y(a),a).\rho = -2\, \mathcal{C}^* \Lambda \mathcal{C} y(a)-z) \qquad (12)$$

where Λ is a canonical isomorphism from $\mathcal{H}$ onto its dual $\mathcal{H}'$.

Proof : see appendix.

Let us see now how this result practically enable us to calculate the gradient of function J with respect to the unknown prameters a . Formula (11) can be rewritten :

$$J'(a).\delta a = \left(\rho, \frac{\partial \Psi}{\partial a}(y(a),a)\delta a\right)_{F'F} \qquad (13)$$

for every $a \in \mathcal{A}_c$ and every $\delta a \in \mathcal{A}$

We shall distinguish two cases :

Case 1 : There is a finite number m of unknown scalar parameters $a_1, a_2, \ldots, a_m$, i.e. $\mathcal{A} = \mathbb{R}^m$.

Then the dual space $\mathcal{A}'$ of $\mathcal{A}$ can be identified to $\mathcal{A} = \mathbb{R}^m$ by the usual scalar product; then formula (13) becomes :

$$J'(a).\delta a = \sum_{j=1}^{m} \gamma_j \delta a_j \qquad (14)$$

for every $\delta a = (\delta a_1, \ldots, \delta a_m) \in \mathbb{R}^m$.

From formula (14) we get the components of the gradient of J with respect to the vector a :

$$\frac{\partial J}{\partial a_1}(a) = \gamma_1, \;\ldots\ldots, \; \frac{\partial J}{\partial a_m}(a) = \gamma_m \qquad (15)$$

Case 2 : There is one unknown parameter function a of the variable u in a bounded domain $D \subset \mathbb{R}^p$.

Variable u can be an independent variable such as space or time (see Applications 1, 2), or a dependent variable, such as the state y of the system (see Application 3). $\mathcal{A}$ is then an infinite-dimensional function space, and generally there is no duality application from $\mathcal{A}$ onto its dual $\mathcal{A}'$. For instance, let us suppose here that

$$\mathcal{A} = \mathcal{C}(D) \qquad (16)$$

Then the gradient of J with respect to a is in $\mathcal{A}'$, which is the space of Radon's measures on D , and which is greater than the space of continuous functions on D !

Fortunately, in the practical applications, it happens that the measure $J'(a)$ is of the form $\gamma(u)du$, where γ is a Lebesgue integrable function on D ; formula (13) becomes then :

$$J'(a).\delta a = \int_D \gamma(u)\delta a(u)du \quad \forall \quad \delta a \in \mathcal{A} \qquad (17)$$

(this is to compare with formula (14) : the discrete sum has been replaced by an integral).

The function $\gamma(u)$, when it exists, is called the functional partial derivative of J with respect to a :

$$\frac{\partial J}{\partial a}(u) = \gamma(u) \qquad \forall\, u \in D \qquad (18)$$

(compare with (15)).

In both cases, Prop . 2 enables us to calculate at one time the gradient of J with respect to a (this gradient being a vector $(\gamma_1, \ldots, \gamma_m)$ or a function $\gamma(u)$) from y, ρ and a with formula (13). So the computation of that gradient requires only two resolutions of systems : Direct State Equation (1), Adjoint State Equation (12), the latter being always linear.

Remark 3 : $\frac{\partial J}{\partial a}(u)$ being generally a less regular function of u than $a(u)$, it is impossible to apply rigourosly a gradient method to the continuous case ! ■

Remark 4 : $\frac{\partial J}{\partial a}(u)$ is the gradient of J with respect to the function a , without any assumptions on the algebraic form of that function a. But if, for practical or physical reasons, one wants to have $a(u)$ under the form :

$$a(u) = \alpha(u, \beta_1, \beta_2, \cdots, \beta_k) \,, \qquad (19)$$

where $\alpha(u, \beta_1, \ldots, \beta_k)$ is a known function of u and k unknown parameters $\beta_1, \beta_2, \cdots, \beta_k$, it is straightforward to compute

$$\frac{\partial J}{\partial \beta_j} \qquad j = 1, 2, \ldots k \quad \text{from } \gamma(u) \; :$$

$$\frac{\partial J}{\partial \beta_j}(\beta) = \int_D \gamma(u) \frac{\partial \alpha}{\partial \beta_j}(u, \beta_1, \ldots, \beta_k)du \quad j=1,2,\ldots,k \; (21)$$

■

Remark 5 : Suppose now that a numerical programm has been written for the identification of a as a function of u , without any specification of the algebraic form of a . At each gradient iteration, the programm updates a by the (discrete analogue of the) formula :

$$a^{n+1}(u) = a^{n}(u) - \rho \frac{\partial J}{\partial a}(a^{n})(u) \quad (22)$$

If now one wants a with the form (19), where α is supposed to be linear with respect to the parameter β (a is a polynomial for instance), the parameters β are to be updated by the formula :

$$\beta^{n+1} = \beta^{n} - \rho \frac{\partial J}{\partial \beta}(\beta^{n}) \quad (23)$$

But, thanks to the linearity of (19) with respect to β , (23) is equivalent to :

$$a^{n+1}(u) = a^{n}(u) - \rho\alpha(u, \frac{\partial J}{\partial \beta}(\beta^{n})) \quad (24)$$

where a^n and a^{n+1} are of the form (19). Comparing (22) to (24) we see that the only modifications to do in the programm are to insert, just after the point where

$$\frac{\partial J}{\partial a}(a_n)(u)$$

has been evaluated :

- the calculation of $\frac{\partial J}{\partial \beta}(\beta^n)$ by formula (21) (quadratures)
- the assignment of $\alpha(n, \frac{\partial J}{\partial \beta}(\beta^n))$ in the memory reserved to $\frac{\partial J}{\partial a}(a^n)(u)$.

The whole rest of the programm remains unchanged. This makes very easy the trial of different algebraic forms of the function a . ∎

It is now time to turn towards the numerical applications, which, I hope, will bring some light on the preceding considerations.

IV. APPLICATIONS

Throughout this paragraph :

$\Omega \subset \mathbb{R}$ or $\mathbb{R}^2$	will be the bounded space domain, with boundary Γ , $x \in \Omega$ being the space variable ,
$]0T[\subset \mathbb{R}$	will be the time interval of integration, $t \in]0,T[$ being the time variable ,
$y \in \mathbb{R}$	will be the dependent variable.

APPLICATION 1 : *Identification of the permeability of an aquifer in the area of Bordeaux.*

We consider here the problem of modeling a static aquifer on a two dimensional (square) space domain

The system equation is :

$$\begin{cases} -\frac{\partial}{\partial x_1}(a(x)\frac{\partial y}{\partial x_1}) - \frac{\partial}{\partial x_2}(a(x)\frac{\partial y}{\partial x_2}) = f(x) \text{ in } \Omega \\ y = g \text{ on } \Gamma \end{cases} \quad (25)$$

where :

- y is water-pressure at point $x \in \Omega$
- $a(x)$ is the unknown (isotropic) permeability of the aquifer at point $x = (x_1,x_2) \in \Omega$
- $f(x)$ is a known function of $x = (x_1,x_2) \in \Omega$ related to the outputs q_j and position of the jth well , $j=1,2,...43$ by :

$$f(x) = \sum_{j=1}^{43} q_j \chi_j(x) \quad (26)$$

where $\chi_j(x)$ is a function characterizing the position and geometry of the jth well:

$$\chi_j(x) = \begin{cases} 0 & \text{if } x \text{ does not belong to the } j^{th} \text{ well,} \\ \text{constant} & \text{if } x \text{ belongs to the } j^{th} \text{ well} \end{cases}$$

the constant being chosen such that :

$$\int_\Omega \chi_j(x)dx = 1$$

- g is the known pressure at the boundary Γ of Ω .

We suppose here that we have a distributed observation, that is, we know a function $z(x)$, which is a measure of the function $y(x)$. In fact, z is deduced by interpolation from the isopiezes shown in Fig. 1. Function g is obtained in the same way. We choose then for cost function associated with a (cf (5)) :

$$J(a) = \int_\Omega (y(x;a) - z(x))^2 \, dx_1 \, dx_2 \quad (27)$$

Let us now settle this problem into the theoretical frame of § III. Replacing in (25) y by $y-\tilde{y}$ (where $\tilde{y}$ is a raising-up of g in Ω), we are lead to a system of the same form than (25), but with $g = 0$.

We suppose :

$$a \in L^\infty(\Omega) \quad , \quad f \in L^2(\Omega) \quad (28)$$

Let the space $L^2(\Omega)$ be identified to its dual space, and H_0^1 be identified to a subspace of $L^2(\Omega)$. Then $L^2(\Omega)$ can be identified to a subspace of the dual space $[H_0^1(\Omega)]'$ of $H_0^1(\Omega)$:

$$H_0^1(\Omega) \subset L^2(\Omega) \subset [H_0^1(\Omega)]' \quad (29)$$

It is now possible to associate, to each function $a \in L^\infty(\Omega)$, a linear operator $A(a)$ from $H_0^1(\Omega)$ into $[H_0^1(\Omega)]'$ by :

$$(A(a)u,v)=\int_\Omega a(x) \sum_{i=1}^{2} \frac{\partial u}{\partial x_i}\frac{\partial v}{\partial x_i}\,dx \quad \forall u,v \in H_0^1(\Omega) \qquad (30)$$

It is then known that a weak formulation of equation (25) (with g=0) is :

$$\begin{cases} \text{find } \; y \in H_0^1(\Omega) \text{ such that :} \\ A(a)y = f \end{cases} \qquad (31)$$

Equation (31) is the state equation. It is of the form (1), where $Y = H_0^1(\Omega)$, $\mathcal{A} = L^\infty(\Omega)$, $F = (H_0^1(\Omega))'$, and

Ψ is the application: $(y,a) \in Y \times \mathcal{A} \to A(a)y \in F$. (32)

The subset $\mathcal{A}_c$ of $\mathcal{A}$ for which equation (31) has one and only one solution y is here :

$$\mathcal{A}_c = \{a \in L^\infty(\Omega) \mid \exists\ \alpha > 0,\ a(x) \geq \alpha > 0 \text{ a.e on } \Omega\} \qquad (33)$$

The subset $\mathcal{A}_{ad}$ of admissible parameters introduced in (7) can be defined here by :

$$\mathcal{A}_{ad} = \{a \in L^\infty(\Omega) \mid a_{Max} \geq a(x) \geq a_{min} > 0 \text{ a.e on } \Omega\} \qquad (34)$$

where a_{Max} and a_{min} are given upper and lower bounds of the permeabilities.
Concerning the observation, we take :

$$\mathcal{H} = L^2(\Omega)\ , \quad \mathcal{C} = \text{canonical injection from } H_0^1(\Omega) \text{ into } L^2(\Omega) \qquad (35)$$

so that the functionnal J defined in (27) coincides with J defined in (5).

We can now apply Property 2 (it is easy to check that its hypothesis are satisfied) :

Proposition 3 : Under assumption (28), the functional J defined in (27) is derivable with respect to function $a \in L^\infty(\Omega)$, and $J'(a)$ is given by :

$$J'(a).\delta a = (A(\delta a)y,p) \qquad \forall a \in \mathcal{A}_c\ , \ \forall a \in \delta\mathcal{A} \qquad (37)$$

where $p \in H_0^1(\Omega)$ is the adjoint state, solution of:

$$A(a)p = -2(y-z) \qquad (38)$$

Using (30) ,(37) becomes :

$$J(a).\delta a = \int_\Omega \delta a(x) \sum_{i=1}^{2} \frac{\partial y}{\partial x_i}\frac{\partial p}{\partial x_i}\,dx \quad \forall\ \delta a \in L^\infty(\Omega) \qquad (39)$$

which proves that the function $\frac{\partial J}{\partial a}(x)$ exists and has for expression :

$$\frac{\partial J}{\partial a}(x) = \sum_{i=1}^{2} \frac{\partial y}{\partial x_i}(x)\frac{\partial p}{\partial x_i}(x) \in L^1(\Omega) \qquad (40)$$

Formula (40) enables us to compute easily $\frac{\partial J}{\partial a}$, when we know y , solution of (31) i.e. weak solution of (25), and p , solution of (38), i.e. weak solution of

$$\begin{cases} \frac{\partial}{\partial x_1}\left(a(x_1 x_2)\frac{\partial p}{\partial x_1}\right) - \frac{\partial}{\partial x_2}\left(a(x_1 x_2)\frac{\partial p}{\partial x_2}\right) = \\ \qquad = -2(y(x_1,x_2)-z(x_1,x_2)) \text{ in } \Omega \\ p = 0 \text{ on } \Gamma \end{cases} \qquad (41)$$

<u>Numerical results</u> : the datas [1] are summarized in Fig. 1.

Ω was discretized by a finite difference grid of 20 X 20 points. The unknownfunction $a(x_1 x_2)$ was discretized over that grid (so there were 400 unknown scalar coefficients). At each gradient step, equations (25) and(41) were solved by classical finite difference schemes. Then the function $\frac{\partial J}{\partial a}(x)$ was evaluated at each node of the grid by

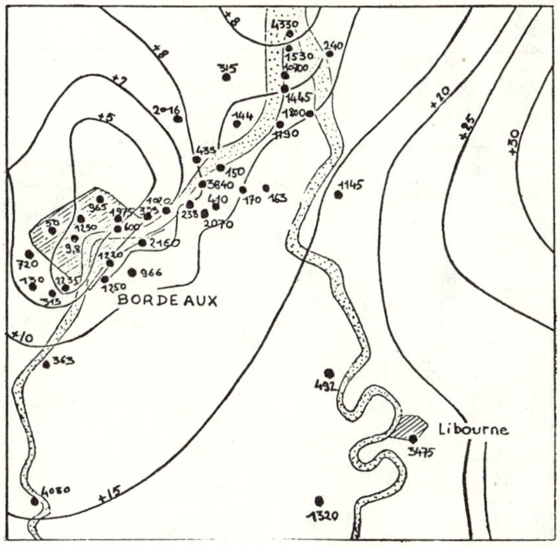

Fig. 1

The Datas of the Aquifer Identification Problem

[1] We thank the "CENTRE D'INFORMATIQUE GEOLOGIQUE et MINIERE" of the "ECOLE DES MINES" in Fontainebleau which supplied us with those datas and authorized us to publish them.

finite difference from equation (40). Then a steepest-descent algorithm was used to update a , with an approximate determination of the optimal gradient step. As we had no idea of the order of magnitude of the permeability a , we, first of all, searched the best constant functions $a(x)$ (using Remark 5) ; starting from $a^{\circ} = 150$, we got $a = 1622$ after 9 gradient iterations and 24s of IBM 360-91.
Then, starting from that function $a(u)$ constant and equal to 1622, we searched the best function a , and we got the function shown in Fig. 2, after 5 gradient iterations and 22 s. of IBM 360-91. The pressure y corresponding ot the transmissivity $a(x_1 x_2)$ of Fig. 2 is shown in Fig. 3.

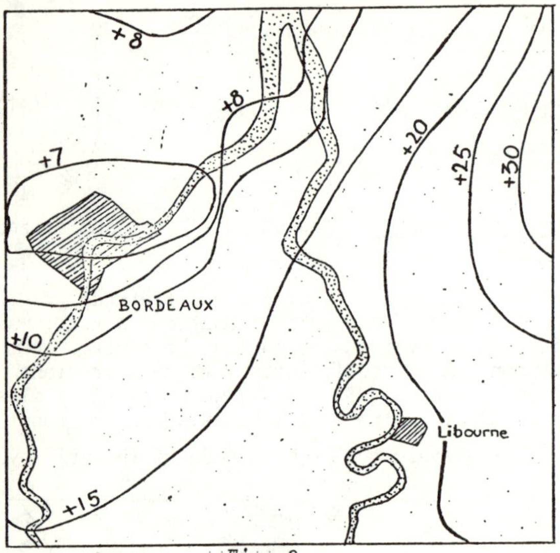

Fig. 2
The function a(x) found with the algorithm.

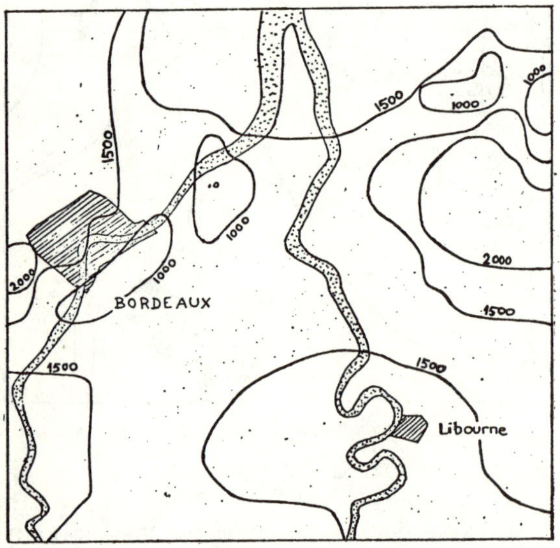

Fig. 3
The isopiezes corresponding to the transmissivities a(x) of Fig. 2.

APPLICATION 2 : *Identification of the permeabi- of a (monophasic) petroleum field.*

Ω is a domain of $\mathbb{R}^2$, shown in Fig. 4. The state equation for the pressure $y(x,t)$ is :

$$\left\{\begin{array}{l} \frac{\partial y}{\partial t} - \sum_{i=1}^{2} \frac{\partial}{\partial x_i}\left(a(x)\frac{\partial y}{\partial x_i}\right) = f(x,t) \text{ in } \Omega\times]0,T[\\ \frac{\partial y}{\partial t} = 0 \quad \text{on } \Gamma \text{ (Neumann boundary conditions)} \\ y(x,o) = y_o(x) \text{ on } \Omega \text{ (Initial conditions)} \end{array}\right. \tag{42}$$

where :

- $a(x)$ is the unknown (isotropic) permeability of the petroleum field,
- $f(x,t)$ is a known function of the form (26), when the q_j are now functions $q_j(t)$ of time,
- $y_o(x,t)$ is the known initial pressure in the field.

Concerning the observations, we have here less information about the pressure $y(x,t)$ than in the former case : only the mean pressure in each of the eleven wells is measured at each instant t, which gives eleven functions $z_j(t)$, $j=1,2,\ldots,11$.

The natural cost function is then :

$$J(a) = \sum_{j=1}^{11}\int_0^T\left[\int_\Omega \chi_j(x)y(x,t;a)-z_j(t)\right]^2 dt \tag{43}$$

As in Application 1, this problem can be settled in the frame of § III, using the variational theory of partial differential equations, and Prop.2 can be applied. For further details concerning those theoretical questions, see [3][4].
The formulas (13) and (12) become :

$$J'(a)\delta a = \int_\Omega\left[\int_0^T \sum_{j=1}^{2}\frac{\partial y}{\partial x_i}\frac{\partial p}{\partial x_i}dt\right]\delta a(x)dx \qquad \forall\ \delta a\in L^\infty(\Omega). \tag{44}$$

(compare with (17) and (39)) where y is the solution of (42) and p (the adjoint state) is solution of :

$$\left[\begin{array}{l} -\frac{\partial p}{\partial t} - \sum_{i=1}^{2}\frac{\partial}{\partial x_i}\left(a(x)\frac{\partial p}{\partial x_i}\right) = -2\sum_{j=1}^{11}\chi_j(x)\left[\int_\Omega \chi_j(x)y(xt)dx - z_j(t)\right] \text{ in } \Omega\times]0,T[\\ \frac{\partial p}{\partial n} = 0 \text{ on } \Gamma \text{ (Neumann conditions)} \\ p(x,T) = 0 \text{ on } \Omega \text{ (final conditions)} \end{array}\right. \tag{45}$$

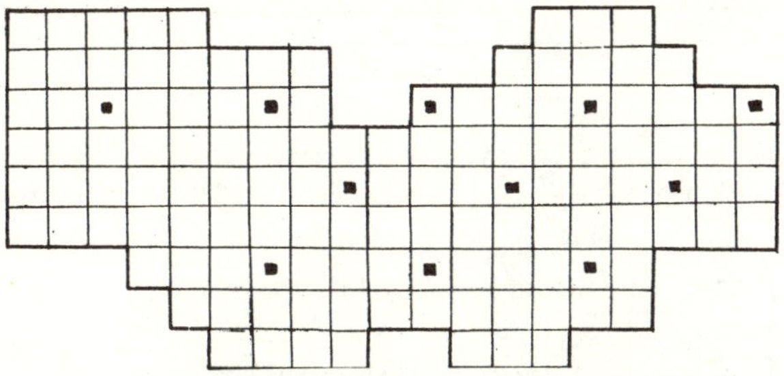

Fig.4 : The space domain and its discretization (the meshes are figurated ; black squares show the position of wells).

So formula (44) shows that there exists in the present case a partial functional derivative $\frac{\partial J}{\partial a}(x)$, and enables us to compute easily from y and p.

Numerical results : Application of this theory to datas supplied by the "INSTITUT FRANCAIS DU PETROLE (IFP)" has been made in the frame of a research contract between IRIA and IFP. We thank the IFP for authorization of publishing the following results. The field in included in a rectangle of about 8 X 16,8 kms, and discretized in 127 meshes (the space step is 884 m). The historic of production (the $q_i(t)$'s) covers 2070 days (the time step is 23 days), and though not figurated here, is known. The initial pressure $y_o(x)$ is known and equals to 482 kg/cm^2. The measures of the pressure $z_i(t)$ at each well are simulated by integrating equation (42) with the "true" permeabilities a(x) shown in Fig. 5. Moreover, the true permeabilities are practically known at each well. The method will allow us to take into account this information very easily.

The initial value of permeabilities was set equal to 200 mdm at each node of the grid, except at the wells, where the exactly known value was imposed.

At each iteration, systems (42) and (45) were solved by a classical finite difference scheme (one system resolution takes 0,561 s on CDC 7600), then $\frac{\partial J}{\partial a}$ was evaluated at each node by (44), then set equal to zero at the nodes where a well was located. The permeabilities obtained after 20 gradient iterations and 85 s of CDC 7600 are shown in Fig.6. The global fit (mean square error) on the pressure in the wells was 0,99 kg/cm^2. The graphics corresponding to the wells having the best (0.44 kg/cm^2) and the worse (1.63 kg/cm^2) fit are plotted in Figures 7 and 8.

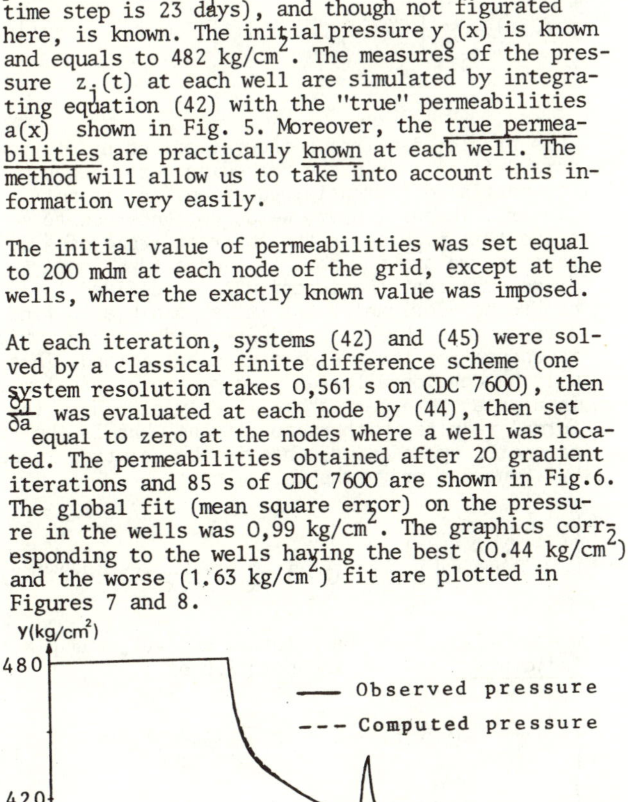

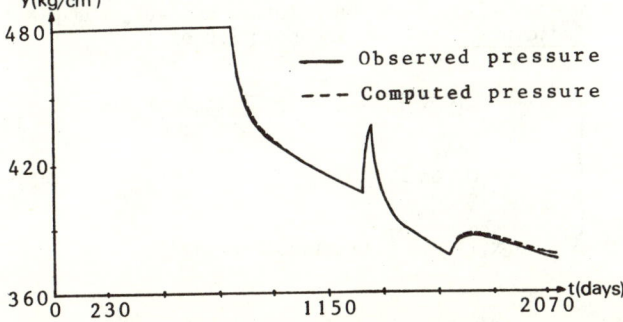

Fig. 7. Observed and computed pressure evolution at well N° 1. (Fit = .44 kg/cm²).

50.	167.	250.	250.	250.	-0.	-0.	-0.	-0.	-0.	-0.	-0.	-0.	300.	300.	300.	-0.	-0.	-0.
50.	180.	180.	250.	250.	100.	30.	30.	-0.	-0.	-0.	-0.	575.	500.	600.	600.	575.	-0.	-0.
50.	180.	180.	180.	250.	100.	160.	30.	-0.	-0.	12.	3.	600.	920.	920.	900.	600.	500.	100.
50.	180.	180.	280.	280.	100.	20.	20.	3.	3.	3.	800.	300.	300.	300.	300.	500.	100.	100.
167.	250.	230.	280.	9.	254.	254.	254.	3.	3.	400.	600.	800.	600.	900.	500.	350.	50.	50.
280.	280.	280.	280.	254.	254.	254.	254.	254.	200.	320.	100.	400.	300.	300.	100.	50.	100.	100.
-0.	-0.	-0.	200.	254.	254.	254.	254.	254.	200.	320.	320.	300.	300.	300.	10.	-0.	-0.	-0.
-0.	-0.	-0.	-0.	250.	254.	254.	254.	254.	100.	200.	320.	320.	300.	300.	10.	-0.	-0.	-0.
-0.	-0.	-0.	-0.	-0.	200.	200.	200.	200.	-0.	-0.	100.	200.	10.	-0.	-0.	-0.	-0.	-0.

Fig. 5
The true permeabilities used in the simulation of the datas.

199.	195.	196.	204.	200.	0.	0.	0.	0.	0.	0.	0.	0.	206.	228.	203.	0.	0.	0.
197.	181.	160.	229.	243.	222.	66.	187.	0.	0.	0.	0.	210.	284.	448.	297.	220.	0.	0.
195.	122.	180.	321.	298.	73.	160.	33.	0.	0.	12.	3.	421.	566.	920.	672.	465.	453.	100.
197.	182.	139.	209.	216.	201.	9.	165.	127.	188.	3.	143.	306.	412.	627.	455.	273.	253.	193.
199.	195.	189.	194.	199.	213.	259.	100.	3.	10.	136.	196.	800.	287.	327.	262.	350.	71.	185.
200.	199.	199.	197.	200.	221.	320.	231.	65.	174.	243.	204.	188.	208.	164.	227.	159.	165.	195.
0.	0.	0.	200.	204.	240.	254.	275.	218.	211.	320.	216.	177.	131.	300.	181.	0.	0.	0.
0.	0.	0.	0.	200.	202.	233.	213.	198.	193.	210.	165.	179.	154.	165.	195.	0.	0.	0.
0.	0.	0.	0.	0.	200.	202.	202.	199.	0.	0.	194.	196.	194.	0.	0.	0.	0.	0.

Fig. 6
The permeabilities obtained after 20 gradient iterations.

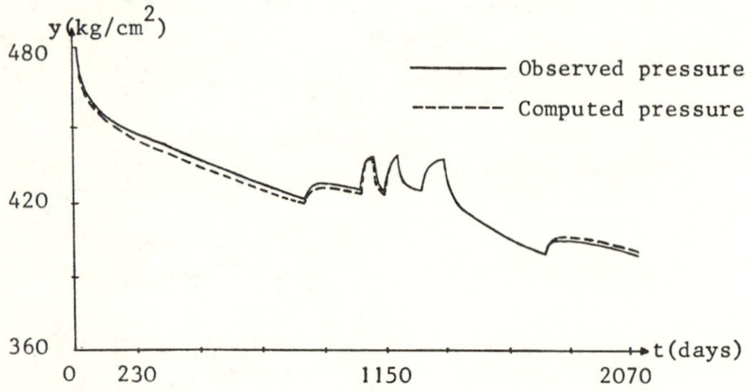

Fig. 8

Observed and Computed Pressure Evolution at well n° 10 (Fit = 1.63 kg/cm²)

Comparing Figures 5 and 6, one sees that the obtained permeabilities $a(x)$ are fairly different from the "exact" one a_o.

This comes from the fact that there is no uniqueness result on a in our problem. In some regions, $a(x)$ has stayed near its initial value of 200 mdm (on the boundary and in the regions where little wells are located) : this means that the permeabilities in those regions have little influence on the measures of pressure and cannot be determined from those measures.

In other regions (where wells with significant outputs are located) one sees that the permeabilities have moved in the right direction, and that the order of magnitude is in accordance with that of the exact coefficients.

APPLICATION 3 : *Identification of a parameter function depending of the solution of an one-dimensional parabolic equation.*

We consider, on the space domain $\Omega =]0,1[$ and the time domain $]0,T[=]0,1[$ the parabolic equation :

$$\begin{cases} \frac{\partial y}{\partial t} - \frac{\partial}{\partial x}\left(a(y)\frac{\partial y}{\partial x}\right) = f(x,t) & \text{in } \Omega\times]0,1[\\ y = g \quad \text{on } \Gamma\times]0,1[& \text{(Dirichlet boundary conditions)} \\ y(x,o) = y_o(x) \quad \text{on } \Omega & \text{(Initial condition)} \end{cases} \qquad (46)$$

The functions f, g and y_o are supposed known, and the function $y \to a(y)$ is to be estimated. (system (46) can be viewed as the equation of an one-dimensional heat diffusion process with a temperature dependent diffusion parameter).

The solution $y(x,t)$ is measured, for each $t \in]0,1[$, at five points $x_1 \ldots x_5$ of the space domain $]0,1[$, giving five functions $z_1(t)..z_5(t)$.

The cost function is then :

$$J(a) = \sum_{j=1}^{5} \int_0^T (y(x_j,t;a) - z_j(t))^2 \, dt \qquad (47)$$

This problem is, from a theoretical point of view, more difficult than the preceding ones, because the partial differential equation (46) is not linear ; in particular, we do not know how to solve this problem in the abstract frame of § III, that is, we are not able to find a functional frame such that the functional J defined in (47) be Frechet-derivable. But it is possible to find a frame into which J is Gâteaux-derivable, which is enough for the use of a gradient method. This will be developed for an n-dimensional problem in a paper to appear. We will only give here the formal results, without specifying what is the parameter space $\mathcal{a}$ and what is the solution space Y. One can prove the following :

$$\begin{cases} \forall \delta a \in \mathcal{a} \\ J'(a).\delta a = \iint_{\Omega}\int_0^1 \delta a(y(x,t))\frac{\partial y}{\partial x}(x,t)\frac{\partial p}{\partial x}(x,t)\, dx dt \end{cases} \qquad (48)$$

where $y(x,t)$ is the solution of (46), and $p(x,t)$ (adjoint state), is the solution of :

$$\begin{cases} -\frac{\partial p}{\partial t} - a(y(x,t))\frac{\partial^2 p}{\partial x^2} = -2\sum_{j=1}^{5} \delta(x-x_j)(y(x_j,t)-z_j(t)) \\ p = 0 \quad \text{on } \Gamma\times]0,1[\\ p(x,T) = 0 \quad \text{(final condition)} \end{cases} \qquad (49)$$

Setting :

$$\varphi(x,t) = \frac{\partial y}{\partial x}(x,t)\,\frac{\partial p}{\partial x}(x,t) \quad \forall x,t \in \Omega \times]0,1[\quad (50)$$

equation (48) can be rewritten :

$$\begin{cases} \forall \delta a \in \mathcal{a} \\ J'(a).\delta a = \int_\Omega \int_0^1 \delta a(y(x,t))\varphi(x,t)\,dxdt \end{cases} \quad (51)$$

Comparing (51) with the general form (17) assumed for $J'(a).\delta a$, we see that (51) does not give us directly the partial functional derivative $\gamma(y)$ of J with respect to $a(y)$, defined by :

$$\begin{cases} \forall \delta a \in \mathcal{a} \\ J'(a).\delta a = \int_{y_m}^{y_M} \delta a(z)\gamma(z)\,dz \end{cases} \quad (52)$$

where y_m and y_M are known lower and upper bounds of the solution y of (46).

As in § III, we will distinguish two cases :

Case 1 : We assume algebraïc form for the function $y \to a(y)$, with only a finite number of unknown parameters $\beta_1, \dots \beta_k$ (see Remark 4, with $u = y$). It is then straightforward to calculate the gradient of J with respect to the vector $\beta = (\beta_1, \dots, \beta_k)$ from (51) :

$$\frac{\partial J}{\partial \beta_j}(\beta) = \int_\Omega \int_0^T \frac{\partial \alpha}{\partial \beta_j}(y(x,t),\beta_1 \dots \beta_k)\varphi(x,t)\,dxdt \quad (53)$$

$$j=1,2 \dots k$$

So we see that, once the function $\varphi(x,t)$ has been computed (which supposes that the two systems (46) and (49) have been solved), the computation of the gradient of J with respect to β requires only k numerical quadratures. Moreover, Remark 5 applies here (with slight modifications).

Case 2 : We do not assume any algebraïc form for the function $y \to a(y)$, so we must find a way of derivating (52) from (51).

Define, for every real number ζ :

$$Q(\zeta) = \{(x,t) \in \Omega \times]0,1[\;|\; y(x,t) \geq \zeta\} \quad (54)$$

and :

$$\eta(\zeta) = \int_{Q(\zeta)} \varphi(x,t)\,dxdt \quad (55)$$

Then formula (51) can be rewritten :

$$\begin{cases} \forall \delta a \in \mathcal{a} \\ J'(a)\,\delta a = \int_{y_m}^{y_M} \delta a(\zeta)\,d\eta(\zeta) \end{cases} \quad (56)$$

which, together with (52) gives :

$$\frac{\partial J}{\partial a}(a)(y) = \gamma(y) = \eta'(y) \quad (57)$$

So we are able to calculate the partial functional derivative of J with respect to the function $y \to a(y)$. Let us see now how we proceed numerically.

NUMERICAL RESULTS : The space domain was discretized into 20 intervals of length .05, and the time interval into 40 steps of length .25. The function f was chosen to be a combination of (discrete) space dual functions at the points $x_1=.1$, $x_2=.3$, $x_3=.5, x_4=.7, x_5=.9$ of $\Omega =]0,1[$.

The initial value was chosen constant, and the "true" function $a_o(y)$ used for simulation of the datas was :

$$a_o(y) = 0.21 - .28\,y + .7\,y^2 \quad (58)$$

The lower and upper bounds of y were chosen as

$$y_m = .3 \qquad y_M = 2 \quad (59)$$

and the interval y_m, y_M was divided into 20 intervals of length Δ :

and the function $a(y)$ was represented on this interval by a continuous piecewise linear function :

$$a(y) = \sum_{j=1}^{20} \left(a_j + \frac{1}{\Delta}(y-y_j)(a_{j+1}-a_j)\right) \chi_j(y) \quad (60)$$

where :

$$\begin{cases} \chi_j(y) = \begin{cases} 1 & \text{if } y \in [y_j, y_{j+1}[\\ 0 & \text{if not.} \end{cases} \\ a_j = \text{discretized value of } a(y_j) \end{cases} \quad (61)$$

The system was then integrated with the "true" value a_o of the function a, using a finite difference scheme of predictor corrector type, and gave the simulated observations $z_j(t)$ for $j = 1,2,\dots, 5$.

To recover the function a(y), we used then a standard steepest descent algorithm with an approximate determination of the optimal gradient step, and projection on a convex polyedra $\mathcal{U}_{ad}$.

We, first of all, searched a(y) under the form of a polynomial of a polynomial of degree 2 (3 unknown scalar parameters), the gradient with respect to those coefficients being computed from (53). As the function a(y) has to be positive, in order that the equation (46) have an unique solution, the convex set $\mathcal{U}_{ad}$ of $\mathbb{R}^3$ was determined by writing that, at every point of discretization $y_i \in]y_m, y_M[$, the polynomial a(y) was positive (21 linear constraints).

The initial value and the functions a(y) obtained after 14 gradient iterations are shown in Figures 9 and 10. After 32 iterations, the obtained function is practically identical to the true function (58). The corresponding measured and computed outputs are shown for well 1 on Figure 11.

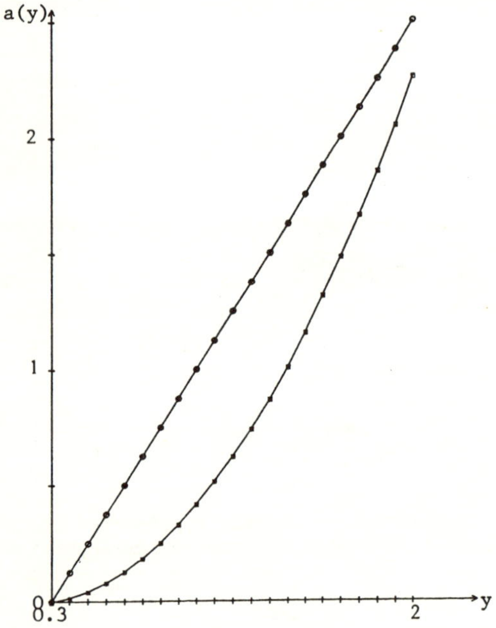

Fig. 9

Initial value of function a(y)

We then searched a(y) under the form (60) of a continuous piecewise linear function (21 unknown scalar parameters).

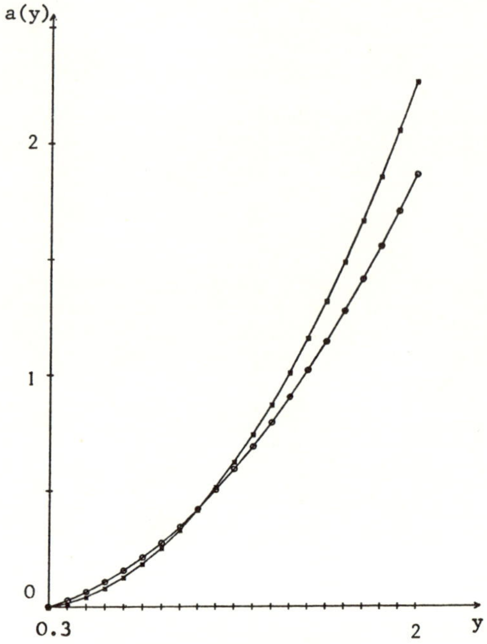

Fig. 10

The polynomial a(y) obtained after 14 gradient iterations.

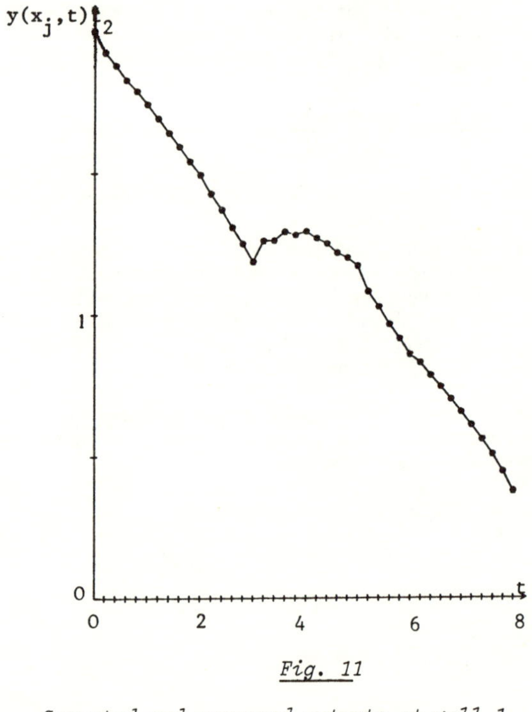

Fig. 11

Computed and measured outputs at well 1.

The derivative $\frac{\partial J}{\partial a_j}$ is computed then from (57) for each j.
This consists numerically in integrating a function (related to φ defined in (50)) on the domain $Q(y_{j+1})-Q(y_j)$ for each j. The programm for that is somewhat tedious to write, but does not take too much time.
We first of all supposed we knew nothing about a but it is a positive function (a is a "free" positive function) :

$$\mathcal{A}_{ad} = \{ a \mid a_j \geqslant 0 \quad j=1,2,\ldots,21 \} \qquad (62)$$

It is straightforward to project on this convex !

The initial value was that of Fig. 9 , and the functions a(y) obtained after 19 gradient iterations, is shown in Fig. 12

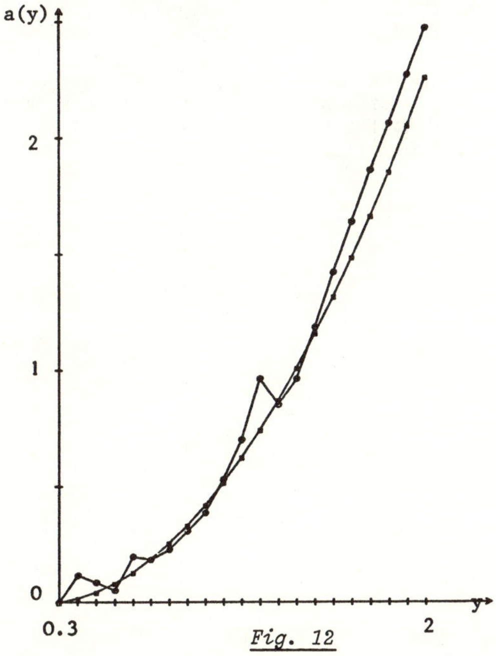

Fig. 12

The function a(y) obtained after 19 iterations (a^o as in Fig.9 , a=free function).

We after that supposed we had some information about the second derivative of a(y) :

$$\mathcal{A}_{ad}=\{a \mid a_j \geqslant 0 \quad j=1,2\ldots21, -m_j \leqslant \frac{a_{j+1}-2a_j+a_{j-1}}{\Delta^2} \leqslant M_j \quad j=2,3,\ldots,20\} \quad (63)$$

(where the m_j and M_j are given positive numbers).

This is interesting because such information is often available from physicists, and we see here that we can take it easily into account. The results corresponding to the initial value of Fig. 9 and different m_j and M_j's are plotted in figures 13 and 14.

The mean time for a gradient iteration is 30 s of CII 10070 (which is around 25 time slower than the CDC 7600).

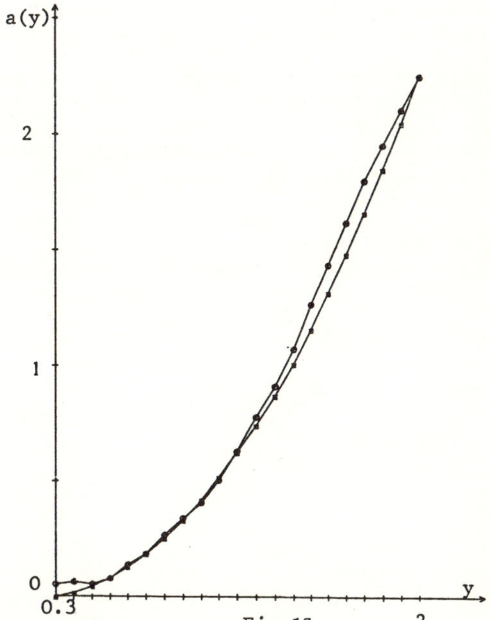

Fig.13

The function a(y) obtained after 10 iterations.

(a^o as in Fig.9, $|a''| \leqslant 5$).

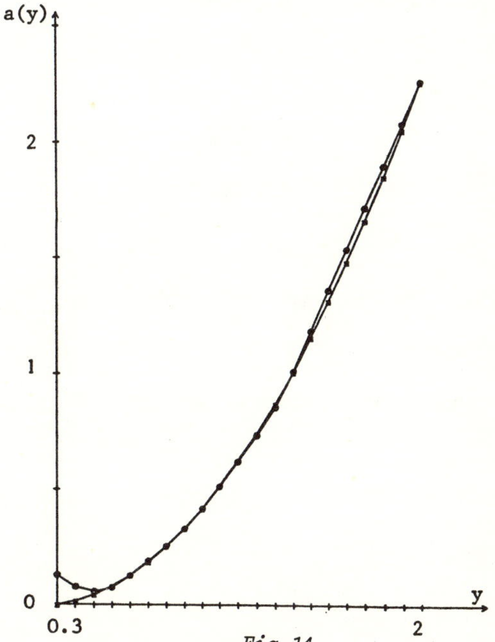

Fig.14

The function a(y) obtained after 9 iterations

(a^o as in Fig.9 , $0 \leqslant a'' \leqslant 5$).

V. CONCLUSIONS

We have tried to convince the reader that, when encountering unknown functions in modelization of systems, it is possible to avoid some algebraic assumptions on the form of those functions, up to the numerical treatment, by the computation of the partial functional derivative.
Even when one is trying to find the function of a given algebraic form, the use of the partial functional derivative as an intermediate, gives a very flexible way of computing the gradient.
The use of an observation operator $\mathcal{C}$, which is required only to be differentiable, enables us to take into account the available measures on the system, without having to make major previous treatment of these datas.
This approach is systematic, and can be appkied to many other fields. Many other examples are studied in [4]. We are now applying this approach to the recovery of sound velocities from seismic datas, and to the modelization of a petroleum field in the case of multiphase flow.

VI. APPENDIX

Proof of Prop. 2 :

Under hypothesis of Prop. 2, the equation (1) defines an application $a \to y$ which is derivable from

$\mathcal{a}_c$ into Y , by implicit function theorem.

the derivative at point $a \in \mathcal{a}_c$ being the application which, to every $\delta a \in \mathcal{a}^c$, makes the corresponding $\delta y \in Y$ defined by ;

$$\frac{\partial\Psi}{\delta y}(y,a).\delta y = -\frac{\partial\Psi}{\delta a}(y,a).\delta a \qquad (64)$$

If we now derivate the function J defined in (5), we get, for the variation J associated with a variation $\delta a \in \mathcal{a}$:

$$\delta J = 2(\mathcal{C}y(a) - z,\ \mathcal{C}\,\delta y)_{\mathcal{H}}$$

i.e. $$\delta J = 2(\Lambda(\mathcal{C}y(a)-z, \mathcal{C}\,\delta y)_{\mathcal{H}'-\mathcal{H}}$$

i.e. $$\delta J = 2(\mathcal{C}^{*}\Lambda(\mathcal{C}y(a)-z-, \delta y)_{Y'-Y}$$

If then $\rho \in F'$ is defined by (12), we get :

$$\delta J = -(\frac{\partial\Psi}{\delta y}(y,a).\rho, \delta y)_{Y'.Y}$$

i.e. $$\delta J = (\rho, -\frac{\partial\Psi}{\delta y}(y,a).\delta y)_{F'F}$$

which, using (64) gives (11) q.e.d. ∎

VII. REFERENCES

[1] LIONS J.L. Contrôle optimal de systèmes gouvernés par des équations aux dérivées partielles. Dunod 1968.

[2] LIONS J.L. Quelques méthodes de résolution de problèmes aux limites non linéaires Dunod Gauthier Villars Paris 1969.

[3] CHAVENT G. Sur une Méthode de résolution du Problème inverse dans les Equations aux dérivées partielles paraboliques. Note CRAS Paris, t.260, Déc. 1969.

[4] CHAVENT G. Deux résultats sur le problème inverse dans les équations aux dérivées partielles du 2ème ordre en t et sur l'unicité de la solution du problème inverse de la diffusion. Note CRAS. Paris t. 270, Janvier 1970.

[5] CHAVENT G. Analyse fonctionnelle et Identification de coefficients répartis dans les équations aux dérivées partielles Thèse d'Etat. Paris 1971.

--- § ---

IDENTIFICATION OF NONLINEAR DISTRIBUTED SYSTEMS BY USING AN INVERSE DESCRIBING FUNCTION METHOD * **

Guy Jumarie

Université du Québec à Montréal, Canada
Laboratoire d'Automatique de l'Université de Lille, France

Abstract. A class of nonlinear distributed systems which exhibit a structure similar to that of the standard closed loop with lumped parameter is considered. After the statement of the main results of a new harmonic linearization method which has been proposed recently for these systems, one describes the identification procedure directly derived, by inversion, from this theoretical approach.

1. OUTLINE OF THE APPROACH

The context of the approach is that of the mathematical inversion of a given model which may be either deterministic or stochastic. Such a tentative assumes that are satisfied some conditions which we can summarize as follows: (i) The model of the system to identify is known, at least approximately. This is the case when the purpose of the identification is to check that the parameters of a given system, previously designed, are lying inside the permissible range. (ii) The inversion of the operators wich describe the model is relatively easy, what is required for a better efficiency of the method regarding both its accuracy and its execution fastness. (iii) This minimal context is improved when the system is input-output stable in such a manner that it provides a better access to the observation of its response when it is subject to the test signal.

1.2 - The model

The class of models considered is diagrammed in

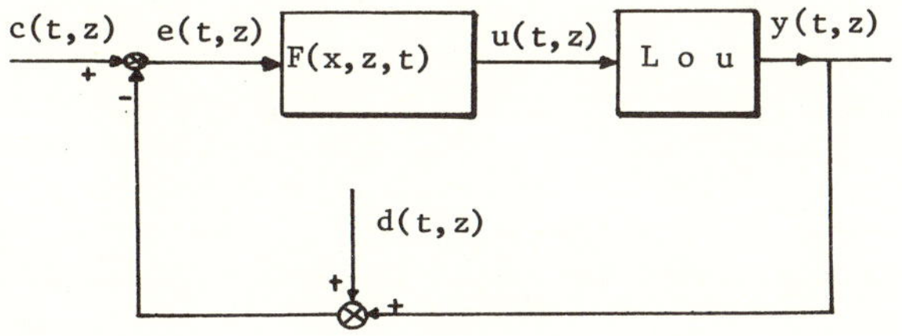

Fig. 1 The general model

* Research supported by National Research Council of Canada.

** Submitted to 3rd IFAC Symposium on Identification and System Parameter Estimation. 1973

Fig.1 where F denotes the nonlinearity which may be lumped of distributed and L represents the linear part which acts on the output u of F. The problem is to identify F and L in some various circumstances.

1.3 - Outline of Identification procedure

Assume that F( ) is subject to the signal

$$e(t,z) = X \sin \Omega z \sin \omega t \qquad (1.1)$$

With the accuracy of the first harmonic in space and time, its response is

$$u(t,z) = N(F,X)\, X \sin\Omega z \sin \omega t,$$

and that of L is

$$y(t,z) = d(t,z) + G_L(i\omega, i\Omega) N(F,X) X \sin\Omega z \sin \omega t \qquad (1.2)$$

where N( ) is the equivalent gain of F, and G_L( ) is the complex gain of L. Assume that d(t,z) is known.

We have the following situations.

(i) ω,Ω are fixed, X is variable. If L is known, $G_L(i\omega,i\Omega)$ N(F,X) yields N(F,X), and F is obtained via the inversion of N(F,X).

(ii) ω,Ω are variable, X is fixed. If F is known, $G_L(i\omega,i\Omega)$ N(F,X) provides $G_L(i\omega,i\Omega)$ and therefore L.

(iii) When both F and L are unknown, a combination of (i) and (ii) can yield F and L with some indetermination under the forme of arbitrary constant terms.

(iv) We now assume that F(x,z,t) is subject to the signal.

$$e(t,z) + n_t(z) = X \sin\Omega z \sin \omega t + n_t(z) \qquad (1.3)$$

where $n_t(z)$ represents a given disturbing noise.

We define the expected value of the nonlinear gain as to be

$$E\{\tilde{u}(t,z)\} = E\{N(F,X)\}\ X \sin\Omega z \sin\omega t \qquad (1.4)$$

and the identification is made via $E\{N(F,X)\}$

This is the basic idea of the approach that we shall expand in the sequel. All the question is to have a suitable representation for the describing function method. After we recall some mathematical datas, we shall identify some specific models. The procedure will be uniform: description of the systems, description of the describing function method, description of the identification procedure.

2. SOME MATHEMATICAL RESULTS

▶▶ *

2.1 Characterisation of a nonlinearity

2.1.1 Input with zero mean value

In the context of the first harmonic accuracy, we consider the response $f(\tilde{e},z) = u(t,z)$ of the nonlinearity $f(x,z)$ subject to the input

$$\tilde{e} = Y \sin\omega t \qquad (2.1)$$
$$Y = X \sin\Omega z \qquad (2.2)$$

We shall successively examine different classes of functions

(i) Class F_{oo}

$F_{oo} = \{f(x,z) \mid f(-x,z) = -f(x,z)\ ;\ f(x,-z) = -f(x,z)\}$

$$u(t,z) \cong Q(Y,z)\ X \sin\omega t \qquad (2.3)$$

with

$$Q(Y,z) \triangleq \frac{4}{\pi XY}\int_0^Y x\, f(x,z)\ \chi(Y,x)\, dx \qquad (2.4)$$

$$\chi(Y,z) = 1/\ (Y^2-x^2)^{\frac{1}{2}}$$

(ii) Class F_{oe}

$F_{oe} = \{f(x,z) \mid f(-x,z) = -f(x,z)\ ;\ f(x,-z) = f(x,z)\}$

$$u(t,z) \cong N(Y,z)\ Y \sin\omega t \qquad (2.5)$$

with

$$N(Y,z) \triangleq \frac{4}{\pi Y^2}\int_0^Y x f(x,z)\ \chi(Y,x)\ dx \qquad (2.6$$

(iii) Class F_{eo}

$F_{eo} = \{f(x,z) \mid f(-x,z) = f(x,z)\ ;\ f(x,-z) = -f(x,z)\}$

$$u(t,z) \cong [M(Y,z)\ /\ 2\,]Y \qquad (2.7)$$

$$M(Y,z) \triangleq \frac{4}{\pi Y}\int_0^Y f(x,z)\ \chi(Y,x)\ dx \qquad (2.8)$$

(iv) Class F_{ee}

$F_{ee} = \{f(x,z) \mid f(-x,z) = f(x,z)\ ;\ f(x,-z) = f(x,z)\}$

$$u(t,z) = [R(Y,z)\ /\ 2\,]\ X \qquad (2.9)$$

$$R(Y,z) \triangleq \frac{4}{\pi X}\int_0^Y f(x,z)\ \chi(Y,x)\ dx \qquad (2.10)$$

(v) $f(x,z)$ arbitrary

We set

$$f_{ee}(x,z) = [f(x,z) + f(-x,z) + f(x,-z) + f(-x,-z)]\ /4 \qquad (2.11)$$

$$f_{eo}(x,z) = [f(x,z) + f(-x,z) - f(x,-z) - f(-x,-z)]\ /4 \qquad (2.12)$$

$$f_{oe}(x,z) = [f(x,z) - f(-x,z) + f(x,-z) - f(-x,-z)]\ /4 \qquad (2.13)$$

$$f_{oo}(x,z) = [f(x,z) - f(-x,z) - f(x,-z) + f(-x,-z)]\ /4 \qquad (2.14)$$

with

$f_{ee} \in F_{ee}$, $f_{eo} \in F_{eo}$, $f_{oe} \in F_{oe}$, $f_{oo} \in F_{oo}$

and

$$f = f_{ee} + f_{eo} + f_{oe} + f_{oo} \qquad (2.15)$$

Therefore $u(t,z)$ via (i) - (iv) under the form

$$u = u_{ee} + u_{eo} + u_{oe} + u_{oo} \qquad (2.16)$$

with the correspondance $u_\alpha \leftrightarrow f_\alpha, \alpha=(e,z,x,o)$.

2.1.2 Input with mean value different from zero

We now assume that the sollication is

$$\tilde{e}_1(t,z) = X_o + X \sin\Omega z \sin\omega t. \qquad (2.17)$$

We define $\phi(X_o,\ x,\ z)$ as to be

$$\phi(X_o,\ x,\ z) \triangleq f(X_o + x,\ z)$$

We can then determine $\phi(X_o,\ u,\ z)$ via (i) - (iv)

For instance, when $f \in F_{oe}$, we have

$$u(t,z) = A(X_o,Y,Z)\ X + B(X_o,Y,Z)\ Y \sin\omega t \qquad (2.18)$$

with

$$A(X_o,Y,z) \triangleq \frac{2}{\pi X}\int_o^Y \phi_{ee}(X_o,x,z)\ \chi(Y,x)dx \qquad (2.19)$$

$$B(X_o,Y,z) \triangleq \frac{4}{\pi Y^2}\int_o^Y x\ \phi_{oe}(X_o,x,z)\ \chi(Y,x)dx \qquad (2.20)$$

2.1.3. Stochastic input

We now assume that $f(x,z)$ is subject to the input

$$e(t,z) + n(t,z\tilde{\omega})$$

where $n(t,z\tilde{\omega})$ denotes a stochastic process depending on the probability parameter $\tilde{\omega} \in \tilde{\Omega}$. The stochastic model so obtained is considered as to be the improvement of the suitable deterministic model. So we refer to the deterministic signal (2.17). As an example, we assume $f \in F_{oe}$ we then have

---

* The text written between the symbols ▶▶...◀◀ may be dropped at first reading without loss of continuity.

$$u(t,z) = E\{A(X_o,Y,z)\}\, X + E\{B(X_o,Y,z)\}\, Y \sin \omega t \qquad (2.21)$$

where E{ } denotes the expected value symbol. For instance, (2.19) yields:

$$E\{A(X_o,Y,z)\} = \frac{2}{\pi X}\int_0^Y E\{\phi_{ee}(X_o,x,z)\} \quad (Y,x)dx \qquad (2.22)$$

2.2 A mathematical result

Let there be the following singular integral equation of Volterra

$$h(X) = \frac{2}{k\,\pi}\int_0^X \frac{x f(x)\, dx}{\sqrt{X^2 - x^2}} \qquad (2.23)$$

in which k is a constant, h(X) is a given function, f(x) is the unknown to determine. The solution of this equation is

$$f(x) = \frac{k}{x\sqrt{\pi}}\,\frac{d}{dx}\int_o^x \frac{Xh(X)\, dX}{\sqrt{x^2 - X^2}} \qquad (2.24)$$

$$= \frac{k}{x\sqrt{\pi}}\left[\frac{h(o)}{x} + \int_o^x \frac{(dh/dX)\, dX}{\sqrt{x^2 - X^2}}\right]$$

for the proof see Annexe .

3. SYSTEM DEFINED BY TRANSFER FUNCTIONS IN SPACE AND TIME

3.1 System and list of symbols

The model of the system considered is diagrammed in Fig.2

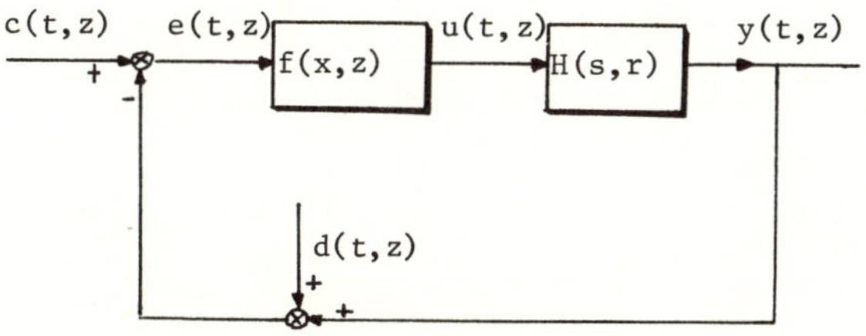

Fig. 2 System represented by transfer function in space and time

We define:

z	one-dimensional distributed parameter
[o,L]	L > o, admissible variation range of z
s	Laplace's variable associated with t
r	Laplace's variable associated with z
H(s,r)	transfer function in time and space
ω	time frequency
Ω	space frequency
Ω_k	$= k\pi/L$, k positive integer
$H(\omega,\Omega)$	absolute value of $H(i\omega,i\Omega)$
$\theta(\omega,\Omega)$	phase of $H(i\omega,i\Omega)$
$H_o(\omega,\Omega)$	$= [H(\omega,\Omega) - H(-\omega,\Omega)]\,/2$
$H_e(\omega,\Omega)$	$= [H(\omega,\Omega) + H(-\omega,\Omega)]\,/2$
$\theta_o(\omega,\Omega)$	$= [\theta(\omega,\Omega) - \theta(-\omega,\Omega)]\,/2$
$\theta_e(\omega,\Omega)$	$= [\theta(\omega,\Omega) + \theta(-\omega,\Omega)]\,/2$
$\lVert v(z)\rVert$	L^2-norm on [0,L]
$\langle v(z)\rangle$	Integral of v(x) on [0,L]

3.2 Characterization of H(s,r).

We describe the output y(t,z) of H(s,r) subject to different inputs u(t,z).

(i) Input $\tilde{u}(t,x) = X \sin\Omega z \sin \omega t$

$$y = X\, H_e(\omega,\Omega) \sin(\Omega z + \theta_o) \sin(\omega t + \theta_e) - X\, H_o(\omega,\Omega) \cos(\Omega z + \theta_o) \cos(\omega t + \theta_e) \qquad (3.1)$$

(ii) Constant input K

$$y = K \lim_{s,r \to o} H(s,r) = K\, H(o,o) \qquad (3.2)$$

(iii) Input X sin ωt

$$y = H(\omega,o) \sin[\omega t + \theta(\omega,o)] \qquad (3.3)$$

(iv) Input X sinΩz

$$y = H(o,\Omega) \sin[\Omega z + \theta(o,\Omega)] \qquad (3.4)$$

3.3 Some theoretical results

3.3.1. Self-oscillations

We define

$$g(x,\Omega) = \Omega\int_o^{\pi/2\Omega} f(x \sin\Omega z, z) \sin\Omega z\, dz \qquad (3.5)$$

Proposition 1.

Assume that $f(x,z) \in F_{oe}$. In the context of the describing function method, we state.

(i) The frequencies (ω_o,Ω_{ko}) of the self-oscillations satisfy the equations.

$$\theta_o(\omega_o,\Omega_{ko}) = (2m + 1)\pi \; ; \; \theta_e(\omega_o,\Omega_{ko}) = 2n\pi$$

or

$$\theta_o(\omega_o,\Omega_{ko}) = 2m\pi \; ; \; \theta_e(\omega_o,\Omega_{ko}) = (2n + 1)\pi$$

(ii) For every admissible oscillating pair (ω_o,Ω_{ko}) a necessary condition in order that there exist an oscillation is that the intersection of the lines C and D,

$$C : x \mapsto g(x,\Omega_{ko})$$

$$D : x \mapsto \pi x/4H(\omega_o,\Omega_{ko})$$

is nonvoid.

(iii) For a given oscillating pair (ω_o,Ω_{ko}), the number of the associated oscillations is less or equal to the number of the intersections of C and D. When this number is odd, there is at least one oscillation . When it is even, either there is no oscillation or there are at least two ones. The magnitude X_o of the oscillations are the solutions of the equations

$$\frac{4}{\pi X_o^2}\int_o^{X_o} x\, g(x,\Omega_{ko})\, \chi(X,x)\, dx = \frac{\pi}{4H(\omega_o,\Omega_{ko})}$$

(iv) Let x_o denote the abcissa of the (C,D) intersection which is immediately less than X_o. This oscillation is stable when the derivative $g_x(x,\Omega_{ko})$ fulfills

$$g_x(x_o,\Omega_{ko}) < \pi/4H(\omega_o,\Omega_{ko})$$

and is unstable when

$$g_x(x_o,\Omega_{ko}) > \pi/4H(\omega_o,\Omega_{ko})$$

3.3.2. *Stability of the equilibrium position*

Proposition 2.
Assume that the following assumptions hold:
(i) The state zero is an equilibrium position.

(ii) There is only one possibly oscillating (not necessarly oscillating !) pair (ω_o,Ω_{ko}). Then the equilibrium zero is stable when

$$4H(\omega_o,\Omega_{ko})\ g_x\ (o,\Omega_{ko})\ /\ \pi < 1$$

and is unstable when

$$4H(\omega_o,\Omega_{ko})\ g_x\ (o,\Omega_{ko})\ /\ \pi > 1$$

Assume now that these equations are reduced to the equality (critical case of Ljapunov). Then the position zero is stable when

$$x\ g_{xx}(o,\Omega_{ko}) < 0,$$

it is unstable when

$$x\ g_{xx}(o,\Omega_{ko}) > 0,$$

the subscript xx denoting the second-order partial derivative w.r.t. x

For the proofs of these propositions see Jumarie (1972b).

3.3.3. Input-output stability
We say that the system considered is input-output stable whenever it is simultaneously L_∞-stable, L_2-stable, quasi-asymptotically stable. We make the following assumptions.

(A1) H(s,r) is stable
(A2) $f(o,z) = 0 \quad \forall\ z \in [o,L]$
(A3) $a(z) \leqslant f(x,z)/x < b(z)\ ,\ z \in [o,L]$

We define

$$g(a,b) \triangleq [\|a\|^2 - \|b\|^2]\ /\ 2\ (\langle a\rangle - \langle b\rangle)$$
$$G(a,b) \triangleq \|a - g(a,b)\| = \|b - g(a,b)\|$$

We state (Jumarie 1972 a)

Proposition 3.
Assume that assumptions (A1) - (A3) *holds. The system considered is input-output stable in the sense above mentionned provided that*

$$\sup_{\substack{\omega \in R \\ \Omega_k \in \{\Omega_k\}}} \left| \frac{H(i\omega,i\Omega_k)}{1 + g(a,b)\ H\ (i\omega,i\Omega_k)} \right| G(a,b) < 1.$$

In the framework of an identification procedure, these theoretical results provide some inquiries on the parameters of the system. ◄◄

3.4 Identification of f(x,z)
3.4.1. Off-line deterministic identification
One excites f(x,z) with the signal $\mathscr{E}(t,z)$ expressed by eqn. (2.1). According to section (2.1), eqn.(2.16), the corresponding outcome u(t,z) is the sum of the outputs $u_{\alpha\beta}(t,z)$, $(\alpha,\beta) = (e,z,x,0)$ of the four components $f_\alpha(x,z)$ of f(x,z), whereby the following identification procedure:
(i) for different values of X, we measure the equivalent gain on the component $u_\alpha(t,z)$.

(ii) The element $f_\alpha(x,z)$ is then given by the inversion (2.24).

For instance, f_{oe} is determined by the inversion

$$f_{oe}(x,z) = \frac{2}{x}\left[\int_o^x \frac{d[Y^2N(Y,z)]/dY}{\sqrt{x^2-Y^2}}\,dY\right] \qquad (3.6)$$

$$Y = X\cos\Omega z$$

3.4.2. On-line deterministic identification

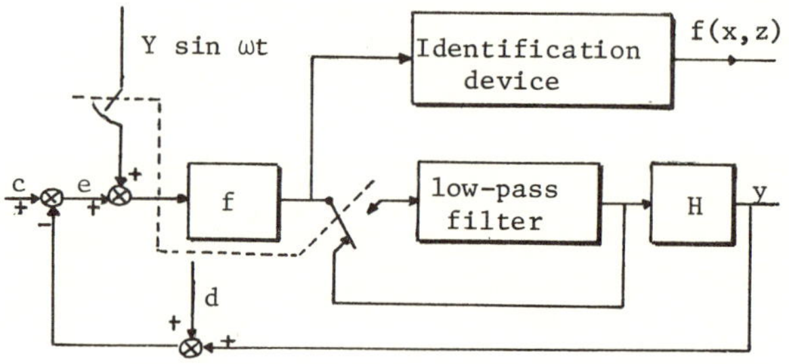

Fig. 3 On-line identification of f(x,z)

(i) We add to the on-line input e(t,z) the identification signal X sinΩz sin ωt with a magnitude X low w.r.t.e(t,z) and a frequency pair (Ω,ω) sufficiently high w.r.t. that of e(t,z).

(ii) f(x,z) is then determined by the inversion of the gain provided by the formulation of section 2.1.2.

In some circumstances we shall be abble to approximate by using the linearization

$$f(e + u) = f(e) + uf_x(e) + o(u^2) \qquad (3.7)$$

so that, in first approximation, the "constant term" of f(e + u) is f(e).

3.4.3 Stochastic identification
The procedure is essentially conserved. We refer to the section 2.1.3. According to its results, we shall identify E{f(x,z)}.

3.4.4 Some comments

(i) According to Proposition 1 we can generate a signal $\tilde{e}(t,z)$ as (2.1) by using a self-oscillating system as that of Fig.2.

(ii) It may happen that, in on-line identification, it is not easy to add to e(t,x) an identification signal which depends explicity on z. In this case, one takes $\tilde{e}(t,z)$ under the form $\tilde{e}$ $\tilde{e} = X \sin \omega t$ and the linearization (3.7) is available.

3.5 Identification of H(s,r)

3.5.1 Off-line identification

(i) We stimulate H(s,r) by the signal $\tilde{e}$ of eqn. (2.1). According to the characterization given by (3.1), when Ω has a fixed value Ω^* and ω is varying, one can identify the functions.

$$\phi_1(\omega,\Omega^*) = \theta_e(\omega,\Omega^*)$$

$$\phi_2(z,\omega,\Omega^*) = X\, H_e(\omega,\Omega^*) \sin [\Omega^* z + \theta_o(\omega,\Omega^*)]$$

$$\phi_3(z,\omega,\Omega^*) = X\, H_o(\omega,\Omega^*) \cos [\Omega^* z + \theta_o(\omega,\Omega^*)]$$

(ii) We now render Ω^* varying in the set $\{\Omega_k\}$ and we obtain θ_e, θ_o, H_e, H_o as defined on the set $\mathbb{R}^+ x\{\Omega_k\}$.

3.5.2 On-line identification

Due to the necessity to operate inside the pass band of H(s,r), this identification is not really possible by using the sinusoidal test signal. Does the approach not fail? Not so much. Indeed we must not forget the context which is that of identifying the variations of the parameters of H(s,r). So, since we can identify (on-line) f(x,z), the comparison of y(t,z) with the theoretical output yields some inquiries on the variations of H(s,r).

3.6 Identification of H(s,r) o f(x,z)

We assume here it is not possible to directly measure the f output u(t,z) so that we have to determine f and H only via e and y.

(i) The identification is off-line

(ii) We stimulate H∘f by the signal (2.1).y(t,z) has the general form

$$y = \frac{R(y,z)}{2} H(o,o)X + \frac{M(Y,z)}{2} H_e(o,\Omega)X$$
$$\sin [\Omega z + \theta(o,\Omega)]$$
$$+ Q(Y,z)\, H(\omega,o)\, X \sin [\omega t + \theta(\omega,o)] +$$
$$+ N(Y,z)\, H_e(\omega,\Omega)\, X \sin (\Omega z + \theta_o) \sin (\omega t+\theta_e)$$
$$- N(Y,z)\, H_o(\omega,\Omega)\, X \cos (\Omega z + \theta_o) \cos (\omega t+\theta_e)$$

(iii) We measure the gain of the different components of this output, for different values of X, ω and Ω. We plot a set of functions which permit to identify H(s,r) and f(x,z).

4. SYSTEM DEFINED BY LAPLACE - GREEN'S KERNELS

4.1 System and list of symbols

The system considered is governed by the eqn.

$$y(s,z) = \int_U G(s,z,\xi)\, \mathcal{L}\{f(e,\xi)\}\, d\xi \qquad (4.1)$$

$$e(t,z) = c(t,z) - y(t,z) - d(t,z)$$

with zero boundary conditions, as shown in Fig. 4.

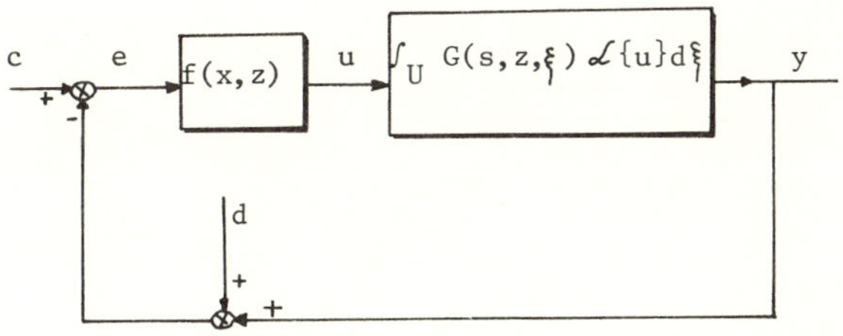

Fig. 4 System defined by Laplace-Green's kernel

We define

x(s,z)	Laplace's transform $\mathcal{L}\{x(t,z)\}$ of x(t,z)
$G(s,z,\xi)$	Laplace-Green's kernels of the system
U	domain of ξ and z variations
B(U)	boundary of U
d(s,z)	output equivalent to the initial conditions different from zero
$G(\omega,z,\xi)$	absolute value of $G(i\omega,z,\xi)$
$\theta(\omega,z,\xi)$	phase of $G(i\omega,z,\xi)$
a(z)x	lower bound of f(x,z)
b(z)x	upper bound of f(x,z)
$K(s,z,\xi)$	resolvent of every Fredholm's equation defined on U with the kernel

$$G(s,z,\xi)/\frac{a(\xi)+b(\xi)}{2}$$

4.2 Characterization of f(x,z)

4.2.1 Input with zero mean value

The characterization of Section 2 is available here. An other one which may be of interest in this precise context of Laplace-Green's kernels is the following.

(i) Let

$\varphi : U \to C,\ z \mapsto \varphi(z)$

a function with real or complex value such that

$$||\varphi(z)||^2 \triangleq \int_U \varphi(z)\, \overline{\varphi(z)}\, dz = 1$$

We define the operator

$$I_\varphi(f) = \int_U f[x\,\varphi(z),z]\, \overline{\varphi(z)}\, dz \qquad (4.2)$$

and we set

$g_\varphi \triangleq I_\varphi(f)$

We consider the reference input

$$\tilde{e}_2(t,z) = X\,\varphi(z) \sin \omega t \qquad (4.3)$$

(ii) Case $f \in F_1$, i.e.:

$F_1 = \{f | f(-x,z) = -f(x,z)\}$

$\tilde{u}(t,z) = N(X,\varphi)\, X\, \psi(z) \sin \omega t + \mathcal{O}(\sin 3\, \omega t)$ (4.4)

where N (X,φ) is given by eqn (2.6) in which we make the substitution

$y \leftarrow X\, \varphi(z),\ f \leftarrow g_\varphi$ (4.5)

(iii) Case $f \in F_2$, i.e.

$F_2 = \{f | f(-x,z) = f(x,z)\}$

$$\tilde{u}\,(t,z) = \frac{M(X,\varphi)}{2}\, X\, \varphi(z) + \mathcal{O}(\cos 2\, \omega t) \qquad (4.6)$$

where $M(X,\varphi)$ is given by eqn. (2.8) in which we make the substitution (4.5)

(iv) Case f arbitrary.

We set

$f_1(x,z) = [f(x,z) - f(-x,z)]/2$

$f_2(x,z) = [f(x,z) + f(-x,z)]/2$

$f(x,z) = f_1(x,z) + f_2(x,z)$

with $f_1 \in F_1$ and $f_2 \in F_2$.

4.2.2 Input with mean value different from zero

We consider the input

$\tilde{e}_3(t,z) = X_0\,(z) + X\,\varphi(z) \sin \omega t$ (4.7)

We define

$f_0(x,z) = [f(X_0 + x,z) - f(X_0 - x,z)]/2$

$f_e(x,z) = [f(X_0 + x,z) + f(X_0 - x,z)]/2$

One has

$\tilde{u}\,(t,z) = A(X_0,X,\varphi)\, X + B(X_0,X,\varphi)\, X\, \varphi(z) \sin \omega t$ (4.8)

where A(.) and B(.) are given by eqns (2.19) and (2.20) in which we make the substitution

$\phi_{oe} \leftarrow I_\varphi\,(f_0),\ \phi_{oe} \leftarrow I_\varphi\,(f_e),\ Y \leftarrow X$

4.3 Characterization of a hysteresis

For the sake of simplicity, we consider here only the zero mean value input (4.3). The reader shall extend easily to other sollicitations. The hysteresis is defined by

$$h(x,z) = \begin{cases} h_1(x,z), & \delta x < 0 \\ h_2(x,z), & \delta x > 0 \end{cases}$$

(i) Odd hysteresis $h \in H_1$

$H_1 = \{h | h_1(x,z) = -h_2(-x,z)\}$

$\tilde{u}\,(t,z) \simeq G(X,\varphi)\, X\, \varphi(z) \sin \omega t + H(X,\varphi)\, X\, \varphi(z) \cos \omega t$ (4.9)

with

$$G(X,\varphi) = \frac{4}{\pi X^2} \int_0^X x\, I_\varphi\, \frac{(h_1+h_2)}{2}\, \chi(X,x)\, dx \qquad (4.10)$$

$$H(X,\varphi) = \frac{4}{\pi X^2} \int_0^X I_\varphi\, \frac{(h_2-h_1)}{2}\, dx \qquad (4.11)$$

(ii) Even hysteresis $f \in H_2$

$H_2 = \{h | h_1(x,z) = h_2(-x,z)\}$

$\tilde{u}(t,z) \simeq \frac{J(X,\varphi)}{2}\, X\, \psi(z) + K(X,\varphi)\, X\, \varphi(z) \sin 2\, \omega t$ (4.12)

J is given by (4.10) with I_ψ instead of x I_φ and

$$K\,(x) = \frac{8}{\pi X^3} \int_0^X I_\varphi\, \frac{(h_2-h_1)}{2}\, dx \qquad (4.13)$$

(iii) Arbitrary hysteresis

f(x,z) is then decomposed in the sum

$h(x,z) = h_0(x,z) + h_e(x,z)$

with

$$h_0(x,z) = \begin{cases} [h_1(x,z) - h_2(-x,z)]/2, & \delta x<0 \\ [h_2(x,z) - h_1(-x,z)]/2, & \delta x>0 \end{cases}$$

$$h_e(x,z) = \begin{cases} [h_1(x,z) + h_2(-x,z)]/2, & \delta x<0 \\ [h_2(x,z) + h_1(-x,z)]/2, & \delta x>0 \end{cases}$$

h_0 and h_e belong respectively to H_1 and H_2 and thus $h(\tilde{e}_2,z)$ is known via (4.9) and (4.12)

These results may be established easily by writing the first term of Fourier's expansion of the response u(t,z) and making the suitable change of variable in the integral so obtained.

4.4 Characterization of the linear part

The characterization of this linear part poses the delicate problem of how to define it. What is more important: the knowledge of G(.) or of its application? Both viewpoints are of interest: the former to design the system, the latter to observe it. It is therefore clear that, if the observation is made to control the results of the synthesis, the two points of view are strongly linked.

4.4.1 Characterization by sinuoïdal response

We define

$G(\omega,z,\xi) \triangleq G(\omega,z,\xi) \exp\,[i\theta(\omega,z,\xi)]$

$= R(\omega,z,\xi) + i\, I(\omega,z,\xi)$

4.4.2 Characterization by series of functions

Let $\psi_i(z,\xi)$ denote a complete base of real orthonormal functions on U x U, i.e.

$$\int_U \int_U \psi_i(z,\xi)\, \psi_j(z,\xi)\, dz\, d\xi = \delta_{ij} \qquad (4.14)$$

where δ_{ij} denotes the Kronecker's symbol. We expand $G(s,z,\xi)$ in this base to obtain

$$G(s,z,\xi) = \sum_0^\infty G_j(s)\, \psi_j(z,\xi) \qquad (4.15)$$

with

$$G_j(s) = \int_U \int_U G(s,z,\xi)\, \psi_j(z,\xi)\, dz\, d\xi \qquad (4.16)$$

$G(s,z,\xi)$ is then characterized by the lumped transfer functions $G_j(s)$.

►►

4.5. Some theoretical results

4.5.1 Self-oscillations

Proposition 4. *Assume* $f(x,y) \in F_1$. *In the context of the describing function method we state.*

(i) The system governed by the eqn (4.1) has an oscillation as

$y(t,z) = X_0\, \varphi(z,\omega_0)\, X \sin \omega_0 t$

only if $\varphi(z,\omega)$ *is an eigenfunction of* $R(\omega,z,\xi)$ *which is orthogonal to* $I(\omega,z,\xi)$ *for* $\omega = \omega_0$. *Let* $1/\lambda(\omega)$ *denote the eigenvalue associated with* $\varphi(z,\omega)$. X_0 *and* ω_0 *comply with the conditions*

$$\varphi[B(U),\omega_0] = 0$$

$$N(X_0,\varphi) = -1/\lambda(\omega_0) \qquad (4.17)$$

with the additional condition that $\lambda(\omega_0)$ *be negative.*

(ii) For every admissible oscillating frequency ω_0, *a necessary condition in order that there exist an oscillation is that the intersection of the lines*

$C: \; x \mapsto h(x,\omega_0)$

$D: \; x \mapsto -x/\lambda(\omega_0)$

is nonvoid

(iii) For a given oscillating frequency ω_0, *the number of the associated oscillations is less or equal to the number of the intersections of C and D. When this number is odd, there is at least one oscillation. When it is even, either there is no oscillation or there are at least two ones.*

(iv) Let x_0 *denote the abcissa of the (C,D) intersection which is immediately less than* X_0. *The oscillation is stable when the derivative* $h_x(x,\omega_0)$ *is such that*

$$h_x(x_0,\omega_0) < -1/\lambda(\omega_0); \qquad (4.18)$$

it is unstable when

$$h_x(x_0,\omega_0) > -1/\lambda(\omega_0). \qquad (4.19)$$

4.5.2 Stability of the equilibrium position

Proposition 5. Assume that the following assumptions hold: (i) The state zero is an equilibrium position. (ii) There is only one possibly oscillating frequency ω_0. *Then the state zero is stable when*

$$h_x(0) < -1/\lambda(\omega_0) \qquad (4.20)$$

and is unstable when

$$h_x(0) > -1/\lambda(\omega_0) \qquad (4.21)$$

In the case where these inequalities are reduced to the equality (Ljapunov critical case) zero is stable when

$$x\, h_{xx}(0) < 0 \qquad (4.22)$$

it is unstable if

$$x\, h_{xx}(0) > 0 \qquad (4.23)$$

The proofs of these propositions are similar to those of Propositions 1 and 2, what is moreover emphasized by the similarity of their statements.

4.5.3 Input-output stability

Definition

(i) The stability concept is the same as that of section 3.3.3, with the norm

$$||x(t,z)||_p = [\textstyle\int_0^\infty \int_U x^p(t,z)\, dt\, dz]^{1/p}$$

(ii) The L_2-norm of a z and ξ dependent function is

$$||k(z,\xi)|| = [\textstyle\int_U \int_U k^2(z,\xi)\, dz\, d\xi]^{1/2}$$

(iii) The L_2-norm of a z-dependent function a (z) is

$$||a(z)|| = [\textstyle\int_U a^2(z)\, dz]^{1/2}$$

(iv) $K(s,z,\xi)$ denotes the resolvent of the Fredholm's eqn.

$$x(s,z) = y(s,z) + \int_U G(s,z,\xi)\, \frac{a(\xi) + b(\xi)}{2}\, x(s,\xi)\, d\xi.$$

In other words, the solution of this eqn. is

$$x(s,z) = y(s,z) + \int_U K(s,z,\xi)\, y(s,\xi)\, d\xi.$$

Assumption

(A4) $G(s,z,\xi)$ is stable for every $(z,\xi) \in \bar{U} \times \bar{U}$.

We state (Jumarie 1972 a)

Proposition 6. Assume that assumptions (A2) - (A4) hold. The system governed by eqn. (4.1) is input-output stable in the sense of section 3.3.3 provided that

$$\sup_{\omega \in R^+} ||K(i\omega,z,\xi)\, G(i\omega,z,\xi)|| . \left|\left| \frac{a(z) - b(z)}{2} \right|\right| < 1 \qquad (4.24)$$

◄◄

4.6 Identification of the non-linearity

The identification procedure given by the characterizations 4.2 and 4.3 is direct. For instance, in an off-line identification, we stimulate the non-linearity by the test signal (4.3). According to the formulae (4.4), (4.6), (4.9) and (4.12) it is easy to test the class of the non-linearity since these expressions are disjoint. For X varying we measure the suitable gain of the response, and the inversion (2.2.4) yields the non-linearity.

For instance, assume that $h \in H_1$, we refer to eqns. (4.9)-(4.11) and we have:

$$I_\varphi \frac{(h_1+h_2)}{2} = \frac{1}{2x} \int_0^x \frac{[X^2 G(X,\varphi) X}{\sqrt{x^2-X^2}} \, dX$$

$$I_\varphi \frac{(h_2-h_1)}{2} = \frac{\pi}{4} \frac{d}{dX} [X^2 H(X,)]$$

whereby the knowledge of I (h_1) and I (h_2). The procedure for on-line identification is similar to that of section 3.4.2 via the formulation of sections 4.2 and 4.3.

Note that here we identify I_φ (h_i) instead of h_i, what agrees with the remark stated in the begining of section 4.4 related to the characterization of G(.).

4.7 Identification of G(s,z,ξ)

We refer to section 4.4.2 and we consider off-line identification. We stimulate the linear part by the signal

$$\tilde{u} \, (t,\psi_j) = \psi_j(z,\xi) \sin \omega t \qquad (4.25)$$

according to eqns. (4.14) and (4.16) the outcome of the linear part is

$$y(t,\psi_j) = G_j(\omega) \sin[\omega t + \theta_j(\omega)] \qquad (4.26)$$

whereby the identification of $G_j(\omega)$ and $\theta_j(\omega)$.

The crucial difficulty of this approach is to generate a signal such that (4.25) Indeed, it may happen that the expression of $\psi_j(z,\xi)$ is not suitable for the realization of (4.25)
In this case, the following remark may be helpful: we come back to Proposition 4 which deals with self-oscillations. In the accuracy of the component $G_j(i\omega)$ we have

$$R_j(\omega,z,\xi) = G_j(\omega) \ \psi_j(z,\xi) \cos \theta_j(\omega)$$
$$I_j(\omega,z,\xi) = G_j(\omega) \ \psi_j(z,\xi) \sin \theta_j(\omega)$$

Then, a self-oscillation, if there are any is such that
(i) The space magnitude $\varphi(z)$, indépendent of ω, is an eigenfunction of $\psi_j(z,\xi)$ with the eigenvalue $1/G_j(\omega) \cos \theta_j(\omega)$.

(ii) The frequency ω_0 is solution of the eqn. $\sin \theta_j(\omega_0) = (2k+1)\pi$.

So, if we can modify the system (it is evident that this is possible only in off-line identification) we can hope the following situation: by changing the linear part of the system so that it becomes

$$H_0(s) \int_U G(s,z,\xi) \mathcal{N}\{u(t,\xi)\} \, d\xi \qquad (4.27)$$

where $H_0(s)$ is known, the system is self-oscillating. So, the inversion of the formulation of Proposition 4 yields the identification procedure.

5. DISCUSSION

The identification method above-explained is only a tentative and there remains to improve it in order that it be more efficient. Furthermore these ameliorations are strongly dependent on the development of the direct describing function method. In the sequel we shall briefly examine some suggestions for further research.
(i) The amplitude response via the first harmonic has the disadvantage of a long measuring time because of the determination of a high number of magnitude responses. A way for avoid this failure is the following, which is a direct extension of the method that we have proposed for lumped parameter systems (Jumarie 1970). It is based upon following property: it is possible to associate with a given function f(x,z) a Tchebitchef's expansion wich converges towards it, i.e.

$$f(x,z) = \sum_{m,n} f_{m,n} \, T_m(x) \, T_n(z) \qquad (5.1)$$

where $T_n(z)$ denotes the Tchebitchef's polynomials.

Furthermore, these coefficients $f_{m,n}$ are direct yielded by Fourrier's expansion of f(X sin Ωz sin ωt). So we measure the coefficients of this last series and we obtain the development (4.12).

(ii) An other disadvantage is the following: it may happen that the system cannot be energized by sinusoïdal oscillations. In this case one can use rectangular waves. Then several harmonics act simultaneously on the system, but in low pass systems the higher harmonics are more strongly damped than the fundamental wave.

(iii) We have to generate waves under the form $\psi(z,\xi) \sin \omega t$
and it may occur that, depending on U, it is not easy to realize $\psi(z,\xi)$. We have shown that in some instances one may operate via self-oscillation. An other way is to stimulate the system only in some points so that the signal appears as being sampled in space. The development of such a theory can be of interest.

(iv) The inversion of the describing function method involves the derivative of the gain w.r.t. the magnitude X. This requires that the sampling period in X is sufficiently small to have a good accuracy in the calculation of this derivative. Note that this disadvantage is mended by the method outlined in (ii).

(v) It its very important to avoid subharmonic phemenon because the corresponding magnitude are generally very large.

(vi) The accuracy of the identification above explained is that of the first harmonic. It is intuitive that different non-linearities can have the same first harmonic. Here again, it is possible to improve the preciseness by using the gains of the higher harmonics. This is a direct extension of the method which has been proposed for lumped parameter systems (Jumarie, 1970).

(vii) While the identification procedure of section 4 does not introduce the dimension of z and the form of U, the method of section 2 clearly mentions the dimension one of z and the range $[0,L]$. Its generalization to a multidimensional parameter distributed in an hyper range $[0,L_1] \times [0,L_2] \times \ldots [0,L_n]$ is direct. For instance for a two-dimensional parameter (z_1,z_2), $f(z,z_1,z_2)$ will be stimulated by the test signal

$$\tilde{e} = X \sin {}^1\Omega\, z_1 \sin {}^2\Omega\, z_2 \sin \omega t$$
$${}^1\Omega \in {}^1\{\Omega\} = \{{}^1\Omega_k | \Omega = k\pi/L_1\}$$
$${}^2\Omega \in {}^2\{\Omega\} = \{{}^2\Omega_k | \Omega = k\pi/L_2\}$$

and the formulation of section 2 will be available with some evident modifications.

6. CONCLUSION

We have proposed an approach to identification of non-linear distributed system via the inversion of a direct method. So doing we think that one can identify suitably only if we have at hand some inquiries regarding the system. To advance in this investigation we need some theoretical results on the controllability and the observability of distributed systems. This problem is presently in infancy but we think that a fruitful area of research is the following: by using a z-discretization, the system is approximated by a multivariable linear system. We develops the theory for this latter and by letting the discretization step tends to zero, we come back to the initial system. We intend to work in this sense.

ANNEXE

Proof of the inversion (2.2)

To begin with, we bear in mind the definition of the arbitrary order derivative of a given function y(x)

(i) When α is negative one set

$$\frac{d^\alpha y(x)}{dx^\alpha} = \frac{1}{\Gamma(-\alpha)} \int_0^x (x-t)^{-\alpha-1} y(t)\, dt \qquad (A.1)$$

(ii) When α is positive different fron an inter, one introduces n = the integer immediately greater than α, and one defines

$$\frac{d^\alpha y(x)}{dx^\alpha} = \frac{d^n}{dx^n} \frac{d^{-n+\alpha}}{dx^{-n+\alpha}} y(x) \qquad (A.2)$$

We note that the change of variable $t = u^2$ in (A.1) yields

$$\frac{d^\alpha y(x)}{dx^\alpha} = \frac{2}{\Gamma(-\alpha)} \int_0^{\sqrt{x}} (x-u^2)^{-\alpha-1} y(u^2)\, u\, du \quad (A.3)$$

This being so, to solve eqn. (2.23) we set $X = \sqrt{W}$ to obtain

$$kh(\sqrt{W}) = \frac{2}{\sqrt{\pi}} \int_0^{\sqrt{W}} \frac{x\, f(x)}{\sqrt{W-x^2}}\, dz$$
$$= \frac{d^{-1/2} f(\sqrt{W})}{d\, W^{1/2}}$$

The inversion yields

$$f(\sqrt{W}) = k \frac{d^{1/2} h(\sqrt{W})}{d\, W^{1/2}}$$
$$= k \frac{d}{dW} \frac{d^{-1/2} h(\sqrt{W})}{d\, W^{-1/2}}$$

We now come back to the definition of the -1/2 - order derivative to obtain

$$f(x) = \frac{2k}{\sqrt{\pi}} \frac{d}{dx^2} \int_0^x \frac{X\, g(X)}{\sqrt{x^2-X^2}}\, dX$$

An the relation

$$\frac{d}{dx^2} = \frac{1}{2x} \frac{d}{dx}$$

yields eqn. (2.24).

REFERENCES

JUMARIE, G. (1970) "On a class of closed loop non linear systems identifications". 2nd IFAC Symposium on "Identification an Process parameter estimation". Paper No. 8.3.

(1972 a) "Deux théorèmes de stabilité globale entrée-sortie pour systèmes non-linéaires à paramètre réparti". C.R. Acad. Sc., t. 275, Série A, pp. 655.

(1972 b) "Une méthode du premier harmonique pour systèmes non-linéaires à paramètre réparti soumis à des sollicitations déterministes ou stochastiques". Intern. Jour. Electronics. To appear Dec. 72 or Jan. 73.

KEY WORDS

Non-linear distributed systems, describing function method, inversion, input-output stability, singular integral equation.

TD-4

COMBINED STATE AND PARAMETER ESTIMATION FOR DISTRIBUTED-PARAMETER SYSTEMS USING DISCRETE OBSERVATIONS

Howard Sherry and David W. C. Shen
The Moore School of Electrical Engineering
University of Pennsylvania
Philadelphia, Pennsylvania
U.S.A.

This paper presents an algorithm for the on-line estimation of the state and parameters of a distributed-parameter system in which measurements are made only at a finite number of spatial locations, as well as at discrete times. Using a weighted least square approach, a sequential algorithm for nonlinear filtering of a continuous-discrete distributed-parameter system has been developed which is directly applicable to the combined state and parameter estimation. The continuous-discrete filter circumvents the difficulty of solving a nonlinear Riccati-like operator equation for continuous time measurements.

1. INTRODUCTION

Extensive work has been done in state and parameter estimation for both linear and nonlinear lumped systems. Similar work for distributed-parameter system has just begun. Therefore it is not surprising to find that most investigators have directed their efforts to the more theoretical aspects of the problem. Very few practical estimation algorithms appear in the literature [1].

Many of the initial efforts extend the Kalman filter to a partial differential equation form consistent with the distributed-parameter system whose state is to be estimated [2] [3] [4]. Others use limiting operations to obtain state estimators from similar results derived for lumped systems [5] [6] [7]. Except for the simplest of problems, very few of the existing estimators can be effectively implemented on-line.

This paper presents on-line algorithms for the sequential estimation of both the state and the parameters of a class of linear and non-linear continuous distributed parameter systems under the conditions that discrete time measurements are made at only a finite number of fixed points in space. The major purpose of this work is to develop an estimation procedure to circumvent many of the computational difficulties inherent in the on-line implementation of the existing methods.

2. COMBINED STATE AND PARAMETER ESTIMATION

Consider a system that is described by the linear partial differential equation

$$\frac{\partial \underline{U}_1(\underline{x},t)}{\partial t} = \mathcal{L}_{\underline{x}}\, \underline{U}_1(\underline{x},t) + \underline{W}_1(x,t) \qquad (\underline{x}\in D) \tag{1}$$

and boundary conditions

$$B_{\underline{x}}\, \underline{U}_1(\underline{x},t) = 0 \qquad (\underline{x}\in\partial D) \tag{2}$$

where $\underline{U}_1$ is the state vector, $\underline{W}$ is a zero mean input disturbance. $\mathcal{L}_{\underline{x}}$ and $\underline{B}_{\underline{x}}$ are linear spatial partial differential operator matrices whose parameters can be functions of both space and time. Assume that one or more parameters of $\mathcal{L}_{\underline{x}}$ are unknown. These parameters are either constants or their Markovian state vector description must be given.

Assume noisy measurements are made at "M" distinct locations in the spatial domain at discrete times. The observation equation is

$$\underline{Y}(\underline{x}_i,t_n) = H_1(\underline{x}_i,t_n)\underline{U}_1(x_i,t_n) + \underline{V}_1(\underline{x}_i,t_n) \tag{3}$$

defined for $i=1,2,\ldots,M$: $x_i \in D$ and $t_o<t_n<t_f$. In equation (3), H_1 is a measurement matrix and $\underline{V}_1$ is a zero-mean measurement noise vector.

The problem is not only the estimation of the state vector $\underline{U}(\underline{x},t)$, but also the identification of the unknown parameters of $\underline{\mathcal{L}}_{\underline{x}}$. In order to solve this combined problem, we extend the method due to Kalman [8] in connection with lumped parameter systems. This procedure has the advantage that it operates in real time. In this method, we adjoin the unknown parameter vector to the original state vector to form a new state vector and at the same time modify the dynamics of the system to account for this new component. For the case when a constant parameter of $\underline{\mathcal{L}}_x$ is unknown, the augmented state and measurement equations are given by

$$\frac{\partial \underline{U}_2(\underline{x},t)}{\partial t} = \begin{bmatrix} \underline{\mathcal{L}}_x \; \underline{U}_1(\underline{x},t) \\ 0 \end{bmatrix} + \begin{bmatrix} \underline{W}_1(\underline{x},t) \\ 0 \end{bmatrix} \tag{4}$$

$$\underline{y}(\underline{x}_i,t_n) = [H_1\;(x_i,t_n)\;0]\;\underline{U}_2\;(\underline{x}_i,t_n) + \underline{V}_1\;(x_i,t_n) \tag{5}$$

where

$$\underline{U}_2(\underline{x},t) = \begin{bmatrix} \underline{U}_1(x,t) \\ \rho \end{bmatrix} \tag{6}$$

and ρ is the unknown parameter.

It is readily seen that the augmented state equation is no longer linear since there are products of the components of $\underline{U}_2$ in $\underline{\mathcal{L}}_x\,\underline{U}_1$. Thus, combined parameter and state estimation in a linear system is essentially a nonlinear state estimation problem.

3. NONLINEAR ESTIMATION USING DISCRETE MEASUREMENTS.

In view of the above discussion, we develop a sequential filter for estimating the state of a general nonlinear distributed-parameter process. This filter is directly applicable to the problem of combined state and parameter estimation. Consider a nonlinear stochastic distributed process described by the equation

$$\frac{\partial \underline{U}(\underline{x},t)}{\partial t} = \underline{F}_x(\underline{U},\underline{x},t) + \underline{W}(\underline{x},t) \qquad (x \epsilon D) \tag{7}$$

defined for $t \geq t_o$ on a spatial domain D; where D is a connected (not necessarily bounded) subset of E^{n_1}, an n_1-dimensional Euclidean space;

$\underline{U}(\underline{x},t)$ is the state vector of the system;

$\underline{W}(\underline{x},t)$ is a zero mean unknown input disturbance;

$\underline{F}_x$ is a nonlinear spatial differential operator whose coefficients may be functions of both time and space.

The spatial boundary conditions are assumed to be homogeneous and given by the equation

$$\underline{B}_x(\underline{U},x,t) = 0 \qquad (\underline{x} \in \partial D) \tag{8}$$

defined for $t > t_o$ on ∂D, the boundary surface of D where

$\underline{B}_x$ is a spatial differential operator which is similar in form to $\underline{F}_x$.

The problem is to estimate the current state of the system, $\underline{U}(\underline{x},t_f)$, given a set of observations made at discrete times, t_j, at M distinct locations in the spatial domain. The observation equation is given by:

$$\underline{Y}(\underline{x}_i,t_j) = \underline{H}(\underline{x}_i,t_j)\underline{U}(\underline{x}_i,t_j) + \underline{V}(\underline{x}_i,t_j) \tag{9}$$

defined for i=1,2,...,M $x_i \epsilon D$ and $t_o \leq t_j \leq t$

where

$\underline{Y}(\underline{x}_i,t_j)$ is the observation made at the ith sensor located at $\underline{x}_i$ at time t_j;

$\underline{V}(\underline{x}_i,t_j)$ is a zero mean disturbance associated with this measurement;

$\underline{H}(x_i,t_j)$ is a matrix that relates the state, $\underline{U}(\underline{x}_i,t_j)$, to the observation $\underline{Y}(\underline{x}_i,t_j)$.

The criterion used to obtain the estimate of the state is chosen to be weighted least square error. If knowledge of the statistics of the random disturbance $\underline{W}$ and $\underline{V}$ is available, this knowledge can be used to advantage in choosing the weighting functions.

To formulate the estimation problem from the least squares point of view, we define the following error criterion.

$$J = \tfrac{1}{2}\int_D [\underline{U}(\underline{x},t_o) - \underline{U}_a(x,t_o)]^T P^{-1}_{o_{\underline{x},\underline{s}}} [\underline{U}(s,t_o) - \underline{U}_a(s,t_o)]d\underline{x}$$
$$+ \tfrac{1}{2}\int_{t_o}^{t_f}\int_D W^T(\underline{x},t)\; Q^{-1}_{\underline{x},\underline{s}}\; W(\underline{s},t)\; d\underline{x}\; dt$$
$$+ \tfrac{1}{2}\sum_{n=1}^{N}\sum_{i=1}^{M} \underline{V}^T(\underline{x}_i,t_n) R^{-1}_{\underline{x}_i,\underline{x}_j}\; \underline{V}(\underline{x}_j,t_n) \tag{10}$$

where $\underline{U}_a(\underline{x},t_o)$ is the apriori estimate of the state at $t = t_o$, $P_{o_{\underline{x},\underline{s}}}$, $Q_{\underline{x},\underline{s}}$ and $R_{\underline{x}_i,\underline{x}_j}$ are linear operators defined by

$$P_{o_{\underline{x},\underline{s}}}\;(\cdot) \triangleq \int_D P_o(\underline{x},\underline{s})\;(\cdot)\;d\underline{s} \tag{11}$$

$$Q_{\underline{x},\underline{s}}(\cdot) \triangleq \int_D Q(\underline{x},\underline{s},t)\;(\cdot)\;d\underline{s} \tag{12}$$

and

$$R^{-1}_{\underline{x}_i,\underline{x}_j}\;(\cdot) \triangleq \sum_{j=1}^{M}[R^{-1}(\underline{x}_i,\underline{x}_j,t_n)\;(\cdot)] \tag{13}$$

$P_o(\underline{x},\underline{s})$, $Q(\underline{x},\underline{s},t)$ and $R^{-1}(\underline{x}_i,\underline{x}_j,t_n)$ are positive definite matrices defined respectively, for all $(x,s) \in DxD$, $(x,s,t) \in DxDx(t_o,t_f)$ and $(x_i,x_j,t_n) \in DxDx\;(t_o,t_f)$.

Notice that equation (10) contains both discrete and continuous functions. Using Dirac delta functions, we will convert this equation to consist only of continuous functions. Let us define

$\delta_1(t-t_n)$ by $\sum_{n=1}^{N} p(t_n) = \int_{t_o}^{t_f} p(t)\ \delta_1(t-t_n)dt$ where $t_o \le t_n \le t_f$ and t_n takes on the N discrete values $t_1, t_2, \ldots, t_N$.

Therefore we can express J as a continuous functional of the form

$$J = G(\underline{U}_o) + \tfrac{1}{2}\int_{t_o}^{t_f}\int_D \underline{W}^T(\underline{x},t)Q^{-1}_{\underline{x},\underline{s}}\ \underline{W}(\underline{s},t)d\underline{x}\ dt + \tfrac{1}{2}\int_{t_o}^{t_f} \underline{V}^T(\underline{x},t)\ \bar{R}^{-1}_{\underline{x},\underline{s}}\ \underline{V}(\underline{s},t)d\underline{x}\ dt \quad (14)$$

$$= G(\underline{U}_o) + \tfrac{1}{2}\langle \underline{W}(\underline{x},t),\ \underline{W}(\underline{s},t)\rangle_{Q^{-1}_{\underline{x},\underline{s}}} + \tfrac{1}{2}\langle \underline{V}(\underline{x},t),\ \underline{V}(\underline{s},t)\rangle_{\bar{R}^{-1}_{\underline{x},\underline{s}}} \quad (15)$$

where

$$G(\underline{U}_o) = \tfrac{1}{2}\int_D [\underline{U}(\underline{x},t_o) - \underline{U}_a(\underline{x},t_o)]^T P^{-1}_{o_{\underline{x},\underline{s}}}\ [\underline{U}(\underline{s},t_o) - \underline{U}_a(\underline{s},t_o)]d\underline{x} \quad (16)$$

$$R^{-1}_{\underline{x},\underline{s}}\ (\cdot) \triangleq \int_D R^{-1}(\underline{x},\underline{s},t)\ (\cdot)ds \quad (17)$$

and

$$\bar{R}^{-1}_{\underline{x},\underline{s}}\ (\cdot) \triangleq R^{-1}_{\underline{x},\underline{s}}\ \delta_1(\underline{x}-\underline{x}_i)\ \delta_1(\underline{s}-x_j)\ \delta_1(t-t_n)\ (\cdot) \quad (18)$$

In order to obtain the least square estimate of the state, we must minimize J with respect to $\underline{U}(\underline{x},t)$, $\underline{W}(\underline{x},t)$ and $\underline{U}_o$ subject to the initial conditions, the boundary conditions and the state equation. Using the Lagrange multipler, $\underline{\lambda}(\underline{x},t)$, to introduce the process as a system constraint, we form the modified performance criterion

$$\bar{J} = J + \langle \underline{\lambda}(\underline{x},t),\ \underline{F}_x(\underline{U},\underline{x},t) + \underline{W}(\underline{x},t)\rangle - \langle \underline{\lambda}(\underline{x},t),\ \frac{\partial \underline{U}(\underline{x},t)}{\partial t}\rangle \quad (19)$$

4. OPTIMAL SMOOTHING VIA THE CANONICAL EQUATIONS

The problem can now be interpreted as a deterministic optimization problem and variational calculus can be applied to solve for the optimal estimator.

Define the Hamiltonian as

$$H[\underline{W},\underline{U},\underline{\lambda},\underline{x},\underline{s},t) = \tfrac{1}{2}\langle \underline{W}(\underline{x},t),\underline{W}(\underline{s},t)\rangle_{Q^{-1}_{\underline{x},\underline{x}}} + \tfrac{1}{2}\langle \underline{Y}(\underline{x},t) - \underline{H}(\underline{x},t)\underline{U}(\underline{x},t),\underline{Y}(\underline{s},t) - \underline{H}(\underline{s},t)\underline{U}(\underline{s},t)\rangle_{\bar{R}^{-1}_{\underline{x},\underline{s}}} + \langle \underline{\lambda}(\underline{x},t),\ \underline{F}_{\underline{x}}(\underline{U},\underline{x},t) + W(\underline{x},t)\rangle \quad (20)$$

A necessary condition for a minimum of $\bar{J}$ is $\delta\bar{J}$, the first variation of $\bar{J}$, must equal zero, i.e.,

$$\delta\bar{J} = \int_{t_o}^{t_f}[\frac{\delta H}{\delta \underline{U}}^T \delta\underline{U} + \frac{\delta H}{\delta \underline{W}}^T + \frac{\delta H}{\delta \underline{\lambda}}\ \delta\underline{\lambda} - \frac{\partial \underline{U}}{\partial t}^T\ \delta\underline{\lambda} - \underline{\lambda}^T\ \delta(\frac{\partial \underline{U}}{\partial t})]dt + \frac{\delta G}{\delta \underline{U}_o}\ \delta\underline{U}_o = 0 \quad (21)$$

(For simplicity we drop the arguments of $\underline{W}$, $\underline{U}$, $\underline{\lambda}$, H, and G).

Integrating the fifth term in equation (21) by parts and rearranging terms, we obtain the following relationships which constitute the necessary conditions for a minimum of $\bar{J}$.

$$\frac{\delta H}{\delta \underline{U}} = -\frac{\partial \underline{\lambda}}{\partial t} \quad (22) \qquad \frac{\delta H}{\delta \underline{\lambda}} = \frac{\partial \underline{U}}{\partial t} \quad (23)$$

$$\frac{\delta H}{\delta \underline{W}} = 0 \quad (24) \qquad \underline{\lambda}(\underline{x},\ t_f) = 0 \quad (26)$$

$$\underline{\lambda}(\underline{x},t_o) + \frac{\delta G}{\delta \underline{U}_o} = 0 \quad (25)$$

Evaluation of the functional derivatives in equations (22) to (25) together with equation (26) lead to a two point boundary value problem for the optimal least square estimate, $\underline{U}^*(\underline{x},t)$, in which the initial state and initial costate are functionally related and the final costate is specified. The Canonical equations are

$$\frac{\partial \underline{U}^*(\underline{x},t)}{\partial t} = \underline{F}_x(\underline{U}^*,\underline{x},t) - \int_D Q(\underline{x},\underline{s},t)\underline{\lambda}(\underline{s},t)d\underline{s} \quad (27)$$

$$\partial\underline{\lambda}(\underline{x},t) = \underline{H}^T(\underline{x},t)\bar{R}^{-1}_{\underline{x},\underline{s}}[\underline{Y}(\underline{s},t) - \underline{H}(\underline{s},t)U^*(\underline{s},t)] - \frac{\partial \underline{F}_x(\underline{U}^*,\underline{x},t)}{\partial \underline{U}^*}\ \underline{\lambda}(\underline{x},t) \quad (28)$$

and the transversality conditions

$$\underline{U}^*(\underline{x},t_o) = -\int_D P_o(\underline{x},\underline{s})\underline{\lambda}(\underline{s},t)ds + \underline{U}_a(\underline{x},t) \quad (29)$$

$$\underline{U}^*(\underline{x},\underline{t}_f) = 0 \quad (30)$$

The solution of this two point boundary value problem that satisfies the spatial boundary conditions is called the optimal smoothing solution of the state for $t \in (t_o,t_f)$, since it makes use of all the measured data over (t_o,t_f). Furthermore, $\underline{U}^*(\underline{x},t_f)$ is also the filtering solution of the problem since it employs only data up to time t_f. This solution, however, cannot be used online. We will therefore utilize the sweep method [9] to convert the two point boundary value to the related initial value problem which can be solved recursively.

5. SEQUENTIAL LEAST SQUARE FILTER

Employing the sweep method, we assume a solution for the filtered estimate, $\underline{\hat{U}}(\underline{x},t)$ of the form

$$\underline{U}^*(\underline{x},t) = -\int_D P(\underline{x},\underline{s},t)\,\underline{\lambda}(\underline{s},t)d\underline{s} + \underline{\hat{U}}(\underline{x},t) \tag{31}$$

In addition, we assume that the nonlinearities of the system are such that if the initial estimate is sufficiently close to the true state, the dynamics of the difference between the state vector and its filtered estimate can be characterized by a set of linear equations.

The differentiation of equation (24) with respect to time yields

$$\frac{\partial \underline{U}^*(\underline{x},t)}{\partial t} = -\int_D \frac{\partial \underline{P}(\underline{x},\underline{s},t)}{\partial t}\,\underline{\lambda}(\underline{s},t)d\underline{s} + \frac{\partial \underline{\hat{U}}(\underline{x},t)}{\partial t} - \int_D \underline{P}(\underline{x},\underline{s},t)\,\frac{\partial \underline{\lambda}(\underline{s},t)}{\partial t}\,d\underline{s} \tag{32}$$

After expanding $\underline{F}_{\underline{x}}(\underline{U}^*,\underline{x},t)$ in a Taylor series about $\underline{\hat{U}}(\underline{x},t)$ and dropping the nonlinear terms, we use the resulting equations to eliminate $\underline{U}^*(\underline{x},t)$ from the canonical equations. The new canonical equations now contain $\underline{\hat{U}}(\underline{x},t)$ and $P(\underline{x},\underline{s},t)$ instead of $\underline{U}^*$. After substituting into equation (32) and rearranging terms, we obtain an equation of the following form:

$$\underline{A}_o + A_1\,\underline{\lambda}(\underline{s},t) = 0 \tag{33}$$

The above equation is satisfied for all $\underline{\lambda}$ by setting $\underline{A}_o$ and $\underline{A}_1$ each equal to zero. Therefore, after using equation (18) to substitute $\bar{R}^{-1}(\underline{x},\underline{s},t)$ and evaluating the integrals in $\underline{A}_o$ and $\underline{A}_1$; we obtain the following two relationships.

$$\frac{\partial \underline{\hat{U}}(\underline{x},t)}{\partial t} = \underline{F}_{\underline{x}}(\hat{U},\underline{x},t) + \sum_{i=1}^{M}\sum_{j=1}^{M} P(\underline{x},\underline{x}_i,t)\underline{H}^T(x_i,t)\cdot R^{-1}(\underline{x}_i,\underline{x}_j,t)\,\delta_1(t-t_n)[\underline{Y}(x_j) - \underline{H}(\underline{x}_j,t)\underline{\hat{U}}(\underline{x}_j,t)] \tag{34}$$

$$\frac{\partial \underline{P}(\underline{x},\underline{s},t)}{\partial t} = (\underline{F}_{\underline{x}})\Big|_{\underline{\hat{U}}(\underline{x},t)} P(\underline{x},\underline{s},t) + P(\underline{x},\underline{s},t)(\underline{F}_s)^T\Big|_{\underline{\hat{U}}(\underline{s},t)} + Q(\underline{x},\underline{s},t) - \sum_{i=1}^{M}\sum_{j=1}^{M} P(\underline{x},\underline{x}_i,t)\underline{H}^T(\underline{x}_i,t)R(\underline{x}_i,\underline{x}_j,t)\,\delta_1(t-t_n)\,\underline{H}(\underline{x}_j,t)\;P(\underline{x}_j,\underline{s},t) \tag{35}$$

which together with the initial conditions

$$\underline{\hat{U}}(\underline{x},t_o) = \underline{U}_o(x,t_o) \tag{36}$$

$$P(\underline{x},\underline{s},t_o) = P(\underline{x},\underline{s}) \tag{37}$$

and the boundary conditions

$$B_{\underline{x}}(\underline{\hat{U}},\underline{x},t) = 0 \tag{38}$$

$$B_{\underline{x}}(\underline{P},\underline{x},\underline{s},t) = 0 \tag{39}$$

$$B_{\underline{s}}(\underline{P},\underline{x},\underline{s},t) = 0 \tag{40}$$

specify an initial value problem for the least square filter. The delta functions in equations (34) and (35) indicate a discontinuity in the filter at the times when measurements are taken. Therefore, at observation time, there are two values of the state of interest at each point in the spatial domain.

Let $\underline{\hat{U}}_-(\underline{x},t_n)$ and $P_-(\underline{x},\underline{s},t_n)$ designate respectively $\underline{\hat{U}}(\underline{x},t)$ and $P(\underline{x},\underline{s},t)$ just prior to the observation made at $t=t_n$. Similarly, let $\underline{\hat{U}}_+(\underline{x},t_n)$ and $P_+(\underline{x},\underline{s},t_n)$ denote $\underline{\hat{U}}(\underline{x},t)$ and $P(\underline{x},\underline{s},t)$ immediately after measurements are taken at $t=t_n$.

We now proceed to develop a sequential scheme for updating the estimated state, $\underline{\hat{U}}(\underline{x},t)$, as time proceeds and new observations become available. Begin by considering the filtering process for $t\neq t_n$. It is clear from equations (34) and (35) that as a result of the Dirac delta function, $\delta,(t-t_n)$, the filter is governed by the following equations between sampling times.

$$\frac{\partial \underline{\hat{U}}(\underline{x},t)}{\partial t} = \underline{F}_{\underline{x}}(\underline{\hat{U}},x,t) \qquad (\underline{x}\epsilon D) \tag{41}$$

$$\frac{\partial \underline{P}(\underline{x},\underline{s},t)}{\partial t} = \underline{F}_{\underline{x}}\Big|_{\underline{\hat{U}}} P(\underline{x},\underline{s},t) + P(\underline{x},\underline{s},t)(\underline{F}_{\underline{s}})^T\Big|_{\underline{\hat{U}}} + \underline{Q}(\underline{x},\underline{s},t) \qquad (\underline{x},\underline{s}\epsilon D) \tag{42}$$

$$B_{\underline{x}}(\underline{\hat{U}},\underline{x},t) = 0 \tag{43}$$

$$B_{\underline{x}}(\underline{P},\underline{x},\underline{s},t) = B_{\underline{s}}(\underline{P},\underline{x},\underline{s},t) = 0 \tag{44}$$

Assuming $P^{-1}(\underline{x},\underline{s},t)$ exists, we obtain the following relationship between $\dfrac{\partial P^{-1}(\underline{x},\underline{s},t)}{\partial t}$ and $\dfrac{\partial P(\underline{x},\underline{s},t)}{\partial t}$

$$\frac{\partial P^{-1}(\underline{x},\underline{s},t)}{\partial t} = P^{-1}(\underline{x},\underline{s},t)\,\frac{\partial P(\underline{x},\underline{s},t)}{\partial t}\,P^{-1}(\underline{x},\underline{s},t) \qquad (45)$$

which after combining with equation (28) yields

$$\frac{\partial P^{-1}(\underline{x},\underline{s},t)}{\partial t} = -P^{-1}(\underline{x},\underline{s},t)\,\underline{F}_x\Big|_{\underline{\hat{U}}} P(\underline{x},\underline{s},t) - (\underline{F}_s)^T\Big|_{\underline{\hat{U}}} P^{-1}(\underline{x},\underline{s},t) - P^{-1}(\underline{x},\underline{s},t)Q(\underline{x},\underline{s},t)P^{-1}(\underline{x},\underline{s},t)$$
$$+ [P^{-1}(\underline{x},\underline{s},t)\sum_{i=1}^{M}\sum_{j=1}^{M} P(\underline{x},\underline{x}_i,t)\underline{H}^T(\underline{x}_i,t)$$
$$R^{-1}(\underline{x}_i,\underline{x}_j,t)\underline{H}(\underline{x}_j,t)P(\underline{x}_j,\underline{s},t)P^{-1}(\underline{x},\underline{s},t)\,\delta_1(t-t_n)$$
$$(\underline{x},\underline{s}\epsilon D) \qquad (46)$$

Since $P^{-1}(\underline{x},\underline{s},t)$ is the inverse of $P(\underline{x},\underline{s},t)$, solving for $P^{-1}(\underline{x},\underline{s},t)$ is equivalent to solving for $\underline{P}(\underline{x},\underline{s},t)$.

Employing a limiting argument to equations (34) (46), we obtain the relationships for the filter when observations are taken, that is, at the measurement time, t_n, $\underline{\hat{U}}_-(\underline{x},t_n)$ and $P_-(\underline{x},\underline{s},t_n)$ are calculated from equations (41) and (42) with $\hat{U}_+(x,t_n-T)$ and $P_+(\underline{x},\underline{s},t_n-T)$ serving as initial conditions. (T represents the time between successive measurements). $\hat{U}_+(\underline{x},t_n)$ and $P_+(\underline{x},\underline{s},t_n)$ are then given by

$$\underline{\hat{U}}_+(\underline{x},t_n) = \underline{\hat{U}}_-(\underline{x},t_n) + \sum_{i=1}^{M}\sum_{j=1}^{M} P_+(\underline{x},\underline{x}_i,t_n)\underline{H}^T(\underline{x}_i,t_n)$$
$$R^{-1}(\underline{x}_i,\underline{x}_j,t_n)[\underline{Y}(\underline{x}_j,t_n) - \underline{H}(\underline{x}_j,t_n)\underline{\hat{U}}_-(x_j,t_n)] \qquad (47)$$

and

$$P_+(\underline{x},\underline{s},t_n) = [P_-^{-1}(\underline{x},\underline{s},t_n) + P_-^{-1}(\underline{x},\underline{s},t_n)\sum_{i=1}^{M}\sum_{j=1}^{M}$$
$$P_-(\underline{x},\underline{x}_i,t_n)\underline{H}^T(\underline{x}_i,t_n)R^{-1}(\underline{x}_i,\underline{x}_j,t_n)\underline{H}(\underline{x}_j,t_n)$$
$$P_-(\underline{x}_j,\underline{s},t_n)P_-^{-1}(\underline{x},\underline{s},t_n)]^{-1} \qquad (48)$$

6. A SIMPLE EXAMPLE OF COMBINED STATE AND PARAMETER ESTIMATION

Consider a situation similar to a one dimensional heat flow problem in which the system equations are:

$$\frac{\partial \underline{U}_1(\underline{x},t)}{\partial t} = \alpha\frac{\partial^2 \underline{U}_1(\underline{x},t)}{\partial x^2} + \underline{W}(\underline{x},t) \qquad 0 < x < 1$$

$$\underline{U}_1(o,t) = \frac{\partial U_1(1,t)}{\partial x} = 0$$

$$Y(x_i,t_n) = U_1(x_i,t_n) + V(x_i,t_n)$$

where $W(\underline{x},t)$ and $V(x_i,t_n)$ are zero mean independent Gaussian disturbances which are white in both x and t. The covariance of $W(x,t)$ and $V(x_i,t_n)$ are assumed to be given by

$$Q(x,s,t) = .5\ \delta(x-s) \text{ and } R(x_i,x_y,t_n) = 1\delta_{ij}$$

respectively. The diffusion constant α is also assumed to be unknown. Therefore, define α as an addition state to be estimated, we obtain the modified system equations

$$\frac{\partial}{\partial t}\begin{bmatrix} U_1(x,t) \\ \alpha \end{bmatrix} = \begin{bmatrix} \alpha\dfrac{\partial^2 U_1(x,t)}{\partial x^2} \\ 0 \end{bmatrix} + \begin{bmatrix} W(x,t) \\ 0 \end{bmatrix}$$

$$Y(x_i,t_n) = [1\ 0]\begin{bmatrix} U_1(x_i)t_n \\ \alpha \end{bmatrix} + V(x_i,t_n)$$

The process is simulated using a uniform distribution of 5.0 for $U_1(x,o)$ and a diffusion constant equal to .5. The initial values of $\hat{U}(x,t)$ and $P(x,s,t)$ are assumed to be given by

$$\underline{\hat{U}}(x,o) = \begin{bmatrix} 10.0 \\ .75 \end{bmatrix} \qquad 0 \le x \le 1$$

and

$$\underline{P}(x,s,o) = \begin{bmatrix} 10.0 & 5 \\ 5 & 3.0 \end{bmatrix} \qquad o \le x,s \le 1$$

Measurements are taken at the following five locations: x=.15,.35,.55,.75,.95, and the observation interval is 0.03 seconds. The results of the simulation are shown in Figures 1-4. Figure 1 shows $U_1(.5,t)$ and $\hat{U}_1(.5,t)$ over the time interval $0 \le t \le .5$ for a typical run. The filtered estimate of α, $\hat{\alpha}$, is a function over the spatial domain. Since α is known to be a constant, $\hat{\alpha}$ is averaged over the spatial domain after each observation. Figure 2 shows the spatial average of α as a function of time. Finally Figures 3 and 4 show the error weighting matrices $P_{11}+(.5,.5,t)$ and $P_{12}+(.5,.5,t)$ respectively.

7. CONCLUDING REMARKS

This paper has developed an algorithm for the on-line estimation of the state and parameters of a class of nonlinear distributed parameter processes in which measurements are taken at a finite number of spatial locations as well as at discrete times. In most industrial processes, a number of transducers are placed at selected points to make all the measurements. Since each transducer covers only a small fraction of the total spatial domain, we can consider the transducers as making measurements of discrete points. A variety of problems are encountered in locating sensors. They are dependent on the type of systems considered as well as the method of control used. The problems of optimal sensor and controller locations is a major one but is beyond the scope of this paper. An important problem for futher research is that where measurements cannot be obtained in the interior of spatial domain D but only as a number of discrete points in ∂D.

In many practical problems because of limitations in computing speed, computer storage and/or the response time of the associated measuring equipment, measurements are taken at discrete times. This situation is particularly true in distributed parameter systems where the continuous-time least square filtering algorithms involve nonlinear partial integro-differential Riccati type equations. In the linear case, it is possible to solve these equations off-line. In the nonlinear case, it is necessary to solve them on-line and this is impracticable. By using discrete time measurements, this paper has provided a useful tool for effectively solving a class of nonlinear estimation problems for distributed parameter systems.

REFERENCES

1. Athans, M.: "Toward a Practical Theory for Distributed Parameter Systems", IEEE Trans. on Automatic Control, April, 1970.
2. Tzafestas, S.G. and J.M. Nightingale: "Optimal Filtering, Smoothing, and Prediction in Linear Distributed-Parameter Systems", Proc. IEEE, Vol. 115, No. 8, August 1968.
3. Tzafestas, S.G. and J.M. Nightingale: "Concerning Optimal Filtering Theory of Linear Distributed Parameter Systems", Proc. IEEE, Vol. 115, No. 11, November, 1968.
4. Tzafestas, S.G. and J.M. Nightingale: "Maximum Likelihood Approach to the Optimal Filtering of Distributed Parameter Systems", Proc. IEEE, Vol. 116, No. 6, June, 1969.
5. Seinfeld, J.H.: "Nonlinear Estimation for Partial Differential Equations", Chemical Engineering Science, Vol. 24, January, 1969.
6. Seinfeld, J.H.: "Estimation in Distributed Parameter Systems", 1970 Symposium on Adaptive Processes, University of Texas at Austin, December, 1970.
7. Pell, T.M. and R. Aris: "Problems in Chemical Reactor Analysis with Stochastic Features", Industrial and Engineering Chemistry Fundamentals, Vol. 9, No. 1, February, 1970.
8. Kalman, R.E.: "New Methods in Witner Filtering Theory", Proc. Sym. Eng. Appl. Random Function Theory and Probability, Wiley, New York, 1963.
9. Gelfand, I.M. and S.U. Fomin: Calculus of Variations, Prentice-Hall, Englewood Cliffs, New Jersey, 1963.

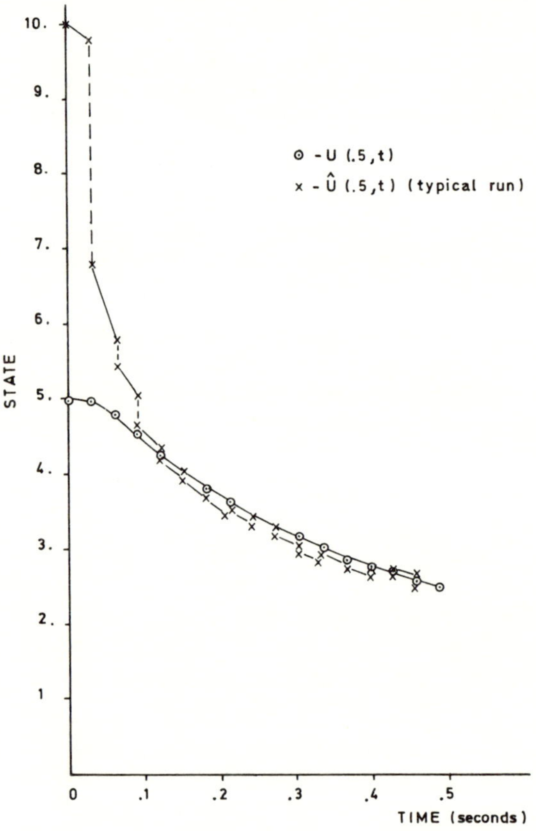

Figure 1 U_1 and $\hat{U}_1$ at midpoint vs. t.

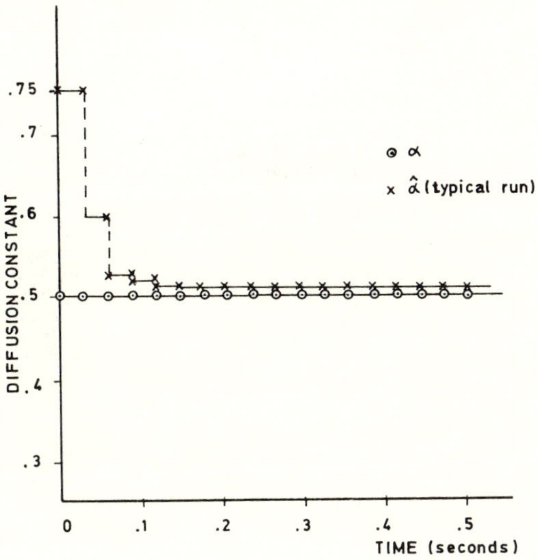

Figure 2 Estimated Space Average of unknown Parameter vs. t.

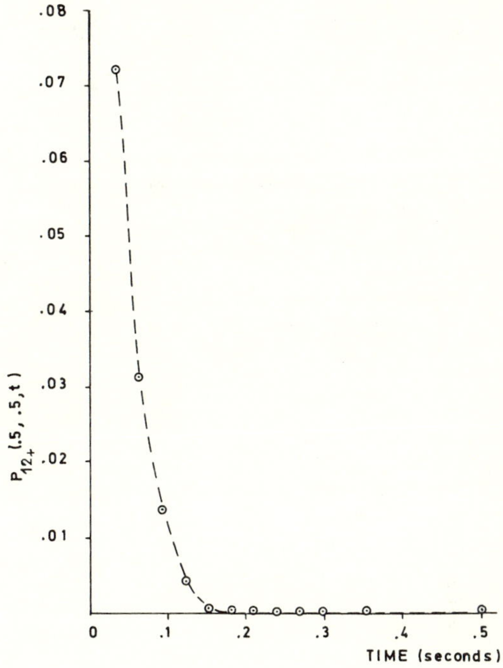

Figure 4 Error Weighting Matrix P_{12+} (.5, .5, t)

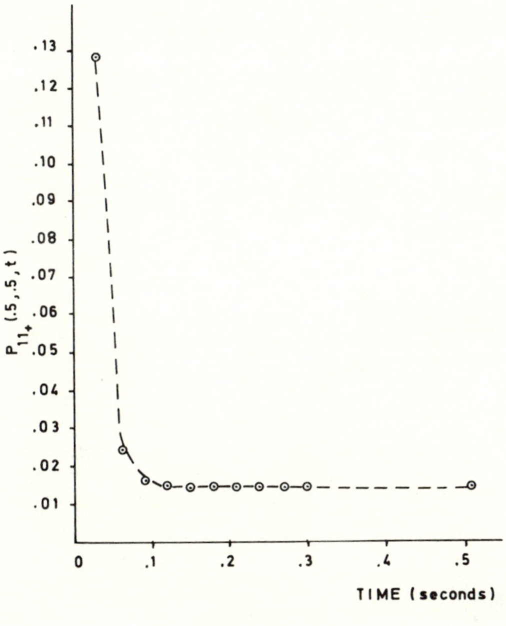

Figure 3 Error Weighting Matrix P_{11_+} (.5, .5, t)

IDENTIFICATION AND STOCHASTIC CONTROL OF NON-DYNAMIC SYSTEMS

A. V. Balakrishnan

Department of System Science
University of California, Los Angeles

We develop a theory of Identification and Stochastic Control of linear Non-Dynamic systems (i.e., Linear Systems whose transfer functions are not necessarily rational) subject to random disturbance as well as measurement error.

1. INTRODUCTION

The class of problems considered is best described with the aid of a block diagram as in figure 1.

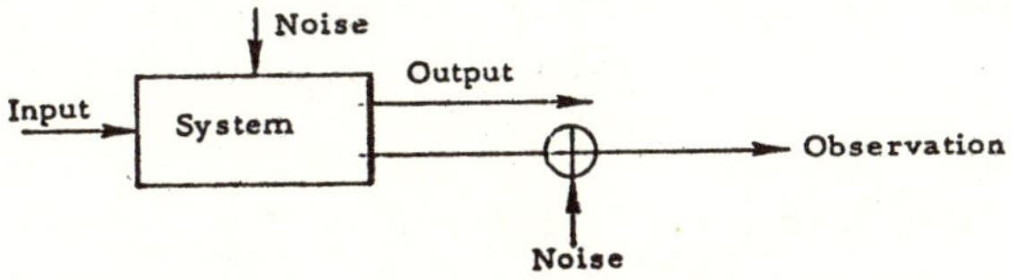

Figure 1

It is a multi-input, multi-output system. The optimization problem is to minimize the effect of the random disturbance at the output by means of an appropriate control input derived from the available observation. The system itself may be partly unknown, and must be identified from the observations also. However it is known [or assumed!] that the system is linear both with respect to the input as well as the external disturbance. The latter is modelled as a white Gaussian noise filtered by a linear system. Thus we can write for the output:

$$y(t;\omega) = s(t;\omega) + n_2(t;\omega) \tag{1.1}$$

where $n_2(t;\omega)$ is white Gaussian noise (of known non-singular spectral density, taken to be the Identity) modelling the observation error, and $s(t;\omega)$ is the response due to the input as well as external disturbance

$$s(t;\omega) = \int_0^t B(t-s)\,u(s)ds + \int_0^t F(t-s)\,n_1(s;\omega)ds \tag{1.2}$$

the (matrix) functions $B(\cdot), F(\cdot)$ representing system or weighting matrices, and $n_1(s;\omega)$ is again white Gaussian with unit spectral density matrix, and $u(\cdot)$ denotes the input (eventually to be a function of the observation). We assume that $B(\cdot)$, $F(\cdot)$ satisfy:

$$\left.\begin{aligned} \|F(t)\| + \|B(t)\| &\leq M_1 \exp - k_1 t \quad t > 0 \\ \|\dot{F}(t)\| + \|\dot{B}(t)\| &\leq M_2 \exp - k_2 t \quad t > 0 \end{aligned}\right\} \tag{1.3}$$

where k_1, k_2 are both > 0. This assumption is made for mathematical simplicity. Note that the condition assures stability. The stochastic control problem in this formulation for the case where the system (i.e., $B(\cdot), F(\cdot)$) is completely known, is to determine a (feedback) control $u(t;\omega)$ based on $y(s;\omega)$, $0 \leq s \leq t$ so as to minimize:

$$\frac{1}{T}\int_0^T E[Q\,s(t;\omega),\, s(t;\omega)]dt + \frac{\lambda}{T}\int_0^T E[u(t;\omega),\, u(t;\omega)]dt \tag{1.4}$$

for fixed T, λ being a fixed positive constant and Q non-negative definite.

By 'non-dynamic' we mean that the functions $B(\cdot), F(\cdot)$ are allowed to be such that the Laplace transforms are not necessarily rational. Our formulation includes a special case (i.e., when the Laplace transforms of $B(\cdot)$ and $F(\cdot)$ are rational) all previously known stochastic control problems of the "linear-quadratic" type. But there are many problems where the generalization is essential. Of particular interest is the case where $F(\cdot)$ does not possess a rational Laplace transform. One application here is to the problem of minimizing the effect of turbulence on an aircraft, the turbulence being modelled using the Von Karman version of the spectral density.

Because the theory is quite new, we have had to confine this presentation to developing the theory with as much detail as space permits. We hope to present numerical calculations in the near future. We point out that the theoretical development is new even where it touches upon the Riccatti equation in a Hilbert space. For one thing, the setting needs to be more general than in [2], and the treatment is quite different. Preliminary results were announced in [3].

2. THE CONTROL PROBLEM

We shall consider the problems of identification and control separately. We begin with the control problem since it is by far the easier one to handle. When $B(\cdot), F(\cdot)$ have rational Laplace transforms, we can easily obtain a (finite dimensional) state space representation for the problem and obtain the optimal feedback control using this representation. This is of course the major 'triumph' of the state space theory. This includes the Bucy-Kalman filtering and the separation principle [1]. In our problem, the functions $B(\cdot)$ and $F(\cdot)$ need not possess rational Laplace transforms and hence it is not possible to exploit the usual procedure. Our first step is to obtain a state-space representation, albeit infinite dimensional [3, 4]. Referring to these for further amplification of system theoretic aspects, we shall simply state and prove the relevant result here.

First let us rewrite (1.1) and (1.2) in the usual 'integrated' version using Wiener processes:

$$Y(t;\omega) = \int_0^t s(\sigma;\omega)d\sigma + W_2(t;\omega) \qquad (2.1)$$

$$s(t;\omega) = \int_0^t B(t-s)u(s;\omega)ds + \int_0^t F(t-s)dW_1(s;\omega) \qquad (2.2)$$

where $W_1(t;\omega)$, $W_2(t;\omega)$ are $m \times 1$ and $n \times 1$ mutually independent Wiener processes respectively, with $W_1(0;\omega) = W_2(0;\omega) = 0$, and each has the unit matrix for the spectral density matrix. To be specific we shall assume

$u(s;\omega)$ is $p \times 1$

$B(\cdot)$ is $n \times p$

$F(\cdot)$ is $n \times m$

Theorem 1

Let $H = L_2[0,\infty]^n$, (L_2-space of $n \times 1$ functions). Let A be the operator defined by $Af = f'$; Domain of $A = [f | f$ is absolutely continuous and $f' \in H]$ (Prime denotes derivative.)

There exist linear transformations B, F mapping respectively the Euclidean spaces E_p, E_m into H such that

$$x(t;\omega) = \int_0^t A\,x(s;\omega)ds + \int_0^t B\,u(s)ds + F\,W_1(t;\omega) \qquad (2.3)$$

where $W_1(t;\omega)$ is a Wiener process, has a unique solution. Moreover $Y(t;\omega)$ in (2.1), (2.2) can then be expressed:

$$Y(t;\omega) = \int_0^t C\,x(s;\omega)ds + W_2(t;\omega) \qquad (2.4)$$

where C is a linear transformation mapping H into E_n. Moreover, B can be taken as: (dot representing derivative):

$$Bu \sim (-1)\sum_0^\infty \dot{B}(s+nL)u, \quad u \in E_r$$

and

$$Fw \sim (-L)\sum_0^\infty \dot{F}(s+nL)w, \quad w \in E_m$$

where L is a fixed positive number, $0 < L < \infty$, and C is then defined by:

$$Cx = \frac{1}{L}\int_0^L x(s)ds$$

Proof

First of all we note that

$$\sum_0^\infty \|\dot{B}(s+nL)u\|$$

converges because of the condition (1, 1). So does similarly

$$\sum_0^\infty \|\dot{F}(s+nL)w\| .$$

Let $S(t)$ denote the semigroup of operators on H defined by

$$S(t)f = g; \quad g(\xi) = f(t+\xi), \quad \xi \geq 0$$

Then A is the infinitesimal generator of $S(t)$ and

$$x(t;\omega) = \int_0^t S(t-\sigma)B\,u(s)ds + \int_0^t S(t-\sigma)F\,dW_1(\sigma;\omega)$$

Moreover

$$Cx(t;\omega) = \int_0^t C\,S(t-\sigma)Bu(\sigma)d\sigma + \int_0^t C\,S(t-\sigma)F\,dW_1(\sigma;\omega)d\sigma$$

while

$$CS(t)Bu = (-1)\int_0^L \sum_0^\infty \dot{B}(\xi+t+nL)u\,d\xi$$
$$= (-1)\sum_0^\infty (B(L+t+nL) - B(t+nL))u$$
$$= B(t)u, \quad t > 0$$

similarly

$$CS(t)\,Fw = F(t)w$$

Hence

$$Cx(t;\omega) = \int_0^t B(t-\sigma)u(\sigma)d\sigma + \int_0^t F(t-\sigma)dW_1(\sigma;\omega) = s(t;\omega)$$

so that (2.4) follows. See [3] for some system theoretic ramifications of this representation.

To prove the main result on the control problem we shall assume that we shall only be dealing with controls linear in the observations. That is to say, $u(t;\omega)$ of the form:

$$u(t;\omega) = \int_0^t H(t;s)dY(s;\omega) \qquad (2.5)$$

The result is undoubtedly true without this restriction on the class of controls; but at the present time, the theory is incomplete in this regard even for the case of finite dimensional state space - see [5]. We may then state our solution to the control problem:

Theorem 2.2

The optimal control of the type (2.5) is given by

$$u_0(t;\omega) = -\frac{B^*}{\lambda}P_C(t)\hat{x}(t;\omega) \qquad (2.6)$$

where $P_C(t)$ is a Nuclear, non-negative definite operator mapping H into H and satisfies:

$$[\dot{P}_C(t)x, y]+[P_C(t)Ax, y]+[P_C(t)x, Ay]+[Cx, QCy]$$

$$-\left[\frac{P_C(t)BB^*P_C(t)}{\lambda}x, y\right] = 0; \; P_C(T)=0 \qquad (2.7)$$

for x, y in (A). Further $\hat{x}(t;\omega)$ is the unique solution of the (stochastic) equation:

$$\hat{x}(t;\omega)=\int_0^t(A-P_f(s)C^*C)\hat{x}(s;\omega)ds+\int_0^t Bu(s;\omega)ds$$

$$+\int_0^t P_f(s)C^*dY(s;\omega) \qquad (2.8)$$

where $P_f(s)$ is also a non-negative definite nuclear operator such that for x in (A*)

$$\dot{P}_f(t)x = AP_f(t)x + P_f(t)A^*x$$

$$+ FF^*x - P_f(t)C^*CP_f(t)x \qquad (2.9)$$

$$P_f(0) = 0$$

Proof

The main ideas are essentially the same as in the finite dimentional case except that allowance must be made for the fact that A is unbounded and that the random variables $x(t;\omega)$ are now no longer finite dimensional. However they are such that

$$E(\|x(t;\omega)\|^2) < \infty$$

and hence there is no need to go to the generality of weak distributions in a Hilbert space. We begin by setting:

$$x_u(t;\omega) = \int_0^t Ax_u(s;\omega)ds + \int_0^t Bu(s;\omega)ds$$

$$\tilde{x}(t;\omega) = x(t;\omega) - x_u(t;\omega)$$

$$= \int_0^t A\tilde{x}(s;\omega)ds + \int_0^t F\,dW_1(\sigma;\omega)d\sigma$$

$$\tilde{Y}(t;\omega) = Y(t;\omega) - \int_0^t Cx_u(s;\omega)ds$$

$$= \int_0^t C\tilde{x}(s;\omega)ds + W_2(t;\omega)$$

Now

$$\tilde{x}(t;\omega) = \int_0^t S(t-s)F\,dW_1(s;\omega)ds$$

and defines a random variable in H for each t with finite variance. (Actually, $\tilde{x}(t;\omega)$ is in the domain of A, since F is.) Further

$$E(\|\tilde{x}(t;\omega)\|^2) = \text{Trace } R(t)$$

where

$$R(t)x = \int_0^t T(\sigma)FF^*T(\sigma)^*x\,d\sigma$$

where

$$[R(t)x, y] = E([\tilde{x}(t;\omega), x]\,[\tilde{x}(t;\omega), y])$$

and we use the notation:

$$R(t) = E[\tilde{x}(t;\omega)\tilde{x}(t;\omega)^*]$$

It is clear that R(t) is nuclear (or trace-class). Next let R denote the covariance operator on $L_2[0, T]^n$ defined by:

$$[Rf, g] = E\left(\int_0^T[f(t), dY(t;\omega)]\cdot\int_0^T[g(t), dY(t;\omega)]\right)$$

Then by Krein's Theorem we can factorize $(I+R)^{-1}$ as:

$$(I+R)^{-1} = (I-\mathscr{L})^*(I-\mathscr{L})$$

where I is the identity operator and $\mathscr{L}$ is Volterra. Let

$$\mathscr{L}f = g; \quad g(t) = \int_0^t L(t;s)\,g(s)ds$$

Then

$$Z(t;\omega) = \tilde{Y}(t;\omega) - \int_0^t ds\int_0^s L(s;\sigma)d\tilde{Y}(\sigma;\omega)$$

is also a Wiener process and the σ-algebra generated by the process $Z(t;\omega)$ is equivalent to that generated by $\tilde{Y}(t;\omega)$, for each t. Since the control $u(t;\omega)$ is of the form:

$$u(t;\omega) = \int_0^t H(t;s)dY(s;\omega)$$

$$= \int_0^t H(t;s)\,d\tilde{Y}(s;\omega) + \int_0^t H(t;s)Cx_u(s;\omega)ds$$

it is readily seen that $u(t;\omega)$ is actually of the form:

$$u(t;\omega) = \int_0^t K(t;s)\,dZ(s;\omega)$$

Here $Z(t;\omega)$ is the so-called "innovation process" which is seen to be definable without reference to the control used. Finally, we also have in turn that the σ-algebra generated by $Y(t;\omega)$ is equivalent to that generated by $Z(t;\omega)$ for each t. Next let:

$$\hat{\tilde{x}}(t;\omega) = E[\tilde{x}(t;\omega)\,|\,\tilde{Y}(s;\omega), \quad 0 \le s \le t]$$

Even though $\tilde{x}(t;\omega)$ is Hilbert-space valued, there is no problem in defining this. One way is to define

$$\hat{\tilde{x}}(t;\omega) = \sum_1^\infty a_k(t;\omega)\,\phi_k$$

where $\{\phi_k\}$ is any orthonormal basis in H and

$$a_k(t;\omega)=E[[\tilde{x}(t;\omega), \phi_k]\,|\,\tilde{Y}(s;\omega), \quad 0\le s\le t]$$

Next for each t, $\hat{\tilde{x}}(t;\omega)$ is in the domain of A, since $\tilde{x}(t;\omega)$ is, and A is closed. Again,

$$\hat{\tilde{x}}(t;\omega) - \int_0^t A\,\hat{\tilde{x}}(s;\omega)ds = H(t;\omega)$$

is a Gaussian Martingale with respect to the growing sigma-algebra generated by $\tilde{Y}(t;\omega)$.

Hence we must have that

$$H(t;\omega) = \int_0^t k_1(s)\, d\tilde{Y}(s;\omega)$$
$$= \int_0^t k_2(s)\, dZ(s;\omega)$$

Let

$$P(t) = E[(\tilde{x}(t;\omega)-\hat{\tilde{x}}(t;\omega))(\tilde{x}(t;\omega)-\hat{\tilde{x}}(t;\omega))^*]$$

Then

$$\lim_{\Delta\to 0} \frac{1}{\Delta} E\left(\int_t^{t+\Delta} dH(\sigma;\omega)\left(\int_t^{t+\Delta} dZ(s;\omega)\right)^*\right) = P(t)C^*$$

and hence

$$H(t;\omega) = \int_0^t P(s)C^*\, dZ(s;\omega)$$

Since the Krein Factorization is unique, and we can directly calculate that

$$\tilde{Y}(t;\omega) - \int_0^t C\hat{\tilde{x}}(s;\omega)ds$$

is also a Gaussian Martingale (Wiener process), it follows that we must have:

$$C\hat{\tilde{x}}(t;\omega) = \int_0^t L(t;s)\, d\tilde{Y}(s;\omega)$$

and

$$Z(t;\omega) = \tilde{Y}(t;\omega) - \int_0^t C\hat{\tilde{x}}(s;\omega)ds$$

Also:

$$\hat{\tilde{x}}(t;\omega) = \int_0^t S(t-\sigma)P(\sigma)C^*\, dZ(\sigma;\omega)$$

Hence also:

$$\hat{\tilde{x}}(t;\omega) = \int_0^t (A-P(s)C^*C)\hat{\tilde{x}}(s;\omega)ds + \int_0^t P(s)C^*d\tilde{Y}(s;\omega)$$

Let

$$\hat{x}(t;\omega) = \hat{\tilde{x}}(t;\omega) + x_u(t;\omega)$$

Then

$$\hat{x}(t;\omega) = E[x(t;\omega)\,|\,Y(s;\omega), \quad 0 \le s \le t],$$

and using the fact that

$$Z(t;\omega) = Y(t;\omega) - \int_0^t Cx(s;\omega)ds$$

we obtain (2.8). To obtain (2.9) we proceed as follows. First of all we note that for $t > 0$,

$$R(t)x = E(\tilde{x}(t;\omega)[\tilde{x}(t;\omega), x])$$

is in the domain of A, since $\tilde{x}(t;\omega)$ is, and A is closed. Hence $AR(t)$ is linear bounded for $t > 0$. Next let

$$R(t) = E[\tilde{x}(t;\omega)\tilde{x}(t;\omega)^*]$$

Then since $\tilde{x}(t;\omega)$ is in the domain of A, and in fact

$$A\tilde{x}(t;\omega) = E[A\tilde{x}(t;\omega)\,|\,Y(s;\omega), \quad 0 \le s \le t]$$

it follows that $AR(t)$ is linear bounded also. Now for x, y in the domain of A^*, we have

$$\frac{d}{dt}[R(t)x, y] = \frac{d}{dt}\int_0^t [F^*T(t-\sigma)^*x, F^*T(t-\sigma)^*y]d\sigma$$
$$= [FF^*x, y]+[R(t)A^*x, y]+[x, R(t)A^*y]$$
$$= [FF^*x+R(t)A^*x+AR(t)x, y]$$

Hence for x in $D(A^*)$:

$$\frac{d}{dt} R(t)x = FF^* + R(t)A^*x + AR(t)x$$

In a similar manner:

$$\frac{d}{dt}[\hat{R}(t)x, y]$$
$$= \frac{d}{dt}\int_0^t [CP_f(\sigma)T(t-\sigma)^*x, CP_f(\sigma)T(t-\sigma)^*y]d\sigma$$
$$= [P_f(t)C^*C\, P_f(t)x, y]+[\hat{R}(t)A^*x, y]+[\hat{R}(t)x, A^*y]$$

leading to

$$\dot{\hat{R}}(t)x = A\hat{R}(t)x + \hat{R}(t)A^*x + P_f(t)C^*C\, P_f(t)x$$

for x in $\mathcal{D}(A^*)$, and since

$$P_f(t) = R(t) - \hat{R}(t)$$

(2.9) follows. Next let us note that since

$$R(t)x = \int_0^t T(\sigma)FF^*T(\sigma)^*x\, d\sigma$$

and it is readily verified (by virtue of (1.3)) that

$$\int_0^\infty \|F^*T(\sigma)^*x\|^2\, d\sigma < \infty,$$

$R(t)$ converges (monotonically) as $t\to\infty$. Since

$$\hat{R}(t) \le R(t)$$

so does $\hat{R}(t)$, and hence

$$P_f(t)x$$

converges as $t\to\infty$. Moreover, the limit $P_f(\infty)$, satisfies:

$$0 = A\,P_f(\infty)x + P_f(\infty)A^*x + FF^*x - P_f(\infty)CC^*\, P_f(\infty)x$$

The problem of optimal control can now be handled as in [4]. The first thing is to show existence of a solution to (2.7). We can clearly find a sequence of operators C_n^* with range in the domain of A^* such that C_n^* converges to C^*.

$$\dot{P}_n(t)x + A^*P_n(t)x + P_n(t)Ax + C_n^*QC_nx - \frac{P_n(t)BB^*P_n(t)x}{\lambda} = 0 \qquad (2.10)$$

$$P_n(T) = 0$$

for every x in $\mathcal{D}(A)$. We show that this has a non-negative definite nuclear solution. For this we exploit the results above. Thus, A^* generates a semigroup, $S(t)^*$, and hence for each x in $\mathcal{D}(A)$:

$$\dot{Q}_n(t)x = A^*Q_n(t)x + Q_n(t)Ax + C_n^*QC_nx - \frac{Q_n(t)BB^*Q_n(t)x}{\lambda}$$

has a non-negative definite nuclear solution with

$$Q_n(0) = 0$$

since C_n^* maps into the domain of A^*. Setting

$$P_n(t) = Q_n(T-t)$$

we have a required kind of solution to (2.10). Moreover we note that

$$\int_0^t T(\sigma)^*C_n^* C_n T(\sigma)x\, d\sigma = Q_n(t)x + \int_0^t \frac{Q_n(\sigma)BB^*Q_n(\sigma)x d\sigma}{\lambda}$$

Since the operators on the right-side are non-negative definite, and the left-side converges to

$$\int_0^t T(\sigma)^* C^*C\, T(\sigma)x\, d\sigma$$

So does each term on the right. Hence $Q_n(\cdot)$ converges strongly, and hence so does $P_n(\cdot)$. Moreover, for x, y in (A), we have:

$$[\dot{P}_n(t)x, y] + [P_n(t)x, Ay] + [P_n(t)Ax, y] + [C_n^*C_nx, y] - \frac{[P_n(t)BB^*P_n(t)x, y]}{\lambda} = 0$$

Taking limits, and setting

$$P_C(t)x = \operatorname{limit}_n P_n(t)x$$

we obtain (2.7) as required. Since the key issue of the existence of $P_c(t)$ satisfying (2.7) has been taken care of, the remainder of the arguments for the optimality of

$$u(t;\omega) = \frac{-B^*}{\lambda} P_C(t)\hat{x}(t;\omega)$$

can be carried out in similar fashion to that in [5].

First let us note that the 'evolution equation' in H:

$$\dot{x}(t) = (A - P_f(t)C^*C)x(t)$$

has a unique solution given by

$$x(t) = \Phi(t)\,x(0)\ ;\ x(0) \in \mathcal{D}(A)$$

This is seen very simply from the fact that the equation can be rewritten as:

$$x(t) = S(t)x(0) + \int_0^t S(t-\sigma)P_f(\sigma)C^*C\, x(\sigma)d\sigma$$

and which in turn can be written as

$$(I - L)\, x(\cdot) = g$$

where

$$g(t) = S(t)\, x(0)$$

and L is a Volterra operator:

$$Lx = y;\ u(t) = \int_0^t S(t-\sigma)P_f(\sigma)C^*C\, x(\sigma)d\sigma$$

But (I - L) has a bounded inverse; in fact

$$x(\cdot) = \sum_0^\infty L^k g$$

or,

$$x(t) = \Phi(t)\, x(0)$$

For properties of $\Phi(\cdot)$ [such as the 'transition' property] see [6].

Let $\mathscr{L}_u$ denote the space of control functions of the form

$$u(t;\omega) = \int_0^t k(t;s)dZ(s;\omega) \qquad (2.11)$$

$u(t;\omega)$ being jointly measurable in t and ω and

$$\int_0^T E[u(t;\omega), u(t;\omega)]dt < \infty$$

Then $\mathscr{L}_u$ is a Hilbert space under the inner-product

$$[u, v] = \int_0^T E[u(t;\omega), v(t;\omega)]dt$$

Now we wish to minimize (1.4) which can be expressed:

$$\int_0^T E[C^*QCx(t;\omega), x(t;\omega)]dt + \lambda[u, u] \qquad (2.12)$$

But

$$E[C^*QC[x(t;\omega)-\hat{x}(t;\omega)], x(t;\omega)-\hat{x}(t;\omega)]]$$

$$= \text{Tr. } C^*QC\ P_f(t)$$

and since $P_f(t)$ is fixed, it is enough to minimize

$$\int_0^T E[C^*QC\ \hat{x}(t;\omega), \hat{x}(t;\omega)]dt + \lambda[u, u] \qquad (2.13)$$

This embodies the 'separation principle'. Next let L denote the linear bounded operator on $\mathscr{L}_u$ defined by

$$Lu = v;\ v(t;\omega) = \int_0^t S(t-\sigma)B(\sigma)u(\sigma;\omega)d\sigma$$

$$w(t;\omega) = \int_0^t S(t-\sigma)\ P_f(\sigma)C^*dZ(\sigma;\omega)$$

Then the cost functional (2.13) can be written as

$$[C^*QC(Lu + w),\ Lu + w] + \lambda[u, u]$$

and has a unique minimum given by

$$u_o = [L^*C^*QCL + \lambda I]^{-1}\ L^*C^*QC\ w$$

or,

$$u_o = -\frac{1}{\lambda}\ L^*C^*QC(Lu_o + w)$$

But

$Lu_o + w$ corresponds to $\hat{x}(t;\omega)$

and

$$L^*C^*QC(Lu_o + w)$$

is the function

$$B^*\ E(\int_t^T S^*(\sigma-t)C^*QC\ \hat{x}(\sigma;\omega)d\sigma\,|\mathscr{B}(t)) \qquad (2.14)$$

(where $\mathscr{B}(t)$ = σ-algebra generated by $Z(s;\omega)$, $0 \leq s \leq t$) provided it is of the form (2.11). We shall now show that it is, by showing that $\hat{x}(\sigma;\omega)$ can be determined appropriately, the optimal control being unique. Thus let us seek a solution of the form:

$$\hat{x}(t;\omega) = \int_0^t (A-H(\sigma))\hat{x}(\sigma;\omega)d\sigma + \int_0^t P_f(\sigma)C^*\ dZ(\sigma;\omega)$$

where $H(\sigma)$ is linear bounded operator mapping H into H for each σ and is continuously strongly differentiable, and $H(0)$ is zero. Then

$$E[\hat{x}(t;\omega)\,|\mathscr{B}(s)] = \Phi(t)\ \Phi(s)^{-1}\ \hat{x}(s;\omega)$$

where

$$\Phi(t)\,x$$

'solves'

$$\dot{x}(t) = (A-H(t))x(t); x(0) = x$$

[See [6] for relevent material on vvolution equations.]

Then substituting into (2.14) we have

$$u_o(t;\omega) = \frac{-B^*}{\lambda}\int_t^T S^*(\sigma-t)C^*QC\Phi(\sigma)\ \Phi(t)^{-1}\hat{x}(t;\omega)d\sigma$$

Let $K(t)$ denote the operator

$$K(t)x = \int_t^T S^*(\sigma-t)C^*QC\ \Phi(\sigma)\ \Phi(t)^{-1}\ x\ d\sigma$$

Then

$$u_o(t;\omega) = \frac{-B^*}{\lambda}\ K(t)\hat{x}(t;\omega)\ ;$$

on the other hand, substituting this into (2.8) we have:

$$\hat{x}(t;\omega) = \int_0^t A\ \hat{x}(\sigma;\omega)d\sigma - \int_0^t \frac{BB^*}{\lambda}\ K(\sigma)\hat{x}(\sigma;\omega)d\sigma + \int_0^t P_f(\sigma)C^*\ dZ(\sigma;\omega)$$

Hence

$$H(t) = \frac{-BB^*}{\lambda}\ K(t)$$

But for x, y in the domain of A

$$\frac{d}{dt}[K(t)x, y]$$

$$= \frac{d}{dt}\int_t^T [QC\Phi(\sigma)\Phi(t)^{-1}x,\ CS(\sigma-t)y]d\sigma$$

$$= -[QC_x, Cy]$$

$$- \int_t^T [QC\Phi(\sigma)\ \Phi(t)^{-1}x, CS(\sigma-t)Ay]d\sigma$$

$$+ \int_t^T [QC\Phi(\sigma)\ \Phi(t)^{-1}(A-H(t))x,\ CS(\sigma-t)y]d\sigma$$

$$= -[C^*QCx, y]$$

$$-[K(t)x, Ay]$$

$$-[K(t)Ax, y]$$

$$+[K(t)\ \frac{BB^*}{\lambda}K(t)x, y]$$

or, in our notation,

$$K(t) = P_C(t)$$

as required, yielding (2.6).

3. THE IDENTIFICATION PROBLEM

Next let us turn to the identification problem. Here we shall assume that the system is unknown except for fixed finite number of parameters in B(·) and F(·). Let us denote the parameter vector by θ, which has thus its range in a subset of a Euclidean space. Let us outline our approach to the problem. We assume that there is a 'true' value θ_o. We assume that the input u(t) is known and that

$$R_u(t) = \lim_{T \to \infty} \frac{1}{T}\int_0^T u(s+t)u(s)^* ds \tag{3.1}$$

exists and is continuous in t. We further assume that (1.3) is satisfied for all θ in a known neighborhood of θ_o.

Let $p(Y;\theta;T)$ denote the probability density functional (Radon-Nikodym derivative of the process $Y(t;\omega)$, $0 \leq t \leq T$ with respect to Wiener measure) for fixed time-interval T. Then we show that, ∇_θ denoting gradient with respect to θ:

$$\lim_{T \to \infty} \frac{1}{T} \nabla_\theta \text{Log } p(Y;\theta_o;T) = 0 \text{ in probability.} \tag{3.2}$$

Further, if the matrix M defined by:

$$\lim_{T \to \infty} \frac{1}{T}(\nabla_\theta \text{Log } p(Y;\theta_o;T))(\nabla_\theta \text{Log } p(Y;\theta_o;T))^* \quad \text{(in pr.)} \tag{3.3}$$

is positive definite, then for all T sufficiently large

$$\frac{1}{T} \nabla_\theta \text{Log } p(Y;\theta;T)$$

has a unique root in a sufficiently small neighborhood of θ_o, and the root converges in probability to θ_o as T goes to infinity. Further, the iteration, (Newton-Raphson):

$$\theta_{n+1} = \theta_n - M_n^{-1}\left(\frac{1}{T}\nabla_\theta \text{Log } p(Y;\theta_n;T)\right) \tag{3.4}$$

where

$$M_n = \frac{1}{T}(\nabla_\theta \text{Log } p(Y;\theta_n;T))(\nabla_\theta \text{Log } p(Y;\theta_n;T))^*$$

can be used for locating the root. It is natural to say that the system is 'identifiable' if M is positive definite.

Hence the main step in the identification problem in our approach is the calculation of the functional $p(Y;\theta;T)$. Here we can make use of the results in Section 3. We have

$$\text{Log } p(Y;\theta;T) = -\frac{1}{2}\left\{\int_0^T [C\hat{x}(t;\omega), C\hat{x}(t;\omega)]dt - 2\int_0^T [C\hat{x}(t;\omega), dY(t;\omega)]\right\} \tag{3.5}$$

where $\hat{x}(t;\omega)$ is determined by:

$$\hat{x}(t;\omega) = \int_0^t (A - P_f(s)C^*C)\hat{x}(s;\omega)ds + \int_0^t P_f(s)C^* dY(s;\omega) + \int_0^t B\, u(s)ds \tag{3.6}$$

and $P_f(\cdot)$ is determined from (2.9). For large T one may use $P_f(\infty)$ in place of $P_f(t)$. The Ito integral in (3.5) may be approximated as:

$$\frac{1}{T}\int_0^T [C\hat{x}(t;\omega), dY(t;\omega)]$$

$$\approx \frac{1}{T}\sum_i [C\hat{x}(t_i;\omega), Y(t_i;\omega) - Y(t_{i-1};\omega)] - \frac{1}{T}\int_0^T \text{Tr. } CP(t)C^* dt \tag{3.7}$$

The finite dimensional results as in [5] go over, in other words. In particular $P_f(\infty)$ can be obtained by the iteration:

$$AP_{n+1} + P_{n+1}A^* + FF^* - P_nC^*CP_n = 0$$

$$P_1 x = \int_0^\infty S(t)\, FF^*\, S(t)^*\, x\, dt$$

4. RELATION TO WIENER FILTERING

Finally it is of interest to note that setting $u(t) \equiv 0$, the results of Section 3 lead to a recursive solution to the Wiener filtering problem, for the case of independent signal and noise, with signal spectrum given by:

$$\left(\int_0^\infty e^{2\pi i f t} B(t)dt\right)\left(\int_0^{\infty} e^{-2\pi i f t} B^*(t)dt\right)$$

and noise spectrum by

$$\left(\int_0^\infty e^{2\pi i f t} F(t)dt\right)\left(\int_0^\infty e^{-2\pi i f t} F^*(t)dt\right)$$

We obtain that:

$$\hat{s}(t;\omega) = E[s(t;\omega) \mid Y(s;\omega),\ 0 \leq s \leq t]$$

is given by

$$\hat{s}(t;\omega) = C\hat{x}(t;\omega)$$

and in particular, as $t \to +\infty$, we have that $\hat{x}(t;\omega)$ is the 'steady-state' solution of

$$\hat{x}(t;\omega) = \int_0^t (A - P_f(\infty)C^*C)\hat{x}(s;\omega)ds + \int_0^t P_f(\infty)C^* dY(s;\omega)$$

REFERENCES

[1] R. S. Bucy and P. D. Joseph: 'Filtering for Stochastic Processes with Applications to Guidance', Interscience, 1968.

[2] A. Bensoussan: 'Filterage Optimal des Systemes Lineares', Dunod, 1970.

[3] A. V. Balakrishnan: "System Theory and Stochastic Optimization", Proceedings NATO Institute, Peter Peregrinns, Ltd, 1972.

[4] A. V. Balakrishnan: 'Introduction to Optimization Theory in a Hilbert Space', Lecture Notes, Springer-Verlag, 1970.

[5] A. V. Balakrishnan: "Stochastic Differential Systems, I", System Science Department, University of California, Los Angeles, 1972.

[6] S. G. Krein: "Linear Differential Equations in a Banach Space", Translations of Mathematical Monographs, Vol. 29, American Mathematical Society, 1972.

ACKNOWLEDGMENT

Research supported in part under AFOSR Grant No. 68-1408, United States Air Force.

IDENTIFICATION OF DISTRIBUTED PARAMETER SYSTEMS USING THE EPSILON METHOD: A NEW ALGORITHM (*)

G. Di Pillo, L. Grippo and M. Lucertini

Centro di Studio dei sistemi di Controllo e Calcolo Automatici del C.N.R., Roma, Italy
Istituto di Automatica, Università di Roma, Italy

In this paper the application of the ε-method to the state estimation and the parameter identification of distributed systems is considered. Using a suitable approximation of the state evolution, a function space algorithm for the minimization of the ε-functional is described. Numerical experiments are reported.

1. INTRODUCTION

The ε-method, which from the theoretical standpoint has been stadied mainly in Balakrishnan (1968 a), (1968 b) and De Julio (1969 a), is a computing technique which extends the penalty function method to identification and optimal control problems.

As regards the application of this technique to distributed parameter systems (DPS), the computation of optimal control laws has been considered in De Julio (1969 b) and Kenneth (1969) and the estimation of unknown parameters in Di Pillo (1973 a). In these applications the minimization of the ε-functional J_ε (cost functional plus penalty terms) was performed by approximating it with a function of several variables obtained using finite-difference methods.

In this paper both the state estimation and the parameter identification of DPS are considered. The cost functional is obtained by taking a finite number of observation points on the spatial domain and considering the mean square observation error in a fixed time interval. An algorithm is proposed for the minimization of J_ε with respect to the state and the unknown parameters, which does not require neither finite-difference approximations nor Ritz-Galerkin procedures. To this aim, the output evolution y(x,t) of the system is approximated by a sum of the type:

$$\tilde{y}(x,t) = \sum_{i=1}^{N} f_i(t)g_i(x) \tag{1}$$

where $f_i(t)$ and $g_i(x)$ belong to proper Sobolev spaces; the minimization of J_ε is then performed with respect to $f_i(t)$, $g_i(x)$ and the unknown parameters on the resulting product space.

A function space technique is proposed which avoids the need of performing the approximate differentiation of $f_i(t)$ and $g_i(x)$. The analytical expression of the gradient of J_ε is derived on the considered space; then a computational scheme based on the conjugate gradient method (CGM) is described. Finally, numerical results are reported which show the effectiveness of the method.

2. PROBLEM FORMULATION

We consider DPS described by an equation of the type:

$$F(t,x,y,y_t,y_x,y_{tt},y_{xx},y_{tx},\underline{\alpha}) = 0 \tag{2}$$

where $t \in [0,T]$, $x \in (0,L)$; y(x,t) is an output variable, supposed to be scalar, $\underline{\alpha} \in EP$ is a vector of unknown parameters and the subscripts denote partial derivatives.

In addition to (2), boundary conditions may be given:

$$G_1(t,x,y,y_t,y_x,y_{tx},\underline{\alpha})\Big|_{x=0} = 0 \tag{3}$$

$$G_2(t,x,y,y_t,y_x,y_{tx},\underline{\alpha})\Big|_{x=L} = 0 \tag{4}$$

The dependence of y on distributed and boundary inputs is embodied in F, G_1, G_2.

The observations are given by:

$$z(x_k,t) = y(x_k,t) + n(x_k,t) \tag{5}$$

in which x_k, k=1,...,M are measurement points and $n(x_k,t)$ is a stationary noise with zero mean.

The problem is that of minimizing the error functional:

$$J = \sum_{k=1}^{M} \int_0^T \{z(x_k,t) - y(x_k,t)\}^2 dt \tag{6}$$

where $y(x_k,t)$ is the solution of (2), (3), (4) for given values of $\underline{\alpha}$ and of the initial conditions.

The application of the ε-method consists in introducing the penalty term

$$\frac{1}{\varepsilon} P = \frac{1}{\varepsilon} \sum_{i=1}^{3} P_i \qquad \varepsilon > 0 \tag{7}$$

$$P_1 = \int_0^T \int_0^L F^2 \, dx\, dt \tag{8}$$

$$P_{i+1} = \int_0^T G_i^2 \, dt \qquad i=1,2 \tag{9}$$

and in minimizing the new functional

$$J_\varepsilon = J + \frac{1}{\varepsilon} P \tag{10}$$

with respect to y(x,t) and $\underline{\alpha}$, and without the constraints (2), (3), (4).

Under suitable hypotheses, as ε goes to zero, the solution of the ε-problem converges to the true solution.

(*) This work was supported by Ente Nazionale Energia Elettrica and Consiglio Nazionale delle Ricerche;

In order to get a numerical solution of the problem, it is necessary to employ some kind of approximation of the functional J_ε.

In previous applications of the ε-method to DPS the functional J_ε has been discretized using finite-differences.

As already said, we assume for the solution y(x,t) an approximation $\tilde{y}(x,t)$ of the type (1) which, by defining:

$$\underline{f}(t) = [f_1(t),\ldots,f_N(t)]^T$$

$$\underline{g}(x) = [g_1(x),\ldots,g_N(x)]^T$$

and denoting by <·,·> the inner product on E^N, can be put in the form:

$$\tilde{y}(x,t) = \langle \underline{f}(t),\underline{g}(x)\rangle \qquad (11)$$

By substituting $\tilde{y}(x,t)$ for y(x,t) in $\tilde{(6)}$-(10) we obtain the approximate ε-functional $\tilde{J}_\varepsilon$:

$$\tilde{J}_\varepsilon = \tilde{J} + \frac{1}{\varepsilon}\tilde{P} = \tilde{J} + \frac{1}{\varepsilon}\sum_{i=1}^{3}\tilde{P}_i$$

with

$$\tilde{J} = \sum_{k=1}^{M}\int_o^T \{z(x_k,t) - \langle \underline{f}(t),\underline{g}(x_k)\rangle\}^2\,dt$$

$$\tilde{P}_1 = \int_o^T\int_o^L \{\tilde{F}(t,x,\underline{f},\underline{f}_t,\underline{f}_{tt},\underline{g},\underline{g}_x,\underline{g}_{xx},\underline{\alpha}\}^2 dxdt \qquad (12)$$

$$\tilde{P}_{i+1} = \int_o^T \{\tilde{G}_i(t,x,\underline{f},\underline{f}_t,\underline{g},\underline{g}_x,\underline{\alpha}\}^2 dt \qquad i=1,2$$

where the subscripts t and x denote differentiation.

Then the minimization of $\tilde{\mathcal{J}}_\varepsilon$ is performed with respect to the vectors $\underline{f}(t)$, $\underline{g}(x)$ and $\underline{\alpha}$ on a suitable Hilbert space.

The expansion (11) is general enough to allow the exact or approximate representation of the solution of several linear and nonlinear partial differential equations (Ames, 1965). For a given N, the above expansion leads to a representation of the solution more accurate then in the case in which for $f_i(t)$ or $g_i(x)$ a predetermined basis is used. However the following developments can be easily adapted to the case in which a family of functions is fixed.

We point out that theoretical foundations for the application of the ε-method to the identification and optimal control of infinite dimensional systems have been given only in the linear case. However, the algorithm described here, based on the approximation (11), can be applied, on a heuristic base, also to nonlinear problems.

3. GRADIENT EXPRESSIONS

Let $W^m(0,S)$ be the Sobolev space of the N-vector functions w(·), absolutely continuous with absolutely continuous derivatives up to the order (m-1), and with $\underline{w}^{(m)}(\cdot)\in L_2^N(0,S)$, endowed with the inner product:

$$\langle \underline{w}_1,\underline{w}_2\rangle_{W^m(0,S)} = \sum_{i=o}^{m-1}\langle \underline{w}_1^{(i)}(0),\underline{w}_2^{(i)}(0)\rangle + \int_o^S \langle \underline{w}_1^{(m)}(s),\underline{w}_2^{(m)}(s)\rangle ds$$

In order to obtain the analytical expression of the gradient of $\tilde{\mathcal{J}}_\varepsilon$, we assume that in (2) both y_{tt} and y_{xx} appear explicitly, so that in (12) both $\underline{f}_{tt}$ and $\underline{g}_{xx}$ will appear explicitly.

In this case we take $\underline{f}(\cdot)\in W^2(0,T)$ and $\underline{g}(\cdot)\in W^2(0,L)$ and minimize $\tilde{\mathcal{J}}_\varepsilon$ on the product space

$$H = W^2(0,T)\times W^2(0,L)\times E^P$$

which is a Hilbert space.

Proposition - Assume that $\tilde{F}$, $\tilde{G}_1$, $\tilde{G}_2$ are continuously differentiable with respect to $\underline{f}$, $\underline{g}$ and their derivatives; denote by

$$\nabla\tilde{J}_\varepsilon = \begin{bmatrix}\underline{\Phi}(t)\\ \underline{\Gamma}(x)\\ \underline{\Psi}\end{bmatrix}$$

the gradient of $\tilde{\mathcal{J}}_\varepsilon$ on H, where $\underline{\Phi}(t)$, $\underline{\Gamma}(x)$, $\underline{\Psi}$ are defined by:

$$\lim_{\lambda\to o}\frac{\tilde{J}_\varepsilon(\underline{f}+\lambda\underline{\phi}) - \tilde{\mathcal{J}}_\varepsilon(\underline{f})}{\lambda} = \langle\underline{\Phi},\underline{\phi}\rangle_{W^2(0,T)}$$

$$\lim_{\lambda\to o}\frac{\tilde{\mathcal{J}}_\varepsilon(\underline{g}+\lambda\underline{\gamma}) - \tilde{\mathcal{J}}_\varepsilon(\underline{g})}{\lambda} = \langle\underline{\Gamma},\underline{\gamma}\rangle_{W^2(0,L)}$$

$$\lim_{\lambda\to o}\frac{\tilde{\mathcal{J}}_\varepsilon(\underline{\alpha}+\lambda\underline{\psi}) - \tilde{\mathcal{J}}_\varepsilon(\underline{\alpha})}{\lambda} = \langle\underline{\Psi},\underline{\psi}\rangle_{E^P}$$

Then, defining for convenience:

$$Q = \sum_{k=1}^{M}\{z(x_k,t) - \langle\underline{f}(t),\underline{g}(x_k)\rangle\}^2 + \frac{1}{\varepsilon}\{\int_o^L \tilde{F}^2\,dx + \tilde{G}_1^2 + \tilde{G}_2^2\}$$

$$R = \frac{1}{\varepsilon}\int_o^T \tilde{F}^2\,dt$$

$$E(x_k) = \int_o^T \{z(x_k,t) - \langle\underline{f}(t),\underline{g}(x_k)\rangle\}^2 dt$$

$$E(x_o) = \frac{1}{\varepsilon}\int_o^T \tilde{G}_1^2\,dt \qquad x_o = 0$$

$$E(x_{M+1}) = \frac{1}{\varepsilon}\int_o^T \tilde{G}_2^2\,dt \qquad x_{M+1} = L$$

it results:

(a) $$\underline{\Phi}(t) = \underline{\Phi}(0) + \underline{\Phi}_t(0)t + \int_o^t\int_o^\tau \underline{\Phi}_{tt}(\eta)d\eta\,d\tau$$

$$\underline{\Phi}(0) = \int_o^T Q_{\underline{f}}\,dt$$

$$\underline{\Phi}_t(0) = \int_o^T \{t\,Q_{\underline{f}} + Q_{\underline{f}_t}\}dt$$

$$\underline{\Phi}_{tt}(t) = (T-t)\int_o^T Q_{\underline{f}}dt + \int_t^T \{Q_{\underline{f}_t}(\tau) - \int_o^\tau Q_{\underline{f}}(\eta)d\eta\}d\tau + Q_{\underline{f}_{tt}}$$

(b) $\underline{\Gamma}(x)=\underline{\Gamma}(0)+\underline{\Gamma}_x(0)x+\int_0^x\int_0^\xi \underline{\Gamma}_{xx}(\chi)d\chi d\xi$

$$\underline{\Gamma}(0) = \sum_{k=0}^{M+1} E_{\underline{g}}(x_k) + \int_0^L R_{\underline{g}}\,dx$$

$$\underline{\Gamma}_x(0) = \sum_{k=1}^{M+1} x_k E_{\underline{g}}(x_k) + E_{\underline{g}_x}(x_o) + E_{\underline{g}_x}(x_{M+1}) + \int_0^L \{x R_{\underline{g}} + R_{\underline{g}_x}\}dx$$

$$\underline{\Gamma}_{xx}(x) = \sum_{k=1}^{M+1} (x_k-x)E_{\underline{g}}(x_k)1(x_k-x) + E_{\underline{g}_x}(x_{M+1}) + (L-x)\int_0^L R_{\underline{g}}dx + \int_x^L\{R_{\underline{g}_x}(\xi) - \int_0^\xi R_{\underline{g}}(\chi)d\chi\}d\xi + R_{\underline{g}_{xx}}$$

(c) $\underline{\Psi} = \int_0^T \underline{Q}_\alpha\, dt$

In the above formulas 1(x) is the unit step function and the derivative of a scalar with respect to a vector, denoted by a subscript, is intended as a column vector.

The proof of this proposition is reported in Di Pillo (1973 b).

Remark - If in (2) y_{tt} does not appear we can assume $\underline{f}(\cdot)\in W^1(0,T)$; the gradient formulas can be simplified in an obvious manner. A similar remark holds for y_{xx}.

4. COMPUTING ALGORITHM

The application of the ε-method requires the minimization of $\tilde{J}_\varepsilon$ for decreasing values of ε.

Using the analytical expression of the gradient it is possible to minimize $\tilde{J}_\varepsilon$ on H by a function space technique. The computations can be organised so as to avoid the approximate differentiation of $\underline{f}(t)$ and $\underline{g}(x)$.

In what follows we describe an iterative computing procedure, based on the CGM, for the numerical solution of each ε-problem.

(a) Set starting values

$$[\underline{f}^o_{tt}(t)\ \underline{f}^o_t(0)\ \underline{f}^o(0)\ \underline{g}^o_{xx}(x)\ \underline{g}^o_x(0)\ \underline{g}^o(0)\ \underline{\alpha}^o] \triangleq \underline{V}^o$$

Set $i = 0$

(b) Compute $\underline{f}^i_t(t), \underline{f}^i(t), \underline{g}^i_x(x), \underline{g}^i(x)$.

(c) Compute

$$[\underline{\Phi}^i_{tt}(t)\ \underline{\Phi}^i_t(0)\ \underline{\Phi}^i(0)\ \underline{\Gamma}^i_{xx}(x)\ \underline{\Gamma}^i_x(0)\ \underline{\Gamma}^i(0)\ \underline{\Psi}] \triangleq \underline{\Lambda}^i$$

(d) Compute the descent direction on H,

$$\underline{d}^i = \begin{bmatrix} \underline{\phi}^i(t) \\ \underline{\gamma}^i(x) \\ \underline{\psi}^i \end{bmatrix}$$

This can be performed by evaluating first

$$[\underline{\phi}^i_{tt}(t)\ \underline{\phi}^i_t(0)\ \underline{\phi}^i(0)\ \underline{\gamma}^i_{xx}(x)\ \underline{\gamma}^i_x(0)\ \underline{\gamma}^i(0)\ \underline{\psi}^i] \triangleq \underline{D}^i$$

by means of the CGM formulas:

$$\underline{D}^o = -\underline{\Lambda}^o\ , \quad \underline{D}^i = -\underline{\Lambda}^i + \beta^{i-1}\underline{D}^{i-1} \qquad i>0$$

$$\beta^{i-1} = \|\nabla\tilde{J}^i_\varepsilon\|^2_H / \|\nabla\tilde{J}^{i-1}_\varepsilon\|^2_H$$

and then carrying out the needed integrations.

(e) Determine the optimal displacement σ^i along the descent direction $\underline{d}^i$

(f) Compute $\underline{V}^{i+1} = \underline{V}^i + \sigma^i\underline{D}^i$, and restart from (b) with $i\leftarrow i+1$.

5. NUMERICAL RESULTS

In the numerical application of the proposed algorithm, we considered the system:

$$\frac{\partial y}{\partial t} = \alpha\frac{\partial^2 y}{\partial x^2} \qquad 0<x<\pi;\ 0\le t\le 1$$

$$y(0,t) = 0 \qquad y(\pi,t) = 0$$

with observations given by (5) in which $n(x_k,t)$ was a stationary white noise uniformly distributed in $[-K, K]$.

We first performed the state estimation, supposing α known; then we considered the combined state and parameter estimation. In both cases, we assumed that the noiseless state evolution y(x,t) was determined by the true values:

$$\alpha = 0.5 \qquad y(x,0) = 0.75\sin x - 0.25\sin 2x$$

so that

$$y(x,t) = 0.75\,e^{-0.5t}\sin x - 0.25\,e^{-2t}\sin 2x$$

In the expansion (1) we assumed $N = 2$.

The minimization of $\tilde{J}_\varepsilon$ was performed on the space $H' = W^1(0,T)\times W^2(0,L)\times E^1$.

The numerical integrations which occur in the algorithm were carried out with discretization steps $\Delta x = \pi/20$, $\Delta t = 1/20$.

The computations were started with $\varepsilon_o = 1$; then the subsequent ε-problems were solved with $\varepsilon_{i+1} = \varepsilon_i/10$.

Case 1: State estimation - Observations were taken at the points $x_k = k\,\pi/4$ $k = 1,2,3$ with noise level given by $K = 0.05$.

The starting values for the first ε-problem were:

$$\underline{f}^o_t(t) = \underline{0}. \qquad \underline{f}^o(0) = [0.75 \quad -0.125]^T$$

$$\underline{g}^o_{xx}(x) = [-0.2\sin x \quad 0.]^T \qquad \underline{g}^o_x(0) = \underline{0}. \qquad \underline{g}^o(0) = \underline{0}.$$

Some numerical results, obtained for $\varepsilon_1 = 0.1$ are reported in fig. 1 and 2. Fig. 1 shows the true initial state, the initial guess and the samples of the estimated initial state; fig. 2 shows $y(\pi/2,t)$, the observations in $x = \pi/2$ and the estimated samples.

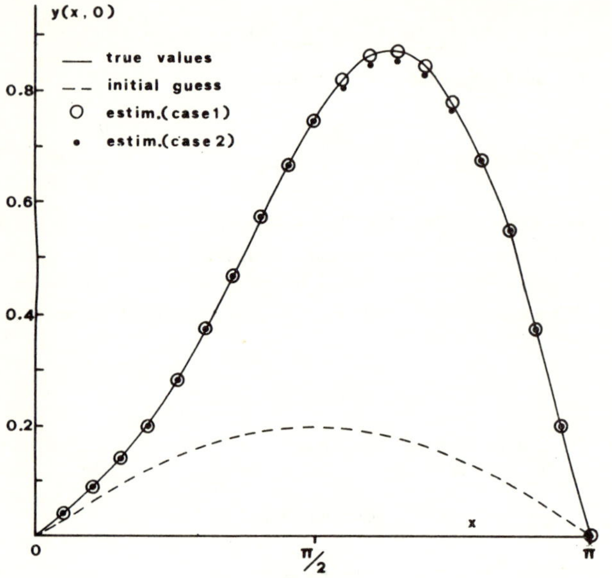

Fig. 1

The estimation errors were negligible also for other values of x and t.

Case 2: Combined state and parameter estimation - In this case, to obtain a good estimate it was necessary to increase the number of measurement points, with respect to the previous case; observations were taken at $x_k = k\,\pi/10$, $k = 1,\ldots,9$.

The noise level and the starting values were the same as in the preceding case, with, in addition $\alpha^o = 0$.

Quite satisfactory results were obtained for $\varepsilon_1=0.1$. Also for this case, the estimated samples of the initial state and of $y(\pi/2,t)$ are shown in fig. 1 and 2

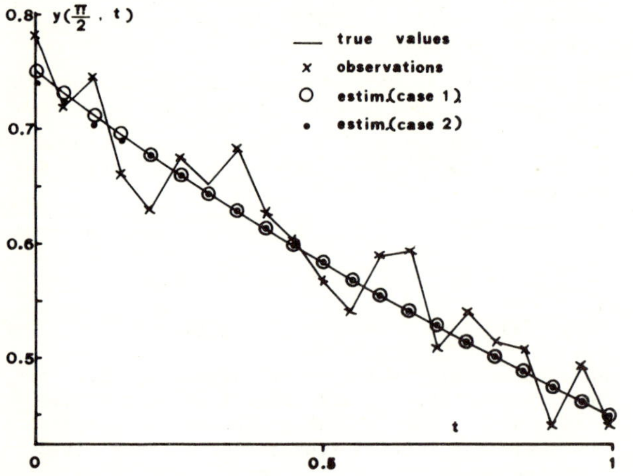

Fig. 2

The estimated value of the parameter was $\alpha = 0.48$.

Further numerical results can be found in Di Pillo (1973 b).

6. CONCLUSIONS

The ε-method appears to be an effective technique in DPS identification. The main advantage consists in avoiding the numerical solution of the system equations.

As we remarked in Di Pillo (1973 a) and as the numerical results reported above confirm, the number of ε-problems which must be solved to get satisfactory results is usually small when dealing with identification and estimation problems.

The algorithm proposed here for the numerical implementation of the ε-method revealed itself, on the basis of the gained experience, more efficient than the algorithm employing a preliminary discretization of the ε-functional performed by a finite-differences method.

REFERENCES

AMES, W.F. (1965) - "Non-linear Partial Differential Equations in Engineering" - Academic Press.

BALAKRISHNAN, A.V. (1968 a) - "On a New Computing Technique in Optimal Control" - SIAM J. Control, Vol. A6, n. 2.

BALAKRISHNAN, A.V. (1968 b) - "A New Computing Technique in System Identification" - J. Computer and System Sciences, Vol. 2, n. 1.

DE JULIO, S. (1969 a) - "On the Optimization of Infinite Dimensional Linear Systems" - In 'Computing Methods in Optimization Problems', Vol. 2, Academic Press.

DE JULIO, S. (1969 b) - "Applicazione del metodo ε alla soluzione di due problemi di ottimizzazione di sistemi a parametri distribuiti". Rendiconti della LXX Riunione Annuale AEI, Rimini (Italy).

DI PILLO, G. and GRIPPO, L. (1973 a) - "Application of the ε-Technique to Distributed Parameter Systems Identification" - J. Opt. Th. and Appl., Vol. 11, n. 1.

DI PILLO, G., GRIPPO, L. and LUCERTINI, M. (1973 b) "Identification of Distributed Parameter Systems using the ε-Method: a New Algorithm" - Rep. n. 3-01. Istituto di Automatica, Università di Roma.

KENNETH, P., SIBONY, M. and YVON, J.P. (1969) - "Penalization Technique for Optimal Control Problems Governed by Partial Differential Equations" - In 'Computing Methods in Optimization Problems', Vol. 2, Academic Press.

CONVERGENCE PROPERTIES OF THE GENERALIZED LEAST SQUARES IDENTIFICATION METHOD

Torsten Söderström

Division of Automatic Control, Lund Institute of Technology, Lund, Sweden

Convergence properties of the generalized least squares method are analyzed. The method can be interpreted as optimization of a likelihood function. The number of local maximum points of the likelihood function is examined. It is shown that this number is influenced by the signal to noise ratio. The theoretical results are illustrated by numerical examples. Plant measurements as well as simulated systems are used.

1. INTRODUCTION

The generalized least squares (GLS) identification method is an extension of the well-known least squares (LS) method, see Clarke (1967), Åström (1971). The basic advantage of the GLS method is the possibility to estimate the parameters for an arbitrary noise spectrum.

Assume that a process (called the system in this paper) is asymptotically stable and given by

$$A^*(q^{-1})\, y(t) = B^*(q^{-1})\, u(t) + v(t) \qquad (1.1)$$

where y(t) is the output at time t, u(t) the input and v(t) is noise. q^{-1} is the backward shift operator and

$$A^*(q^{-1}) = 1 + a_1 q^{-1} + \ldots + a_n q^{-n}$$

$$B^*(q^{-1}) = b_1 q^{-1} + \ldots + b_n q^{-n}$$

For simplicity the following conventions are used. e(t) is always denoting white noise. Its variance $E\, e(t)^2$ is denoted by σ^2. S denotes the signal to noise ratio.

In the following it will be assumed that the noise v(t) can be expressed as

$$v(t) = H^*(q^{-1})\, e(t)$$

where $H^*(q^{-1})$ is an asymptotically stable filter of finite order.

The purpose of an identification is to fit a mathematical model to given data. The LS method corresponds to the model

$$\hat{A}^*(q^{-1}) y(t) = \hat{B}^*(q^{-1}) u(t) + \varepsilon(t) \qquad (1.2)$$

and the parameter estimates are obtained from the unique minimum of the loss function (N denotes the number of data)

$$W = \frac{1}{2N} \sum_{t=1}^{N} \varepsilon^2(t)$$

If the noise v(t) is white this method will give consistent estimates, Åström (1968). If v(t) is not white the parameter estimates will in general converge to false values with probability one as the number of data tends to infinity.

If the correlation function of v(t) (or equivalently $H^*(q^{-1})$) is known it is possible to filter data with $H^*(q^{-1})^{-1}$ so that

$$v^F(t) = A^*(q^{-1}) y^F(t) - B^*(q^{-1}) u^F(t) \qquad (1.3)$$

is white noise. The LS method will then give consistent estimates if $u^F(t)$ and $y^F(t)$ are used.

However, the assumption that $H^*(q^{-1})$ is known is not realistic. The idea of the GLS method, Clarke (1967), is to estimate $H^*(q^{-1})$ as well as $A^*(q^{-1})$ and $B^*(q^{-1})$ in an iterative way. The main steps are the following.

1. Estimate $A^*(q^{-1})$ and $B^*(q^{-1})$ by the LS method.
2. Estimate $H^*(q^{-1})$ by fitting an autoregression to the residuals.
3. Filter data. Make a new LS estimation of $A^*(q^{-1})$ and $B^*(q^{-1})$ using the filtered data and repeat from 2.

In this paper two versions of the GLS method are discussed. They differ in the way of filtering and are both described in Clarke (1967). In version 1 always the original data are filtered. This version corresponds to the model Söderström (1972).

$$\hat{A}^*(q^{-1}) y(t) = \hat{B}^*(q^{-1}) u(t) + \frac{1}{\hat{C}^*(q^{-1})} \varepsilon(t) \qquad (1.4)$$

In version 2 the filters are applied to the already filtered data. This version corresponds to a model of the form

$$\hat{A}^*(q^{-1}) y(t) = \hat{B}^*(q^{-1}) u(t) + \frac{1}{\prod_{i=1}^{\infty} \hat{C}_i^*(q^{-1})} \varepsilon(t) \qquad (1.5)$$

The results presented in this paper can be found in a more detailed form in Söderström (1972), which contains proofs of all theorems in the paper. The main part of the paper will deal with the first version.

2. PRELIMINAIRIES

The concept of persistently exciting signals is used in the following theorems. See Åström (1971) and Ljung (1971) for a definition and discussions.

The following lemma gives a valuable interpretation of the first version of GLS.

Lemma 1. The first version of GLS is a minimization of the loss function

$$W(\hat{\theta}) = \frac{1}{2N} \sum_{t=1}^{N} \varepsilon^2(t)$$

$$\hat{\theta} = [\hat{a}_1, \ldots \hat{a}_n, \hat{b}_1, \ldots \hat{b}_n, \hat{c}_1, \ldots \hat{c}_n]^T$$

$$\varepsilon(t) = \hat{C}^*(q^{-1})[\hat{A}^*(q^{-1})y(t) - \hat{B}^*(q^{-1})u(t)]$$

by a relaxation method.

Further if

$$v(t) = \frac{1}{C^*(q^{-1})} e(t) \tag{2.1}$$

with e(t) white gaussian noise the GLS is the maximum likelihood method applied to (1.1).

Remark 1.: It follows from Caines (1971) that if (2.1) holds the estimates have nice asymptotic properties, they are e.g. consistent, asymptotically efficient and asymptotically normal.

Remark 2.: $W(\hat{\theta}_k)$, k = 1,2 ... being the number of iterations, form a decreasing bounded sequence which implies convergence. Possible bounded limits must be stationary points of $W(\hat{\theta})$. It is shown in Söderström (1972) that they must be minimum points if "stability" is required. (By a stable limit point is understood that a start sufficiently close to the point will imply convergence to the point). Thus it is of great interest to know if the loss function has a unique local minimum or not.

Remark 3.: Note that the convergence can be very slow. Close to a minimum point $\hat{\theta}_k$ will converge linearly.

Clearly W is a polynomial in $\hat{a}_1 \ldots \hat{c}_n$ where the coefficients are different sample covariances. An analysis of W and especially the local minimum points of this function must therefore be done in a probabilistic setting. In order to do the analysis reasonable asymptotic theory will be used.

In the following some assumptions are made

i) $u(t) = u_1(t) + G^*(q^{-1})e_1(t)$

$u_1(t)$ is deterministic and periodic. $G^*(q^{-1})$ is a stable filter of finite order.

$e_1(t)$ is white noise.

One of the terms may vanish.

ii) $e_1(t)$ and e(t) (u(t) and v(t)) are independent.

Using standard ergodic theory it can be seen, Söderström (1972), that these assumptions imply that $W(\hat{\theta})$ has a limit $V(\hat{\theta})$ for all $\hat{\theta}$ with probability one as N tends to infinity. Further

$$V(\hat{\theta}) = \frac{1}{2} E\left[\frac{\hat{C}^*(q^{-1})}{A^*(q^{-1})}\left\{\hat{A}^*(q^{-1})\, B^*(q^{-1}) - A^*(q^{-1})\, \hat{B}^*(q^{-1})\right\}u(t)\right]^2 + E\left[\frac{\hat{A}^*(q^{-1})\hat{C}^*(q^{-1})}{A^*(q^{-1})}\, v(t)\right]^2 \tag{2.2}$$

where $E\,u_1^2(t)$ denotes $\lim_{N\to\infty} \frac{1}{N}\sum_{t=1}^{N} u_1^2(t)$.

In the next section the number of local minimum points of $V(\hat{\theta})$ is examined.

3. MAIN RESULTS

In order to simplify the analysis a bit only "interesting" values of $\hat{\theta}$ will be considered. The following assumptions are made.

i) $\hat{A}^*(z)$ has all zeros outside the circle $|z| = r > 1$

ii) $\hat{C}^*(z)$ has all zeros outside the circle $|z| = r > 1$

iii) $\hat{b}_i$ is bounded $1 \leq i \leq n$.
r is some number close to 1.

The theoretical results presented in this paper require an infinite number of data. For practical purposes it is interesting to know if the results hold with "good approximation" for a finite number of data. In order to examine this problem simulations as well as identifications of plant measurements were made. All computations were carried out on a UNIVAC 1108. In Söderström (1972) a description of the used programs and other simulation results can be found.

Test of order has been performed in the following way, compare Bohlin (1970). When the order is increased by one, the decrease of the loss function is ΔV. The model of the lower order is accepted (ΔV is not significant) if the test quantity t = NΔV/(3 V) is small. Since 3 t is asymptotically χ^2 distributed under similar conditions it may be appropriate to accept the lower order model if t is less than 3. A straight forward application of such a test of order would in general result in more complex models for the actual processes. However, the order of the models are not unreasonable.

It is worth mentioning that the main purpose of the identifications of the plant measurements was to investigate the possible existence of more than one local minimum of the loss function. It is to be noted that for a "false" minimum point the estimated covariance matrix has statistically a dubious meaning.

The results of the identifications are compared with models obtained with the "ordinary" maximum likelihood model

$$\hat{A}^*(q^{-1})y(t) = \hat{B}^*(q^{-1})u(t) + \hat{C}^*(q^{-1})\varepsilon(t) \quad (3.1)$$

The results of the identifications of the plant measurementsare illustrated by plotted signals and normalized covariance functions. The following signals are plotted: the input $u(t)$, the output $y(t)$, the model output $y_m(t) = \hat{B}^*(q^{-1})/\hat{A}^*(q^{-1})\, u(t)$, the model error $e_m(t) = y(t) - y_m(t)$ and the residuals $\varepsilon(t)$. The plotted covariance functions are $r_\varepsilon(\tau)$, $r_{\varepsilon u}(\tau) = E\varepsilon(t)u(t+\tau)$ and $r_{e_m u}(\tau)$. The 5 % confidence intervals are given by dashed lines. The time is given in sampling periods in all the diagrams.

Large signals to noise ratios

The following theorem gives conditions for a unique minimum.

Theorem 1 - Let the system (1.1) of order n be controllable and the input $u(t)$ persistently exciting of order 2n. Assume that the order of the model is n. Consider parameter estimates in Ω, an arbitrary compact set.

Then there is a constant S_o such that if $S_o \leq S < \infty$ then the loss function (2.2) has exactly one stationary point in Ω. This point is a local minimum and satisfies.

$$\begin{aligned} \hat{a}_i &= a_i + O(1/S), S\to\infty \quad i = 1 \dots n \\ \hat{b}_i &= b_i + O(1/S), S\to\infty \quad i = 1 \dots n \\ \hat{c}_i &= \bar{c}_i + O(1/S), S\to\infty \quad i = 1 \dots n \end{aligned} \quad (3.2)$$

where $\bar{C}^*(q^{-1}) = 1 + \bar{c}_1 q^{-1} + \dots + \bar{c}_n q^{-n}$ and $(\bar{c}_1, \dots, \bar{c}_n)$ is the minimum point of

$$[E\, \hat{C}^*(q^{-1})v(t)]^2 \quad (3.3)$$

The different assumptions have the following meanings.

i) The restriction on the input signal is very natural. If the condition is not satisfied the loss function may have several global minimum points.

ii) The study of only parameter estimates in Ω is motivated before.

iii) The restriction on the signal to noise ratio is crucial as is shown in Theorem 3.

iv) The assumption of controllability is essential. If the system is not controllable, there is a factor in common between $A^*(q^{-1})$ and $B^*(q^{-1})$. Equation (1.1) can be divided by this factor, obtaining a controllable system of lower order than the original and with another correlation of the noise. If the system is not controllable, it is thus equivalent to regard the order of the model as higher than the order of the (controllable part of the) system. This situation is treated in

Theorem 2 - Let the system (1.1) be controllable and of order n.

Assume that the order of the model is n+k, $k > 0$ and that $u(t)$ is persistently exciting of order 2n+k. Consider parameter estimates in Ω, an arbitrary compact set. Then there is a constant S_o such that if $S_o \leq S < \infty$.

i) All local minimum points of the loss function (2.2) fulfil

$$\hat{A}^*(q^{-1}) = A^*(q^{-1})L^*(q^{-1}) + o(1),\ S \to \infty \quad (3.4)$$

$$\hat{B}^*(q^{-1}) = B^*(q^{-1})L^*(q^{-1}) + o(1),\ S \to \infty \quad (3.5)$$

where $L^*(q^{-1}) = 1 + \ell_1 q^{-1} + \dots + \ell_k q^{-k}$.

Further $L^*(q^{-1})$ and $\hat{C}^*(q^{-1})$ fulfil

$$L^*(q^{-1}) = \bar{L}^*(q^{-1}) + o(1),\ S \to \infty \quad (3.6)$$

$$\hat{C}^*(q^{-1}) = \bar{C}^*(q^{-1}) + o(1),\ S \to \infty \quad (3.7)$$

where $(\bar{\ell}_1, \dots, \bar{\ell}_k, \bar{c}_1, \dots, \bar{c}_{n+k})$ is a stationary point of

$$V_3(\ell_1, \dots, \ell_k, c_1, \dots, c_{n+k}) =$$

$$= [E\, L^*(q^{-1})\hat{C}^*(q^{-1})v(t)]^2 \quad (3.8)$$

The matrix of second order derivatives of V_3 in $(\bar{\ell}_1, \dots, \bar{c}_{n+k})$ must be positive definite or positive semidefinite.

ii) If the matrix of second order derivatives of V_3 in $(\bar{\ell}_1, \dots, \bar{c}_{n+k})$ is positive definite, then there exists a unique local minimum point of the form (3.4) - (3.7) and the terms $o(1)$ can be replaced by $O(1/S)$. Further the matrix V" is positive definite in this point.

Note that all minimum points have the property

$$\frac{\hat{B}^{*}(q^{-1})}{\hat{A}^{*}(q^{-1})} = \frac{\hat{B}^{*}(q^{-1})}{\hat{A}^{*}(q^{-1})} + o(1), \quad S \to \infty$$

Example 1

The data used in this example are plant measurements from a nuclear reactor in OECD Halden Reactor Project in Norway. The input is reactivity created by control rod movement and the output is the nuclear power. For a description of the experiment see Olsson (1973). ML identifications using the model (3.1) have been performed by Carlsson (1972). The system contains a direct term which is easily handled by shifting the data.

The test quantity for comparing the models of order 1 and 2 is 11.4, while the value is 1.1 when the models of orders 2 and 3 are compared. Thus the order two seems to be good. Two minimum points of $W(\theta)$ were found for this order.

The results of the identification is given in table 1 and figures 1 - 3.

Parameter	Model 1	Model 2	Model in Carlsson (1972)
$\hat{a}_1$	-0.18±0.06	-0.91±0.02	-0.89±0.02
$\hat{a}_2$	0.07±0.02	-0.01±0.02	-0.01±0.02
$\hat{b}_o$	2.41±0.14	2.38±0.14	2.43±0.12
$\hat{b}_1$	5.80±0,21	4.01±0.19	4.02±0.17
$\hat{b}_2$	-1.08±0.38	-5.83±0.16	-5.68±0.15
$\hat{c}_1$	-0.79±0.07	-0.06±0.04	Comparison is
$\hat{c}_2$	-0.12±0.06	0.06±0.03	impossible
$W\cdot10^4$	3.41	3.47	3.45
$\hat{\sigma}\cdot10^2$	2.61	2.63	2.63
Static gain	7.95	7.07	7.50

Table 1 - Parameter estimates from identification of the nuclear reactor data.

It is seen from the figures that the differences between the models are small. Further ($\hat{a}_1$, $\hat{a}_2$, $\hat{c}_1$, $\hat{c}_2$) of model 1 is close to ($\hat{c}_1$, $\hat{c}_2$, $\hat{a}_1$, $\hat{a}_2$) of model 2. In fact, both models as well as the model in Carlsson (1972) may be simplified to a first order system

$$y(t) = 2.4(1 + 2.6q^{-1})u(t) + \frac{1}{1 - 0.9q^{-1}}e(t) \quad (3.9)$$

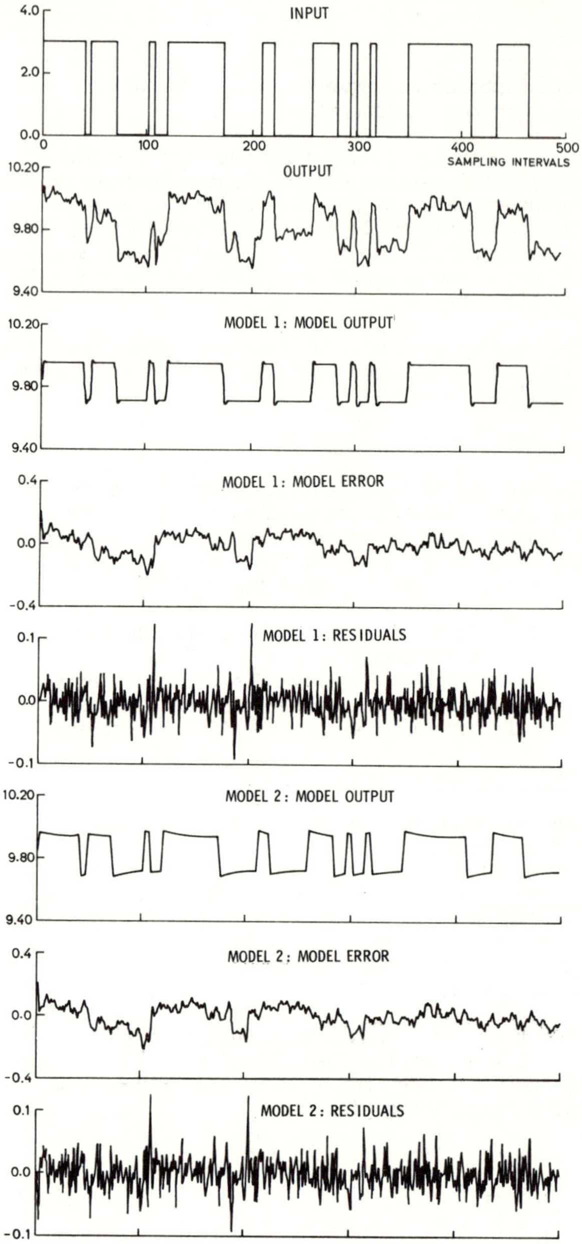

Fig 1. Models of the nuclear reactor. The input is given in digital units and the other variables in MW. The sampling period is 2 seconds.

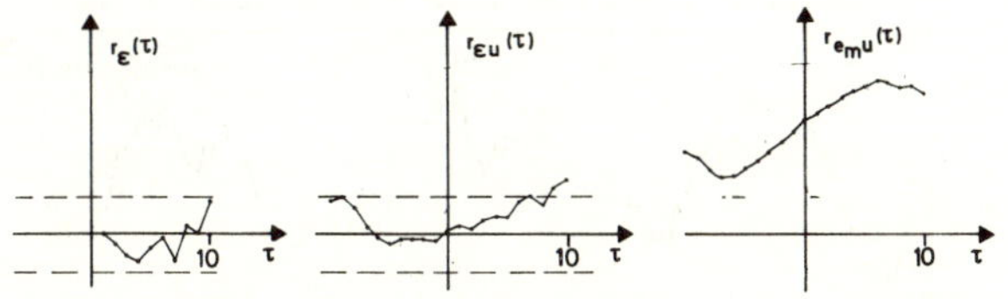

Fig 2. Normalized sample covariance functions for the nuclear reactor, model 1.

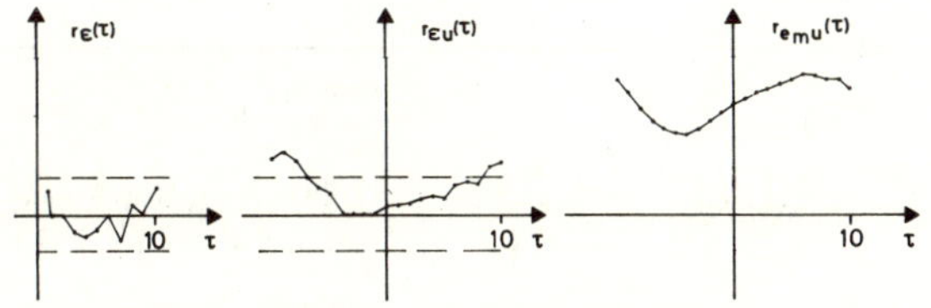

Fig 3. Normalized sample covariance functions for the nuclear reactor, model 2.

if approximate factors in common and small zeros are omitted.

An identification of a first order model gave the result:

$$y(t) = \frac{2.396 + 6.234q^{-1}}{1 - 0.00012q^{-1}} u(t) + \frac{1}{1 - 0.918q^{-1} + 0.0001q^{-2}} e(t)$$

and $\lambda = 2.66 \cdot 10^{-2}$ which differs just a little from the simplified model.

Since the two models do not differ very much it is impossible to call any of them the "best" or most "correct" one.

If (3.9) is an adequate description of the dynamical behaviour of the process then it is expected with Theorem 2 in mind, that there will be (at least) two different but equivalent models of second order. The models obtained by identification are in fact close to these expected models. Of course, this is a loose discussion according to the assumption that (3.9) describes the system adequately enough.

Small signal to noise ratios

When the signal to noise ratio is low it turns out that the loss function V in general has no unique minimum point. In order to describe the situation a concept called the "noise condition" (NC) is introduced.

Definition 1.

The noise $v(t) = H^*(q^{-1})e(t)$ fulfils the "noise condition" (NC) if there exist at least two different pairs of polynomials $\hat{A}_1^*(q^{-1})$, $\hat{C}_1^*(q^{-1})$ and $\hat{A}_2^*(q^{-1})$, $\hat{C}_2^*(q^{-1})$, such that

$$V_2(\hat{a}_1 \ldots \hat{a}_n, \hat{c}_1 \ldots \hat{c}_n) =$$

$$= E\left[\frac{\hat{A}^*(q^{-1})\hat{C}^*(q^{-1})H^*(q^{-1})}{A^*(q^{-1})} e(t)\right]^2$$

has local minimum points with positive definite matrices of second order derivatives in $(\hat{a}_{11} \ldots \hat{a}_{1n}, \hat{c}_{11} \ldots \hat{c}_{1n})$ and $(\hat{a}_{21} \ldots \hat{a}_{2n}, \hat{c}_{21} \ldots \hat{c}_{2n})$.

An analysis of (NC) and when it is fulfilled can be found in Söderström (1972). For example it is sufficient that V_2 has a minimum point with V_2'' positive definite and $\hat{A}^*(q^{-1}) \neq \hat{C}^*(q^{-1})$.

Theorem 3 - Assume that u(t) is persistently exciting of order n and that v(t) fulfils (NC). Then there is a number $S_1 > 0$ such that $0 < S \leq S_1$ implies that the loss function $V(\hat{\theta})$ has more than one local minimum point. Two of them are of the form

$$\hat{A}^*(q^{-1}) = \hat{A}_i^*(q^{-1}) + O(S) \qquad S \to 0$$
$$\hat{C}^*(q^{-1}) = \hat{C}_i^*(q^{-1}) + O(S) \qquad S \to 0 \qquad i = 1,2$$

with $\hat{A}_i$ and $\hat{C}_i$ from def. 1.

Bohlin (1971) has given results, which can be used to test if an estimate is the true maximum likelihood estimate. The test quantity involves sample covariances of $\varepsilon(t)$ and u(t). If, however, the noise level is high this method cannot be used successfully in the case described here. The minimum points of the loss function will give residuals $\varepsilon_1(t)$ and $\varepsilon_2(t)$ satisfying $\varepsilon_1(t) - \varepsilon_2(t) = O(S^{1/2})$ so that also all possible test quantities will differ just a little if S is small.

It can be argued that this fact is not so serious. When S is small identification is a hard task and it is difficult to get accurate models at all. However, in the examples below S is not unreasonably small. From the sample covariance functions no clear conclusions can be done.

Example 2

The system is a binary distillation column. The data have been received from Dr B.J. Williams, who has performed the measurements at National Physical Laboratory, London. Results of maximum likelihood identifications are reported in Gustavsson (1969). The input signal is the reflux ratio and the output signal is the top product composition. The test quantity for comparing models of orders 2 and 3 is 36. Since the ML identification indicates a model of order 2 as reasonable, this order was considered in spite of the great value of the test quantity.

For second order models two minimum points of the loss function were found. The results from the identification are given in Table 2 and Figures 4 - 6.

From Table 2 it is seen that $\hat{C}^{*}(q^{-1})$ of model 2 is very close to $\hat{A}^{*}(q^{-1})$ of model 1. With Theorem 3 in mind this is not astonishing.

The model from Gustavsson (1969) is very close to the model 1, which means that the noise can be well modelled as

$$v(t) = \hat{C}^{*}_{ML}(q^{-1})e(t)$$

as well as

$$v(t) = \frac{1}{\hat{C}^{*}_{GLS}(q^{-1})} e(t)$$

with e(t) white noise.

The values of the static gain indicate that model 1 gives the best description of the process. Also from the lower value of loss function at the corresponding minimum point, it can be expected that this model is to be preferred.

The plots of the results are a nice illustration of the expected differences.

From Figures 5 - 6 it is noted that the residuals are most white for model 2 and most uncorrelated with the input for model 1. That means it is hard (or impossible) to choose the "best" model from these figures.

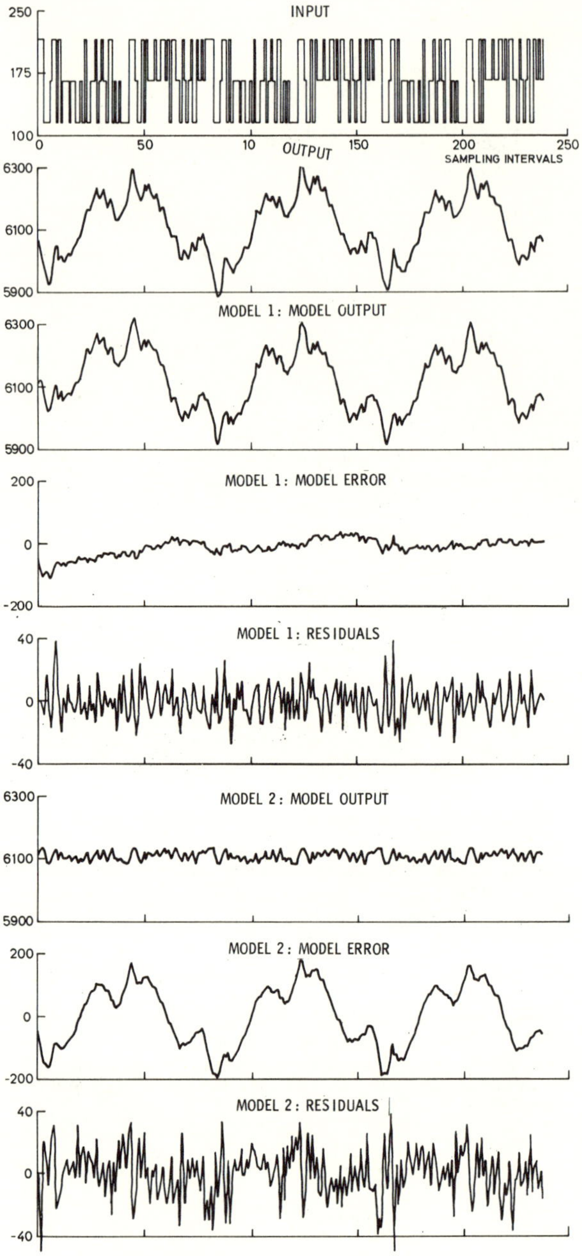

Fig 4. Models of the distillation column. Digital units are used. The sampling period is 96 seconds.

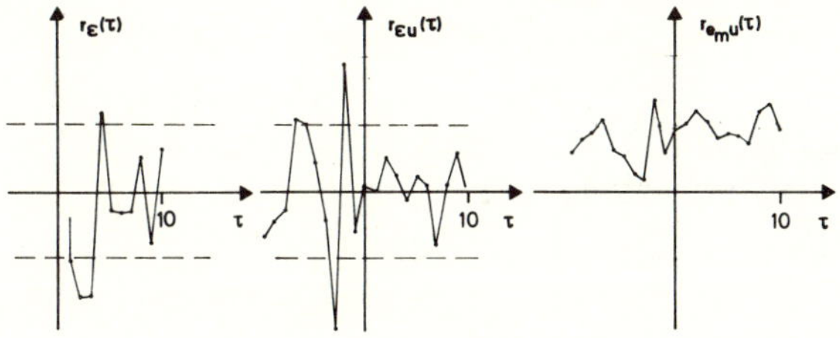

Fig 5. Normalized sample covariance functions for the distillation column, model 1.

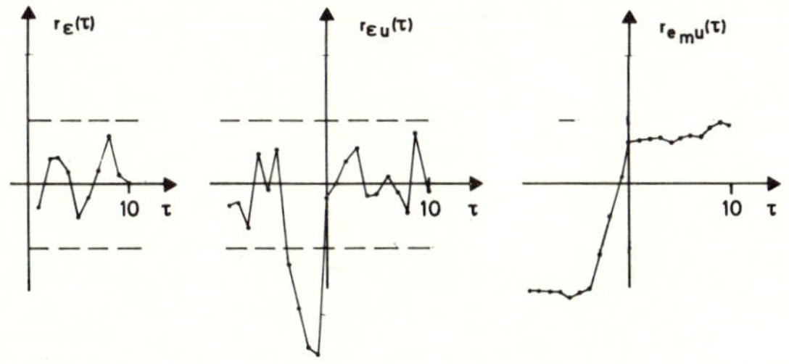

Fig 6. Normalized sample covariance functions for the distillation column, model 2.

Parameter	Model 1	Model 2	Corresp. ML model in Gustavsson (1969)
$\hat{a}_1$	-1.53 ± 0.02	0.19 ± 0.08	-1.54 ± 0.02
$\hat{a}_2$	0.55 ± 0.02	-0.02 ± 0.05	0.55 ± 0.02
$\hat{b}_1$	0.24 ± 0.02	0.37 ± 0.02	0.23 ± 0.02
$\hat{b}_2$	-0.62 ± 0.02	0.22 ± 0.03	-0.60 ± 0.02
$\hat{c}_1$	0.82 ± 0.06	-1.51 ± 0.08	Comparison
$\hat{c}_2$	0.41 ± 0.06	0.54 ± 0.07	impossible
W	69.04	164.37	71.68
$\hat{\sigma}$	11.75	18.13	11.97
Static gain	-18.77	0.507	-22.46

Table 2 - Parameter estimates from GLS identification of the distillation column data.

Example 3

The system is a laboratory heat diffusion process at the Division of Automatic Control, Lund Institute of Technology.

Identification results using the ML model (3.1) as well as a short description of the process are given in Leden (1971).

The test quantity for comparing models of orders 4 and 5 has the value 190. Since the ML identification indicates a model of order 4 as reasonable, this order was considered in spite of the great value of the test quantity.

The loss function turns out to have (at least) two minimum points for fourth order models. The results are presented in Table 3 and Figures 7 - 9.

	Model 1	Model 2	Corresponding model in Leden (1971)
$\hat{a}_1$	-2.43 ± 0.04	-1.24 ± 0.05	-2.96 ± 0.002
$\hat{a}_2$	1.88 ± 0.11	0.61 ± 0.09	3.27 ± 0.005
$\hat{a}_3$	-0.37 ± 0.10	-0.52 ± 0.07	-1.61 ± 0.005
$\hat{a}_4$	-0.07 ± 0.03	0.35 ± 0.03	0.30 ± 0.002
$\hat{b}_1\cdot10^3$	0.15 ± 0.06	-0.64 ± 0.07	0.0
$\hat{b}_2\cdot10^3$	-0.02 ± 0.12	-0.34 ± 0.08	0.13 ± 0.01
$\hat{b}_3\cdot10^3$	-0.84 ± 0.13	-0.71 ± 0.07	-0.42 ± 0.03
$\hat{b}_4\cdot10^3$	1.51 ± 0.08	-0.50 ± 0.08	0.79 ± 0.01
$\hat{c}_1$	1.39 ± 0.05	-0.66 ± 0.05	Comparison
$\hat{c}_2$	1.28 ± 0.05	-1.06 ± 0.06	is impossible
$\hat{c}_3$	0.96 ± 0.05	0.22 ± 0.05	
$\hat{c}_4$	0.39 ± 0.03	0.50 ± 0.05	
W	$2.16\cdot10^{-7}$	$5.13\cdot10^{-2}$	$0.92\cdot10^{-7}$
$\hat{\sigma}$	$0.658\cdot10^{-3}$	$1.013\cdot10^{-3}$	$0.428\cdot10^{-3}$
Static gain	0.2528	-0.0100	0.2440

Table 3 - Parameter estimates from GLS identification of the heat rod.

The theoretical value of the static gain is 0.25, which indicates that model 1 is superior to model 2. This fact is very much confirmed by the plotted signals.

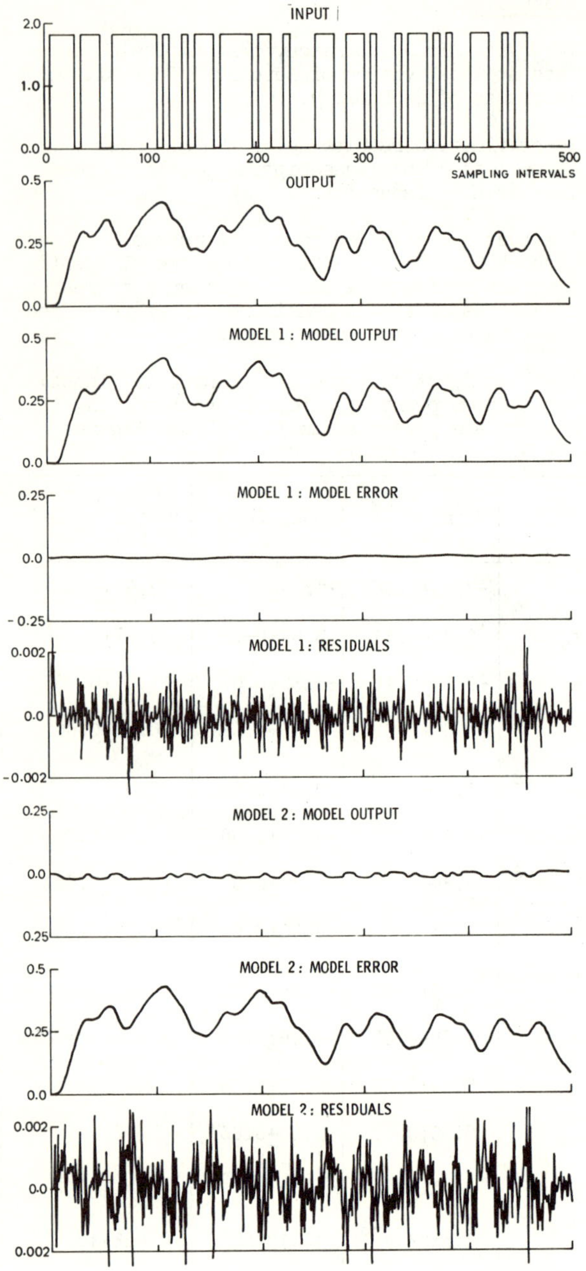

Fig 7. Models of the heat-rod process. All variables are given in °C. (Constants are added to the input, the output and the model outputs). The sampling period is 10 seconds.

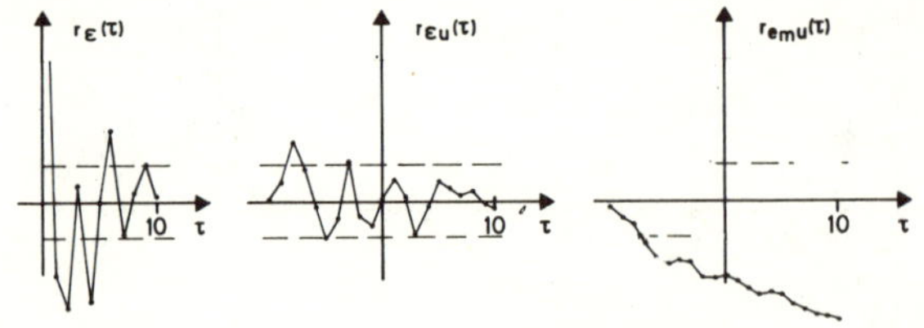

Fig 8. Normalized sample covariance functions for the heat-rod model 1.

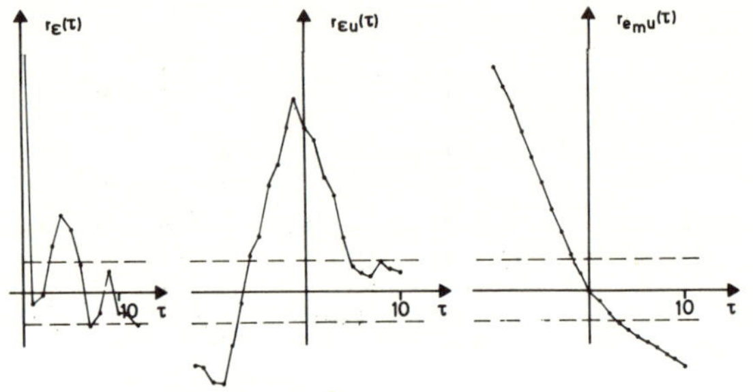

Fig 9. Normalized sample covariance functions for the heat-rod model 2.

A comparison with plots of the ML model (given in Leden (1971)) shows small difference between that model and model 1.

In the model 2 the output is "interpreted" as mainly due to noise.

From Figures 8 - 9 it is seen that the residuals are not white in any of the two models. The input signal and the residuals are considerably more correlated for the second model.

The second version of GLS

Now an example illustrating the possible behaviour of the second version of GLS is described. The question of convergence of this version under suitable conditions cannot be answered easily, and it has not been studied by the author.

The following case will be taken into consideration. The system and the model are both of first order. The iteration is started with the LS estimate of a and b. Conditions for convergence in the next step are examined. If the estimated operator $\hat{C}^*(q^{-1}) \equiv 1$ then the next estimation of $\hat{a}$ and b will give the same result as before.

The interesting equations are thus

$$\hat{c} = -\frac{r_\varepsilon(1)}{r_\varepsilon(0)} = 0 \tag{3.10}$$

$$\varepsilon(t) = (1+\hat{a}q^{-1})y(t)-\hat{b}q^{-1}u(t) \tag{3.11}$$

$$\begin{bmatrix} r_y(0) & -r_{yu}(0) \\ -r_{yu}(0) & r_u(0) \end{bmatrix} \begin{bmatrix} \hat{a} \\ \hat{b} \end{bmatrix} = \begin{bmatrix} -r_y(1) \\ r_{yu}(1) \end{bmatrix} \tag{3.12}$$

Example - Consider the system

$$(1 + aq^{-1})y(t) = u(t) + v(t),$$

$$v(t) = \frac{1}{1 + cq^{-1}} e(t)$$

where u(t) is white noise. There is a number $S_o > 0$ such that if $0 < S \leq S_o$ then (3.10) has two solutions with respect to a, which satisfy

$a = O(S), \qquad S \to 0$

$a = - c + O(S), \qquad S \to 0$

Calculations which prove the assertion can be found in Söderström (1972).

Examples 4 - 6

The following examples illustrate that the second version of the GLS method can give false values of the estimates.

All the systems were generated by the equation

$$(1 + aq^{-1})y(t) = bq^{-1}u(t) + \frac{1}{1 + cq^{-1}} e(t)$$

The number of samples was 500. The input signal was a PRBS with amplitude 1.0, a good approximation of white noise. The results of the identifications confirm the theory. $\sigma_{\hat{c}}$ denotes the estimated standard deviation of $\hat{c}$.

Example	Parameters	σ	$a(\hat{a})$	$b(\hat{b})$	$c(\hat{c})$	$\sigma_{\hat{c}}$
4	true values	100.0	-0.5	1.0	0.5	
	expected estimates		0.0	1.0	0.0	0.045
	obtained estimates		-0.01	0.98	0.007	0.045
5	true values	100.0	0.0	1.0	-0.8	
	expected estimates		-0.8	1.0	0.0	0.045
	obtained estimates		-0.79	1.08	-0.026	0.045
6	true values	1.2	-0.7	1.0	0.9	
	expected estimates		0.09	1.0	0.0	0.045
	obtained estimates		0.16	0.94	0.043	0.045

Table 4 - Identification results with the second version of GLS.

4. CONCLUSIONS

Some essential properties of the generalized least squares (GLS) method for identification of dynamical systems are summarized below. Part of the material is well-known.

- The GLS method can be interpreted as a maximum likelihood method when suitable assumptions of the structure of the equations governing the system are made. The estimation of the correlation of the noise can be done in some different ways, and every way to do it corresponds to some structure of the system equations. The GLS method is then a special minimization algorithm applied to a corresponding loss function, which is a sum of squared residuals. The GLS method will then have nice asymptotic properties.
- The GLS method is an uncomplicated extension of the least squares (LS) method. Besides a program for LS identification only programs for administration and filtering are needed.
- The GLS method gives a very slow convergence

close to a minimum point of the loss function. The method is thus inappropriate if great accuracy is required.

o Applied to nice data the GLS method gives good results comparable with the results of the more complicated "ordinary" maximum likelihood method. The required conditions of the data are weaker than for the simpler LS method. For a sufficiently high value of the signal to noise ratio it can be shown theoretically that the loss function has only one local minimum point.

o The loss function corresponding to the GLS method may have more than one minimum point. In this case the result of the GLS identification depends on the start values of the parameter estimates. The existence of several minimum points can be shown theoretically for low signal to noise ratios. In practice it can happen also for reasonable values of this ratio. It is not always easy without intimate knowledge of the actual process to decide which of the models that will be the "best" or most "correct".

ACKNOWLEDGEMENTS

The author wants to thank his thesis supervisor prof. K.J. Åström who proposed this problem and provided valuable guidance for the solution. In particular he pointed out the possibility of interpreting the GLS procedure as a special maximization of the likelihood function. Many stimulating discussions with tekn. lic. Ivar Gustavsson are also gratefully acknowledged. He is grateful for the measurements, which were supplied to the Div. of Aut. Control, Lund Institute of Technology by OECD Halden Reactor Project, Dr B.J. Williams, London and tekn. lic. Bo Leden, Lund. The project was partially supported by the Swedish Board for Technical Development under contracts 71-50/U 33 and 72 - 202/U 137.

REFERENCES

Åström, K.J. (1968).
Lectures on the Identification Problem - the Least Squares Method. Report 6806, Division of Automatic Control, Lund Institute of Technology.

Åström, K.J. and Eykhoff, P. (1971).
System Identification - A Survey. Automatica 7, 123 - 162.

Bohlin, T. (1970).
On the Maximum Likelihood Method of Identification. IBM J. Res. and Dev., 14, No 1, 41 - 51.

Bohlin, T. (1971).
On the Problem of Ambiguities in Maximum Likelihood Identification. Automatica 7, 199 - 210.

Caines, P.E. (1971).
The Parameter Estimation of State Variable Models of Multivariable Linear Systems. Control Systems Centre Report No. 146, The University of Manchester, Institute of Science and Technology.

Carlsson, S. (1972).
Maximum Likelihood Identification of Reactor Dynamics from Multivariable Experiments. (In Swedish). Master Thesis, Division of Automatic Control, Lund Institute of Technology.

Clarke, D.W. (1967).
Generalized Least Squares Estimation of the Parameters of a Dynamic Model. 1st IFAC Symposium on Identification in Automatic Control Systems. Prague.

Gustavsson, I. (1969)
Identification of Dynamics of a Distillation Column. Report 6916, Division of Automatic Control, Lund Institute of Technology.

Leden, B. (1971)
Identification of Dynamics of a One Dimensional Heat Diffusion Process. Report 7121, Division of Automatic Control, Lund Institute of Technology.

Ljung, L. (1971)
Characterization of the Concept of "Persistently Exciting" in the Frequency Domain. Report 7119, Division of Automatic Control, Lund Institute of Technology.

Olsson, G. (1973).
Modelling and Identification of Nuclear Power Reactor Dynamics from Multivariable Experiments. Preprints 3rd IFAC Symp. Identification and system parameter estimation, Hague. June, 1973.

Söderström, T. (1972).
On the Convergence Properties of the Generalized Least Squares Identification Method. Report 7228, Division of Automatic Control, Lund Institute of Technology.

TWO STAGE LEAST SQUARES ESTIMATORS AND THEIR RECURSIVE APPROXIMATIONS

R.N. Pandya,
Communication Engineering Dept.,
Government Engineering College,
Jabalpur, M.P., India.

B. Pagurek,
Division of Systems Engineering,
Carleton University,
Ottawa, Ontario, Canada.

Two Stage Least Squares (2SLS) estimators for the various sets of parameters associated with a single output, linear stochastic difference equation model of known order are presented. Recursive (on-line) approximations to these Two Stage Least Squares estimators are formulated. It is shown that the basic structure of these recursive approximations fits most of the on-line estimators that are presently available for this model.

1. INTRODUCTION

Linear Simulations Equation models are frequently employed in econometrics for representing the behaviour of economic systems. This multivariable model can be written in the form:

$$A\underline{y}(k) = B\underline{u}(k) + \underline{e}(k) \qquad (1.1)$$

where k is the discrete time index and

$\underline{y}(k)$ = G × 1 vector of endogenous variables,
$\underline{u}(k)$ = K × 1 vector of predetermined variables,
$\underline{e}(k)$ = G × 1 vector of random errors,

A = G × G nonsingular matrix of parameters,
B = G × K matrix of parameters.

The endogenous variables (outputs) are determined within the system and are in general, correlated with the errors. The predetermined variables may include the exogenous variables (inputs) that are determined outside the systtem and are statistically independent of the errors for every time index k; and some lagged endogenous variables (past outputs) that are statistically independent of the present and future errors. For this paper, it will be assumed that $\underline{u}(k)$ include only the exogenous variables.

The error vector $\underline{e}(k)$ is zero mean and for a given time index k, the errors across the system (from one equation to the other) are correlated with a constant covariance matrix. However, within an individual equation, the errors are not correlated from one time period to the other. Since the coefficient matrix A in (1.1) is nonsingular, we have:

$$\underline{y}(k) = A^{-1}B\underline{u}(k) + A^{-1}\underline{e}(k) \qquad (1.2)$$

This constitutes the so called 'Reduced Form' model, which is characterised by the fact that all the endogenous variables are specified in terms of the exogenous variables alone. A single equation (say the first) in the system (1.1) may be written as:

$$y_1(k) = b_{11}u_1(k)+b_{12}u_2(k)+..+b_{1J}u_J(k) - a_{12}y_2(k)-\ldots-a_{1H}y_H(k)+e_1(k) \qquad (1.3)$$

where $H \leq G$ and $J \leq K$, which implies that some of the elements of A and B in (1.1) are zero.

If the number of unknown parameters in the first structural equation (1.3) is $p = J+H-1$, then the equation is said to be 'unidentified' if $K<p$; it is 'just-identified' if $K=p$; and it is 'over-identified' if $K>p$. It is the last case that is of interest to us in this paper.

The Ordinary Least Squares (OLS) estimator for the structural parameters $(b_{11},b_{12},\ldots,b_{1J},-a_{12},-a_{13},\ldots,-a_{1H})$ in (1.3) is not consistent because the regressors $y_2(k),y_3(k),\ldots,y_H(k)$ are correlated with the errors $e_1(k)$. This correlation results from the fact that these regressors are determined in the subsequent equations of the system, whose error terms are correlated with $e_1(k)$.

The Two Stage Least Squares (2SLS) estimator, proposed by Basmann [1957] and Theil [1958], uses the following two stages for obtaining consistent estimates of the structural parameters:

Stage 1: Use OLS to fit the 'reduced form model for the endogenous variables $y_2(k)$, $y_3(k)$, ..., $y_H(k)$ in (1.3). Since the reduced form model employes only the exogeneous variables as explanatory variables, the OLS estimates of the reduced form parameters are consistent.

Stage 2: In model (1.3), replace the endogeneous variables $y_2(k), y_3(k), \ldots\ldots, y_H(k)$ by their fitted (estimated) values from Stage 1; and apply OLS to the resulting model, to get the required estimates of the structural parameters.

Needless to say, the foregoing discussion of Linear Simultaneous Equation models and the 2SLS technique is rather brief and formal. For a detailed description and rigorous analysis of these topics, see Dhrymes [1970], [1972].

The model to be considered in this paper is the single output, n^{th} order, linear stochastic difference equation model having the general canonic form:

$$A(q^{-1})y(k) = B(q^{-1})u(k) + C(q^{-1})e(k) \qquad (1.4)$$

where $A(q^{-1})$, $B(q^{-1})$ and $C(q^{-1})$ are polynomials in the backward shift operator q^{-1} (with $a_o = c_o = 1$); $[u(k); y(k)]$ represents the measured input-output pairs, and $e(k)$ is a sequence of identically and independently distributed (i.i.d.) random variables with zero mean and constant variance λ^2. It is further assumed that the characteristic roots of $A(q^{-1})$ and $C(q^{-1})$ lie inside the origin centered unit circle and that the input sequence $u(k)$ is determined outside the system, so that its members are statistically independent of the errors $e(k)$ for every k. Equation (1.4) expands into:

$$y(k) = b_o u(k) + b_1 u(k-1) + \ldots + b_n u(k-n) - a_1 y(k-1) - \ldots - a_n y(k-n) + v(k) \qquad (1.5)$$

where the consolidated error term $v(k)$ is the moving average process:

$$v(k) = e(k) + c_1 e(k-1) + \ldots + c_n e(k-n) \qquad (1.6)$$

$$e(k) \sim \text{i.i.d. } (0, \lambda^2) \qquad (1.7)$$

The OLS estimator for the structural or process paramters $(b_o, b_1, \ldots, b_n, -a_1, \ldots, -a_n)$ in the above model is not consistent because this model also contains a set of regressors viz. $y(k-1), y(k-2), \ldots, y(k-n)$, that are correlated with the error $v(k)$. The correlation in this case is due to the simultaneous presence of serially correlated errors and lagged outputs in the model.

In this paper an attempt has been made to answer the following questions:

1. Since the inconsistency of OLS in the canonical model (1.5) and the single equation (1.3) of the Simultaneous Equation model (1.1) can be traced to the same source - is it possible to formulate a 2SLS estimator for the process parameters in the model under study?

2. The design of a suitable control strategy requires consistent estimates of the process as well as the noise parameters $(c_1, c_2, \ldots, c_n)$. To this end, is it possible to suitably reinterpret the principle behind the 2SLS technique, so that based on this extended definition, 2SLS estimators may be formulated for (a) the noise parameters when the process parameters or their consistent estimates are known; and (b) the process as well as the noise parameters?

3. Is it possible to formulate recursive (on-line) approximations to the 2SLS estimators proposed in 1. and 2. above? If so, how are they related to the prevailing on-line estimators for the corresponding sets of parameters? The motive behind taking up the last question is to investigate the possibility of providing a unified structure for the various on-line estimators that are currently available for this model.

2. ESTIMATION OF PROCESS PARAMETERS

2.1 2SLS Estimator: The 2SLS estimator for the process parameters in model (1.5)-(1.7) will now be considered. The first stage in the 2SLS procedure requires that a 'reduced form' model be fitted for those explanatory variables that are correlated with the error. For the model being considered, an obvious choice for the reduced form model is the 'Impulse Response' model for the system. Thus (1.4) may be put as:

$$y(k) = \frac{B(q^{-1})}{A(q^{-1})} u(k) + \frac{C(q^{-1})}{A(q^{-1})} e(k) \qquad (2.1)$$

$$\text{where } x(k) = \frac{B(q^{-1})}{A(q^{-1})} u(k) \qquad (2.2)$$

$$\text{and } w(k) = \frac{C(q^{-1})}{A(q^{-1})} e(k) \qquad (2.3)$$

are respectively, the noise-free process outputs and the errors associated with the impulse response model.

Theoretically, the impulse response representation would involve an infinite number of parameters. In practice, however, the sequence can be terminated after a finite number of terms under the stability assumption on $A(q^{-1})$. The decision regarding the number of terms to be retained may be based on priliminary step response tests on the system. Thus the impulse response model may be specified as:

$$y(k)=\gamma_o u(k)+\gamma_1 u(k-1)+\ldots+\gamma_n u(k-n)+\ldots$$
$$\ldots+\gamma_{s-1}u(k-s+1)+w(k) \qquad (2.4)$$

where the number of parameters is s. For (s+N) pairs of input-output data, the structural model (1.5) may be written as:

$$\underline{y} = M\underline{\beta}+\underline{v} = [M_1 \vdots M_2]\begin{bmatrix}\underline{\beta}_1\\ \hline \underline{\beta}_2\end{bmatrix} + \underline{v} \qquad (2.5)$$

$$\underline{y} = [y(s+1)\ y(s+2)\ldots\ldots y(s+N)]' \qquad (2.6)$$

$$\underline{\beta} = [\underline{\beta}_1' \vdots \underline{\beta}_2']' = [b_o b_1 \ldots b_n \vdots -a_1 - \ldots -a_n]' \quad (2.7)$$

The partitioning of the design matrix M in (2.5) is meant to emphasize the fact that elements of M_1 consist of input (or exogenous) variables only, whereas the elements of M_2 consist of lagged outputs that arecorrelated with the element of $\underline{v}$.

In the vector-matrix notation the impulse response model (2.4) is:

$$\underline{y} = U\underline{\gamma}+\underline{w} = [M_1 \vdots U_2]\underline{\gamma} + \underline{w} \qquad (2.8)$$

where $\underline{y}$ is the same as in (2.6), $\underline{\gamma}$ is the (s×1) vector of impulse response coefficients and U is the associated (N×s) design matrix whose first (n+1) columns are common with M, as indicated by the partitioning. The reduced form errors (elements of $\underline{w}$) are zero mean and are, in general, serially correlated. Since the matrix U contains only the input variables, we also have:

$$E(U'\underline{w}) = \underline{0} \qquad (2.9)$$

Assuming a full rank for U, the OLS estimator for $\underline{\gamma}$ is given by

$$\hat{\underline{\gamma}} = (U'U)^{-1}U'\underline{y} \qquad (2.10)$$

in which the unbiasedness (for finite sample size) and consistency of $\hat{\underline{\gamma}}$ is assured by (2.9). However, the estimates are not minimum variance because the errors w(k) are serially correlated. They will be minimum variance iff. $c_1 = a_1$, which implies that w(k) = e(k).

Since the number of process parameters to be estimated in (1.5) is p = (2n+1); it may be noted that the model will be 'unidentified' if s < p, it will be 'just-identified' if s = p and will be 'over-identified' if s > p. For the present discussion, it will be assumed that the model is over-identified.

The 2SLS estimator for the vector of process parameters $\underline{\beta}$ can, therefore, be implimented in practice, by executing the following steps sequentially:

1. For the impulse response model (2.8), estimate the impulse response parameters $\underline{\gamma}$, using the OLS estimator (2.10).

2. Obtain the fitted or estimated values of y(k) from

$$\hat{\underline{y}} = U\hat{\underline{\gamma}} \qquad (2.11)$$

It may be noted that these also represent the estimates of the noise-free process outputs x(k) as defined in (2.2) and we shall make use of this fact in the latter part of this section.

3. In the structural model (2.5) replace the lagged outputs on the right hand side i.e. the elements of the sub-matrix M_2, by their fitted values from 2 above to give

$$\underline{y} = [M_1 \vdots \hat{M}_2]\begin{bmatrix}\underline{\beta}_1\\ \hline \underline{\beta}_2\end{bmatrix} + \underline{v} + \hat{W}_2\underline{\beta}_2 \qquad (2.12)$$

where $\hat{W}_2 = (M_2 - \hat{M}_2)$ is the matrix of reduced form residuals.

4. Use OLS on model (2.12) to give the 2SLS estimator for $\underline{\beta}$:

$$\hat{\underline{\beta}}(2SLS) = \begin{bmatrix}\hat{\underline{\beta}}_1\\ \hat{\underline{\beta}}_2\end{bmatrix} = \begin{bmatrix}M_1'M_2 & M_1'\hat{M}_2\\ \hat{M}_2'M_1 & \hat{M}_2'M_2\end{bmatrix}^{-1}\begin{bmatrix}M_1'\underline{y}\\ \hat{M}_2'\underline{y}\end{bmatrix} \qquad (2.13)$$

The OLS estimator for $\underline{\gamma}$ (and M_2) in the first stage and that for $\underline{\beta}$ in the second stage can be combined to give a single expression for the 2SLS estimator. This consolidated expression has been derived in the Appendix and is given by:

$$\hat{\underline{\beta}}(2SLS) = [M'U(U'U)^{-1}U'M]^{-1}M'U(U'U)^{-1}U'\underline{y} \qquad (2.14)$$

which is in a form that is often found in econometrics e.g. see Dhrymes [1970]. It is not difficult to reduce (2.13) to the following form that is also frequently used:

$$\hat{\underline{\beta}}(2SLS)=\begin{bmatrix} M_1'M_1 & M_1'M_2 \\ M_2'M_1 & (M_2'M_2-\hat{W}_2'\hat{W}_2) \end{bmatrix}^{-1}\begin{bmatrix} M_1'\underline{y} \\ (M_2-\hat{W}_2)'\underline{y} \end{bmatrix} \tag{2.15}$$

which is equivalent to the form:

$$\hat{\underline{\beta}}(2SLS)=\begin{bmatrix} M_1'M_1 & M_1'M_2 \\ \hat{M}_2'M_1 & \hat{M}_2'M_2 \end{bmatrix}^{-1}\begin{bmatrix} M_1'\underline{y} \\ \hat{M}_2'\underline{y} \end{bmatrix} \tag{2.16}$$

Derivation of the above forms and proof of consistency may be found in Pandya [1972].

It is important to note that the form (2.16) implies, that the 2SLS estimator (2.13) is equivalent to the Instrumental Variable (IV) estimator in which the matrix $(M_1 \mid \hat{M}_2)$ is used as an IV matrix in model (2.5) - $\hat{M}_2$ being obtained through an impulse response fit for M_2. This equivalence of the 2SLS estimator to an IV estimator is also apparent from (2.14).

Remark: If the model is 'just-identified' (s=p), so that the matrix (U'M) in (2.14) is p×p, and also if this matrix has a nonsingular limit, then the 2SLS estimator reduces to:

$$\hat{\underline{\beta}}(2SLS) = (U'M)^{-1}U'\underline{y} \tag{2.17}$$

which is the so called 'Indirect Least Squares' estimator for a just-identified model. If (2.17) is used on an over-identified model, the estimates are consistent, though much less efficient (asymptotically) as compared to the 2SLS estimator. For comparative results, see Pandya [1972].

Remark: If the input sequence u(k); k=1,2,3,... is such that it satisfies the relation:

$$\underset{N\to\infty}{\text{p.lim}} \left[\frac{1}{N} U'U\right] = \alpha I \tag{2.18}$$

then the 2SLS estimator (2.14) simplifies to:

$$\hat{\underline{\beta}}(2SLS) = (M'UU'M)^{-1}M'UU'\underline{y} \tag{2.19}$$

In other words, if the input sequence is obtained either as samples from a white noise process or as a suitably designed Pseudo Random Binary Sequence (PRBS), for large sample size, the impulse response coefficients may be estimated as the simple average:

$$\hat{\underline{\gamma}} = \left(\frac{1}{\alpha N}\right) U'\underline{y} \tag{2.20}$$

This result is extensively used in correlation experiments and leads to a drastic reduction in computation time and storage. If necessary, therefore, a large number of impulse response coefficients may be used, with nominal increase in the computational lead.

Remark: Mayne [1967] has used a form of 2SLS procedure for estimating the process parameters, in which suitably lagged outputs have been included in the reduced form model. By doing this, not only is the computational advantage mentioned in the preceeding remark lost, but as Pandya [1972] has found, there is little improvement in the efficiency of the estimates.

2.2 Recursive Approximation to $\hat{\underline{\beta}}(2SLS)$:

As mentioned earlier, the fitted values of y(k); k = 1,2,3,.... obtained in the first stage of the 2SLS procedure are also the estimates of the noise-free process outputs x(k); k = 1,2,3,..... Hence we may say that $\hat{\underline{\beta}}(2SLS)$ is obtained by using OLS on the model:

$$y(k)=b_0u(k)+b_1u(k-1)+\ldots+b_nu(k-n) - a_1\hat{x}(k-1)-\ldots-a_n\hat{x}(k-n)+\hat{w}(k) \tag{2.21}$$

Alternatively, the equivalence of the 2SLS estimator (2.13) and the IV estimator (2.16) implies that $\hat{\underline{\beta}}(2SLS)$ are obtained by using $\hat{x}(k)$; k = 1,2,3,.... as instrumental variables in model (1.5) or by using the IV matrix

$$\hat{X}' = (M_1 \mid \hat{M}_2) = (M_1 \mid \hat{X}_2)$$

on the vector matrix model (2.5).

Both, Wong [1967] and Young [1967] have derived an exact recursive form for an IV estimator. One of the important advantages of the recursive form is that it calls for the instrumental variables sequentially – one at a time (same being true for the input-output data pairs) and the complete set need not be known apriori.

To reiterate therefore, consistent estimates of the process parameters β can be obtained (recursively or otherwise), if consistent estimates of $x(k)$; $k = 1,2,3,..$ are available. However, the converse is also true i.e. if consistent estimates of β are available, then consistent estimates of $x(k)$ can be obtained from the recursive relation (difference equation):

$$\hat{x}(k)=\hat{b}_0 u(k)+\hat{b}_1 u(k-1)+\ldots+\hat{b}_n u(k-n) - \hat{a}_1\hat{x}(k-1)-\ldots-\hat{a}_n\hat{x}(k-n) \quad (2.22)$$

in which a knowledge of the exact initial conditions is not critical, because the process is assumed to be stable.

The recursive 2SLS|IV estimator for β and the recursive estimator for $x(k)$ may be 'bootstrapped' together to give the following recursive approximation to $\hat{\beta}(2SLS)$:

$$\tilde{\underline{\beta}}(k+1)=\tilde{\underline{\beta}}(k)+\frac{C(k)\underline{z}(k+1)[y(k+1)-\underline{m}'(k+1)\tilde{\beta}(k)]}{1+\underline{m}'(k+1)C(k)\underline{z}(k+1)}$$

$$C(k+1)=C(k)-\frac{C(k)\underline{z}(k+1)\underline{m}'(k+1)C(k)}{1+\underline{m}'(k+1)C(k)\underline{z}(k+1)}$$

$C(k)$ = error covariance matrix

$$\underline{m}(k+1)=[u(k+1)..u(k-n+1)y(k)..y(k-n+1)]'$$

$$\underline{z}(k+1)=[u(k+1)..u(k-n+1)\tilde{x}(k)..\tilde{x}(k-n+1)]'$$

$$\tilde{x}(k) = \underline{z}'(k)\,\tilde{\underline{\beta}}(k)$$

In practice, the initial conditions for the algorithm may be chosen as:

$$\tilde{\underline{\beta}}(o)=\underline{0};\ C(o)=\alpha I(\alpha>>1);\ \underline{z}(o)=\underline{m}(o)$$

which implies a recursive OLS starter with a minimum data set.

But for certain ad-hoc changes, the on-line IV estimators proposed by Wong [1967] and Young [1967] have the same basic structure as the simple recursive approximation to 2SLS estimator presented here. Extensive tests, using artificially generated data as well as experimental data, have been carried out on the simple recursive algorithm. No problem with convergence were ever encountered and on the average it seems to have an edge over the existing on-line estimators, as far as computational simplicity and accuracy is concerned.

2.3 <u>Results</u>: Test results on data generated by a second order model are given here. Average estimates over 20 different noise sequences have been tabulated together with the RMS error.

Pertinent details of the simulation experiment are given below:

<u>Noise</u>: Normally and independently distributed with zero mean and unit variance.

<u>Input</u>: PRBS of unit amplitude, clock time of one sampling period and basic length of 127 samples.

<u>No. of Trials</u>: 20

<u>No. of Data</u> : 1,000

<u>No.of impulse response parameters</u>: 25

The results are given in Table 2.1.

3. ESTIMATION OF NOISE PARAMETERS

3.1 <u>2SLS Estimator</u>: In model (1.5)-(1.7), if the process parameters are known, the problem of estimating the noise parameters $(c_1,c_2,\ldots,c_n)$ is equivalent to estimating the parameters of the moving average process:

$$v(k)=e(k)+c_1e(k-1)+\ldots+c_ne(k-n) \quad (3.1)$$

$$v(k)=C(q^{-1})e(k);\ C(q^{-1})=1+c_1q^{-1}+\ldots..+c_nq^{-n} \quad (3.2)$$

$e(k)\sim$ i.i.d.$(0,\lambda^2)$

in which the only known variables are $v(k)$; $k = 1,2,3,\ldots..$

Formulation of a 2SLS estimator for the parameters of the above moving average model requires a reinterpretation of the 2SLS technique. To this end, we bring in the concept of 'structural' and 'incidental' parameters introduced by Neyman [1948]. From the discussion of the 2SLS estimator for the process parameters β and its recursive approximation, presented in the preceeding section, it may be inferred that the associated model (1.5) is characterised by the presence of two sets of unknown parameters - a fixed number of structural parameters viz. the process parameters β, and a set of incidental parameters viz. the noise-free process outputs $x(k)$; $k = 1,2,3,\ldots\ldots.$ whose size increases as the number of

ESTIMATOR	$\hat{\beta}_1$	$\hat{\beta}_2$	$\hat{\beta}_3$	$\hat{\beta}_4$	$\hat{\beta}_5$
True values	1.000	1.000	1.000	1.500	-0.700
2SLS	1.023	1.017	0.975	1.499	-0.698
(RMS Errors)	(0.035)	(0.033)	(0.070)	(0.010	(0.007)
2SLS (Rec.)	1.023	1.016	0.973	1.500	-0.699
(RMS Errors)	(0.035)	(0.032)	(0.071)	(0.010)	(0.008)
IV (Wong)	1.023	1.014	0.969	1.502	-0.702
(RMS Errors)	(0.035)	(0.032)	(0.072)	(0.010)	(0.009)
OLS	1.022	1.128	1.152	1.388	-0.598
(RMS Errors)	(0.033)	(0.135)	(0.165)	(0.110)	(0.100)

Table 2.1: Comparative Results for the 2SLS Estimator and some Recursive Estimators for the Process Parameters.

measurements is increased. Presence of these ever increasing number of incidental parameters, renders the model unidentifiable unless some alternate model or additional information is available for estimating the incidental parameters seperately.

Thus the 2SLS technique may be interpreted as a procedure in which, the indicental parameters are estimated in the first stage by applying OLS to a suitable model. In the second stage, the unknown incidental parameters in the given structural model are replaced by their estimates from stage 1 and OLS is applied to the resulting model, to give the estimates of the required structural parameters. Under this interpretation, the requirement that a reduced form model having a pre-specified structure, be available, has been dropped. Also, the incidental parameters in a given model may occur either implicitly or explicitly.

Coming back to the moving average model (3.1), it is apparent that the noise parameters $(c_1, c_2, \ldots, c_n)$ can be estimated by OLS, if consistent estimates of $e(k)$; $k = 1,2,3,\ldots$ are available. For this model, therefore, the choice of structural and incidental parameters is obvious because the incidental parameters $e(k)$ in this model are explicit. The 2SLS estimator for the noise parameters may be formulated as follows:

Stage 1: Estimate Incidental Parameters

Convert the moving average process (3.2) into the infinite order auto regressive process;

$$e(k) = \left[\frac{1}{C(q^{-1})}\right] v(k) = \Gamma(q^{-1}) v(k) \tag{3.3}$$

which may be terminated after a suitable number of terms under the stability assumption on $C(q^{-1})$, to give

$$e(k) = v(k) + \gamma_1 v(k-1) + \ldots + \gamma_s v(k-s) \tag{3.4}$$

$$v(k) = -\gamma_1 v(k-1) - \gamma_2 v(k-2) - \ldots - \gamma_s v(k-s) + e(k) \tag{3.5}$$

$$\underline{v} = V\gamma + \underline{e} \tag{3.6}$$

in which γ can be estimated by the OLS estimator:

$$\hat{\gamma} = (V'V)^{-1} V'\underline{v} \tag{3.7}$$

and the vector of incidental parameters $\underline{e}$ by the residuals:

$$\hat{\underline{e}} = (\underline{v} - V\hat{\gamma}) \tag{3.8}$$

Stage 2: Estimate Structural Parameters

Substitute the estimates $\hat{e}(k)$; $k = 1,2,3,\ldots$, obtained in Stage 1 for the unknown incidental parameters in (3.1) to give:

$$v(k) = c_1\hat{e}(k-1) + \ldots + c_n\hat{e}(k-n) + \eta(k) \tag{3.9}$$

$$\underline{v} = \hat{E}\underline{\delta} + \underline{\eta}, \quad \underline{\delta} = [c_1 c_2 \ldots c_n]' \tag{3.10}$$

where $\eta(k)$ represent the effective errors after such a substitution. The vector of noise parameters δ may now be estimated by the OLS estimator:

$$\hat{\underline{\delta}} = (\hat{E}'\hat{E})^{-1}\hat{E}'\underline{v} \quad (3.11)$$

and the variance of e(k) by:

$$\hat{\lambda}^2 = \frac{1}{M}\Sigma\,[\,\hat{e}(k)\,]^2 \quad (3.12)$$

3.2 Recursive Approximation to $\underline{\delta}$(2SLS)

Using arguments similar to those used in the preceeding section, the following recursive approximation can be obtained for the 2SLS estimator for the noise parameters:

$$\tilde{\underline{\delta}}(k+1)=\tilde{\underline{\delta}}(k) + \frac{S(k)\tilde{\underline{\varepsilon}}(k+1)\;[v(k+1)-\tilde{\varepsilon}'(k+1)\tilde{\delta}(k)]}{1+\tilde{\varepsilon}'(k+1)S(k)\tilde{\underline{\varepsilon}}(k+1)}$$

$$S(k+1)=S(k)-\frac{S(k)\tilde{\underline{\varepsilon}}'(k+1)\tilde{\underline{\varepsilon}}(k+1)S(k)}{1+\tilde{\varepsilon}'(k+1)S(k)\tilde{\varepsilon}(k+1)}$$

$S(k) = (\tilde{E}'\tilde{E})^{-1}$ at time k.

$$\tilde{\varepsilon}(k+1) = [\tilde{e}(k)\tilde{e}(k-1)\ldots\tilde{e}(k-n+1)]'$$

$$\tilde{e}(k) = v(k) - \tilde{\underline{\varepsilon}}'(k)\tilde{\underline{\varepsilon}}(k)$$

$$\tilde{\lambda}^2(k)= \left(\frac{k-1}{k}\right)\lambda^2(k-1) + \frac{1}{k}\,[e(k)]^2$$

The above algorithm is essentially the same as the on-line estimator for noise parameters proposed by Young [1970]. It involves a 'bootstrapping' operation in so far as the recursive estimator for $\underline{\delta}$ is used for updating the recursive estimator for e(k), and vice versa.

In the foregoing discussion it was assumed that the process parameters were known exactly. The analysis, however, is still valid if consistent estimates of the process parameters (instead of their true values) are available. This would imply that instead of v(k), it is their consistent estimates that are used for estimating the noise parameters.

3.3 Results: The model used in section (2.3) was also used for testing the above two estimators for the noise parameters. The number of trials and the number of data points were also the same. The number of terms in the auto regressive series (3.5) were taken to be 16. Results are given in Table 3.1.

ESTIMATOR	$\hat{\delta}_1$	$\hat{\delta}_2$	$\hat{\lambda}^2$
True Values	-1.000	0.200	1.000
2SLS	-1.000	0.196	0.993
RMS Errors	(0.037)	(0.049)	(0.035)
2SLS (Rec.)	-0.982	0.167	1.003
RMS Errors	(0.046)	(0.072)	(0.034)

Table 3.1: Results for the 2SLS estimator and its Recursive Approximation, for the Noise Parameters

4. ESTIMATION OF PROCESS AS WELL AS NOISE PARAMETERS

If the process parameters and the noise parameters are to be estimated together in the given model:

$$A(q^{-1})y(k)=B(q^{-1})u(k)+C(q^{-1})e(k) \quad (4.1)$$

$$\begin{aligned} y(k)=&b_0u(k)+b_1u(k-1)+..+b_nu(k-n)\\ &- a_1y(k-1)-..-a_ny(k-n)+e(k)+\\ &c_1e(k-1)+..+c_ne(k-n)\end{aligned}$$

$$\underline{y}= [M\,\vdots\,E]\begin{bmatrix}\underline{\beta}\\ \underline{\delta}\end{bmatrix} +\underline{e} =[M\,\vdots\,E]\underline{\theta}+\underline{e} \quad (4.2)$$

we can again define the vector of structural parameters as θ and the incidental parameters as e(k); k = 1,2,3,..... The 2SLS estimator for this case may be implimented as follows:

Stage 1. Estimate Incidental Parameters

From (4.1) we have

$$e(k)=\frac{A(q^{-1})}{C(q^{-1})}\,y(k) - \frac{B(q^{-1})}{C(q^{-1})}\,u(k) \quad (4.3)$$

$$e(k) = G(q^{-1})y(k) - H(q^{-1})u(k) \quad (4.4)$$

$$G(q^{-1})y(k) = H(q^{-1})u(k)+e(k) \quad (4.5)$$

where $G(q^{-1})$ and $H(q^{-1})$ are polynomials of infinite order. Since $C(q^{-1})$ has all its roots inside the unit circle, these may be truncated after a suitable number of terms to give:

$$y(k)=\gamma_o u(k)+\gamma_1 u(k-1)+..+\gamma_r u(k-r) - \gamma_{r+1} y(k-1)-..-\gamma_s y(k-r)+e(k) \quad (4.6)$$

where $s = 2r+1$, or $\underline{y} = M^*\underline{\gamma}+\underline{e}$ (4.7)

Thus $\underline{\gamma}$ can be estimated from the OLS estimator

$$\hat{\underline{\gamma}} = (M^{*\prime}M^*)^{-1}M^{*\prime}\underline{y} \quad (4.8)$$

and the vector of incidental parameters from the residuals:

$$\hat{\underline{e}} = (\underline{y}-\hat{\underline{y}}) = (\underline{y}-M^*\hat{\underline{\gamma}}) \quad (4.9)$$

Stage 2. Estimate Structural Parameters

Estimates of the incidental parameters obtained in Stage 1 may be substituted in the structural model to give:

$$\underline{y} = [M \vdots \hat{E}] \begin{bmatrix} \underline{\beta} \\ \underline{\delta} \end{bmatrix} + \hat{\underline{e}} \quad (4.10)$$

and the structural parameters estimated by the OLS estimator:

$$\hat{\underline{\theta}} = \begin{bmatrix} M'M & M'\hat{E} \\ \hat{E}'M & \hat{E}'\hat{E} \end{bmatrix}^{-1} \begin{bmatrix} M'\underline{y} \\ \hat{E}'\underline{y} \end{bmatrix} \quad (4.11)$$

The underlying technique is made more explicit, if (4.11) is written as:

$$\hat{\underline{\beta}} = (M'M)^{-1}M'(\underline{y}-\hat{E}\hat{\underline{\delta}}) = (M'M)^{-1}M'\hat{\underline{y}} \quad (4.12)$$

$$\hat{\underline{\delta}} = (\hat{E}'\hat{E})^{-1}\hat{E}'(\underline{y}-\hat{M}\hat{\underline{\beta}}) = (\hat{E}'\hat{E})^{-1}\hat{E}'\hat{\underline{v}} \quad (4.13)$$

which have to be solved simultaneously for $\hat{\underline{\beta}}$ and $\hat{\underline{\delta}}$ and the observation vectors $\hat{\underline{y}}$ and $\hat{\underline{v}}$ respectively correspond to the modified observations:

$$\hat{y}(k) = y(k)-\hat{c}_1\hat{e}(k-1)-..-\hat{c}_n\hat{e}(k-n) \quad (4.14)$$

$$\hat{v}(k) = y(k)+\hat{a}_1 y(k-1)+..+\hat{a}_n y(k-n) -\hat{b}_o u(k)-\hat{b}_1 u(k-1)-..-\hat{b}_n u(k-n) \quad (4.15)$$

It may be remarked that Astrom [1966] also uses (4.3) for estimating the errors e(k), whose sum of squares is minimised in the Maximum Likelihood (MLE) method. The estimation of e(k) is implicit in the Maximum Likelihood estimator, whereas in the 2SLS method, discussed here, we use these estimates in a more explicit manner.

As in the previous two sections, the recursive approximation to the above 2SLS estimator can be formulated by using the recursive form of the OLS estimator (4.11), in which the elements of $\hat{E}$ (estimates of the incidental parameters) are furnished, one at a time, by the recursive relation:

$$\tilde{e}(k)=y(k)-[u(k)u(k-1)..u(k-n)y(k-1)...y(k-n)e(k-1)...e(k-n)]\tilde{\underline{\theta}}(k) \quad (4.16)$$

This recursive estimator was originally proposed by Panuska [1969] and has been discussed in detail by Pandya [1972]. Panuska [1969] has proved the mean square convergence of this algorithm using the theory of stochastic approximations. However, his proof depends on the critical assumption that the recursive estimator for e(k) is stable.

Results: 2SLS estimator for $\underline{\theta}$ and its recursive approximation were tested on a second order model. The polynomials $G(q^{-1})$ and $H(q^{-1})$ were terminated after 10 terms and the total number of data pairs was 400. The remaining details were the same as in section 3.3 and 2.3.

5. CONCLUSIONS

Starting from the Two Stage Least Square (2SLS) estimator that is used in econometric models, 2SLS estimators have been proposed for the Process parameters, the Noise Parameters and Process + Noise parameters for Astrom's canonical model. It is shown, that in terms of the structural parameters η (say) and the incidental parameters $\xi(k)$; $k = 1,2,3,...$ the underlying principle of the 2SLS technique may be represented by Fig. 5.1 below:

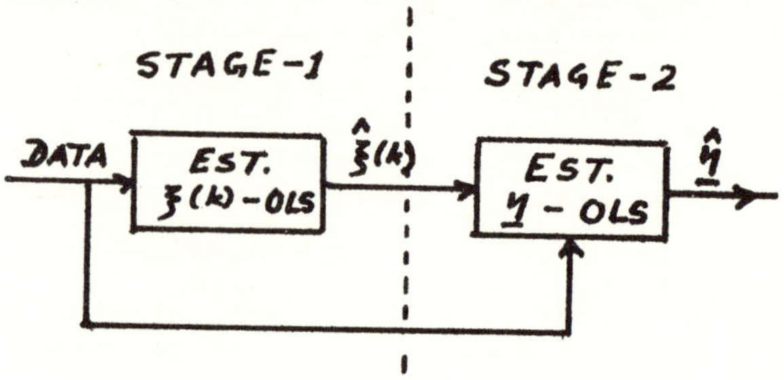

Fig. 5.1: Basic Structure of 2SLS Estimator.

ESTIMATOR	$\hat{\theta}_1$	$\hat{\theta}_2$	$\hat{\theta}_3$	$\hat{\theta}_4$	$\hat{\theta}_5$	$\hat{\theta}_6$	$\hat{\lambda}^2$
True Values	1.800	-0.820	1.000	0.500	-1.000	0.500	1.000
2SLS	1.803	-0.824	1.035	0.441	-1.007	0.501	0.940
RMS Errors	(0.014)	(0.014)	(0.064)	(0.100)	(0.049)	(0.076)	(0.089)
2SLS (Rec.)	1.805	-0.826	1.045	0.449	-0.980	0.411	0.979
RMS Errors	(0.033)	(0.022)	(0.069)	(0.114)	(0.071)	(0.140)	(0.980)

Table 4.1: Results for the 2SLS estimator and its Recursive Approximation, for the complete parameter set.

Except for the estimation of process parameters, two matrix inversion are generally required in the 2SLS technique. In the former case, the matrix inversion associated with the estimation of incidental parameters may be avoided by proper input signal design. In practice, therefore, it is more convenient to use seperate 2SLS estimators for the process and noise parameters rather than to use a single 2SLS estimator for all the parameters.

Due to the dynamic nature of the model under consideration, it is possible to formulate recursive approximations to the 2SLS estimators for the various sets of parameters. These recursive (on-line) estimators have the 'bootstrap' structure shown in Fig. 5.2.

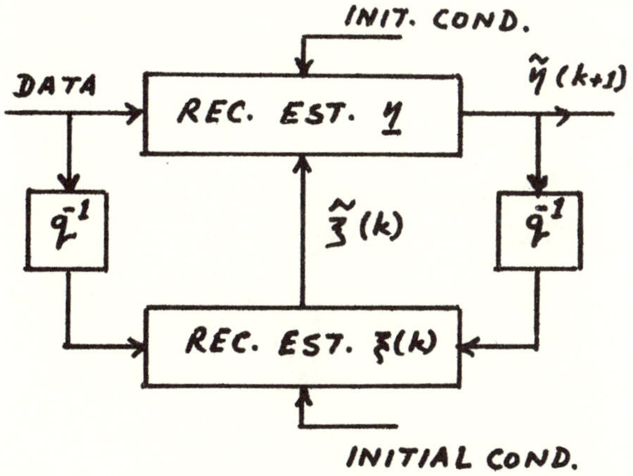

Fig. 5.2: Basic Structure of the Recursive Approximation to 2SLS Estimator (Bootstrap Estimators).

No general proof of convergence for such Bootstrap estimators is available. Though tests using simulated and practical data indicate that they are stable, a rigorous proof of their convergence is certainly desirable.

Most of the on-line estimators that are currently available for the model under study conform to the basic structure (Fig. 5.2) that has been proposed in this paper. This may be considered as a step towards classification of these on-line estimators under a single heading viz. Bootstrap estimators or Recursive Approximations to 2SLS estimators.

6. REFERENCES

Astrom, K.J. and Bohlin, T., (1966) 'Numerical Identification of Linear Dynamic Systems from Normal Operating Records', in Theory of Self Adaptive Systems, edited by P.H. Hammond (New York: Plenum Press).

Basmann, R.L., (1957), 'A Generalised classical method of linear estimation of coefficients in a structural equation', Econometrica, 25, 77.

Dhrymes, P.J., (1970), Econometrics - Statistical Foundations and Applications (New York: Harper and Rowe).

Dhrymes, P.J., (1972), 'Simultaneous Equations Inference in Econometrics', IEEE Trans. AC-17, 427.

Mayne, D.Q., (1967), 'A Method for estimating Discrete Time Transfer Functions', I.E.E. Conference on Computers in Control, University of Bristol, Paper No. C-2.

Neyman, J. and Scott, E.L., (1948), 'Consistent estimates based on partially consistent observations', Econometrica, 16, 1.

Pandya, R.N., (1972), 'A Class of Bootstrap Estimators for Identification of Linear Discrete Time Models', Technical Report No. SE-72-3, Division of Systems Engineering, Carleton University, Ottawa, Canada.

Panuska, V., (1969), 'An adaptive recursive least square identification algorithm,' Proceedings, Eight IEEE Symp. Adaptive Processes, IV 6-1.

Theil, H., (1958), 'Economic Forecaste and Policy' (Amsterdam: North Holland Publishing Co.)

Wong, K.Y. and Polak, E., (1967), 'Identification of Linear Discrete Time Systems using the Instrumental Variable Method', IEEE Trans., AC-12, 707.

Young, P.C., (1967), 'An Instrumental Variable Method for Real Time Identification of Noisy Processes', Fourth IFAC Congress, Warsaw, Paper No. 26.6.

Young, P.C., (1970), 'An Extension to the Instrumental Variable Method for the Identification of a Noisy Dynamic Process' Dept. of Engineering, University of Cambridge, Technical Note CN|70|1.

APPENDIX

A composite expression for the 2SLS estimator for the process parameters may be derived under the assumption that s is much larger than p. Considering the partitioned form of the model (2.5)

$$\underline{y} = M_1\underline{\beta}_1 + M_2\underline{\beta}_2 + \underline{v},$$

replacement of the elements of M_2 by their estimates, obtained through the reduced form fit, is equivalent to the model

$$\underline{y} = M\underline{\beta}_1 + U(U'U)^{-1}U'M_2 + \underline{v} + \hat{W}_2\underline{\beta}_2$$

However matrix U has the form

$$U = [M_1 \,\vdots\, U_2] \rightarrow M_1 = U \begin{bmatrix} I_{n+1} \\ \hline 0 \end{bmatrix}$$

so that $U(U'U)^{-1}U'M_1 = M_1$ and

$$\underline{y} = U(U'U)^{-1}U'M\underline{\beta} + \underline{v} + \hat{W}\underline{\beta}$$

where $\hat{W} = (0 \,\vdots\, \hat{W}_2) = (M - \hat{M})$

which gives the OLS estimator for $\underline{\beta}$ as

$$\hat{\underline{\beta}}(2SLS) = [M'U(U'U)^{-1}U'M]^{-1}M'U(U'U)^{-1}U'\underline{y}$$

where the error term $\hat{W}\underline{\beta}$ does not enter into the normal equations for $\hat{\underline{\beta}}(2SLS)$ because

$$M'U(U'U)^{-1}U'\hat{W}\underline{\beta} = \underline{0}$$

ON THE ESTIMATION OF THE TRANSFER FUNCTION PARAMETERS OF PROCESS- AND NOISE DYNAMICS USING A SINGLE-STAGE ESTIMATOR

J.L. Talmon* and A.J.W. van den Boom
University of Technology, E.E. Department,
Eindhoven, the Netherlands

This paper deals with the bias aspects of least squares estimators. The frequently used generalized least squares technique does not suffer from bias problems but requires a comparitively large number of noise parameters to be estimated. This is due to the fact that the noise is modelled as an autoregressive series. In this paper an updating scheme is proposed, which enables the modelling of the noise filter as a mixed autoregressive moving-average series. This type of modelling of the noise characteristics has the important advantage that in principle a smaller number of parameters has to be estimated. Moreover the process- and noise parameters are determined using a single-stage estimator instead of separated estimators as applied in the generalized least squares scheme. It turns out that the proposed estimator is of a very general nature. Practical results on simulated data show the efficiency and the applicability of the method presented.

CONTENTS

1. INTRODUCTION

It is well known that least squares estimators can lead to biased results if the residuals are correlated cf. Åström & Eykhoff (1971). The generalized least squares technique, proposed by Clarke (1967) and in iterative form by Hastings-James and Sage (1969), filters the observed signals in such a way that the resulting residuals become uncorrelated.
For the implementation of such a filter, knowledge of the characteristics of the unfiltered residuals is necessary.
Usually this knowledge is obtained by estimating the autoregressive parameters of these residuals. This way of modelling the residuals may become cumbersome with respect to the number of parameters involved. In general, for a complete description a mixed autoregressive moving-average type of modelling will need less parameters. This type of representation of the noise is more adequate because the noise dynamics are treated in the same way as the process itself, i.e. by estimation of both numerator- and denominator terms of the transfer function of a linear, time invariant system.
Clarke and Hastings-James/Sage used separated estimators for the process parameters and noise parameters.
By rewriting the matrix form of the describing difference equation of the whole system a very general estimator can be found, which estimates both process parameters and noise parameters. This estimator is suited for mixed autoregressive moving-average, pure moving-average or pure autoregressive modelling of both process- and noise parameters. It can be written in a way which permits updating, without the need of recursion like Clarke's estimator.

2. PROCESS AND NOISE DESCRIPTION

Consider a discrete process P, for which the relation between input and output can be described by the following linear, time-invariant difference equation:

$$x_k + \sum_{i=1}^{q} a_i x_{k-i} = \sum_{i=1}^{p} b_i u_{k-i} \qquad (1)$$

Let the output x_k be disturbed by an independent, zero mean additive noise signal n_k (see fig. 1), viz.:

$$y_k = x_k + n_k \qquad (2)$$

where y_k is the observable disturbed output signal of the process.

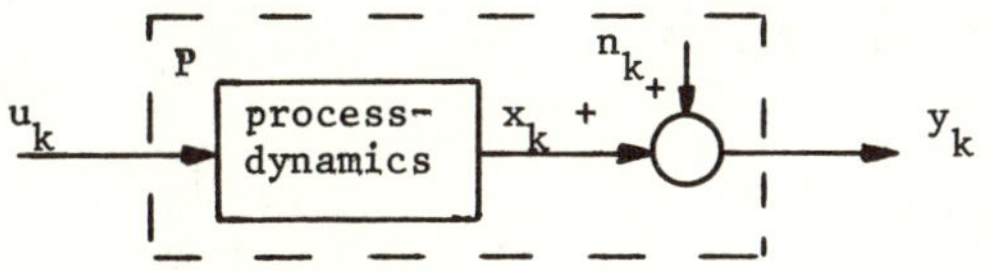

figure 1.

Combining (1) and (2):

$$y_k + \sum_{i=1}^{q} a_i y_{k-i} = \sum_{i=0}^{p} b_i u_{k-i} + e_k \qquad (3)$$

where e_k is the equation error:

$$e_k = n_k + \sum_{i=1}^{q} a_i n_{k-i} \qquad (4)$$

*J.L. Talmon is now with the Institute of Medical Physics TNO, Utrecht, the Netherlands

Define the shift operator z:

$$u_{k-j} = z^{-j}u_k \qquad (5)$$

Rewrite (3) and (4) in terms of shift polynomials

$$\{1 + A(z^{-1})\}y_k = \{b_o + B(z^{-1})\}u_k + e_k \qquad (6)$$

$$e_k = \{1 + A(z^{-1})\}n_k \qquad (7)$$

with:

$$\left.\begin{aligned} A(z^{-1}) &= a_1 z^{-1} + \dots + a_q z^{-q} \\ B(z^{-1}) &= b_1 z^{-1} + \dots + b_p z^{-p} \end{aligned}\right\} \qquad (8)$$

To guarantee the stability of the process P, the roots of $1 + A(z^{-1})$ are assumed to lie inside the unit circle.

In section 3.1. we will prove that in general a least-squares estimate of $A(z^{-1})$ and $B(z^{-1})$ will be asymptotically biased unless their residual e_k is a sample of a white noise sequence.

In order to arrive at a residual with white noise properties, we have to model the noise properly. Therefore we will consider in detail the description of the noise.

We assume that the noise n_k can be described as a filtering of a well behaved, zero mean white noise signal ξ_k:

$$n_k + \sum_{i=1}^{r} d_i n_{k-i} = \xi_k + \sum_{i=1}^{s_o} g_i \xi_{k-i} \qquad (9)$$

or equivalently:

$$n_k = \frac{1 + G(z^{-1})}{1 + D(z^{-1})}\,\xi_k \qquad (10)$$

with:

$$\begin{aligned} D(z^{-1}) &= d_1 z^{-1} + \dots + d_r z^{-r} \\ G(z^{-1}) &= g_1 z^{-1} + \dots + g_{s_o} z^{-s_o} \end{aligned} \qquad (11)$$

Again, the roots of $1 + D(z^{-1})$ are assumed to lie inside the unit circle. Combining (7) and (10) leads to:

$$e_k = \frac{1 + C(z^{-1})}{1 + D(z^{-1})}\,\xi_k \qquad (12)$$

with:

$$\begin{aligned} 1 + C(z^{-1}) &= \{1 + A(z^{-1})\}\{1 + G(z^{-1})\} = \\ &= 1 + c_1 z^{-1} + \dots + c_s z^{-s} \end{aligned} \qquad (13)$$

and $s = s_o + q$

In (12), e_k is described as a sample of a mixed autoregressive moving-average time series. By defining:

$$1 + C'(z^{-1}) = \frac{1 + C(z^{-1})}{1 + D(z^{-1})} \qquad (14)$$

so that:

$$e_k = \{1 + C'(z^{-1})\}\xi_k \qquad (15)$$

e_k is described as a pure moving-average series, and by defining:

$$1 + D'(z^{-1}) = \frac{1 + D(z^{-1})}{1 + C(z^{-1})} \qquad (16)$$

so that:

$$e_k = \frac{1}{1 + D'(z^{-1})}\,\xi_k \qquad (17)$$

e_k is described as a sample of a pure autoregressive series.

Åström, Bohlin and Wensmark (1965) have proved that each discrete linear process can be described by the following equation:

$$\begin{aligned} \{1 + A^*(z^{-1})\}y_k = \{b_o^* + B^*(z^{-1})\}u_k &+ \\ + \{1 + C(z^{-1})\}\xi_k& \end{aligned} \qquad (18)$$

By combining (6) and (12):

$$\begin{aligned} \{1 + A(z^{-1})\}y_k = \{b_o + B(z^{-1})\}u_k &+ \\ + \frac{1 + C(z^{-1})}{1 + D(z^{-1})}\,\xi_k& \end{aligned} \qquad (19)$$

we can find the relation between the description of (18) and (19):

$$\left.\begin{aligned} 1 + A^*(z^{-1}) &= \{1 + A(z^{-1})\}\{1 + D(z^{-1})\} \\ b_o^* + B^*(z^{-1}) &= \{b_o + B(z^{-1})\}\{1 + D(z^{-1})\} \end{aligned}\right\} \qquad (20)$$

If we use the process-description of (18) we see easily that more parameters are involved then in equation (19).

Other possible modifications of (19) can be found by making use of (14) or (16).

As the polynomials $1 + C'(z^{-1})$ and $1 + D'(z^{-1})$ may have, in principle, an infinite length, we conclude that the description (19) has the attractive property of minimum parameter representation.

3. LEAST SQUARES ESTIMATION OF THE PROCESS- AND/OR NOISE PARAMETERS

3.1. only process parameters estimated

Suppose that the sequences u_k and y_k with $k = 1, \dots, N$ are available. If we assume that $q \geq p$, which is not essential, we can write:

$$\underline{y} = \Omega(u,y)\underline{b}' + \underline{e} \qquad (21)$$

with:

$$\begin{aligned} \underline{y}^T &= [y_{q+1}, \dots, y_N] \\ \underline{e}^T &= [e_{q+1}, \dots, e_N] \\ \underline{b}'^T &= [b_o, \dots, b_p, -a_1, \dots, -a_q] \end{aligned} \qquad (22)$$

$$\Omega(u,y) = \begin{bmatrix} u_{q+1} \cdots & u_{q+1-p} & y_q & \cdots y_1 \\ \cdot & \cdot & \cdot & \cdot \\ \cdot & \cdot & \cdot & \cdot \\ \cdot & \cdot & \cdot & \cdot \\ u_N \cdots & u_{N-p} & y_{N-1} & \cdots y_{N-q} \end{bmatrix} \tag{23}$$

To find a least squares estimator $\underline{\beta}'$ of the parameters $\underline{b}'$ define $\underline{\varepsilon} = \underline{y} - \Omega(u,y)\underline{\beta}'$ and minimize $\underline{\varepsilon}^T\underline{\varepsilon}$ with respect to $\underline{\beta}'$. This results in (Deutsch (1965), Goldberger (1964)):

$$\underline{\beta}' = [\Omega^T(u,y)\Omega(u,y)]^{-1}\Omega^T(u,y)\underline{y} \tag{24}$$

Note that we assume here the order of the process to be known. We can write (24) in a way, which permits updating (Talmon (1971)).

$$\left.\begin{aligned} \underline{\beta}'_{k+1} &= \underline{\beta}'_k - P_k\underline{\omega}_{k+1}\left[1 + \underline{\omega}^T_{k+1}P_k\underline{\omega}_{k+1}\right]^{-1} \cdot \\ &\quad \cdot \left[\underline{\omega}^T_{k+1}\underline{\beta}'_k - y_{k+1}\right] \\ P_{k+1} &= P_k - P_k\underline{\omega}_{k+1}\left[1 + \underline{\omega}^T_{k+1}P_k\underline{\omega}_{k+1}\right]^{-1}\underline{\omega}^T_{k+1}P_k \end{aligned}\right\} \tag{25}$$

where $\underline{\beta}'_k$ is the estimate of $\underline{b}'$ after k samples;

$$\underline{\omega}^T_k = [u_k, \ldots, u_{k-p}, y_{k-1}, \ldots, y_{k-q}]$$

$$P_k = [\Omega^T_k(u,y)\Omega_k(u,y)]^{-1}$$

and: (26)

$$\Omega_k(u,y) = \begin{bmatrix} u_{q+1} \cdots & u_{q+1-p} & y_{q-1} \cdots & y_1 \\ \cdot & \cdot & \cdot & \cdot \\ \cdot & \cdot & \cdot & \cdot \\ \cdot & \cdot & \cdot & \cdot \\ u_k \cdots & u_{k-p} & y_{k-1} \cdots & y_{k-q} \end{bmatrix}$$

The asymptotic properties of this estimator can be found by taking the probability limit of (24) and using Slutsky's theorems cf. Goldberger (1964)

$$\begin{aligned} \underset{N\to\infty}{\text{plim}}\ \underline{\beta}' &= \underset{N\to\infty}{\text{plim}}\ [\{\Omega^T(u,y)\Omega(u,y)\}^{-1}\Omega^T(u,y)\underline{y}] = \\ &= \underline{b}' + [\underset{N\to\infty}{\text{plim}}\{\tfrac{1}{N}\ \Omega^T(u,y)\Omega(u,y)\}]^{-1} \cdot \\ &\quad \cdot \underset{N\to\infty}{\text{plim}}[\tfrac{1}{N}\ \Omega^T(u,y)]\underline{e} \end{aligned} \tag{27}$$

Now $\underline{\beta}'$ will be an asymptotically unbiased estimate if the following two conditions are fulfilled:

I $\underset{N\to\infty}{\text{plim}}\ [\frac{1}{N}\ \Omega^T(u,y)\Omega(u,y)]$ is nonsingular

II $\underset{N\to\infty}{\text{plim}}\ [\frac{1}{N}\ \Omega^T(u,y)\underline{e}| = \underline{0}$

Condition I assures that the measured signals contain enough information to make the estimation possible.
Condition II is fulfilled if $e_k(k=1, \ldots, N)$ is a white noise sequence. As $e_k = \{1+A(Z^{-1})\}n_k$, the sequence n_k $(k=1, \ldots, N)$ must be noise, which is derived from white noise by a filter, having the same autoregressive parameters as the process. In general, this will not be the case, and the estimator (24) will yield biased results. To overcome this unwanted property we define a weighted least squares estimator:

$$\underline{\beta}' = [\Omega^T(u,y)R\Omega(u,y)]^{-1}\Omega^T(u,y)R\underline{y} \tag{28}$$

Following Eykhoff (1967):

$$R = F^TF \tag{29}$$

F being a lower triangular matrix.
Substituting (29) and (21) in (28):

$$\underline{\beta}' = \underline{b}' + [\{F\Omega(u,y)\}^T\{F\Omega(u,y)\}]^{-1}\{F\Omega(u,y)\}^T F\underline{e} \tag{30}$$

If the sequence $\underline{e}^* = F\underline{e}$ is uncorrelated with $\Omega^*(u,y) = F\Omega(u,y)$ then $\underset{N\to\infty}{\text{plim}}\ \frac{1}{N}\ \Omega^{*T}(u,y)\underline{e}^* = \underline{0}$ and the estimate is asymptotically unbiased. This is the case if $\underline{e}^*$ is a white noise sequence, i.e. F is a "noise whitening" filter.
In: (31)

$$\underline{e}^*_k = \{1 + F(z^{-1})\}e_k = \{1 + F(z^{-1})\}.\frac{1 + C(z^{-1})}{1 + D(z^{-1})}\xi_k$$

we see that $1 + F(z^{-1})$ is the inverse equation error filter. This estimator coincides with the Markov estimator where $R = \Sigma^{-1}$; Σ being the covariance matrix of the equation error.
This Markov estimate is only applicable if proper a priori knowledge $(\Sigma, D'(z^{-1}))$ is available. In many practical situations this will not be the case.

3.2. Both process- and noise parameters estimated

When a priori knowledge of the noise is lacking, we have to estimate the noise parameters. Clarke (1967) suggested an explicit method, where he estimates the autoregressive parameters $\underline{\delta}'$ of the approximated equation error sequence $\hat{e}_k$ with the estimator:

$$\underline{\delta}' = -(\hat{E}^T\hat{E})^{-1}\hat{E}^T\hat{\underline{e}} \tag{32}$$

The sequence $\hat{e}_k$ $(k = 1, \ldots, N)$ is generated by using the estimate of (24): (33)

$$\hat{e}_k = y_k - (u_k, \ldots, u_{k-p}, y_{k-1}, \ldots, y_{k-q})\underline{\beta}'$$

where:

$$\left.\begin{aligned} \hat{\underline{e}}^T &= [\hat{e}_{q+1}, \ldots, \hat{e}_N] \\ \hat{E} &= \begin{bmatrix} \hat{e}_q & \cdots\ \hat{e}_{q+1-r'} \\ \cdot & \cdot \\ \cdot & \cdot \\ \cdot & \cdot \\ \hat{e}_{N-1} & \cdots\ \hat{e}_{N-r'} \end{bmatrix} \\ \underline{\delta}'^T &= [\delta'_1, \ldots, \delta'_{r'}] \end{aligned}\right\} \tag{34}$$

With this estimate $\underline{\delta}'$, the sequences u_k and y_k ($k = 1, \ldots, N$) are filtered:

$$\left.\begin{aligned} u_k^* &= u_k + \delta_1 u_{k-1} + \ldots + \delta_{r'} u_{k-r'} \\ y_k^* &= y_k + \delta_1 y_{k-1} + \ldots + \delta_{r'} y_{k-r'} \end{aligned}\right\} \quad (35)$$

and a least squares estimate $\underline{\beta}^*$ is found:

$$\underline{\beta}^* = [\Omega^T(u^*, y^*)\Omega(u^*, y^*)]^{-1}\Omega^T(u^*, y^*)\underline{y}^* \quad (36)$$

which can be considered as an approximate version of (30). Whith this $\underline{\beta}^*$, a new improved sequence $\hat{e}_k$ can be generated and also a new estimate $\underline{\delta}'$, etc.

The main characteristics of Clarke's method are:

1) removal of the asymptotical bias;
2) separated estimators for the process- and noise parameters;
3) use of explicit (one shot) estimators and consequently the need for iteration of the solution as mentioned above;
4) the use of an autoregressive model for the noise.

Hastings-James and Sage (1969) proposed an implicit (updating) scheme, approximating Clarke's scheme. Their estimator has the attractive properties that intermediate estimates become available and that no recursion on the solution is needed.

4. THE EXTENDED MATRIX METHOD

4.1. The general form

In this section we propose a single stage estimator which can be used for estimation of process- and noise parameters together. This estimator can be implemented such that both numerator and denominator parameters of the transfer functions of process and noise-process are estimated. But, if wanted, other types of process- and noise modelling (e.g. autoregressive noise parameters) are applicable.

If we rewrite (19), we obtain:

$$\{1 + A(z^{-1})\}y_k = \{b_o + B(z^{-1})\}u_k - D(z^{-1})e_k + C(z^{-1})\xi_k + \xi_k \quad (37)$$

In matrix notation:

$$\underline{y} = \Omega(u, y, \xi, e)\begin{bmatrix} \underline{b} \\ -\underline{a} \\ \underline{c} \\ -\underline{d} \end{bmatrix} + \underline{\xi} \quad (38)$$

and:

$$\Omega(u,y,\xi,e) = [U|Y|\Xi|E] = \left[\begin{array}{ccc|ccc|ccc|ccc} u_{q+1} & \cdots & u_{q+1-p} & y_q & \cdots & y_1 & \xi_q & \cdots & \xi_{q+1-s} & e_q & \cdots & e_{q+1-r} \\ \vdots & & \vdots & \vdots & & \vdots & \vdots & & \vdots & \vdots & & \vdots \\ u_N & \cdots & u_{N-p} & y_{N-1} & \cdots & y_{N-q} & \xi_{N-1} & \cdots & \xi_{N-s} & e_{N-1} & \cdots & e_{N-r} \end{array}\right] \quad (39)$$

It should be noted that the equation error $\underline{\xi}$ is a white noise signal and that the order of both process and noise-process are given.

If this latter knowledge is lacking, a suitable test of order should be used, which is able to determine the order of process and noise-process separately, cf. Van den Boom and Van den Enden (1973) for descriptions where $p = q$ and $r = s$.

An asymptotically unbiased estimator

$$\underline{\beta}^{+T} = [\underline{\beta}^T, -\underline{\alpha}^T, \underline{\gamma}^T, -\underline{\delta}^T] \text{ of } [\underline{b}^T, -\underline{a}^T, \underline{c}^T - \underline{d}^T]$$

is given by:

$$\underline{\beta}^+ = [\Omega^T(u,y,\xi,e)\,\Omega(u,y,\xi,e)]^{-1}\Omega^T(u,y,\xi,e)\underline{y} \quad (40)$$

as:

$$\underset{N\to\infty}{\text{plim}}\,[\underline{\beta}^+] = \begin{bmatrix} \underline{b} \\ -\underline{a} \\ \underline{c} \\ -\underline{d} \end{bmatrix} + \underset{N\to\infty}{\text{plim}}\left[\tfrac{1}{N}\Omega^T(u,y,\xi,e)\Omega(u,y,\xi,e)\right]^{-1} \cdot$$

$$\cdot\ \underset{N\to\infty}{\text{plim}}\left[\tfrac{1}{N}\Omega^T(u,y,\xi,e)\xi\right] = \begin{bmatrix} \underline{b} \\ -\underline{a} \\ \underline{c} \\ -\underline{d} \end{bmatrix} \quad (41)$$

Because we do not know the elements of Ξ and E, we replace them by their estimates:

$$\left.\begin{aligned} \hat{e}_k &= y_k - \underline{\omega}_k^T \begin{bmatrix} \underline{\beta}_k \\ -\underline{\alpha}_k \end{bmatrix} \\ \hat{\xi}_k &= \hat{e}_k - \underline{\nu}_k^T \begin{bmatrix} \underline{\gamma}_k \\ -\underline{\delta}_k \end{bmatrix} \end{aligned}\right\} \quad k = q+1, \ldots, N \quad (42)$$

with:

$$\left.\begin{aligned} \underline{\omega}_k^T &= [u_k, \ldots, u_{k-p}, y_{k-1}, \ldots, y_{k-q}] \\ \underline{\nu}_k^T &= [\hat{\xi}_{k-1}, \ldots, \hat{\xi}_{k-s}, \hat{e}_{k-1}, \ldots, \hat{e}_{k-r}] \end{aligned}\right\} \quad (43)$$

The scheme becomes then:

$$\underline{\beta}^+ = [\Omega^T(u,y,\hat{\xi},\hat{e})\Omega(u,y,\hat{\xi},\hat{e})]^{-1}\Omega^T(u,y,\hat{\xi},\hat{e})\underline{y} \quad (44)$$

Because intermediate estimates of $\underline{\beta}^+$ are needed for the generation of $\hat{\underline{e}}$ and $\hat{\underline{\xi}}$ we will use an updating estimator:

$$\underline{\beta}^+_{N+1} = \underline{\beta}^+_N - P_N\underline{\omega}^+_{N+1}\left[\rho + \underline{\omega}^{+T}_{N+1}P_N\underline{\omega}^+_{N+1}\right]^{-1} \cdot$$

$$\cdot \left[\underline{\omega}^+_{N+1}\underline{\beta}^+_N - y_{N+1}\right] \quad (45a)$$

$$P_{N+1} = \frac{1}{\rho}\left[P_N - P_N\underline{\omega}^{+}_{N+1}\{\rho + \underline{\omega}^{+T}_{N+1} P_N\underline{\omega}^{+}_{N+1}\}^{-1}\underline{\omega}^{+T}_{N+1}P_N\right] \tag{45b}$$

with:

$$\underline{\omega}^{+T}_{N+1} = [u_{N+1}, \ldots, u_{N+1-p}, y_N, \ldots, y_{N+1-q}, \hat{\xi}_N, \ldots, \hat{\xi}_{N+1-s}, \hat{e}_N, \ldots, \hat{e}_{N+1-r}] \tag{46}$$

being the last row of Ω_{N+1}:

$$\Omega_{N+1} = \begin{bmatrix} \Omega_N \\ \underline{\omega}^{+T}_{N+1} \end{bmatrix} \tag{47}$$

For small N, we will have poor estimates of the parameters and consequently bad $\hat{e}$ and $\hat{\xi}$. For this reason, the influence of new input-output pairs is made relatively greater than the influence of old measurements.
This is done by choosing as loss function:

$$V = \sum_{i=q+1}^{N} \rho^{N-i}(y_i - \underline{\omega}^{+T}_i \underline{\beta}^{+}_i)^2 \tag{48}$$

resulting in the updating scheme of (45).
For the initialization of the estimation, the following procedure is applied:

1) Define an initial square P-matrix whose dimensions equal the total number of parameters to be estimated. By choosing $P_{init} = aI$, a being large we obtain rapid convergence cf. Lee (1964)

$$P_{init} = \begin{bmatrix} 10^6 & & 0 \\ & \ddots & \\ 0 & & 10^6 \end{bmatrix}$$

2) Fill $\underline{\omega}^{+}_{q+1}$ with input-output samples:

$$\underline{\omega}^{+}_{q+1} = [u_{q+1}, \ldots, u_{q+1-p}, y_q, \ldots, y_1, \underbrace{0, \ldots 0}_{\text{for } \hat{\xi}}, \underbrace{0, \ldots, 0}_{\text{for } \hat{e}}]$$

3) Apply (45) for a first estimate $\underline{\beta}^{+}_{q+1}$ and generate $\hat{e}_{q+2} = y_{q+2} - \underline{\omega}^{+T}_{q+2}\underline{\beta}^{+}_{q+2}$
4) Form

$$\underline{\omega}^{+T}_{q+2} = [u_{q+2}, \ldots, u_{q+2-p}, y_{q+1}, \ldots, y_2, 0, \ldots 0, 0, \ldots, 0]$$

and apply (45) for the second estimate $\underline{\beta}^{+}_{q+2}$ and generate $\hat{e}_{q+2} = y_{q+2} - \underline{\omega}^{+T}_{q+2}\underline{\beta}^{+}_{q+2}$
5) Proceed as in 4) until r samples of $\hat{e}$ are available.
6) Substitute these samples in the $\underline{\omega}^{+}$ vector:

$$\underline{\omega}^{+T}_{q+r+1} = [u_{q+r+1}, \ldots, u_{q+r+1-p}, y_{q+r}, \ldots, y_{1+r}, 0, \ldots, 0, \hat{e}_{q+r}, \ldots, \hat{e}_{q+1}]$$

and apply (45) to obtain an estimate $\underline{\beta}^{+}_{q+r+1}$ of $\underline{a}$, $\underline{b}$ and $\underline{d}$.
7) Generate $\hat{e}_{q+r+1}$ as usual and also $\hat{\xi}_{q+r+1}$ using the available samples $\hat{e}$ and the estimate of $\underline{d}$ and substitute them in $\underline{\omega}^{+}_{q+r+2}$.
8) Apply (45) and generate $\hat{e}_{q+r+2}$. For the generation of $\hat{\xi}_{q+r+2}$ the previous approximated sample $\hat{\xi}_{q+r+1}$ can be used.
9) In the following iterations still more samples $\hat{\xi}$ become available resulting in a complete filling of the $\underline{\omega}^{+}$ vector.

4.2. Special cases.

In the previous section the general form of the extended matrix estimator is given. In this general case the noise $\underline{e}$ was modelled as an autoregressive moving-average series leading to a minimal parameter description.
Two special cases of the estimator can be distinguished if we model the noise $\underline{e}$ as either a pure autoregressive cf. Smets (1970) or a pure moving average series cf. Young (1968). As mentioned before these noise descriptions may have relatively large numbers of parameters.
The estimators for these types of modelling can easily be found if in (46) $\underline{\omega}$ is adapted.
For the pure autoregressive case the part with the samples $\hat{\xi}$ is removed and the dimension r of $\underline{\delta}$ is enlarged while for the pure moving-average case the part with the samples $\hat{e}$ is removed and the dimension s of $\underline{\gamma}$ is enlarged.

5. PRACTICAL RESULTS

To test the schemes given in section 4, an Algol-60 program was written which permits estimation with the extended matrix method and its special cases. In this program the simulated process input data and the noisy output data were generated. The process chosen was the same as given by Åström, Bohlin and Wensmark (1965):

$$(1 - 1.5z^{-1} + 0.7z^{-2})y_k = (z^{-1} + 0.5z^{-2})u_k + e_k \tag{50}$$

and the noise process was:

$$(1 - 0.5z^{-1})e_k = (1 + 0.3z^{-1})\xi_k \tag{51}$$

ξ_k being a white noise signal with a rectangular amplitude distribution between $-\lambda$ and $+\lambda$.
The input u_k was a filtered sequence:

$$u_k = \eta_k - 0.5u_{k-1} \tag{52}$$

where η_k was a white noise sequence with a rectangular amplitude distribution between -1 and + 1.
A block diagram of the whole set-up is given in figure 2:

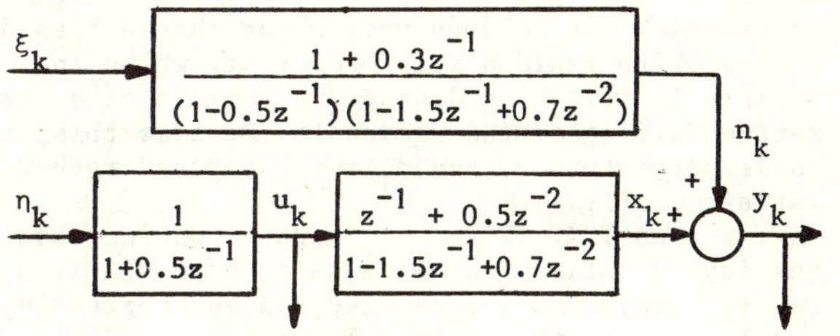

figure 2.

If the white input signal η_k has a power of σ_η^2 then the power of the output, σ_x^2, is given by

$$\sigma_x^2 = \frac{\sigma_\eta^2}{2\pi j} \oint_{|z|=1} H(z)H(z^{-1}) \frac{dz}{z} \quad (53)$$

so the ratio $\frac{\sigma_x^2}{\sigma_\eta^2}$ becomes:

$$\frac{\sigma_x^2}{\sigma_\eta^2} = \frac{1}{2\pi j} \oint_{|z|=1} H(z)H(z^{-1}) \frac{dz}{z} \quad (54)$$

Åström, Jury & Agniel (1970) give a fast method to calculate this integral.
For the given combination of process and input filter, $\sigma_x^2/\sigma_\eta^2 = 8.75$ and for the

noise filter $\sigma_n^2/\sigma_\xi^2 = 43.6$ (if $\lambda = 1$)

We define the signal to noise ratio:

$$\frac{S}{N} = \frac{\sigma_x^2}{\sigma_\eta^2} \quad (55)$$

The program permits a change in weighting factor:

$$\rho_{k+1} = (1 - \Delta\rho)\,\rho_k + \Delta\rho \quad (56)$$

In figure 4 and table 1 the results with different types of estimated noise models are given. The input- output data of all examples except X and XI are identical.
example I no noise estimated
example II noise estimated as a mixed autoregressive moving-average series
example III, IV, V noise estimated as a pure autoregressive series
example VI, VII, VIII, IX noise estimated as a pure moving average series.
The signal to noise ratio S/N was -7 dB ($\lambda = 1$). The data weighting parameters were chosen as $\rho = 0{,}9913$ and $\Delta\rho = 0{,}001$.
In example X and XI the signal to noise ratio was -19dB ($\lambda = 4$) and -31dB ($\lambda = 16$) respectively and in example XII the influence of the choice of $\Delta\rho$ is shown.
In this table the mean values and standard deviations of 10 independent runs of 1000 samples each are given.
In example I it becomes very clear that a bias due to non white residuals is occurring, while in example II this bias disappears because of a correct modelling of the residuals. Besides this, the noise parameters c_1 and d_1 are estimated rather satisfactorily.
In examples III, IV and V, there is an increasing quality of estimation. Because $d_3' = -0.072$ is a comparatively small parameter, an autoregressive model of order two is useful.

In examples VI, VII, VIII, IX there is also a slight increase of quality of estimation. Here a fourth order moving-average model gives approximately comparable results with a mixed autoregressive moving-average model of first order.
As a general conclusion we can state that the bias of the estimates using the special models is very dependent on the values of the noise parameters, while in the case of the general noise model this bias is independent on the values of the noise parameters.
The results for increasing S/N (example II, X and XI) show the good properties of the method chosen. Globally, the behaviour of the variance of the estimates for different S/N is shown in fig. 3 (logarithmic scales)

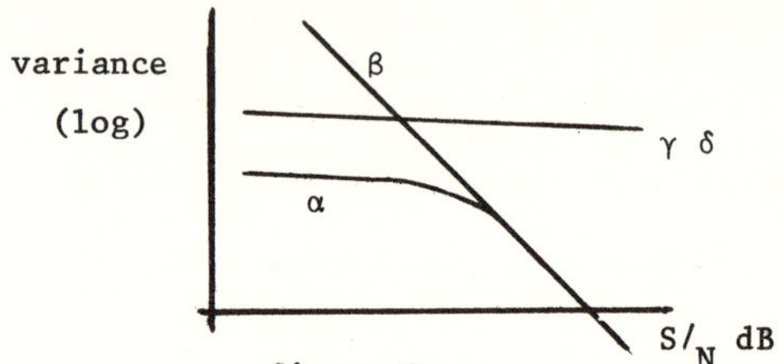

figure 3.

This can be explained as follows:
The variance of all estimates is, roughly, proportional to:

$$\frac{1}{N} \cdot \frac{\text{power of disturbances on signal}}{\text{power of measured signal}}$$

For the parameter β we find:

$$\text{var } \beta :: \frac{\sigma_\xi^2}{N\sigma_u^2}$$

which increases with increasing σ_ξ^2
For the parameters γ and δ we find:

$$\text{var } \delta :: \frac{\sigma_\xi^2}{N\sigma_e^2} \quad \text{and var } \gamma :: \frac{\sigma_\xi^2}{N\sigma_\xi^2}$$

which is independent of σ_ξ^2 as σ_e^2 is linearly dependent of σ_ξ^2
For the parameter α we find:

$$\text{var } \alpha :: \frac{\sigma_\xi^2}{N\sigma_y^2} = \frac{\sigma_\xi^2}{N(\sigma_x^2 + \sigma_n^2)}$$

If we distinguish two cases:

a) $\sigma_n^2 << \sigma_x^2$ then $\text{var } \alpha :: \frac{\sigma_\xi^2}{N\sigma_x^2}$

which is increasing with σ_ξ^2
b) $\sigma_n^2 >> \sigma_x^2$ then $\text{var } \alpha :: \frac{\sigma_\xi^2}{N\sigma_n^2}$

which is independent of σ_ξ^2 as σ_n^2 is linearly dependent of σ_ξ^2.

EXAMPLE	S/N dB	$\Delta\rho$	α_1	α_2	β_o	β_1	β_2	γ_1	δ_1	δ'_1	δ'_2	δ'_3	γ'_1	γ'_2	γ'_3	γ'_4
I	-7	0.001	-1.689	0.846	0.012	0.998	0.410									
			0.013	0.014	0.040	0 073	0.064									
II	-7	0.001	-1.508	0.703	0.002	1.000	0.504	0.308	-0.482							
			0.042	0.034	0.046	0.051	0.059	0.092	0.085							
III	-7	0.001	-1.481	0.718	-0.011	0.981	0.509			-0.700						
			0.041	0.029	0.040	0.051	0.065			0.051						
IV	-7	0.001	-1.512	0.696	0.005	0.998	0.507			-0.770	0.200					
			0.034	0.033	0.049	0.055	0.062			0.068	0.066					
V	-7	0.001	-1.511	0.710	0.010	1.008	0.503			-0.783	0.232	-0.069				
			0.037	0.034	0.052	0.057	0.060			0.046	0.067	0.031				
VI	-7	0.001	-1.602	0.767	-0.014	0.995	0.465						0.677			
			0.020	0.018	0.033	0.040	0.053						0.053			
VII	-7	0.001	-1.552	0.726	0.005	0.997	0.487						0.741	0.297		
			0.034	0.031	0.048	0.050	0.058						0.073	0.056		
VIII	-7	0.001	-1.542	0.707	0.007	1.009	0.498						0.779	0.346	0.151	
			0.030	0.028	0.052	0.056	0.059						0.046	0.065	0.059	
IX	-7	0.001	-1.514	0.706	0.003	1.006	0.495						0.790	0.362	0.175	0.077
			0.038	0.033	0.055	0.069	0.068						0.047	0.071	0.096	0.058
X	-19	0.001	-1.515	0.706	0.005	0.969	0.517	0.311	-0.473							
			0.057	0.045	0.202	0.218	0.220	0.094	0.104							
XI	-31	0.001	-1.516	0.707	-0.013	0.863	0.576	0.312	-0.471							
			0.057	0.045	0.806	0.870	0.883	0.097	0.106							
XII	-7	0.025	-1.538	0.721	-0.005	0.985	0.475	0.294	-0.458							
			0.022	0.025	0.038	0.060	0.045	0.073	0.071							
true parameters			-1.500	0.700	0.000	1.000	0.500	0.300	-0.500	-0.800	0.240	-0.072	0.800	0.400	0.200	0.100

table 1 results of estimation with the extended matrix method.

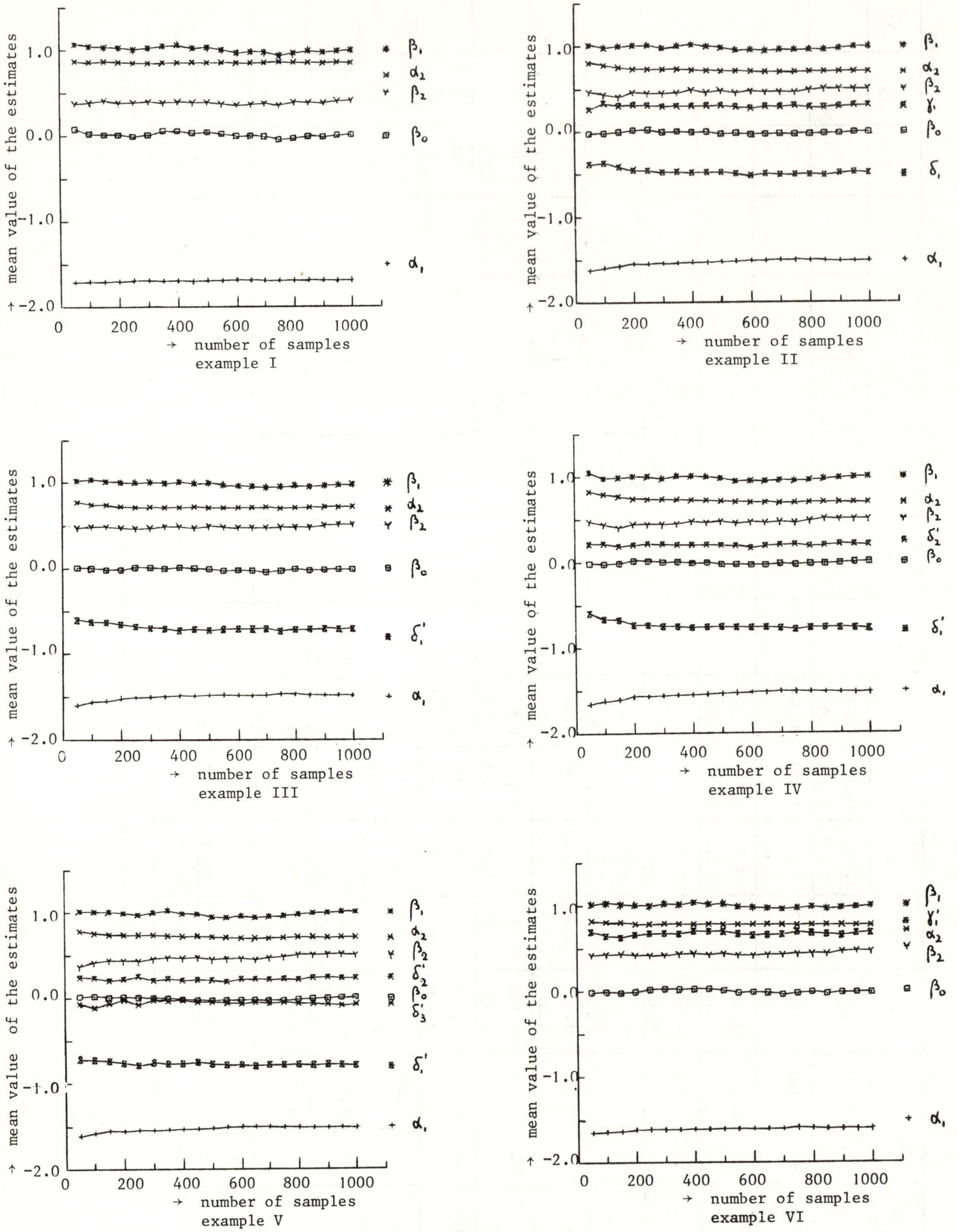
mean value of the estimates
1.0
0.0
-1.0
-2.0
0
200
400
600
800
1000
→ number of samples
example I
β1
α2
β2
β0
α1
example II
β1
α2
β2
γ1
β0
δ1
α1
example III
β1
α2
β2
β0
δ1'
α1
example IV
β1
α2
β2
δ2'
β0
δ1'
α1
example V
β1
α2
β2
δ2'
β0
δ3'
δ1'
α1
example VI
β1
γ1'
α2
β2
β0
α1

figure 4

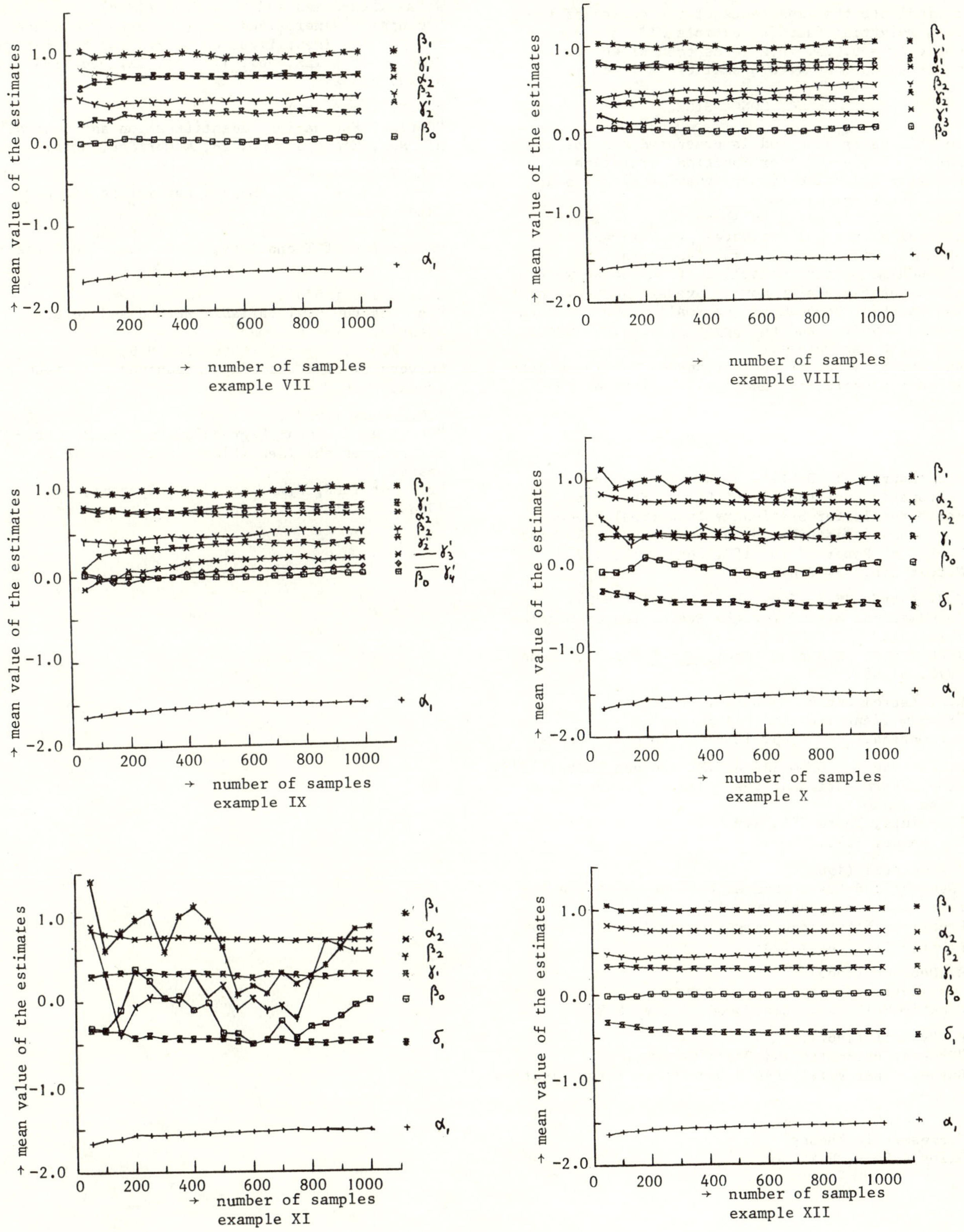
mean value of the estimates
number of samples
example VII
example VIII
example IX
example X
example XI
example XII

figure 4

To indicate the importance of the choice of the data weighting factors, example XII is given where $\Delta\rho = 0.025$. This leads to a slower parameter adjustment with a smaller variance.

6. CONCLUSIONS

In this paper a method is presented for the estimation of the transfer function parameters of both process- and noise filter dynamics of a linear time invariant system.
The scheme presented is shown to be of a very general nature and estimates process- and noise dynamics with a single-stage estimator.
The scheme permits modelling of the noise with a mixed autoregressive moving average series. This is important because of the small number of parameters involved and the physical significance of the noise parameters.
An extensive set of estimations on simulated data shows the usefullnes and applicability of the method.

LITERATURE

K.J. Åström, T. Bohlin, S. Wensmark (1965). "Automatic Construction of Linear Stochastic Dynamic Models for Stationary Industrial Processes with Random Disturbances using Operating Records" Technical Paper TP 18, 150, June 1965, I.B.M. Nordic Lab., Sweden.

K.J. Åström, E.I. Jury, R.G. Agniel (1970). "A Numerical Method for the Evaluation of Complex Integrals" IEEE Trans. on Aut. Control, AC-15, no. 4, August 1970, p. 468.

K.J. Åström and P. Eykhoff (1971). "System Identification - A Survey" Automatica, vol. 7, p. 123-162, Pergamon Press.

A.J.W. van den Boom and A.W.M. van den Enden (1973) "The Determination of the Order of Process- and Noise Dynamics" Preprints, Third IFAC Symposium on Identification, The Hague, 1973.

D.W. Clarke (1967). "Generalized-Least Squares Estimation of the Parameters of a Dynamic Model" Preprints First IFAC Symposium on Identification, paper 3-17, Prague 1967.

R. Deutsch (1965). "Estimation Theory" Prentice Hall Inc.; Englewood Cliffs, N.Y.

P. Eykhoff (1967). "Process Parameter and State Estimation" Survey Paper First IFAC Symposium on Identification Prague,

A.S. Goldberger (1964). "Econometric Theory" Wiley & Sons, N.Y.

R. Hastings-James and M.W. Sage (1969). "Recursive Generalized Least Squares Procedure for On-Line Identification of Process Parameters" Proc. of the IEE, vol. 116, no. 12, Dec. 1969, p. 2057-2062.

R.C.K. Lee (1964). "Optimal Estimation, Identification and Control" MIT Res. Monograph no. 28, Cambridge, Mass.

J.L. Talmon (1971). "Approximated Gauss-Markov Estimators and Related Schemes" T.H. Report 71-E-17, Febr. 1971, Eindhoven University of Technology, Eindhoven, The Netherlands.

A.J. Smets (1970). "The Instrumental Variable Method and Related Identification Schemes" T.H. Report 70-E-15, November 1970, Eindhoven University of Technology, Eindhoven, The Netherlands.

P.C. Young (1968). "The Use of Linear Regression and Related Procedures for the Identification of Dynamic Processes" Proc. of IEEE, Symp. on Adapt. Systems, dec. 1968.

THE DELTA-LMS ADAPTIVE ALGORITHM

John Eric Timm

Stanford University
Stanford, California

A new least-mean-squares (LMS) adaptive algorithm, the delta-LMS adaptive algorithm, is developed in this paper. Using a two step iterative process, this new algorithm solves a specific convergence problem that occurs in LMS algorithms when the desired response and the input signals have large mean values.

The properties of this new algorithm are illustrated by identification examples. These examples are computer simulations in which tapped delay lines are used to model time invariant, asymptotically stable systems.

1. INTRODUCTION

The delta-least-mean-squares (LMS) adaptive algorithm, which is developed in this paper, solves a convergence problem of existing LMS adaptive algorithms. This problem occurs in system identification problems when a tapped delay line (i.e., a transversal filter) is used to model systems that have large mean valued input and output signals.

LMS adaptive algorithms are recursive, steepest descent methods for finding an optimum weight vector that minimizes mean square error. These methods were developed by Widrow and Hoff [1]. Nagumo and Noda [2] applied LMS algorithms to system identification problems and gave convergence results. Mendel [3], [4] extended these results by considering the time-varying and the measurement noise cases.

A system identification problem is described in Section 2. A tapped delay line is used to model an unknown system that may be linear or nonlinear. The convergence problem that arises when existing LMS algorithms are used to attempt to find the optimum weight vector for this tapped delay line is discussed in Section 3. This is the convergence problem that provides motivation for the development of the delta-LMS algorithm. This new algorithm is given in Section 4. Section 5 contains simulation results and the conclusions are given in Section 6.

In this paper column vectors are represented by underlined small letters, matrices by capital letters, and scalars by small letters. The transpose of a matrix is denoted by T, the expected value of a vector by $E[\cdot]$, and the Euclidean norm of a vector by $\|\cdot\|$.

2. THE SYSTEM IDENTIFICATION PROBLEM

Figure 1, a block diagram of an unknown system and a tapped delay line, represents the system identification problem considered in this paper. An LMS algorithm may be used to iteratively attempt to find values for the weights of the tapped delay line that minimize the mean square error between the system and tapped delay line ouputs and thereby solve the identification problem.

The unknown system is assumed to be asymptotically stable and time-invariant. If nonlinear, it is assumed that for the range of inputs considered, this unknown system can be approximated by a linear system and a bias offset. The scalar input u_j at time j to the system is a stationary random variable, and thus, the system output d_j is a random variable.

The tapped delay line model is used because no *a priori* knowledge of the form or order of the system equations is required. This particular tapped delay line may be used to model linear systems which are subject to disturbances or nonlinear systems which can be approximated by a linear system and a bias offset. It is of the form considered in reference [5] and it differs from those considered in references [2] - [4] because of the addition of a bias weight w_0.

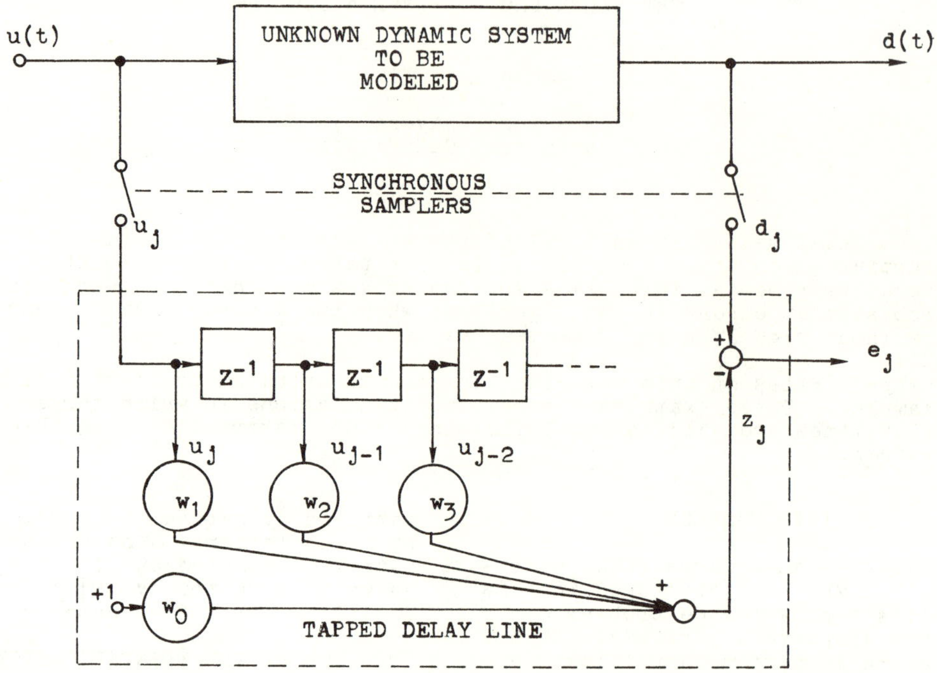

Figure 1. A System Identification Problem

The output of the tapped delay line model, z_j , a randomvariable, is given by

$$z_j = w_0 + \underline{\omega}^T \underline{X}_j \quad (1)$$

where the n dimensional dynamic, tapped delay line input vector $\underline{X}_j$ has components

$$X_{i,j} = u_{j-i+1} \text{ , } i = 1, 2, \ldots , n$$

and where $\underline{\omega}$ is the dynamic weight vector. Equation (1) may also be written as the product of two augmented (i.e., n+1 dimensional) vectors

$$z_j = \underline{w}^T \underline{x}_j \quad (2)$$

where the first component of $\underline{x}_j$, $x_{0,j}$, is equal to one and the remaining n components are

$$x_{i,j} = X_{i,j} \text{ , } i = 1, 2, \ldots , n$$

and where

$$\underline{w}^T = [\, w_0 \, \underline{\omega}^T \,] .$$

Since the error, $e_j = d_j - z_j$, is a random variable the mean square error function is used in this paper. The optimum weight vector $\underline{w}^*$ (which minimizes the mean square error and thereby solves the identification problem) is a constant vector because of the assumptions made, that is, a stationary input and a time-invariant system.

LMS adaptive algorithms may be used to attempt to find $\underline{w}^*$ (see [1] - [7] for the mathematical properties of these algorithms). These algorithms are based on the method of steepest descent, but use an unbiased estimate, $-e_j \underline{x}_j$, in place of the gradient of the mean square error. They do not require the a priori knowledge of the auto-correlation matrix

$$R_{\underline{xx}} = E[\underline{x}_j \underline{x}_j^T]$$

or the cross-correlation matrix

$$P_{d\underline{x}} = E[d_j \underline{x}_j]$$

and the matrix inversion that would be required if $\underline{w}^*$ was obtained by solving the Wiener-Hopf equation

$$\underline{w}^* = R_{\underline{xx}}^{-1} P_{d\underline{x}} \, . \quad (3)$$

They may also be used to solve the identification problem during the real time operation of the unknown system.

In Section 3 a form of LMS algorithm, called the alpha-LMS algorithm in this paper, will be used to study the convergence problem. The alpha-LMS algorithm iteratively attempts to find $\underline{w}^*$ by using

$$\underline{w}_{j+1} = \underline{w}_j + (\alpha/\|\underline{x}_j\|^2)e_j\underline{x}_j . \quad (4)$$

When $\underline{x}_j$ satisfies appropriate conditions (e.g., $\underline{x}_j$ is a random variable with a positive definite auto-correlation matrix) and $0 < \alpha < 2$, there exist theorems which state that $\underline{w}_j$ as found by (4) will converge to $\underline{w}^*$. The fraction of error corrected at each j is specified by α, and when appropriate conditions are satisfied, the fastest convergence occurs for $\alpha = 1$. However, if the minimum mean square error is nonzero, $\underline{w}_j$ will not converge to $\underline{w}^*$ but will have some excess mean square error. To minimize this excess mean square error, α should be chosen less than or equal to one. Reference [7] contains a detailed discussion of these properties.

3. THE CONVERGENCE PROBLEM

An LMS algorithm is implemented on a digital computer and used to attempt to find $\underline{w}^*$ for the tapped delay line of Figure 1. Convergence will not occur when u_j has large mean values even if the conditions of a convergence theorem are satisfied. This will be illustrated by a simple example.

Figure 2 is a block diagram of a system

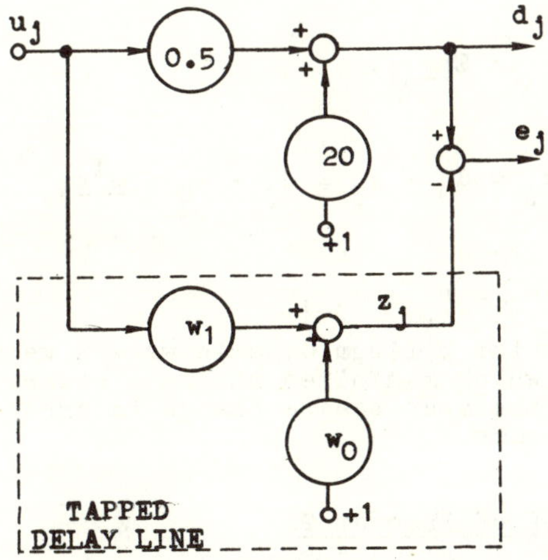

Figure 2. A Simple System

consisting of a gain g_1 of 0.5 and an output bias g_0 of 20 and a tapped delay line model. The output of this two weight tapped delay line is given by (2) with

$$\underline{x}_j^T = [1 \ u_j].$$

The alpha-LMS algorithm is used to attempt to find $\underline{w}^*$ with u_j having various mean values. The alpha-LMS algorithm satisfies the following convergence theorem of reference [7] that is stated without proof.

Theorem 1.

Let

$$d_j = \underline{g}^T\underline{x}_j \quad (5)$$

where $\underline{x}$ is a vector random process such

$$P[x_{1,j}x_{\ell,k}] = P[x_{1,j}]P[x_{\ell,k}]; \ j\neq k \text{ or } \ell\neq 1$$

and R_{xx} is positive definite. Then $\underline{w}^*$ which minimizes the mean square error,

$$E[e_j^2] = E[(d_j - \underline{w}^T\underline{x}_j)^2] ,$$

is

$$\underline{w}^* = \underline{g} . \quad (6)$$

Furthermore, when the alpha-LMS algorithm,

$$\underline{w}_{j+1} = \underline{w}_j + (\alpha/\|\underline{x}_j\|^2)e_j\underline{x}_j , \quad (7)$$

is used with $0 < \alpha < 2$,

$$\lim_{j\to\infty} E[\|\underline{w}_j - \underline{w}^*\|] = 0 .$$

With an initial weight vector of $\underline{w}_0 = \underline{0}$ and an $\alpha = 1$, (7) is used to attempt to find $\underline{w}^*$. When the inputs--stationary, independent random variables--were uniformly distributed in the range (-5,5) convergence of the weights to the system parameters occurred in ten iterations. However, when the input range was changed to (10,20) no apparent convergence occurred even after 300 iterations. This lack of convergence occurred even with other values of α satisfying the sufficient conditions of Theorem 1 (e.g., 0.1 and 1.9). Thus, factors other than the conditions of Theorem 1 need to be considered.

The reason for this lack of convergence can be determined by considering the shape of the mean square error surface. As LMS algorithms use an unbiased estimate of the gradient of the mean square error surface, the shape of this surface will affect the convergence of these algorithms.

When LMS adaptive algorithms are used to minimize mean square error, the eigenvalues of the auto-correlation matrix for $\underline{x}_j$ determine the shape of the mean square error surface. When the input range is (10,20) for the above example, the eigenvalues of the auto-correlation matrix are 0.0356 and 234, and thus define a highly elliptical mean square error surface. If an adaptive algorithm is used on such a highly elliptical surface, the resulting convergence rate will be extremely slow. In fact, since adaptive algorithms are implemented on digital computers which have finite word lengths, convergence to $\underline{w}^*$ will fail to occur when the inputs have large mean values.

A possible solution to this convergence problem is scaling the bias weight to make the mean square error surface less elliptical. However, this would entail knowing the statistics of $\underline{u}_j$ and then scaling the correlation matrix to reduce the ratio of the largest to the smallest eigenvalue. Another, simpler, solution is to use the delta-LMS algorithm which is given in the next section.

4. THE DELTA-LMS ALGORITHM

The delta-LMS algorithm solves the convergence problem given in Section 3 by using a two step iterative process. First, a new estimate for the dynamic weight vector is found by using the alpha-LMS algorithm to minimize the change in error squared. Second, using this new estimate for the dynamic weight vector, the new estimate for the bias weight is found by implementing the alpha-LMS algorithm to minimize the error squared.

The dynamic weight vector is found by minimizing the change in error squared. The change in error is given by

$$\Delta e_j = e_j - e_{j-1} \ . \tag{8}$$

Defining

$$\Delta d_j = d_j - d_{j-1}$$

and

$$\Delta \underline{X}_j = \underline{X}_j - \underline{X}_{j-1}$$

and using (1), (8) becomes

$$\Delta e_j = \Delta d_j - \underline{\omega}^T \Delta \underline{X}_j \tag{9}$$

which is a function of the dynamic weight vector but not of the bias weight. By using the alpha-LMS algorithm, the $\underline{\omega}^*$ which minimizes the mean square change in error is found by using

$$\underline{\omega}_{j+1} = \underline{\omega}_j + (\alpha / \|\Delta \underline{X}_j\|^2) \Delta e_j \Delta \underline{X}_j \tag{10}$$

with $0 < \alpha < 2$.

The bias weight is found by minimizing the error squared. The error is given by

$$e_j = (d_j - \underline{\omega}_{j+1}^T \underline{X}_j - w_{0,j}) \tag{11}$$

where $\underline{\omega}_{j+1}$ is obtained from (10). Using the alpha-LMS algorithm, the w_0^* which minimizes the mean square error is found by using

$$w_{0,j+1} = w_{0,j} + \alpha e_j \tag{12}$$

with $0 < \alpha < 2$.

The statistics of $\underline{X}_j$ may affect the optimum weight vector as found by the delta-LMS algorithm. Theorem 2 gives sufficient conditions for this weight vector to equal the theoretical optimum as found by (3), the Wiener-Hopf equation.

Theorem 2.

Let

$$d_j = \underline{g}^T \underline{x}_j \tag{13}$$

or expanding

$$d_j = g_0 + \underline{g}'^T \underline{X}_j \ . \tag{14}$$

Let $\underline{X}_j$ be a stationary, vector random process with a positive definite auto-correlation matrix, $R_{\underline{XX}}$, and let

$$m_d = E[d_j]$$

and

$$\underline{m}_{\underline{X}} = E[\underline{X}_j].$$

If

$$e_j = d_j - z_j = d_j - w_0 - \underline{\omega}^T \underline{X}_j \tag{15}$$

and

$$\Delta e_j = e_j - e_{j-1} \tag{16}$$

then the optimum dynamic weight vector $\underline{\omega}^*$ which minimizes the mean square error and the mean square change in error is the same.

Proof of Theorem 2.

First, $\underline{\omega}^*$ which minimizes the mean square error is found by taking the gradient of $E[e_j^2]$ with respect to $\underline{\omega}$ and w_0 , setting it equal to zero, and solving to obtain

$$\underline{\omega}^* = (R_{\underline{X}\underline{X}} - \underline{m}_{\underline{X}}\underline{m}_{\underline{X}}^T)^{-1}(P_{d\underline{X}} - m_d\underline{m}_{\underline{X}}) \quad . \tag{17}$$

Using

$$P_{d\underline{X}} = g_0\underline{m}_{\underline{X}} + R_{\underline{X}\underline{X}}\underline{g}'$$

and

$$m_d = g_0 + \underline{g}'^T\underline{m}_{\underline{X}}$$

in (17) gives

$$\underline{\omega}^* = \underline{g}' \quad . \tag{18}$$

Second, ω^* which minimizes the mean square change in error is found by taking the gradient of $E[\Delta e_j^2]$ with respect to $\underline{\omega}$, setting it equal to zero, and solving to obtain

$$\underline{\omega}^* = R^{-1}_{\Delta\underline{X}\Delta\underline{X}} P_{\Delta d\Delta\underline{X}} \quad . \tag{19}$$

Using

$$P_{\Delta d\Delta\underline{X}} = R_{\Delta\underline{X}\Delta\underline{X}}\underline{g}'$$

in (19) gives

$$\underline{\omega}^* = \underline{g}' \quad . \tag{20}$$

Since (18) and (20) are identical, the theorem is proved.

When the dimension of $\underline{w}$ is less than the dimension of $\underline{g}$, the delta-LMS algorithm will yield different values than those which would be obtained by minimizing the mean square error. If $\underline{\omega}^*$ is defined as the optimum dynamic weight vector obtained by minimizing the mean square error and $\underline{\omega}^*_\delta$ is defined as the optimum delta-LMS dynamic weight vector, then the difference is

$$\underline{s}^* = \underline{\omega}^*_\delta - \underline{\omega}^* \quad . \tag{21}$$

The following theorem shows how the statistics of $\underline{x}_j$ affect $\underline{s}^*$.

Theorem 3.

Let

$$d_j = \underline{g}^T\underline{x}_j \tag{22}$$

where

$$\underline{x}_j^T = [1 \;\; \underline{X}_j'^T \;\; \underline{X}_j''^T]$$

and $\underline{X}_j'$ and $\underline{X}_j''$ are stationary, vector random processes and have positive definite auto-correlation matrices, or expanding (22)

$$d_j = g_0 + \underline{g}'^T\underline{X}_j' + \underline{g}''^T\underline{X}_j'' \quad . \tag{23}$$

If

$$z_j = w_0 + \underline{\omega}^T\underline{X}_j' \tag{24}$$

then

$$\underline{s}^* = \Big(R^{-1}_{\Delta\underline{X}'\Delta\underline{X}'}R_{\Delta\underline{X}'\Delta\underline{X}''} - (R_{\underline{X}'\underline{X}'} - \underline{m}_{\underline{X}'}\underline{m}^T_{\underline{X}'})^{-1}(R_{\underline{X}'\underline{X}''} - \underline{m}_{\underline{X}'}\underline{m}^T_{\underline{X}''})\Big)\underline{g}'' \quad .$$

Proof of Theorem 3.

As obtained in Theorem 2 (i.e., use (17) but substitute $\underline{X}_j'$ for $\underline{X}_j$)

$$\underline{\omega}^* = (R_{\underline{X}'\underline{X}'} - m_{\underline{X}'}m^T_{\underline{X}'})^{-1} \cdot (P_{d\underline{X}'} - m_d\underline{m}_{\underline{X}'}) \quad . \tag{25}$$

Substituting

$$m_d = g_0 + \underline{g}'^T\underline{m}_{\underline{X}'} + \underline{g}''^T\underline{m}_{\underline{X}''}$$

and

$$P_{d\underline{X}'} = g_0\underline{m}_{\underline{X}'} + R_{\underline{X}'\underline{X}'}\underline{g}' + R_{\underline{X}'\underline{X}''}\underline{g}''$$

in (25) yields

$$\underline{\omega}^* = g' + (R_{\underline{X}'\underline{X}'} - m_{\underline{X}'}m^T_{\underline{X}'})^{-1} \cdot (R_{\underline{X}'\underline{X}''} - \underline{m}_{\underline{X}'}\underline{m}^T_{\underline{X}''})g'' \quad . \tag{26}$$

The optimum weight vector for the delta-LMS algorithm is given by

$$\underline{\omega}^*_\delta = R^{-1}_{\Delta\underline{X}'\Delta\underline{X}'}P_{\Delta d\Delta\underline{X}'} \tag{27}$$

(i.e., obtain (27) as (19) but substitute $\underline{X}'$ for $\underline{X}$).

Substituting

$$P_{\Delta d\Delta\underline{X}'} = R_{\Delta\underline{X}'\Delta\underline{X}'}\underline{g}' + R_{\Delta\underline{X}'\Delta\underline{X}''}\underline{g}''$$

in (27) yields

$$\underline{\omega}^*_\delta = \underline{g}' + R^{-1}_{\Delta\underline{X}'\Delta\underline{X}'}R_{\Delta\underline{X}'\Delta\underline{X}''}\underline{g}'' \quad . \tag{28}$$

By using (26) and (28) in (21) the proof of the theorem is completed.

In practice the unknown system will contain feedback, and thus have in effect an infinitely long tapped delay line. By choosing a long enough tapped delay line, $\underline{s}^*$ as given by Theorem 3 will be negligible and the delta-LMS algorithm will yield the same weight vector as could be obtained by directly minimizing the mean square error.

The delta-LMS algorithm converges when u_j has large mean values because the correlation matrix of $\Delta \underline{X}_j$ which determines the shape of mean square change in error surface does not include the eigenvalue arising from the bias weight input. This can be illustrated by considering the correlation matrices for $\underline{x}_j$ and $\Delta \underline{X}_j$. If u_j, the input to the unknown system and the tapped delay line satisfies

$$E[u_j] = m \text{ , for all } j$$

$$E[u_j^2] = \xi^2 \text{ , for all } j$$

and

$$E[u_j u_i] = E[u_j]E[u_i] \text{ , } j \neq i$$

then the correlation matrix for

$$\underline{x}_j^T = [\,1 \;\; X_j^T\,]$$

is

$$R_{\underline{xx}} = \begin{bmatrix} 1 & m & - - - - - m \\ m & \xi^2 & m^2 \\ | & & \\ m & m^2 & \xi^2 \end{bmatrix} \qquad (29)$$

and defining

$$\sigma^2 = (\xi^2 - m^2)$$

the correlation matrix for $\Delta \underline{X}_j$ is

$$R_{\Delta \underline{X} \Delta \underline{X}} = \begin{bmatrix} 2\sigma^2 & -\sigma^2 & 0 & \dots & 0 \\ -\sigma^2 & 2\sigma^2 & -\sigma^2 & \dots & 0 \\ 0 & -\sigma^2 & 2\sigma^2 & \dots & 0 \\ \dots & & & & \\ 0 & 0 & 0 & \dots & 2\sigma^2 \end{bmatrix} . \qquad (30)$$

Since (30) does not include the bias term arising from the bias weight, for large mean values it will define a better error surface.

The value of α also affects the excess mean square error for the delta-LMS algorithm. As with the alpha-LMS algorithm, the alphas should be chosen less than one, and how much less depends on the acceptable amount of excess mean square error. However, if α is too small compared to the quantization width of the computer, then the correction for $\underline{w}_j$ may be rounded off to zero and convergence will not occur.

5. EXAMPLES

Examples which demonstrate the characteristics and advantages of the delta-LMS algorithm are given in this section. These example are computer simulations performed on a Hewlett-Packard 2116 digital computer, and they consist of system identification problems in which a tapped delay line is used to model an unknown, time-invariant system. Three different discrete systems are considered. The system inputs were stationary, independent random variables that were uniformly distributed in some range (except when correlated inputs were used). For these examples the initial weight vector was assumed to be zero.

The first example is the system given in Figure 2 (discussed in Section 3). Convergence to the minimum mean square error of zero occurred within ten iterations when the delta-LMS algorithm was used and the inputs were uniformly distributed in the range (10,20). In contrast, when the alpha-LMS algorithm was used with the same input range, no convergence occurred after 300 iterations. Further, when the inputs were distributed in the range (-10,10), the convergence of the delta-LMS algorithm (convergence occurred in five iterations) was superior to that of the alpha-LMS algorithm.

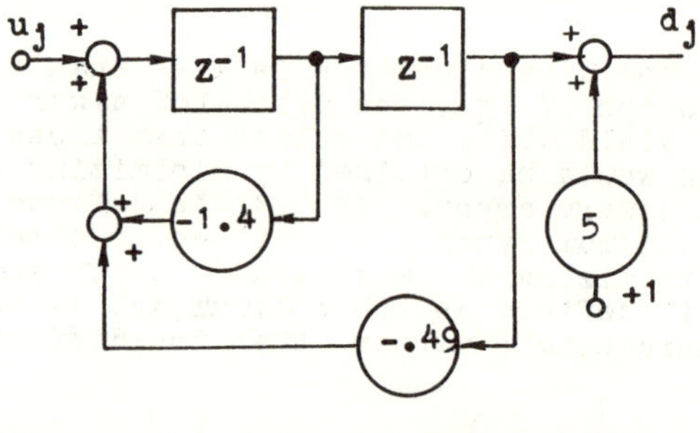

Figure 3. A Second Order System

The second system studied is shown in Figure 3. This second order system was chosen to illustrate the applicability of using the delta-LMS algorithm to find the parameters of a system which requires a long tapped delay line model. Although the correlation matrix for $\Delta \underline{X}_j$ does not include the eigenvalue arising from the bias weight, the eigenvalue ratio may be poor for long tapped delay lines. A tapped delay line of 25 dynamic weights and a bias weight was used to model this system. With inputs uniformly distributed in the range (0,20) and alphas of 0.7 the tapped delay line weight vector converged to the optimum weight vector.

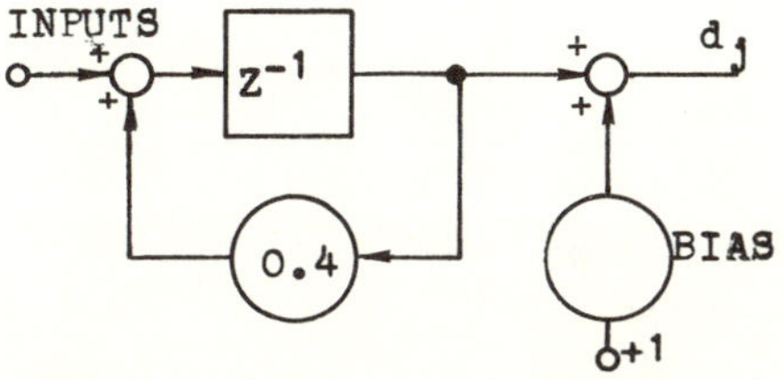

Figure 4. A First Order System

The third system studied is shown in Figure 4. This system was used to compare the performance of the delta-LMS and the alpha-LMS adaptive algorithms and to test the delta-LMS algorithm with correlated inputs, various values for alphas, and zero mean, white measurement noise.

Table 1. Delta-LMS Algorithm Results for System of Figure 4.
($\alpha = 0.5$, 500 iterations)

Input Range	System Bias	w_0	w_1	w_2	w_3	w_4	w_5	w_6	w_7
(-10,10)	0	-0.004	-0.002	0.998	0.399	0.158	0.062	0.027	0.007
(-10,10)	5	5.010	0.000	0.999	0.397	0.157	0.060	0.022	0.006
(10,30)	5	5.500	0.000	0.999	0.397	0.157	0.061	0.022	0.006

Table 2. Alpha-LMS Algorithm Results for System of Figure 4.
($\alpha = 0.5$, 500 iterations)

Input Range	System Bias	w_0	w_1	w_2	w_3	w_4	w_5	w_6	w_7
(-10,10)	0	0.000	0.000	1.000	0.400	0.160	0.066	0.025	0.011
(-10,10)	5	3.500	-0.084	0.929	0.318	0.098	0.030	-0.008	0.006
(10,30)	5	0.020	0.053	1.029	0.428	0.195	0.103	0.081	0.091

Tables 1 and 2 show the weight vectors for the system of Figure 4 found by using the delta-LMS algorithm and the alpha-LMS algorithm. These weight vectors are given for the time $j = 500$. Enough weight vectors were used so that the minimum obtainable mean square error was effectively zero. As expected, when the inputs had large mean values (i.e., the input range was (10,30)) and there was a system bias, the alpha-LMS algorithm failed to converge while the delta-LMS algorithm converged.

The delta-LMS algorithm converged when correlated inputs were applied to the system of Figure 4. These correlated inputs were generated by passing zero-mean, white noise through the filter

$$(1 - K^2)^{\frac{1}{2}}(1 + Kz^{-1})^{-1}$$

and then adding the desired scaling factor and bias. Typical values of K used were 0, 0.5, and 0.95. Convergence of the tapped delay line weights to values which minimize mean square error occurred for all these values of K when a tapped delay line of seven dynamic weights and a bias weight was used. As the degree of correlation in the input signal increased (i.e., larger values of K were used) the rate of convergence decreased. Also, when a tapped delay line of three dynamic weights and a bias was used, convergence occurred. For this shorter tapped delay line the minimum mean square error was nonzero, and the optimum weight vector as given by the delta-LMS algorithm was different than that given by minimizing the mean square error.

The delta-LMS algorithm was tested with various values for the alphas. The smaller values of alpha decreased the excess mean square error. Also, as the alphas were decreased below one, the rate of adaption decreased. However, it was experimentally observed that too small a value for the alphas resulted in no convergence since the correction to w_j was apparently below the quantization level of the digital computer.

The effect of noise on the delta-LMS algorithm was also tested. The rate of convergence decreased and the excess mean square error was increased when the algorithm was tested with various values of zero mean measurement noise.

6. CONCLUSIONS

A new adaptive algorithm, the delta-LMS algorithm, is presented in this paper. It solves a convergence problem that occurs when existing LMS algorithms are used to attempt to find the tapped delay line model weights for physical systems which have large mean valued inputs and outputs. These are the typical operating conditions of many systems. The delta-LMS algorithm should also prove useful for solving similar convergence problems in the fields of pattern recognition and adaptive signal filtering.

The properties of this algorithm are derived and are illustrated by examples.

REFERENCES

1. B. Widrow and M.E. Hoff, Jr., "Adaptive switching circuits," IRE WESCON Conv. Rec., pt. 4, pp. 96-104, 1960.

2. J. Nagumo and A. Noda, "A learning method for system identification," IEEE Trans. on Automatic Control, vol. AC-12, pp. 282-287, June 1967.

3. J.M. Mendel, "Gradient, error-correction identification algorithms," Information Sciences, vol. 1, pp. 23-42, December 1968.

4. ___________, "Gradient identification for linear systems," Adaptive, Learning and Pattern Recognition Systems: Theory and Applications, eds. J.M. Mendel and K.S. Fu (New York: Academic Press, 1970) pp. 209-242.

5. B. Widrow, "Adaptive model control applied to real-time blood-pressure regulation," Pattern Recognition and Machine Learning, ed. K.S. Fu (Plenum Press, 1971) pp. 310-324.

6. B. Widrow, "Adaptive filters I: fundamentals," Stanford Electronics Laboratories, Stanford, Calif., Tech. Rept. SEL-66-126 (TR No. 6764-6), December 1966.

7. J. Timm, "The delta-least-mean squares adaptive algorithm with applications," Ph.D. dissertation, Stanford University, Stanford, Calif., November 1972.

ON THE IDENTIFICATION OF LINEAR DISCRETE TIME SYSTEM MODELS USING THE INSTRUMENTAL VARIABLE METHOD

Brian Finigan - Ian H. Rowe

Department of Electrical Engineering
University of Toronto
Ontario, Canada

The identification of linear discrete time system models using the instrumental variable method requires the generation of an instrumental matrix sequence. A method of using a time invariant, linear, discrete time system model to generate an instrumental matrix sequence is discussed.

INTRODUCTION

Identification of linear discrete time system models using the instrumental variable approach was discussed by Wong [2]. A method was proposed by Wong for generating a sequence of matrices which approximated an optimal instrumental matrix sequence. The method involved using a time varying linear discrete system model. However, no analysis guaranteeing that the matrix sequence generated is an instrumental matrix sequence was given. This paper discusses a method of using a time invariant, linear, discrete time system model to generate a sequence of matrices. Moreover, it is shown that the sequence generated is an instrumental matrix sequence.

In section I, notation for system models' input-output data, and, the definitions of instrumental matrix sequences, instrumental variables and instrumental variable estimates are given. In section II, a statement of the problem to be considered is given and the main theorem, Theorem A, is quoted. Section III is devoted to the proof of Theorem A and section IV gives an example showing how Theorem A can be used to generate an instrumental matrix sequence.

I NOTATION AND DEFINITIONS

Notation

Consider the system model SM_i^1 represented by the nth order difference equation

$$y_i(t) + \sum_{k=1}^{n} a_{ik}y_i(t-k) = \sum_{k=0}^{n} b_{ik}u(t-k) \qquad \text{(I-1)}$$

The Z transform representation is

$$Y_i(z) = \frac{\sum_{k=0}^{n} b_{ik}z^{n-k}}{z^n + \sum_{k=1}^{n} a_{ik}z^{n-k}} U(z) \qquad \text{(I-2)}$$

1 Unless stated otherwise, the subscript i is assumed to belong to the set {i:i=0,1,2,...}

Remark: The system representation used by Wong [2] has the Z transform given by

$$Y_i(z) = \frac{z^n + \sum_{k=1}^{n} b_{ik}z^{n-k}}{\sum_{k=0}^{n} a_{ik}z^{n-k}} U(z) \qquad \text{(I-3)}$$

This representation cannot be used for the frequently occuring system model which has more poles than zeros. Thus the representation (I-1) is preferred by the authors.

Define $M_N(n,\xi) \equiv$

$$\begin{bmatrix} \xi(n-1) & \cdots & \xi(0) & u(n) & \cdots & u(0) \\ \xi(n) & \cdots & \xi(1) & u(n+1) & \cdots & u(1) \\ \vdots & & \vdots & \vdots & & \vdots \\ \xi(n+N-2) & \cdots & \xi(N-1) & u(n+N-1) & \cdots & u(N-1) \end{bmatrix}_{N\times(2n+1)} \qquad \text{(I-4)}$$

$M_N^p(n,\xi) \equiv$

the pth column of the matrix $M_N(n,\xi)$ (I-5)

$$\underline{\theta}_i \equiv [-a_{i1},\ldots,-a_{in},b_{i0},\ldots,b_{in}]'_{(2n+1)\times 1} \qquad \text{(I-6)}$$

where the prime denotes transpose. Letting t have the values n, n+1,...,n+N-1, N equations are obtained from equation (I-1) which can be written as

$$M_N^1(n+1,y_i) = M_N(n,y_i)\,\underline{\theta}_i \qquad \text{(I-7)}$$

For the system model SM_0 let

$$y_0^*(t) = y_0(t) + v(t) \qquad \text{(I-8)}$$

Defining

$$V_N(n) \equiv \begin{bmatrix} v(n) & v(n-1) & & v(0) \\ v(n+1) & v(n) & & v(1) \\ \cdot & \cdot & & \cdot \\ \cdot & \cdot & & \cdot \\ \cdot & \cdot & & \cdot \\ v(n+N-1) & v(n+N-1) & \dots & v(N-1) \end{bmatrix}_{Nx(n+1)} \quad \text{(I-9)}$$

$$\underline{\theta}_0^* \equiv [1, a_{01}, a_{02}, \dots, a_{0n}]' \quad \text{(I-10)}$$

equation (I-7) can be rewritten as

$$M_N^1(n+1,y_0^*) = M_N(n,y_0^*)\underline{\theta}_0 + V_N(n)\underline{\theta}_0^* \quad \text{(I-11)}$$

Equation (I-1) can also be written in a vector form as follows.

Define

$$\underline{x}_i(t) \equiv [y_i(t), y_i(t-1), \dots, y_i(t-n+1)]' \quad \text{(I-12)}$$

$$A_i \equiv \left[\begin{array}{cccc|c} -a_{i1}, & -a_{i2}, & \dots, & -a_{in-1} & -a_{in} \\ \hline & & & & 0 \\ & I_{(n-1)x(n-1)} & & & \cdot \\ & & & & \cdot \\ & & & & 0 \end{array}\right]_{(nxn)} \quad \text{(I-13)}$$

$$\underline{B}_{ij} \equiv [b_{ij}, 0, \dots, 0]'_{(nx1)} \quad \text{(I-14)}$$

Then equation (I-1) can be written as

$$\underline{x}_i(t) = A_i\underline{x}_i(t-1) + \sum_{k=0}^{n} \underline{B}_{ik}u(t-k) \quad \text{(I-15)}$$

The impulse response of the vector representation is given by

$$\underline{H}_i(t) \equiv \begin{cases} 0 & \dots \quad t<0 \\ \sum_{k=0}^{t} A_i^{t-k}\underline{B}_{ik} & \dots \quad 0\le t\le n \\ A_i^{t-n}\underline{h}_i & \quad n<t \end{cases} \quad \text{(I-16)}$$

$$\text{where } \underline{h}_i \equiv \underline{H}_i(n) = \sum_{k=0}^{n} A_i^{n-k}\underline{B}_{ik} \quad \text{(I-17)}$$

<u>Definitions</u>

Consider the system model SM_0 represented by equation (I-1) with the subscript $i = 0$.

Suppose that only the corrupted output $y_0^*(t)$ given by equation (I-8) is available for observation. Given an input sequence $\{u(t)\}_{t=0}^{\infty}$, a sequence of $Nx(2n+1)$ matrices $\{Z_N\}_{N\ge 2n+1}^{\infty}$ is an instrumental matrix sequence if the following probability limits satisfy

$$P\lim_{N\to\infty} \frac{1}{N} Z_N' V_N(n) = 0$$
$$P\lim_{N\to\infty} \frac{1}{N} Z_N' M_N(n,y_0) \quad M, \ M \text{ is nonsingular.} \quad \text{(I-18)}$$

The elements of the matrices Z_N, $N\ge 2n+1$, are called instrumental variables.

The instrumental variable estimate, $\hat{\underline{\theta}}_0(N)$, of the unknown parameter $\underline{\theta}_0$, computed on the basis of $N+n-1$ observations of $y_0^*(t)$ is

$$\hat{\underline{\theta}}_0(N) = \left[\frac{1}{N} Z_N' M_N(n,y_0^*)\right]^+ \left[\frac{1}{N} Z_N' M_N^1(n+1,y_0^*)\right] \quad \text{(I-19)}$$

$$N = 2n+1, 2n+2, \dots$$

where the superscript + denotes the pseudoinverse. Wong [2] showed that $\hat{\underline{\theta}}_0(N)$ is a consistent estimate of $\underline{\theta}_0$.

II STATEMENT OF THE PROBLEM

Consider the nth order system model SM_0 represented by equation (I-1). Suppose that only the input $u(t)$ and the corrupted output $y_0^*(t)$ (see equation (I-8)) are available for measurement. The purpose of this paper is to show that given SM_0 is stable[2] and has no pole-zero cancellation, then for a particular class of inputs, and certain weak conditions on $v(t)$, the signal corrupting the output $y_0(t)$, any stable n^{th} order system model SM_i, represented by equation (I-1), which has no pole-zero cancellation can be used to generate a matrix sequence which is an instrumental matrix sequence. This result is stated formally in Theorem A.

For ease of reference, the following conditions are listed.

Conditions C: C1) SM_i, represented by equation (I-1), is a stable, n^{th} order, linear model. C2) let $\underline{U}(t) \equiv [u(t+n), u(t+n-1), \dots, u(t)]'_{(n+1)x1}$,

2 SM_i stable implies the eigenvalues of A_i lie inside the unit circle in the z plane.

where $\{u(t)\}_{t=0}^{\infty}$ satisfies the condition

$\lim_{N\to\infty} \frac{1}{N} \sum_{t=0}^{N-1} \underline{U}(t)\underline{U}'(t) = I$; C3) $\{u(t)\}_{t=0}^{\infty}$ is the input for both SM_0 and SM_i; C4) $|u(t)|$ is uniformly bounded; C5) $y_0^*(t) = y_0(t) + v(t)$ where $\{v(t)\}_{t=0}^{\infty}$ is a sample from a zero mean, stationary noise process whose covariance function $r_v(\tau)$ satisfies $\lim_{\tau\to\infty} \tau r_v(\tau) = 0$; C6) $\underline{\theta}_i$ is statistically independent of the noise $v(t)$; C7) both SM_0 and SM_i have no pole-zero cancellation.

Theorem A

Given conditions C1) to C6), then $\{M_N(n,y_i)\}_{N\geq 2n+1}^{\infty}$ is an instrumental matrix sequence <=> C7) is true.[3]

The next section is devoted to the proof of Theorem I.

III THEOREMS AND PROOFS

In this section, theorems required for the proof of Theorem A are given and then finally a proof of Theorem A is given.

Define

$$S_i \equiv \sum_{t=n}^{\infty} H_i(t)\underline{H}_0'(t) = \sum_{t=0}^{\infty} A_i^t \underline{h}_i \underline{h}_0' A_0'^t \qquad \text{(III-1)}$$

$$H_i \equiv [\underline{h}_i, A_i\underline{h}_i, \ldots, A_i^{n-1}\underline{h}_i] \qquad \text{(III-2)}$$

Notation: $R(Q)$ denotes the range of the matrix Q; $N(Q)$ denotes the null space of Q; ϕ denotes the empty set; $A \supset B$ denotes the set A contains the set B; $(\forall \eta \in A)\exists$(property 1)$\ni$(property 2) is to be read "for all η, η an element of the set A, there exists property 1 such that property 2 is true".

Theorem 1

Given condition C1), then a) S_i exists and is the unique solution to the equation

$$S_i - A_i S_i A_0' = \underline{h}_i \underline{h}_0' \quad \text{and}$$

b) $$S_i = \sum_{t=0}^{j-1} A_i^t \underline{h}_i \underline{h}_0' A_0'^t + A_i^j S A_0'^j \;, \quad (j=1,2,\ldots)$$

Proof: a) See Theorem 3 in Lancaster [1].

b) $$S_i = \sum_{t=0}^{\infty} A_i^t \underline{h}_i \underline{h}_0' A_0^t$$

$$= \sum_{t=0}^{j-1} A_i^t \underline{h}_i \underline{h}_0' A_0'^t + A_i^j \Big(\sum_{t=0}^{\infty} A_i^t \underline{h}_i \underline{h}_0' A_0'^t \Big) A_0'^j \qquad \text{Q.E.D}$$

Theorem 2

$(A_i, \underline{h}_i)$ is a controllable pair <=> SM_i has no pole-zero cancellations. (Note that by definition, the pair $(A_i, \underline{h}_i)$ is controllable <=> H_i is full rank).

Proof: Let an impulse be applied to SM_i resulting in the output sequence $\{y_i(t)\}_{t=0}^{\infty}$ and corresponding vector sequence $\{\underline{x}_i(t)\}_{t=n-1}^{\infty}$. Assuming zero initial conditions, then $H_i = [\underline{x}_i(n), \underline{x}_i(n+1), \ldots, \underline{x}_i(2n-1)]$. i) Suppose SM_i has pole-zero cancellation. Then $\exists(\lambda \in R^n)\ni(\forall t \geq n)(\langle\underline{\lambda}, \underline{x}_i(t)\rangle = 0)$[4] => $\underline{\lambda}' H_i = 0$ => H_i is singular.[5] Thus H_i full rank => SM_i has no pole-zero cancellation. ii) Suppose H_i is not full rank. Then $\exists(\underline{\lambda} = [\lambda_1, \lambda_2, \ldots, \lambda_n]')$ $\ni(\sum_{j=1}^{n} \lambda_j A_i^{j-1} \underline{h}_i = 0)$ => $(\forall k \geq 0)\ (\sum_{j=1}^{n} \lambda_j A_i^{k+j-1} \underline{h}_i = 0)$ => $(\forall t \geq n)\ (\langle\underline{\lambda}, \underline{x}_i(t)\rangle = 0)$ => SM_i has pole-zero cancellation. Thus SM_i has no pole-zero cancellation => H_i is full rank. Q.E.D.

Theorem 3

Assume condition C1). Let both $(A_0, \underline{h}_0)$ and (A_i, h_i) be controllable pairs. Then S_i is full rank.

Proof:

For simplicity, the subscript i is dropped from A_i and $\underline{h}_i$, and F and $\underline{g}$ are defined by $F \equiv A_0'$, $\underline{g} \equiv \underline{h}_0$. Thus equation (III-1) can be written as

$$S = \sum_{t=0}^{\infty} A^t \underline{h}\, \underline{g}' F^t \qquad \text{(T3-1)}$$

The proof of this theorem is based mostly on contradiction arguments. For clarity, the proof is divided into sections. In section I, a basic property (given by equation (T3-2)) is developed and two lemmas are given. Referring to equation (T3-1), F and A can be either singular or nonsingular. Section II considers the case in which F is nonsingular, (A may or may not be nonsingular) and the case in which A is nonsingular, (F may or may not be nonsingular). Both these cases lead to S being nonsingular. Section III considers the case in which both F and A are singular. For F, $a_{0n-1} \neq 0$ or $a_{0n-1} = 0$ and for A, $a_{in-1} \neq 0$ or $a_{in-1} = 0$ (see equation (I-13)). Thus section III is

3 The symbol <=> denotes equivalence.

4 $\langle\underline{\eta}, \underline{\xi}\rangle$ denotes the inner product of the vectors $\underline{\eta}$ and $\underline{\xi}$.

5 The symbol => denotes implication.

divided into sections i) and ii). Section i) considers the case in which $a_{0n-1} \neq 0$, (a_{in-1} may or may not be zero) and the case in which $a_{in-1} \neq 0$, (a_{0n-1} may or may not be zero). Both these cases lead to S being nonsingular. Section ii) considers the case in which both $a_{0n-1} = 0$ and $a_{in-1} = 0$ and it is shown that this case leads to S being nonsingular.

I Theorem 1 => $S = \sum_{t=0}^{n-1} A^t \underline{h}\, \underline{g}' F^t + A^n S F^n$. Thus, $S - A^n S F^n = [\underline{h}, A\underline{h}, \ldots, A^{n-1}\underline{h}][\underline{g}, F'\underline{g}, \ldots, F'^{n-1}\underline{g}]'$. Since $(A,\underline{h})$ and $(F,\underline{g})$ are controllable pairs =>

$$S - A^n S F^n \text{ is nonsingular} \qquad (T3\text{-}2)$$

Define the set ψ as follows

$$\psi \equiv \{\underline{q} \varepsilon R^n : \underline{q} \notin N(S), \underline{q} \varepsilon R(F), F\underline{q} \varepsilon N(AS)\}$$

Lemma 1 Let 1) $\Omega = \{\underline{r} \varepsilon R^n : \underline{r} \neq 0, \underline{r} \varepsilon R(F), \underline{r} \varepsilon N(S)\} \neq \phi$ and 2) $(\forall \underline{r} \varepsilon R(F)) \exists (\underline{q} \varepsilon R(F)) \ni (F\underline{q} = \underline{r})$. Then $\psi \neq \phi$.

Proof: Suppose $\psi = \phi$. Let $\underline{r} \varepsilon \Omega$. Assumption 2) => for $(j=1,2,\ldots,n)$, $\exists (\underline{q}_j \varepsilon R(F)) \ni (F\underline{q}_1 = \underline{r}, F\underline{q}_j = \underline{q}_{j-1}, 2 \leq j \leq n)$. $\psi = \phi$ => $\underline{q}_1 \varepsilon N(S)$. $\psi = \phi$ and $\underline{q}_1 \varepsilon N(S)$ => $\underline{q}_2 \varepsilon N(S)$. Iterating this logic => $\underline{q}_n \varepsilon N(S)$. Noting that $F^n \underline{q}_n = \underline{r}$ => $S\underline{q}_n - A^n S F^n q_n = 0$ => a contradiction to (T3-2). Q.E.D.

Lemma 2 Suppose 1) $(\forall \underline{r} \varepsilon R(F)) \exists (\underline{q} \varepsilon R(F)) \ni (F\underline{q} = \underline{r})$. Then $\psi \neq \phi$ => S is full rank.

Proof: Suppose $\psi \neq \phi$. Then $\exists (\underline{q}_1 \varepsilon R^n) \ni (\underline{q}_1 \notin N(S), \underline{q}_1 \varepsilon R(F), F\underline{q}_1 \varepsilon N(AS))$.

Assumption 1) => for $(j=2,3,\ldots,n) \exists (\underline{q}_j \varepsilon R^n) \ni (F\underline{q}_j = \underline{q}_{j-1})$.

Theorem 1 => $S\underline{q}_1 - ASF\underline{q}_1 = S\underline{q}_1 = \underline{h}\langle \underline{g}, \underline{q}_1 \rangle \neq 0$ => $R(S) \supset R(\underline{h})$. Also $S\underline{q}_2 - A^2 S F^2 \underline{q}_2 = S\underline{q}_2 - A^2 S F \underline{q}_1 = S\underline{q}_2 = \underline{h}\langle \underline{g}, \underline{q}_2 \rangle + A\underline{h}\langle \underline{g}, \underline{q}_1 \rangle$.

But $\langle \underline{g}, q_1 \rangle \neq 0$ and $\underline{h}$ and $A\underline{h}$ are linearly independent => $S\underline{q}_2 \neq 0$ and $R(S) \supset R([\underline{h}, A\underline{h}])$. Iterating this logic on $S\underline{q}_j - A^j S F^j \underline{q}_j$, $(j=3,4,\ldots,n)$ => $R(S) \supset R([\underline{h}, A\underline{h}, \ldots, A^{n-1}\underline{h}] = R^n$ => S is full rank. Q.E.D.

II Suppose A1) S is singular and A2) F is nonsingular. Since Lemma 1 => $\psi \neq \phi$, Lemma 2 => S is full rank => a contradiction to A1). As a result, F nonsingular => S is nonsingular. Similarly, using S', (see equation (T3-1)) it can be shown that A nonsingular => S is nonsingular.

III Suppose A1) S is singular, A3) F is singular and A4) A is singular.

Lemma 3 Assume A1), A3) and A4). Let $\underline{z} \varepsilon N(S)$. Then $\langle \underline{g}, \underline{z} \rangle = 0$ and $F\underline{z} \varepsilon N(AS)$.

Proof: Theorem 1 => $S\underline{z} - ASF\underline{z} = -ASF\underline{z} = \underline{h}\langle \underline{g}, \underline{z} \rangle$. But $(A,\underline{h})$ is a controllable pair so A4) => $\underline{h} \notin R(A)$ => $\langle \underline{g}, \underline{z} \rangle = 0$ and $ASF\underline{z} = 0$. Q.E.D.

i) Suppose A5) $a_{0n-1} \neq 0$

Lemma 4 Assume A3) and A5). Then $(\forall \underline{r} \varepsilon R(F)) \exists (\underline{q} \varepsilon R(F)) \ni (F\underline{q} = \underline{r})$.

Proof: Let $\underline{r} \varepsilon R(F)$. Noting that A3) => $a_{0n} = 0$, $\underline{r} \varepsilon R(F)$ => $\underline{r} = [r_1, r_2, \ldots, r_{n-1}, 0]'$. But, $a_{0n-1} \neq 0$ => the $(n-1)x(n-1)$ matrix obtained by omitting the nth row and the nth column of F is nonsingular => $\exists (\underline{q} \varepsilon R(F)) \ni (F\underline{q} = \underline{r})$. Q.E.D.

Lemma 5 Assume A1), A3), A4) and A5). Then $\psi \neq \phi$.

Proof: Let $\underline{z} \varepsilon N(S)$ and $\underline{z} \neq 0$. Lemma 3 => $F\underline{z} \varepsilon N(AS)$. Lemma 4 => $\exists (\underline{r} \varepsilon R(F)) \ni (F\underline{r} = F\underline{z})$. Note if $F\underline{z} = 0$, then $S\underline{z} - A^n S F^n \underline{z} = 0$ => a contradiction to (T3-2). Thus $F\underline{z} \neq 0$ and $\underline{r} \neq 0$. Now consider $S\underline{r} - ASF\underline{r} = S\underline{r}$. If $S\underline{r} \neq 0$, $\psi \neq \phi$ and the proof is complete. If $S\underline{r} = 0$, Lemma 4 and Lemma 1 => $\psi \neq \phi$. Q.E.D.

Since Lemma 5 => $\psi \neq \phi$, Lemma 4 and Lemma 2 => S is full rank => a contradiction to A1). As a result, A singular, F singular and $a_{0n-1} \neq 0$ => S is nonsingular. Similarly, using S', it can be shown that A singular, F singular and $a_{in-1} \neq 0$ => S is nonsingular.

ii) Suppose A6) $a_{0n-1} = 0$ and A7) $a_{in-1} = 0$.

Lemma 6 Assume A3), A4), A7) and suppose $\exists (j,\underline{z}) \ni ((j=1,2,\ldots), \underline{z} \varepsilon R^n$, and $-A^{j+1} S F^{j+1} \underline{z} = A^j \underline{h}\langle \underline{g}, F^j \underline{z} \rangle)$. Then $\langle \underline{g}, F^j \underline{z} \rangle = 0$.

Proof: Let $-A^{j+1} S F^{j+1} \underline{z} = A^j \underline{h}\langle \underline{g}, F^j \underline{z} \rangle$. Suppose $\langle \underline{g}, F^j \underline{z} \rangle \neq 0$. Then $A^j \underline{h} \varepsilon R(A^j (AS))$. But, (A,h) is controllable so A4) => $\underline{h} \notin R(A)$. Thus $\exists (\underline{\xi} \varepsilon R^n) \ni (\underline{\xi} \varepsilon N(A), (\underline{h} + \underline{\xi}) \varepsilon R(A))$. But, A7) => $R(A) \supset N(A)$. As a result, $\underline{\xi} \varepsilon R(A)$ and $(\underline{h} + \underline{\xi}) \notin R(A)$ => a contradiction. Thus $\langle \underline{g}, F^j \underline{z} \rangle = 0$. Q.E.D.

Let $\underline{z} \varepsilon N(S)$. Theorem 1 => $S - A^j S F^j = \sum_{t=0}^{j-1} A^t \underline{h}\underline{g}' F^t$, $(j=1,2,\ldots)$. Thus $S\underline{z} - A^j S F^j \underline{z} = -A^j S F^j \underline{z} = \sum_{t=0}^{j-1} A^t \underline{h}\underline{g}' F^t \underline{z} = \sum_{t=0}^{j-1} A^t \underline{h}\langle \underline{g}, F^t \underline{z} \rangle$. For j=1,

Lemma 3 => $\langle \underline{g}, \underline{z} \rangle = 0$. For j=2, $-A^2SF^2\underline{z} = \underline{h}\langle \underline{g}, \underline{z} \rangle + A\underline{h}\langle \underline{g}, F\underline{z} \rangle = Ah\langle \underline{g}, F\underline{z} \rangle$. Lemma 6 => $\langle \underline{g}, F\underline{z} \rangle = 0$. Iterating the same logic for (j=3,4,...) it can be shown that $\langle \underline{g}, F^n\underline{z} \rangle = 0$ => $S\underline{z} - A^n h \underline{g}' F^n \underline{z} = 0$ => contradiction of (T3-2). As a result, F singular, A singular, $a_{0n-1} = 0$ and $a_{in-1} = 0$ => S is nonsingular.

Summarizing, section II shows that F nonsingular => S is nonsingular and A nonsingular => S is nonsingular. Section III shows that F singular and A singular => S is nonsingular. Q.E.D.

Remark: Note A_i, is defined in equation (I-13). Thus this theorem is stated and proved for the special case of A_i being in the companion form. For a change of basis T_i in (I-15), it is straight forward to show that the resulting S_i^* satisfies $S_i^* = T_i^{-1} S_i (T_0')^{-1}$. Thus S_i^* is nonsingular <=> S_i is nonsingular and theorem 3 is valid for the more general case for which A_i is not in companion form.

Theorem 4

Given conditions C1) to C4), then
$\lim_{N\to\infty} \frac{1}{N} M_N'(n,y_i) M_N(n,y_0)$ is nonsingular
<=> S_i is nonsingular.

Proof: Let $\underline{\omega}_i(t) \equiv [y_i(t-1), y_i(t-2), \ldots, y_i(t-n), u(t), \ldots, u(t-n)]'$

It is a straight forward calculation to show that

$$\lim_{N\to\infty} \frac{1}{N} M_N'(n,y_i) M_N(n,y_0) =$$

$$\lim_{N\to\infty} \frac{1}{N} \sum_{t=n}^{N+n-1} \underline{\omega}_i(t) \underline{\omega}_0'(t) = \qquad \text{(III-4)}$$

$$\left[\begin{array}{c|cccc} \sum_{t=0}^{\infty} \underline{H}_i(t)\underline{H}_0'(t) & 0 & \underline{H}_i(0) & \underline{H}_i(1) \ldots & \underline{H}_i(n-1) \\ \hline 0 & & & & \\ \underline{H}_0'(0) & & & & \\ \underline{H}_0'(1) & & I_{(n+1)\times(n+1)} & & \\ \vdots & & & & \\ \underline{H}_0'(n-1) & & & & \end{array}\right]$$

$$\sim \left[\begin{array}{c|ccc} \sum_{t=n}^{\infty} \underline{H}_i(t)\underline{H}_0'(t) & 0 & \cdots & 0 \\ \hline 0 & & & \\ \vdots & & I_{(n+1)\times(n+1)} & \\ 0 & & & \end{array}\right]_{(2n+1)\times(2n+1)}$$

where A~B denotes that the matrix B can be obtained from the matrix A by a sequence of elementary transformations. Q.E.D.

Theorem 5

Let $\{Z_N\}_{N\geq 2n+1}^{\infty}$ be a sequence of Nx(2n+1) matrices whose elements are uniformly bounded. If ($\forall N \geq 2n+1$), the elements of Z_N are statistically independent of the elements of $V_N(n)$, and if $\{v(t)\}_{t=0}^{\infty}$ satisfies C5), then
$P \lim_{N\to\infty} \frac{1}{N} Z_N' V_N(n) = 0$.

Proof: See Theorem 2 in Wong [2].

Theorem A

Given conditions C1) to C6), then $\{M_N(n,y_i)\}_{N\geq 2n+1}^{\infty}$ is an instrumental matrix sequence <=> C7) is true.

Proof: i) Suppose $\{M_N(n,y_i)\}_{N\geq 2n+1}^{\infty}$ is an instrumental matrix sequence. Then referring to equation (III-4), $\nexists(\lambda \in R^{2n+1}) \ni (\forall t \geq \mu)(\langle \underline{\lambda}, \underline{\omega}_i(t) \rangle = 0)$ or $\langle \underline{\lambda}, \underline{\omega}_0(t) \rangle = 0)$ => C7) is true.

ii) Conditions C1), and C3) to C6) and Theorem 5 =>

$$P \lim_{N\to\infty} \frac{1}{N} M_N'(n,y_i) V_N(n) = 0 \qquad \text{(TA-1)}$$

Suppose C7) is true. Then Theorem 2 => $(A_i, \underline{h}_i)$ and $(A_0, \underline{h}_0)$ are controllable pairs. Thus Theorem 3 => S_i is nonsingular =>
$\lim_{N\to\infty} \frac{1}{N} M_N'(n,y_i) M_N(n,y_0)$ is nonsingular (TA-2)
(by Theorem 4).
(TA-1) and (TA-2) => $\{M_N(n,y_i)\}_{N\geq 2n+1}^{\infty}$ is an instrumental matrix sequence Q.E.D.

IV APPLICATION

Generation of an instrumental matrix sequence using a time invariant, stable linear system.

Let SM_0 be a 1st order system model given by

$$Y_0(z) = \frac{b_0 z + b_1}{z + a_1} U(z)$$

where $b_0 \neq 0$ and $|a_1| < 1$. No pole-zero cancellation requires $a_1 b_0 \neq b_1$. Define $\underline{\theta}_0$ by $\underline{\theta}_0 = [-a_1, b_0, b_1]'$. Let $y_0^*(t) = y_0(t) + v(t)$. Suppose the elements of $\underline{\theta}_0$ are unknown and that only the corrupted output $y_0^*(t)$ is available for measurement. Let SM_1 be a 1st order system model given by

$$Y_1(z) = \frac{z+1}{z} U(z)$$

Define $\underline{\theta}_1 = [0,1,1]'$. Let $\{u(t)\}_{t=0}^{\infty}$ satisfy

$$\lim_{N\to\infty} \frac{1}{N} \sum_{t=0}^{n-1} \begin{bmatrix} u(t+1) \\ u(t) \end{bmatrix} [u(t+1), u(t)] = \begin{bmatrix} 1 & 0 \\ 0 & 1 \end{bmatrix}$$

where $|u(t)|$ is uniformly bounded. If $\{v(t)\}_{t=0}^{\infty}$ satisfies condition C5) then Theorem A => SM_1 can be used to generate an instrumental matrix sequence for the purpose of generating a consistent estimate of $\underline{\theta}_0$.

For this example $(i = 0,1)$

$$M_N(1,y_i) = \begin{bmatrix} y_i(0) & u(1) & u(0) \\ y_i(1) & u(2) & u(1) \\ \cdot & \cdot & \cdot \\ \cdot & \cdot & \cdot \\ \cdot & \cdot & \cdot \\ y_i(N-1) & u(N) & u(N-1) \end{bmatrix}$$

$$M_N(2,y_i) = [y_i(1), y_i(2) \ldots, y_i(N)]'$$

$$\underline{x}_i(t) = y_i(t)$$

$$A_0 = -a_1 \quad , \quad A_1 = 0$$

$$b_{00} = b_0 \quad , \quad b_{10} = 1$$

$$b_{01} = b_1 \quad , \quad b_{11} = 1$$

$$\underline{H}_0(t) \begin{cases} 0 & t<0 \\ b_0 & t=0 \\ -a_1 b_0 + b_1 = \underline{h}_0 & t=1 \\ -a_1^{t-1}(-a_1 b_0 + b_1) & t \geq 1 \end{cases}$$

$$\underline{H}_1(t) \begin{cases} 0 & t<0 \\ 1 & t=0 \\ 1 & t=1 \\ 0 & t \geq 1 \end{cases}$$

Since $\underline{\theta}_1$ is statisically independent of $v(t)$,

$$\operatorname{P\,lim}_{N\to\infty} \frac{1}{N} M_N'(1,y_1) V_N(1) = 0 \qquad \text{(IV-1)}$$

$$\lim_{N\to\infty} \frac{1}{N} M_N'(1,y_1) M_N(1,y_0) = \left[\begin{array}{c|cc} \sum_{t=0}^{\infty} \underline{H}_1(t)\underline{H}_0(t) & 0 & \underline{H}_1(0) \\ \hline 0 & 1 & 0 \\ \underline{H}_0(0) & 0 & 1 \end{array}\right]$$

$$= \left[\begin{array}{c|cc} b_0 + (-a_1 b_0 + b_1) & 0 & 1 \\ \hline 0 & 1 & 0 \\ b_0 & 0 & 1 \end{array}\right] \sim \left[\begin{array}{c|cc} -a_1 b_0 + b_1 & 0 & 0 \\ \hline 0 & 1 & 0 \\ 0 & 0 & 1 \end{array}\right]$$

Recall, no pole-zero cancellation => $a_1 b_0 \neq b_1$. Thus $\lim_{N\to\infty} \frac{1}{N} M_N'(1,y_1) M_N(1,y_0)$ exists and is non-singular. (IV-2)

Define $\hat{\underline{\theta}}_0(N)$ by

$$\hat{\theta}_0(N) = \left[\frac{1}{N} M_N'(1,y_1) M_N(1,y_0^*)\right]^{+} \left[\frac{1}{N} M_N'(1,y_1) M_N^1(2,y_0^*)\right]$$

$$N = 3,4,\ldots$$

(IV-1) & (IV-2) => $\{M_N(1,y_1)\}_{N>3}^{\infty}$ is an instrumental matrix sequence. As a result, $\hat{\underline{\theta}}_0(N)$ is a consistent estimate of $\hat{\underline{\theta}}_0$.

V CONCLUSIONS

Consider the nth order, time invariant, linear, discrete time, system model SM_0. To generate a consistent estimate of the parameter $\underline{\theta}_0$ (see equations (I-1) and (I-6)) using the instrumental variable method, an instrumental matrix sequence is required. Suppose SM_0 is stable and has no pole-zero cancellation. Let $\{M_N(n,y_1)\}_{N \geq 2n+1}^{\infty}$

(see equation (I-4)) be generated by *any* stable, n^{th} order, time invariant, linear, discrete time, system model SM_i with no pole-zero cancellation. If the input sequence, $\{u(t)\}_{t=0}^{\infty}$, satisfies C2) to C4) and the noise sequence, $\{v(t)\}_{t=0}^{\infty}$, satisfies C5), then it has been shown that $\{M_N(n,y_i)\}_{N\geq 2n+1}^{\infty}$ is an instrumental matrix sequence.

REFERENCES

[1] Lancaster, P. (1970); Explicit Solutions of Linear Matrix Equations; SIAM Review; 12, 544.

[2] Wong, K.Y., Polak, E. (1967); Identification of Linear Discrete Time Systems Using the Instrumental Variable Approach; IEEE Trans. Auto. Control; AC-12, 707.

ACKNOWLEDGEMENT

The authors wish to thank the National Research Council of Canada for their research support under Grant number A5276.

MAXIMUM LIKELIHOOD IDENTIFICATION OF ÅSTRÖM MODEL BY QUASILINEARIZATION

Borivoje P. Furht
Institute "Boris Kidric" - Vinca, POB 522
Belgrade, Yugoslavia

This paper presents an application of quasilinearization technique to the identification of linear stochastic model, proposed by Åström. Interpreting maximum likelihood method as finding the coefficients of the prediction model, an iterative algorithm is derived using quasilinearization. The algorithm proposed is very simple for computer application and can be implemented in on-line identification.

1. INTRODUCTION

Consider linear discrete-time time invariant one input one output model suggested by Åström

$$A(z^{-1})y(t) = B(z^{-1})z^{-k}u(t) + \lambda C(z^{-1})e(t) \qquad (1.1)$$

where
u and y are system input and output, respectively,
$\{e(t)\}$ is sequence of independent normal N(0,1) random variables,
λ is noise standard deviation,
z^{-1} is delay operator defined as

$$y(t) = z^{-1}y(t+1) \qquad (1.2)$$

$A(z^{-1})$, $B(z^{-1})$ and $C(z^{-1})$ are polynomials of z^{-1}

$$A(z^{-1}) = 1 + a_1 z^{-1} + \ldots + a_n z^{-n}$$

$$B(z^{-1}) = b_1 z^{-1} + \ldots\ldots + b_n z^{-n} \qquad (1.3)$$

$$C(z^{-1}) = 1 + c_1 z^{-1} + \ldots + c_n z^{-n}$$

n and k are known system order and system delay, respectively.

This model is very convenient for the synthesis of the minimum mean-squares control law (Åström, 1967).

The task of the identification is the estimation of (3n+1) unknown parameters a_i, b_i, c_i (i=1,...n) and λ, on the basis of a finite sequence of N input/output observations.

Åström (1965) has implemented maximum likelihood method to the identification of unknown parameters in the model (1.1), with an assumption that the distribution of e is gaussian. The gradient technique has been proposed for the minimization of the likelihood function. The generalized least-squares method (Clark, 1967) can also be implemented to the identification of the model (1.1). Then, process parameters a_i and b_i and noise parameters c_i are estimated separately. Peterka (1970) has derived "tally" estimate where process parameters are obtained in one step. Panuska (1968) has applied stochastic approximation method to the identification the model proposed.

Comparative analysis of these methods with respect to noise influence, accuracy and rate of convergence has been done by Furht (1972).

A new estimation algorithm is proposed in the present paper. Maximum likelihood criterion is used. The quasilinearization technique is applied for the minimization of the loss function. Same technique has been used by Schulz (1968) for the identification of Z-transfer function in the restricted case where system output is corrupted with additive white noise.

2. MAXIMUM LIKELIHOOD

Maximum likelihood method is based on the maximization of the likelihood function. It is equivalent to the minimization of the negative logarithm of the likelihood function

$$F(\theta,\lambda) = -\log L(\theta,\lambda) = \frac{1}{2\lambda^2}\sum_{t=1}^{N} \varepsilon^2(t) + \frac{N}{2}\log\lambda + \frac{N}{2}\log 2\pi \qquad (2.1)$$

where

$$\varepsilon(t) = C^{-1}(z^{-1})A(z^{-1})y(t) - C^{-1}(z^{-1})B(z^{-1})z^{-k}u(t) \qquad (2.2)$$

The minimization of F can be performed separately. First, the minimization of the loss function

$$V(\theta) = \sum_{t=1}^{N} \varepsilon^2(t) \qquad (2.3)$$

with respect to θ gives the parameter estimates $\hat{\theta}$. The error $\varepsilon(t)$ is linear in the parameters a_i and b_i, but non-linear in c_i. The gradient Newton-Raphson procedure has been proposed for the minimization of the loss function V (Åström, 1965). The minimization with respect to λ can be performed analytically

$$\hat{\lambda}^2 = \frac{1}{N}\min_{\theta} V(\theta) \qquad (2.4)$$

The maximum likelihood criterion (2.3) can be interpreted as a "model adjustment" technique. If one defines

$$\varepsilon(t) = y(t) - v(t) \qquad (2.5)$$

where v(t) is the model output, the prediction model in the form

$$v(t)=C^{-1}(z^{-1})B(z^{-1})z^{-k}u(t)-C^{-1}(z^{-1})[A(z^{-1})-C(z^{-1})]y(t) \tag{2.6}$$

is obtained. Now, the identification constits of finding out the parameters of the prediction model (2.6), such that the loss function

$$V(y,v) = \sum_{t=1}^{N}[y(t)-v(t)]^2 = \sum_{t=1}^{N}\varepsilon^2(t) \tag{2.7}$$

is as small as possible.

3. QUASILINEARIZATION

In this paper the non-linearity in the parameters c_i of the prediction model is handled by means of quasi-linearization technique. An iterative algorithm can be developed assuming that the estimates from the i-th iteration are available. By linearizing the form

$$C(z^{-1})v(t)=B(z^{-1})z^{-k}u(t)-[A(z^{-1})-C(z^{-1})]y(t) \tag{3.1}$$

about the solution from the i-th iteration, the expression for the (i+1) th estimates may be obtained. The linearization yields

$$C^i(z^{-1})v^i(t)+[C^{i+1}(z^{-1})-C^i(z^{-1})]v^i(t)+C^i(z^{-1})[v^{i+1}(t) - v^i(t)] = B^{i+1}(z^{-1})z^{-k}u(t)-[A^{i+1}(z^{-1})-C^{i+1}(z^{-1})]y(t) \tag{3.2}$$

In this relation $v^{i+1}(t)$ is a linear approximation to the v(t) that results from the (i+1) th iteration. Solving (3.2) for $v^{i+1}(t)$ and introducing in (2.5) one obtains the expression for the error on the (i+1) th iteration

$$\varepsilon^{i+1}(t) = y(t)-v^i(t)+A^{i+1}(z^{-1})\frac{y(t)}{C^i(z^{-1})} - B^{i+1}(z^{-1})z^{-k}\frac{u(t)}{C^i(z^{-1})} - C^{i+1}(z^{-1})\frac{y(t)-v^i(t)}{C^i(z^{-1})} \tag{3.3}$$

Note that the error is linear in the parameters a_i, b_i and c_i from (i+1) th iteration.

Filtering process inputs and otuputs and model outputs through the digital filter $1/C^i(z^{-1})$

$$\tilde{y}^i(t) = y(t)/C^i(z^{-1})$$
$$\tilde{u}^i(t) = u(t)/C^i(z^{-1}) \tag{3.4}$$
$$\tilde{v}^i(t) = v^i(t)/C^i(z^{-1})$$

the error becomes

$$\varepsilon^{i+1}(t) = y(t) - v^i(t) + A^{i+1}(z^{-1})\tilde{y}^i(t) - B^{i+1}(z^{-1})z^{-k}\tilde{u}^i(t) - C^{i+1}(z^{-1})[\tilde{y}^i(t)-\tilde{v}^i(t)] \tag{3.5}$$

Introducing the following notation

$$M^i(t) = [-\tilde{y}^i(t-1)\ldots-\tilde{y}^i(t-n)\vdots\tilde{u}^i(t-k-1)\ldots\tilde{u}^i(t-k-n)\vdots \tilde{y}^i(t-1)-\tilde{v}^i(t-1)\ldots\tilde{y}^i(t-n)-\tilde{v}^i(t-n)] \tag{3.6}$$

$$w^i(t) = y(t)-v^i(t)+\tilde{v}^i(t) \tag{3.7}$$

and

$$\theta = \mathrm{col}[a_1,\ldots a_n:b_1,\ldots b_n:c_1,\ldots c_n] \tag{3.8}$$

the error takes the following form

$$\varepsilon^{i+1}(t) = w^i(t) - M^i(t)\theta^{i+1} \tag{3.9}$$

Substituting (3.9) into the loss function (2.3) and minimizing it with respect to θ one obtains the (i+1)th estimates

$$\hat{\theta}^{i+1} = \left[\sum_{t=k+n+1}^{N} M^{i^T}(t)M^i(t)\right]^{-1}\left[\sum_{t=k+n+1}^{N} M^{i^T}(t)w^i(t)\right] \tag{3.10}$$

The iterative algorithm consits of the following steps:

a. Find an initial value for the parameter estimates.
b. Calculate the outputs from the prediction model using (2.6).
c. Filter input/output observations and model outputs through the digital filter using (3.4).
d. Estimate new parameters using (3.10).
e. Repeat steps b, c and d until the parameter estimates converge.

4. INITIAL ESTIMATES

It is possible to show that the algorithm converges with quadratic properties and quite fast to the global minimum of the loss function if the initial parameter estimates is chosen adequately.

Assuming $c_i=0$, V becomes a quadratic function of a and b and the algorithm gives the least-squares estimate a^o and b^o of a and b in one step. The initial estimates for the iteration procedure are then taken as $\theta^o=\mathrm{col}[a^o, b^o, 0]$. These initial estimates do not ensure the convergence in all cases. On the other hand, if the convergence exists, the number of iterations can be too great. Hence, it is also necessary to find good initial estimates for the parameters c_i.

The following procedure gives the initial estimates for the parameters c_i. The procedure is based on the error analysis and it is similar to the one proposed by Clark (1967). In the first step, the equivalent error is computed on the basis of the least-squares estimates for the parameters a_i and b_i.

$$\hat{\eta}(t)=C(z^{-1})\,\varepsilon(t)=A^o(z^{-1})y(t)-B^o(z^{-1})z^{-k}u(t) \tag{4.1}$$

Noise parameters are estimated in the next step. If one defines

$$\frac{1}{C(z^{-1})} \cong D(z^{-1}) = 1+d_1z^{-1}+\ldots+d_nz^{-n} \tag{4.2}$$

it follows that the auto-regression model for $\hat{\eta}(t)$ is given in the form

$$D(z^{-1})\,\hat{\eta}(t) = \varepsilon(t) \tag{4.3}$$

The conventional least-squares method gives

$$\hat{d} = \left[\sum_{t=n+1}^{N} S^T(t)S(t)\right]^{-1}\left[\sum_{t=n+1}^{N} S^T(t)\,\hat{\eta}(t)\right] \qquad (4.4)$$

where

$$S(t) = \left[-\hat{\eta}(t-1), -\hat{\eta}(t-2), \ldots -\hat{\eta}(t-n)\right] \qquad (4.5)$$

and

$$d = \mathrm{col}\left[d_1, d_2, \ldots . d_n\right] \qquad (4.6)$$

Having the parameters d_i the initial estimates for c_i are approximatively computed from (4.2).

5. EXPERIMENTAL RESULTS

In order to illustrate the mentioned theoretical considerations some experiments have been done by simulation on a general purpose digital computer.

5.1. Example 1

The second-order process has been simulated:

$$y(t) = \frac{1.0z^{-1}+0.5z^{-2}}{1-1.5z^{-1}+0.7z^{-2}}u(t)+\lambda\frac{1-1.0z^{-1}+0.2z^{-2}}{1-1.5z^{-1}+0.7z^{-2}}e(t) \qquad (5.1)$$

A pseudo random binary sequence with amplitudes +1 and -1 has been used to excite the process. The discrete white noise e(t) has been gaussian with variance 1. The number of input/output observations has been 400. The identification of unknown parameters has been performed by the original maximum likelihood method using Newton-Raphson procedure and by the algorithm proposed in this paper. The procedure proposed for finding the initial estimates has been used for both identification algorithms. Table 1 allows the comparison of the results obtained using these two methods.

Table 1

	TRUE	NEWTON-RAPHSON	QUASILINE-ARIZATION
a_1	-1.5	-1.520	-1.525
a_2	0.7	0.711	0.714
b_1	1.0	0.974	0.972
b_2	0.5	0.471	0.462
c_1	-1.0	-1.023	-1.039
c_2	0.2	0.218	0.217
λ	1.0	1.027	1.032
V		422.12	422.43
number of iterations		6	6

It has been shown that the algorithm proposed gives parameter estimates near the true maximum likelihood estimates, but the computing time is smaller and the computer application is much simpler. Namely, a gradient routine involves computation of the gradient V_θ of V with respect to the parameters as well as the matrix of second partial derivatives $V_{\theta\theta}$. The number of iterations for both procedures is equal.

5.2. Example 2

To enable the comparison with other identification methods the parameters of two processes proposed by Isermann (1972) have been estimated by the algorithm. The structure of the processes is given in the form

$$y(t) = \frac{B(z^{-1})}{A(z^{-1})}u(t) + \lambda\frac{F(z^{-1})}{G(z^{-1})}e(t) \qquad (5.2)$$

The processes are second- and third-order, respectively, and the noise filter F/G has the same form for both processes

$$\frac{F(z^{-1})}{G(z^{-1})} = \frac{0.0114z^{-1}}{1-1.027z^{-1}+0.264z^{-2}} \qquad (5.3)$$

Using the approximation

$$\frac{A(z^{-1})F(z^{-1})}{G(z^{-1})} \cong C(z^{-1}) \qquad (5.4)$$

where $C(z^{-1})$ is a polynomial of finite order, the structure (5.2) is reduced to the Åström form (1.1) and the algorithm proposed can be implemented to the identification of the process parameters a and b.

In these experiments the number of input/output observations has been 700. The input signal has been pseudo random binary sequence with amplitudes +0.5 and -0.5. The identification of unknown parameters has been performed for two noise/input signal ratios: 0.1 and 0.2, and with 5 different noise data sets. The mean values of the parameters are presented in tables 2 and 3. Their standard deviations are given for the second-order process immediately below the averages.

Table 2
Second-order process

	TRUE	ESTIMATE $\eta = 0.1$	ESTIMATE $\eta = 0.2$
a_1	-1.425	-1.431	-1.428
		0.0114	0.0247
a_2	0.496	0.500	0.512
		0.0114	0.0257
b_1	-0.102	-0.101	-0.100
		0.0031	0.0067
b_2	0.173	0.171	0.168
		0.0040	0.0081
c_1		-0.413	-0.409
c_2		-0.205	-0.185
λ		0.058	0.117
V		2.382	9.508
$\delta_{\Sigma 1}$		0.0454	0.0960
$\delta_{\Sigma 2}$		0.0107	0.0242
δ_q		0.0496	0.1433
number of iterations		3	4

Table 3
Third-order process

	TRUE	ESTIMATE $\eta = 0.1$	ESTIMATE $\eta = 0.2$
a_1	-2.084	-2.000	-2.095
a_2	1.422	1.319	1.469
a_3	-0.316	-0.297	-0.362
b_1	0.018	0.018	0.017
b_2	0.010	0.011	0.011
b_3	-0.006	-0.0066	-0.0146
c_1		-0.994	-1.093
c_2		0.091	0.157
λ		0.056	0.113
V		2.204	8.878
$\delta_{\Sigma 1}$		0.180	1.445
$\delta_{\Sigma 2}$		0.052	0.027
δ_g		0.154	2.030
number of iterations		6	10

The errors given in the tables have the following definitions

$$\delta_{\Sigma 1} = \left[\sum_{i=1}^{p} \left(\frac{\Delta\theta_i}{\theta_i} \right)^2 \right]^{\frac{1}{2}} \tag{5.5}$$

$$\delta_{\Sigma 2} = \left[\frac{\sum_{i=1}^{p} (\Delta\theta_i)^2}{\sum_{i=1}^{p} \theta_i^2} \right]^{\frac{1}{2}} \tag{5.6}$$

with: $\Delta\theta_i = \theta_i - \hat{\theta}_i$, $i=1,2,\ldots p$,
θ_i : exact parameter values,
$\hat{\theta}_i$: parameter estimates,
p : number of all parameters to be identified.

$$\delta_g = \left[\frac{\quad}{\Delta g^2(k)/g^2(k)} \right]^{\frac{1}{2}} \tag{5.7}$$

with: $\Delta g(k) = g(k) - \hat{g}(k)$
$g(k)$: exact weighting function
$\hat{g}(k)$: weighting function, resulting from the estimated parameters.

The figure 1 presents V as a function of number of iterations for the second- and third-order process, for noise/input signal ratio 0.1.

6. CONCLUSION

The algorithm proposed has been shown to be effective for the identification of Åström model (1.1). The procedure for finding the initial estimates enable fast convergence. The algorithm is simple for computer application. Accuracy is high and the estimates are near the true maximum likelihood estimates obtained by gradient procedures, but the computing time is much smaller. Also, it is possible to find out a recursive procedure which enables the application of the algorithm in on-line identification (Furht, 1973).

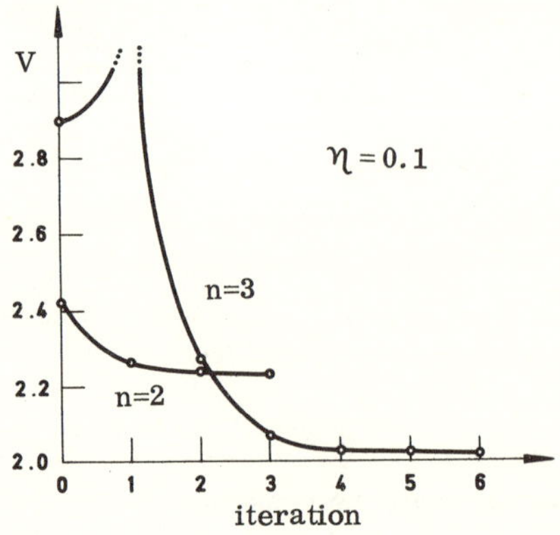

Fig.1 Convergence of the loss function

REFERENCES

Åström, K. J. (1967). "Computer control of a paper machine-an application of linear stochastic control theory", IBM J. Res. Dev. 11, 289.

Åström, K. J. and Bohlin, T. (1965). "Numerical identification of linear dynamics systems from normal operating records", IFAC symp. Theory of self-adaptive control systems, Teddington.

Åström, K. J. and Eykhoff, P. (1970, 1971). "System identification-a survey", 2nd IFAC symposium on identification and process parameter estimation and Automatica, Vol. 7, 123.

Clark, D. W. (1967). "Generalized-least-squares estimation of a dynamic model", IFAC symposium on identification in automatic control systems, Prague, paper 3.17.

Furht, B. P. (1973). "A new recursive procedure for on-line linear stochastic system identification", ETAN Conf. Yugoslavia (in serbo-croatien).

Furht, B. P. (1972). "Comparative analysis of methods for linear stochastic system identification", M. Sc. Thesis, Electrical Faculty, University of Belgrade, Yugoslavia (in serbo-croatien).

Isermann, R. (1972). "Test-cases for evalution and comparison of different identification and parameter-estimation methods", University Stuttgart.

Panuska, V. (1968). "A stochastic approximation method for identification of linear systems using adaptive filtering", Proc. JACC, Michigan, 1014.

Peterka, V. and Halouskova, A (1970). "Tally estimate of Åström model for stochastic systems", 2nd IFAC symposium on identification and process parameter estimation, Prague, paper 2.3.

Sage, A. P. and Melsa, J. L. (1971). "System identification", Academic Press, New York, Ch. 6.

Schulz, E. R. (1968). "Estimation of pulse transfer function parameters by quasilinearization", IEEE Trans. on Aut. Control AC-13, 424.

THE DESIGN OF MODEL REFERENCE PARAMETER ESTIMATION SYSTEMS USING HYPERSTABILITY THEORIES

Chang-Chieh Hang
Inter-Institute of Engineering Control
University of Warwick,
Coventry, England

The implementation of Landau's hyperstable model reference parameter estimation scheme is found to give difficulty due to the series derivative compensator introduced to ensure stability. This paper examines the application of the "state variable filters" technique to avoid the signal differentiation problem associated with Landau's scheme. The merits of the resultant design are then compared with the generalized equation error design. Multi-dimensional and nonlinear systems are also considered.

INTRODUCTION

Process parameter estimation using an adjustable model can be designed by either the response error (RE) method or the generalized equation error (GEE) method [6]. The latter has been most widely used on account of the assured stability and the ease of implementation. This paper examines the RE method derived by Landau using Popov's hyperstability theorems [4], [5]. Although Landau's design is guaranteed to be globally stable for all inputs, it requires the measurement of the output derivatives. A solution avoiding this problem by using the "state variable filters" is introduced in this paper. Linear single dimensional systems are considered in detail and comparisons with the GEE design are made. Extensions to multi-dimensional and nonlinear systems are discussed subsequently.

LINEAR SINGLE DIMENSIONAL SYSTEMS

1. Statement of the Problem

Given a linear time-invariant plant with transfer function

$$\frac{N_p(s)}{D_p(s)} = \frac{\sum_{j=0}^{n-1} b_{pj}\, S^j}{S^n + \sum_{j=0}^{n-1} a_{pj}\, S^j} \tag{1}$$

and a model with transfer function

$$\frac{N_m(s)}{D_m(s)} = \frac{\sum_{j=0}^{n-1} b_{mj}\, S^j}{S^n + \sum_{j=0}^{n-1} a_{mj}\, S^j} \tag{2}$$

the estimation problem is to determine the design laws for adjusting the parameters b_{mj} and a_{mj} so that the error $e_1(t)$ between the plant output $\theta_p(t)$ and the model output $\theta_m(t)$ is reduced to zero asymptotically. It will be assumed that the input signal is active enough so that $e_1 \to 0$ implies $b_{mj} \to b_{pj}$ and $a_{mj} \to a_{pj}$. This "identifiability" condition is identical to that of the GEE method and has been discussed in detail by Young [9]. Further a generalised error $V_1(t)$ can be obtained by processing the error $e_1(t)$ with a linear compensator $Z_1(s) = \sum_{i=0}^{\ell} z_i s^i$. The function and design of $Z_1(s)$ will be discussed later.

2. The Stable Design Rule

Using a state space formulation the system dynamics are described by :

plant
$$\dot{\underset{\sim}{y}}_p = A_p \underset{\sim}{y}_p + B_p \underset{\sim}{u} \tag{3}$$
$$\theta_p = C\, \underset{\sim}{y}_p \tag{4}$$

model
$$\dot{\underset{\sim}{y}}_m = A_m(t) \underset{\sim}{y}_m + B_m(t)\, \underset{\sim}{u} \tag{5}$$
$$\theta_m = C\, \underset{\sim}{y}_m \tag{6}$$

error
$$\underset{\sim}{e} = \underset{\sim}{y}_p - \underset{\sim}{y}_m \tag{7}$$
$$v_1 = D\, \underset{\sim}{e} \tag{8}$$

where y and u are in phase variable (canonical) forms, $C = [1, 0, \ldots, 0]$, $D = [1, z_1, \ldots, z_\ell]$ so that A and B are in companion forms :

$$A = \begin{bmatrix} 0 & 1 & 0 & \cdots & \\ 0 & 0 & 1 & & \\ \vdots & & & \ddots & 1 \\ -a_0 & -a_1 & \cdots & & -a_{n-1} \end{bmatrix}, \quad B = \begin{bmatrix} 0 & \cdots\cdots & 0 \\ \vdots & & \vdots \\ b_0 & b_1 \cdots & b_{n-1} \end{bmatrix}.$$

With the assumptions that the pair (A_p, B_p) is completely controllable and that the pair (C, A_p) is completely observable, the following theorem has been verified by Landau using the Hyperstability theory of Popov [4],[5].

THEOREM: Sufficient and partially necessary conditions in order that the parameter adjustments are asymptotically hyperstable are

(i) the transfer function $Z_1(s).C.(sI-A_p)^{-1}$ be strictly positive real;

(ii) the parameter adjustment laws are

$$\dot{a}_{mj} = -\left[\alpha_j v_1 y_{mj} + \gamma_j \frac{d}{dt}(\alpha_j v_1 y_{mj})\right] \tag{9}$$

$$\dot{b}_{mj} = \beta_j v_1 u_j + \delta_j \frac{d}{dt}(\beta_j v_1 u_j) \tag{10}$$

where the α_j, γ_j, β_j, δ_j are constants.

These two conditions constitute a Stable Response Error (SRE) method. It can be shown that condition i) is satisfied if $Z_1(s)/D_p(s)$ is positive real and hence the order ℓ of the compensator $Z_1(s)$ should satisfy $(n-1)<\ell<(n+1)$. In condition ii), the feedforward gains γ_j, δ_j are selected to provide damping to the transient response of e_1 [1] when the adaptive gains α_j, β_j or the input signal are very large.

3. Estimation with Input and Output Records

The design equations (9), (10) can be implemented if the plant states are accessible and v_1 obtained using equations (7), (8). In most practical cases, however, only the input u_1 and output θ_p are available. The application of the "state variable filters" (SVF) technique [6] to implement the SRE design laws is investigated in the following.

Referring to Fig.1a the input and output signals are filtered by two SVF each having a (n-1) order transfer function $1/N_f(s)$. Since the plant is time-invariant, commutation is allowed and one obtains

$$\theta_{pf}(s) = \frac{N_p(s)}{D_p(s)} u_{f1}(s) \tag{11}$$

Replacing the input and output states of the plant in equations (3)~(8) by corresponding filtered values, the stable design laws of equations (9), (10) become

$$\dot{a}_{mj} = -\left[\alpha_j v_{f1} y_{mj} + \gamma_j \frac{d}{dt}(\alpha_j v_{f1} y_{mj})\right] \tag{12}$$

$$\dot{b}_{mj} = \beta_j v_{f1} u_{fj} + \delta_j \frac{d}{dt}(\beta_j v_{f1} u_{fj}) \tag{13}$$

$$\text{where } v_{f1}(s) = Z_1(s)\, e_{f1}(s) \tag{14}$$

$$e_{f1}(s) = \theta_{pf}(s) - \theta_m(s) \tag{15}$$

Now since u_f and the derivatives of θ_{pf} and θ_m are available from the SVF and the model, equations (12)~(15) can be implemented without using any differentiators. Fig. 2 shows the complete estimation scheme of a second order plant. The required compensator there is $Z_1(s) = z_0 + z_1 s$, $z_0 \geq 0$, $z_1 > 0$, and $z_0/z_1 < a_{p1}$.

In the above we have assumed that the order (ℓ) of Z_1 is (n-1). When ℓ is n or n+1, the order of the SVF can be increased accordingly. In certain cases we may be able to choose $N_f(s) = Z_1(s)$. The scheme then becomes that shown in Fig.1b where only one SVF is required.

4. Noisy Measurements

So far we have assumed noise-free measurements and the plant is linear, time-invariant while the model and plant have the same orders. The relaxation of all these assumptions have been examined by Landau [4] who shows that the SRE method is still valid for most practical cases.

It can be shown [2] that the SRE method yields asymptotically unbiased estimates if the output measurement noise has a zero mean. This is in contrast with the GEE method where the estimates of all a_{mj} are asymptotically biased. If the so called Instrumental Variable (IV) [8],[10] is used to remove the noise biasing effect, the assured global stability of the GEE design will be lost. A further advantage of the SRE method is that it requires less analogue hardware than the GEE method. A comparison is shown in Table 1. Included in the comparison is the multiple GEE method [6] which has the advantage of rapid convergence over the GEE method. However its implementation requires prohibitive hardware as indicated and the noise biasing effect is still retained. The best compromise is the SRE method which has the properties of global stability, least hardware, adjustable damping and unbiased estimates. The difficulty in using this method is in the design of the compensator to satisfy the Popov's positive real condition. The economy of the SRE method over other response error methods has been demonstrated previously by Parks [7].

EXTENSIONS

The use of the SRE method to identify multi-dimensional and nonlinear systems will now be discussed. Detailed proofs and simulated examples will be found in ref. [2].

1. Multivariable Systems

A wide class of multivariable systems is the multi-input, single-output system as shown in Fig. 3. Most multi-output systems can be decomposed into a number of these uncoupled, single-output sub-systems. Now if only the inputs u and the output θ_p are measurable, the GEE method is not applicable. If the output and each input are filtered by the SVF as shown, direct application of Landau's Theorem [4] will yield the following hyperstable design laws :

(i) $Z_1(s).C.(sI-A_p)^{-1}$ be strictly positive real;

$$\text{(ii)}\ \dot{a}_{imj}(t) = -\left[\alpha_{ij} v_{f1} y_{imj} + \gamma_{ij} \frac{d}{dt}(\alpha_{ij} v_{f1} y_{imj})\right] \tag{16}$$

$$\dot{b}_{imj}(t) = \beta_{ij} v_{f1} u_{ifj} + \delta_{ij} \frac{d}{dt}(\beta_{ij} v_{f1} u_{ifj}) \tag{17}$$

Here again the SVF and the phase variable forms facilitate implementation without using any differentiators. An examination of condition (i) shows that it requires each $Z_1(S)/D_{ip}(S)$ to be positive real. Hence a restriction of this method is that the difference among the orders of $D_{ip}(S)$ cannot be more than two [3] in order that a possible $Z_1(S)$ exists.

2. Nonlinear Systems

Consider a class of nonlinear systems with the following state equations :

$$\dot{\underset{\sim}{y}}_p = A_p \underset{\sim}{y}_p + B_p u + A_p' F(\underset{\sim}{y}_p, \underset{\sim}{u}) \tag{18}$$

where F( ) represents a single-valued nonlinear function the form of which is assumed known.

If the SVF chosen approximates a transportation lag the commutation of nonlinear terms is allowed. The design using the SRE method is then straight forward as one can treat the nonlinear terms as additional inputs to the estimation model. Thus

$$\dot{\underset{\sim}{y}}_m = A_m \underset{\sim}{y}_m + B_m \underset{\sim}{u}_f + A_m' F(\underset{\sim}{y}_{pf}, \underset{\sim}{u}_f) \qquad (19)$$

The disadvantage of this approach is that the estimates of a_{mj}' will be asymptotically biased when the noise at output is significant. Replacing $F(\underset{\sim}{y}_{pf}, u_f)$ in equation (19) by $F(\underset{\sim}{y}_m, \underset{\sim}{u}_f)$ will remove the bias at the expense of the global stability assurance. Here the advantage over the GEE method is the economy since no extra instrumentation is needed to generate IV signals.

CONCLUDING REMARKS

Other real case examples (time-varying parameters and lower order adjustable model) have been analysed by Landau [4]. The use of the SVF technique in these cases presents no problem. The SRE method via hyperstability theory is found very attractive since the estimates obtained are asymptotically unbiased and the hardware involved is relatively little. The stability and performance of such designs have also been confirmed using digital simulations and reported in [2]. A typical result of the example in Fig. 2 is shown here in Fig. 4.

The major disadvantage of the SRE method is that the upper bound of the unknown parameters must be known a priori so that the linear compensator can be designed. Possibilities of relieving this problem are under investigation.

REFERENCES

1. Gilbart, J.W., Monopoli, R.V. and Price, C.F. (1970), "Improved Convergence and Increased Flexibility in the Design of Model Reference Adaptive Control Systems", IEEE 9th Symp. on Adaptive Processes, iv.3.1.

2. Hang, C.C. (1973), Ph.D. dissertation to be submitted to Department of Engineering, University of Warwick, England.

3. Kuo, F.F. (1966), Network Analysis and Synthesis, Wiley, Chapter 10.

4. Landau, I.D. (1970), "Hyperstability and Identification", IEEE 9th Symp. on Adaptive Processes, x.4.1.

5. ——— (1972), "A Generalization of the Hyper-Stability Conditions for Model Reference Adaptive Systems", IEEE Trans. Autom. Control, 17, 246.

6. Lion, P.M. (1967), "Rapid Identification of Linear and Nonlinear Systems", AIAA Journal, 10, 1835.

7. Parks, P.C. (1967), "Stability Problems of Model Reference and Identification Systems", IFAC Symp. on Identification, Prague, 5.4.

8. Rucker, R.A. (1963), "Real Time Systems Identification in the Presence of Noise", Proc. Western Electronics Convention, 2.3.

9. Young, P.C. (1968), "Identification Problems Associated with the Equation Error Approach to Process Parameter Estimation", Proc. 2nd Asilomar Conf. on Circuits and Systems, 416.

10. ——— (1970), "An Instrumental Variable Method for Real-Time Identification of a Noisy Process", Automatica, 6, 271.

Method \ Hardware		Integrators	Multipliers
SRE	Fig. 1a	5n-2 (8)	4n (8)
	Fig. 1b	4n-1 (7)	4n (8)
GEE		4n (8)	4n (8)
GEE and IV network		6n (12)	6n (12)
Multiple GEE (Ref. [6])		8n-2 (14)	$8n^2$ (32)

Table 1. Hardware Comparison (bracketed numbers refer to a 2nd order case)

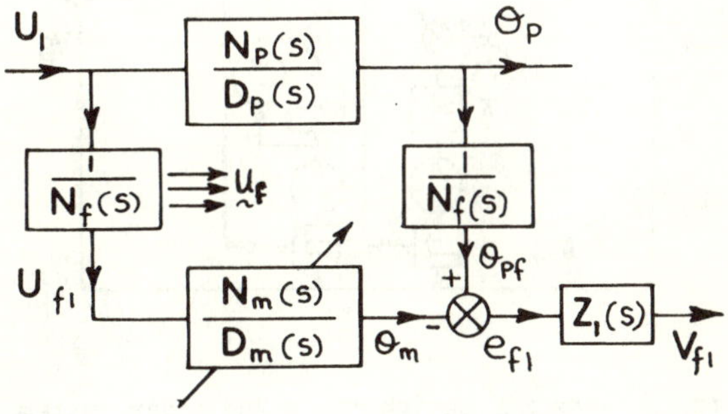

Fig. 1a

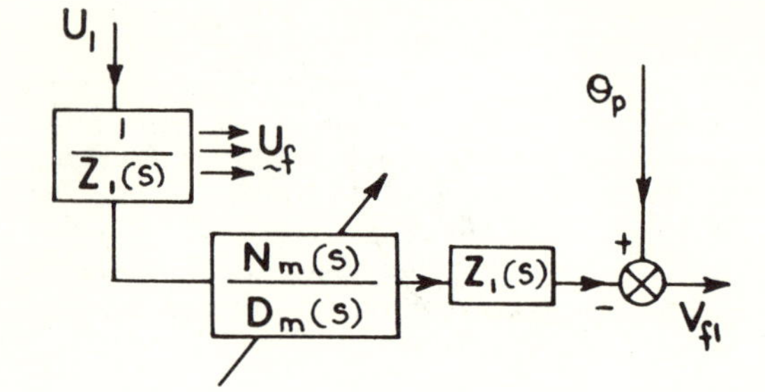

Fig. 1b

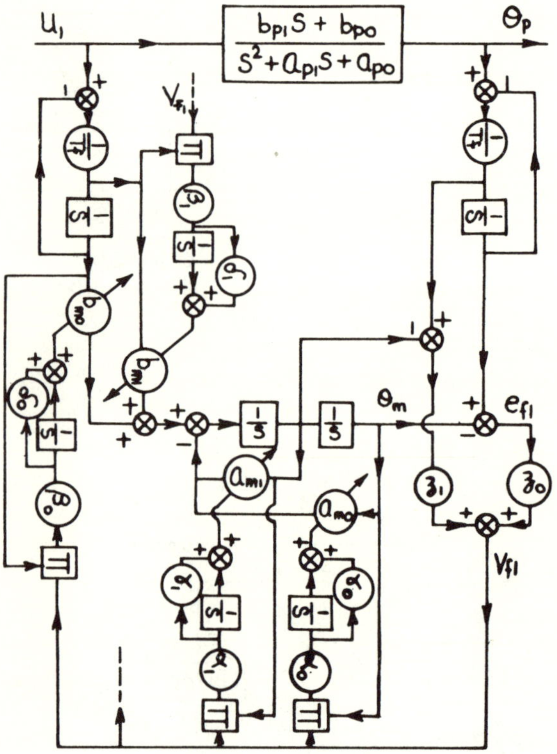

Fig. 2 The SRE design for a 2nd order system

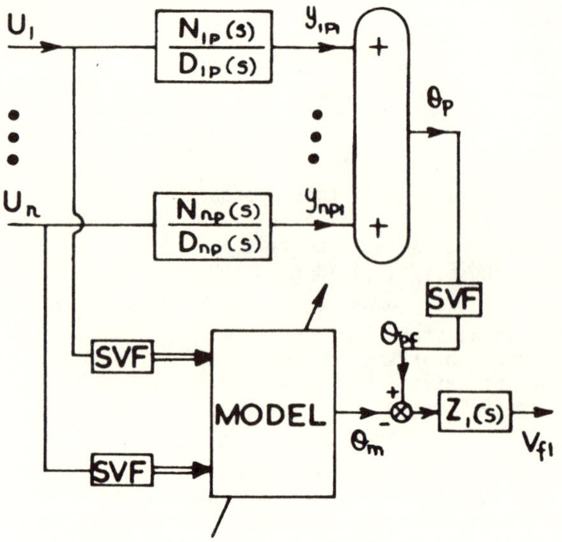

Fig. 3 The multi-input system

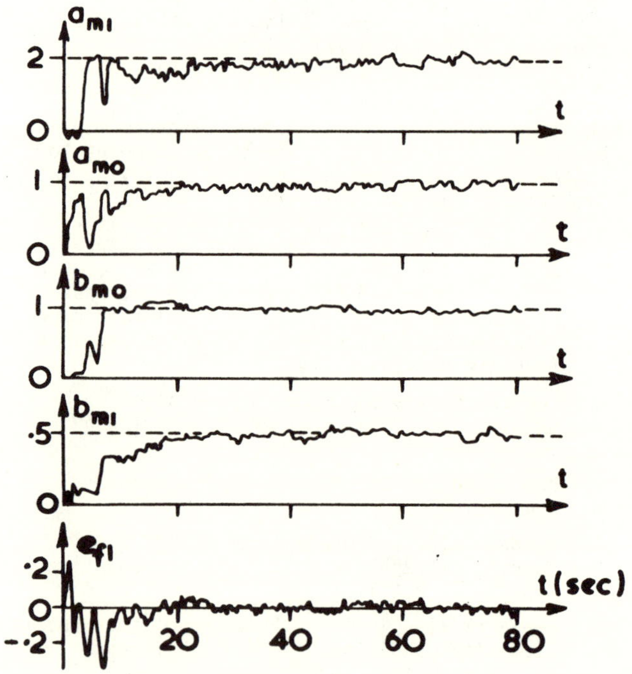

Fig. 4 Identification results for a linear system.
u_1 = P.R.B.S. Noise-signal ratio = 0.1.
---- True values. $T_f = 0.5$, $Z_1(s) = 0.5 + s$.
$\alpha_1 = 10$, $\alpha_0 = \beta_0 = 2$, $\beta_1 = 0.5$, all $\gamma = \delta = 0$.

CONVERGENCE PROPERTIES IN LINEAR PARAMETER TRACKING SYSTEMS

S.J. Merhav and E. Gabay

Department of Aeronautical Engineering
Technion-Haifa, ISRAEL.

The ensemble averaged convergence process of the parameter vector is investigated in terms of criterion surface geometry both for scalar and vector error criteria. The ratio between the smallest and largest diameter of this elliptic surface, termed "elongation", is given by the ratio between largest and smallest eigenvalues of a covariance matrix which defines the differential equation of the parameter vector. Almost always elongation and time of convergence increase with the order of the system to be identified. Factorization of numerator and denominator polynomials separates dominant modes and permits their rapid identification.

1. BACKGROUND AND STATEMENT OF THE PROBLEM

Let S be a linear time-invariant system. x(t) is a scalar band limited input and y(t) the corresponding output. The parameters of S can be identified by means of a model M. Let R(p) and Q(p) ($p^i = d^i/dt^i$) represent rational functions in p operating on x(t) and y(t) respectively. A scalar error e(t) can be defined:

$$e(t) = Q(p)y(t) - R(p)x(t) \qquad (1)$$

The choice of R(p) and Q(p) decides the type of M. If the structure of S is known, and following notation of ref. [5] ,it can be described by $N(p,\underline{b})/D(p,\underline{a})$ where $N(p,\underline{b}) \triangleq \sum_m b_j p^j$, $D(p,\underline{a}) \triangleq p^n + \sum_{n-1} a_i p^i$. Choosing $Q(p)=H_D(p)D(p,\underline{\alpha})$ and $R(p)=H_N(p)N(p,\underline{\beta})$, and setting $H_N(p)=H_D(p)=H(p)$ then if $\underline{\alpha}=\underline{a}$ and $\underline{\beta}=\underline{b}$, $e(t)\equiv 0$. Denote $\underline{c} \triangleq \mathrm{Col}(a_o, a_1 \ldots a_{n-1}, b_o, b_1 \ldots b_m)$, $\underline{\Gamma} \triangleq \mathrm{Col}(\alpha_o, \alpha_1, \ldots \alpha_{n-1}, \beta_o, \beta_1, \ldots \beta_m)$ and $y_k \triangleq H(p)p^k y$ ($k=0,1\ldots n$) and $x_h = H(p)p^h x$, ($h=0,1,\ldots m$) the sensitivity vector $\underline{w} \triangleq \mathrm{Col}(y_o, y_1 \ldots y_{n-1}, x_o, x_1 \ldots x_m)$, is independent of $\underline{\Gamma}$. Thus,

$$e = y_n + \underline{w}^T \underline{\Gamma} = y_n + \sum_{i=o}^{n-1} \alpha_i y_i + \sum_{j=o}^{m} \beta_j x_j \qquad (2)$$

Systems of this type are known as Linear parameter trackers [1] . The generation of e with additive noise n(t) is shown in heavy lines in Fig. 1. $\underline{\Gamma}$ and $\underline{w}$ are of dimension q=m+n+1. For n(t)=0 we have:

Condition 1: If all y_k and x_h are linearly independent, then $e \equiv 0$ if, and only if $\underline{\Gamma} = \underline{c}$.

Definition 1: A system S for which $e \equiv 0$ can be fulfilled within a finite interval of time is completely identifiable. In accordance with Eq. (2) the two-norm $||e||_2 = \{\int e^2(t)dt\}^{1/2}$ is a measure of the distance $d(\underline{\Gamma},\underline{c})$ in parameter space.

Definition 2: A model in which $e \equiv 0$ can be fulfilled, is a "zero-distance model". If not all y_k and x_h are linearly independent, $e \equiv 0$ can be satisfied with $\underline{\Gamma} \neq \underline{c}$. Thus, zero distance is a necessary but not sufficient condition for complete identifiability. In practice, $e \equiv 0$ cannot be completely satisfied. This may be due to: 1) Extraneous noise: $n(t) \neq 0$. 2) Filter mismatch:

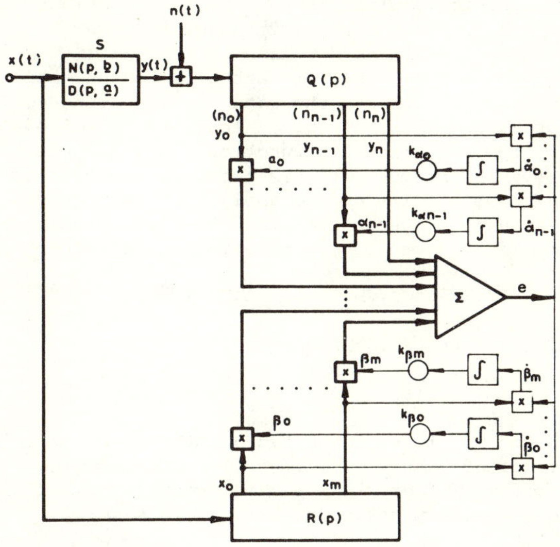

Figure 1. Generation of e and implementation of Eq. 5.

$H_D(p) \neq H_N(p)$. 3) Model mismatch: The dimension of M is smaller than the dimension of S. as a result, a residual error e_r will persist.

Define the quadratic criterion function

$$J = \frac{1}{2} e^2 \qquad (3)$$

Applying a steep descent law for any component of $\underline{\Gamma}$, we have:

$$\dot{\alpha}_k = -k_{\alpha k} \frac{\partial J}{\partial \alpha_k} = -k_{\alpha k} e \frac{\partial e}{\partial \alpha_k} = -k_{\alpha k} e\, y_k$$
$$\dot{\beta}_h = -k_{\beta h} \frac{\partial J}{\partial \beta_h} = -k_{\beta h} e \frac{\partial e}{\partial \beta_h} = -k_{\beta h} e(-x_h) \qquad (4)$$

Implementation of Eqs. (4) is indicated by the thin lines in Fig. 1. Substituting e from Eq. (2) and writing out Eqs. (4) for all components of $\underline{\Gamma}$, leads to:

$$\dot{\underline{\Gamma}} = -KA\underline{\Gamma} - K\underline{u} \qquad (5)$$

K is a diagonal gain matrix with constant

elements $k_{\alpha o}\ldots k_{\alpha n-1}, k_{\beta o}\ldots k_{\beta m}$. $A \triangleq \underline{w}\underline{w}^T$, which is a qxq symmetrical time varying matrix, is positive semi-definite, being zero only at isolated points [1].

$\underline{u} = \mathrm{Col}(y_n y_o, y_n y_1 \ldots y_n y_{n-1}, y_n x_o \ldots y_n x_m)$. For $t \to \infty$, $\underline{\Gamma} = \underline{\Gamma}_\infty$ and $\dot{\underline{\Gamma}} = 0$. In a zero-distance system $\underline{\Gamma}_\infty = \underline{c}$. Thus, from Eq. (5), $KA\Gamma_\infty = K\underline{u}$ so that $\underline{u} = A\Gamma_\infty$. Denoting $\underline{\gamma} = \underline{\Gamma} - \underline{\Gamma}_\infty$, substituting into Eq. (5) and rearranging we have:

$$\dot{\underline{\gamma}} = - KA\underline{\gamma} \qquad (6)$$

The proof of the complete and asymptotic stability of Eq. (6) is given in [5] and [7]. From the definition of e in Eq. (2) and the criterion function, it is clear that $\underline{\Gamma}_\infty$ is unique. Uniqueness and complete asymptotic stability are necessary conditions for the convergence of $\underline{\Gamma}$ to $\underline{\Gamma}_\infty$. They do not, however, establish quantitative relationships between the rate of convergence of $\underline{\Gamma}$, structure and order of S, input x(t), noise n(t), type of criterion function J and descent law K. The purpose of this paper is to establish such relationships and to apply them so that the convergence properties of the model for stationary random inputs can be quantitatively predicted.

2. AVERAGE CONVERGENCE PROCESS.

Define the criterion function

$$I = \frac{1}{2} E(e)^2 > 0 \qquad (7)$$

in which E is the expectation. Taking ensemble averages of both sides of Eq. (4) interchanging the order of averaging and differentiation and with previously defined notations we have:

$$\frac{d}{dt} E(\underline{\Gamma}) = - KE(\underline{w}e) \qquad (8)$$

In the equilibrium state $E(\underline{\Gamma})$ = const. so that Eq. (8) yields

$$E(\underline{w}\, e) = 0 \qquad (9)$$

In nonzero distance systems, the solution of Eqs. (9), after substitution of e yields the best estimate of $\underline{\Gamma}$. In zero distance systems it is fulfilled by $e \equiv 0$, i.e., $\underline{\Gamma} = \underline{c}$. Denoting $E(\underline{\Gamma}) \triangleq \hat{\underline{\Gamma}}$, $E(\underline{w}\,\underline{w}^T) \triangleq \bar{A}$ and $E(y_n \underline{w}) \triangleq \bar{\underline{u}}$, it is now shown that $\hat{\underline{\Gamma}}$ is determined by the differential equation:

$$\dot{\hat{\underline{\Gamma}}} = - K\bar{A}\hat{\underline{\Gamma}} - K\bar{\underline{u}} \qquad (10)$$

$K\bar{A}$ determines the convergence properties $\hat{\underline{\Gamma}}$ and $\bar{\underline{u}}$ determines $\hat{\underline{\Gamma}}_\infty$. Substitution from Eq. (2) into Eq. (8), and expansion of the right hand side results in:

$$\frac{d}{dt} E(\Gamma) = - KE(\underline{w}\,\underline{w}^T\underline{\Gamma}) - KE(y_n \underline{w}) \qquad (11)$$

Assume that a solution $E(\underline{\Gamma})$ exists and express Eq. (11) in the following form:

$$\frac{d}{dt} E(\underline{\Gamma}) = - KP(t)E(\underline{\Gamma}) - KE(y_n \underline{w}) \qquad (12)$$

P(t) is a qxq time varying matrix independent of $\underline{\Gamma}$. From Eqs. (8) and (12), $E(\underline{w}\,\underline{w}^T\underline{\Gamma}) = P(t)E(\underline{\Gamma})$ implying $P(t) = E[(\underline{w}\,\underline{w}^T)|\underline{\Gamma}]$. But since $\underline{w}$ is independent of $\underline{\Gamma}$ it follows that $E[(\underline{w}\,\underline{w}^T)|\underline{\Gamma}] = E(\underline{w}\,\underline{w}^T)$. Consequently $P(t) = \bar{A}$ which confirms Eq. (10). In Eq. (10), in the steady state, $\dot{\hat{\underline{\Gamma}}} = 0$ and the steady state solution of $\underline{\Gamma}$ is $\hat{\underline{\Gamma}}_\infty = - \bar{A}^{-1}\bar{\underline{u}}$. In analogy with Eq. (6) we have:

$$\dot{\hat{\underline{\gamma}}} = - K\bar{A}\hat{\underline{\gamma}} \qquad (13)$$

Eq. (13) expresses the convergence process of the ensemble average $\hat{\underline{\gamma}}$. It is entirely determined by $K\bar{A}$. Expanding Eq. (7) one finds $\underline{\Gamma}^T\bar{A}\underline{\Gamma} > 0$. Consequently:

1. $\bar{A}$, which is symmetrical is also P.D.
2. Since K is P.D. and since the product of two P.D. matrices is a P.D. matrix [5], $K\bar{A}$ is also P.D.
3. Consequently all the eigenvalues of $-K\bar{A}$ have negative real parts, and Eq. (13) is asymptotically and completely stable.
4. If all elements in K are equal (steepest descent), $K\bar{A}$ retains its symmetry. Thus, all the eigenvalues of $- K\bar{A}$ are negative and real.
5. On solving Eq. (13) for steepest descent it is readily shown that any of the q components of $\underline{\gamma}$ can have at most q-1 overshoots.

3. GEOMETRICAL INTERPRETATION.

Relations between I (Eq. (7)) and $\bar{A}$, are found from:

$$e = y_n + \underline{w}^T\hat{\underline{\Gamma}}$$
$$0 = y_n + \underline{w}^T\underline{c} \qquad (14)$$

Substracting, e is:

$$e = \underline{w}^T(\hat{\underline{\Gamma}} - \underline{c}) = \underline{w}^T\hat{\underline{\gamma}} \qquad (15)$$

Consequently, I is given by:

$$I = \frac{1}{2} \hat{\underline{\gamma}}^T\bar{A}\hat{\underline{\gamma}} \qquad (16)$$

These are closed surfaces in the q = m+n+1 dimensional parameter space defined by the components of $\hat{\underline{\gamma}}$. Since $\bar{A}$ is P.D. and symmetrical these surfaces are, concentric ellipsoids centered at the origin $\hat{\underline{\gamma}} = 0$. Considerable insight can be obtained by studying a two dimensional system S(q=2) in which e.g., a_o and b_o are to be identified and for which $\bar{A}$ is:

$$\bar{A} = \begin{pmatrix} \overline{y_o y_o} & \overline{y_o x_o} \\ \overline{x_o y_o} & \overline{x_o x_o} \end{pmatrix} \qquad (17)$$

Denoting $a_o - \hat{\alpha}_o \triangleq \hat{\alpha}$ and $b_o - \hat{\beta}_o \triangleq \hat{\beta}$, so that $\mathrm{Col}(\hat{\alpha}, \hat{\beta}) = \hat{\underline{\gamma}}$; with these notations, expansion of Eq. (16) yields:

$$I = \overline{y_o^2}\,\hat{\alpha}^2 + 2\,\overline{x_o y_o}\,\hat{\alpha}\,\hat{\beta} + \overline{x_o^2}\,\hat{\beta}^2 \qquad (18)$$

which is a family of concentric ellipses. Following

the geometrical interpretation of quadratic forms given in [5], it follows that the principal axes, r_1, r_2 of the unit ellipse are equal to the inverses of the corresponding eigenvalues $1/\lambda_1$, $1/\lambda_2$ of $\bar{A}$. The smaller eigenvalue represents the correspondingly slow descent along the major axis of the elliptic criterion surface. The ratio between the largest and smallest eigenvalues $\varepsilon=\lambda_{max}/\lambda_{min}$, will be referred to as the *elongation* of the criterion surface.

Conclusions:

1. To obtain rapid convergence to the origin, the elongation of the criterion surface must be reasonably small.
2. On criterion surfaces with large elongation, $\hat{\underline{\gamma}}$ will tend to settle for long periods of time on wrong values which strongly depend on the initial conditions $\underline{\gamma}_o$.
3. The convergence properties of the system are entirely determined by $\bar{A}$. K decides the direction of the descent path.
4. A steep descent path with carefully chosen elements in K may "shoot" the descent path towards the origin and appear to provide good convergence properties [6]. However, different initial conditions may reverse the situation so that convergence does not seem to be achieved at all.

For large elongations, i.e. $\lambda_1 \to 0$, the solutions for $\hat{\underline{\gamma}} = \text{Col}\ (\hat{\alpha},\hat{\beta})$ lie on the major axis which is a subspace of the plane $(\hat{\alpha},\hat{\beta})$. If r eigenvalues vanish, the solution lies in an r dimensional subspace including the origin. In general none of the diameters of the ellipsoid are aligned with any of the parameter axes. Consequently, even if one eigenvalue vanishes, none of the components of $\hat{\underline{\gamma}}$ converge rapidly to their correct value. These conclusions are experimentally verified by the recordings shown in Fig. 2.

4. EFFECT OF MODEL MISMATCH AND NOISE

Let $\hat{\underline{\Gamma}}_\infty$ be the steady state value of $\underline{\Gamma}$ which minimizes $E(e^2)$ in an imperfectly matched model. The residual error due to this mismatch is:

$$e_{rm} = y_n + \underline{w}^T\hat{\underline{\Gamma}}_\infty \qquad (19)$$

The contribution of n(t) to the residual error is (see Fig. 1)

$$e_{rn} = n_n + \sum_{i=o}^{n-1} n_i\hat{\alpha}_{i\infty} + \sum_{j=o}^{m} 0 \cdot \hat{\beta}_{j\infty} = n_n + \underline{n}^T\hat{\underline{\Gamma}}_\infty. \qquad (20)$$

From Eq.(20) the last m components of $\underline{n}$ are identically zero. The total residual error e_r is $e_r = e_{rm} + e_{rn}$.

From Eqs. (4) we have for $k_{\alpha k} = k_{\beta h} = k$, i.e, $kI = K$, (I is the identity matrix)

$$\dot{\underline{\Gamma}} = -k\underline{w}e \qquad (21)$$

Thus,

$$\underline{\Gamma}(t) = -k\int_o^t w(\tau)e(\tau)d\tau \qquad (22)$$

In the steady-state, as $t \to \infty$, $\underline{\Gamma}(t)$ can be expressed by

$$\underline{\Gamma}(t) = \hat{\underline{\Gamma}}_\infty + \tilde{\underline{\Gamma}}(t) \qquad (23)$$

The second term is the zero-mean stationary parameter-noise vector. Substituting Eq.(23) into Eq. (2) we have:

$$e(t)=y_n+\underline{w}^T[\hat{\underline{\Gamma}}_\infty+\tilde{\underline{\Gamma}}(t)]+n_n+\underline{n}^T[\hat{\underline{\Gamma}}_\infty+\tilde{\underline{\Gamma}}(t)]=e_{rm}+e_{rn}+$$
$$+(\underline{w}^T+\underline{n}^T)\tilde{\underline{\Gamma}}(t) \qquad (24)$$

Substituting Eq. (23) and (24) into Eq.(21) we have:

$$\dot{\tilde{\underline{\Gamma}}}(t)+k(\underline{w}\,\underline{w}^T+\underline{w}\,\underline{n}^T)\tilde{\underline{\Gamma}}(t) = -k\underline{w}(e_{rm}+e_{rn}) \qquad (25)$$

For the perfect match condition, $e=y_n+\underline{w}^T\underline{c} = 0$ is substracted from Eq.(19) so that:

$$e_{rm} = \underline{w}^T(\hat{\underline{\Gamma}}_\infty - \underline{c}) \triangleq \underline{w}^T\hat{\underline{\gamma}}_\infty \qquad (26)$$

Substituting Eq. (26) into Eq. (25), we have:

$$\dot{\tilde{\underline{\Gamma}}}(t)+k(\underline{w}\,\underline{w}^T+\underline{w}\underline{n}^T)\tilde{\underline{\Gamma}}(t) = -k(\underline{w}\,\underline{w}^T)\hat{\underline{\gamma}}_\infty - k\,\underline{w}e_{rn} \qquad (27)$$

Using $p^i = d^i/dt^i$, Eq. (27), with $\underline{w}\,\underline{n}^T$ neglected, can be written as follows:

$$\tilde{\underline{\Gamma}}(t) = -k[pI+k(\underline{w}\,\underline{w}^T)]^{-1}[(\underline{w}\,\underline{w}^T)\hat{\underline{\gamma}}_\infty + \underline{w}e_{rn}] \qquad (28)$$

I is a qxq identity matrix.

The qxq matrix $(\underline{w}\,\underline{w}^T)$ can be shown to have the following properties:

1) $\text{adj}[pI + k(\underline{w}\,\underline{w}^T)](\underline{w}\,\underline{w}^T) \equiv p^{q-1}(\underline{w}\,\underline{w}^T)$

2) $\text{adj}[pI + k(\underline{w}\,\underline{w}^T)]\underline{w} \equiv p^{q-1}\underline{w}$

3) $\det[pI + k(\underline{w}\,\underline{w}^T)] \equiv p^q + kp^{q-1}\text{tr}(\underline{w}\,\underline{w}^T)$

Using these properties, Eq. (28) can be expressed in the operational form.

$$\tilde{\underline{\Gamma}}(t) = -\frac{k(\underline{w}\,\underline{w}^T)\hat{\underline{\gamma}}_\infty}{p+k\text{tr}(\underline{w}\,\underline{w}^T)} - \frac{k\,\underline{w}\,e_{rn}}{p+k\text{tr}(w\,w^T)} \qquad (29)$$

The first term in Eq. (29) represents parameter noise $\tilde{\underline{\Gamma}}_m$ due to model mismatch and the second term represents parameter noise $\tilde{\underline{\Gamma}}_n$ due to input disturbance. Although Eq. (29) cannot be regarded as a solution, the following observations can be made:

1. Each component of $\tilde{\underline{\Gamma}}(t)$ is governed by a first order time varying differential equation.
2. $\tilde{\underline{\Gamma}}_m(t)$ and $\tilde{\underline{\Gamma}}_n(t)$ both increase monotonically with k but become independent of it for large values.
3. Increasing the input x, i.e. $\underline{w}$, or k has the same effect on $\tilde{\underline{\Gamma}}_m$. Small values of k and/or $\underline{w}$ will tend to decrease $\tilde{\underline{\Gamma}}_m$ but will cause long-term divergence.
4. For a given k and e_{rn}, $\tilde{\underline{\Gamma}}_n$ decreases as x, i.e. $\underline{w}$ is increased. In other words a large input supresses the parameter noise $\tilde{\underline{\Gamma}}_n$.

5. EFFECT OF ORDER OF THE SYSTEM

Reconsider $\bar{A} \triangleq E(\underline{w}\,\underline{w})^T$. Its elements have the following properties: $\overline{y_i y_j} = 0$; i-j, odd (i,j=0...n-1), $\overline{x_k x_h} = 0$; k - h, odd (k,h=0...m). Therefore, for the general case defined in Eq. (2), $\bar{A}$ with e.g. n odd and m even has the following form:

$$\begin{bmatrix} \overline{y_o y_o} & 0 & \cdot\cdot & \overline{y_o y_{n-1}} & \overline{y_o x_o} & \cdots & \overline{y_o x_m} \\ 0 & \overline{y_1 y_1} & \cdot\cdot & 0 & \overline{y_1 x_o} & \cdots & \overline{y_1 x_m} \\ \overline{y_2 y_o} & 0 & \cdot\cdot & \overline{y_2 y_{n-1}} & \overline{y_2 x_o} & \cdots & \overline{y_2 x_m} \\ \cdot & & & & & & \\ \overline{y_{n-1} y_o} & 0 & & \overline{y_{n-1} y_{n-1}} & \overline{y_{n-1} x_o} & \cdots & \overline{y_{n-1} x_m} \\ \overline{x_o y_o} & \overline{x_o y_1} & \cdot\cdot & \overline{x_o y_{n-1}} & \overline{x_o x_o} & 0 & \overline{x_o x_m} \\ \overline{x_1 y_o} & \overline{x_1 y_1} & \cdot\cdot & \overline{x_1 y_{n-1}} & 0 & \overline{x_1 x_1} & \overline{x_1 x_m} \\ \cdot & \cdot & & \cdot & \cdot & \cdot & \cdot \\ \overline{x_m y_o} & \overline{x_m y_1} & \cdot\cdot & \overline{x_m y_{n-1}} & \overline{x_m x_o} & 0\ \cdot & \overline{x_m x_m} \end{bmatrix} \tag{30}$$

A system S in which only the denominator parameters are to be identified is represented by the square part of $\bar{A}$ within the dashed lines. First consider a system with denominator dynamics only, $S=S_n$ defined by $\prod_o^{n-1} p_i [\prod_o^{n-1}(s+p_i)]^{-1} = N(s)/D(s)$. p_i are real or complex L.H.P. poles. x(t) is assumed to be white noise filtered by 1/(1+s) and having unity variance. The poles p_i will be denoted by $p_i \triangleq p\mu_i$ where $\mu_o = 1$ and $\mu_i \geq 1 (i=1,...n)$ are the scattering factors of p_i. Thus, $p_o = p$, is the dominant mode normalized with respect to the input bandwidth. Thus, the output $Y_o(s)$ is $Y_o(s) = p^n \prod_o^{n-1}\mu_i[(s+1)\prod_o^{n-1}(s+p\mu_i)]^{-1}$. For a given filter H(s), all $\overline{y_i y_j}$ (i,j=1...n-1) in $\bar{A}$ are determined. Consider three ranges of p. CASE I: p >>1; CASE II: p<<1; CASE III: p≈1. Irrespective of μ_i, it can be demonstrated that the last diagonal term $\overline{y^2_{n-1}}$ grows rapidly with n for case I and diminishes rapidly for Case II. The characteristic equation of $\bar{A}$ can be expressed in the form: $c(\lambda) = (\overline{y^2_o}-\lambda)(\overline{y^2_1}-\lambda)....(\overline{y^2_{n-1}}-\lambda) - f(\lambda) = 0$. $f(\lambda) \neq 0$ is due to the off-diagonal terms $\overline{y_i y_j} \neq 0$. It is easily verified that for Case I, $(\overline{y_i y_j})_{max} << \overline{y^2_{n-1}}$ and for case II, $(\overline{y_i y_j})_{max} << \overline{y^2_o}$. Due to this it can be demonstrated that for case I and Case II, on substituting $\lambda = \overline{y^2_{n-1}}$ and $\lambda = \overline{y^2_o}$ respectively into $c(\lambda)$, $f(\overline{y^2_{n-1}})$ and $f(\overline{y^2_o})$ are negligibly small. Therefore, from $c(\lambda)$, λ_{max} and λ_{min} are given to a very good approximation by $\lambda_{max} \cong \overline{y^2_{n-1}}$ and $\lambda_{min} \cong \overline{y^2_o}$ for case I and by $\lambda_{max} \cong \overline{y^2_o}$, $\lambda_{min} \cong \overline{y^2_{n-1}}$ for case II. The elongations therefore are $\varepsilon = \overline{y^2_{n-1}}/\overline{y^2_o}$, (p>>1) and $\varepsilon = \overline{y^2_o}/\overline{y^2_{n-1}}$, (p<<1).

Conclusion: ε in an n-dimensional parameter space increases rapidly with n if the dominant mode p of S_n is either small or large in comparison with the input bandwidth. In case III, if all μ_i are close to unity, ε grows slowly with the dimension n. However, large scattering of the poles of S_n involves a comparatively rapid growth of ε with n. The following numerical example demonstrates these characteristic trends for S_1, S_2, S_3 and S_4: $\mu_1 = 1,2,4,8...$ Case I. p = 10, Case II. p = 0.1, Case III. p=1.

The resulting elements of $\bar{A}$ and corresponding eigenvalues and elongations are given in Table I. The numerical results indicate :

1. The trend of rapid growth of ε with n.
2. λ_{max} of $\bar{A}$ is nearly equal to the largest power component which is either $\overline{y^2_{n-1}}$ for case I or $\overline{y^2_o}$ for case II. For case I, the large elongations give rise to extremely large values of y_{n-1}. For a given input level of x(t) this will cause saturation of active computing elements if analog equipment is used. For case II, Table I demonstrates that λ_{min} in all cases is extremely small. Eq.(29) indicates that increasing k may lead to excessive parameter noise.

Consequently large elongation, whether due to p<<1 or to p>>1 indicates the fundamental difficulty in obtaining rapid convergence of $\hat{\Gamma}$. If, for example, $\overline{y^2_o} = 25v^2$, is an upper allowable limit then, e.g. for S_4, p=0.1, k=1, Table I shows: $\lambda_{min} = 25/7000 = 3.6 \times 10^{-3}$ 1/sec, indicating an average convergence time constant of τ = 280sec! In contrast, for p=1 and the same $\overline{y^2_o}$, $\lambda_{min} = 25/33 = 0.8$ 1/sec, or τ = 1,2 sec. If numerator is included, the system is described by: $S^m_n = K\prod_o^{m-1}(s+z_j)[\prod_o^{n-1}(s+p_i)]^{-1}$. The corresponding $\bar{A}$ is now the complete matrix in Eq. (30). The following additional points regarding ε may be observed:

1. If the D.C.gain $K_o = K \prod_o^{m-1} z_j / \prod_o^{n-1} p_i$ is either $K_o >> 1$ or $K_o << 1$, the elongation will increase.
2. If any component variable x_j is proportional to some y_i, λ_{min} vanishes since $\bar{A}$ becomes singular. This situation will occur e.g. if a pole in S^m_n cancels a zero. The resulting singularity of $\bar{A}$ and infinite time of convergence of the parameter vector is analogous to the uncontrollability of the state vector in linear dynamic systems with pole zero cancellation [7].

6. GENERALIZED ERROR CRITERION.

A vector equation error $\underline{e}$, can be defined consisting of components $e_k = \hat{e}_{k-1}$, k = 1...r; r≤q. $e_k = y_{n+k} + \sum_{n-1} \alpha_i y_{i+k} + \sum_m \beta_j x_{j+k}$. A generalized criterion surface based on $\underline{e}$ is

$$J = \frac{1}{2}(\underline{e}^T R \underline{e}) \tag{31}$$

R is a diagonal weighting matrix with elements $\rho_o...\rho_q$. Sufficient conditions for convergence of Γ with this generalized criterion surface are presented in detail in [1]. Since for the statistical approach presented here, $\bar{A}$ contains all the information regarding convergence, it will be formulated for Eq. (31). From e_k we define $\underline{w}_r \triangleq Col(y_r, y_{1+r} \cdots y_{n-1+r}, x_r, \cdots x_{m+r})$, (r=0,1,2..) Defining the (r+1) xq matrix W,

$$W \triangleq \begin{pmatrix} y_o & y_1 & \cdots & y_{n-1} & x_o & \cdots x_m \\ y_1 & y_2 & \cdots & y_n & x_1 & \cdots x_{n+1} \\ \vdots & \vdots & & \vdots & \vdots & \vdots \\ y_r & y_{1+r} & \cdots & y_{n+r} & x_r & x_{m+r} \end{pmatrix} \tag{32}$$

we have $\underline{e} = W\underline{\Gamma} + \underline{y}_n$ where $\underline{y}_n = Col(y_n\ y_{n+1} \cdots y_{n+r})$. Applying the steepest decent law:

$$\frac{d}{dt}\underline{\Gamma} = -k\frac{\partial J}{\partial \underline{\Gamma}} = -k\left(\frac{\partial \underline{e}}{\partial \underline{\Gamma}}\right)^T R\underline{e} \qquad (33)$$

Substituting $\underline{e}=W\underline{\Gamma}+\underline{y}_n$ and $\partial\underline{e}/\partial\underline{\Gamma} = W^T$ into Eq. (33) we have:

$$\frac{d}{dt}\underline{\Gamma} = -kW^T RW\underline{\Gamma} - kW^T R y_n \qquad (34)$$

Defining $I = (1/2)E[\underline{e}^T R\underline{e}]$ and in complete analogy with Eq. (10),

$$\hat{\underline{\Gamma}} = -k\bar{A}^r\hat{\underline{\Gamma}} - k\bar{u}^r \qquad (35)$$

$\bar{A}^r$ is the covariance matrix derived from an r+1 dimensional vector and $\bar{u}^r$ a corresponding covariance vector. Developing $\bar{A}^r=E[W^T RW]$, we find that the structure of $\bar{A}^r$ is identical to that of $\bar{A}=\bar{A}^o$ given in Eq. (30) and that each element a^r_{ij} in it is given by $(a_{ij})_r = a_{ij}+\rho_1 a_{i+1,j+1}+\cdots+\rho_r a_{i+r,j+r}$. Denoting the covariance matrix corresponding to each component e_k of $\underline{e}$ by A_K having the elements $a_{i+k,j+k}$, we also have:

$$\bar{A}^r = \bar{A} + \rho_1\bar{A}_1+\cdots+\rho_k\bar{A}_k \qquad (36)$$

All the basic properties, symmetry and positive-definiteness are retained in $\bar{A}^r$ and therefore, all the results obtained in the foregoing sections remain unchanged.

Results:

1. For $S=S_n$ (denominator dynamics only) and $\underline{p}_o<<1$ the generalized criterion does not alter $\bar{A}^o$ significantly since higher derivatives y_{i+K} diminish and all $\bar{A}_k<<\bar{A}^o$.
2. For $S = S^m_n$, and $p\geq 1$, the implementation of the generalized error criterion will, as a rule, cause increased elongation since the derivatives y_{i+k} grow rapidly.

This is demonstrated in Table II by numerical examples for a system $S^m_n = S^1_2$, with four parameters: gain, zero, two poles. They describe human operator transfer functions [8]. The results demonstrate that a generalized vector error criterion function indeed increases λ_{min}, indicating more rapid convergence. This, however, is achieved at the expense of even larger elongation.

A sample of the convergence process corresponding to the first row of Table II is illustrated in Fig. 2. It shows the projections of $\underline{\gamma}$ in the $(\alpha'_o,\beta'_1);(\alpha'_1,\beta'_1)$ plane. $\alpha'_o,\alpha'_1,\beta'_1$ denote components of $\underline{\gamma}$. The large elongation and the inclination of the axes of the ellipsoid are clearly indicated. This explains the slow convergence of all the components of $\underline{\Gamma}$. This difficulty can be overcome if $N(s,\underline{\beta})$ and $D(s,\underline{\alpha})$ are factored so that S is identified in terms of p_i and z_i. Factorization, reveals the significant modes which converge rapidly and isolates them from the rest. The effectiveness of this factorization is illustrated in Fig. 3. It shows a sample recording corresponding to the first row of Table II. p'_o and p'_1 denote the deviations of p_o and p_1 from their equilibrium values. The extremely rapid and smooth convergence of the dominant pole p_o is clearly seen.

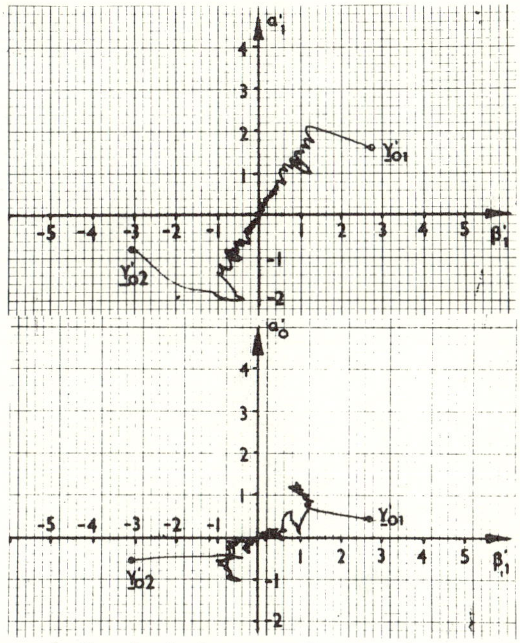

Figure 2. Smaple Convergence Procees for example of 1st Row Table II.

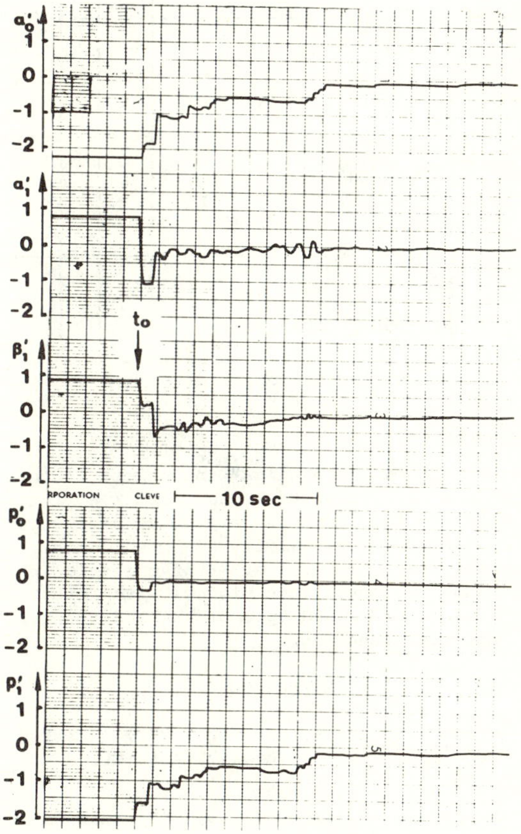

Figure 3. Sample Convergence Process in time domain for 1st Row Table II.

TABLE I*: NUMERICAL VALUES OF A AND ε FOR UNITY VARIANCE AND UNITY BANDWIDTH STATIONARY INPUT.

	p	ELEMENTS OF $\overline{y_i y_j}$ in $\bar{A}_n$						EIGENVALUES				ε
		00	11	22	33	02	13	λ_0	λ_1	λ_2	λ_3	
S_1	0.1	4.5-2	-	-	-	-	-	4.55-2	-	-	-	-
	1.0	2.5-1	-	-	-	-	-	2.5 -1	-	-	-	-
	10	4.5-1	-	-	-	-	-	4.5 -1	-	-	-	-
S_2	0.1	3.3-2	5.0-4	-	-	-	-	3.3-2	4.9-4	-	-	6.7+1
	1.0	2.2-1	1.1-1	-	-	-	-	2.2-1	1.0-1	-	-	2.0+0
	10	4.5-1	2.9+0	-	-	-	-	4.5-1	2.9+0	-	-	6.5+0
S_3	0.1	3.1-2	3.0-4	2.9-5	-	-3.3-8	-	3.1-2	3.3-4	2.5-5	-	1.2+3
	1.0	2.1-1	9.3-2	2.5-1	-	-9.3-2	-	2.5-1	1.1-1	7.1-2	-	3.6+0
	10	4.5-1	2.67+0	3.5+2	-	-2.7+0	-	4.2-1	2.7+0	3.5+2	-	8.3+2
S_4	0.1	3.0-2	2.96-4	1.9-5	5.6-6	-2.7-4	-1.9-5	3.0-2	3.0-4	1.6-5	4.4-6	7.0+3
	1.0	2.1-1	8.9-2	2.2-1	2.2+0	-8.9-2	-2.6-1	6.7-2	1.2-1	3.0-1	2.2+0	3.3+0
	10	4.5-1	2.62+0	3.1+2	2.5+5	-2.6+0	-3.1+2	4.2-1	2.2+0	3.2+2	2.5+5	6.0+6

TABLE II*: NUMERICAL VALUES OF ε FOR SYSREMS S_2^1 WITH GENERALIZED ERROR CRITERIA.

Input Spectrum	H(s)	S_2^1	R			λ_1	λ_2	λ_3	λ_4	ε
			ρ_0	ρ_1	ρ_2					
$\frac{1}{4-s^2}$	$\frac{10^4}{(s+10^4)}$	$\frac{s+2}{(s+0.11)(s+4.5)}$	1	0	0	3.2+1	6.6-1	4.4-2	4.4-4	7.2+3
			1	1	1	9.3+2	3.9+1	6.0-2	6.5-3	1.44+4
			1	1	1	8.9+3	9.7+2	3.1-1	4.1-2	2.2+5
$\frac{1}{16-s^2}$	$\frac{10^4}{(s+10)^4}$	$\frac{s+3}{(s+0.2)(s+11)}$	1	0	0	7.8+1	1.0+1	7.8-2	1.55-3	5.0+3
			1	1	0	2.9+2	9.4+1	1.6-1	5.3-2	5.6+3
			1	1	1	3.2+4	3.2+2	7.4+1	2.5-1	1.4+5
$\frac{1}{4-s^2}$	$\frac{10^4}{(s+10)^4}$	$\frac{s+0.3}{(s+3)(s+9.5)}$	1	0	0	1.4+2	1.4+1	2.5-1	1.8-1	7.7+2
			1	1	0	4.6+3	1.5+2	2.3+1	7.8-1	5.9+3
			1	1	1	2.5+5	6.1+3	2.3+4	1.0+1	2.0+4

* ± r following numerical values in Tables denote x $10^{\pm r}$.

Conclusions:

The concept of "elongation" introduced in this paper has been shown to be an effective figure of merit of the convergence process of the parameter vector. Its validity has been demonstrated both for scalar and vector error criteria in relating the convergence process with input statistics, order and type of S and the type of criterion function.

REFERENCES.

1. Åström K.J. and Eykhoff P. (1971), "System Identification - a Survey", Automatica Vol.7 p.123.

2. Athans M. and Falb P.(1966), "Optimal Control" Chapter 4, McGraw-Hill (Book).

3. Bellman R.(1960) "Introduction to Matrix Analysis" Chapter 4, McGraw-Hill (Book).

4. Ibid, Chapter 1.

5. Hofmann L.G., Lion P.M. and Best J. (1966), "Theoretical and Experimental Research on Parameter Tracking Systems" NASA CR-452

6. Hsia T.C. and Vilmolvanich, V.(1969)" An on Line Technique for System Identification", IEEE Trans. Vol. AC-14 No. 1 p.92.

7. Lion P.M. (1966) "Rapid Identification of Linear and Nonlinear Systems", Preprint JACC. p. 605.

8. McRuer D.T. and Krendel E.S. (1959) "The Human Operator as a Servo System Element" J. Franklin Inst. Vol. 267 p. 511.

TM-9 751

A METHOD FOR DISCRETE SYSTEM IDENTIFICATION

Raymond Clarke

Inter-University Institute of Engineering Control
University of Warwick, Coventry, CV4 7AL, England

John W. Green

Oldham College of Technology
Oldham, England

A method is presented for the estimation of discrete system parameters. It is derived from a maximum descent law and compares favourably with contemporary techniques. Sensitivity models are not required and stability can be ensured by correct choice of the adaptive gain.

INTRODUCTION

The techniques developed using the method of stable maximum descent, (Green, 1969), (Walker 1972) for the estimation of continuous system parameters (Clarke, 1971 and 1973) may be extended to the estimation of discrete or sampled data system parameters.

Stable maximum descent relies upon the formation of a vector of error states between the process and the adaptive model in which the model parameters such as $\tilde{a}$ are adjusted according to a maximum descent law:

$$\dot{\tilde{a}}_{ij} = - \frac{\partial e_m}{\partial \tilde{a}_{ij}} \qquad (1)$$

with e_m representing a quadratic function of error states and their derivatives :

$$e_m = \tfrac{1}{2}||C_1 e + C_2 \dot{e}||^2 \qquad (2)$$

By choosing C_1 and C_2 in a certain manner (Green, 1969) an asymptotically stable adjustment law is produced. The responses of the adaptive parameters compare favourably with those obtained by contemporary techniques (Parks, 1966), (Gilbart, 1971) and in the ideal case can be made to adapt at arbitrarily fast rates by increasing the adaptive loop gain.

The discrete version of this method which is described in this paper, incurs limit to the adaptive loop gain but this is a typical problem encountered in discretizing any continuous system. Nevertheless it is felt that the discrete version has many useful applications.

METHOD

It is assumed that the discrete process can be represented by the state equation

$$x(k + 1) = [I + h\ A]x(k) + Bu(k) \qquad (3)$$

Only changes in the A matrix will be considered in this analysis, although the method is readily extendable to more complicated systems. Thus the adaptive model is represented by the state equation

$$y(k + 1) = [I + h\ \tilde{A}]y(k) + Bu(k) \qquad (4)$$

In consequence the following error state equation can be formed by subtracting (2) from (1)

$$e(k + 1) = [I + h\ \tilde{A}]e(k) + h\ W\ x(k) \qquad (5)$$

where $W = A - \tilde{A}$ (6)

Each adaptive parameter is adjusted by a maximum descent law :

$$\tilde{a}_{ij}(k + 1) = \tilde{a}_{ij}(k) - h \frac{\partial e_m}{\partial \tilde{a}_{ij}(k)} \qquad (7)$$

in which : $e_m = \tfrac{1}{2}||C_1 e(k) + C_2\ e\ (k + 1)||^2$ (8)

so that :

$$\tilde{a}_{ij}(k + 1) = \tilde{a}_{ij}(k) - h\left[C_1 e(k) + C_2 e(k + 1)\right]^T * \left[\frac{\partial\ (C_1 e(k))}{\partial \tilde{a}_{ij}(k)} + \frac{\partial\ (C_2 e(k + 1)}{\partial \tilde{a}_{ij}(k)}\right] \qquad (9)$$

The presence of the sensitivity terms on the right side of (9) appear to make the adaptive law difficult to implement. However the terms can be simplified by considering the differentiation of (5) with respect to $\tilde{a}_{ij}(k)$.

i.e.
$$\frac{\partial\ (e(k + 1))}{\partial \tilde{a}_{ij}(k)} = \frac{\partial\ ((I + hA)e(k)}{\partial \tilde{a}_{ij}(k)} + \frac{h \partial\ W}{\partial \tilde{a}_{ij}(k)} x(k) \qquad (10)$$

C_2 is now choosen as gI, in which g represents adaptive loop gain, and is used to premultiply every term in (9) :

i.e.
$$\frac{\partial\ (C_2\ e(k + 1))}{\partial \tilde{a}_{ij}} = \frac{\partial\ (C_2(I + hA)e(k))}{\partial \tilde{a}_{ij}} + C_2\ h \frac{\partial W}{\partial \tilde{a}_{ij}(k)} x(k) \qquad (11)$$

by choosing C_1 such that

$$C_1 = -C_2[I + h\tilde{A}] \qquad (12)$$

(10) shows that

$$\frac{\partial\,(C_2\, e(k+1))}{\partial \widetilde{a}_{ij}(k)} + \frac{\partial\, C_1 e(k)}{\partial \widetilde{a}_{ij}(k)} =$$

$$C_2 h \frac{\partial W}{\partial \widetilde{a}_{ij}(k)} x(k) \qquad (13)$$

and simplifies the adaptive law (9) to

$$\widetilde{a}_{ij}(k+1) = \widetilde{a}_{ij}(k) - h\left[C_1 e(k) + C_2 e(k+1)\right]^T * \left[C_2\, h \frac{\partial W}{\partial \widetilde{a}_{ij}(k)} x(k)\right] \qquad (14)$$

which forms the basis of an algorithm to obtain estimates of the elements of A.

STABILITY ANALYSIS

It is the aim of this analysis to show that in the steady state the adaptive controllers produce

$$Wx(k) \equiv 0 \qquad (15)$$

so that (5) becomes

$$e(k+1) = \left[I + h\widetilde{A}\right] e(k) \qquad (16)$$

This means that the convergence of the error state variables to zero depends upon $\widetilde{A}$, and therefore A, which in turn determines the ultimate speed of performance. Thus in any application it is necessary to ensure that the range of A will keep within satisfactory limits.

To show that the conditions for (15) are successfully maintained, (8), is re-written in terms of the parameter error by assuming model changes can be made to operate much faster than process changes so that

$$\omega_{ij}(k+1) - \omega_{ij}(k) = -(\widetilde{a}_{ij}(k+1) - \widetilde{a}_{ij}(k))$$

and when this is applied to (8) it produces

$$\omega_{ij}(k+1) = \omega_{ij}(k) - h\left[C_1 e(k) + C_2 e(k+1)\right]^T \left[C_2 h \frac{\partial W}{\partial \omega_{ij}(k)} x(k)\right] \qquad (17)$$

Through the use of (5) and (12) this becomes

$$\omega_{ij}(k+1) = \omega_{ij}(k) - h^3\left[C_2 W x(k)\right] \left[C_2 \frac{\partial W}{\partial \omega_{ij}(k)} x(k)\right] \qquad (18)$$

For the simplest case under consideration here $\frac{\partial W}{\partial \omega_{ij}(k)}$ is represented by a square matrix of order n with zero in every position except that corresponding to $\omega_{ij}(k)$ where unity is written When (18) is applied to all $\omega_{ij}(k)$ operating simultaneously the following error parameter state equation is produced

$$\underline{\omega}(k+1) = \left[I - h^3 g^2\, \underline{X}\right] \underline{\omega}(k)$$

in which, for the largest case of n^2 adaptive parameters, $\underline{X}$ is a square matrix of order n^2 with a set of n sub-matrices $x(k)\, x(k)^T$ lying symmetrically along the leading diagonal and zero everywhere else. e.g. for n = 2

$$\underline{X} = \begin{bmatrix} X_1^2(k) & X_1(k)X_2(k) & 0 & 0 \\ X_1(k)X_2(k) & X_2^2(k) & 0 & 0 \\ 0 & 0 & X_1^2(k) & X_1(k)X_2(k) \\ 0 & 0 & X_1(k)X_2(k) & X_2^2(k) \end{bmatrix}$$

The eigenvalues of X are easily obtained as

$$\lambda_i = \sum_{p=1}^{n} X^2 p(k) \quad \text{for } i = 1, 2 \ldots\ldots n \qquad (21)$$

$$\lambda_j = 0 \quad \text{for } j = n+1, n+2 \ldots\ldots n^2 \qquad (22)$$

Thus with reference to $\left[h^3 g^2 \underline{X}\right]$ which is part of (19) this has n non zero eigenvalues each equal to

$$h^3 g^2 \sum_{p=1}^{n} X^2 p(k)$$

Therefore by imposing the limiting condition

$$h^3 g^2 \sum_{p=1}^{n} X^2 p(k) < 2 \qquad (23)$$

the eigenvalues of $\left[I - h^3 g^2 X\right]$ will all lie within the unit circle and guarantee that (15) is maintained in the steady state. In order to have a rule for selecting a satisfactory adaptive loop gain (23) is rearranged as

$$g^2 < \frac{2}{h^3 \sum_{p=1}^{n} X^2 p(k)} \qquad (24)$$

CONCLUSIONS

The success of this method depends upon choosing an error measure, $e_m = \frac{1}{2}\|C_1 e(k) + C_2 e(k+1)\|^2$ with C_1 and C_2 obeying the rule $C_1 = -C_2[I + h\widehat{A}]$ The ultimate speed of operation depends upon the adaptive loop gain, $g < \dfrac{2}{h^3 \sum_{p=1}^{n} X^2 p(k)}$ and A of the process.

The method has been used successfully to identify the parameters of several simulated systems and also of a pilot heat-exchanger plant. These results will be made available for discussion.

List of Principal Symbols :

$A = \{a_{ij}\}$, parameter characteristic matrix of order n

$\widetilde{A} = \{\widetilde{a}_{ij}(k)\}$, model characteristic matrix of order n

e_m = error measure of performance

C_1, C_2 = square matrices of order n, associated with e_m

u(k) = process and model input at kth instant

B = driving matrix

x(k), y(k), e(k) = process, model and error state vectors respectively at kth instant

h = discretigation interval

$W = \{\omega_{ij}(k)\}$, parameter error matrix

$\underline{\omega}(k)$ = column vector of all elements in W

$\underline{X}$ = square matrix of order n^2, associated with x(k)

λ = eigenvalue

REFERENCES

CLARKE, R., GREEN, J.W., NELSON, J.C.C. (1971) The application of hierarchical control strategy to a parallel flow heat exchanger. 4th U.K.A.C. Control Convention, Manchester.

CLARKE, R. (1973) "Identification of industrial systems", Ph.D. Thesis. To be submitted to Dept. of Elec. & Electron. Eng., Leeds University.

GILBART J.W., MONOPOLI, R.V., and PRICE, C.F. (1970). Improved Convergence and Increased Flexibility in the design of model reference adaptive control systems. I.E.E. 9th Symposium on Adaptive Processes, IV.3.1.

GREEN, J.W. (1969) Adaptive control of multiloop speed control systems with particular reference to the Ward Leonard System. Ph.D. Thesis. Dept. of Elec. & Electron. Eng., Leeds University.

PARKS, P.C. (1966) Liapunov redesign of model ref adaptive control systems. I.E.E.E. Trans. Vol. AC-11 No. 3, 362.

WALKER, P.H., and GREEN, J.W. (1972) "Controller design for linear multivariable systems" J. of Inst. of measurement and control. Vol. 5, No. 5, 188.

SYSTEM IDENTIFICATION USING STOCHASTIC APPROXIMATION

Naresh K. Sinha

Department of Electrical Engineering and
Communications Research Laboratory
McMaster University
Hamilton, Ontario,
Canada.

This paper discusses the application of stochastic approximation to the identification of linear systems. The parameters of a discrete-time model are first determined in a recursive manner from the samples of the observed input and output data contaminated with noise using a new algorithm for stochastic approximation proposed recently. From this model, the parameters of the corresponding continuous-time system are then estimated. Simulation results are presented indicating that the proposed scheme for identification converges much faster than the existing methods based on stochastic approximation, provided that the signal-to-noise power ratio is not less than one.

INTRODUCTION.

Application of stochastic approximation to system identification was proposed by Kirvatis and Fu (1) and several algorithms for this purpose have since been developed by Saridis and Stein (2,3), Saridis, Nikolir and Fu (4), Holmes (5), and Neal and Bekey (6). The main advantages of this approach, compared to conventional methods, such as maximum likelihood estimation are:

(i) only a small interval of data needs processing.

(ii) only simple computations are required, even when the actual functional dependence of the regression functions on the parameters of interest is non-linear.

(iii) a priori knowledge of the process statistics is not necessary, nor is the detailed knowledge of the functional relationship between the desired parameters and the observed data.

The main difficulty with this method has been that the algorithms proposed by the various authors converge rather slowly to the true value, although convergence to the correct value in the mean-square sense and with probability one has been proved. Even the latest "improved" algorithm proposed by Saridis and Lobbia (7) requires more than 100 iterations to give a normalized error of the order of 10% in examples of first- and second-order systems given by these authors.

In a recent paper (8) Sinha and Griscik have presented a new algorithm for stochastic approximation which converges much faster than the two general algorithms proposed earlier by Fu, Nikolic, Chien and Wee (9). The objective of this paper is to discuss the application of this new algorithm to the identification of the parameters of a discrete-time system from samples of the observed input-output data contaminated with noise. The method can easily be extended to on-line identification of continuous systems (10).

First, the algorithms for stochastic approximation will be discussed briefly, followed by application to the identification of linear discrete-time systems. Results of simulation will then be presented.

Algorithms for Stochastic Approximation.

A general algorithm for stochastic approximation of the form

$$\hat{x}_{n+1} = \hat{x}_n + \gamma_{n+1} \left\{ f(\underline{r}_{n+1}) - x_n \right\} \quad ,n = 1,2,\ldots \quad (1)$$

is said to be of the Dvoretzky (11) type, where

$\hat{x}_n$ = n^{th} estimate of x

r_n = n^{th} observation taken

$\underline{r}_{n+1} = \left[r_1 \; r_2 \; \ldots . \; r_n \; r_{n+1} \right]^T$,

superscript T denoting transposition,

$f(\underline{r}_{n+1})$ = scalar functional of $\underline{r}_{n+1}$, and

γ_n = gain sequence

The first algorithm in [9] uses

$$f_1(\underline{r}_n) \quad (2)$$

where the observations $r_n = x + \zeta_n$, (3)

ζ_n being a random noise sequence of zero mean value

and the gain sequence $\gamma_n = \frac{1}{n+\alpha}$ (4)

where α is an arbitrary constant. The value of α is taken as zero in the absence of a priori statistics, otherwise, for best results it depends upon the variance of the noise and the initial value of the expected mean square error.

The second algorithm in [9] uses

$$f_2(\underline{r}_n) = \frac{1}{n}\sum_{i=1}^{n} r_i \quad (5)$$

where the samples r_n are of the same form as in (3), but the gain sequence is given by

$$\gamma_n = \frac{n}{n(n+1)/2 + \alpha} \quad (6)$$

For the best convergence, α is again dependent upon a priori statistics, but may be taken as zero in the absence of such information.

The new algorithm in [8] proposes that

$$f_3(\underline{r}_n) = |R_{r_n}(\ell)|^{\frac{1}{2}} \quad (7)$$

where $R_{r_n}(\ell)$ is an estimate of the antocorrelation function of the samples r_1, r_2, ... r_n, defined as

$$\hat{R}_{r_n}(\ell) = \frac{1}{n-}\sum_{i=1}^{n-} r_i \; r_{i+\ell} \quad (8)$$

and ℓ and n are sufficiently large. In most practical cases, it is sufficient to make ℓ =10 and n=20 or more.

It has been shown in [8] that the algorithm f_3 converges much faster than f_1 and f_2 with the γ-sequence given by either

$$\gamma_n = \frac{1}{n} \quad (9)$$

or

$$\gamma_n = 1 - \frac{1}{n} \quad (10)$$

<u>Application to System Identification:</u>

Consider a linear discrete-time system described by the difference equation

$$a_o u_i + a_1 u_{i-1} + \ldots + a_n u_{i-n} = c_i + b_1 c_{i-1} + \ldots + b_n c_{i-n} \quad (11)$$

where $u_i = u(iT)$ is the input sequence, $c_i = c(iT)$ is the output sequence and T is the sampling interval. The determination of the 2n+1 parameters $a_o, a_1, \ldots, a_n, b_1, b_2, \ldots, b_n$ is straightforward in the absence of noise, as these can determined by solving p=2n+1 linear simultaneous equations in terms of the input-output date. However, due to noise, instead of c_i, measurements are made of y_i, defined as

$$y_i = c_i + v_i \quad (12)$$

where v_i is the output noise sequence. Although not much may be known statistically about the noise sequence, it is customary to consider it as an uncorrelated zero-mean sequence.

To use the general stochastic algorithm of equation (1) for system identification, one would have to determine each of the 2n+1 parameters recursively. The observation, vector $\underline{r}_i$, then, corresponds to the observation of parameter vector obtained by inverting a matrix of the type

$$A_i \underline{r}_i = \underline{z}_i \quad (13)$$

where $A_i \triangleq$

$$\begin{bmatrix} u_i & u_{i-1}\cdots u_{i-n} & -y_{i-1} & - & -y_{i-2}\cdots -y_{i-n} \\ u_{o+1} & u_i \;\cdots u_{i-n+1} & -y_i & & -y_{i-1}\cdots -y_{i-n+1} \\ \vdots & & & & \\ u_{1+p-1} & u_{1+p-2}\cdots u_{i-n+p-1} & -y_{-i+p-2} & & -y_{i+p-3}\cdots -y_{i-n+p-1} \end{bmatrix} \quad (14)$$

$$z_i \triangleq \left[y_i \; y_{,+1} \cdots y_{i+p-1}\right]^T \quad (15)$$

and $\underline{r}_i$ is the ith observation of the parameter vector

$$\underline{x} \triangleq \left[a_o \; a_1 \ldots a_n \; b_1 \; b_2 \ldots b_n\right]^T \quad (16)$$

As more input and output data is obtained, one may delete the first row of A_i, and add another row at the bottom to obtain the matrix A_{i+1}. By inverting the matrix A_{i+1}, the next observation vector $\underline{r}_{i+1}$ is obtained as

$$\underline{r}_{i+1} = A_{i+1}^{-1} \; \underline{z}_{i+1} \quad (17)$$

In practice, the inverse of the matrix A_{i+1} can be obtained from the inverse of A_i by using a simple algorithm discussed in reference [12]. Thus, following only one matrix inversion, the observation vectors $\underline{r}_i'$, $\underline{r}_{i+1}$, ..., are obtained for use in the stochastic approximation algorithms.

However, these observation vectors do not satisfy the condition of equation (3), in that the effect of noise in (17) will cause the resulting sequence ζ_n to have a non-zero mean. As a result, the application of the Dvoretzky-type algorithms will lead to biased estimates.

This difficulty can be overcome by using a "bootstrap" technique in conjunction with the above algorithms [13]. A simple way of doing this is to estimate the mean value of the noise sequence $\underline{N}_n$, and subtracting from the output sequence. One can use the current estimate $\underline{x}_n$ to estimate the vector noise sequence $\underline{N}_n$ as

$$\underline{N}_n = A_n\underline{x}_n - \underline{z}_n \tag{18}$$

and use these samples to determine the sample-mean of $\underline{N}_n$. These can now be subtracted from $\underline{z}_{n+1}$ and the corresponding terms in A_{n+1} to determine $\underline{r}_{n+1}$ which nearly satisfies equation (3), and should lead to an estimate which may be considered unbiased.

Determination of the continuous-time system transfer function: The determination of the continuous-time system transfer function from the discrete-time model has been discussed in detail in a recent paper [10]. There it has been shown that if the sampling rate is such that $|p_k T| < 0.5$, where T is the sampling period and $|p_k|$ is the system pole farthest from the origin of the s-plane, then the bilinear z-transformation, described below, provides good results besides being computationally efficient.

$$z = \frac{2 + sT}{2 - sT} \tag{19}$$

$$s = \frac{2}{T}\left(\frac{z-1}{z+1}\right) \tag{20}$$

This requirement on the sampling period may be simply interpreted as keeping the sampling frequency above the Nyquist rate in terms of the impulse response of the system.

Results of Simulation: To test the above procedure, results of two simulations will be presented. The first example was that of a first-order discrete-time system, represented by

$$c_{k+1} = a_1 u_k - b_1 c_k \tag{21}$$

$$y_k = c_k + v_k \tag{22}$$

The constants a_1 and b_1 were chosen as 0.7 and 0.5, respectively, for the simulation, and u_k and v_k were uncorrelated white-noise sequences of zero mean, and variance one. The input sequence u_k and the noisy output sequence y_k were then used for identification of the parameters a_1 and b_1 using the process discussed earlier. The stochastic approximation algorithms selected for this example were the older algorithm f_1 and the new algorithm f_3, with the gain sequence $\gamma_n = \frac{1}{n}$ for each case.

With the first algorithm, the estimates of the parameters converged to within 5 per cent of the correct values in about 250 iterations, whereas, the new algorithm, with =10, caused convergence to within 5 per cent in about 70 iterations. In each case, the output sequence was corrected for obtaining unbiased estimates, as discussed earlier.

The second example simulated was that of the second-order system described by

$$c_{k+1} = a_1 u_k + a_2 u_{k-1} - b_1 c_k - b_2 c_{k-1} \tag{23}$$

and

$$y_k = c_k + v_k \tag{24}$$

The values of the parameters selected for the simulation were

$$a_1=0.5,\quad a_2=0.2,\quad b_1=-0.8 \text{ and } b_2=0.15 \tag{25}$$

The input sequence $\{u_k\}$ and the output noise sequence $\{v_k\}$ were again taken as uncorrelated random sequences of zero mean and unit variance.

This time the estimates took a little longer to converge to within 5% of the correct values; about 300 iterations with the first algorithm and 90 iterations with the new algorithm for stochastic approximation. The computer-time required for the second example was much longer, due to the inversion of a 4x4 matrix required at each iteration after correction in the values of the output sequence.

Conclusion: In this paper, the application of a new algorithm for stochastic approximation to the identification of linear discrete-time systems has been discussed. A bootstrap method for correcting the effect of noise has been proposed so that the biassing effect is considerably reduced. Two examples of simulation indicate convergence at rates much faster then obtained by precious authors. Further investigations are being carried out on the effect of increased output noise.

It may be added that although only single-input single-output systems have been considered in this paper, extension to multivariable systems is straightforward.

Acknowledgement: Most of this work was done while the author was a Visiting Professor with the Information Systems Laboratory, Department of Electrical Engineering, Stanford University; Stanford, California. The author likes to express his gratitude for being allowed to use all the facilities at Stanford.

References

[1] Kirvatis, K. and Fu, K.S.: "Identification of nonlinear systems by stochastic approximation", Proc. 1966 Joint Automatic Control Conference, pp. 225.

[2] Saridis, G.N. and Stein. G.: "Stochastic approximation algorithms for linear discrete-time identification", IEEE Trans. on Automatic Control, vol. AC-13, October 1968, pp. 515-523.

[3] Saridis, G.N. and Stein, G.: "A new algorithm for linear system identification", IEEE Trans. on Automatic Control, vol. AC-13, October 1968, pp. 592-594.

[4] Saridis, G.N., Nikolic, Z.Z. and Fu, K.S.: "Stochastic approximation algorithms for system identification, estimation and decomposition of mixtures," IEEE Trans. on System Science and Cybernetics, vol. SSC-5, January 1969, pp. 8-15.

[5] Holmes, J.K.: "Two stochastic approximation procedures for identifying linear systems", IEEE Trans. on Automatic Control, vol. AC-14, June 1969, pp. 292-295.

[6] Neal, C.B. and Bekey, G.A.: "Estimation of the parameters of sampled-data systems by means of stochastic approximation", International Journal of Systems Science, vol. 1, 1970, pp. 63-74.

[7] Saridis, G.N. and Lobbia, R.N.: "Parameter identification and control of linear discrete-time systems", IEEE Trans. on Automatic Control, vol. AC-17, February 1972, pp. 52-60.

[8] Sinha, N.K. and Griscik, M.P.: "A Stochastic Approximation Method", IEEE Trans. on Systems, Man and Cybernetics, vol. SMC-1, October 1971, pp. 338-344.

[9] Fu, K.S., Nicolic, Z.Z., Chien, T.Y. and Wee, W.G.: "On the stochastic approximation and related learning techniques", Purdue University, Lafeyette, Ind., Rep. TR-EE66, April 1966.

[10] Sinha, N.K.: "Estimation of transfer function of continuous system from sampled data", Proc. IEE, vol. 119, May 1972, pp. 612-614.

[11] Dvoretzky, A.: "On stochastic approximation", Proc. 3rd Berkeley Symposium on Mathematical Statistics and Probability, vol. 1, Los Angeles: University of California Press, 1956, pp. 39-56.

[12] Sinha, N.K. and Lawton, J.S.: "Unbiased estimates for on-line system identification", Proc. IEEE, vol. 59, November 1971, pp. 1631-1632.

[13] Rowe, I.H.: "A bootstrap method for statistical estimation of model parameters", International Journal of Control, vol. 12, November 1971, pp. 721-738.

EQUIVALENT FILTER IDENTIFICATION FROM AUTOCOVARIANCE ESTIMATES

Krister Gerdin
Chalmers University of Technology, Division of Automatic Control
Gothenburg, Sweden[1]

From an experiment on a time-independent Gaussian process with rational spectrum the equivalent (shaping) filter is identified. Measurement data are first reduced to an autocovariance estimate and then filter parameter estimates, asymptotically equivalent to Maximum Likelihood estimates, are computed. Experimental results from simulated time series are given.

1. INTRODUCTION

The presented method sets out to estimate the parameters a_i, c_i in the stochastic, time-independent difference equation

$$y(t)+a_1y(t-1)+\ldots+a_ry(t-r)=e(t)+c_1e(t-1)+\ldots+c_re(t-r) \quad (1)$$

from a measured sequence $y(t)$, $t=1,\ldots,N$. $e(t)$ is supposed to be a sequence of independent, normal random variables, $Ee(t)=0$ and $Ee^2(t)=\lambda^2=\mu$; μ is also to be estimated. The problem can also be formulated as the identification of the equivalent (or shaping) filter $G(z)=C(z)/A(z)$ with

$$A(z)=1+a_1z^{-1}+\ldots+a_rz^{-r} \quad (2)$$

$$C(z)=1+c_1z^{-1}+\ldots+c_rz^{-r} \quad (3)$$

z^{-1} is the unit delay operator. Confining the zeros of $z^rA(z)$ and $z^rC(z)$ to the region inside the unit circle means no restriction here. Measurement data are first reduced by forming an autocovariance function estimate for $y(t)$ and then an approximative likelihood function is minimized. Some ideas in this context emanate from Gerdin (1970). A somewhat similar approach was early suggested by Durbin (1960). The interesting approach in Zetterberg(1969) should be mentioned in this context; this paper may also be referred to for a more extensive literature revue. In certain respects the present method is very close to the method by Åström (1966). The main difference is the use of autocovariances instead of unreduced time series as input to the estimation procedure. Of course this data reduction implies less efficient estimates. Why then "spend blood and tears" on a method like this? Justification is given by the facts that

- the method is fast and can be used to search the parameter space for initial parameter vector estimates to be used in full data utilizing methods
- it can contribute towards a deeper understanding of the information content in autocovariance estimates.

2. OUTLINE OF THE METHOD

We will work with a slightly different model than above (eq. (1)):

$$A(z)y(t)=(\lambda/\lambda')C(z)\varepsilon(t), \quad \varepsilon(t)=(\lambda'/\lambda)e(t) \quad (4)$$

The constant λ' and a new polynomial $A'(z)$,

$$A'(z)=1+a_1'z^{-1}+\ldots+a_r'z^{-r}, \quad (5)$$

are uniquely (apply e.g. the Schur-Cohn stability criterion) defined by the identity $(\mu'=(\lambda')^2)$

$$(\mu'/\mu)A(z^{-1})A(z)=A'(z)C(z^{-1})+A'(z^{-1})C(z). \quad (6)$$

From an experiment we have available the sequence $y(t)$, $t=1,\ldots,N$, and we then form the autocovariance function estimate

$$\phi_y(k)=(1/N)\sum_{t=1}^{N-|k|} y(t)y(t+|k|), \quad k=0,\pm1,\ldots,\pm M \quad (7)$$

Although $\varepsilon(t)$, $t=1,\ldots,N$, is not available we can formally introduce $\phi_\varepsilon(k)$, in the sequel for simplicity denoted $\phi(k)$:

$$\phi(k)=(1/N)\sum_{t=1}^{N-|k|} \varepsilon(t)\varepsilon(t+|k|), \quad k=0,\pm1,\ldots,\pm M \quad (8)$$

In section 3 it will be shown that $\phi(k)$ can be computed from $\phi_y(k)$ according to (9), (10), (11):

$$C(z)\phi^+(k)=A'(z)\phi_y(k), \quad k=-M+r,\ldots,M \quad (9)$$

$$\phi^+(-M+i-1)=\psi_i, \quad i=1,\ldots,r \quad (10)$$

$$\phi(k)=\phi^+(k)+\phi^+(-k). \quad (11)$$

It cannot generally be recommended here to set the initial values ψ_i to zero; they must be estimated together with the parameters. (9)-(11) are exactly valid if and only if $\{\varepsilon(t)=0/t=-r+1,\ldots,0; t=N+1,\ldots,N+r\}$ (that is to say if the initial and final states in the inverse filtering relation $e(t)=(A(z)/C(z))y(t)$ are zero). (9)-(11) are thus true asymptotically in N and this is the first approximation of the method.

In section 4 the loss function

$$L=-(1-(2/N))\ln(\phi(0)/\mu')+(\phi(0)/\mu')+(1/\mu')^2\sum_{k=1}^{M}(N/(N-k))\phi^2(k) \quad (12)$$

is motivated. L is (apart from a certain constant) an approximation to the likelihood function of the vector $\underline{\phi}_y$ and this is the second and last approximation involved in the method. L is minimized with respect to a_i', c_i, ψ_i through the relation (9)-(11) between on one side $\phi(k)$ and on the other side the parameters and the estimated $\phi_y(k)$; this minimization is discussed in section 5. It can be seen from the results in the sequel that the parameter estimates will be ML estimates for $N\to\infty$, $M\to\infty$, $M/N\to0$.

---

[1]Now at Fiskeby AB, Skärblacka, Sweden

As a last step μ and the a_i's are computed from (6) by spectral factorization of the right hand member. Several working numerical procedures for spectral factorization of polynomials are available. In the numerical experiments described later a method by Vostrý (1973) was used; it is a development of the method given in Vostrý (1972). The method is iterative with very fast convergence and is easy to implement on a digital computer.

3. DIFFERENCE EQUATION FOR AUTOCOVARIANCE ESTIMATES

Equations (6), (9)-(11) are to be derived. First, introduce

$$Y(z)=(1/N)\sum_{t=1}^{N} y(t)z^{-t},\quad \hat{\Phi}_y(z)=\sum_{k=-N}^{N} \hat{\phi}_y(k)z^{-k} \tag{13}$$

To simplify the next deduction, assume $y(t)$ defined and equal to zero for $t<1$, $t>N$ (the result is not dependent upon this assumption). Then

$$\hat{\phi}_y(k)=(1/N)\sum_{t=1}^{N} y(t)y(t+k), \tag{14}$$

$$\hat{\Phi}_y(z)=\sum_{k=-N}^{N} \hat{\phi}_y(k)z^{-k}=(1/N)\sum_{k=-N}^{N}\sum_{t=1}^{N} y(t)y(t+k)z^{-k}=$$
$$=(1/N)\sum_{t=1}^{N} y(t)z^{t}\sum_{k=-N}^{N} y(t+k)z^{-(t+k)}=(1/N)Y(z^{-1})Y(z). \tag{15}$$

Correspondingly, $\hat{\Phi}(z)=(1/N)\mathcal{E}(z^{-1})\mathcal{E}(z)$. Now, disregarding initial and final states,

$$\mathcal{E}(z^{\pm1})=(\lambda'/\lambda)\{A(z^{\pm1})/C(z^{\pm1})\}\, Y(z^{\pm1}) \tag{16}$$

and consequently

$$\hat{\Phi}(z)=(\mu'/\mu)\{A(z)A(z^{-1})/C(z)C(z^{-1})\}\hat{\Phi}_y(z). \tag{17}$$

Choosing μ' properly, we finally have

$$\frac{\mu'}{\mu}\,\frac{A(z)A(z^{-1})}{C(z)C(z^{-1})}=\frac{A'(z)}{C(z)}+\frac{A'(z^{-1})}{C(z^{-1})} \tag{18}$$

and with $\hat{\Phi}^{+}(z)=\{A'(z)/C(z)\}\hat{\Phi}_y(z)$, eq. (6), (9)-(11) are obviously true.

4. DISTRIBUTION PROPERTIES OF AUTOCOVARIANCE ESTIMATES FOR A PURELY RANDOM TIME SERIES

Let us discuss the density function $f_a(\underline{x})$ of the vector $\underline{\phi}_a=((N/\mu')\hat{\phi}(0),(N/(\sqrt{N-1}\mu'))\hat{\phi}(1),\dots,(N/(\sqrt{N-M}\mu'))\hat{\phi}(M))$. Directly from the definition (8) it is found that if

$$f(\underline{x})=f_{\chi^2,N}(x_0)\cdot f_{N(0,1)}(x_1)\cdot\ldots\cdot f_{N(0,1)}(x_M) \tag{19}$$

then $f_a(\underline{x})$ and $f(\underline{x})$ have the same first and second order moments. $f_{\chi^2,N}$ is the χ^2 density function, N degrees of freedom, and is the exact density of $(N/\mu')\hat{\phi}(0)$. $f_{N(0,1)}$ is the normal density function, mean zero and variance one, and is the asymptotic (in N) density of $(N/(\sqrt{N-k}\,\mu')\hat{\phi}(k)$. (Cochran (1934) and Anderson (1941), (1942)). As a matter of fact, $f_a\to f$ as $N\to\infty$ (M fixed) which is shown for reference in the theorem at the end of this section. L (12) is formed from the likelihood function corresponding to f. If the ψ_i were zero, a linear transformation would carry $\underline{\phi}$ to $\underline{\phi}_y$. Asymptotically in N and M, $M/N\to 0$, then, minimizing L is equal to minimizing the likelihood function of $\underline{\phi}_y$ as the effect of the ψ_i being zero and of the Jacobian becomes neglectable. Not very much effort has been spent in finding out whether there are better approximations (in some sense) than L.

<u>Theorem</u>: $f_a\to f$, $N\to\infty$ (M fixed)

The proof of the theorem is sketched very briefly; numbers within parenthesis refer to sections in Kendall (1969), (1967), (1968). For simplicity, standardize to $\mu'=1$.

$$f_{\chi^2,N}(x_0)\to(1/\sqrt{2N})f_{N(0,1)}((x_0-N)/\sqrt{2N}),\ N\to\infty \tag{16.3}$$

Introduce the serial covariance

$$\hat{\phi}_c(k)=\frac{1}{N}\Big\{\sum_{t=1}^{N-|k|}\varepsilon(t)\varepsilon(t+|k|)+\sum_{t=N-|k|+1}^{N}\varepsilon(t)\varepsilon(t+|k|-N)\Big\}$$
$$k=0,\pm1,\dots,\pm M$$

$\underline{\phi}_c=(\hat{\phi}_c(0),\dots,\hat{\phi}_c(M))\to\underline{\phi}=(\hat{\phi}(0),\dots,\hat{\phi}(M))$ in probability and thus also in distribution. $\hat{\phi}_c(k)$ can be written as a quadratic form, $\hat{\phi}_c(k)=\underline{\varepsilon}^{T}A_k\underline{\varepsilon}$, and then the cumulant-generating function of $\underline{\phi}_c$ is

$$\Phi_c(t_0,\dots,t_M)=-(1/2)\ln\{\det\{I-2it_0A_0-\dots-2it_MA_M\}\}. \tag{15.12}$$

The matrix inside brackets is circulant and thus analytically computable (Rosenblatt (1962)). An investigation of the cumulants confirms the proposition. Q.E.D.

5. MINIMIZATION OF THE LIKELIHOOD FUNCTION

L is to be minimized with respect to

$$\underline{\theta}=(\theta_1,\dots,\theta_{3r})=(a_1',\dots,a_r',c_1,\dots,c_r,\psi_1,\dots,\psi_r) \tag{20}$$

and μ'. Derivation with respect to μ' gives immediately

$$\mu'=\Big\{\hat{\phi}(0)+\sqrt{\hat{\phi}^2(0)+8(1-(2/N))\sum_{k=1}^{M}(N/(N-k))\hat{\phi}^2(k)}\Big\}\Big/\{2(1-(2/N))\} \tag{21}$$

$\hat{\phi}(k)$ is found by solving (9). Because of the nonlinear nature of the minimization problem some iterative method must be used; a steepest descent method and the Newton-Raphson method have been successfully applied. Of course the problem of local minima may be troublesome. For reference, first and second order derivatives are listed below.

<u>5.1. First order derivatives</u>

$$\frac{\partial L}{\partial\theta_i}=\frac{1}{\mu'^2}\Big\{\Big(\frac{\mu'}{\hat{\phi}(0)}\Big)\Big[\hat{\phi}(0)-\Big(1-\frac{2}{N}\Big)\mu'\Big]\frac{\partial\hat{\phi}(0)}{\partial\theta_i}+2\sum_{k=1}^{M}\frac{N}{(N-k)}\hat{\phi}(k)\frac{\partial\hat{\phi}(k)}{\partial\theta_i}\Big\} \tag{22}$$

By implicit, partial derivation in (9), difference equations with solutions $\partial\hat{\phi}^{+}(k)/\partial\theta_i$ are gained and then $\partial\hat{\phi}(k)/\partial\theta_i=\partial\hat{\phi}^{+}(k)/\partial\theta_i+\partial\hat{\phi}^{+}(-k)/\partial\theta_i$.

$$\left.\begin{aligned} &C(z)\partial\phi^+(k)/\partial a_1' = \phi_y(k-1) \\ &\partial\phi^+(k)/\partial a_{i+1}' = \partial\phi^+(k-1)/\partial a_i' + \\ &\qquad +\phi_y(-M+r-i-1)\,[1/C(z)]_{k-(-M+r)} \end{aligned}\right\} k=-M+r,\ldots,M \quad (23)$$

$$\partial\phi^+(k)/\partial a_i' = 0,\ k=-M,\ldots,-M+r-1$$

$$\left.\begin{aligned} &C(z)\partial\phi^+(k)/\partial c_1 = -\phi^+(k-1) \\ &\partial\phi^+(k)/\partial c_{i+1} = \partial\phi^+(k-1)/\partial c_i - \\ &\qquad -\phi^+(-M+r-i-1)\,[1/C(z)]_{k-(-M+r)} \end{aligned}\right\} k=-M+r,\ldots,M \quad (24)$$

$$\partial\phi^+(k)/\partial c_i = 0,\ k=-M,\ldots,-M+r-1$$

$$\left.\begin{aligned} &C(z)\partial\phi^+(k)/\partial\psi_i = 0 \\ &\partial\phi^+(k)/\partial\psi_{i+1} = \partial\phi^+(k-1)/\partial\psi_i - \\ &\qquad -c_{r-i}\,[1/C(z)]_{k-(-M+r)} \end{aligned}\right\} k=-M+r,\ldots,M \quad (25)$$

$$\partial\phi^+(k)/\partial\psi_i = \delta(k-(-M+i-1)),\ k=-M,\ldots,-M+r-1$$

Here $[1/C(z)]_k$ denotes the coefficient of z^{-k} in the result of the long division $1/C(z)$.

5.2. Second order derivatives

$$\frac{\partial^2 L}{\partial\theta_i\partial\theta_j} = \frac{1}{\mu'^2}\Big\{\Big(\frac{\mu'}{\phi(0)}\Big)\Big[\phi(0)-\Big(1-\frac{2}{N}\Big)\mu'\Big]\frac{\partial^2\phi(0)}{\partial\theta_i\partial\theta_j} + \Big(\frac{\mu'}{\phi(0)}\Big)^2\Big(1-\frac{2}{N}\Big)\frac{\partial\phi(0)}{\partial\theta_i}\frac{\partial\phi(0)}{\partial\theta_j} + 2\sum_{k=1}^{M}\frac{N}{(N-k)}\Big[\phi(k)\frac{\partial^2\phi(k)}{\partial\theta_i\partial\theta_j} + \frac{\partial\phi(k)}{\partial\theta_i}\frac{\partial\phi(k)}{\partial\theta_j}\Big]\Big\} \quad (26)$$

$$\frac{\partial^2\phi(k)}{\partial a_i'\partial a_j'} = \frac{\partial^2\phi(k)}{\partial\psi_i\partial\psi_j} = \frac{\partial^2\phi(k)}{\partial a_i'\partial\psi_j} = 0$$

$$C(z)\partial^2\phi^+(k)/\partial c_1^2 = -2\partial\phi^+(k-1)/\partial c_1;\ k=-M+r+1,\ldots,M$$

$$\begin{aligned} &\partial^2\phi^+(k)/\partial c_{i+1}^2 = \partial^2\phi^+(k-2)/\partial c_i^2 + \qquad k=-M+r+i+1,\ldots,M \\ &+2\phi^+(-M+r-i-1)\,[1/C^2(z)]_{(k-i-1)-(-M+r)} \end{aligned} \quad (27)$$

$$\begin{aligned} &\partial^2\phi^+(k)/\partial c_i\partial c_{j+1} = \partial^2\phi^+(k-1)/\partial c_i\partial c_j + \qquad k=-M+r+i, \\ &+\phi^+(-M+r-j-1)\,[1/C^2(z)]_{(k-i)-(-M+r)} \qquad \ldots,M;\ i\le j \end{aligned}$$

$$\partial^2\phi^+(k)/\partial c_i\partial c_j = 0;\ k=-M,\ldots,-M+r+i-1$$

$$\left.\begin{aligned} &C(z)\partial^2\phi^+(k)/\partial a_1'\partial c_1 = -\partial\phi^+(k-1)\partial a_1' \\ &\partial^2\phi^+(k)/\partial a_{i+1}'\partial c_1 = \partial^2\phi^+(k-1)/\partial a_i'\partial c_1 - \\ &-\phi_x(-M+r-i-1)\,[1/C^2(z)]_{(k-1)-(-M+r)} \end{aligned}\right\} k=-M+r+1,\ldots,M \quad (28)$$

$$\partial^2\phi^+(k)/\partial a_i'\partial c_{j+1} = \partial^2\phi^+(k-1)/\partial a_i'\partial c_j;\ k=-M+r+j+1,\ldots,M$$

$$\partial^2\phi^+(k)/\partial a_i'\partial c_j = 0;\ k=-M,\ldots,-M+r+j-1$$

$$\left.\begin{aligned} &C(z)\partial^2\phi^+(k)/\partial c_1\partial\psi_1 = -\partial\phi^+(k-1)/\partial\psi_1 \\ &\partial^2\phi^+(k)/\partial c_1\partial\psi_{j+1} = \partial^2\phi^+(k-1)/\partial c_1\partial\psi_j - \\ &-\delta(j+1-r)\,[1/C(z)]_{k-(-M+r)} + \\ &+c_{r-j}\,[1/C^2(z)]_{(k-i)-(-M+r)} \\ &\partial^2\phi^+(k)/\partial c_{i+1}\partial\psi_j = \partial^2\phi^+(k-1)/\partial c_i\partial\psi_j - \\ &-\delta(i+j-r)\,[1/C(z)]_{k-(-M+r)} \end{aligned}\right\} k=-M+r,\ldots,M \quad (29)$$

$$\partial^2\phi^+(k)/\partial c_i\partial\psi_j = 0,\ k=-M,\ldots,-M+r-1$$

6. A NUMERICAL EXPERIMENT

Sampled independent and normally distributed values from an analog noise generator (PACE 201 A) were fed into a computer, normalized to N(0, 1) by means of the mean and variance estimates for the whole experiment and then filtered by $(1+0.5z^{-1})/(1-0.5z^{-1})$. Autocovariances and parameter estimates were then calculated for 10 time series with N=500 and M=10, 30, 50. The results are given in table 1. It may be interesting to

M =	10	30	50	10	30	50	10	30	50
	$-a_1$	$-a_1$	$-a_1$	c_1	c_1	c_1	λ	λ	λ
	.502	.490	.504	.455	.488	.490	1.108	1.153	1.174
	.542	.540	.535	.610	.577	.587	1.026	1.062	1.081
	.549	.535	.517	.491	.501	.497	0.976	1.024	1.072
	.414	.384	.379	.554	.578	.567	1.008	1.042	1.061
	.435	.423	.412	.458	.471	.477	1.007	1.040	1.059
	.470	.438	.439	.519	.532	.559	1.072	1.120	1.165
	.515	.495	.481	.562	.584	.564	0.943	0.969	1.004
	.388	.402	.363	.558	.556	.586	1.043	1.066	1.104
	.540	.534	.540	.537	.539	.536	1.009	1.032	1.069
	.522	.551	.544	.431	.395	.401	0.981	1.014	1.053
$\hat{m}$ =	.488	.479	.471	.518	.522	.526	1.017	1.052	1.084
$\hat{\sigma}_1$=	.056	.063	.068	.057	.059	.059	0.048	0.053	0.052
$\hat{\sigma}_2$=	.056	.063	.071	.057	.061	.062	0.049	0.072	0.097

Table 1. Results from numerical experiment.

compare the experimental parameter deviations here with the theoretical lower bounds (Cramér-Rao) on the standard deviations for a full data approach (Åström (1967)): $\sigma_a=\sigma_c=0.048$, $\sigma_\lambda=0.032$. $\hat{\sigma}_1$ and $\hat{\sigma}_2$ in the table are calculated around $\hat{m}$ and the known exact parameter values resp.. Striking observations are also the good estimate quality for such a small number of shifts as 10 and the obvious bias in λ for larger M. No definite convergence problems were encountered and in most cases it was possible to start from $a_1=c_1=\psi_1=0$, using the steepest descent method.

7. REFERENCES

1. Anderson, R.L. (1941): Serial Correlation in the Analysis of Time Series. Unpublished thesis, Library, Iowa State College, Ames, Iowa.

2. Anderson, R.L. (1942): Distribution of the Serial Correlation Coefficient. Ann. Math. Stat. 13 no 1 p. 1.

3. Åström, K-J & Bohlin, T. (1966): Numerical Identification of Linear Systems from Normal Operating Records. In: Hammond, P.H. (ed): Theory of Self-Adaptive Control Systems. Plenum Press, London.

4. Åström, K-J (1967): On the Achievable Accuracy in Identification Problems. In: Preprints 1st IFAC Symposium on Identification in Automatic Control Systems. Academia, Prague.

5. Cochran, W.G. (1934): Distribution of Quadratic Forms in a Normal System with Applications to the Analysis of Covariance. Proc. Camb. Phil. Soc. 30 p. 178.

6. Durbin, J. (1961): Efficient Fitting of Linear Models for Continuous Stationary Time Series from Discrete Data. Bull. Inst. Stat. 38 p. 273.

7. Gerdin, K. (1970): Markov Modelling via Correlation Analysis Applied to Identification of a Paper Machine in the Absence of Test Signal. In: Preprints 2nd IFAC Symposium on Identification and Process Parameter Estimation. Academia, Prague.

8. Kendall, M.G. & Stuart, A. (1969),(1967),(1968) The Advanced Theory of Statistics. London (Griffin & Co) 1 (3rd ed), 2 (2nd ed), 3 (2nd ed).

9. Vostrý, Z. (1972): Numeričeskij Metod Spektralnoj Faktorizacii Polynomov (in Russian). Kybernetika 8 no 4 p. 323.

10. Vostrý, Z. (1973): Personal communication. To be published in part 2 of a thesis 'Synthesis of Linear Discrete Control with Respect to a Quadratic Performance Index for Systems Having Different Number of Inputs and Outputs'. Inst. Inf. Theory Aut., Prague.

11. Zetterberg, L.H. (1969): Estimation of Parameters for a Linear Difference Equation with Application to EEG Analysis. Math. Biosc. 5 p. 227.

DETERMINATION OF CHANGES IN THE PROPERTIES OF RANDOM PROCESSES BY THE BAYES METHOD

Laimutis Telksnys

Institute of Physics and Mathematics
Academy of Sciencies of the Lithuanian SSR,
Vilnius, USSR

The problem of change detection in the properties of random processes by their single realizations of the finite length is considered. A constructive method to get the estimations of the values of arguments, optimal according to the Bayes criterion, in the presence of which the properties of random processes take change, is described. A method of realization of the necessary calculations by means of computers is presented. Examples are given.

1. INTRODUCTION

The problem of change detection in the properties of random processes comes forth while solving the questions of automatic control and change detection in the functioning conditions of technical devices and biological organisms, while processing the results of various tests and measurements.

The methods of determination of the most probable changes in the properties of random processes is presented in the papers by Telksnys [1,2], Kligienė [3]. Let's note, however, that in practice we need not only such estimates. In a number of cases different loss functions are taken and therefore it is required to determine the changes in the properties so that the risk to err were minimal. It's a pity, there are no methods, acceptable for their practical realization, to solve this Bayes task. That's why let's consider a constructive method of solving the task, mentioned above.

2. THE STATEMENT OF THE PROBLEM

Let a random process have a finite frequency band and can be represented by a random sequence $X_i(N)$ $(i=n_0+1,\dots,n_2)$ described by the equation of the following type

$$X_i(N)=\begin{cases}-\sum_{j=1}^{p_1} a_j^{(1)}X_{i-j}+b^{(1)}V_i\ (i=n_0+1,\dots,N)\\ -\sum_{j=1}^{p_2} a_j^{(2)}X_{i-j}+b^{(2)}V_i\ (i=N+1,\dots,n_2),\end{cases} \quad (1)$$

where N is a number of the component (random or unknown) of the random sequence $X_i(N)$ $(i=n_0+1,\dots,n_2)$, in the presence of which it changes its properties; X_i $(i=n_0+1,\dots,n_0+p_1)$ is a random sequence with a zero expectation and a given correlation function $K_1=[k_{ij}^{(1)}]$ $(i,j=n_0+1,\dots,n_0+p_1)$, $k_{ij}^{(1)}=M\{X_{i+p_1}X_{j+p_1}\}$ $(i,j=n_0+1,\dots,n_0+p_1)$; V_i $(i=n_0+1,\dots,n_2)$ is a sequence of independent Gaussian random variables with a zero expectation and a unitary variance; the parameters $a_j^{(k)}$ $(j=1,\dots,p_k;\ k=1,2)$ are such that the roots $z_{k1},\dots,z_{kp_k}$ $(k=1,2)$ of the characteristic equations $z_k^{p_k}+a_1^{(k)}z_k^{p_k-1}+\dots+a_{p_k}^{(k)}=0$ $(k=1,2)$ are within the unit circles $|z_{ki}|<1$ $(i=1,\dots,p_k;\ k=1,2)$.

Note that the random sequence $X_i(N)$ $(i=n_0+1,\dots,n_2)$ is Gaussian under the condition that the value of the random parameter N is fixed, i.e. $N=n$.

We have got the realization $x_i(N=n)$ of the random sequence $X_i(N)$ at the finite interval $(i=n_0+1,\dots,n_2)$. Both a functional form of the conditional probability density function $p(\bar{x}|n)$, $\bar{x}=\{x_i(n)\}$ $(i=n_0+1,\dots,n_2)$ and the a priori probability density function $\alpha(n)$ of the random parameter N are known. If the parameter N is unknown, it is assumed that at the interval $n_0+1\le n\le n_2$ $c=$ const. A loss function is given $\ell(n,\tilde{n})$. n -is the component in the presence of which the change in the properties really took place. $\tilde{n}$ is an estimate (not optimal) of the value n.

It is urgent to find such a value n^* of the component that the average risk of making an error were minimal when the loss function is given.

3.SOLVING OF THE PROBLEM

It is possible to determine n^* by the minimum of the conditional average risk, i.e.

$$n^* = \arg\min_{\tilde{n}\in\mathcal{N}} R(\tilde{n}), \qquad (2)$$

where $R(\tilde{n}) = M\{\ell(n,\tilde{n})|\bar{x}\} = \sum_{n=n_o+1}^{n_2} \omega(n|\bar{x})\ell(n,\tilde{n})$.

In formula (2) $\omega(n|\bar{x})$ is an a posteriori probability density function

$$\omega(n|\bar{x}) = c(\bar{x})\alpha(n)f(\bar{x}|n), \qquad (3)$$

where $f(\bar{x}|n)$ is a likelihood function, corresponding to the conditional probability density function $p(\bar{x}|n)$ of the random sequence $X_i(N=n)$ $(i=n_o+1,\dots,n_2)$;

$$c^{-1}(\bar{x}) = \sum_{n=n_o+1}^{n_2} \alpha(n)f(\bar{x}|n). \qquad (4)$$

In order to get the estimates n^* with the acceptable for practice accuracy we must deal with rather long realizations. The number of components of random sequences $x_i(n)$ $(i=n_o+1,\dots,n_2)$ often compiles thousands and even more. Under such conditions calculation of n^* by formula (2) is extremely cumbersome and labour-consuming. It can't be realized even with the help of computers. In order to eliminate these difficulties let's consider the ways of transformation of expression (2).

4.TRANSFORMATION OF THE EXPRESSIONS

For the purpose of eliminating the difficulties in computing the likelihood function $f(\bar{x}|n)$, included by expression (2), let's present the conditional probability density function as follows:

$$p(\bar{x}|n) = p(x_{n_o+1},\dots,x_{n_o+p_1})p(x_{n_o+p_1+1},\dots,x_n)\times \times p(x_{n+1},\dots,x_{n_2}), \qquad (5)$$

where $p(x_{n_o+1},\dots,x_{n_o+p_1}) =$

$$= (2\pi)^{-\frac{p_1}{2}}(\det K_1)^{-\frac{1}{2}}\exp\Big\{-\frac{1}{2}\sum_{i,j=n_o+1}^{n_o+p_1} \varkappa_{ij}x_ix_j\Big\},$$

$\det K_1$ is a determinant of the matrix

$$K_1 = [k_{ij}^{(1)}] \ (i,j = n_o+1,\dots n_o+p_1),$$

$$k_{ij}^{(1)} = M\{X_{i+p_1}X_{j+p_1}\}, \quad K_1^{-1} = [\varkappa_{ij}]$$

$(i,j = n_o+1,\dots,n_o+p_1)$;

$p(x_{n_o+p_1+1},\dots,x_n) =$

$$= (2\pi D_I)^{-\frac{n-n_o-p_1}{2}}\prod_{i=n_o+p_1+1}^{n}\exp\Big\{-\frac{(x_i-M_I)^2}{2D_I}\Big\},$$

$$M_I = M\{X_i|x_{i-1},\dots,x_{i-p_1}\},$$

$$D_I = M\{X_i|x_{i-1},\dots,x_{i-p_1} - M_I\}^2$$

$(i = n_o+p_1+1,\dots,n)$;

$p(x_{n+1},\dots,x_{n_2}) =$

$$= (2\pi D_{II})^{-\frac{n_2-n}{2}}\prod_{i=n+1}^{n_2}\exp\Big\{-\frac{(x_i-M_{II})^2}{2D_{II}}\Big\},$$

$$M_{II} = M\{X_i|x_{i-1},\dots,x_{i-p_2}\},$$

$$D_{II} = M\{X_i|x_{i-1},\dots,x_{i-p_2} - M_{II}\}^2$$

$(i = n+1,\dots,n_2)$.

For the further simplification of calculations to get the estimates n^* let's consider two situations. The first situation is when the constants have no influence on determining n^* from $R(\tilde{n})$. The second situation is when the constants should be taken into account in $R(\tilde{n})$

In the first situation n^* could be calculated according to the following formula:

$$n^* \simeq \arg\min_{\tilde{n}\in\mathcal{N}} \sum_{n=n_o+1}^{n_2} \ell(n,\tilde{n})\exp\beta_4(n|\bar{x}), \qquad (6)$$

where

$$\beta_4(n|\bar{x}) = \begin{cases} \beta_3(n|\bar{x}) - \beta_{max}, \text{ if } \beta_3(n|\bar{x}) - \beta_{max} \geq -c \\ -\infty, \text{ if } \beta_3(n|\bar{x}) - \beta_{max} < c, \end{cases} \qquad (7)$$

$$\beta_3(n|\bar{x}) = \ln[\alpha(n)f(\bar{x}|n)],$$

$$\beta_{max} = \max_{n_o+1\leq n\leq n_2} \beta_3(n|\bar{x}).$$

In formula (7) c should be chosen large enough. In most cases in practice (for example, when $n_2 - n_o$ does not exceed 10^5) it's enough to choose c =40.

In the second situation n^* ought to be calculated according to the formula:

$$n^* \simeq \arg\min_{\tilde{n}\in\mathcal{N}} \sum_{n=n_o+1}^{n_2} \ell(n,\tilde{n})\Big[\sum_{s=n_o+1}^{n_2}\exp\beta_5(n,s|\bar{x})\Big]^{-1}, \qquad (8)$$

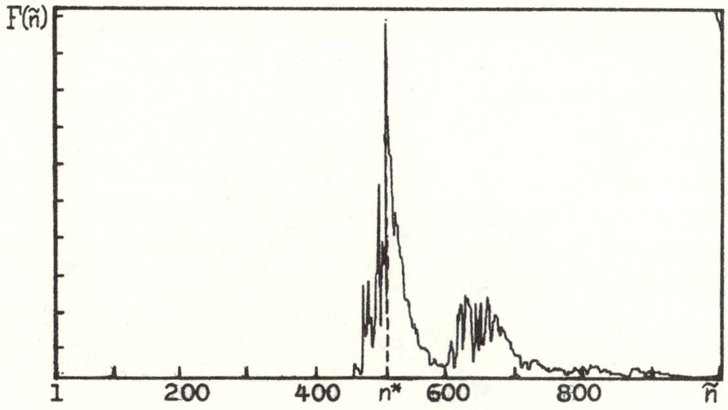

Fig.1

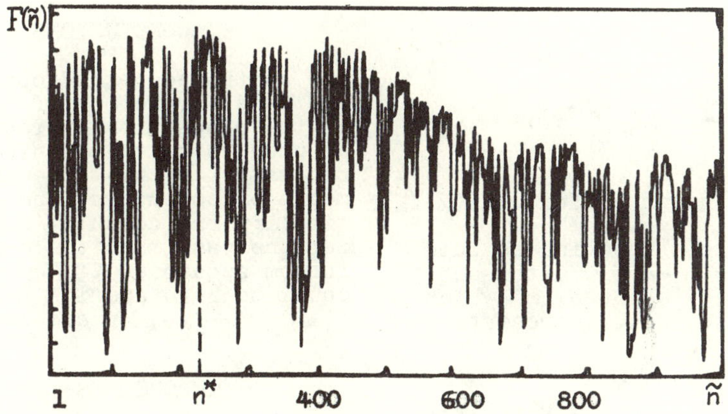

Fig.2

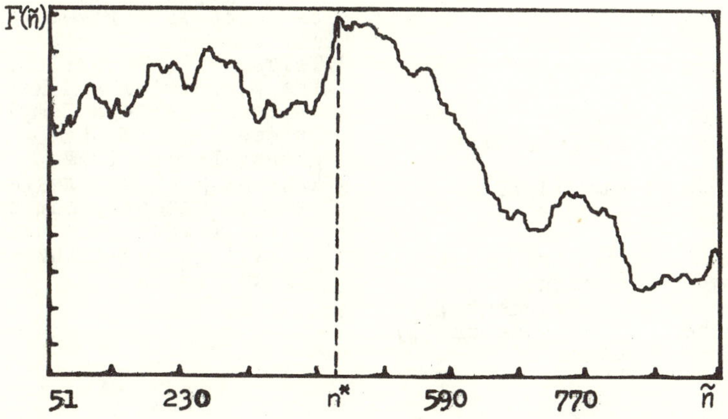

Fig.3

where

$$\beta_5(n,s|\bar{x}) = \begin{cases} \beta_3(s|\bar{x}) - \beta_3(n|\bar{x}), \\ \text{if } c \geq \beta_3(s|\bar{x}) - \beta_3(n|\bar{x}) \geq -c \\ \infty, \text{ if } \beta_3(s|\bar{x}) - \beta_3(n|\bar{x}) > c \\ -\infty, \text{ if } \beta_3(s|\bar{x}) - \beta_3(n|\bar{x}) < -c. \end{cases} \quad (9)$$

We shall illustrate how it is possible and worth while to use the above mentioned formulas.

5. EXAMPLES

Situation 1. Let a loss function have the following form

$$\ell(n,\tilde{n}) = \begin{cases} 1, |n-\tilde{n}| > a \\ 0, |n-\tilde{n}| \leq a. \end{cases} \quad (10)$$

Then taking into account that $\sum_{n=n_0+1}^{n_2} \omega(n|\bar{x}) = 1$ we can write:

$$n^* \simeq \arg\max_{n_0+a+1 \leq \tilde{n} \leq n_2-a} \sum_{n=\tilde{n}-a}^{\tilde{n}+a} \exp \beta_4(n|\bar{x}), \quad (11)$$

where $\beta_4(n|\bar{x})$ is given by formula (7).

Situation 2. Let a loss function be quadratic, i.e. $\ell(n,\tilde{n}) = (n-\tilde{n})^2$. With such a loss function, as it is known, n^* which minimizes the average risk, represents the conditional expectation:

$$n^* = \sum_{n=n_0+1}^{n_2} n\,\omega(n|\bar{x}). \quad (12)$$

Then in accordance with expression (9), the estimate n^* should be calculated by the formula:

$$n^* \simeq \sum_{n=n_0+1}^{n_2} n \left[\sum_{s=n_0+1}^{n_2} \exp \beta_5(n|\bar{x}) \right]^{-1}. \quad (13)$$

Note, that formulas (11) and (13) are convenient for calculations by computers.

We shall conclude with a numerical example.

Let $X_i(N=n)$ $(i = n_0+1, \ldots, n_2)$ represent a realization of the random sequence $X_i(N)$, where $n_2 - n_0 = 1000$ components, and which is described by expression (1), when $p_1 = p_2 = 1$. The loss function is described by formula (10).

In the sequel we give the results of experiments performed by computers when the change in the properties really took place when $n = 500$.

In fig.1 a diagram of the function $F(\tilde{n}) = \sum_{n=\tilde{n}-a}^{\tilde{n}+a} \exp \beta_4(n|\bar{x})$ is presented (see formula (11)), when $a_1^{(1)} = -0{,}9048$, $a_1^{(2)} = -0{,}8187$, $b^{(1)} = 0{,}0363$, $b^{(2)} = 0{,}033$, and the parameter $a = 0$ in the loss function (10). The component n^*, in the presence of which the random sequence x_i $(i = 1, \ldots, 1000)$ changes its properties, was defined by the maximum of the function $F(\tilde{n})$. In this experiment it is obtained that $n^* = 500$.

In fig.2 a diagram of the function $F(\tilde{n})$ is presented, when $a_1^{(1)} = -0{,}9048$, $a_1^{(2)} = -0{,}9025$, $b^{(1)} = 0{,}0363$, $b^{(2)} = 0{,}0362$ and $a = 0$. In this case $n^* = 230$.

When the parameter of the loss function $a = 50$, the estimate $n^* = 435$. A diagram of the function $F(\tilde{n})$ for this case is shown in fig. 3.

6. CONCLUSIONS

The mentioned method (optimal according to the Bayes criterion) of change detection in the properties of random processes can be used to solve practical problems. The necessary calculations can be realized by means of computers. The given methods to simplify the colculations can also be used for the case when the expectation of the random sequence X_i $(i = n_0+1, \ldots, n_2)$ is not equal to zero and its parameters $a_j^{(k)}$, $b^{(k)}$ $(j = 1, \ldots, p_k;\ k = 1,2)$ are unknown or random.

REFERENCES

1. Telksnys L., Černiauskas V. Determination of the Moments of Time of a Change in Statistical Properties of Random Process. IFAC Symposium 1970 on Identification and Process Parameter Estimation. Prague, 1970.
2. Telksnys L. Determination of Changes in the properties and Recognition of Random Processes with a Complicated Structure. Proceedings of the 5th IFAC World Congress, Paris, 1972.
3. Клигене Н. Определение момента изменения свойств авторегрессивной последовательности. Техническая кибернетика. Материалы XXII научно-технической конференции. Каунас, 1972.

'ANALOGUE MODEL IDENTIFICATION OF SYSTEM DYNAMIC ERROR TRANSFER FUNCTIONS'

Professor D.R. Towill, U.W.I.S.T. Cardiff, U.K.

A method based on frequency domain sensitivity functions is presented in flow diagram form suitable for human operator implementation in analogue model experiments in which the error between system and model performance is displayed as a Lissajous figure. The method is successfully applied to determine the dynamic error transfer function of a simulated sixth order radar control system. In deriving the flow diagram, considerable use is made of the fact that many high order systems are adequately described by a coefficient plane model.

1. INTRODUCTION

Interactive system identification is now feasible in laboratories possessing an on-line computer plus suitable software for aiding the experimenter in such matters as choosing appropriate stimuli, evaluating performance criteria, and providing guidelines for further iterative steps necessary for establishing an optimum process model[1]. Where such facilities are not available due to economic or geographical reasons, a suitable transfer function describing system performance may be obtained using an analogue model in which the coefficients are adjusted until there is good agreement between model and system behaviour to selected stimuli. Two major problems then arise. Firstly, there is difficulty in obtaining a consistent high order model (this is also a difficulty with other methods[2]), and secondly a high level of technical skill is required of the human operator involved in the experimental work so that success depends on familiarity with the test method and the ability to learn about the system as the experiment proceeds.

Fortunately it is possible to describe the dynamic error performance of a wide variety of tracking type systems by a particular form of low order model with normalised coefficients occupying a relatively small part of the coefficient plane[3], even when the systems are high order and non-linear. This paper exploits this fact by generating a standard set of model gain and phase sensitivity functions leading to a simple flow chart aid for the human operator, thus providing a basic identification technique valid for tracking type systems. Some results are given for the in-situ modelling of a simulated sixth order radar control system.

2. PREVIOUS WORK ON ANALOGUE MODEL METHODS

The initial reception to analogue modelling techniques was extremely enthusiastic since the possibility of direct transfer function determination seemed on the horizon. Various papers on specific instruments then appeared[4][5], but only recently has the operator aiding problem been tackled from a fundamental viewpoint. A method for checkout of a known type of system based only on phase information has been developed[6], and tested on a sample of simulated systems[7]. Some general guidelines for identification have appeared together with proof that a quasi-linear model established in-situ may be valid in both time and frequency domains[8], but the skill level required in identification remained high.

The system under test may be described for a set of experimental conditions by the transfer function,

$$T(s) = K\left[\frac{1+b_1 s+b_2 s^2+\ldots\ldots b_q s^q}{1+a_1 s+a_2 s^2+\ldots\ldots a_n s^n}\right]\ldots\ldots(1)$$

Since many high order systems can be described by a third order zero velocity lag model[3], the analogue model to be identified may be written,

$$T(s) = K\left[\frac{1+A_1 s}{1+A_1 s+A_2 s^2+A_3 s^3}\right]\ldots\ldots(2)$$

and reasonable equivalence is sought over the frequency range 20% to 200% of system bandwidth between the performance of model and system since this ensures satisfactory prediction of dynamic errors[3]. For the experimental work described in this paper, the analogue model was used in the parallel mode as shown in fig. 1, with the error between model and system displayed as a Lissajous figure. At any test frequency, a departure of the display from a closed line at 45° represents an error in model matching. A phase error is seen as the line becoming an ellipse, and a gain error tilts the ellipse. Time domain stimuli available for the test include step, ramp, and parabolic excitation, supplemented by a range of spot frequencies. If required, some noise rejection capability is achieved by filtering model and system responses identically prior to display.

3. MODEL SENSITIVITIES

If S_{α}^{T} is defined as $\left(\frac{dT}{d\alpha}\cdot\frac{\alpha}{T}\right)$ where α is a system parameter, it is well known[9] that the model gain and phase sensitivities are given by, respectively,

$$S_{\alpha}^{|T|} = R_e \cdot S_{\alpha}^{T} \quad \ldots\ldots(3)$$

$$\phi \cdot S_{\alpha}^{\phi} = I_m \cdot S_{\alpha}^{T} \quad \ldots\ldots(4)$$

Since K in equation (2) is determined from a steady state test, only A_1, A_2, and A_3 remain to be determined from frequency domain measurements. Table I lists the sensitivity functions for the three coefficients.

4. GENERATING A STANDARD SET OF SENSITIVITY FUNCTIONS

Equation (2) may be transformed into the coefficient plane model[10],

$$T(s) = K \cdot \frac{1 + c(s/\omega_0)}{1 + c(s/\omega_0) + b(s/\omega_0)^2 + (s/\omega_0)^3} \quad \ldots(5)$$

Because many systems may be modelled with coefficients lying in a relatively small part of the coefficient plane, their gain and phase sensitivity functions are unlikely to be very different, and may be approximated by using just one typical (b,c) combination. A suitable combination is b=c=2, the Butterworth array, with sensitivity functions as shown in fig. 2. A flow chart will now be derived for identification purposes.

5. SETTING UP AN IDENTIFICATION FLOW CHART

A reasonable way to proceed is to select a number of discrete frequencies and grade the sensitivity functions into high, medium, and low categories based on an arbitrary set of levels[6]. To start

TABLE I.

MODEL SENSITIVITY FUNCTIONS.

$S_{A_1}^{T}$	$\dfrac{-A_1 s^3 (A_2 + A_3 s)}{1 + 2A_1 s + (A_1^2 + A_2)s^2 + (A_1 A_2 + A_3)s^3 + A_1 A_3 s^4}$
$S_{A_2}^{T}$	$\dfrac{-A_2 s^2}{1 + A_1 s + A_2 s^2 + A_3 s^3}$
$S_{A_3}^{T}$	$\dfrac{-A_3 s^3}{1 + A_1 s + A_2 s^2 + A_3 s^3}$

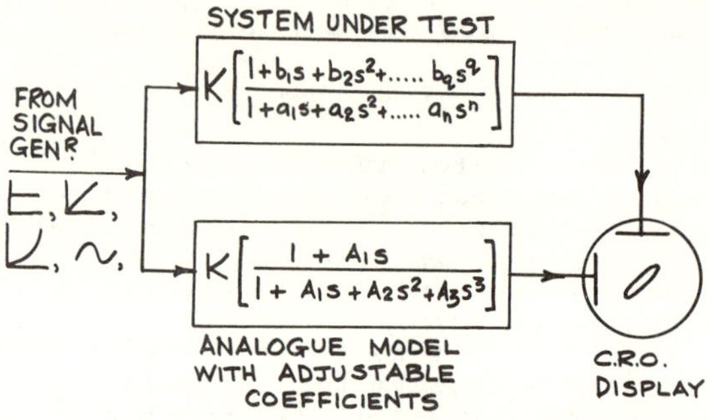

FIG. 1. ANALOGUE MODEL USED IN THE PARALLEL MODE.

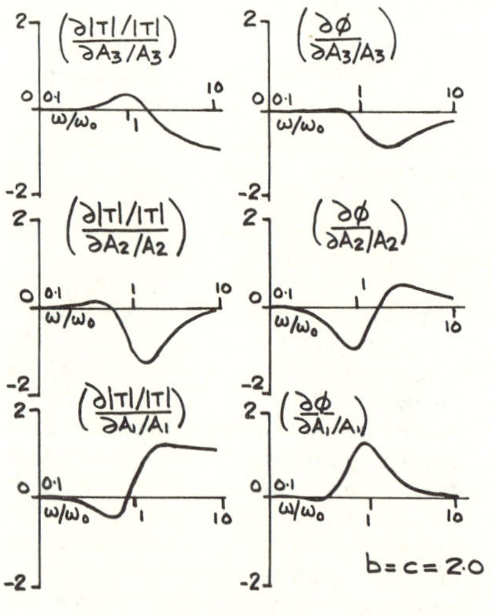

FIG. 2. GAIN AND PHASE SENSITIVITY PLOTS FOR THE MODEL.

$$T(s) = \left[\frac{1 + c(s/\omega_0)}{1 + c(s/\omega_0) + b(s/\omega_0)^2 + (s/\omega_0)^3}\right]$$

the identification procedure, at least one frequency measurement must be highly sensitive to a particular coefficient. When the sensitivity functions of fig.2. are so graded, it is apparent that at a test frequency of $0.6\omega_0$ a phase only error is most likely due to A_2 being in error, whilst if gain and phase are simultaneously in error, the likely cause is an error in A_1. At a test frequency $1.5\omega_0$, a combined phase and gain error is likely to be due to an error in A_1, and a phase only error is likely to be due to an error in A_3. Finally, any remaining errors at $3.0\omega_0$ are removed by adjusting A_3. The iterative identification procedure may be summarised in flow diagram form as shown in fig.3., the sequence in which the measurements are made assuming considerable importance.

6. AN APPLICATION

The identification procedure of fig.3. has been applied to the determination of a third order zero velocity lag model of a simulated two zero, six pole radar control system previously studied in detail in reference 11. In the absence of any theoretical model of the system under test it is reasonable to set $\omega_0 = \omega_{90°}$ for the purpose of this identification scheme. Since b=c=2, is the starting point for the analogue model, the initial settings of A_1, A_2, and A_3 are now known and implemented. Since conventional oscillators provide pre-determined discrete frequencies, the nearest convenient frequencies to $0.60\omega_0$, $1.50\omega_0$, and $3.00\omega_0$ are used. In this experiment, the frequencies used were 0.3Hz, 0.9Hz and 1.4Hz. Only two stages of iteration proved necessary, the final Lissajous figures for a wide frequency scan being shown in fig.4. It is considered that the match between system and model is highly acceptable for practical purposes especially considering the significant difference between the system and model orders.

Fig.5. shows the model pole-zero array for the initial model setting and for the two stages of iteration. The final values are b=1.87,c=4.75, showing that the identification scheme is successful in rapidly modelling a system with coefficients quite different from the initial model settings.

7. CONCLUSIONS

An on-line analogue model based identification scheme has been successfully applied to a simulated sixth order radar control system. The flow chart provided in the paper enables the human operator

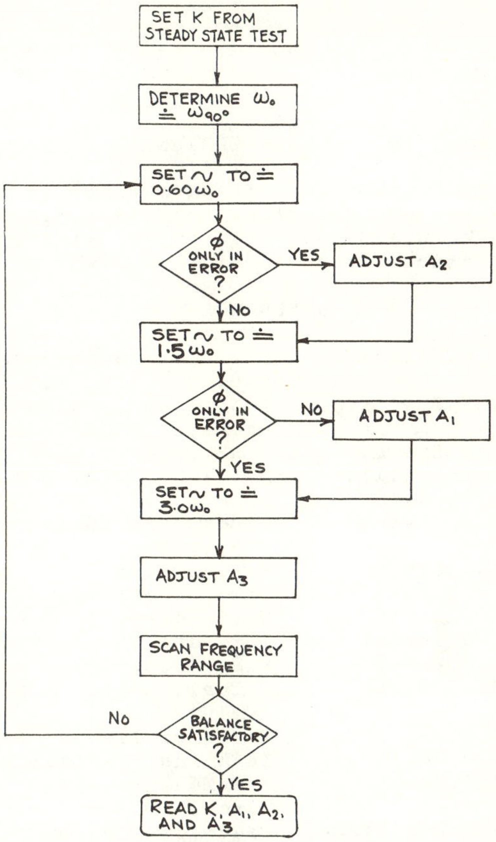

FIG.3. MODEL COEFFICIENT ADJUSTMENT CHART.

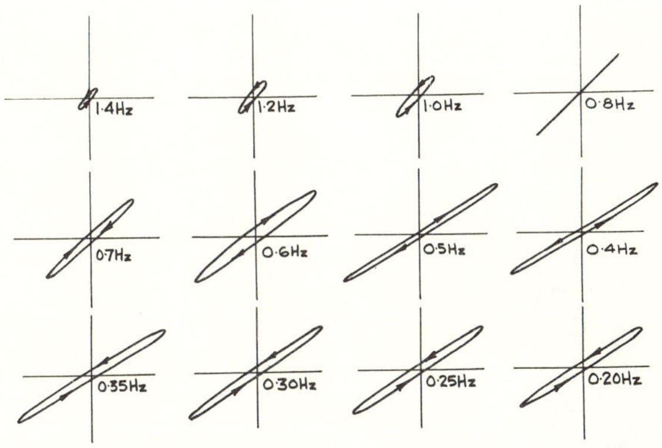

FIG.4. LISSAJOUS FIGURES FOR FINAL THIRD ORDER MODEL BALANCING SIXTH ORDER RADAR CONTROL SYSTEM.

to rapidly establish an adequate model by selecting appropriate test frequencies and relating Lissajous error displays to the model coefficients.

8. ACKNOWLEDGEMENTS

The simulation used in this paper is part of a project undertaken at U.W.I.S.T. by Jim Heard and sponsored by the Ministry of Defence (Procurement Executive).

REFERENCES

1. D.C.Williams — Electronics Letters, V.6, No. 15, July 1970
2. L.W.Taylor — Proc. JACC, Atlanta, Georgia, 1970
3. J.M.Brown, D.R.Towill, P.A.Payne — IERE Paper No. 1426/IC56, Jan. 1972
4. R.W.Levinge — Electronic Engineering V.37, 1965, p.218
5. C.A.Cox — Systems and Communications, V.2, No.2, 1966, p.18
6. D.R.Towill and K.J.Baker — Int. Jnl. Prod. Res. V.8, No.3, 1970, p.285
7. K.J. Baker — M.Eng.Thesis, UWIST, 1971
8. J.E.D.Kirby, D.R.Towill, K.J.Baker — To be published. Proc. IEEE Instrumentation and Measurement, March, 1973
9. J.J.DiStefano, A.R.Stubberud, I.J.Williams — Feedback and Control Systems, McGraw Hill, 1967, p. 168
10. D.R.Towill — Transfer Function Techniques for Control Engineers. Iliffe Book Co. 1970, Chapter 5
11. Z.Mehdi — Ph.D.Thesis,UWIST,1971

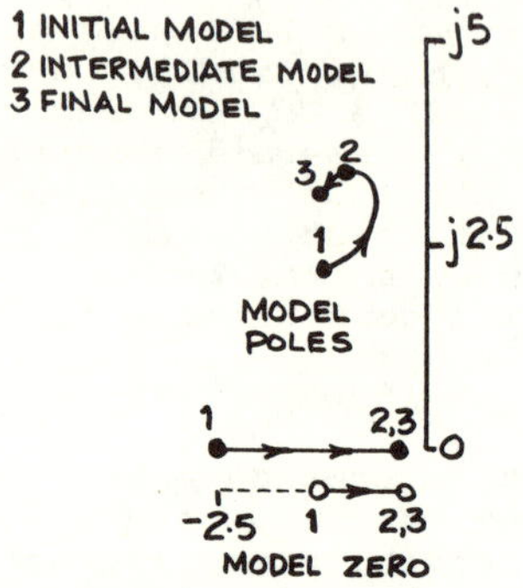

FIG.5. MODEL POLES AND ZEROS AT VARIOUS STAGES OF ITERATION.

IDENTIFICATION OF VERY NOISY DYNAMIC PROCESSES USING MODELS WITH FEW PARAMETERS

Rolf Isermann
University of Stuttgart
7000 Stuttgart, W.-Germany

SUMMARY

The variances of the parameter estimates of a dynamic process can be reduced by using models with as many equal parameters as possible. This is shown for continuous processes by using the Cramér-Rao inequality. A very simple method for identifying a dynamic process with three unknown parameters is also given.

1. INTRODUCTION

In the problem of the identification of processes with large noise-to-signal ratio a commom question arises: how to save the measuring time and still getting a sufficient description of the process dynamics. It will be shown that for higher order processes this can be done by assuming a simpler model with fewer different parameters which replaces the more exact model with many more different parameters. this approach takes into account a loss of information about detailed process dynamics.

For linear processes the following simplifications which lead to only small errors in the description of the closed loop behavior are investigated, Isermann (1971,b):

a. Omission of small time constants
b. Replacement of small time constants by a time delay
c. Replacement of different time constants by equal time constants.

These simplifications hold for relatively large parameter variations so that it very often seems possible to reduce the order of the process model (a.or b.) and use models with a reduced number of different parameters (c.).

2. THE CRAMER-RAO LOWER BOUND FOR PARAMETER ESTIMATION OF CONTINUOUS PROCESSES

Consider a linear continuous process Åström (1967),

$$A(p)\ y(t) = B(p)u(t) + C(p)e(t) \qquad (1)$$

where

$$A(p) = p^m + a_{m-1}p^{m-1} + \ldots + a_1 p + a_o$$
$$B(p) = b_m p^m + b_{m-1}p^{m-1} + \ldots + b_1 p + b_o \qquad (2)$$
$$C(p) = p^m + c_{m-1}p^{m-1} + \ldots + c_1 p + c_o$$

and $py(t)$ denotes $dy(t)/dt$.

The residuals $e(t)$ are assumed to be white gaussian noise with conditional probability density $p(e|\underline{\theta})$, where $\underline{\theta}$ is the parameter vector consisting of all a_i, b_i, c_i.

Introducing bandlimited continuous white noise with limit frequency ω_ℓ and approximating its autocorrelation function (acf) $\phi_{ee}(\omega)$ by the acf of a discrete white noise, Isermann (1972), yields

$$p(e|\theta) = \left[\frac{1}{\sigma_e\sqrt{2\pi}}\right]^{T\omega_\ell} \exp\left[-\frac{\omega_\ell}{2\sigma_e^2}\int_o^T e^2(t)dt\right] \qquad (3)$$

and the logarithm of the likelihood function

$$L = \ln p(e|\underline{\theta})$$
$$= -\frac{\omega_\ell}{2\sigma_e^2}\int_o^T e^2(t)dt - T\omega_\ell \ln\sigma_e - \frac{T\omega_\ell}{2}\ln 2\pi. \qquad (4)$$

The lower bound of the covariances of an unbiased parameter estimate $\hat{\underline{\theta}}$ is given by the CRAMÉR-RAO-inequality

$$E\{(\hat{\underline{\theta}}-\underline{\theta})(\hat{\underline{\theta}}-\underline{\theta})^T\} \geq \underline{J}^{-1} = \left[-E\{\delta^2 L/\delta\theta_i\delta\theta_j\}\right]^{-1}. \qquad (5)$$

Using Parseval's formula the expectation of the second derivatives of L is

$$E\{\frac{\delta^2 L}{\delta\theta_i\delta\theta_j}\} = -\frac{T_M\omega_\ell}{\sigma_e^2}E\{\frac{\delta e(t)}{\delta\theta_i}\cdot\frac{\delta e(t)}{\delta\theta_j}\}$$
$$= -\frac{T_M\omega_\ell}{2\pi\sigma_e^2}\int_{-\infty}^{\infty}\lim_{T\to\infty}\frac{1}{T}\frac{\delta}{\delta\theta_i}e_T(i\omega)\frac{\delta}{\delta\theta_j}e_T(-i\omega)d\omega. \qquad (6)$$

Compare Åström (1967). If $u(t)$ and $e(t)$ are statistically independent, one gets e.g.

$$E\{\frac{\delta^2 L}{\delta a_i^2}\} = -\frac{T_M\omega_\ell}{2\pi\sigma_e^2}\int_{-\infty}^{\infty}\left[\left|\frac{B(i\omega)}{C(i\omega)A(i\omega)}\frac{\delta A(i\omega)}{\delta a_i}\right|^2\phi_{uu}(\omega) + \left|\frac{1}{A(i\omega)}\frac{\delta A(i\omega)}{\delta a_i}\right|^2\phi_{ee}(\omega)\right]d\omega. \qquad (7)$$

The information matrix is

$$\underline{J} = \begin{bmatrix} J_{aa} & J_{ab} & J_{ac} & 0 \\ J_{ba} & J_{bb} & 0 & 0 \\ J_{ca} & 0 & J_{cc} & 0 \\ 0 & 0 & 0 & J_{dd} \end{bmatrix} \qquad (8)$$

with $J_{aa} = -E\{\delta^2 L/\delta a_i^2\}; J_{ab} = -E\{\delta^2 L/\delta a_i \delta b_i\}$; etc.; $J_{dd} = -E\{\delta^2 L/\delta \sigma_e^2\}$

For i=1 the inversion of the information matrix is

$$J^{-1} = \begin{bmatrix} \frac{J_{cc}}{Q} & -\frac{J_{ab}J_{cc}}{QJ_{bb}} & -\frac{J_{ac}}{Q} & 0 \\ \frac{J_{ab}J_{cc}}{QJ_{bb}} & \frac{1}{J_{bb}} & 0 & 0 \\ -\frac{J_{ac}}{Q} & 0 & \frac{J_{aa}J_{bb}-J_{ab}^2}{QJ_{bb}} & 0 \\ 0 & 0 & 0 & \frac{2\sigma_e^2}{T_M \omega_\ell}\left(1-\frac{J_{cc}J_{ab}^2}{QJ_{bb}}\right) \end{bmatrix} \quad (9)$$

where $Q = J_{aa}J_{cc}-J_{ac}^2$. Notice that

$$\sigma_{\Delta b}^2 = 1/J_{bb} \text{ and } \sigma_{\Delta a}^2 \approx 1/J_{aa} \text{ if } J_{ac}^2/J_{cc} << J_{aa}. \quad (10)$$

3. THE EFFECT OF USING MODELS WITH FEW PARAMETERS

As an example the process model

$$G_1(s) = \frac{B(s)}{A(s)} = \frac{b}{(a_1+s)(a_2+s)\dots(a_m+s)} \quad (11)$$

$$C(s) = a_2(a_1+s)(a_3+s)\dots(a_m+s)$$

is taken. It is assumed that in certain ranges of a_i's, $G_1(s)$ can be approximated by

$$G_2(s) = \frac{b}{(a+s)^m} \approx G_1(s) \quad (12)$$

(e.g., for m=5 with $a_1/a_m \leq 1/5$ the above approximation in a closed loop with PID-controller gives a relative r.m.s.error of $y(t) \leq 10\%$, Isermann (1971,b)).

If all the parameters $a_2, a_3, \dots a_m$ are assumed to be known exactly and only a_1 has to be estimated, J_{a1a1} is independent of all the other members $J_{a2a2}, \dots, J_{amam}$. Then one obtains from (5), (9), (10)

$$\sigma_{\Delta a1}^2 \approx J_{a1a1}^{-1} = -1/E\{\delta^2 L/\delta a_1^2\}. \quad (13)$$

Therefore one needs (7). For $G_1(s)$ given by (11),

$$P_{11} = \frac{B(i\omega)}{C(i\omega)A(i\omega)} \frac{\delta A(i\omega)}{\delta a_1} = \frac{b}{a_2(a_1+s)^2(a_3+s)\dots(a_m+s)} \approx \frac{b}{a_2(a+s)^m} \quad (14)$$

$$P_{12} = \frac{1}{A(i\omega)} \frac{\delta A(i\omega)}{\delta a_1} = \frac{1}{a_1+s} \approx \frac{1}{a+s} \quad (15)$$

and for $G_2(s)$ given by (12),

$$P_{21} = \frac{B(i\omega)}{C(i\omega)A(i\omega)} \frac{\delta A(i\omega)}{\delta a} = m\frac{b}{a(a+s)^m} \quad (16)$$

$$P_{22} = \frac{1}{A(i\omega)} \frac{\delta A(i\omega)}{\delta a} = m\frac{1}{a+s} \quad (17)$$

Introducing (14)-(17) in (7) and assuming $a_1 \approx a$, and $a_2 \approx a$, it follows that

$$\sigma_{\Delta a}^2/\sigma_{\Delta a_1}^2 = E\{\delta^2 L/\delta a_1^2\}/E\{\delta^2 L/\delta a^2\} \approx 1/m^2 \quad (18)$$

If $a_2, a_3, \dots, a_n$ are not assumed known exactly, $\sigma_{\Delta a_1}^2$ will increase so that

$$\sigma_{\Delta a}/\sigma_{\Delta a_1} > 1/m. \quad (19)$$

The same holds for the b_i parameters. This means that the lower bound of a parameter's standard deviation can be reduced by >1/m if m equal parameters are assumed. Therefore the measuring time can be reduced by $>1/m^2$. These results were obtained by assuming that the original parameters, which are replaced by equal parameters, are not too different.

Also, the variance of the output variable, caused by variances of the estimated parameters, decreases. For small parameter variations with Y(s) = G(s)U(s) one gets, b=1,

$$\Delta Y_1(s) = -(a+s)^{-(m+1)} U(s) \sum_{i=1}^{m} \Delta a_i \text{ for } G_1(s) \quad (20)$$

$$\Delta Y_2(s) = -m(a+s)^{-(m+1)} U(s)\ \Delta a \text{ for } G_2(s) \quad (21)$$

If only a_1 has to be estimated and $a_2, a_3, \dots a_m$ are known exactly, (20), (21) and (18) lead to

$$E\{\Delta Y_2^2(s)\}/E\{\Delta Y_1^2(s)\} = m^2 E\{\Delta a^2\}/E\{\Delta a_1^2\} = 1. \quad (22)$$

If all parameters $a_1, a_2, \dots, a_m$ have to be estimated, then

$$E\{\Delta Y_2^2(s)\}/E\{\Delta Y_1^2(s)\} = m^2 E\{\Delta a^2\}/E\{(\sum_{i=1}^{m} \Delta a_i)^2\} < 1. \quad (23)$$

For example, if the Δa_i are assumed to be statistically independent (in general they are not) and $a_1 \approx a_2 \approx \dots \approx a_m$

$$E\{\Delta Y_2^2(s)\}/E\{\Delta Y_1^2(s)\} \approx 1/m. \quad (24)$$

Consequently it is advantageous for the identification of very noisy processes to use models with as many equal parameters as possible. Such models can be obtained by

- Replacement of small time constants in the nominator of a transfer function by a time delay
- Shifting of poles to cancel zeros
- Replacement of remaining different poles by equal poles.

Details are given in Isermann (1971,b). Notice that in all cases the sum of all the time constants has to remain approximately constant.

Example for model simplification:

$$\frac{(1+13s)}{(1+17.5s)(1+15s)(1+10s)(1+2s)(1+1s)}$$

$$\approx \frac{(1+13s)}{(1+17.5s)(1+15s)(1+10s)} e^{-3s}$$

$$\approx \frac{1}{(1+18.5s)(1+11s)} e^{-3s} \approx \frac{1}{(1+15.75s)^2} e^{-3s}.$$

But the assumption of such simplified models with few different parameters in general does not lead also to a simplification of the estimation procedure. In the case of the introduction of a time delay, the estimation procedure in general becomes even more complicated, Åström, Eykhoff(1970), van den Bos(1969), especially for on-line identification. For this reason an on-line identification method for estimation of three unknown parameters is regarded briefly which uses Fourier analysis and results in very simple algorithms also in the presence of a time delay. The simplicity of the algorithms is obtained by choosing a special input signal, and by first identifying parts of a nonparametric model and then using simple relationships to determine the parameters.

4. IDENTIFICATION OF A MODEL WITH THREE UNKNOWN PARAMETERS USING FOURIER ANALYSIS

It is assumed that a linearizable continuous process with continuous input $u(t)$ and output $y(t)$, fig.1., can be described by a

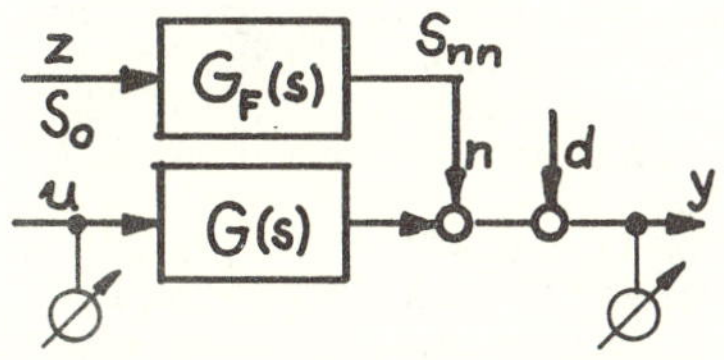

Fig. 1. Dynamic continuous process. u, input; y, measurable output; z, white noise with density S_o; n, noise; d, drift.

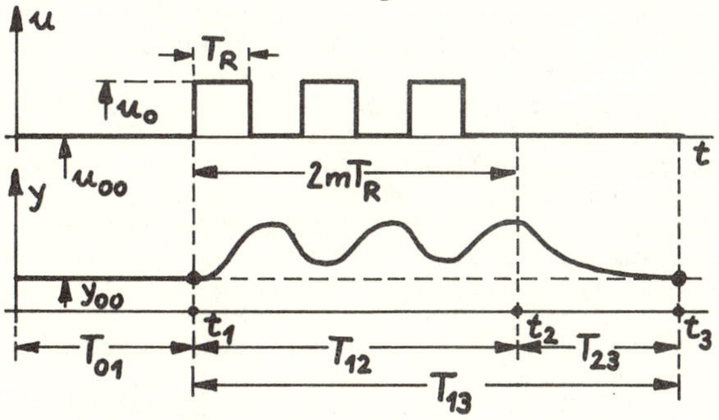

Fig. 2. Input and output variables.

model with three unknown parameters. For example the model could be

$$G(s) = \frac{Y(s)}{U(s)} = \frac{K}{(1+Ts)^n} e^{-T_D s}, \qquad (25)$$

where n is known, and K, T and T_D have to be estimated or

$$G(s) = \frac{Y(s)}{U(s)} = \frac{K}{T_2^2 s^2 + T_1 s + 1} \qquad (26)$$

where K, T_1 and T_2 have to be estimated.

In the following the model eq.(25), a simplified model for lowpass processes, is used for the description of the identification procedure. A sequence of m rectangular pulses is chosen as testsignal, fig.2. By taking appropriate pulse width T_R this sequence gives strong excitation in the middle frequency region, for which the identified model has to be most accurate if the model is used for calculating the closed loop behavior, Isermann (1971,c). Taking rectangular pulses also an effective estimation of the gain K is possible.

4.1. Evaluation method for a low pass process

First two values of a nonparametric model, the frequency response

$$G(i\omega) = \frac{Y(i\omega)}{U(i\omega)} = \frac{\int_{t_1}^{t_3} [y(t)-y_{oo}] \exp(-i\omega t)\,dt}{\int_{t_1}^{t_3} [u(t)-u_{oo}] \exp(-i\omega t)\,dt} \qquad (27)$$

are determined for $\omega=0$ and $\omega_\nu=\pi/T_R$. The reference value u_{oo} is always known, but the reference value y_{oo} in general has to be identified before applying eq.(27) by averaging

$$y_{oo} = \frac{1}{T_{01}} \int_0^{t_1} y(t)\,dt. \qquad (28)$$

The gain K is easily obtained by

$$K = G(0) = \int_{t_1}^{t_3} [y(t)-y_{oo}]\,dt \Big/ \int_{t_1}^{t_3} [u(t)-u_{oo}]\,dt. \qquad (29)$$

Using rectangular pulses one can get relatively good estimates of the gain, especially for large pulse width $T_R \geq T_s$, T_s settling time, as was shown in, Isermann (1972).

The time constant T can be obtained from

$$T = \frac{1}{\omega_\nu} \left[\left[\frac{K}{|G(i\omega_\nu)|} \right]^{\frac{2}{n}} - 1 \right]^{\frac{1}{2}} \qquad (30)$$

and the delay time T_D from

$$T_D = \frac{1}{\omega_\nu} \left[\varphi(\omega_\nu) - n \text{ arc tg} \omega_\nu T \right] \qquad (31)$$

which both follow from eq.(25). $\varphi = \arg G(i\omega_\nu)$.

The pulse width T_R of the input signal is chosen as to minimize the variance of T which is caused by the noise n(t).

For low pass processes this minimum in general is relatively flat and is located within $0.6 \geq |G(i\omega)| \geq 0.2$, so that $\omega_\nu = \omega_o = \pi/T_R$ can be selected from $|G(i\omega_\nu)| \approx 0.35$.

The information about the process dynamics with this method is obtained by evaluation of the response function for one strong excited medium frequency and for zero frequency. For obtaining very simple algorithms not all frequencies are taken for evaluation and therefore it cannot be expected that the parameter estimates are efficient.

4.2. Selection of evaluation time T_{13} of the pulse response

An error Δy_{oo} in estimating the reference value y_{oo}, eq.(28), causes an error

$$\Delta Y(i\omega) = \mathcal{F}\{\Delta y_{oo}\} = -\int_0^{T_{13}} \Delta y_{oo}\exp(-i\omega t')dt'$$

$$= -\int_0^{\infty} \Delta y_{ooR}(t')\exp(-i\omega t')dt'$$

$$= -\Delta y_{oo}T_{13}\left[\frac{\sin\frac{\omega T_{13}}{2}}{\frac{\omega T_{13}}{2}}\right]e^{-i\frac{\omega T_{13}}{2}} \qquad (32)$$

with $t' = t - t_1$ and $\Delta y_{ooR}(t')$ as a rectangular pulse with pulse width T_{13}. An error Δy_{oo} has no influence on $Y(i\omega)$ if $\sin(\omega T_{13}/2) = 0$. This is fulfilled for

$$T_{13} = 2\pi\alpha/\omega \qquad \alpha = 1,2,3,\ldots \qquad (33)$$

With $\omega_o = \pi/T_R$ this leads to

$$T_{13} = 2\alpha_o T_R \qquad \alpha_o = 1,2,3,\ldots \qquad (34)$$

Selecting the evaluation time T_{13} due to eq.(34) errors Δy_{oo} do not cause errors in determing $G(i\omega_o)$.

4.3. Influence of noise n(t)

If stochastic noise $n(t)$ with $E\{n(t)\}=0$ is contaminating the output variable, consistent estimates of all three parameters are obtained. For the process

$$G(s) = K\,e^{-T_D s}/(1+Ts)^n; \quad G_F(s) = 1/(1+T_F s) \qquad (35)$$

$$\overline{n^2(t)} = S_o/2T_F\,; \; \eta = \sqrt{\overline{n^2(t)}}\,/u_o \qquad (36)$$

the standard deviations have been calculated similar to Isermann (1971a and 1972):

$$\sigma_{\delta K} = \sqrt{E\left\{\left(\frac{\Delta K}{K}\right)^2\right\}} = \frac{\sqrt{2T_F T_{13}}}{KmT_R}\sqrt{1+\frac{T_{13}}{T_{o1}}}\cdot\eta \qquad (37)$$

$$\sigma_{\delta T}(\omega_o) = \sqrt{E\left\{\left(\frac{\Delta T}{T}\right)^2\right\}}$$

$$\leq \frac{\pi\,(1+\omega_o^2T^2)\,\sqrt{T_F T_{13}}}{|G(i\omega_o)|\,\sqrt{2}\,n\,\omega_o^2T^2\sqrt{1+\omega_o^2T_F^2}\;mT_R}\,\eta \qquad (38)$$

$$\sigma_{\delta TD}(\omega_o) = \sqrt{E\left\{\left(\frac{\Delta T_D}{T}\right)^2\right\}} \leq \frac{n}{1+\omega_o^2T^2}\sigma_{\delta T}(\omega_o). \qquad (39)$$

4.4. Comparison with the Cramér-Rao lower variance bound

For model (35) with $n=4, T_D=0$ and $b=K/T^4$; $a=1/T$; $c=a$ eq.(1) becomes

$$(a+s)^4y(s) = bu(s) + c(c+s)^3e(s). \qquad (40)$$

A PRBS input is used with clock interval λ. Then eq.(7) and (8) lead to

$$\left.\begin{aligned}
J_{aa} &= \frac{T_M}{T}\left[\frac{15\pi}{48}\frac{K\lambda T^2}{T_F\eta^2} + 8\pi\,T^2\right]; \\
J_{bb} &= \frac{T_M}{T}\frac{3\pi}{128}\frac{\lambda T^8}{T_F\eta^2}\,; J_{cc} = \frac{T_M}{T}\left[\frac{15\pi}{2}T^2 + \omega_\ell T^3\right] \\
J_{ab} &= -\frac{T_M}{T}\frac{15\pi}{198}\frac{\lambda T^5}{T_F\eta^2}\,; J_{ac} = -8\pi T^2\left(\frac{T_M}{T}\right); \\
J_{dd} &= \left[\pi/\overline{n^2(t)}\right]\cdot\left[T_M/T_F\right].
\end{aligned}\right\} \qquad (41)$$

The variances $\sigma^2_{\delta a}$, $\sigma^2_{\delta b}$ and $\sigma^2_{\delta c}$ can be calculated with eq.(9). Then one obtains

$$\sigma^2_{\delta T} = \sigma^2_{\delta a}\,; \quad \sigma^2_{\delta K} = \sigma^2_{\delta b} + 8\sigma^2_{\delta ba} + 4\sigma^2_{\delta a} \qquad (42)$$

Table 1 shows the results. [1)]

Table 1. Comparison of the parameter variances for the example: $G(s)=K/(1+Ts)^4$; $K=1$; $T=T_F=100s$; $\eta=0.25$; $T_M=140T$; $T_{13}/T_{o1}=2$; $m=10$ pulses.

Identific. method	Testsign. parameter	$\sigma_{\delta K}$	$\sigma_{\delta T}$	$\sigma_{\delta TD}$
Three param. identific.	$T_R=3.93T$	0.147	≤ 0.195	≤ 0.475
Cramér-Rao lower bound with PRBS	$\lambda_1=0.7T$	0.132	0.018	-
	$\lambda_2=3.0T$	0.070	0.010	-

References:

ÅSTRÖM,K.J. (1967). On the achievable accuracy in identification problems. IFAC-Symposium on Identification,Prague.

ÅSTRÖM,K.J.and Eykhoff,P. (1971). System identification - a survey. IFAC-Automatica, pp. 123.

VAN DEN BOS,A. (1970). Estimation of linear system coefficients from noisy responses to binary multifrequency test signals. Prepr. IFAC-Symposium on Identification, Prague.

DEUTSCH,R. (1965). Estimation theory. Prent. Hall, Englewood Cliffs.

GUPTA,S.C. (1965). Performance of process models. JACC, Session X, Paper 5.

ISERMANN,R. (1971,a). Experimentelle Analyse der Dynamik von Regelsystemen -Identifikat. I. University textbook Nr.515, Bibliographisches Institut Mannheim,Germany.

ISERMANN,R.(1971,b). Theoretische Analyse der Dynamik industrieller Prozesse-Identifikation II. University textbook Nr.764. Bibliographisches Institut Mannheim,Germany.

ISERMANN,R. (1971,c).Required accuracy of linear time invariant mathematical models of controlled elements. IFAC-Automatica, Vol.7 ,p.333.

ISERMANN,R. (1972). Identification of the static behavior of very noisy dynamic processes. Preprints IEEE-Conference on Cybernetics and Society, Washington D.C.

1) Also for another example this method gives good results. See: Isermann,R. et al.: Evaluation and comparison for six on-line identification methods.... IFAC-Symposium on Identification 1973.

SIMULTANEOUS IDENTIFICATION OF THE DYNAMIC SYSTEM AND RESTORATION OF AN INPUT SIGNAL

Celestinas Paulauskas

Institute of Physics and Mathematics
Academy of Sciences of the Lithuanian SSR
Vilnius, USSR

The approach to identification of a linear dynamic system and restoration of its input signal is considered in the case when only an output signal is observed and the input signal is an unobserved sequence of unknown pulses of an arbitrary form and final duration with a nonzero interval between the pulses.

In the problem of system identification (Rajbman(1970), Åström (1971)) as a rule it is assumed that realizations of input and output signals are available. In some problems (determination of parameters of the vocal tract (Flanagan (1965)), determination of parameters of the mechanical system with rotating parts (Pavlov (1971)) etc.) input signals are quite unobservable. The problem without essential assumptions on an input signal and the system makes no sense at all. Let us assume that the system is linear, the form of its operator with unknown parameters is known; a nonobserved input signal is a sequence of unknown pulses of an arbitrary form and final duration with a nonzero interval between the pulses.

The model of the system is

$$y_n = \sum_{i=0}^{n} h_{n-i} x_i + \varepsilon_n, \; n = 0,1,\dots,K, \qquad (1)$$

where $x_n, y_n, h_n, \varepsilon_n$, $n = 0,1,\dots,K$, are input, output, weighting function of the system, and observing noise, respectively. z-transformation for (1) (Kuzin (1962), Burrus (1970)) is

$$Y(z) = \frac{C(z)X(z) + B(z)}{A(z)} + \mathcal{E}(z) \qquad (2)$$

where $C(z) = \sum_{j=0}^{\ell} c_j z^{-j}$, $B(z) = \sum_{j=0}^{p-1} b_j z^{-j}$, $A(z) = \sum_{j=0}^{p} a_j z^{-j}$. $B(z)$ takes into account nonzero initial states. When (2) is considered in the time domain we have a matrix equation

$$\begin{bmatrix} b \\ 0 \end{bmatrix} + \begin{bmatrix} e^\circ \\ e \end{bmatrix} = \begin{bmatrix} Y^\circ & X^\circ \\ Y & X \end{bmatrix} \begin{bmatrix} a^\circ \\ c \end{bmatrix} \qquad (3)$$

where $b = [b_0, b_1, \dots, b_{p-1}]'$, $a^\circ = [a_0, a_1, \dots, a_p]'$, $c = [c_0, c_1, \dots, c_\ell]'$, $\bar{e} = [e^\circ \mid e]'$ - $(K+1)$-vector of errors; Y°, Y - $(p \times p)$-matrix and $(p \times (K-p))$-matrix composed from y_n, $n = 0,1,\dots,K$, respectively; X°, X - $(\ell \times p)$-matrix and $(\ell \times (K-p))$-matrix composed from x_n, $n = 0,1,\dots,K$, respectively.

Solution of (3) is impossible in the general case. Under the assumptions made earlier at the interval between the pulses $\bar{X} = [X^\circ \mid X]' \equiv 0$. When (3) is composed from the values of the output sequence at that interval we have

$$\begin{bmatrix} b \\ 0 \end{bmatrix} + \begin{bmatrix} e^\circ \\ e \end{bmatrix} = \begin{bmatrix} Y^\circ \\ Y \end{bmatrix} [a^\circ]. \qquad (4)$$

Let $\bar{e} = [e^\circ \mid e]' \equiv 0$. Then the value of vector of parameters a can be obtained from the lower part of (4), i.e.

$$Y a^\circ = 0, \quad a_0 = 1 \qquad (5)$$

where Y - $((p+1) \times p)$-matrix, a_0 - component of the vector a°. The initial state vector b is obtainable from

$$b = Y^\circ a^\circ. \qquad (6)$$

When $\bar{e} \neq 0$ the estimate $\hat{a}$ for the vector a can be obtained by the least-square method (Linnik (1958)). As far as according to (3)

$$e = Y a^\circ \qquad (7)$$

or

$$e = \tilde{y} + \tilde{Y} a \qquad (8)$$

where $a^\circ = [1 \mid a]'$, $\tilde{y} = [y_p, y_{p+1}, \dots, y_K]'$, $Y = [\tilde{y} \mid \tilde{Y}]$ then

$$\hat{a} = -[\tilde{Y}'\tilde{Y}]^{-1} \tilde{Y}' \tilde{y}. \qquad (9)$$

In order to obtain $\hat{a}$ from (9) it is worthy to apply a recurrent algorithm using the pseudoinversion (Lee (1964)), Sinha (1970)).

For the purpose of obtaining the estimate $\hat{b}$ for the vector b equation (4) is transformed like this:

$$\begin{bmatrix} b \\ 0 \end{bmatrix} + \bar{e} = A \bar{y} \qquad (10)$$

where $\bar{y} = [y^\circ \mid \tilde{y}]' = [y_0, y_1, \dots, y_p, \dots, y_K]'$, A - $((K+1) \times (K+1))$ -matrix which consist of $a_0, a_1, \dots, a_p$. Substituting $\bar{e} = A\bar{\varepsilon}$ instead of (10) we have

$$\bar{\varepsilon} = \bar{y} - A^{-1} [b \mid 0]'. \qquad (11)$$

Denoting $A^{-1} = M$ (11) can be expressed by

$$\bar{\varepsilon} = \bar{y} - [M_1 \vdots M_0][\beta \vdots 0]' = \bar{y} - M_1 \beta \tag{12}$$

hence

$$\hat{\beta} = [M_1' M_1]^{-1} M_1' \bar{y} \tag{13}$$

Irreducibility of fraction (Kapitonenko (1972))

$$V(z) = \frac{B(z)}{A(z)} \tag{14}$$

is the necessary and sufficient condition under which it is possible to obtain the estimate for the vector a by applying a single trajectory y_n, $n = 0, 1, \ldots$ caused by nonzero initial states. If this is the case its resultant will be nonzero (Mishina (1962)):

$$R(A, B) \neq 0. \tag{15}$$

The value of $R(A, B)$ depends on β and a. Condition (15) must exist when a is arbitrary. Condition (15) is guaranteed when

$$\beta = [0, 0, \ldots, 0, \beta_{p-1}]', \ \beta_{p-1} \neq 0. \tag{16}$$

(16) is equivalent to

$$y_{p-1} \neq 0 \tag{17}$$

The interval where $\bar{X} \equiv 0$ can be found by the minimum of Gram's determinant (Sobakin (1972)) which can be obtained from Y (Linnik (1958)):

$$G = \det [Y'Y]. \tag{18}$$

In the general case $G \geqslant 0$. When the rows of the matrix Y are linear dependent $G = 0$. It occurs when Y is formed from y_n, $n = 0, 1, \ldots$, which correspond to the interval where $\bar{X} \equiv 0$. In the general case the Gram's determinant has a nonnegative minimal value G_{min} at this interval.

z - transformation of the weighting function h_n, $n = 0, 1, \ldots, K$ in the general case is

$$H(z) = \frac{C(z)}{A(z)}. \tag{19}$$

Since only $\hat{a}$ in $A(z)$ is known and C in $C(z)$ is unknown in the general case, it is impossible to get the estimate $\hat{H}(z)$ and as well as the estimate $\hat{h}_n$, $n = 0, 1, \ldots, K$. It is possible in a particular, important for practice case when $C(z) \equiv 1$. Then

$$\hat{h}_n = -\sum_{j=1}^{p} \hat{a}_j \hat{h}_{n-j} + \begin{cases} 1, & n = 0, \\ 0, & n \geqslant 1 \end{cases} \tag{20}$$

what corresponds to the system described by

$$y_n + \sum_{j=1}^{p} a_j y_{n-j} = x_n, \ n = 0, 1, \ldots \tag{21}$$

Having (20) or (21) we can state a number of independent problems (determination of the inverse model of the system, restoration of the input sequence etc.).

Under practical testing of the abovementioned method it was found that the increasing of the method accuracy for systems of the higher order (3 and more) is possible by using the method to the one- two-difference of the output sequence. By it removes the filtering influence of input pulses to the system's weighting function. At this of course it is necessary a compromisse among the errors conditioned both by nonuniform intensivity of spectral components of input sequence and by observing noise.

The method is illustrated by a simple example. In fig. 1 fragments of normalized realizations of the output sequence y_n, $n = 0, 1, \ldots, 175$ and of its one-difference Δy_n, $n = 0, 1, \ldots, 175$ are presented. The sequence of isosceles triangle pulses is used as an input signal x_n, $n = 0, 1, \ldots, 175$. The system corresponded to the three-formant vocal tract (Flanagan (1965)) is described by 6th order recurrent equation with the parameters $a_1 = -4.5589628$, $a_2 = 9.3399734$, $a_3 = -11.417961$, $a_4 = 8.9499893$, $a_5 = -4.2021560$, $a_6 = 0.89064788$. Values of Gram's determinant G_n and values of parameters $a_1, a_2, \ldots, a_6$ are obtained at the intervals contained 42 values of the Δy_n, $n = 0, 1, \ldots, 175$ and moved along the time axis with one step. They are plotted in fig. 1 and 2. There are obtained $\hat{a}_1 = -4.55822754$, $\hat{a}_2 = 9.33837891$, $\hat{a}_3 = -11.4157714$, $\hat{a}_4 = 8.94653321$, $\hat{a}_5 = -4.19934082$, $\hat{a}_6 = 0.889709473$ corresponded to the $G_{min} = 0.0\ldots 0 \times 10^{-19}$. An input signal restorated from the output signal realization y_n, $n = 0, 1, \ldots, 175$ by mean of the inverse model with the obtained parameters $\hat{a}_1, \hat{a}_2, \ldots, \hat{a}_6$ is identical to the actual signal.

Also satisfactory results are obtained at the solution of the certain testing examples and the real vocal tract identification problems in the case when the systems are described by equations to 10th order.

References

1. Astrom K.J., Eykhoff P. (1971), System identification - a survey. Automatica, 7, No 2, p. 123.
2. Burrus Ch.S., Parks Th.W. (1970), Time Domain Design of Recursive Digital Filters. IEEE Trans. Audio and Electroacoustics, 18, No 2, p. 137.
3. Flanagan J.L. (1965). Speech Analysis, Synthesis and Perception. Springer-Verlag, Berlin a.o.
4. Капитоненко В.В., Саввин А.Б. (1972), Восстанавливающие эксперименты в задачах идентификации, Автоматика и телемеханика, № 1, стр. 155.
5. Кузин Л.Т. (1962). Расчет и проектирование дискретных систем управления. Машиностроение, Москва.

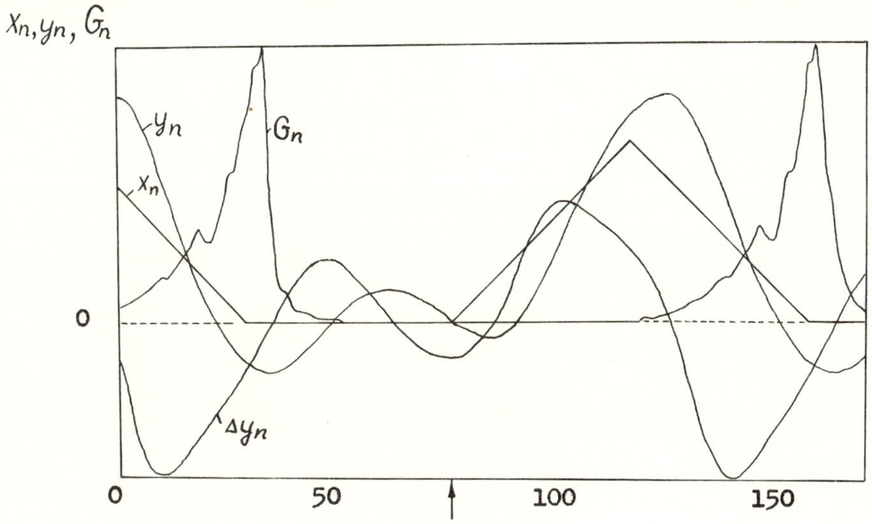

Fig. 1

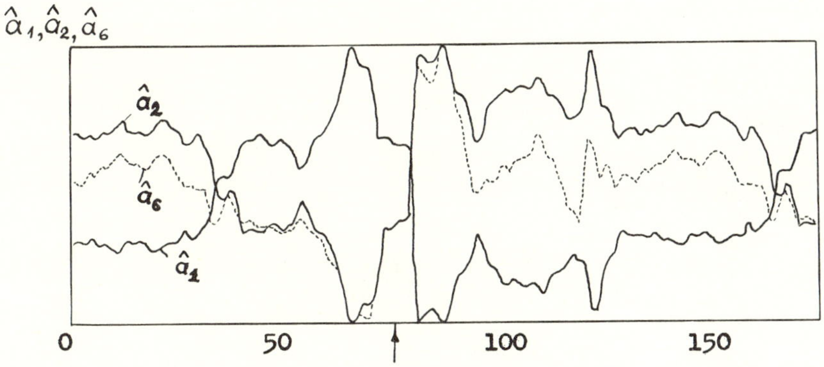

Fig. 2

6. Lee R.C.K. (1964). Optimal Estimation, Identification, and Control. The MIT Press, Cambridge. Mass.

7. Линник Ю.В. (1958). Метод наименьших квадратов и основы обработки наблюдений Физматгиз, Москва.

8. Мишина А.П., Проскуряков И.В.(1962). Высшая алгебра. Физматгиз, Москва.

9. Павлов Б.В. (1971). Акустическая диагностика механизмов. Машиностроение, Москва.

10. Sihha N.K., Pille W. (1970), Online system identification using matrix pseudoinverse. Electr. Letters, 6, No 15, 453.

11. Райбман Н.С. (1970). Что такое идентификация. Наука, Москва.

12. Собакин А.Н. (1972), Об определении формантных параметров голосового тракта по речевому сигналу с помощью ЦВМ. Акустический журнал, 18, № 1, стр. 106.

AN IDENTIFICATION TECHNIQUE FOR MULTI-INPUT MULTI-OUTPUT SYSTEMS

A.Y.Bilal and M.F.Hassan
Electronic and Communication Department
Faculty of Engineering, Cairo University
Giza-Egypt.

A method for real time identification of multi-input multi-output linear time invariant systems is presented. Weighting sequence models are experimentaly determined from input-output records using delayed data information. Both noise free and noise corrupted signals are considered, and expressions for the variances of the estimators are derived. The proposed computational algorithms have the property that errors usually present in numerical deconvolution techniques are greatly reduced.

1-INTRODUCTION

Identification of multidimensional systems is developed in terms of impulse response functions. First, two schemes, namely, a "one column" and an "all column" identification are developed and discussed. Both yield very accurate results. The "one column" identification can be considered as an extension of the work done by Sage(1971) concerning the problem of point estimation of impulse response functions for single input, single output systems. In the "all column" identification, however, our technique uses a special scheme of delayed data information at the input and specifies certain instances for the proper output measurements. In the second part of the paper the input and output data assumed to be contaminated by noise and estimators for both identification schemes are derived. It is shown that using the "all column" technique a significant reduction in the estimator standard deviation can be obtained.

2-NOISE FREE ANALYSIS

(2-1) All Column Identification:

We are concerned with a model configuration with an accessible set of n-input signals $\{w_1(t), w_2(t), \ldots w_n(t)\}$, m-output signals $\{y_1(t), \ldots y_m(t)\}$ and are required to identify a set of nxm impulse response functions $h_{pk}(t)$, $p=1,\ldots m; k=1,\ldots n$, in the finite time interval $[0,T]$. Assuming linearity and time invariance, the model can be described by:

$$y_p(t)=\sum_{k=1}^{n}\int_0^t h_{pk}(\tau).w_k(t-\tau)\, d\tau \quad \ldots\ldots\ldots(1)$$

As the above problem statement is a bit ambitious, it is necessary to simplify it in order to make it computationaly treatable. For this reason, the impulse response functions will be approximated by their midpoint values in each interval Δ, where $N\Delta=T$, that is:

$$h_{pk}(t)=h_{pk}\left(\frac{2i+1}{2}\Delta\right),\quad i\Delta\leq t<(i+1)\Delta \quad \ldots\ldots\ldots(2)$$

This leads to nxmxN unknowns to be estimated, assuming that the input and output data are experimentaly measured at specified sampling instances. The ordinary numerical deconvolution techniques usually applied to solve one dimensional problems (n=m=1) result in an inaccurate method of identification mainly due to error accumulation, especially if the time of interest is quite large. In order to overcome these difficulties, we will use a special scheme of delayed data test signals. First, we apply an input sequence of functions $\{w_1(t), \ldots, w_k(t), \ldots, w_n(t)\}$ in such a manner that $w_1(t)$ is not delayed, $w_2(t)$ is delayed behined $w_1(t)$ by (Δ/n), $w_k(t)$ is delayed by $(k-1)\Delta/n$ and all the outputs are measured at the set of time instances $\{(\Delta-\Delta/n), \Delta, (2\Delta-\Delta/n), 2\Delta, \ldots, (N\Delta-\Delta/n), \Delta N\}$. Therefore, the input sequence is given as:

$$\{w_1(t), w_2(t-\Delta/n), \ldots, w_k[t-(k-1)\Delta/n], \ldots, w_n[t-(n-1)\Delta/n]\}.$$

where we define:

$$w(t-\tau)=0 \quad \text{for } t<\tau$$
$$\quad\; =w(t) \quad \text{for } t\geq\tau$$

The experiment is to be repeated n-times with the outputs measured at the set of sampling instances described as above, but with an input sequence of functions given by $\{w_1[t-(n-1)\Delta/n], w_2(t), w_3(t-\Delta/n), \ldots, w_n[t-(n-2)\Delta/n]\}$ for the second experiment, and so on. Let:

$y_p^k(j\Delta) \triangleq$ the output $y_p(t)$ measured at the instant $j\Delta$ during the experiment number k, $(k=1,\ldots,n)$ in which $w_k(t)$ is not delayed.

The sequence of output $y_p^k(j\Delta)$ and $y_p^{k+1}(j\Delta-\Delta/n)$ will be used for the identification of $h_{pk}[(2j-1)\Delta/2]$. One can easily show that:

$$y_p^k(j\Delta)-y_p^{k+1}(j\Delta-\Delta/n)-\sum_{i=1}^{j-1}\Delta h_{pk}[(2i-1)\Delta/2]\times\{$$

$$w_k[(2j-1)\Delta/2-i\Delta+\Delta]-w_k[(2j-1)\Delta/2-i\Delta]\}$$
$$=\Delta h_{pk}[(2j-1)\Delta/2].\ w_k(\Delta/2)\ldots\ldots\ldots\ldots(3)$$

If $w_k(t)$ is a unit step function, equation (3) reduces to:

$$y_p^k(j\Delta)-y_p^{k+1}(j\Delta-\frac{\Delta}{n})=\Delta h_{pk}(\frac{2j-1}{2}\Delta)\ldots\ldots(4)$$

where: $j=1,..,N$; $p=1,..,m$ and $k=1,..,n$

The last result shows that using step functions as test signals, the identification of $h_{pk}(t)$ at the instant $t=(2j-1)\Delta/2$ is independent of its previous values; a fact of paramount importance from the point of view of the accuracy of identification.

(2-2) One Column Identification :

In this experiment, in order to identify the one column $h_{pk}(t)$, $p=1,..,m$ only the input $w_k(t)$ is applied while keeping the other inputs zero and measuring the outputs at the sampling instances $\{\Delta, 2\Delta, .., N\Delta\}$, from which we have:

$$y_p(j\Delta)-\Delta\sum_{i=1}^{j-1}h_{pk}(\frac{2i-1}{2}\Delta)w_k(\frac{2j-1}{2}\Delta)$$
$$=h_{pk}(\frac{2j-1}{2}\Delta).w_k(\frac{\Delta}{2})\ldots\ldots\ldots\ldots\ldots\ldots(5)$$

With unit step functions as test signals equation (5) reduces to :

$$y_p(j\Delta)-\Delta\sum_{i=1}^{j-1}h_{pk}(\frac{2i-1}{2}\Delta)=h_{pk}(\frac{2j-1}{2}\Delta).\Delta.(6)$$

In the material that follows we will discuss identification in the presence of noise and will derive computational algorithms for the necessary estimators for both identification schemes.

3- NOISE PRESENT ANALYSIS

As mentioned before, the best conditions of accuracy are achieved with step functions as test signals. Therefore, we will adopt this condition for our noise present situations. In the analysis, both inputs and outputs are assumed to be contaminated with random noise signals which are ergodic, independent and with zero mean. Input noise signals are also assumed to be independent on the test signal sources.

(3-1) All Column Identification :

We define:

$$z_p^k(t)=y_p^k(t)+v_p(t)\quad;\ \text{and}$$
$$q_p^k(j\Delta)=z_p^k(j\Delta)-z_p^{k+1}(j\Delta-\frac{\Delta}{n})$$

Where $v_p(t)$ is the noise at the pth output channel at any instant of time t. Assuming $u_i(t)$ to be the input noise at i^{th} input terminal, it is not difficult to show that :

$$q_p^k(\Delta)=\Delta h_{pk}(\frac{\Delta}{2})+\sum_{i=1}^{n}\left[\Delta u_i(\frac{\Delta}{2})-(\Delta-\frac{\Delta}{n})u_i(\frac{\Delta}{2}-\frac{\Delta}{2n})\right]\times h_{pi}(\frac{\Delta}{2})+v_p(\Delta)-v_p(\Delta-\frac{\Delta}{n})=\Delta\hat{h}_{pk}(\frac{\Delta}{2})=\Delta h_{pk}(\frac{\Delta}{2})+\Delta\epsilon_{pk}(\frac{\Delta}{2})\ldots\ldots\ldots\ldots(7)$$

Where $\hat{h}_{pk}$ denotes the estimated value of h_{pk} with an error ϵ_{pk} given by :

$$\epsilon_{pk}(\frac{\Delta}{2})\triangleq(\frac{1}{\Delta})\left\{\sum_{i=1}^{n}\left[\Delta u_i(\frac{\Delta}{2})-(\Delta-\frac{\Delta}{n})u_i(\frac{\Delta}{2}-\frac{\Delta}{2n})\right]h_{pi}(\frac{\Delta}{2})+v_p(\Delta)-v_p(\Delta-\frac{\Delta}{n})\right\}\ldots\ldots\ldots(8)$$

Similarly :

$$\epsilon_{pk}(\frac{3\Delta}{2})\triangleq(\frac{1}{\Delta})\left\{\sum_{i=1}^{n}\left[\Delta u_i(\frac{3\Delta}{2})-\Delta u_i(\frac{3\Delta}{2}-\frac{\Delta}{n})\right]h_{pi}(\frac{\Delta}{2})+\Delta u_i(\frac{\Delta}{2})-(\Delta-\frac{\Delta}{n})u_i(\frac{\Delta}{2}-\frac{\Delta}{2n})\ h_{pi}(\frac{3\Delta}{2})+v_p(2\Delta)-v_p(2\Delta-\frac{\Delta}{n})\right\}\ldots\ldots(9)$$

In matrix notation we finally have :

$$\mathcal{E}_{pk}=\sum_{i=1}^{n}U_iH_{pi}+(1/\Delta)V_p\ \ldots\ldots\ldots(10)$$

Where :

$$\mathcal{E}_{pk}'=\left[\epsilon_{pk}(\frac{\Delta}{2})\quad\epsilon_{pk}(\frac{3\Delta}{2})\ \ldots\ldots\ \epsilon_{pk}\left\{\frac{(2N-1)}{2}\Delta\right\}\right]$$
$$H_{pi}'=\left[h_{pi}(\frac{\Delta}{2})\quad h_{pi}(\frac{3\Delta}{2})\ \ldots\ldots\ h_{pi}\left\{\frac{(2N-1)}{2}\Delta\right\}\right]$$
$$V_p'=\left[\left\{v_p(\Delta)-v_p(\Delta-\frac{\Delta}{n})\right\}\ \left\{v_p(2\Delta)-v_p(2\Delta-\frac{\Delta}{n})\right\}\ \ldots\ldots\ \left\{v_p(N\Delta)-v_p(N\Delta-\frac{\Delta}{n})\right\}\right]$$

$U_i=\left[u_{rs_i}\right]$ where

$$u_{rs}=0\quad\text{for } s>r$$
$$=u_i(\frac{\Delta}{2})-\frac{n-1}{n}u_i(\frac{\Delta}{2}-\frac{\Delta}{2n})\quad\text{for } s=r$$
$$=u_i(\frac{2r-1-2s+2}{2}\Delta)-u_i(\frac{2r-1-2s+2}{2}\Delta-\frac{\Delta}{n})\quad\text{for } s<r$$

and

X' denotes the transpose of the matix X

Equation(10) represents the error propagation matrix in the identification of the impulse response eliments $h_{pk}(t)$ which have zero expected value since :

$$E\{\mathcal{E}_{pk}\}=\sum_{i=1}^{n}E\{U_i\}H_{pi}+(\frac{1}{\Delta})\ E\{V_p\}=0\ldots\ldots(11)$$

Where E( ) denotes the expected value of( ). Now from equation (7) we get :

$$Q_p^k=H_{pk}+\mathcal{E}_{pk}=\hat{H}_{pk}\ \ldots\ldots\ldots\ldots\ldots(12)$$

where :

$$Q_p^{k'} = \left[q_p^k(\Delta) \quad q_p^k(2\Delta) \quad \ldots\ldots \quad q_p^k(N\Delta)\right]$$

Expectation of equation (12) gives :

$$E\{Q_p^k\} = E\{H_{pk} + \mathcal{E}_{pk}\} = H_{pk}$$

Therefore :

$$H_{pk} = \lim_{M\to\infty} \frac{1}{M}\sum_{j=1}^{M} Q_{p_j}^k = \lim_{M\to\infty} \hat{H}_{pk_M} \quad \ldots\ldots\ldots(13)$$

Where $\hat{H}_{pk_M}$ is the estimated value of H_{pk} obtained by repeating the identification experiment M times.

The variance of the estimator can be obtained as follows :

$$\sigma^2_{\hat{H}_{pk_M}} = E\{(H_{pk_M}-H_{pk})^2\} = E\{(H_{pk_M}-H_{pk})'(H_{pk_M}-H_{pk})\} = (1/M^2)E\Big\{\sum_{j=1}^{M}\mathcal{E}'_{pk_j}\mathcal{E}_{pk_j} + \sum_{i\neq j}\sum \mathcal{E}'_{pk_j}\,\mathcal{E}_{pk_i}\Big\}$$

In the average,the cotribution of the second term is zero,therefore :

$$\sigma^2_{\hat{H}_{pk_M}} = (1/M^2)\sum_{j=1}^{M} E\{\mathcal{E}'_{pk_j}\,\mathcal{E}_{pk_j}\} \quad \ldots\ldots\ldots(14)$$

Using matrix equation (10) we have :

$$\mathcal{E}'_{pk_j}\,\mathcal{E}_{pk_j} = \sum_{i=1}^{n} H'_{pi_j}U'_{i_j}U_{i_j}H_{pi_j} + (1/\Delta^2)V'_{p_j}V_{p_j}$$

+terms that will vanish in the average.

One can easily show that :

$$\sigma^2_{\hat{H}_{pk_M}} = \left(\frac{1}{M}\right)\Big\{\sum_{i=1}^{n} H'_{pi}\Phi_{u_i}H_{pi} + (2/\Delta^2)\left[\sigma^2_{v_p} - \phi_{v_p}\left(\frac{\Delta}{n}\right)\right]\Big\} \quad \ldots\ldots\ldots(15)$$

Where σ_x and $\phi_x(\tau)$ are the standard deviation and auto-correlation of the noise x respectively, and :

$$\Phi_{u_i} = [\gamma_{rs}] \quad ; \text{ where } \gamma_{rs} = \gamma_{sr}$$

Where for r=s

$$\gamma_{rs} = \gamma_{sr} = \left[\frac{2N-1-2r+2}{N} + \frac{(n-1)^2}{n^2.N}\right]\sigma^2_{u_i} - \left(\frac{2}{N}\right)\left[(N-r).\phi_{u_i}\left(\frac{\Delta}{n}\right) + \frac{n-1}{n}\phi_{u_i}\left(\frac{\Delta}{2n}\right)\right]$$

for r<s

$$\gamma_{rs} = \left(\frac{n-1}{nN}\right)\Big\{\phi_{u_i}\left[(s-r)\Delta - \frac{\Delta}{2n}\right] - \phi_{u_i}\left[(s-r)\Delta + \frac{\Delta}{2n}\right]\Big\} + \left(\frac{2N-1-2s+2}{N}\right)\phi_{u_i}[\Delta(s-r)] - \left(\frac{N-s+1}{N}\right)\phi_{u_i}\left[(s-r)\Delta - \frac{\Delta}{n}\right] - \left(\frac{N-s}{N}\right)\phi_{u_i}\left[(s-r)\Delta + \frac{\Delta}{n}\right] \quad \ldots\ldots\ldots(16)$$

(3-2) One Column Identification:

Following the previous notation for the output and input noise we define:

$$z_p(t) = y_p(t) + v_p(t) \quad \ldots\ldots\ldots(17)$$

$z_{pk}(t) \triangleq$ the output $z_p(t)$ when the input $w_k(t)$ is applied,the other inputs being zero.

Therefore we have:

$$z_{pk}(\Delta) = \Delta h_{pk}\left(\frac{\Delta}{2}\right) + \Delta\epsilon_{pk}\left(\frac{\Delta}{2}\right) = \Delta\hat{h}_{pk}\left(\frac{\Delta}{2}\right) \quad \ldots\ldots(18)$$

where,the error $\epsilon_{pk}(\frac{\Delta}{2})$ is given by :

$$\epsilon_{pk}\left(\frac{\Delta}{2}\right) = \sum_{i=1}^{n} h_{pi}\left(\frac{\Delta}{2}\right)u_i\left(\frac{\Delta}{2}\right) + \frac{1}{\Delta}v_p(\Delta) \quad \ldots\ldots(19)$$

Also :

$$z_{pk}(2\Delta) = \Delta\hat{h}_{pk}\left(\frac{\Delta}{2}\right) + \Delta\hat{h}_{pk}\left(\frac{3\Delta}{2}\right) = \Delta h_{pk}\left(\frac{\Delta}{2}\right) + \Delta\epsilon_{pk}\left(\frac{\Delta}{2}\right) + \Delta h_{pk}\left(\frac{3\Delta}{2}\right) + \Delta\epsilon_{pk}\left(\frac{3\Delta}{2}\right)$$

where:

$$\epsilon_{pk}\left(\frac{\Delta}{2}\right) + \epsilon_{pk}\left(\frac{3\Delta}{2}\right) = \sum_{i=1}^{n}\Big\{h_{pi}\left(\frac{\Delta}{2}\right)u_i\left(\frac{3\Delta}{2}\right) + h_{pi}\left(\frac{3\Delta}{2}\right) \times u_i\left(\frac{\Delta}{2}\right)\Big\} + (1/\Delta)v_p(2\Delta) \quad \ldots\ldots(20)$$

From equation (19) and (20) we have:

$$\epsilon_{pk}\left(\frac{3\Delta}{2}\right) = \sum_{i=1}^{n}\Big\{h_{pi}\left(\frac{\Delta}{2}\right)\left[u_i\left(\frac{3\Delta}{2}\right) - u_i\left(\frac{\Delta}{2}\right)\right] + h_{pi}\left(\frac{3\Delta}{2}\right) \times u_i\left(\frac{\Delta}{2}\right)\Big\} + \left(\frac{1}{\Delta}\right)\left[v_p(2\Delta) - v_p(\Delta)\right] \quad \ldots\ldots(21)$$

Similarly,as followed before,we get:

$$\mathcal{E}_{pk} = \sum_{i=1}^{n} U_iH_{pi} + (1/\Delta)V_p \quad \ldots\ldots(22)$$

where:

$$\mathcal{E}'_{pk} = \left[\epsilon_{pk}\left(\frac{\Delta}{2}\right) \quad \epsilon_{pk}\left(\frac{3\Delta}{2}\right) \quad \ldots\ldots \quad \epsilon_{pk}\left(\frac{2N-1}{2}\Delta\right)\right]$$

$$H'_{pi} = \left[h_{pi}\left(\frac{\Delta}{2}\right) \quad h_{pi}\left(\frac{3\Delta}{2}\right) \quad \ldots\ldots \quad h_{pi}\left(\frac{2N-1}{2}\Delta\right)\right]$$

$$V'_p = \left[\{v_p(\Delta)\} \quad \{v_p(2\Delta) - v_p(\Delta)\} \quad \ldots\ldots \quad \{v_p(N\Delta) - v_p(N\Delta - \Delta)\}\right]$$

$U_i = [u_{rs_i}]$;where $u_{rs_i} = 0$ for $s > r$

$u_{rs_i} = u_i(\frac{\Delta}{2})$ for $r=s$ and for $s<r$

$$u_{rs_i} = u_i\left(\frac{2r-1-2s+2}{2}\Delta\right) - u_i\left(\frac{2r-1-2s}{2}\Delta\right)$$

Equation (22) gives :

$$E\{\mathcal{E}_{pk}\} = \sum_{i=1}^{n} E\{U_i\}H_{pi} + \left(\frac{1}{\Delta}\right)E\{V_p\} = 0 \quad \ldots\ldots(23)$$

Let :

$$Z'_{pk} = \left[z_{pk}(\Delta) \quad z_{pk}(2\Delta) \quad \ldots\ldots \quad z_{pk}(N\Delta)\right]$$

$$\hat{H}'_{pk} = \left[\hat{h}_{pk}\left(\frac{\Delta}{2}\right) \quad \hat{h}_{pk}\left(\frac{3\Delta}{2}\right) \quad \ldots\ldots \quad \hat{h}_{pk}\left(\frac{2N-1}{2}\Delta\right)\right]$$

and the NxN matrix $W = [w_{ij}]$

$w_{ij} = 0$ for $j > i$

$= 1$ for $j \leq i$

Therefore :

$$Z_{pk} = \Delta W\ \hat{H}_{pk} = \Delta W \left[H_{pk} + \mathcal{E}_{pk}\right] \quad \ldots\ldots\ldots(24)$$

From this equation one can get :

$$H_{pk} = E\left\{\frac{1}{\Delta} W^{-1} Z_{pk}\right\} = \lim_{M\to\infty} \frac{1}{M} \sum_{i=1}^{M} \frac{1}{\Delta} W^{-1} Z_{pk_i}$$
$$= \lim_{M\to\infty} \hat{H}_{pk_M} \quad \ldots\ldots\ldots\ldots\ldots\ldots\ldots(25)$$

Where $\hat{H}_{pk_M}$ is the estimated value of H_{pk} based on M experiments.

The variance of the estimator can be obtained using equation (22) of the error propagation matrix, giving:

$$\sigma^2_{\hat{H}_{pk_M}} = \frac{1}{M}\left\{\sum_{i=1}^{n} H'_{pi} \Phi_{u_i} H_{pi} + \frac{1}{\Delta^2}\left[\frac{2N-1}{N} \sigma^2_{v_p} - \frac{2N-2}{N} \times \phi_{v_p}(\Delta)\right]\right\} \quad \ldots\ldots\ldots\ldots(26)$$

$$\Phi_{u_i} = \left[\gamma_{rs_i}\right] \quad ;\text{where } \gamma_{rs} = \gamma_{sr}$$

$$\gamma_{rs} = (2N-1-2r+2)\sigma^2_{u_i} - (2N-2-2r+2)\,\phi_{u_i}(\Delta) \qquad \text{for } r=s$$

$$\gamma_{rs} = -(N-s+1)\phi_{u_i}(s\Delta - r\Delta - \Delta) + (2N-1-2s+2)\times \phi_{u_i}(s\Delta - r\Delta) - (N-s)\phi_{u_i}(s\Delta - r\Delta + \Delta) \qquad \text{for } s > r$$

By comparing equation(16) and (26) we can see that the estimator using "all column" identification scheme is more accurate than the corresponding "one column identification scheme.The accuracy of estimation improves as the number of accessible inputs increases.It appears from the above two equations,that approximately 10 $\log_{10} n$ dbs improvement has been achieved in the estimator variances.

4-EXPERIMENTAL RESULTS

The algorithms (4) and (6) were used to identify a model with 2-inputs and 2-outputs having $h_{11}=\exp(-t)$, $h_{12}=\exp(-2t)$, $h_{21}=\exp(-3t)$ and $h_{22}=\exp(-4t)$ without noise using the I.C.L. 1905E digital computer available at Cairo University. Then,algorithms (13) and (25) were applied for the identification of the same model using pseudo random noise sources at 1% of the test signal level and $\sigma^2=0.0001$ at both the inputs and outputs.The experimental results are given in figures (1) and (2).Excellent correlation with the theoretical investigations was achieved.

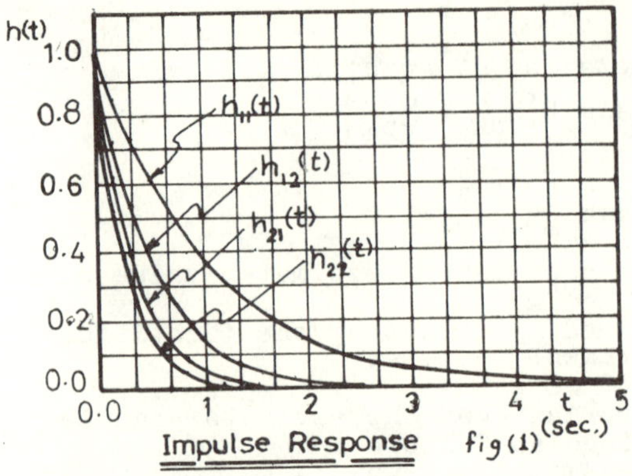

Impulse Response fig (1)

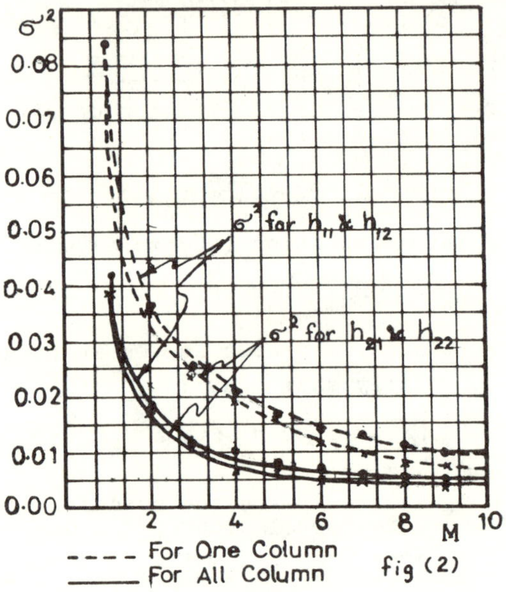

fig (2)

5-REFERENCES

1- Bell,D. and Griffin,A.W.J. Modern Control Theory and Computing McGraw-Hill, 1969.

2- Cohem,W.,Rault,A.and Welingasten,H. A New Identification Algorithm for Linear Discreat Time Systems IFAC symposium 1970, Czhchoslovakia.

3- Davenport,W.B. and Root,W.L. An Introduction to The Theory of Random Signals and Noise McGraw-Hill 1958 .

4- Sage,A.P. and Melsa,J.L. System Identification Academic Press 1971.

5- Sage,A.P. and Melsa,J.L. Estimation Theory with Application to Communication and Control McGraw-Hill 1971.

NUMERICAL ALGORITHMS FOR THE IDENTIFICATION OF SYSTEMS, VIA SPLINES*

Arvind Caprihan

Programa de Engenharia de Sistemas e Computação
COPPE/UFRJ, Rio de Janeiro, Brasil

We consider the identification of linear systems as a problem involving a solution of an ill-posed linear equation. General condition are given for the regularization of this problem. A numerical solution is obtained in the subspace of cubic splines.

* This work was supported by Banco Nacional do Desenvolvimento Econômico

1. INTRODUCTION

Let X and Y be Hilbert spaces with inner products $(.,.)_X$ and $(.,.)_Y$ respectively, $T(X,Y)$ is the space of linear continuous operators from X to Y. The norms in all the Hilbert spaces we consider will be derived from the inner products. It will be assumed that the system output $f \in Y$, can be written as

$$Ax = f, \tag{1}$$

where $A \in T(X,Y)$, it is determined by the input to the system $x \in X$. Due to errors in observation, we observe $g \in Y$, where

$$|| f - g ||_Y \leq \delta , \tag{2}$$

for some $\delta > 0$. Our objective is to give a reasonable scheme to estimate x from the knowledge of g.

Our identification procedure requires that A be a linear continuous operator. The system itself may in general be linear or non-linear, time invariant or time-varying. In particular a linear time invariant system can be identified by this scheme.

For any operator $A \in T(X,Y)$, its range will be denoted by $R(A)$, its null space by $N(A)$. $N(A)^{\perp}$ will mean the orthogonal complement of $N(A)$ in X.

In majority of identification problem, the operator A that arises has the following properties

a) $N(A)$ may not be empty (3)

b) $R(A)$ is not closed

Since we are only given $g \in Y$, we may try to solve

$$A\,x = g \tag{4}$$

(3) implies that the solution of (4) is an ill-posed problem. By this we mean that either g may not be in the $R(A)$ or there may not be a unique solution to (4) or the solution may not depend continuously on g.

In order to resolve these difficulties we have to change our notion of a solution. We construct a well-posed problem, the solution of which will be the estimate of the unknown system. This process is called regularization. Tihonov [1] and Ivanov [2] have given methods for regularization based on Approximation Theory. Our method is similar to Tihonov [1]. Franklin [3] has used stochastic estimation for regularization. Parzen [4] has discussed the similarity between the two approaches. Mosca [5] has given solutions based on reproducing kernel Hilbert spaces. Root [6] has considered questions of identifiability in an abstract framework. The results given here are related to those given in Caprihan -de Figueiredo [7].

We obtain the solution $x \in X$ by solving a well-posed least squares problem. In the beginning we give a weak set of conditions for this regularization. The second part of the paper gives a numerical scheme for its solution. Error bounds have also been obtained when the solution is restricted to the subspace of cubic splines.

2. REGULARIZATION

Let Z be a Hilbert space with the inner product $(. , .)_Z$. We are also given an initial guess $\bar{x} \in X$, for the unknown system. The optimal estimate of the unknown system will be an $x \in X$ which minimizes

$$J(x,\bar{x},g) = \lambda \; ||Ux - U\bar{x}||_Z^2 + ||Ax - g||_Y^2 , \tag{5}$$

where λ is some positive constant and $U \in T(X,Z)$.

U is chosen so that the above minimization problem has a unique solution. It also serves the role of a smoothing operator. We are trying to find an $x \in X$ which is smooth and whose deviation from G is less. The choice of λ depends on δ. We give conditions for chosing U.

A re-definition of x and g imply that (5) is equivalent to finding an $x \in X$, which minimizes

$$J(x,g) = \lambda \, ||Ux||_Z^2 + ||Ax - g||_Y^2 \tag{6}$$

A Hilbert space W is defined to be the cartesian product of Z and Y. Thus $(z,y) = w \in W$, where $z \in Z$ and $y \in Y$. The inner product in W is defined by

$$(w_1,w_2)_W = \lambda(z_1,z_2)_Z + (y_1,y_2)_Y \tag{7}$$

If we define an operator $L \in T(X,W)$ by

$$Lx = (Ux,Ax) \quad , \; x \in X \; , \tag{8}$$

then (6) is equivalent to

$$J(x,g) = ||Lx - h||_W^2 \quad , \tag{9}$$

where

$$h = (0,g) \tag{10}$$

There exists a unique $x \in X$ which minimizes (9) if $R(L)$ is closed and $N(L)$ is the zero subspace. The next theorem shows how to choose U so that the above two conditions are met.

Theorem 1. Let U satisfy the properties:

a) $R(U)$ is closed

b) $N(U) \cap N(A) = \theta_X$, where θ_X is the null subspace of X. (11)

c) $||Ax||_Y \geq m \, ||x||_X$, for all $x \in N(U)$ and where m is some positive constant.

Then it follows that

a) $R(L)$ is closed

b) $N(L) = \theta_X$ (12)

c) $||Lx||_W \geq c \, ||x||_X$, for all $x \in X$ and where c is some positive constant.

3. SOLUTION IN A PRESCRIBED SUBSPACE

Let $X_n \subset X$ be a finite-dimensional subspace with basis functions $\{\phi_i\}$, $i = 1,\dots n$. Define

$$B = \text{matrix} \; \{(U\phi_i \, , U\phi_j)_Z\}, \; i,j = 1,\dots n$$

$$H = \text{matrix} \; \{(A\phi_i \, , A\phi_j)_Y\}, \; i,j = 1,\dots n \tag{13}$$

$$a = \text{col.} \; \{ \, (A\phi_i \, , g)_Y \} \, , \; i = 1,\dots n$$

Theorem 2. The unique element $x \in X_n$, which minimizes $J(x,g)$ is given by

$$x = \sum_{i=1}^{n} b_i \, \phi_i$$

where b_i, $i=1, \dots n$ are solved from

$$(\lambda B + H)b = a \tag{14}$$

with

$$b = \text{col} \, (b_1, b_2, \dots b_n)$$

Thus the optimum $x \in X_n$ can be calculated from (14). It has the following interesting property.

Lemma 1. Let $x \in X_n$ be given by (14) and let $\bar{x} \in X$ minimize $J(x,g)$, then

$$(L^* L \bar{x} - L^* L x, \; y)_X = 0 \quad , \tag{15}$$

for all $y \in X_n$. L^* is the adjoint of L.

Since L is a bounded operator, there exists a constant α , such that

$$||L x||_W \leq \alpha || x ||_X \tag{16}$$

The above lemma can be used to prove Lemma 2.

Lemma 2. If P_n is the orthogonal projection operator from X to X_n and if x and $\bar{x}$ are given as in Lemma 1 then

$$|| \bar{x} - x ||_X \leq \alpha \, c^{-1} \, || \bar{x} - P_n x ||_X \tag{17}$$

The above theorem also proves the convergence of x to $\bar{x}$ with increasing n if $X_n \subset X_{n+1}$ and so on.

4. CUBIC SPLINE SUBSPACE

We will assume that $\bar{x}$ as defined above lies in C^m (0,T) with $m \geq 4$. By C^m (0,T), we mean the class of all functions x(t) defined on (0,T) which possess an absolutely continuous $(n-1)^{th}$ derivative on (0,T) and whose n^{th} derivative is in L^2 (0,T) . Let $\pi : 0<h<2h<\ldots<Nh = T$ be a uniform partition of the interval (0,T) with mesh size h. We define Sp^2 (π), the cubic spline interpolation space to be the collection of all piecewise-polynomial functions w(t) defined on (0,T) such that w(t) is a polynomial of degree 3 in each interval $[ih,ih+h]$, i=0,...N-1 and such that $w(t) \in C^2$ (0,T).

Let $\phi(t)$ be zero outside the interval(-2h,2h) and within that interval be defined by

$$\phi(t) = \begin{cases} (t+2h)^3 & t \in (-2h,-h) \\ h^3+3h^2(t+h)+3h(t+h)^2-3(t+h)^3 & t \in (-h,0) \\ h^3+3h^2(h-t)+3h(t-h)^2-3(t-h)^3 & t \in (0,h) \\ (2h-t)^3 & t \in (h,2h) \end{cases} \tag{18}$$

Then the basis functions of $Sp^2(\pi)$ are given by ϕ_i, i=-1,0,1,... N,N+1, where

$$\phi_i(t) = \phi(t-ih) \text{ , if } t \in (0,T) \quad i=-1,0,1,\ldots N+1$$
$$= 0 \quad \text{otherwise} \tag{19}$$

The following result is a special case of results of Jerome-Varga [8]. We define $D = \frac{d}{dt}$

Theorem 3. Let $\bar{x}(t) \in C^m(0,T)$, $m \geq 4$, let $\{\pi_i\}_{i=1}^{\infty}$ be any sequence of uniform partitions of 0,T with mesh size h_i and $\lim_{i\to\infty} h_i = 0$. Also let $w_i(t)$ be the unique interpolation of $\bar{x}(t)$ in $Sp^2(\pi_i)$. In other words,

$$D^k\bar{x}(0) = D^k w_i(0) \qquad k = 0,1$$
$$D^k\bar{x}(T) = D^k w_i(T) \qquad k = 0,1 \tag{20}$$

and

$$\bar{x}(jh_i) = w_i(jh_i) \qquad j = 0,\ldots N$$

Then

$$||\bar{x}-w_i||_{L^2} \leq K_1 \, ||D^4\bar{x}||_{L^2} \, h_i^4 \tag{21}$$

for some constant K_1 independent of h_i.

Theorem 3 and Lemma 2 imply the next theorem.

Theorem 4. If $X = L^2(0,T)$ and $\bar{x}$ and x are given as in Lemma 1 and if $X_n = Sp^2(\pi)$, then

$$||\bar{x} - x||_X \leq K_2 \, ||D^4 \bar{x}||_X \, h^4 \tag{22}$$

5. FURTHER APPROXIMATIONS

Before we can calculate the optimum $x \in Sp^2(\pi)$ we have to calculate the entries of matrices B and H and of the vector a. Since we are looking for smooth solutions, the operator U can be taken to be D^2. In this case entries of B can be analytically calculated. It has the following properties,

$$(U\phi_i,U\phi_j)_Z = (U\phi_{N-i} \text{ , } U\phi_{N-j})_Z$$

$$(U\phi_i,U\phi_j)_Z = 0 \quad \text{if } |i-j| = 2$$
$$= 6h^3 \quad \text{if } |i-j| = 3$$
$$= 0 \quad \text{if } |i-j| > 3 \tag{23}$$

$$(U\phi_i,U\phi_i)_Z = 12h^3 \quad \text{if } i = -1 \text{ or } N+1$$
$$= 48h^3 \quad \text{if } i = 0 \text{ or } N$$
$$= 84h^3 \quad \text{if } i = 1 \text{ or } N-1$$
$$= 96h^3 \quad \text{otherwise}$$

$$(U\phi_i,U\phi_{i+1})_Z = -18h^3 \text{ if } i = -1 \text{ or } N$$
$$= -36h^3 \text{ if } i = 0 \text{ or } N-1$$
$$= -54h^3 \text{ otherwise}$$

It will be possible to analytically calculate terms like $(A\phi_i \text{ , } A\phi_j)_Y$ and $(g, A\phi_i)_Y$ only in very special cases. In general some numerical quadrature schemes will have to be used for their calculation. Let these approximate calculations of H and a be $\bar{H}$ and $\bar{a}$ respectively. Let $\hat{x}$ be the corresponding approximation to x. Thus

$$\hat{x} = \sum_{i=1}^{n} \bar{b} \, \phi_i \text{ ,}$$

where

$$(\lambda B + \bar{H}) \, \bar{b} = \bar{a} \tag{24}$$

and

$$\bar{b} = \text{col} (\bar{b}_1,\bar{b}_2,\ldots \bar{b}_n)$$

The convergence of $\hat{x}$ to $\bar{x}$ depends on the quadrature schemes used. We can prove the following

Theorem 5. Suppose the approximation $\bar{H}$ and $\bar{a}$ satisfy the properties

a) $(\bar{H})_{ij} = (H)_{ij} + O(h^p)$

b) $(\bar{a})_i = (a)_i + O(h^q)$ (25)

c) $|| \lambda Bb + \bar{H}b||_{R^n} \geq \beta \, ||b||_{R^n}$, where β is some positive constant.

Then

$$|| \bar{x} - \hat{x} ||_X \leq K_3 h^r , \quad (26)$$

where

$$r = \text{minimum} \ (p - \frac{3}{2} , q - \frac{1}{2} , 4) \quad (27)$$

and K_3 is some constant independent of h.

6. CONCLUSIONS

None of the theorems have been proved here. Theorem 1 is the one that needs proof. All other lemmas and theorems are either straightword or well-known. They in combination give a convergent scheme for identification of systems. The results of numerical experimentation will be presented at the conference.

The approach presented here can be used for systems which obey the following input-output relationship.

$$f(v) = \sum_{i=1}^{K} \int_0^T .. \int_0^T x_i(v,s_1,..,s_i)u(s_1)..u(s_i)ds_1..ds_i$$

(28)

where $f(v)$ is the output of the system for the input $u(s)$.

REFERENCES

1. A.N.Tihonov, "Solution of Incorrectly Formulated Problems and the Regularization Method" , Soviet Math. Dokl.4,(1963), 1035 - 1038.

2. V.Ivanov, "On Linear Problems which are not well Posed", Soviet Math. Dokl.3, (1962), 981 - 983.

3. J.N.Franklin, "Well-Posed Stochastic Extension of Ill-posed Linear Problems", J.Math.Anal. Appl. 31, (1970), 682 - 716.

4. E.Parzen, "On the Equivalence Among Time Series Parameter Estimation, Approximation Theory and Optimal Control Theory", Proc. 5th Annual Princeton Conference on Information Sciences and Systems, (1971), 1 -5.

5. E.Mosca, "On a Class of Ill-Posed Estimation Problem and a Related Gradient Iteration", IEEE Trans. on Automatic Control, Vol. AC-17, (1972) 459 - 465.

6. W.L.Root, "On the Structure of a Class of System Identification Problems", Automatica, Vol. 7, Nº 2, (1971), 219 - 232.

7. A.Caprihan and R.J.P. de Figueiredo, " The Generalized Smoothing Spline with Application to Signal Reconstruction and System Identification", (Submitted to S.I.A.M. J. of Numerical Analysis).

8. J.W. Jerome and R.S.Varga, "Generalization of Spline Functions with Applications to Nonlinear Boundary Value and Eigenvalue Problems", Theory and Applications of Spline Functions, edited by T.N.E. Greville, Academic Press , (1969), 103 - 156.

APPLICATION OF CUBIC SPLINES TO SYSTEM IDENTIFICATION

M. Shridhar, Assistant Professor
N. A. Balatoni, Graduate Student

Department of Electrical Engineering
University of Windsor
Windsor, Ontario, Canada

The application of cubic splines to the identification of lumped and distributed systems is considered. A general procedure for identification of systems described by a set of first-order differential equations is developed. Numerical examples of system identification for both lumped and distributed systems where the observations are subject to additive measurement noise, are presented.

1. INTRODUCTION

The identification of system parameters from a knowledge of experimental observations on the system is of interest to engineers concerned with the application of modern control techniques to industrial processes. With the increasing availability of low-priced digital computers, implementation of real-time adaptive-optimal control strategies is now proving feasible.

Techniques currently being used, Bellman [2], Collins [3], Lanczos [5], involve the derivation of a discrete-time equivalent of a continuous system and the use of recursive least-squares estimation procedures for obtaining the system parameters. The discrete-time equivalent of continuous systems is obtained either by application of Prony's method or by finite difference approximation of the differential equations that describe the system under study. It has been shown that these methods yield accurate estimates of system parameters, only if the experimental data is available with excessive accuracy. It is easy to see that for industrial processes where random disturbances are always present, high accuracies in experimental observations are not easily attainable.

The proposed method deals with the application of cubic splines to system identification. It is shown that fitting experimental data with a known function such as "cubic spline" leads to a finite difference approximation of the continuous system, that is convergent and which in addition possesses the "best approximation" property. Due to the smoothing properties of the cubic spline, accurate estimates of system parameters are obtained even when the observed data is known only approximately.

2. MATHEMATICAL SPLINE

The spline function is a piecewise polynomial defined on a mesh $\Delta: a = x_0 \leq x_1 \leq \cdots \leq x_N = b$. In the ensuing discussion only cubic splines will be considered.

2.1. The Cubic Spline

The cubic spline, $S_\Delta(x)$, is a piecewise cubic function defined on a mesh Δ with the following properties.

i) $S_\Delta(x)$ is continuous together with the first and second derivatives on [a,b]

ii) $S_\Delta(x)$ is a cubic in each sub-interval $x_{i-1} \leq x \leq x_i$ $(i = 1,2,\ldots,N)$

iii) $S_\Delta(x)$ satisfies the following equation

$$S_\Delta(x_i) = y(x_i) \quad i = 0,1,2,\ldots,N$$

where $y(x)$ is the function being approximated by $S_\Delta(x)$.

Under the above conditions $S_\Delta(x)$ is said to be a spline with respect to the mesh Δ, interpolating to the values y_i at the mesh points.

In any interval $[x_{j-1}, x_j]$, the cubic spline satisfies

$$S''_\Delta(x) = M_{j-1}\frac{x_j - x}{h_j} + M_j\frac{x - x_{j-1}}{h_j} \tag{2.1}$$

where $M_j = S''_\Delta(x_j)$, $j = 0,1,2,\ldots,N$ and $h_j = x_j - x_{j-1}$.

The continuity conditions for $S'_\Delta(x)$ at the mesh points leads to the following relation:

$$\frac{h_j}{6}M_{j-1} + \frac{h_j + h_{j+1}}{3}M_j + \frac{h_{j+1}}{6}M_{j+1} = \frac{y_{j+1} - y_j}{h_{j+1}} - \frac{y_j - y_{j-1}}{h_j}$$

$$i = 1, 2, \ldots, N-1 \tag{2.2}$$

Similarly it can be shown that

$$h_{j+1}m_{j-1} + 2(h_j + h_{j+1})m_j + h_j m_{j+1} = 3\frac{y_j - y_{j-1}}{h_j}h_{j+1} + 3\frac{y_{j+1} - y_j}{h_{j+1}}h_j$$

$$j = 1, 2, \ldots, N-1 \tag{2.3}$$

where $m_j = S'_\Delta(x_j)$ $j = 0, 1, 2, \ldots, N$.

The quantities M_j and m_j $(j = 0,1,2,\ldots,N)$ are uniquely determined if suitable end conditions are specified on the function $y(x)$ being interpolated. Typically those are $S'_\Delta(a) = y'_o$ and $S'_\Delta(b) = y'_N$.

The spline function has the best approximation and minimum curvature properties that can be used to advantage in the identification of system parameters from experimental data. This is discussed in the next section.

For a complete description of the properties of splines the reader is referred to the text "The Theory of Splines" by Ahlberg et al. [1].

3. SYSTEM IDENTIFICATION USING CUBIC SPLINES

The identification of physical systems described by a set of first-order ordinary differential equations will be considered in this section. The model for the dynamic system is assumed to be as follows:

$$\dot{\underline{X}}(t) = A\,\underline{X}(t) + B\,U(t) + \Gamma\,w(t) \tag{3.1}$$

$$y(t) = C^T\,\underline{X}(t) + v(t) \tag{3.2}$$

where A, B, Γ and C are n x n, n x 1, n x 1 and n x 1 matrices, $\underline{X}(t)$ is the state vector, $y(t)$ and $U(t)$ are measurable and $w(t)$ and $v(t)$ are independent white Gaussian random noise with zero means.

It is desired to obtain the best estimate of A and B under the following assumptions:

1) The input U(t) can be expressed as the solution of a homogeneous system of first-order equations. This implies that U(t) may be expressed as a sum of exponentials.

2) The system satisfies observability conditions.

The system of equations defined by (3.1) can then be reduced to a homogeneous set by defining additional state-variables to represent U(t). Thus a new system is defined as follows

$$\dot{\underline{X}} = \Phi\underline{X} + \Gamma\,w(t) \tag{3.3}$$

$$y(t) = C^T\,\underline{X}(t) + v(t) \tag{3.4}$$

where Φ is the augmented matrix to be estimated.

The identification procedure involves the following two steps:

1) The estimation of the state-vector using discrete Kalman filtering, given an initial approximation to the matrix Φ.

2) The derivation of an improved estimate for Φ using the cubic spline fit.

3.1. Cubic Spline Identification

3.1.1. Identification of Lumped Systems

Let $\underline{S}_\Delta(t)$ be a vector of cubic splines defined on an uniform mesh Δ: $0 = t_o < t_1 < t_2 \ldots\ldots < t_N = T$ $(h = t_{i+1} - t_i,\ i = 0,1,2,\ldots,N-1)$. The cubic spline approximation to the solution of equation (3.3) may be expressed as

$$\underline{X}(t) \simeq \underline{S}_\Delta(t) \tag{3.5}$$

where $\underline{S}_\Delta(t_i) = \underline{X}(t_i) \triangleq \underline{X}_i$

Substituting equation (3.5) in equation (3.3) one obtains

$$\dot{\underline{S}}_\Delta(t) = \Phi\underline{S}_\Delta(t) + \underline{\varepsilon}(t) \tag{3.6}$$

where the noise terms and the error due to approximations have been included in the vector $\underline{\varepsilon}(t)$.

At $t = t_i$, $i = 0, 1, 2, \ldots, N$

$$\dot{\underline{S}}_\Delta(t_i) = \Phi\underline{S}_\Delta(t_i) + \varepsilon(t_i) \tag{3.7}$$

Multiplying equation (3.7) by the coefficients of m_j in equation (2.3) for $i = j-1$, j and $j+1$ respectively, and adding these, one obtains

$$h\dot{\underline{S}}_\Delta(t_{j-1}) + 4h\,\dot{\underline{S}}_\Delta(t_j) + h\,\dot{\underline{S}}_\Delta(t_{j+1})$$

$$\Phi\,[\underline{S}_\Delta(t_{j-1}) + 4\,\underline{S}_\Delta(t_j) + \underline{S}_\Delta(t_{j+1})] + \underline{\varepsilon}_1$$

Using equations (2.3) and (3.5) and simplifying, the above equation reduces to

$$\frac{\underline{X}_{j+1} - \underline{X}_{j-1}}{2h} = \frac{1}{6}\,\Phi\left[\underline{X}_{j-1} + 4\underline{X}_j + \underline{X}_{j+1}\right] + \underline{\varepsilon}_1 \tag{3.8}$$

Let $\underline{Y}_j = \underline{X}_{j+1} - \underline{X}_{j-1}$

$$\underline{Z}_j = \frac{h}{3}\left[\underline{X}_{j-1} + 4\underline{X}_j + \underline{X}_{j+1}\right]$$

Then

$$\underline{Y}_j = \Phi\,\underline{Z}_j + \underline{\varepsilon}_1 \tag{3.9}$$

If the state variables $\underline{X}_j$ can be measured or estimated then equation (3.9) may be solved by recursive least-squares procedure for the unknown matrix Φ, column by column without any inversion [6].

In the case when the state-vector cannot be measured, the well known Kalman filtering procedure [4] may be used to estimate the state-vector.

3.1.2. Identification of Distributed Systems

In this section the spline identification procedure for a system described by a linear parabolic partial differential equation will be developed. The extension to the general case is straightforward.

Let the distributed systems be represented by

$$\frac{\partial y(x,t)}{\partial t} - a\frac{\partial^2 y(x,t)}{\partial x^2} = f(x,t), \qquad (3.10)$$

$$\Omega: \{0 \leqslant x \leqslant 1\}, \quad \Gamma = \{x: x=0 \text{ and } x=1\}$$
$$0 \leqslant t \leqslant T$$
$$\Sigma: \quad \Gamma \times (0,T)$$

Initial condition:

$$y(x,0) = u_2(x), \quad x \in \Omega$$

Boundary condition:

$$y(S,t) = u_1(S,t), \quad S, T \in \Sigma$$

Measurable data: (1) $Z(x_j,t) = y(x_j,t) + w_j(t)$ $j = 0, 1, \ldots\ldots, N$ and $\Delta x = x_j - x_{j-1}$ where $w_j(t)$ is a Gaussian zero mean white noise process.

(2) $f(x, t)$

It is desired to obtain an estimate of 'a' given the observation $Z(x_j,t)$.

Let the cubic spline approximation to the solution of equation (3.10) be given by

$$y(x,t) \simeq \sum_{j=0}^{N} y(x_j,t)\, S_{\Delta j}(x) \qquad (3.11)$$

where $S_{\Delta j}(x)$, $j = 0, 1, \ldots\ldots, N$ are cubic splines defined as a uniform mesh Δ_x: $0 = x_o < x_1 < \ldots\ldots < x_N = 1$ with

$$S_{\Delta j}(x_i) = \delta_{ji} \qquad \text{(Kroneeker's delta)}$$

Substituting (3.11) in equation (3.10), one obtains after simplification,

$$\sum_{j=0}^{N} y'(x_j)\, S_{\Delta j}(x) - a\sum_{j=0}^{N} y(x_j)\, S''_{\Delta j}(x) = f(x,t) + \varepsilon_1$$

where $y'(x_j) = \frac{\partial y}{\partial t}(x_j)$.

At $x = x_i$, this reduces to

$$y'(x_i) - a\sum_{j=0}^{N} y(x_j)\, S''_{\Delta}(x_i) = f(x_i,t) + \varepsilon_1 \qquad (3.12)$$

Multiplying equation (3.12) by the coefficients of M_j in equation (2.2) for $j = i-1$, i and $i+1$ respectively and adding, one obtains after simplification

$$\frac{1}{6}\left[y'(x_{i-1}) + 4y'(x_i) + y'(x_{i+1})\right] - a\left[\frac{y(x_{i-1}) - 2y(x_i) + y(x_{i+1})}{(\Delta x)^2}\right] =$$

$$\frac{1}{6}\left[f(x_{i-1}) + 4f(x_i) + f(x_{i+1})\right] + \varepsilon_2$$

$$i = 1, 2, \ldots\ldots, N-1 \qquad (3.13)$$

The system of equations defined by (3.13) may be expressed in the vector notation as

$$A\,\dot{\underline{Y}} = a\cdot B\,\underline{Y} + A\,\underline{f} + \underline{r} + \underline{\varepsilon} \qquad (3.14)$$

where A is a (N-1) x (N-1) matrix.

$$A = \frac{1}{6}\begin{bmatrix} 4 & 1 & 0 & & & \\ 1 & 4 & 1 & & & \\ & \ddots & \ddots & \ddots & & \\ & & & 1 & 4 & 1 \\ & & & & 0 & 1 & 4 \end{bmatrix}$$

$$B = \frac{1}{(\Delta x)^2}\begin{bmatrix} -2 & 1 & 0 & & & \\ 1 & -2 & 1 & & & \\ & \ddots & \ddots & \ddots & & \\ & & & 1 & -2 & 1 \\ & & & & 0 & 1 & -2 \end{bmatrix}$$

$$\underline{Y} = \left[y(x_1,t),\ y(x_2,t),\ \ldots\ldots,\ y(x_{N-1},t)\right]^T$$

$$\underline{f} = \left[f(x_1,t),\ f(x_2,t),\ \ldots\ldots,\ f(x_{N-1},t)\right]^T$$

$$\underline{r} = \begin{bmatrix} \frac{1}{6} f(x_o,t) - \frac{1}{6} y'_o + \frac{a}{(\Delta x)^2} y_o \\ 0 \\ 0 \\ \cdot \\ \cdot \\ 0 \\ \frac{1}{6} f(x_N,t) - \frac{1}{6} y'_N + \frac{a}{(\Delta x)^2} y_N \end{bmatrix}$$

where

$$y_o = y(x_o,t)$$
$$y_N = y(x_N,t)$$

Premultiplying equation (3.14) by A^{-1} and by redefining the variables one obtains

$$\dot{\underline{Y}} = a\,Q\,\underline{Y} + T\,\underline{U} \qquad (3.15)$$

The above systems of first-order differential equations may be solved for the unknown parameter 'a' by the techniques developed in the previous section. It should be noted that the vector $\underline{Y}$ is measurable.

4. DISCUSSION

An examination of equation (3.8) reveals that the cubic approximation to the solution of equation (3.3) results in a vector difference equation. It is further noted that the left hand side of (3.8) is indeed a finite-difference approximation of the vector $\dot{\underline{X}}$. One might then be tempted to ask what the relative advantages of cubic spline approximation are over the conventional finite-difference approximation of a system of ordinary differential equations. The answer to this question may be deduced if one observes the right hand side of equation (3.8). This term which represents smoothing of the state vector, results in an improved estimate of the matrix Φ. Further, the "best approximation" and "minimum curvature" properties of spline functions results in the smoothing of the derivatives at the mesh points.

While the statistical consistency of the estimate is not discussed here, computational results presented in the next section indicate the feasibility of the above procedure. Since a finite-difference equation results when cubic spline approximation is used, recursive identification is made possible; this feature is important for adaptive control of systems and for the identification of time varying systems.

The use of higher-order splines (eg. quintic splines) is recommended if additional smoothing is required or when systems of order higher than 2 are to be identified.

5. COMPUTATIONAL RESULTS

The computational schemes generated in the earlier sections are illustrated by considering the following examples:

5.1. 4th Order Model

$$\begin{bmatrix}\dot{x}_1\\ \dot{x}_2\\ \dot{x}_3\\ \dot{x}_4\end{bmatrix} + \begin{bmatrix}-10 & 1 & 0 & 0\\ -35 & 0 & 1 & 0\\ -50 & 0 & 0 & 1\\ -24 & 0 & 0 & 0\end{bmatrix}\begin{bmatrix}x_1\\ x_2\\ x_3\\ x_4\end{bmatrix} + \begin{bmatrix}0\\0\\0\\1\end{bmatrix}\sqrt{1700}\ \sin(t)$$

$$y = [1\ \ 0\ \ 0\ \ 0]\ \underline{X} + \nu$$

The identification involved the estimation of the 1st column of the A-matrix. It was assumed that no input noises were present. The peak value of the input was chosen so that the steady state output had a peak value of 1.

Eight different noise levels ($\sigma^2 = 10^{-7}, 10^{-6}, \ldots, 10^{-1}, 1$) were used in this experiment. An initial estimate of the parameters was obtained by solving a conventional finite-difference approximation of the system for the unknown parameters. The iterations were continued until the successive estimates changed by less than 1 percent. The computational results are shown in Figure 1. It is seen that the accuracy of the estimates decrease as the noise variance is increased beyond 10^{-1}. Since the initial estimates were different in each case, it is felt that a more accurate estimate of the effects of noise level may be obtained by using the same initial estimate.

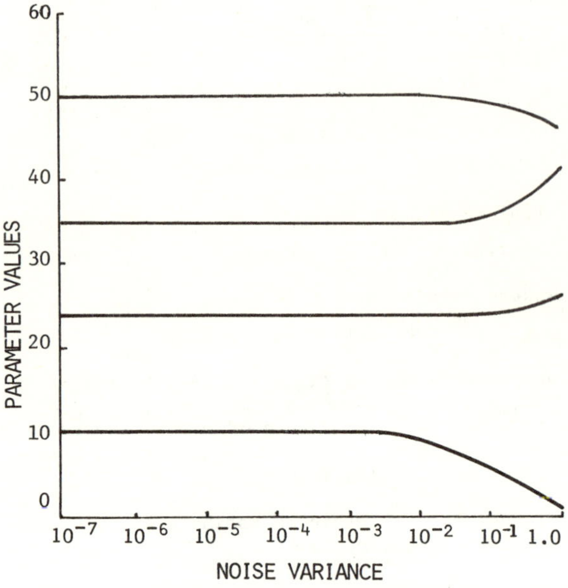

FIGURE 1: Effect of Noise Variance on Parameter Estimation.

The rate of convergence was not appreciably affected by the noise level. The results are displayed in Figure 2.

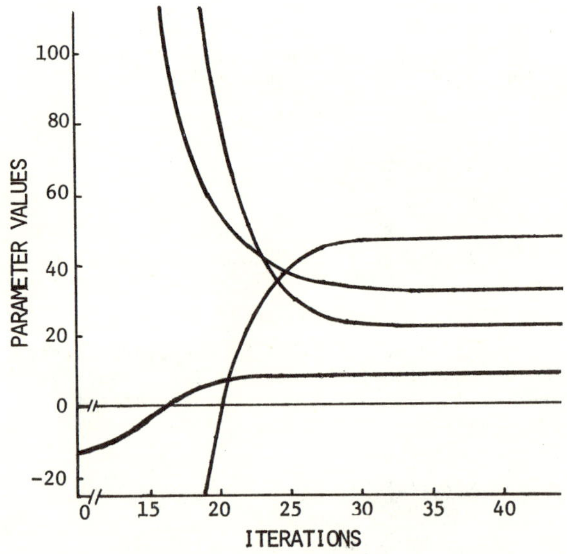

FIGURE 2: Parameter Convergence as a Function of the Number of Iterations.

5.2. Distributed System

The model for the distributed system was assumed to be

$$\frac{\partial U}{\partial t} = a \frac{\partial^2 U}{\partial x^2} + f(x,t)$$

$$a = 1.50$$

$$f(x,t) = 5.0$$

B.C. $U(0,t) = U(\pi,t) = 0,\ t > 0$

$U(x,0) = 0$

observation $Z(x_j,t) = U(x_j,t) + w_j(t)$

$j = 0, 1, 2, \ldots\ldots, N.$

The effects of output noise level are displayed in Figure 3. The identification accuracy is found to decrease when noise variance increases above 10^{-2}. The rate of convergence was found to be smaller when compared to the case of lumped systems (Figure 4).

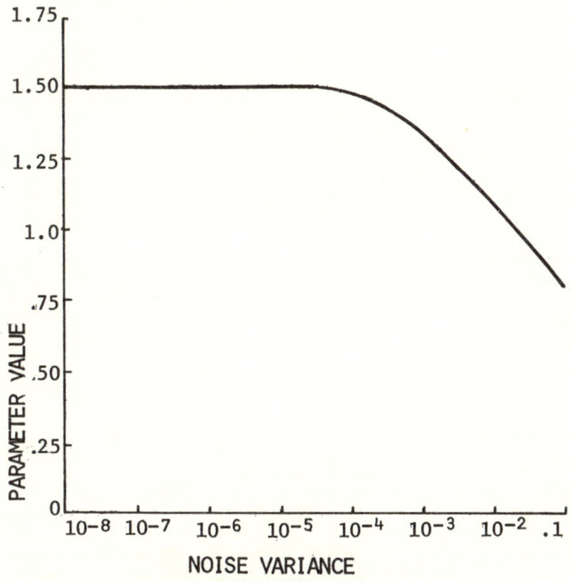

FIGURE 3: Effect of Noise Variance on Parameter Identification

6. CONCLUSION

(1) The use of cubic splines in the identification of both lumped and distributed parameter systems has been shown to be feasible.

(2) The extension of spline identification procedure to a system described by a set of first order differential equations has been shown to result in a feasible general procedure for system identification.

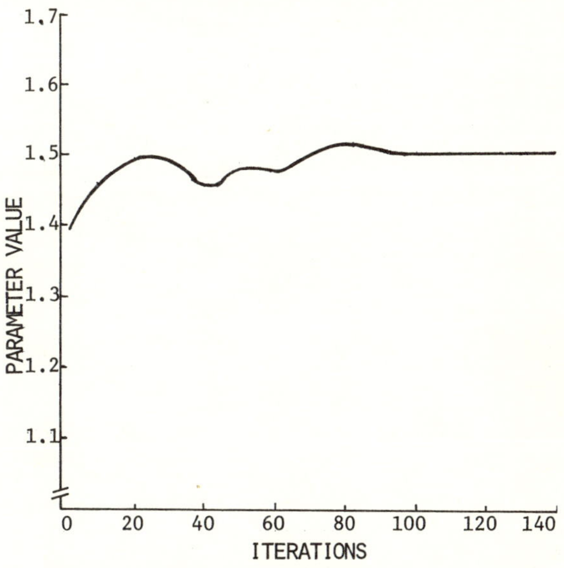

FIGURE 4: Parameter Convergence as a Function of the Number of Iterations.

(3) Since the spline technique results in a difference equation, on-line identification of system parameter and state-vector estimation has become feasible.

7. REFERENCES

[1] Ahlberg, J.H., Nilson, E.N., Walsh, J.L., *"The Theory of Splines and Their Applications"*, Academic Press, New York, (1967).

[2] Bellman, R., Roth, R.S., *"The Use of Splines with Unknown End Points in the Identification of Systems"*, J. Math. Anal.and Appl.34, (1971), p.26.

[3] Collins, P.L., Khatri, H.C., *"Identification of Distributed Parameter Systems Using Finite Differences"*, Trans. ASME, J. Basic Engineering, (1969), p.239.

[4] Kalman, R.E., *"New Methods and Results in Linear Prediction and Estimation Theory"*, Tech. Report 61-I, RIAS, (1961).

[5] Lanczos, C., *"Applied Analysis"*, Prentice-Hall, Englewood Cliffs, N.J., (1956).

[6] Lee, R.C.K., *"Optimal Estimation, Identification and Control"*, M.I.T. Press, Cambridge, Mass., Research Monograph No.28, (1964).

SOME ESTIMATION METHODS FOR NONLINEAR DISCRETE TIME IDENTIFICATION

Cs.Bányász
Computer and Automation Institute
Hungarian Academy of Sciences
Budapest
Hungary

R.Haber L.Keviczky
Technical University, Budapest
Automatic Control Department
Budapest
Hungary

In our paper some parameter estimation methods of simple discrete-time systems consisted of a no-memory nonlinearity followed by a linear dynamic system are considered. The polynomial form is used as a static nonlinearity and the impulse transfer function model as a discrete time linear dynamics. Comparisons of some known model structures are performed and the deterministic description is supplemented by noise-structures used at the linear discrete-time systems. The possibility of identification of the Hammerstein model is investigated in detail in case of quadratic polynomial form and different noise situations. Over and above the well-known iterative and noniterative methods the extension of generalized least-squares and maximum likelihood method - used at the linear systems - for nonlinear case is presented. These methods are supported by simulation results.

INTRODUCTION

Nonlinear systems have many different types. Since there is no simple mathematical method for the description of different structures /and probably will not be in the future, too/ only special nonlinear systems' parameter estimation tasks have been solved with success. In the same way as the impulse response for linear systems the Volterra series expansion means a non-parametric system description for a wide class of nonlinear systems. During parametrizing this linear form, approximating the series of "twice infinite size" /infinite in the time and in order of the expansion/, a function relation - linear in parameters - can be obtained which simplifies the identification procedure

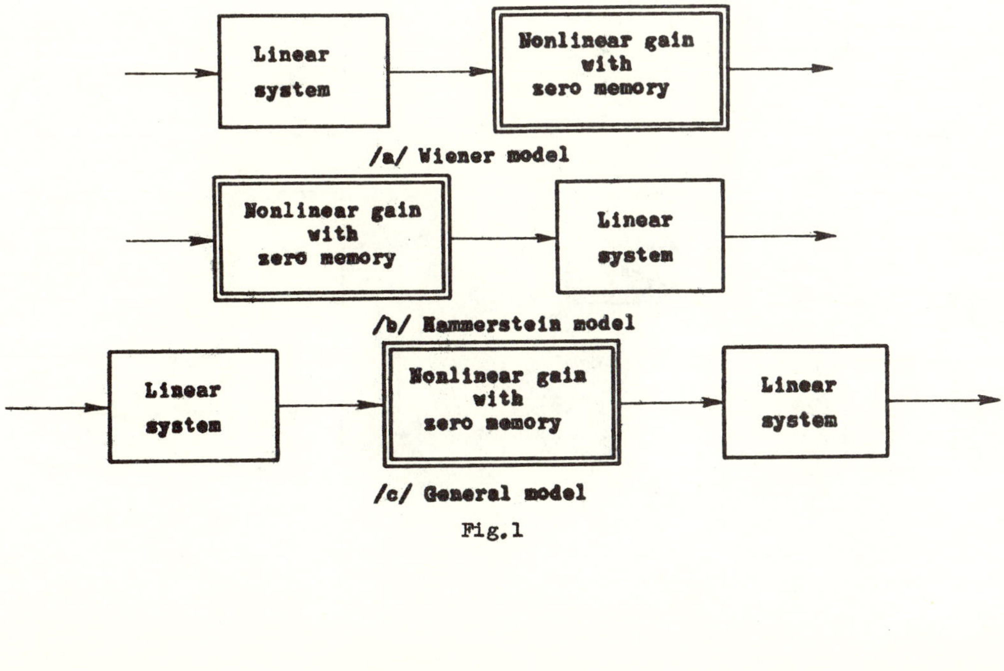

Fig.1

considerably. Unfortunately, for many practical cases the number of necessary parameters is too large;c.f. Aström /1970/.

For special classes of nonlinear systems where the linear dynamics and the nonlinearity-without-memory can be separated the methods worked out for the identification of linear discrete-time systems can be extended very simply or they can be applied with some changing. Three most well-known forms of such models - the Wiener, the Hammerstein and the general one - are shown in Fig.1;c.f. Goldberg /1971/.

In this paper the Hammerstein model is detailed in case of second order polynomial form. If the nonlinearity and the linear dynamics are separable then the operator corresponding to the Hammerstein model is the special case of Uryson operator;c.f. Narendra /1966/. Considering a discrete-time system, assuming second order polynomial as a nonlinearity and using the impulse transfer function the Hammerstein model is shown in Fig.2. The equation of the no-memory-nonlinearity is

$$v/t/ = r_o+r_1u/t/+r_2u^2/t/ \qquad /1/$$

and the difference equation for the discrete-time system is /assuming unit sampling time/

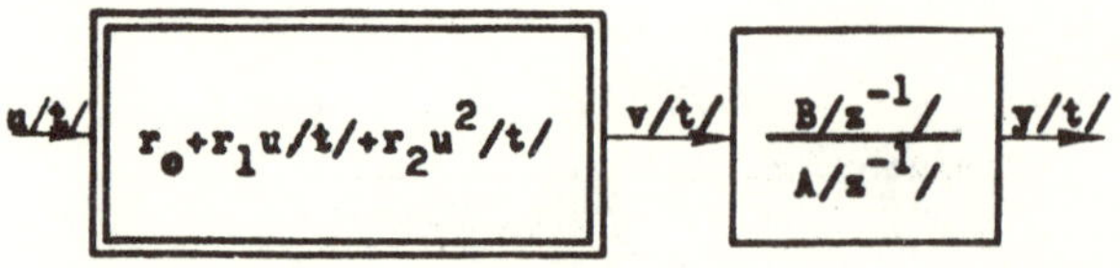

Fig.2

$$A/z^{-1}/\ y/t/ = B/z^{-1}/\ v/t/ \qquad /2/$$

where

$$A/z^{-1}/ = 1+a_1z^{-1}+\ldots+a_nz^{-n} \qquad /3/$$

and

$$B/z^{-1}/ = 1+b_1z^{-1}+\ldots+b_mz^{-m}\ ;\ m \leqq n\ . \qquad /4/$$

/Here z^{-1} is a backward shift operator : $z^{-1}u/t/=u/t-1/$. Notice that the first term of the polynomial $B/z^{-1}/$ is taken unit in order to make the description surely unambigous./ On the basis of /1/ and /2/ the relationship between the input and output is :

$$y/t/ = \sum_{i=o}^{m} b_i v/t-i/ - \sum_{i=1}^{n} a_i y/t-i/ =$$

$$= r_o \sum_{i=o}^{m} b_i + r_1 \sum_{i=o}^{m} b_i u/t-i/ + r_2 \sum_{i=o}^{m} b_i u^2/t-i/ -$$

$$- \sum_{i=1}^{n} a_i y/t-i/ \qquad . \qquad /5/$$

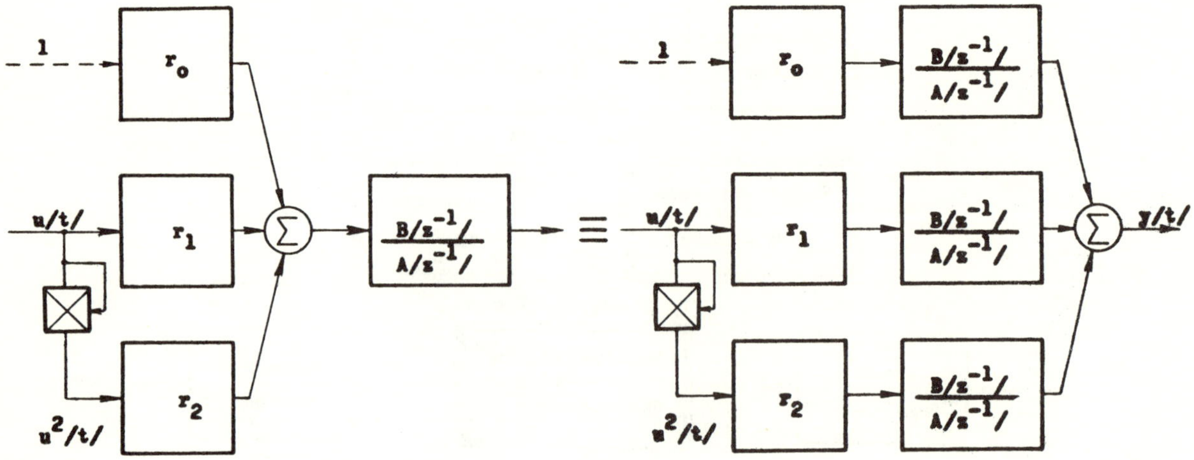

Fig.3

This equation corresponds to a "multiple input" single output system formally, as it is illustrated in Fig.3. Introducing the /n+2m+3/×1 vector $\underline{g}$:

$$\underline{g}/t/=[1,u/t/,\dots,u/t-m/,u^2/t/,\dots,u^2/t-m/,$$
$$-y/t-1/,\dots,-y/t-n/]^T \qquad /6/$$

and the parameter vector $\underline{p}$:

$$\underline{p}=[r_o^{*},r_1,r_1b_1,\dots,r_1b_m,r_2,r_2b_1,\dots,r_2b_m,$$
$$a_1,\dots,a_n]^T \quad , \qquad /7/$$

where

$$r_o^{*}=r_o\ /1+\sum_{i=1}^{m} b_i/ \quad , \qquad /8/$$

and the equation /5/ can be given as a scalar product

$$y/t/=\underline{g}^T/t/\ \underline{p} \quad . \qquad /9/$$

/Here T means the transposition./ Thus we get a system equation having linearity in parameters, the vector $\underline{p}$, however, has redundant elements, since

$$b_i=\frac{p_{2+i}}{p_2}=\frac{p_{3+m+i}}{p_{3+m}} \quad ; \quad 1\leqq i\leqq m \quad . \qquad /10/$$

On the basis of Fig.3 we can attain to the generalization of the Hammerstein model if different transfer functions are assumed in every channel of the "multiple input" system. Fig.4 shows such a system /after reducing to a common denominator/. If

$$B_1/z^{-1}/=1+b_{11}z^{-1}+\dots+b_{1m_1}z^{-m_1} \quad ; \quad m_1\leqq n \qquad /11/$$

and

$$B_2/z^{-1}/=1+b_{21}z^{-1}+\dots+b_{2m_2}z^{-m_2} \quad ; \quad m_2\leqq n \qquad /12/$$

then it can be seen easily the output signal of the system is also linear in parameters and corresponds to /9/ if the following /n+m_1+ +m_2+3/× 1 vectors $\underline{g}$ and $\underline{p}$ are applied :

$$\underline{g}/t/=[1,u/t/,\dots,u/t-m_1/,u^2/t/,\dots,u^2/t-m_2/,$$
$$-y/t-1/,\dots,-y/t-n/]^T \qquad /13/$$

$$\underline{p}=[r_o^{*},r_1b_{11},\dots,r_1b_{1m_1},r_2,r_2b_{21},\dots,r_2b_{2m_2},$$
$$a_1,\dots,a_n]^T \qquad /14/$$

Here

$$r_o^{*}=r_o\ /1+\sum_{i=1}^{n} a_i/ \quad . \qquad /15/$$

In practical applications the measurement noises must be taken into accunt. The majority of methods - failing sufficient apriori information - demand to measure the input signal without noise. At the output, however, a linear noise model having rational spectrum is assumed according to the practical experiences, this situation is presented in Fig.5.

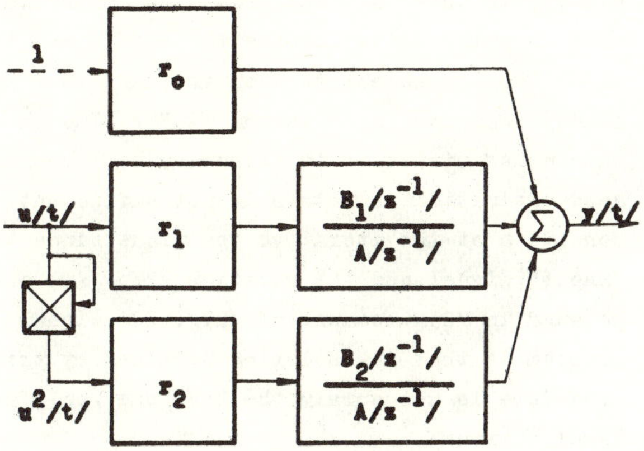

Fig.4

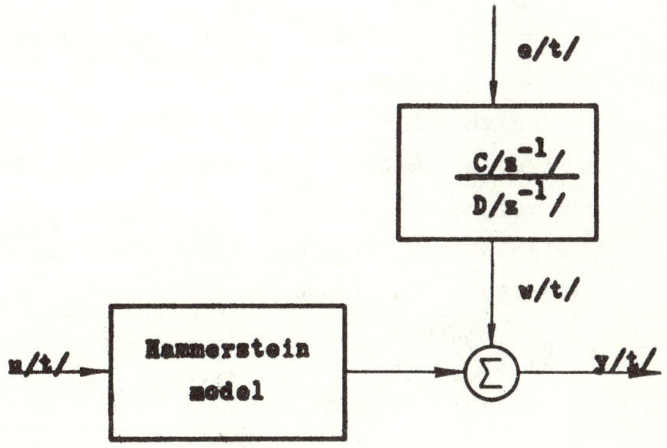

Fig.5

Here

$$C/z^{-1}/=1+c_1z^{-1}+\dots+c_kz^{-k} \quad ; \quad k\leqq \ell \qquad /16/$$

and

$$D(z^{-1})=1+d_1z^{-1}+\ldots+d_\ell z^{-\ell} \quad . \qquad (17)$$

It is assumed that the source noise $e(t)$ is normally distributed white noise with variance one and independent of $u(t)$.

The system to be identified is considered as a structurally stable one with constant parameters and it is assumed that every root of polynomial $z^kC(z^{-1})$ lies inside the unit circle. In this paper the possibilities of the off-line parameter estimation are considered, i.e. when we have N values of $\{u(t),y(t)\}$, where $u(t)$ is a "persistently and sufficently exciting" signal;c.f. Åström /1970/.

ITERATIVE ESTIMATION TECHNIQUE

Narendra /1966/ suggested an iterative technique for the identification of the Hammerstein model. Arrange the N values $u(t)$ and $y(t)$ $(t=1,2,..,N)$ to $N\times 1$ vector $\underline{u}$ and $\underline{y}$ and define the vector $\underline{v}$ similarly. Furthermore, introduce the vectors

$$\underline{p}_1 =[\ r_o,r_1,r_2]^T \quad , \qquad (18)$$

$$\underline{g}_1(t) =[\ 1,u(t),u^2(t)]^T \quad , \qquad (19)$$

and

$$\underline{p}_2 =[\ b_o,b_1,\ldots,b_m,a_1,a_2,\ldots,a_n]^T \quad , \qquad (20)$$

$$\underline{g}_2(t) =[\ v(t),v(t-1),\ldots,v(t-m), -y(t-1),-y(t-2),\ldots,-y(t-n)]^T \qquad (21)$$

The determination of the parameters consists of the following steps :

/a/ Assume that $B(z^{-1})/A(z^{-1}) \equiv 1$, thus $\hat{v}_1(t)=y(t)$, where ^ refers to the estimated value.

/b/ A simple least-squares /LS/ estimation is made for $\underline{p}_1$ on the basis of $\underline{u}$ and $\hat{\underline{v}}_1$, i.e.

$$\hat{\underline{p}}_1 = (\underline{\underline{G}}_1^T \underline{\underline{G}}_1)^{-1} \underline{\underline{G}}_1^T \underline{v}_1 \qquad (22)$$

Here

$$\underline{\underline{G}}_1 = \begin{bmatrix} \underline{g}_1^T(1) \\ \vdots \\ \underline{g}_1^T(N) \end{bmatrix} \qquad (23)$$

/c/ Estimate $v(t)$ again :

$$\hat{v}_2(t) = \underline{g}_1(t)\ \hat{\underline{p}}_1 \quad . \qquad (24)$$

/d/ Make a LS estimation for $\underline{p}_2$ on the basis of $\hat{\underline{v}}_2$ and $\underline{y}$:

$$\hat{\underline{p}}_2 = (\underline{\underline{G}}_2^T \underline{\underline{G}}_2)^{-1} \underline{\underline{G}}_2^T \underline{y} \quad , \qquad (25)$$

where

$$\underline{\underline{G}}_2 = \begin{bmatrix} \underline{g}_2^T(1) \\ \vdots \\ \underline{g}_2^T(N) \end{bmatrix} \qquad (26)$$

Naturally, now $\hat{v}_2$ is taken in $\underline{g}_2(t)$ instead of v.

/e/ After these, estimate $v(t)$ from $y(t)$ by $\hat{\underline{p}}_2$:

$$\hat{v}_1(t)= y(t)+\sum_{i=1}^{n}\hat{a}_i y(t-i) - \frac{1}{\hat{b}_o}\sum_{i=1}^{m}\hat{b}_i v_1(t-i) \quad . \qquad (27)$$

Here the unanimity is guaranteed with division by $\hat{b}_o$. The iteration is continued from /b/ up to a sufficient accuracy or a given iteration number. If we have some apriori information about the parameters of dynamic or nonlinear part of the system then we can enter into the iteration at the other point, as well, according to the meaning. It is to be remarked that other iterative algorithms can also be formed similarly to this; c.f. Haber /1972/. During the identification from noisy measurements it must be taken into account that - in this algorithm - the necessary condition of unbiased parameter estimation is $C(z^{-1})\equiv 1$ and $D(z^{-1})\equiv A(z^{-1})$. Namely, in this case the "equation error" in /25/ is white noise /equals to $e(t)$ / and the LS estimation coincides with the maximum likelihood /ML/ one ;c.f. Åström /1970/. Though this noise structure is necessary to the unbiased estimation, but this is not sufficient condition at the iterative technique since the "input" signal and the equation error are correlated in /25/ because of /27/. Additional problem is that the solution obtained by this procedure is not surely the best one /only a local minimum/.

DIRECT /NONITERATIVE/ ESTIMATION TECHNIQUE

Hsia /1968/ and Chang /1971/ suggested a direct method instead of the iterative technique for a special case and for the Hammerstein model, respectively. It is shown in the previous section that both the simple and the generalized Hammerstein model can be described by equation /9/, linear in parameters. This means that in case of $C/z^{-1}/\equiv 1$ and $D/z^{-1}/=A/z^{-1}/$ the ordinary LS estimation gives an unbiased estimation of $\underline{p}$, i.e.

$$\hat{\underline{p}} = /\underline{\underline{G}}^T \underline{\underline{G}}/^{-1} \underline{\underline{G}}^T \underline{y} \qquad /28/$$

Here

$$\underline{\underline{G}} = \begin{bmatrix} \underline{g}^T/1/ \\ \vdots \\ \underline{g}^T/N/ \end{bmatrix} \qquad /29/$$

This **method** gives considerable saving in computation time compared to the iterative technique. In every case it is to be investigated which the better estimation of $\hat{b}_i$ is because the parameter vector /7/ in the linearized equation of the simple Hammerstein model has redundant elements, see /10/. This can be performed by the investigation of covariance matrix $/\underline{\underline{G}}^T\underline{\underline{G}}/^{-1}$ or by comparison of squared sums of residuals. Estimation techniques can be elaborated for other noise structures, too, applying the methods used at the identification of linear discrete-time systems. In the next section the generalized-least-squares /GLS/ method,c.f. Clarke /1967/ and the ML method,c.f. Åström /1965/ are extended for the Hammerstein model.

GENERALIZED-LEAST-SQUARES METHOD

The GLS method suggested by Clarke is actually a special case of Aitken estimation for the linear discrete-time systems. The estimation procedure consists of the following steps; c.f. Clarke /1967/ :

/a/ An ordinary LS estimation for the parameters of /9/ according to /27/.

/b/ Computation of the residuals /equation errors/

$$f/t/ = y/t/ - \underline{g}^T/t/\, \hat{\underline{p}} \quad ; \ t=1,2,\ldots,N \qquad /30/$$

by the estimated parameters of $\hat{\underline{p}}$.

/c/ Assume that the equation of noise model is

$$f/t/ = \underline{q}^T/t/\, \underline{h} \qquad /31/$$

i.e. linear in parameters, where

$$\underline{q}/t/ = [\, f/t-1/, f/t-2/, \ldots, f/t-s/]^T \qquad /32/$$

and

$$\underline{h} = [\, h_1, h_2, \ldots, h_s]^T \quad . \qquad /33/$$

Thus the LS estimation of parametervector $\underline{h}$ can be determined on the basis of values f/t/ computed according to /30/ :

$$\hat{\underline{h}} = /\underline{\underline{F}}^T \underline{\underline{F}}/^{-1} \underline{\underline{F}}^T \underline{f} \qquad /34/$$

where

$$\underline{\underline{F}} = \begin{bmatrix} \underline{q}^T/1/ \\ \vdots \\ \underline{q}^T/N/ \end{bmatrix} \qquad /35/$$

and vector $\underline{f}$ contains the N values of f/t/. The equation /31/ is valid only in that case when $C/z^{-1}/\equiv 1$ and $D/z^{-1}/=A/z^{-1}/\, H/z^{-1}/$, where

$$H/z^{-1}/ = 1+h_1 z^{-1}+\ldots+h_s z^{-s} \quad . \qquad /36/$$

Thus the GLS method gives an asymptotically unbiased estimation. In general, the noise structure assumed above approximates other noise structures, too, very well. The approximation is the better, the condition

$$H/z^{-1}/ \cong D/z^{-1}/ \big/ [A/z^{-1}/\, C/z^{-1}/]$$

is realized the more exactly ;c.f. Söderström /1972/.

/d/ Compute the filtered values multiplying the both sides of equation /9/ by $H/z^{-1}/$:

$$u_1^F/t/=\hat{H}/z^{-1}/\, u/t/ \ ; \ u_2^F/t/=\hat{H}/z^{-1}/\, u^2/t/$$
$$y^F/t/=\hat{H}/z^{-1}/\, y/t/ \qquad /37/$$

For the generalized Hammerstein model the system equation with the filtered values is

$$y^F/t/ = \underline{g}_F^T/t/\, \underline{p}_F \quad , \qquad /38/$$

where

$$g_F/t/=[1^F, u_1^F/t/, \ldots, u_1^F/t-m_1/, u_2^F/t/, \ldots$$

$$\ldots, u_2^F/t-m_2/, -y^F/t-1/, \ldots, -y^F/t-n/]^T \qquad /39/$$

and

$$p_F = [r_o^*, r_1, \ldots, r_1 b_{1m_1}, r_2, \ldots, r_2 b_{2m_2},$$

$$a_1, \ldots, a_n]^T = p \qquad /40/$$

Here

$$1^F = /1 + \sum_{i=1}^{s} h_i/ \qquad . \qquad /41/$$

/e/ Constituting vector y_F from values $y^F/t/$ and matrix

$$G_F = \begin{bmatrix} g_F^T/1/ \\ \vdots \\ g_F^T/N/ \end{bmatrix} \qquad /42/$$

the GLS estimation of p_F is :

$$\hat{p} = \hat{p}_F = /G_F^T G_F/^{-1} G_F^T y_F \qquad /43/$$

The procedure is continued from point /b/. /Notice that the second filtering method of Clarke can also be applied but it has worse convergence behaviours in general;c.f. Clarke /1967/, Söderström /1972/./

MAXIMUM LIKELIHOOD METHOD

Methods investigated above assume that the noise model is a special case of general one shown in Fig.5 and reduce the essentially nonlinear estimation problem to a linear one in parameters via "quasilinearization". Åström /1965/ has developed the ML method for linear discrete-time systems, in case of $D/z^{-1}/=A/z^{-1}/$ solving the nonlinear estimation problem and has given its computation technique./Since the system model and noise model can be reduced to common denominator in every case so condition $D/z^{-1}/ = A/z^{-1}/$ is not too severe./ Let us investigate the application of ML method for the Hammerstein model.

Residual $\varepsilon/t/$ is defined by equation

$$C/z^{-1}/\ \varepsilon/t/=A/z^{-1}/\ y/t/ - r_o^* - B_1/z^{-1}/\ u/t/ -$$

$$- B_2/z^{-1}/\ u^2/t/ \quad , \qquad /44/$$

and it is the estimation of source noise $\lambda e/t/$ shown in Figs.4 , 5.

The logarithm of likelihood function becomes

$$L = -\frac{1}{2\lambda^2}\sum_{t=1}^{N}\varepsilon^2/t/ - N\log\lambda + \text{const.} \qquad /45/$$

Maximizing this function is equivalent to minimizing the loss function :

$$V/p/ = \frac{1}{2}\sum_{t=1}^{N}\varepsilon^2/t/ \qquad /46/$$

The ML estimation of λ can be obtained by $\hat{p}$ belonging to the minimum of the loss function:

$$\hat{\lambda} = \frac{2}{N} V/\hat{p}/ \qquad /47/$$

Now p has the following form :

$$p = [r_o^*, b_{1o}, b_{11}, \ldots, b_{1m_1}, b_{2o}, b_{21}, \ldots, b_{2m_2},$$

$$a_1, \ldots, a_n, c_1, \ldots, c_k]^T \qquad /48/$$

and

$$b_{1o} = r_1 \quad ; \quad b_{2o} = r_2$$

according to /14/. The ML estimate is consistent, asymptotically normal and efficient under mild conditions ;c.f. Åström /1965/.

In general combined Gauss-Newton and Newton--Raphson algorithms are used to minimize the loss function :

$$\hat{p}_{k+1} = \hat{p}_k - \alpha[V_{pp}/\hat{p}_k/]^{-1} V_p/\hat{p}_k/ \quad , \qquad /49/$$

where V_p is the gradient of $V/p/$ and V_{pp} is the matrix of second order partial derivatives. Since during the minimum seeking every root of $z^k C/z^{-1}/$ must lie inside the unit circle so the task has restriction. In the vicinity of illegitimate region the factor α has a role whose computation can be made according to different strategies;c.f. Bányász/1971/. The computation of derivatives is detailed in the Appendix. Failing apriori information the seeking is started from a LS estimation. The globality of obtained minimum has to be controlled by restarting the seeking from other initial points. /Notice that Goldberg /1971/ has published a procedure which is very near to the idea of Åström method. He elaborated a program for the general model of Fig.1 in case of $C/z^{-1}/ \equiv \equiv D/z^{-1}/ \equiv 1$ and minimized the loss function by conjugate gradient method./

It can be seen easily from these relationships that equation /44/ corresponds to a triple input, single output linear discrete-time system formally, where $u_1/t/\equiv 1$, $u_2/t/=u/t/$ and $u_3/t/=u^2/t/$. /This, of course, can be extended for higher order nonlinearities./ The parameters to be estimated are the coefficients of polynomials $C/z^{-1}/, A/z^{-1}/, B_1/z^{-1}/, B_2/z^{-1}/$ and $B_3/z^{-1}/$. Here $B_1/z^{-1}/$ must be assumed zero order. Thus, if somebody has a program for the ML identification of a multiple input, single output system then this program can be made easily suitable for identification of the generalized Hammerstein model, as well.

Both the GLS method and the ML one have been considered for the generalized Hammerstein model. This is needed because first we have to be convinced of that fact whether the approximate separability, i.e. condition /10/, is valid. If it exists then the task must be solved under restriction /10/ in order to ensure the unanimity. We get a possible simple solution of this problem if we introduce equation - equivalent to /9/ -

$$y/t/ = \underline{g}_3^T/t/\ \underline{p}_3 \qquad /50/$$

instead of /9/, where

$$\underline{p}_3 = [\ r_o^{*}, r_1, r_1 b_1, \ldots, r_1 b_m, r_2, a_1, \ldots, a_n]^T \qquad /51/$$

and

$$\underline{g}_3/t/ = [\ 1, u/t/, u/t-1/, \ldots, u/t-m/, x/t/, -y/t-1/, \ldots, -y/t-n/]^T \qquad /52/$$

where

$$x/t/ = u^2/t/ + \sum_{i=1}^{m} b_i u^2/t-m/ \qquad /53/$$

Since both the GLS method and ML one contain iterative procedures so values b_i can be computed always between the iteration cycles on the basis of /10/ to constitute x/t/. In this way the parameter vector becomes unambiguous and convenient to the simple separable Hammerstein model.

SIMULATION RESULTS

The programs of every algorithm mentioned in this paper have been made. Their operations and estimation properties, the necessary computing times have been investigated and compared by several simulation examples. The results of these investigations and the practical applications will be presented in our lecture. Now we give the estimated parameter values of the fol-

	$C(z^{-1})$	λ	r_1	$r_1 b_{11}$	r_2	$r_2 b_{21}$	a_1	a_2	h_1 / c_1	h_2 / c_2	b_{11}	b_{21}	$\hat{\lambda}$	MSE
			2	1	1	0.5	-1.5	0.7						
LS	(I)	0.5	1.952	0.971	1.009	0.499	-1.517	0.713	-	-	0.497	0.494	-	0.472
GLS			1.952	0.967	1.010	0.498	-1.518	0.714	0.034	-	0.496	0.493	-	0.480
GLS			1.971	1.025	1.002	0.509	-1.500	0.700	-0.526	-0.086	0.520	0.508	-	0.157
ML			1.961	1.009	1.002	0.509	-1.509	0.705	0.502	-	0.515	0.508	0.534	0.391
ML			1.973	1.025	1.005	0.511	-1.504	0.701	0.568	0.236	0.520	0.508	0.519	0.355
LS	(II)	1.5	1.903	1.346	0.955	0.635	-1.411	0.618	-	-	0.707	0.664	-	1.459
GLS			1.926	1.135	0.995	0.553	-1.484	0.687	0.488	-	0.589	0.555	-	0.406
GLS			1.926	1.088	1.003	0.532	-1.496	0.697	0.633	0.115	0.565	0.531	-	0.384
ML			1.895	1.117	0.987	0.547	-1.493	0.694	-0.492	-	0.589	0.554	1.607	0.427
ML			1.916	1.115	0.995	0.546	-1.492	0.694	-0.579	0.239	0.548	0.577	1.511	0.377
LS	(III)	3.0	1.829	2.110	0.943	0.898	-1.142	0.380	-	-	1.153	0.952	-	3.951
GLS			1.837	1.470	0.978	0.659	-1.433	0.649	0.774	0.363	0.800	0.674	-	1.167
ML			1.846	1.246	1.018	0.538	-1.481	0.686	-0.994	0.198	0.676	0.528	3.066	1.029
LS	(IV)	3.0	1.851	2.639	0.917	1.109	-0.923	0.175	-	-	1.426	1.209	-	5.109
GLS			1.974	1.422	1.003	0.657	-1.409	0.625	0.997	0.498	0.720	0.655	-	1.368
ML			1.879	1.196	0.979	0.565	-1.488	0.694	-1.519	0.747	0.637	0.577	3.067	0.467

lowing simple Hammerstein model in a table for different polynomials $C/z^{-1}/$ and N=500.

$r_o \equiv 0$; $r_1 = 2$; $r_2 = 1$; $B/z^{-1}/=1+0.5z^{-1}$

and $A/z^{-1}/=D/z^{-1}/=1-1.5z^{-1}+0.7z^{-2}$. The noise/signal ratio at the output was 30 percent which is a large value enough. The input signal was a random sequence with normal distribution, zero mean and variance one. In the Table the value

$$MSE = \sqrt{\frac{1}{N-1} \sum_{t=1}^{N} [y_o/\hat{p}/ - y_o/p/]^2}$$

is also given which is the mean square error. Here y_o is the system's output without noise. The following polynomials $C/z^{-1}/$ were applied : in case /I/ $C/z^{-1}/= \frac{1}{1-0.6z^{-1}}$; in /II/ $C/z^{-1}/= \frac{1}{1+0.6z^{-1}}$; in /III/ $C/z^{-1}/=1-z^{-1}+0.2z^{-2}$; in /IV/ $C/z^{-1}/=1-1.5z^{-1}+0.7z^{-2}$. In the Table the noniterative /LS/, the GLS and the ML methods are compared, too.

CONCLUSIONS

In this paper it has been shown that the identification technique of a special nonlinear system, the Hammerstein model, is very near to the methods elaborated for linear systems. We have reviewed the published estimation procedures, pointed out to the relationships and suggested to extend the GLS method and ML one for nonlinear case. We have given the necessary formulas to the application. The usefulness of the suggested methods is supported by simulation results. Further extensive investigations are needed to study in practical cases whether the Hammerstein model describes suitably the real nonlinear dynamic systems.

In our opinion the suggested methods may have most extensive application on that fiels - at this moment - where it is to be investigated whether some process can be described by means of linear discrete-time model or not in the vicinity of the working point until the input signal is changing within a given region. In case of strong nonlinear character the linear approximation is allowed only under less changing of input signal.

Methods applied for the identification of the Hammerstein model can be used at the adaptive extremum control, as well. Think about that the on-line estimation can be performed by the GLS method, the adaptive system model necessary to the dual control, however, can be produced by the Hammerstein model.

APPENDIX

The computations necessary to the ML identification of the Hammerstein model are made on the basis of Gustavsson's results /1969/ obtained at the identification of multiple input, single output systems.

Differentiating /46/ we get

$$\frac{\partial V}{\partial p_i} = \sum_{t=1}^{N} \varepsilon/t/ \frac{\partial \varepsilon/t/}{\partial p_i} , \qquad /A1/$$

and

$$\frac{\partial^2 V}{\partial p_i \partial p_j} = \sum_{t=1}^{N} \frac{\partial \varepsilon/t/}{\partial p_i} \frac{\partial \varepsilon/t/}{\partial p_j} + \sum_{t=1}^{N} \varepsilon/t/ \frac{\partial^2 \varepsilon/t/}{\partial p_i \partial p_j} . \qquad /A2/$$

By differentiation of /44/ we get

$$\left.\begin{aligned}
C/z^{-1}/ \frac{\partial \varepsilon/t/}{\partial a_i} &= z^{-i} y/t/ \ ; \ i=1,\dots,n \\
C/z^{-1}/ \frac{\partial \varepsilon/t/}{\partial b_{1i}} &= -z^{-i}u/t/ \ ; \ i=0,\dots,m_1 \\
C/z^{-1}/ \frac{\partial \varepsilon/t/}{\partial b_{2i}} &= -z^{-i}u^2/t/; \ i=0,\dots,m_2 \\
C/z^{-1}/ \frac{\partial \varepsilon/t/}{\partial r_o^*} &= -1 \\
C/z^{-1}/ \frac{\partial \varepsilon/t/}{\partial c_i} &= -z^{-i}\varepsilon/t/ \ ; \ i=1,\dots,k
\end{aligned}\right\} \qquad /A3/$$

Differentiating the last equation of /A3/ once more :

$$C/z^{-1}/ \frac{\partial^2 \varepsilon/t/}{\partial a_i \partial c_j} = -z^{-i-j+1} \frac{\partial \varepsilon/t/}{\partial a_1}$$

$$C/z^{-1}/\, \frac{\partial^2 \varepsilon/t/}{\partial b_{1i}\partial c_j} = -z^{-i-j}\, \frac{\partial \varepsilon/t/}{\partial b_{1o}}$$

$$C/z^{-1}/\, \frac{\partial^2 \varepsilon/t/}{\partial b_{2i}\partial c_j} = -z^{-i-j}\, \frac{\partial \varepsilon/t/}{\partial b_{2o}}$$

$$C/z^{-1}/\, \frac{\partial^2 \varepsilon/t/}{\partial r_o^{*}\,\partial c_j} = -z^{-j}\, \frac{\partial \varepsilon/t/}{\partial r_o^{*}} \qquad /A4/$$

$$C/z^{-1}/\, \frac{\partial^2 \varepsilon/t/}{\partial c_i\,\partial c_j} = -2\, z^{-i-j+1}\, \frac{\partial \varepsilon/t/}{\partial c_1}$$

The remaining second order derivatives are zeros. Here the equality

$$\frac{\partial \varepsilon/t/}{\partial a_i} = z^{-i+1}\, \frac{\partial \varepsilon/t/}{\partial a_1} = \frac{\partial \varepsilon/t-i+1/}{\partial a_1}\, ; \qquad /A5/$$

$$i \leqq t+1$$

is valid which is easy to see. /This is used for the other derivatives, too./

It is advisable to compute the residuals of equation /44/ in the following state variable form :

$$\underline{x}_1/t/= \begin{bmatrix} -c_1 & 1 & 0 & \dots & 0 \\ -c_2 & 0 & 1 & \dots & 0 \\ \vdots & \vdots & \vdots & \ddots & \vdots \\ -c_{k-1} & 0 & 0 & \dots & 1 \\ -c_k & 0 & 0 & \dots & 0 \\ 0 & 0 & 0 & \dots & 0 \\ \cdot & \cdot & \cdot & \dots & \cdot \\ 0 & 0 & 0 & \dots & 0 \end{bmatrix} \underline{x}_1/t-1/+ \begin{bmatrix} a_1 \\ a_2 \\ \vdots \\ \cdot \\ \cdot \\ \cdot \\ \cdot \\ a_n \end{bmatrix} y/t-1/-$$

$$- \begin{bmatrix} b_{11} \\ b_{12} \\ \vdots \\ b_{1m_1} \\ 0 \\ \vdots \\ 0 \end{bmatrix} u/t-1/- \begin{bmatrix} b_{1o} \\ 0 \\ \vdots \\ 0 \\ 0 \\ \vdots \\ 0 \end{bmatrix} u/t/- \begin{bmatrix} b_{21} \\ b_{22} \\ \vdots \\ b_{2m_2} \\ 0 \\ \vdots \\ 0 \end{bmatrix} u^2/t-1/- \begin{bmatrix} b_{2o} \\ 0 \\ \vdots \\ 0 \\ 0 \\ \vdots \\ 0 \end{bmatrix} u^2/t/+$$

$$+ \begin{bmatrix} 1 \\ 0 \\ \vdots \\ 0 \end{bmatrix} y/t/- \begin{bmatrix} r_o^{*} \\ 0 \\ \vdots \\ 0 \end{bmatrix} \qquad /A6/$$

and

$$\varepsilon/t/ = [\,1,0,\dots,0\,]\ \underline{x}_1/t/ \qquad /A7/$$

Similarly to the residuals the gradient vector can also be computed by state-variables. According to the first equation of /A3/

$$\underline{x}_2/t/ = \begin{bmatrix} -c_1 & -c_2 & \dots & -c_{k-1} & -c_k & 0 & \dots & 0 \\ 1 & 0 & \dots & 0 & 0 & 0 & \dots & 0 \\ 0 & 1 & \dots & 0 & 0 & 0 & \dots & 0 \\ \vdots & \vdots & \ddots & \vdots & \vdots & \vdots & \ddots & \vdots \\ 0 & 0 & & 1 & 0 & 0 & \dots & 0 \end{bmatrix} \times$$

$$\times\, \underline{x}_2/t-1/ \;+ \begin{bmatrix} 1 \\ 0 \\ \vdots \\ 0 \end{bmatrix} y/t/ \qquad /A8/$$

and we get

$$\frac{\partial \varepsilon/t/}{\partial a_1} = x_{21}/t/\ ;\ \frac{\partial \varepsilon/t/}{\partial a_2} = x_{21}/t-1/ = x_{22}/t/, \qquad /A9/$$

and so on.

Similarly to /A3/ the other equations can also be represented simply by state-variables. So the computation of derivatives and gradient becomes considerably simple.

From the residuals and their derivatives by parameters, the gradient $\underline{V}_p/\underline{p}/$ and the matrix of approximate second order derivatives can already be computed very simply. /See the first term of right side of /A2/./ To compute the matrix of exact second order derivatives the equations /A4/ must be taken into account, as well. Since /A4/ corresponds to /A3/ formally the second order derivatives can be produced by such state-variable equations as /A8/. It is worth mentioning that there are only so many state variables in one equation of /A4/ as different values /i+j/ can be constituted from the indexes of parameters. This statement - similarly to /A8/ - simplifies the computation of exact second order derivatives, too.

REFERENCES

Åström, K.J. and T. Bohlin /1965/. Numerical Identification of Linear Dynamic Systems from Normal Operating Records, IFAC, Teddington.

Åström, K.J., T. Bohlin and S. Wensmark /1965/. Automatic Construction of Linear Stochastic Dynamic Models for Industrial Processes with Random Disturbances Using Operating Records,Report,IBM Nordic Laboratory,Stockholm.

Åström,K.J. and P.Eykhoff /1970/. System Identification - A survey,IFAC,Prague.

Bányász,Cs. /1971/. Investigation of Discrete Identification Methods by Digital Computer,Ph. D.Thesis,Automation Research Institute of the Hungarian Academy of Sciences.

Chang,F.H.I. and R.Luus /1971/. A Noniterative Method for Identification Using Hammerstein Model,IEEE Trans.Aut.Control,AC-16,/Oct./,464.

Clarke,D.W. /1967/. Generalized-Least-Squares Estimation of the Parameters of a Dynamic Model, IFAC,Prague.

Goldberg,S. and A.Durling /1971/. A Computational Algorithm for the Identification of Nonlinear Systems,Journ.Frankl.Inst.,/June/,427.

Gustavsson,I. /1969/. Parametric Identification of Multiple Input, Single Output Linear Dynamic Systems,Report 6907,Lund Institute of Technology,Division of Automatic Control.

Haber,R. /1972/. Identification of Nonlinear Discrete Processes.Master Thesis,Technical University of Budapest,Automatic Control Department.

Hsia,T.C. /1968/. Least Square Method for Nonlinear Discrete System Identification,2nd Asilomar Conf.Circ. and Syst.,423.

Keviczky,L. and Cs.Bányász /1973/. Identification of Linear Discrete Time Systems,Measurement and Automation,Hungary,/To be published/.

Narendra,K.S. and P.G.Gallmann /1966/. An Iterative Method for the Identification of Nonlinear Systems Using a Hammerstein Model,IEEE Trans.Aut.Control,AC-11,/July/,546.

Söderström,T. /1972/. On the Convergence Properties of the Generalized Least Squares Identification Method,Report 7228,Lund Institute of Technology,Division of Automatic Control.

IDENTIFICATION OF POWER BLOCKS AS NON-LINEAR DYNAMIC OBJECTS

Vytautas Kaminskas, Antanas Nemura

Institute of Physical and Technical Problems of Energetics, Lithuanian Academy of Sciences, Kaunas, USSR

Problems of on-line determining consumption functions of power blocks are considered. Algorithms for computing parameter estimates of dynamic models are derived, and a stability test for non-stationary intervals of power block characteristics is presented. Identification algorithms are checked by the method of statistical simulation. Results for 300MW power blocks suggest an expedient applicability of the method.

1. INTRODUCTION

For optimal control of power systems and power stations, mathematical models are indispensible of the relations of relative fuel consumption and active loading. Performance characteristics of power blocks are dependable on conditions of performance, wear, maintenance a.o. This makes on-line identification of power blocks important. Power block can be described mathematically by Hammerstein operator. Estimates of Hammerstein operators have been considered by Narenda [5] , Gluchov [3] , Chang [2] a.o.

2. MATHEMATICAL MODEL OF POWER BLOCK

An observed k-th value of fuel consumption z_k at time instant $t=k\Delta t$ can be described as

$$Z_k = \sum_{j=0}^{m} w_j f(x_{k-j}) + N_k \qquad (2.1)$$

where $f(x)$ - stationary consumption function, describing the relation of fuel consumption and actual load in normal performance, w_j - weight factors accounting for the dynamics of the process, N_k - noise due to unaccounted disturbances and observation errors. To have no effect on performance, the weight factors should satisfy the following condition:

$$\sum_{j=0}^{m} w_j = 1, \qquad (2.2)$$

Series expansion of consumption function into linearly independent functions $\varphi_i(x)$ of the form

$$f(x) = \sum_{i=0}^{n} a_i \varphi_i(x)$$

leads to equation

$$Z_k = a^T \Phi_k w + N_k, \qquad (2.3)$$

where

$$a^T=(a_0,a_1,\dots,a_n),\quad w^T=(w_0,w_1,\dots,w_m)$$

corresponding vector parameters,

$$\Phi_k = \|\varphi_i(x_{k-j})\|,\quad i=\overline{0,n},\quad j=\overline{0,m}$$

- matrix of order $(n+1)\times(m+1)$. Identification consists of constructing algorithms to compute estimates $\hat{a}$ and $\hat{w}$ of vector parameters a and w from observations $(x_0, x_1, \dots, x_s)$ and $(z_0, z_1,\dots, z_s)$, $s>2m+n+2$ of actual loading and fuel consumption respectively. It is also necessary to find quasi-stationary intervals that the estimates apply to.

3. ESTIMABILITY OF PARAMETERS

A non-singular Fisher-type information matrix

$$\mathcal{J}_{a,w} = M[\nabla_{a,w} \log p(z|a,w,x)] \times [\nabla_{a,w} \log p(z|a,w,x)]^T \qquad (3.1)$$

is considered a necessary and sufficient for estimability condition of parameters a and w . For noise of normal distribution $N\sim\mathcal{N}(0, \sigma_N^2)$ matrix (3.1) becomes

$$\mathcal{J}_{a,w} = \frac{1}{\sigma_N^2}\begin{Vmatrix} A^TA & A^TB \\ B^TA & B^TB \end{Vmatrix}, \qquad (3.2)$$

where

$$A=\|c^T\Phi_k^T\|,\quad B=\|a^T\Phi_k\|,\quad k=\overline{m,s}$$

matrices of order $(s-m)\times(n+1)$ and $(s-m)\times(m+1)$, respectively. But then the following statement is true:
1. If functions $\varphi_i(x)$, $i=\overline{0,n}$ are linearly independent and the rank of matrix

$$\Phi=\|\varphi_i(x_k)\|,\ k=\overline{m,s},\ i=\overline{0,n}$$

equals the number of coordinates of parameter $a\ (r(\Phi)=n+1)$, and 2. if the rank of matrix

$$X=\|x_{k-j}\|,\ j=\overline{0,m},\ k=\overline{m,s}$$

equals the number of weight factors $(r(X)=m+1)$, then information matrix (3.2) is non-singular for all values of a and w, except the origin of coordinates $(a=0,\ w=0)$.
Indeed, as confirmed by Rao [6], determinant of matrix (3.2) is

$$|J_{a,w}|=\sigma_N^{-2(n+m+2)}|A^TA|\times \times|B^TB-B^TA(A^TA)^{-1}A^TB|. \quad (3.3)$$

With assumptions taken, matrices A^TA and $B^TB-B^TA(A^TA)^{-1}A^TB$ are positive definite in the whole range of parameters a and w, except the origin of coordinates. Then (3.3) confirms our statement, because $|J_{a,w}|>0$ S_0 in the whole range of parameters a and w except the origin of coordinates $(a=0,\ w=0)$. The rank of matrix X presents the maximum number of weight factors w_j estimable from observation of object variables $(x_0, x_1,\dots,x_s)$ and $(z_0, z_1,\dots,z_s)$.

4. PARAMETER ESTIMATION BY THE METHOD OF LEAST SQUARES

In this method estimates $\hat{a}$ and $\hat{w}$ are determined to satistfy the conditions

$$\hat{a},\hat{w}:\ Q(\hat{a},\hat{w}|z,x)=\min_{a,w\in W} Q(a,w|z,x), \quad (4.1)$$

$$Q(a,w|z,x)=\sum_{k=m}^{s}(z_k-a^T\Phi_k w)^2, \quad (4.2)$$

$$W=\{w:\ d^Tw=1\}, \quad (4.3)$$

where $d^T=(1,1,\dots,1)$ row-vector of order $(m+1)$.
The extreme value problem (4.1) is equivalent to a specific case of Lagrange problem

$$\begin{cases}\nabla_{a,w}Q(a,w|z,x)-\lambda\nabla_{a,m}(d^Tw)\\ d^Tw-1=0\end{cases} \quad (4.4)$$

where λ - Lagrange multiplier. To solve the set of equations (4.4) the following iterative algorithms is applied:

$$\begin{cases}\hat{a}_\ell=(A_\ell^TA_\ell)^{-1}A_\ell^Tz\\ \hat{w}_\ell=(B_\ell^TB_\ell)^{-1}(B_\ell^Tz-\lambda_\ell d),\\ \lambda_\ell=\dfrac{d^T(B_\ell^TB_\ell)^{-1}B_\ell^Tz-1}{d^T(B_\ell^TB_\ell)^{-1}d}\\ \ell=1,2,\dots\end{cases} \quad (4.5)$$

where

$$A_\ell=\|\hat{w}_{\ell-1}^T\Phi_k^T\|,\ B_\ell=\|\hat{a}_\ell^T\Phi_k\|,\ k=\overline{m,s}$$

$$z^T=(z_m, z_{m+1},\dots,z_s)$$

ℓ - number of iteration.
For significance test of parameter coordinates a and w the value of statistics is found [6]

$$F=\frac{s-2m-n-2}{q}\ \frac{R_1^2-R_0^2}{R_0^2}, \quad (4.6)$$

where

$$R_0^2=\min_{a,w\in W}Q(a,w|z,x), \quad (4.7)$$

$$R_1^2=\begin{cases}\min\limits_{a\in H_0,w\in W}Q(a,w|z,x),\\ \text{if } H_0: Ca=0\\ \min\limits_{a,w\in W\cap H_0}Q(a,w|z,x),\\ \text{if } H_0: Ca=0\end{cases} \quad (4.8)$$

residual sums of squares, q - number of coordinates of a and w tested for significance, C = matrix of order $q\times(n+1)$ or $q\times(m+1)$, with all zero elements in each row except a single unit to represent the coordinate being tested. If $F>F_\alpha$, the coordinates are significant. The critical value of F_α is chosen to satisfy $P\{F>F_\alpha\}=\alpha$ from Fisher distribution tables with degrees of freedom q and $s-2m-n-2$ and significance level α. If $F<F_\alpha$, the coordinates may be considered zero. As the result of the significance test, the necessary number of factors a_i is found, to reflect the degree of non-linearity of consumption function $f(x)$,

and the number of weight factors w_j. Note that the decision on significance is in this case approximate, estimates $\hat{a}$ and $\hat{w}$ being non-linear with respect to z. Empirical values of non-linearity by Beale [1] suggest a possible application of linear estimations theory to the case.
Respective confidence regions for consumption function $f(x)$ and for weight factor w are

$$R_f=\left\{f(x): \hat{f}(x)-t_\alpha\hat{\sigma}_f(x)\le f(x)\le\hat{f}(x)+ +t_\alpha\hat{\sigma}_f(x)\right\} \quad (4.9)$$

$$R_w=\left\{w:(\hat{w}-w)^T B_*^T B_*(\hat{w}-w)\le(m+1)\sigma_N^2F_\alpha\right\} \quad (4.10)$$

where

$$\hat{\sigma}_f^2(x)=\hat{\sigma}_N^2\varphi^T(x)(A_*^TA_*)^{-1}\varphi(x), \qquad \hat{\sigma}_N^2=\frac{R_0^2}{s-2m-n-2} \quad (4.11)$$

are corresponding dispersion estimates,

$$\hat{f}(x)=\hat{a}^T\varphi(x),\quad \varphi^T(x)=(\varphi_0(x),\varphi_1(x),\dots,\varphi_n(x)),$$

$$A_*=\|\hat{w}^T\Phi_k^T\|,\quad B_*=\|\hat{a}^T\Phi_k\|,\quad k=\overline{m,s}$$

Constant t_α is chosen from Student distribution tables to satisfy $P\{|t|>t_\alpha\}=\alpha$ with degrees of freedom $s-2m-n-2$ and significance level α, and constant is chosen from Fisher distribution tables to satisfy $P\{F>F_\alpha\}=\alpha$ with degrees of freedom $(m+1)$ and $s-2m-n-2$. Finding confidence regions (4.9) and (4.10) is analogic to the case of linear estimation.

5. PARAMETER ESTIMATION BY THE METHOD OF MAXIMUM POSTERIOR PROBABILITY DENSITY

This method treats parameters a and w as random variables of prior probability density $p(a)$ and $p(w)$, respectively. Estimates $\hat{a}$ and $\hat{w}$ are found for the conditions

$$\hat{a},\hat{w}:\ p(\hat{a},\hat{w}|z,x)=\min_{a,w\in W}p(a,w|z,x), \quad (5.1)$$

$$p(a,w|z,x)=\frac{p(a,w)\ p(z|a,w,x)}{\int p(a,w)p(z|a,w,x)\,da\,dw} \quad (5.2)$$

where $p(a,w)$ - prior probability density of common a and w distribution $p(z|a,w,x)$ - conditional probability density function of the output signal z. We assume a, w and N_k independent and of normal distribution with parameters of distribution

$$a\sim\mathcal{N}(\bar{a},V_a),\ w\sim\mathcal{N}(\hat{w},V_w),\ N_k\sim\mathcal{N}(0,\sigma_N^2) \quad (5.3)$$

where $\bar{a},\bar{w}$ and V_a, V_w are mean values and covariance matrices respectively. Then solving extreme value problem (5.1) with the application of Lagrange multiplier leads to the following iterative algorithm:

$$\begin{cases}\hat{a}_\ell=(\sigma_N^2V_a^{-1}+A_\ell^TA_\ell)^{-1}(\sigma_N^2V_a^{-1}\bar{a}+A_\ell^Tz),\\ \hat{w}_\ell=(\sigma_N^2V_w^{-1}+B_\ell^TB_\ell)^{-1}(\sigma_N^2V_w^{-1}\bar{w}+B_\ell^Tz-\lambda_\ell d),\\ \lambda_\ell=\dfrac{d^T(\sigma_N^2V_w^{-T}+B_\ell^TB_\ell)^{-1}-1}{d^T(\sigma_N^2V_w^{-1}+B_\ell^TB_\ell)^{-1}d}\times\\ \qquad\times[(\sigma_N^2V_w^{-1}\bar{w}+B_\ell^Tz)-1],\\ \ell=1,2,\dots\end{cases} \quad (5.4)$$

or we may write alternatively

$$\begin{cases}\hat{a}_\ell=\bar{a}+\Pi_\ell A_\ell^T(z-A_\ell\bar{a}),\\ \hat{w}_\ell=\bar{w}+\Lambda_\ell[B_\ell^T(z-B_\ell\bar{w}-\lambda_\ell d],\\ \lambda_\ell=\dfrac{d^T[\bar{c}+\Lambda_\ell B_\ell^T(z-B_\ell\bar{w})-1}{d^T\Lambda_\ell d}\\ \ell=1,2,\dots\end{cases} \quad (5.5)$$

where

$$\Pi_\ell=(\sigma_N^2V_a^{-1}+A_\ell^TA_\ell)^{-1},\ \Lambda_\ell=(\sigma_N^2V_w^{-1}+B_\ell^TB_\ell)^{-1} \quad (5.6)$$

It may be seen from equations (4.5), (5.4) or (5.6), that the least squares algorithm (4.5) presents a specific case of algorithms (5.4) or (5.5). For $V_a=0$ and $V_a^{-1}=0$ (5.4), (5.5) and (4.5) become identical. This corresponds to the case of variance of parameters a and w tending to infinity. Thus for a limited number of observations on object variables, estimates based on prior information are more accurate than the least squares estimates. The number of observations increasing, weight of prior information decreases, estimates of algorithms (4.5), (5.4) and (5.5) are assimptotically equal.

6. DETERMINING QUASI-STATIONARY INTERVALS OF BLOCK VARIABLES

Length β of quasi stationary intervals is determined by fitting new observations

$$(x_{s+\beta+1}, x_{s+\beta+2}, \dots, x_{s+\beta+h}),$$

$$(z_{s+\beta+1}, z_{s+\beta+2}, \dots, z_{s+\beta+h}),$$

into the mathematical model of the first sampling. For stationarity test 4 the following statistics is found:

$$F = \frac{s+h-4m-2n-4}{n+m+2} \; \frac{R_1^2 - R_0^2}{R^2} \tag{6.1}$$

where

$$R_0^2 = \min_{a,w\in W} \sum_{k=m}^{s} (z_k - a^T\Phi_k w)^2 + \min_{a,w\in W} \sum_{l=s+m+\beta}^{s+\beta+h} (z_l - a^T\Phi_l w)^2, \tag{6.2}$$

$$R_1^2 = \min_{a,w\in W} \left\{ \sum_{k=m}^{s} (z_k - a^T\Phi_k w)^2 + \sum_{l=s+m+\beta}^{s+\beta+h} (z_l - a^T\Phi_l w)^2 \right\} \tag{6.3}$$

residual sums of squares. If inequality $F > F_\alpha$ is valid, with F_α chosen for $P\{F > F_\alpha\} = \alpha$ with $(n+m+2)$ and $(s+h-4m-2n-4)$ degrees of freedom, then the new model with confidence factor no longer fits into the previous one. The test is analogic to linear estimation theory, and its applicability in our case is based on non-linearity measure by Beale [1].

7. EXPERIMENTAL RESULTS

Quantitative results of convergence for algorithms (4.5) and (5.4) hane been found by the method of statistical simulation with an electronic computer. A descrete autoregressive first order process was used as input sequence. Covariance matrices V_a and V_w were chosen as

$$V_a = \| 6^2_{a_i} \delta_{ij} \|, \quad i,j = \overline{0,n}$$
$$V_w = \| 6^2_{w_k} \delta_{k,l} \|, \quad k,l = \overline{0,m} \tag{7.1}$$

where $6_{a_i} = 0{,}1\bar{a}_i$, $6_{w_k} = 0{,}1\bar{w}_k$ and $\bar{a}_i$ and $\bar{w}_k$ - mean factors. Fig. 1 represents the relation of the mean square error

$$6 = \sum_k [f(x_k) - \hat{f}(x_k)]^2 \Big/ \sum_k [f(x_k)]^2 \tag{7.2}$$

and the number of iterations l for different ratios $6_N^2 / 6_y^2$, where $y_k = a^T\Phi_k w$ actual output signal. The experiment proves fast convergence and sufficient stability of algorithms (4.5) and (5.4) in disturbance.

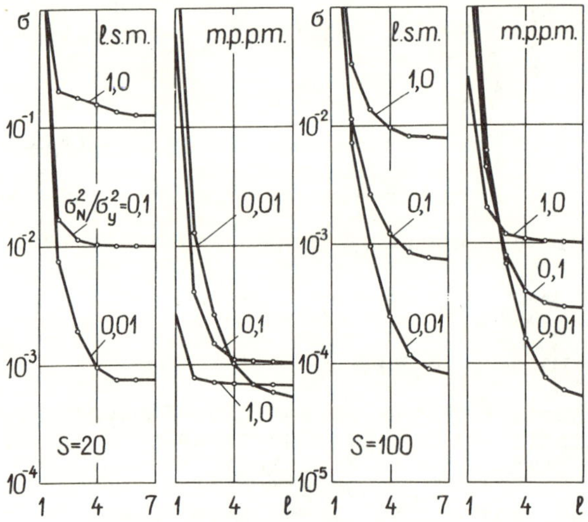

Fig. 1

The least squares algorithm (4.5) was applied for on-line determination of fuel consumption of a 300 MW power block, from data covering 18 hours of performance with a 3 min discrete interval on time axis. The on-line identification resulted in the second-third order polynomial expression of the stationary consumption function in the normal range of performance. To account for the dynamics, 11 or 12 weight factors should be used.

Adequacy check of mathematical model (2.1) revealed a three times smaller residual variance $6_N^2 = R_0^2/(s-2m-n-2)$ as compared to a static model of the block ($z_k = f(x_k) + N_k$). The identification method is also applicable to other non-linear dynamic objects which can be described by Hammerstein operators.

REFERENCES

1. E.M. Beale (1960). Confidence Regions in Nonlinear Estimations. J. Royal Stat. Soc. vol B 22 N 1 pp. 47-75.
2. F.H.I. Chang, R. Luus (1971). Noniterative Method for Identification Using Hammerstein Model. IEEE Trans. Autom. Control, vol AC - 16 N 5 pp. 464-468.
3. Л.В. Глухов (1967). Восстановление характеристик одного нелинейного звена по входному и выходному сигналу. АН СССР Автоматика и телемеханика № II, ст.ст.82-90.
4. В.А. Каминскас, А.А. Немура (1972). Об определении нестационарности одного вида оператора динамических объектов. Б, т.I(68), ст.ст.I07-III. Труды АН Лит. ССР.
5. K.S. Narenda, P.G. Gollman (1966), AN Iterative Method for the Identification of Nonlinear Systems Using a Hammerstein Model. IEEE Trans. Autom. Control. vol AC - 11, pp. 546-560
6. С.Р. Рао (I968). Линейные статистические методы и их применения. Перевод с английского. М, "Наука".

THE IDENTIFICATION OF PROCESSES WITH DIRECTION-DEPENDENT DYNAMIC RESPONSES

K. R. Godfrey

Inter-University Institute of Engineering Control
University of Warwick, Coventry, CV4 7AL, England

P.A.N. Briggs
Control Engineering Division
Warren Spring Laboratory, Gunnels Wood Road,
Stevenage, Herts., SG1 2BX, England

Many processes have dynamic responses which are dependent on the direction in which the response variable is moving. The effects of such non-linear behaviour on the weighting function model of the process obtained by crosscorrelation and on the difference equation model obtained by a generalized least-squares procedure are determined theoretically for a process with first-order dynamics perturbed with pseudo-random binary signals. The theory is confirmed by results from a hybrid computer simulation, and computer-simulated results for processes with second-order dynamics are also presented. Experimental work on two pilot-scale processes with direction-dependent dynamics is reported, and further examples from the literature are examined.

INTRODUCTION

The implementation of a control system on an industrial process involves four stages, namely identification to obtain a model of the process, simulation of the model, design of the control system, and final tuning of the control system on the process. This paper is concerned with the first stage, and in particular with statistical models obtained by analysing the input-output records of the process.

Although in a few cases it is possible to obtain adequate models from the analysis of input-output records with the process in normal operation (Rake 1970), it is usually necessary to use a test signal to perturb those inputs of interest. For some processes, adequate models may be obtained from simple step response tests (Godfrey 1970), but when a substantial amount of noise is present, either in the process or in the measurement devices, some form of statistical analysis must be used to obtain the models. The most frequently-used perturbation signals are those based on pseudo-random binary sequences, when the process model can be a weighting function obtained by crosscorrelation or a difference-equation model obtained by more complex computation, such as generalised least-squares estimation (Clarke 1967).

Experimental work on an industrial process nearly always has to be carried out while the plant is in production, and in order not to produce off-specification material, the amplitudes of the perturbation signals have to be kept small. This results in a model which should be a linear approximation to the process, so that linear system control theory can be used in the control design. If the control system is required to operate over a wide range of throughputs, a series of these approximately linear models is obtained about several operating points. Although general non-linear modelling techniques have been and are being developed (Ream 1970, Barker 1972), these are complex and the accuracy of estimation of their parameters is particularly susceptible to noise.

One of the most persistent non-linearities, even when the perturbation signal amplitude is small, is that in which the dynamics of the process are different according to whether the variable under investigation is increasing or decreasing. This situation often occurs in industry and two well-known examples are a temperature loop operating well above ambient temperature where, owing to heat losses, an increase in the variable by a given amplitude tends to take longer than a decrease of the same amplitude; and a steam drum where "flashing off" of steam causes pressure to fall quite slowly when increasing the flow out of the drum, but when decreasing the flow out, flashing off does not occur and pressure rises more rapidly.

This direction-dependent non-linearity does not present too much of a problem in step response measurements, where the responses in both directions can be measured. Controller settings based on these responses can then be made direction-dependent or, more simply, designed for the slower dynamics, accepting a somewhat downgraded performance in the opposite direction. The situation is not so simple when a pseudo-random binary signal is used to perturb the process and it is the purpose of this paper to establish the effects of this type of non-linearity on the weighting-function and difference-equation models, both of which assume linearity of the process.

THEORY

Process with first-order dynamics

Consider a noise-free first-order process with input x and output y described by the equations:-

$$T_U \dot{y} + y = x, \quad \dot{y} \text{ positive} \tag{1A}$$

$$T_D \dot{y} + y = x, \quad \dot{y} \text{ negative} \tag{1B}$$

For a first-order process, y responds immediately to a change of sign in x, so that if the input signal is a pseudo-random binary signal (p.r.b.s) with levels ±1, and clock-pulse interval Δt

$$\operatorname{sgn}(\dot{y}) = \operatorname{sgn}(x) \qquad (2)$$

The situation is therefore described by the difference equation :

$$y_t = \alpha_{t-1}\, y_{t-1} + (1-\alpha_{t-1})\, x_{t-1}\,, \qquad (3)$$

$$\text{where } \alpha_{t-1} = a + bx_{t-1} \qquad (4A)$$

$$a = \tfrac{1}{2}(\exp(-\Delta t/T_U) + \exp(-\Delta t/T_D)) \qquad (4B)$$

$$b = \tfrac{1}{2}(\exp(-\Delta t/T_U) - \exp(-\Delta t/T_D)) \qquad (4C)$$

When corresponding expressions for y_{t-1}, y_{t-2},... to that of equation(3)for y_t are summed, it is found (Godfrey 1972) that for stable processes, (i.e. with a < 1 and b < 1) y_t can be expressed as a non-linear function of the past inputs $x_{t-\ell}$:

$$y_t = K \sum_{p=o}^{v} \prod_{\ell=1}^{p} (a + b\, x_{t-\ell}) - \frac{1-a}{b} \qquad (5)$$

$$\text{where } K = \frac{(1-a)^2}{b} - b \qquad (6)$$

In deriving equation (5), the integer v is taken as large, but less than the number of digits in the p.r.b.s. The right-hand side of equation (5) may be split up into constant, first-order and higher-order terms.

(i) Constant term

$$W_o = K(1 + a + a^2 + \ldots) - \frac{1-a}{b} \qquad (7)$$

(ii) First-order terms

$$W_1 = K \frac{b}{1-a}(x_{t-1} + ax_{t-2} + a^2x_{t-3} + a^3x_{t-4} + a^4x_{t-5} + \ldots. \qquad (8)$$

(iii) Second-order terms

$$\begin{aligned} W_2 = K \frac{b^2}{1-a} (& x_{t-1}x_{t-2} + ax_{t-2}x_{t-3} + a^2x_{t-3}x_{t-4} + \\ & + a^3x_{t-4}x_{t-5} + \ldots.. \\ & + ax_{t-1}x_{t-3} + a^2x_{t-2}x_{t-4} + \\ & + a^3x_{t-3}x_{t-5} + \ldots.. \\ & + a^2x_{t-1}x_{t-4} + a^3x_{t-2}x_{t-5} + \ldots. \\ & + a^3x_{t-1}x_{t-5} + \ldots.. \\ & + \ldots\ldots..) \end{aligned} \qquad (9)$$

(iv) Third-order terms

$$\begin{aligned} W_3 = K. \frac{b^3}{1-a}(& x_{t-1}x_{t-2}x_{t-3} + ax_{t-2}x_{t-3}ax_{t-4} + \\ & + a^2x_{t-3}x_{t-4}x_{t-5} + \ldots.. \\ & + ax_{t-1}x_{t-2}x_{t-4} + a^2x_{t-2}x_{t-3}x_{t-5} + \ldots \\ & + ax_{t-1}x_{t-3}x_{t-4} + a^2x_{t-2}x_{t-4}x_{t-5} + \ldots \\ & + a^2x_{t-1}x_{t-2}x_{t-5} + \ldots. \\ & + a^2x_{t-1}x_{t-3}x_{t-5} + \ldots. \\ & + a^2x_{t-1}x_{t-4}x_{t-5} + \ldots. \\ & + \ldots\ldots.) \end{aligned} \qquad (10)$$

(v) Fourth-order terms

$$\begin{aligned} W_4 = K. \frac{b^4}{1-a}(& x_{t-1}x_{t-2}x_{t-3}x_{t-4} \\ & + ax_{t-2}x_{t-3}x_{t-4}x_{t-5} + \ldots. \\ & + ax_{t-1}x_{t-2}x_{t-3}x_{t-5} + \ldots.. \\ & + ax_{t-1}x_{t-2}x_{t-4}x_{t-5} + \ldots.. \\ & + ax_{t-1}x_{t-3}x_{t-4}x_{t-5} + \ldots.. \\ & + \ldots\ldots\ldots.) \end{aligned} \qquad (11)$$

(vi) Fifth-order terms

$$\begin{aligned} W_5 = K. \frac{b^5}{1-a}(& x_{t-1}x_{t-2}x_{t-3}x_{t-4}x_{t-5} + \ldots\ldots \\ & + \ldots\ldots..) \end{aligned} \qquad (12)$$

(vii) Higher-order terms

The pattern of the formation of each set of successively higher-order terms is evident, and it is seen that they are multiplied by factors involving b, b^2, b^3, ... The factor b is a measure of the difference in dynamics between positive-going and negative-going directions. Higher powers of b tend to zero quite rapidly, so that higher-order terms become small.

For a process with additive (high-frequency) noise, equation (2) will no longer hold exactly. Nevertheless, the above theoretical treatment will still represent the process to a large extent with the noise causing, as is usual in correlation experiments, an increase in the variance of the crosscorrelation estimates.

Process with higher-order dynamics

It is possible to split up the transfer function of a process with nth-order dynamics into partial fractions and to regard the process response as the sum of n first-order responses. At first sight, therefore, it appears possible to extend the above theory in this way, but unfortunately, equation (2) no longer applies to the higher-order process, for which the time of the change of sign in $\dot{y}$ following a change of sign in x is variable. The higher-order case does not appear to be amenable to exact analytical treatment along the above lines, although it would be expected intuitively that a result bearing some resemblance to that for the first-order case will be obtained.

An alternative approach for higher-order systems, based on earlier work on input transducers with direction-dependent responses (Godfrey 1966) is currently being investigated and it is hoped to report on this at the Conference.

COMPARISON OF DIFFERENT CLASSES OF PSEUDO-RANDOM BINARY SIGNALS

For most industrial processes, the difference in dynamics in the two directions will not be very large, so that $|\underline{b}|$ will be substantially less than $\underline{a}$, so that we shall only be concerned with the contributions of W_o, W_1 and W_2.

The expression (equation (7)) for W_o is independent of the class of p.r.b.s. used. When the response signal is crosscorrelated with the p.r.b.s., this term gives rise as usual to a d.c. offset when a signal based on a maximum-length sequence or on a quadratic residue code is used and zero when a signal based on an inverse-repeat sequence is used (Godfrey 1969).

The first-order terms (equation (8)) give rise to an impulse response in the crosscorrelation function, starting at lag one, which has a coefficient of $\frac{Kb}{1-a}$ and which decreases by a factor of $\underline{a}$ at each sampling interval. This is independent of the class of p.r.b.s. used. For the inverse-repeat sequence, this response is repeated, with its sign reversed, one half period later, as usual.

By contrast, the effects of the second-order terms depend on the internal structure of the p.r.b.s. used. The effects for the three most commonly-used classes of p.r.b.s. are considered below.

Maximum-length sequences (m-sequences)

Some simplification of equation (9) is possible by making use of the shift-and-multiply property (Zierler 1959), corresponding to the shift-and-add property in the (0,1) field:

$$x_{t-\alpha}x_{t-\beta} = -x_{t-u} \qquad (13)$$

Defining a lag k_1 such that

$$x_{t-\ell}x_{t-\ell-1} = -x_{t-\ell-k_1}, \qquad (14)$$

it may be seen that

$$\begin{aligned} x_{t-\ell}\,x_{t-\ell-2} &= x_{t-\ell}x_{t-\ell-1}x_{t-\ell-1}x_{t-\ell-2} \\ &= x_{t-\ell-k_1}\,x_{t-\ell-k_1-1} \\ &= -x_{t-\ell-2k_1} \end{aligned} \qquad (15)$$

In general,

$$x_{t-\ell}\,x_{t-\ell-2^p} = -x_{t-\ell-2^p k_1} \qquad (16)$$

Similarly defining a lag k_3 such that

$$x_{t-\ell}\,x_{t-\ell-3} = -x_{t-\ell-k_3} \qquad (17)$$

it follows that

$$x_{t-\ell}\,x_{t-\ell-6} = -x_{t-\ell-2k_3} \qquad (18)$$

and so on. Each shift on the left-hand side of these equations from 1 to $2^n - 2$, where n is the number of stages in the shift register producing the m-sequence, produces a different lag on the right-hand side (Tsao 1964).
The lag k_1, k_3, ... are dependent on the particular m-sequence used and are different for different m-sequences of the same period.

Thus, the pattern of lag pairs in equation (9) may be replaced by single lags:-

$$\begin{aligned} W_2 = -K.\frac{b^2}{1-a}\,(&x_{t-k_1-1} + ax_{t-k_1-2} + a^2x_{t-k_1-3} + \\ &+ a^3x_{t-k_1-4} + \ldots\ldots \\ &+ ax_{t-2k_1-1} + a^2x_{t-2k_1-2} + \\ &+ a^3x_{t-2k_1-3} + \ldots\ldots \\ &+ a^2x_{t-k_3-1} + a^3x_{t-k_3-2} + \ldots\ldots \\ &+ a^3x_{t-4k_1-1} + \ldots\ldots \\ &+ \ldots\ldots\ldots\ldots) \end{aligned} \qquad (19)$$

It may be seen that the second-order terms contribute to the crosscorrelation function a series of impulse responses, each decaying by a factor of $\underline{a}$ at each sampling interval and oriented with respect to the contribution of the first-order terms according to the sign of -b. The first starts at lag (k_1+1) and is smaller than the first-order term by a factor of b, and successive ones start at lag $(2k_1+1)$ (smaller by a factor ab), at lag (k_3+1) (smaller by a factor a^2b), at lag $(4k_1+1)$ (smaller by a factor a^3b) and so on. The second-order terms do in fact contribute an inpulse response starting at every lag, but these get successively smaller by a factor of $\underline{a}$. Since a < 1 for stable processes, only the first few terms are significant, and the second-order effects consist of a few sharp discontinuities in the crosscorrelation function. Due to the periodicity of the input signal, any of the transients which have not died away by lag (2^n-1), i.e. the period of the sequence, will be continued from lag zero onwards.

Inverse-repeat sequences

For inverse-repeat sequences, the fact that the

second half of the sequence is the negative of the first half means that the contribution to the crosscorrelation function of the second-order term is nil.

Quadratic Residue Codes

Quadratic residue codes do not possess the shift-and-multiply property, so that the effects of second-order terms on the crosscorrelation function do not follow a similar pattern to that for maximum-length sequences. Several short quadratic residue codes have been examined and no obvious pattern emerges, but the significant contributions of the second-order terms are far more numerous than for m-sequences of comparable period, and could easily be taken for the effects of high-frequency noise.

As noted at the start of this Section, in the industrial process dynamics case, $\underline{a}$ is generally substantially larger than $|\underline{b}|$, so that the effects of W_3, W_4, W_5, ... on the cross-correlation function can be ignored. A different case has been considered by Moore in connection with experiments on a gas turbine engine (Moore 1970), in which the input transducer had direction-dependent dynamics. Moore considered the limiting case when the dynamics are instantaneous in one direction and exponential in the other. This was considered in some earlier work (Godfrey 1966) but with the limitation that the exponential time-constant was sufficiently small for the transient to have been largely completed within one clock-pulse interval. The input-transducer error with this limitation removed produces a particularly interesting result when an m-sequence input is used, and this is considered in more detail in the next section.

INPUT-TRANSDUCER ERROR WITH AN m-SEQUENCE INPUT

When the input-transducer is instantaneously acting in one direction and exponential in the other direction, then from equation (4B) and (4C), $a = |b|$. In this case, the d.c. term and the main peak on the crosscorrelation function are still given by equations (7) and (8) respectively, while the first additional peak still comes from the W_2 term (equation (9)) and is less than the main peak by a factor of $\underline{b}$. This peak will be referred to as the additional peak of magnitude 1. The difference between this case and the industrial process dynamics case considered above is that the further additional peaks from the W_2 term now have additional peaks of the same magnitude from the higher-order terms.

To show this, it is necessary to consider further the shift-and-multiply property. As noted above,

$$x_{t-1}\, x_{t-\delta} = -x_{t-\lambda}\ , \quad \delta > 1 \tag{20}$$

but in addition,

$$x_{t-1}\, x_{t-\gamma}\, x_{t-\delta} = +x_{t-\nu}\ , \quad \delta > \gamma > 1 \tag{21}$$

$$x_{t-1}\, x_{t-\beta}\, x_{t-\gamma}\, x_{t-\delta} = -x_{t-\sigma}, \quad \delta>\gamma>\beta>1 \tag{22}$$

$$x_{t-1}\, x_{t-\alpha}\, x_{t-\beta}\, x_{t-\gamma}\, x_{t-\delta} = +x_{t-\omega}, \quad \delta>\gamma>\beta>\alpha>1 \tag{23}$$

and so on (Zierler 1959). Also, provided that $(\delta-1)$ is less than $(n-1)$, where n is the number of stages in the shift register, all the resulting lags λ, ν, σ, ω are different (Tsao 1964). Further, if the separation $(\delta-1)$ equals n, each lag λ, ν, σ, ω is still different except for the combination given by the feedback polynomial itself which gives $-x_{t-1}x_{t-1} = -1$ on the right-hand side. If the lag separation $(\delta-1)$ exceeds n, then resultant lags on the right-hand side start repeating, although as noted above, each value of λ in equation (20) is distinct up to $(\delta-1) = (2^n-2)$. The expressions for the higher-order terms W_3, W_4, W_5... may be reduced in a similar manner to the W_2 term, with single values of x, appropriately delayed replacing the triple, quadruple, quintuple ... combinations of equations (10), (11), (12)...

Thus, there are now two additional peaks of magnitude 2, one from the W_2 term, a factor of $\underline{a}$ smaller than the first additional peak and one from the W_3 term, a factor of $\underline{b}$ smaller. The W_2 term again gives one additional peak of magnitude 3, smaller than the first additional peak by a factor of a^2, while W_3 contributes two peaks of magnitude 3, smaller by a factor of ab and W_4 contributes one peak of magnitude 3, smaller by a factor of b^2. The pattern is now emerging, and there are eight additional peaks of magnitude 4, one from W_2, smaller than the first additional peak by a factor of a^3, three from W_3, smaller by a factor of a^2b, three from W_4, smaller by a factor of ab^2 and one from W_5 smaller by a factor of b^3. In this case, the dynamics of the transducer should not be very long in comparison with the clock-pulse interval, so that higher powers of a will tend to zero fairly rapidly. Hence in most practical cases, peaks of magnitude 5 and above should be negligible compared with the main peak and even magnitude 4 peaks should be very small.

The direction of the additional peaks depends on the sign of b, and if it is negative, all the additional peaks are positive due to the alternation of signs in equations (20) to (23). When b is positive additional peaks derived from W_2, W_4,... are negative, so that half the additional peaks of magnitude greater than one are negative.

This theory confirms the simulation results obtained by Moore (1970) who also noted that the positions of the additional peaks of magnitude p are the same as the backward shift in the m-sequence from the end of the run of n ones to the ends of runs of (n-p) ones or minus ones in the sequence (for $p < (n-1)$). This is readily confirmed when considering the shifting operation of equations (20) to (23) relative to the run of n ones for values of $(\delta-1)$ between 1 and $(n-1)$, noting the inequalities which must be obeyed for each equation. If the run of length $(n-p)$ is a run of ones, the additional peak in the crosscorrelation function is always positive, while if it is a run of minus ones, the additional peak is negative if b is positive. This argument can be extended to include the case of $(\delta-1) = n$, for which the $(2^{\delta-2}-1)$ additional peaks of magnitude $(\delta-1)$ will be in positions corresponding to the backward shift in the m-sequence from the end of the run of n ones to the ends of runs of no ones and no minus ones. Here, a run of no ones means anywhere within a run of minus ones, while a run of no minus ones means anywhere within a run of ones. (As noted above, one additional peak is excluded when $(\delta-1) = n$, since one of the combinations of equations (21), (23)... corresponding to the feedback polynomial produces -1). For $(\delta-1)>n$, the additional peaks start overlapping, but they will generally be negligibly small.

COMPUTER SIMULATION WORK

We return now to the industrial process case, for which the magnitude of b is considerably less than a, so that the effects of third-order and higher-order terms may be neglected.

Confirmation of the theory for a process with first-order dynamics

A first-order process with time constant $2\Delta t$ in the -1 to +1 direction and $4\Delta t$ in the +1 to -1 direction has been simulated on a hybrid computer, and input-output crosscorrelation functions computed (digitally) for a number of different perturbation signals. The process was made noise-free to facilitate interpretation of the results. For the process simulated, a = 0.693 and b = -0.087.

The crosscorrelation function obtained when using a 31-digit m-sequence, generated from a five-stage shift register with feedback from stages 3 and 5, is shown in Fig. 1A. The main peak has been normalised to 1 and the d.c. level has been removed. For this sequence, $k_1 = 14$, $2k_1 = 28$, $k_3 = 5$, $4k_1 = 25$ and $k_5 = 3$. Pulse responses of the same shape as the first are therefore expected to start at lag 15 with amplitude, relative to the main peak, of -b (=0.087), at lag 29 with relative amplitude -ab(=0.060), at lag 6 with relative amplitude $-a^2b$ (=0.042), at lag 26 with relative amplitude $-a^3b$ (=0.029), at lag 4 with relative amplitude $-a^4b$ (=0.020) and so on. This is confirmed in Fig. 1A.

Fig. 1B shows the crosscorrelation function when the perturbation signal was the 31-digit quadratic residue code with a plus one in the 31st position. It may be seen that the effects of the second-order terms are now distributed over the whole crosscorrelation function and are very similar in appearance to the effects of noise.

The crosscorrelation function resulting from using a 62-digit inverse-repeat sequence, obtained by inverting every other digit of the 31-digit m-sequence used above, is shown in Fig. 1C. Here, the sawtooth off-peak pattern of the sequence autocorrelation function and the sharp discontinuity in the impulse response of the first-order process combine to produce a small amplitude sawtooth pattern in the input-output crosscorrelation function (Godfrey 1969), and this has been removed in Fig. 1C. Only the first half of the crosscorrelation function is

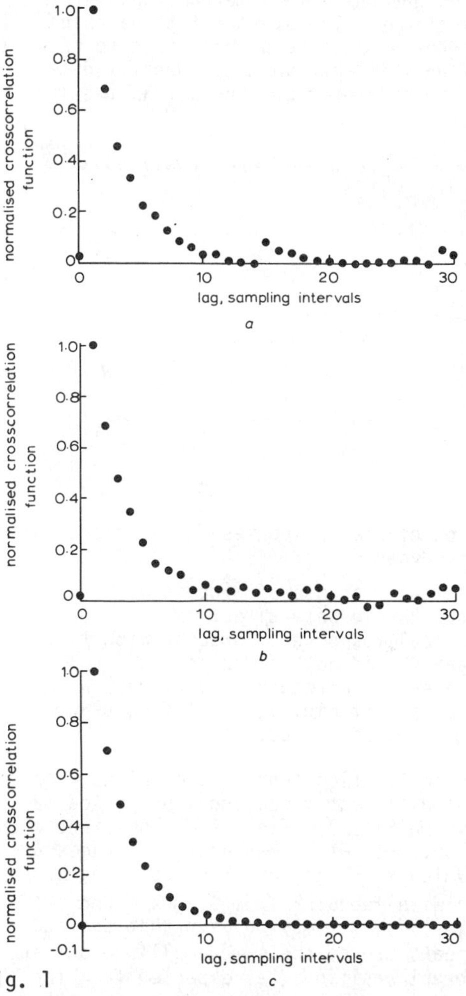

Fig. 1

Crosscorrelation functions for process with 1st-order dynamics (time constant $2\Delta t$ in -1 to +1 direction and $4\Delta t$ in +1 to -1 direction)

Perturbation signals based on

a 31-digit m-sequence, feedback from stages 3 and 5

b 31-digit quadratic residue code

c 62-digit inverse-repeat sequence

shown, as the second half is the negative of the first half. It may be seen that the peaks due to the second-order terms disappear, as expected. The third-order terms are very small in amplitude and the first additional peak due to these, of relative amplitude b^2 (=0.0076), is only just discernible at delay 24.

In Fig. 2 the step response obtained by integrating the pulse response of Fig. 1C is shown and is compared with the step response of two linear processes, one with time-constant $2\Delta t$ in both directions and one with time constant $4\Delta t$ in both directions. The steady-state gains of all three processes have been normalised to one. It may be seen that the dynamics identified tend towards the shorter time-constant as the lag increases.

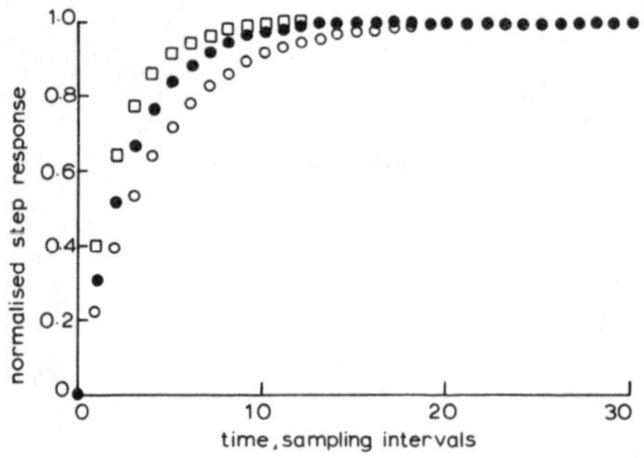

Fig. 2

Comparison of step responses for processes with 1st-order dynamics

□- theoretical response of process with time constant $2\Delta t$ in both directions
○- theoretical response of process with time constant $4\Delta t$ in both directions
●- integrated pulse response of process with dynamics as for Fig. 1, identified using 62-digit inverse-repeat

The crosscorrelation functions obtained when using two different m-sequences of period 127 are shown in Fig. 3. Fig. 3A was obtained when using a sequence with feedback from Stages 6 and 7 (for which k_1 = 121) and Fig. 3B using a sequence with feedback from Stages 4 and 7 (for which k_1 = 97). It may be seen that the additional peaks are of the same amplitude but are in different positions, as expected from the theory.

Simulation results for processes with second-order dynamics

The crosscorrelation function for a process with two time-constants, each of $4\Delta t$, in the -1 to +1 direction and two time-constants, each of $2\Delta t$, in the +1 to -1 direction is shown in Fig. 4 for a 127-digit m-sequence with feedback from stages 4 and 7 and in Fig. 5 for a 62-digit inverse-repeat sequence. The difference in dynamics has deliberately been made rather larger than would usually be encountered in practice, to illustrate the effects of the non-linearity clearly. It may be seen from Fig. 4 that extra peaks due to second-order terms start in the same positions as for the first-order process, but they are not now all of the same shape. As before, the inverse-repeat sequence (Fig. 5) produces a smoother crosscorrelation function, although not quite as smooth as for the first-order process, because the effects of the third-order terms are a little larger with the greater difference in dynamics between the two directions. The step response obtained by integrating the curve of Fig. 5 again tends towards the step response of the faster linear process as the lag increases.

EXPERIMENTAL WORK ON PILOT-SCALE PROCESSES

Experiments designed to illustrate the effects of direction-dependent dynamics in practical situations have been carried out on two pilot-scale processes.

Reactor section of an acetone plant

The first experiment was carried out on the reactor section of a pilot-scale chemical process for the production of acetone from isopropanol (Smith 1966) which, until recently, was operational at Warren Spring Laboratory. A block diagram of this section is shown in Fig. 6. Liquid feed of isopropanol azeotrope (88% by weight isopropanol/water mixture) enters the vaporiser, together with some hydrogen to prevent carbonisation of the catalyst, and the resulting vapour is further heated in the preheater by the reactor products before entering the reactor itself. Both the reactor and vaporiser are heated by a flow of hot gas and a by-pass valve (V2) is used to control the flow of heating gas through the reactor. Earlier experimental work in connection with a hybrid computer simulation study (Davies 1972) of this section of the plant had indicated the possibility of obtaining unequal dynamics in the two directions of the temperature responses of the reactants when changing the heating gas flow through the reactor due to heat losses from the system.

For the present experiment, the perturbation signal, derived from a 127-digit m-sequence with feedback shift register connections from Stages 4 and 7 and with a clock-pulse interval of 1 min., was applied to the heating gas by-pass valve V2, and the response of the product temperature within the reactor was measured. The crosscorrelation function between these two variables is shown in Fig. 7, from which it may be seen that there are clear discontinuities starting at lags 48, 68 and 98. These are in the same positions as the main discontinuities of Fig. 4, thus indicating the presence of unequal dynamic responses in the pilot plant. It would have been interesting to record the step responses in each direction also, but unfortunately pressure of experimental work on the plant at the time (immediately prior to its closure) was too great for this to be done.

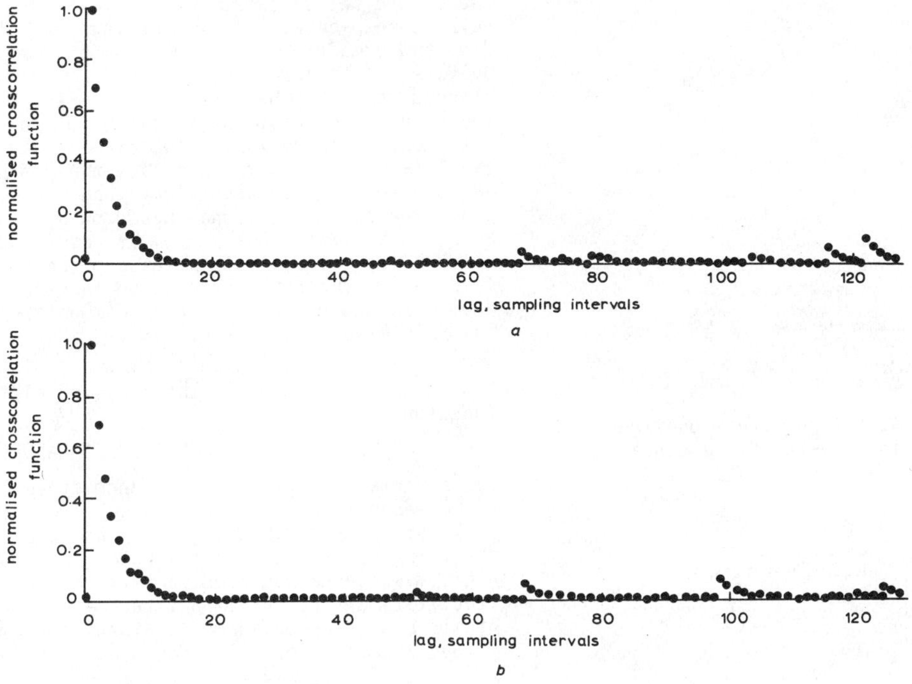

Fig. 3
Process as for Fig. 1, but using perturbation signals based on 127-digit m sequences

a Feedback from stages 6 and 7

b Feedback from stages 4 and 7

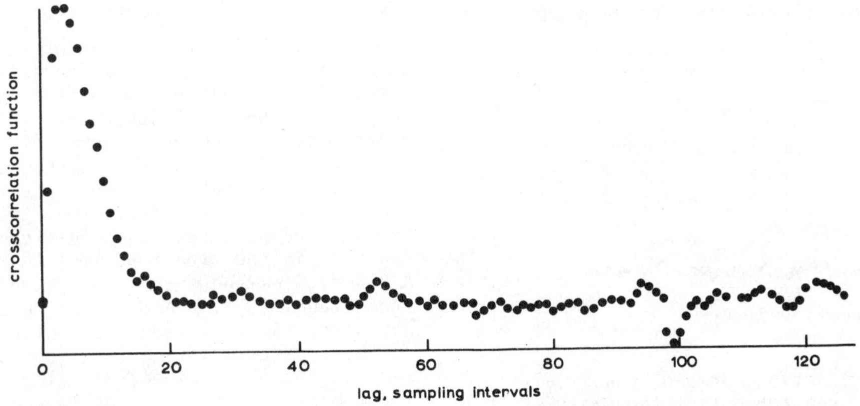

Fig. 4
Crosscorrelation functions for process with 2nd-order dynamics (two time constants each of $4\Delta t$ in the - 1 to +1 direction, and two time-constants each of $2\Delta t$ in the +1 to -1 direction)
The perturbation signal is based on a 127-digit m sequence with feedback from stages 4 and 7

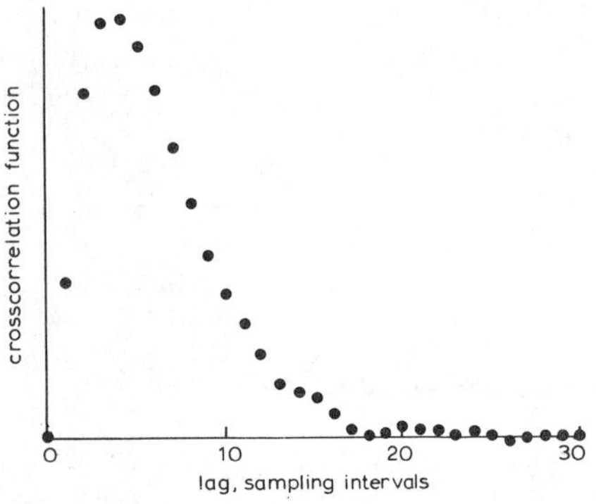

Fig. 5

Process as for Fig. 4, but perturbation signal based on a 62-digit inverse-repeat sequence.

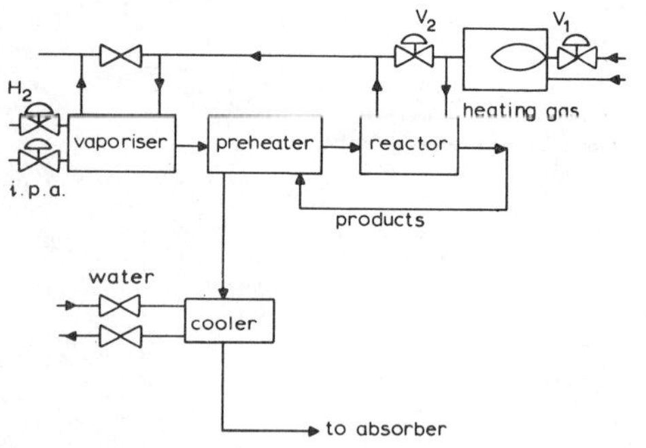

Fig. 6

Reactor section of pilot-scale acetone plant

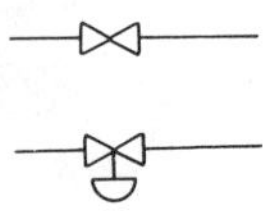

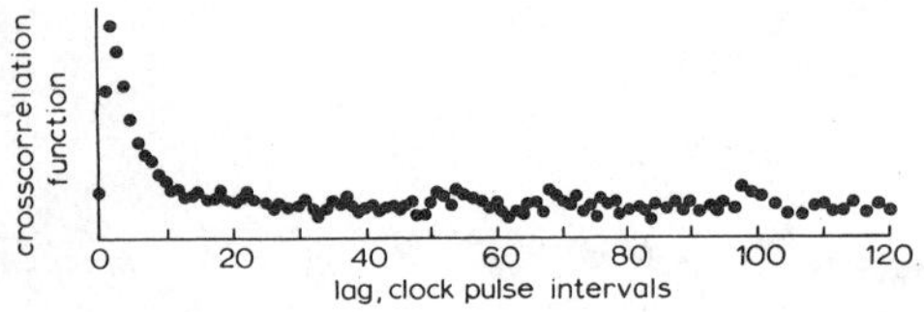

Fig. 7

Crosscorrelation function between bypass-valve (V_2) position and reactor-outlet temperature

The perturbation signal is based on a 127-digit m sequence, feedback from stages 4 and 7, and a clock-pulse interval of 1 min

Steam raising plant

The second set of experiments investigated the relationship between feed flow into the steam drum and pressure within the drum of the Oxford University steam-raising rig (Clarke 1973), a block diagram of which is shown in Fig. 8. It was suspected that this relationship had slightly different dynamics in the two directions due to the "flashing off" of steam, but the difference was too small to be seen clearly on step responses. When a perturbation signal based on a 31-digit m-sequence with feedback from Stages 2 and 5 and clock-pulse interval of 2s was applied to the feed valve, the crosscorrelation function with drum pressure output of Fig. 9 was obtained. For this m-sequence, $(k_1 + 1) = 19$, $(2k_1 + 1) = 6$ and $(k_3 + 1) = 30$, and quite distinct discontinuities are evident in the crosscorrelation function of these delays.

FURTHER EXAMPLES FROM THE LITERATURE

The effects of direction-dependent dynamic responses have been evident in some recent applications described in the literature. In the first of these, a signal based on a 1023-digit m-sequence was used to perturb the reactivity of the DRAGON nuclear reactor at the Atomic Energy Establishment, Winfrith (Cummins 1968a, 1968b). For this sequence, $k_1 = 76$, and subsidiary peaks starting at lag 77 are evident on the crosscorrelation functions. Unfortunately, the graphs are plotted only for small lags, so that it is not possible to see whether there are also peaks starting at lags $(2k_1 + 1)$, etc.

In a second application, a molecular beam was modulated with a 255-digit m-sequence (Hirschy 1971), for which $k_1 = 58$ and here a small-amplitude peak starting at lag 59 is evident. The noise level on the results is too high for any other subsidiary peaks to be resolved.

From a quite different field of study, the cross correlation between pseudo-random variations of extracellular calcium concentration and the response of peak isometric tension of an isolated heart muscle has been measured (Toll 1973). Here, the (static) dose response curve is severely non-linear and cannot be taken as linear even over the small amplitude range of the applied perturbations; this causes the dynamics in the two directions to be different. A 255-digit m-sequence signal with feedback from stages 2, 3, 4 and 8, for which $k_1 = 25$, was used. Sharp discontinuities are evident on the crosscorrelation function starting at lags 26 and 51 as expected. (The correlation functions are not shown for lags greater than 60 for clarity of presentation). A further set of experimental results was obtained using a 510-digit inverse-repeat sequence and in these the discontinuities disappear, as expected from the theory above.

The final application was concerned with the measurement of speed response of a gas turbine engine (Moore 1970), but here the dynamics of

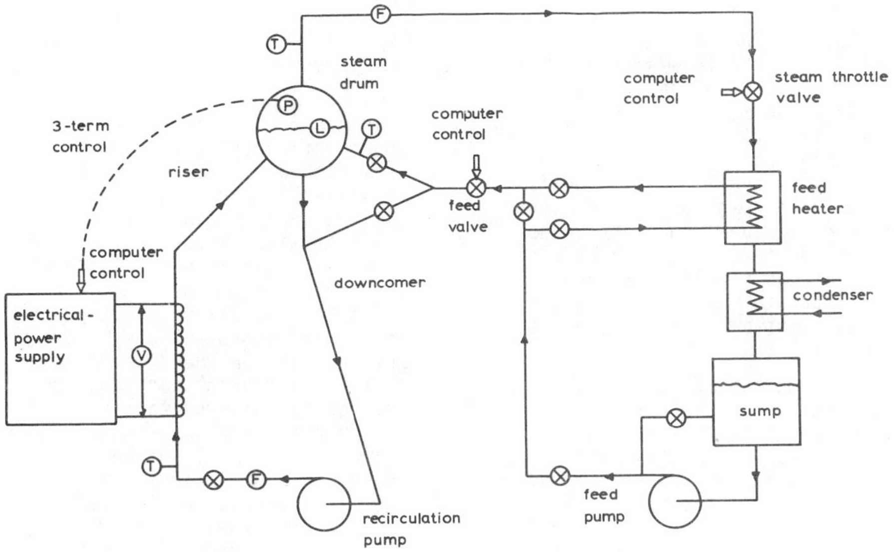

Fig. 8

Block diagram of Oxford University steam-raising plant

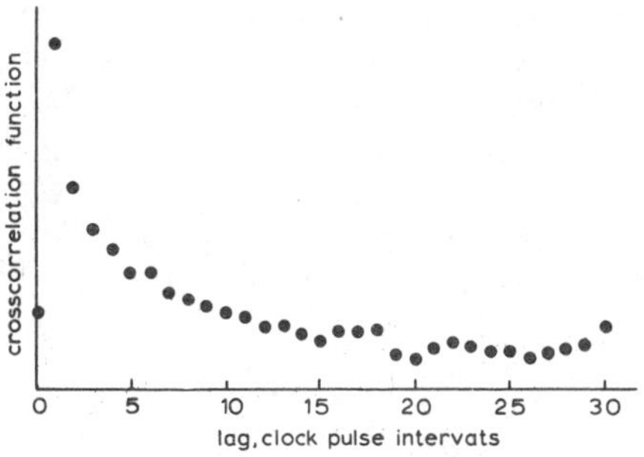

Fig. 9

Crosscorrelation function between feed-valve position and drum pressure

The perturbation signal is based on a 31-digit m-sequence, feedback from stages 2 and 5, and a clock pulse interval of 2s

of the input-transducer were direction-dependent while the engine itself was thought to be linear over the small range of perturbation signals applied. Provided that the input-transducer dynamics are fast with respect to the dynamics of the overall system, the additional peaks on the input-output cross-correlation function are of the same shape as the main peak (Godfrey 1966), so that they can easily be mistaken for an effect in the process itself. For this reason, Moore has referred to them as echoes. In the gas-turbine engine results, at a low clock-pulse rate, only one additional peak, starting at lag (k_1 + 1) occurred, but as the clock-pulse rate was increased, further additional peaks occurred, starting in the positions expected from the theory on input-transducer error in the present paper.

THE EFFECTS ON DIFFERENCE-EQUATION MODELS

The input-output records from the computer simulation runs and pilot-plant experiments described above have also been used to compute low-order difference-equation models by generalised least-sequares estimation (Clarke 1967). For the experiments using inverse-repeat sequences the pulse responses of the difference-equation models are in very good agreement with the crosscorrelation function estimates of the weighting functions, a first-order difference equation model being used for the results of Fig. 1C and a second-order difference equation model for the results of Fig. 5. This is to be expected because both models are based on least-squares estimation from the same data.

For the m-sequence tests, the two models are in good agreement, the pulse response of the difference-equation model being similar to the main peak of the weighting-function model. The discontinuities in the crosscorrelation function caused by the second-order non-linear terms are not modelled by the difference equation model, as expected, since the latter is only of low order. For this reason, any small discontinuities on the main peak itself caused by these second-order terms are not modelled by the difference-equation models, but the departures from the crosscorrelation functions are quite small. The differences are in fact too small for the pulse responses of the difference-equation models to be shown on the same Figures as the corresponding crosscorrelation functions without getting very confusing graphs.

CONCLUSIONS

The effects on weighting-function models obtained by crosscorrelation and on difference-equation models obtained by generalised least-squares estimation of unequal dynamics in the up and down directions - a situation frequently occurring in practice - have been examined for perturbation signals based on pseudo-random binary signals. The theory has been developed for a first-order process, for which it has been found that the initial main peak tends towards the shorter of the two time-constants as the lag increases; this is rather unfortunate because any controller based on the results must be able to cope with the longer dynamics.

For an m-sequence input, the effects on the crosscorrelation function of second-order terms assume a pattern which is quite systematic, being based on the shift-and- multiply property, and unless the dynamics in the two directions are very different, there are comparatively few additional peaks with significant amplitude. When an input signal based on a quadratic residue code is used, the effects of second-order terms are distributed over the whole crosscorrelation function with no obvious pattern and could very easily be taken for the effects of noise. If an inverse-repeat sequence is used there are no additional peaks due to second-order terms and only the main peak together with the effects of higher odd-order terms, which are extremely small, is obtained.

When the dynamics in the two directions are very different, as might occur in an input transducer, for example, a very interesting result is obtained when a p.r.b.s. based on an m-sequence is used. Here, if the transition in one direction is instantaneous while in the other direction it is exponential, additional peaks in the cross-correlation function are more numerous, and take on certain amplitudes, there being 2^0 peaks of magnitude 1, 2^1 peaks of magnitude 2 (smaller than magnitude 1) and so on. The positions of the peaks are related to the backward shift of runs of plus ones and minus ones from the run of n ones in the sequence.

Extension of the theory to higher-order processes is not straightforward, although an alternative approach is currently being investigated. Intuitively, a result along similar lines to that for the first-order process is expected, and this is indeed what has been obtained in computer-simulated results for several processes with second-order dynamics. The discontinuity starting at lag $(k_1 + 1)$ is very evident when a perturbation signal based on an m-sequence is used, and this is also true for the pilot-plant experiments.

The pulse responses of difference-equation models obtained by generalised least-squares estimation are in good agreement with the main peaks of the corresponding correlation functions, as expected. The difference-equation models, being of low order, naturally do not model the small discontinuities in the weighting function due to the second-order terms when signals based on m-sequences are used.

In this paper, no attempt has been made to model the non-linearities by using higher-order correlation functions and the analysis has been restricted to examining the effects on the more familiar linear models of the process. This is because the work has been carried out with industrial applications in mind, for which complex models are too intractable for control system design. In fact, if an m-sequence is used, the dynamics in both directions for a first-order process can be evaluated from the extra peaks on the crosscorrelation function but in most practical cases these will be masked by noise so that it will not be possible to do this with any accuracy. Frequently the experimenter will not know whether the direction-dependent non-linearity is significant, but if an m-sequence input is used, it should be possible to find this out by examining whether there is an extra peak in the crosscorrelation function in the region immediately following $(k_1 + 1)$ sampling intervals. It is not possible to give an analytical expression for k_1 and it has to be found by using the shift-and-multiply property. Values of $(k_1 + 1)$ for several short m-sequences are given in Table 1.

TABLE 1 - VALUES OF $(k_1 + 1)$ FOR SOME SHORT m-SEQUENCES

Number of stages, n	Length of m-sequence (2^n-1)	Feedback from stages	$(k_1 + 1)$
3	7	2, 3	6
4	15	3, 4	13
5	31	3, 5	15
5	31	1,3,4,5	14
5	31	1,2,3,5	13
6	63	5, 6	59
7	127	4, 7	98
7	127	6, 7	122

REFERENCES

BARKER, H.A., OBIDEGWU, S.N. and PRADISTHAYON, T. (1972). Performance of antisymmetric pseudo-random signals in the measurement of 2nd-order volterra kernels by crosscorrelation. Proceedings I.E.E., 119(3), pp. 353-362.

CLARKE, D.W. (1967). Generalised least-squares estimation of the parameters of a dynamic model. Paper 3.17 to IFAC Symposium on Identification in Automatic Control Systems, Prague, June 1967.

CLARKE, D.W., DYER, D.A.J., HASTINGS-JAMES, R., ASHTON, R.P. and EMERY, J.P. (1973) Identification and control of a pilot-scale boiling rig. Third IFAC Symposium on Identification and System Parameter Estimation, The Hague, June 1973.

CUMMINS, J.D. (1968a). Parameter estimation from DRAGON high-temperature gas-cooled reactor dynamic experiments. UK Atomic Energy Establishment, Winfrith, Report AEEW-R 571.

CUMMINS, J.D. (1968b) Further optimization studies of experimental dynamic responses measured on the HTGC DRAGON reactor. UK Atomic Energy Establishment, Winfrith, Report AEEW-R 603.

DAVIES, J. and FOWLES, A.J. (1972). Hybrid modelling of a chemical reactor. Proc. 2nd. International Symposium on Chemical Reaction Engineering, Amsterdam, 2-4 May 1972, pp. B8.63-B8.74.

GODFREY, K.R., EVERETT, D. and BRYANT, P.R. (1966). Input-transducer errors in binary crosscorrelation experiments - 2. Proceedings I.E.E., 113(1), pp. 185-189.

GODFREY, K.R. (1969). The theory of the crosscorrelation method of dynamic analysis and its application to industrial processes and nuclear power plant. Measurement and Control, 2(5), pp. T65-T72.

GODFREY, K.R. and SHACKCLOTH, B. (1970) Dynamic modelling of a steam reformer and the implementation of feedforward/feedback control. Measurement and Control, 3(5), pp. T65-T72.

GODFREY, K.R. and BRIGGS, P.A.N. (1972). The identification of processes with direction-dependent dynamic responses. Proceedings I.E.E., 119(12), pp. 1733-1739.

HIRSCHY, V.L. and ALDRIDGE, J.P. (1971). A crosscorrelation chopper for molecular beam modulation. Rev. Sci. Instrum., 42(3), pp. 381-383.

MOORE, D.J. (1970). Error correction applied to dynamic analysis. Rolls-Royce Ltd., Bristol Engines Division, Report EER/5033/70.

RAKE, H. (1970). Correlation analysis of blast-furnace operating data. Paper 11.1 to IFAC Symposium on Identification and Process Parameter Estimation, Prague, June 1970.

REAM, N. (1970). Non-linear identification using inverse-repeat m-sequences. Proceedings I.E.E., 117(1), pp. 213-218.

SMITH, W., FOWLES, A.J., MILES, J.E.P. and RAY, D.J. (1966). Design and construction of a system for the computer control of a complex chemical plant. Chem. Engr., 201, pp. CE.188 - CE.192.

TSAO, S.H. (1964). Generation of delayed replicas of maximal-length linear binary sequences. Proceedings I.E.E., 111, pp. 1803-1806.

TOLL, M. (1973). Systems analysis of inotropic interventions in isolated heart muscle. Ph.D. Thesis, University of London, January 1973.

ZIERLER, N. (1959). Linear recurring sequences. J. Soc. Indust. Appl. Math., 7(1), pp.31-48.

LEAST-SQUARES ESTIMATION OF RESTRICTED PARAMETERS IN PHYSICAL AND CHEMICAL MODELS

H. van Houwelingen, J. Petiet, C. Schweigman.
Akzo Research Laboratories, Corporate Research
Arnhem, The Netherlands.

The least-squares estimation of parameters in physical and chemical models may well be done by a modified Gauss-Newton method. A extension of this method allowing the parameters to lie within bounds is described. Moreover, a method to compute confidence regions in case the parameters are bounded is illustrated.

INTRODUCTION

The problem of computing the least-squares estimation of parameters in non-linear models on the basis of a series of experimental data can be formulated as a minimization problem of the form:

$$\text{Minimize: } F(x) = \sum_{i=1}^{m} f_i^2(x) \qquad (1)$$

where $f_i(x)$, $i=1,2,\ldots,m$ are non-linear functions in the variables or parameters $x=(x_1,x_2,\ldots,x_n)$

The function $f_i(x)$ represents the deviation ε_i between the ith-observation and its expected value. The errors ε_i are assumed to have zero means and the variance - covariance matrix of the errors ε_i is assumed to be equal to $I\sigma^2$, I being the identity matrix and σ^2 the common unknown variance of the errors. If the variance-covariance matrix has the form $V\sigma^2$, where V is a known (positive definite) matrix the same minimization problem may be obtained.
For the construction of the confidence regions it is further assumed that the errors ε_i are normally distributed.
The number of squares functions m is assumed to be greater than or equal to the number of variables n: $m \geq n$.
Besides, the functions $f_i(x)$ are assumed to be differentiable and their derivatives to be continuous and bounded.

In parameter-estimation problems the parameters may frequently not be chosen freely because the mathematical model is not defined for all values of the parameters or because on physical or chemical grounds it is known that the parameters have to lie within certain bounds.
E.g. parameters may be concentrations or reaction rate constants which have to be positive.
For parameter-estimation problems the type of restrictions:

$$a_i \leq x_i \leq b_i \quad i=1,2,\ldots,n \qquad (2)$$

where a_i and b_i are constants, is in practice very important.

This paper describes both an efficient method to handle constraints of type (2) as well as some statistical aspects of the estimation of parameters, especially the construction of confidence regions in cases where (2) has to be satisfied.

MINIMIZATION METHOD

To minimize functions of the special type (1) efficient methods have been published e.g. by: Box [3], Hartley [6], Marquardt [9, 10], Powell [11].
In this paper the following modification of Gauss-Newton's method will be applied.

Given an iteration point x, a new point $x + \xi$ is generated as follows:

$$\xi = -H^{-1} \nabla F \qquad (3)$$

where:
(i) ∇F is the gradient of $F(x)$
(ii) H^{-1}, the inverse of the $n * n$ matrix $H=2\,G'G$ and G', the transposed of the $m * n$ matrix G with elements:

$$G_{i,j} = \frac{\delta f_i}{\delta x_j} \quad i=1,2,\ldots,m;\ j=1,2,\ldots,n.$$

It will be assumed that the matrix H is not singular.

If $F(x+\xi) < F(x)$, the point $x + \xi$ is accepted as a new iteration point.
If $F(x+\xi) \geq F(x)$ and $\xi \neq 0$, a value of τ can be computed such, that $0 < \tau < 1$ and $F(x+\tau\xi) < F(x)$. Now the point $x+\tau\xi$ is accepted as a new iteration point.

The handling of the constraints (2) may without loss of generality be illustrated by considering the constraints:

$$x_i \geq 0 \ ,\ i=I \qquad (4)$$

where the index set I is defined by:
$I= \{1,2,\ldots,n\}$

If a point x satisfies (4), it is called feasible. If the function F(x) given by (1) is minimal in a feasible point x, it should be true that:

$$x_i \frac{\delta F}{\delta x_i} = 0 \qquad i \in I$$
$$\frac{\delta F}{\delta x_i} \geq 0 \qquad i \in I \tag{5}$$

For convex functions F the above conditions are sufficient to guarantee that F(x) is a global minimum; see e.g. Kunzi [8].
To deal with the restrictions (4) for the minimization method described above, we introduce the following index sets:

$\underline{N}(x)$: $\{i \in I \mid x_i = 0\}$
$\overline{N}(x)$: $\{i \in I \mid x_i > 0\}$
$\nu(x) \subset I$

and a corresponding n-dimensional vector $\zeta(\nu;x)$ given by:

$$\sum_{j \in \nu(x)} H_{ij}\, \zeta_j(\nu;x) = -\frac{\delta F}{\delta x_i} \qquad i \in \nu(x)$$
$$\zeta_i(\nu;x) = 0 \qquad i \notin \nu(x),\ i \in I \tag{6}$$

Thus $\zeta_i(\nu;x)$, $i \in \nu(x)$ is defined as the Gauss-Newton increment if only the coordinates x_i, $i \in \nu(x)$ are taken into consideration.
The elements H_{ij} are defined in (3).

The choice of $\nu(x)$ is the key to the proposed method and is based on the conditions in (5).
The choice is as follows:

Step 1: Investigation of the conditions in (5)

If $N_1(x) := \{i \in \underline{N}(x) \mid \frac{\delta F}{\delta x_i} \geq 0\}$

then $\nu(x) := \{i \in I \mid i \notin N_1(x)\}$

Step 2: Computation of the new direction $\zeta(\nu,x)$ with the aid of (6)

Step 3: Is the new point feasible?

If $N_2(x) := \{i \in \underline{N}(x) \mid \frac{\delta F}{\delta x_i} < 0 \cap \zeta_i(\nu;x) < 0\}$

then $\nu_1(x) := \{i \in \nu(x) \mid i \notin N_2(x)\}$

$\nu(x) := \nu_1(x)$

Now go back to "step 2", unless $N_2(x)$ is empty, in which case "step 4" should be executed.

Step 4: Crossing new boundaries
The increment vector $\zeta(\nu;x)$ corresponding to the last index set $\nu(x)$ is called $\xi(x)$.

Now define the index set $\Gamma : \{i \in \overline{N}(x) \mid x_i + \xi_i < 0\}$

and compute $\gamma = \min_{i \in \Gamma} \frac{x_i}{|\xi_i|}$.

If Γ is empty, $\gamma := 1$.

Step 5: Determination of the new iteration point
If $F(x+\gamma\xi) < F(x)$, then the point $x+\gamma\xi$ is accepted as the new iteration point.
If $F(x+\gamma\xi) \geq F(x)$ and x no solution point then a value of τ can be computed such that $0<\tau<1$ and $F(x+\tau\gamma\xi) < F(x)$.
Now the point $x+\tau\gamma\xi$ is accepted as a new iteration point.

This algorithm has the property that the increment $\xi(x)$ is zero, if, and only if (5) is satisfied; see Schweigman [12].
On the basis of this property, convergence of the iteration process above can be proved for a wide class of conditions.

EXAMPLE

The above described algorithm has been succesfully applied to a large number of physical and chemical problems, containing up to 50 parameters. An example illustrating the proposed method to minimize (1), given the constraints (2), concerns the interpretation of an X-ray diagram of a mixture of some materials.
The model used is:

$$y = \theta_1\{1+(\frac{u-\theta_2}{\theta_3})^2\}^{-\theta_4} + \theta_5\{1+(\frac{u-\theta_6}{\theta_7})^2\}^{-\theta_8}$$

$$+R\ \frac{\theta_1\theta_3}{\theta_{10}}\ \frac{\Gamma(\theta_4-0.5)\Gamma(\theta_{11})}{\Gamma(\theta_{11}-0.5)\Gamma(\theta_4)}\ \{1+(\frac{u-\theta_9}{\theta_{10}})^2\}^{-\theta_{11}}$$

$$+\ \theta_{12}+\theta_{13}u$$

where: y is the dependent variable
u is the independent variable
R is a given constant

$$\Gamma(m) = \int_0^\infty t^{m-1} e^{-t}\, dt$$

and $\theta_1, \theta_2, \ldots, \theta_{13}$ are the parameters to be computed.

The restrictions on the parameters $\theta_1, \theta_2, \ldots, \theta_{13}$ are of the form (2) with:
$75 \leq \theta_1 \leq 125$, $20 \leq \theta_2 \leq 20.8$, $.75 \leq \theta_3 \leq .95$, $1.55 \leq \theta_4 \leq 1.85$, $5 \leq \theta_5 \leq 35$, $20.8 \leq \theta_6 \leq 21.2$, $.5 \leq \theta_7 \leq 1.3$, $.2 \leq \theta_8 \leq .86$, $23.4 \leq \theta_9 \leq 23.8$, $.01 \leq \theta_{10} \leq 1.7$, $.51 \leq \theta_{11} \leq 2.1$, $1.9 \leq \theta_{12} \leq 2.1$, $-.01 \leq \theta_{13} \leq .01$.

These constants are partly based on physical grounds and partly on the fact that the above model is not defined for all values of the parameters.

The function $F(\theta)$ to be minimized is of the form (1):

$$F(\theta) = \sum_{i=1}^{101} \left[y_i - y(u_i;\theta)\right]^2$$

where:

θ is $(\theta_1, \theta_2, \ldots, \theta_{13})$

y_i, the i-th measured value of the dependent variable y.

u_i, the i-th measured value of the independent variable u.

Table I gives the course of the iteration process It shows that in almost each iteration point the iteration process continues on one or more boundaries. However, the solution point is an interior point.

It should be noted that without the restrictions above Gauss-Newton's method failed.

CONFIDENCE REGIONS

Various methods are known to construct a confidence region for the unrestricted parameters in a non-linear model. Hartley [5] gives an exact confidence region, which is in general difficult to construct. Box and Cox [2]found an approximate confidence region, which makes use of the contours of the likelihood function (see below). Mezaki and Kittzell [7] give an approximate confidence region, closely related to the Box-region, though generally larger. When the number of observations increases, the regions tend to be identical.
All these regions have the disadvantage that in general they have a complex form. In fact we should like to have such a region that for each parameter a confidence interval can be derived in a simple way. This means a rectangular (conservative) region. Dunn [4] gives such a region for the mean of a multivariate normal distribution with known variances, which can also be used for the parameters in a linear model. The disadvantage of this method is that for highly correlated parameters it gives an underestimate of the true confidence. In that case, however, reparametrization [1] of the model in order to reduce the correlation between the parameters may be considered.

The following presents a method as a possible approach to construct a confidence region, similar to the conservative region mentioned above, for the parameters in a linear model when one or more of the parameters have a lower or an upper bound or both. For application to non-linear models a linear approximation is used.

The method will be described for the case of two parameters θ_1 and θ_2, of which θ_1 has a lower bound:
$\theta_1 \geq 0$. It can easily be extended for more parameters or rectrictions.

Consider the model

$$\eta = \eta\,(u_1, u_2, \ldots, u_k;\ \theta_1, \theta_2)$$

where η is a function of known form, θ_1 and θ_2 are the unknown parameters and $u_1, u_2, \ldots, u_k$ are the independent variables.
Suppose that experiments are performed for n settings $u_{1j}, u_{2j}, \ldots, u_{kj}$, each with actual observation:

$y_j = \eta_j + \varepsilon_j$ $(j=1,2,\ldots,n)$, where the ε_j are a set of n independent errors having a normal distribution with zero mean and unknown variance σ^2.
Using the actual observations, a likelihood function $f(\theta_1,\theta_2)$ of the parameters can be found. This function is the joint density function of the errors ε_j in which the observed values y_j $(j=1,2,\ldots,n)$ are substituted so that it is a function of θ_1 and θ_2 only.
We define:

$\hat{\theta}_i$: unrestricted estimate of θ_i $(i=1,2)$

$\hat{\theta}_i^*$: restricted estimate of θ_i $(i=1,2)$ $(\hat{\theta}_i^* \geq 0)$

s_i: estimated standard error of $\hat{\theta}_i$ $(i=1,2)$

ρ: correlation coefficient between $\hat{\theta}_1$ and $\hat{\theta}_2$

ν: degrees of freedom of residual sum of squares $\sum_{j=1}^{n} [y_j - \eta(u_{1j},u_{2j},\ldots,u_{kj};\hat{\theta}_1,\hat{\theta}_2)]^2$

Suppose

$$h(\theta_1,\theta_2) = \frac{1}{1-\rho^2}\left[\left(\frac{\theta_1-\hat{\theta}_1}{s_1}\right)^2 - 2\rho\left(\frac{\theta_1-\hat{\theta}_1}{s_1}\right)\left(\frac{\theta_2-\hat{\theta}_2}{s_2}\right) + \left(\frac{\theta_2-\hat{\theta}_2}{s_2}\right)^2\right]$$

Then:

$$f(\theta_1,\theta_2) = \frac{\Gamma(1+\frac{\nu}{2})}{\nu\pi s_1 s_2\sqrt{1-\rho^2}\;\Gamma(\frac{\nu}{2})}\left[1 + \frac{h(\theta_1,\theta_2)}{\nu}\right]^{-(\frac{\nu}{2}+1)}$$

If $\theta_1 \geq \theta_2$ the conditional likelihood function is:

$$g(\theta_1,\theta_2|\theta_1 \geq 0) = cf(\theta_1,\theta_2) \text{ if } \theta_1 \geq 0$$
$$= 0 \text{ if } \theta_1 < 0$$

$$\text{Here } c^{-1} = \int_0^\infty \int_{-\infty}^\infty g(\theta_1,\theta_2|\theta_1 \geq 0)\, d\theta_1\, d\theta_2 \tag{7}$$

The marginal likelihood functions are

$$g_1(\theta_1|\theta_1>0) = \int_{-\infty}^\infty g(\theta_1,\theta_2|\theta_1 \geq 0)\, d\theta_2 \text{ if } \theta_1 \geq 0; \quad = 0 \text{ if } \theta_1 < 0 \tag{8}$$

$$g_2(\theta_2) = \int_0^\infty g(\theta_1,\theta_2|\theta_1 \geq 0)\, d\theta_1 \tag{9}$$

If $1-\alpha$ is the confidence level, then for each of the parameters θ_1 ($\theta_1>0$) and θ_2 a confidence interval is constructed with confidence level $\sqrt{1-\alpha}$. In constructing a confidence interval for θ_1 two cases should be distinguished:

i) $\int_0^{\hat{\theta}_1^*} g_1(\theta_1|\theta_1\geq 0)d\theta_1 \geq \frac{1}{2}\sqrt{1-\alpha}$. Here a symmetric interval on $\hat{\theta}_1^*$ is constructed by determining k_1 such that

$$\int_{\hat{\theta}_1^*-k_1}^{\hat{\theta}_1^*+k_1} g_1(\theta_1|\theta_1\geq 0)d\theta_1=\sqrt{1-\alpha} \qquad (10)$$

ii) $\int_0^{\hat{\theta}_1^*} g_1(\theta_1|\theta_1\geq 0)d\theta_1 < \frac{1}{2}\sqrt{1-\alpha}$. In this case the value $\theta_1=0$ is taken as lower bound for the interval and k_1 is determined such that

$$\int_0^{\hat{\theta}_1^*+k_1} g_1(\theta_1|\theta_1\geq 0)d\theta_1=\sqrt{1-\alpha} \qquad (11)$$

A confidence interval for θ_2 is obtained by constructing a symmetric interval on $\hat{\theta}_2^*$ i.e. k_2 is determined such that

$$\int_{\hat{\theta}_2^*-k_2}^{\hat{\theta}_2^*+k_2} g_2(\theta_2)d\theta_2 = \sqrt{1-\alpha} \qquad (12)$$

The rectangle

$\hat{\theta}_1^*-k_1 < \theta_1 < \hat{\theta}_1^* + k_1$ (or $0 < \theta_1 < \hat{\theta}_1^* + k_1$)

$\hat{\theta}_2^*-k_2 < \theta_2 < \hat{\theta}_2^* + k_2$

then form a confidence region for the restricted parameters θ_1 and θ_2, for which the confidence level is approximately $1-\alpha$.

The integrals (7) - (9) or integralequations (10) - (12) can be evaluated or solved numerically.

Figure 1 illustrates the method for the linear model

$\eta = \theta_1 u_1 + \theta_2 u_2$, in which $\theta_1 \geq 0$.

It gives:

1) the confidence ellipse for the unresticted parameters with a confidence level of 0.95 (—).
2) the conservative confidence region for the unrestricted parameters with a confidence level of $\geq$ 0.95 (-.-.-.-)
3) the confidence region for restricted parameters ($\theta_1 \geq 0$) constructed according to the method discussed above, with a confidence level of approximately 0.95 (-----).

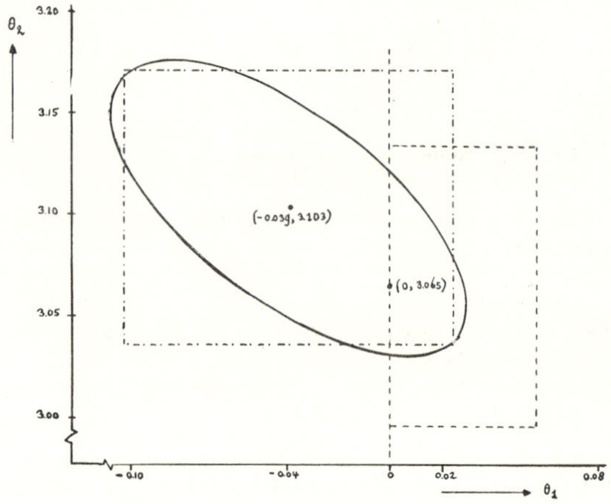

REFERENCES

1. Box (1960), Fitting empirical data, Ann. N.Y. Acad. Sci., 86, p. 792.
2. Box and Cox (1964), An analysis of transform. J. of the Royal Stat. Soc. B, 26, p. 211.
3. Box (1958), Use of stat. methods in the elucidation of basic mechanisms, Bull. Inst. Intern, de Statistique 36, p. 215.
4. Dunn (1958), Estimation of the means of dependent variables, Ann. Math. Stat,29.p.1095.
5. Hartley (1964), Exact confidence regions for the parameters in non-linear regression laws, Biometrika, 51, p. 347.
6. Hartley (1961), The modified Gauss-Newton method for the fitting of non-linear regression functions by least squares, Technometrics, 3, p.269.
7. Mezaki and Kittrel (1967), Parametric sensitivity in fitting non-linear kinetic models, Ind. and Eng. Chem.,59, p. 63.
8. Kunzi, Tzschach, Zehnder (1966) Mathematische optimierung, p. 61.
9. Marquardt (1963), An algorithm for the least squares estimation of non-linear parameters, I. Soc. Ind. Appl. Math., 11, p. 431.
10. Marquardt (1970), Generalized inverses, Ridge regression, Biased Linear Estimation and Nonlinear estimation, Technometrics, 12, p. 591.
11. Powell (1965), A method for minimizing a sum of squares of non-linear functions without calculating derivatives, Comp. J. 7, p. 303.
12. Schweigman, Non-linear programming: Handling linear and non-linear constraints; to be published.

Table I: Iteration process of the example
(numbers in rectangulars are boundary values)

step	θ_1	θ_2	θ_3	θ_4	θ_5	θ_6	θ_7	θ_8	θ_9	θ_{10}	θ_{11}	θ_{12}	θ_{13}	$F(\theta)$
0	100.00	20.40	.850	1.710	20.00	21.00	.900	.560	23.60	.750	.550	2.00	-.0010	$.495*10^5$
1	99.19	20.40	.851	1.724	20.51	20.98	.921	.586	23.60	.751	.552	2.10	-.0017	$.492*10^5$
2	93.35	20.39	.870	1.850	24.24	20.87	1.021	.701	23.64	.722	.568	2.10	-.0041	$.441*10^5$
3	87.79	20.37	.848	1.850	27.95	20.80	1.005	.791	23.66	.724	.592	2.10	-.0099	$.388*10^5$
4	87.77	20.37	.848	1.850	27.95	20.80	1.005	.792	23.66	.724	.592	2.10	-.0100	$.387*10^5$
5	86.53	20.37	.842	1.850	27.95	20.80	1.078	.860	23.67	.722	.606	2.10	-.0100	$.352*10^5$
6	78.36	20.33	.803	1.850	35.00	20.80	.869	.860	23.68	.723	.659	2.10	-.0100	$.283*10^5$
7	75.00	20.31	.784	1.850	35.00	20.80	.871	.860	23.70	.792	.750	2.10	-.0100	$.193*10^5$
8	75.00	20.30	.750	1.850	35.00	20.80	.906	.860	23.71	.907	.881	2.10	-.0100	$.129*10^5$
9	75.00	20.25	.750	1.850	35.00	20.95	.974	.860	23.74	1.267	1.357	2.10	-.0100	$.357*10^4$
10	75.00	20.27	.764	1.850	35.00	21.03	1.005	.640	23.75	1.589	1.986	2.10	-.0100	$.408*10^3$
11	76.07	20.29	.792	1.850	31.67	21.11	1.232	.781	23.76	1.571	2.042	2.10	-.0100	$.527*10^2$
12	76.10	20.29	.791	1.850	32.11	21.11	1.238	.800	23.75	1.575	2.045	2.12	.0084	$.483*10^2$
13	76.11	20.28	.791	1.850	32.13	21.11	1.236	.799	23.75	1.575	2.045	2.01	.0084	$.483*10^2$
14	76.10	20.28	.784	1.827	32.08	21.11	1.237	.805	23.75	1.573	2.040	2.05	.0080	$.482*10^2$

ON-LINE ESTIMATION OF NON-LINEAR PROCESS PARAMETERS

A McN Jackson† and D W T Rippin†
Power-Gas Limited — Eidgenossische Technische Hochschule
London, England — Zurich, Switzerland

An on-line sequential parameter estimation technique has been developed and applied to an experimental laboratory scale chemical reactor system. The estimator is based upon a gradient method of minimising the sum of squared errors, the objective being to cater for non-linear systems. The gradient vector is multiplied by a modifying matrix to give a parameter adjustment vector, a suitable fraction of which is used to update the model parameters. Starting with steepest descent convergence given by a unit matrix, successive refinement of the modifying matrix leads to second order hill-climbing and eventually to an exact correspondence with linear sequential least-squares filters.

1. INTRODUCTION

The estimation of system state variables and parameters by the use of sequential least-squares filters [1], [2], [3], [4] , is an efficient method of extracting information from the data input, providing that the system is linear. When system non-linearities are severe, however, the performance of a 'once-through' filter is degraded. Additionally, there are many situations in which data may be accumulated at a faster rate than is necessary for keeping slowly varying parameters adequately updated. Thus the efficiency of the estimation procedure, in terms of the quantity of data utilized, is not of prime importance.

Under these circumstances, a simple estimation procedure may be employed, which avoids the matrix inversion required by least-squares filters, but which nevertheless converges upon least-squares estimates of slowly varying parameters.

Such an estimation procedure has been developed from hill-climbing techniques [5] .

2. THE HILL-CLIMBING ESTIMATOR

The estimator operates upon successive sets of n observations to produce a series of adjustments to the system parameter estimate vector, $\hat{\underline{b}}$. Parameter dynamics may be evaluated externally to the estimator, so the following treatment is confined to the estimation of constant or slowly varying deviations from the predicted trajectories. The processing of each set of data is called an 'adjustment cycle'. Thus :

$$\hat{\underline{b}}_k = \hat{\underline{b}}_{k-1} - h\hat{S}^{-1}\underline{s}_k \qquad (1)$$

where the subscript k denotes quantities generated during the k'th adjustment cycle. The vector $\underline{s}_k$ contains the first derivatives, with respect to the parameters, of the sum of squares surface generated by the n observations, evaluated at the previous estimate $\hat{\underline{b}}_{k-1}$.

† Formerly of Department of Systems Engineering, University of Lancaster, Lancaster, England.

$$s_{j,k} = \frac{\partial s_k}{\partial b_j} = 2\sum_{i=1}^{n} e_{i,k}\frac{\partial e_{i,k}}{\partial b_j} \ , \ j=1---K \qquad (2)$$

The parameter sensitivities may be obtained analytically or by perturbation.

Assuming for the moment that $\hat{S}$ is a unit matrix, then in an off-line model fitting context, equation (1) represents a steepest descent method of minimising the sum of squares: h being an adjustment gain, chosen to give suitable step sizes. In this case a single set of stored observations is re-used for each iteration (adjustment cycle).

For on-line operation, input data storage is eliminated. Each observation is processed and discarded as it becomes available, and the components of the gradient vector are accumulated over n observations according to (2). After updating the parameter estimates, the next adjustment cycle is performed using fresh data.

The disadvantages of the basic steepest descent method are; oscillatory, and hence slow, progress on highly elliptical surfaces; and the lack of any direct indication of a suitable value for the adjustment gain. To overcome these problems, the second order, scale invariant formulation is used in which $\hat{S}$ is a matrix of estimates of the second derivatives of the sum of squares surface. Ideally, the second derivatives of a non-quadratic function should be evaluated at the minimum of the sum of squares surface, at which position the following approximation may be used:

$$\hat{s}_{ij} = \frac{\partial^2 s}{\partial b_i \partial b_j} = 2\sum_{h=1}^{m}\frac{\partial e_h}{\partial b_i}\cdot\frac{\partial e_h}{\partial b_j} \ , \ i=1---K, j=i---K \qquad (3)$$

Whilst this may be an unrealistic approach for an off-line model fitting procedure, it is reasonable to assume that, during the commissioning of an on-line process application, sufficient information can be generated to obtain an estimate, $\hat{S}$, which is characteristic of the region of operation of the estimator. In this way, the effects of parameter correlation and scaling can be at least reduced if not removed entirely.

3. THE RESPONSE OF THE ESTIMATOR

For the moment, it is convenient to assume that n, the number of observations in a set, and the distribution characteristics of the system inputs, are such that they have little effect upon the shape and position of each cycles' individual sum of squares surface. Otherwise, this source of variation may be related to an equivalent system output measurement noise.

The response of the estimator is analysed assuming a quadratic form for the sum of squares surface. This is equivalent to assuming a linear model and provides a basis for comparing the performance with non-linear systems. On a quadratic surface, the product $\hat{S}^{-1}\underline{s}_k$ is an estimate of the error or displacement, $\underline{x}_k$, of the previous estimate $\hat{\underline{b}}_{k-1}$, from the 'actual' parameter values, $\underline{b}^*$:

$$\underline{x}_k = \hat{\underline{b}}_{k-1} - \underline{b}^* \qquad (4)$$

$$\hat{\underline{x}}_k = \hat{S}^{-1}\underline{s}_k = \underline{x}_k - \underline{a}_k \qquad (5)$$

where $\underline{a}$ represents the effects of measurement error [5]. Relating all random sources to a single equivalent system output measurement error n_o, the following relationship holds:

$$E\left[\underline{a}.\underline{a}^T\right] = 2S^{-1}E\left[n_o^2\right] \qquad (6)$$

Defining $\hat{\underline{b}}(k)$ as an estimate based solely upon the k'th observation set, then:

$$\hat{\underline{b}}(k) = \hat{\underline{b}}_{k-1} - \hat{\underline{x}}_k \qquad (7)$$

$$\begin{aligned}\hat{\underline{b}}_k &= \hat{\underline{b}}_{k-1} - h\,\hat{\underline{x}}_k \\ &= (1-h)\,\hat{\underline{b}}_{k-1} + h\,\hat{\underline{b}}(k) \\ &= (1-h)^k\hat{\underline{b}}_o + h\sum_{i=0}^{k-1}(1-h)^i\hat{\underline{b}}(k-i) \\ &= \phi^k\hat{\underline{b}}_o + (1-\phi)\sum_{i=0}^{k-1}\phi^i\hat{\underline{b}}(k-i)\end{aligned} \qquad (8)$$

where $\phi = 1 - h$. The estimate $\hat{\underline{b}}_k$ is a weighted average of all previous estimates $\hat{\underline{b}}(k-i)$.

In terms of the estimate errors:

$$\begin{aligned}\underline{x}_{k+1} &= \phi\underline{x}_k + (1-\phi)\,\underline{a}_k \\ &= \phi^k\underline{x}_1 + (1-\phi)\sum_{i=0}^{k-1}\phi^i\underline{a}_{k-i}\end{aligned} \qquad (9)$$

The deterministic response, ie with zero $\underline{a}$, is an exponential decay of the initial estimate errors for $|\phi| < 1$. Putting k equal to infinity in (9) leads to the 'equilibrium' response:

$$E\left[\underline{x}.\underline{x}^T\right] = \frac{1-\phi}{1+\phi}E\left[\underline{a}.\underline{a}^T\right] \qquad (10)$$

The parameter estimate variances tend to zero as ϕ tends to 1. However, a value of ϕ close to 1 corresponds to an adjustment gain, h, close to zero which restricts the ability of the estimator to respond to changes in the system parameters.

4. SELECTING THE ADJUSTMENT GAIN

Although, in a non-linear situation, the prime consideration may be to achieve stability by using a low adjustment gain, investigation of the optimum linear response gives a useful indication of the influence of the gain upon the estimator performance.

Two objectives have been investigated: 1) to produce minimum variance parameter estimates after N cycles, and 2) to minimise the mean square prediction error over an operating period of N cycles. In many practical applications, the parameter estimates would be used continuously for control or optimization and so the second objective would be more relevant, as it seeks to reduce large deviations rapidly.

The details of the analyses are given in [5] : the principal results are presented here.

Defining $R^2 = E\left[x_{i,1}^2\right]/\text{var}\left[a_i\right]$, minimum $E\left[x_{i,N+1}^2\right]$ is given by :

$$h \simeq \frac{\ln(4NR^2)}{2N-1} \quad \text{for small h} \qquad (11)$$

This result applies distinctly to each parameter. As the same value of R^2 will not necessarily apply to each parameter, the scalar adjustment gain, h, could be replaced by a diagonal matrix containing the individual optimum adjustment gains. However, unless widely differing values are returned by (11), this is not worthwhile as the performance is relatively insensitive to the value of h, once the correct order of magnitude has been found.

An analytical solution for the minimum overall mean square error has been found for the special case where $E\left[\underline{x}_1.\underline{x}_1^T\right] = R^2E\left[\underline{a}.\underline{a}^T\right]$. This condition is met if the set of initial estimates is derived from data obtained under conditions similar to those under which the estimator operates. For this case, the single optimum gain is given by:

$$h = \left[(R^2/N) + \tfrac{1}{4}(R^2/N)^2\right]^{\frac{1}{2}} - \tfrac{1}{2}(R^2/N) \qquad (12)$$

The general case may be solved numerically by minimising the general expression for the mean square error with respect to the individual adjustment gains. Again, the performance function is flat near the optimum. For example, when $R^2/N=1$, a value of 0.62 is returned by (12), but the increase in the mean square error is less than 10% within the range 0.4 to 0.9. This range of optimum gains is given by a range of R^2/N from 0.27 to 3.2, ie more than a decade. It is, therefore, reasonable to select a single gain based on an 'average' value of R^2/N.

The approximations used to arrive at the above equations were tested by comparison with a high speed repetitive analogue/logic simulation of equation (9). Multiple sampling and correlation of a gaussian noise generator were used to obtain the noise series $\underline{a}$, and low pass filters were used to display continuously the performance mean and variance. The results agreed with the analytical predictions over the wide ranges of R^2 and N investigated.

5. PERFORMANCE WITH A NON-LINEAR SYSTEM

The work on the laboratory scale reactor system involved the estimation of catalyst activity parameters. The plant was essentially noise free, and the catalyst activity changed relatively slowly compared with the 3-4 days duration of an experimental run. Thus measurement errors and initial parameter estimate errors were introduced artificially. This allowed direct comparison of a plant run with a digital computer simulation run utilizing identical noise series. Once a satisfactory correspondence between plant and simulation had been established with a relatively small number of runs, the large amount of data needed to characterise a stochastic situation was obtained by simulation.

Only two parameters were available for estimation, both exhibiting non-linear behaviour. The shape of the sum of squares surface was such that, at the extremes of the range of parameter values considered, positive displacements of one of the parameters (b_2) were under-estimated, and negative displacements over-estimated, by a factor of approximately 3 in both cases. Hence, with an adjustment gain of 1, the noise free response to a negative displacement was a large overshoot of the minimum sum of squares followed by a slow recovery from the positive displacement side.

The net effect of this behaviour was to produce a rapid increase in the mean square error and the parameter variances at high adjustment gains. At lower gains the simulation results converged upon the theoretical performance with a linear system. This tended to favour low values of the adjustment gain, particularly when the probability distribution of initial displacements of b_2 was biassed towards the negative side. Even so, the performance with the optimum linear gain was only slightly below the best performance achieved, when the measurement noise was high.

The parameter estimate variances were tabulated for a grid of values of individual adjustment gains, h_1 and h_2. This showed conclusively that the estimate variance of one parameter was independent of the gain used to estimate the other parameter. Thus, in spite of the severe non-linearity of the system, the correlation between parameters was completely removed.

6. RELATIONSHIP WITH LEAST-SQUARES FILTERS

Instead of using a fixed, a priori estimate $\hat{S}$, it is feasible to estimate the second derivative matrix within the on-line algorithm, although at the expense of introducing a matrix inversion into the procedure. The advantages are that the variation of the sum of squares surface from cycle to cycle can be taken into account, and that the estimate can be updated if long term movement of the plant parameters changes the shape of the surface.

The summations of equation (3) may be performed over the n points of an observation set to produce an estimate $\hat{S}(k)$. As $\hat{\underline{b}}_k$ converges upon the system parameters, this estimate will converge upon S(k). If n is small, some or all of the $\hat{S}(k)$ may be singular, ie the parameters may be unobservable within a single adjustment cycle. In this case, a weighted average estimate can be used:

$$\hat{S}_k = (1-g)\,\hat{S}_{k-1} + g\,\hat{S}(k)$$

$$= (1-g)^k \hat{S}_o + g \sum_{i=0}^{k-1} (1-g)^i \hat{S}(k-i) \qquad (13)$$

which may be compared with equation (8).

The standard linear regression equation may be reformulated for perturbations about a reference state:

$$\hat{\underline{b}} = [G^T G]^{-1} G^T \dot{\underline{y}} \qquad (14)$$

where the elements of G are $\frac{\partial y_i}{\partial b_j}$.

In an iterative situation, the reference state may be identified with the current parameter estimates, so that $\dot{\underline{y}}$ becomes the prediction error vector, and $\hat{\underline{b}}$ corresponds to $-\hat{\underline{x}}$. Thus :

$$\hat{\underline{x}}_k = -[G_k{}^T G_k]^{-1} G_k^T \underline{e}_k \qquad (15)$$

Inspection reveals that, apart from a factor of 2 which cancels when the product is taken, $G_k^T e_k$ corresponds to $-s_k$ and $G_k^T G_k$ to $\hat{S}(k)$. An exponentially weighted sequential least-squares filter, incorporating n observations at a time, has the following basic form:

$$\hat{\underline{b}}_k = \hat{\underline{b}}_{k-1} + P_k G_k^T \underline{e}_k \tag{16}$$

$$P_k^{-1} = cP_{k-1}^{-1} + G_k^T G_k$$

$$= c^k P_o^{-1} + \sum_{i=0}^{k-1} c^i G_{k-i}^T G_{k-i} \tag{17}$$

For large k, the influence of the initial estimates $\hat{S}_o$ and P^{-1} may be ignored. Then if $g = 1-c$, comparison of (13) and (17) shows that:

$$2gP_k^{-1} = \hat{S}_k \tag{18}$$

Thus, equating g and h, which gives the same weighting function for both $\hat{\underline{b}}_k$ and $\hat{S}_k$, produces identity between the hill-climbing estimator using $\hat{S}_k$, and a basic sequential least-squares filter with an exponential weighting coefficient of 1-h.

The variance of an element of P_k^{-1} is related to the variance of an element of $G_k^T G_k$ by the factor $(1-c)^2$. For values of c close to 1, and assuming stationary statistics for G, P_k will become ostensibly constant, once the starting transients have died away. Under these conditions there will be little difference between using a constant matrix, and continuously updating and inverting P_k^{-1}.

7. CONCLUSIONS

It seems likely that, for many long term process applications, a 'least-squares' filter may be replaced by a computationally simpler 'low-squares' filter which produces estimates in the vicinity of the least-squares estimates. Analysis and simulation indicate that the reduction in performance is slight.

Where the non-linearities of a system intrude significantly upon the response of an on-line estimator, the least-squares basis of a linear filter is invalidated. Either simulation or operational testing is required to determine if the computational complexities of a least-squares filter bring any additional benefits over the use of a hill-climbing estimator.

1 BRAY, J.W., HIGH, R.J., McCANN, A.P. and JEMMESON, H. "On-line Model Making for a Chemical Plant" Trans. SIT 17 (3) 59 and 65 (Sep 1965).

2 COGGAN, G.C. and NOTON, A.R.M. "Optimal Discrete Time Recursive State and Parameter Estimation in Chemical Engineering" Symposium "Current Trends with Computers in Chemical Engineering" IChE Nottingham (21 April 1969).

3 GAVALAS, G.R. and SEINFELD, J.H. "Sequential Estimation of States and Kinetic Parameters in Tubular Reactors with Catalyst Decay" CES 24 (4) 625 (April 1969).

4 GOLDMANN, S.F. and SARGENT, R.W.H. "Applications of Linear Estimation Theory to Chemical Processes: a Feasibility Study" CES 26 (10) 1535 (October 1971).

5 JACKSON A.McN. "On-line Parameter Estimation of Non-linear Systems" PhD thesis, University of Lancaster (1971).

FREQUENCY DOMAIN IDENTIFICATION OF NONLINEAR SYSTEMS

Alex. B. Gardiner
Paisley College of Technology
Paisley, Renfrewshire, U.K.

Dynamic systems with single-valued nonlinearities can be represented in cascade or multi-dimensional parallel form. The latter can be identified using time domain correlation techniques, but the resulting model is extremely clumsy and cannot, as yet, be reduced to the neater cascade form. A frequency domain identification technique is introduced which does allow the simple cascade form to be determined.

1. INTRODUCTION

Open loop dynamic systems containing single valued nonlinearities can be represented in cascade form (figure 1) or parallel form (figure 2). In the time domain, the cascade to parallel transformation is readily accomplished, George (1959), but the inverse operation does not seem possible, especially if the parallel multi-dimensional impulse responses only exist as a matrix of values as obtained from a multi-dimensional correlation experiment. This work shows how the cascade description can be determined using frequency domain testing and analysis.

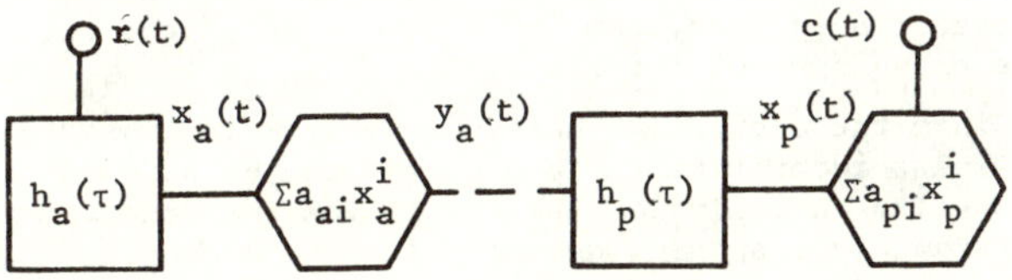

Figure 1 Cascade nonlinear representation

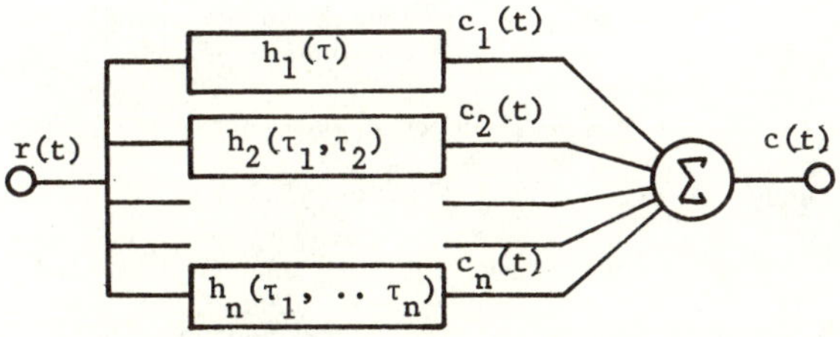

Figure 2 Parallel nonlinear representation

2. SYSTEM REPRESENTATION

2.1. Time domain

The cascade representation is a neat analytical description but it is the parallel form that, to date, has been the more useful in analysis and identification. The fourth power cascade system shown in figure 1 can be described by the four channel parallel system shown in figure 2, each parallel channel representing one power of nonlinearity. Using functional notation, the input/output relationship can be described by

$$c(t) = L[r(t)] = L_1[r(t)] + L_2[r(t)] + L_3[r(t)] + L_4[r(t)] \quad (1)$$

Each operator in this equation represents a single nonlinear channel so has the property that

$$L_n[g.r(t)] = g^n L_n[r(t)] \quad (2)$$

where g is a scalar. The total output for input g.r(t) is then

$$L[g.r(t)] = g.L_1[r(t)] + g^2.L_2[r(t)] + g^3.L_3[r(t)] + g^4.L_4[r(t)] \quad (3)$$

Therefore, by testing at different input levels, the output of each parallel channel, $L_n[r(t)]$ can be determined experimentally, Gardiner (1966 and 1973). Correlation analysis would then allow the system impulse responses, $l_n(\tau_1, \dots \tau_n)$, to be determined independent of the correlation of other channels.

2.2. Frequency Domain

Although most nonlinear identification has utilised time domain techniques, transformation can be made to the multidimensional Laplace domain, George (1959), i.e.

$$L_n(s_1,\dots s_n) = \int\dots\int l_n(t_1,\dots t_n)\exp(s_1t_1 + \dots + s_nt_n)dt_1\dots dt_n \quad (4)$$

However, it is the frequency domain representation that is required, not the Laplace domain representation. For steady state responses in linear systems, jω is equivalent to s, but this requires amplification in the multidimensional case. By putting the input sinusoid in its exponential form

$$A \cos \omega t = \frac{A}{2} e^{j\omega t} + \frac{A}{2} e^{-j\omega t} \quad (5)$$

George has shown that the steady state frequency responses are obtained from the multidimensional Laplace transform by replacing the s's by jω or

$-j\omega$ and multiplying by an appropriate factor which depends on the channel power and the harmonic being considered. The sum of $j\omega$'s and $-j\omega$'s determines which harmonic is being considered. For example, in the 3rd power channel $L_3(s_1,s_2,s_3)$, the 1st harmonic (fundamental) frequency response to an input A sin ωt is given by

$$L_3(j\omega) = \frac{3A^3}{4} L_3(s_1,s_2,s_3);\; s_1=s_2=j\omega,\; s_3 = -j\omega$$

$$= \frac{3A^3}{4} L_3(j\omega,j\omega,-j\omega)$$

$$= \frac{3A^3}{4} L_3((j\omega)_2.\ (-j\omega)_1) \qquad (6)$$

The 3rd harmonic would be given by

$$L_3(j3\omega) = \frac{A^3}{4} L_3(s_1,s_2,s_3);\quad s_1=s_2=s_3=j\omega$$

$$= \frac{A^3}{4} L_3((j\omega)_3,\ (-j\omega)_0) \qquad (7)$$

In general, the $(n-r)^{th}$ harmonic of the nth power channel is given by

$$L_n(j(n-r)\omega) = \sum_{r=o}^{m} \binom{n}{r}\frac{A^n}{2^{n-1}} \cdot L_n((j\omega)_{n-r},(-j\omega)_r)$$

$$+ \text{(for even n)} \binom{n}{\frac{n}{2}} \frac{A^n}{2^n} L_n((j\omega)_{n/2},\ (-j\omega)_{n/2}) \qquad (8)$$

where m is the integer $(n-2)/_2$ or $(n-1)/_2$.

If a nonlinear system can be described in functional notation as in equation 1 and transformed into the Laplace domain, equation 8 allows the calculation of the frequency responses of all the harmonics. An identification procedure, however, must work from the frequency response back to system description, provided that the frequency responses can be measured.

3. HARMONIC FREQUENCY RESPONSE

3.1. Harmonic Responses

Investigation of the form of multidimensional Laplace transfer functions shows that the s-terms are either additive or multiplicative, e.g. a cubic term and a 1st order filter would give

$$H_3(s_1,s_2,s_3) = \frac{b}{(s_1+s_2+s_3+a)} \qquad (9)$$

when the cubic is before the filter

$$\text{or } J_3(s_1,s_2,s_3) = \frac{b}{(s_1+a)} \cdot \frac{b}{(s_2+a)} \cdot \frac{b}{(s_3+a)} \qquad (10)$$

when the cubic is after the filter. Combinations of the two forms also exist.

The multiplicative terms show the frequency dependent nonlinear behaviour of the first harmonic since one root is multiplied by its conjugate, e.g. with $s_1=s_2=-s_3=j\omega$ and input A cos ωt,

$$J_3(j\omega) = \frac{3A^3}{4} \cdot \frac{b^3}{(j\omega+a)(j\omega+a)(-j\omega+a)}$$

$$= \frac{3A^3}{4} \left|\frac{b}{j\omega+a}\right| \frac{b^2}{\omega^2+a^2} \exp(j \arg(^{b/}j\omega+a)) \qquad (11)$$

The highest harmonic of each power channel, however, always exhibits a linear type response with frequency, e.g.

$$H_3(j3\omega) = \frac{A^3}{4} \frac{b}{j3\omega + a} \qquad (12)$$

$$\text{and } J_3(j3\omega) = \frac{A^3}{4} \frac{b^3}{(j\omega+a)}3 = \frac{A^3}{4} \frac{27b^3}{j3\omega+3a)}3 \qquad (13)$$

3.2. Equivalent Linear Harmonic Networks

In general, from equation 8,

$$L_n(jn\omega) = \frac{A^n}{2^{n-1}} L_n((j\omega)_n) \qquad (14)$$

Since s is now always positive, this indicates that the frequency response of the highest harmonic of each channel can be considered as being that of a linear network preceded by a harmonic generator, an nth-power harmonic generator being a system which gives out $\frac{A^n}{2^{n-1}}$ cos $n\omega t$ for an input of A cos ωt.

Since the equivalent harmonic network is linear, linear identification procedures can be used to develop an analytic expression from the measured frequency response, assuming that this can be measured.

3.3. Measurement of the harmonic responses

From equations 3 and 8, the output for a sinusoid of frequency ω can be written with terms indicating the harmonic and the power contributing that harmonic, e.g. a fourth power example with input g_1A cos ωt has the output

$$C(g_1A \cos \omega t) = g_1 \cdot c_1(j\omega) + g_1^3 \cdot c_3(j\omega) + g_1^2 \cdot c_2(j2\omega) + g_1^4 \cdot c_4(j2\omega) + g_1^3 \cdot c_3(j3\omega) + g_1^4 \cdot c_4(j4\omega) \qquad (15)$$

With input g_2 A cos ωt, each component of the output changes amplitude depending on the channel producing it. Extracting each harmonic at each level allows the highest harmonic of each channel, $c_n(jn\omega)$, to be determined, e.g. if $c(j2\omega)_1$ and $c(j2\omega)_2$ are the measured 2nd harmonics,

$$c(j2\omega)_1 = g_1^2\, c_2(j2\omega) + g_1^4\, c_4(j2\omega) \qquad (16)$$

$$c(j2\omega)_2 = g_2^2\, c_2(j2\omega) + g_2^4\, c_4(j2\omega) \qquad (17)$$

Solving these simultaneous equations gives the 2nd harmonic of the 2nd power channel.

The choice of the level factors, g, depends on measurement noise considerations and this is discussed in the reference cited, Gardiner (1973).

In this way, the frequency responses of the equivalent linear networks of an Nth power system can be obtained by measuring the harmonics at $^N/2$ levels over the frequency range of measurement.

4. CASCADE SYSTEM DESCRIPTION

4.1. Significance of the equivalent harmonic networks

The channel harmonic frequency responses characterise the parallel description of the nonlinear system. In the multidimensional impulse response, the confusing and redundant data for identification purposes is in the cross product terms which cannot be separated from the wanted power terms, but in the frequency domain the redundant data is in the lower harmonics of each channel and they can be separated from the required data in the highest harmonic of each power channel. The cascade description of the processes can be determined by comparing the higher power equivalent linear networks with the linear network. The nature of the comparison depends on the frequency base to which the harmonic output responses are taken. They can either be plotted in terms of the harmonic frequency, $n\omega$, or the system input frequency, ω. The former would seem the obvious choice but the latter results in an easier analysis. If $L_n(jn\omega)_\omega$ is obtained by plotting $L_n(jn\omega)$ or a function of ω, rewriting equations 12 and 13 in terms of $j\omega$ gives

$$H_3(j3\omega)_\omega = \frac{A^3}{4} \; \frac{b}{3(j\omega+{}^a/_3)} \tag{18}$$

and
$$J_3(j3\omega)_\omega = \frac{A^3}{4} \; \frac{b^3}{j\omega+a)^3} \tag{19}$$

Equation 18 shows that the power before the dynamics apparently causes the pole (and zero) values to be divided by the order of the nonlinearity and equation 19 shows that the power after the dynamics raises the order of the filter by the order of the power. In this way it can be deduced which part of the dynamics are before or after power terms.

Analysis of all equivalent harmonic networks allows the complete cascade description of the process to be determined.

4.2. Rules for Determining the Cascade System Description

1. Measure the frequency response of each harmonic at $^N/2$ levels over the frequency range of interest.

2. Combine the frequency responses at each level to determine the frequency response of the highest harmonic of each channel.

3. Treat each response as linear and identify the equivalent harmonic networks from the plots of output response vs input frequency.

4. Compare the linear and harmonic responses. Poles and zeros whose values have changed (always by an integer amount) are after the nonlinearity, the magnitude of change giving the power of nonlinearity they follow. Poles and zeros which are duplicated are before nonlinearities, the power of which is shown by the number of duplications.

4.3. A 4th power example

A nonlinear system was simulated on an analogue computer and the responses of the four measurable equivalent harmonic networks determined. The harmonic responses are shown in figure 3.

The linear network is characterised by poles at 6π and 30π rad/sec., the 2nd power network by double poles at 6π and 30π rad/sec, a pole at 3π rad/sec and a zero at the origin, the cubic by a triple pole at 30π rad/sec and poles at 3π and 6π rad/sec and the 4th power by a quadruple pole at 30π rad/sec and a double pole at 3π rad/sec.

The smaller poles of the 4th power network show that the 6π filter is between two 2nd power terms and the quadruple poles at 30π shows that the 30π filter is before any nonlinearity. This is confirmed by the 2nd power network which, because of the presence of the zero, must consist of two parallel paths, one with all dynamics before the squarer (giving double poles at 6π and 30π) and one with the squarer after the 30π filter but before the 6π filter.

The cubic network shows that all the cubic terms are after the 30π filter but that there is a squarer before a 6π filter and another 6π filter with no nonlinearity before or after it. But this suggests only a 2nd power nonlinearity so the cubic must be formed by the multiplication of a linear signal (passing through the 30π filter then the 6π one) with a 2nd power signal (passing through the 30π filter, being squared then passing through the 6π filter.) This is consistent with two second power polynomial nonlinearities separated by a 6π filter and preceded by a 30π filter. By arbitrarily assuming that the linear gain of each nonlinearity is unity, all gain factors can be determined from the gains of the various networks.

5. CONCLUSIONS

The frequency responses of the highest harmonic of each power channel behave in a linear manner with frequency and, if they can be measured, the responses can be analysed using linear techniques to give the equivalent harmonic networks which are related to the linear network by the relative positions of the dynamics and nonlinearities.

A knowledge of the possible relationships allows the cascade description of the nonlinear system to be determined. This is a compact description of the process which can be expanded into the parallel form using functional analysis techniques.

The harmonic responses can be determined by performing the frequency response measurements at $^{N}/2$ levels, where N is the overall power of nonlinearity. Although it can be done, this is a cumbersome procedure and there are problems in extracting the harmonics which are attenuated with frequency much faster than the fundamental. Digital signal processing may be better than the hybrid techniques used so far in the tests, and work is continuing to find an easier measurement procedure to match the easy analysis technique.

Since the technique was shown to be possible by using the Laplace transform of the Volterra functional expansion, it is expected that the multi-dimensional Fourier transforms of a measured multidimensional impulse response will give the required harmonic frequency responses.

6. REFERENCES

GARDINER, A B (1966): 'Elimination of the Effects of nonlinearities on process crosscorrelations'. Electronics Letters, 2, p164.

GARDINER, A B (1973): "identification of processes containing single valued nonlinearities', IJC, 17.

GEORGE, D A (1959): 'Continuous nonlinear systems', MIT Technical Report 355.

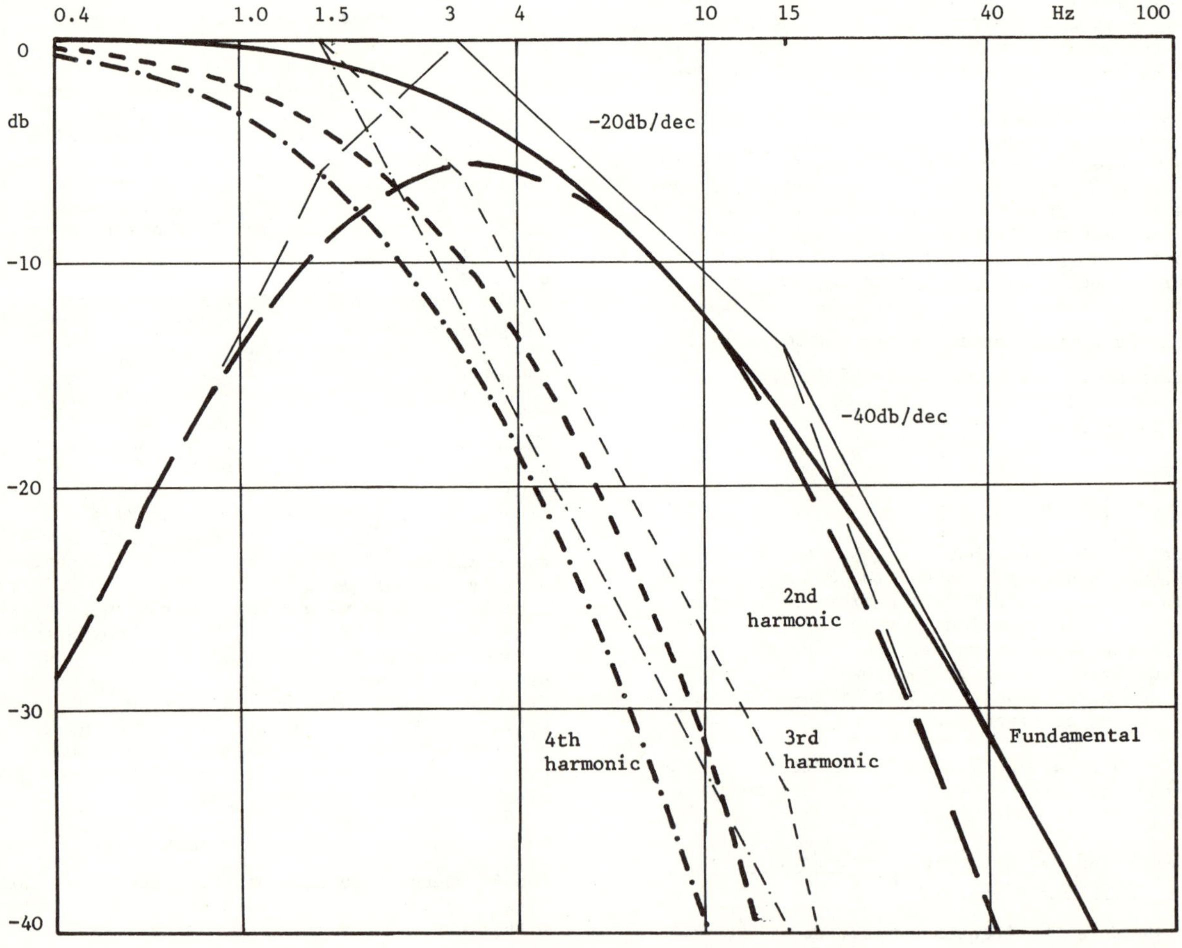

Figure 3 Equivalent network amplitude responses

APPLICATION OF THREE-LEVEL-PSEUDORANDOM-SIGNALS FOR PARAMETER ESTIMATION OF NONLINEAR SYSTEMS

R. Krempl

Ruhr-Universität Bochum, Germany

In this paper three methods are proposed for parameter estimation of nonlinear systems. The nonlinear systems are approximated by a linear part and a nonlinear characteristic of the degree 2n. For the complete estimation of such a model only measurements with n different signal levels are needed. Finally a test result is presented to show the effectivenes of the proposed methods.

1. INTRODUCTION

On identification of linear systems there exists a great number of theoretical publications as well as practical application. Estimation methods for nonlinear systems using pseudorandom signals however are not developed to such an extent so far. In the following chapters some methods are shown which allow the estimation of nonlinear models with different structures. However the application of these methods is restricted to testing of time-invariant systems. As test-signals three-level-pseudo-random signals are used. For the approximation models are used consisting of a linear part and a nonlinear gain.

2. HAMMERSTEIN-MODEL

The so called Hammerstein-model [1,2] consists of an amplifier with a nonlinear characteristic which is followed by a linear system (Fig.1). The linear part is supposed to have no integrative action. The characteristic of the nonlinear part is unique:

$$Q(x) = a_1 x + a_2 x^2 + a_3 x^3 \ldots \tag{1}$$

The output of the model can be written:

$$y(t) = \int_0^\infty g(\tau) Q(x(t-\tau))\, d\tau \,. \tag{2}$$

Crosscorrelating output and input-signals one gets:

$$\phi_{xy}(\tau_1) = a_1 \int_0^\infty g(\tau)\, \phi_{xx}(\tau_1 - \tau)\, d\tau + a_3 \int_0^\infty g(\tau) \cdot \phi_{xxxx}(0,0,\tau_1 - \tau)\, d\tau + \ldots \tag{3}$$

In equation (3) all the integrals with even coefficients vanish, because for 3-level pseudo-random signals ("t.p.r.s.") the following expression is valid:

$$\phi_{xxx}(\tau_1, \tau_2) = \phi_{xxxxx}(\tau_1, \tau_2, \tau_3, \tau_4) = 0 \tag{4}$$

Furthermore, for t.p.r.s with a signal amplitude A the equation (5) hold:

$$\phi_{xxxx}(0,0,\tau) = A^2\, \phi_{xx}(\tau), \tag{5}$$

$$\phi_{xxxxxx}(0,0,0,0,\tau) = A^4\, \phi_{xx}(\tau).$$

Equation (3) then becomes:

$$\phi_{xy}(\tau_1) = \sum_{i=1}^{n} a_{2i-1} A^{2i-2} \int_0^\infty g(\tau)\, \phi_{xx}(\tau - \tau_1)\, d\tau \,. \tag{6}$$

If the settling time is less then half of the sequence-length of the input signal the following approximation is permitted:

$$\phi_{xy}(\tau_1) \approx K \sum_{i=1}^{n} a_{2i-1} A^{2i} g(\tau_1). \tag{7}$$

The value of K depends on the sequence length of the input signal only. For the determination of the even coefficients of the characteristic the output signal is correlated with:

$$x_m(t) = x^2(t) - \frac{1}{N\Delta t} \int_0^{N\Delta t} x^2(t)\, dt \,. \tag{8}$$

Using the properties of higher-order autocorrelation functions [3] the following is valid:

$$\phi x_m y(\tau_1) = \sum_{i=1}^{n} a_{2i} A^{2i-2} \int_0^\infty g(\tau) \cdot \phi x_m xx(\tau - \tau_1, \tau - \tau_1)\, d\tau \approx \sum a_{2i} A^{2i+2} \hat{K}\, g(\tau_1) \tag{9}$$

Solving equation (7) and (9), one gets the coefficients of the nonlinear characteristic and the impulse-response function respectively.

2. CHARACTERISTIC WITH HYSTERESIS

For further investigations the nonlinear amplifier is assumed to have hysteresis. The width of the hysteresis being constant along the whole characteristic (Fig.3). Concerning the linear part the assumption made in the previous chapter is still valid. If the testing signal is a t.p.r.s., the output signal can be divided into two parts:

$$v(t) = v_1(t) + bv_2(t) \tag{10}$$

This subdivision is carried out in order to estimate the hysteresis-property of the system. The signal $v_1(t)$ changes its value in the same rhytm as the input-signal, its amplitude-values being e.g. $\bar{A}_I$, 0 and $\bar{A}_I$. The signal $v_1(t)$ is approximated by a polynomial of the form $\sum_1 \bar{a}_i x^i$. This polynomial is not identical with the polynomial of the characteristic (Fig.3). The second part of (10) $bv_2(t)$ represents the value of the signal v(t), if the input changes to zero:

$$v_2(t) = \left[1 - \left(\frac{x(t)}{A}\right)^2\right] \text{sign}\left[x(t-\Delta t) + v_2(t-\Delta t)\right]. \tag{11}$$

The value Δt means the shift-register clock-pulse interval. For the crosscorrelation function between input- and output-signal it can be written:

$$\phi_{xy}(\tau_1) \approx \sum_1^n \bar{a}_{2i-1} A^{2i} K g(\tau_1) + bA \cdot \int_0^\infty g(\tau) \phi_{x^* v_2}(\tau - \tau_1) d\tau \; ; \; x(t) = Ax^*(t) \; . \tag{12}$$

The calculation of the impulse-response function and of the polynomial-coefficients respectively, is not yet possible, because the right-hand integral is still unknown.

A fictive signal $\hat{x}(t)$ is defined, which has the following properties:

$$\phi_{x^* v_2}(t) = \phi_{\hat{x} x^*}(t) \; ; \; \phi_{\hat{x} x^* x^*}(t) = 0 \; . \tag{13}$$

The correlation between the output signal y(t) and the signal $\hat{x}(t)$ yields:

$$\phi_{\hat{x}y}(\tau_1) = \sum_{i=1}^n \bar{a}_{2i-1} A^{2i-1} \int_0^\infty g(\tau) \phi_{\hat{x}x^*}(\tau - \tau_1) d\tau + b \int_0^\infty g(\tau) \phi_{\hat{x}v_2}(\tau - \tau_1) d\tau \; . \tag{14}$$

By measuring with two different amplitudes and subtracting the crosscorrelation functions a value for the unknown integral of equation (12) can be given:

$$\int_0^{\tau_m} g(\tau) \phi_{x^* v_2}(\tau - \tau_1) d\tau = \int_0^{\tau_m} g(\tau) \phi_{\hat{x}x^*}(\tau - \tau_1) d\tau =$$

$$= K \frac{\phi_{\hat{x}y}(\tau_1)_I - \phi_{\hat{x}y}(\tau_1)_{II}}{\int_0^{\tau_m} \left(\phi_{x^*y}(\tau)_I - \phi_{x^*y}(\tau)_{II}\right) d\tau} = C(\tau_1) \; . \tag{15}$$

For a great value τ_1 the first term of equation (12) decays to zero ($g(\tau_1) \approx 0$; $\phi_{xy}(\tau_1) \neq 0$).

$$b \approx \frac{\phi_{xy}(\tau_1)}{A \cdot C(\tau_1)} \; , \quad \tau_1 \gg 0 \; . \tag{16}$$

From equation (12) the coefficients $\bar{a}_i$ of the hysteresis characteristic can be estimated as well:

$$\tau_m \sum_1^n \bar{a}_{2i-1} A^{2i} \int_0^{\tau_m} g(\tau_1) d\tau_1 = \frac{1}{K} \left[\cdot \int_0^{\tau_m} \phi_{xy}(\tau_1) d\tau_1 - bA \int_0^{\tau_m} C(\tau_1) d\tau_1 \right] \approx \sum_{i=1}^n \bar{a}_{2i-1} A^{2i} \tag{17}$$

For the estimation of the even coefficients $\bar{a}_i$ the output signal y(t) is correlated with the signal $x_m(t)$:

$$\phi_{x_m y}(\tau_1) = \sum_{i=1}^n \bar{a}_{2i-1} A^{2i-2} \int_0^\infty g(\tau) \phi_{x_m x}(\tau - \tau_1) d\tau + \sum_{i=1}^n \bar{a}_{2i} A^{2i-2} \int_0^\infty g(\tau) \phi_{x_m xx}(0, \tau - \tau_1) d\tau + b \int_0^\infty g(\tau) \phi_{x_m v_2}(\tau - \tau_1) d\tau \; . \tag{18}$$

Calculating the crosscorrelating function and taking it into account the properties of t.p.r.s. respectively the following equations hold:

$$\phi_{xx_m}(\tau) \equiv 0; \quad \phi_{x_m v_2}(\tau) \equiv 0 \tag{19}$$

Integrating with respect to τ_1:

$$\sum_{i=1}^n \bar{a}_{2i} A^{2i-2} \approx \frac{1}{K} \int_0^\tau \phi_{x_m y}(\tau_1) d\tau_1 \; . \tag{20}$$

From the equtions (16), (17) and (20) the coefficients $\bar{a}_i$ and the height of the hysteresis can be estimated. The impulseresponse function follows from equation (12):

$$g(\tau_1) \approx \frac{1}{K} \frac{\phi_{xy}(\tau_1) - b \cdot A \cdot C(\tau_1)}{\sum_{i=1}^n \bar{a}_{2i-1} A^{2i}} \tag{21}$$

The coefficients $\bar{a}_i$ describe a characteristic which passes through the points I,II,..(Fig.2). Knowing these points and the height of the hysteresis it is possible to calculate the coefficients a_i and the width b_1 of the hysteresis. For the complete estimation of a model, consisting of a characteristic of the degree 2n in connection with a linear part, measurements with n different signal levels are needed.

4. SIMPLIFIED WIENER-MODEL [1,2]

This model consists of a linear part followed by a nonlinear gain (Fig. 2). The assumptions made in previous chapters are still valid for the nonlinear characteristic and the linear part. The crosscorrelation function can be written as:

$$\phi_{xy}(\tau_1) = \sum_{i=1}^n a_{2i-1} \cdot \phi_{xv}{}^{2i-1} \; , \tag{22}$$

$$\phi_{xv}{}^{2i}(\tau_1) \equiv 0 \; . \tag{23}$$

Carrying out measurements with different signal amplitudes, the crosscorrelation function is:

$$\phi_{xy}(\tau_1)_{I,II,..} = A^2_{I,II,..} \cdot a_1 \phi_{xv}^{**}(\tau_1) + A^4_{I,II,..} \cdot a_3 \cdot \phi^{*}_{x(v)}{}^{*}3(\tau_1) + a_5 \cdots \qquad (24)$$

or

$$\underline{\phi_{xy}(\tau)} = \underline{A} \cdot \begin{bmatrix} a_1 \ \phi_{xv}^{**}(\tau) \\ a_3 \ \phi^{*}_{x(v)}{}^{*}3(\tau) \\ \vdots \end{bmatrix} ; \qquad (25)$$

$$a_1 \phi_{xv}^{**}(\tau) = \frac{|A_1(\tau)|}{|A|} . \qquad (26)$$

The crosscorrelation function ϕ_{xv}^{**} can be calculated in a simple way:

$$\phi_{xv}^{**}(\tau) = \frac{1}{A^2} \int_0^{\infty} g(\vartheta) \ \phi_{xx}(\tau - \vartheta) \ d\vartheta \approx K g(\tau); \qquad (27)$$

$$g(\tau) \approx \frac{1}{a_1 K} \frac{|A_1(\tau)|}{|A|} \quad ; \quad \int_0^{\tau_m} \phi_{xv}^{**}(\tau) \ d\tau \approx K . \qquad (28)$$

Knowing the impulse-response function, all the other crosscorrelation functions $\phi^{*}_{x(v)}{}^{*}i(\tau)$ can be calculated. From the set of equations (25) and an analogous set all the coefficients can be calculated. To determine the characteristic of a polynomial of the degree 2n and the impulse-response function of the linear part only n different input levels are necessary.

5. EXPERIMENTAL RESULTS

For the proving of these methods a number of experiments have been carried out. The systems have been simulated with an analog computer. The further data processing was done seperately using a digital computer. Fig. 4 and 5 shows the results obtained using the method described in chapter 3 for the system F(s) = 1/(1+3.8s)(1+4.6s) connected with a nonlinearity. In a similar way experiments have been carried out, with a noise signal added to the output. The results of these experiments are also satisfying.

REFERENCES

[1] S. Goldberg an A. Durling, "A Computational Algortithm for the identification of Nonlinear Systems", Jorn. of the Franklin Inst. Vol. 291 pp, 427-97, June 1971.

[2] K.S.Narendra, P.G. Gallmann and J.H.Chang, "Identification of Nonlinear Systems using Gradient and Iterative Methods" Technical report No CT-4, Dunham Lab. Yale Univ., New. Haven Conn. Aug. 1966.

[3] W.D.T.Davies, "System Identification für Self-Adaptive Control" J. Wiley and Sons, London 1970.

[4] K.R.Godfrey, "Three-Level m Sequences", Electr. Letters, Vol. 2 pp. 241-242, July 1966.

[5] A.B. Gardiner, "Elimination of the Effect of Nonlinearities" on Process Crosscorrelation", Electr. Letters Vol. 2, pp. 164-165, May 1966.

[6] D.W.Clarke and K.R.Godfrey, "Simultaneous Estimation of first and Second Derivatives of Cost Functions", Electr. Letters Vol. 2, pp. 338-339, Sept. 1966.

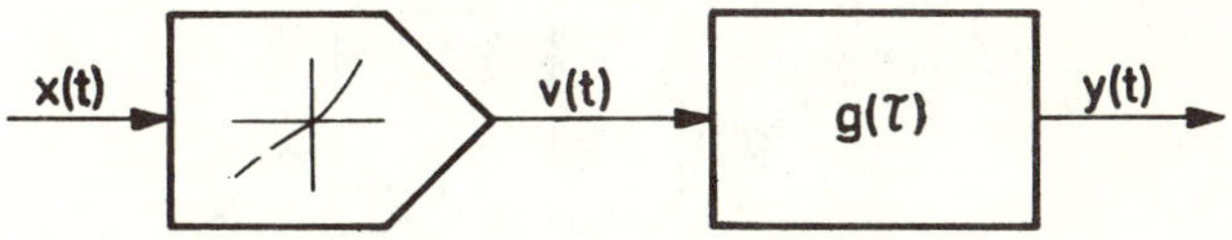

Fig. 1 Hammerstein-model

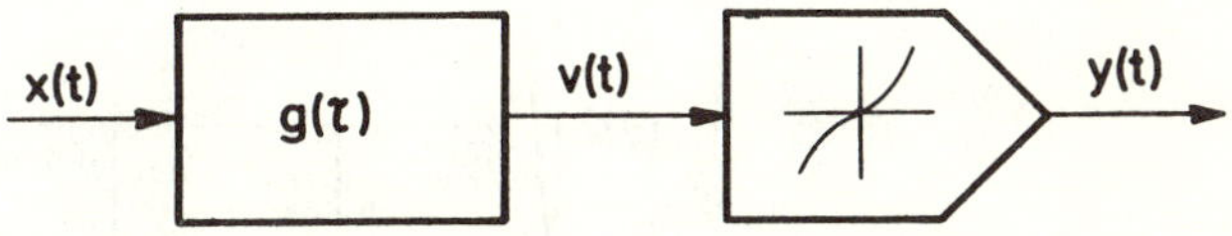

Fig. 2 Simplified Wiener-model

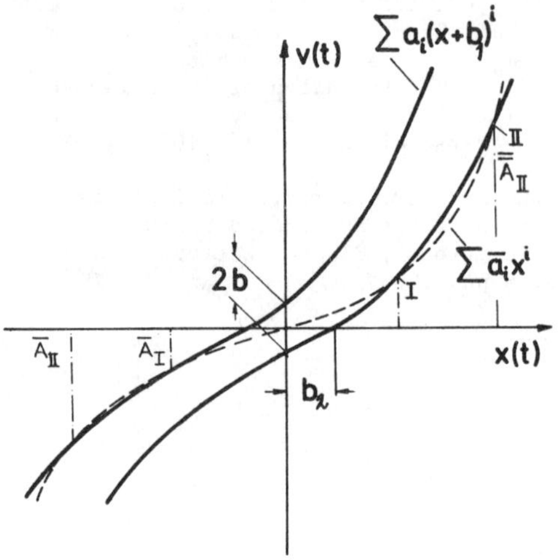

Fig. 3 Characteristic with hysteresis

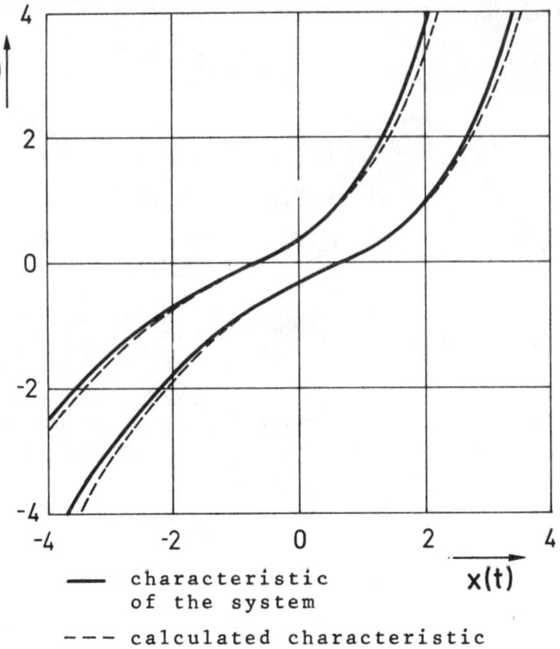

Fig. 4 Experimental result: nonlinear characteristic (test-signal levels: A_I=1.5, A_{II}=2.5, A_{III}=3.5)

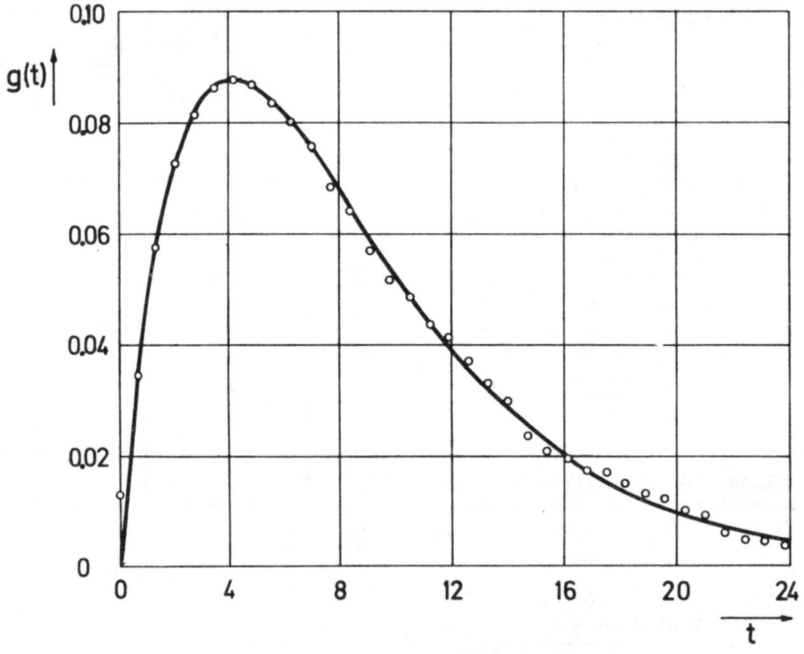

Fig. 5 Impulse-response function (N=242, Δt=0.7sec)

IDENTIFICATION OF NONLINEAR SYSTEMS THROUGH THE UTILISATION OF DISCONTINUOUS ORTHOGONAL FILTERS IN APPLYING A MULTILEVEL PSEUDORANDOM SIGNAL

Ľubomír Šutek, Miloslav Varga

Institute of Technical Cybernetics
Slovak Academy of Sciences
Bratislava, Czechoslovakia

The contribution concisely describes the method of estimating the parameters of a model that approximates the nonlinear no-memory portion of the Hammerstein-type nonlinear memory system by discrete orthogonal Rademacher functions and the linear memory portion of the nonlinear memory system by discrete orthogonal Haar functions, applying a multilevel inversely repeated pseudorandom test signal.

1. INTRODUCTION

The functional description applied to characterize nonlinear systems offered a compact and unified classification but its large-scale application encounters technical difficulties. The twolevel pseudorandom input signal allowed the estimation of kernels of the Volterra series [Kadri, 1971] with insufficient accuracy and with an ambiguity for higher kernel dimensions [Kichatov, 1967] .

Wiener's approach based on the application of orthogonal functionals [Wiener, 1958] provided for the development of practically usable methods at least for special input signals [Roy, 1962].

In this contributions the procedure of estimating the parameters of a Hammerstein-type model [Gallman, 1966] will be shown, approximating the nonlinear no-memory portion of the system by orthogonal Rademacher functions and the linear memory portion by orthogonal Haar functions. The parameters of the model are made up by the product of the development coefficients of the characteristics for both parts of the system. The use of a multilevel inversely repeated pseudorandom test signal [Ream, 1970] allowed to remove interaction between the model parameters and to set up a course suitable for practical use.

2. HAMMERSTEIN'S MODEL OF NONLINEAR SYSTEMS

The identification method referred to requires a nonlinear system model made up of a nonlinear no-memory gain followed by a linear memory system, Fig. 1

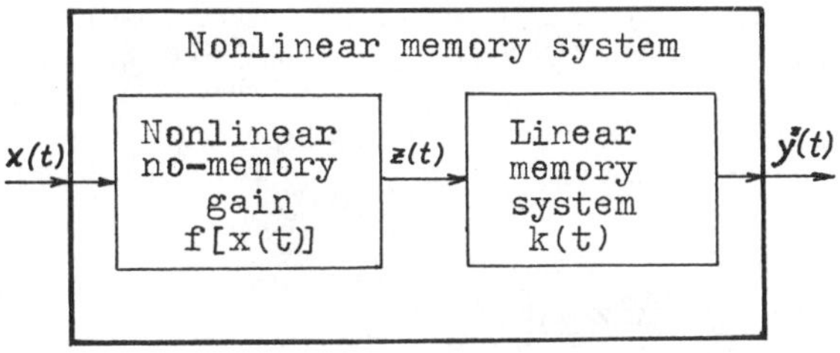

Fig. 1

This Hammerstein model (H-model)[Narendra, 1966] of the nonlinear system with constants independent of time is described by the relation

$$y^*(t) = \int_0^\infty k(\tau).f[x(t-\tau)]\,d\tau \quad , \qquad (1)$$

where $k(t)$ is the response to the impulse of the linear memory portion of the nonlinear memory system, $f(x)$ is the nonlinear characteristic of the nonlinear no-memory portion of the nonlinear memory system.

The principle of the suggested identification method for the H-model of nonlinear systems consists in approximating the characteristics of both portions of the nonlinear memory system by the system of Rademacher's and Haar's discrete orthogonal functions.

For the sake of conciseness the definition of the Rademacher and Haar functions and their interesting properties, known in mathematics for a relatively long time but practically utilized only of late

[Root, 1971] ,[Šutek, 1970] , will not be introduced [Alexits, 1961] , [Soboľ, 1969].

Let the linear memory system $k(t)$ be approximated by Haar's orthogonal functions* $h_i(t)$, thus

$$k(t) = \sum_{i=1}^{\infty} a_i h_i(t) \tag{2}$$

and let the nonlinear no-memory system $f(x)$ be approximated by Rademacher's orthogonal functions $r_j(x)$, thus

$$f(x) = \sum_{j=1}^{\infty} b_j r_j(x) , \tag{3}$$

where the coefficients of development a_i, b_j are determined by relations

$$a_i = \int_0^L k(t).h_i(t)dt , \tag{4}$$

$$b_j = \sum_{i=i_{min}}^{i_{max}} f[x(i)].r_j[x(i)] . \tag{5}$$

The form of relation (5) is due to the properties of the Rademacher functions. In relation (4) L is the final settling time of the system and, at the same time, of the model and in relation (5) $\langle x(i_{min}), x(i_{max})\rangle$ is the considered variation range of the input signal $x(t)$ and, at the same time, the approximation range of characteristic $f(x)$.

The block diagram of the model that approximates the characteristics of both portions of the nonlinear memory system by the Rademacher and Haar orthogonal functions is on Fig. 2 in which the coefficients of model $c_{i,j}$, which are to be determined are presupposed to have the form

$$c_{i,j} = a_i.b_j = \sum_{i=i_{min}}^{i_{max}} \int_0^L k(\tau).h_i(\tau).f[x(i)].r_j[x(i)]d\tau . \tag{6}$$

---

*The transition from dual to simple indexing of the Haar functions is to[Soboľ].

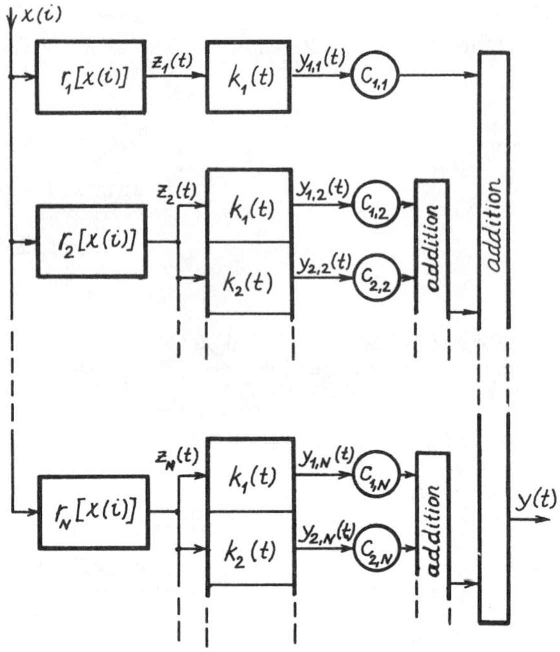

Fig. 2

3. DESCRIPTION OF THE METHOD OF ESTIMATING THE MODEL PARAMETERS

The outputs of the nonlinear no-memory portions of the model (Fig. 2) are determined by relations

$$z_j(t) = r_j[x(t)] \tag{7}$$

and the outputs of the linear memory portions for the jth nonlinear no-memory gain are described by relations

$$y_{i,j}(t) = \int_0^L h_i(\tau).r_j[x(t-\tau)]d\tau . \tag{8}$$

The model output is determined by relation

$$y(t) = \sum_{i=1}^{N}\sum_{j=1}^{N} c_{i,j} \int_0^L h_i(\tau).r_j[x(t-\tau)]d\tau \tag{9}$$

To determine coefficients $c_{i,j}$ characterizing the unknown nonlinear memory system let us calculate the mathematical expectation of the output product of the unknown system $y^*(t)$ and of output $y_{i,j}(t)$ of the model (correlation method) hence for coefficient $c_{i,j}$ we shall have

$$E\{y^*(t).y_{i,j}(t)\} = \int_0^L \int_0^L k(\tau).h_i(\nu).$$
$$E\{f[x(t-\tau)].r_j[x(t-\nu)]\}\, d\tau.d\nu \quad . \tag{10}$$

To determine the mathematical expectation on the right-hand side of relation (10) let us presume that the input test signal x(t) has the following properties

a) finite number of values (levels),
b) change of values x(t) being possible only at discrete points $t = i.\Delta T$, ΔT = const., i = 0,1,2,...,n. Denote $x(i\,\Delta T) = x(i)$,
c) uniform probability density $p[x(i)]$ on interval $\langle x(i_{min}), x(i_{max})\rangle$,
d) it is stationary,
e) the values are statisticaly independent, hence

$$p[x(i),\ x(i+j)] = p[x(i)].p[x(i+j)]$$

for $i \neq j$ and for $i = j$ it will be

$$p[x(i),\ x(i+j)] = p[x(i)].\delta[x(i)-x(i-j)] \tag{11}$$

By applying these presumptions relation (10) will have the form

$$E\{y^*(t).y_{i,j}(t)\} =$$
$$= \begin{cases} 0 & \text{for } \tau \neq \nu \\ \sum_{i=i_{min}}^{i_{max}} \int_0^L k(\tau).h_i(\tau).f[x(i-\tau)].r_j[x(i-\tau)] \\ p[x(i-\tau)]\,d\tau & \text{for } \tau = \nu \ . \end{cases} \tag{12}$$

This relation is in conformity with relation (6) , thus the mathematical expectation of the output product of the unknown nonlinear system $y^*(t)$ and of the output of model $y_{i,j}(t)$ is proportional to the value of coefficient $c_{i,j}$ which is the product of the development coefficients a_i, b_j, of the characteristics of the non-linear no-memory and the linear memory system. The values of coefficients $c_{i,j}$ characterize the unknown nonlinear system and are the result of the suggested method.

4. PROPERTIES AND THE GENERATION OF THE INVERSELY REPEATED MULTILEVEL PSEUDORANDOM SEQUENCE

The procedure of estimating the parameters of the suggested model requires an input signal in form of a dicrete "white" noise with uniform probability density. The accuracy of estimating the statistical characteristics of the current modes of generating such a signal depends on the length of observation. The abridgement of this period without affecting the attained accuracy of estimation of the characteristics has led to the application of a multilevel inversely-repeated pseudorandom sequence. The block diagram of generating this sequence is in Fig. 3.

The multilevel pseudorandom sequence u(i), the generation and properties of which are known [Davies, 1971], constitutes as a product with sequence w(i) the output sequence x(i) . The sequence w(i) is the product of the pseudorandom sequence $v_n(i)$ and of sequence $(-1)^i$. The GI block is the generator of the feed impulses with step ΔT, n is the number of flip-flop circuits of the shift register, setting up through feedback the mod 2 pseudorandom sequence $v_n(i)$.

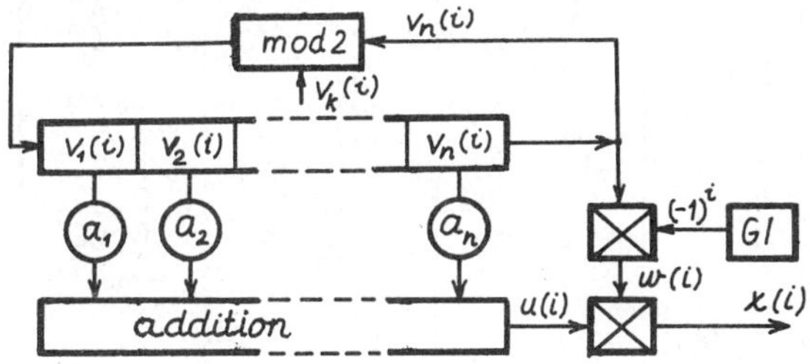

Fig. 3

Sequence x(i) has the length of period $2(2^n-1)\Delta T$ with uniform probability density (zero mean value) and with the correlation function $R_x(\tau\Delta T)$ given by the relation

$$R_x(\tau\Delta T) = \Big\{\Big[\sum_{k=1}^{n}\sum_{l=1}^{n} a_k a_l R_v[(\tau-l+k)\ \Delta T] +$$
$$+ (2^n-1)^{-1}\sum_{j=1}^{n} a_j\Big].KR[\tau \bmod (2^n-1)] - (2^n-1)^{-1}$$
$$\sum_{j=1}^{n} a_j\Big\}\Big\{(-1)^{\tau}\{[1+(2^n-1)^{-1}\}.KR[\tau \bmod (2^m-1)] -$$
$$-(2^m-1)^{-1}]\Big\} \ , \tag{13}$$

where KR is the Kronecker symbol, $R_v(\tau)$ is the correlation function of signal $v_n(i)$.

5. VERIFICATION OF THE METHOD

The suggested method was verified by simulation on the GIER digital computer and with results sufficiently convenient for practical application. As an example the Table introduces the values of coefficients the accurate ones and those determined by the method for the nonlinear memory system with the characteristics of both portions according to Fig. 4, approximating with 16 points, with percentage values of the $\Delta\sigma$ effect of parameter estimation error upon the overall value of the mean quadratic deviation of the system and the model.

At present this method is being verified by a real process(glass tank with temperature dynamics in dependence on the gas--air amount and ratio and on the amount and composition of the glass former)and data collection being under way.

Coef.	Values		$\Delta\sigma$ %
	accurate	determined	
$c_{2,1}$	3,217	3,2604	14,760
$c_{2,2}$	0,03535	0,03563	0,095
$c_{2,3}$	-1,2	-1,2094	3,189
$c_{2,4}$	0,4	0,37795	6,979
$c_{2,5}$	-1,4142	-1,4253	3,763
$c_{2,6}$	0,14142	0,1425	0,376
$c_{2,7}$	0,3536	0,3563	0,924
$c_{2,8}$	0,1414	0,1425	0,376
$c_{2,9}$	-0,2	-0,1763	6,977
$c_{2,10}$	-2,0	-1,9906	3,148
$c_{2,11}$	-0,4	-0,3780	6,981
$c_{2,12}$	0,4	0,4283	11,280
$c_{2,13}$	0,2	0,2267	10,168
$c_{2,15}$	0,1	0,1259	10,960
$c_{2,16}$	0,1	0,1259	10,960

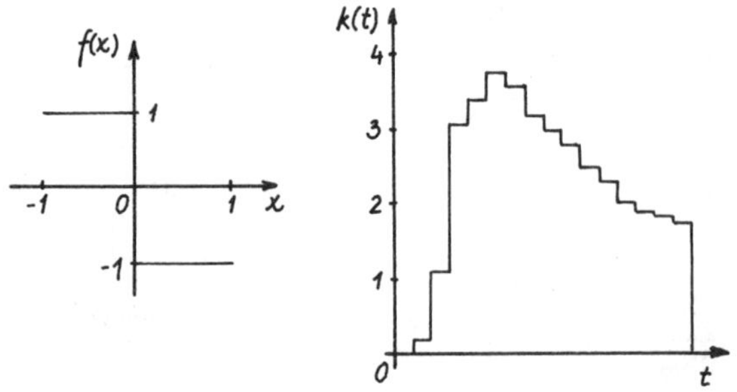

Fig. 4

6. CONCLUSION

With regard to the use of Rademacher's and Haar's orthogonal functions along with the generation of multilevel pseudorandom sequences, description is given of a method suitable not only for computer software but also for special single-purpose facilities devised to estimate the parameters of a model based on the approximation of the no-memory nonlinear and the memory linear portion of the unknown nonlinear memory system by means of Rademacher's and Haar's orthogonal functions.

REFERENCES

Alexits,G. 1961 : Convergence Problems of Orthogonal Series. Akadémia Kiadó, Budapest, 1961.

Davies,A.C. 1971 : Properties of Waveforms Obtained by Nonrecursive Digital Filtering of Pseudorandom Binary Sequences. IEEE Trans. Comp., Vol. C-20, p. 270, 1971.

Kadri, F.L. 1971 : Nonlinear Plant Identification by Crosscorrelation. IFAC Symposium, Györ, 1971.

Kichatov,Yu.F., Tenengolts,G.M., Dynkin, V.N. 1967 : Nonlinear System Identification with Pseudorandom Multilevel Sequences. IFAC Symposium, Prague, 1967.

Gallman,P.G., Narendra,K.S. 1966 : An Iterative Method for the Identification of Nonlinear Systems Using a Hammerstein Model. IEEE Trans. AC, p. 546, 1966.

Ream,N. 1970 : Nonlinear Identification Using Inverse-repeat m-sequences. Proc. IEEE, Vol. 117, 1970.

Root,W.L. 1971 : On the Structure of a Class of System Identification Problems. Automatica, Vol. 7, p. 219, 1971.

Roy,R.J., DeRusso,P.M. 1962 : A Digital Orthogonal Model for Nonlinear Processes with Two-Level Inputs. IRE Tranc. AC, p. 93, 1962.

Sobol,J.M. 1969 : Mnogomernyje kvadraturnyje formuly i funkcii Chaara. Nauka, Moskva, 1969.

Šutek,Ĺ., Varga,M. 1970 : On the Approximation of Systems by a Set of Incontinuous Orthogonal Functions and the Impact of Properties of the Input Signal upon Attainable Accuracy. IFAC Symposium, Prague, 1970.

Wiener,N. 1958 : Nonlinear Problems in Random Theory. Techn. Press MIT, New York, 1958.

AN IMPROVED PERFORMANCE CRITERION FOR PSEUDORANDOM SEQUENCES IN THE MEASUREMENT OF 2-ND. ORDER VOLTERRA KERNELS BY CROSSCORRELATION

F. L. Kadri and J. D. Lamb

Dynamic Analysis Group,
University of Wales Institute
of Science and Technology,
Cardiff, U.K.

SUMMARY

A property of pseudorandom sequences is introduced which shows that the general patterns of the anomalies associated with the multidimensional autocorrelation functions are the same, but shift as the multidimensional delay point moves along a line parallel to the diagonal. This property is used to develop a criterion to describe the performance of pseudorandom binary sequences as test signals for finding the two-dimensional crosscorrelation function of a nonlinear system. The criterion is discussed and illustrated by an example.

I- INTRODUCTION

The behaviour of a linear system can be described completely by its impulse response, and in practice there are many ways of measuring it, such as subjecting the system to an impulse, differentiating the the response to a step, indirectly from harmonic response analysis or by crosscorrelation. The Volterra kernels describe the behaviour of continuous nonlinear systems just as the impulse response does for linear systems ; here the output (y) is expressed as a series of multidimensional convolution terms of the input (x) and the Volterra kernels (k_i), viz.

$$y(t)=\int k_1(\tau_1)x(t-\tau_1)d\tau_1+\int\int k_2(\tau_1,\tau_2)x(t-\tau_1)x(t-\tau_2)d\tau_1 d\tau_2+..+... \qquad (1)$$

where the integration is performed between $+\infty$ and $-\infty$ throughout, unless stated otherwise. Notice that linear systems contain only the first order convolution term of the series in eq. (1).

One of the increasingly popular methods of deriving the impulse response of linear systems is by crosscorrelating the output with delayed pseudo-white noise input, this method can be adapted to nonlinear systems by crosscorrelating the output with delayed pseudo-white noise, (Lee and Schetzen {1965}). In fact, pseudorandom sequences have been used to find the two-dimensional crosscorrelation function of nonlinear systems (Hooper and Gyftopoulos {1966}), such sequences when used as a substitute for white noise greatly reduce the computational effort, but there"still a need for much longer run times than those required for linear systems, the reason for this is the existance of 'anomalies', or nonzero values, in the fields of the multidimensional autocorrelation functions. These were found to be due to the deterministic characteristics of pseudorandom sequences (Barker and Pradisthayon {1970}).

Most recently Barker et al.{1972}proposed a criterion to describe the performance of pseudorandom sequences for measuring the two-dimensional crosscorrelation function of nonlinear systems. In the present paper, it is demonstrated that the patterns of the anomalies associated with the multidimensional autocorrelation functions stay the same, but shift as the multidimensional delay point moves along a line parallel to the diagonal i.e. the line passing through the origin and being equidistant from all axes. As a result, simpler patterns of error are obtained when finding the two-dimensional crosscorrelation function of nonlinear systems, which makes feasible the ready selection, or rejection, of test signal characteristics. This property is used to develop a performance criterion for PseudoRandom Binary Sequences (PRBS) which is illustrated and compared with the approach of Barker et al.{1972} , and the advantages of the present approach placed in evidence.

II- A PROPERTY OF MULTIDIMENSIONAL AUTOCORRELATION FUNCTIONS OF PSEUDORANDOM SEQUENCES

A property is hereby introduced, where the patterns of the anomalies associated with the multidimensional autocorrelation function stay the same, but shifts as the multidimensional delay point moves along a line parallel to the diagonal. An algebraic statement of this property follows.

Theorem

$$\phi_n(i_1,i_2..i_h,J_1+c,J_2+c..J_m+c)=\phi_n(i_1-c,i_2-c..i_h-c,J_1,J_2..J_m)\cdot(2)$$

where

ϕ_n = the n-dimensional autocorrelation function of a pseudorandom sequence.

and $h+m = n$, where

$$1 \leqslant h,m \leqslant n-1$$

and $i_1,i_2..i_h,J_1,..J_m$ = time delays (No. of clock periods)

also c= integer multiple of the clock period generating the sequence.

The multidimensional delay point in R.H.S. of eq. (2), for example, has coordinates $J_1,..J_m$.

Proof

Consider the definition of the n-dimensional autocorrelation function of a pseudorandom sequence (Barker et al. {1972}). Applying the definition to L.H.S. of eq. (2) ;

$$\phi_n = \frac{1}{N}\sum_{i=0}^{N-1} x_{i-i_1}..x_{i-i_h}x_{i-J_1-c}..x_{i-J_m-c} \qquad (3)$$

where x_i is the i-th element of the sequence and N is the sequence length. Since pseudorandom sequences repeat after every N element, the limits of the summation can be altered so that ;

$$\phi_n = \frac{1}{N}\sum_{i=-c}^{N-1-c} x_{i-i_1}..x_{i-i_h}x_{i-J_1-c}..x_{i-J_m-c} \qquad (4)$$

Now, substituting $i' = i-c$ in eq. (4),

$$\phi_n = \frac{1}{N}\sum_{i'=0}^{N-1} x_{i'-(i_1-c)}..x_{i'-(i_h-c)}x_{i'-J_1}..x_{i'-J_m}$$

$$= \phi_n(i_1-c,..i_h-c,J_1,..J_m) \quad . \quad . \quad . \quad . \quad . \quad (5)$$

which proves the theorem.

III-A PERFORMANCE CRITERION FOR PRBS AS A TWO-DIMENSIONAL CROSSCORRELATION FUNCTION TEST SIGNAL

The major consequence of the property introduced in section II is that when finding multidimensional crosscorrelation functions along lines parallel to the diagonal, the patterns of error obtained will be simpler than if they were found in the more usual way, i.e. along lines parallel to a time axis (Kadri {1972}). On this basis, a performance criterion can be found which describes the performance of a pseudorandom sequence for two dimensional crosscorrelation function estimation. This results in figures of merit (Q(1),Q(2)..Q(c)..), where Q(c) corresponds to a line at a time-displacement (c-clock periods) from the diagonal.

Barker et al. {1972} give a comprehensive criterion to describe the performance of pseudorandom sequences in the measurement of two-dimensional crosscorrelation function of nonlinear systems. This describes the performance of a pseudorandom sequence as a set of figures of merit $(R_1,R_2..R_r..)$. A simplified definition of R_r is that "it is the largest square area containing the delay point (J_1,J_2) in the i_1-i_2 plane of $\phi_4(i_1,i_2,J_1,J_2)$ where there are less than r 'anomalies'". Here, a criterion is suggested for PRBS where a simple definition of it (Q(c)) is that "it is the largest square area corresponding to the longest path of the delay point along a line parallel to and at a time-displacement of c clock periods from the diagonal in the i_1-i_2 plane of ϕ_n where there are no anomalies". In fig. 1, ϕ_4 is shown for delay point (0,1) moving parallel to the diagonal for the four stage m-sequence PRBS (10011).

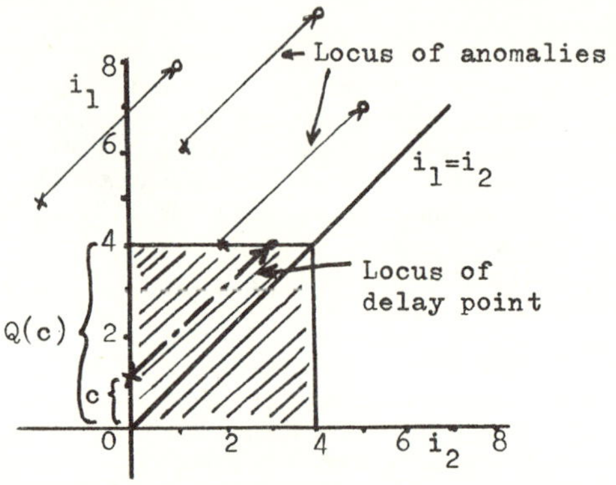

X Anomalies for delay (0,1)

o Anomalies for delay (3,4)

Fig. 1) The performance criterion Q(c), c=1, for m-sequence PRBS (10011).

A direct comparison between the two criteria is valid for R_1 only for PRBS, this will be demonstrated by finding a relationship between R_1 and Q(c).

Assume it is required to find Q(c), $c \leqslant R_1$. Considering the definition of R_1, and since a line parallel to the diagonal and at a distance c-clock periods from it passes through the square $R_1 x R_1$ for a time of (R_1-c) clock periods,(see fig. 2), the following relationship holds ;

$$Q(c) \geqslant R_1-c$$
or
$$Q(c)+c \geqslant R_1 \quad . \quad . \quad . \quad . \quad . \quad . \quad (6)$$

This relationship serves as a useful check on the derivation of Q(c).

The criterion has been applied to sequences derived by Barker et al. {1972},as these sequences have optimum characteristics for their lengths. The results for c = 1 to 10 are presented in Table 1.

The Q's in this table indicate the boundary of the area with no anomalies; this accounts for the one clock period difference when applying eq.(6).

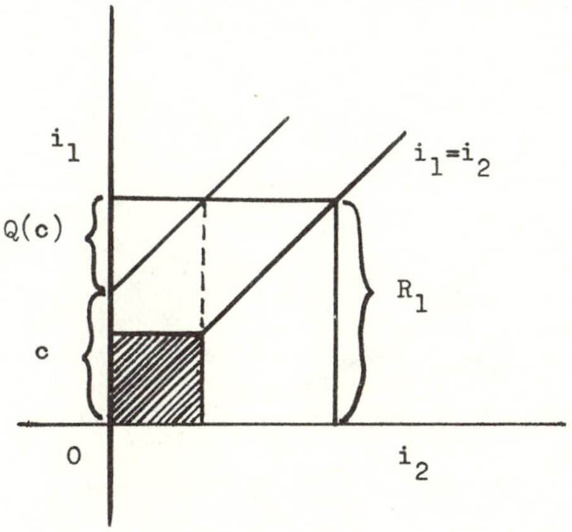

Fig. 2) The relationship between R_1 and Q(c).

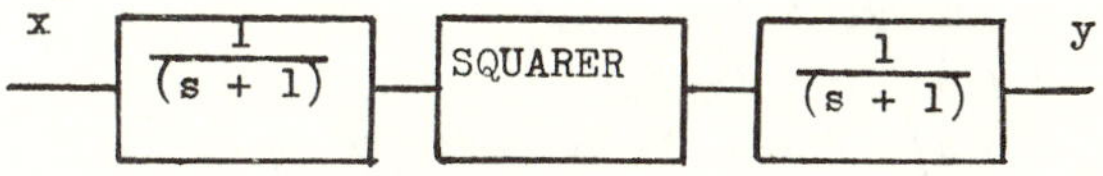

Fig. 3) Representation of a nonlinear system.

SEQUENCE	CONJUGATE SEQUENCE	R_1	Q(1)	Q(2)	Q(3)	Q(4)	Q(5)	Q(6)	Q(7)	Q(8)	Q(9)	Q(10)
100101011	110101001	12	12	13	16	18	9	14	12	15	17	20
1010000111	1110000101	18	26	17	22	26	15	20	27	21	23	26
10111111011	11011111101	26	38	32	29	46	28	22	21	34	42	26
111101011101	101110101111	35	40	56	33	51	61	47	67	44	38	50

Table 1) Values of R_1 and Q(1) to Q(10) for some optimal sequences .

IV- EXAMPLE

The nonlinear system of fig. 3 has been considered to find the expected two-dimensional crosscorrelation function along lines parallel to the diagonal($\Phi_2(\tau_J,\tau_c)$). Multidimensional Laplace transforms are used (Kadri {1972}), which leads to the expression ;

$$\Phi_2(s_J,s_c) = \frac{2A^2}{(s_J+1)(s_J+2)\quad(s_c+1)} \tag{7}$$

where A= equivalent impulse of PRBS autocorrelation function , s_J and s_c correspond to the time delay along the line and the time displacement of the line from the diagonal respectively.

A digital simulation package consisting of (a) PRBS generator, (b) representation of the system of fig. 3, and (c) the two-dimensional crosscorrelation function computational procedures, has been developed. The results of using the (101110101111) sequence is presented in fig. 4. The PRBS signal corresponding to this sequence had ±1 v (amplitude) and 10 Hz. (clock speed), the graphs shown are for c=1,2,3 and 4 clock periods, for 60 integer increments of the clock period for each curve. Q(c) is indicated on each graph to show the way the criterion describes the performance of the PRBS used.

V- DISCUSSION

In the overall investigation, other sequencs were tried on the digital model , and on an analogue model where readily possible. Also other systems were tested and all the results indicated that the criterion may be used with a variable degree of success. Space limitations preclude a complete reporting of this work here, but the example described is considered to be useful for the reasons ;

(a) The system is simple yet representative.
(b) The particular test sequence has an optimum predicted performance for its length as judged by the related criterion (Barker et al. {1972}).
(c) The sequence is the longest tried, hence it should exhibit a minimum evidence of the anomalies on the two-dimensional crosscorrelation function.

The multidimensional crosscorrelation functions of feedback systems also may be found using the above approach and with fuller use of multidimensional Laplace transforms, this has the consequence of isolating the nonlinear part

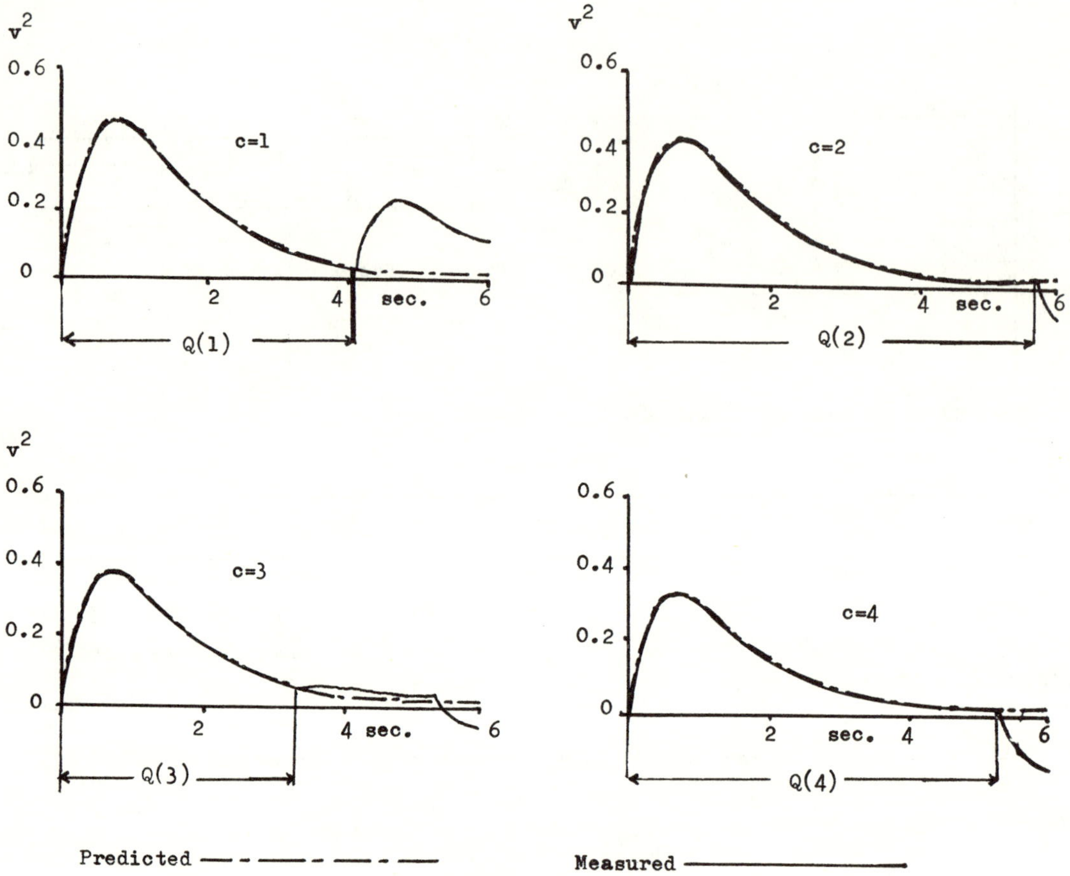

Fig. 4) Expected and measured two-dimensional crosscorrelation function for c=1,2,3 and 4 using PRBS characterized by (101110101111).

entirely from the linear part of the response which would have many usful applications such as fault finding and improved identification techniques.

REFERENCES

BARKER, H.A. and PRADISTHAYON, T.,"High Order Autocorrelation Functions of Pseudorando Signals Based on m-Sequences", Proc IEE, Vol. 117, No. 9, pp 1857, Sep. 1970.

BARKER, H.A., OBEDEGWU, S.N. and PRADISTHAYON,T., "Performance of Antisymmetric Pseudorandom Signals in the Measurement of 2-nd Order Volterra Kernels by Crosscorrelation", Proc IEE, Vol. 119, No. 3, pp 353, March 1972.

HOOPER, R.J. and GYFTOPOULOS, E.P.,"On the Measurement of Characteristic Kernels of a Class of Nonlinear Systems", USAEC Conference Report No. 660206, 1966.

LEE, Y.W. and SCHETZEN, M.,"Measurement of the Kernels of a Nonlinear System by Crosscorrelation", Int. J. of Control, Vol. II, No. 3, pp 237, Sep. 1965.

KADRI, F.L.,"The Dissociation of Variables ; A Transform Technique for the Determination of Volterra Kernels in i-Space", XIX International Congress on Nuclear and Aerospace Electronics, Rome, March 1972.

IDENTIFICATION BY PSEUDOSENSITIVITY FUNCTIONS AND QUASILINEARIZATION

M. R. Mataušek and M. D. Milovanović
Institute "Boris Kidrič" - Vinča
Beograd, Yugoslavia

The paper treats the identification of n o n l i n e a r deterministic multivariable systems by pseudosensitivity functions, from the point of view of quasilinearization. The p s e u d o quasilinearization procedure is introduced and the pseudosensitivity functions, defined for linear systems in Vušković (1970) and Bingulac (1973), are given the broader sense. The cases when the pseudosensitivity functions lead to the noniterative (one-shot) identification algorithms are defined. A detailed analysis is also performed in order to distinguish between the Gauss-Newton algorithm and an algorithm based on the modified quasilinearization, Baird (1969) and Baird (1970). Finally, combining the modified and pseudo quasilinearization, an algorithm with improved convergency characteristics is proposed.

1. STATEMENT OF THE PROBLEM

Let the structure of the model be known exept for a number of unknown stationary parameters q and unknown initial conditions x(0), i.e. let the dynamic behaviour of a process be described by the nonlinear vector differential equation

$$\dot{x} = f(x,q,u)\ , \quad x(0) = x_o \tag{1}$$

and the mathematical model be given by

$$\dot{x}^k = f(x^k,q^k,u)\ , \quad x^k(0) = x_o^k\ , \tag{2}$$

where $x^k(0) \neq x(0)$ and $q^k \neq q$. Dimensions of the column vectors - state vector x(t), input vector u(t) and unknown parameter vector q, are n, $m \leq n$ and r, respectively. f is a known n-dimensional column vector function.

Writing equation (2) as

$$\begin{Bmatrix}\dot{x}^k\\ \dot{q}^k\end{Bmatrix} = \begin{Bmatrix} f(x^k,q^k,u)\\ 0\end{Bmatrix},\quad \begin{Bmatrix}x^k(0)\\ q^k(0)\end{Bmatrix} = \begin{Bmatrix}x_o^k\\ q^k\end{Bmatrix}, \tag{3}$$

and linearizing equations (3) in the vicinity of q^k and $v^k(t)$, where at present $v^k(t)$ is an undefined n-dimensional column vector, it is obtained

$$\begin{Bmatrix}\dot{x}_L^k\\ \dot{q}_L^k\end{Bmatrix} = \begin{Bmatrix} f_x(v^k,q^k,u) & f_q(v^k,q^k,u)\\ 0 & 0\end{Bmatrix}\begin{Bmatrix}x_L^k - v^k\\ q_L^k - q^k\end{Bmatrix} + \begin{Bmatrix} f(v^k,q^k,u)\\ 0\end{Bmatrix}. \tag{4}$$

Here, f_x and f_q are (n x n) and (n x r) matrices, with elements $f_{xij} = \partial f_i(v^k,q^k,u)/\partial x_j$, $i,j = 1,2,\ldots,n$ and $f_{qij} = \partial f_i(v^k,q^k,u)/\partial q_j$, $i = 1,2,\ldots,n$, $j = 1,2,\ldots,r$.

Given the initial conditions $x_L^k(0)$ and $q_L^k(0)$, the solution of the system (4) is

$$x_L^k(t) = S_{11}^k(t)x_L^k(0) + S_{12}^k(t)q_L^k(0) + C_1^k(t),\quad q_L^k(t) = q_L^k(0) \tag{5}$$

where

$$\dot{S}_{11}^k = f_x(v^k,q^k,u)S_{11}^k\ ,\quad S_{11}^k(0) = I\ , \tag{6}$$

$$\dot{S}_{12}^k = f_x(v^k,q^k,u)S_{12}^k + f_q(v^k,q^k,u)\ ,\quad S_{12}^k(0) = 0, \tag{7}$$

$$\dot{C}_1^k = f_x(v^k,q^k,u)(C_1^k - v^k) + f(v^k,q^k,u) - f_q(v^k,q^k,u)q^k,\quad C_1^k(0) = 0\ . \tag{8}$$

S_{11}^k and S_{12}^k are (n x n) and (n x r) matrices, C_1^k is a n-dimensional column vector and I is a unit matrix.

Supposing that the observational values of the state coordinates $x_i(t_j)$, $0 \leq t_j \leq \tau$, $j = 1,\ldots,N$, are given

$$x_i(t_j) = x_{ij}\ ,\quad i = 1,2,\ldots,s \leq n,\quad sN > (n+r), \tag{9}$$

some iterative, $k = 1,2,\ldots,$ identification procedures will now be investigated.

Starting with the relation

$$v^k(t) = x_L^{k-1}(t), \tag{10}$$

finding the unknown initial conditions $x_L^k(0)$ and $q_L^k(0)$ so that $x_L^k(t)$ satisfies the given boundary conditions (9) and proceeding in the same manner for k=1, 2, ..., the quasilinearization (QL) procedure applied to the solution of identification problems, see e.g. Lavy (1965), is obtained.

It will be shown in Sec. 2 that starting with the relation

$$v^k(t) = x^k(t)\ , \tag{11}$$

i.e. applying the modified quasilinearization (MQL) procedure, Baird (1969), to the solution of identification problems, the Gauss-Newton (GN) algorithm, see e.g. Marquardt (1963), is obtained.

In Sec. 3 a noniterative pseudo quasilinearization (PQL) algorithm is formulated, supposing that all state coordinates are measurable and linearizing equations (3) in the vicinity of x(t), i.e. starting with the relation

$$v^k(t) = x(t). \tag{12}$$

For a case when some state coordinates are not measurable it is shown that a combination of the modified and pseudo quasilinearization leads to an algorithm with improved convergency characteristics.

In what follows, notation

$$S_{ij}^k = W_{ij}^k, \quad C_1^k = P_1^k \quad \text{when} \quad v^k(t) = x^k(t),$$
$$S_{ij}^k = Z_{ij}^k, \quad C_1^k = G_1^k \quad \text{when} \quad v^k(t) = x(t), \tag{13}$$

is used. The elements of matrices W_{11}^k and W_{12}^k are sensitivity functions of the system (1) in respect to the initial conditions x(0) and parameters q. By the elements of matrices Z_{11}^k and Z_{12}^k the pseudosensitivity functions of the nonlinear multivariable system (1) are defined.

2. ITERATIVE IDENTIFICATION ALGORITHM

The algorithm is based on the modified quasilinearization, i.e. $v^k(t) = x^k(t)$. Let the equations (2), (6), (7) and (8) be solved starting with the initial conditions $x^k(0) \neq x(0)$, $q^k(0) \neq q$. Then, according to the relations (5) and (13)

$$x_L^k(t) = W_{11}^k(t)x_L^k(0) + W_{12}^k(t)q_L^k(0) + P_1^k(t). \tag{14}$$

The unknown initial conditions $x_L^k(0)$ and $q_L^k(0)$ are to be determined so that $x_L^k(t)$ satisfies the given boundary conditions (9). In order to simplify the presentation the following boundary conditions will be used

$$x_L^k(j\,\Delta t) = x(j\,\Delta t), \quad j = 1, \ldots, N, \tag{15}$$

where Δt is the sampling period.

Let $(nN \times (n+r))$ matrix $\{K_{1W}\ K_{2W}\}$ and nN-dimensional column vector D_P be defined by the relations

$$\left\{K_{1W}\ K_{2W}\right\} = \begin{Bmatrix} W_{11}^k(\Delta t) & W_{12}^k(\Delta t) \\ \vdots & \vdots \\ W_{11}^k(N\Delta t) & W_{12}^k(N\Delta t) \end{Bmatrix},$$

$$D_P = \begin{Bmatrix} x(\Delta t) & - P_1^k(\Delta t) \\ \vdots & \vdots \\ x(N\Delta t) & - P_1^k(N\Delta t) \end{Bmatrix}. \tag{16}$$

Then, from the relations (14) and (15), it follows

$$\left\{K_{1W}\ K_{2W}\right\} \begin{Bmatrix} x_L^k(0) \\ q_L^k(0) \end{Bmatrix} = D_P. \tag{17}$$

After the pseudo inversion, see e.g. Penrose (1956), equation (17) yields

$$\begin{Bmatrix} x_L^k(0) \\ q_L^k(0) \end{Bmatrix} = \left\{\left\{K_{1W} K_{2W}\right\}^T \left\{K_{1W} K_{2W}\right\}\right\}^{-1} \left\{K_{1W} K_{2W}\right\}^T D_P. \tag{18}$$

Setting

$$x^k(0) = x_L^{k-1}(0), \quad q^k = q_L^{k-1}(0), \tag{19}$$

the iterative identification algorithm MQL is obtained.

Using the procedure applied to form the relations (24) to (27) and according to the relations (19), it is easy to show that

$$W_{11}^k(t)x_L^{k-1}(0) + W_{12}^k(t)q_L^{k-1}(0) = x^k(t) - P_1^k(t). \tag{20}$$

Evaluating P_1^k from the relation (20) and substituting in D_P, relation (18), it follows

$$\begin{Bmatrix} \Delta x_L^k(0) \\ \Delta q_L^k(0) \end{Bmatrix} = \left\{\left\{K_{1W} K_{2W}\right\}^T \left\{K_{1W} K_{2W}\right\}\right\}^{-1} \left\{K_{1W} K_{2W}\right\}^T R, \tag{21}$$

where

$$R = \begin{Bmatrix} x(\Delta t) & - x^k(\Delta t) \\ \vdots & \vdots \\ x(N\Delta t) & - x^k(N\Delta t) \end{Bmatrix}, \quad \begin{Bmatrix} \Delta x_L^k(0) \\ \Delta q_L^k(0) \end{Bmatrix} = \begin{Bmatrix} x_L^k(0) - x_L^{k-1}(0) \\ q_L^k(0) - q_L^{k-1}(0) \end{Bmatrix}. \tag{22}$$

From the relations (19) and the second of the relations (22) it follows

$$x^k(0) = x^{k-1}(0) + \Delta x_L^{k-1}(0), \quad q^k = q^{k-1} + \Delta q_L^{k-1}(0). \tag{23}$$

Relations (21), (23) are in fact the Gauss-Newton (GN) algorithm.

It should be mentioned that one could obtain the Gauss-Newton algorithm directly, without the transformations (18) and (20), but starting with the perturbation format of the modified quasilinearization, Baird (1970). However, the modified quasilinearization format containing the particular solution $P_1^k(t)$ has been used here because, as it will be shown in Sec. 3, the vector $P_1^k(t)$, i.e. the vector $G_1^k(t)$, plays the crucial role in the formulation of noniterative algorithms for the nonlinear systems identification.

3. NONITERATIVE IDENTIFICATION ALGORITHMS

These algorithms are based on the pseudo quasilinearization, i.e. $v^k(t) = x(t)$. The following definitions are used: linear system - f(x, q, u) linear function in both x and q; system linear-in-parameters - f(x, q, u) nonlinear function in x but linear in q; completely nonlinear system - f(x, q, u) nonlinear function in both x and q.

Multiplying the equation (6) by the unknown initial condition vector x(0) and the equation (7) by the unknown parameter vector q and adding so obtained equations, taking into account the relations (12) and (13), one obtaines

$$\dot{E}_1 = f_x(x, q^k, u)E_1 + f_q(x, q^k, u)q, \quad E_1(0) = x(0), \tag{24}$$

$$E_1(t) = Z_{11}^k(t)\,x(0) + Z_{12}^k(t)q. \tag{25}$$

Subtracting the equation (8) from the equation (1) it follows

$$\dot{E}_2 = f_x(x,q^k,u)E_2 + f(x,q,u) - f(x,q^k,u) +$$
$$+ f_q(x,q^k,u)q^k, \quad E_2(0) = x(0) , \qquad (26)$$

$$E_2(t) = x(t) - G_1^k(t). \qquad (27)$$

It is obvious from the equations (24), (26) and the relations (25), (27) that, in systems where

$$f(x,q,u) - f(x,q^k,u) = f_q(x,q^k,u)(q - q^k) , \qquad (28)$$

i.e. in systems linear-in-parameters q, the following relation holds

$$Z_{11}^k(t)x(0) + Z_{12}^k(t)q = x(t) - G_1^k(t) . \qquad (29)$$

If the (nN x (n+r)) matrix $\{K_{1Z} \; K_{2Z}\}$ and the nN-dimensional column vector D_G are formed according to the relations (16), i.e. substituting $W_{11}^k(t)$, $W_{12}^k(t)$ and $P_1^k(t)$ with $Z_{11}^k(t)$, $Z_{12}^k(t)$ and $G_1^k(t)$ respectively, it follows from the relation (29)

$$\{K_{1Z} \; K_{2Z}\} \begin{Bmatrix} x(0) \\ q \end{Bmatrix} = D_G . \qquad (30)$$

After the pseudo inversion, equation (30) yields

$$\begin{Bmatrix} x(0) \\ q \end{Bmatrix} = \left\{\{K_{1Z} K_{2Z}\}^T \{K_{1Z} K_{2Z}\}\right\}^{-1} \{K_{1Z} K_{2Z}\}^T D_G . \qquad (31)$$

Relation (31) defines, for systems linear-in-parameters q, the noniterative identification algorithm PQL.

It is obvious from the relation (28) that for completely nonlinear systems it is not possible to formulate a noniterative algorithm based on the pseudosensitivity functions.

According to the relations (20), (21), (13), (31), the noniterative Gauss-Newton algorithm - pseudo Gauss--Newton (PGN) algorithm - can be formulated only when the following relation holds

$$Z_{11}^k(t)x^k(0) + Z_{12}^k(t)q^k = x^k(t) - G_1^k(t) . \qquad (32)$$

Using the procedure applied to form the relations (24) to (27), it can easily be shown that relation (32) holds only for systems where

$$f(x,q^k,u) - f(x^k,q^k,u) = f_x(x,q^k,u)(x - x^k) . \qquad (33)$$

This means, however, that the noniterative Gauss--Newton algorithm can only be formulated in the case of linear systems. The only exeption are the systems where $f(x,q,u)$ is of the form $f_L(x,u) + f_N(x,q,u)$, $f_L(x,u)$ being a linear function in x, $f_N(x,q,u)$ being a nonlinear function in x but linear in q, and if initial guess for unknown parameters is $q^o = 0$.

Suppose now that some coordinates of the state vector x(t) are not measurable. If unmeasurable coordinates do not appear in the partial derivative $f_q(x,q^k,u)$, what means that thay do not appear in the terms containing the unknown parameters q, the relations (28), (29) are still valid and the unknown parameters can be determined after the first iteration. For linear systems $\dot{x} = Ax + Bu$, this means that, if $x_a(t)$, $x_b(t)$ and $x_c(t)$ are measurable coordinates of the state vector x(t), the unknown parameters can be determined after the first iteration when they appear in the a-th, b-th and c-th column of the matrix A.

If the unmeasurable coordinates of the state vector x(t) do appear in the partial derivative $f_q(x,q^k,u)$, the unknown parameters cannot be determined after the first iteration. Numerical examples show, however, that in this case, by combining the pseudo and modified quasilinearization a new algorithm with improved convergency characteristics can be obtained. In this combined algorithm the equations (3) are linearized in the vicinity of the vector $v^k(t)$, whose coordinates consist of the measurable coordinates of the state vector x(t) and of those coordinates of the vector $x^k(t)$ which correspond to the unmeasurable coordinates of the state vector x(t).

4. NUMERICAL EXAMPLES

Three numerical examples will be presented and discussed, illustrating the identification of a linear system, a system linear-in-parameters, Lavy (1965), and a completely nonlinear system

$$\dot{x}_1 = a_{11}x_1 + a_{12}x_2 + b_{11}u, \quad \dot{x}_2 = a_{21}x_1 + a_{22}x_2 + b_{22}u; \qquad (34)$$

$$\dot{x}_1 = x_2, \quad \dot{x}_2 = -q_1x_2 - q_2\sin x_1 ; \qquad (35)$$

$$\dot{x}_1 = x_2, \quad \dot{x}_2 = -q_1x_2 - \sin(q_2x_1) . \qquad (36)$$

The following abbreviations, in addition to the ones already adopted, will be used: PMQL to denote the combined quasilinearization algorithm and KGN to denote the combined Gauss-Newton algorithm. In these algorithms the coordinate $v_1^k(t) = x_1(t)$ is taken from the system to be identified, i.e. from the model with true values of the initial conditions and parameters, while coordinate $v_2^k(t) = x_2^k(t)$ is taken from the model. Let it be mentioned here that it would also be interesting to examine the use of the equation $\dot{x}_2^k = f_2(x_1, x_2^k, q_1^k, q_2^k)$ for generating $v_2^k(t) = x_2^k(t)$, when identifying the systems (35) and (36) by the PMQL algorithm.

Linear system identification is illustrated by Tables 1 and 2, refering to the case when the boundary value problem is solved only for the coordinate $x_1(t)$.

Table 1

$x_1(0)$	$x_2(0)$	a_{11}	a_{21}	a_{12}	a_{22}	b_{11}	b_{22}	k
1.000	0.000	-1.000	0.000	0.000	0.000	2.000	0.000	
1.000	4.200	-0.500	-1.000	4.000	-2.000	3.000	2.000	0
algorithm KGN								
	0.000	-1.000	0.000	4.000	-2.000	2.000	0.000	1

Table 2

$x_1(0)$	$x_2(0)$	a_{11}	a_{21}	a_{12}	a_{22}	b_{11}	b_{22}	k
1.000	3.000	-1.000	-2.000	3.000	-4.000	2.000	3.000	
1.000	4.200	-0.500	-1.000	4.000	-2.000	3.000	2.000	0
algorithm KGN								
	2.277	-1.000	-1.518	3.952	-4.000	2.000	2.277	5

Identification of the system linear-in-parameters is illustrated by Tables 3, 4 and 5, while the identification of the completely nonlinear system is illustrated by Table 6. The results presented in the Tables 3, 4 and 6 are obtained by solving the boundary value problem on the interval $0 \leq t \leq 4$, in seventeen points $t_j = (j-1)\Delta t$, $j=1,2,\ldots,17$, $\Delta t = 0.25$ for the coordinate $x_1(t)$ and only at the point $t_1 = 0$ for the coordinate $x_2(t)$. Such distribution of points makes possible the direct comparison with the results obtained using the algorithm QL, Lavy (1965). The values in the Table 5 are obtained by solving the boundary value problem for both coordinates in sixteen points $t_j = (j-1)\Delta t$, $j=2,3,\ldots,17$, $\Delta t = 0.25$.

In all the tables, when $k=0$, the initial guesses for initial conditions and parameters are given. If some initial conditions for $k = 0$ are equal to their true values and are not addapted during the iterations, their values are not indicated in the tables for $k \neq 0$.

Table 1 shows that after the first iteration the order of the unknown system is identified too, i.e. that the result of the identification is $x_1^1(t) = x_1(t)$ and $x_2^1(t) = 0$. Analysing the results presented in Table 2, it can be shown that the result of the identification is $x_1^5(t) = x_1(t)$ and $x_2^5(t) = cx_2(t)$, i.e. that the coordinate $x_2(t)$ is identified exept for an unknown multiplier c.

The algorithm GN, when applied to identify the linear system (34) with the initial guesses given in Tables 1 and 2, diverges.

No additional comment concerning Tables 3 to 6 is necessary, exept that the algorithm PMQL in Table 6 converges to a local minimum of the criterion, Lavy (1965), $J = (x_2(0) - x_2^k(0))^2 + \sum_{r=o}^{16} (x_1(r\Delta t) - x_1^k(r\Delta t))^2$.

Table 3

$x_1(0)$	$x_2(0)$	q_1	q_2	k
1.500	0.000	2.000	2.000	
1.500	0.000	0.500	5.000	0
algorithm PQL				
1.500	0.000	2.000	2.000	1
algorithm PMQL				
1.500	-0.000	1.999	1.999	5
algorithm MQL				
1.496	-0.001	1.851	1.855	5
1.500	-0.000	2.000	2.000	7
algorithm QL (Lavy (1965))				
1.500	0.000	2.000	2.000	5

Table 4

$x_1(0)$	$x_2(0)$	q_1	q_2	k
1.500	0.000	2.000	2.000	
1.500	0.000	4.000	4.000	0
algorithm PQL				
1.500	0.000	2.000	2.000	1
algorithm PMQL				
1.488	-0.002	0.148	0.311	1
1.500	-0.000	2.000	2.000	6
algorithm MQL - divergent				
1.472	-0.006	0.227	0.036	1
1.430	-0.097	20.32	1.105	2
1.503	0.000	-1747.	-87.11	3

Table 5

$x_1(0)$	$x_2(0)$	q_1	q_2	k
1.50	0.00	2.000	2.000	
1.50	0.00	0.500	5.000	0
algorithm PGN				
		0.897	2.347	1
		1.999	2.000	3
algorithm KGN				
		1.620	1.799	3
		1.999	2.000	5
algorithm GN				
		1.691	1.766	5
		2.000	1.999	7

Table 6

$x_1(0)$	$x_2(0)$	q_1	q_2	k
1.500	0.000	2.000	1.000	
1.500	0.000	4.000	2.000	0
algorithm PQL				
1.499	-0.000	2.684	1.374	1
1.500	0.000	2.000	1.000	5
algorithm PMQL				
1.468	-0.098	-1.966	-2.149	9
1.467	-0.098	-1.990	-2.145	14
algorithm MQL - divergent				
1.515	0.003	27.57	0.670	1
1.516	0.001	-2462.	-79.71	2

ACKNOWLEDGEMENT

The autors express their thanks to Mr. M. D. Mladenović for his help in numerical analysis of nonlinear systems identification.

REFERENCES

Baird, C.A., Jr. (1969). Modified Quasilinearization Technique for the Solution of Boundary-Value Problems for Ordinary Differential Equations. J. Optimization theory appl., Vol. 3, No. 4, p. 227.

Baird, C.A., Jr. (1970). Quasilinearization and the Methods of Finite Difference and Initial Values. J. Optimization theory appl., Vol. 6, No. 4, p. 320.

Bingulac, S.P. and Djorović, M. (1973). On the Generation of Sensitivity and Pseudosensitivity Functions for Multivariable Systems. To be presented at III IFAC symp. sensitivity, adaptivity, optimality. Ischia, Italy, June 18-21.

Lavy, A. and Strauss, J.C. (1965). Parameter Identification in Continuous Dynamic Systems. IEEE international convention record, Symp. autom. control, syst. sci., cybernetics, human factors, pt. VI.

Marquardt, D.W. (1963). An Algorithm for Least--Squares Estimation of Nonlinear Parameters. J. Soc. ind. appl. math., Vol. 11, No. 2, p. 431.

Penrose, R. (1956). On Best Approximate Solutions of Linear Matrix Equations, Proc. Cambridge Philos. Soc., Vol. 52, Part 1, p. 17.

Vušković, M.I., Bingulac, S.P. and Djorović, M. (1970). Application of the Pseudosensitivity Functions in Linear Dynamic System Identification. IFAC symp. Identification and process parameter estimation, Prague, paper 2.5.

IDENTIFICATION OF MULTIVALUED PARAMETERS IN A CLASS OF NOISY DYNAMIC PROCESSES

R. E. Klein
Associate Professor of Mechanical Engineering
Department of Mechanical and Industrial Engineering
University of Illinois at Urbana-Champaign
Urbana, Illinois
U.S.A.

The problem considered in this paper is that of nonlinear system identification in which an ordinary differential equation model is sought to represent a physical system whose description is initially unknown. An equation error and steep descent method is used to produce numerical values for coefficients which characterize unknown quantities in the differential equations. The classes of dynamic processes are those described by nonlinear, time invariant, lumped parameter models where the nonlinearities may be of a class having multivalued characteristics. The method requires knowledge of the existing system input, or an augmented input, in addition to knowledge of the system output. *A priori* information concerning the order of the unknown system model and the arguments of the nonlinear functions is assumed available.

1. INTRODUCTION

Despite the research effort devoted to the identification problem and the array of methods that have been devised, the problem of identifying nonlinear dynamic systems is still largely incomplete. Many of the existing identification techniques require extensive *a priori* knowledge of the unknown system, special tests, and lengthy computation.

Kohr [1] presents a method for identifying processes having a class of multivalued nonlinear functions. The method by Kohr employs a technique whereby constants and functions in an equation error model may be identified provided any nonlinear term present may be written as a function of a single argument. The procedure provides plots of the functions versus their arguments. This method is reasonably effective, however, it is slow, manually implemented, and requires complete comtrol of the system input and relatively noise-free measurements. The Kohr method has worked for some multivalued nonlinear functions such as backlash.

The nonlinear identification procedure proposed in this paper is an extension of methods developed by Kohr [1], Hoberock and Kohr [2], Lion [3], Miller [4], Sprague and Kohr [5], Butler and Bohn [6], and others. An equation error model and steep descent parameter adjustment are basic features of the method. As higher derivatives of measured states are normally not acessible from on-line data, the necessary derivatives are estimated through the use of linear filters operating on the system input and output, respectively. The new feature of the method is the means used to represent unknown multivalued nonlinear functions which appear in the equation error model. Once the region of the multivalued function is known, the argument range is partitioned into a suitable set of small segments. A simple series expansion of the function is then assumed within each interval and for each argument direction, respectively. In this manner, the multivalued function may be characterized, in each interval, by a small number of coefficients. The sets of coefficients from these elementary expansions are found by employing a steep descent adjustment procedure to minimize the square of the equation error. The result of the identification process is a set of parameters and expansion coefficients which describe an approximation for the differential equation model.

2. PRELIMINARIES

Numerous identification procedures make use of a so-called "equation error" model. This model is a partially known differential equation into which are substituted measured derivatives of the input and output. Because the equation is only partially known, it will not balance, in general, and an equation error or residual will result. This error is operated upon in real time in the identification process to supply estimates of the quantities unknown in the differential equation. The equation error model method requires derivatives of the system input and output which must be evaluated from measured input-output data. Because of the impracticability of differentiating noise corrupted signals, a set of linear filters is used to obtain approximations of the necessary derivatives.

The linear filters, or derivative filters, are devices which are constructed from linear operators, and which have an order equal to or greater than the order of the highest derivative desired. For identification of nonlinear systems, it has been shown by Kohr [1] that the filter operator must commute with the nonlinear operators in the unknown system's differential equation. One suitable linear filter is the transport lag or delay time operator. The convolution theorem can be used to show that this operator, the transport lag, commutes with a broad class of nonlinear operators.

A transport lag cannot be simulated exactly by a finite number of lumped parameter elements; however, ITAE standard forms have been shown to be satisfactory approximations. The coefficients for these forms are given in a variety of standard sources. Consequently, it is a straightforward matter to design and implement a linear derivative filter by use of a polynomial-type transfer function which approximates a pure transport lag for frequencies up to some specified filter frequency or bandwidth.

With input and output derivatives available, in the equation error model the error becomes an algebraic function of the unknown parameters and the system input-output derivatives. Furthermore, partial derivatives of the equation error, with respect to the unknown parameters, are algebraic functions of the system input-output derivatives and can be directly and continually evaluated from the measured data. Thus, it is possible for the identification method to use the equation error approach to achieve continuous parameter adjustment, such as with analog computer components, rather than a pure iterative procedure.

3. THE BASIC IDENTIFICATION METHOD

In the context of system identification, the word "identification" will be restricted here to mean specifically the process of finding a suitable differential equation model for a physical system of unknown identity. The identification process is to be accomplished on the basis of measured input-output data and it is assumed that the system input can be selected or augmented by the experimenter. The identification process will be regarded as complete when the difference between the system and model response to identical, arbitrary inputs is small compared to the response of either.

It is assumed that sufficient *a priori* information is available about the system to write a differential equation for it as

$$a_n x^n + \cdots + f_i(x^j)\, x^i + \cdots + f_o(x^k)\, x^o = u^o + \cdots + g_i(u^\ell)\, u^i + \cdots g_m(u^q)\, u^m \tag{3.1}$$

where

$$x^i = \frac{d^i x}{dt^i} \tag{3.2}$$

a_n is a constant, and the f's and g's are functions of the indicated arguments.

In Eq. (3.1), x^o denotes the output from the system and u^o denotes the input to the system. The synbol n denotes the highest derivative of the system output appearing in Eq. (3.1), and m denotes the highest derivative of the system input.

By use of the equation error method, the error E(t) is formed from Eq. (3.1) by introducing estimates of the unknown quantities with a hat symbol (^) and constant coefficient estimates by the Greek letter α, and transposing, leads to

$$E = \alpha_n x^n + \cdots + \hat{f}_i(x^j)\, x^i + \cdots + \hat{f}_o(x^k)\, x^o - u^o - \cdots - \hat{g}_i(u^\ell)\, u^i - \cdots - \hat{g}_m(u^q)\, u^m \tag{3.3}$$

The major difficulty in identifying nonlinear systems is posed by the unknown functions in Eq. (3.1). In order to find these functions completely and directly requires finding an infinity of point pairs, that is, the entire curve for each of the functions. Considerable difficulty is encountered, however, when such an approach is used on systems whose descriptions contain more than a single unknown function. Consequently, identification methods which utilize direct and brute-force computation of functions are most useful for identification of simple classes of systems.

An alternate method to the direct computation of functions for the model is to characterize the unknown functions, in some appropriate manner, by a finite number of parameters. The method developed in this paper uses series expansion representation of the functions within a suitable set of partition intervals. Values for the unspecified parameters in these series are found in the identification process.

The partitioning procedure is used to assure an adequate approximation to most well-behaved functions by use of a low-ordered polynomial. This procedure is identical to using a single series of polynomials except that the point of expansion is varied from preselected point to preselected point as the function argument changes. This method obviates the need for searching for a high-order set of orthogonal functions which is valid for a broad class of functions over the entire function range.

The expansion technique may be developed by considering an arbitrary term from Eq. (3.1), such as $f_k(x^\ell)x^k$. For convenience, it can be assumed that $\ell = k$, then the term selected can be written as

$$h(x)x = f_k(x^k)x^k \tag{3.4}$$

where h(x)x is the function to be expanded.

Now, the range of x can be divided into an arbitrary number of equal intervals with one partition at x = 0. At any given time x will be in a particular interval which can be denoted by the index i. Within this interval, the function h(x) can be expanded in a power series as

$$h(x)x = h_i(x)x = (a_{k1i} + a_{k2i}\,\Delta x_i + a_{k3i}\,\Delta x_i^2 + \cdots + a_{kri}\,\Delta x_i^{r-1} + \cdots)x^k \tag{3.5}$$

where

$$\Delta x_i \triangleq x - x_i \tag{3.6}$$

and where x_i is a preselected, fixed point in the ith interval.

At the start of the identification process, the expansion coefficients in Eq. (3.4) will be unknown so that $\hat{E}$ will not be zero except over time intervals of zero measure. This fact is exploited by employing a steep descent adjustment procedure to find the required coefficients. This is done by forming the monotonically increasing function of $\hat{E}$

$$J = \frac{1}{2}\hat{E}^2 \tag{3.7}$$

Formal application of steep descent rules [7] to the positive definite function J provides adjustment procedures for individual parameters. For a typical parameter, say α, from the equation error, the steep descent adjustment equation is

$$\dot{\alpha} = -k\,\frac{\partial J}{\partial \alpha} = -k\hat{E}\,\frac{\partial \hat{E}}{\partial \alpha} \tag{3.8}$$

where k is a positive constant.

If the input u^o is other than a constant, the equation error $\hat{E}$ will be nonzero except over only subsets in time of zero measure unless the coefficients have assumed their correct values, and thus, ultimate convergence will occur. Also, because of the positive definite nature of J, one may start the coefficient adjustment process without any initial condition estimates.

On the basis of the development above, it is possible to outline a procedure for identifying a broad class of nonlinear systems which satisfy the assumptions given. A block diagram indicating the computational scheme is shown in Fig. 1.

4. EXAMPLE APPLICATIONS--SIMULATED

Consider the identification of a second-order system whose differential equation contains a deadzone term and a saturation term with hysteresis. The differential equation of this example with a multivalued nonlinearity is

$$a_2 x^2 + f_1(x^1)\,x^1 + f_o(x^o)\,x^o = u^o \tag{4.1}$$

where $f_1(x^1)\,x^1$ represents a deadzone characteristic given by

$$f_1(x^1)\,x^1 = 0 \qquad |x^1| \le 0.5$$

$$f_1(x^1)\,x^1 = 1 - 0.5\ \mathrm{sign}\,(x^1), \quad |x^1| > 0.5 \tag{4.2}$$

and $f_o(x^o)\,x^o$ represents a saturation characteristic, with hysteresis, given by

$$f_o(x^o)\,x^o = 0.5, \qquad |x^o| \ge 1.0 \tag{4.3}$$

and for $|x^o| < 1.0$, by

$$f_o(x^o)\,x^o = 0, \qquad |x^o| < 0.5$$

$$f_o(x^o)\,x^o = x^o - 0.5\ \mathrm{sign}\,(x^o) \tag{4.4}$$

$$0.5 \le |x^o| < 1.0$$

if x^o has most recently been in the region $x^o < -1$ as opposed to the region $x^o > 1$; correspondingly, $f_o(x^o)\,x^o$ is given by

$$f_o(x^o)\,x^o = x^o, \quad |x^o| < 0.5$$

$$f_o(x^o)\,x^o = 0.5\ \mathrm{sign}\,(x^o),\ 0.5 \le |x^o| < 1.0 \tag{4.5}$$

if x^o has most recently been in the region $x^o > 1$, as opposed to the region $x^o < -1$.

The coefficient a_2 was assumed to be equal to unity and the functions f_1 and f_o were assumed to be unknown. A sinusoidal signal, u = sin t, was selected as an input. Derivative filters were chosen to be third-order, standard ITAE filters with a filter frequency of 10.0 radians per second. The interval length for each variable was chosen as 0.25 and skew-symmetry was assumed for both functions.

This example was carried out using an IBM 360/75 digital computer. Results of the computation are shown in Figs. 2 and 3 by plots of actual and estimated function values. In this particular example, the singlevalued nonlinear term $f_1(x^1)\,x^1$ and the multivalued nonlinear term $f_o(x^o)\,x^o$ are composed of straight-line segments. The smoothing or roundness of the approximating functions near the points of discontinuity was due primarily to errors induced by the filter which only approximates a pure transport operator. Selection of the partition locations to coincide with the discontinuities in the "unknown" function improved the identification accuracy to some extent. Other values of partition size have been used which still produced good, but somewhat more rounded results at the points of function discontinuity.

5. APPLICATION TO POPULATION DYNAMICS

There exist broad classes of problems in engineering, biology, and economics, where the system dynamics involves nonlinear differential equations with multivalued nonlinearities. One such problem is that of brown lemming (*Lemmus trimucronatus*) population dynamics. This section of the paper is devoted to the application of the identification method to the lemming population dynamics problem.

Many hypotheses have been proposed to explain lemming population cycles in the Arctic. Although the ecosystems change, few show the temporal regularity exhibited by Arctic communities in which brown lemmings are the principal herbivore. Although lemmings are the subject of many legends (which are primarily false), it has been documented that lemmings undergo dramatic and cyclic population changes (with amazing regularity for all of recorded history). (Krebs [8] gives an excellent history and account of lemming population dynamics.) Based on all reasonable information available, lemming population dynamics are described by a first-order differential equation of the form

$$x^{1} + f_{o}(x^{o}) = 0 \qquad (5.1)$$

where $f_o(x^o)$ is a multivalued nonlinearity taking on both positive and negative values.

The multivalued function $f_o(x^o)$ is caused by a variety of factors, not all known or understood, such as:

1. Avian predators,
2. Competition for food supply,
3. Changes in litter size due to crowding, weather, nutrition, etc.,
4. Breeding changes due to stress,
5. Disease due to stress,
6. Weather, and
7. Genetic changes due to inbreeding in low population densities.

It is possible to reconstruct a record of $x^o(t)$ from data available from Krebs [8] and others. The limit cycle behavior of $x^o(t)$ is such that it is repetitive, excepting small changes due to various disturbances, allowing one to assume that the system is basically that of a limit cycling nonlinear process. Consequently, the input u(t) is assumed to be zero.

Research is in progress to explore further the identification of the function $f_o(x^o)$ responsible for the limit cycle population response of lemmings. The equation error method is ideal as it requires only on-line data for identification. Knowledge of the multivalued function $f_o(x^o)$ will enable one to explore the biological causes of population cycles in lemmings.

6. Electromechanical System

Nonlinearities such as backlash present in instrumentation play a central role in the degradation of performance in mechanical systems. The example considered in this section consists of an electromechanical position control system. The system considered uses a power amplifier to produce a current i(t) which drives a torque motor having the transfer function

$$\frac{T(t)}{i(t)} = \frac{k_m}{\tau_m D + 1} \qquad (6.1)$$

where $D = d/dt$, k_m = motor gain, and τ_m = the motor time constant. The torque T(t) then acts on the rotational inertia to be positioned producing an acceleration $\ddot{\theta}_o$ such that

$$\frac{\ddot{\theta}_o}{T(t)} = \frac{1}{I} \qquad (6.2)$$

where I is the angular moment of inertial of the load. Now, the double integration of $\ddot{\theta}_o$ yields the angular position of the load θ_o.

The feedback information consists of both position and velocity feedback by means of a gear-driven angular potentiometer and a tachometer. Consequently, the motor voltage, out of a summing amplifier and a power amplifier, is

$$i(t) = u(t) - k_D\dot{\theta}_o - k_a f(\theta_o) \qquad (6.3)$$

where u(t) is the system command input, k_D is the tachometer feedback gain, k_a is the position feedback gain, and $f(\theta_o)$ represents a backlash element with unity slope and backlash separation of δ radians. In the physical model tested, the backlash separation was measured, by direct testing, to be 0.12 radians.

The electromechanical positioning system described above is thus governed by the following differential equation

$$\dddot{\theta}_o + \left(\frac{1}{\tau_m}\right)\ddot{\theta}_o + \left(\frac{k_m}{I\tau_m}\right)k_D\,\dot{\theta}_o + \left(\frac{k_m k_a}{I\tau_m}\right) f(\theta_o) = u(t) \qquad (6.4)$$

Owing to the third-order nature of the system and the presence of the backlash element in the angular potentiometer's gear train, it is a straightforward matter to select appropriately

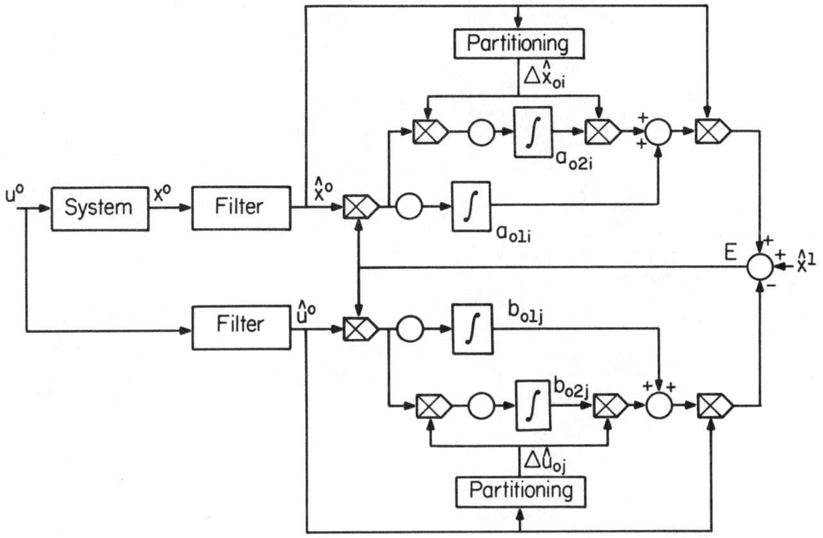

Figure 1 Block Diagram for Equation Error Identification

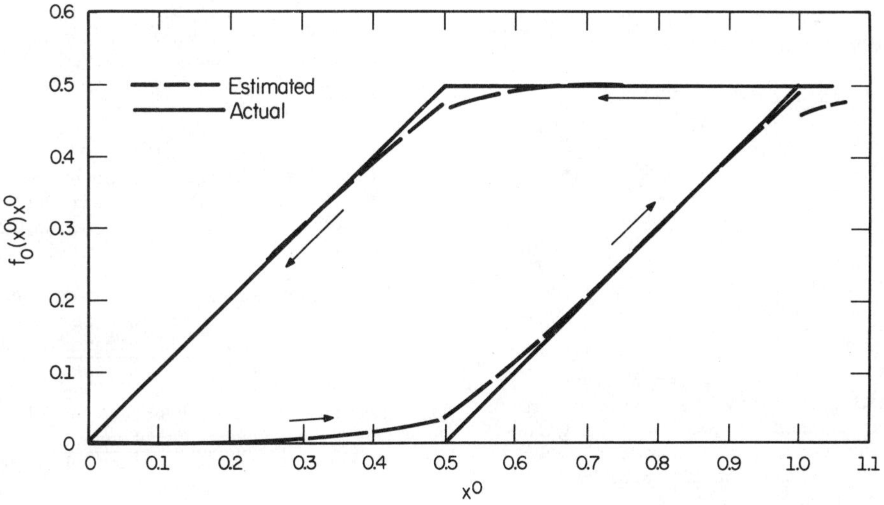

Figure 2 Actual and Estimated Values of $f_o(x^o)\ x^o$

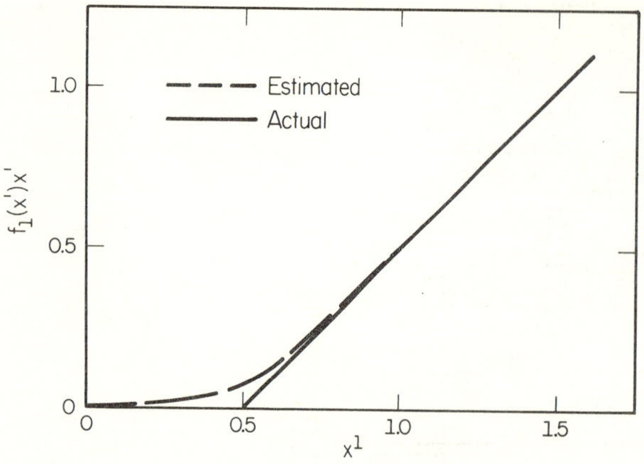

Figure 3 Actual and Estimated Values of $f_1(x^1)x^1$

high enough gains k_m and k_a, provided k_D is suffienciently small, to cause the system to display limit cycle behavior. Also, the value of the time constant, τ_m, was assumed known.

Equation (6.4) now contains two terms, one of them being a multivalued nonlinearity, which can be assumed to be unknown. The objective of the identification procedure is then the identification of these terms. The method outlined in the paper was applied to the above problem, first with a computer simulated version, and then with real on-line data. The identification algorithm worked well in both cases. The value of δ was identified as 0.124 and 0.109 in the simulated and real cases, respectively. The slope of the backlash ($k_m k_a / I\tau_m$) was identified within 6 percent and 11 percent, respectively.

7. SUMMARY AND REMARKS

The method outlined for the identification of multivalued nonlinearities in ordinary differential equations represents an extension and outgrowth of previous works in the area. (See References 1 through 6.) *A priori* information concerning the presence of a multivalued nonlinearity allows one to implement an equation error identification procedure which will identify a broad class of multivalued nonlinearities. The following remarks are in order concerning the method and its application.

7.1 Remark. The steep descent gains employed with the technique outlined have been arrived at in a trial and error process. Too low of a gain resulted in excessive convergence times; too high of a gain resulted in instability of the identification algorithm.

7.2 Remark. The order and bandwidth of the transport lag filter approximations can be prescribed as outlined in Sprague and Kohr [5] which gives techniques for the singlevalued case.

7.3 Remark. The multivalued identification method described herein is limited to piecewise continuous nonlinear elements. Hence, an important class of functions is not admissible, *i.e.*, those having step discontinuities, such as relays and on-off switches. The limitation lies in the linear derivative filters which will pass frequencies only below some upper frequency cutoff.

7.4 Remark. Discontinuous multivalued functions could be identified, provided the unknown function was an argument of a measured function, thus obviating the need for derivative filters.

REFERENCES

1. Kohr, R. H., "A Method for the Determination of a Differential Equation Model for Simple Nonlinear Systems," IEEE Trans. on Electronic Computers, Vol. EC-12, No. 4, 1963, pp. 394-400.

2. Hoberock, L. L. and Kohr, R. H., "An Experimental Determination of Differential Equations to Describe Simple Nonlinear Systems," ASME Trans., Series D, Vol. 89, No. 2, June 1967, pp. 393-398.

3. Lion, P. M., "Rapid Identification of Linear and Nonlinear Systems," Joint Automatic Conference Preprints, August 1966, pp. 605-615.

4. Miller, B. J., "A Theoretical and Analog Study of a Steep Descent Coefficient Computer for Process Analysis and Adaptive Control," Ph.D. Thesis, Electrical Engineering Department, The Ohio State University, Columbus, Ohio, 1962.

5. Sprague, C. H. and Kohr, R. H., "The Use of Piecewise Continuous Expansions in the Identification of Nonlinear Systems," Trans. of ASME, Vol. 91, Series D, No. 2, June 1969, pp. 179-184.

6. Butler, R. E. and Bohn, E. V., "An Automatic Identification Technique for a Class of Nonlinear Systems," IEEE Trans. on Automatic Control, Vol. AC-11, 1966, pp. 292-296.

7. Tompkins, C. B., "Methods of Steep Descent," Chap. 18, Modern Mathematics for the Engineer, Beckenbach, E. F., Editor, McGraw-Hill Book Co., New York, 1956, 448-479.

8. Krebs, C. J., "The Lemming Cycle at Baker Lake, Northwest Territories, during 1959-62," Arctic Inst. North Amer. Tech. Paper No. 15: pp. 1-104.

LIMIT IDENTIFIABILITY OF CONTROL SYSTEMS

Jaromír Štěpán

Institute of Information Theory and Automation
Czechoslovak Academy of Sciences
Prague, Czechoslovakia

The present control systems theory is based upon classical (absolutely accurate) mathematics. The paper is devoted to an analysis of the usability of theoretical methods from the viewpoint of practical accuracy of the measurement of signals or of data (as a rule to two and exceptionally to more decimal places). The considerations set out from a "backward analysis" - from the conception of "limit identifiability of control systems". On limiting assumptions, the question is treated whether at a given accuracy of the measurement of the signals there exists to a system of the n-th order a "sufficiently" accurate model of the same order. A discussion is concerned with the connection between the limit identifiability of control systems and the Wiener-Kolmogorov theory of a linear filter and the Kalman concept of the control systems observability.

1. INTRODUCTION

The present control systems theory sets out from "absolute" accuracy of the measurement of signals or data. Only on this assumption it is namely possible to establish simply a connection with classical (absolutely accurate) mathematics and to utilize its perfectly developed apparatus. In practical tasks, however, it is not possible to measure with absolute accuracy, but only with a finite accuracy to two and only exceptionally even to more decimal places. Consequently, only approximate models of control systems can be obtained. This fact unfavourably influences and often prevents the utilization of theoretical methods in the solution of practical tasks.

There arises the basic question whether at the given level of accuracy of measurement it is possible to identify a "sufficiently" accurate model of the control system from the measured data. The "sufficient" accuracy of the model depends on its final utilization in tasks of synthesis or decision. In the present work we shall be concerned with the identification (or identifiability) of control systems in connection with the ascertainment of the degree of stability of control loops set up from the systems under investigation. We shall also treat the question whether at a given "experimental" accuracy there exists to the system of n-th order a model of the same order. The question of the existence of such a model can be answered by means of the null hypothesis applied to the "limit" standard deviation of the estimates of the coefficients of the transfer functions, or of differential equations. The term "limit" standard deviation will here denote the lower bound of the standard deviations of the estimates of the coefficients from the viewpoint of measurement accuracy, of the shape of the input signals, the constants of the controllers in feedbacks and, not in the last place, from the viewpoint of the functional properties of the systems. A task formulated in this way shall be termed problem of "limit identifiability of control systems" (LICS).

2. BASIC RELATIONS

We shall be concerned with single-input-single-output systems for deterministic input signals and for null initial conditions. The regression functions or, in other words, the substitute responses of the systems will be considered in the form

$$^{s}\bar{y}(t) = \sum_{i=0}^{\bar{n}} \bar{a}_i \, {}^{s}v^{(i)}(t) + \sum_{i=1}^{\bar{m}} \bar{b}_i \, {}^{s}\hat{y}^{(i)}(t)\,, \qquad (2.1)$$

where the Laplace transforms $^{s}\bar{y}(t)$, $^{s}\hat{y}^{(i)}(t)$ and $^{s}v^{(i)}(t)$ are given by the relations

$$\mathcal{L}\{^{s}\bar{y}(t)\} = \frac{\left(1+\sum_{i=1}^{\bar{m}} \bar{b}_i p^i\right) S(p)}{\sum_{i=0}^{\bar{n}} \bar{a}_i p^i} = \frac{[1+\bar{M}(p)]S(p)}{\bar{N}(p)} \qquad (2.2)$$

$$\mathcal{L}\{^{s}\hat{y}^{(i)}(t)\} = \frac{p^i S(p)}{N(p)}\,; \quad \mathcal{L}\{^{s}v^{(i)}(t)\} = \frac{p^i S(p)}{N^2(p)} \qquad (2.3)$$

The Laplace transform of an ideal response is

$$\mathcal{L}\{{}^s y(t)\} = \frac{\left(1+\sum_{i=1}^{m} b_i p^i\right) S(p)}{\sum_{i=0}^{n} a_i p^i} = \frac{[1+M(p)]S(p)}{N(p)} \qquad (2.4)$$

We assume that the poles and the zeros are distinct.

Moreover we shall use the substitute responses of closed control loops set up from systems of the type (2.4) and an ideal proportional controller with the gain coefficient K_j. For $b_1 = b_2 = \dots = b_m = 0$ we obtain

$${}^s\overline{y}_j(t) = \sum_{i=0}^{\overline{n}} \overline{a}_i \, {}^s v_j^{(i)}(t) \,, \qquad (2.5)$$

where

$$\mathcal{L}\{{}^s v_j^{(i)}(t)\} = \frac{K_j p^i S(p)}{[N(p)+K_j]^2} \,; \quad \mathcal{L}\{{}^s\overline{y}_j(t)\} = \frac{K_j S(p)}{\overline{N}(p)+K_j}$$

With the subscript j we indicate that we have here to do with substitute responses corresponding to the j-th gain coefficient K_j. For the gain coefficient $K_j \longrightarrow K_k$ (K_k is the critical gain coefficient) we use the subscript k instead of the subscript j.

The transfer function S(p) characterizes the shape of the input signal. In the identification we do not know an exact description of the input signal. The uncertainty of the estimates of the coefficients of the transfer function S(p) will increase the uncertainty of the estimates of the coefficients $\overline{a}_i$ and $\overline{b}_i$. Therefore, we direct our attention at the limit of the currently employed input signals u(t), i.e. at step and impulse ($\mathcal{L}\{{}^0u(t)\} = {}^0S(p) = \frac{{}^0m_j}{p}$; ${}^1S(p) = {}^1m_j$). By the superscript 0 (or maybe 1) on the left side on top we shall denote responses to a step (or to an impulse) and further the respective variances or the standard deviations of the estimates of the coefficients.

The method of determining the LICS is a "backward" analysis. In the regression functions (2.1) and (2.5) we consider the modified sensitivity functions $\frac{\partial \, {}^s y^*(t)}{\partial a_i} = {}^s v^{(i)}(t)$ or, maybe $\frac{\partial \, {}^s y_j(t)}{\partial a_i} = {}^s v_j^{(i)}(t)$ corresponding to the ideal responses ${}^s y(t)$ or ${}^s y_j(t)$. Thus we assume the knowledge of the transfer function of the system in the same way as we assume the knowledge of the accurate solution in the analysis of the stability of numerical processes [Isaacson (1966)].

The discrete regression functions follow directly from the functions (2.1) and (2.5). For instance, according to (2.5) we obtain

$${}^s\overline{y}_j(t_z) = \sum_{i=0}^{\overline{n}} \overline{a}_i \, {}^s v_j(t_z) \;; \quad (z = 1,2,\dots,r) \qquad (2.6)$$

The LICS we ascertain on the basis of the standard deviations of the estimates of the coefficients obtained by means of a continuous linear estimator, for which we assume the knowledge of the sensitivity functionals $\| {}^s v_j^{(i)} \|$. First let us put down the relations for a discrete estimator for the simplest case with the regression function (2.6) (for $\overline{n} = n$). For the points of the measured responses will hold

$${}^s\tilde{y}_j(t_z) = {}^s y_j(t_z) + n(t_z) \;; \quad (z = 1,2,\dots,r) \,, \qquad (2.7)$$

where $n(t_z)$ is the Gaussian stationary and ergodic noise. We shall assume that the measurement errors of the responses of the systems as well as of the responses of the closed control loops can be approximated by a noise with Gaussian distribution (Par. 3). We set out from the regression function (2.6). The least squares linear unbiased estimate of $\boldsymbol{a}$ is defined as the vector $\overline{\boldsymbol{a}}$ which minimizes the quadratic risk function

$$R(\boldsymbol{a}) = ({}^s\tilde{\boldsymbol{y}}_j - {}^s\boldsymbol{G}_j \boldsymbol{a})^T ({}^s\tilde{\boldsymbol{y}}_j - {}^s\boldsymbol{G}_j \boldsymbol{a}) \,, \qquad (2.8)$$

where ${}^s\tilde{\boldsymbol{y}}_j^T = \left| {}^s\tilde{y}_j(t_1) \quad {}^s\tilde{y}_j(t_2) \; \dots \; {}^s\tilde{y}_j(t_r) \right|$

$$\boldsymbol{a}^T = \left| a_0 \; a_1 \; \dots\dots \; a_n \right|$$

$${}^s\boldsymbol{G}_j = \begin{vmatrix} {}^s v_j^{(0)}(t_1) & {}^s v_j^{(1)}(t_1) & \dots\dots & {}^s v_j^{(n)}(t_1) \\ {}^s v_j^{(0)}(t_2) & {}^s v_j^{(1)}(t_2) & \dots\dots & {}^s v_j^{(n)}(t_2) \\ \vdots & & \ddots & \vdots \\ {}^s v_j^{(0)}(t_r) & \dots & \dots & {}^s v_j^{(n)}(t_r) \end{vmatrix}$$

The vector of the estimates and the pertinent covariance matrix are given by the following relations

$$\bar{a} = \left({}^s G_j^T \, {}^s G_j\right)^{-1} {}^s G_j^T \, {}^s\tilde{y}_j \qquad (2.9)$$

$$\hat{\Psi} = E\{(\bar{a}-a)(\bar{a}-a)^T\} = \left({}^s G_j^T \, {}^s G_j\right)^{-1} {}^s G_j^T \, E\{nn^T\} \, {}^s G_j \left({}^s G_j^T \, {}^s G_j\right)^{-1} \qquad (2.10)$$

For independent noise $E\{nn^T\} = I\sigma^2$ (where I is identity matrix) we can write

$$\bar{\Psi} = \sigma^2 \left({}^s G_j^T \, {}^s G_j\right)^{-1} \qquad (2.11)$$

The relations determining the LICS are essentially simplified if we set out from the continuous linear estimator. According to Štěpán (1971), we give the resultant relations for a continuous linear estimator as a limit case of the discrete estimator given by the relations (2.8) to (2.11).

We set out from one response and from the limit error of the measurement δ_L ($\delta_L^2 = T\sigma_L^2$). The relation for the limit standard deviation of the measurement σ_L we derive in Par. 3. The vector of the estimates of the coefficients $\bar{a}_i$ is given analogously to the relation (2.9)

$$\bar{a} = \left[\int_0^T \left({}^s H_j \, {}^s H_j^T\right) dt\right]^{-1} \int_0^T \left({}^s H_j \, {}^s\tilde{y}_j\right) dt \,, \qquad (2.12)$$

where

$${}^s H_j^T = \left| {}^s v_j^{(0)}(t) \,,\; {}^s v_j^{(1)}(t) \ldots \; {}^s v_j^{(n)}(t) \right| .$$

The pertinent covariance matrix is given analogously to the relation (2.11).

$$\Psi = \delta_L^2 \left[\int_0^T \left({}^s H_j \, {}^s H_j^T\right) dt\right]^{-1} = \delta_L^2 \, {}^s_j A^{-1} \,. \qquad (2.13)$$

We emphasize that the limit transition from a discrete to a continuous estimator must be performed over a number of discrete responses $q \longrightarrow \infty$. The distance between neighbouring points on one measured discrete response is selected so that the respective noise is independent.

The variances or the relative standard deviations of the estimates of the coefficients $\bar{a}_i$ are given by the relations

$${}^s var_j(\bar{a}_i) = {}^s D_j(\bar{a}_i) = \delta_L^2 \, {}^s_j \alpha^{ii} \,, \qquad (2.14)$$

$${}^s s_{ij} = \frac{\left[{}^s D_j(\bar{a}_i)\right]^{\frac{1}{2}}}{a_i} \,; \qquad (2.15)$$

where the elements ${}^s_j\alpha^{ii}$ of inverse matrix ${}^s_j A^{-1}$ are calculated from

$${}^s_j\alpha^{ih} = \frac{\left|{}^s_j A_{hi}\right|}{\left|{}^s_j A\right|} \,, \qquad (2.16)$$

where $\left|{}^s_j A_{hi}\right|$ is complement of element ${}^s_j\alpha_{hi}$ in matrix ${}^s_j A$ and $\left|{}^s_j A\right|$ is the determinant of matrix ${}^s_j A$. Elements ${}^s_j\alpha_{hi}$ of the matrix ${}^s_j A$ are determined by relation

$${}^s_j\alpha_{hi} = \left({}^s v_j^{(h)} \,,\; {}^s v_j^{(i)}\right) = \int_0^T {}^s v_j^{(h)}(t) \, {}^s v_j^{(i)} \, dt \,. \qquad (2.17)$$

The mentioned linear estimator gives efficient estimates, i.e. estimates with the smallest possible variance. In the Cramér-Rao inequality there holds the sign of equality. This follows from the proposition of Cramér (1958): If an efficient estimate $\bar{a}$ of a exists, the likelihood equation will have a unique solution equal to $\bar{a}$. Likelihood function $\ell(a)$ has in the considered case according to (2.8) the form

$$\ell(a) = C \exp\left\{-\tfrac{1}{2}\left({}^s\tilde{y}_j - {}^s G_j a\right)^T \left({}^s\tilde{y}_j - {}^s G_j a\right)\right\} \qquad (2.18)$$

Hence

$$\log_e \ell(a) = \log_e C - \tfrac{1}{2}\left({}^s\tilde{y}_j - {}^s G_j a\right)^T \left({}^s\tilde{y}_j - {}^s G_j a\right) \qquad (2.19)$$

The constant C is independent of the coefficients a_i, and thus the likelihood equation

$$\nabla_a \log_e \ell(a) = 0 \,, \qquad (2.20)$$

(where $\nabla_a^T = \left|\frac{\partial}{\partial a_0} \; \frac{\partial}{\partial a_1} \cdots \frac{\partial}{\partial a_n}\right|$) leads to the conditions that minimize the risk function (2.8) with respect to a.

3. MEASUREMENT ACCURACY

In the introduction let us indicate at least in general the significance of the term accuracy of measurement. The accuracy of measurement is characterized first by the precision of the measurement, i.e. by the fineness of the graduation of the scale of the measured signal and secondly by systematic errors. An absolutely accurate measurement is consequently an absolutely precise measurement, which means that the scale of the measured signal is divided into graduations $\varepsilon \longrightarrow 0$ and the systematic errors are zero.

In the control systems theory, the question of accuracy of data was neither asked nor answered. For this reason, we resort to papers on geodesy, This field has more than hundred years experiences in the application of statistics. Of significance for further considerations is first of all the statement on practical measurement accuracy according to Böhm (1964). For a measured parameter $\tilde{\ell}\,[\,E(\tilde{\ell}) = \ell\,]$ and the errors of measurement $\varepsilon_i\,[\,E(\varepsilon_i) = 0\,]$ there holds the limit relation

$$\lim_{n\to\infty} \frac{1}{n} \sum_{i=1}^{n} \tilde{\ell}_i = \ell \qquad (3.1)$$

only on the assumption of absolute accuracy of measurement.The validity of the relation (3.1) is limited by the precision of measurement. If e.g. ℓ is a certain length dimension and the parameter $\tilde{\ell}_i$ are measured in millimetres, then we obtain the "ideal" value of ℓ again only with an accuracy of millimetres even if the measurement will be infinite times repeated.In the sense of the ergodic theory,this proposition holds also for dynamic processes. From this viewpoint we must conceive the relations for the continuous estimator in Par. 2.

In the literature on control,the authors set out as a rule from the unrestricted validity of the limit (3.1). Statistical methods are here interpreted as "perfect" methods by means of which it is possible to attain arbitrarily accurate results on the only assumption: The set of measured data must be "sufficiently" large, or the time of observation of the measured signal must be "sufficiently" long.

In the wide range of problems of measurement accuracy we shall be concerned with the key problem of quantization the measured data to the given number of decimal places. We set out from the demand of the same information contents of the rectangular error distribution and the Gaussian (normal error) distribution with a standard deviation σ_L according to Bell (1968).The error due to quantizing the measuring scale of the points of responses ${}^s\tilde{y}(t_z) = {}^s\tilde{y}_z$ is determined by the probability distribution function

$$f({}^s\tilde{y}_z) = 0 \quad \text{when} \quad {}^s\tilde{y}_z < {}^s y_z - \Delta, \qquad {}^s\tilde{y}_z > {}^s y_z + \Delta \qquad (3.2)$$

$$f({}^s\tilde{y}_z) = \frac{1}{2\Delta} \quad \text{when} \quad {}^s y_z - \Delta < {}^s\tilde{y}_z < {}^s y_z + \Delta$$

The information content in natural units is given by relation

$$I = \int_{-\infty}^{+\infty} f({}^s\tilde{y}_z)\, \log_e f({}^s\tilde{y}_z)\, d\,{}^s\tilde{y}_z = \log_e \frac{1}{2\Delta} \qquad (3.3)$$

For the Gaussian probability distribution with the standard deviation σ_L we obtain the information content

$$I_G = \int_{-\infty}^{+\infty} \left\{ \frac{1}{\sigma_L (2\pi)^{\frac{1}{2}}} \exp\left[-\frac{({}^s\tilde{y}_z - {}^s y_z)^2}{2\sigma_L^2} \right] \right\} \left[\log_e \frac{1}{\sigma_L (2\pi)^{\frac{1}{2}}} - \frac{({}^s\tilde{y}_z - {}^s y_z)^2}{2\sigma_L^2} \right] d\,{}^s\tilde{y}_z = \log_e \frac{1}{\sigma_L (2\pi e)^{\frac{1}{2}}} \qquad (3.4)$$

From the condition $I = I_G$ we obtain

$$\sigma_L = \frac{2\Delta}{(2\pi e)^{\frac{1}{2}}} \qquad (3.5)$$

For quantization the measured data to N decimal places $(2\Delta = 10^{-N})$ we obtain

$$\sigma_L = \frac{10^{-N}}{(2\pi e)^{\frac{1}{2}}} \qquad (3.6)$$

The suitability of the given model will be discussed in Par. 5.

4. DEFINITION OF CONTROL SYSTEMS IDENTIFIABILITY

We shall limit ourselves to systems with a constant in the numerator of the transfer function. The further mentioned definitions as well as the linear continuous estimator from Par. 2 can be expanded also for systems with responses of the type (2.1).

Definition 4.1: The control system described by the transfer function $F_s(p) = 1/\sum_{i=0}^{n} a_i p^i$ with distinct poles is identifiable for the limit measuring accuracy according to the relation (3.5) if all coefficients a_i are identifiable.

The identifiability of the coefficients a_i shall be judged according to the null hypothesis applied to the relative standard deviation of the estimates of the coefficients $\bar{a}_i$.

Definition 4.2: The coefficient a_i is identifiable for the limit accuracy of measurement according to the relation (3.5) $(\delta_L^2 = T\sigma_L^2)$,if holds

$$^s s_{ij} = \frac{[^sD_j(\bar{a}_i)]^{\frac{1}{2}}}{a_i} \leq 1 \qquad (4.1)$$

For Gaussian noise the values $^s s_{ij} = 1$ indicate limits $\pm a_i$ round the ideal value of coefficient a_i within which the estimate will lie with the probability P = 68 %.

The identifiability of control system at given measurement accuracy will vary in dependence on the set of input signals $u(t) \in U$ and on the set of constants of the controllers in feedback $F_R \in R$. From the viewpoint of both practice and theory we must know the "best possible", or more briefly said, the "limit" identifiability of the system for the "limit" signal $u^*(t)$ from some suitably selected subset U_1 and the "limit" controller F_R^* from some suitably selected subset R_1. In the selection of the subsets U_1 and R_1 as well as in deriving the limit input signal $u^*(t)$ and the limit controller F_R^* we are limited by both the mathematical apparatus and the viewpoints of practical usability.

5. LIMIT IDENTIFIABILITY OF ONE CLASS OF CONTROL SYSTEMS

In this paragraph we shall attempt to show the possibilities of using the given conception of LICS. We shall not ascertain the limit identifiability of some control system, but we shall try to find simple relations determining the identifiability of one entire subclass of control systems. We shall be concerned with the subclass of systems Φ_1 of a practically important class of systems Φ with real distinct poles, which can be described by the transfer function ($a_0 = 1$)

$$F_s(p) = \frac{1}{\prod_{k=1}^{n}(T_k p+1)} = \frac{1}{1+\sum_{i=1}^{n} a_i p^i} \qquad (5.1)$$

The pertinent transfer function of the closed control loop for the change of the command variable and for an ideal proportional controller with the gain coefficient K_j had the form

$$F_{wj}(p) = \frac{K_j}{\sum_{i=1}^{n} a_i p^i + 1 + K_j} \qquad (5.2)$$

Consequently, we shall seek an LICS of the type (5.1) in dependence on the subset of gain coefficients of the controllers in feedback $K_j \in R_1$.

We shall define the subclass of control systems Φ_1. We shall be concerned only with control systems of the type (5.1), where the limit identifiability of control systems is determined only by the identifiability of the coefficient a_n. For the relative standard deviations of the estimates of the coefficients of the subclass Φ_1 must hold

$$^s s_{nj} \geq {^s s_{ij}} \qquad (i = n-1,\ n-2, \dots 1, 0) \qquad (5.3)$$

5.1 - Limiting Assumptions

The square of the norms of ideal responses $^s y_j(t)$ pertinent to transfer function of the type (5.2) can be written in the form (for $a_{0j}=1+K_j$)

$$\|^s y_j\|^2 = a_{0j}^2 \|^s v_j^{(0)}\|^2 + 2a_{0j}\sum_{i=1}^{n} a_i (^s v_j^{(0)}, {^s v_j^{(i)}}) + \sum_{i=1}^{n}\sum_{h=1}^{n} a_i a_h (^s v_j^{(i)}, {^s v_j^{(h)}}) \qquad (5.4)$$

Let us consider the simplest possible case - orthogonal estimates of $\bar{a}_i$. Only the diagonal elements of the matrices $^sG_j^T\, {^sG_j}$ or ^sA_j are nonzero. The square of the relative standard deviations $^s s_{ij}^2$ is then given according to (2.15) by

$$^s s_{ij}^2 = \frac{\delta_L^2}{a_i^2 \|^s v_j^{(i)}\|^2} \qquad (5.5)$$

The relative standard deviations of the estimates of the coefficients $\bar{a}_i$ are inversely proportional to the norms of the corresponding components $a_i \|^s v_j^{(i)}\|$. The relations (5.4) and (5.5) clearly show that the relative standard deviations of the estimates of the coefficients depend:

1) on the magnitude of the norm of the useful signal
2) on the functionals of the sensitivity $\|^s v_j^{(i)}\|$.

ad 1) In practice we can measure only signals with a finite amplitude. For instance, for the unit channel of output signals we obtain

$$0 < {}^{s\sim}\!y_j(t) < 1 \qquad \text{for } t \in (0,T) \qquad (5.6)$$

This condition is difficult to incorporate into the treatment. For this reason, we set out from an approximate fulfilment of the condition (5.6) - from a constant ratio of the "mean" amplitudes of the output signal and the noise

$$\frac{\|{}^s y_j\|}{\delta_L} = \text{konst} \qquad (5.7)$$

The input signal will be selected so as to correspond to

$$\|{}^s y_j\|^2 = a_1 \qquad (5.8)$$

The adequacy of this value follows from the magnitude of the squares of the norms of the responses of the systems of the type (5.1) to a unit step. These responses satisfy the condition (5.6) for an arbitrary value of the coefficient a_1. For the most favourable case for the dead-time lag control system $F_S(p) = 1/e^{-T_d p} = 1/1 + T_d p + \frac{1}{2} T_d^2 p^2 + \cdots$ we obtain $\|{}^0 y\|^2_{T_d} = T_d = a_1$. The further text will consequently for the greater part not be concerned with unit steps or impulses.

The square of the measuring error $\delta_L^2 = 2 a_1 \sigma_L^2$ we select according to Štěpán (1970), and by means of (3.6) we obtain

$$\frac{\|{}^s y_j\|}{\delta_L} = \frac{1}{\sqrt{2}\,\sigma_L} = (\pi e)^{\frac{1}{2}}\, 10^N \qquad (5.9)$$

The fulfilment of the condition (5.9) by the methods setting out from the Wiener-Kolmogorov theory will be discussed in Par. 6.

ad 2) By the fulfilment of the condition (5.9), the sensitivity functionals $\|{}^s v_j^{(i)}\|$ vary essentially only in dependence on the gain coefficient K_j. The ascertainment of the gain coefficient K_j^*, for which the identifiability of the coefficients a_n is most favourable, is the subject of the following paragraphs.

As far as the dependent estimates of the coefficients are concerned, the situation is more complicated. Decisive is again the magnitude of the useful signal. However, we must set out from the ratio of the relative standard deviations, as will be shown in the further text.

5.2 - Systems of Third Order

The relations derived in this paragraph hold for systems with a transfer function

$$F_S(p) = \frac{1}{1 + a_1 p + a_1^2 \varkappa p^2 + a_1^3 \varkappa \rho p^3} \qquad (5.10)$$

where $\varkappa = \frac{a_2}{a_1^2}$ and $\rho = \frac{a_3}{a_1 a_2}$.

We set out from the papers of Štěpán (1970) and (1972 b). The regression functions pertinent to the transfer function (5.2) for input signals, step and impulse have the form ($\bar{n} = n$)

$${}^1\bar{y}_j(t) = {}^1\eta\left[\bar{a}_{0j}\, v_j^{(0)}(t) + \sum_{i=1}^{n} \bar{a}_i\, v_j^{(i)}(t)\right] \qquad (5.11)$$

$${}^0\bar{y}_j(t) = {}^0\eta\left[\bar{a}_{0j}\, v_j^{(-1)}(t) + \sum_{i=1}^{n} \bar{a}_i\, v_j^{(i-1)}(t)\right] \qquad (5.12)$$

For systems of third order it is possible to derive according to Newton (1957) analytical relations for the squares of the norms $\|v_j^{(i)}\|^2$ for the integration interval $T = \infty$ in the relation (2.17). Difficulties in fulfilment of the relation (5.9) even in the region of the gain coefficient $K_j \in (0{,}5 K_j ; K_k)$ (K_k is the critical gain coefficient) are obviated by introducing the ratios of the relative standard deviations. Decisive for the identifiability of the coefficient a_3 are first of all the ratios ${}^1G_{3j}$ and ${}^0G_{3j}$

$${}^1G_{3j} = \left(\frac{{}^1 s_{3j}}{{}^1 s_{1j}}\right)^2 = \frac{a_1^2 \|v_j^{(1)}\|^2}{a_3^2 \|v_j^{(3)}\|^2} = \frac{\varkappa + a_{0j}\rho^2}{a_{0j}\rho^2(1 + a_{0j}\varkappa)} \qquad (5.13)$$

$${}^0G_{3j} = \left(\frac{{}^0 s_{3j}}{{}^0 s_{1j}}\right)^2 = \frac{(1 - a_{0j}\rho)^3 + a_{0j}^3\rho^3 + a_{0j}\varkappa}{a_{0j}^3 \rho^2 (\varkappa + \rho)} \qquad (5.14)$$

In the case of ${}^0G_{3j}$ we can neglect in the decisive region $K_j \in (0{,}5 K_j ; K_k)$ the dependence of the estimates of $\bar{a}_{0j}$ and $\bar{a}_1$.

Now we analyse the dependence of the ratios ${}^1G_{3j}$ and ${}^0G_{3j}$ on a_{0j} ($a_{0j} = 1 + K_j$)

$$\frac{d\,{}^1G_{3j}}{d\,a_{0j}} = -\frac{\varkappa\left(1+a_{0j}^2\varrho^2\right)+2a_{0j}\varkappa^2}{a_{0j}^2\varrho^2\left(1+a_{0j}\varkappa\right)^2} \qquad (5.15)$$

$$\frac{d\,{}^0G_{3j}}{d\,a_{0j}} = -\frac{3\left(1-a_{0j}\varrho\right)^2+2a_{0j}\varkappa}{a_{0j}^4\varrho^2\left(\varkappa+\varrho\right)} \qquad (5.16)$$

For stable control processes must hold $\varkappa > 0$; $\varrho > 0$; $a_{0j} > 0$ and $a_{0j}\varrho < 1$. The derivative in both cases is always negative. For $K_j \to K_k$; $a_{0j} \to a_{0k}$ and $a_{0k}\varrho = 1$ we obtain

$$\lim_{K_j\to K_k} {}^1G_{3j} = {}^1G_{3k} = 1 \;;\; \lim_{K_j\to K_k} {}^0G_{3j} = {}^0G_{3k} = 1 \qquad (5.17)$$

Hence holds

$${}^1G_{3j} \geqslant 1 \quad ; \quad {}^0G_{3j} \geqslant 1 \qquad (5.18)$$

The identifiability of the coefficient a_3 depends further on the ratios ${}^1s_{3j}/{}^1s_{2j}$; ${}^1s_{3j}/{}^1s_{0j}$ or ${}^0s_{3j}/{}^0s_{2j}$; ${}^0s_{3j}/{}^0s_{0j}$. For $\frac{\varkappa}{\varrho} > 1{,}5$ can be shown that holds

$$\left(\frac{{}^1s_{3j}}{{}^1s_{2j}}\right)^2+\left(\frac{{}^1s_{3j}}{{}^1s_{0j}}\right)^2 \geqslant \frac{2\varkappa}{\varrho} \quad \text{or} \left(\frac{{}^0s_{3j}}{{}^0s_{2j}}\right)^2+\left(\frac{{}^0s_{2j}}{{}^0s_{2j}}\right)^2 \geqslant \frac{2\varkappa}{\varrho} \qquad (5.19)$$

The limit identifiability of the coefficient a_3 of systems of the type (5.10) for $\frac{\varkappa}{\varrho} \geqslant 1{,}5$ is determined by the "most favourable" value of the gain coefficient $K_j^* \to K_k$. On the mentioned assumptions, the identifiability of a_3 does not depend on the form of the input signals, because there holds

$${}^1G_{3k} = {}^0G_{3k} = \frac{a_1^2}{a_3^2}\,\frac{1}{\omega_k^4} \;;$$

where ω_k is the critical frequency.

The calculation of the "limit" relative standard deviation of the estimate of the coefficient a_3 sets out from the relations (5.8), (5.9), (5.19) and

$$\frac{1}{{}^1s_{0k}^2}+\sum_{i=1}^{3}\frac{1}{{}^1s_{ik}^2} = 2a_1\delta_L^{-2}$$

$${}^1s_{3k}^* = {}^0s_{3k}^* = s_3^* = \left(\frac{\varkappa+\varrho}{\varrho}\right)^{\frac{1}{2}}\frac{\delta_L}{\|{}^sy_j\|} = \left(\frac{\varkappa+\varrho}{\varrho}\right)^{\frac{1}{2}}\frac{10^{-N}}{(\pi e)^{\frac{1}{2}}} \qquad (5.20)$$

The intervals of the dimensionless parameters $\varkappa$ and ϱ are calculated from the Euler inequalities [Mařík (1964)]. For the coefficients of the polynomials with only real roots holds

$$a_i^2 \geqslant a_{i+1}a_{i-1}\left(1+\frac{1}{i}\right)\left(1+\frac{1}{n-i}\right) \qquad (5.21)$$

According to the denominator of the transfer function (5.1) we obtain for $n = \infty$

$$a_1^2 \geqslant 2a_2 \implies \varkappa \leqslant 0{,}5$$

$$a_2^2 \geqslant 1{,}5\,a_1a_3 \implies \varkappa \geqslant 1{,}5\varrho \;;\; \varrho \leqslant 0{,}33 \qquad (5.22)$$

Of significance from the viewpoint of practical tasks are only cases with $a_{0k} \leqslant 50$; $\varrho \geqslant 0{,}02$ i.e. intervals of $\varkappa$ and ϱ

$$\varkappa \in (0{,}03\,;\,0{,}5) \;;\; \varrho \in (0{,}02\,;\,0{,}33) \qquad (5.23)$$

For the intervals $\varkappa$ and ϱ, according to (5.23), and for $\frac{\varkappa}{\varrho} \geqslant 1{,}5$ we calculate the interval of the "limit" relative standard deviation

$$s_3^* \in (0{,}54.10^{-N}\,;\,1{,}9.10^{-N}) \qquad (5.24)$$

Now we shall also demonstrate the connection of the LICS of the type (5.10) with the calculation of the critical gain coefficient of control loop. For the relative standard deviation of the estimate of the coefficient $a_{ok} = 1 + K_k$ we can derive from the relation $a_{0k} = \frac{a_1a_2}{a_3}$

$$s(\bar{a}_{0k}) = \left(\sum_{i=1}^{3}s_i^{*2}\right)^{\frac{1}{2}} , \qquad (5.25)$$

where the asterisk denoted "limit" relative standard deviations. Since there holds (5.3), decisive for the deviation $s(\bar{a}_{0k})$ is first of all the deviation s_3^* . If there holds $s_3^* > 1$, there will hold according to (5.25) $s(\bar{a}_{0k}) > 1$. From this follows: If the control system of the type (5.10) is not identifiable in the sense of Par.4, at the given level of measuring accuracy, the calculation of the critical gain coefficient or the algebraic criterions of the stability from the coefficients of the model will lose its sense.

5.3 - Systems of Higher Orders

In this paragraph we extend the considerations of Par. 5.2 to systems of higher orders of the type (5.1). The judging of the identifiability of the coefficients $a_n (n > 3)$ will be based on the ratio

$$\frac{{}^1G^n_{(n-2)j}}{{}^1G^n_{nj}} = \frac{a_n^2 \,\| v_{jn}^n \|^2 \, a_{n-4}^2 \,\| v_{jn}^{(n-4)} \|^2}{a_{n-2}^4 \,\| v_{jn}^{(n-2)} \|^4} \; ; \qquad (5.26)$$

where the index n (only for $n > 3$) in ${}^1G^n_{nj}$ at right on top, and in $\| v_{jn}^{(i)} \|$ at right below denotes the order of the respective system. We assume that the ratios ${}^1G^n_{ij}$ or ${}^0G^n_{ij}$ are given only by the diagonal elements of the matrix ${}^s_jA$ [Štěpán (1972 b)]. Now we show that holds

$$\frac{{}^1G^n_{(n-2)j}}{{}^1G^n_{nj}} \geqslant \frac{{}^1G^n_{(n-2)k}}{{}^1G^n_{nk}} \quad \text{or} \quad \frac{{}^0G^n_{(n-2)j}}{{}^0G^n_{nj}} \geqslant \frac{{}^0G^n_{(n-2)k}}{{}^0G^n_{nk}} \, . \qquad (5.27)$$

The sensitivity functionals $\| v_{jn}^{(i)} \|$ are linked by the Schwarz-Bunjakovsky inequality

$$\| v_{jn}^{(n)} \| \; \| v_{jn}^{(n-4)} \| \geqslant \left| \left(v_{jn}^{(n)} , \; v_{jn}^{(n-4)} \right) \right| = \| v_{jn}^{(n-2)} \|^2 \qquad (5.28)$$

The sign of equality holds for $K_j \to K_k$. For the coefficients a_i, which are independent of the gain coefficient, follows the validity of the inequalities (5.27) from the relation

$$\frac{\| v_{jn}^{(n)} \|^2 \;\; \| v_{jn}^{(n-4)} \|^2}{\| v_{jn}^{(n-2)} \|^4} \geqslant \frac{\| v_{kn}^{(n)} \|^2 \;\| v_{kn}^{(n-4)} \|}{\| v_{kn}^{(n-2)} \|^4} = 1 \qquad (5.29)$$

The limit identifiability of the coefficient a_n of systems of the type (5.1) is again obtained for the critical gain coefficient K_k. The identifiability of the coefficient a_n is independent of the shape of the input signals.

By means of the relations (5.26),(5.29) and the Euler inequalities (5.21) for the coefficients a_i we can put down the inequalities

$${}^1G^n_{nk} = {}^0G^n_{nk} = G^n_{nk} \geqslant \lambda_{n-3}^2 \, \lambda_{n-2}^4 \, \lambda_{n-1}^2 \, G^n_{(n-2)k} \geqslant \cdots$$

$$\cdots \geqslant \lambda_2^2 \, (\lambda_3 \lambda_4 \ldots \lambda_{n-2})^4 \, \lambda_{n-1}^2 \; G^n_{3k} \; , \qquad (5.30)$$

where

$$\lambda_i = \left(1 + \frac{1}{i} \right) \left(1 + \frac{1}{n-i} \right) \; . \qquad (5.30)$$

Thus we can calculate from ratio (for odd n)

$$\left(\frac{{}_n s^*_{nk}}{{}_n s_{3k}} \right)^2 = G^n_{nk} \; G^n_{(n-2)k} \cdots \; G^n_{5k} \qquad (5.31)$$

the lower bound of the limit relative standard deviations ${}_n s^*_{nk}$. We shall assume that the ratio of the vectors of the estimates of even and odd coefficients is approximately given by the relation (5.19) and that we can put down ${}_n s_{3k} \doteq s^*_3$. Moreover, we must correct the expressions G^n_{ik} with regard to the variations of the critical frequency [Štěpán (1972b)].

Table 5.1 summarizes the lower bound of the limit relative standard deviations of the estimates of the coefficients a_n of the systems of 5th, 7th and 9th order in dependence on the number of measured decimal places.

Table 5.1

	2	3	4	5
s^*_{5k}	1.10^{-1}	1.10^{-2}	1.10^{-3}	1.10^{-4}
	$3,7.10^{-1}$	$3,7.10^{-2}$	$3,7.10^{-3}$	$3,7.10^{-4}$
s^*_{7k}	3,5	$3,5.10^{-1}$	$3,5.10^{-2}$	$3,5.10^{-3}$
	12,5	1,25	$1,25.10^{-1}$	$1,25.10^{-2}$
s^*_{9k}	190	19	1,9	$1,9.10^{-1}$
	680	68	6,8	$6,8.10^{-1}$

The upper or, maybe, the lower value was calculated from the lower or the upper limit of the interval s^*_3 according to (5.24). The bold line separates the systems which are no more identifiable according to the relation (4.1). Table 5.1 may seem to be pessimistic. The results given in the literature, e.g. Åström (1967) are still more unfavourable. Here we must bear in mind that the LICS is based upon the possibly most favourable assumptions. Let us summarize at least the most favourable ones:

1) We consider a linear continuous estimator.
2) The mentioned procedure is not subject to problems concerning the

stability of numerical processes. Practical tasks necessarily lead to ill-conditioned matrices.

3) The identification based on the responses of control loops acting near the stability limits is an open question.
4) The amplitudes of input signals, or of signals at the output of the controllers were not limited.
5) Difficulties in the fulfilment of the conditions (5.8) and (5.9) for the gain coefficient $K_j \rightarrow K_k$.

In conclusion we discuss the model of the measuring errors from Par. 3. From the considerations in Par. 5 and first of all from the relations (5.4) and (5.8) follows that the demand to identify the noise cannot be realized. The selection of the Gaussian noise in connection with the quantization of the data is justified. It has offered the possibility of a simple use of the null hypothesis, and at the same time it has shown a new conception of the relation of the control tasks to the problem of rounding errors [Isaacson (1966), Wilkinson (1965)].

In this paragraph we have shown on one of the practically most important classes of control systems that at the present state of measuring technique it is in most cases not possible to obtain sufficiently accurate estimates of the coefficients a_i pertinent to the transfer functions. It appears that it will be necessary to find new approaches that would approximate the control theory to actual conditions - to limited accuracy of the measurement of signals, and thus also more closely to practical problems.

6. THE CONNECTION OF LICS WITH SOME IDENTIFICATION METHODS

We discuss some assumptions in Par. 2 and Par. 5.1. This relates first of all to the question of using deterministic input signals. In the case of identification methods based on the Wiener-Kolmogorov theory it is possible by means of measuring the signals satisfying the condition (5.6) to obtain a more favourable ratio between signal and noise then given by the relation (5.9) [Herles (1970)]. We consider the limit case of white noise to which an impulse corresponds in the region of deterministic signals. The ratio of the norms of the signal and the noise for a system of first order is given exactly (for systems of higher orders of the type (5.1) approximately) by

$$\frac{\|{}^1y\|}{\delta_L} = \left(\frac{\pi e}{2}\right)^{\frac{1}{2}} \frac{10^N}{a_1} \qquad (6.1)$$

The ratio (6.1) depends on the magnitude of the coefficient a_1 [Štěpán (1971)]. In the case of equal measuring accuracy it is possible to obtain better results for systems with a coefficient $a_1 \ll 1$. In this sense we must restrict the validity of the results given in Par. 5. In this way we can at least partly explain the successes of the Wiener-Kolmogorov theory in the field of radar technique ($a_1 \ll 1$) as well as the failures in the identification of industrial systems ($a_1 > 1$).

To the theory of Kalman-Bucy filter is linked the concept of control systems observability. In considered case the control systems observability is connected to the linear dependence of sensitivity functions ${}^s v_j^{(i)}(t)$ in the sense of absolutely accurate mathematics (in the sense of approximation of functions ${}^s y_j(t)$). The response ${}^s y_j(t)$ is approximable with substitute function (2.5) if for the Gram's determinants of the sensitivity matrices from the relations (2.10) and (2.13) holds

$$|{}^sG_j^T\, {}^sG_j| > 0 \quad ; \quad |{}^s_jA| > 0 \qquad (6.2)$$

When the given Gram's determinants are equal zero, then the pertinent sensitivity functions are linear dependent. Thus it is demanded that for variances, or for the relative standard deviations of the estimates holds (for $a_i \neq 0$)

$${}^sD_j(\bar{a}_i) < \infty \quad ; \quad {}^s s_{ij} < \infty \qquad (6.3)$$

Control systems of the type (5.1) are "approximable" or observable throughout. On the other hand only a small part of control systems of the type (5.1) is identifiable according to (4.1) for measuring accuracy in practice, i.e. for the interval $N \in (2,5)$ (Table 5.1).

7. REFERENCES

Åström K.J. (1967). On the achievable accuracy in identification problems. Proc. IFAC symp. Identification in Automatic Control Systems, Prague, paper 1.8

Åström K.J. and Eykhoff P. (1970). System identification. Proc. IFAC symp. Identification and Process Parameter Estimation, Prague, survey paper

Bell D.A. (1968) Information theory. London, Pitman

Böhm J. (1964). The smoothing calculus (in czech). Prague, SNTL

Cramér H.(1958). Mathematical methods of statistics. Princeton,Prin.univ.press

Deutsch R. (1965). Estimation theory. Englewood Cliffs, N.J., Prentice Hall

Herles, V. (1970). Limit of precision in identification procedures. Proc. IFAC symp. Identification and Process Parameter Estimation, Prague,paper 9.8

Isaacson E. and Keller H.B. (1966).Analysis of numerical methods. N.Y.Wiley

Mařík J. (1964). On polynomials with real roots (in czech). Časopis pro pěstování matematiky,89,5

Newton G.C. and Gould L.A.,Kaiser J.F. (1957). Analytical design of linear feedback controls. N.Y.Wiley

Štěpán J.(1967).The identification of control systems and the limit of stability. Proc.IFAC symp. Identification in Automatic Control Systems,Prague, paper 1.11

Štěpán J. (1969). The applicability limitation of identification methods.Kybernetika,5,97

Štěpán J. (1970).The uncertainty problem in system identification.Proc.IFAC symp. Identification and Process Parameter Estimation,Prague,paper 1.8

Štěpán J.(1971). Some problems of system identification.Kybernetika,7,133

Štěpán J.(1972a).Messgenauigkeit und Regelungstheorie.Regelungstechnik,20, 203

Štěpán J.(1972b).The usability of control systems theory and the measurement accuracy.Prague,Institute of Information Theory and Automation,Report (in czech)

Wilkinson J.H.(1965).The algebraic eigenvalue problem.Oxford,Clarendon Press.

ON THE IDENTIFIABILITY OF LINEAR DYNAMICAL SYSTEMS

Keith Glover
Decision and Control Sciences Group
Department of Electrical Engineering
Massachusetts Institute of Technology
Cambridge, Massachusetts 02139, U.S.A.

Jan C. Willems
Mathematisch Instituut
Rijksuniversiteit Groningen
Groningen
The Netherlands

Consider the situation in which the unknown parameters of a stationary linear system may be parametrized by a set of unknown parameters. The question thus arises of when such a set of parameters can be uniquely identified on the basis of observed data. This problem is considered here both in the case of input and output observations and in the case of output observations in the presence of a white noise input. Conditions for local identifiability are derived for both situations and a sufficient condition for global identifiability is given for the former situation, i.e., when simultaneous input and output observations are available.

1. INTRODUCTION

Much work has recently been devoted to the problem of finding canonical forms for linear systems in state space form. (e.g., Popov (1972), Mayne (1972)). Thus it is well known that global canonical forms exist only in the single input and in the single output cases. Otherwise, a finite combinatorial number of such forms are needed to represent all minimal n-dimensional systems. Consequently, before identification can proceed an appropriate (non-global) parametrization must usually be determined.

In this paper we will consider the identification of systems described by the linear differential or difference equations:

$$\frac{dx}{dt}(t) = Ax(t) + Bu(t)$$

$$y(t) = Cx(t) + Du(t)$$

or

$$x(k+1) = Ax(k) + Bu(k)$$

$$y(k) = Cx(k) + Du(k)$$

where $x \in R^n$; $u \in R^p$; $y \in R^r$, $A \in R^{n\times n}$, $B \in R^{n\times p}$, $C \in R^{r\times n}$.

In practical identification problems such equations may often be postulated on the basis of a priori knowledge on the structure and physics of the system, with the elements of the matrices A, B, C and D, either zero, known physical constants, or certain known functions of unknown parameters. Thus if the unknown parameters are denoted by $\alpha \in \Omega \subset R^q$, then the system matrices may be written as $A(\alpha)$, $B(\alpha)$, $C(\alpha)$ and $D(\alpha)$, where $A: R^q \rightarrow R^{n\times n}$, $B: R^q \rightarrow R^{n\times p}$, $C: R^q \rightarrow R^{n\times r}$, and $D: R^q \rightarrow R^{r\times p}$.

A natural question which arises in the context of such identification problems is whether or not these unknown parameters, $(A,B,C,D)(\alpha)$, can be identified from observations of the system?

Two situations will be considered here. In Section 2 input and output observations will be permitted, and in Section 3 the system is assumed to be driven by white noise and only output observations are allowed. Stochastic aspects of this problem have been considered by, for example, Tse (1972).

2. IDENTIFIABILITY FROM INPUT/OUTPUT OBSERVATION

In this section it will be assumed that *both* the input and the output of a linear time invariant system are observed and that the input is persistently exciting (as defined for example by Åström and Bohlin (1966)). That is, there are sufficient assumptions so that the transfer function of the system can be identified from input and output observations. Consider now the following definition:

<u>Definition 1</u>:
Let $(A,B,C,D)(\alpha): \Omega \subset R^q \rightarrow R^{n(n+p+r)+pr}$ be a parametrization of the system matrices (A,B,C,D). This parametrization is said to be *locally identifiable* at $\hat{\alpha} \in \Omega$ if there exists an $\varepsilon > 0$ such that

(i) $||\alpha-\hat{\alpha}|| < \varepsilon$, $||\beta-\hat{\alpha}|| < \varepsilon; \alpha,\beta \in \Omega$ and

(ii) $C(\alpha)(Is-A(\alpha))^{-1}B(\alpha) + D(\alpha)$

$$= C(\beta)(Is-A(\beta))^{-1}B(\beta) + D(\beta)$$

for all $s \in \mathbb{C}$ (= complex plane)

imply $\alpha = \beta$ ∎

In other words, in the neighborhood of $\hat{\alpha}$, there are no two systems with distinct parameters which will have the same transfer function. This definition is similar to the definition of "*non-degeneracy*" as given by Kalman (1966).

One possible approach to this problem is to consider the Markov parameters $H(\alpha) = (C(\alpha)B(\alpha), \ldots, C(\alpha)A^{2n-1}(\alpha)B(\alpha))$ and to check whether as a function of α it is one-to-one at $\hat{\alpha}$.

Research supported by NASA under Contract No. NGL-22-009-124 with the Electronics Systems Laboratory of the Massachusetts Institute of Technology, Cambridge, Massachusetts 02139 U.S.A.

A local sufficient condition for this to be the case is that the Jacobian of $H(\alpha)$ at $\hat{\alpha}$ should be of full rank. This condition is in principle not difficult to verify but is from a computational point of view inefficient and cumbersome. If however the system $(A,B,C,D)(\alpha)$ is assumed to be minimal (i.e., reachable and observable) then the following much simpler result may be obtained.

Theorem 1

Let $(A,B,C,D)(\alpha):\Omega \subset R^q \to R^{n(n+p+r)+pr}$ (with Ω an open set in R^q) *be a* C' (i.e., continuously differentiable on Ω) *parametrization of the system matrices* (A,B,C,D), *and let* $(A,B,C,D)(\alpha)$ *be minimal. Then* $(A,B,C,D)(\alpha)$ *is locally identifiable at* $\hat{\alpha}$ *if and only if there exists an* $\varepsilon > 0$ *such that* $(TA(\alpha)T^{-1}, TB(\alpha), C(\alpha)T^{-1}, D(\alpha)) \neq (A,B,C,D)(\beta)$ *for all* $T\varepsilon GL(n)=\{T|\det T \neq 0\}$ *and* $\alpha,\beta\varepsilon N_\varepsilon(\hat{\alpha}) = \{\alpha|\ ||\alpha-\hat{\alpha}|| < \varepsilon\}$.

A sufficient condition for this is that $\det(X'X)\neq 0$ where

$$\hat{X}=\begin{bmatrix} I \otimes A'(\hat{\alpha})-A(\hat{\alpha}) \otimes I \\ I \otimes B'(\hat{\alpha}) \\ -C(\hat{\alpha}) \otimes I \\ 0 \end{bmatrix}, \quad M = \begin{bmatrix} \frac{\partial a_{ij}}{\partial\alpha}(\hat{\alpha}) \\ \vdots \\ \frac{\partial b_{ij}}{\partial\alpha}(\hat{\alpha}) \\ \vdots \\ \frac{\partial c_{ij}}{\partial\alpha}(\hat{\alpha}) \\ \vdots \\ \frac{\partial d_{ij}}{\partial\alpha}(\hat{\alpha}) \end{bmatrix}$$

$$X = [\hat{X}, M]$$

The indices of M are such that the elements of A, B, C and D are listed by rows. (see Proof). $\otimes$ denotes Kronecker product. ∎

Proof: The first part follows immediately from the facts that minimal systems from an open set in parameter space, and that any two equivalent minimal systems are related by a similarity transformation, $(A,B,C,D) \overset{T}{\to} (TAT^{-1},TB,CT^{-1},D)$ (see Brockett, (1970)).

The sufficient condition is proved as follows. Let $F:GL(n)\times\Omega\times\Omega \to R^{n(n+p+r)+pr}$ be defined by

$$F(T,\alpha,\beta) = (TA(\alpha)T^{-1}-A(\beta), TB(\alpha)-B(\beta), C(\alpha)T^{-1}-C(\beta), D(\alpha)-D(\beta))$$

We will use the implicit function theorem to show that the equation $F(T,\alpha,\beta) = 0$ has a unique solution in some neighborhood of $(I,\hat{\alpha},\hat{\alpha})$. Clearly F is a C' function, and it will now be shown that

$$X = \frac{\partial F}{\partial(T,\alpha)}(I,\hat{\alpha},\hat{\alpha})$$

$$\lim_{h\to 0} \frac{1}{h}(F(I+h\delta T,\hat{\alpha},\hat{\alpha})-F(I,\hat{\alpha},\hat{\alpha}))$$

$$= \lim_{h\to 0} \frac{1}{h}((I+h\delta T)A(\hat{\alpha})(I+h\delta T)^{-1}-A(\hat{\alpha}), (I+h\delta T)B(\hat{\alpha})-B(\hat{\alpha}), C(\hat{\alpha})(I+h\delta T)^{-1}-C(\hat{\alpha}))$$

$$= (\delta TA(\hat{\alpha})-A(\hat{\alpha})\delta T,\ \delta TB(\hat{\alpha}), -C(\hat{\alpha})\delta T)$$

and $\lim_{h\to 0} \frac{1}{h}(F(I,\hat{\alpha}+h\delta\hat{\alpha},\hat{\alpha})-F(I,\hat{\alpha},\hat{\alpha}))$

$$= \left(\left(\frac{\partial a_{ij}}{\partial\alpha}(\hat{\alpha})\delta\alpha\right), \left(\frac{\partial b_{ij}}{\partial\alpha}(\hat{\alpha})\delta\alpha\right), \left(\frac{\partial c_{ij}}{\partial\alpha}\delta\alpha\right), \left(\frac{\partial d_{ij}}{\partial\alpha}\delta\alpha\right)\right)$$

Now order the components of the domain of F as the vector, $\mathrm{col}(t_{11},\ldots,t_{1n},t_{21},\ldots,t_{2n},\ldots, t_{n1},\ldots,t_{nn}, \alpha_1,\ldots,\alpha_q)$ and the components of the range of F as the vector $\mathrm{col}(a_{11},\ldots,a_{1n},\ldots, a_{n1},\ldots,a_{nn},b_{11},\ldots,b_{1p},\ldots,b_{n1},\ldots,b_{np},c_{11},\ldots, c_{1n},\ldots,c_{r1},\ldots,c_{rn},d_{11},\ldots,d_{1p},\ldots,d_{r1},\ldots,d_{rp})$. It is easily verified that under this ordering the linear equations in δT and $\delta\alpha$ given above correspond precisely to the matrix X given in the statement of Theorem 1. The remainder of the proof proceeds by considering 3 different cases.

(i) Case $q = n(p+r)+pr$

Then $F:D \subset R^{n(n+p+r)+pr} \to R^{n(n+p+r)+pr}$ and $\det \frac{\partial F}{\partial(T,\alpha)}(I,\hat{\alpha},\hat{\alpha}) \neq 0$

Hence the implicit function theorem implies that there exist neighborhoods, $U_1 \subset R^q$ of $\hat{\alpha}$ and W of $(I,\hat{\alpha})$, such that for any $(T,\alpha)\varepsilon W$ and $\beta\varepsilon U_1$, the equation $F(T,\alpha,\beta) = 0$ has a unique solution, $(T,\alpha) = g(\beta)$. Further there exist neighborhoods V_1 of I and V_2 of $\hat{\alpha}$ such that $V_1 x V_2 \subset W$ and $g^{-1}(V_1 x V_2) = U_2 \subset U_1$. Hence for any $T\varepsilon V_1$, $\alpha\varepsilon V_2$, $\beta\varepsilon U_2$ the equation $F(T,\alpha,\beta) = 0$ has a unique solution. However we require that $F(T,\alpha,\beta) = 0$ has a unique solution for all $T\varepsilon GL(n)$, $\alpha,\beta\varepsilon N_\varepsilon(\hat{\alpha})$. In order to prove this, we proceed by contradiction. Assume therefore that for all $\varepsilon > 0$ there exists α_ε, β_ε $\varepsilon N_\varepsilon(\hat{\alpha}) \subset V_2 \cap U_2$ and $T_\varepsilon \notin V_1$ such that $F(T_\varepsilon,\alpha_\varepsilon,\beta_\varepsilon) = 0$. This implies that $T_\varepsilon = [M'(\beta_\varepsilon)M(\beta_\varepsilon)]^{-1}M'(\beta_\varepsilon)M(\alpha_\varepsilon)$ where $M(\alpha) = [C'(\alpha),A'(\alpha)C'(\alpha),\ldots,A'^{n-1}(\alpha)C'(\alpha)]'$. Therefore T_ε is continuous in $(\alpha_\varepsilon,\beta_\varepsilon)$ in a neighborhood of $(\hat{\alpha},\hat{\alpha})$ by the observability of $(A(\hat{\alpha}),C(\hat{\alpha}))$, and hence $||T_\varepsilon-I||$ can be made arbitrarily small by choosing ε sufficiently small, which contradicts the fact $T_\varepsilon \notin V_1$. This completes the proof for the case $q = n(p+r)+pr$.

(ii) If $q > n(p+r)+pr$ then there are too many degrees of freedom and no identifiability possible

(iii) If $q<n(p+r)+pr$, then $\det(X'X) \neq 0$ implies the nullspace of X consists of the zero vector only. Hence there exists $k=n(p+r)+pr-q$ vectors in $R^{n(n+p+r)+pr}$, say $\{m_i\}$, $i=1,2,\ldots,k$ such that $\det[X,m_1,\ldots m_k] \neq 0$. The remainder of the proof follows exactly as in Case (i). ∎

Theorem 1 gives a straightforward test for local identifiability and is significantly simpler than the methods based on the so-called *information matrix*. Disadvantages of the concept of local identifiability are that the nominal values, $\hat{\alpha}$, must be known and the size of the neighborhood of $\hat{\alpha}$ is in general not easily found. However, if a parametrization is locally identifiable for all $\alpha\epsilon\Omega$, then one may conclude that identification algorithms will be locally well-behaved but may converge to one of several solutions depending on the initial estimates and on the actual data received. It is thus desirable to attempt to generalize the result of the theorem to global identifiability.

Definition 2

Let $(A,B,C,D)(\alpha):\Omega \subset R^q \rightarrow R^{n(n+p+r)+pr}$ be a parametrization of the system matrices (A,B,C,D). This parametrization is said to be *globally identifiable* if

(i) $C(\alpha)(Is-A(\alpha))^{-1}B(\alpha)+D(\alpha)=$

$C(\beta)(Is-A(\beta))^{-1}B(\beta)+D(\beta)$ for all $s\epsilon\mathbb{C}$.

and

(ii) $(A,B,C,D)(\alpha)$ is minimal imply that $\alpha=\beta$. ■

Condition (ii) could be deleted in the above definition but then the definition would be very restrictive since most useful parametrizations admit multiple representations of non-minimal systems.

Local identifiability for all $\alpha\epsilon\Omega$ will in general not imply global identifiability. For example the parametrization, $A(\alpha)=\begin{pmatrix}1 & 0\\ 1+\alpha & 2\end{pmatrix}$, $B(\alpha)=\begin{pmatrix}1+\alpha\\ 1+3\alpha\end{pmatrix}$; $C(\alpha) = (2-\alpha,4-3\alpha)$;and $D(\alpha) = 0$, can be shown to be locally identifiable for all $\alpha\epsilon R$ but is not globally identifiable since $\alpha=0$ and $\alpha=1$ lead to distinct minimal representations of the same transfer function. The problem of concluding global results based on local conditions is a difficult problem in global analysis (see Palais (1959)).

The following proposition gives a sufficient condition for global identifiability, when the parametrization is affine, (i.e., a linear map plus an offset).

Proposition 1:

An affine parametrization $(A,B,C,D)(\alpha)$: $\Omega \subset R^q \rightarrow R^{n(n+p+r)+pr}$, *is globally identifiable if* $\det[Y'(\alpha,\beta)Y(\alpha,\beta)] \neq 0$ *for all* $\alpha,\beta\epsilon\Omega$, *where*

$$Y(\alpha,\beta)=\begin{bmatrix} Z(\alpha,\beta) & 0 & M\\ 0 & Z(\beta,\alpha) & -M\end{bmatrix}$$

$$Z(\alpha,\beta)=\begin{bmatrix} I\otimes A'(\alpha) - A(\beta)\otimes I\\ I\otimes B'(\alpha)\\ -C(\beta)\otimes I\\ 0\end{bmatrix}$$

and M is given in Theorem 1.

Proof: Since we are only concerned with minimal systems global identifiability is implied if the equations, $TA(\alpha) = A(\beta)T$, $TB(\alpha) = B(\beta)$, $C(\alpha) = C(\beta)T$, $D(\alpha) = D(\beta)$, have a unique solution for all $\alpha,\beta\epsilon\Omega$ and $T\epsilon GL(n)$. Let q_1 and q_2 be the vectors formed by listing respectively the elements of $(T-I)$ and $(T^{-1}-I)$ by rows. Then it is easily verified that the above equations are equivalent to $[q_1,q_2,\alpha-\beta]Y'(\alpha,\beta) = 0$, since $(A,B,C,D)(\alpha)$ is affine. Therefore since $\det(Y'(\alpha,\beta)Y(\alpha,\beta)) \neq 0$ the nullspace of $Y(\alpha,\beta) = N(Y(\alpha,\beta)) = \{0\}$ and the result is thus verified. ■

Remarks: 1. The condition is not necessary since $(\bar{q}_1,\bar{q}_2,\bar{\alpha}-\bar{\beta})\epsilon N(Y(\hat{\alpha},\hat{\beta}))$ does not imply that $(I+\bar{Q}_1)^{-1} = I+\bar{Q}_2$, $\bar{\alpha}=\hat{\alpha}$ and $\bar{\beta}=\hat{\beta}$, which is required for a system not to be globally identifiable.

2. A somewhat more restrictive sufficient condition for global identifiability is that $N(Z(\alpha,\beta), M) = \{0\}$ for all $\alpha,\beta\epsilon\Omega$. We remark that this condition is in fact satisfied by the canonical forms of Luenberger (1967) and Popov (1972).

3. IDENTIFIABILITY FROM OUTPUT OBSERVATION

In this section we will consider the identifiability of a continuous time linear stationary system under the following assumptions.

A1. The input $u(t)$ is not observed directly, but is assumed to be a white noise process with $E(u(t)u'(\tau)) = I\delta(t-\tau)$.

A2. The matrix A is asymptotically stable, (i.e., the eigenvalues of A are strictly in the left half plane).

A3. The system has reached steady state when the observations begin (i.e., the output process $y(t)$ is a stationary process).

A4. The system to be identified is globally minimal, i.e., the dimension of the state is less than or equal to that of any other system having the same output spectral density when driven by white noise (see Anderson (1967)).

Under these assumptions the most information that may be obtained from the output observations is the output spectral density, $\phi(s) = G(s)G'(-s)$, where $G(s) = C(Is-A)^{-1}B + D$. This motivates the following definition.

Definition 3

Let $(A,B,C,D)(\alpha):\Omega\subset R^q \rightarrow R^{n(n+p+r)+pr}$ be a parametrization of the system matrices (A,B,C,D). This parametrization is said to be *locally identifiable from its output spectral density* at $\hat{\alpha}\epsilon\Omega$ if there exists an $\varepsilon > 0$ such that

(i) $||\alpha-\hat{\alpha}|| < \varepsilon$, $||\beta-\hat{\beta}|| < \varepsilon$, $\alpha,\beta\epsilon\Omega$.

and

(ii) $G(s,\alpha)\, G'(-s,\alpha) = G(s,\beta)G'(-s,\beta)$ for all $s\epsilon\mathbb{C}$.

imply $\alpha = \beta$.

where $G(s,\alpha) = C(\alpha)(Is-A(\alpha))^{-1}B(\alpha) + D(\alpha)$ ■

A solution to this problem could be obtained by looking at the mapping from α into the power series expansion of $\phi(s)$ about $s = \infty$, and showing that this mapping is locally one-to-one. Such an approach would give computationally complex conditions. However, there exists a characterization of all globally minimal solutions to the spectral factorization problem. This is summarized in the following lemma:

Lemma 1

Let (A,B,C,D) and (F,G,H,J) be the system matrices of two globally minimal continuous time systems satisfying conditions (A1) and (A2). Then these systems have the same output spectral density function if and only if there exist matrices $T\in GL(n)$ and $P=P'$ such that

$$TAT^{-1}=F,\ CT=H,\ AP + PA' = BB'-TGG'T'$$
$$PC' = BD'-TGJ',\ DD' = JJ'.$$

The proof of Lemma 1 is a straightforward consequence of Lemma 2 in Anderson (1967). ∎

The local identifiability problem considered in this section thus reduces to verifying whether or not the following equations have the unique solution $\alpha = \beta$, $T=I$, $P=0$, for all $\alpha,\beta\in N_\varepsilon(\hat{\alpha})$.

$$P = P',\ TA(\alpha)T^{-1} = A(\beta),\ C(\alpha)T = C(\beta),$$
$$A(\alpha)P + PA'(\alpha) = B(\alpha)B'(\alpha)-TB(\beta)B'(\beta)T',$$
$$PC'(\alpha) = B(\alpha)D'(\alpha) - TB(\beta)D'(\beta),$$
$$D(\alpha)D'(\alpha) = D(\beta)D'(\beta).$$

The following theorem can be proved in an analogous manner to the proof of Theorem 1, using the implicit function theorem to show that (β,T,P) is a unique function of α.

<u>Theorem 2</u>

Let $(A,B,C,D)(\alpha):\Omega \subset R^q \to R^{n(n+p+r)+pr}$ *(with Ω an open set in R^q) be a C' parametrization of the system matrices* (A,B,C,D) *of a continuous time system satisfying* (A1)-(A4). *Then this parametrization is locally identifiable from its output spectral density at $\hat{\alpha}\in\Omega$, if the following linear equations in* $(\delta B,\delta D,\delta T,\delta P,\delta\beta)$, *have a unique solution (i.e., zero).*

(i) $\delta P = \delta P'$

(ii) $(\hat{A}\delta P+\delta T\hat{B}\hat{B}'+\delta B\hat{B}')+(\hat{A}\delta P+\delta T\hat{B}\hat{B}'+\delta B\hat{B}')' = 0$

(iii) $\delta P\hat{C}' = -(\delta T\hat{B}\hat{D}' + \delta B\hat{D}' + \hat{B}\delta D')$

(iv) $\hat{D}\delta D' + \delta D D' = 0$

(v) $(\hat{A}\delta T-\delta T\hat{A},\delta B,\ \hat{C}\delta T,\delta D)$
$$=\left(\left(\frac{\partial a_{ij}}{\partial\alpha}(\hat{\alpha})\delta\beta\right),\left(\frac{\partial b_{ij}}{\partial\alpha}(\hat{\alpha})\delta\beta\right),\left(\frac{\partial c_{ij}}{\partial\alpha}(\hat{\alpha})\delta\beta\right),\left(\frac{\partial d_{ij}}{\partial\alpha}(\hat{\alpha})\delta\beta\right)\right)$$

where $(\hat{A},\hat{B},\hat{C},\hat{D}) = (A,B,C,D)(\hat{\alpha})$. ∎

We remark that the condition of Theorem 2 may be restated as a non-zero determinant condition for a matrix of dimension less than or equal to $[\frac{n}{2}(3n+2p+1)+pr]$.

A completely analogous argument can be followed through for discrete time systems. In this case, the linear equations which must have a unique solution to imply local identifiability from the output spectral density are:

(i)' $\delta P = \delta P'$

(ii)' $\hat{A}\delta P\hat{A}'-\delta P= -(\delta T\hat{B}\hat{B}'+\delta B\hat{B}'+\hat{B}\delta B'+\hat{B}\hat{B}'\delta T')$

(iii)' $A\delta PC' = -(\delta T\hat{B}\hat{D}'+\delta B\hat{D}' + \hat{B}\delta D')$

(iv)' $\hat{C}\delta P\hat{C}' = - \hat{D}\delta D' - \delta D\hat{D}'$

(v)' Same as (v)

4. CONCLUSIONS

In this paper we have presented some tests for the identifiability of parametrizations of stationary linear dynamical systems. These conditions should be of great value in situations where sufficient a priori knowledge is available so that state space equations can be written down with relatively few unknown parameters (i.e., $\leq n(p+r)+pr$)).

An open problem, presently under investigation, is that of finding weaker sufficient conditions for global identifiability.

5. REFERENCES

B.D.O. Anderson, (1967) "The Inverse Problem of Stationary Covariance Generation", J. of Statistical Physics, Vol. <u>1</u>

K.J. Åström and T. Bohlin,(1966), "Numerical Identification of Linear Dynamic Systems from Normal Operating Records", Paper IFAC Symp.Theory of Self-adaptive Control Systems, Teddington, England. In <u>Theory of Self-Adaptive Control Systems</u>,(Ed. P.H. Hammond), Plenum Press, New York

R.W. Brockett,(1970), <u>Finite Dimensional Linear Systems</u>, John Wiley.

R.E. Kalman,(1966) "On Structural Properties of Linear, Constant, Multivariable Systems", Paper 6.A, Proceedings of Third IFAC Congress, London

D.G. Luenberger,(1967) "Canonical Forms for Linear Multivariable Systems" <u>IEEE Trans. on Auto. Control</u>, Vol. <u>AC-12</u>, pp. 290-293.

D.Q. Mayne (1972) "A Canonical Model for Identification of Multivariable Linear Systems", <u>IEEE</u> Trans. Automatic Control, Vol. <u>AC-17</u>, pp. 728-729

R.S. Palais,(1959) "Natural Operations on Differential Forms", <u>Trans. Amer. Math. Soc</u>., Vol. <u>92</u>, pp. 125-141

V.M. Popov,(1972) "Invariant Description of Linear, Time-Invariant Controllable Systems" SIAM J. Control, Vol. <u>10</u>, No. 2, pp. 252-264

E. Tse and J. Anton,(1972),"On the Identifiability of Parameters" IEEE Trans. on Automatic Control, Vol. <u>AC-17</u>, No. 5, pp. 637-645

IDENTIFIABILITY OF LINEAR DYNAMICAL SYSTEMS

Hans E. Berntsen

Jens G. Balchen

Division of Automatic Control
University of Trondheim, Norway

In this paper we consider methods for examining if the parameters in a dynamical model are identifiable. We want our methods to be used in a Computer Aided Design - system. A criterion for identifiability of linear dynamical system based on frequency response is presented.

1. INTRODUCTION

One of the problems in identifying dynamical systems is to combine apriori knowlegde with experimental data. We often use a model of the form:

$$\dot{x} = f(x,\beta,u)$$

$$\beta = 0 \tag{1}$$

$$y = g(x,\beta)$$

where

x = n dimensional state vector
β = p dimensional parameter vector
u = r dimensional control vector
y = m dimensional measurement vector

We wish to find numerical values for β such that the model (1) gives a response which is a good approximation to that of the real system. For this task we have available apriori information about β, e.g. design data, and experimentally obtained measurement data. Since not all of the parameters characterizing the system are observable from the measurement vector, we are forced to decide which parameters to determine from design data and which to determine from experiments.

If we decide to determine only a few parameters from the experimental data we must expect rather large deviations between system- and modelresponse. If on the other hand we try to determine too many parameters or some unfortunate combinations of parameters from experimental data we may experience the divergence of the parameter estimation algorithm because not all the parameters are identifiable. Bellmann (1970).

In this paper we will discuss some possibilities for determining whether a combination of parameters which have been chosen is identifiable. We are particularly interested in methods which can be implemented in a Computer Aided Design - system.

If there is a one to one relationship between a chosen parameter vector β and the response of the model we will define the model as being identifiable.

If we, in order to achieve a one to one relationship as above, must specify that β is near β_0 we define the model as being locally identifiable near β_0.

2. OBSERVABILITY CRITERION BASED ON TAYLOR SERIES EXPANSION.

The model (1) can be rewritten:

$$\tilde{x} = \begin{bmatrix} x \\ \beta \end{bmatrix} \tag{2}$$

$$\dot{\tilde{x}} = \tilde{f}(\tilde{x}, u) \quad , \quad y = \tilde{g}(\tilde{x}) \tag{3}$$

Using this form the linear model

$$\dot{x} = A(\beta)x + B(\beta)u \quad , \quad y = D(\beta)x \tag{4}$$

will become a nonlinear model. Thus if we want to determine the identifiability of the model by investigating if the augmented state vector $\tilde{x}$ is observable, then we need a criterion for the observability of a nonlinear system. This is not simple. Linearizing the model (3) and investigating the observability of the state of this linearized model will lead to a too restrictive criterion. A simple model of the type given in (4):

$$\dot{x} = \beta_1 x + \beta_2 u \quad , \quad y = x \tag{5}$$

will using linearization, turn out to be non-observable, but is in fact observable if $u \neq 0$.

Schoenwandt (1970) has presented a method for investigating the observability of a nonlinear system of the form (3) by expanding:

$$y(t) = \sum_{i=0}^{\infty} \frac{1}{i!} \overset{(i)}{y}(t) \Big|_{t=t_0} \cdot (\Delta t)^i \tag{6}$$

where

$$\Delta t = t - t_0$$

and

$\dot{y}, \ddot{y}, \ldots, y^{(i)}$ = time derivatives $y(t)$.

It is seen that the time derivatives of $y(t)$ are determined by $\tilde{x}(t_0)$, $u(t_0)$, $\dot{u}(t_0)$, $\ddot{u}(t_0)$, ... such that

$$Y = \begin{bmatrix} y \\ \dot{y} \\ \ddot{y} \\ \cdot \\ \cdot \end{bmatrix} = \Psi(\tilde{x}(t_0), u(t_0), \dot{u}(t_0), \ddot{u}(t_0), \ldots) \quad (7)$$

If the relationship between $x(t_0)$ and Y is one to one with the given control vector, then $\tilde{x}(t_0)$ is observable. It is, however, difficult to determine this relationship. We therefore linearize (7) around some point $\tilde{x}(t_0) = \tilde{x}_0$ in the augmented state space:

$$\delta Y = \frac{\partial}{\partial \tilde{x}(t_0)} \Psi(\tilde{x}(t_0), u(t_0), \ldots) \Big|_{\tilde{x}(t_0)=\tilde{x}_0} \delta\tilde{x}$$

$$= F(\tilde{x}_0)\delta\tilde{x} \quad (8)$$

If and only if $\operatorname{rank}(F(\tilde{x}_0)) = \dim \tilde{x} = n + p$ then $\tilde{x}(t_0)$ is locally observable near $\tilde{x}_0$.

It seems reasonable to refer to the matrix F above as the observability matrix. We observe that when we compute F we need the time derivatives of $u(t)$. This is inconvenient because it is often difficult to tell what values to assign to u, $\dot{u}$, $\ddot{u}$, ...

In principle F, determined according to (8), will be a matrix of dimension $(\infty) \times (n + p)$. To determine local observability we need, however, only establish so many rows in F that $\operatorname{rank}(F) = n + p$.

If the model is of the form (4) with the matrices A, B, and D as linear functions of β, F can be determined numerically by means of a standard computer program. Homstvedt (1971).

For the linear system:

$$\dot{x} = Ax + Bu, \quad y = Dx \quad (9)$$

we get

$$Y = \begin{bmatrix} y \\ \dot{y} \\ \ddot{y} \\ \cdot \\ \cdot \end{bmatrix} = \begin{bmatrix} Dx \\ DAx + DBu \\ DA^2x + DABu + DB\dot{u} \\ \cdot \\ \cdot \end{bmatrix} \quad (10)$$

$$F^T = \left[D^T \vdots A^TD^T \vdots (A^T)^2D^T \vdots \ldots \right] \quad (11)$$

Since we know (theorem of Hamilton-Cayley) that

$$\operatorname{rank}\left[D^T \vdots A^TD^T \vdots (A^T)^2D^T \vdots \ldots \right]$$
$$= \operatorname{rank}\left[D^T \vdots A^TD^T \vdots \ldots \vdots (A^T)^{n-1}D^T \right]$$

we are led to the usual criterion for observability of a linear model.

3. IDENTIFIABILITY CRITERION BASED ON FREQUENCY RESPONSE

Rather than expanding $y(t)$ in the time domain as above, we now establish the equivalent relationships in the frequency domain. For the system given in (4) we obtain

$$y(j\omega) = D(\beta)\left[j\omega I - A(\beta)\right]^{-1}B(\beta)u(j\omega)$$

$$= Z(j\omega,\beta)u(j\omega) \quad (12)$$

Here the transfer matrix $Z(j\omega,\beta)$ is a $m \times r$ matrix in which the elements are complex functions of β and ω, represented by its real and imaginary components. If there is a one to one relationship between β and Z such that

$$Z(j\omega,\beta_1) = Z(j\omega,\beta_2) \Rightarrow \beta_1 = \beta_2 \quad (13)$$

then we define the model as being identifiable. Determining whether the condition of (13) is fulfilled is not simple. Similar to that done in (8) we will therefore linearize the transfer matrix Z around $\beta = \beta_0$:

$$\delta Z = \{\delta z_{ij}\} = Z(j\omega,\beta) - Z(j\omega,\beta_0)$$

$$\approx \frac{\partial}{\partial\beta} Z(j\omega,\beta) \Big|_{\beta=\beta_0} \delta\beta \quad (14)$$

where $\delta\beta = \beta - \beta_0$

If (14) yields a one to one relationship between δZ and $\delta\beta$ when β is near β_0 such that (13) is satisfied when β_1 and β_2 are near β_0 we shall say that the model is locally identifiable near β_0. We then have to investigate the 3-dimensional $m \times r \times p$ array

$$\frac{\partial}{\partial\beta} Z(j\omega,\beta) \Big|_{\beta=\beta_0}$$

whose elements are complex functions in β_0 and ω. (14) is reformulated such that we get a real vector equation:

$$\delta\varepsilon(j\omega,\delta\beta) \stackrel{\Delta}{=} \left[\operatorname{Re}\delta z_{11}, \operatorname{Im}\delta z_{11}, \operatorname{Re}\delta z_{21}, \operatorname{Im}\delta z_{21} \ldots \operatorname{Re}\delta z_{m1}, \operatorname{Im}\delta z_{m1}, \operatorname{Re}\delta z_{12}, \operatorname{Im}\delta z_{12} \ldots \operatorname{Re}\delta z_{mr}, \operatorname{Im}\delta z_{mr}\right]^T \quad (15)$$

We may now write (14) as:

$$\delta\varepsilon(j\omega,\delta\beta) = F(j\omega,\beta_0)\delta\beta \tag{16}$$

where F is a (2·m·r) x p matrix. It is simpler to investigate if (16) has a unique solution for $\delta\beta$ than if (14) has one. We now assume that the control vector u contains power at the distinct frequencies $\omega = \omega_1,\omega_2,\ldots,\omega_s$. Using these s different frequences in (16) we get s equations which we collect in the form:

$$\delta\mathcal{E} = \begin{bmatrix} \delta\varepsilon(j\omega_1,\delta\beta) \\ \delta\varepsilon(j\omega_2,\delta\beta) \\ \cdot \\ \cdot \\ \delta\varepsilon(j\omega_s,\delta\beta) \end{bmatrix} = \begin{bmatrix} F(j\omega_1,\beta_0) \\ \cdot \\ \cdot \\ \cdot \\ F(j\omega_s,\beta_0) \end{bmatrix} \cdot\delta\beta = \tilde{F}(\beta_0)\cdot\delta\beta \tag{17}$$

We have now arrived at an expression which is analogous to (8) and we must require that the identifiability matrix $\tilde{F}$ is such that $\mathrm{rank}(\tilde{F}) = p$ in order that the model shall be locally identifiable near β_0. In order to establish local identifiability the control vector need only contain power at as many frequencies as neccesary to make $\mathrm{rank}(\tilde{F}) = p$.

If the system matrices A, B and D are linear functions of β, $\tilde{F}$ can easily be determined using a standard computer program.

An alternative way of testing the rank of $\tilde{F}$ is to use the property that $\tilde{F}$ has rank p if and only if $\tilde{F}^T\tilde{F}$ is nonsingular.

4. DEGREE OF IDENTIFIABILITY

We may wish to have a quantitative measure of the degree of identifiability. We apply different weights on the components of (17) by introducing a uncertainty vector w such that:

$$\delta\mathcal{E} = \tilde{F}\delta\beta + w \tag{18}$$

We then define a diagonal matrix W in which w_{jj} is the inverse of the weight applied to $\delta\mathcal{E}_j$. Analogous to the results of linear minimum variance estimation we then get:

$$\delta\hat{\beta} = (\tilde{F}^T W^{-1}\tilde{F})^{-1}\tilde{F}^T W^{-1}\delta\mathcal{E} \tag{19}$$

In (19) we may interpret $(F^T W^{-1} F)^{-1} = S^{-1}$ as a covariance matrix (uncertainty matrix).

The degree of local identifiability near β_0 must be correlated with the degree of linear independence between the columns of S. If the columns of S are linearly dependent the model is not identifiable near β_0 and the degree of identifiability is zero. If the columns of S are orthogonal we will define that the degree of identifiability is 1.0. We use the well known Hadamard inequality [1]:

$$|\det(S)| \leq \prod_{i=1}^{p} \left(\sum_{j=1}^{p} s_{ij}^2\right)^{\frac{1}{2}} \tag{20}$$

In (20) the to sides are equal if the columns of S are orthogonal.

$$\kappa \overset{\Delta}{=} \left(\frac{|\det(S)|}{\prod_{i=1}^{p}\left(\sum_{j=1}^{p} s_{ij}^2\right)^{\frac{1}{2}}}\right)^{1/p} \tag{21}$$

κ in (21) may be used as a convenient measure of local degree of identifiability. Also the trace of the matrix S^{-1} may be a helpful parameter when we are studying the local identifiability of the model.

$$\left(\mathrm{trace}(A) = \sum_{i=1}^{n} a_{ii}\right)$$

By investigating the degree of local identifiability for a number of points in the space spanded by β we may get an idea of the identifiability of the model and thereby the possibilities of succeeding in estimating β.

5. THE LOCAL IDENTIFIABILITY OF PARAMETERS IN NON-LINEAR MODELS.

To examine the local identifiability of a model that is nonlinear in the state with the help of the frequency method seems a little suspicious, but we will consider an idea.

We linearize the model (1) around a steady state x_0, u_0:

$$\delta\dot{x} = A(x_0,u_0,\beta)\delta x + B(x_0,u_0,\beta)\ \delta u$$

$$\delta\dot{y} = D(x_0,\beta)\delta x \tag{22}$$

We compute the identifiability matrix $\tilde{F}(x_0,u_0,\beta_0)$. If the real system is varying over a greater part of the state space, we may linearize the model in several different possible (but not neccesarily actual) steady states $x_1,x_2,\ldots x_q$ and create an augmented identifiability matrice $\bar{F}$.

$$\bar{F} = \begin{bmatrix} \tilde{F}(x_1,u_1,\beta_0) \\ \tilde{F}(x_2,u_2,\beta_0) \\ \cdot \\ \cdot \\ \tilde{F}(x_q,u_q,\beta_0) \end{bmatrix}$$

If the model is not locally identifiable around β_0, then $\mathrm{rank}(\bar{F}) < p$. On the other hand, if $\mathrm{rank}(\bar{F}) = p$, then this, for the lack of a better criterion, may be taken as clearly indicating that the model is identifiable.

6. NUMERICAL ASPECTS OF THE FREQUENCY CRITERION

Numerical experiments have been performed with the frequency criterion for identifiability of linear models 4 . The methods seem to be useful. A model of order 10, with 20 parameters has been examined and distinction between an identifiable and non-identifiable parameter vector β has been established. Due to numerical errors it very often happens that $\det S \neq 0$ even if the model is not identifiable. Computing the "covariance" matrix $V = S^{-1} = (\tilde{F}^T W^{-1} \tilde{F})^{-1}$ in such a case is possible with a low degree of accuracy. The matrix V may then give us an indication of which parameters that could be included in β in order to make the model identifiable. An example that illustrates this follows:

Example:

$$\dot{x} = Ax + Bu \quad , \quad y = Dx$$

Apriori values of the matrices

$$A = \begin{bmatrix} -0.38 & 0.38 & 0 \\ 0.2 & -2 & 2 \\ 0 & 0.3 & -0.7 \end{bmatrix} , \quad B = \begin{bmatrix} 0.2 \\ 0 \\ 0 \end{bmatrix} , \quad D = \begin{bmatrix} 1 & 0 & 0 \\ & & \\ 0 & 1 & 0 \end{bmatrix}$$

It is assumed that $a_{12} = -a_{11}$.
We use the following frequences when computing the observability matrix $\tilde{F}$:

$$\omega = 0.01,\ 0.1,\ 0.2,\ 0.5,\ 1,\ 2,\ 5,\ 10,\ 20,\ 50.$$

We choose W = the identity matrix. The local identifiability near the apriori values of the model is examined.

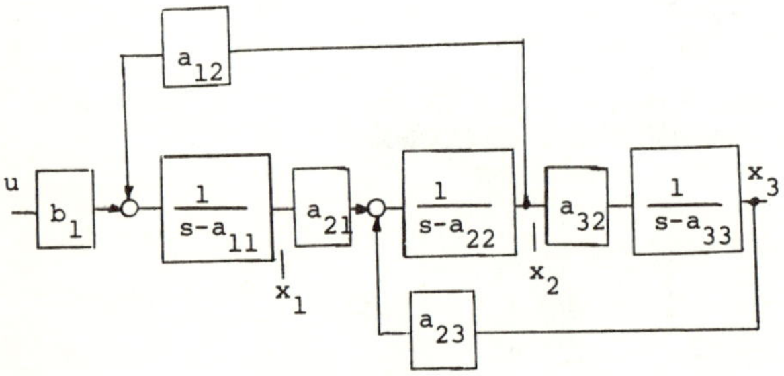

Fig. 1. The linear model.

Case 1.

$$\beta = \begin{bmatrix} a_{12} & a_{22} & b_1 & a_{23} & a_{32} \end{bmatrix}^T$$

The following results were found.

$$\kappa = 1.8 \cdot 10^{-5} \quad , \quad \text{trace } (V) = 10^{19}$$

$$V = \begin{bmatrix} 1.38 & 0.62 & 0.48 & 5.0 & 0.75 \\ 0.62 & 1070 & -4.14 & -1650 & -108 \\ 0.48 & -4.14 & 0.23 & 7.0 & 0.68 \\ 15.4 & -1400 & 9.94 & 10^{18} & -1.5 \cdot 10^{18} \\ -0.9 & -145 & 0.25 & -1.5 \cdot 10^{18} & 2 \cdot 10^{17} \end{bmatrix}$$

The fact that the computed V is non-symetric indicates a poor numerical accuracy. The results clearly indicates that the model is not identifiable with the chosen β and also that a_{23} and a_{32} not can be determined simultaneously by experimental data. This is also seen from fig. 1.

Case 2.

$$\beta = \begin{bmatrix} a_{12} & a_{22} & b_1 & a_{23} \end{bmatrix}^T$$

$$\kappa = 0.03 \quad , \quad \text{trace (V)} = 6512$$

$$V = \begin{bmatrix} 1.38 & 0.62 & 0.48 & 9.3 \\ 0.62 & 1070 & -4.1 & -2360 \\ 0.48 & -4.1 & 0.23 & 11.6 \\ 9.3 & -2360 & 11.6 & 5440 \end{bmatrix}$$

The result tells us that the model is identifiable. The numerical values of V indicates that to estimate both a_{22} and a_{23} may give a poor accuracy.

Case 3.

$$\beta = a_{12}\ a_{22}\ b_1{}^T$$

$$\kappa = 0.12 \quad , \quad \text{trace (V)} = 45$$

$$V = \begin{bmatrix} 1.36 & 4.66 & 0.46 \\ 4.66 & 43.1 & 0.89 \\ 0.46 & 0.89 & 0.21 \end{bmatrix}$$

REFERENCES

1. E.F.Beckenbach and R. Bellman, (1965), "Inequalities", Springer-verlag, Berlin, p64.

2. R.Bellman and K.J.Åstrøm, (1970), "On Structural Identifiability", Mathematical Biosciences 7, pp 329-339.

3. H.Berntsen, (1972), "Parameter- and state estimation", Thesis. Division of Automatic Control, University of Trondheim, Norway. (In norwegian).

4. H.Berntsen, (1973), "Identifiability of linear dynamical systems", Report no. 73-12-W. Division of Automatic Control, University of Trondheim, Norway.

5. R.G.Brown, (1966), "Not Just Observable, But How Observable?" Proc. 1966 National Electronic Conf., Vol. 22, pp 709-714.

6. G.Homstvedt, (1971), "Parameterestimation using Kalmanfilter." Thesis, Division of Automatic Control, University of Trondheim, Norway. (In norwegian)

7. U.Schoenwandt, (1970), "On observability of nonlinear systems". 2.nd. IFAC symposium in Prague 15-20 june 1970, part II, 8.2.

TEST SIGNALS AND NONLINEAR OBSERVABILITY

K. S. P. Kumar

Department of Electrical Engineering, University of Minnesota
Minneapolis, Minnesota 55455 U.S.A.

The success of state and parameter estimation algorithms depends strongly on the observability of the system under consideration. For linear systems, the known input has no effect on observability. For nonlinear systems, however, the observability property critically depends on the applied input signal. A proper choice of the input test signal may render an unobservable nonlinear system observable. When this is established, estimation algorithms may be constructed to provide the required estimates. The role of test signals on the observability of nonlinear systems is introduced here via examples.

1. INTRODUCTION

The qualitative property of observability of a system is important for various reasons. In linear systems, this property can be established by a variety of tests (Chen 1970). It is well known that a known input to the linear system has no effect on the observability property. The observability property of the noise free linear system plays an important role in the behavior of the well known Kalman-Bucy algorithm for state estimation (Jazwinski 1969).

For nonlinear systems, the observability property is a function of the input to the system. While some inputs may cause the system to be unobservable, some other input may make it observable. This is much like the stability property of a nonlinear system. It is known, for example, that some inputs quench limit cycle oscillations while others do not. It then remains to examine the methods by which a suitable test signal can be selected to make a nonlinear system observable. When the observability property is established, we can hope that an estimation algorithm will provide useful estimates.

2. DEFINITIONS

Consider the system described by

$$\dot{x} = f(t,x,u) \qquad x(t_o) = x_o \tag{1}$$

$$y = h(t,x) \tag{2}$$

where x is the n-vector of states, u is the r-vector of inputs and y is the m-vector of observations. Denote by $K \; R^n$ the set of admissible initial states of (1). Let $[t_o,t_1]$ be the time interval of interest.

Solutions of (1) corresponding to an initial state x_o and an input u are denoted by $\phi(t;x_o,u)$. It is assumed that each solution of (1) has the uniqueness property and is defined on $[t_o,t_1]$. The observation corresponding to x_o and u is denoted by $y(t;x_o,u)$. It is clear that

$$y(t;x_o,u) \equiv h(t,\phi(t;x_o,u)), \; t\varepsilon[t_o,t_1].$$

<u>Definition 1</u>. For a given input u, the state $x_o\varepsilon K$ is said to be <u>observable</u> with respect to K on $[t_o,t_1]$ if there does not exist another $z_o\varepsilon K$ such that $y(t;x_o,u) \equiv y(t;z_o,u)$ for all $t\varepsilon[t_o,t_1]$.

<u>Definition 2</u>. It follows that $x_o\varepsilon K$ is unobservable with respect to K on $[t_o,t_1]$ if there exists another $z_o\varepsilon K$ such that

$$y(t;x_o,u) \equiv y(t;z_o,u) \text{ for all } t\varepsilon[t_o,t_1].$$

<u>Definition 3</u>. Suppose K is open in R^n. For an input u, $x_o\varepsilon K$ is said to be <u>locally observable</u> on $[t_o,t_1]$ if there exists a neighborhood $N(x_o)$ such that x_o is observable with respect to $N(x_o)$ on $[t_o,t_1]$.

<u>Definition 4</u>. $x_o\varepsilon K$ is <u>locally unobservable</u> on $[t_o,t_1]$ if x_o is unobservable with respect to every $N(x_o)$ on $[t_o,t_1]$.

3. PRIOR RESULTS

Some necessary conditions and sufficient conditions for observability have been reported in (Griffith 1971). This involves constructing a differential question satisfied by the observation function, showing that the solutions of this differential equation have the uniqueness property and finally establishing that two observation functions have the same initial state. In the work of (Schoenwandt 1970), a power series expansion of the observation was used. A one-to-one map between the set of initial states and the corresponding set of coefficients in the series expansion was shown to be a necessary and sufficient condition for observability. The work of (Fitts 1970) gave sufficient conditions for observability of nonlinear discrete systems. Other works of relevance are (Roitenberg 1970, de Figueiredo 1970, Seinfeld 1972, Dessau 1972).

In using any of the above approaches to input design, a convenient way to start is to assume a polynomial whose coefficients can be adjusted until observability is established. In so doing, one must not ignore the stability of the system. This is a difficult task since nonlinear systems with forcing functions are involved. An attempt to guarantee stability under observability is discussed in (Kostyukovskii 1970). Much remains to be done in exploiting this idea.

4. EXAMPLES

The following examples demonstrate how observability property can be influenced by a proper choice of the input test signal.

Example 1. $\dot{x} = x^2 + u \qquad x(0) = x_o < 0$

$$y = \sin x$$

Consider $u(t) \equiv 0$. The solution is

$$\phi(t;x_o) = x_o/(1-x_o t); \quad x_o < 0$$

Thus, $y(t;5\pi/2) = y(t;x_o)$ for $t > 0$ only if there is a point $x_o \neq -5\pi/2$ for which

$$x_o/(1-x_o t) = [\tfrac{-5\pi}{2}/(1 + 5\pi/2t)] + 2n\pi$$

for all $t \geq 0$ for some n, $n = 0,1,\ldots$

No such point exists and hence $x_o = -5\pi/2$ is observable with respect to R' on [0,T] for any $T > 0$ and $u \equiv 0$.

Now consider $u(t) = -\pi^2(e^{-6\pi} + 9/4)$ and $x_o=-\pi/2$. Thus

$$\phi(t;x_o) = (\tfrac{3\pi}{2} + x_o)e^{-3\pi t} - \tfrac{3\pi}{2}; \quad x_o < 0.$$

It is verified that $y(t;-5\pi/2) - y(t;-\pi/2) = 0$ for all t. Thus, $x_o = -5\pi/2$ is unobservable with respect to R' on $[0,\infty)$. A suitable change in u(t) has made an observable state unabservable.

Example 2. $x + 3\dot{x} + 2x + ae^{-bt}x^3 = u(t)$

$$y = x$$

where a and b are constants. Let $x_1 = x$, $x_2 = \dot{x}_1$, $x_3 = ae^{-bt}$ and $x_4 = b$. Then, the above system can be written as

$$\dot{x}_1 = x_2$$

$$\dot{x}_2 = -2x_1 - 3x_2 - x_1^3 x_3 + u(t)$$

$$\dot{x}_3 = -x_3 x_4$$

$$\dot{x}_4 = 0$$

$$y = x_1$$

Note that $x_o = (x_{10}, x_{20}, a, b)$.
Let $u(t) \equiv 0$. Then, the necessary condition of (Griffith 1971) can be used to show that states of the form $(0,0,x_{30} \neq 0,\ x_{40} \neq 0)$ are unobservable with respect to R^4 on $[0,\infty)$. Letting u(t) be an analytic function on $[0,\infty)$, but not identically zero, the previously cited paper allows us to conclude that the unobservable states now become observable with respect to R^4 on $[0,\infty)$.

Example 3. $\dot{x}_1 = x_2 + u$

$$\dot{x}_2 = x_2 + x_1^2$$

$$y_1 = x_2 + x_1^2$$

$$y_2 = x_2^2$$

In the paper (Dessau 1972), the author shows that a choice of u=0 renders the system unobservable whereas a choice of u = -1 renders it observable. The theorems of (Griffith 1971) corroborate this conclusion.

5. CONCLUSIONS

The observability property of a nonlinear system depends on the input signal in contrast with that of a linear system. By a proper choice of the input, unobservable states can be made observable. A systematic method for carrying out this program consists in assuming an input polynomial with undetermined coefficients and adjusting these coefficients until observability can be established. Such a program necessarily has to be worked out for specific systems. Currently research is going on towards this aim where the systems being considered are those that exhibit mild nonlinearities as well as strong nonlinearities.

6. REFERENCES

Chen, C.T. (1970), Introduction to Linear System Theory, Holt, Rinehart and Winston, Inc., New York.
de Figueiredo, R.J.P., L. W. Dyer and A.N. Netrawali (1970), Estimation of States and Parameters of Nonlinear Dynamical Systems by Stochastic Approximation Methods, Proc. Symp. on Nonlinear Estimation Theory and Its Applications, 196.
Dessau, H.R. (1972), Dynamic Linearization and Ω-observability of Nonlinear Systems, J. Math. Anal. and Applications 40, 409.
Fitts, J.M. (1970), On the Observability of Nonlinear Systems with Applications to Nonlinear Regression Analysis, Proc. Symp. on Nonlinear Estimation Theory and Its Applications, 128.
Jazwinski, A.H. (1970), Stochastic Processes and Filtering Theory, Academic Press, New York.
Kostyukovskii, Y.M.L. (1970), An Algorithm of Sequential Estimation of the State of a Nonlinear System and of Lyapunov's Functions, Automation and Remote Control 6, 867.
Griffith, E.W., and K.S.P. Kumar (1971), On the Observability of Nonlinear Systems, J. Math. Anal. and Applications 35, 135.
Roitenberg, Y.Y. (1970), Observability of

Nonlinear Systems, J. SIAM Control 8, 338.
Seinfeld, J.H. and M. Hwang (1972), Observability of Nonlinear Systems, J. Optimization Theory and Applications 10, 67.
Schoenwandt, U. (1970), On Observability of Nonlinear Systems, 2nd IFAC Symposium on Identification and Process Parameter Estimation.

TREATMENT OF TIME VARYING STOCHASTIC BIAS IN RECURSIVE FILTERING

Mahendra M. Shah

Department of Electrical Engineering
University of Nairobi
P.O. Box 30197
Nairobi, Kenya

Abstract

The problem of estimating the state x(i) of a linear system in the presence of a time-varying stochastic bias vector b(i) is considered via the Supplemented Partitioning Approach (S.P.A.) to filtering. The motivation of the S.P.A. method is to use the partitioning philosophy to reduce the computational difficulties associated with computer storage, computing time and computational errors in the application of Kalman-Bucy filtering theory to state-augmented systems. In the S.P.A. method two effectively decoupled Kalman-Bucy filters are constructed for the estimation of the system state and the estimation of the bias vector respectively. The interaction between the system state vector and the bias vector is accounted for by supplementing the system, bias and measurement noise vectors as well as the respective noise covariance matrices. These supplementing terms are derived from linear transformation of the system, bias and measurement equations and from the concept of innovations noise sequences.

1. INTRODUCTION

Since the appearance of Kalman and Bucy's filtering results, [1] and [2], considerable progress has been made in the practical implementation aspects of this theory. This has resulted in extensive applications of Kalman-Bucy filtering theory to aerospace, internal navigation and industrial systems as well as economic systems.

Kalman-Bucy filtering theory assumes an accurate model of the system, measurement and noise parameters. In many practical systems only an approximate knowledge of these parameters is available and if care is not taken then the Kalman-Bucy filter based on these parameters may lead to large estimation errors. The usual approach to reduce these estimation errors is to use the method of state-augmentation or extended Kalman-Bucy filtering theory, Kalman [3], Aoki [4], to account for the system, measurement and noise uncertainties. These uncertainties are referred to as bias terms. However, this approach of state augmentation leads to an increase in the dimension of the overall system and as a result of this severe computational difficulties may arise. This occurs especially in systems where the dimension of the system state (n) plus the dimension of the bias vector (r), i.e. (n + r), is significantly larger than the system state vector (n). The main computational difficulties that arise are computing time requirement (of order $(n + r)^3$) and computing storage requirement (or order $(n + r)^2$). Furthermore the numerical accuracy of the estimates is a function of the number of computations in the solution of the extended Kalman-Bucy filter equations and hence in the case of large (n + r) significant numerical errors may arise.

The above mentioned computational difficulties are nonlinear functions of the extended system dimension and one obvious approach to overcome the computational difficulties would be to decouple the estimation of the system state vector from that of the estimation of the bias vector. This approach forms the basis of the method developed by Friedland [5]; however Friedland only considered the case of constant bias $\dot{b} = 0$ and suggested that results could be readily extended to the case of time-varying bias $\dot{b} = Cb$.

2. DISCRETE TIME FILTERING

Statement of the Problem

Given the system and bias equation,

$$x(i + 1) = A(i)x(i) + B(i)b(i) + u(i) \quad (1)$$

$$b(i + 1) = C(i)b(i) + v(i) \quad (2)$$

and the measurement equations,

$$y(i) = M(i)x(i) + N(i)b(i) + w(i) \quad (3)$$

Where i is discrete time instant

$x(i)$ n-th order system state vector
$A(i)$ n x n system transition matrix.
$b(i)$ r-th order bias state vector.
$B(i)$ n x r matrix
$u(i)$ n-th order system noise vector.
$C(i)$ r x r bias transition matrix.
$v(i)$ r-th order bias noise vector
$y(i)$ m x 1 measurement vector.
$M(i)$ mxn system measurement matrix.
$N(i)$ mxr bias measurement matrix.
$w(i)$ m-th order measurement noise vector.

$u(i)$, $v(i)$ and $w(i)$ are assumed to be uncorrelated zero mean white noise sequences of covariance,

$$E[u(i)\ u'(j)] = Q(i)\delta_{ij}$$

$$E[v(i)\ v'(j)] = R(i)\delta_{ij}$$

$$E[w(i)\ w'(j)] = S(i)\delta_{ij} \tag{4}$$

Let us define the new state.

$$z(i) = \begin{bmatrix} x(i) \\ b(i) \end{bmatrix} \tag{5}$$

In this case the system-bias dynamic equations and the measurement equations can be expressed as

$$z(i+1) = F(i)\ z(i) + G(i)\ \xi(i) \tag{6}$$

$$y(i) = L(i)\ z(i) + w(i) \tag{7}$$

where,

$$F(i) = \left[\begin{array}{c|c} A(i) & B(i) \\ \hline 0 & C(i) \end{array}\right] \begin{array}{l} \updownarrow n \\ \updownarrow r \end{array} \quad (\text{columns: } n,\ r)$$

$$G(i) = \left[\begin{array}{c} I \\ \hline I \end{array}\right] \begin{array}{l} \updownarrow n \\ \updownarrow r \end{array}$$

$$L(i) = \left[\begin{array}{c|c} M(i) & N(i) \end{array}\right] \updownarrow m \quad (\text{columns: } n,\ r)$$

$$\text{and } E[\xi(i)\ \xi'(j)] = \left[\begin{array}{c|c} Q(i) & 0 \\ \hline 0 & R(i) \end{array}\right] \begin{array}{l} \updownarrow n \\ \updownarrow r \end{array} = \bar{Q}(i) \tag{8}$$

The filtering problem is to find the optimal estimate $\hat{x}(i/i)$ in the presence of bias $b(i)$ and based on the measurements $y(i)$. Hence $\hat{x}(i/i)$, the a-posteriori estimate is defined as the state estimate based on measurements $[y(1), y(2), ---, y(i)]$ and similarly for the a-priori estimate $\hat{x}(i/i-1)$. The above method of combining the state and bias equations is known as the state-augmented method, [3], and in this case the application of extended Kalman-Bucy filter equations for the optimal estimate of $z(i)$, given the sequence of measurements $[y(1), y(2), \ldots.\ y(i)]$ results in the following filter equations,

$$\hat{z}(i+1/i) = F(i)\ \hat{z}(i/i) \tag{9}$$

$$\hat{z}(i/i) = \hat{z}(i/i-1) + K(i)\ [y(i) - L(i)\ \hat{z}(i/i-1)] \tag{10}$$

where $K(i)$, the filter gain matrix is evaluated from the following set of matrix recursive equations,

$$K(i) = P(i)\ L'(i)[L(i)\ P(i)\ L'(i) + S(i)]^{-1} \tag{11}$$

$$P(i+1) = F(i)\ P^*(i)\ F'(i) + \bar{Q}(i) \tag{12}$$

$$P^*(i) = [I - K(i)\ L(i)]\ P(i)\ [I - K(i)\ L(i)]' + K(i)\ S(i)\ K'(i) \tag{13}$$

Here, $P(i)$ = a-priori error covariance matrix

$$= E[\tilde{z}(i/i-1)\ \tilde{z}'(i-1)] \tag{14}$$

and $P^*(i)$ = a-posteriori error covariance matrix

$$= E[\tilde{z}(i/i)\ \tilde{z}'(i/i)] \tag{15}$$

Now if $P(0)$ and $z(0)$ are assumed to be known then the above set of extended Kalman-Bucy filter equations, (9) to (13), can be solved recursively and from these the system state and bias estimates $\hat{x}(i/i)$, $\hat{x}(i/i-1)$, $\hat{b}(i/i)$ and $\hat{b}(i/i-1)$ can be obtained.

Computational Aspects of the Extended Kalman-Bucy Filter : equations (9) to (13)

It is well known, Aoki [4], that the Kalman-Bucy filter for an n-th dimensional system with an m-th dimensional measurement vector requires the following number of multiplications for each iteration of the filter recursive equations:

$= 4n^3 + n^2 + nm(3n + 2m + 2)$ multiplications per iteration (16)

In a digital computer the multiplication operation takes the bulk of the computing time and hence the above equation gives an idea of the computing time in the solution of the Kalman-Bucy filter equation.

Therefore the computing time for the solution of the extended Kalman-Bucy filter equations, (9) to (13), will be:

$= 4(n + r)^3 + (n + r)^2 + (n + r)m$

$(3n + 3r + 2m + 2)$ mulitplications per iteration (17)

Consider a numerical example:

Example 1:

For a chemical absorption tower, Shah [6],

$n = 6 \quad m = 2 \quad r = 2$

The computation time for the solution of the extended Kalman-Bucy filter equation is of the order 2592 multiplications per iteration.
If there was no bias i.e. $r = 0$ then the computation time is of the order 1188 multiplications per iteration. However, if on the other hand the dimension of the bias vector is comparable to the dimension of the state vector, i.e. for example $r = 6$, then the computation time is of the order 8064 multiplications per iteration. Hence this example shows that in examples with a significant number of bias terms a serious computational difficulty may arise. In addition to the large computing time requirement, computer storage requirement and errors will also be significantly increased since these latter quantities are also nonlinear functions of the extended system dimension.

Since the computational requirements for the solution of the recursive filter equations vary as nonlinear functions of the extended system dimension, it may be possible to overcome the computational difficulties by partitioning the extended system into two subsystems I and II, namely, the dynamical system and the bias system. In the next section the S.P.A. method is applied to the problem of state estimation in the presence of time-varying stochastic bias. A Kalman-Bucy filter is constructed for the estimation of states of the dynamical system, subsystem I, and the bias is accounted for by obtaining an estimate of the bias from a second Kalman-Bucy filter for the bias system, subsystem II.

3. SUPPLEMENTED PARTIONING APPROACH (S.P.A.)

The system and measurement equation (1) to (3) are considered in the partitioned form as follows,

Subsystem I

Dynamical System and Measurement Equations:

$$x(i + 1) = A(i)\, x(i) + \bar{u}^*(i) \qquad (1)$$

$$\bar{y}_x(i) = M(i)\, x(i) + \bar{w}_x^*(i) \qquad (18)$$

Subsystem II

Bias System and Measurement Equations:

$$b(i + 1) = C(i)\, b(i) + v(i) \qquad (2)$$

$$\bar{y}_b(i) = N(i)\, b(i) + \bar{w}_b^*(i) \qquad (19)$$

where

$$\bar{u}^*(i) = B(i)\, b(i) + u(i) \qquad (20)$$

$$\bar{w}_x^*(i) = N(i)\, b(i) + w(i) \qquad (21)$$

$$\bar{w}_b^*(i) = M(i)\, x(i) + w(i) \qquad (22)$$

Considering the "total" noise disturbances $\bar{u}^*(i)$, $\bar{w}_x^*(i)$ and $\bar{w}_b^*(i)$

If these terms could be considered as white noise sequences then it would be straightforward to construct two independent Kalman-Bucy filters for Subsystem I and II. However examination of the terms in equation (20) to (22) clearly shows that terms such as $B(i)\, b(i)$, $N(i)\, b(i)$ and $M(i)\, x(i)$ are clearly not white noise sequences.

This difficulty may be overcome by reformulating the problem such that the parts of the "total" disturbances which are white noises are emphasised.

Problem Reformulation: The S.P.A. Method

The subsystem I and II equations (1), (18), and (2), (19) are modified by adding and substracting certain terms to the r.h.s. as follows:

Subsystem I

$$x(i+1) = A(i)\,x(i) + B(i)\,\hat{b}(i/i) + u^*(i) \qquad (23)$$

$$y_x(i) = M(i)\,x(i) + w_x^*(i) \qquad (24)$$

where

$$u^*(i) = u(i) + B(i)\,\tilde{b}(i/i) \qquad (25)$$

$$w_x^*(i) = w(i) + N(i)\,\tilde{b}(i/i-1) \qquad (26)$$

$$\tilde{b}(i/i) = b(i) - \hat{b}(i/i) \qquad (27)$$

and

$$\tilde{b}(i/i-1) = b(i) - \hat{b}(i/i-1) \qquad (28)$$

Here $y_x(i)$ is referred to as the observation process for the Subsystem I.

Subsystem II

$$b(i+1) = C(i)\,b(i) + v(i) \qquad (2)$$

$$y_b(i) = N(i)\,b(i) + w_b^*(i) \qquad (29)$$

where

$$w_b^*(i) = w(i) + M(i)\,\tilde{x}(i/i-1) \qquad (30)$$

$$\tilde{x}(i/i-1) = x(i) - \hat{x}(i/i-1) \qquad (31)$$

and $y_b(i)$ is the observation process for the bias system, Subsystem II.

Consider Subsystem I

The Kalman-Bucy filtering theory can be applied to Subsystem I - equation (23) and (24) if $u^*(i)$ and $w_x^*(i)$ can be treated as white noise sequences and if $\hat{b}(i/i)$ is assumed known and hence can be treated as a deterministic input to the filters. In Appendix A it is shown that $u^*(i)$, $w_x^*(i)$ and $w_b^*(i)$ can be approximated as white noise sequences of known covariance; discussion and derivation of this result is deferred to Appendix A and here we quote the final results:

Lemma I

(a) $u^*(i)$ is a white noise sequence of covariance,

$$E[u^*(i)\,u^{*\prime}(i)] = E[u(i)] + B(i)\,E[b(i/i)\,\tilde{b}'(i/i)]\,B'(i)$$

$$= Q(i) + B(i)\,P_{bb}^*(i)\,B'(i) \qquad (32)$$

(b) $w_x^*(i)$ is a white noise sequence of covariance,

$$E[w_x^*(i)\,w_x^{*\prime}(i)] = E[w(i)\,w'(i)] + N(i)\,E[\tilde{b}(i/i-i)\,\tilde{b}'(i/i-1]\,N'(i)$$

$$= S(i) + N(i)\,P_{bb}(i)\,N'(i) \qquad (33)$$

(c) $w_b^*(i)$ is a white noise sequence of covariance,

$$E[w_b^*(i)\,w_b^{*\prime}(i)] = E[w(i)\,w(i)] + M(i)\,E[\tilde{x}'(i/i-1)\,\tilde{x}(i/i-1)]\,M'(i)$$

$$= S(i) + M(i)\,P_{xx}(i)\,M'(i) \qquad (34)$$

Here $P_{bb}^*(i)$ and $P_{bb}(i)$ are the a-posteriori and the a-priori error covariance matrices respectively of the bias state $b(i)$ and $P_{xx}(i)$ is the a-priori error covariance matrix of the system state $x(i)$. Using the result of Lemma I, Kalman-Bucy filtering theory is applied to Subsystem I:

Kalman-Bucy Filter: Subsystem I

$$\hat{x}(i+1/i) = A(i)\,\hat{x}(i/i) + B(i)\,\hat{b}(i/i) \qquad (35)$$

$$\hat{x}(i/i) = \hat{x}(i/i - 1) + K_{xx}(i)[y_x(i) - M(i)\,\hat{x}(i/i - 1)] \quad (36)$$

The filter gain matrix $K_{xx}(i)$ is computed from,

$$K_{xx}(i) = P_{xx}(i)\,M'(i)[M(i)\,P_{xx}(i)\,M'(i) + S(i) + N(i)\,P_{bb}(i)\,N'(i)]^{-1} \quad (37)$$

$$P_{xx}(i + 1) = A(i)\,P^*_{xx}(i)\,A'(i) + Q(i) + B(i)\,P^*_{bb}(i)\,B'(i) \quad (38)$$

$$P^*_{xx}(i) = [I - K_{xx}(i)\,M(i)]\,P_{xx}(i)\,[I - K_{xx}(i)\,M(i)]' + K_{xx}(i)\,[S(i) + N(i)\,P_{bb}(i)\,N'(i)]\,K'_{xx}(i) \quad (39)$$

Given the initial dynamic system characteristics,,

$$x(t_o) = x(i_o)$$

$$P_{xx}(t_o) = P_{xx}(i_o)$$

where i_o is the time instant at $t = t_o$, The above set of matrix recursive equations (37) to (39) can be solved to obtain the subsystem I filter gain $K_{xx}(i)$ and hence the dynamical system state estimates,

$\hat{x}(i + 1/i)$ and $\hat{x}(i/i)$ from equations (35) (36), provided $\hat{b}(i/i)$ is known at each iteration. The bias estimate $\hat{b}(i/i)$ can be obtained from Subsystem II, which is solved "in parallel" with the solution of Subsystem I.

Note that for Subsystem I the term $B(i)\,\hat{b}(i/i)$ is treated as an external deterministic disturbance, Kalman [1].

Similary for Subsystem II:

Referring to the bias system and measurement equations (2) and (29) and using Lemma 1, the Kalman-Bucy filter equations for the bias system, subsystem II are:

$$\hat{b}(i + 1/i) = C(i)\,\hat{b}(i/i) \quad (40)$$

$$\hat{b}(i/i) = \hat{b}(i/i - 1) + K_{bb}(i)\,[y_b(i) - N(i)\,\hat{b}(i/i - 1)] \quad (41)$$

$$K_{bb}(i) = P_{bb}(i)\,N'(i)[N(i)\,P_{bb}(i)\,N'(i) + S(i) + M(i)\,P_{xx}(i)\,M'(i)]^{-1} \quad (42)$$

$$P_{bb}(i + 1) = C(i)\,P^*_{bb}(i)\,C'(i) + R(i) \quad (43)$$

$$P^*_{bb}(i) = [I - K_{bb}(i)\,N(i)]\,P_{bb}(i)\,[I - K_{bb}(i)\,N(i)]' + K_{bb}(i)\,[S(i) + M(i)\,P_{xx}(i)\,M'(i)]\,K'_{bb}(i) \quad (43)$$

Given the initial bias system characteristics,

$$b(t_o) = b(i_o)$$

and

$$P_{bb}(t_o) = P_{bb}(i_o)$$

where i_o is the time instant at $t = t_o$ the above set of equations (42) to (44) can be solved and hence the bias estimates $\hat{b}(i + 1/i)$ and $\hat{b}(i/i)$ can be obtained from equations (40) and (41). The a-priori state estimate $\hat{x}(i/i - 1)$ is available from the Subsystem I filter.

In the above S.P.A. method the estimation of the state $x(i)$ in the presence of a bias $b(i)$ has been obtained via the partitioning approach such that the estimation of $x(i)$ is essentially decoupled from that of $b(i)$. This can be seen from the steps involved in solving the filtering equations for Subsystem I and II:

Steps in the Supplemented Partitioning Approach

Given the initial system and bias chara-characteristics

$x(i_o)$, $b(i_o)$, $P_{xx}(i_o)$ and $P_{bb}(i_o)$

Calculation of the filtering gains for Subsystem I and Subsystem II

Subsystem I	Subsystem II
1. Solve equation (37) for $K_{xx}(i_o)$	1. Solve equation (42) for $K_{bb}(i_o)$
2. Solve equation (39) for $P^*_{xx}(i_o)$	2a. Solve equation (44) for $P^*_{bb}(i_o)$
3a. Solve equation (38) for $P_{xx}(i_o + 1)$	3a. Solve equation (43) for $P_{bb}(i_o + 1)$
3b. Form term $[M(i_o + 1)$ $P_{xx}(i_o + 1)$ $M'(i_o + 1)]$ and transfer to filtering algorithm for Subsystem II	3b. Form term $[N(i_o + 1)$ $P_{bb}(i_o + 1)$ $N'(i_o + 1)]$ and transfer to filtering algorithm for Subsystem I
4. Go to step 1, updating time $i_o \rightarrow i_o + 1$	4. Go to step 1, updating time $i_o \rightarrow i_o + 1$

The output of the Subsystem I, recursive equations (37) and (39), gives $K_{xx}(i)$ and that of Subsystem II, recursive equations (42) to (44), gives $K_{bb}(i)$ at time instant i and these are now used to produce the output of the two filters as follows:

5. Solve equation (36) for $\hat{x}(i/i)$	5a. Solve equation (41) for $\hat{b}(i/i)$
	5b. Form term $B(i)$ $\hat{b}(i/i)$ and transfer to filtering algorithm for Subsystem I
6a. Solve equation (35) for $\hat{x}(i + 1/i)$	6a. Solve equation (40) for $\hat{b}(i + 1/i)$
6b. Form term $M(i + 1)$ $\hat{x}(i + 1/i)$ and transfer to filtering algorithm Subsystem II.	6b. Form term $N(i + 1)$ $\hat{b}(i + 1/i)$ and transfer to filtering algorithm Subsystem I.
OUTPUT	OUTPUT
State estimate $\hat{x}(i/i)$ in the presence of bias $\hat{b}(i/i)$.	Bias estimate $\hat{b}(i/i)$ (if required)

The complete procedure for the application of the S.P.A. method for the above Subsystems I and II is depicted in Figures 1 and 2. Figure 1 shows the partitioning of the system and measurements and the relevant coupling terms, and Figure 2 shows the S.P.A. recursive solutions of the filter gain-matrices and the respective filters for Subsystem I and II together with the relevant supplementing terms.

Computational Aspects of the S.P.A. Filtering Method

The S.P.A. filters for the two Subsystems I and II, as developed in the above section, are basically two Kalman-Bucy filters with the interactions between the two Subsystems being accounted for by the supplementing terms in equations(35) to (39) and equations(40) to (44).

Consider the Filter for Subsystem I. Equations (35) to (39):

This is an n-th dimensional system with an m-th dimensional measurement vector. The recursive solution of the S.P.A. filter basically requires the following number of multiplications per iteration:

$$[4n^3 + n^2 + nm(3n + 2m + 2)]$$

multiplications per iteration (16)

In addition to this the evaluation of the supplementing terms in equations (35) to (39) requires:

System Coupling (Bias) Correction Terms

One matrix multiplication of (n x r) by (r x r)

One matrix multiplication of (n x r) by (r x n)
One matrix-vector multiplication of (n x r) by (r x 1)

Measurement Coupling (Bias) Correction Terms

One matrix multiplication of (m x r) by (r x r)
One matrix multiplication of (m x r) by (r x m)
One matrix-vector multiplication of (m x r) by (r x 1)

Assume that in multiplying a matrix (n x m) by a matrix (m x r), the number of multiplication operation is nmr. Hence the number of additional mulitplications from the above is,

$$= [nr^2 + n^2r + nr + mr^2 + m^2r + mr]$$

multiplications per iteration (45)

From equations (16) and (45), the total number of multiplications, i.e. order of the computing time, for the recursive solution of the S.P.A. filter for Subsystem I is,

$$= [4n^3 + n^2 + nm(3n + 2m + 2) + nr(r + n + 1) + mr(r + m + 1)]$$

multiplications per iteration (46)

In a similar manner for Subsystem II, the computing time for the recursive solution of the S.P.A. filter is of the order,

$$= [4r^3 + r^2 + rm(3r + 2m + 2) + mn(n + m + 1)]$$ multiplications per iteration (47)

Note that in equation (47) some terms are missing as compared to equation (46). This is because of the dynamics of the bias system, equation (2), in that the system state does not directly affect the bias dynamics.

Consider the numerical example as considered in the case of the extended Kalman Bucy filter.

Case 1

$n = 6, \quad m = 2, \quad r = 2.$

The two S.P.A. filters for subsystem I and II require 1508 mulitplications per iteration.

Case 2

$n = 6, \quad m = 2, \quad r = 0$

In this case there is only one S.P.A. filter and this is the Kalman-Bucy filter for Subsystem I.

Case 3

$n = 6, \quad m = 2, \quad r = 6.$

The two S.P.A. filters for Subsystem I and II require 3060 multiplications per iteration.

The above results show that the use of the S.P.A. method leads to a considerable reduction in the number of computations, i.e. computing time, as compared to the state-augmented or extended Kalman-Bucy filter method. The computer storage required will also be reduced in the S.P.A. method since all computer operations involve smaller dimensional matrices.

In the S.P.A. method it has been assumed that all the components of the measurement vector y(i) contains system state as well as bias terms. In some practical systems it may be that some measurements are bias-free and in this case it is not necessary to process the bias-free measurements through the S.P.A. filter for the bias system. This in turn will lead to a further reduction in the computer requirements of the S.P.A. method. Also note that if it is not desired to use inverse computer routines as required in equations (37) and (42), then this can be done by using the method of sequential processing of measurements, Shah [6].

A Numerical Error Aspect of the S.P.A. Method

In many practical systems the system and noise parameters vary slowly with time and hence the dynamics of the bias systems will be slow. A result of this is that the range of eigen-values of the state-augumented system will be large. It is well known, Vaughan [9], that in systems of this type, i.e. large range of eigen-values, the solution of the extended Kalman-Bucy filter recursive equations runs into computational error difficulties in that the essential information is coded in numbers of widely differing magnitude. Now since the computer calculations are done with words of finite length, the essential information coded in small numbers is lost and cannot be recovered in the process of computer additions of very large and very small numbers. In the

S.P.A. method the dynamic system (fast) and the bias system (slow) are considered separately and hence the above mentioned difficulty is avoided. This aspect has been observed in the application of the S.P.A. method to a nuclear power plant prototype, Atary and Shah [7], where the S.P.A. results are similar and in some cases "better" than the optimal Kalman-Bucy results.

4. CONCLUSIONS AND APPLICATIONS

In the application of the S.P.A. method to the problem of state estimation in the presence of time varying stochastic bias the overall system is considered as two interacting Subsystems I and II, namely, the dynamic system (state x) and the bias system (state b). By using the method of linear transformation the Subsystem I and II system and measurement noise sequences are reformulated as "total" disturbances. These "total" disturbances are composed of the actual system and measurement noise sequences plus correction terms to account for the interactions between Subsystems I and II. The concept of innovations processes, Kailath [10], is used to show that the "total" disturbances can be approximated as white noise sequences, the covariances of which are derived. In this derivation the approximation is that the correlation between the state estimate errors of Subsystem I and II are small. Due to this approximation the S.P.A. results would be expected to be "suboptimal" as compared to the optimal Kalman-Bucy filter results. However, the loss in performance due to this approximation is compensated by the increase in numerical accuracy. The S.P.A. method is particularly useful in systems with a large range of eigenvalues where the extended Kalman-Bucy filter runs into computational error difficulties.

The computational requirements, i.e. computer storage and computing time of the S.P.A. method are considerably less than the extended Kalman-Bucy filter method since the computations in the latter method involves large dimensional matrices whereas in the former method involves a sequence of computations with smaller dimensional matrices. There are many applications in inertial guidance, orbit determination problems, industrial process and economic models in which the system state vector as well as the corresponding bias vector is large dimensional. In such applications the S.P.A. method is particularly useful since computers in these applications are used for numerous other purposes besides as control system design aides and hence it is important to reduce the computational requirements.

In this paper the discrete time case has been treated; however the results can readily be extended to the continuous time case.

REFERENCES

[1] R.E. Kalman, "A new approach to linear filtering and prediction problems," Trans. ASME, J. Basic Engrg., Ser.D, Vol.82, pp. 35-45, March 1960.

[2] R.E. Kalman and R.S. Bucy, "New results in linear filtering and prediction theory," Trans. ASME, J. Basic Engrg., Ser. D., Vol.83, pp. 95-108, March 1961.

[3] R.E. Kalman, "New Methods in Wiener Filtering Theory," in Proc. 1st Symp on Engineering Applications of Random Function Theory and Probability, J.L. Bogdanoff and F. Kozin, Eds. New York, 1963, pp. 270-388.

[4] M. Aoki, "Optimization of Stochastic Systems," Mathematics in Science and Engineering, Vol. 32, Academic Press, New York, 1967.

[5] B. Friedland, "Treatment of Bias in Recursive Filtering," IEEE Trans. on Automatic Control, Vol. AC-14, No.4, pp.359-367, August 1969.

[6] M.M. Shah, "Supplemented Partioning Approach to Filtering," submitted to Trans. on Automatic Control for publication.

[7] J. Atary and M.M. Shah, ""Modeling and Analytical Control System design of a complete nuclear power plant prototype," Proceedings of the 5th World Congress of IFAC, Part I, Session 6-1, pp 1-13, Paris, June 1972.

[8] P.D. Joseph, "Sub-Optimal Linear Filtering," 10C 9321.4-653, TRW Space Technology Labs., Redando Beach, Calif., December 1963.

[9] D.R. Vaughan, "A negative exponential solution for the linear optimal regulator problem", J.A.C.C. pp. 717-725, 1968.

[10] T. Kailath, "An innovations Approach to Least-Squares Estimation Part I: Linear Filtering in Additive White Noise," IEEE Trans. on Automatic Control, Vol. AC-13, No.6, pp. 646-655, December 1968.

APPENDIX A

On the approximation of "Total" System and Measurement Disturbances as White Noise Sequences

"Total" system disturbances

$$u^*(i) = u(i) + B(i)\,\tilde{b}(i/i) \qquad (A1)$$

"Total" Measurement disturbances

$$w_x(i) = w(i) + N(i)\,\tilde{b}(i/i-1)$$

$$w_b(i) = w(i) + M(i)\,\tilde{x}(i/i-1) \qquad (A2)$$

On the right hand side of the above equations (A1) and (A2), the noise terms $u(i)$ and $w(i)$ are uncorrelated zero mean white noise sequences. The estimation error term $B(i)\,\tilde{b}(i/i)$, $N(i)\,\tilde{b}(i/i-1)$ and $M(i)\,\tilde{x}(i/i-1)$ are all gaussian since the estimation errors in the Kalman-Bucy filtering theory has a gaussian distribution, Kalman [1].

In the S.P.A. method the main point is whether anything can be said about the "Whiteness" of these terms. In order to investigate this aspect let us first consider the concept of innovations process.

Rewriting the reformulated system bias and measurement equations (23), (2), (24) and (29):

$$x(i+1) = A(i)\,x(i) + B(i)\,\hat{b}(i/i) + u(i) + B(i)\,\tilde{b}(i/i) \qquad (A3)$$

$$b(i+1) = C(i)\,b(i) + v(i) \qquad (A4)$$

$$y_x(i) = M(i)\,x(i) + w(i) + N(i)\,\tilde{b}(i/i-1) \qquad (A5)$$

$$y_b(i) = N(i)\,b(i) + w(i) + M(i)\,\tilde{x}(i/i-1) \qquad (A6)$$

Consider Subystem I:

The observation process is,

$$y_x(i) = M(i)\,x(i) + w(i) + N(i)\,\tilde{b}(i/i-1) \qquad (A5)$$

The new observation $\eta(i)$ brought into the system at time i is:

$$\begin{aligned}\eta(i) &= y_x(i) - M(i)\,\hat{x}(i/i-1)\\ &= M(i)\,\tilde{x}(i/i-1) + N(i)\,\tilde{b}(i/i-1) + w(i)\\ &= [M(i)\ N(i)]\begin{bmatrix}\tilde{x}(i/i-1)\\ \tilde{b}(i/i-1)\end{bmatrix} + w(i)\\ &= L(i)\,\tilde{z}(i/i-1) + w(i) \qquad (A7)\end{aligned}$$

It is well known, Kailath [10] and Shah [6], that the innovations sequence $\eta(i)$ is a white noise sequence of covariance,

$$\begin{aligned}E[\eta(i)\,\eta'(j)] &= \{S(i) + L(i)\,E[\tilde{z}(i/i-1)\,\tilde{z}'(i/i-1)]\,L'(i)\}\delta_{ij}\\ &= [S(i) + L(i)\,P_{zz}(i)\,L'(i)]\delta_{ij} \qquad (A8)\end{aligned}$$

Let us consider the "partial" innovations sequence $y_x(i)$:

$$y_x(i) = M(i)\,x(i) + w(i) + N(i)\,\tilde{b}(i/i-1) \qquad (A5)$$

For Subsystem I, the "total" measurement disturbance is

$$w_x^*(i) = w(i) + N(i)\,\tilde{b}(i/i-1) \qquad (A9)$$

Therefore,

$$E[w_x^*(i)\,w_x^{*\prime}(j)] = E[w(i)\,w'(j)] + E[w(i)\,\tilde{b}'(j/j-1)]\,N'(j) + N(i)\,E[\tilde{b}(i/i-1)\,w'(j)]$$

$$+ N(i)\, E[\tilde{b}(i/i-1)\, \tilde{b}'(j/j-1)]\, N'(j) \qquad (A10)$$

Case i>j

The second term on the right hand side of the above expression is zero since future noise is independent of past signal. Combining the third and fourth terms,

$$= E\left\{N(i)\, \tilde{b}(i/i-1)[w(j) + N(j)\, \tilde{b}(j/j-1)]'\right\}$$

$$= E\left\{N(i)\, \tilde{b}(i/i-1)[w(j) + N(j)\, \tilde{b}(j/j-1) + M(j)\, \tilde{x}(j/j-1)]' - N(i)[\tilde{b}(i/i-1)\, x'(j/j-1)]\, M'(j)\right\}$$

$$= N(i)\, E[\tilde{b}(i/i-1)\, \eta'(j)] - N(i)\, E[\tilde{b}(i/i-1)\, \tilde{x}'(j/j-1)]\, M'(i) \qquad (A11)$$

By the orthogonal projections theorem, Kalman [1], the first term in the above expression is zero and if the error correlation between the dynamic system and the bias system is weak then the second term is also approximately zero.

Hence equation (A10) becomes,

$$E[w_x^*(i)\, w_x^{*\prime}(j)] = E[w(i)\, w'(j)] = 0 \qquad (A12)$$

Similarly we can show this to be true for the case $i < j$.

Case i = j

$$E[w_x^*(i)\, w_x^{*\prime}(i)] = N(i)\, E[\tilde{b}(i/i-1)\, \tilde{b}'(i/i-1)]\, N'(i) + E[w(i)\, w'(i)] + N(i)\, E[\tilde{b}(i/i-1)\, w'(i)] + E[w(i)\, \tilde{b}'(i/i-1)]\, N'(i) \qquad (A13)$$

Now at time i, the measurement noise $w(i)$ is independent of $\tilde{b}(i/i-1)$ since the estimate $\hat{b}(i/i-1)$ is based on measurements up to time $(i-1)$.

Hence equation (A13) becomes

$$E[w_x^*(i)\, w_x^{*\prime}(i)] = E[w(i)\, w'(i)] + N(i)\, E[\tilde{b}(i/i-1)\, \tilde{b}'(i/i-1)]\, N'(i) = S(i) + N(i)\, P_{bb}(i)\, N'(i) \qquad (A14)$$

Similarly for Subsystem II it can be shown that the "partial" innovations process,

$$w_b^*(i) = w(i) + M(i)\, \tilde{x}(i/i-1) \qquad (A15)$$

has a covariance of,

$$E[w_b^*(i)\, w_b^{*\prime}(j)] = [S(i) + M(i)\, P_{xx}(i)\, M'(i)]\, \delta_{ij} \qquad (A16)$$

Now let us consider the "total" system disturbance,

$$u^*(i) = u(i) + B(i)\, \tilde{b}(i/i)$$

$$= u(i) + B(i)[b(i) - \hat{b}(i/i)] \qquad (A1)$$

Substituting from equation (41),

$$u^*(i) = u(i) + B(i)\left\{b(i) - \hat{b}(i/i-1) - K_{bb}(i)[M(i)\, \tilde{x}(i/i-1) + N(i)\, \tilde{b}(i/i-1) + w(i)]\right\}$$

$$= u(i) + B(i)\left\{[I - K_{bb}(i)\, N(i)]\, \tilde{b}(i/i-1) - K_{bb}(i)\, w(i) - K_{bb}(i)\, M(i)\, \tilde{x}(i/i-1)\right\} \qquad (A17)$$

Using arguments similar to those for $w_x^*(i)$, it can be shown that for cases i > j and i < j,

$$E[u^*(i)\, u^{*\prime}(j)] = E[u(i)\, u'(j)] + B(i)\, K_{bb}(i)\, E[w(i) w'(j)]\, K'_{bb}(i)\, B'(j) + \text{Terms in } \{\tilde{b}(i/i-1),\ \tilde{x}(i/i-1) \text{ and } \tilde{b}(j/j-1),\ \tilde{x}(j/j-1)\} = 0 \quad (A18)$$

and for case i = j,

$$E[u^*(i)\, u^{*\prime}(i)] = Q(i) + B(i)\, P^*_{bb}(i)\, B'(i) + E[\overline{\tilde{x}(i/i-1), \tilde{b}(i/i-1)}] \quad (A19)$$

where $\overline{\tilde{x}(i/i-1), \tilde{b}(i/i-1)}$ represents terms involving products of $\tilde{x}(i/i-1)$ and $\tilde{b}(i/i-1)$ and

$$P^*_{bb}(i) = [I - K_{bb}(i)\, N(i)]\, E[\tilde{b}(i/i-1)\, \tilde{b}'(i/i-1)][I - K_{bb}(i)\, N(i)]' + K_{bb}(i) \{ E[w(i)\, w'(i)] + M(i)\, E[\tilde{x}(i/i-1)\, \tilde{x}'(i/i-1)]\, M'(i) \} K'_{bb}(i)$$

$$= [I - K_{bb}(i)\, N(i)]\, P_{bb}(i)\, [I - K_{bb}(i)\, N(i)]' + K_{bb}(i)\, [S(i) + M(i)\, P_{xx}(i)\, M'(i)]\, K'_{bb}(i) \quad (A20)$$

Therefore as an approximation if the correlation between $\tilde{x}(i/i-1)$ and $\tilde{b}(i/i-1)$ is ignored, then

$$E[u^*(i)\, u^{*\prime}(i)] = Q(i) + B(i)\, P^*_{bb}(i)\, B'(i) \quad (A21)$$

Hence $u^*(i)$, $w^*_x(i)$ and $w^*_b(i)$ can be treated as approximate white noise sequences of known covariance. The crucial step in the formulation of the S.P.A. is this approximation of "total" disturbances as white noise sequences. Numerical application studies, Shah [6] and Atary and Shah [7], has shown that inspite of this approximation the S.P.A. results are "near" optimal since the expected performance loss due to the approximation is compensated by the increased accuracy of the computations.

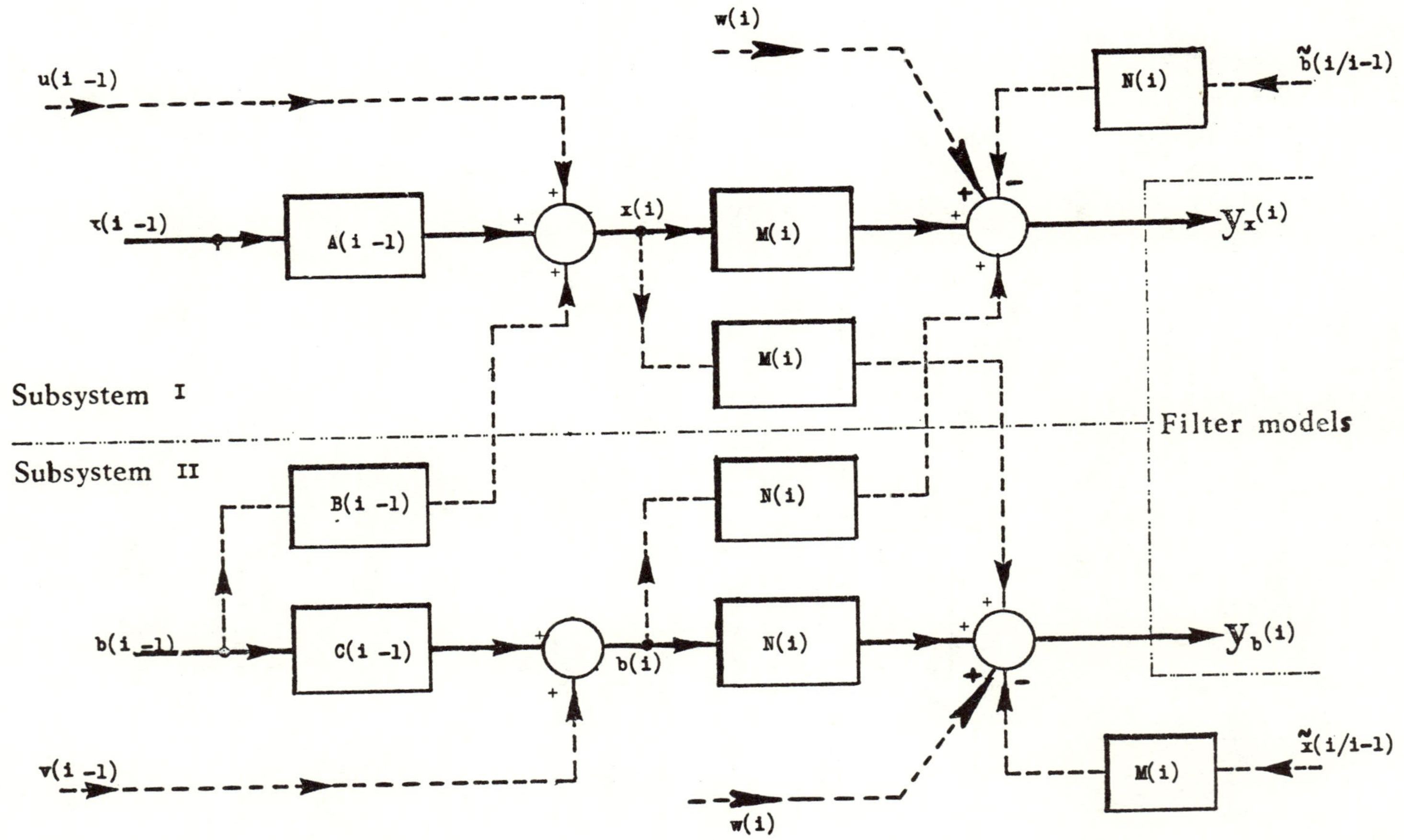

Fig. 1.- Partitioned system model for states estimation by the SPA method
(In the presence of Time-Varying Stochastic Bias)

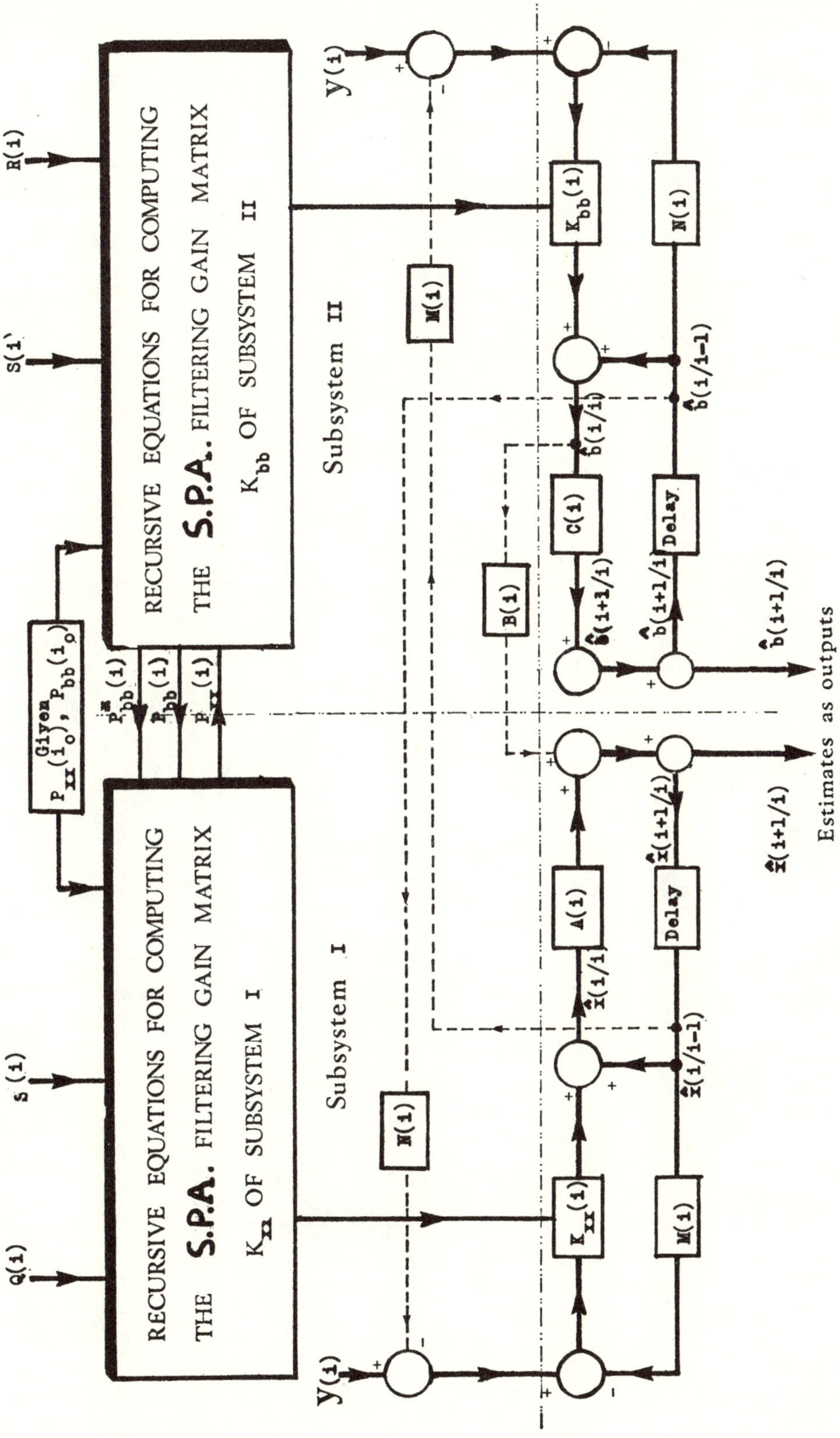

Fig. 2.- Supplemented Partitioning Approach to Filtering (SPA method)
(In the presence of Time-Varying Stochastic Bias)

NONLINEAR FILTERING ERROR-ANALYSIS ALGORITHM

Antonín Vaněček

INORGA Institute for Automation
Prague 11806, Czechoslovakia

Error-analysis algorithm of the continuous first-order conditional-mean nonlinear filter (extended Kalman filter) is given. Simulation results for a simple case are presented.

INTRODUCTION

Nonlinear filtering (on-line nonlinear identification) can solve wide spectrum of problems, e.g., state estimation, on-line parameter identification, and sequential processing of measured data for parameter identification. Nonlinear filtering nevertheless introduces the problems of its own, being virtually infinite dimensional. To reduce the dimensionality of the problem, many approximate filters have been derived and/or used. The approximation can be invalidated because of its errors. We shall be concerned with error-analysis of one probably most frequently used nonlinear filter.

ANALYSIS

We shall begin with the Itô-type models of the plant (table 2, equations (A,C)) and of the measurement, (B,C). In the plant model the increment of the state vector dx(t) is dependent on the current state x(t), on time t, and on additive disturbance $\omega(t)$. Input disturbance $\omega(t)$ is a vector Wiener -Lévy process and the integral of an input white noise process w(t). Similarly for the measurement model with additive measurement noise $\nu(t)$. Stochastic vector differential equations (A,B) are to be interpreted via the Itô integral, see e.g. GICHMAN (1968), SAGE (1971). For our purpose we shall introduce the problem of nonlinear filtering as a problem of continuous estimation of the first two moments of the state, the history of the measured output $Z(t)=(z(\tau)| 0\leqslant \tau<t)$ being given. The exact equations for the first two moments are, KUSHNER (1967), SAGE (1971):

$$d\hat{x}(t)=\hat{f}(x(t),t)dt+E((x(t)-\hat{x}(t))h^T(x(t),t)| Z(t))\Psi_v^{-1}(t)(z(t)-\hat{h}(x(t),t))dt, \qquad (1)$$

$$dV_{\tilde{x}}(t)+d\hat{x}(t)d\hat{x}^T(t)=E(f(x(t),t)(x(t)-x(t))^T|Z(t)d\tau+E((x(t)-x(t))f^T(x(t),t)|Z(t))dt+E(G(x(t),t)\Psi_w(t)G^T(x(t),t)|Z(t))dt+E((x(t)-\hat{x}(t))(x(t)-\hat{x}(t))^T(h(x(t),t)-\hat{h}(x(t),t))^T\Psi_v^{-1}(x)(z(t)-h(x(t),t)|Z(t))dt. \qquad (2)$$

Nevertheless the equations (1,2) together with the initial conditions do not determine the evolution of the both moments. E.g., for computation of $f(\hat{x}(t),t)$ at time dt, the knowledge of $\hat{x}(0), V_{\tilde{x}}(0)$ does not suffice because $f(\cdot,\cdot)$ is a nonlinear function. For computing the first two moments the knowledge is necessary of all the higher moments, so the problem is infinite dimensional. Known solutions of (1,2) are based on approximations, on either the stochastic linearization of $f(\cdot,\cdot)$, $G(\cdot,\cdot)$, $h(\cdot,\cdot)$ or the Taylor expansion of $f(\cdot,\cdot)$, $G(\cdot,\cdot)$, $h(\cdot,\cdot)$ in the vicinity of estimated state. We shall use the second approach, and sketch the derivation of the most simple approximation as a byproduct, but especially that of the errors of the mentioned approximation. As a consequence of that the error-variance $V_{\tilde{x}}(\cdot)$ is a matrix function, we shall use the tensor calculus. The necessary relations are summed in table 1.

$A=\mathcal{B}c$	$a_{ij}=\sum_s b_{ijs}c_s$
$A=b\mathcal{C}$	$a_{ij}=\sum_s b_s c_{sij}$
$\mathcal{A}=\partial B/\partial c$	$a_{ijk}=\partial b_{ij}/\partial c_k$
$\mathcal{A}=\mathcal{B}\odot C$	$a_{ijk}=\sum_s b_{ijs}c_{sk}$
$\mathcal{A}=B\odot\mathcal{C}$	$a_{ijk}=\sum_s b_{is}c_{sjk}$

Table 1. Tensor symbolism

Now we can define the sensitivities of the estimated state and error-variance, $S_x(t)$, $\mathcal{S}_V(t)$, see table 2, equations (K,M). The derivations are taken in q.m. For the problems of existence see GICHMAN (1968), SUNAHARA (1971). In the following, we shall use the notorious Lagrange theorem for vector function f(•) of vector argument $x=[x_1 x_2 \dots]^T$:

$$f(x)=f(x_0)+\begin{bmatrix} \left.\dfrac{\partial f^T(x)}{\partial x_1}\right|_{x={}^1\eta} \\ \left.\dfrac{\partial f^T(x)}{\partial x_2}\right|_{x={}^2\eta} \\ \text{-----------} \end{bmatrix}(x-x_0) \quad (3)$$

$\|x-{}^1\eta\| < \|x-x_0\|$, $\|x-{}^2\eta\|<\|x-x_0\|$, ..., and x belongs to the vicinity of x_0, where the f(•) is differentiable. We shall denote [•] as $\partial f(\eta(t),t)/\partial\eta(t)$.

Now using the lagrangean approach and technical assumptions

$$E(x(t)-\hat{x}^*(t)\,|\,Z(t))=0 \quad (4)$$

$$E((x(t)-\hat{x}^*(t))(x(t)-\hat{x}^*(t))^T(x(t)-\hat{x}^*(t)) \frac{\partial h^T(\eta(t),t)}{\partial\eta(t)}\Psi_v^{-1}(t)(z(t)-h(\hat{x}^*(t),t)\,| \; Z(t))\, dt=0 \quad (5)$$

we obtain the equations (O,P,Q,R) of the table 2. (The $\eta(t)$'s should be generally different; in following, we shall see reasons for which our simplified notation is possible.) Our equations (O,P,Q,R) under the assumptions (4,5) are exact solutions of a problem of the estimation for the first two conditional moments, nevertheless, we do not know the $\eta(t)$'s, that is, we do not know at what points we have to continuously linearize. To solve this remaining problem, we introduce the nominal conditions $\eta(t)=\hat{x}(t)$, $\hat{x}^*(t)=\hat{x}(t)$, $V^*_{\tilde{x}}(t)=V_{\tilde{x}}(t)$ and obtain the notorious first-order conditional-mean nonlinear filter, equations (D,E,F,G) of the table 2. Using our sensitivities defined in (K,M), we obtain our error-analysis algorithm (H,I,J,L,N).

Table 2. Analysed algorithm

First-order conditional-mean nonlinear filter (extended Kalman filter) error-analysis algorithm

Plant model	$dx(t)=f(x(t),t)dt+G(x(t),t)\,d\omega(t)$, $x(0)=\alpha$, $d\omega(t)=w(t)dt$	(A)
Measurement model	$dy(t)=h(x(t),t)\,dt+d\nu(t)$, $dy(t)=z(t)dt$, $d\nu(t)=v(t)dt$	(B)
Prior statistics	$\mu_w(t)=0$, $V_\omega(t,\tau)=\Psi_w(t)\min(t,\tau)$; $\mu_v(t)=0$, $V_\nu(t,\tau)=\Psi_v(t)\min(t,\tau)$; $V_{\omega\nu}(t,\tau)=0$, $V_{x\omega}(0,t)=0$, $V_{x\nu}(t,\tau)=0$	(C)
Innovation	$I(t)=z(t)-h(\hat{x}(t),t)$	(D)
Gain	$L(t)=V_{\tilde{x}}(t)\dfrac{\partial h^T(\hat{x}(t),t)}{\partial\hat{x}(t)}\Psi_v^{-1}(t)$	(E)
Filter algorithm	$d\hat{x}(t)=f(\hat{x}(t),t)dt+L(t)I(t)dt$, $\hat{x}(0)=\hat{\alpha}$	(F)
Error-variance algorithm	$dV_{\tilde{x}}(t)=\Big[\dfrac{\partial f(\hat{x}(t),t)}{\partial\hat{x}(t)}V_{\tilde{x}}(t)+V_{\tilde{x}}(t)\dfrac{\partial f^T(\hat{x}(t),t)}{\partial\hat{x}(t)}-L(t)\dfrac{\partial h(\hat{x}(t),t)}{\partial\hat{x}(t)}V_{\tilde{x}}(t)+G(\hat{x}(t),t)\Psi_w(t)G^T(x(t),t)\Big]dt$, $V_{\tilde{x}}(0)=\beta$	(G)
Pertubation of estimated state	$\hat{x}^*(t)-\hat{x}(t)=S_x(t)\delta\eta(t)+o_1(\delta\eta(t))$	(H)
Pertubation of error-variance	$V^*_{\tilde{x}}(t)-V_{\tilde{x}}(t)=\mathcal{S}_v(t)\delta\eta(t)+o_2(\delta\eta(t))$	(I)
Pertubation argument estimation	$\|\delta\eta(t)\|=k\,\|V_{\tilde{x}}^{1/2}(t)\|$	(J)
	$S_x(t)\triangleq\left.\dfrac{\partial\hat{x}^*(t)}{\partial\eta(t)}\right|_{nom}$	(K)
Estimated state sensitivity	$dS_x(t)=\Big[\dfrac{\partial f(\hat{x}(t),t)}{\partial\hat{x}(t)}-L(t)\dfrac{\partial h(\hat{x}(t),t)}{\partial\hat{x}(t)}\Big]\cdot S_x(t)dt+\dfrac{\partial h^T(\hat{x}(t),t)}{\partial\hat{x}(t)}\Psi_v^{-1}(t)I(t)\cdot\mathcal{S}_v(t)dt+\Big[V_{\tilde{x}}(t)\odot\dfrac{\partial^2 h(\hat{x}(t),t)}{\partial x^2(t)}\Big]\Psi_v^{-1}(t)I(t)\,dt$, $S_x(0)=0$	(L)
	$\mathcal{S}_v(t)\triangleq\left.\dfrac{\partial V^*_{\tilde{x}}(t)}{\partial\eta(t)}\right|_{nom}$	(M)
Error-variance sensitivity	$d\mathcal{S}_v(t)=\Big[\dfrac{\partial G(\hat{x}(t),t)}{\partial\hat{x}(t)}\odot\Psi_w(t)G^T(\hat{x}(t),t)+G(\hat{x}(t),t)\Psi_w(t)\odot\dfrac{\partial G^T(\hat{x}(t),t)}{\partial\hat{x}(t)}\Big]\odot S_x(t)dt+\Big[\dfrac{\partial f(\hat{x}(t),t)}{\partial\hat{x}(t)}+\dfrac{\partial f^T(\hat{x}(t),t)}{\partial\hat{x}(t)}-L(t)\dfrac{\partial h(\hat{x}(t),t)}{\partial\hat{x}(t)}-\dfrac{\partial h^T(\hat{x}(t),t)}{\partial\hat{x}(t)}L(t)\Big]\odot\mathcal{S}_v(t)dt+\Big[V_{\tilde{x}}(t)\odot\dfrac{\partial^2 f^T(\hat{x}(t),t)}{\partial x^2(t)}+\dfrac{\partial^2 f(\hat{x}(t),t)}{\partial x^2(t)}\odot V_{\tilde{x}}(t)-L(t)\odot\dfrac{\partial^2 h(\hat{x}(t),t)}{\partial x^2(t)}\odot V_{\tilde{x}}(t)-V_{\tilde{x}}(t)\odot\dfrac{\partial^2 h^T(\hat{x}(t),t)}{\partial x^2(t)}\odot L^T(t)\Big]dt$, $\mathcal{S}_v(0)=0$	(N)
Langrangean innovation	$I^*(t)=z(t)-h(\hat{x}^*(t),t)$	(O)
Lagrangean gain	$L^*(t)=V^*_{\tilde{x}}(t)\dfrac{\partial h^T(\eta(t),t)}{\partial\eta(t)}\Psi_v^{-1}(t)$	(P)
Lagrangean filter algorithm	$d\hat{x}^*(t)=f(\hat{x}^*(t),t)dt+L^*(t)I^*(t)dt$, $\hat{x}^*(0)=\hat{\alpha}$	(Q)
Lagrangean error-variance algorithm	$dV^*_{\tilde{x}}(t)=\Big[\dfrac{\partial f(\eta(t),t)}{\partial\eta(t)}V^*_{\tilde{x}}(t)+V^*_{\tilde{x}}(t)\dfrac{\partial f^T(\eta(t),t)}{\partial\eta(t)}-L^*(t)\dfrac{\partial h(\eta(t),t)}{\partial\eta(t)}V^*_{\tilde{x}}(t)+G(\eta(t),t)\Psi_w(t)G^T(\eta(t),t)\Big]dt$, $V^*_{\tilde{x}}(0)=\beta$	(R)

Let us notice that the first order conditional-mean nonlinear filter (D,E,F,G) is identical with the extended Kalman filter which can be obtained from the classical Kalman filter, KALMAN (1961), using quasilinearization approach. Let us mention some examples of the use of the extended Kalman filter. JAZWINSKI (1970, p.277) refers that "the extended Kalman filter is commonly used in orbit determination". WISMER (1972) refers on application of the extended Kalman filter to continuous carbon estimation as a key to optimum basic oxygen furnace control and states that this extension was successfully used after unsuccessful attempts with linear estimation. CHOQUETTE (1970) refers on the use of the extended Kalman filter as a part of a multivariable direct digital control of chemical reactor and states that the theoretical basis for the extension is missing. The situation which motivated our analysis precises further MORTENSEN (1972): "It is beginning to become clear that theoretical nonlinear filtering and extended Kalman filtering are really emerging as two separate subject areas, with two separate schools of workers, and two separate bodies of results and folklore".

Interchanging (3) by the Taylor expansion with the Lagrange remainder, we obtain the analysis of higher-order conditional-mean nonlinear filters. Our analysis being rather formal ignored the observability problem: we will touch this problem in the next part. The sensitivity equations (L,N) increase significantly the dimensionality of the problem: it is supposed that (L,N) will be used just for investigation prior the proper nonlinear filtering. The investigation of stochastic diferential equations is possible -- at least in principle -- purely analytically. GICHMAN (1968) refers of analytical investigation of the moments of the solution, of the stability and boundeness of the solutions. We take as being feasible for us the investigation via simulation.

SIMULATION

We shall consider the simplest scalar model still demonstrating the behaviour of sensitivities. The plant model

$$dx(t) = -\,x(t)dt + d\omega(t) \qquad (6)$$

and the measurement model

$$dy(t) = x(t)dt + \varepsilon x^2(t)dt + d\nu(t) \qquad (7)$$

This particular use of the plant with linear dynamics is not really restrictive, VANĚČEK (1972). Let us note that output transformation $x(t) \mapsto x(t) + \varepsilon x^2(t)$ $(\varepsilon > 0)$ is injective ~~for $x(t) > -1/2\varepsilon$, but not for smaller $x(t)$.~~

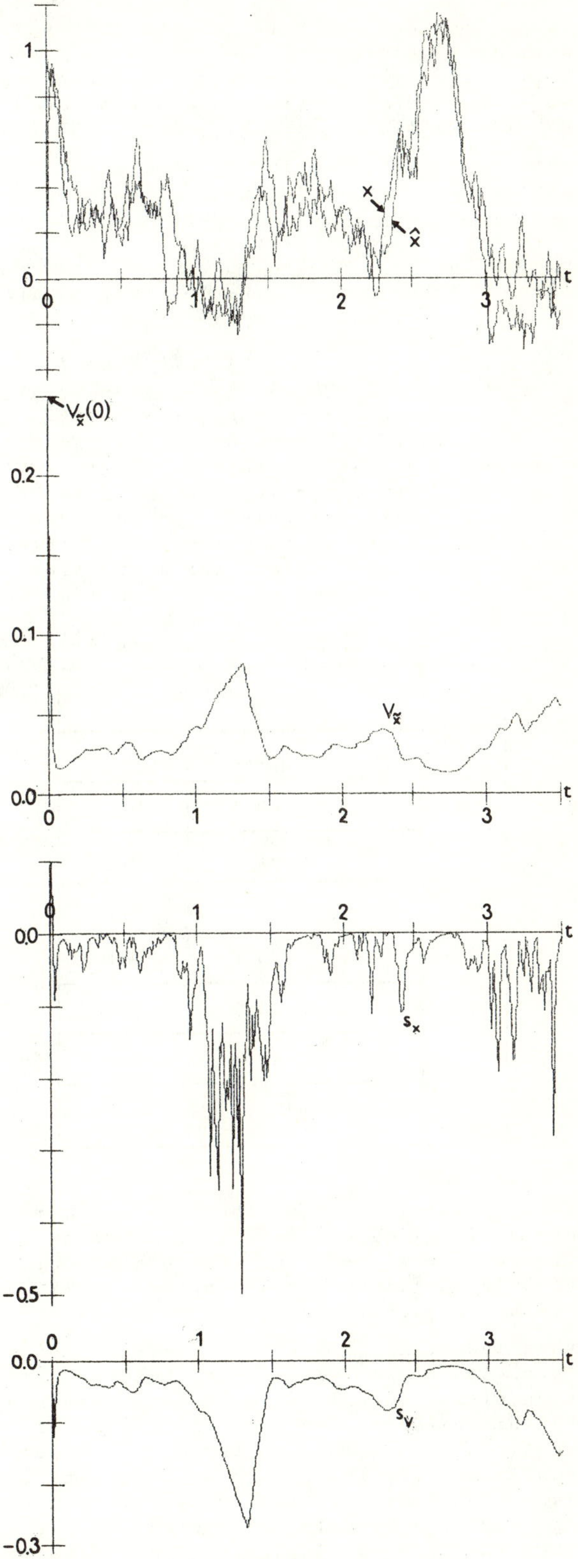

Figure 1. A sample of simulation

In simulations we shall be concerned only with $(x(t),\varepsilon)$ obeying the mentioned observability condition. To transform the continuous white noises w(t), v(t) to discrete ones we use the relations $\Psi_w = T\delta_w^2, \Psi_v = T\delta_v^2$, SAGE (1971), where T is the sampling period. For digital simulation, the white gaussian noise generator and the 4th order Runge-Kutta algorithm were used. To characterize the conditions of simulations, the ratio of nonlinear and linear components of power of output signal $NL/L \triangleq \delta_{\varepsilon x 2}^2/\delta_x^2$, and the ratio of power of output signal and output noise $S/N \triangleq \delta_{x+\varepsilon x 2}^2/\delta_v^2$, were introduced. The results of one realization -- the one with largest errors -- are shown in figure 1. First of all we can see that the nonlinear filtering was stable, at least in given interval. The instantaneous errors grow with decreasing instanteneous power of signal but they are recoverable. The q.m. relative errors for a spectrum of S/N, NL/L indices are given in the table 3.

Table 3. Estimation errors

S/N	NL/L %	s_x %	$s_v/\sqrt{V_{\tilde{x}}}$ %
1	0	0.0	0.0
	5	-0.5	-9.3
	20	-1.4	-16.5
	100	-5.2	-30.3
4	0	0.0	0.0
	5	-0.3	-7.0
	20	-0.8	-12.8
	100	-3.3	-26.6

It is interesting that the relative q.m. error of estimated state (estimated by s_x %) is one order lower that the relative q.m. error of estimated error-variance (estimated by $s_v/\sqrt{V_{\tilde{x}}}$%). The reason for this may be that the estimation of the first moment is corrected by the second one, but the estimation of the second moment is not corrected by the third one. We take the errors of investigated nonlinear filtering to be suprisingly good. This is so especially in the case when nonlinear filtering is used for the separable problem of control, i.e., when the control is derived just from the estimated state and not from the higher moments.

LITERATURE

I.I.Gichman, A.V.Skorochod (1968): Stochastic differential equations. (In Russian). Naukova dumka, Kiev.

P.Choquette, A.R.M.Noton, C.A.G.Watson (1970): Remote computer control of an industrial process. Proc.IEEE 58, 10.

A.H.Jazwinski (1970): Stochastic processes and filtering theory. Acad.Pr., N.Y.

R.E.Kalman, R.S.Bucy (1961): New results in linear filtering and prediction theory. Tr.ASME, J.Bas.Eng.83D, 95.

H.J.Kushner (1967): Dynamical equations for optimal nonlinear filtering. J.Diff. Equations 3, 179.

R.E.Mortensen (1972), IEEE Tr.Aut.Control AC-17, 184.

A.P.Sage, J.L.Melsa (1971): Estimation theory with applications to communications and control. McGraw-Hill, N.Y.

Y.Sunahara, Y.Sawaragi, K.Ohnimato (1971): On the estimation of stochastic sensitivity for non-linear dynamical systems in state estimation problems. Int.J.Control 13, 1083.

A.Vaněček (1972): On-line nonlinear identification. (In Czech.) Res.Rep.INORGA Inst.Autom.,Prague.

A.Vaněček (1972'): A simple decomposition of a non-linear Volterra system. Int.J. Control 16, 1197.

D.A.Wismer, C.H.Wells(1972): A modern approach to industrial process control, Automatica 8, 117.

A SEQUENTIAL METHOD FOR THE ESTIMATION OF PARAMETERS IN NONLINEAR MODELS OF MULTIVARIABLE SYSTEMS

Manuel Mendes and Christian De Polignac
German-French Institut Laue-Langevin
Rue des Martyrs ,GRENOBLE, France

The extended Kalman-Filter Equations have been applied in the estimation of parameters of non-stationary nonlinear models of impulse response and input functions from multivariable systems. Several computational aspects are exposed on simulated examples. Also a program has been developpped for the practical applications on the analysis and data reduction of physical spectra in neutron and x-ray experiments.

1. INTRODUCTION

When data with strong noisy smearing are measured methods for smoothing and filtering become necessary, for example, in the evaluation of impulse-response and input functions of nonstationary plants, in the correction of smearing effects caused by instrument functions on physical x-ray and neutron scattering data, in the image processing.

The problem of estimation of parameters in the following model will be considered:

$$\underline{y}(t_i) = \int_{-\infty}^{t_i} \underline{G}(t_i,t',\underline{p}_1).\underline{u}(t',\underline{p}_2)dt' + \underline{v}(t_i) \quad (1)$$

which represents a non-stationary multivariable plant. $\underline{y}(t_i)$ is the m×1 vector of measured variables (*output*), $\underline{u}(t,\underline{p}_2)$ the r×1 vector of *input* functions, $\underline{v}(t_i)$ the vector of the *measurement* noise and $\underline{G}(t,t',\underline{p}_1)$ the m×r matrix of *impulse--response functions* of the plant. The parameter vectors $\underline{p}_1$ and $\underline{p}_2$ are to be estimated from all the available measurements $\underline{y}(t_i)$ i=1,...,N . It should be noted that these measurements don't need to be uniformly spaced.

Two typical situations can be treated: a)-$\underline{p}_2$ is known (*the input functions are given*) so that only $\underline{p}_1$ must be evaluated; this is the general situation of *IDENTIFICATION* in Control Theory; b)-$\underline{p}_1$ is known,i.e., given the impulse-response Matrix $\underline{G}(t,t',\underline{p}_1)$, the input functions are to be estimated (*for ex. in the correction of smearing effects in physical spectra*). Naturally the more general situation of unknown $\underline{p}_1$ and $\underline{p}_2$ is also possible and of great interest.

Several ways of solving the problem are now possible: a)-If no analytical (physical) models are available, general assumptions are necessary. For example step function models have been utilized by LEVIN [1] . The authors generalized this method in another work [2] . MENDES and DELESTRE [3] have also recently utilized spline-models; b)- As a result of the previous procedures or of other " a-priori" knowledge, the engineer is able to make analytical model assumptions so that the fit of the corresponding parameters is desired. *This case has been studied in this work.*

2.DESCRIPTION OF THE METHOD

2.1- MAXIMUM LIKELIHOOD ESTIMATIONS

The first approach to be considered is the *Maximum-Likelihood method.* It results from Eq.(1):

$$\underline{y}(t_i) = \underline{g}_i(\underline{p},t_i) + \underline{v}(t_i) \quad (2)$$

and for all the measurements

$$\underline{y} := [\underline{y}^T(t_1),\ldots,\underline{y}^T(t_N)]^T = \underline{g}(\underline{p}) + \underline{v} \quad (3)$$

and, if the observations are gaussian distributed $\underline{v} \rightarrow N(\underline{0},\underline{V})$, the Max.Lik. estimation is obtained from:

$$\operatorname*{Max}_{\underline{p}}\{L(\underline{y}=\underline{y}_{obs};\underline{p})\} \rightarrow \operatorname*{Min}_{\underline{p}}\{\|\underline{y}_{obs}-\underline{g}(\underline{p})\|^2_{\underline{V}^{-1}}\} \quad (4)$$

where L(...) is the *likelihood -function.* This is the general situation of *non linear curve fitting*, where a non quadratic (in the parameter vector $\underline{p}$) function must be minimized (maximized). The solution can be obtained by utilization of different gradient techniques, but normally these methods are very cumbersome and not adequate for "real-time" utilizations.

Because the Max.Lik. estimations are asymptotically efficient, the CRÁMER-RAO bound gives an excellent idea of the best (minimal) variances, which can be obtained for the estimations. It can be therefore highly desirable to calculate this bound in order to study "off-line" the dependence of estimation on the parameters of the measurement scheme (input parameters, quality of sensors, etc). This has been done from in another work |4| , in practical cases of neutron-scattering. It is known that

$$\mathrm{VAR}\{\hat{\underline{p}}\} := \underline{P} \geqq \underline{I}^{-1}$$

where $\underline{I}$ is the FISHER Information matrix, given by

$$\underline{I} = \varepsilon\{[\nabla_{\underline{p}} \log p(\underline{y}/\underline{p})][\nabla_{\underline{p}} \log p(\underline{y}/\underline{p})]^T\} =$$

$$= \nabla^T_{\underline{p}} [\underline{g}(\underline{p})] \cdot \underline{V}^{-1} \cdot \nabla_{\underline{p}} [\underline{g}(\underline{p})] \tag{5}$$

for the special case of Eq.(4). The minimization of a function of $\underline{I}^{-1}$ often solves the important problem of *optimal design of experiments* (*for ex input variables*).

2.2- SEQUENTIAL ESTIMATION METHODS

For several reasons, sequential algorithms have been considered: a)-The treatment of the information of each measurement $\underline{y}(t_i)$ in a sequential way allows for on-line implementations (for ex., by means of data acquisition by small computers connected in real time mode with a central procesor); b)-Time variable parameters and different types (correlated or uncorrelated, known or unknown,...) of disturbances can be easily treated.

The following general system of equations is now considered:

$$\underline{p}(k+1) = \underline{f}_p(\underline{p}(k),k) + \underline{w}(k) \tag{6a}$$

$$\underline{y}(k) = \int_{-\infty}^{t_k} \underline{G}(t',t_k,\underline{p}_1(k)).\underline{u}(\underline{p}_2(k),t')dt' + \underline{v}(k)$$

$$= \underline{g}(\underline{p}(k),k) + \underline{v}(k) \tag{6b}$$

where $\underline{y}(k) \equiv \underline{y}(t_k)$ and $\underline{p} := |\underline{p}_1^T \,\vdots\, \underline{p}_2^T|^T$. Eq.(6a) gives the time evolution of the parameters and is supposed to be known (*state equations*). Often it is sufficient to consider linear relations

$$\underline{p}(k+1) = \underline{\emptyset}(k).\underline{p}(k) + \underline{w}(k)$$

where the matrix $\underline{\emptyset}(k)$ is known or itself dependent on unknown time constant parameters, which can also be treated in the general formalism. The disturbances $\underline{w}(k)$ are representative for the unavoidable uncertainties of the model. Eq.(6b) (*the output eq. of the model*) is strong non linear on the parametervector (*modelstate*) to be estimated.

The application of the extended KALMAN Filter equations on the System eq.(6) results in:

$$\hat{\underline{p}}(k+1/k) = \underline{f}_p(\hat{\underline{p}}(k),k)$$

$$\tilde{\underline{P}}(k+1/k) = \underline{\Phi}(k).\tilde{\underline{P}}(k).\underline{\Phi}^T(k) + \underline{W}(k)$$

$$\underline{K}(k+1) = \tilde{\underline{P}}(k+1/k)\underline{G}^T(k+1)[\underline{G}(k+1)\tilde{\underline{P}}(k+1/k). \tag{7}$$
$$.\underline{G}^T(k+1) + \underline{V}(k+1)]^{-1}$$

$$\hat{\underline{p}}(k+1) = \hat{\underline{p}}(k+1/k) + \underline{K}(k+1)[\underline{y}(k+1) - \underline{g}(\hat{\underline{p}}(k+1/k),k+1)]$$

$$\tilde{\underline{P}}(k+1) = [\underline{E} - \underline{K}(k+1)\underline{G}(k+1)]\tilde{\underline{P}}(k+1/k)$$

with the notations:

$$\hat{\underline{p}}(k+1) := \varepsilon\{\underline{p}(k+1)/\underline{Y}(0 \to k+1)\} ;$$

$$\hat{\underline{p}}(k+1/k) := \varepsilon\{\underline{p}(k+1)/\underline{Y}(0 \to k)\} ; \tag{8}$$

$$\tilde{\underline{P}}(k+1) := \varepsilon\{(\underline{p}(k+1)-\hat{\underline{p}}(k+1))(\underline{p}(k+1)-\hat{\underline{p}}(k+1))^T\};$$

The matrixes $\underline{\Phi}(k)$ and $\underline{G}(k)$ result from the linearization of the model equations and are defined as:

$$\underline{\Phi}(k) := \nabla_{\underline{p}} \underline{f}_p(\underline{p},k)\Big|_{\underline{p}=\hat{\underline{p}}(k)} \tag{9a}$$

$$\underline{G}(k+1) := \nabla_{\underline{p}} \underline{g}(\underline{p},k+1)\Big|_{\underline{p}=\hat{\underline{p}}(k+1/k)} = \tag{9b}$$

$$= \int_{-\infty}^{t_{k+1}} \nabla_{\underline{p}}[\underline{G}(t',t_{k+1},\underline{p}_1)\underline{u}(\underline{p}_2,t')]dt'\Big|_{\underline{p}=\hat{\underline{p}}(k+1/k)}$$

Until now the model (6) was exclusively considered supposing that *no other system description forms were possible*. This is often the situation in chemical and other industrial plants. Neverthless it is interesting to compare the computational efficiency of Eq.(7) in systems, where an alternative state description exists . In this case the classical system of equations are considered:

$$\underline{x}(k+1) = \underline{f}_x(\underline{x}(k),\underline{p}(k),\underline{u}(k),k) + \underline{w}_x(k)$$

$$\underline{p}(k+1) = \underline{f}_p(\underline{p}(k),k) + \underline{w}_p(k) \tag{10}$$

$$\underline{y}(k) = \underline{h}(\underline{x}(k),\underline{p}(k),k) + \underline{v}(k)$$

The extended KALMAN Filter equations are now applied for the estimation of the augmented state $\underline{x}^*(k) := [\underline{x}^T(k) \,\vdots\, \underline{p}^T(k)]^T$. Therefore it is well known that the *state augmentation* results in increasing computation time and memory. On the other hand the computation of Eq(9b) *is independent from the order of the system*, so that when the input-output representation of the system Eq(1) is sufficient for control purposes, *there is apparently no advantage if the state representation is introduced.*

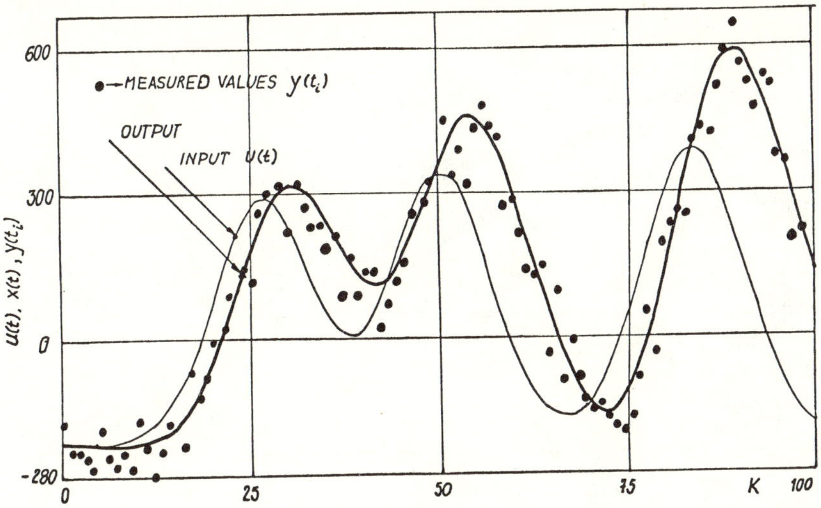

FIGURE 1

3. COMPUTATIONAL ASPECTS

A program for practical applications of Filter-equations (7) has been tested on different numerical examples.

The numerical integration of Eq(6b),(9b) allows for the treatment of several classes of problems: a)-The input function can be either defined by step functions with variable sampling interval, or continuously; b)-In the case of constant parameters no restriction on the spacing of the measurements is present; c)-When the parameter vector $\underline{p}$ of the input variables are to be estimated, the impulse response functions are often the result of previous deconvolution procedures, and therefore given in the form of step-,spline-models,...

As it is typical for nonlinear problems several cases of *filter divergence* have been observed. They were mainly caused by *systematical model de-del deviations*, so that judicious choice of the variance matrix $\underline{W}(k)$ (system disturbances) was often enough in order to obtain better results. Otherwise the application of *iterative filter equations* was especially efficient to compensate for bad initial values of the parameters: when at the time t_{k+1} a first estimation $\underline{\hat{p}}^1(k+1)$, $\underline{\tilde{P}}^1(k+1)$ is computed it is possible to use the 1-step smoothing equations to obtain the values $\underline{\hat{p}}(k/k+1)$, $\underline{\tilde{P}}(k/k+1)$. Eq(7) started again with these new initial values yields $\underline{\hat{p}}^2(k+1)$ $\underline{\tilde{P}}2(k+1)$. Several iterations are time consuming,but the experience shows a great improvement already at the second iteration.

EXAMPLES:

A simple 1-input-1-output system of 1st order with the impulse-response $g(t)=K.\exp(-t/T)$ was simulated. K=1.0 is a constant unknown parameter and T is supposed to be unknown and time-variable. Fig.1 shows simulation results, with given input variable, output sampling interval $T_s=0.3$[sec] and variance values $\sigma_v^2(k)=2500$. The true parameter T follows an exponential with

$$T(k+1) = \alpha.T(k)$$

with $\alpha=1.01$ and $T(0)= 1.0$[sec] .

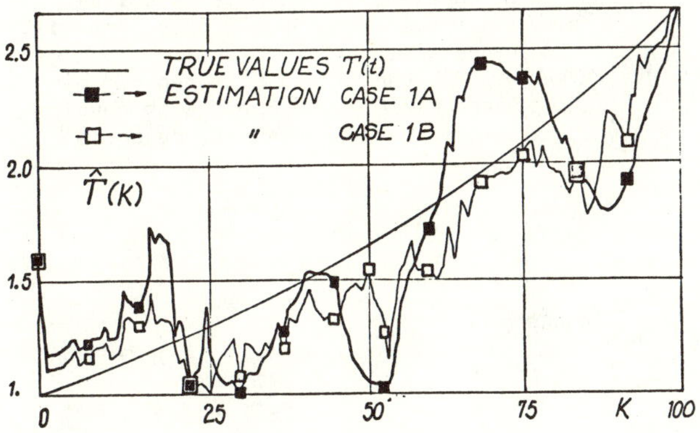

FIGURE 2a

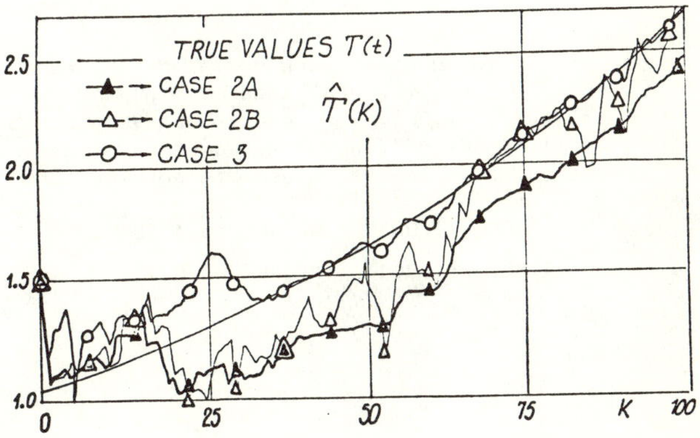

FIGURE 2b

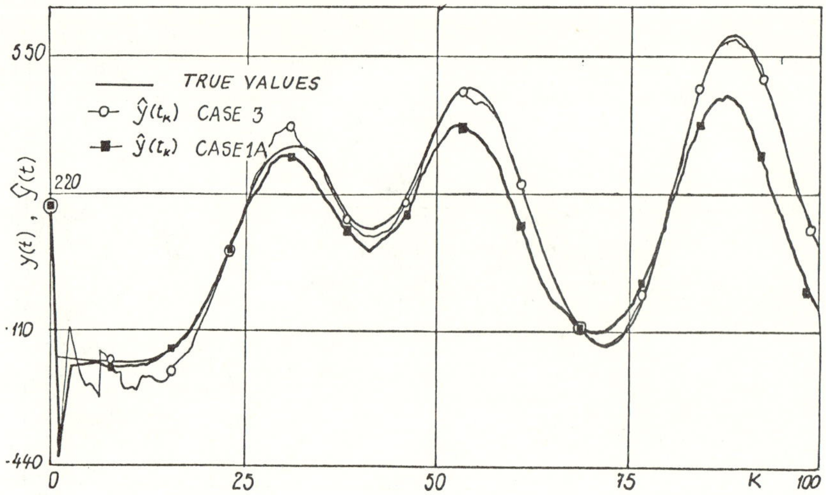

FIGURE 3

Several cases have been simulated:

CASE 1 : The parameter T(k) was supposed to be constant and unknown, a wrong value $\alpha=1.0$ was considered. A first simulation CASE 1A with W(k)=Diag(0.,0.) showed great distortion in the estimations of the output variable $\hat{y}(t)$, Fig.3. At the same time the parameter estimation $\hat{T}(k)$ is greatly biased, Figure 2a. In a second simulation CASE 1B with W(k)=Diag(0.,0.01) the results are improved.

CASE 2 : Now a correct equation for the evolution of T(k) was considered, by setting $\alpha=1.01$ the true value. In CASE 2A W(k) is equal Diag(0.,0.) and the estimations can be biased. In CASE 2B with W(k)=Diag(0.,0.01) again systematic deviations are corrected. (Fig.2b).

CASE 3 :Finally the parameter α was also considered as unknown with the following state equations

$$K(k+1) = K(k) \quad +w_1(k)$$

$$T(k+1) = \alpha(k)T(k) \quad +w_2(k)$$

$$\alpha(k+1) = \alpha(k) \quad +w_3(k)$$

The state equations are now also nonlinear Fig.2b shows the evolution of $\hat{T}(k)$ for W(k) =Diag(0.,0.,0.). The results are very comparable to Case 2. For the two other parameters the following values are obtained:

$$\hat{K}(100) = 1.008 \qquad \sigma_k^2(100)=0.1.10^{-2}$$

$$\hat{\alpha}(100) = 1.011 \qquad \sigma_\alpha^2(100)=0.8.10^{-6}$$

4. CONCLUSIONS

A sequential algorithm has been developped for the estimation of unknown parameters in the impulse -response functions and/or input variables . It is very adequate for on line situations and for the treatment of time-variable parameters.Otherwise the order of the system has no influence in the computing time. The extended Kalman filter equations turned out to be of great flexibility in the treatment of this class of problems.

5. REFERENCES

[1] LEVIN 1960: Optimum estimation of impulse response in the presence of noise.IRE Trans.Cir.Th.,CT-7,1,pp 50-56

[2] MENDES a. POLIGNAC: Recursive Bayes deconvolution in Physical Experiments. Acta Cryst. 1973, A 29, pp 1-9.

[3] MENDES a. DELESTRE: Sequential deconvolution in physical experiments by splines. 1st Int.Conf.on Spec. Lines, Knoxville Tennessee, 1972.

[4] MENDES a. POLIGNAC: Neue Methoden für die Entschmierung von Spectra in physikalischen Experimenten (to be publ.)

DESIGN OF STATE ESTIMATOR FOR SYSTEMS ORIGINALLY DESCRIBED BY INPUT-OUTPUT RELATION

Y. Funahashi and K. Nakamura

Automatic Control Laboratory, Nagoya University
Fro-cho, Chikusa-ku, Nagoya, Japan

Generally, the identification method gives an input-output relation of the system rather than a state space description. In this paper, we construct an optimal filter for systems which are described by an input-output equation. It is shown that the steady-state optimal estimator, which is obtained simply from the system parameters, gives an exact value of the system state.

1. INTRODUCTION

The problem of state estimation for dynamical systems is of fundamental importance in the design of optimal control systems. For a class of systems whose state space description is given beforehand, there are considered two kinds of approach : the Luenberger observer [1,2] and the Kalman filter [3]. If no system equations are available, one must determine them from observations of the input and output variables. In the identification procedures, it is unrealistic to separate system disturbance from output measurement noise. Therefore the system equation directly resulted from the identification procedures is rather an input-output relation than a state description.

The main purpose of this paper is to propose an approach to the state estimation problems for linear discrete-time dynamical systems whose description is given in terms of inpur-output relation. In section 2 we develop a canonical input-output equation for stochastic linear systems, indicating a relationship with the familiar state description. In section 3 we derive a canonical state-space description from the canonical input-output relation and design an optimal state estimator. It is shown that the optimal steady-state filter gives an exact value of the system state and the optimal steady-state estimator gain is simply given by parameters of the canonical input-output equation, without solving the Riccati-type equation. For the case of noise free, the steady-state can be shown to give a dead-beat deterministic observer.

2. CANONICAL INPUT-OUTPUT DESCRIPTION FOR STOCHASTIC LINEAR SYSTEMS

In this section we derive a difference equation relating to input-,output-variables and stochastic disturbances. This form is a canonical one for parameter identification of multivariable systems [4,5,6].

A linear constant multivariable discrete-time system subjected to colored noise can be modelled by the following state space equation :

$$S_1 : \begin{aligned} x(t+1) &= Fx(t) + Gu(t) + v_1(t) \\ y(t) &= Hx(t) + v_2(t) \end{aligned} \qquad (1)$$

where t takes integer values. The state vector x, which is an augmented vector composed of the physical system and of colored noise process, has dimension n . The input u and output y have dimensions of r and m respectively. The sequences $\{v_1(t)\}$ and $\{v_2(t)\}$ are independent equally distributed random vectors with zero mean and covariances R_1 and R_2 . The matrix H is assumed, for simplicity, to have rank of m . It is assumed that the noise free system :

$$\begin{aligned} x(t+1) &= Fx(t) + Gu(t) \\ y(t) &= Hx(t) \end{aligned} \qquad (2)$$

is completely observable with observability index p . Then there exist n linearly independent row vectors in the matrix Ω_p .

$$\Omega_p = \begin{bmatrix} H \\ HF \\ \vdots \\ HF^{p-1} \end{bmatrix} . \qquad (3)$$

In order to derive an input-output equation for the system S_1 , we must exclude the state $x(t)$ from (1). From (1) we get :

$$\begin{aligned} y(t) &= Hx(t) + v_2(t) \\ y(t+1) &= HFx(t) + HGu(t) + Hv_1(t) + v_2(t+1) \\ &\cdots\cdots \\ y(t+p-1) &= HF^{p-1}x(t) + HF^{p-2}Gu(t) + \ldots \\ &\quad + HGu(t+p-2) + HF^{p-2}v_1(t) + \ldots \\ &\quad Hv_1(t+p-2) + v_2(t+p-1) . \end{aligned} \qquad (4)$$

As the system is observable, we can calculate $x(t)$ using some n equations selected from (4). The choice of n from $m \times p$ equations in (4) is not unique. We chose the n independent row vectors in Ω_p in the following order :

$$h_1F^{p-1},\ h_2F^{p-1},\ \ldots,\ h_1F^{p-2},\ h_2F^{p-2},\ \ldots,\ h_mF^{p-2},$$
$$\ldots,\ h_1,\ h_2,\ \ldots,\ h_m. \qquad (5)$$

Using the n equations corresponding to the independent rows chosen above, we calculate $x(t)$ and substitute it into

$$y(t+p) = HF^p x(t) + HF^{p-1}Gu(t) + \ldots + HGu(t+p-1)$$
$$+ HF^{p-1}v_1(t) + \ldots + v_2(t+p). \qquad (6)$$

and rearranging terms we obtain an input-output description for S_1 :

$$y(t+p) = -A_1y(t+p-1) - A_2y(t+p-2) - \ldots - A_py(t)$$
$$+B_1u(t+p-1) + B_2u(t+p-2) + \ldots + B_pu(t)$$
$$+D_1v_1(t+p-1)+D_2v_1(t+p-2) + \ldots+D_pv_1(t)$$
$$+v_2(t+p) + A_1v_2(t+p-1) +\ldots +A_pv_2(t) \qquad (7)$$

or by introducing the shift operator defined by $z^{-1}y(t) = y(t-1)$,

$$A(z^{-1})y(t) = B(z^{-1})u(t) + D(z^{-1})v_1(t) + $$
$$+A(z^{-1})v_2(t) \qquad (8)$$

where

$$A(z^{-1}) = I_m + z^{-1}A_1 + z^{-2}A_2 + \ldots + z^{-p}A_p$$
$$B(z^{-1}) = z^{-1}B_1 + z^{-2}B_2 + \ldots + z^{-p}B_p$$
$$D(z^{-1}) = z^{-1}D_1 + z^{-2}D_2 + \ldots + z^{-p}D_p \qquad (9)$$

The structure of A_i matrices is such that if a certain, say k-th ; column in the A_i matix is equal to zero, then also the k-th columns in the matrices A_{i+1} , A_{i+2}, ..., A_p are all equal to zero. Define p_k as the least integer such that the above statement holds. Then the order of the system and the observability index are related with $p_1, p_2, \ldots, p_m$ [6].

$$n = p_1 + p_2 + \ldots + p_m , \qquad (10)$$
$$p = \max\{ p_1, p_2, \ldots, p_m \} , \qquad (11)$$

The disturbance model is

$$v(t) = D(z^{-1})\ v_1(t) + A(z^{-1})v_2(t). \qquad (12)$$

The separation of $v_1(t)$ and $v_2(t)$ based on the observations of input and output variables has been found impossible. It is necessary to resort to spectral factorization of discrete-time processes. Notice that the autocorrelation of $v(t)$

$$R(\tau) = E[\ v(t)\ v^T(t-\tau)\ ]$$

is zero for $\tau = p+1, p+2, \ldots$. We propose the disturbance model as

$$v(t) = C(z^{-1})e(t)$$
$$= e(t) + C_1e(t-1) + \ldots + C_pe(t-p) \qquad (13)$$

Where $\{ e(t) \}$ is an m-vector of independent random process of normal distribution with zero mean value and covariance matrix R. The number of parameters in R and $C(z^{-1})$ is $m(m+1)/2+pm^2$. They are to be identified or determined from the followling equations.

$$R + C_1RC_1^T + C_2RC_2^T + \ldots +C_pRC_p^T = R(0)$$
$$C_1R + C_2RC_1^T + \ldots +C_pRC_{p-1}^T =R(1)$$
$$C_2R + C_3RC_1^T+\ldots+C_pRC_{p-2}^T = R(2)$$
$$\vdots$$
$$C_pR = R(p) \qquad (14)$$

In (14) C_i can be chosen to satisfy the stability requirement of the filter $C(z^{-1})$ [7]. This implies that C_i are chosen so that the roots of

$$|C(z^{-1})| = 0 \qquad (15)$$

lie inside the unit circle.

Combining (8) and (13) we have finally the canonical input-output description for the discrete-time stochastic system S_1 :

$$S_2 : A(z^{-1})y(t) = B(z^{-1})u(t) + C(z^{-1})e(t). \qquad (16)$$

In the following we will denote by a^k_{ij}, b^k_{ij} and c^k_{ij} the (k,j) element of A_i, B_i and C_i, respectively.

3. CANONICAL STATE SPACE DESCRIPTION

In this section we show that there exists a canonical state space description for the linear stochastic system.

Let us define the new state variables as follows:

$$\xi_{j1}(t+1) = -\sum_{i=1}^{m} a^j_{1i}\xi_{i1}(t) + \xi_{j2}(t) + $$
$$+ \sum_{i=1}^{r} b^j_{1i}u_i(t) + \sum_{i=1}^{m}\gamma^j_{1i}e_i(t)$$

$$\ldots\ldots$$

$$\xi_{jp_j-1}(t+1) = -\sum_{i=1}^{m} a^j_{p_j-1\,i}\xi_{i1}(t) + \xi_{jp_j}(t) + $$
$$+\sum_{i=1}^{r} b^j_{p_j-1\,i}u_i(t) + \sum_{i=1}^{m}\gamma^j_{p_j-1\,i}e_i(t)$$
$$\xi_{jp_j}(t+1) = -\sum_{i=1}^{m} a^j_{p_ji}\xi_{i1}(t)+ \sum_{i=1}^{r} b^j_{p_ji}u_i(t)+$$
$$+\sum_{i=1}^{m}\gamma^j_{p_ji}e_i(t),\ j=1,\ldots, m \qquad (17)$$

where

$$\gamma^j_{ki} = c^j_{ki} - a^j_{ki} \quad i=1,\ldots,m\ ,\ j=1,\ldots,m\ ,$$
$$k=1,\ldots,p_i \qquad (18)$$

The output variables are given by

$$y_i(t) = \xi_{i1}(t) + e_i(t) \qquad i=1,..,m \tag{19}$$

Writing (17) and (19) in matrix notation we have the canonical state space description for S_2 :

$$S_3 : \quad \xi(t+1) = A\,\xi(t) + Bu(t) + \Gamma e(t)$$
$$y(t) = H\,\xi(t) + e(t) \tag{20}$$

where

$$\xi(t) = \begin{bmatrix} \xi_{11}(t) \\ \xi_{12}(t) \\ \cdot \\ \cdot \\ \xi_{1p_1}(t) \\ \xi_{21}(t) \\ \cdot \\ \cdot \\ \xi_{mp_m}(t) \end{bmatrix} \qquad A = \begin{bmatrix} A^{11} & A^{12} & \cdots\cdots & A^{1m} \\ A^{21} & A^{22} & \cdots & \\ \cdot & & & \\ \cdot & & & \\ A^{m1} & \cdots\cdots & & A^{mm} \end{bmatrix}$$

$$A^{ii} = \begin{bmatrix} -a^i_{1i} & 1 & 0 & 0 & \cdots \\ -a^i_{2i} & 0 & 1 & 0 & \cdots \\ \cdots\cdots & & & & \\ -a^i_{p_i i} & 0 & 0 & \cdots & 0 \end{bmatrix} \qquad A^{ij} = \begin{bmatrix} -a^i_{1j} & 0 & \cdots \\ -a^i_{2j} & 0 & \cdots \\ \cdots & & \\ -a^i_{p_i j} & 0 & \cdots \end{bmatrix} \quad i \neq j$$

$$B = \begin{bmatrix} B^1 \\ B^2 \\ \cdot \\ \cdot \\ B^m \end{bmatrix} \qquad B^i = \begin{bmatrix} b^i_{11} & b^i_{12} & \cdots & b^i_{1r} \\ \cdot & & & \\ \cdot & & & \\ b^i_{p_i 1} & \cdots & & b^i_{p_i r} \end{bmatrix}$$

$$\Gamma = \begin{bmatrix} \Gamma^1 \\ \cdot \\ \cdot \\ \Gamma^m \end{bmatrix} \qquad \Gamma^i = \begin{bmatrix} \gamma^i_{11} & \cdots & \gamma^i_{1m} \\ \cdots & & \\ \gamma^i_{p_i 1} & \cdots & \gamma^i_{p_i m} \end{bmatrix}$$

$$H = \begin{bmatrix} 1 & 0 & 0 & 0 & 0 & 0 & \cdots & & & \\ 0 & 0 & 0 & \cdot & 1 & 0 & \cdots & & & \\ \cdot & \cdot & \cdot & \cdot & \cdot & \cdot & \cdot & \cdot & & \\ \cdot & \cdot & \cdot & \cdot & \cdot & \cdot & \cdot & \cdot & 1 & 0 & 0 \end{bmatrix} \tag{21}$$
$$\underbrace{\qquad}_{p_1} \quad \underbrace{\qquad}_{p_2} \quad \cdots \quad \underbrace{\qquad}_{p_m}$$

4. DESIGN OF STATE ESTIMATOR

Now we will design an optimal state estimator. Denote by $\hat{\xi}(t) = \hat{\xi}(t/t-1)$ the value of estimate of $\xi(t)$ based on the measurements $\{\, y(t-1), y(t-2), \ldots, u(t-1), u(t-2), \ldots \}$. In the next theorem the covariance matrix R is assumed to be positive definite, which be relaxed afterwards.

Theorem 1 : The optimal state estimator for the system S_3 is given by the relation

$$\hat{\xi}(t+1) = A\,\hat{\xi}(t)+Bu(t)+K(t)[y(t)-H\,\hat{\xi}(t)] \tag{22}$$

with $\hat{\xi}(0) = E\{\xi(0)\}$, and the filter gain $K(t)$ is specified by the relations

$$K(t)=[AP(t)H^T+\Gamma R][HP(t)H^T+R] \tag{23}$$

$$P(t+1)=AP(t)A^T+\Gamma R\Gamma^T-K(t)[AP(t)H^T+\Gamma R]^T \tag{24}$$

Proof: Define

$$\tilde{\xi}(t) = \xi(t) - \hat{\xi}(t)$$
$$P(t) = E\,\{\tilde{\xi}(t)\,\tilde{\xi}^T(t)\}$$

Since the structure of the estimator is assumed as (22), the error equation becomes

$$\tilde{\xi}(t+1)=[A-K(t)H]\,\tilde{\xi}(t)+[\Gamma-K(t)]e(t) \,. \tag{25}$$

The corariance matrix of (25) is

$$P(t+1)=[A-K(t)H]P(t)[A-K(t)H]^T+ +[\Gamma-K(t)]R[\Gamma-K(t)]^T \,. \tag{26}$$

Minimization of $P(t+1)$ with respect to $K(t)$ gives the optimal gain $K(t)$ as (23) and (24).

Next, we will consider the steady-state filter. Assuming that $\lim P(t)$ exists, we obtain the matrix equality from (23) and (24).

$$P=APA^T+\Gamma R\Gamma^T-[APH^T+\Gamma R][HPH^T+\Gamma]^{-1}[APH^T+\Gamma R]^T. \tag{27}$$

$P=0$ is one of the solutions of (27) and for $P=0$ the optimal gain is

$$K = \Gamma \,.$$

Hence, if the model with the structure (16) is known or can be obtained either by the transformation from (1) described in section 2 or through some identification procedures, then there is no need to solve a Riccati-type equation in order to obtain the steady-state estimator. We will examine in the next theorem the steady-state estimator whose gain is Γ .

Theorem 2 Steady-state state estimator :

$$\hat{\xi}(t+1)=A\,\hat{\xi}(t)+Bu(t)+\Gamma[y(t)-H\,\hat{\xi}(t)] \tag{28}$$

$\hat{\xi}(t)$ converges to the true value of $\xi(t)$ with probability one. The rate of convergence is determined by the matrix $A-\Gamma H$, whose elements are those of $C_1, C_2, \ldots, C_p$ in (13) .

Proof : Estimation error resulting from (28) is

$$\tilde{\xi}(t+1) = [A - \Gamma H]\,\tilde{\xi}(t) \tag{29}$$

From (21),

$$A - \Gamma H = \begin{bmatrix} -c^1_{11} & 1 & 0 & 0 & & -c^1_{1m} & 0 & & \\ -c^1_{21} & 0 & 1 & 0 & & -c^1_{2m} & 0 & & \\ \cdots & & & & & \cdots & & & \\ -c^1_{p_1 1} & 0 & 0 & . & 0 & -c^1_{p_1 m} & 0 & & \\ \vdots & & & & & & & & \\ -c^m_{11} & & & & & -c^m_{1m} & 1 & 0 & .. \\ -c^m_{21} & & 0 & & & -c^m_{2m} & 0 & 1 & 0\ .. \\ \cdots & & & & & \cdots & & & \\ -c^m_{p_m 1} & & & & & -c^m_{p_m m} & 0 & \cdots & 0 \end{bmatrix} \qquad (30)$$

As

$$|\lambda I - (A - \Gamma H)| = |C(\lambda^{-1})|$$

and the stability assumption of (15), all the eigenvalues of $(A-\Gamma H)$ are less than one in absolute value.

$$\tilde{\xi}(t) = [A-\Gamma H]^t \, \tilde{\xi}(0) \qquad (31)$$

converges to zero with probability one.

Concerning the prediction of output , we have the next theorem.

<u>Theorem 3</u> The optimal output predictor is given by

$$y(t+1/t) = H \, \hat{\xi}(t+1) \ . \qquad (32)$$

If the steady-state state estimator given by (28) is used, the covariance matrix of steady-state output prediction error is equal to R.

The invertibility of covariance matorix R of noise process $\{e(t)\}$ is assumed in Theorem 1. It is needed to guarantee the convergence of the optimal state estimator. We will show that this condition can be relaxed in the steady-state estimator. As is evident in the proof of Theorem 2, the estimation error of steady-state estimator does not depend on the noise process $\{e(t)\}$. Hence steady-state gain Γ is irrelevant of singularity of R.

<u>Comment 1</u> Even if covariance matrix R is singlar, Theorem 2 is valid as well.

<u>Comment 2</u> As an extreme case we consider the situation where there is no stochastic disturbance and measurement noise . From (12) to (14),

$$R = 0$$

$$C_i = 0, \ i = 1, . \ . \ . \ , p. \qquad (34)$$

The state equation corresponding to (20) becomes

$$\xi(t+1) = A\,\xi(t) + Bu(t),$$
$$y(t) = H\,\xi(t), \qquad (35)$$

and the steady-state gain is, from Γ in (28), given by

$$K_d = \begin{bmatrix} -a^1_{11} & -a^1_{12} & \cdots & -a^1_{1m} \\ \cdots & & & \\ -a^1_{p_1 1} & -a^1_{p_1 2} & \cdots & -a^1_{p_1 m} \\ -a^2_{11} & -a^2_{12} & \cdots & -a^2_{1m} \\ \cdots & & & \\ -a^2_{p_2 1} & -a^2_{p_2 2} & \cdots & -a^2_{p_2 m} \\ \cdots & & & \\ -a^m_{p_m 1} & -a^m_{p_m 2} & \cdots & -a^m_{p_m m} \end{bmatrix} \qquad (36)$$

From Theorem 2, the state estimator is

$$\hat{\xi}(t+1) = A\,\hat{\xi}(t) + Bu(t) + K_d[y - H\,\hat{\xi}(t)] \qquad (37)$$

The estimation error in this case has a nice property.

$$\tilde{\xi}(t+1) = [A-K_dH]\,\tilde{\xi}(t)$$

$$= \begin{bmatrix} \begin{matrix} 0\,1\,0\,0 \\ 0\,0\,1\,0 \\ \cdots \\ 0\,0\,0\,0 \end{matrix} & 0 & 0 \\ 0 & & 0 \\ 0 & 0 & \begin{matrix} 0\,1\,0\,0 \\ 0\,0\,1\,0\ .. \\ \cdots\ 1 \\ 0\,0\,0\ .\ .\ 0 \end{matrix} \end{bmatrix}$$

The matrix $[A-K_dH]$ has a special structure such that

$$[A - K_dH]^p = 0 \qquad (39)$$

where p is defined to be the observability index of the system (35). Hence

$$\tilde{\xi}(t) = 0. \quad \forall t \geq p \qquad (40)$$

and we have thus obtained the dead-beat state estimator for multivariable deterministic systems.

5. CONCLUSION

By starting from the canonical input-output description for stochastic linear systems, we have developed a simple design procedure for the steady-state optimal state estimator. Under this frame-work, the steady-state optimal estimators are constructed to estimate the state of a linear discrete-time system on the bases of both deterministic and noisy measurements. For a deterministic system, the steady-state estmator was shown to become a dead-beat observer.

REFERENCES

1. Luenberger, D. G. : Observing the state of a linear system, IEEE Trans. Military Electronics, vol.MIL-8, 74-80, April 1964
2. ——— : Observers for multivariable systems, IEEE Trans. Automatic Control, vol.AC-11, 190-197, April 1966
3. Kalman, R. E. : A new approach to linear filtering and prediction problems, Trans. ASME, J. Basic Engrg., Ser. D, vol.82, 35-45, March 1960
4. Rowe, I.H.: A statistical model for the identification of multivariable stochastic systems, IFAC Symp.on Mult. Cont. Sys. , Düsseldorf, Oct., 1968
5. Aström, K.J.: System identification — A survey, IFAC, 2nd Symp. on Identification and Process Parameter estimation, Prague, June 1970
6. Valis, J. : On-line identification of Multivariable linear systems of unknown structure from input output data, ibid.
7. Motyka, P.R. and Cadzow, J.A. : The factorization of discrete-process spectral matrices, IEEE Trans. Automatic Control, vol.AC-12, 698-707, Dec.1967

THE NON-LINEAR DYNAMIC OBJECTS PARAMETERS ESTIMATION BASED ON THE SENSITIVITY-ALGORITHM

Rouban Anatoly Ivanovitch

The Institute of Automatized Control Systems and Radioelectronics, Tomsk, U S S R

By using common approach based on application of linearization method and sensitivity equations, the retrospective algorithms of simultaneous observation and identification problems solution are obtained for objects described by usual differential equations (D.E.), the difference equations, partial D.E. as well as the multipoint bound non-linear problems solution algorithms. The square type tendency of algorith is shown. The more complex identification problems solution possibility is approached. The adaptive sensitivity algorithm is obtained.

INTRODUCTION

The linearization method application idea was proposed by Sokolov (1961) for minimization of any functionals dependent on unknown parameters. This approach includes iteration method by Gauss-Newton (so called non-linear estimations method) based on error value minimization. The authors of this paper deal with the linearization method application for unknown D.E. parameters calculation. The paper of Medler (1969), Rouban (1969), Petrov (1970), Gorodetsky (1971) and et all are devoted to this problem. The problem of only usual D.E.coefficients calculation is mainly solved in these papers. Since on building the functioning algorithm it is necessary to use the main results of sensitivity theory (sensitivity equation solution and obtaining), they began to call linearization algorithm as sensitivity one. Petrov's paper(1970) is rich of ideas. Sensitivity algorithm is applied for usual D.E. coefficients identification as well as for outer affects one, for observation problem solution and the regions of this algorithm possible application are noted in this paper.

The results of more complex observation and identification problems solution as well as the solution of bound problems are given in this paper.

1. SIMULTANEOUS SOLUTION OF OBSERVATION AND IDENTIFICATION PROBLEM

Suppose mathematical model structure of object is the usual non-linear D.E.system with even right parts

$$\frac{dy}{dt} = f(y,\alpha,t),\ y(0)= y_o, \qquad (1.1)$$

where y is n-component vector-column of stationary coefficients which characterize dynamic features of object, its structure and outer affects.

m-component vector - column

$$\eta(t) = \eta(y,\alpha,t) \qquad (1.2)$$

of coordinates possible to be measured is given on y coordinates. It is necessary to calculate y_o initial conditions vectors as well as the vectors of α coefficients according to $\eta^*(t)$ measurements of $\eta(t)$ vector at $t\in[0,t_1]$; the observation and identification problem is being solved simultaneously [8].

In all cases, even for linear D.E., the measurements depend on parameters in non-linear way and these dependences are linearized in each iteration in sensitivity algorithm:

$$\overset{1+I}{\eta}(t)\approx \overset{1+I}{\bar{\eta}}(t) = \overset{1}{\eta}(t) + \Gamma^1_I(t)\, U^1_I(t)\,\Delta \overset{1+I}{y_o} + (\Gamma^1_I(t)\, U^1_2(t)+\Gamma^1_2(t))\Delta \overset{1+I}{\alpha} \qquad (1.3)$$

Here: 1 is iteration number;

$$\Delta \overset{1+I}{y}(t) = \overset{1+I}{y}(t) - \overset{1}{y}(t);$$

$$\Delta \overset{1+I}{y_o}= \overset{1+I}{y_o}- \overset{1}{y_o},\ \Delta\overset{1+I}{\alpha} = \overset{1+I}{\alpha} - \overset{1}{\alpha}$$

$$\Gamma^1_I(t) = \frac{\partial\eta(y^1,\alpha^1,t)}{\partial y} \qquad (1.4)$$

$$\Gamma^1_2(t) = \frac{\partial\eta(y^1,\alpha^1,t)}{\partial\alpha},$$

$y^1(t)$ and sensitivity function matrix

$$U^1_I(t) = \frac{\partial y^1(t)}{\partial y_o},\ U^1_2(t) = \frac{\partial y^1(t)}{\partial\alpha} \qquad (1.5)$$

are found from D.E. system.

$$\frac{dy^1}{dt} = f(y^1,\alpha^1,t),\ y^1(0) = y^1_o$$

$$\frac{dU_I^l}{dt} = j_I^l(t)\, U_I^l,\quad U_I^l(0) = I,$$
$$\frac{dU_2^l}{dt} = j_I^l(t)\, U_2^l - j_2^l(t),\quad U_2^l(0) = 0, \qquad (1.6)$$

which are solved simultaneously as a rule. In (1.6) $j_I(t)$, $j_2(t)$ are Yakoby's matrixes

$$j_I^l(t) = \frac{\partial f(y^l,\alpha^l,t)}{\partial y},$$
$$j_2^l(t) = \frac{\partial f(y^l,\alpha^l,t)}{\partial \alpha}. \qquad (1.7)$$

Unknown vector of parameter dynamic increments

$$\Delta\theta^{l+I} = \begin{pmatrix}\Delta y_0^{l+I}\\ \Delta\alpha^{l+I}\end{pmatrix} \qquad (1.8)$$

is calculated from criterion

$$\int_0^{t_I} (\eta^*(t) - \bar{\eta}^{l+I}(t))^T Q(t) (\eta^*(t) - \bar{\eta}^{l+I}(t))\, dt = \min, \qquad (1.9)$$

i.e.

$$\Delta\theta^{l+I} = \left(\int_0^{t_I} (A^l)^T Q A^l dt\right)^+ \int_0^{t_1} (A^l)^T Q (\eta^* - \eta^l) dt \qquad (1.10)$$

Here: T is transpose operation, Q(t) is given as positevely definite matrix of weight functions,

$$A^l = (\Gamma_I^l U_I^l;\ \Gamma_I^l U_2^l + \Gamma_2^l), \qquad (1.11)$$

+ is calculation operation of pseudoinverse matrix.

If measurements are made in descrete moments of time, summation according to the respective descretes will take place instead of integration in equations(1.9) and (1.10)

According to the increments θ^{l+I} we find the following approximation for unknown parameters

$$\theta^{l+I} = \theta^l + \gamma^l \Delta\theta^{l+I},\quad l = 0,1,2,\ldots \qquad (1.12)$$

where $0 < \gamma^l \leq I$ scalar introduction enables to extend sensitivity algorithm convergent regions [4,13].

The algorithm of quasilinearization is close to sensitivity algorithm by idea and universality. Their comparison is given below.

Firstly, the sensitivity algorithm is more economic in calculation aspect at other equal calculation difficulties. Let unknown parameter number include n_1 ($n_1 \leq n$) of initial conditions and r of D.E. coefficients. Then at the sensitivity algorithm application the order of D.E.system to be solved (Koshi problem) is by $n(n - n_1 + 1)$ value more than in quasilinearization algorithm during (l+I) iteration.

Secondly, both algorithms have speed of convergent of one and the same order. If mathematical model of objects (1.1) and measuring unit (1.2) describes processes in them in full, then

$$\eta^{l+I}(t) \xrightarrow[l\to\infty]{} \eta^*(t) \qquad (1.13)$$

and sensitivity algorithm gives a second-order convergent [7,8] of model parameters θ^l to θ, i.e.

$$\|\theta^{l+I} - \theta\| \leq \delta \|\theta^l - \theta\|^2,\ \delta < 1, \qquad (1.14)$$

in some regions of θ parameters real values. Here $\|\ \|$ is a norm. When condition (1.13) is not true, sensitivity algorithm leads to a first-order convergent process[2].

Sensitivity algorithm gave high speed of convergent and was stable good enough as to additive errors of measurement for numerical examples: parameters calculations of non-linear pendulum motion equation [7,8], the auto-controlled vessel motion [7], chemical reactions kinetics equations [7,11,12] , and some others.

If the object model is descrete, the parameters calculation is made according to scheme described above and some results are given in paper [12]. The numerical examples demonstrate high speed of sensitivity algorithm of convergent.

2. DISTRIBUTED OBJECTS IDENTIFICATION

Sensitivity-algorithm is successfully applied for distributed objects identification at partial observation of variables. These results can be found in papers [7,10].

Assume the object is described by partial D.E. system of the following type

$$\frac{\partial y}{\partial t} = f(y,z,t,\tau,\alpha), \frac{\partial z}{\partial \tau} = \varphi(y,z,t,\tau,\alpha) \quad (2.1)$$

or

$$\frac{\partial y}{\partial t} = f\left(t,\tau,\alpha, y, \frac{\partial y}{\partial \tau}, \frac{\partial^2 y}{\partial \tau^2}, \ldots\right), \quad (0 \le t \le t_I, \ 0 \le \tau \le \tau_I) \quad (2.2)$$

with given bound conditions, which also depend on unknown parameters α. Denote observed values multiplicity of object variables by $\{\eta^*(t_i, \tau_j) = \eta^*_{ij}\}$ and multiplicity of foretold values by

$$\{\eta_{ij} = \eta[y(t_i, \tau_j), z(t_i, \tau_j), \alpha]\}, \quad (2.3)$$

where $\eta[y,z,\alpha]$ is a vector-column of known changes made by measuring device. It is necessary to determine such a group of parameters α that multiplicity $\{\eta^*_{ij}\}$ and $\{\eta_{ij}\}$ respective elements should differ from one another as less as it is possible, for example, in criterion aspect

$$I(\alpha) = \|\eta^* - \eta\|^2 = \sum_{i,j} (\eta^*_{ij} - \eta_{ij})^T Q_{ij} (\eta^*_{ij} - \eta_{ij}) = \min. \quad (2.4)$$

The problem solution scheme is the same but sensitivity equations for distributed models [10] are of a particular interest.

3. MULTIPOINT NON-LINEAR BOUNDARY-VALUE PROBLEMS SOLUTION

It is necessary to find non-linear vector DE solution

$$\frac{dx}{dt} = f(x), \quad (3.1)$$

with given multipoint bound conditions

$$g(x(t_1), x(t_2), \ldots, x(t_n)) = 0, \quad 0 \le t_1 \le t_2 \le \ldots \le t_n \le b. \quad (3.2)$$

Unknown initial conditions x_o DE (3.1) are added in each iteration on the base of sensitivity algorithm [9]:

$$x_o^{l+I} = x_o^l + \gamma^l \, x_o^{l+I}, \quad x_o^{l+I} = - A^+ g^l, - \quad (3.3)$$

where $g^l = g(x^l(t_1), \ldots, x^l(t_n))$,

$$A = \sum_{i=I}^{n} G_i^l U^l(t_i), \quad G_i^l = \frac{dg^l}{dx(t_i)},$$

U^l is sensitivity matrix by x_o.

This algorithm was formerly applied by Paul (1969) for two-point bound problems solution.

By approaching real initial conditions x_o algorithm gives a second-order convergent [9], but the calculations are much easier as compared to quazilinearization algorithm.

4. IDENTIFICATION COMPLEX PROBLEMS STATING

1) Dynamic object and measuring device model structure is given in the form of equations:

$$\frac{dy}{dt} = f(y,\alpha,\beta, t), \quad (4.1)$$

$$F(y_o, t_o(\alpha,\beta,\gamma), \alpha,\beta,\gamma) = 0 \quad (4.2)$$

$$\varphi(\alpha,\beta) = 0, \quad (4.3)$$

$$\eta(t) = \eta(y,\alpha,\beta, t) \quad (4.4)$$

Independent parameters here are α, γ and (4.3) is possible to be solved with regard to β, i.e. $\left|\frac{\partial \varphi}{\partial \beta}\right| \ne 0$. Initial value y_o is connected with unknown parameters α, β, γ and with initial moment of time t_o by unevident relation (4.2) solvable with regard to y_o, t_o being known function $t_o(\alpha,\beta,\gamma)$ of α, β, γ, parameters. It is necessary to calculate the parameters according to $\eta^*(t)$ vector measurements.

The calculation idea according to sensitivity algorithm is the same but the sensitivity equations are more complex. The production of such equations is of a particular interest since the results obtained in these cases have more common character as compared to those of mentioned in literature [3].

2) When the right parts of DE (4.1) are uneven functions, the sensitivity algorithm idea is also applied but it is necessary to apply the results and ideas of Gorodetsky (1971) to produce sensitivity equations.

3) Object model can be given in the form of differential difference equations (DDE) at some other conditions considered above. Now much will depend on chosen way of DDE solution in identification problem solution. For example, if the steps method is chosen, the DDE system solution is just the same as the DE with uneven right parts system solution, and

we come to the case 2 **considered** above.

4) identification of non-stationary objects. Consider the object expressed by usual DE

$$\frac{dz}{dt} = f_1(z, \beta(t), t), \; z(0) = z_0, \qquad (4.5)$$

$$\eta(t) = \eta_1(z(t), \beta(t), \alpha_2, t)$$

Here $\beta(t)$, α_2 are vectors of unknown variables and constant parameters respectively. Following Tsipkin (1972) we find $\beta(t)$ as DE solution

$$\frac{d\beta(t)}{dt} = f_2(z(t), \beta(t), \alpha_1, t), \quad \beta(0) = \beta_0, \qquad (4.6)$$

in which β_0, α_1 are unknown constant parameters. Uniting (4.5) and (4.6) we come to problem § 1.

5. IDENTIFICATION ADAPTIVE ALGORITHMS

Common idea of identification adaptive algorithms production for determined concentrated objects follows from Tsipkin's paper (1972) if quality function is chosen in a form of

$$I = \frac{1}{2}\int_0^t (\eta^*(\tau) - \eta(\tau))^T Q(\tau)(\eta^*(\tau) - \eta(\tau)) \cdot d\tau, \qquad (5.1)$$

where $Q(\tau)$ is known weight matrix reflecting static character of measurement errors. Then estimatores $\hat{y}(t)$, $\alpha(t)$ for phase coordinates of $y(t)$ object and α parameters meet the system of equations

$$\frac{d\hat{y}}{dt} = f(\hat{y}, \alpha(t), t) + U_1(t)\frac{d\alpha(t)}{dt}, \; \hat{y}(0) = y_0,$$

$$\frac{d\alpha}{dt} = -\gamma(t)\, V^T Q\, (\eta^* - \eta), \; \alpha(0) = \alpha_0,$$

$$V = \frac{\partial \eta}{\partial y} U_1 + \frac{\partial \eta}{\partial \alpha}, \; (t) = (y, (t), t),$$

$$U_1(t) = \frac{\partial y(t)}{\partial \alpha}, \qquad (5.2)$$

This system of equation **enables** to **make** estimations more precise constantly with new information provision $\eta^*(t)$. The matrix optimal value calculation $\gamma(t)$ demands $\hat{y}(\tau)$, $\eta^*(\tau)$ to be stored in mind at $\tau \in [0, t]$. This causes difficulties at algorithm realization (5.2). That is why the suboptimal algorithms are being worked out and $\gamma(t)$ (or its analogue) is calculated only on base of current information in these algorithms. One of these algorithms is obtained on base of linearization method application.

$$\frac{d\hat{y}}{dt} = f(\hat{y}, \alpha(t), t) + U_1(t)\frac{d\alpha}{dt}, \; \hat{y}(0) = y_0,$$

$$\frac{d\alpha}{dt} = D(t) V^T Q(\eta^* - \eta), \; \alpha(0) = \alpha_0,$$

$$\frac{dD}{dt} = -DV^T QVD, \; D(0) = D_0. \qquad (5.3)$$

LITERATURE

1. Bellman R., Kalaba R. (1968). Quasilinearization and nonlinear boundary-value problems. Mir Publishers, Moscow.
2. Gavourin M.K., Farforovskaya Yu.B. (1966) About One Method of Finding out the Squares Sum Minimum. J.Comp.Math.and Math.Physics, 6, N 6, 1094.
3. Gorodetsky V.I., Zakharin F.M., Rosenvasser E.N., Yusupov R.M. (1971) Sensitivity Theory Methods in Automic Control. "Energia", Leningrad.
4. Medler Charles L., Hsu Chih-Chi (1969) An algorithm for nonlinear parameter identification. IEEE Trans.Aut.Cont., 14, N 6, 726.
5. Paul R.J.A., Legge C.G. (1969) Direct-sensitivity method of solving boundary-value problems in optimal control studies. Proc. IEE, 116, N 2, 273.
6. Petrov B.N., Kroutko P.D. (1970) Sensitivity Theory Application in Automatic Control Problems. USSR As Proc., Technical Cybernetics, N 2, 202.
7. Rouban A.I. (1969). Some Problems of Mathematical Description of Dynamic Objects. The Candidate Dissertation Tomsk State University.
8. Rouban A.I. (1971) Non-Linear Dynamic Objects Observation and Identification Algorithms. USSR AS Proc., Technical Cybernetics, N 3, 205.
9. Rouban A.I. (1971) Sensitivity Algorithm Application for Non-Linear Boundary Problems Solution. In Collection "Theory of Invariation and Theory of Sensitivity in Automatic Control", part 3, Kiev.
10. Rouban A.I. (1971) Distributed Dynamic Objects Identification on Base of Sensitivity Algorithm. USSR As Proc., Technical Cybernetics N 6, 191.
11. Rouban A.I. (1972) Chemical Reactors Identification on Base of Linearization Method Application. In Collection "Modelling Chemical Processes and Reactors", v.4, part 1, 92.
12. Rouban A.I. (1972) Descrete Dynamic Systems Identification on Base of Linearization Method Application, Automatics and Computing Technique, N 6.
13. Sokolov S.N., Silin I.N. (1961) Finding out Functionals Minimums by Linearization Method. Preprint IINR, D-810, Doubna.
14. Tsipkin Ja.Z. (1972) Dynamic Adaptation Algorithms. Automatics and Telemechanics, N 1.

A SEQUENTIAL METHOD FOR SYSTEM IDENTIFICATION IN HIERARCHICAL STRUCTURE*

N. J. Smith**
IBM Federal Systems Division
Huntsville, Alabama

A. P. Sage
Information and Control Sciences Center
SMU Institute of Technology
Dallas, Texas 75222

This paper is concerned with the application of hierarchical system theory to the identification problem. Specifically, the equations associated with a given identification problem are recast such that they may be decomposed into infimal subproblems of system identification which can be coordinated using hierarchical systems theory. The maximum a posteriori approach to system identification is taken. This leads to a two point boundary value problem solution of which determines optimum state ahd parameter estimates and estimates of any unknown prior statistics. Invariant imbedding is used to resolve this two point boundary value problem such that a recursive or sequential solution to the identification problem is obtained. Several examples indicate the use of the identification algorithms.

INTRODUCTION

Hierarchical techniques [1,2] are currently being applied to many areas of modern system theory such as optimization, state estimation, and identification [3,4]. The concept of structuring a system into subsystems and coordinating these infimal units with a supremal controller allows much versatility and is especially useful in systems of high dimensionality.

Probably the most restrictive feature of hierarchical system theory is its inherently iterative nature. For optimization this is a necessary part of the solution since optimality is desired over the entire time interval of interest. For estimation and identification problems, it seems reasonable to seek a sequential coordination technique which will allow hierarchical estimation problems to be solved in real time. This work presents such a sequential coordination technique. The maximum a posteriori (MAP) approach to estimation theory [5-7] is taken and the resulting two point boundary value problem (TPBVP) is cast into a hierarchical structure. The infimal TPBVP's are solved using invariant imbedding [7] and coordinated sequentially. This technique allows state estimation and parameter identification to be carried out on-line in a hierarchical system.

When estimation and identification algorithms are actually implemented, it is usually assumed that the noise parameters are known exactly. This assumption is rather restrictive and for some problems is unacceptable. For example, in the modeling of urban and societal systems, noise sources are often difficult to identify and the prior determinationof their moments may not be feasible. Hence, the sequential identification algorithms would be applicable to a far wider class of systems if they did not require exact knowledge of noise parameters.

Two previous approaches to this problem are related to the problem considered here. Sage and Wakefield [8] use a maximum likelihood technique to determine the estimator gain directly. Sage and Husa [9] identify the noise moments to obtain adaptive estimation algorithms. Both these results are incorporated into the sequential hierarchical identification algorithms and several examples are solved to demonstrate their use.

*This research was supported by the Air Force Office of Scientific Research under contract no. F44620-68-C-0023 and the National Science Foundation under Grant GK33348.

**Formerly with the Information and Control Sciences Center, SMU Institute of Technology, Dallas, Texas 75222.

HIERARCHICAL DECOMPOSITION

Cost function development for MAP identication is well known [5-7] and results in the need to minimize

$$\bar{J} = \frac{1}{2}\|x(t_0) - \mu_x(t_0)\|^2_{V_x^{-1}(t_0)} + \frac{1}{2}\int_{t_0}^{t_f}\{\|z(t)-h[x(t),t]\|^2_{R^{-1}(t)} + \|w(t)\|^2_{Q^{-1}(t)}\}dt \quad (1)$$

where t_f is fixed, subject to the differential equation equality constraint representing the message model dynamics

$$\dot{x}(t) = \bar{f}[x(t),t] + w(t) \quad , \quad E\{x(t_0)\} = \mu_{x_0} \quad (2)$$

The continuous-time observation equation is assumed to be

$$z(t) = h[x(t),t] + v(t) \quad (3)$$

The plant and measurement noises are assumed to be white uncorrelated zero mean gaussian sequences with $\mathrm{Cov}\{w(t),w(\tau)\}=Q(t)\delta_D(t-\tau)$ and $\mathrm{Cov}\{v(t),v(\tau)\}=R(t)\delta_D(t-\tau)$, where δ_D is the Dirac delta function.

We may now apply hierarchical system theory to this optimization problem. From [4] it follows that the infimal problem is: minimize wrt $w_i(t)$

$$J_i = \frac{1}{2}\|x_i(t_0)-\mu_{x_{i0}}\|^2_{V^{-1}_{x_{i0}}} + \int_{t_0}^{t_f}\{\|z_i-h_i[x_i,t]\|^2_{R_i^{-1}(t)} + \frac{1}{2}\|w_i(t)\|^2_{Q_i^{-1}(t)} + \beta_i'(t)\pi_i(t) - \sum_{j\neq i}^{N}\beta_j'(t)g_{ji}(x_i)\}dt \quad (4)$$

subject to

$$\dot{x}_i(t) = f_i[x_i,\pi_i,t] + w_i(t) \quad (5)$$

and with

$$z_i(t) = h_i[x_i,t] + v_i(t) \quad (6)$$

where $\pi_i(t)$ represents the effects of the other infimal units upon infimal unit i. Correlated noises may cause difficulties in decomposing the cost functional since Q(t) and R(t) may not be block diagonal. We now address this problem briefly. Correlation between plant and measurement noises may be handled by standard techniques [6,p.280]. For plant noise correlation among the subsystems, we may define a new noise input $w(t)=G(t)w_I(t)$ where $G(t)G'(t)=Q(t)$ and $\mathrm{Cov}\{w_I(t),w_I(\tau)\}=I\delta_D(t-\tau)$. For measurement noise correlation, we define $\bar{z}(t)=A(t)z(t)$ where A(t) is chosen so that the performance functional may be properly decomposed. Then $\pi_i(t)$ is a function only of the other states so that

$$\pi_i(t) = g_i[x_j(t)] \quad , \quad j \neq i \tag{7}$$

Applying the maximum principle [7] yields

$$\hat{w}_i(t) = -Q_1(t)\lambda_i(t) \tag{8}$$

$$-\frac{\partial H_i}{\partial \hat{x}_i} = \dot{\lambda}_i(t) = \frac{\partial h_i'[\hat{x}_i,t]}{\partial \hat{x}_i(t)} R_i^{-1}(t)\{z_i-h_i[\hat{x}_i,t]\} - \frac{\partial f_i'[x_i,\pi_i,t]}{\partial \hat{x}_i(t)}\lambda_i(t)+\frac{\partial}{\partial \hat{x}_i}\sum_{j\neq i}^{N}\beta_j'(t)g_{ji}(\hat{x}_i) \tag{9}$$

with H_i the Hamiltonian for the *i*th problem in (4)-(6)

$$\lambda_i(t_0) = -V_{x_{i0}}^{-1}[\hat{x}_i(t_0) - \mu_{x_{i0}}] \quad , \quad \lambda_i(t_f)=0 \tag{10}$$

Thus, Eqs. (5) and (8) through (10) specify the TPBVP which the *i*th infimal unit must solve. Strictly speaking, the solution of this TPBVP yields the fixed interval smoothing estimate of x(t).

The invariant imbedding technique will be used to solve for $\hat{x}(t)$. The important feature of the invariant imbedding approach is that the solution is sequential. Thus because of the way the TPBVP is solved, we will obtain the filter estimate of the states and any unknown parameters which are to be identified. If we use sequential coordination, then the overall hierarchical identification procedure will be sequential and may be performed on-line.

The basic idea of invariant imbedding is to replace a specific problem with a more general one. Then once the general problem is solved, the solution to the specific one is also obtained. Interestingly enough, it is often easier to solve the general problem than the specific one. We begin by imbedding $\lambda_i(t_f)$; that is, we let

$$\lambda_i(t_f) = C \tag{11}$$

where C can take on any value including C = 0, and t_f is now variable rather than fixed. Now assume that $\hat{x}_i(t_f)$ can be expressed as a function of C and t_f, say

$$\hat{x}_i(t_f) = r(C,t_f) \tag{12}$$

Now for convenience, let α and σ represent the right sides of Eqs. (5) and (9) respectively. Then following [5-7] a partial differential equation may be derived for $r(C,t_f)$

$$\frac{\partial r(C,t_f)}{\partial t_f} + \frac{\partial r(C,t_f)}{\partial C}\sigma[r,C,t_f] = \alpha[r,C,t_f] \tag{13}$$

This is a vector partial differential equation and its general solution is not known. Hence, we seek an approximate solution

$$r(C,t_f) = \hat{x}_i(t_f) - P_i(t_f)C \tag{14}$$

This approximation will lead to differential equations for $\hat{x}_i(t)$ and $P_i(t)$ which will yield the filter estimate of $x_i(t)$. We now substitute this approximate solution into Eq. (13). Expanding α and σ in a Taylor series about C = 0 and retaining only terms up to first order in C yields the desired result. Equating like powers of C and replacing t_f by the running time variable t produces differential equations for $\hat{x}_i(t)$ and $P_i(t)$ as summarized in Table 1. Thus using the invariant imbedding procedure to solve the MAP identification problem yields a sequential solution to the infimal optimization problems. If these infimal solutions are coordinated using a sequential coordination technique then the overall hierarchical identification process will be sequential and may be carried out in real time. We now turn our attention to the coordination process.

SEQUENTIAL COORDINATION--COORDINATION FOR IMPROVEMENT

The determination of proper values for the coordination variables $\beta(t)$ and $\pi(t)$ is the central problem of hierarchical system theory. It is through proper manipulation of these variables by the supremal controller that the infimal units meet the overall system objectives. The basic objective of coordination is to insure that for the final solution the interconnection constraint of Eq. (7) is satisfied, for only then can we claim that the infimal solutions also solve the original problem.

The coordination technique used here will be based upon the Prediction Principle of Mesarovic, *et al.* [1]. Using this coordination method, the supremal predicts a value for the interconnection variable $\pi(t)$ and supplies a value for $\beta(t)$ to the infimal units. Each infimal unit then solves its problem and g[x(t)] is computed. The Prediction Principle states that the overall problem has been solved if $\pi(t) = g[x(t)]$. That is, a value of $\beta(t)$ has been found which allows the supremal to correctly predict the value of $\pi(t)$.

The coordination variables $\pi(t)$ and $\beta(t)$ play basically different roles in the coordination process. $\pi(t)$ is more closely associated with the infimal units and their optimization problems. On the other hand, $\beta(t)$ is more closely associated with the supremal unit and Pearson [2] has shown that we actually wish to maximize the overall system Hamiltonian wrt $\beta(t)$. This is a very constrained maximization problem, however, taking place only over those solutions which minimize the infimal costs. This is not as inconsistent as it may seem since we would expact the cost for those solutions satisfying the interconnection constraint to be greater than the cost for those solutions which do not.

The first requirement for on-line coordination is the ability to propagate the coordination variables in time; that is, we need differential equations describing the coordination variables. The second requirement is that these differential equations propagate in such a way that the Hamiltonian is extremized, or more realistically, in such a way that the Hamiltonian extremization is improved. This suggests a two step, "predictor-corrector" coordination process:

- Predict the value of the coordination variables using an approximation to the derivative.
- Correct this value using Hamiltonian extremization.

These two steps will now be discussed in detail. Let Δt represent the computational step size. Then a linear approximation to the derivative is given by

$$\dot{\beta}(t) = \frac{1}{\Delta t}\,[\beta(t) - \beta(t-\Delta t)] \tag{15}$$

The coordination variables may then be propagated by

$$\beta_p(t+\Delta t) = \beta(t) + \dot{\beta}(t)\Delta t \tag{16}$$

where the subscript p denotes the predicted value of the variable. Similar expressions may be written for π and π_p. We also wish to adjust the coordination variables so that the Hamiltonian is minimized. Hence, the supremal updates the predicted values of the coordination varibles as

$$\beta(t) = \beta_p(t) + K_\beta \frac{\partial H(t)}{\partial \beta(t)} \tag{17}$$

$$\pi(t) = \frac{1}{2}\,(\pi_p(t) + g[x(t)]) \tag{18}$$

$\beta(t)$ is updated using a one step gradient technique while $\pi(t)$ is updated using an averaging process. However, Eq. (7) was used in place of Eq. (20) with similar results. This method for predicting $\pi(t)$ was motivated by the equality method of Gunizy and Sage [4]. Another logical choice would be to predict $\pi(t)$ by minimizing the infimal Hamiltonians but this would require solving for $\lambda(t)$. The complete sequential identification process is detailed in the flow diagram presented in Fig. 1. This coordination process allows real-time, on-line solutions if the infimal solutions are also sequential.

UNKNOWN NOISE MOMENTS

It is convenient to assume that plant and measurement noise moments are precisely known when estimation algorithms are actually implemented. Then if estimation accuracy is unacceptable, some attempt is made to obtain more accurate values for these moments in the hope that these new values will produce acceptable results.

To accomplish identification of the prior statistics, the results of Sage and Husa [9] will be incorporated into the sequential hierarchical identification algorithms. Sage and Wakefield use a maximum likelihood technique to treat the case of unknown plant noise covariances. Using this approach, the estimator gain is determined directly by maximizing a likelihood function. The resulting algorithms do not require knowledge of the plant noise covariance but do require the measurement noise moments. If these moments are unknown or random, they may be adaptively identified using the covariance matching techniques of Sage and Husa [9]. Alternatively, it would appear feasible to incorporate any of the existing sequential algorithms for the identification of prior statistics [10,11] into the hierarchical framework.

The alteration of the sequential hierarchical identification algorithms to incorporate these results is straightforward. Each infimal is assumed to be solving a separate problem and the coordination variables are treated as known inputs for these problems Table 1 also presents the sequential hierarchical identification algorithms for unknown noise moments.

COMPUTATIONAL RESULTS

A simple example will now be solved to demonstrate the use of the invariant imbedding approach to system identification in hierarchical structures. Consider the linear second order system

$$\dot{x}_1 = -a_1x_1 + x_2 + w_1 \quad ; \quad \dot{x}_2 = x_1 - a_2x_2 + w_2$$

$$z_1 = x_1 + v_1 \quad ; \quad z_2 = x_2 + v_2$$

with $\mu_{x_1}(t_0) = 5.5$, $V_{x_1}(t_0) = 1.0$, and $\mu_{x_2}(t_0) = 4.5$, $V_{x_2}(t_0) = 1.0$. The parameters a_1 and a_2 are unknown and we wish to determine their respective values based upon the observations z_1 and z_2.

We adjoin the unknown parameters as new states and introduce interconnection variables π_1 and π_2 to decompose the system into two infimal units.

Infimal Unit 1

$$\dot{x}_{11} = -x_{12}x_{11}+\pi_1+w_{11} \quad ; \quad \dot{x}_{12} = w_{12} \quad ; \quad z_2 = x_{21}+v_2$$

Infimal Unit 2

$$\dot{x}_{21} = \pi_2-x_{22}x_{21}+w_{21} \quad ; \quad \dot{x}_{22} = w_{22} \quad ; \quad z_2 = x_{21}+v_2$$

The plant and measurement noises are assumed to be white, zero mean, gaussian processes with $Q_{11} = Q_{22} = 1$, $Q_{12} = Q_{21} = 0$, and $R_1 = R_2 = 0.01$. The parameters to be identified, a_1 and a_2, represented by x_{12} and x_{22}, are assumed to be samples from a gaussian population with a mean of 2.0 and a variance of 1.0. This is not to imply that a_1 and a_2 are random. This is for convenience in making initial estimates and physically represents prior knowledge of the approximate values of a_1 and a_2.

To generate an initial trajectory for $\pi(t)$, the system was simulated using the mean values of a_1, a_2, $x_{11}(0)$ and $x_{21}(0)$ with no noise. The initial guess for $\pi(t)$ was then taken as $\pi_1(t) = x_{21}(t)$ and $\pi_2(t) = x_{11}(t)$. This is a reasonable choice of $\pi(t)$ since it uses all the available information. Choosing $\beta(t) = 0$ means that the infimal units are initially uncoupled.

The infimal identification problems are solved using the invariant imbedding algorithms of Table 1. The coordination is performed sequentially as detailed in Fig. 1. The results of this identification process are shown in Fig. 2. As may be noted from the figure, the correct values of a_1 and a_2 were determined very accurately. The important advantage of the sequential coordination is that the identification can be performed sequentially so that it may be carried out on-line.

The example will now be solved using the results for unknown measurement noise moments. Specifically, the sequential identification algorithms will be used as before, but the measurement noise moments will be identified using the measurement noise moment algorithms of Table 1. The results of identifying a_1 and a_2 are presented in Fig. 3 and the results of identifying the noise moments for Infimal Unit 2 are shown in Fig. 4. The results for Infimal Unit 1 were comparable. As may be seen from the figures, the results are now comparable to those for the case where the measurement noise moments were known.

The example will now be solved as though the plant noise covariance is unknown. The results of identifying a_1 and a_2 are presented in Fig. 4. The results of identifying the estimator gains for Infimal Unit 2 are presented in Fig. 5. The gains begin tracking the optimal steady state value quickly. The identification time for a_1 and a_2 is increased somewhat as might be expected.

References

[1] Mesarovic, M. D., *et al.*, Theory of Hierarchical Multilevel Systems, Academic Press, 1970.

[2] Pearson, J. D., "Dynamic Decomposition Techniques," in D. Wismer (ed.) Optimization Methods for Large Scale Systems, McGraw-Hill, 1971.

[3] Noton, A. R., "A Two-Level Form of the Kalman Filter," IEEE Trans. Auto. Control, April 1971.

[4] Guinzy, N. J. and A. P. Sage, "System Identification in Large Scale and Hierarchical Structures," Computers & Electrical Engineering, 1973.

[5] Sage, A. P. and J. L. Melsa, System Identification, Academic Press, 1971.

[6] Sage, A. P. and J. L. Melsa, Estimation Theory with Applications to Communications and Control, McGraw-Hill, 1971.

[7] Sage, A. P., Optimum Systems Control, Prentice Hall, 1968.

[8] Sage, A. P. and C. D. Wakefield, "Maximum Likelihood Identification of Time Varying and Random System Parameters," Int. J. Control, July 1972.

[9] Sage, A. P. and G. W. Husa, "Adaptive Filtering with Unknown Prior Statistics," Proc. JACC, 1969.

[10] Sage, A. P. and G. W. Husa, "Algorithms for Sequential Adaptive Estimation of Prior Statistics," Proc. 8th Sym. on Adaptive Processes, 1969.

[11] Mehra, R. K., "Approaches to Adaptive Filtering," IEEE Trans. Auto. Con., October 1972.

TABLE 1

System Model	$\dot{x}_i(t) = f_i[x_i, \pi_i, t] + w_i(t)$
Observation Model	$z_i(t) = h_i[x_i, t] + v_i(t)$
Partial Gain Model	$\dot{\gamma}_i(t) = A_i(t)\gamma_i(t) + \zeta_i(t)$
Gain Model	$K_i(t) = T_{1i}n_i(t) + T_{2i}\gamma_i(t)$
Prior Statistics	$E\{x_i(t_o)\} = \mu_{x_{io}}$, $Var\{x_i(t_o)\} = V_{x_{io}}$ $E\{\gamma_i(t_o)\} = \mu_{\gamma_{io}}$, $Var\{\gamma_i(t_o)\} = V_{\gamma_{io}}$ $E\{v_i(t)\} = \mu_{v_i}(t)$, $E\{w_i(t)\} = E\{\zeta_i(t)\} = 0$ $Cov\{w_i(t), w_i(\tau)\} = Q_i(t)\delta_D(t-\tau)$ $Cov\{v_i(t), v_i(\tau)\} = R_i(t)\delta_D(t-\tau)$ $Cov\{\zeta_i(t), \zeta_i(\tau)\} = S_i(t)\delta_D(t-\tau)$ All noise sources are uncorrelated. $Q_i(t)$, $R_i(t)$, $\mu_{v_i}(t)$ are random or unknown.
Innovation	$\tilde{z}_i(t) = z_i(t) - h_i[\hat{x}_i, t] - \hat{\mu}_{v_i}(t)$
Definitions	$\beta(t)$ and $\pi(t)$ are coordination variables. $F_i(t) \triangleq \frac{\partial f_i[\hat{x}_i, \pi_i, t]}{\partial \hat{x}_i(t)}$, $H_i(t) \triangleq \frac{\partial h_i[\hat{x}_i, t]}{\partial \hat{x}_i(t)}$
Partial Gain Estimator	$\dot{\hat{\gamma}}_i(t) = A_i(t)\hat{\gamma}_i(t) + P'_{x\gamma_i}(t)H'_i(t)\hat{R}_i^{-1}(t)\tilde{z}_i(t)$
Gain Algorithm	$\hat{K}_i(t) = T_{1i}n_i(t) + T_{2i}\hat{\gamma}_i(t)$
Filter Algorithm	$\dot{\hat{x}}_i(t) = f_i[\hat{x}_i, \pi_i, t] + P_{x_i}(t)H'_i(t)\hat{R}_i^{-1}(t)\tilde{z}_i(t)$ $+ \hat{K}_i(t)\tilde{z}_i(t) + P_{x_i}(t)\frac{\partial}{\partial \hat{x}_i}(\sum_{j \neq i}^{N} \beta'_j(t) g_{ji}(\hat{x}_i))$
Measurement Noise Moment Algorithms	$\dot{\hat{\mu}}_{v_i}(t) = \frac{1}{t}\{ z_i(t) - h_i[\hat{x}_i, t] \}$ $\dot{\hat{R}}_i(t) = \frac{1}{t}\tilde{z}_i(t)\tilde{z}'_i(t)$
Variance Algorithms	$\dot{P}_{x_i}(t) = P_{x_i}(t)\{F_i(t) - \hat{K}_i(t)H_i(t)\}'$ $+ \{F_i(t) - \hat{K}_i(t)H_i(t)\}P_{x_i}(t)$ $+ \{T_{2i}P'_{x\gamma_i}(t) + P_{x\gamma_i}(t)T'_{2i}\}\tilde{z}_i(t)$ $+ P_{x_i}(t)\frac{\partial}{\partial \hat{x}_i}\{H'_i(t)\hat{R}_i^{-1}(t)\tilde{z}_i(t)\}P_{x_i}(t)$ $+ P_{x_i}(t)\frac{\partial}{\partial \hat{x}_i}(\frac{\partial}{\partial \hat{x}_i} \sum_{j \neq i}^{N} \beta'_j(t) g_{ji}(\hat{x}_i))P_{x_i}(t)$ $\dot{P}_{x\gamma_i}(t) = \{F_i(t) - \hat{K}_i(t)H_i(t)\}P_{x\gamma_i}(t)$ $+ P_{x\gamma_i}(t)A'_i(t) + T_{2i}P_{\gamma_i}(t)\tilde{z}_i(t)$ $+ P_{x_i}(t)\frac{\partial}{\partial \hat{x}_i}\{H'_i(t)\hat{R}_i^{-1}(t)\tilde{z}_i(t)\}P_{x\gamma_i}(t)$ $\dot{P}_{\gamma_i}(t) = A_i(t)P_{\gamma_i}(t) + P_{\gamma_i}(t)A'_i(t) + S_i(t)$ $+ P'_{x\gamma_i}(t)\frac{\partial}{\partial \hat{x}_i}\{H'_i(t)\hat{R}_i^{-1}(t)\tilde{z}_i(t)\}P_{x\gamma_i}(t)$
Initial Conditions	$\hat{x}_i(t_o) = \mu_{x_{io}}$, $P_{x_i}(t_o) = 0$, $P_{x\gamma_i}(t_o) = 0$ Initial Conditions for measurement noise moments are best guess values. Initial conditions for γ_i are determined from the steady state value of the linear filter gain using best guess values for noise parameters.

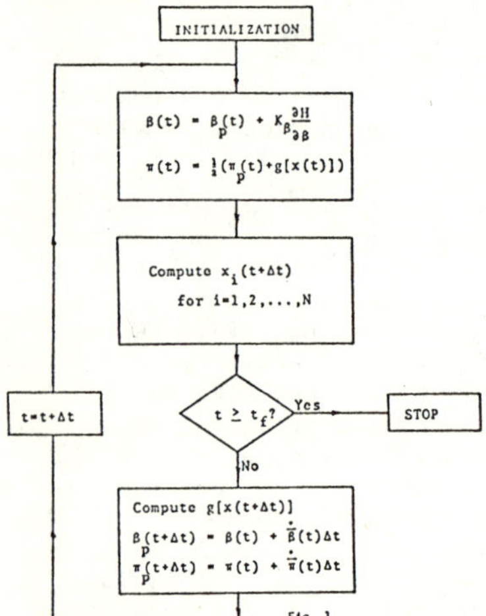

Fig. 1

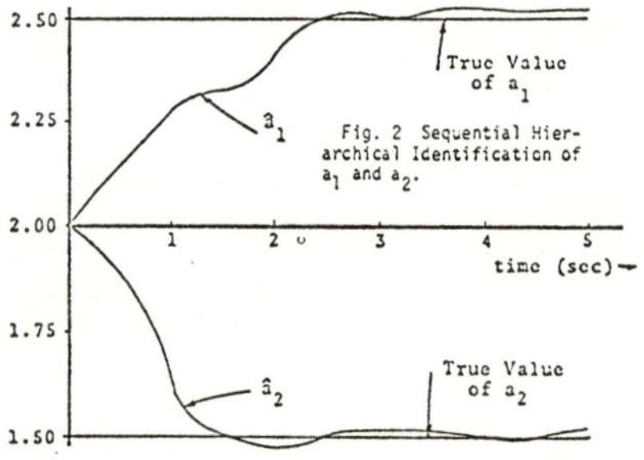

Fig. 2 Sequential Hierarchical Identification of a_1 and a_2.

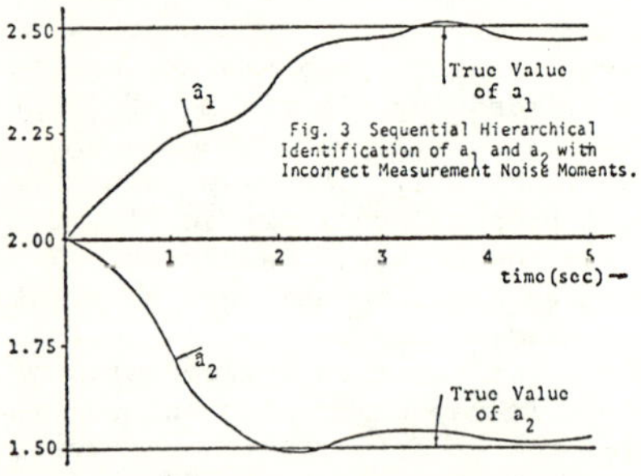

Fig. 3 Sequential Hierarchical Identification of a_1 and a_2 with Incorrect Measurement Noise Moments.

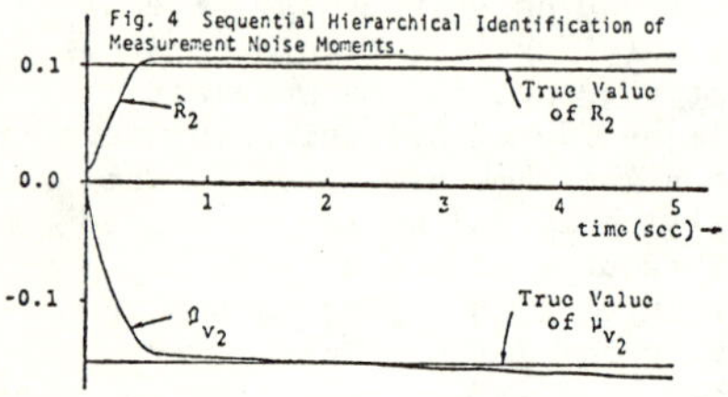

Fig. 4 Sequential Hierarchical Identification of Measurement Noise Moments.

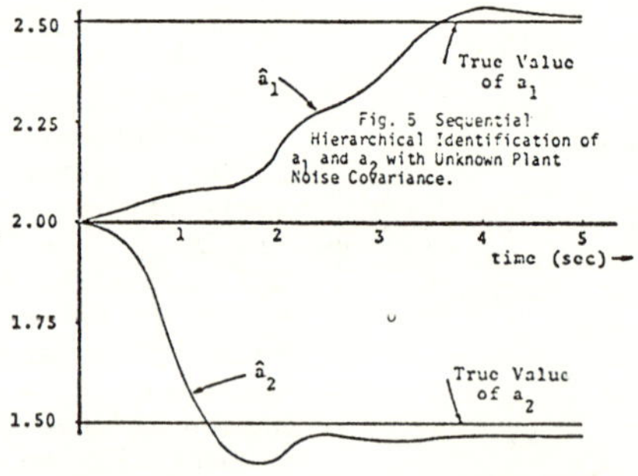

Fig. 5 Sequential Hierarchical Identification of a_1 and a_2 with Unknown Plant Noise Covariance.

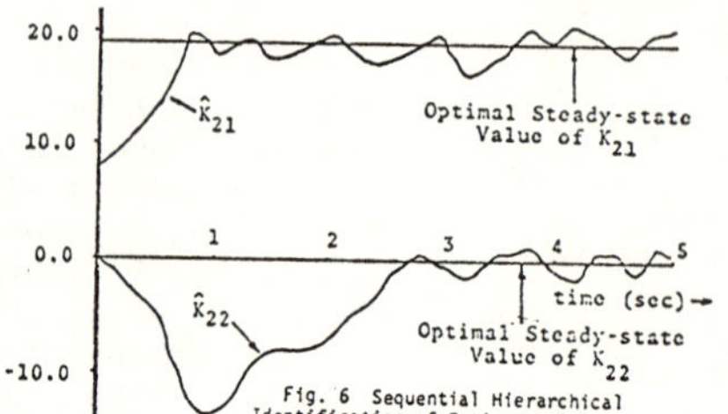

Fig. 6 Sequential Hierarchical Identification of Estimator Gains.

ON BIAS REDUCTION IN FILTERING PROBLEMS WITH NON-LINEAR MEASUREMENTS BY FIRST ORDER ITERATIVE METHODS*

HUYNH huu Thanh
Office National d'Etudes et de Recherches Aérospatiales (O.N.E.R.A.)
92320 Châtillon - France

The paper presents a study on first order filtering methods for estimation problems with measurements which are non-linear functions of the state. A qualitative interpretation of the linearized, the extended and the iterated Kalman filters is first given in the scalar case. The superiority of the iterated filters for bias reduction, as compared with the classical filters, is then proved when the measurement noise is very small relative to the estimate error or for a large number of measurements. These results are illustrated by a numerical simulation of the attitude restitution of a spinning satellite.

INTRODUCTION

For a large class of estimation problems (restitution of satellite attitude, orbit, rocket trajectory,...) the linear Kalman-Bucy filter (1961) has been used successfully, by linearizing the dynamic equations or the measurements equations along a nominal path.

Nevertheless, in some conditions, for instance in presence of large non-linearities, the classical linearized or extended Kalman filter does not give a reliable convergence of the estimates. More sophisticated methods have to be considered to overcome the difficulties encountered. One of them, called the iterated Kalman filter, which remains a first order method, seems to be appropriate for reducing the non-linear effects and to provide results equivalent to those given by second order methods - Denham (1966) - Jazwinski (1970) Wishner et al. (1969) - Mehra (1971).

According to experimental results, the iterated Kalman method is particularly useful with respect to the extended Kalman method when the measurement noise is small and the estimate error still remains large. Denham (1966) - Jazwinski (1970).

This paper tries to justify, in a theoretical way, the superiority of two iterated filters with respect to classical filters for reducing the non-linear effects in some particular cases. The study begins with a qualitative explanation of the Kalman filters behaviour. An illustration is also given by numerical results of a satellite attitude restitution.

GEOMETRICAL INTERPRETATION OF FIRST ORDER ITERATIVE FILTERS

The Kalman filtering equations are first written in a suitable form, which allows for an interesting geometrical interpretation.

Filtering equations

We consider, in this study, a non evolutive system and, to simplify the analysis, only the scalar case is discussed, except for the numerical example.

The measurement, at discrete time t_k, is :

$$y_k = h(x) + \eta_k \qquad k = 1,2,\ldots \tag{1}$$

the state x is a Gaussian random variable with a priori mean $\hat{x}_0$ and variance Γ_0 ; $h(x)$ is a non-linear function of the state x ; η_k is the "Gaussian white noise" independant of x, with zero mean and variance R

The estimation problem consists in computing the conditional mean $E(x/Y_N)$ and the conditional variance $cov(x/Y_N) = E\{(x - E(x/Y_N))^2/Y_N\}$ where Y_N represents the available measurement set $\{y_1, y_2, \cdots, y_N\}$.

Let us recall first the linearized or extended Kalman filter.

The linear Kalman-Bucy (1961) filter is applied to this problem by linearizing the measurement equations around a set of nominal values $\bar{x}_0, \bar{x}_1, \cdots, \bar{x}_k \ldots$. The approximate values $\hat{x}_k$ and Γ_k of the conditional mean and conditional variance are computed from the recursive equations :

$$\left.\begin{aligned}
&\Delta\hat{x}_k = \bar{x}_{k-1} - \bar{x}_k + \Delta\hat{x}_{k-1} + K_k(\Delta y_k - H_{k-1}\Delta\hat{x}_{k-1})\\
&K_k = \Gamma_{k-1}H_{k-1}(\Gamma_{k-1}H_{k-1}^2 + R)^{-1}\\
&\Gamma_k = \Gamma_{k-1} - \Gamma_{k-1}^2 H_{k-1}^2(\Gamma_{k-1}H_{k-1}^2 + R)^{-1} = \frac{\lambda_{k-1,k-1}}{1+\lambda_{k-1,k-1}}\Gamma_{k-1}\\
&\text{with : } \Delta\hat{x}_k = \hat{x}_k - \bar{x}_k, \quad \Delta y_k = y_k - h(\bar{x}_{k-1})\\
&\qquad H_{k-1} = [\partial h/\partial x]_{x=\bar{x}_{k-1}}, \quad \lambda_{k-1,k-1} = R/\Gamma_{k-1}H_{k-1}^2
\end{aligned}\right\} \tag{2}$$

The linearized Kalman filter corresponds to nominal values $\bar{x}_0, \bar{x}_1, \ldots \bar{x}_k \ldots$ equal to a constant (usually the a priori estimate $\hat{x}_0$) the extended Kalman filter corresponds to a linearization around the previous estimate.

The iterated Kalman filters, which were proposed by Denham and Pines (1966), try to improve the linearization of the measurement equations by a change of the nominal values $\bar{x}_0, \bar{x}_1, \ldots \bar{x}_k \ldots$

The local iterated filter consists in performing I iterations at each instant $t_1, t_2 \ldots t_k$. At each iteration the linearization is performed around the estimate obtained from the previous iteration. The corresponding equations are :

$$\left.\begin{aligned}
&\hat{x}_k^i = \hat{x}_{k-1} + K_k^i\left[y_k - h(\hat{x}_k^{i-1}) - H_k^{i-1}(\hat{x}_{k-1} - \hat{x}_k^{i-1})\right]\\
&K_k^i = \Gamma_{k-1}H_k^{i-1}\left[\Gamma_{k-1}(H_k^{i-1})^2 + R\right]^{-1} \qquad i = 1,2,..I
\end{aligned}\right\} \tag{3}$$

$$\text{with} \qquad \hat{x}_k^0 = \hat{x}_{k-1}, \quad H_k^{i-1} = \left[\frac{\partial h}{\partial x}\right]_{x=\hat{x}_k^{i-1}}$$

* *This work was supported in part by the European Space Research and Technology Center (ESTEC) under contracts nº 1179/70 HP and nº 1617/72 SK. The autor wishes to acknowledge the contribution of Miss F. Depasse, Drs C. Aumasson and J.P. Marec.*

The estimate and its variance at instant t_k are :

$$\left.\begin{aligned} &\hat{x}_k = \hat{x}_k^I \\ &\Gamma_k = \Gamma_{k-1} - (\Gamma_{k-1} H_k^{I-1})^2 \left[\Gamma_{k-1}(H_k^{I-1})^2 + R\right]^{-1} = \frac{\lambda_{k-1,I-1}}{1+\lambda_{k-1,I-1}}\,\Gamma_{k-1} \\ &\text{with} \quad \lambda_{k-1,I-1} = R/\Gamma_{k-1}(H_k^{I-1})^2 \end{aligned}\right\} \quad (4)$$

The global iterated filter consists in iterating on a large number of measurements. At each run, the linearization is done around the last estimate of the previous iteration and the estimation process starts again with the a priori mean $\hat{x}_0$ and variance Γ_0.

The corresponding equations are then, for the i th iteration :

$$\left.\begin{aligned} &\Delta\hat{x}_k^i = \Delta\hat{x}_{k-1}^i + K_k^i(\Delta y_k^i - \bar{H}^i \Delta\hat{x}_{k-1}^i) \\ &K_k^i = \Gamma_{k-1}^i \bar{H}^i(\Gamma_{k-1}^i(\bar{H}^i)^2 + R)^{-1} \\ &\Gamma_k^i = \Gamma_{k-1}^i - (\Gamma_{k-1}^i \bar{H}^i)^2(\Gamma_{k-1}^i(\bar{H}^i)^2 + R)^{-1} \end{aligned}\right\} \quad (5)$$

$$k = 1,2,\ldots,N \;;\; i = 1,2,\ldots$$

With $\Delta\hat{x}_k^i = \hat{x}_k^i - \bar{x}^i$, $\bar{x}^1 = \hat{x}_0$, $\bar{x}^i = \hat{x}_N^{i-1}$ $(i>1)$

$$\hat{x}_0^i = \hat{x}_0 \,,\quad \Gamma_0^i = \Gamma_0 \,,\quad \Delta y_k^i = y_k - h(\bar{x}^i) \,,\quad \bar{H}^i = \left[\frac{\partial h}{\partial x}\right]_{\bar{x}^i}$$

Interpretation of the Kalman filter

Let us define two values x_k and $\tilde{x}_k$ of the state x as follows : (see fig. 1)

$$y_k = h(x_k) \quad , \quad y_k - h(\bar{x}_{k-1}) = H_{k-1}(\tilde{x}_k - \bar{x}_{k-1})$$

x_k is the state value relative to the measurement y_k, $\tilde{x}_k$ can be considered as an approximation of this value, obtained by replacing the curve $h(x)$ by its tangent at $\bar{x}_{k-1}$.

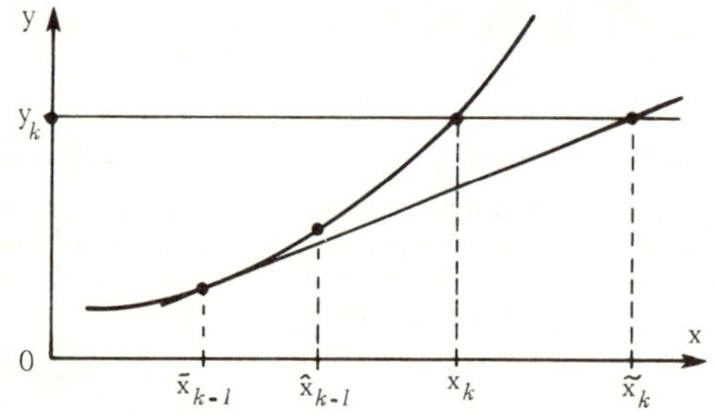

Fig. 1 - Interpretation of the Kalman filter.

Taking into account the definition of $\tilde{x}_k$, the equations (2) become :

$$\left(\frac{1}{\Gamma_{k-1}} + \frac{H_{k-1}^2}{R}\right)\hat{x}_{k-1} = \frac{1}{\Gamma_{k-1}}\hat{x}_{k-1} + \frac{H_{k-1}^2}{R}\tilde{x}_k$$

$$\frac{1}{\Gamma_k} = \frac{1}{\Gamma_{k-1}} + \frac{H_{k-1}^2}{R}$$

If the measurement was linear in terms of x, we should have $\tilde{x}_k \equiv x_k$ and the estimate $\hat{x}_k$ could be considered as the barycenter of the previous estimate $\hat{x}_{k-1}$, and the estimate x_k resulting from the measurement y_k, with masses respectively equal to the inverse of the mean square errors $E[(\hat{x}_{k-1}-x)^2/Y_{k-1}] = \Gamma_{k-1}$ and $E[(x_k - x)^2/Y_{k-1}] = R/H_{k-1}^2$

In the actual case, the linearization of the measurement equation consists in replacing x_k by its approximate value $\tilde{x}_k$. It is obvious that this approach is justified only if the nominal value $\bar{x}_{k-1}$ (around which the linearization is performed) is close to the value x_k.

Interpretation of the iterated Kalman filter

We only give the interpretation for the local iterations case. The case of global iterations could be deduced easily.

Let us consider, as previously, a value $\tilde{x}_k^i$ defined by the equation : (see fig. 2)

$$y_k - h(\hat{x}_k^{i-1}) = H_k^{i-1}(\tilde{x}_k^i - \hat{x}_k^{i-1})$$

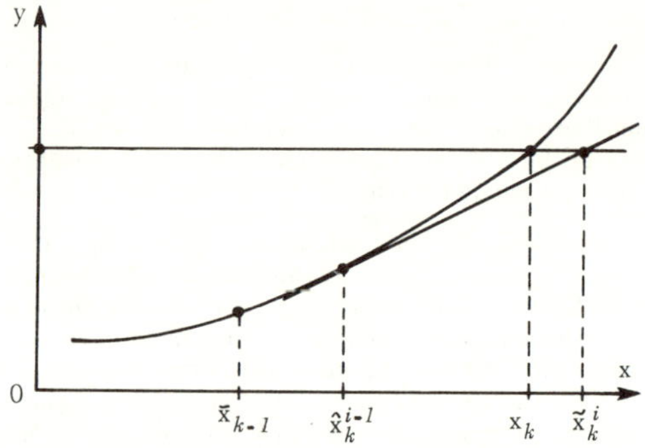

Fig. 2 - Interpretation of the iterated Kalman filter.

The filter equations (3) can be written then :

$$\left(\frac{1}{\Gamma_{k-1}} + \frac{(H_k^{i-1})^2}{R}\right)\hat{x}_k^i = \frac{1}{\Gamma_{k-1}}\hat{x}_{k-1} + \frac{(H_k^{i-1})^2}{R}\tilde{x}_k^i$$

An interpretation can be given similar to the one presented for the Kalman filter, the difference is that the approached value $\tilde{x}_k$ is replaced by $\tilde{x}_k^i$ which is assumed to be closer to the actual value x_k than $\tilde{x}_k$.

The successive iterations then consists in trying to improve the knowledge of the state x_k corresponding to the actuel measurement y_k.

Let us remind that the (local) iterated Kalman filter has also a probabilistic interpretation.

With the assumptions that $\hat{x}_{k-1}$ is the conditional mean $E(x/Y_{k-1})$ with a Gaussian probability density and conditional variance Γ_{k-1}, it is known that the iterated Kalman filter is a modified version of the Newton-Raphson method for finding the maximum likelihood estimate for the conditional probability $p(x/Y_k)$ (see Jazwinski (1970), page 350).

THEORETICAL STUDY OF BIAS REDUCTION IN QUADRATIC CASE

Two particular cases are considered in the following : reduction of the bias by the local and global iterated Kalman filters. Assumptions are made about the ratio between the measurement noise and the estimate variance, and the non-linear function $h(x)$ is supposed to be quadratic.

Bias reduction by local iterations

We consider here only the effect of the iterations on one measurement.

Let us assume that $\hat{x}_{k-1}$ is the conditional mean, with conditional variance Γ_{k-1}, and derive the bias of the estimate in the extended Kalman filter.

Expanding the quantity Δy_k from equation (2) the expression for the difference $\varepsilon_k = \hat{x}_k - x$ can be obtained :

$$\varepsilon_k = \frac{R}{\Gamma_{k-1} H_{k-1}^2 + R}\,\varepsilon_{k-1} + \frac{\Gamma_{k-1} H_{k-1}}{\Gamma_{k-1} H_{k-1}^2 + R}\left(\frac{1}{2} H'' \varepsilon_{k-1}^2 + \eta_k\right)$$

where H'' is the second derivative of $h(x)$.

The expectation and the actual mean square of the estimation error ε_k can be then computed :

$$\left.\begin{aligned} E(\varepsilon_k / Y_{k-1}) &= \frac{1}{1+\lambda_{k-1,k-1}} \cdot \frac{1}{2}\beta_{k-1}\Gamma_{k-1} \\ P_k = E(\varepsilon_k^2 / Y_{k-1}) &= \underbrace{\frac{\lambda_{k-1,k-1}}{1+\lambda_{k-1,k-1}}\Gamma_{k-1}}_{\Gamma_k} + \frac{1}{(1+\lambda_{k-1,k-1})^2}\cdot\frac{3}{4}\beta_{k-1}^2\Gamma_{k-1}^2 \end{aligned}\right\} \quad (6)$$

with $\lambda_{k-1,k-1} = R/\Gamma_{k-1}H_{k-1}^2$, $\beta_{k-1} = H''/H_{k-1}$

and $E(\varepsilon_{k-1}^4 / Y_{k-1}) = 3\left[E(\varepsilon_{k-1}^2 / Y_{k-1})\right]^2 = 3\,\Gamma_{k-1}^2$

The actual mean square error P_k is composed of one term relative to the measurement error (Γ_k) and one term relative to the bias of the estimate.

We see that the bias term is larger than the computed variance Γ_k when the ratio $\lambda_{k-1,k-1}$ is very small compared to $\beta_{k-1}^2\Gamma_{k-1}$, that is for instance when the noise variance is very small, but not the variance Γ_{k-1}, nor the non-linear coefficient β_{k-1}.

In this particular case, let us compute the bias and the actual mean square error of the (local) iterated Kalman filter, for a large number of iterations.

If the sequence $\{\hat{x}_k^i\}$ of the process given in (3) converges, the limit value $\hat{x}_k^\infty$ satisfies the equation :

$$\hat{x}_k^\infty = \hat{x}_{k-1} + \Gamma_{k-1} H_k^\infty \left[\Gamma_{k-1}(H_k^\infty)^2 + R\right]^{-1}\left[y_k - h(\hat{x}_k^\infty) - H_k^\infty(\hat{x}_{k-1} - \hat{x}_k^\infty)\right]$$

Expanding the expression $y_k - h(\hat{x}_k^\infty)$ and taking into account the relation $H_k^\infty = H + H''(\hat{x}_k^\infty - x)$ the following equation is obtained :

$$\left[(1+\beta\varepsilon_k^\infty)^2 + \lambda_{k-1}\right]\beta\varepsilon_k^\infty = \lambda_{k-1}\beta\varepsilon_{k-1} + (1+\beta\varepsilon_k^\infty)\cdot\frac{1}{2}(\beta\varepsilon_k^\infty)^2 + (1+\beta\varepsilon_k^\infty)\frac{\beta\eta_k}{H} \quad (7)$$

with $\varepsilon_k^\infty = \hat{x}_k^\infty - x$; $\lambda_{k-1} = \frac{R}{\Gamma_{k-1}H^2}$; $\beta = \frac{H''}{H}$; $H = \left(\frac{\partial h}{\partial x}\right)_x$

The exact computation of the expectation and the mean square of the limit value error ε_k^∞ is tedious and would require the computation of higher order moments.

Assuming that $\beta\eta_k/H$ is very small, equation (7) gives, to first order in $\beta\varepsilon_k^\infty$:

$$\varepsilon_k^\infty \simeq \lambda_{k-1}\varepsilon_{k-1} + \eta_k/H$$

Neglecting the third and higher order terms of $\beta\varepsilon_k^\infty$ in (7) and replacing second order terms by the above expression, we get an approximate solution of this equation :

$$(1+\lambda_{k-1})\,\varepsilon_k^\infty \simeq \lambda_{k-1}\varepsilon_{k-1} - \frac{\beta}{2}\left[3\lambda_{k-1}^2\varepsilon_{k-1}^2 + \frac{\eta_k^2}{H}\right] + (1 - 2\lambda_{k-1}\beta\varepsilon_{k-1})\frac{\eta_k}{H}$$

It is now possible to compute the expectation of the estimate error ε_k^∞. After some simplifications we get, at first order in λ_{k-1} :

$$E(\varepsilon_k^\infty / Y_{k-1}) \simeq -\frac{1}{2}\lambda_{k-1}\beta\Gamma_{k-1}$$

With an expression of the solution equation (7) to third order terms with respect to λ_{k-1}, in a similar way as above, the mean square error of the estimate can be deduced, at second order in λ_{k-1}:

$$E\left[(\varepsilon_k^\infty)^2 / Y_{k-1}\right] \simeq \frac{\lambda_{k-1}\Gamma_{k-1}}{1+\lambda_{k-1}} + \frac{15}{4}\lambda_{k-1}^2\beta^2\Gamma_{k-1}^2$$

Compared with the bias of the extended Kalman filter (6) the bias is here of first order with respect to λ_{k-1} and the actuel mean square is, to first order in λ_{k-1}, the same as the computed variance given by the Kalman equation (2) or (3).

The (local) iterated Kalman filter provides better results than the extended Kalman filter when the above assumptions on λ_{k-1} and $\beta\eta_k/H$ are valid, that is when :

- either the noise variance R is very small meanwhile the estimate variance Γ_{k-1} is not ;
- or the non-linearity, defined by $\beta = H''/H$, is very small while the estimate variance Γ_{k-1} is very large.

Let us notice also that, with the assumption λ_{k-1} very small, the convergence of the iterations of the iterated Kalman filter could be proved.

Let us described briefly the scheme of the proof. If $u_i = \beta(\hat{x}_k^i - \hat{x}_k^\infty)$, the recurrent relation for the sequence $\{u_i\}$ is given by :

$$u_i = f(u_{i-1})u_{i-1} \text{ with } f(u_{i-1}) = \frac{\frac{u_{i-1}}{2}(1+u_{i-1}) - \lambda_{k-1}u_0}{(1+u_{i-1})^2 + \lambda_{k-1}}$$

with the assumption $1 + u_0 > 0$, it can be shown then :

$$|f(u_{i-1})| \leqslant \tau < 1 \qquad i = 2, 3, \ldots$$

from which we conclude the convergence of the sequence.

When the assumptions given above are not justified, it is difficult to establish the superiority of the iterated Kalman filter with respect to the extended Kalman filter. It may occur that the iterated filter produces a bias larger than the extended filter - Breakwell (1969).

Bias reduction by global iteration

When the measurement noise is not small, iterations on a large number of measurements can improve the estimate, in a way similar to the local iterated filter.

The expression of the bias from the linearized Kalman filter is first derived below. With $\varepsilon_k = \hat{x}_k - x$, let us expand the equation (2). The final estimate error ε_N (on N measurements) can be expressed in terms of the a priori error ε_0 :

$$\varepsilon_N = \frac{\Gamma_N}{\Gamma_0}\varepsilon_0 + N\,\Gamma_N\frac{H_0}{R}\cdot\frac{1}{2}H''\varepsilon_0^2 + \frac{\Gamma_N H_0}{R}\sum_{j=1}^{N}\eta_j$$

$$\Gamma_N = \frac{\Gamma_0}{1 + N\,\Gamma_0 H_0^2/R} \qquad H_0 = \left(\frac{\partial h}{\partial x}\right)_{x=\hat{x}_0}.$$

The linearized Kalman filter provides a bias when the number of measurements becomes infinite :

$$b_{LK} = \lim_{N\to\infty}\varepsilon_N = \frac{1}{2}\,\frac{H''}{H_0}\,\varepsilon_0^2$$

noticing that the term $\frac{1}{N}\sum_{j=1}^{N}\eta_j$ tends towards zero with $1/N$

On the other hand, the final estimate error $\varepsilon_N^i = \hat{x}_N^i - x$ at the ith iteration of the global iterated Kalman filter, can be derived in terms of the a priori error ε_0 and the final estimate error of the previous iteration :

$$\varepsilon_N^i = \frac{\Gamma_N^i}{\Gamma_0}\varepsilon_0 + N\frac{\Gamma_N^i\bar{H}^i}{R}\cdot\frac{1}{2}H''(\varepsilon_N^{i-1})^2 + \frac{\Gamma_N^i\bar{H}^i}{R}\sum_{j=1}^{N}\eta_j$$

$$\Gamma_N^i = \frac{\Gamma_0}{1 + N\,\Gamma_0(\bar{H}^i)^2/R}$$

With $\beta = H''/H$, $\bar{H}^i = H(1+\beta\,\varepsilon_N^{i-1})$, for a large measurement set we have approximatively :

$$\varepsilon_N^i \simeq \frac{1}{2}\cdot\frac{\beta\,(\varepsilon_N^{i-1})^2}{1+\beta\,\varepsilon_N^{i-1}}$$

With $1+\beta\varepsilon_0 > 0$, the following inegalities can be shown :

$$0 \leq \varepsilon_N^i < \frac{1}{2}\,\varepsilon_N^{i-1} < \dots < \frac{1}{2^{(i-1)}}\,\varepsilon_N^1$$

At each iteration, except for the first one, the final bias is reduce with respect to the bias of the previous iteration at least by a factor two.

APPLICATION TO THE ATTITUDE RECONSTITUTION OF SPINNING SATELLITES

The extended and the (local) iterated Kalman filters have been applied to the attitude reconstitution of spinning satellites for which the non-linear effects were significant. The illustration on the bias reduction in the second method is given here. Let us describe the estimation problem.

It consists in restituting both components of the angular momentum (right ascension α_H and declination δ_H in an equatorial frame) of an ESRO satellite (GEOS) during the transfer phase Huynh et al. (1973).

The attitude sensor system is composed with a sun sensor and a dual-beam infrared earth sensor. The sun sensor provides the difference of occurrence time between the pulses delivered by two slits inclined one to each other. One earth sensor beam delivers a pulse when it enters the earth infrared disk, or gets out of it. The earth sensor measurements are the difference of occurrence times between these pulses and one reference pulse from the sun sensor.

The measurement noise variance is small (mean square root less than 0.2 degree) with respect to a priori variances of the angular momentum components (mean square root about 4 degrees). On the other hand, the earth sensor measurements are strongly non-linear functions in terms of the angular momentum at the altitude where the sensors operate (near the apogee).

A large number of numerical simulation cases have been performed. In the estimation program, several kinds of errors have also been estimated (systematics, setting errors) in order to improve the attitude reconstitution. The state vector presents as many as 13 components.

When a priori discrepancies remain small (less than 1 degree for the angular momentum components) the extended Kalman filter performs well enough, as can be seen on figure 3. That means, in particular, for all cases considered, that the estimate accuracy is consistent with the computed root mean square.

For large a priori discrepancies (about 4 degrees for the angular momentum components) the bias becomes significant compared to the computed variance (see fig. 4). Iterations on first measurements set allow a considerable reduction of this bias, and provide then an estimate whose accuracy corresponds to the computed variance.

CONCLUSION

In this paper geometrical aspects of the iterated Kalman filter are presented.

Then, theoretical study proves that the local iterated Kalman filter is particularly useful when the noise variance is very small relative to the estimate variance, and the non-linearities are not too small, or when the second order terms are small while the estimate variance is very large. The global iterated Kalman filter provides also a bias reduction for a large number of measurements.

Lastly numerical results of an attitude restitution of a spinning satellite illustrates the improvement due to the use of the iterated Kalman method compared with the extended Kalman method.

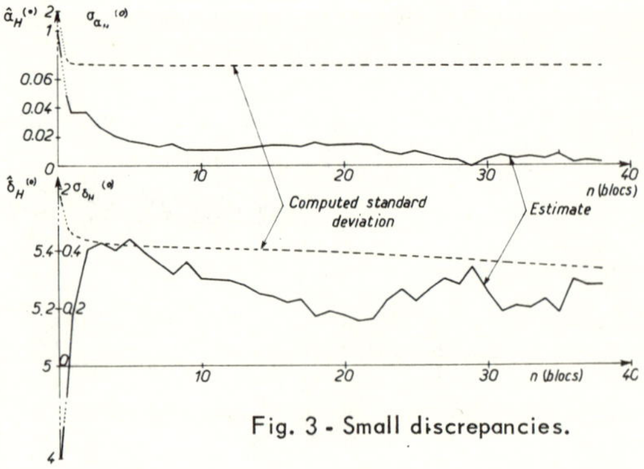

Fig. 3 - Small discrepancies.

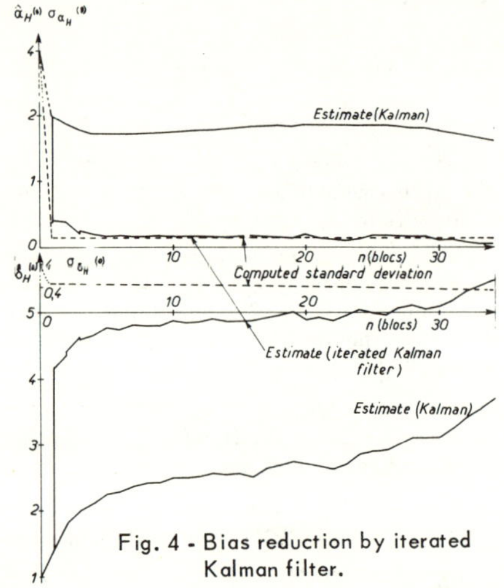

Fig. 4 - Bias reduction by iterated Kalman filter.

REFERENCES

Breakwell J.V. (1969) - Slighthy non-linear estimation with noisy data. AIAA Paper n° 60-839, AIAA Guidance, Control and Flight Mechanics Conference.

Denham W.F. and Pines S. (1966) - Sequential estimation when measurement function nonlinearity is comparable to measurement error. AIAA J., vol n° 6, p. 1071-1076.

Huynh H.T., Depasse F., Aumasson C. (1973) - Restitution d'attitude de GEOS. Contrat ESTEC/ ONERA 1617/12 SK.

Jazwinski A.H. (1970) - Stochastic processes and filtering theory. Academic Press.

Kalman R.E., Bucy R.S. (1961) - New results in linear filtering and prediction theory. J. of Basic Eng., Trans. of the ASME, vol. 83, p. 95-107.

Mehra R.K. (1971) - A comparison of several non-linear filters for reentry vehicle tracking. IEEE Trans. on Automatic Control, vol. AC-16, n° 4.

Wishner R.P., Tabaczynski J.A., Athans M. (1969) - A comparison of three non-linear filters. Automatica, vol. 5, p. 487-496, Pergamon Press.

APPLICATION OF DIFFERENT STATISTICAL TESTS FOR THE DETERMINATION OF THE MOST ACCURATE ORDER OF THE MODEL IN PARAMETER ESTIMATION

H. Unbehauen, Doz. Dr.-Ing.
B. Göhring, Dipl.-Ing.
University of Stuttgart, W. Germany

From the variety of parameter estimation methods used today for the identification of stochastically disturbed control systems, the direct least squares method and the maximum likelihood method are critically compared in this paper with regard to the accuracy of the estimated parameters of the model. Additionally, in order to determine the most accurate order of the model, seven structure testing methods are applied. These testing methods are thoroughly investigated with regard to their efficiency and reliability. Before the actual parameter estimation, the possible orders of the model are essentially limited by using the determinant ratio testing method. These different structure testing methods are compared in connection with the application of the maximum likelihood method and the direct least squares method. The investigated control system was simulated on an analogue computer, because an accurate judgement of the efficiency of these testing methods under different and extreme working conditions can only be made with an exact knowledge of the real process. With the experience and results obtained in this investigation, rules for the application of structure testing methods are formulated. Because of the large number of cases investigated, the probability of the general validity of these rules is high.

1. INTRODUCTION

In the field of control engineering the task of the identification often consists of determining a discrete linear time-invariant mathematical model of a stochastically disturbed control system from the measured input and output signals. The identification itself includes the determination of the parameters (step 1) as well as the structure (step 2) of the model. For this step by step modelling, in step 1 statistical parameter estimation methods, and in step 2 a number of different testing methods for determining the most accurate order of the model, are appropriate. Since numerous estimation and testing methods exist today, the user often has to decide which of these methods is the most appropriate. At the present time, several, to some extend excellent, survey papers on statistical parameter estimation methods have been published [1 to 6], but only a few papers give a critical comparison [7] or show the connexions and similarities between these methods [8]. Apart from a few papers about single structure testing methods[7,9,10] studies are missing which compare critically these methods under different working conditions with regard to their efficiency and reliability. Moreover, there are no papers containing rules for the application of these testing methods in connection with the parameter estimation methods. Therefore this paper deals thoroughly with the investigation of these above stated problems.

2. PARAMETER ESTIMATION METHODS

Many apparently different parameter estimation methods can be shown to have a lot of similarities and relations [8]. Therefore the parameter estimation methods can be classified into two groups. Those which are based on the principle of the "least squares methods" and those which are not. The "direct least squares method" as well as the "maximum likelihood method" can be considered as typical representatives for each group. Therefore these two parameter estimation methods in connection with the investigation of different structure testing methods are used in this paper. In both cases the stochastically disturbed process (Fig. 1) is described by the same structure of the model as shown in Fig.2.

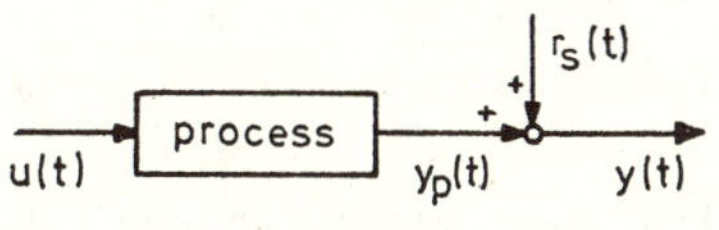

Fig. 1
Block diagram of a stochastically disturbed process

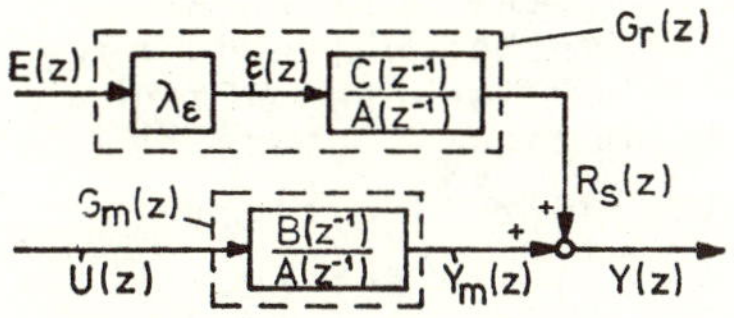

Fig. 2
Block diagram of the model for a stochastically disturbed process

Since both estimation methods are only different in the partial model $G_r(z)$ for the stochastic disturbance signal r_s and in the numerical treatment, their principle is briefly shown in the following section.

2.1. The direct least squares method (DLS-method)

The model for the DLS-method is given by

$$A(z^{-1})Y(z)-B(z^{-1})U(z)=\mathcal{E}(z)=\lambda_\varepsilon E(z) \qquad (1)$$

with the polynomials

$$A(z^{-1})=1+a_1z^{-1}+\ldots+a_nz^{-n} \qquad (2a)$$

$$B(z^{-1})=b_1z^{-1}+\ldots+b_nz^{-n}. \qquad (2b)$$

According to Fig. 2 it follows $C(z^{-1})=1$.

By transforming of Eq.(1) in the discrete time domain and by defining the column vectors $\underline{y}$ (output-signal), $\underline{\varepsilon}$ (error signal), $\underline{p}$ (parameter vector) and the data matrix $\underline{M}$ [11] we get the vector expression

$$\underline{y}(N) = \underline{M}(N)\underline{p}(N) + \underline{\varepsilon}(N) \qquad (3)$$

with

$$\underline{p}(N) = (a_1a_2\ldots a_n,\ b_1b_2\ldots b_n)^T$$

$$\underline{y}(N) = (y(n+1)y(n+2)\ldots y(n+N))^T$$

$$\underline{\varepsilon}(N) = (\varepsilon(n+1)\varepsilon(n+2)\ldots\varepsilon(n+N))^T.$$

The data matrix $\underline{M}(N)$ in Eq.(3) contains the measured values of the input and output signals u(k) and y(k) for k = n, n+1, n+2, ...,n+N-1. The minimization of the performance index $I = 1/2\ \hat{\underline{\varepsilon}}^T\hat{\underline{\varepsilon}}$ of the estimated model error $\hat{\underline{\varepsilon}}$ leads to the equation

$$\underline{M}^T(N)\cdot\underline{M}(N)\cdot\underline{p}(N) = \underline{M}^T(N)\cdot\underline{y}(N). \qquad (4a)$$

From this equation we get the direct analytical solution for the parameter vector

$$\underline{p}(N) \equiv \hat{\underline{p}}(N) = \left[\underline{M}^T(N)\cdot\underline{M}(N)\right]^{-1}\cdot\underline{M}(N)\cdot\underline{y}(N). \qquad (4b)$$

If the model errors are $(0,\lambda_\varepsilon)$-normally distributed and if their variance $\lambda_\varepsilon^2\cdot\underline{I}$ is known, the standard deviation λ_{p_i} of the parameters p_i can be calculated by

$$\lambda_{p_i}^2(N) = \lambda_\varepsilon^2\cdot\left[\underline{M}^T(N)\cdot\underline{M}(N)\right]_{ii}^{-1}. \qquad (5)$$

2.2. Maximum likelihood method (ML-method)

This method represents the best known parameter estimation method which is not based on the direct least squares principle. The model of the ML-method is given by

$$A(z^{-1})y(z)-B(z^{-1})U(z)=C(z^{-1})\cdot\mathcal{E}(z), \qquad (6)$$

where $A(z^{-1})$ and $B(z^{-1})$ are defined by Eqs.(2a) and (2b) and

$$C(z^{-1}) = 1+c_1z^{-1}+\ldots+c_nz^{-n}. \qquad (2c)$$

The aim of this method is to calculate the maximum value of the likelihood function L or, if the model error $\underline{\varepsilon}$ is normally distributed, the minimum value of the performance index $I = 1/2\cdot\hat{\underline{\varepsilon}}^T\hat{\underline{\varepsilon}}$. This can be done by means of the iterative optimization algorithm

$$\underline{p}(i+1)=\underline{p}(i)-\beta\left[\underline{I}_{pp}(\underline{p}(i))\right]^{-1}\cdot\underline{I}_p(p(i)), \qquad (7)$$

where $\underline{I}_p$ represents the gradient, $\underline{I}_{pp}$ the Hesse-matrix of I, β is a weighting factor and $\underline{p}=(a_1\ldots a_n,b_1\ldots b_n,c_1\ldots c_n)^T$ the wanted parameter vector.
The large expense of the numerical treatment of Eq.(7) is justified because the inverse of matrix $\underline{I}_{pp}$ can simultaneously be used for the calculation of

$$\lambda_{p_i}^2(N) = \lambda_\varepsilon^2\left[\underline{I}_{pp}(N)\right]_{ii}^{-1} \qquad (8)$$

and hence for the standard deviation λ_{p_i}.
If λ_ε is not known, but normally distributed, then it can be estimated from

$$\hat{\lambda}_\varepsilon = \left[2\cdot I(N)/N\right]^{1/2}. \qquad (9)$$

3. STRUCTURE TESTING METHODS

The parameters of model structures of different order n can be estimated by the above described methods. To limit the possible model orders and to determine the most accurate order, the following described parameter testing methods can be used.

3.1. Determinant ratio test

This test, proposed by Woodside [12], is only used to limit the number of possible model orders before starting the parameter estimation. For this testing method the (2n)-dimensional vector

$$\underline{h}_r(n)=(u(k-1)y(k-1)\ldots u(k-n)y(k-n))^T$$

is defined for a fixed order n and with this we calculate the (2n, 2n)-dimensional matrix

$$\underline{H}_r(n) = \frac{1}{N}\sum_{k=n+1}^{n+N}\underline{h}_r(n)\cdot\underline{h}_r^T(n),$$

which has the following properties

$$\det \underline{H}_r(n)\begin{Bmatrix}>\\=\end{Bmatrix}0 \text{ for } n\begin{Bmatrix}\leq\\>\end{Bmatrix}n_o, \qquad (10)$$

where n represents the momentary order of the model and n_o the exact order of the real process. The determinant ratio

$$DV(n) = \det\underline{H}_r(n)/\det\underline{H}_r(n+1) \qquad (11)$$

can now be calculated for succeeding model orders $n=1,2,\ldots,n_{max}$. If the value of the determinant ratio DV(n) shows a distinct increase compared with the previous value DV(n-1), then n corresponds approximately to n_o.

3.2. Condition number test

This testing method is based on the idea to get a measure for the condition of the inverse of a (m,m)-dimensional matrix $\underline{A}(m)$. As a possible measure for this a condition number

$$BZ(n) = \|\underline{A}(m)\|\cdot\|\underline{A}^{-1}(m)\| \quad (12)$$

is defined, where n is the order of the model and the norm $\|\underline{A}(m)\|$ is the maximal sum of a row in $\underline{A}(m)$ [13, 14]. If the chosen order of the model was too high, then the inverses of the matrices in the DLS-method and the ML-method $M^T(N)\cdot M(N)$ respectively I_{pp} can become singular. In this case, BZ(n) will be very high or theoretically infinit. It was shown in several investigations [13, 15] that for the practical application the value 10^7 can be considered to be an upper critical limit for BZ(n). Under these conditions BZ(n) can be taken to indicate that the order of the model is too high.

3.3. Polynomial test

This testing method is mainly used to investigate whether the polynomials $A(z^{-1})$, $B(z^{-1})$ and $C(z^{-1})$ defined by Eqs.(2a,b,c), have common roots. On the other hand, this method indicates whether some roots remain approximately invariable for increasing orders of the model. Such roots can be defined as characteristic. Furthermore the stability of the whole model can be investigated by studying the roots of $A(z^{-1})$. Since the convergence of the employed Bairstow method [16] depends largely on the condition of the investigated polynomials, the convergence behaviour for several models can also be used to estimate the correct order.

3.4. Test for independence

The assumption of nearly all parameter estimation methods, namely that the model errors ε are uncorrelated, is checked by this method for increasing order n. If ε is not autocorrelated then the autocorrelation function,

$$R_{\varepsilon\varepsilon}(\tau_N)=\frac{1}{N}\sum_{k=1}^{N}\varepsilon(k)\cdot\varepsilon(k+\tau_N) \quad (13)$$

$$(\tau_N=0,1,..,10)$$

becomes zero for all values $\tau_N \neq 0$, where τ_N is normalized on the sampling rate T.

3.5. Test for normality

The test for normality is used to check the statistical distribution of a population by taking N samples of the model error ε(k) (k=1,2,...,N). For this, the discrete probability distribution function

$$P[(\varepsilon(k) \text{ for } k=1,2,..,N)<\varepsilon_i]=P_i \quad (14)$$

$$(i=0,1,..,i_{max})$$

is calculated, where the reference value $\varepsilon_i=\varepsilon_{min}+i\cdot\Delta\varepsilon$ $(i=0,1,...,i_{max})$ is permanently varied and the associated probability value P_i is calculated. This value specifies the cumulative frequency of the model errors which are smaller than the given upper limit ε_i. Thus $\varepsilon_{min}\leq\varepsilon_i\leq\varepsilon_{max}$ corresponds to $0\leq P_i\leq 100\%$.

If the calculated values of P_i (in %) are plotted over the values of ε_i on probability paper, on which the ordinate scale is graduated according to the area under a normal distribution function, then the connecting line of all values of P_i must, in the case of a normal distribution, be a straight line. In this case, the average value can also be checked and is zero if the condition

$$P[(\varepsilon(k) \text{for } k=1,2,..,N)<(\varepsilon_i=0)]\equiv P_i=50\% \quad (15)$$

is satisfied. If the disturbance signal r_s of the process to be identified is known to be normally distributed, then this testing method can simultaneously be applied to select only these model orders, for which the probability distribution function fits a straight line.

3.6. Statistical F-test

A rough estimation of the order of the model can often be made by using the apriori knowledge of the investigated control system. This can be formulated by a so-called zero-hypothesis [7,9,10, 17,18]

$$H_0: a_i=b_i=c_i\left\{\begin{matrix}\neq\\=\end{matrix}\right\}0 \text{ for } \left\{\begin{matrix}0\leq i\leq n\\ i>n\end{matrix}\right\}. \quad (16)$$

This hypothesis can be tested by calculating the performance index I_2 for $n_2=n+l$ (l=1,2,3,..,9), which should not be very smaller than the performance index I_1 for the accurate supposed order $n_1=n$ of the model. For the numerical treatment the test measure

$$t_g(n) = \frac{I_1-I_2}{I_2}\cdot\frac{N-n_2}{n_2-n_1} \quad (17)$$

is defined. If the errors ε(k) of the model are normally distributed and if both variables I_1 and I_2, having the χ^2-distribution, have the same standard deviation, then the random variable t_g has an F-distribution with $f_1=n_2-n_1$ and $f_2=N-n_2$ degrees of freedom, Fig. 3. For a level of significance $\alpha\leq1\%$ or $\leq5\%$ and with N>200 the tabulated reference value $t_\alpha\approx3{,}0$ or $\approx2{,}6$ is obtained [17,18]. For $t_g<t_\alpha$ the probability of I_2 not being essentially smaller than I_1 is higher than the permitted value of α. This means that the zero hypothesis H_0 is valid and that $n_1=n$ is the correct

estimated order $\hat{n}_o$ of the model.

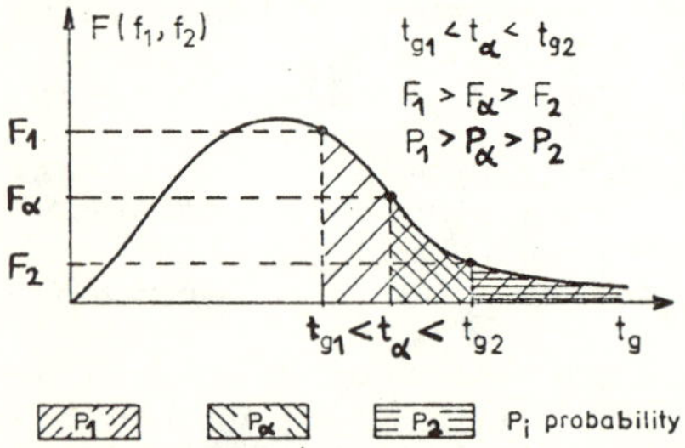

Fig. 3
Diagram of the F-distribution

In the case $t_g > t_\alpha$ the correct estimated order of the model is $n_2 = \hat{n}_o$. Instead of H_o the alternative hypothesis

$$H_1: a_i = b_i = c_i \begin{Bmatrix} \neq \\ = \end{Bmatrix} 0 \text{ for } \begin{Bmatrix} 0 \leq i \leq n+1 \\ i > n+1 \end{Bmatrix} \quad (18)$$

is then valid.

3.7. Test of signal errors

A simple, but very effective method for testing the structure of a model is given by a comparison of the characteristic time responses with the exact signals for all investigated model orders. For this comparison the output signal $\hat{y}_m(k)$, the step response $\hat{h}_m(k)$ and the impulse response $\hat{g}_m(k)$ of the estimated model are appropriate. Furthermore the signal errors

$$\hat{e}_y(k) = y(k) - \hat{y}_m(k), \quad (19a)$$

$$\hat{e}_h(k) = h(k) - \hat{h}_m(k), \quad (19b)$$

$$\hat{e}_g(k) = g(k) - \hat{g}_m(k), \quad (19c)$$

can be calculated and compared using the exact signals $y(k)$, $h(k)$ and $g(k)$ of the real process. An optical comparison of the error between y and $\hat{y}_m$ is often difficult. Therefore Eq.(19a) is used for the numerical calculation. Then the results are plotted for different orders of the model. The signals h and g can be compared graphically with $\hat{h}_m$ and $\hat{g}_m$ respectively and the best fitting one then determines the order n of the model.

It should be mentioned that still more statistical testing methods could be applied in connection with the parameter estimation methods described above, e.g. the probability ratio test, the test of standard deviations of the parameters, the test for unnecessary parameters, the test of errors in the frequency domain, the test of stationarity etc. However, in this paper only the most efficient ones are applied. Furthermore it should be noticed that a single test can cause incorrect decisions but the application of several different testing methods will specify the most accurate order of a model.

4. THE INVESTIGATED PROCESS

The structure testing methods described before are applied in this paper in connection with the DLS- and ML-method on an analogous simulated control system of second order ($n_o = 2$) with the continuous transfer function

$$G(s) = \frac{1,0+0,2s+0,01s^2}{1,0+4,5s+1,0s^2} . \quad (20)$$

The discrete transfer function

$$G(z) = \frac{B(z^{-1})}{A(z^{-1})} = \frac{0,01+0,0004z^{-1}-0,0024z^{-2}}{1-1,6296z^{-1}+0,6376z^{-2}} = \frac{0,01z^2+0,0004z-0,0024}{(z-0,977)(z-0,653)} \quad (21)$$

is obtained after the application of the exact z transformation with a sampling time T = 100ms.

This control system to be identified is characterized by an aperiodic behaviour, and by the fact that $b_o \neq 0$ (i.e. $h(0) \neq 0$) and that the values of the other b_i-parameters are very small. Consequently, the gain factor $K_s = \sum_{i=0}^{n} b_i / \sum_{i=0}^{n} a_i = 1,0$ is about 1/10 lower than in nearly all other examples described in literature till now. Furthermore, one of the roots of the characteristic polynomial $A(z^{-1})$ is situated very near to the stability margin. Consequently, this system is rather difficult to identify and provides the best conditions to check the parameter estimation as well as the structure testing methods with regard to their efficiency. A pseudo-random binary signal, a m-sequence with $N_u = 7,15,21,63$, has been chosen as the input signal u. The additional noise signal r_s is $(0,\lambda_{r_s})$-normally distributed, where λ_{r_s} is varied within the region of $0 \leq \lambda_{r_s} \leq 1,0$ in very small steps.

5. COMPARISON OF THE RESULTS

5.1. Results of the parameter estimation methods

From the measured output and input signals models of different order were estimated by the DLS- and ML-method under diverse working conditions, e.g. an additional noise r_s with diverse levels $\lambda_s = r_{s\,eff}/u_{eff}$ $(0 \leq \lambda_s \leq 1)$, a linear drift d_1 $(0 \leq d_1 \leq 0,02$ V/s) and a linear drifting noise were superposed on the output signal y_P. The accuracy of the models can be best judged with the estimated parameters $\hat{a}_i$ and $\hat{K}_s$.

System without noise and drift: Under the working conditions described in Fig. 4a the ML-method gives very precise results

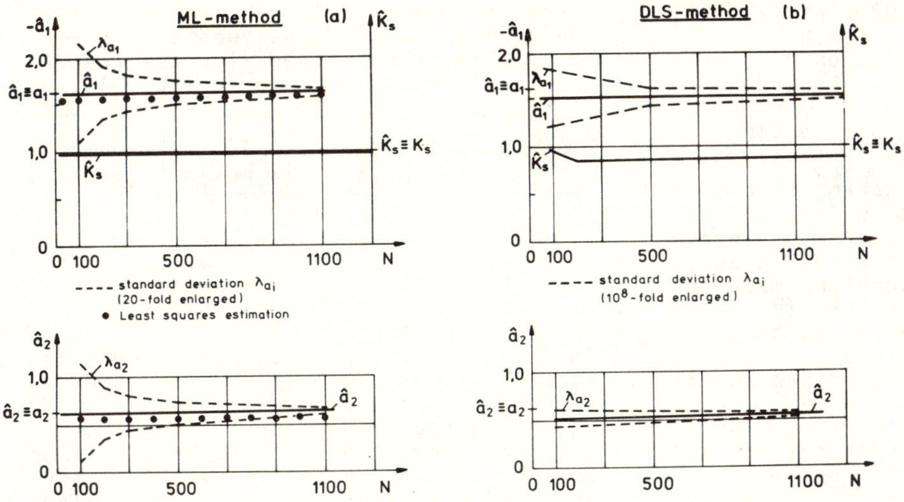

Fig. 4

Influence of number of data N on the estimated parameters $\hat{a}_i$ and the gain factor $\hat{K}_s$ for the ML-method (a) and the DLS-method (b) (T=100ms, N_u=31, Δt_u=1s, n=2, N=1, 2,...,1300)

for the order $\hat{n}=n_o=2$ and the parameters $\hat{a}_i$ ($\hat{a}_1$=-1,625, $\hat{a}_2 \cong$ 0,63) as well as for the gain factor $\hat{K}_s$=1,003 using only N=62 pairs of measuring data. The high accuracy of $\hat{K}_s$ can be obtained for all model orders $\hat{n} \geq n_o$. Only the model of first order is of no use because $\hat{K}_s$ becomes nearly 100% too large. Under the same conditions the DLS-method also gives stable, but essentially more unprecise results for $\hat{n}=n_o=2$ as Fig. 4b shows.

The sampling time T=100ms and N=62 yield a measuring time T_{mes}=6,2s which corresponds about 1,5-times to the dominating time constant T_1=4,4s or 0,25-times to the settling time T_a=20s of the investigated process. No better results in $\hat{a}_i$, $\hat{b}_i$ and $\hat{K}_s$ can be obtained by using more measuring data up to N=1100. Only the standard deviations λ_{p_i} of the parameters, corresponding to Eqs.(5) and (8) decrease insignificantly (see Fig. 4). But this fact is not so important for the quality of the model. In order to get similar good estimates with the DLS-method as with the ML-method, the sampling time T must be chosen sufficiently small. However, if T cannot be sufficiently decreased, the ML-method can be applied with the condition that T_{mes} is large enough.

System with noise: The above statements concerning the influence of the different variables on the ML-method are still valid if the output of the process is disturbed by a noise signal r_s. To get the same accuracy as for the system without noise disturbances, the minimal quantity of measuring data has to be increased, e.g. for the noise level λ_s=0,05, N=800. For increasing noise level it is better to take a longer measuring time than a smaller sampling time.

In the case of the DLS-method the estimated parameters of the model and consequently the gain factor become worse for increasing λ_s. This tendency cannot at all be influenced by N_u. Smaller values of T yield only a little bit better values of the parameters $\hat{a}_i$, but the gain factor $\hat{K}_s$ becomes very poor. However, better results can be obtained with a higher amplitude of the PRBS input signal.

System with drift: In the investigated region of the linear drift d_1 ($0 \leq d_1 \leq 0{,}02$ V/s) the accuracy of $\hat{a}_i$ and $\hat{K}_s$ decreases rapidly with increasing values of d_1 for the ML-method. This is valid for all sampling times T=20,40 and 100ms. However, if T_{mes} and T are reduced, then the estimated value $\hat{K}_s$ will become essentially better for $d_1 \leq 0{,}005$ V/s, but at the same time the estimated parameters $\hat{a}_i$ will become poor. This is also valid for the DLS-method. With both methods the time behaviour of the model can be very much improved by a smaller sampling time T, but the accuracy of the estimated parameters decreases.

System with noise and drift: The ML-method gives poor results for the parameters $\hat{a}_i$ and $\hat{K}_s$ for large values of T and T_{mes} with increasing drift. Wrong results are obtained for d_1=0,02 V/s and λ_s>0,2. Decreasing the sampling time from 100ms to 40ms or 20ms gives quite better results for $\hat{K}_s$.

In the case of the DLS-method the estimated parameters $\hat{a}_i$ are nearly independant of the drift for T=100ms and $\lambda_s \geq 0{,}05$, however, they are similarly as unprecise as the value of $\hat{R}_s$. A reduction of T does give better results.

5.2. Results of the structure testing methods

In the comparison of the parameter estimation methods, conducted so far, the assumption has been made that the order $n = n_o = 2$ of the model is known in advance. This assumption cannot be made in most practical cases. Therefore in the following section we investigate how precise, the order of the model can be estimated with the above described structure testing methods.

Determinant ratio test: Under the working conditions described in Fig. 5

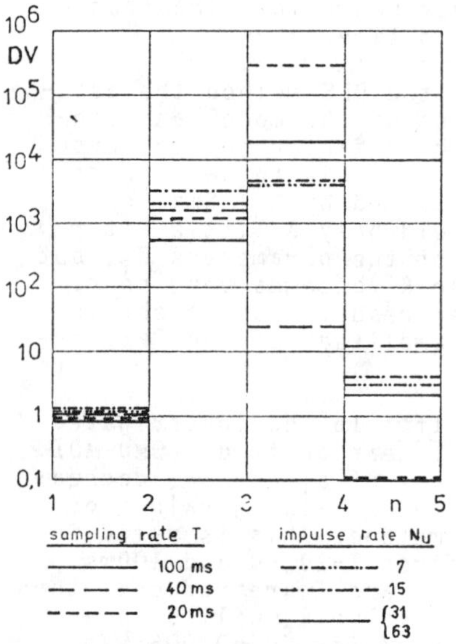

Fig. 5

Influence of model order n on the determinant ratio DV (Δt_u = 1s, N = 1100)

this test gives the correct model order $\hat{n}_o = n_o = 2$ independent from N_u, because from all calculated values of DV(n) only DV(2) shows a considerable step. This step is not so distinct for a system with noise. For a system with drift ($0 \leq d_1 \leq 0{,}02$V/s) the correct order of the model can be determined independent of T. For a system with drift and a high noise level λ_s this test still indicates the correct order $\hat{n}_o = 2$, but only an insignificant ascent of the DV(n)-value can be observed for n = 2.

On the basis of this preliminary examination of $\hat{n}_o$ with 120 sets of different data it seems sufficient to apply all the following testing methods only for the orders n = 1 until 4 of the model.

Condition number test: This test provides the lowest value of the condition number BZ(n) for the ML-method with n = 1 under the conditions described in Table 1.

n	1	2	3	4
BZ	$3{,}38\cdot10^3$	$6{,}13\cdot10^6$	$4{,}89\cdot10^7$	$3{,}24\cdot10^8$

Table 1

Influence of model order n on the condition number BZ for the ML-method (T = 100ms, N_u = 31, Δt_u = 1s, N = 1100)

For n = 2 a step of about 10^3 occurs and the respective value of BZ(n) becomes very near to the critical limit value 10^7. For n = 3 and 4 the calculated values of BZ(n) exceed this limit. From this it follows that the estimated value of the order of the model must be $\hat{n}_o = 1$. Also, this wrong estimation cannot be influenced by varying the quantity N_u.

The results are similar for a system with noise. However, with a sampling time T = 100ms the results for $n \geq 3$ don't exceed the critical limit value. The wrong estimated order $\hat{n}_o = 1$ cannot be improved by varying N_u or by decreasing T. Also for a system with drift this test provides a wrong estimated model order $\hat{n}_o = 1$. A system with drift and noise shows the same results.

By applying different numerical techniques for the DLS-method the most accurate solution $\hat{\underline{p}}(N)$ has been obtained from Eq. (4a) by the Cholesky-technique. However, in this case the inverse matrix of $\underline{M}^T(N)\cdot\underline{M}(N)$ is not known explicitely. Therefore the order of the model cannot be checked by this testing method.

Polynomial test: Under the conditions described in Table 2 the roots of $A(z^{-1})$ for all orders n = 1,2,3,4 lie inside the unit circle for the ML-method. Furthermore Table 2a shows for $n \geq n_o = 2$ that these roots appear repeatedly and that they remain nearly constant. Comparing these roots with those of the polynomials $B(z^{-1})$ and $C(z^{-1})$ it can be seen that all the three polynomials have n-2 common roots for every model order. These common roots are unnecessary. The remaining two roots of each of these polynomials for n=2,3 and 4 are identical and therefore characteristic for the investigated system. From this behaviour we derive on the correct order $\hat{n}_o = 2$ of the system.

Also by the DLS-method and under the same conditions the order of the model $\hat{n}_o = n_o = 2$

as well as the characteristic roots can be estimated exactly (Table 2b).

n	1		2		3		4	
	Re	Im	Re	Im	Re	Im	Re	Im
$A(z^{-1})$	0,994	0	0,977	0	0,977	0	0,977	0
	–	–	0,648	0	0,650	0	0,650	0
	–	–	–	–	-0,879	0	-0,868	0
	–	–	–	–	–	–	-0,175	0
$B(z^{-1})$	-0,337	0	0,774	+0,873	0,772	+0,876	0,772	+0,875
	–	–	0,774	-0,873	0,772	-0,876	0,772	-0,875
	–	–	–	–	-0,854	0	-0,841	0
	–	–	–	–	–	–	-0,180	0
$C(z^{-1})$	-0,533	0	0,495	+0,186	0,491	+0,172	0,494	+0,179
	–	–	0,495	-0,188	0,491	-0,172	0,494	-0,179
	–	–	–	–	-0,870	0	-0,859	0
	–	–	–	–	–	–	-0,183	0

(a)

characteristic and non-redundant roots; redundant roots

(b)

n	1		2		3		4	
	Re	Im	Re	Im	Re	Im	Re	Im
$A(z^{-1})$	–	–	0,978	0	0,977	0	0,977	0
	–	–	0,551	0	0,586	0	0,600	0
	–	–	–	–	-0,517	0	-0,418	+0,556
	–	–	–	–	–	–	-0,418	-0,556
$B(z^{-1})$	–	–	>100	0	>100	0	>100	0
	–	–	2,549		-1,150	0	-2,177	0
	–	–	–	–	8,170	0	-0,056	+1,203
	–	–	–	–	–	–	-0,056	-1,203

Table 2

Influence of model order n on the roots of the polynomials $A(z^{-1})$, $B(z^{-1})$ and $C(z^{-1})$ for the stability test for the ML-method (a) and the DLS-method (b) (T = 100ms, N_u = 31, Δt_u = 1s, N = 1100; exact roots of the A-Polynom; 0,9768 and 0,6527)

For a system with noise the roots of the characteristic polynomial $A(z^{-1})$ lie inside the unit circle for all orders n = 1, 2,3,4 by applying the ML-method. It is, however, more difficult to separate the estimated roots into characteristic and common ones. But only those common roots which do not exist for different orders of the model are unnecessary. The DLS-method with λ_s<0,2 still gives good results for the larger root of the characteristic polynomial (≈0,9). The estimation of the correct order of the model is not possible with this testing method.

For a system with drift d_1<0,01 V/s the polynomial test gives very good results using the ML-method. If d_1≥0,01 V/s, then for the same orders the model will become instable, or the numerical Bairstow-technique diverges or too many common roots occur. Out of this follows a variety of model orders, from which the correct estimation value $\hat{n}_o$ can be derived only very seldom. In the case of the DLS-method this testing method gives the correct order of the model $\hat{n}_o$=n_o=2 as well as the correct estimation of the two characteristic roots for d_1<0,01 and T = 100ms. Although the Bairstow-technique does not fail for d_1≥0,01 V/s and T = 100ms, the correct order of the model cannot be found.

For a system with noise and drift, the polynomial test, when used with the ML-method, gives similar results for small λ_s and high values of T and T_{mes} as for a system without noise and drift. With increasing λ_s, once again it is, however, difficult to separate the characteristic and common roots in order to estimate the correct order of the model $\hat{n}_o$. If T_{mes} and T are diminuished at the same time, then, as for a system without noise and drift, this testing method gives explicitly the correct order $\hat{n}_o$=2 of the model and the correct roots.

For the DLS-method, under the same working conditions, the polynomial test supplies only one correct characteristic root of the polynomial $A(z^{-1})$ which is nearly independent of T and d_1 with λ_s<0,2. Thus we get the wrong order $\hat{n}_o$=1 of the model. If λ_s≥0,2, this test fails completely.

<u>Test for independence</u>: Under the working conditions mentioned in Fig. 6 the order n = 1 can be excluded with the ML-method (a). For all other orders the condition $R_{\varepsilon\varepsilon N}(\tau_N \neq 0)$ = 0 is satisfied best for $\hat{n}_o$=2. The same result is valid for the DLS-method (b) and is independent of N_u and T.

For a system with noise this test prooves for all orders that the model errors are uncorrelated when used with the ML- and DLS-methods. The order $\hat{n}_o$ = 2 can be found as the smallest possible of all correct orders.

For a system with drift, this test used with the ML-method gives, in most cases, similar good results for d_1<0,01 V/s. It fails completely, however, for d_1≥0,01V/s. When used with the DLS-method, this test also proves the independency of the model errors, but never gives a correct estimate of the order.

For a system with noise and drift, this test used with the ML-method proves for all orders n the independency of the mo-

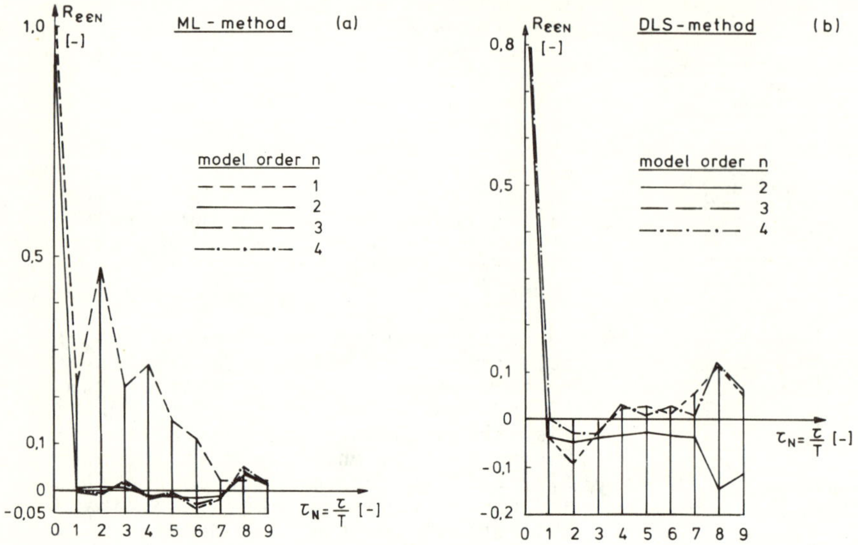

Fig. 6

Influence of model order n on the normalized autocorrelation function $R_{\varepsilon\varepsilon N}(\tau_N)$ for the test of independence of the model error ε for the ML-method (a) and the DLS-method (b) (T = 100ms, N_u = 31, Δt_u=1s, N=1100)

del errors and also the correct order $\hat{n}_o$ = 2, independent of λ_s. The efficiency of this test to estimate the correct model order n_o can be increased if T is decreased.

Test for normality: Under the conditions mentioned in Fig. 7

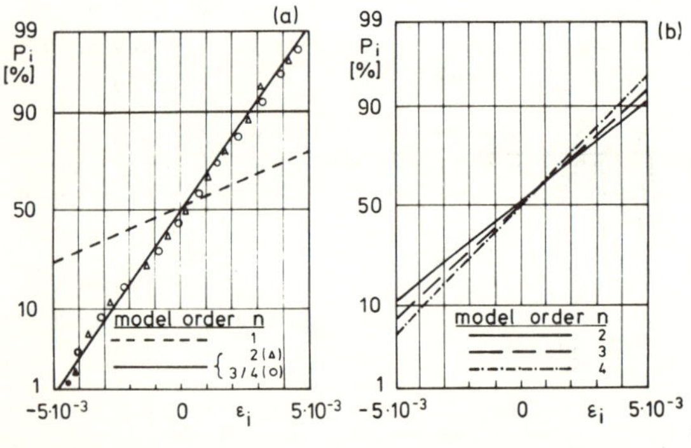

Fig. 7

Influence of model order n on the probability distribution function P_i for the test of normality of the model error ε for the ML-method (a) and the DLS-method(b) (T = 100ms, N_u= 31, Δt_u = 1s, N = 1100)

the order n=1 can be definitely excluded for the ML-method (a) because the connecting line of all P_i values is a straight line within the region of $30 \leq P_i \leq 70\%$. In contrast, for $n \geq n_o = 2$ within the whole region of $1 \leq P_i \leq 99\%$ one gets straight lines, which differ slightly from one another. The correct estimate $\hat{n}_o$ = 2 is found by the fact, that for this order the value P_i for ε_i = 0 lies nearest to 50 %. Similar good results are obtained with the DLS-method, Fig. 7b.

If a system with noise and/or drift requires testing, it is also possible to prove for the ML- and DLS- methods that the model errors are normally distributed and have zero mean. The correct order n_o itself can never be determined but, at least, the number of orders n limited.

Statistical F-test: Under the conditions mentioned in Table 3 the order n=1 can,

t_g \ n_1/n_2	1/2	1/3	1/4	2/3	2/4	3/4
(a) ML	7370	3701	2463	3,12	1,57	0,018
(b) DLS	-	-	-	52,63	40,70	26,3
t_α [α=1%]	4,63	3,4	2,8	4,63	3,4	4,63

Table 3

Comparison of the test measure t_g with the tabulated value t_α (α=1%) for the F-test for the ML-method (a) and the DLS-method (b) (T=100ms,N_u=31,Δt_u=1s,N=1100)

for the ML-method (a) definitely be excluded, because $t_g(1) \gg t_\alpha$ is valid for α=1%. For n = n_o= 2 the condition $t_g(n) < t_\alpha$ is fulfilled for all N_u and T values.

For a noisy system with or without drift this test also gives for all N_u and T values the correct estimate $\hat{n}_o$= 2. It fails, however, for a system with drift only. In contrast, the F-test used with the DLS-method for a system without noise and drift, Table 3b, and for all other conditions always gives the wrong model order.

Test of signal errors: Under the conditions mentioned in Fig. 8

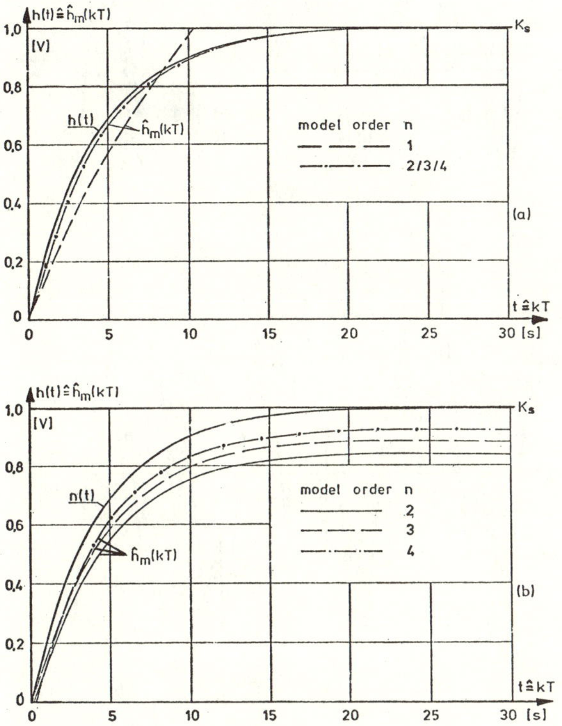

Fig. 8

Analog measured and estimated step response $h(t)$ and $\hat{h}_m(kT)$ for the ML-method (a) and the DLS-method (b).
(T = 100ms, N_u = 31, Δt_u = 1s, N = 1100)

one can see directly that the step response $\hat{h}_m(kT)$, estimated with the ML-method (a), already describes the control systems very well for $n = \hat{n}_o = 2$. In contrast, the error $\hat{e}_h(kT)$ always becomes smaller for the DLS-method (b) with increasing n. The same results are obtained for the impulse response $\hat{g}_m(kT)$ and the error $\hat{e}_g(kT)$. Comparing these results with Fig. 9 one can see that the output signal $\hat{y}_m(kT)$ (b), calculated with the PRBS input signal $u(t)$ (a), simulates for the ML-method as well as for the DLS-method the analogous measured output signal $y(t)$ (b), already quite good for $n = 2$. Furthermore one can definitely exclude $n = 1$ with the aid of the signal error $\hat{e}_y(kT)$, Fig.(9c and 9d), calculated for the ML-method. Thus the correct estimate is $\hat{n}_o = 2$.

6. CRITICAL REVIEW OF THE PARAMETER ESTIMATION AND STRUCTURE-TEST-METHODS

Parameter estimation methods: For a system with drift the ML-method gives more accurate and especially more stable $\hat{a}_i$- and $\hat{K}_s$-values, nearly independent of N_u and T for $n = n_o = 2$, than the DLS-method. The accuracy of $\hat{K}_s$ can, for the DLS-method, be increased by reducing T or increasing n but only at the expense of the accuracy of the $\hat{a}_i$ parameters.

For a system with noise the ML-method also gives very accurate $\hat{a}_i$ and $\hat{K}_s$ values. To obtain the same accuracy as for systems without noise one has only to increase T_{mes}. In contrast, the DLS-method gives worse results with increasing λ_s.

For a system with drift, other measure conditions prove to be better for the ML-method. Decreasing T and T_{mes} simultaneously gives more accurate $\hat{K}_s$ but also worse $\hat{a}_i$ values. This effect appears still more distinct for the DLS-method.

For a system with noise and drift, the ML-method still gives useful results for small T and T_{mes} values. The DLS-method, however, fails almost completely.

With these results and through the large number of data sets, we can definitely say that the ML-method under all practical plant conditions gives much more accurate models than the DLS-method.

Structure testing methods

Determinant ratio test: The determinant ratio test proves to be very efficient under nearly all working conditions. With this preliminary test it is possible to reduce the possibilities of model orders to $n_{max} = 4$, consequently saving a lot of computing time.

Condition number test: The condition number test which was only applied to the ML-method failed completely under all working conditions.

Polynomial test: With this test, together with the ML-method, it is possible to estimate the exact order $\hat{n}_o = 2$ as well as the roots of the polynomial $A(z^{-1})$ and to prove the stability of the model. Similar good results gives the DLS-method for a system with or without drift. This test, however, fails for a system with noise, when used with the DLS-method.

Test for independence: With the test for independence, together with both parameter estimation methods for a system without noise and drift for $n = 2$ to 4, it is

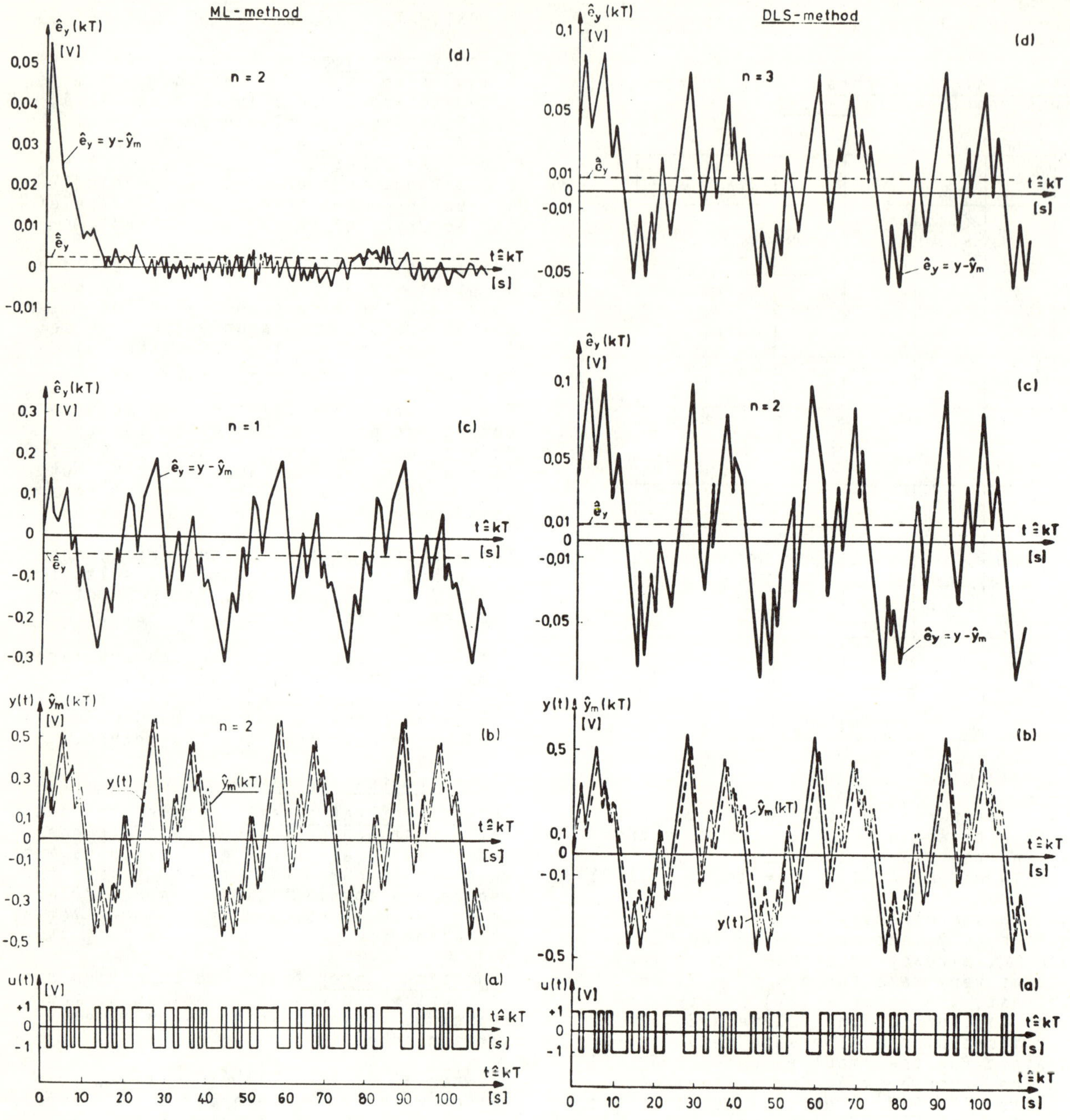

Fig. 9

Analog measured input u(t) (a) and output signal y(t) (b), estimated output signal $\hat{y}_m$(kT) (b) and signal error $\hat{e}_y$(kT)(c) and (d) for the ML-method and DLS-method.
(T = 100ms, N_u = 31, Δt_u = 1s, N = 1100)

always possible to prove the independency of the model errors and to estimate the correct order $\hat{n}_o$ = 2. This test is less efficient if a system with noise or drift is investigated. It only fails when used with the DLS-method for a system with drift.

Test for normality: With the test for normality, together with the ML- and DLS-methods, it is possible to prove that the model errors are normally distributed under all working conditions and have zero mean. The model order itself cannot be definitely determined but at least the possible orders can be limited further.

Statistical F-test: The statistical F-test, together with the ML-method, gives the correct order under all working conditions except for a system with drift. This test fails altogether in connection with the DLS-method.

Signal error test: By means of several characteristic time responses, it is possible to estimate, together with the ML-method, the correct order $\hat{n}_o = 2$, and to simulate the dynamic behaviour very accurately. In contrast, the model order cannot be, when the signal error test is used, found together with the DLS-method, although the dynamic behaviour is as good as for the ML-method.

Summarizing, we can say that the model order $\hat{n}_o$ can be estimated more accurately when these structure testing methods are used in conjunction with the ML-method than when they are used in conjunction with the DLS-method. Because the model parameters, estimated before with the ML-method, are, without exception, better than those with the DLS-method, it can be concluded that the quality and efficiency of the structure testing methods, Table 4

No	structure test		order $\hat{n}_o$ ML	order $\hat{n}_o$ DLS	quality of test
1	determinant ratio test		$1 \le \hat{n}_o \le 4$		very good
2	condition number test		1	-	bad
3	polynom test		2	2	very good
4	independence test		2	2	good
5	normality test		2	2	good
6	statistical F-test		2	≥ 4	very good
7	signal error test of	output signal $\hat{e}_y$	2	2	very good
		step response $\hat{e}_h$	2	4	very good
		impulse response $\hat{e}_g$	2	4	very good

Table 4

Estimated order $\hat{n}_o$ for the ML- and DLS-method and the quality of the investigated structure testing methods.

depend largely on the accuracy of the results obtained before with the parameter estimation method.

7. RULES FOR THE STRUCTURE TESTING METHODS

With regard to practical applications we formulate the following rules:

a) Determinant ratio test: Compute the determinant ratio DV(n) according to Eq. (11) for succeeding orders n. Choose that order n as $\hat{n}_o$ for which the associated DV(n) value increases considerably for the first time.

b) Condition number test: Compute the condition number BZ(n) according to Eq. (12) for succeeding orders n. Choose the smallest order n as $\hat{n}_o$ for which the associated BZ(n) value is either the smallest one or still does not increase considerably or does not exceed the upper limit value of 10^7.

c) Polynomial test: Compute the roots of the polynomials according to Eqs. (2a) to (2c) for succeeding orders n. Choose that order n as $\hat{n}_o$ for which the pairs of polynomials A and B or A and C have no common roots and all the roots of A lie inside the stable region. Furthermore all roots which hardly change for increasing n can be regarded as characteristic for the system.

d) Test for independence: Compute the autocorrelation function $R_{\varepsilon\varepsilon}$ according to Eq. (13) for succeeding orders n. Choose that order n as $\hat{n}_o$ for which the condition $R_{\varepsilon\varepsilon}(\tau_N \neq 0)=0$ is best satisfied.

e) Test for normality: Compute the discrete probability distribution function according to Eq. (14) for succeeding orders n. Choose that order n as $\hat{n}_o$ for which the connection line of all P_i values is a straight line within the region of $10 \le P_i \le 90\%$ and for which also the condition according to Eq. (15) is satisfied.

f) Statistical F-test: Compute the test measure $t_g(n)$ according to Eq. (17) for succeeding orders n. Choose the smallest order n as $\hat{n}_o$ for which $t_g(n) < t_\alpha$ is valid.

g) Test of signal errors: Compute the discrete signals $\hat{y}_m(k)$, $\hat{h}_m(k)$ and $\hat{g}_m(k)$ as well as the error $\hat{e}_y(k)$ according to Eq. (19) for succeeding orders n. Choose that order n as $\hat{n}_o$ for which the time responses of the above mentioned signals best fit the measured curves and/or the error signal $\hat{e}_y$ is smallest.

The practical value of these testing methods does not only depend on the quality that we have assigned to them but also on the purpose and the further application of the estimated model. Nevertheless, one must consider that one testing method alone can effect wrong decisions, but different methods together determine the correct model order with quite high accuracy.

8. CONCLUSION

In this paper, a critical comparison of different structure testing methods for the estimation of the correct model order has been made. The testing methods were thoroughly investigated in connection with the maximum likelihood and direct least squares method with regard to their efficiency and reliability. With the obtained results and experiences, the value of the testing methods was judged and finally rules for practical applications were formulated.

Remark: This paper was supported by the "Deutsche Forschungsgemeinschaft" within the project Un 25/1+3.

References

[1] Aström, K.J. and Eykhoff, P.: System identification, a survey report. IFAC-Symposium, Prag (1970), Paper 0.1.

[2] Balakrishnan, A.V. and Peterka, V.: Identification in automatic control systems. 4th IFAC-Congress, Warschau (1969), Survey paper 9.

[3] Cuenod, M. and Sage, A.P.:Comparison of some methods for process identification. IFAC-Symposium, Prag (1967), Paper 0.1.

[4] Eykhoff, P.: Process parameter and state estimation. IFAC-Symposium, Prag (1967), Paper 0.2.

[5] Eykhoff, P.: Systems modelling and identification. 3rd IFAC-Congress, London (1966), Survey paper.

[6] Niemann, R.E.: A review of process identification and parameter estimation techniques. Int.J.Control 13 (1971), pp. 209-264.

[7] Gustavsson, I.: Comparison of different methods for identification of industrial processes.Automatica 8 (1972), pp.127-142.

[8] Unbehauen, H. and Göhring, B.: Modellstrukturen and numerische Methoden für statistische Parameterschätzverfahren zur Identifikation von Regelsystemen. Regelungstechnik und Prozeßdatenverarbeitung 21 (1973), (in print).

[9] Gustavsson, I.: Parametric identification of time series. Report 6803. Lund Institute of Technology, Lund, (1968).

[10] Gustavsson, I.: Identification of dynamics of a distillation column. Report 6916. Lund Institute of Technology, Lund, (1969).

[11] Unbehauen, H.: Übersicht über Methoden zur Identifikation (Erkennung) dynamischer Systeme. Regelungstechnik und Prozeßdatenverarbeitung 21 (1973), No.1 (in print).

[12] Woodside, C.H.: Estimation of the order of linear systems. Automatica 7 (1971), pp. 727-733.

[13] Wilkinson, J.H.: Rundungsfehler. Springer-Verlag, Berlin (1969).

[14] Isaacson, E. and Keller, H.B.: Analysis of numerical methods. John Wiley, New York (1966).

[15] Kant, D. and Winkler, D.: Numerische Lösung schlecht konditionierter Gleichungssysteme auf dem Prozeßrechner. Regelungstechnik und Prozeß-Datenverarbeitung 19 (1971), No.4, pp.145-149.

[16] Kuo, S.S.: Numerical methods and computers. Addison Wesley, Massachusetts (1965).

[17] Fisz, M.: Wahrscheinlichkeitsrechnung und mathematische Statistik. VEB-Verlag, Berlin (1970).

[18] Waerden, B.L. van der: Mathematische Statistik Springer-Verlag,Berlin(1965).

Notation

u, y	input/output signal	[V]
N	number of data	
r_s	noise signal	[V]
λ_s	noise level	[-]
d_1	linear drift	[V/s]
ε	model error	[V]
a_i, b_i, c_i	parameters of the model	
K_s	gain factor	[V/V]
n and n_o	order of model and system	
T	sampling rate	[ms]
T_a	settling time	[s]
T_{mes}	measuring time	[s]
τ_N	time delay	[-]
T_u	time period	[s]
Δt_u	bit interval	[s]
N_u	number of bit intervals (impulse rate)	
$DV(n)$	determinant ratio	
$BZ(n)$	condition number	
$R_{\varepsilon\varepsilon N}$	autocorrelation function normalized on $R_{\varepsilon\varepsilon}(0)$	[-]
P_i	probability distribution function	[%]
$t_g(n)$ and t_α	test and tabulated value	
$\hat{y}_m$	model output	[V]
$\hat{h}_m$	step response	[V]
$\hat{g}_m$	impulse response	[V]
$\hat{e}_y, \hat{e}_h, \hat{e}_g$	error of signals	[V]

THE DETERMINATION OF THE ORDER OF PROCESS- AND NOISE DYNAMICS

A.J.W. van den Boom and A.W.M. van den Enden
University of Technology, E.E. Department
Eindhoven, the Netherlands

In most papers on parameter estimation schemes the order of the process under study is assumed to be known a-priori. In many practical situations, however, this information is lacking. Consequently simple methods for the determination of the order are necessary.
In this paper a description is given of five tests based on respectively: 1) the behaviour of the error function, 2) whiteness of the residuals (correlation function), 3) statistical independency of loss functions, 4) the behaviour of the determinant, and 5) the pole-zero cancellation effect.
Some of these methods are known (pole-zero cancellation, test of whiteness, F-test), the others are rearranged in such a way that a better discrimination of the order is obtained. The method based on the behaviour of the error function is extended in such a way that the order of both process- and noise dynamics are estimated separately.
An extensive set of simulations of the Åström-process is presented in order to elucidate the comparison of the different methods. It turns out that an acceptable determination of the order is possible even with a signal-to-noise ratio at the process output of -15dB.

CONTENTS

1. INTRODUCTION

Although today's literature on parameter estimation schemes is huge, relatively little attention has been paid towards the problems of the determination of the order. Usually, the order is assumed to be known beforehand.
Woodside (1970) suggested three very interesting tests of order, based on the product moment matrix of the measurables. His methods are attractive because no estimations of the parameters have to be performed as a phase in the determination of the order. He is making use, however, of a priori information concerning the covariance matrix of the noise and consequently does not meet the problem of determining the order of the noise. Åström (1968) describes a statistical test (F-test) which is, theoretically, applicable for Gaussian measurables.
This test has been used by Gustavsson (1969) for constructing models of process-and noise dynamics. However, the noise was modelled as a pure moving-average series and the order of process-and noise dynamics were taken equal. Recently, Chow (1972) proposed a method for the determination of the orders of the moving-average and autoregressive parts separately of a mixed autoregressive moving-average series with noisy observations.
In this paper we will summarize and compare different methods suitable for the determination of the orders of process and, if applicable, noise dynamics. This work is partially based on an introductory study by Van der Sommen (1971)

2. OUTLINE OF THE PROBLEM

Consider a linear time invariant discrete process:

$$\left(1 + \sum_{i=1}^{q} a_i z^{-i}\right) y_k = \left(b_o + \sum_{i=1}^{q} b_i z^{-i}\right) u_k + e_k \qquad (1)$$

where u_k and y_k are the input- and disturbed output signals respectively.
Define the equation error e_k as a linear filtering of a zero mean white noise ξ_k:

$$e_k = \frac{1 + \sum_{i=1}^{s} c_i z^{-i}}{1 + \sum_{i=1}^{s} d_i z^{-i}} \xi_k \qquad (2)$$

which is a mixed autoregressive moving-average type of description.
Combine (1) and (2):

$$y_k = \frac{b_o + \sum_{i=1}^{q} b_i z^{-i}}{1 + \sum_{i=1}^{q} a_i z^{-i}} u_k + \frac{1 + \sum_{i=1}^{s} c_i z^{-i}}{\left(1 + \sum_{i=1}^{s} d_i z^{-i}\right)\left(1 + \sum_{i=1}^{q} a_i z^{-i}\right)} \xi_k \qquad (3)$$

Define:

$$x_k = \frac{b_o + \sum_{i=1}^{q} b_i z^{-i}}{1 + \sum_{i=1}^{q} a_i z^{-i}} u_k \qquad (4)$$

and:

$$n_k = \frac{1 + \sum_{i=1}^{s} c_i z^{-i}}{\left(1+ \sum_{i=1}^{s} d_i z^{-i}\right)\left(1+ \sum_{i=1}^{q} a_i z^{-i}\right)} \xi_k \qquad (5)$$

Now x_k and n_k can be interpreted as the undisturbed output signal of the process and the corrupting output noise respectively.
As no knowledge is available of the quantities q and s we have to estimate these orders from the noisy observations y_k and the input signal u_k.
Note that no detailed a priori knowledge of the statistical properties of the noise is assumed to be available. The need for the determination of s is motivated by the fact that for obtaining consistent estimates by the generalized least squares method, proper modelling of the noise is essential; cf. Talmon and Van den Boom (1973).
Before describing the different tests, it is useful to give explicitly the notation used in the following sections.
Starting from (1) and (2) we define:

$$\underline{y}^T = [y_{q+1}, y_{q+2}, \ldots, y_{q+N}] \quad (6)$$

$$\underline{b}'^T = [b_o, \ldots, b_q, -a_1, \ldots, -a_q] \triangleq [\underline{b}^T, -\underline{a}^T] \quad (7)$$

$$\Omega(u,y) = [U|Y] = \left[\begin{array}{ccc|ccc} u_{q+1} & \cdots & u_1 & y_q & \cdots & y_1 \\ \vdots & & \vdots & & & \vdots \\ u_{q+N} & \cdots & u_N & y_{q+N-1} & \cdots & y_N \end{array}\right] \quad (8)$$

$$\underline{e}^T = [e_{q+1}, \ldots, e_{q+N}] \quad (9)$$

where N is the number of samples.
Rewriting (1) in matrix-notation:

$$\underline{y} = \Omega(u,y)\underline{b}' + \underline{e} \quad (10)$$

A least squares estimate $\underline{\beta}'$ of $\underline{b}'$ is:

$$\underline{\beta}' = [\Omega^{*T}\Omega^*]^{-1}\Omega^{*T}\underline{y} \quad (11)$$

where $\underline{\beta}'^T \triangleq (\underline{\beta}^T, -\underline{\alpha}^T)$ has dimension $2\hat{q}+1$; $\Omega^*(u,y)$ is a $N\times(2\hat{q}+1)$ matrix and $\hat{q}$ is the model order. Now $\underline{\beta}'$ is only consistent if $\underline{e}$ is white. Analogous to (10) we can rewrite (2) in matrix vector notation:

$$\underline{e} = \Xi\underline{c} - E\underline{d} + \underline{\xi} \quad (12)$$

with:

$$E = \begin{bmatrix} e_q & \cdots & e_{q-s+1} \\ \vdots & & \vdots \\ e_{q+N-1} & \cdots & e_{q-s+N} \end{bmatrix} \quad (13)$$

$$\Xi = \begin{bmatrix} \xi_q & \cdots & \xi_{q-s+1} \\ \vdots & & \vdots \\ \xi_{q+N-1} & \cdots & \xi_{q-s+N} \end{bmatrix} \quad (14)$$

$$\underline{\xi}^T = [\xi_{q+1}, \ldots, \xi_{q+N}] \quad (15)$$

$$\underline{d}^T = [d_1, \ldots, d_s] \quad (16)$$

$$\underline{c}^T = [c_1, \ldots, c_s] \quad (17)$$

Where s represents the unknown order of the noise filter. Combining (10) and (12) yields:

$$\underline{y} = \Omega(u,y,\xi,e)\underline{b}'' + \underline{\xi} \quad (18)$$

with:

$$\Omega(u,y,\xi,e) = [U|Y|\Xi|E] \quad (19)$$

$$\underline{b}''^T = [\underline{b}^T, -\underline{a}^T, \underline{c}^T, -\underline{d}^T] \quad (20)$$

A least squares estimate $\underline{\beta}''$ of $\underline{b}''$ is:

$$\underline{\beta}'' = [\Omega^{*T}(u,y,\hat{\xi},\hat{e})\Omega^*(u,y,\hat{\xi},\hat{e})]^{-1}\Omega^{*T}(u,y,\hat{\xi},\hat{e})\underline{y} \quad (21)$$

with:

$$\Omega^*(u,y,\hat{\xi},\hat{e}) = [U^*|Y^*|\hat{\Xi}^*|\hat{E}^*] \quad (22)$$

$$\underline{\beta}''^T = [\underline{\beta}^T, -\underline{\alpha}^T, \underline{\gamma}^T, -\underline{\delta}^T] \quad (23)$$

where U^* and Y^* are $N\times(\hat{q}+1)$ and $N\times\hat{q}$ matrices respectively containing the measurables; $\hat{\Xi}^*$ and $\hat{E}^*$ are both $N\times\hat{s}$ matrices with $\hat{s}$ being the order of the noise model.
As $\underline{e}$ and $\underline{\xi}$ are not available, we have to estimate them by making use of old estimates of process- and noise parameters yielding $\hat{\underline{e}}$ and $\hat{\underline{\xi}}$ and consequently $\hat{E}^*$ and $\hat{\Xi}^*$ cf. Talmon and Van den Boom (1973).

3. DESCRIPTION OF THE EXAMPLE

In the following section results will be given of order tests applied on the following combination of discrete process- and noise dynamics:

$$\text{process: } y_k - 1.5y_{k-1} + 0.7y_{k-2} = u_{k-1} + 0.5u_{k-2} + e_k \quad (24)$$

$$\text{noise: } e_k - 0.5e_{k-1} = \xi_k + 0.3\xi_{k-1} \quad (25)$$

The process used is the same as proposed by Åström (1968), while the noise filter used is the same as proposed by Talmon and Van den Boom (1973).
The signals u_k and ξ_k are white noise sequences with a rectangular distribution between (-1,+1) and $(-\lambda,+\lambda)$ respectively.
The input signal u_k and the disturbed output signal y_k are available for estimation (500 samples).
The signal-to-noise ratios of the undisturbed signal x_k and the additive noise n_k are for:

$\lambda = \frac{1}{4}$ S/N = 8,3 dB
$\lambda = 1$ S/N = - 3,7 dB
$\lambda = 4$ S/N = -15,7 dB

The parameter estimation method used is the aforementioned extended matrix method.

4. DESCRIPTION OF THE TESTS

4.1. The behaviour of the error function.

Let us study the behaviour of the error function $V_1 = \frac{1}{N}\hat{\underline{e}}^T\hat{\underline{e}}$ as it is related to the order of the estimates.

$$\hat{\underline{e}} = \underline{y} - \Omega^*(u,y)\underline{\beta}' = \Omega(u,y)\underline{b}' - \Omega^*(u,y)\underline{\beta}' + \underline{e} \triangleq \triangleq \underline{m} + \underline{e} \quad (26)$$

$$V_1 = \frac{1}{N}\hat{\underline{e}}^T\hat{\underline{e}} = \quad (27)$$

$$= \frac{1}{N}\{\underline{m}^T\underline{m} + 2\underline{b}'^T\Omega^T(u,y)\underline{e} - 2\underline{\beta}'^T\Omega^{*T}(u,y)\underline{e} + \underline{e}^T\underline{e}\}$$

In order to gain insight, assume that $\underline{e}$ is white noise and take the probability limit of V_1 using (8):

$$\underset{N\to\infty}{\text{plim}}[V_1] = \underset{N\to\infty}{\text{plim}}[\frac{1}{N}\underline{m}^T\underline{m}] + 2\underline{b}'^T \underset{N\to\infty}{\text{plim}}\left[\frac{1}{N}\begin{bmatrix}U^T\\X^T+\mathcal{N}^T\end{bmatrix}\underline{e}\right] -$$

$$- 2\underset{N\to\infty}{\text{plim}}[\underline{\beta}'^T]\underset{N\to\infty}{\text{plim}}\left[\frac{1}{N}\begin{bmatrix}U^{*T}\\X^{*T}+\mathcal{N}^{*T}\end{bmatrix}\underline{e}\right] + \quad (28)$$

$$+ \underset{N\to\infty}{\text{plim}}[\frac{1}{N}\underline{e}^T\underline{e}]$$

where $\mathcal{N}^*$ is a $N\times\hat{q}$ matrix containing the disturbing noise cf. equation (5).

Now, as $\underline{e}$ is independent of $\underline{u}$ and $\underline{x}$, and $\underline{e}$ is independent of the rows of $\mathcal{N}^*$ due to its white character, we find:

$$\underset{N\to\infty}{\text{plim}}[V_1] = \underset{N\to\infty}{\text{plim}}[\frac{1}{N}\underline{m}^T\underline{m}] + \underset{N\to\infty}{\text{plim}}[\frac{1}{N}\underline{e}^T\underline{e}] \quad (29)$$

Let us consider the first term of the right-hand part of (29). Therefore we take into account the asymptotic properties of the estimate $\underline{\beta}'$, cf. equation (11):

$$\underset{N\to\infty}{\text{plim}}[\beta_i] = \begin{cases} \neq b_i & \text{if } \hat{q}<q \text{ due to the truncation effect.} \\ = b_i & \text{if } \hat{q}=q \\ = b_i & i\leq q \\ = 0 & i>q \end{cases} \Big\} \text{if } \hat{q}>q \quad (30)$$

$$\underset{N\to\infty}{\text{plim}}[\alpha_i] = \begin{cases} \neq a_i & \text{if } \hat{q}<q \text{ due to truncation} \\ = a_i & \text{if } \hat{q}=q \\ = a_i & i\leq q \\ = 0 & i>q \end{cases} \Big\} \text{if } \hat{q}>q \quad (31)$$

This leads to:

$$\underset{N\to\infty}{\text{plim}}[\frac{1}{N}\underline{m}^T\underline{m}] = \begin{cases} > 0 & \text{for } \hat{q}<q \\ = 0 & \text{for } \hat{q}\geq q \end{cases} \quad (32)$$

Consequently, the behaviour of the error function V_1 is significantly changing in the neigbourhood of q, cf. figure 1 curve a.

In practical situations we are dealing with data sequences of finite length, with a non white equation error $\underline{e}$ and with estimates of $\underline{e}$, so the behaviour of V_1 will be less pronounced and more noisy both depending on the signal to noise ratio and the colouring of $\underline{e}$, cf. figure 1 curve b.

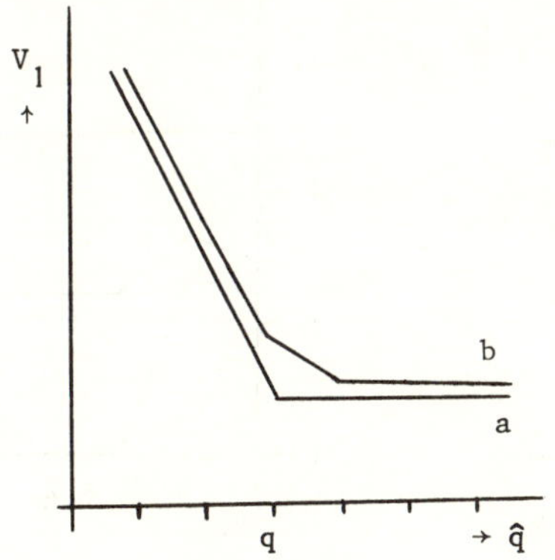

figure 1.

For the determination of the order of the noise dynamics s we can use similar reasonings. Therefore let us consider first the following undisturbed autoregressive moving-average process cf. equation (12):

$$\underline{e} = \Xi\underline{c} - E\underline{d} + \underline{\xi} \quad (33)$$

The residual can be defined as:

$$\tilde{\underline{\xi}} \triangleq \underline{e} + E^*\underline{\delta} - \Xi^*\underline{\gamma} = - E\underline{d} + \Xi\underline{c} + E^*\underline{\delta} - \Xi^*\underline{\gamma} + \underline{\xi}$$

$$\triangleq \underline{r} + \underline{\xi} \quad (34)$$

where E^* and Ξ^* are $N\times\hat{s}$ matrices and $\underline{\delta}$ and $\underline{\gamma}$ are vectors of dimensions $\hat{s}$, $\hat{s}$ being the order of the model.

$$\text{Define: } V_2 = \frac{1}{N}\tilde{\underline{\xi}}^T\tilde{\underline{\xi}} \quad (35)$$

Now:

$$\underset{N\to\infty}{\text{plim}}[V_2] = \underset{N\to\infty}{\text{plim}}[\frac{1}{N}\{\underline{r}^T\underline{r}+2\underline{r}^T\underline{\xi}+\underline{\xi}^T\underline{\xi}\}] =$$

$$= \underset{N\to\infty}{\text{plim}}[\frac{1}{N}\underline{r}^T\underline{r}] + \underset{N\to\infty}{\text{plim}}[\frac{1}{N}\underline{\xi}^T\underline{\xi}] \quad (36)$$

because $\underline{r}$ and $\underline{\xi}$ are uncorrelated due to the white character of $\underline{\xi}$.

As:

$$\underset{N\to\infty}{\text{plim}}[\gamma_i] = \begin{cases} \neq c_i & \text{if } \hat{s}<s \\ = c_i & \text{if } \hat{s}=s \\ = c_i & i\leq s \\ = 0 & i>s \end{cases} \Big\} \text{if } \hat{s}>s \quad (37)$$

and:

$$\underset{N\to\infty}{\text{plim}}[\delta_i] = \begin{cases} \neq d_i & \text{if } \hat{s}<s \\ = d_i & \text{if } \hat{s}=s \\ = d_i & i\leq s \\ = 0 & i>s \end{cases} \Big\} \text{if } \hat{s}>s \quad (38)$$

we find that:

$$\underset{N\to\infty}{\text{plim}}[\frac{1}{N}\underline{r}^T\underline{r}] = \begin{cases} >0 & \text{for } \hat{s}<s \\ =0 & \text{for } \hat{s}\geq s \end{cases} \quad (39)$$

Similar to V_1, the behaviour of V_2 is qualitatively changing in the neighbourhood of the correct order of the model. In practical situations, we only have approximated values of the signals $\underline{e}$ and $\underline{\xi}$, leading to a less ideal behaviour of V_2 comparable to curve b of figure 1.

By observing the qualitative character of V_1 and V_2 simultaneously we are able to determine the orders of process- and noise dynamics separately.

In figure 2 the quantity log $V_1(\hat{q})$ shows a change in behaviour for $\hat{q}=2$, for different noise filter orders ($\hat{s}=0,1,2$).

In figure 3 the quantity log $V_2(\hat{s})$ with $\hat{q}=2$ leads to $\hat{s}=1$. Also, by observing V_1 and V_2 we are able to distinguish the correct orders of process- and noise dynamics separately. It is obvious that V_2 is also a function of $\hat{q}$. This gives the possibility to use V_2 as an indicator for the order of the process itself. In figure 4 it can be seen that $\hat{q}=2$ looks a very reasonable estimate for the given λ.

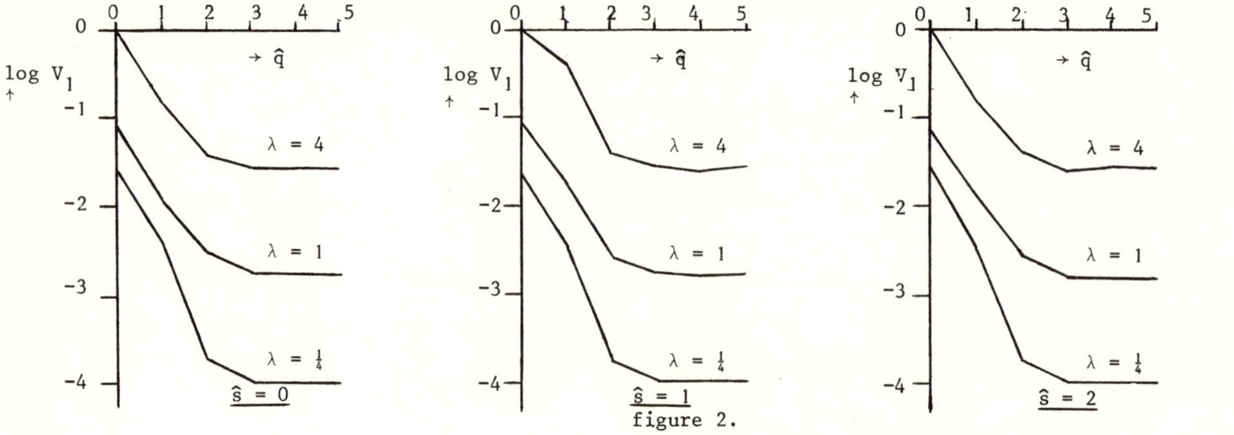

figure 2.

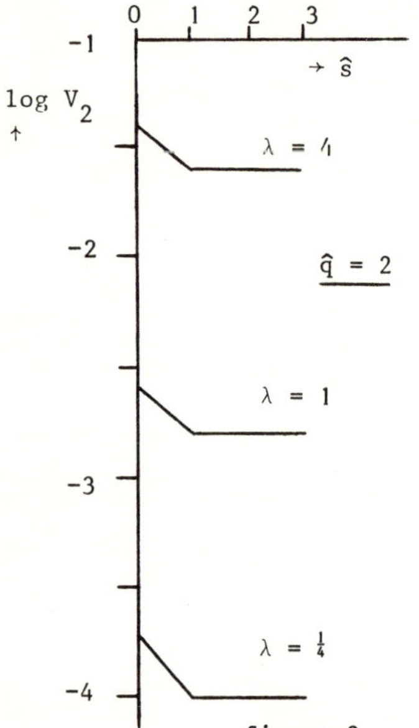

figure 3.

4.2. Whiteness of residuals.

This method is based on the fact that by proper modelling of process- and noise dynamics $\hat{\xi}$ can be seen as an acceptable estimate of the white noise ξ. Therefore, testing $\hat{\xi}$ for its statistical properties gives us an indication of the quality of the modelling.

If we compute:

$$\tilde{\Psi}_{\hat{\xi}\hat{\xi}}(\tau) = \frac{1}{N}\sum_{i=1}^{N} \hat{\xi}_{\ell+i}\hat{\xi}_{\ell+i+\tau}$$

we can test if this can be accepted as the autocorrelation function of a white noise sequence, taking into account the limited number of samples used in the computation. Analogous to Laning and Battin (1956) it can be shown that the variance in the correlation function $\tilde{\Psi}(\tau)$ of a gaussian white signal on basis of calculation with only N samples is:

$$\mathrm{var}\{\tilde{\Psi}(\tau)\} = \frac{\Psi^2(0)}{N} \qquad \tau \neq 0 \tag{40}$$

If we define a normed correlation function:

$$\tilde{R}(\tau) \triangleq \frac{\tilde{\Psi}(\tau)}{\tilde{\Psi}(0)} \tag{41}$$

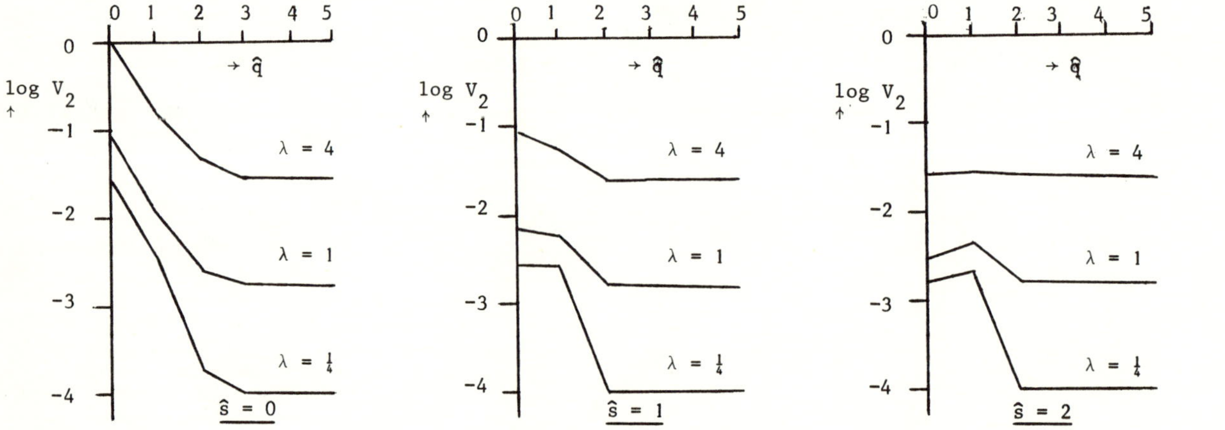

figure 4

then: $\operatorname{var}\{\tilde{R}(\tau)\} \simeq \frac{1}{N}$ (42)

In figure 5a $\tilde{R}_{\hat{\xi}\hat{\xi}}(\tau)$ is plotted for $\hat{s}=1$ and $\hat{q}=0,1,2$ respectively ($\lambda=\frac{1}{4}$). If we calculate the variance $\hat{\sigma}^2(\tilde{R}_{\hat{\xi}\hat{\xi}})$ using $\tilde{R}_{\hat{\xi}\hat{\xi}}(\tau)$, $\tau = 1,\dots 9$ we can check if this $\hat{\sigma}^2$ is below the theoretical limit of (42). In figure 5b $\log \hat{\sigma}^2(\tilde{R}_{\hat{\xi}\hat{\xi}})$ is plotted for different orders of the noise model. It is obvious that $\hat{q}=3$ is the selected process order for $\hat{s}=0$, which can be explained if we rewrite equations (24) and (25):

$$(1-1.5z^{-1}+0.7z^{-2})y_k = (z^{-1}+0.5z^{-2})u_k + \frac{1}{1-0.8z^{-1}+0.24z^{-2}+\dots}\xi_k \quad (43)$$

It looks reasonable that $1-0.8z^{-1}$ is a good approximation of:

$1-0.8z^{-1} + 0.24z^{-2} - 0.072z^{-3}$ .... leading to

$$(1-1.5z^{-1} + 0.7z^{-2})(1-0.8z^{-1})y_k =$$
$$= (z^{-1} + 0.5z^{-2})(1-0.8z^{-1})u_k + \xi_k$$

For $\hat{s}=1$ the method selects $\hat{q}=2$ which is correct. For $\hat{s}=2$ order 0 is accepted by the method. The whiteness of residuals test is not able to distinguish between process- and noise dynamics.

It is the authors' opinion that this method is very suited to be used as a check on the method based on minimization of the loss function. The two tests can be considered as each other's complement as the loss function test takes into account the autocorrelation function of the residuals with zero delay while the whiteness test considers delays unequal to zero.

4.3. The F-Test.

Åström (1968) describes a test based on the statistical independency of the quantities $V_2(\hat{q}_2)$ and $V_2(\hat{q}_1)-V_2(\hat{q}_2)$, where $V_2(\hat{q}_1)$ and $V_2(\hat{q}_2)$ are loss functions based on models with order $\hat{q}_1$ and $\hat{q}_2$ respectively and normal residuals. If $\hat{q}_2>\hat{q}_1\geq q$, then $V_2(\hat{q}_2)$ and $V_2(\hat{q}_1)-V_2(\hat{q}_2)$ are independent random variables with χ^2 distributions and $N-(2\hat{q}_2+1)$ and $2(\hat{q}_2-\hat{q}_1)$ degrees of freedom respectively. The test quantity:

$$t = \frac{V_2(\hat{q}_1) - V_2(\hat{q}_2)}{V_2(\hat{q}_2)} \cdot \frac{N - (2\hat{q}_2+1)}{2(\hat{q}_2-\hat{q}_1)} \quad (44)$$

has an $F(N-2\hat{q}_2-1,\ 2\hat{q}_2-2\hat{q}_1)$ distribution. For a risk level of 5% and $N > 100$ we test for $t < 3$. In table 1 results are given for $\lambda = \frac{1}{4}$, with $N = 200$ (samples 301-500). Note that we still are dealing with uniformly distributed random variables.

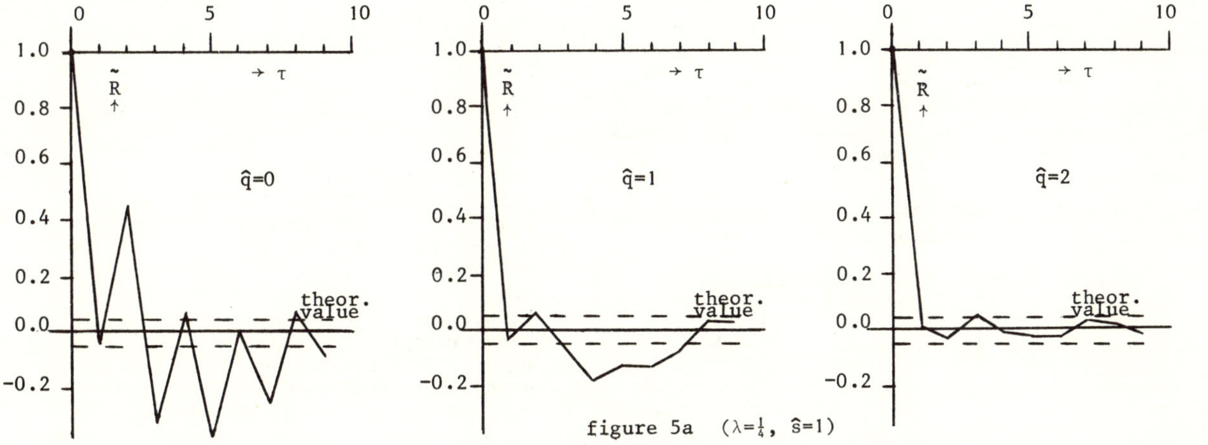

figure 5a ($\lambda=\frac{1}{4}$, $\hat{s}=1$)

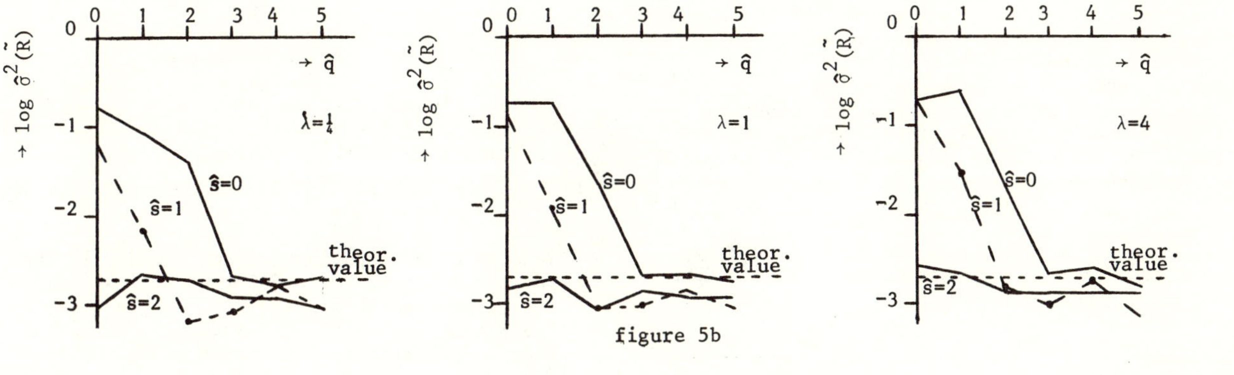

figure 5b

In practice we found no difference in results when testing with normal distributed noise. For $\hat{s}=0$ we choose a third order process, although the increase in order from 4 to 5 gives some improvement. For $\hat{s}=1$ we select $\hat{q}=2$ although the same problem arises at the increase of model order from 3 to 4. For $\hat{s}=2$ we have problems in selecting process order 2. Here also order 3 and 5 give some improvement. As a general impression the authors can state that, compared with the other tests, the F-test yields less impressing results.
It should be well realized that a slight decrease of the errorfunction is qualified by the F-test as a significant improvement. If $V_2(\hat{q}_2)=0.96\ V_2(\hat{q}_1)$ then $t \approx 4$, which is significant!
An effect that may be important is the fact that we are generating our residuals during the iterative estimation scheme, resulting in residuals with poorer quality in the beginning of the procedure.

$\hat{s}$	$\hat{q}_1 \backslash \hat{q}_2$	1	2	3	4	5
0	0	610	6955	8088	6004	4963
	1	-	1859	1653	1091	846
	2	-	-	73.4	36.3	26.4
	3	-	-	-	0.00	2.09
	4	-	-	-	-	4.17
1	0	11.9	1352	883	687	544
	1	-	2402	1176	814	604
	2	-	-	-0.97	1.77	1.17
	3	-	-	-	4.55	2.25
	4	-	-	-	-	0.00
2	0	-26.2	718	506	375	320
	1	-	1988	1049	692	553
	2	-	-	6.09	3.02	4.45
	3	-	-	-	0.00	3.47
	4	-	-	-	-	6.94

Table 1 $(\lambda=\frac{1}{4})$

4.4. The behaviour of the determinant.

This method is based on the fact that the rank of the $N\times(2\hat{q}+1)$ matrix $\Omega^*(u,x)$ is $\min.(q+\hat{q}+1,\ 2\hat{q}+1)$, cf. Lee (1964) p. 97. Consequently, testing the $(2\hat{q}+1, 2\hat{q}+1)$ square matrix $\frac{1}{N}\Omega^{*T}(u,x)\Omega^*(u,x)$ for singularity gives the desired order. In noisy cases cases only $\Omega^*(u,y)$ is available so $\frac{1}{N}\Omega^{*T}(u,y)\Omega^*(u,y)$ has to be tested for near singularity. This test can be performed without executing the estimation algorithm. Of course $\frac{1}{N}\Omega^{*T}(u,y)\Omega^*(u,y)$ has only near singularity properties if the signal to noise ratio is not too small.
Now in equation (45) the ranks of Q_{11}^* and Q_{22}^* give the desired information of the order of process- and noise dynamics respectively. It is obvious that the rank of Q_{22}^* can not be determined without estimating the process- and noise parameters, as these are necessary for the generation of $\hat{\underline{e}}$ and $\hat{\underline{\xi}}$.
Consequently we are forced to combine this type of test with the parameter estimation procedure.

$$\frac{1}{N}\Omega^{*T}(u,y,\hat{\xi},\hat{e})\Omega^*(u,y,\hat{\xi},\hat{e}) = \tag{45}$$

$$= \frac{1}{N}\left[\begin{array}{cc|cc} U^{*T}U^* & U^{*T}Y^* & U^{*T}\hat{\Xi}^* & U^{*T}\hat{E}^* \\ Y^{*T}U^* & Y^{*T}Y^* & Y^{*T}\hat{\Xi}^* & Y^{*T}\hat{E}^* \\ \hline \hat{\Xi}^{*T}U^* & \hat{\Xi}^{*T}Y^* & \hat{\Xi}^{*T}\hat{\Xi}^* & \hat{\Xi}^{*T}\hat{E}^* \\ \hat{E}^{*T}U^* & \hat{E}^{*T}Y^* & \hat{E}^{*T}\hat{\Xi}^* & \hat{E}^{*T}\hat{E}^* \end{array}\right] = \left[\begin{array}{c|c} Q_{11}^* & Q_{12}^* \\ \hline Q_{21}^* & Q_{22}^* \end{array}\right]$$

As during the estimation:

$$\left[\frac{1}{N}\Omega^{*T}(u,y,\hat{\xi},\hat{e})\Omega^*(u,y,\hat{\xi},\hat{e})\right]^{-1} \triangleq \left[\begin{array}{c|c} C_{11}^* & C_{12}^* \\ \hline C_{21}^* & C_{22}^* \end{array}\right] \tag{46}$$

is available, it is worthwhile to investigate if the matrices C^* are useful for order determination. As $Q_{21}^{*T} = Q_{12}^{*T}$ we find:

$$C_{11}^* = \left[Q_{11}^* - Q_{12}^* Q_{22}^{*-1} Q_{12}^{*T}\right]^{-1} \tag{47}$$

$$C_{22}^* = \left[Q_{22}^* - Q_{12}^{*T} Q_{11}^{*-1} Q_{12}^*\right]^{-1} \tag{48}$$

Already we assumed that the input signal $\underline{u}$ is independent of the disturbing noise $\underline{\xi}$, so for large N we find:

$$Q_{12}^* \approx \frac{1}{N}\left[\begin{array}{c|c} 0 & 0 \\ \hline Y^{*T}\hat{\Xi}^* & Y^{*T}\hat{E}^* \end{array}\right] \tag{49}$$

Thus:

$$C_{11}^* \approx \left\{Q_{11}^* - \left[\begin{array}{c|c} 0 & 0 \\ \hline 0 & M \end{array}\right]\right\}^{-1} \tag{50}$$

$$Q_{11}^* \approx C_{11}^{*\,-1} + \left[\begin{array}{c|c} 0 & 0 \\ \hline 0 & M \end{array}\right] \tag{51}$$

where M is a submatrix which is only important when the signal-to-noise ratio is small. This will be indicated later in this section. If we neglect M for good signal-to-noise ratios, then C_{11}^* can be used as an indicator for the order of the process. Now, if Q_{11}^* is near singular, it has some eigenvalues which are small compared to the others. As the eigenvalues of C_{11}^* are - approximately - the inverses of the eigenvalues of Q_{11}^*, C_{11}^* also has eigenvalues which are strongly different in magnitude. This also indicates near singularity, which also occurs if $\hat{q}>q$.
Now define the relative determinant of C_{11}^*:

$$\text{rel.det.}C_{11}^* = \frac{\det C_{11}^*}{\text{maximum content of } C_{11}^*} \tag{52}$$

The use of the relative determinant is important because the determinant itself is a function of the amplitude of the signals:

$$\det Q = f(\text{amplitude}^{\,2(2\hat{q}+1)}) \tag{53}$$

In fig. 6 the behaviour of rel.det.C_{11}^* is shown for $\hat{s}=0,1,2$ leading to $\hat{q}=2$ for $\lambda=\frac{1}{4},1$ and an doubtful determination $\hat{q}=2$ for $\lambda=4$. In practice, we found that the relative determinant of C_{22}^* is an useful indicator for the order of the noise, cf. fig. 7.

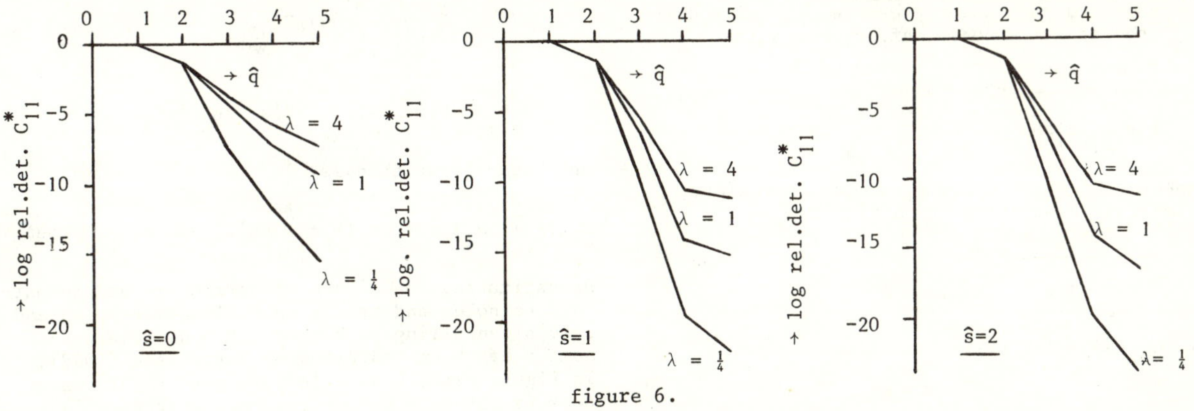

figure 6.

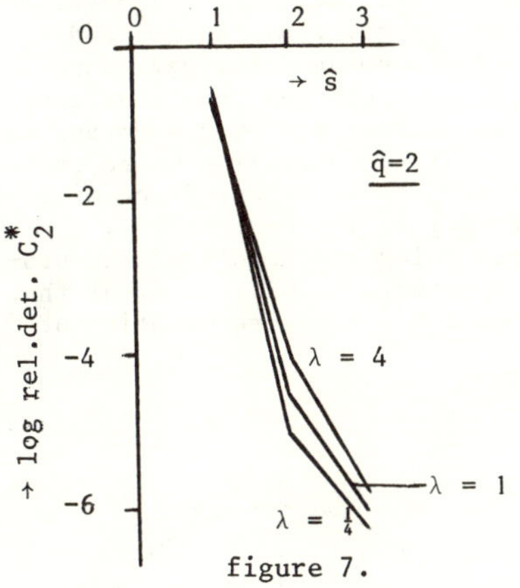

figure 7.

As det Q^*_{11} is small if $\hat{q}>q$ the quantity det C^*_{11} will be large. However as the maximum content of C^*_{11} is increasing more rapidly than det C^*_{11} the relative determinant of C^*_{11} will decrease for $\hat{q}>q$. The amount of increase of det C^*_{11} can be defined by:

$$\Delta C^*_{11} = \frac{\det.C^*_{11}(\text{order } i)}{\det.C^*_{11}\ (\text{order } i-1)} \tag{54}$$

Now by observing where a remarkable increase occurs the order of the process can be found, cf. figure 8. In this figure an increase of C^*_{11} can be observed for $\hat{q}=3$. The increase is dependent on the signal to noise ratio and is limited by the matrix M which has been neglected cf. equation (51). If the signal to noise ratio is larger ($\lambda={}^1/_{64}$) this increase is more apparent cf. fig. 8

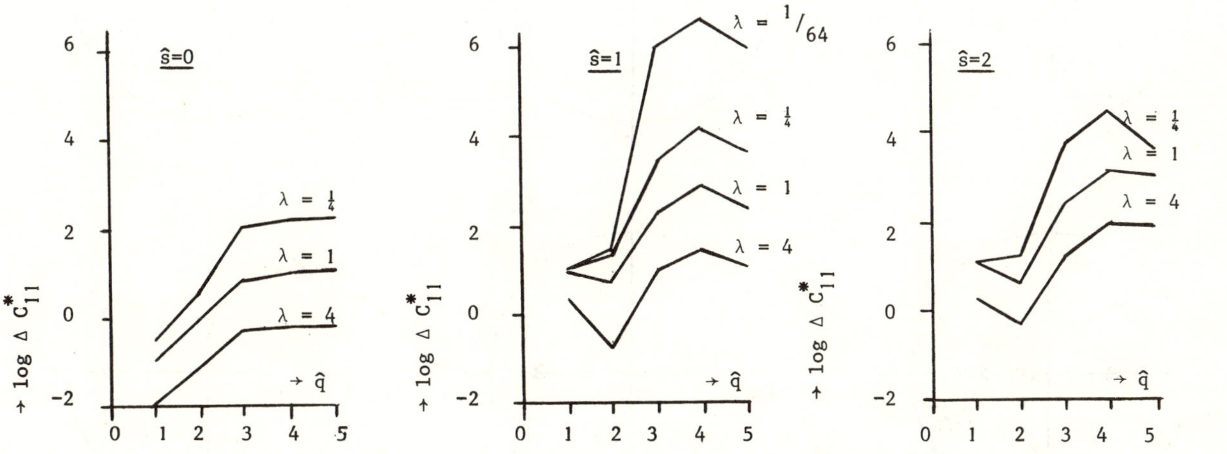

figure 8.

4.5. The pole-zero cancellation effect.

Let us choose a pure autoregressive type of modelling of the additive noise, cf. equation (2):

$$e_k = \frac{1}{1 + \sum_{i=1}^{\infty} d_i' z^{-i}} \xi_k \tag{55}$$

Dependent on the values of d_i' we can approximate this infinite development by a finite sum:

$$e_k = \frac{1}{1 + \sum_{i=1}^{s'} d_i' z^{-i}} \xi_k \tag{56}$$

Combining (56) with (1):

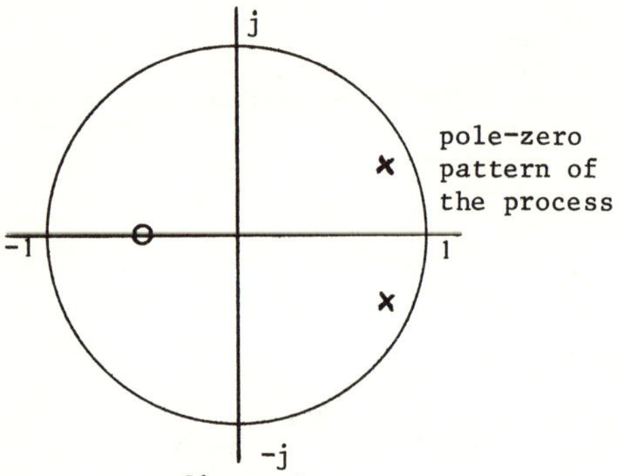

figure 9

$$(1+\sum_{i=1}^{q} a_i z^{-i})(1+\sum_{i=1}^{s'} d_i' z^{-i})y_k =$$

$$= (b_o+\sum_{i=1}^{q} b_i z^{-i})(1+\sum_{i=1}^{s'} d_i' z^{-i})u_k + \xi_k \tag{57}$$

which can be written as:

$$(1+\sum_{i=1}^{q+s'} a_i' z^{-i})y_k = (b_o+\sum_{i=1}^{q+s'} b_i' z^{-i})u_k + \xi_k \tag{58}$$

By estimating the a' and b' parameters and comparing the poles and zero's we will notice poles and zero's cancelling each other if $\hat{q}>q$. This phenomena has been indicated by Gustavsson (1968). In figure 9 the true pole-zero pattern of the process is given in the unit circle of the z-plane. In figure 10 the pole-zero cancellation effect is shown for $\lambda = \frac{1}{4}$. It is clear that the pole-zero pattern of the process is built up for $\hat{q}=1$ and $\hat{q}=2$, and remains hardly unchanged for $\hat{q}\geq 2$. Moreover, for $\hat{q}\geq 3$, cancelling pole-zero combinations are added. It is obvious that the process-order can be determined as $\hat{q}=2$. If the signal-to-noise ratio is small (e.g. $\lambda=4$) the cancellation effect will be less pronounced, cf. figure 11. Furthermore, the bias in the estimated poles and zero's of the process will be more important. It will be clear that this method is suited for order determination of the process only.

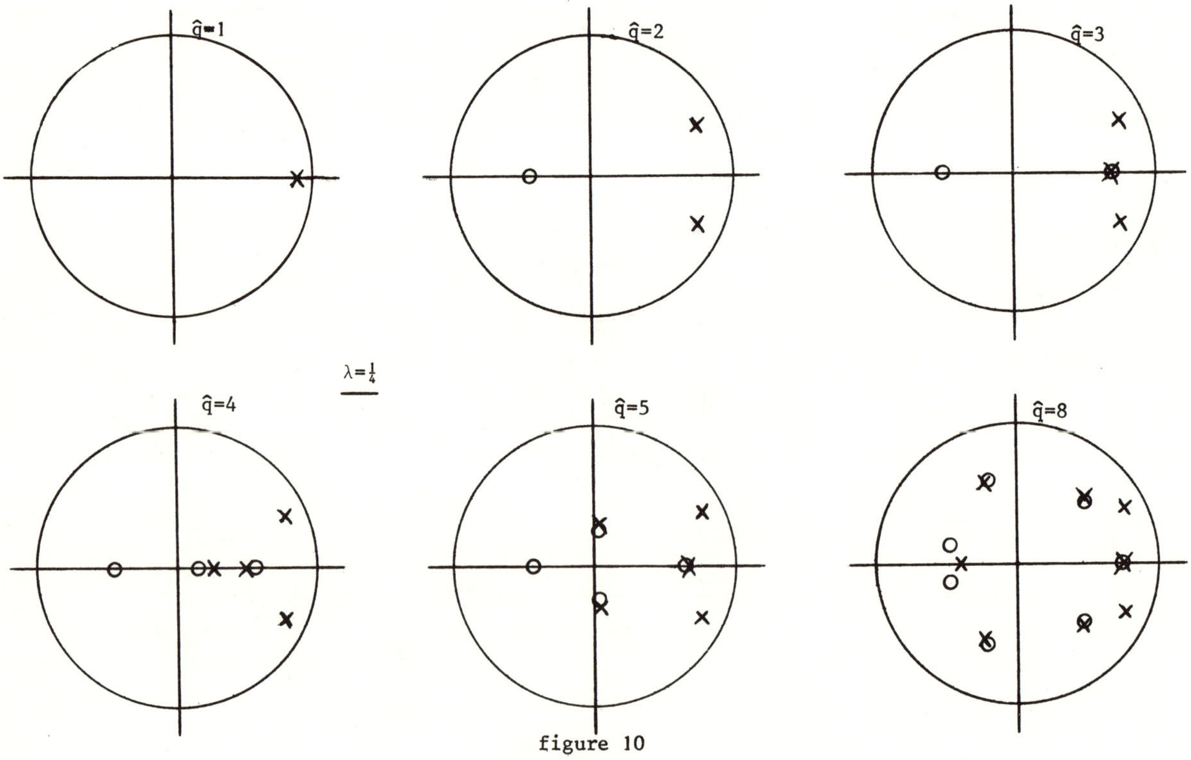

figure 10

The noise is modelled as a pure autoregressive series which has the principal drawback of requiring more parameters for a complete description. As during the order determination by means of the other methods, estimates of the parameters of the process become available for different $\hat{q}$ and $\hat{s}=0$, it is rather simple to implement this method as an additonal check on the results of the other methods.

5. COMPARISON OF THE TESTS

In table 2 the main results of the seven tests described in the previous section are grouped together. It is important to note that these figures have been obtained from a single set of input-output data in order to facilitate a useful comparison. The tests can be divided into three groups:

Group I:
Tests for the determination of the order of process and noise dynamics separately:
Ia V_1 with V_2
Ib V_2
Ic C_{11} with C_{22}

Group II:
Tests for the determination of the order of process dynamics only:
II - pole zero cancellation.

Group III:
Tests for the determination of an acceptable combination of process- and noise order.
IIIa residuals
IIIb F-test

It is obvious that the tests of group I are preferable if the orders of process- and noise dynamics have to be determined separately. The tests of group II and III are useful as an extra verification of the results of the tests of group I. It is the authors' experience that order determination should be performed using the different tests in parallel. In most practical situations this will very well be possible as the quantities necessary for the different tests become available during the estimation.
It can also be noticed in table 2 that the quantitative results of the different tests are approximately equal except those of the F-test, which are inferior.
Furthermore, the V_1, V_2 combination gives somewhat better results.
The main conclusion to be stated is that an acceptable test of order of process- and noise dynamics separately could be achieved even for a signal to noise ratio at the process output of -15.7dB ($\lambda=4$)!

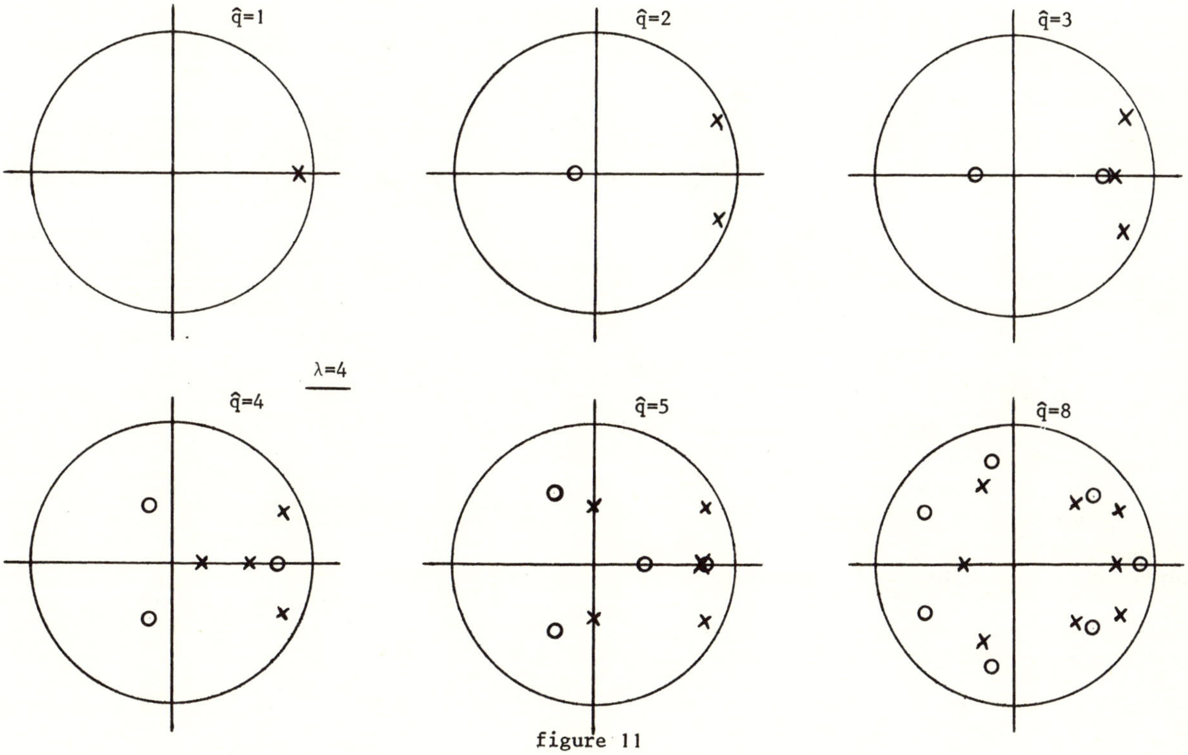

figure 11

	$\lambda = \frac{1}{4}$		$\lambda = 1$		$\lambda = 4$		group
test	$\hat{q}$	$\hat{s}$	$\hat{q}$	$\hat{s}$	$\hat{q}$	$\hat{s}$	
V_1	$\hat{s}=0$, $\hat{s}=1$, $\hat{s}=2$ → $\hat{q}=2$ ↘		$\hat{s}=0$, $\hat{s}=1$, $\hat{s}=2$ → $\hat{q}=2$ ↘		$\hat{s}=0$, $\hat{s}=1$, $\hat{s}=2$ → $\hat{q}=2$ ↘		Ia
V_2		$\hat{q}=2 \rightarrow \hat{s}=1$		$\hat{q}=2 \rightarrow \hat{s}=1$		$\hat{q}=2 \rightarrow \hat{s}=1$	Ib
V_2	$\hat{s}=0$, $\hat{s}=1$, $\hat{s}=2$ → $\hat{q}=2$ ↗		$\hat{s}=0 \rightarrow \hat{q}=2$ $\hat{s}=1 \rightarrow \hat{q}=2$ $\hat{s}=2 \rightarrow \hat{q}=?$ ↗		$\hat{s}=0 \rightarrow \hat{q}=2$ $\hat{s}=1 \rightarrow \hat{q}=2$ $\hat{s}=2 \rightarrow \hat{q}=?$ ↗		
relative determinant C_{11}	$\hat{s}=0$, $\hat{s}=1$, $\hat{s}=2$ → $\hat{q}=2$ ↘		$\hat{s}=0 \rightarrow \hat{q}=?$ $\hat{s}=1 \rightarrow \hat{q}=2$ $\hat{s}=2 \rightarrow \hat{q}=2$ ↘		$\hat{s}=0 \rightarrow \hat{q}=?$ $\hat{s}=1 \rightarrow \hat{q}=2(?)$ $\hat{s}=2 \rightarrow \hat{q}=2(?)$ ↘		Ic
relative determinant C_{22}		$\hat{q}=2 \rightarrow \hat{s}=1$		$\hat{q}=2 \rightarrow \hat{s}=1$		$\hat{q}=2 \rightarrow \hat{s}=1$	
pole-zero cancellation	$\hat{q}=2$		$\hat{q}=2$		$\hat{q}=2$		II
residuals	$\hat{s}=0 \rightarrow \hat{q}=3$ $\hat{s}=1 \rightarrow \hat{q}=2$ $\hat{s}=2 \rightarrow \hat{q}=0$		$\hat{s}=0 \rightarrow \hat{q}=3$ $\hat{s}=1 \rightarrow \hat{q}=2$ $\hat{s}=2 \rightarrow \hat{q}=0$		$\hat{s}=0 \rightarrow \hat{q}=3$ $\hat{s}=1 \rightarrow \hat{q}=2$ $\hat{s}=2 \rightarrow \hat{q}=0(?)$		IIIa
F-test	$\hat{s}=0 \rightarrow \hat{q}=3(5)$ $\hat{s}=1 \rightarrow \hat{q}=2(4)$ $\hat{s}=2 \rightarrow \hat{q}=2(3)(5)$		$\hat{s}=0 \rightarrow \hat{q}=3(5)$ $\hat{s}=1 \rightarrow \hat{q}=2(4)$ $\hat{s}=2 \rightarrow \hat{q}=2(3)(5)$		$\hat{s}=0 \rightarrow \hat{q}=3(5)$ $\hat{s}=1 \rightarrow \hat{q}=2(4)$ $\hat{s}=2 \rightarrow \hat{q}=2(3)(5)$		IIIb

Table 2

6. CONCLUSION

In this paper different tests for the determination of the order in connection with identification problems are compared. Simulations of the Åström process show the different properties of the tests described.
Two types of tests, the behaviour of the error function and the behaviour of the determinants are shown to be suitable for determining the order of process- and noise dynamics separately. These tests are compared in qualitative and quantitave behaviour with other tests e.g. the F-test, the test of whiteness of the residuals and the pole-zero cancellation effect.
Simulated results show that an acceptable test of order of process- and noise dynamics separately is possible even at a signal-to-noise ratio of-15.7dB.

LITERATURE

K.J. Åström (1968). Lectures on the Identification Problem - The Least Squares Method. Report 6806, Lund Institute of Technology, Sweden.

J.C. Chow (1972). On Estimating the Orders of an Autoregressive Moving-Average Process with Uncertain Observations. IEEE, AC-17, p. 707

I. Gustavsson (1969). Maximum Likelihood Identification of the Ågesta Reactor and Comparison with the Results of Spectral Analysis. Report 6903, Lund Institute of Technology, Sweden.

J.H. Laning and R.H. Battin (1956). Random Processes in Automatic Control. McGraw-Hill, New York.

R.C.K. Lee (1964). Optimal Estimation, Identification, and Control. Research Monograph no. 28, MIT Press, Cambridge, Mass.

F.L.M. van der Sommen (1971). The Determination of the Order of Linear Systems (in Dutch) M.Sc. report, E.E. Dept., University of Technology, Eindhoven, the Netherlands

J.L. Talmon and A.J.W. van den Boom (1973). On the Estimation of the Transfer Function Parameters of Process- and Noise Dynamics using a Single-Stage Estimator, Preprints Third IFAC Symposium on Identification, The Hague.

C.M. Woodside (1970). Estimation of the Order of Linear Systems. Preprints Second IFAC Symposium on Identification, Prague.

AN APPLICATION OF REALIZATION THEORY TO IDENTIFICATION OF MULTIVARIABLE PROCESS*

Katsuhisa FURUTA

Tokyo Institute of Technology
Oh-Okayama, Meguro-ku, Tokyo, JAPAN

An identification algorithm for a linear multivariable system without the knowledge of its order is presented, which gives a set of equivalent systems to the plant. By introducing the notion of ε-practically minimal realization, they are reduced to an equivalent system to the plant with the minimal dimension in the practical sense.

1. INTRODUCTION

Identification is defined to determine, on the basis of input and output, an equivalent system to a plant within a specified class of systems. When the structure of the plant is known, an equivalent system is determined by estimating parameters in the model. Therefore, parameter estimation has been focused in the identification problem, and there have been proposed many parameter identification methods based on the statistical estimation theory [1][2][3].

In the practical identification problems, however, the structure of the plant is not known *a priori*. When the plant is linear, the order of the transfer function is not available before the identification. And the determination of the order is the most important problem in the identification. In the identification based on statistical estimation theory, the determination of the order of plant has been achieved by changing the order of plant until the appropriate order being obtained, and the parameters are estimated for the assigned order each time [4]. In this case, the definition of the order is ambiguous and it may be applicable only for the identification of a single variable system, besides it requires a lot of calculation.

Recently, the system theory, especially the ideas of controllability and observability, showed that only the completely observable and completely controllable subsystem of a plant can be identified from input and output, and the so-called order of the plant is the dimension of this subsystem [5]. The minimal realization from the input and output has been recognized to be a proper model for the identification [6]. But there are few works on the realization from input and output contaminated by noise.

In this paper, the identification problem is reconsidered based on the realization theory. And the identification method of the order of the plant using data contaminated by noise is presented. In the proposed method, the model with the arbitrary order larger than that of plant is chosen, and a class of systems which are equivalent to the plant is identified. The properties of the class are discussed and one of their subsystems which are completely controllable and completely observable is derived. This subsystem gives the equivalent external relation to the plant and is the optimum model of the plant. In the case when this subsystem is neither controllable nor observable from the practical view points, its practically minimal realization is derived by defining ε-practical controllability and ε-practical observability.

* This research is in part supported by Matsunaga Science Foundation

2. IDENTIFICATION PROBLEM

It is considered to identify a multivariable system, when its order is not known *a priori*. The procedure is developed based on the identification method of single variable system proposed by the author [7]. The input $u(t)(\in R^m)$ and the output $y(t)(\in R^p)$ of the multivariable system with m inputs and p outputs are measured by $v(t)$ and $z(t)$ contaminated by noises $m(t)$ and $n(t)$

$$v(t) = u(t) + m(t) \quad (1a)$$
$$z(t) = y(t) + n(t) \quad (1b)$$

where t denotes the time t, and $m(t)$ and $n(t)$ are assumed ergodic random vectors such as

$$E(m(t)) = 0,\quad E(n(t)) = 0 \quad \forall t$$
$$E(m(t)m^T(\tau)) = qI_m\delta_{t\tau},\quad E(n(t)n^T(\tau)) = rI_p\delta_{t\tau},$$
$$E(m(t)n^T(\tau)) = 0$$

The input-output relation of the multivariable discrete system is characterized by the pulse transfer function

$$R(z,l) = \frac{1}{q(z,l)}[\Gamma_1 z^{-1} + \Gamma_2 z^{-2} + .. + \Gamma_l z^{-l}] \quad (2)$$

where

$$q(z,l) = 1 + a_1 z^{-1} + \,..\, + a_l z^{-l}$$

$$\Gamma_i = \begin{bmatrix} \gamma_{i1}^T \\ \vdots \\ \gamma_{ip}^T \end{bmatrix} \text{ is } p\times m \text{ matrix, } \gamma_{ij} \text{ is m-vector,}$$

and z^{-1} is the unit time shift operator.
The identification problem is to find the transfer function $R(z,l)$ with appropriate l minimizing the ctiterion function, when input data $\{v(t), t=0,..,N\}$ and output data $\{z(t), z=0,..,N\}$ are given. The order of the multivariable system is not related to l explicitly.

In this paper, the quadratic sum of the equation error is employed as the criterion function J,

$$J= \frac{1}{N-l+1} \sum_{t=l}^{N} \|\varepsilon(t)\|^2 \quad (3)$$

where

$$\varepsilon(t)= y(t)+\sum_{i=1}^{l} a_i y(t-i)-\sum_{i=1}^{l}\Gamma_i u(t-i) \quad (4)$$

Letting

$$\phi= (a_1,a_2,..,a_l,\gamma_{11}^T,\gamma_{12}^T,..,\gamma_{lp}^T)^T \quad (5)$$

$$L_{t-1}^{l}= [z(t-1), ..,z(t-l),V_{t-1}, ..,V_{t-l}] \quad (6)$$

where

$$V_{t-i}= \begin{bmatrix} -v^T(t-i) & 0^T \dots & 0^T \\ 0^T & \ddots & 0^T \\ \vdots & \ddots & \\ 0^T \dots & 0^T & -v^T(t-i) \end{bmatrix} : p\times mp \text{ matrix} \quad (7)$$

then $\varepsilon(t)$ is

$$\varepsilon(t)= [[z(t),L_t^l]-[n(t),n(t-1),..,M_{t-1}]]\binom{1}{\phi} \quad (8)$$

where

$$M_{t-i}= \begin{bmatrix} -m^T(t-i) & 0^T \dots & 0^T \\ 0^T & \ddots & 0^T \\ \vdots & \ddots & \\ 0^T \dots & 0^T & -m^T(t-i) \end{bmatrix} : p\times mp \text{ matrix} \quad (9)$$

Because of ergodicity, the criterion function J can be expressed for $N\to\infty$ as

$$J= \left\|\begin{matrix}1\\ \phi\end{matrix}\right\|^2_{[Q_N^l-C]} \quad N\to\infty \quad (10)$$

where

$$Q_N^l= \frac{1}{N+1-l} \sum_{t=l}^{N} \begin{bmatrix} z^T(t) \\ L_t^{lT} \end{bmatrix} [z(t),L_t^l] \quad (11)$$

$$C= \begin{bmatrix} rp & 0 & \dots & & 0 \\ 0 & rp & & & \vdots \\ \vdots & & q & & \vdots \\ \vdots & & & \ddots & 0 \\ 0 & \dots & \dots & 0 & q \end{bmatrix} \quad (12)$$

By the positive semidefiniteness of J, the symmetric matrix Q_N^l-C is found positive semidefinite. Instead of infinite N, eq.(10) may exist for sufficiently large N.

Decomposing Q_N^l-C into the following form

$$Q_N^l-C= [\ p_N^l \ \vdots\ \Pi_N^l\ ] \quad (13)$$

the following theorem is obtained.

THEOREM 1

When observed input $\{v(t), t=0,..\}$ *and output* $\{z(t), t=0,..\}$ *of a plant, which are contaminated by random vectors with zero means and variances* qI *and* rI, *are given, they can be the observed input and output of the system with the pulse transfer function of the form* $R(z,l)$ *iff*

$$-p_N^l \in \text{Range } (\Pi_N^l) \quad \textit{for large } N \quad (14)$$

where p_N^l *and* Π_N^l *are defined in eq.*(13).

PROOF

(Necessity) If the given input and output data are those of $R(z,l)$, the equation error defined by eq.(4) will be zero and the criterion function J becomes zero for the optimum choice of $\phi=\phi°$, i.e.,

$$\left\|\begin{matrix}1\\ \phi°\end{matrix}\right\|^2_{Q_N^l-C} = 0 \quad (15)$$

Since the symmetric matrix Q_N^l-C is positive semi-definite, there exists $(Q_N^l-C)^{1/2}$. From eq.(15),

$$\|(Q_N^l-C)^{1/2}\binom{1}{\phi°}\|^2= 0 \Rightarrow (Q_N-C)^{1/2}\binom{1}{\phi°}= 0 \quad (16)$$

Multiplying $(Q_N-C)^{1/2}$,

$$(Q_N-C)\binom{1}{\phi°}= [p_N^l, \Pi_N^l]\binom{1}{\phi°}= 0 \quad (17)$$

therefore

$$-p_N^l= \Pi_N^l \phi° \quad (18)$$

Eq.(18) shows eq.(14).
(Sufficiency) It is obvious by following the proof of necessity reversely. QED

From eq.(18), the following lemma is given.

LEMMA 1

When the given input and output data are satisfying eq.(14), *they are the measured input and output of the system with the transfer function* $R(z,l)$ *and the parameters* ϕ *in the following set are satisfying those of* $R(z,l)$.

$$\{\phi\}= -\Pi_N^{l+} p_N^l + \text{Null } (\Pi_N^l) \text{ for large } N \quad (19)$$

where Π_N^{l+} *is the pseudoinverse of* Π_N^l *and* Null (Π_N^l) *is the null space of* Π_N^l.

PROOF

Eq.(19) is directly derived from eq.(18).

Lemma 1 tells that if the given input and output are those of the system with the transfer function of the form $R(z,l)$, the identified system is not unique if Null (Π_N^l) is not $\{0\}$. Any system with $R(z,l)$ having the parameters ϕ of an element in $\{\phi\}$ can be the equivalent system to the plant, and the equivalent system is not unique.

REMARK: Eq.(14) and eq.(19) exist for sufficiently large N, and for small N, it is necessary to decide whether they can be regarded to be satisfied or not. One of such judgement can be done by evaluating the eigenvalues of Q_N^l-C; if any of the eigenvalues of Q_N^l-C are nearly equal zeros, eq.(14) is considered to be satisfied in the practical sense. And in this case $(1,\phi^T)^T$ is the eigenvector of Q_N^l-C corresponding to nearly zero eigenvalue.

3. PRACTICAL MINIMAL REALIZATION

The identified system with the given l may not be unique, since the set of systems with parameters given by eq.(19) are equivalent to the plant. In this section, the minimally dimensional system equivalent to the plant is derived from the practical point of view.
The system with the transfer function $R(z,l)$ can be realized by the observable standard form

$$\Sigma : \begin{array}{l} x(t+1)= Fx(t)+Gu(t) \qquad (20a) \\ y(t)= Hx(t) \qquad (20b) \end{array}$$

where

$$F=\begin{bmatrix} 0 & \cdots & 0 & -a_l I \\ I & & \vdots & \vdots \\ 0 & \ddots & 0 & \vdots \\ \vdots & \ddots & & \vdots \\ 0 & \cdots 0 & I & -a_1 I \end{bmatrix} \qquad G=\eta^{-1}\begin{bmatrix}\Gamma_l \\ \vdots \\ \Gamma_l\end{bmatrix}$$

$$H=\eta[0,..0,I]$$

where $x(\in R^n)$ is the state of the system

The controllability matrix W and observability matrix M are defined as

$$W=[G, FG,.., F^{n-1}G]$$
$$M=[H^T, F^TH^T,..,(F^T)^{n-1}H^T]$$

The ε-practically controllability and observability are defined as [9].

DEFINITION 1

The state x of the system Σ is said to be ε-practically controllable (observable) if x is an element in the subspace spanned by the eigenvectors of $WW^T(MM^T)$ corresponding to the eigenvalues larger than ε, where W and M denote controllability and observability matrices.

When ε is zero, ε-practical controllability becomes controllability in the ordinary definition. The ε-practically minimal realization is defined as [9].

DEFINITION 2

The system represented by the states which are ε-practically controllable and observable is said to be ε-practically minimal realization.

The observable standard form of the system Σ can be completely ε-practically observable by choosing η appropriately. When η is appropriately fixed, the system Σ is minimally realized in ε-practical sense by the Theorem 2.

THEOREM 2

When the system Σ (F,G,H) is completely ε-practically observable, its ε-practically minimal realization is achieved by choosing

$$\bar{F}_0=S^TFS \quad (21a) \qquad \bar{G}_0=S^TG \quad (21b) \qquad \bar{H}_0=HS \quad (21c)$$

where

$$S=[w_1, ..,w_{n_0}] \qquad (22)$$

$w_i(i=1,..,n_0)$ *are normal eigenvectors of* WW^T *corresponding to the eigenvalue larger than* ε. *The* ε-*practically identified* $\Sigma_0(\bar{F}_0,\bar{G}_0,\bar{H}_0)$ *has the transfer function*

$$R_0(z)=\bar{H}_0(Iz-\bar{F}_0)^{-1}\bar{G}_0 \qquad (23)$$

THEOREM 3

The transfer function of the ε-practically observable standard system $\Sigma=(F,G,H)$ is related to that of ε-practically minimal realization $(\bar{F}_0, \bar{G}_0,\bar{H}_0)$ by

$$H(Iz-F)^{-1}G=\bar{H}_0(Iz-\bar{F}_0)^{-1}\bar{G}_0+O(\varepsilon^{1/2}) \qquad (24)$$

PROOF

The system (F,G,H) can be expressed by changing the coordinates in the statespace by

$$\bar{x}=[w_1, ..,w_{n_0},w_{n_0+1}, ..,w_n]^{-1}x \qquad (25)$$

where $w_i(i=1, ..,n)$ is the normal eigenvector of WW^T, and $\{w_{n_0+1}, ..,w_n\}$ are the normal eigenvectors corresponding to eigenvalues less than ε. Then the system (F,G,H) has the equivalent system

$$\bar{x}(t+1)=\begin{bmatrix}\bar{F}_0, & \bar{F}_{12} \\ 0 & \bar{F}_{22}\end{bmatrix}\bar{x}(t)+\begin{bmatrix}\bar{G}_0 \\ \bar{G}_2\end{bmatrix}u(t) \qquad (26a)$$

$$y(t)=[\bar{H}_0,\bar{H}_2]\bar{x}(t) \qquad (26b)$$

where $(\bar{F}_{22},\bar{G}_2,\bar{H}_2)$ is ε-practical uncontrollable. From the fact that the transfer function of the system (F,G,H) is equal to that of eq.(26),

$$\begin{aligned} R(z,l)= H(Iz-F)^{-1}G=\bar{H}_0(Iz-\bar{F}_0)^{-1}\bar{G}_0 \\ +\bar{H}_0(Iz-\bar{F}_0)^{-1}\bar{F}_{12}(Iz-\bar{F}_{22})^{-1}\bar{G}_2 \\ +\bar{H}_2(Iz-\bar{F}_{22})^{-1}\bar{G}_2 \end{aligned} \qquad (27)$$

from Cayley-Hamilton theorem

$$(Iz-\bar{F}_{22})^{-1}\bar{G}_2=[\bar{G}_2,..,(\bar{F}_{22})^{n-n_0-1}\bar{G}_2]\begin{bmatrix}A_1(z) \\ \vdots \\ A_{n-n_0}(z)\end{bmatrix} \qquad (28)$$

$$\|(Iz-\bar{F}_{22})^{-1}\bar{G}_2\|^2\leq\varepsilon\left\|\begin{matrix}A_1(z) \\ \vdots \\ A_{n-n_0}(z)\end{matrix}\right\|^2 \qquad (29)$$

From eq.(27) and eq.(29), eq.(24) is shown.

QED

If ε is chosen as

$$\nu_{n_0}>\varepsilon>\nu_{n_0+1} \qquad (30)$$

where

$$\frac{\sum_{i=1}^{n_0}\nu_i}{\sum_{i=1}^{n}\nu_i} \cong 1 \qquad (31)$$

and $\{\nu_i\}$ are the eigenvalues of WW^T with $\nu_1 \geq \nu_2 \geq \ldots \geq \nu_n$, then for the ordinary input, the output of the system (F_0, G_0, H_0) and the system (F,G,H) can be regarded as equivalent. Thus the minimal realization of Σ can be achieved. Introducing ε-practically minimal realization, the order of the system may be considered to be determined by eliminating estimation error in ϕ.

4. EXAMPLE

To illustrate the proposed identification procedure, the identification of the second order system is achieved.

Example 1. When the input-output data of the system with transfer function

$$G(z^{-1}) = \frac{0.75z^{-1} + 0.25z^{-2}}{1 - 0.85z^{-1} + 0.5z^{-2}}$$

are measured contaminated by noise with the variance 0.0025, and its order is not known, the identification is achieved by assuming 5th order system, then the identified parameters are

$$\phi^T = (1.184575,\ 0.8326572,\ 0.3899577,\ 0.07999564,\ 0.5310771,\ 0.7624783,\ 1.280106,\ 1.060826,\ 0.3320592,\ -0.2566071)$$

Eigenvalues of MM^T for the standard observable form with $\eta=1$ are 6.512, 3.4514, 1.2058, 0.3161, 0.1167. W for the observable canonical form of $R(z,5)$ obtained is

$$W = \begin{bmatrix} -0.2566071E\ 00 & -0.4049347E\ 00 & -0.2001590E\ 00 & 0.1089500E\ {-01} & 0.1353162E\ 00 \\ 0.3320592E\ 00 & -0.3176020E\ 00 & -0.4350845E\ 00 & -0.1985178E\ 00 & 0.3127756E\ {-01} \\ 0.1060826E\ 01 & 0.3472491E{-01} & -0.4645741E\ 00 & -0.4270845E\ 00 & -0.9915824E\ 01 \\ 0.1280106E\ 01 & 0.4259427E\ 00 & -0.2790973E\ 00 & -0.4474923E\ 00 & -0.2149269E\ 00 \\ 0.7624783E\ 00 & 0.3768925E\ 00 & -0.2051491E\ {-01} & -0.2547959E\ 00 & -0.1456673E\ 00 \end{bmatrix}$$

The eigenvalues of WW^T are 4.543988, 0.6718120, 0.2066063×10^{-2}, 0.2151366×10^{-3}, 0.8898509×10^{-4}. When 0.1 is chosen as ε, the given standard observable form with $\eta=1$ is ε-practically observable and the normal eigenvectors w_1, w_2 of WW^T corresponding to eigenvalues larger than ε give

$$S^T = \begin{bmatrix} -0.1340404E\ 00 & 0.1824156E\ 00 & 0.5647459E\ 00 & 0.6839069E\ 00 & 0.4026054E\ 00 \\ 0.5536032E\ 00 & 0.6562795E\ 00 & 0.3562065E\ 00 & -0.1672\text{[illegible]}02E\ 00 & -0.3285753E\ 00 \end{bmatrix}$$

From eq.(21), ε-practically minimal realization $\{\overline{F}_0, \overline{G}_0, \overline{H}_0\}$ of $R(z,5)$ is identified as

$$\overline{F}_0 = \begin{bmatrix} 0.2529890, & 1.045469 \\ -0.3248198, & 0.5783122 \end{bmatrix} \quad \overline{G}_0 = \begin{bmatrix} 1.876516 \\ -0..0.90404 \end{bmatrix}$$

$$\overline{H}_0 = (0.4026054,\ -0.3285753)$$

The pulse transfer function of this system is

$$H_0(Iz - F_0)^{-1}G_0 = \frac{0.759z^{-1} - 0.242z^{-2}}{1 - 0.8313z^{-1} + 0.485z^{-2}}$$

Examples show that the order of a single variable system can be determined by the proposed procedure.

5. CONCLUSION

An identification procedure for a multivariable linear system is presented, which can be used without the knowledge of the structure (order) of the system. In the proposed procedure, if the order of the pulse transfer function of the model is chosen sufficiently large, a set of systems is identified to be equivalent to the plant. By taking the minimal realization of the equivalent systems, especially in the practical sense, the appropriate unique model is derived. The definitions of ε-practical observability, controllability and minimal realization are given, which may be used in the applied fields. The procedure shows that the attentions must be paid on the system structure in the identification.

The author appreciates the comments given by Professor R.E.Kalman of University of Florida for the earlier version of the paper and the computational help of Dr.J-S.Ha of R.O.K.Marchant College.

REFERENCES

[1] K.J.Astrom, P.Eykhoff "Identification and Process Parameter Estimation" Preprint of 2nd IFAC Symp. on Identification (1970)

[2] A.V.Balakrishnan, V.Peterka "Identification in a Automatic Control Systems" Survey paper at 4th Congress of IFAC (1969)

[3] A.P.Sage, J.L.Melsa "System Identification" Academic Press (1971)

[4] C.M.Woodside "Estimation of the order of linear systems" IFAC Symp. on Identification 1.7 (1970)

[5] R.R.Kalman et al. "Topics in Mathematical System Theory" McGraw Hill (1969)

[6] M.A.Budin "Minimal Realization of Discrete Linear Systems from Input-Output Observations" IEEE AC-16 pp.395-401 (1971)

[7] K.Furuta, J.G.Paquet "On the Identification of Time-Invariant Discrete Process" IEEE-AC, vol. 15, No.1, pp.153-155 (1970)

[8] H.A.Antosiewicz "Linear Control Systems Controllability" in "Functional Analysis and Optimization" Academic Press (1966)

[9] K.Furuta "The Determination of the order of Model in Practical Identification" Proc. in The 3rd Symp. on Nonlinear Estimation Theory and Its Applications, pp.74-77 (1972)

[10] J.Rissanen "Estimation of Parameters in Multi-Variate Random Processes" (to appear)

CANONICAL STRUCTURES IN THE IDENTIFICATION OF MULTIVARIABLE SYSTEMS

Roberto P. Guidorzi
Istituto di Automatica
University of Bologna
40136 Bologna - Italy

The role of state-space canonical (block-companion) representations in the structural and parametric identification of linear multivariable systems is investigated. Unitary algorithms for the identification of these systems are then proposed. Their main advantages are that they can operate on quite general input/output sequences and lead (with remarkable saving of computational effort) to canonical state-space representations; moreover the initial state of the system needs not to be zero or known and can be easily estimated. A consistent extension to the case of uncorrelated zero-mean additive noise is also reported.

I. INTRODUCTION

According to the classical definition given by Zadeh |1|, identification is the determination, on the basis of input and output, of a system within a specified class of systems, to which the system under test is equivalent.
This definition is quite general and allows many degrees of freedom in the practical formulation of the identification problem since the class of the models to be identified and the class of input signals are not specified and also the meaning of "equivalent" is often referred to a suitable optimality criterion.
For the practical usefulness of an identification procedure the following considerations should be taken into account:

1) The obvious choice for the class of the models is given by state-space representations since almost all the modern available control theory regards these representations.

2) The class of the input signals should not be restricted to such particular inputs as impulse functions, step functions, colored or white noise, sinusoidal signals, P.R.B.S. etc., since these inputs can seldom be applied to real processes. It is on the contrary desirable to use normal operating records of the plant.

The choice of a state-space, i.e. parametric, model implies the preliminary determination of the system order; it is in fact well known |2| that parametric models can give large errors when the order of the model does not agree with the order of the process. Another practically relevant requirement regards the minimality (or at least a strong limitation) of the number of the significant parameters of the model and this means the choice of a suitable canonical (or quasi-canonical)form. Moreover the simplicity of the link between the selected canonical state-space representation and the set of input/output equations used for the identification (the initial data are input/output sequences) heavily conditions the required amount of computation.
The previous problems have been completely solved for single-input single-output or multi-input single-output systems |2|. Troubles appear when multivariable i.e. multi-input multi-output systems are considered since an unitary approach satisfying previous requirements can not be found in the literature.
The purpose of this paper is to bridge this gap by exploring a class of state-space canonical models linked by particularly simple relations to input/output differential descriptions that can be directly identified from the input/output sequences. The resulting identification procedure is similar to those for single-input single-output systems and can be summarized in the following steps:

1) Structural identification, i.e. determination by direct treatment of the input/output sequences of a set of indexes completely describing the structure (not only the order) of a canonical model for the process.

2) Parametric identification of the subsystems (one for every output component) of an input/output description of the system with the prviously identified structure.

3) Deduction, from the identified input/output representation, of the state-space model.

<u>Comparison with previous literature</u>

A fundamental contribution in this field can be found in the paper by Ho and Kalman |3|; in this work a state-space model is deduced from an infinite sequence of Markov parameters. This procedure to obtain a parametric representation from

a nonparametric one has been successively extended by Gueguen |4| to obtain realizations of input/output differential descriptions and by Gerth |5| who considers the construction of the Markov parameters from input/output sequences. Interesting results have also been given by Tether |6| who studies the equivalence between finite sequences of Markov parameters and state-space models and by Ackermann |7| who gives a canonical realization by the use of an input/output description independently developed by Bonivento and Guidorzi |8|. Gopinath |9| assumes as initial knowledge of the system general input/output sequences; by direct elaboration of the data a set of input/output equations is then obtained. The procedure is however based on the knowledge of a suitable data selector matrix corresponding to a well-defined structure of the equations and an efficient method for the construction of this matrix is not given. The approach by Gopinath is also the basis of a recent paper by Budin |10| who improves the computational requirements and points out a general procedure leading to a minimal but not canonical realization.
The basic difference between the approach presented in this paper and the previous ones lies in the structural identification of the system, i.e. in the direct determination from the input/output sequences of a set of indexes like those described in a dual context in |11| and |12|, which permit a decomposition of the system into independently observable subsystems that can be separately identified. This first step greatly simplifies the remaining ones which are similar to those for single-input single-output systems. A further bonus is that the initial state of the system needs not to be zero or known and can be easily estimated; this feature is useful when only short data sequences are available.

II. ORGANIZATION OF THE PAPER

The class of linear discrete multivariable systems considered can be represented by the usual state-space equations

$$\mathbf{x}(k+1) = \mathbf{F}\mathbf{x}(k) + \mathbf{G}\mathbf{u}(k) \qquad (2.1a)$$

$$\mathbf{y}(k) = \mathbf{H}\mathbf{x}(k) \qquad (2.1b)$$

where $\mathbf{x}$ belongs to the n-dimensional state-space $\mathscr{X}$, $\mathbf{u}$ to the r-dimensional input space $\mathscr{U}$ and $\mathbf{y}$ to the m-dimensional output space $\mathscr{Y}$. The initial data are quite general syncronous input/output sequences i.e. normal operating records of $\mathbf{y}$ and $\mathbf{u}$. The paper is organized as follows.
In section III certain proprieties of canonical (block-companion) state-space representations are stated and the structural and parametric equivalence between this class of models and input/output difference descriptions of the type

$$\mathbf{P}(z)\,\mathbf{y}(k) = \mathbf{Q}(z)\,\mathbf{u}(k) \qquad (2.2)$$

where $\mathbf{P}(z)$ and $\mathbf{Q}(z)$ are polynomial matrices in z (z^{-1} is the unitary delay operator) is established. In section IV the structural identification of equations (2.1) is considered; the subsequent parametric identification of an input/output model of type (2.2) follows in section V. In section VI the algorithms given in section IV and in section V are extended to the case of data corrupted by additive noise with zero mean. In section VII the derivation of two classes of canonical state-space models from the identified input/output description is discussed and in section VIII some concluding remarks are reported.

III. CANONICAL INPUT/OUTPUT STRUCTURES OF MULTIVARIABLE SYSTEMS

Consider the discrete completely observable multivariable system represented by eq. (2.1) and let

$$\mathbf{H} = \begin{bmatrix} \mathbf{h}_1^T \\ \mathbf{h}_2^T \\ \vdots \\ \mathbf{h}_m^T \end{bmatrix} . \qquad (3.1)$$

Construct then the vector sequences

$$\begin{matrix} \mathbf{h}_1 , & \mathbf{F}^T\mathbf{h}_1, \mathbf{F}^{T^2}\mathbf{h}_1 \ldots . \\ \vdots & \\ \mathbf{h}_m , & \mathbf{F}^{'}\mathbf{h}_m \ldots\ldots\ldots \end{matrix} \qquad (3.2)$$

and select them in the following order

$$\mathbf{h}_1, \mathbf{h}_2 \ldots \mathbf{h}_m, \mathbf{F}^T\mathbf{h}_1 \ldots \mathbf{F}^T\mathbf{h}_m, \mathbf{F}^{T^2}\mathbf{h}_1 \ldots . \qquad (3.3)$$

retaining a vector $\mathbf{F}^{T^s}\mathbf{h}_i$ in (3.3) if and only if it is independent from all previously selected ones and all the vectors $\mathbf{F}^{T^j}\mathbf{h}_i$ $(0 \le j < s)$ have already been selected. Let $\nu_1, \nu_2, \ldots, \nu_m$ be the numbers of vectors selected from the first, second,..., m-th sequence in (3.2); the vectors $\mathbf{F}^{T^{\nu_j}}\mathbf{h}_j$ are therefore linearly dependent from previously selected ones. Because of the complete observability of the system it follows that $\nu_1 + \nu_2 + \ldots + \nu_m = n$. It is also valid the following lemma.

<u>Lemma 1</u> - *The set of integers* $\nu_1, \ldots, \nu_m$ *completely defines the structure of the couple* $(\mathbf{F}, \mathbf{H})$ *and is invariant with respect to a change of coordinates and with respect to the class of feedbacks* $\mathbf{F} + \mathbf{K}\mathbf{H}$ *where* $\mathbf{K}$ *is a generic* $(n \times m)$ *matrix.*

The proof directly follows from the content of |11|,|12| by means of obvious duality considerations. Let now construct the nonsingular matrix

$$\mathbf{T}^T = \{\mathbf{h}_1, \mathbf{F}^T\mathbf{h}_1 \ldots \mathbf{F}^{T(\nu_1-1)}\mathbf{h}_1 | \ldots | \mathbf{h}_m \ldots \mathbf{F}^{T(\nu_m-1)}\mathbf{h}_m\} \qquad (3.4)$$

and operate the change of coordinates leading to the new state vector $\mathbf{w} = \mathbf{T}\mathbf{x}$. The new equations of the system are

$$\mathbf{w}(k+1) = \mathbf{A}\mathbf{w}(k) + \mathbf{B}\mathbf{u}(k) \qquad (3.5)$$

$$\mathbf{y}(k) = \mathbf{C}\mathbf{w}(k) \qquad (3.6)$$

and the matrices $\mathbf{A}$, $\mathbf{B}$ and $\mathbf{C}$ exibit the following structures:

$$\mathbf{A} = \mathbf{T}\;\mathbf{F}\;\mathbf{T}^{-1} = \{\mathbf{A}_{ij}\} \qquad (3.7)$$

$(i,j=1,2,\ldots,m)$ where the $(\nu_i \times \nu_i)$ matrices $\mathbf{A}_{ii}$ show the companion structure (3.8) and the $(\nu_i \times \nu_j)$ matrices $\mathbf{A}_{ij}$ are of the type (3.9).

$$\mathbf{A}_{ii} = \begin{bmatrix} 0 & \\ \vdots & \mathbf{I}_{\nu_i-1} \\ 0 & \\ a_{ii,1} & \cdots\; a_{ii,\nu_i} \end{bmatrix} \qquad (3.8)$$

$$\mathbf{A}_{ij} = \begin{bmatrix} 0 & \cdots & \cdots & 0 \\ \vdots & & & \vdots \\ 0 & \cdots & \cdots & 0 \\ a_{ij,1} & \cdots & a_{ij,\nu_j} & 0..0 \end{bmatrix} \qquad (3.9)$$

$$\mathbf{C} = \mathbf{HT}^{-1} = \begin{bmatrix} 1 & 0 & \cdots & \cdots & \cdots & \cdots & \cdots & 0 \\ 0 & \cdots & 0\,1\,0 & \cdots & \cdots & \cdots & \cdots & 0 \\ \cdot & \cdots & \cdots & \cdots & \cdots & \cdots & \cdots & \cdot \\ 0 & \cdots & \cdots & \cdots & \cdots & 0\,1\,0 & \cdot & 0 \end{bmatrix} \qquad (3.10)$$
$$\uparrow_{1} \qquad \uparrow_{(\nu_1+1)} \qquad \uparrow_{(\nu_1+..+\nu_{m-1}+1)}$$

$$\mathbf{B} = \mathbf{TG} = \begin{bmatrix} \mathbf{b}_1^T \\ \vdots \\ \mathbf{b}_n^T \end{bmatrix} = \begin{bmatrix} b_{11} & \cdots & b_{1r} \\ \vdots & & \vdots \\ b_{n1} & \cdots & b_{nr} \end{bmatrix} . \qquad (3.11)$$

For a proof of the canonical structure of the obtained couple $(\mathbf{A},\mathbf{C})$ see (in a dual context) Luenberger |13| or Chen |14|.

Remark 1 - Because of the order followed in the selection of the vectors (3.2) and of the consequent structure of $\mathbf{T}$, in every matrix $\mathbf{A}_{ij}$ at most ν_i+1 elements are non identically zero if $j<i$, ν_i if $j>i$.

Remark 2 - The dependent vectors $\mathbf{F}^{T\nu_i}\mathbf{h}_i$ are a linear combination, with coefficients given by the i-th significant row of $\mathbf{A}$, of previously selected ones i.e.

$$\mathbf{F}^{T\nu_i}\mathbf{h}_i = a_{i1,1}\mathbf{h}_1 + \,..\, + a_{im,1}\mathbf{h}_m + a_{i1,2}\mathbf{F}^T\mathbf{h}_1 + \,..$$

The original system has been decomposed in this way into m interconnected subsystems; a direct consequence of the structure of the matrices $\mathbf{A}$ and $\mathbf{C}$ is given by the following lemma.

Lemma 2 - *The j-th subsystem in the canonical representation $(\mathbf{A},\mathbf{B},\mathbf{C})$ is completely observable from the j-th component of the output.*

An input/output difference equation of the type (2.2) will now be deduced from the triple $(\mathbf{A},\mathbf{B},\mathbf{C})$. Remarkable features of this description are that it exibits the structure of the system and that also the parameters of the triple $(\mathbf{A},\mathbf{B},\mathbf{C})$ can be easily obtained from it (and *viceversa*).

Let $\mathbf{w}(k)=(w_{k,1}\; w_{k,2}\; ..\; w_{k,n})$, $\mathbf{y}(k)=(y_{k,1}\; \cdots\; ..\; y_{k,m})$ and $\mathbf{u}(k)=(u_{k,1}\; \cdots\cdots\; u_{k,r})$.

Consider then the j-th subsystem; because of the complete observability of its state from the j-th output component and of the particularly simple structure of the matrices $\mathbf{A}$ and $\mathbf{C}$ it is easy to derive from eq. (3.5) and (3.6) the following expressions:

$$w_{k,(\nu_1+..+\nu_{j-1}+1)} = y_{k,j}$$

$$w_{k,(\nu_1+..+\nu_{j-1}+2)} = z\, y_{k,j} - \mathbf{b}^T_{(\nu_1+..+\nu_{j-1}+1)}\mathbf{u}(k)$$

$$w_{k,(\nu_1+..+\nu_{j-1}+3)} = z^2 y_{k,j} - \mathbf{b}^T_{(\nu_1+..+\nu_{j-1}+2)}\mathbf{u}(k) - \mathbf{b}^T_{(\nu_1+..+\nu_{j-1}+1)}\, z\, \mathbf{u}(k)$$

$$\cdots\cdots\cdots\cdots\cdots\cdots\cdots\cdots\cdots\cdots \qquad (3.12)$$

$$w_{k,(\nu_1+..+\nu_j)} = z^{\nu_j-1} y_{k,j} - \mathbf{b}^T_{(\nu_1+..+\nu_j-1)}\mathbf{u}(k) - \,\cdots\, - \mathbf{b}^T_{(\nu_1+..+\nu_{j-1}+1)}\, z^{\nu_j-2}\,\mathbf{u}(k)$$

It must be remembered that here and in the following z^{-1} is the unitary delay operator; the notation $z^2\,\mathbf{u}(k)$ is therefore equivalent to $\mathbf{u}(k+2)$ and so on. By writing eq. (3.12) for j=1,..,m the whole state vector can be expressed as

$$\mathbf{w}(k) = \mathbf{V}(z)\,\mathbf{y}(k) - \mathbf{WZ}(z)\,\mathbf{u}(k) \qquad (3.13)$$

where

$$\mathbf{V}(z) = \begin{bmatrix} 1 \cdots 0 \\ z \quad \vdots \\ \vdots \quad \vdots \\ z^{\nu_1-1} \quad \vdots \\ 0 \quad 0 \\ \vdots \quad \vdots \\ 0 \quad 1 \\ \vdots \quad z \\ \vdots \quad \vdots \\ 0 \cdots z^{\nu_m-1} \end{bmatrix} \;(3.14) \qquad \mathbf{W} = \begin{bmatrix} 0 \cdots\cdots\cdots\cdots 0 \\ \mathbf{b}_1^T \;\; 0 \cdots\cdots\cdots 0 \\ \vdots \;\; \ddots \qquad \vdots \\ \mathbf{b}^T_{\nu_1-1} \cdots \mathbf{b}^T_1 \;\; 0..0 \\ \hline \vdots \qquad \vdots \\ \hline 0 \cdots\cdots\cdots\cdots 0 \\ \mathbf{b}^T_{n-\nu_m+1} \;\; 0 \cdots\cdots 0 \\ \vdots \;\; \ddots \qquad \vdots \\ \mathbf{b}^T_{n-1} \cdots \mathbf{b}^T_{n-\nu_m+1} \;\; 0.0 \end{bmatrix} \;(3.15)$$

$$n \times r(\nu_M - 1)$$

$$\mathbf{Z}(z) = \begin{bmatrix} \mathbf{I} \\ z\,\mathbf{I} \\ \vdots \\ z^{\nu_M-1}\,\mathbf{I} \end{bmatrix} . \;(3.16) \qquad \nu_M = \max_i(\nu_i)$$

The substitution of (3.13) in (3.5) leads to the input/output description

$$\{(z\mathbf{I}-\mathbf{A})\,\mathbf{V}(z)\}\,\mathbf{y}(k) = \{(z\mathbf{I}-\mathbf{A})\mathbf{WZ}(z)+\mathbf{B}\}\,\mathbf{u}(k) . \qquad (3.17)$$

In representation (3.17) however only the ν_1-th, $(\nu_1+\nu_2)$-th, .. , n-th equations are significant; the remaining ones are simple identities. The significant equations in (3.17) can therefore be written in the form

$$\mathbf{P}(z)\,\mathbf{y}(k) = \mathbf{Q}(z)\,\mathbf{u}(k) \qquad (3.18)$$

where

$$\mathbf{P}(z) = \begin{bmatrix} p_{11}(z) & \cdots & p_{1m}(z) \\ \vdots & & \vdots \\ p_{m1}(z) & \cdots & p_{mm}(z) \end{bmatrix} \qquad (3.19)$$

$$\mathbf{Q}(z) = \begin{bmatrix} q_{11}(z) & \cdots & q_{1r}(z) \\ \vdots & & \vdots \\ q_{m1}(z) & \cdots & q_{mr}(z) \end{bmatrix} \qquad (3.20)$$

The polynomials of $\mathbf{P}(z)$ can be immediately obtained from (3.17); in fact it stands

$$p_{ii}(z) = z^{\nu_i} - a_{ii,\nu_i} z^{\nu_i-1} - \ldots - a_{ii,2}\, z - a_{ii,1} \qquad (3.21)$$

$$p_{ij}(z) = -a_{ij,\nu_{ij}} z^{\nu_{ij}-1} - \ldots - a_{ij,2} z - a_{ij,1} \quad . \tag{3.22}$$

The polynomials of $\mathbf{Q}(z)$ are obtained by computing the right side of expression (3.17); by simple passages it follows

$$q_{ij}(z) = \beta_{(\nu_1+..+\nu_i),j} z^{\nu_i-1} + .. + \beta_{(\nu_1+..+\nu_{i-1}+2),j} z + \beta_{(\nu_1+..+\nu_{i-1}+1),j} \tag{3.23}$$

where the coefficients $\beta_{i,j}$ are the elements of the matrix

$$\bar{\mathbf{B}} = \mathbf{M}\mathbf{B} = \begin{bmatrix} \beta_{11} & \cdots\cdots & \beta_{1r} \\ \vdots & & \vdots \\ \vdots & & \vdots \\ \beta_{n1} & \cdots\cdots & \beta_{nr} \end{bmatrix} \tag{3.24}$$

and the matrix $\mathbf{M}$ is given by

$$\mathbf{M} = \left[\begin{array}{ccccc|c|ccccc}
-a_{11,2} & -a_{11,3} & \cdot\cdot & -a_{11,\nu_1} & 1 & \cdot\cdot & -a_{1m,2} & \cdot\cdot & -a_{1m,\nu_{1m}} & & 0 \\
-a_{11,3} & -a_{11,4} & \cdots & / & & \cdot\cdot & -a_{1m,3} & \cdot\cdot & / & & \cdot \\
\cdot & & / & & & \cdot\cdot & \cdot & / & & & \cdot \\
\cdot & / & & & & \cdot\cdot & -a_{1m,\nu_{1m}} & & & & \cdot \\
-a_{11,\nu_1} & 1 & & & & \cdot\cdot & 0 & & & & \cdot \\
1 & & & & & \cdot\cdot & 0 & & & & 0 \\
\hline
\vdots & \vdots & \vdots & \vdots & \vdots & \vdots & \vdots & \vdots & \vdots & \vdots & \vdots \\
\hline
-a_{m1,2} & \cdot & \cdot & -a_{m1,\nu_{m1}} & 0 & \cdot\cdot & -a_{mm,2} & \cdot\cdot & -a_{mm,\nu_m} & & 1 \\
-a_{m1,3} & \cdot & \cdot & 0 & & \cdot\cdot & -a_{mm,3} & \cdot\cdot & & / & \\
\cdot & & / & & & \cdot\cdot & \cdot & & / & & \\
-a_{m1,\nu_{m1}} & 0 & & & & \cdot\cdot & -a_{mm,\nu_m} & 1 & & & \\
0 & & & & & \cdot\cdot & 1 & & & &
\end{array}\right] \tag{3.25}$$

<u>Remark 3</u> - It is useful to observe that (see also remark 1) the polynomials in $\mathbf{P}(z)$ and $\mathbf{Q}(z)$ satisfy to the following relations:

$$\begin{array}{ll}
\deg\{p_{ii}(z)\} > \deg\{p_{ij}(z)\} & \text{for } i>j \\
\deg\{p_{ii}(z)\} \geq \deg\{p_{ij}(z)\} & \text{for } j<i \\
\deg\{p_{ii}(z)\} > \deg\{p_{ji}(z)\} & \text{for } i \neq j \\
\deg\{p_{ii}(z)\} > \deg\{q_{ij}(z)\} & .
\end{array}$$

The equivalence between the class of canonical state-space representations (3.5)-(3.6) and the input/output description (3.18) has been thus established.

<u>Remark 4</u> - The set of indexes $\nu_1, \ldots, \nu_m$ can be immediately deduced (by inspection) indifferently from the knowledge of $\mathbf{A}$ or of $\mathbf{P}(z)$; also from the parametric standpoint $\mathbf{A}$ and $\mathbf{P}(z)$ are equivalent. Matrix $\mathbf{C}$ can be directly written if $\nu_1, \ldots, \nu_m$ are known.

<u>Remark 5</u> - To obtain the matrix $\mathbf{Q}(z)$ from the knowledge of the couple $(\mathbf{A}, \mathbf{B})$ it is first necessary to construct the matrix $\mathbf{M}$ (that can be written by direct inspection of $\mathbf{A}$); the elements of $\bar{\mathbf{B}} = \mathbf{M}\mathbf{B}$ are then the coefficients of the polynomials $q_{ij}(z)$.

<u>Remark 6</u> - To obtain the matrix $\mathbf{B}$ from the knowledge of $\mathbf{P}(z)$, $\mathbf{Q}(z)$ it is first necessary to construct (by direct inspection) the matrices $\mathbf{M}$ and $\bar{\mathbf{B}}$; matrix $\mathbf{B}$ is then given by $\mathbf{B} = \mathbf{M}^{-1}\bar{\mathbf{B}}$. Note that $\mathbf{M}$ is always nonsingular because of its structure (in fact $\det(\mathbf{M})=1$).

<u>Example</u> - Let

$$\mathbf{A} = \left[\begin{array}{ccc|cc} 0 & 1 & 0 & & \\ 0 & 0 & 1 & & \\ -2 & -3 & -3 & 2 & -1 \\ \hline & & & 0 & 1 \\ -1 & 2 & 1 & -1 & -2 \end{array}\right] \quad \mathbf{B} = \begin{bmatrix} 1 & 0 \\ 0 & 1 \\ -1 & 0 \\ 0 & 1 \\ 1 & 1 \end{bmatrix} \quad \mathbf{C} = \begin{bmatrix} 1 & 0 & 0 & 0 & 0 \\ 0 & 0 & 0 & 1 & 0 \end{bmatrix}$$

Then $\nu_1 = \nu_M = 3$, $\nu_2 = 2$; the coefficients of the polynomials $q_{ij}(z)$ are the elements of the matrix $\bar{\mathbf{B}}$ given by

$$\bar{\mathbf{B}} = \begin{bmatrix} \beta_{11} & \beta_{12} \\ \beta_{21} & \beta_{22} \\ \beta_{31} & \beta_{32} \\ \beta_{41} & \beta_{42} \\ \beta_{51} & \beta_{52} \end{bmatrix} = \left[\begin{array}{ccc|cc} 3 & 3 & 1 & 1 & 0 \\ 3 & 1 & & 0 & \\ 1 & & & & \\ \hline -2 & -1 & 0 & 2 & 1 \\ -1 & 0 & & 1 & \end{array}\right] \begin{bmatrix} 1 & 0 \\ 0 & 1 \\ -1 & 0 \\ 0 & 1 \\ 1 & 1 \end{bmatrix} = \begin{bmatrix} 2 & 4 \\ 3 & 1 \\ -1 & 0 \\ -1 & 2 \\ -1 & 1 \end{bmatrix}$$

The input/output description is therefore

$$\begin{bmatrix} z^3+3z^2+3z+2 & z-2 \\ -z^2-2z+1 & z^2+2z+1 \end{bmatrix} \mathbf{y}(k) = \begin{bmatrix} -z^2+3z+2 & z+4 \\ -z-1 & z+2 \end{bmatrix} \mathbf{u}(k) .$$

IV. STRUCTURAL IDENTIFICATION

<u>Definition 1</u> - *The structural identification of a multivariable system is the determination of the set of integers $\nu_1, \ldots, \nu_m$ defining the structure of the couple $(\mathbf{A}, \mathbf{C})$ from input/output sequences without the intermediate construction of a parametric model.*

The structural identification will be performed by taking advantage (remark 4) of the fact that this information concerning a canonical state-space representation is also evidentiated in the input/output difference model (3.18).

Let consider the s-th equation in the set (3.18), i.e.

$$\sum_{i=1}^{m} p_{s,i}(z)\, y_{k,i} = \sum_{i=1}^{m} q_{s,i}(z)\, u_{k,i} \tag{4.1}$$

that, on the basis of the relations (3.21), (3.22) and (3.23), can also be written in the form

$$y_{(k+\nu_s),s} = \sum_{i=1}^{m} \sum_{j=1}^{\nu_{si}} a_{si,j}\, y_{(k+j-1),i} + \sum_{i=1}^{r} \sum_{j=1}^{\nu_s} \beta_{(\nu_1+..+\nu_{s-1}+j),i}\, u_{(k+j-1),i} \tag{4.2}$$

(in (4.2) $\nu_{ss} = \nu_s$). Consider now the matrix of input/output data given by

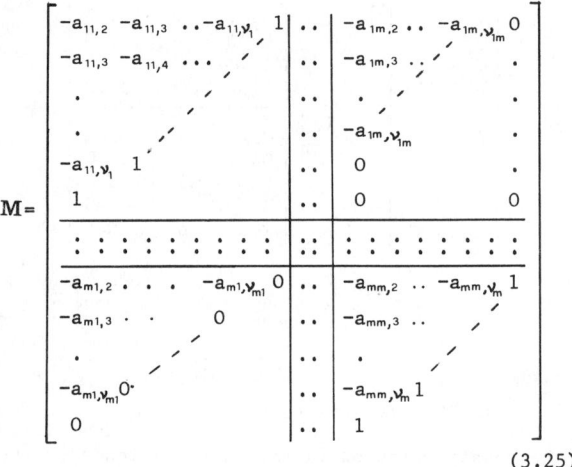

$$\left[\begin{array}{ccc|c|ccc|cc|c|cc} y_{k,1} & y_{k+1,1} & \cdot\cdot & \cdot\cdot & y_{k,m} & y_{k+1,m} & \cdot\cdot & u_{k,1} & \cdot\cdot & \cdot\cdot & u_{k,r} & \cdot\cdot \\ y_{k+1,1} & y_{k+2,1} & & & y_{k+1,m} & y_{k+2,m} & & u_{k+1,1} & & & u_{k+1,r} & \\ \cdot & \cdot & & & \cdot & \cdot & & \cdot & & & \cdot & \\ \cdot & \cdot & & & \cdot & \cdot & & \cdot & & & \cdot & \\ y_{k+N,1} & \cdot & \cdot\cdot & \cdot\cdot & y_{k+N,m} & \cdot & \cdot\cdot & u_{k+N,1} & \cdot & \cdot\cdot & u_{k+N,r} & \cdot\cdot \end{array}\right] =$$

$$= \left[\mathbf{y}_1(k) \;\; \mathbf{y}_1(k+1) \ldots | \ldots | \mathbf{y}_m(k) \ldots | \mathbf{u}_1(k) \ldots | \ldots | \mathbf{u}_r(k) \ldots \right] \tag{4.3}$$

Equation (4.2) shows that the dependence relations among the vectors of (4.3) are the same (taking into account also the input) as those among the vectors (3.2) (see remark 2). This propriety allows the determination of $\nu_1, \ldots, \nu_m$ by selecting the vectors (4.3) according to the same selection plan considered in section 3; the vectors of (4.3) will be therefore selected in the following order

$$\mathbf{y}_1(k) \ldots \mathbf{y}_m(k) \;\; \mathbf{u}_1(k) \ldots \mathbf{u}_m(k) \;\; \mathbf{y}_1(k+1) \ldots \mathbf{y}_m(k+1) \ldots \tag{4.4}$$

A vector is retained if and only if it is independent from previously selected ones; when a dependent vector $\mathbf{y}_s(k+\nu_s)$ is found all the remaining vectors belonging to the same submatrix will also be dependent so that their test is unnecessary.

The selection ends when a dependent vector has been found in every output submatrix; the numbers of vectors selected from these submatrices will be $\nu_1, \ldots, \nu_m$.

Remark 7 - The integer N in (4.3) must be large enough ($N > n + r\nu_M$) in order to permit the selection of the necessary number of independent vectors; in other words the input/output sequences must be of sufficient length to permit the complete structural identification of the system. Usually the number of available data is many times the system order so that the previous condition is largely fulfilled.

Remark 8 - For the structural (and subsequent parametric) identification the input sequence must satisfy well-known conditions |2|, i.e. must "excite all the modes" of the system. Also this requirement can be easily met when the length of the data is large with respect to the system order.

It is certainly not advisable to carry out the structural identification directly on the vectors of (4.3) because of the large amount of storage necessary; moreover the required storage would be a function of N. Since for every matrix, $\mathbf{D}$, it stands $\mathrm{rank}(\mathbf{D}) = \mathrm{rank}(\mathbf{D}^T\mathbf{D})$ a more useful algorithm is the following.

Algorithm for the structural identification

Let

$$\mathbf{L}_i(\mathbf{y}_j) = \left[\mathbf{y}_j(k) \;\; \mathbf{y}_j(k+1) \;\ldots\; \mathbf{y}_j(k+i-1) \right] \tag{4.5}$$

$$\mathbf{L}_i(\mathbf{u}_j) = \left[\mathbf{u}_j(k) \;\; \mathbf{u}_j(k+1) \;\ldots\; \mathbf{u}_j(k+i-1) \right]. \tag{4.6}$$

Then matrix (4.3) taking δ_1 vectors in the first submatrix, δ_2 vectors in the second, etc. can be written as

$$\mathbf{R}(\delta_1, \delta_2, \ldots, \delta_{m+r}) = \{ \mathbf{L}_{\delta_1}(\mathbf{y}_1) \;\ldots\; \mathbf{L}_{\delta_m}(\mathbf{y}_m) | \mathbf{L}_{\delta_{m+1}}(\mathbf{u}_1) \;\ldots\; \mathbf{L}_{\delta_{m+r}}(\mathbf{u}_r) \}. \tag{4.7}$$

Define now the product $\mathbf{R}^T\mathbf{R}$ as

$$\mathbf{S}(\delta_1, \ldots, \delta_{m+r}) = \mathbf{R}^T(\delta_1, \ldots, \delta_{m+r}) \, \mathbf{R}(\delta_1, \ldots, \delta_{m+r}) \tag{4.8}$$

$\mathbf{S}(\delta_1, \ldots, \delta_{m+r})$ is therefore a square matrix whose dimension is given by $\delta_1 + \ldots + \delta_{m+r}$. Construct now the sequence of increasing-dimension matrices

$$\mathbf{S}(2,1,\ldots,1) \;\; \mathbf{S}(2,2,\ldots,1) \;\ldots\; \mathbf{S}(2,2,\ldots,2) \ldots \tag{4.9}$$

and select from (4.9) nonsingular ones. When a singular matrix is found, one of the indexes ν_i is determined; the procedure ends (all remaining matrices are singular) when all m indexes are determined.

Remark 9 - If two adjacent matrices in sequence (4.9) are considered, all the elements of the first are present in the subsequent one that can thus be obtained by computing only a limited number of terms.

Remark 10 - Let $\mathbf{S}(\mu_1, \mu_2, \ldots, \mu_{m+r})$ be a singular matrix in (4.9) and let μ_i be the index increased by one with respect to the previous (nonsingular) matrix in the sequence. Then $\nu_i = \mu_i - 1$ while the indexes ν_{ij} are given by $\nu_{ij} = \mu_j$, $(j=1, \ldots, m)$ $(j \neq i)$; these indexes establish the number of nonzero elements in the submatrices $\mathbf{A}_{ij}$.

V. PARAMETRIC IDENTIFICATION

The parametric identification will be performed on the input/output description (3.18); the parameters to be identified are therefore the coefficients of the polynomials $p_{ij}(z)$ and $q_{ij}(z)$. The computation of the parameters is carried out through m independent steps; in each step the parameters of the polynomials belonging to one row of $\mathbf{P}(z)$ and to the corresponding row of $\mathbf{Q}(z)$ are determined.

Let consider the selection of the vectors in matrix (4.3); when a dependent vector $\mathbf{y}_s(k+\nu_s)$ is found, it is linked to previously accepted vectors by eq. (4.2) or (4.1) so that it is possible to determine the coefficients of $p_{si}(z)$ $(i=1, \ldots, m)$ and of $q_{sj}(z)$ $(j=1, \ldots, r)$. To this purpose let define the vector of parameters

$$\gamma_s = (a_{s1,1} \ldots a_{s1,\nu_{s1}} | \ldots | a_{sm,1} \ldots a_{sm,\nu_{sm}} | \beta_{(\nu_1+\ldots+\nu_{s-1}+1),1} \ldots \beta_{(\nu_1+\ldots+\nu_s),1} | \ldots | \beta_{(\nu_1+\ldots+\nu_{s-1}+1),r} \ldots \beta_{(\nu_1+\ldots+\nu_s),r}) \tag{5.1}$$

and, for simplicity of notation, let

$$\mathbf{R}_s = \mathbf{R}(\nu_{s1}, \nu_{s2}, \ldots, \nu_s, \ldots, \nu_{sm}) \tag{5.2}$$

$$\mathbf{S}_s = \mathbf{S}(\nu_{s1}, \nu_{s2}, \ldots, \nu_s, \ldots, \nu_{sm}) \;. \tag{5.3}$$

By using the full length of the available data

γ_s can be obtained by means of the least-squares estimator

$$\gamma_s = (R_s^T R_s)^{-1} R_s^T y_s(k+\nu_s) = S_s^{-1} R_s^T y_s(k+\nu_s). \quad (5.4)$$

Remark 11 - Note that, during the structural identification, ν_s is determined by recognizing the singularity of the matrix $S(\nu_{s1},..,\nu_s+1,..,\nu_{sm})$; this matrix contains both S_s and the vector $R_s^T y_s(k+\nu_s)$. The additional computations regard the inversion of S_s and its product with the vector $R_s^T y_s(k+\nu_s)$.

Remark 12 - The maximal number of parameters that are identified in a single step is given by $n + r\nu_M$. This is therefore also the maximal order of the matrix to be inverted.

VI. STRUCTURAL AND PARAMETRIC IDENTIFICATION IN THE PRESENCE OF NOISE

In this section the algorithms previously considered are extended to the noisy case. It will be assumed here that the input and output sequences are corrupted by an additive uncorrelated noise with zero mean; the noisy components of the input and output vectors will be denoted with

$$y^*_{k,j} = y_{k,j} + d(y_{k,j}) \quad (6.1)$$

$$u^*_{k,j} = u_{k,j} + d(u_{k,j}) \quad . \quad (6.2)$$

In (6.1),(6.2) and in the following the starred quantities are noisy ones. In this case it is well-known |15||16| that the least-squares formula (5.4) leads to a biased estimate of γ_s also for $N\to\infty$. The origin of the bias in (5.4) can however be easily detected and eliminated if the statistics of the noise are known. In fact, from the expressions (5.3) and (4.8) of S_s, since the noise has zero mean and is not correlated with the input/output sequences of the system, it follows that |17|

$$\operatorname*{plim}_{N\to\infty} \frac{1}{N} S^*_s = \lim_{N\to\infty} \frac{1}{N} S_s + N(d_s) \quad (6.3)$$

where $N(d_s)$ is the covariance matrix of the noise vector

$$d_s=\{d(y_{k,1})..d(y_{k+\nu_{s1},1})|...|d(u_{k,r})..d(u_{k+\nu_{s1},r})\}.$$

It stands also

$$\operatorname*{plim}_{N\to\infty} \frac{1}{N} R_s^{T*} y_s^*(k+\nu_s) = \lim_{N\to\infty} \frac{1}{N} R_s^T y_s(k+\nu_s) + n_s \quad (6.4)$$

where

$$n_s = \lim_{N\to\infty} \frac{1}{N} \{ \sum_{i=k}^{N} (d(y_{i,1})d(y_{i+\nu_s,s})),.., \sum_{i=k}^{N} (d(y_{i+\nu_s,s}) d(y_{i+\nu_{s1}-1,1}))|..|.. \sum_{i=k}^{N} (d(u_{i+\nu_s-1,r}) d(y_{i+\nu_s,s}))\} \quad (6.5)$$

If the statistics of the noise are known it is therefore possible, for N large enough, to obtain a consistent estimation of the quantities S_s and $R_s^T y_s(k+\nu_s)$ by means of the expressions

$$\hat{S}_s = S^*_s - N\, N(d_s) \quad (6.6)$$

$$\hat{R}_s^T \hat{y}_s(k+\nu_s) = R_s^{T*} y_s^*(k+\nu_s) - n_s \,. \quad (6.7)$$

The "compensated" least-squares estimator

$$\hat{\gamma}_s = \hat{S}_s^{-1} \hat{R}_s^T \hat{y}_s(k+\nu_s) \quad (6.8)$$

gives therefore a consistent estimate, $\hat{\gamma}_s$, of the parameters vector γ_s.
The identification of the system structure will, similarly, be performed on the sequence of the matrices $\hat{S}(\mu_1,...,\mu_{m+r})$.

It can be noted that, when the same amount of uncorrelated zero-mean noise is added to the input/output sequences and nonoverlapping sets of data are used, then $N(d_s) = \sigma^2 I$ where σ^2 is the variance of the noise so that

$$\operatorname*{plim}_{N\to\infty} \frac{1}{N} S^*_s = \lim_{N\to\infty} \frac{1}{N} S_s - N\sigma^2 I \quad (6.9)$$

$$\operatorname*{plim}_{N\to\infty} \frac{1}{N} R_s^{T*} y_s^*(k+\nu_s) = \lim_{N\to\infty} \frac{1}{N} R_s^T y_s(k+\nu_s) \quad (6.10)$$

and the estimator (6.8) is given by

$$\hat{\gamma}_s = \{ S^*_s - N\sigma^2 I)^{-1} R_s^{T*} y_s^*(k+\nu_s). \quad (6.11)$$

It is also important to note that, in this case |17||18|, an estimate of $N\sigma^2$ is given by the least eigenvalue of the symmetrical matrix $S(\mu,\mu,...,\mu)$ for $\mu>\nu_M$ so that a consistent estimate of the structure and of the parameters of the system can be obtained starting from a given upper bound to the order of the system. The same technique can be applied when different amounts of noise are present on the various inputs and outputs by performing a previous scaling on the data; the ratio of the different noises must however be known.

VII. FROM INPUT-OUTPUT TO STATE-SPACE MODELS

After the structural and parametric identification of the system carried out according to the algorithms described in the sections 4,5 and 6, an input/output description of the type (3.18) has been obtained. It is then necessary to obtain, with a minimal effort, a canonical state-space representation. A first obvious choice regards the construction of a state-space model of the type (3.5)-(3.11). This involves the following steps:

Algorithm 1

1) Write directly matrix $\mathbf{A}$ from $\mathbf{P}(z)$ (the elements of $\mathbf{A}$ are the coefficients of the polynomials in $\mathbf{P}(z)$). Write directly matrix $\mathbf{C}$ from the knowledge of $\nu_1,\ldots,\nu_m$.

2) Write directly the matrices $\mathbf{M}$ (the elements of $\mathbf{M}$ are those of $\mathbf{A}$, arranged according to (3.25)) and $\bar{\mathbf{B}}$ (the elements of $\bar{\mathbf{B}}$ are the coefficients of the polynomials in $\mathbf{Q}(z)$).

3) Compute matrix $\mathbf{B}$, given by $\mathbf{B} = \mathbf{M}^{-1}\bar{\mathbf{B}}$.

Once the triple $(\mathbf{A},\mathbf{B},\mathbf{C})$ has been obtained it is easy to estimate also the initial state of the system by means of relation (3.13). In (3.13) the only matrix that requires a parametric knowledge of the system for its construction is $\mathbf{W}$ which is constructed with the rows of the input distribution matrix $\mathbf{B}$ arranged as shown in (3.15).

The development of a second class of canonical state-space models obtainable from $\mathbf{P}(z)$ and $\mathbf{Q}(z)$ requires the previous recalling of some theoretical results on the realization of input/output difference descriptions. To avoid excessive heaviness to the paper some results are now only stated; their proof can however be found in |19|.

Let consider the difference input/output description of a multivariable system given by

$$\Lambda(z)\,\tilde{y}(k) = \mathbf{Q}(z)\,u(k) \tag{7.1}$$

where $\Lambda(z)$ and $\mathbf{Q}(z)$ are $(m\times m)$ and $(m\times r)$ matrices whose entries, $\lambda_{ij}(z)$ and $q_{ij}(z)$ are polynomials in z satisfying to the following conditions:

1) The polynomials $\lambda_{ii}(z)$ are monic.

2) Are valid the relations

$$\deg\{\lambda_{ii}(z)\} > \deg\{\lambda_{ij}(z)\} \quad \text{for } i\neq j$$
$$\deg\{\lambda_{ii}(z)\} > \deg\{\lambda_{ji}(z)\} \quad \text{for } i\neq j$$
$$\deg\{\lambda_{ii}(z)\} > \deg\{q_{ij}(z)\}\ .$$

Write also

$$\lambda_{ii}(z) = z^{\nu_i} - \alpha_{ii,\nu_i} z^{\nu_i-1} - \ldots - \alpha_{ii,2} z - \alpha_{ii,1} \tag{7.2}$$

$$\lambda_{ij}(z) = -\alpha_{ij,\nu_{ij}} z^{\nu_{ij}-1} - \ldots - \alpha_{ij,2} z - \alpha_{ij,1} \tag{7.3}$$

$$q_{ij}(z) = \beta_{(\nu_1+\ldots+\nu_i),j}\, z^{\nu_i-1} + \ldots + \beta_{(\nu_1+\ldots+\nu_{i-1}+2),j}\, z + \beta_{(\nu_1+\ldots+\nu_{i-1}+1),j} \tag{7.4}$$

$$n = \nu_1 + \nu_2 + \ldots + \nu_m\ , \quad \nu_M = \max_i(\nu_i)\ .$$

It is then possible to state the following lemma.

Lemma 3 |19| - *An equivalent state-space description of the system (7.1) is given by the triple* $(\tilde{\mathbf{A}},\tilde{\mathbf{B}},\tilde{\mathbf{C}})$ *where*

$$\tilde{\mathbf{A}} = \{\tilde{\mathbf{A}}_{ij}\} \quad (7.5) \qquad \tilde{\mathbf{A}}_{ii} = \begin{bmatrix} 0 \ldots\ldots\ldots 0 & \alpha_{ii,1} \\ & \alpha_{ii,2} \\ I_{\nu_i-1} & \vdots \\ & \alpha_{ii,\nu_i} \end{bmatrix} \tag{7.6}$$

$(i,j=1,..m)$

$$\tilde{\mathbf{A}}_{ij} = \begin{bmatrix} 0 & \cdots & 0 & \alpha_{ij,1} \\ \vdots & & & \vdots \\ \vdots & & & \alpha_{ij,\nu_{ij}} \\ 0 & \cdots & 0 & 0 \end{bmatrix} \tag{7.7}$$

$$\tilde{\mathbf{B}} = \begin{bmatrix} \tilde{b}_1^T \\ \vdots \\ \tilde{b}_n^T \end{bmatrix} = \begin{bmatrix} \beta_{11} & \cdots & \beta_{1r} \\ \vdots & & \vdots \\ \beta_{n1} & \cdots & \beta_{nr} \end{bmatrix} \tag{7.8}$$

$$\tilde{\mathbf{C}} = \begin{bmatrix} 0\ldots0\ 1\ 0 \ldots\ldots\ldots\ldots\ldots\ldots\ldots\ 0 \\ 0\ldots\ldots\ldots\ldots0\ 1\ 0\ \ldots\ldots\ 0 \\ \ldots\ldots\ldots\ldots\ldots\ldots\ldots\ldots\ldots\ldots \\ 0\ \ldots\ldots\ldots\ldots\ldots\ldots\ldots\ldots\ 0\ 1 \end{bmatrix} \tag{7.9}$$

$$\qquad\uparrow\nu_1 \qquad\quad \uparrow\nu_1+\nu_2 \qquad\quad \uparrow n$$

In lemma 3 the word "equivalent" means that the input/output model (7.1) and the state-space model (7.5)-(7.9) have the same external behavior i.e. their output is identical for every input sequence applied with null initial conditions. The link between nonzero initial conditions in (7.1) and the initial state in (7.5)-(7.9) is given by the equation

$$\tilde{w}(k) = \tilde{\mathbf{M}}\,\mathbf{V}(z)\,\tilde{y}(k) - \tilde{\mathbf{W}}\mathbf{Z}(z)\,u(k) \tag{7.10}$$

where $\mathbf{V}(z)$ and $\mathbf{Z}(z)$ are given by (3.14) and (3.16) respectively, $\tilde{\mathbf{M}}$ is given by expression (3.25) (with the obvious substitution of the parameters $a_{ij,k}$ with $\alpha_{ij,k}$) and $\tilde{\mathbf{W}}$ is given by

$$\tilde{\mathbf{W}} = \left[\begin{array}{c} \tilde{b}_2^T \ \ldots\ldots\ldots\ldots\ \tilde{b}_{\nu_1}^T\ 0\ldots0 \\ \vdots \qquad\qquad\qquad \vdots \\ \tilde{b}_{\nu_1}^T \qquad\qquad\qquad \vdots \\ 0\ \ldots\ldots\ldots\ldots\ldots\ldots\ 0 \\ \hline \vdots \qquad\qquad\qquad \vdots \\ \hline \tilde{b}_{n-\nu_m+2}^T \ldots\ \tilde{b}_n^T\ 0\ldots\ldots0 \\ \vdots \qquad\qquad\qquad \vdots \\ \tilde{b}_n^T \qquad\qquad\qquad \vdots \\ 0\ \ldots\ldots\ldots\ldots\ldots\ldots\ 0 \end{array}\right]_{(n\times r\nu_M)} . \tag{7.11}$$

By comparing the proprieties of the polynomials in the identified model $\mathbf{P}(z)$, $\mathbf{Q}(z)$ and in the couple $\Lambda(z)$, $\mathbf{Q}(z)$ of (7.1) it follows that the only propriety of $\Lambda(z)$ that is not shared by $\mathbf{P}(z)$ is the condition $\deg\{\lambda_{ii}(z)\} > \deg\{\lambda_{ij}(z)\}$ for $j<i$. In fact in $\mathbf{P}(z)$ the degree of $p_{ii}(z)$ can be also equal to the degree of $p_{ij}(z)$ when $j<i$; since however the degree of $p_{ij}(z)$ is at most equal to the degree of $p_{ii}(z)$ and $p_{ii}(z)$ is monic it is very simple to determine (see |19|) a triangular $(m\times m)$ real-valued matrix $\mathbf{K}$ such that the product $\mathbf{P}(z)\mathbf{K}$ satisfies to the same conditions as $\Lambda(z)$.

$$\mathbf{K} = \begin{bmatrix} 1 & & & & \\ c_{21} & 1 & & & \\ \vdots & & \ddots & & \\ c_{m-1,1} & \cdots & & 1 & \\ c_{m1} & \cdots\cdots & & c_{m,m-1} & 1 \end{bmatrix} \quad (7.12)$$

It is then possible to write the identified model in the form

$$\mathbf{P}(z)(\mathbf{K}\,\mathbf{K}^{-1})\ \mathbf{y}(k) = \mathbf{Q}(z)\ \mathbf{u}(k) \quad (7.13)$$

or also, taking $\mathbf{\Lambda}(z)=\mathbf{P}(z)\mathbf{K}$, $\tilde{\mathbf{y}}(k)=\mathbf{K}^{-1}\ \mathbf{y}(k)$,

$$\mathbf{\Lambda}(z)\ \tilde{\mathbf{y}}(k) = \mathbf{Q}(z)\ \mathbf{u}(k) \quad (7.14)$$

$$\mathbf{y}(k) = \mathbf{K}\,\tilde{\mathbf{y}}(k)\ . \quad (7.15)$$

It is thus possible to state the following algorithm.

Algorithm 2

1) Write directly the matrix $\tilde{\mathbf{B}}$ from $\mathbf{Q}(z)$ (the elements in $\tilde{\mathbf{B}}$ are the coefficients of the polynomials in $\mathbf{Q}(z)$).

2) Determine a triangular matrix $\mathbf{K}$ such that the polynomials in every row of $\mathbf{P}(z)\mathbf{K}$ have lower degree than diagonal ones. Set $\mathbf{\Lambda}(z)=\mathbf{P}(z)\mathbf{K}$ and write matrix $\tilde{\mathbf{A}}$ (whose elements are the coefficients of the polynomials in $\mathbf{\Lambda}(z)$).

3) Write, from the elements of $\mathbf{K}$, the output distribution matrix which is given by

$$\tilde{\mathbf{C}} = \begin{bmatrix} 0 \ldots 0\ 1\ 0 \ldots\ldots\ldots\ldots\ldots 0 \\ 0 \ldots 0\ c_{21} \ldots 0\ 1\ 0 \ldots\ldots 0 \\ \ldots\ldots\ldots\ldots\ldots\ldots\ldots\ldots\ldots \\ 0 \ldots 0\ c_{m1} \ldots 0\ c_{m2} \ldots\ldots 1 \end{bmatrix}. \quad (7.16)$$

The triple $(\tilde{\mathbf{A}},\tilde{\mathbf{B}},\tilde{\mathbf{C}})$ is a state-space realization of the identified input/output model. By the use of eq. (7.10), the initial state is given by

$$\tilde{\mathbf{w}}(k) = \tilde{\mathbf{M}}\,\mathbf{V}(z)\,\mathbf{K}^{-1}\ \mathbf{y}(k) - \tilde{\mathbf{W}}\,\mathbf{Z}(z)\ \mathbf{u}(k)\ . \quad (7.17)$$

Remark 13 - The obtained state-space models are, because of their structure, completely observable. Their complete reachability (controllability if the dynamical matrix is nonsingular) is however not assured since non reachable (non controllable) initial states could be nonzero; in these cases the obtained model can be reduced by means of standard techniques |20|. This consideration is not relevant when the amount of available input/output data is large with respect to the system order.

At first sight algorithm 1 could look simpler than algorithm 2; in fact algorithm 1 is really simpler *to deduce* than algorithm 2, not for what concerns the required number of operations. Moreover the number of operations actually required by algorithm 2 depends on the number of polynomials $p_{ij}(z)$ on the left side of the main diagonal of $\mathbf{P}(z)$ whose degree is equal to the degree of $p_{ii}(z)$; when the condition $\deg\{p_{ii}(z)\} > \deg\{p_{ij}(z)\}$ for $j<i$ is met by $\mathbf{P}(z)$ algorithm 2 *does not require any computation* ($\mathbf{K}=\mathbf{I}$) and permits to write directly the triple $(\tilde{\mathbf{A}},\tilde{\mathbf{B}},\tilde{\mathbf{C}})$ from $\mathbf{P}(z)$, $\mathbf{Q}(z)$. A further feature of algorithm 2 is that it does not require the inversion of a $(n \times n)$ matrix like algorithm 1 but only a single-step reduction on a $(m \times m)$ polynomial matrix and can thus prove advantageous when n is large.

The following simple numerical example can clarify the whole identification procedure and also permit a comparison between the algorithms proposed for the last step.

A numerical example

The input/output sequences of a system with one input and two outputs are given by

$u(0)=1$, $u(1)=2$, $u(2)=4$, $u(3)=5$, $u(4)=-5$, $u(5)=-12$, $u(6)=15$, $u(7)=50$, $u(8)=10$, $u(9)=-60$, $u(10)=30$, $u(11)=0$, $u(12)=0$

$\mathbf{y}(0)=(0\ 0)^T$, $\mathbf{y}(1)=(0\ 0)^T$, $\mathbf{y}(2)=(1\ 0)^T$, $\mathbf{y}(3)=(0\ 2)^T$, $\mathbf{y}(4)=(2\ -1)^T$, $\mathbf{y}(5)=(5\ 4)^T$, $\mathbf{y}(6)=(-3\ -9)^T$, $\mathbf{y}(7)=(5\ 10)^T$, $\mathbf{y}(8)=(-25\ 8)^T$, $\mathbf{y}(9)=(56\ -7)^T$, $\mathbf{y}(10)=(-3\ 10)^T$, $\mathbf{y}(11)=(-27\ -4)^T$, $\mathbf{y}(12)=(40\ 5)^T$

Step 1 - Structural Identification

Taking N=9 the sequence of matrices

$\mathbf{S}(2,1,1)$, $\mathbf{S}(2,2,1)$ ... $\mathbf{S}(3,2,2)$, $\mathbf{S}(3,3,2)$...

is constructed. The first singular matrix found is $\mathbf{S}(3,3,2)$ so that $\nu_2=3-1=2$. This matrix is deleted from the sequence that continues as

$\mathbf{S}(3,2,3)$, $\mathbf{S}(4,2,3)$, ...

The second singular matrix found is $\mathbf{S}(4,2,3)$ so that $\nu_1=\nu_M=3$. Since m=2 the structural identification is terminated.

Step 2 - Parametric Identification

The input/output equations that can consequently be written are:

$$(-a_{21,3}\,z^2-a_{21,2}\,z - a_{21,1})y_{k,1} + (z^2-a_{22}z - a_{22,1})\,y_{k,2} = (\beta_{5,1}z + \beta_{4,1})\ u_{k,1}$$

$$(z^3-a_{11,3}z^2-a_{11,2}z - a_{11,1})\,y_{k,1} + (-a_{12,2}z - a_{12,1})\,y_{k,2} = (\beta_{3,1}z^2 + \beta_{2,1}z + \beta_{1,1})\ u_{k,1}$$

The associated parameters, obtained by means of the estimator (5.4), are given by

$$\gamma_2=(a_{21,1}\ a_{21,2}\ a_{21,3}|a_{22,1}\ a_{22,2}|\beta_{4,1}\ \beta_{5,1})^T=$$

$= (1\ 0\ -1|1\ -2|1\ 0)^T$

$\gamma_1=(a_{11,1}\ a_{11,2}\ a_{11,3}|a_{12,1}\ a_{12,2}|\beta_{1,1}\ \beta_{2,1}\ \beta_{3,1})^T=$

$=(1\ 0\ -1|-1\ 1|1\ 0\ 0)^T$

and consequently

$$\mathbf{P}(z)=\begin{bmatrix} z^3+z^2-1 & -z+1 \\ z^2-1 & z^2+2z-1 \end{bmatrix} \qquad \mathbf{Q}(z)=\begin{bmatrix}1\\1\end{bmatrix}$$

Step 3 - Construction of a state-space model

Algorithm 1 : The matrices $\mathbf{A}$ and $\mathbf{C}$ can be immediately written and are given by

$$\mathbf{A}=\left[\begin{array}{ccc|cc} 0 & 1 & 0 & & \\ 0 & 0 & 1 & & \\ 1 & 0 & -1 & -1 & 1 \\ \hline & & & 0 & 1 \\ 1 & 0 & -1 & 1 & -2 \end{array}\right] \qquad \mathbf{C}=\begin{bmatrix} 1 & 0 & 0 & 0 & 0 \\ 0 & 0 & 0 & 1 & 0 \end{bmatrix} .$$

The matrices $\mathbf{M}$ and $\bar{\mathbf{B}}$ are given by

$$\mathbf{M}=\left[\begin{array}{ccc|cc} 0 & 1 & 1 & -1 & 0 \\ 1 & 1 & & 0 & \\ 1 & & & & \\ \hline 0 & 1 & 0 & 2 & 1 \\ 1 & & & 1 & \end{array}\right] \qquad \bar{\mathbf{B}}=\begin{bmatrix}1\\0\\0\\1\\0\end{bmatrix}$$

The inverse of $\mathbf{M}$ is then computed and the matrix $\mathbf{B}$ is given by

$$\mathbf{B}=\mathbf{M}^{-1}\bar{\mathbf{B}}=\begin{bmatrix} 0 & 0 & 1 & 0 & 0 \\ 0 & 1 & -1 & 0 & 0 \\ 1 & -1 & 0 & 0 & 1 \\ 0 & 0 & -1 & 0 & 1 \\ 0 & -1 & 3 & 1 & -2 \end{bmatrix}\begin{bmatrix}1\\0\\0\\1\\0\end{bmatrix}=\begin{bmatrix}0\\0\\1\\0\\1\end{bmatrix} .$$

The initial state, computed by means of (3.13) is given by $\mathbf{w}(0)=(0\ 0\ 1\ 0\ 0)^T$.

Algorithm 2 : Matrix $\tilde{\mathbf{B}}$ can be immediately written and is given by

$$\tilde{\mathbf{B}}=\begin{bmatrix}1\\0\\0\\1\\0\end{bmatrix} .$$

Since $\deg\{p_{21}(z)\}=\deg\{p_{22}(z)\}$ it is necessary to reduce $\mathbf{P}(z)$ by subtracting its second column from the first. It follows

$$\mathbf{K}=\begin{bmatrix} 1 & \\ -1 & 1 \end{bmatrix} \qquad \mathbf{\Lambda}=\mathbf{P}(z)\mathbf{K}=\begin{bmatrix} z^3+z^2+z-2 & -z+1 \\ -2z & z^2+2z-1 \end{bmatrix}$$

It follows

$$\tilde{\mathbf{A}}=\begin{bmatrix} 0 & 0 & 2 & 0 & -1 \\ 1 & 0 & -1 & 0 & 1 \\ 0 & 1 & -1 & 0 & 0 \\ 0 & 0 & 0 & 0 & 1 \\ 0 & 0 & 2 & 1 & -2 \end{bmatrix} \qquad \tilde{\mathbf{C}}=\begin{bmatrix} 0 & 0 & 1 & 0 & 0 \\ 0 & 0 & -1 & 0 & 1 \end{bmatrix} .$$

The initial state, computed by means of (7.17) is given by $\tilde{\mathbf{w}}(0)=(1\ 0\ 0\ 0\ 0)^T$.

VIII. CONCLUDING REMARKS

In this paper a unitary approach to the structural and parametric identification of linear multivariable systems has been presented; this approach heavily relies on the structural proprieties of block-companion state-space representations for multivariable systems. Some early results on these proprieties had been obtained by the author in the first half of 1969 |21|; in 1970 an identification procedure based on the same state-space representations has been independently proposed by Velis |22| who, however, does not consider the structural identification. In the past two years some applications of the procedure to both simulated and real processes have been made and are reported in |17||23||24|; the obtained results are satisfactory for what concerns both the obtained models and the computational effort required. The whole procedure can be easily extended, in an entirely obvious way, to systems where an algebraical link between the input and the output is present.

REFERENCES

|1| Zadeh, L.A. (1962). From circuit theory to system theory. Proc. IRE, 50, 856-865.

|2| Åström, K.J. and Eykhoff, P. (1970). System identification, a survey. IFAC symp. Identification and process parameter estimation, Prague, survey paper.

|3| Ho, B.L. and R.E. Kalman (1966). Effective construction of linear state-variable models from input/output functions. Regelungstechnik, 14, 545-548.

|4| Gueguen, C.J. (1971). An algebraic algorithm for reducing a differential system to a state form. Second IFAC symp. on multivariable control systems, paper 1.2.2, Duesseldorf.

|5| Gerth, W. (1971). On the construction of low order linear state-variable models from measured data. Second IFAC symp. on multivariable control systems, paper 1.2.1, Duesseldorf.

|6| Tether, A.J. (1970). Construction of minimal linear state-variable models from finite input-output data. IEEE Trans. autom. control, AC-15, 427-436.

|7| Ackermann, J.E. and R.S. Bucy (1971). Canonical minimal realization of a matrix of impulse response sequences. Inform. Contr, 19, 224-231.

|8| Bonivento, C. and R. Guidorzi (1971). Canonical input-output description of linear multivariable systems. Ricerche di Automatica, 2, 72-83.

|9| Gopinath, B. (1969). On the identification of linear time-invariant systems from input-output data. Bell Syst. Tech. J., 48, 1101-1113.

|10| Budin, M.A. (1971). Minimal realization of discrete linear systems from input-output observations. IEEE Trans. autom. control, AC-16,395-401.

|11| Brunovsky, P. (1970). A classification of linear controllable systems. Kybernetika, 3, 173-187.

|12| Kalman, R.E. (1971). Kronecker invariants and feedback. Conf. on ordinary differential equations, NRL Math. research center.

|13| Luenberger, D.L. (1967). Canonical forms for linear multivariable systems. IEEE Trans. autom. control, AC-12, 290-293.

|14| Chen, C.T. (1970). Introduction to linear system theory. Holt, Rinehart and Winston, New York.

|15| Baender, R.G. and F.W. Smith (1968). A note on the statistical estimation of a hyperplane. IEEE Trans. autom. control, AC-13, 591.

|16| Koopmans, T. (1953). Identification problems in economic model construction. J. Wiley and Sons, New York.

|17| Bonivento, C. and Guidorzi, R. (1971). Parametric identification of linear multivariable systems. Joint autom. control conf., St. Louis, paper 5-C6.

|18| Woodside C.M. (1970). Estimation of the order of linear systems. IFAC symp. on identification and process parameter estimation, Prague, paper 1.7.

|19| Guidorzi, R. and G. Marro (1972). Partial identification of large-scale systems. Allerton conf. on circuit and system theory, Urbana, paper I-B2.

|20| Mayne, D.Q. (1968). Computational procedure for the minimal realization of transfer function matrices. Proc. IEE, 115, 1363-1368.

|21| Guidorzi, R. et alia (1969). Studio di metodi numerici per l'identificazione dei sistemi dinamici. Tech. Rept. Centro Calcoli e Servomeccanismi, Univ. of Bologna.

|22| Valis, J. (1970). On-line identification of multivariable linear systems of unknown structure from input-output data. IFAC symp. on identification and process parameter estimation, Prague, paperl.5.

|23| Dal Bianco, A., Galli, P and R. Guidorzi (1971). Identification of chemical reactors from input-output data. IEEE Conf. on decision and control, paper F.5.8.

|24| Bonivento C. and R. Guidorzi (1972). Application of an identification method to distillation columns. Princeton conf. on information sciences and systems, paper A.6.5.

MINIMAL REALIZATION AND APPROXIMATION OF LINEAR SYSTEMS FROM NORMAL OPERATING RECORDS

Alain Barraud - Philippe de Larminat

Laboratoire d'Automatique de l'Ecole Nationale Supérieure de Mécanique
Nantes, Loire-Atlantique, France

A method is proposed to identify a minimal order state model from the input-output data of a multivariable discrete process.
This realization is directly given as a canonical form, without any treatment of a non minimal order model. From a deterministic point of view, an identifiability condition is mentioned. Experimental results are reported.

1. INTRODUCTION

The minimal realization problem, when starting from input output data, has not so far been very much reported : Budin (1971), de Larminat (1972), except when using an intermediate representation such as impulse responses or transfer function matrix, (see for instance Gerth (1971)).
The algorithm, here proposed, is developed in a similar view as that of Ackermann (1971) for the realisation from impulse responses.
In § 2, we present a condition allowing to identify an observable and controllable process from input-output data. The observable canonical form of Wonham and Johnson (1969) is recalled in § 3. The identification algorithm is derived in § 4, and the experimental results are given in § 5.

2. STATEMENT OF THE PROBLEM, HYPOTHESIS

Let a process (P) and its input-output observations (Fig. 1):

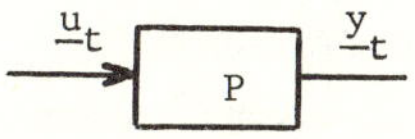

$$\xrightarrow{\underline{u}_t} \boxed{P} \xrightarrow{\underline{y}_t} \quad (t = 1, 2, 3...)$$

Fig. 1

$$\underline{u}'_t = [u^1_t \, u^2_t \ldots u^p_t] \,, \quad \underline{y}'_t = [y^1_t \, y^2_t \ldots y^r_t]$$

$\underline{u}'$ is for transposition of $\underline{u}$.
By assumption, (P) is defined by a set of <u>completely observable and controllable</u> n order state equations.

$$\underline{x}_{t+1} = \mathcal{A} \, \underline{x}_t + \mathcal{B} \, \underline{u}_t$$
$$\underline{y}_t = \mathcal{C} \, \underline{x}_t$$

$\underline{x}$: dim n $\underline{u}$: dim p $\underline{y}$: dim r

<u>Identifiability condition</u> :
It is easily shown that, if $(\mathcal{A}, \mathcal{B})$ is controllable, for any initial state $\underline{x}_1$ and for any integer $q > n+1$, it is possible to find an integer T and an input sequence $\{\underline{u}_t\}$ such as the following matrix W is full rank :

$$W = \begin{bmatrix} \underline{x}'_1 & \underline{u}'_1 & \underline{u}'_2 & \cdots & \underline{u}'_q \\ \underline{x}'_2 & \underline{u}'_2 & \underline{u}'_3 & \cdots & \underline{u}'_{1+q} \\ \vdots & \vdots & \vdots & & \vdots \\ \underline{x}'_{t+1} & \underline{u}'_{t+1} & \underline{u}'_{t+2} & \cdots & \underline{u}'_{t+q} \\ \vdots & \vdots & \vdots & & \vdots \\ \underline{x}'_{T+1} & \underline{u}'_{T+1} & \underline{u}'_{T+2} & \cdots & \underline{u}'_{T+q} \end{bmatrix} \quad \begin{array}{c} (t = 1, 2, \ldots, T \\ T > q + n) \end{array}$$

We assume in the remainder of this paper that (P) is observable and controllable, and that W is full rank.
Note that these properties are unchanged under a regular state transformation. Hence they are related to the process and not to a particular state representation $(\mathcal{A}, \mathcal{B}, \mathcal{C})$.

3. OBSERVABLE CANONICAL FORM OF WONHAM AND JOHNSON WITH ONE-DIRECTIONAL COUPLINGS

Since the couple $(\mathcal{A}, \mathcal{C})$ is by assumption observable there exists a regular transformation T :

$$\underline{x}_t = T \underline{\chi}_t$$

$$A = T \mathcal{A} T^{-1} \qquad B = T \mathcal{B} \qquad C = \mathcal{C} T^{-1}$$

such as A, B, C are given in the following form :

$$A = \begin{bmatrix} A^1 & 0 & \cdots & 0 \\ D^i & A^i & & 0 \\ \vdots & & \ddots & \\ & & D^r & A^r \end{bmatrix} \qquad B = \begin{bmatrix} B_1 \\ \hline B^i \\ \hline B^r \end{bmatrix}$$

$$C = \begin{bmatrix} E^1 & & 0 \\ \hline & E^i & \\ \hline & & E^r \end{bmatrix}$$

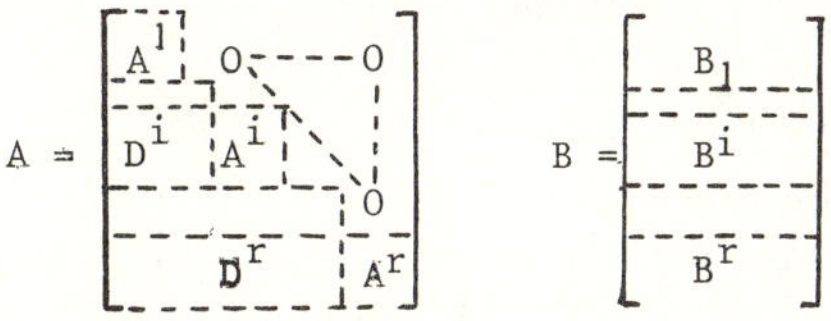

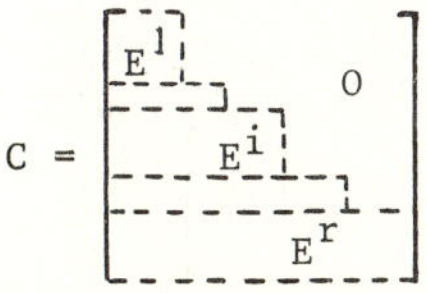

B is constituted by r blocks of $n_i \times p$ dimension:

$$B^i = \begin{bmatrix} b^{1i}_1 & b^{2i}_1 & \cdots & b^{pi}_1 \\ \vdots & \vdots & & \vdots \\ b^{1i}_{n_i} & b^{2i}_{n_i} & \cdots & b^{pi}_{n_i} \end{bmatrix} \triangleq \begin{bmatrix} B^i_1 \\ \hline \vdots \\ \hline B^i_{n_i} \end{bmatrix}$$

A is constituted by blocks D^i and by r companion matrices of $n_i \times n_i$ dimension :

$$A^i = \begin{bmatrix} 0 & & 0 & a_1^i \\ 1 & & & a_2^i \\ & \ddots & 0 & \vdots \\ 0 & & 1 & a_{n_i}^i \end{bmatrix}$$

Define : $m_i = n_1 + n_2 + \dots + n_{i-1}$ $(m_1 \triangleq 0)$

The two following cases are then to be distinguished :

a) $n_i \neq 0$: Then

$$D^i = \begin{bmatrix} c_1^i & c_2^i & \dots & c_{m_i}^i \\ 0 & 0 & \dots & 0 \\ 0 & 0 & \dots & 0 \end{bmatrix} \triangleq \begin{bmatrix} C^i \\ \hline 0 \\ 0 \end{bmatrix}$$

and $E^i = [0 \dots 01]$ (dim. $1 \times m_i$)

b) $n_i = 0$: Then no blocks A_i, B_i, D_i exist, and $E^i = [c_1^i \; c_2^i \dots c_{m_i}^i]$

It is important to note that for given $(\mathcal{A}, \mathcal{B}, \mathcal{C})$, the above form (A,B,C) is unique.

4. IDENTIFICATION ALGORITHM

From the input output data, the following data matrix Y is defined (q is given and is a priori greater than the unknown integer n).

$$Y = \begin{bmatrix} \underline{y}_1' & \underline{y}_2' & \dots & \underline{y}_q' & \vdots & \underline{u}_1' & \dots & \underline{u}_q' \\ \cdot & \cdot & & \cdot & \vdots & \cdot & & \cdot \\ \underline{y}_{t+1}' & \underline{y}_{t+2}' & \dots & \underline{y}_{t+q}' & \vdots & \underline{u}_{t+1}' & \dots & \underline{u}_{t+q}' \\ \cdot & \cdot & & \cdot & \vdots & \cdot & & \cdot \\ \underline{y}_{T+1}' & \underline{y}_{T+2}' & \dots & \underline{y}_{T+q}' & \vdots & \underline{u}_{T+1}' & \dots & \underline{u}_{T+q}' \end{bmatrix}$$

By assumption, the unknown matrix W, associated to these data, is full rank.

The $\underline{x}_t^i$ and $\underline{z}_t^i$ vectors are defined as :

$$[\underline{x}_t^{1'} \dots \underline{x}_t^{i'} \dots \underline{x}_t^{r'}] \triangleq \underline{x}_t \quad \text{and} \quad z_t^{i'} \triangleq [\underline{x}_t^1 \dots x_t^{i-1'}]$$

From the special form of (A,B,C), the following expressions (1), (2) can be verified :

$$\underline{x}_{t+1}^i = \begin{bmatrix} -a_2^i & & -a_{n_i}^i & 1 & \vdots & -B_2^i & & -B_{n_i}^i & 0 \\ & & 1 & 0 & \vdots & & & 0 & \\ -a_{n_i}^i & & & & \vdots & -B_{n_i}^i & & & \\ 1 & 0 & \dots & 0 & \vdots & 0 & \dots & & 0 \end{bmatrix} \begin{bmatrix} y_{t+1}^i \\ \vdots \\ y_{t+n_i}^i \\ \hline \underline{u}_{t+1} \\ \vdots \\ \underline{u}_{t+n_i} \end{bmatrix} \quad (1)$$

$$y_{t+n_i+1}^i = C^i \underline{z}_{t+1}^i + a_1^i y_{t+1}^i + \dots + a_{n_i}^i y_{t+n_i}^i + B_1^i \underline{u}_{t+1} + \dots + B_{n_i}^i \underline{u}_{t+n_i} \quad (2)$$

From the assumption that W is full rank, it is possible to derive that for any $\nu_i \leqslant n_i$, the following Y_{ν_i} matrix is full rank. Moreover the equation (2) implies that the same matrix is not full rank for $\nu_i > n_i$

$$Y_{\nu_i} = \begin{bmatrix} \underline{z}_1^{i'} & \vdots & y_1^i & \dots & y_{\nu_i}^i & \vdots & \underline{u}_1' & \dots & \underline{u}_{\nu_i}' \\ \cdot & \vdots & \cdot & & \cdot & \vdots & \cdot & & \\ \underline{z}_{t+1}^{i'} & \vdots & y_{t+1}^i & \dots & y_{t+\nu_i}^i & \vdots & \underline{u}_t' & \dots & u_{t+\nu_i}' \\ \cdot & \vdots & \cdot & & \cdot & \vdots & \cdot & & \\ \underline{z}_{T+1}^{i'} & \vdots & y_{T+1}^i & \dots & y_{T+\nu_i}^i & \vdots & \underline{u}_T' & \dots & \underline{u}_{T+\nu_i}' \end{bmatrix}$$

So, the identification principle is straightforward.

- For i = 1, and for any integer $\nu_1 \leqslant q$, the Y_{ν_1} matrices are immediately obtained from the data matrix Y, since there is no $\underline{z}_1$ $(m_1 = 0)$.
n_1 is then identified by observing the rank of Y_{ν_1} for increasing values of ν_1.
Hence the system of the equations (2) where i=1, t=0,1,...T, has a unique solution which gives the coefficients of A^1 and B^1.
After computation of this solution, equation(1) gives $\underline{x}_t^1$ from the input output knowledge.
- For i = 2, we knowing that $\underline{z}^2 = \underline{x}^1$, it is then possible to construct the Y_{ν_2} matrices which allow to determine n_2. The next step is to identify the A_2, B_2, D_2 coefficients (or the E_2 alone if $n_2 = 0$).
- The same procedure is carried on until i = r.
This result can be extended to the case where the data are obtained from differents records. Consequently, the method is convenient for minimal realization from impulse responses matrix.

4. IMPLEMENTATION

For each stage (i = 1, 2, ... r), a mono-output system is identified, taking into account the previously computed states. Hence any classical identification method may be used.
Our program presents the main following characteristics :

1°) n_1, $n_2 \dots n_r$ and ABC are first computed by linear regression (equation error minimization)

2°) For A and C so obtained, we determine the inital state $\underline{x}_o$ and the matrix B which minimize the output error (quadratic problem in $\underline{x}_o$ and B).

- Experimental results

On Fig. 2 and 3, the results obtained from the test case proposed by Dr. Richalet. From computation time reasons, we increased the sample period by replacing for all signals each batch of ten points running by their mean value.

Since it is impossible to compare the coefficients of different dimension models, real and simulated output are plotted (Fig.2) and the unit step response matrices are given (Fig. 3) in the two following cases :

a) For a non minimal model : three partial multi-input, mono-output models, each of dimension 5 (light lines).

b) For a minimal model of dimension 8, given in canonical form : $n_1 = 3$, $n_2 = 2$, $n_3 = 3$ (heavy lines).

CONCLUSION

We presented, with a deterministic point of view, a minimal realization algorithm from input out-put data. This method was shown efficient for the determination of a reduced order model, in a real case where, obviously, the linearity and finite order assumptions are not fulfilled.

On the other hand, since the canonical form used here is unique, conclusions simular to that of Caines (1972) and Mayne (1972), concerning generalization of the maximum likelihood method, could be obtained.

REFERENCES

J.E. Ackermann, R.S. Bucy (1971) "Canonical minimal realization of a matrix of impulse response sequence". Information and Control, Vol.19, pp. 224-231.

M.A. Budin (1971) "Minimal realization of discrete linear systems from input-output observations". IEEE Vol.A.C. 16-5, oct. pp. 395-401

P.E. Caines (1972) "Unique state variables models for linear systems". Int.J.Control. Vol.16, n° 5, pp. 939-944.

W. Gerth, W. Mitars (1971) "On the construction of low order linear state variable models from measured data." IFAC Düsseldorf, paper 1.2.1.

C.D. Johnson (1969) "A unified canonical form for controllable and uncontrollable linear dynamical systems". Preprints JACC, june.

Ph. de Larminat (1972) "Un algorithme de réalisation minimale à partir de données d'entrée-sortie". C.R.A.S. Paris, t. 275 (24 juillet) pp. 315-318.

D.Q. Mayne (1972) "A canonical model for identification of multivariable systems" IEEE, Vol.AC 16-5, oct. pp. 728-729.

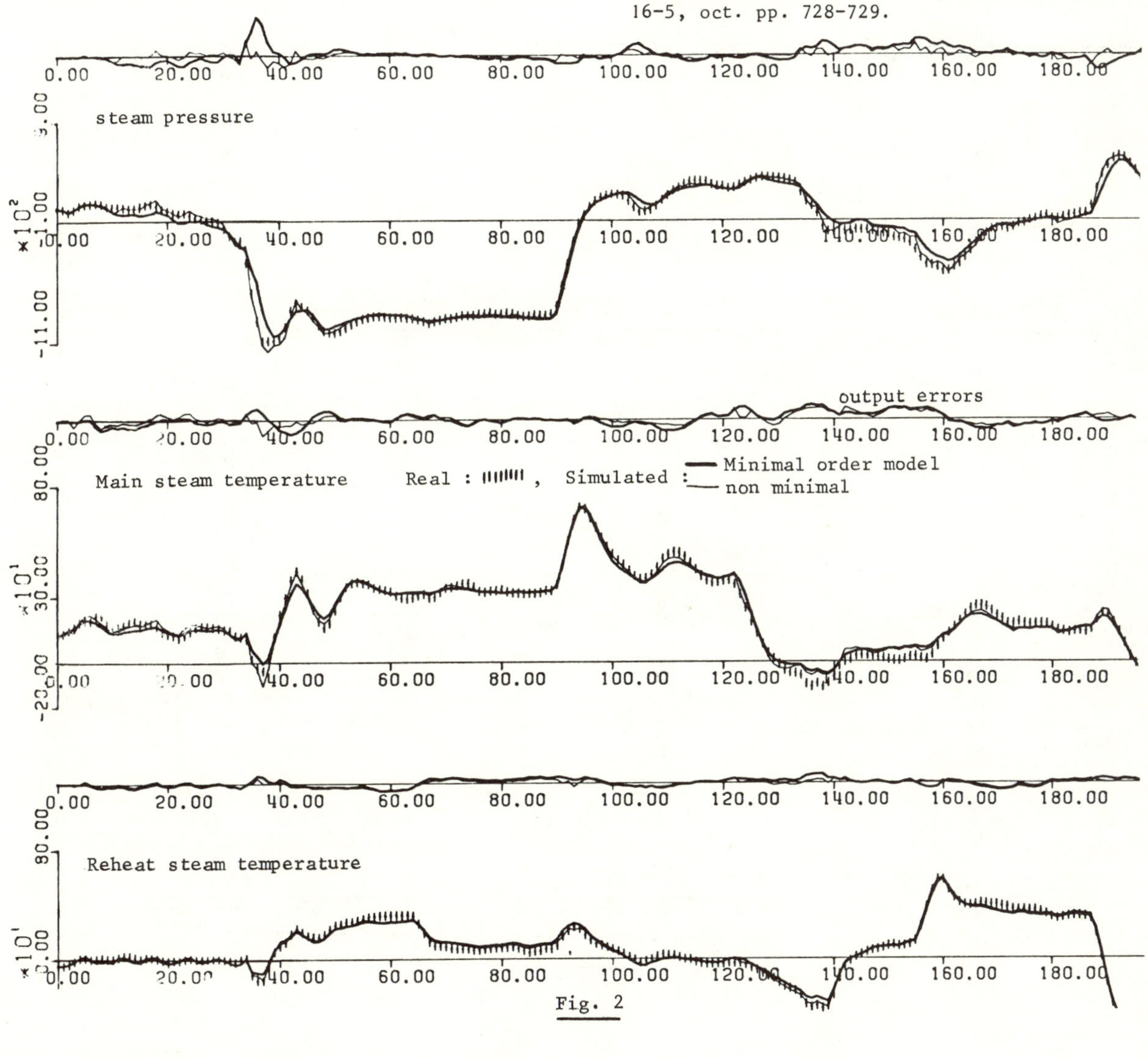

Fig. 2

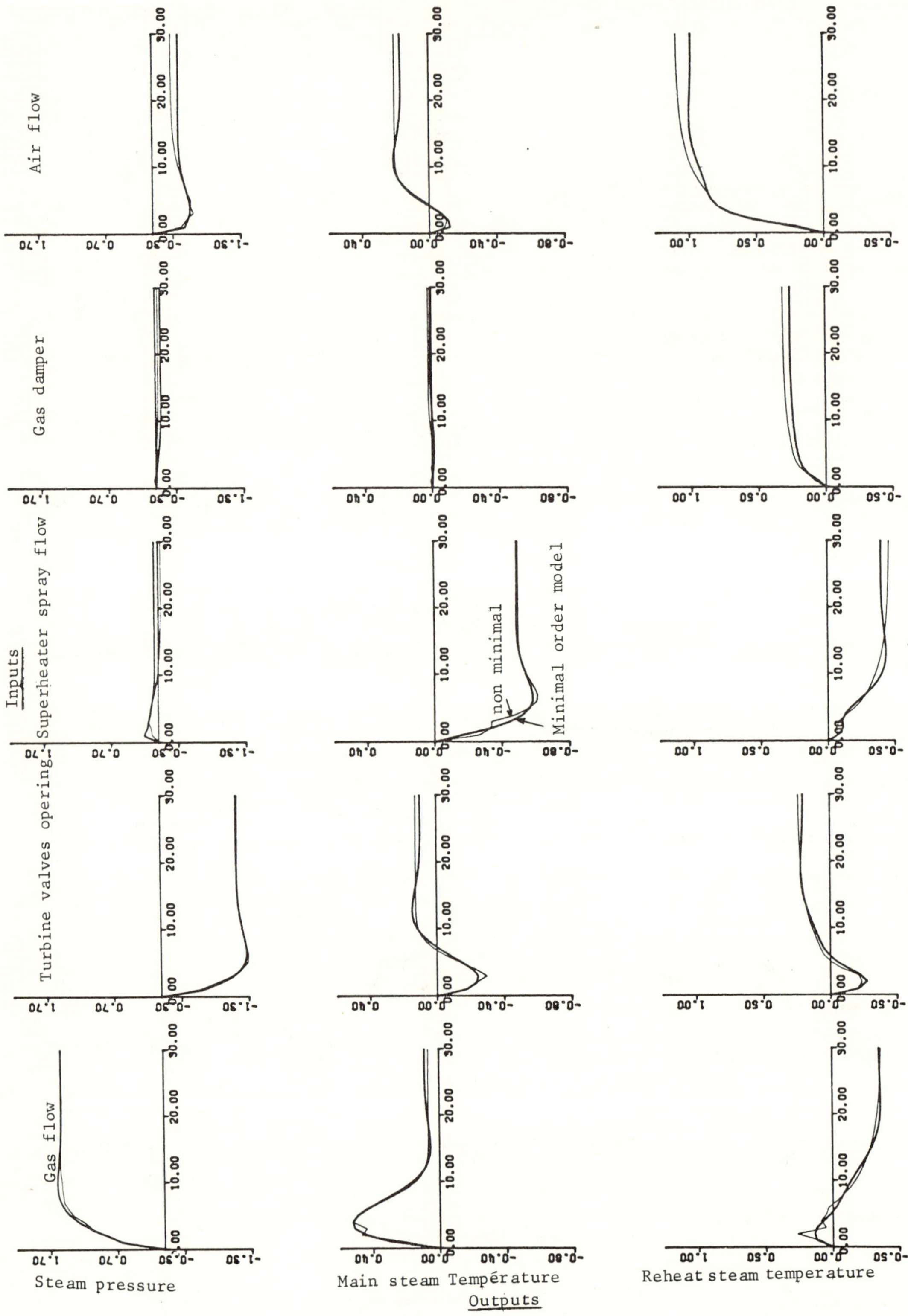

Fig. 3 - Test case B : Unit step responses

MULTIDIMENSIONAL INDUSTRIAL PLANT IDENTIFICATION

A.A. Dorofeyuk, A.D. Kasavin, I.Sh. Torgovitsky
Institute of Control Science (Automation and Telemechanics)
Central Institute of Complex Automation
Moscow, USSR

The paper is concerned with a comprehensive identification procedure to be applied to complex industrial plants with a number of input variables and an "input-output" relation of a type unknown in advance. According to the procedure, informative input variables are selected by extremum grouping algorithms. To construct a static model, piecewise approximation algorithms are developed. A possible way to use the procedure for study of the plant dynamic properties is outlined. The results of three practical solutions are cited.

1. INTRODUCTION

The functioning of an industrial plant is characterized by a number of input variables (raw material composition and quality, technology parameters). The plant performance is evaluated in terms of an output variable (productivity or production quality index). The identification of such a plant is started, as a rule, with constructing a static model.

The core of most methods to solve the problem which are known in the literature is the assumption that "informative" input variables whose effect on the plant output parameter is the greatest and "input-output" relation type are known a priori (Lukomskiy 1961, Rajbman 1966, Tsypkin 1968, Aizerman 1970). When analyzing a specific plant, however, such information is often unavailable. In this situation the comprehensive identification method the paper is concerned with seems to be useful. This method can also allow for plant dynamic properties.

2. PROBLEM STATEMENT

Let us distinguish two subproblems in the problem of plant identification, the search for informative input variables and restoration of the "input-output" relation on the knowledge of these variables. Let us first see how informative variables can be found.

If the number of input parameters is not large, then finding informative inputs is not, as a rule, too difficult. In fact, there are methods of full or shortened enumeration of every possible subspace from the initial set of input variables. For each of these subspaces some performance characteristic of a function restored in a given subset is calculated, for example, residual dispersion of an output parameter (Rajbman 1966, Dorofeyuk 1970). As a result, a subspace is chosen where the performance index value is extremal, for example, the subspace, where residual dispersion is minimal.

However, in many practical problems there are tens or even hundreds of input variables. As a rule, to decrease the number of input variables used to restore the function is a task of paramount importance for such problems, because at high dimensionality a statistically reliable plant model is feasible only when vast experimental data are available. In these problems ordinary enumeration algorithms are clearly of no avail. Further on the problem of reducing the number of input variables is stated. In solving it we maintain in this reduced number of variables the essential information from input parameters readings.

Usually the operation of any industrial plunt is dictated by a small number of "causes" or "factors". They are temperature conditions, chemical reaction kinetics and raw material characteristics, plant operating conditions, the operator's work features, etc. However, in most cases these factors either cannot be measured at all or are extremely difficult to measure. Therefore many indirect parameters to a certain extent characterizing the factors are measured.

If we managed to find the values of factors themselves on the knowledge of indirect parameters and to select those parameters which in a sense are "the nearest" to the factors, then we should solve two problems at once. On the one hand, technologists would acquire valuable information on the causes influencing the plant operation and, on the other hand, the number of input parameters to be measured could be reduced greatly.

In the paper the problem is stated as a problem of parameters extremal grouping (Braverman 1970), i.e. as follows:

Let there be n initial input parameters $x^1, \dots, x^n$. It is necessary to obtain a decomposition of the parameters into k groups $A_1, \dots, A_k$ such that all the parameters which are the "nearest" to each other are brought into the same group. Besides, variables-factors $f_1, \dots, f_k$ must be found such that inside each group the parameters are as close to the associated factor as possible. In other words, an extremal value of some functional

$$J_1 = \sum_{j=1}^{k} \sum_{x^i \in A_j} F\{R(f_j, x^i)\} \qquad (1)$$

must be provided in terms of decomposition and factor selecting. In the expression (1) $R(u,v)$ is some measure of proximity between variables u and v, $F\{\cdot\}$ is some convex function selected in advance.

Usually a correlation coefficient $\rho(u,v)$ between variables u and v is chosen as $R(u,v)$, and the part of $F\{\cdot\}$ is played by functions of the form $F\{z\} = z^2$ or $F\{z\} = |z|$. Depending on the selection of $F\{\cdot\}$ the functional (1) takes the form

$$J_1 = \sum_{j=1}^{k} \sum_{x^i \in A_j} [\rho(f_j, x^i)]^2 \qquad (2)$$

or

$$J_1 = \sum_{j=1}^{k} \sum_{x^i \in A_j} |\rho(f_j, x^i)|$$

Indeed, there are functionals of a form other than (1) whose extremization (in terms of both decomposition and factor selection) agrees with the above statement of the problem. For instance, extremization of a functional of the form

$$J^* = J_1 - \sum_{\substack{\ell, j = 1 \\ \ell \neq j}}^{k} \sum_{x^i \in A_\ell} F\{R(f_j, x^i)\}$$

ensures both that the requirement of mutual proximity of variables and proximity to their factor in each group and the requirement of minimal proximity between factors of different groups are met. The resultant decomposition may prove better for the identification problem at hand than the one obtained from (1).

Once the extremal grouping has been completed, a specified number of parameters nust be selected (e.g. one) from each group that are the nearest to the factor associated with this group. The resultant set of input variables $x^1, \ldots, x^k$ is assumed to be the desired one.

Now let us consider the problem of obtaining the relation between values of y and the values of the selected input variables $x^1, \ldots, x^k$ in static operation. The plant is assumed to operate as a functional transformer $y = F(x)$ where the function $F(x)$ is unknown in advance. The fixed values of the vector of input variables $x = \{x^1, \ldots, x^k\}$ appear in agreement with the generally unknown probability distribution density $p(x)$, $x \in X$.

On the available observations it is required to obtain an approximation $\tilde{F}(x)$ of the function $F(x)$ so that the specified criterion $\mathcal{I}(y, \tilde{F}(x))$ would take an extremal value.

The part of $\mathcal{I}(y, \tilde{F}(x))$ is usually played by the value of residual dispersion y relative to $\tilde{F}(x)$, or a functional of the form

$$\mathcal{I} = \int_X [y - \tilde{F}(x)]^2 p(x)\, dx \tag{3}$$

In most existing methods for solution of this problem (see e.g. Lukomskiy 1961, Rajbman 1966, Tsypkin 1968, Aizerman 1970) the form of the function $\tilde{F}(x, \alpha)$ is selected with an accuracy to the unknown parameter vector α which is to be determined.

Because often there is no information on the form of the function $F(x)$ the approximation quality chiefly depends on the selection of $\tilde{F}(x, \alpha)$.

As a rule, $\tilde{F}(x, \alpha)$ is a segment of a series of expansion in terms of a certain fixed system of the functions $\{\varphi_i(x)\}$

$$\tilde{F}(x, \alpha) = \sum_{j=1}^{r} \alpha_j \varphi_j(x) \tag{4}$$

Recent years saw new methods of function restoration (see the References in Kasavin 1972) which will be conditionally termed piecewise approximation methods. Their chief difference from the methods mentioned above consists in that the system of the function in (4) is not fixed in a certain sense. The basic idea of these methods is as follows. The region where the function $F(x)$ is determined is decomposed into a certain number of subregions and the function is approximated separately in each. Then, if in each of the subregions the function $F(x)$ is of sufficiently simple form, then simple functions would do for approximation.

Let us consider the problem of obtaining the function $y = F(x)$ with (3) as a performance criterion. Select the approximating function $\tilde{F}(x)$ in the form

$$\tilde{F}(x) = \sum_{j=1}^{r} \varepsilon_j(x) \tilde{F}_j(x, \alpha_j) \tag{5}$$

where $\{\tilde{F}_j(x, \alpha_j), j = 1, \ldots, r\}$ specified functions determined with an accuracy to the vectors of the parameters $\{\alpha_j, j = 1, \ldots, r\}$.

The functions $\varepsilon_j(x)$ in (5) is determined as

$$\varepsilon_j(x) = \begin{cases} 1, & \text{if } x \in X_j \\ 0, & \text{if } x \bar{\in} X_j \end{cases} \tag{6}$$

where $\{X_j, j = 1, \ldots, r\}$ a certain decomposition of the space X into of nonintersecting regions, i.e. $X=\bigcup_{j=1}^{\tau} X_j$, $X_i \cap X_j = \varnothing$ for $i \neq j$, τ – a fixed number.

The functional (3) with (5) and (6) can be represented as

$$J = \int_X [y - \sum_{j=1}^{\tau} \varepsilon_j(x) \tilde{F}_j(x, \alpha_j)]^2 p(x) dx$$

or

$$J = \sum_{j=1}^{\tau} \int_{X_j} [y - \tilde{F}_j(x, \alpha_j)]^2 p(x) dx . \quad (7)$$

The problem of constructing a function $\tilde{F}(x)$ minimizing the functional (3) consists in selection of regions $\{X_j, j=1,\ldots, \tau\}$ and determining the vectors $\{\alpha_j, j=1,\ldots,\tau\}$ such that would ensure a minimal value of the functional (7).

If the number of refions τ in (7) is not known in advance, then we could consider a functional whose minimization would yield both optimal estimates of vectors of the coefficients $\{\alpha_j\}$ and decomposition $\{X_j\}$ of the space X and the optimal number of regions τ . This functional differs from (7) in an additional term in the right-hand part which allows for reliability of the calculated estimates of approximating function coefficients (Dorofeyuk 1972).

3. SOLUTION ALGORITHMS

As follows from Sect. 2, the problem of plant identification must be solved in two stages. The first stage is to use the method of extremum grouping for informative input variables search. At present there are a number of algorithms to solve this problem (Braverman 1970). As an example, let us consider the following iteration procedure, providing a mayimum value of the functional (2).

Let the decomposition of $x^1,\ldots,x^n$ variables into $A_1^p,\ldots,A_k^p$ groups be performed at the p-th step of the iteration. The factors f_q^p are obtained for each group of variables A_q^p by the formula

$$f_q^p = \frac{\sum_{x^i \in A_q^p} \alpha_i x^i}{\sqrt{\sum_{x^i, x^j \in A_q^p} \alpha_i \alpha_j \rho(x^i, x^j)}}$$

where $\rho(x^i,x^j)$ is the correlation coefficient of x^i and x^j variables; α_j are the components of eigenvector of a matrix $R_q^p=\{\rho(x^i,x^j)\}, x^i,x^j \in A_q^p$ corresponding to its greatest eigen-number. Then the (p + 1)-th division of $A_1^{p+1},\ldots,A_q^{p+1}$ is built by the rule: the variable x^i is associated with the group A_q^{p+1} if

$$[\rho(x^i, f_q^p)]^2 \geqslant [\rho(x^i, f_\nu^p)]^2 , \quad \nu = 1,\ldots,k \quad (8)$$

In the case when there are two or more factors and a variable x^i such that for these factors and this variable (8) becomes an equality, the variable x^i is associated with one of the corresponding groups arbitrarily.

The algorithm under consideration converges to the maximum (which may be local) of the functional (2) because at whatever factors $f_1^{p+1},\ldots,f_k^{p+1}$ the functional (2) does not diminish at each iteration step.

Once the groups of variables $A_1,\ldots,A_k$ and the factors $f_1,\ldots,f_k$ have been found, a variable whose correlation coefficient with the factor f_i is maximal is chosen out of each group A_i .

The resultant set of input variables, is subjected to either a full or shortened enumeration (Dorofeyuk 1970) or without any further treatment it is used as a set of informative variables at the second stage.

As follows from Sect. 2, at this stage the identification problem is formulated as a problem of minimizing the functional (7). Obtain an adaptive algorithm to solve the problem. (To simplify the discussion, the case when the number of regions in the functional (7) $r=2$ is consider).

As is generally known the necessary condition for the functional (7) to be minimal is that its first variation must be equal to zero:

$$\delta J(X_1, X_2, \alpha_1, \alpha_2) = 0 \ .$$

This condition can be represented in the form of a system of equations:

$$\int\limits_{X_j} [y - \tilde{F}_j(x,\alpha_j)] \nabla_{\alpha_j} \tilde{F}_j(x,\alpha_j) p(x) dx = 0, \quad j = 1, 2,$$

$$\Phi(x,y) = [y - \tilde{F}_1(x,\alpha_1)]^2 - [y - \tilde{F}_2(x,\alpha_2)]^2 = 0, \quad x \in \Lambda \qquad (9)$$

(Λ is a piesewise-smooth boundary surface between the regions X_1 and X_2) and to solve it let us use the procedures of stochastic approximation type:

$$\alpha_1[n+1] = \alpha_1[n] + \gamma_1[n+1] (y[n+1] - \tilde{F}_1(x[n+1], \alpha_1[n])) \nabla_{\alpha_1} \tilde{F}_1(x[n+1], \alpha_1[n]),$$

$$\alpha_2[n+1] = \alpha_2[n]$$

if in learning at the (n + 1)-th algorithm step there arises a situation $(x = x[n+1], y = y[n+1])$ such that

$$\Phi(x[n+1], y[n+1]) = (y[n+1] - \tilde{F}_1(x[n+1], \alpha_1[n]))^2 - (y[n+1] - \tilde{F}_2(x[n+1], \alpha_2[n]))^2 < 0 \ .$$

Otherwise $(\Phi(x[n+1], y[n+1]) \geq 0)$

$$\alpha_1[n+1] = \alpha_1[n],$$

$$\alpha_2[n+1] = \alpha_2[n] + \gamma_2[n+1](y[n+1] - \tilde{F}_2(x[n+1], \alpha_2[n])) \nabla_{\alpha_2} \tilde{F}_2(x[n+1], \alpha_2[n]).$$

The initial conditions for this procedure may be the composition $\{X_j^0\}$ and coefficients $\{\alpha_j^0\}$ of approximating functions obtained by automatic classification methods (Dorofeyuk 1970).

Since the expression (9) explicitly includes the output variable y, the values of which become known only in learning (i.e. in the simulation process), it is impossible to use this decision rule before learning is completed. To eliminate this disadvantage, the boundary Λ between the regions X_1, X_2 associated with the decision rule (9) must be obtained by the procedure of pattern recognition with external supervision using the sign of the expression $\Phi(x[n+1], y[n+1])$ as an indication that the point $x[n+1]$ belongs to some definite region X_i. Algorithms of the potential functions method (Aizerman, 1970) for example can be used to act as such a procedure.

It must be noted that if (9) generates the regions X_1 and X_2 each of which is a merger of several isolated subregions, then to restore the boundary Λ some "stepped" classification procedure must be used (Dorofeyuk, 1971).

A particular case is of interest when it is known in advance that a function $y = F(x)$ to be approximated is convex and the planes are chosen to act as $\tilde{F}_j(x, \alpha_j)$ i.e. $\{\tilde{F}_j(x,\alpha_j) = (c_j, x) + a_j, j = 1,2\}$. Then the condition (9) may be replaced by the equation

$$(c_2 - c_1, x) + a_2 - a_1 = 0 \quad (x \in \Lambda)$$

which does not include y. In such a problem the coefficients of the equation of the boundary Λ between the regions X_1 and X_2 are defined in terms of the coefficients of the approximating functions $\tilde{F}_1(\cdot)$ and $\tilde{F}_2(\cdot)$. This case and a corresponding adaptive algorithm are considered in Dorofeyuk (1970), Opoytsev (1970).

It must be noted that the adaptive algorithm, proposed in the paper, for solution the problems of piecewise approximation can be used to restore the static response of a plant, whose output parameter is measured with an interference having zero mathematical expectation and limited dispersion. In this case the regression surface can be easily shown to be piecewise approximated, which is the desired characteristic of a "noisy" plant.

Kasavin (1972) gives a brief review of other algorithms to solve the problem of piecewise approximation, which differ from the one considered in this paper in the procedures by which the space X is divided into the regions $\{X_j\}$.

In conclusion of the Section note that if the selected informative variables $x^1, \ldots, x^k$ influence the output variable y in different ways in different regions X_i of the space X $(\{x^1, \ldots, x^k\} = x \in X)$ then this will be felt in the results of the piecewise approximation algorithm, i.e. in the obtained values of α_{ij} $(i=1,\ldots,k, j=1,\ldots,r)$.

4. PLANT DYNAMICS

A static model of a plant is the simplest one in the sense that the output value $y(t)$ at time t is considered to be functionally related to a vector of input parameters values at the same time $x(t) = \{x^1(t), \ldots, x^k(t)\}$. But often such a model does not give an idea of the specific nature of plant functioning because $y(t)$ can depend on the input parameters values at some fixed times $t-\tau$ preceding t. For example, in the case of a plant with a "pure" delay $y(t) = F[x(t-\tau)]$. In a more common case $y(t)$ can depend on a finite set $\{x(t-\tau_1), \ldots, x(t-\tau_n)\}$, i.e.

$$y(t) = F(x(t-\tau_1), \ldots, x(t-\tau_n)),$$

$$0 \leq \tau_1 < \tau_2 < \ldots < \tau_n \leq \tau_m$$

where τ_m is a "memory depth" of a plant, maximum from among all input variables. The form of the function $F(\cdot)$ is not known in advance. Then under identification problem we shall understand finding the values $\tau_1, \ldots, \tau_n$ and the estimate $\tilde{F}(\cdot)$ of the function $F(\cdot)$ using input and output signal samples.

It is easily seen that a problem thus stated reduces to the identification problem for a static plant described with a greater number of input variables.

If plant "memory depth" τ_m is known, then breaking up the segment $[0, \tau_m]$ into, for example, equal intervals it is possible to choose variables $x(t-\tau_1), \ldots, x(t-\tau_n)$ allowed for in model construction by the above procedure of informative variable search. The estimate of τ_m can be found, for instance, from cross correlative and cross dispersion functions between the output $y(t)$ and inputs $x^i(t)$ plant variables. To find the approximation function $\tilde{F}(\cdot)$ a piecewise approximation algorithms can be used.

5. EXPERIMENTAL RESULTS

The adaptive algorithm of piecewise approximation proposed in this paper was simulated on a digital computer. The parabola $y = 1 - x^2$ was piecewise linearly approximated by two straight lines. Samples of the argument $(-1 < x_j < +1)$ and of the corresponding values $y_j = 1 - x_j^2$ of a fixed length $(j = 1, \ldots, N)$ were generated. The values of x_j were given by an evenly distributed random number generator. Each of the samples was treated with the algorithm repeatedly. Two straight lines were taken as the initial conditions, the first of them passed via points (x_1, y_1), (x_2, y_2) of the sample obtained and the second passed via points (x_3, y_3), (x_4, y_4).

As the result of the simulation it was found that the algorithm converges with a proper selection of the sequence type $\gamma_i(n)$ $(i=1,2)$ when initial conditions are specified in this way. The form of $\gamma_i(n)$ influences the convergence rate of the algorithm. Satisfactory results were obtained $\gamma_i(n_i) = \frac{A}{B+n_i}$ where A and B are the constants, chosen in advance and n_i $(i=1,2)$ is the number of the points which took part in correcting the i-th straight line coefficients before an n-th step of the algorithm $(n_1 + n_2 = n-1)$.

The comprehensive identification procedure proposed in this paper was used to investigate and construct a static model of the ethyline polimerization process in a tubular high pressure reactor.

The initial statistical data were a totality of 240 points in an eighteen-dimensional space of input variables with the corresponding values of an output variable, the product quality. All the data were obtained in static operation. The use of one extremal grouping algorithm resulted in dividing the initial 18 variables into 7 groups which then were found interpretable in physical terms. Out of each group only one variable was picked up that was the nearest to the corresponding group factor. Finally, two the most informating variables were selected out of these seven variables by means of a shortened enumeration method of (Dorofeyuk 1970). Then in the space of these variables a piecewise-constant model was constructed by one of the piecewise approximation methods (Dorofeyuk 1970). The space of the informative variables was divided into five subregions. The model constructed allowed one to reduce the prediction dispersion of an output variable eight times as compared with prediction by the mean value.

An analogous procedure was used in quality control of single items. In the space of five informative variables a piecewise constant model was constructed by using 75 points. As a result, the five-dimensional space was divided into five subregions. Residual dispersion of an output parameter for the constructed model was 7.6 times less than entire dispersion of an output parameter and 2.5 less than residual dispersion of a linear regression model.

Besides, piecewise approximation method was used to predict the reliability characteristics of electronic devices. To construct a piecewise-stepping predicting function, 120 points in a nine-dimensional space were used. The space was divided into four regions, inside each of them either the upper or the lower guaranteed boundary of a device service life was set. The performance of prediction function obtained was checked in two hundred new devices, which were not used in the model construction. The prediction was correct in 82 per cent of cases which is approximately 1.5 times better than in other models constructed for the problem in question (including regression models).

In conclusion it should be noted that a similar procedure proved efficient in a number of economics and medical diagnostics problems (Rozin 1972, German 1972).

REFERENCES

(1) Aizerman M.A., Braverman E.M., Rozonoer L.I. (1970). Metod potentsial'nych funktsiy v teorii obucheniya mashin. "Nauka", Moskva.

(2) Braverman E.M. (1970). Metody ekstremal'noy gruppirovki parametrov i zadacha vydeleniya sushchestvennykh factorov. "Avtomatica i telemekhanika", No.1, p.123.

(3) Dorofeyuk A.A., Kasavin A.D., Torgovitsky I.Sh. (1970). Application of Automatic Classification Methods to Process Identification in Industry. Preprints of papers of 2-nd IFAC Symposium on Identification and Process Parameter Estimation. Prague, paper 4.1.

(4) Dorofeyuk A.A. (1971). Algoritmy avtomaticheskoy klassifikatsii(obzor) "Avtomatica i telemekhanika", No.12, p.78.

(5) Dorofeyuk A.A., Kasavin A.D., Zhivov N.P., Torgovitsky I.Sh. (1972) Identifikatsiya mnogorezhimnykh promyshlennykh ob'ektov. "Problemy sistemotekhniki", "Sudostroenie", Leningrad, vyp. 1, p.106.

(6) German V.A., Muchnik I.B. (1972) Primenenie algoritmov teorii raspoznavaniya obrazov dlya approksimatsii funktsii po ee znacheniyam v sluchayno vybrannykh tochakakh. "Materialy XXII nauchno-tekhnicheskoy konferentsii", Kaunas, IV, p.68.

(7) Kasavin A.D. (1972). Adaptivnye algoritmy kusochnoy approksimatsii v zadache identifikatsii. "Avtomatika i telemekhanika", No.12, p.98.

(8) Lukomskiy Ya.I. (1961) Teoriya korrelyatsii i ee primenenie k analizu proizvodstva. "Gosstatizdat", Moskwa.

(9) Opoytsev (1970) Optimal'nye razbieniya v zadachakh identifikatsii i raspoznavaniya obrazov. V sb. "Identifikatsiya". Doklady II Vsesoyuznogo soveshchaniya po statisticheskim metodam teorii upravleniya (Tashkent, 1970), "Nauka", Moskwa, p.193.

(10) Rajbman N.S., Chadeev V.M. (1966) Adaptivnye modely v sistemakh upravleniya. "Sovetskoe radio", Moskwa.

(11) Rozin B.B. (1972) Dvukhurovnevye diskretno-nepreryvnye modeli ekonomicheskikh pokazateley. V sb. "Raspoznavanie obrazov i regressionnyy analiz v ekonomicheskikh issledovaniyakh. Novosibirsk, p.5.

(12) Tsypkin Ya.Z. (1968) Adaptatsiya i obuchenie v avtomaticheskikh sistemakh. "Nauka", Moskwa.

OPTIMAL MODEL BUILDING

J. Gordesch - P.P.Sint
Computing Center, University of Vienna
Vienna, Austria

An algorithm is developed which builds models,optimal in a threefold way: optimal reproduction of data, guaranteed by common methods of estimation; simplicity and feasibility, ensured by automatic construction of the model with possible man-machine interaction; usefulness in explanation and projection, aimed at by using clustering techniques for variables and items.

A model is defined as a simplified mapping of reality. The reality and its picture may be either physical or mental. We confine ourselves to mental pictures only, i.e. the model will be constructed with the help of abstract concepts from pregiven data.

A model should be optimal in an at least threefold sense:
a) optimal reproduction of data;
b) simplicity and feasibility (also entailing low costs etc.);
c) usefulness in gaining additional information (explanation, projection).

To a certain degree these three requirements exclude each other. Optimal reproduction of data is achieved by the usual statistical estimation procedures, i.e. the minimization of a risk function. Certainly the feasibility of the model is promoted by automation of the process of constructing the model. The complexity of the process, however, enhances a restriction of automation and leads to a man-machine interaction. Especially the problem of explanation and similar tasks is difficult to handle automatically: the heuristics involved are too complicated for rigid algorithms and should be left to man.

An algorithm for the automatic construction of a certain class of models is developed in the following and summarized in the flowchart below (Fig. 1).

1. DATA BASE

The data are supposed to be in the form of values of variables (binary, multistate, numeric) measured on well-distinguished items of a population.

2. FEATURE EXTRACTION

Optional introduction of artificial variables (e.g. forming index numbers, transformation of variables as taking logarithms).

3. STRUCTURAL GRAPH

A graph is defined over the set of all variables, where the nodes of the graph represent the variables and the edges the relations between the variables. A model is, at this stage, an enumeration of selections of elementary paths of the graph. The enumeration of even a moderate number of elementary paths is a hopeless proposition, hence a severe simplification of the graph will be necessary. This may be achieved either by the fusion of nodes (cf. 2 feature extraction, 6 clustering of variables, and 8 reduction) or by (external) removement of paths (relations).

4. EXTERNAL INFORMATION

Exclusion of variables and relations according to the preknowledge of the scientist. At this stage it is also possible to give the knowledge concerning the third requirement for an optimal model. Thus the variables may be divided into predictor variables (e.g. causes) and criterion variables (e.g. effects). The criterion variables represent the wanted additional information, being calculated from the predictor variables. In the sequel we distinguish between the case where the above division is provided (treated first) and where not.

5. CLUSTERING OF ITEMS

In many cases the variables concerning subgroups may follow similar laws, and it may be worthwile to divide the population into these subgroups. A whole battery of such clustering methods has been developed (for a survey see JARDINE-SIBSON and LERMAN, 1971). Different models may be built for each cluster. Example: a population is divided into groups (clusters) with similar somatic attributes.

The steps 4 and 5 may also be fused resulting in a two way clustering.

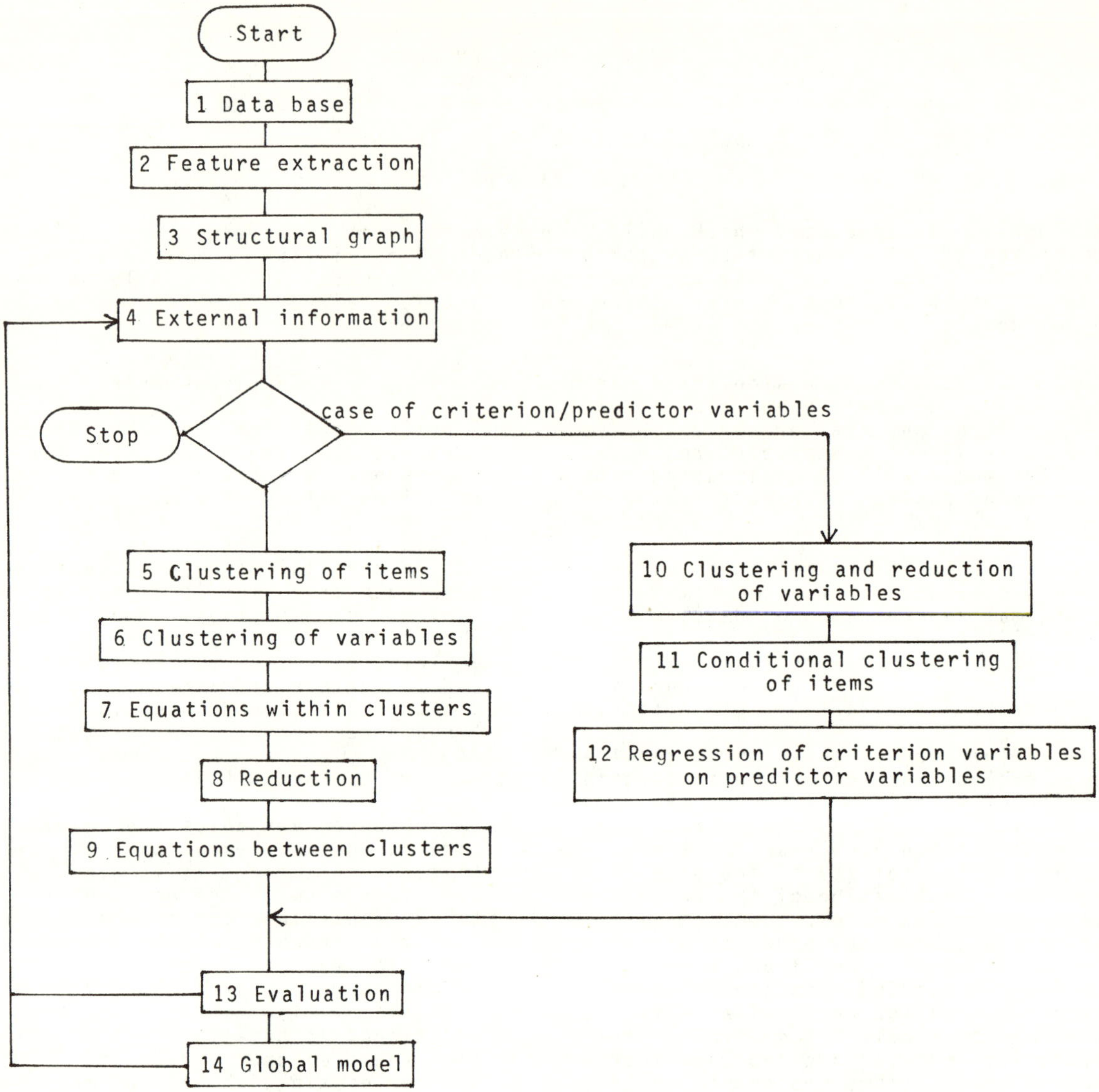

Fig. 1: Flow chart of model building algorithm

6. CLUSTERING OF VARIABLES

Several variables may bear the same information on the items or may, at least, be approximately equivalent. This leads to the proposal to cluster the variables, and condense the variables in one cluster to one or at least fewer variables than contained in the cluster. Example: measurements on human beings may be divided into groups giving proportional results (the measurements on the left and the right hand side may even be nearly identical).

7. EQUATIONS OF VARIABLES IN CLUSTERS

Analyzing the similarities in the model leads to the construction of regression equations between variables within the cluster. Because all the variables in one cluster are equivalent, orthogonal regression is recommended (estimation of the parameters in an equation $a_0+\Sigma a_i f_i(x_i)=0$, cf. MALINVAUD, 1966).

8. REDUCTION-VARIABLE SELECTION OR FACTOR EXTRACTION

Now we reduce the space of variables by choosing one variable out of each cluster as representative. This may be done either arbitrarily or by some optimization process (e.g. variable nearest to the centre of the cluster). An alternative method may be the concentration of all the variables in the cluster into one or more factors.

9. EQUATIONS BETWEEN CLUSTERS

We may form an equation immediately with the aid of variables obtained in step 8. A more sophisticated approach is to start by determining a total ordering of the clusters. For this purpose we use the capacity of each selected variable to predict all the other selected variables. Now we search a path which has its first node in the first cluster, its second node in the second cluster, etc.,and which is of minimal length in the sense of a "distance" function between the variables (e.g. $1-|r|$, r correlation coefficient, or some "distance" derived from measures of association). These variables are included in a regression equation the coefficients of which are estimated by the method of orthogonal regression (cf. MALINVAUD, 1966). Several algorithms exist for finding a path of minimal length in a directed labelled graph (see e.g. ALBRECHT, 1969; the method of dynamic programming: BELLMANN, 1957). It is, however, necessary to find not merely the best policy but also the second best, the third best, etc., the k-th best:

$$a_0^{(j)} + \sum_{i=1}^{n} a_i^{(j)} f_i(x_i) = 0,$$

$j=1,2,\ldots,k$ number of subpolicy(equation),
$i=1,2,\ldots,n$ number of variable

$a_i^{(j)} \neq 0$ for one and only one element x_i in each cluster.

For when we determine an exact solution ($j=1$) to the mathematical problem of optimization, we may have only an approximate solution of the real problem. If a mathematically optimal solution is unsatisfactory from the viewpoint of reality, we can either find a suboptimal solution ($j=2,3,\ldots,k$) in hopes that it will be better from the viewpoint of reality, or reformulate the mathematical problem (see 13 and 4).

10. CLUSTERING AND REDUCTION

Seperate clustering of predictor and criterion variables and replacement of variables in each cluster by a selected variable or a factor (optional).

11. CONDITIONAL CLUSTERING OF ITEMS

While traditional clustering methods use all variables indiscriminately in the criterion of assessing the quality of a classification, conditional clustering uses only the criterion variables. On the other hand, the classes may be found with the help of the predictor variables alone. Example: healthy people are classified into classes for which similar medical diagnostic methods in the case of illness promise good results. Several conditional cluster methods result in classes of items for which the relationships (cf.12) of predictor and criterion variables are similar. For a more detailed description of the method see the Appendix, and SINT, 1971.

12. REGRESSION OF CRITERION VARIABLES ON PREDICTOR VARIABLES

At this stage the parameters of the system of equations

$$y_i = \sum_j a_{ij} x_j$$

i number of criterion variable,
j number of predictor variable,
are estimated by a method described by GOLUB, 1965.

13. EVALUATION

This step yields the evaluation according to our three requirements of optimality, particularly the optimal reproduction of data, and prepares the basis for external evaluation.

14. GLOBAL MODEL

Finally the various submodels constructed are combined to one global model, which may still be reformulated (see 4).

The performance of the model depends essentially on the quality of the clustering process. A system of programs is in preparation; parts of it, a subroutine control system (SCS) developed by N. Winterleitner, and programs for various algorithms, are already working.

For the special problem of causal models see GORDESCH, 1972.

APPENDIX: CONDITIONAL CLUSTERING

Traditional clustering methods try to subdivide the space of items into regions of "similar" elements. The location of an item is given by a vector of the values of its attributes.
Assuming k attributes x_i we describe an item e by a vector

$$e = (x_1, x_2, \ldots, x_k),$$

and a non-hierarchical clustering method is a mapping

$$C: e \rightarrow n, \quad n \in N,$$

where N is a nominal scale, conveniently a finite set of integers. In addition,

the variables are used as criterion to decide on the quality of the clustering achieved. The method of Edwards and Cavalli-Sforza (1965) e.g. searches for clusters for which the sum of variances within the clusters will be minimized.
In conditional clustering we assume that there are two different types of variables such that an element is described by a vector of k predictor variables and l criterion variables,

$$e=(x_1,x_2,..x_k;y_1,y_2...y_l)=(x,y).$$

While the process of assigning an element to a cluster is done by the predictor variables, i.e.

$$CC:x \rightarrow n,\ n \in N,$$

the quality of this clustering is evaluated by the criterion variables. The clustering of the elements by the predictor variables is subject to the condition that the clusters should be homogeneous in the criterion variables. Classical cluster analysis searches for densely populated regions in the given attribute space, whereas conditional cluster analysis aims at the prediction of the values of the criterion variables, which may be achieved by searching for dense regions in the criterion space and studying the corresponding regions in the predictor space (see Fig.2). It may even happen that regions separated in the predictor space correspond to similar elements in the criterion space.
In traditional cluster analysis, the criterion used (e.g. minimum within cluster variance) determines the method of partitioning the underlying space. The method of subdividing the predictor space, however, is not necessarily related to the criterion for evaluating the resulting clustering. Each method of pattern recognition may be used in this process. One method of determining the clusters - a rather restricted one - searches for dense regions in the predictor space which correspond to dense regions in the criterion space. These methods find clusters for which the relations between predictor variables and criterion variables are similar. Hence they are especially useful for our algorithm of model construction. Regions not connected in the predictor space will be separated in this case, one of the features we could not take account for in the above model because of the necessary simplicity and feasibility.

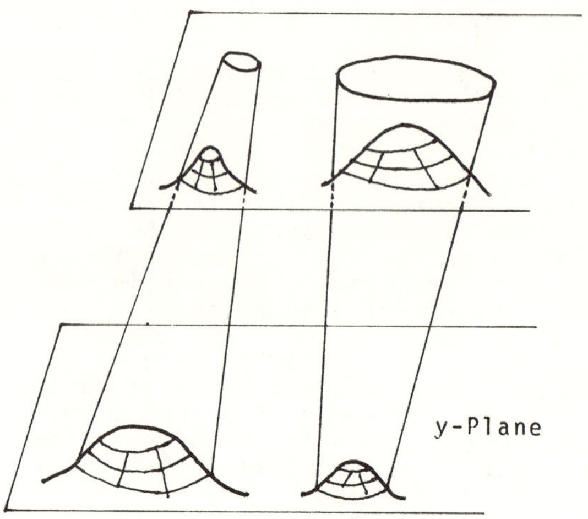

Fig. 2: Conditional Clustering

REFERENCES

ALBRECHT,R.-VISOTSCHNIG,E.: ALGOL-Prozeduren zu den modifizierten Algorithmen nach Minty und Moore. Computing 4(1969), 76-81.
BELLMAN,R.: Dynamic Programming. Princeton Univ.Press: New Jersey 1957.
EDWARDS,A.W.F.,and CAVALLI-SFORZA,L.L.: A method for cluster analysis. Biometrics 21 (1965), 362-375.
GOLUB,G.: Numerical methods for solving linear least squares problems. Numerische Mathematik 7 (1965),3,206-216.
GORDESCH,J.: Multivariate Verfahren in den Sozial- und Wirtschaftswissenschaften. Physica: Würzburg 1972.
JARDINE,N.-SIBSON,R.: Mathematical Taxonomy. Wiley: London 1971.
LERMAN, I.C.: Les bases de la classification automatique. Gauthier-Villars: Paris 1970.
MALINVAUD,E.: Statistical Methods of Econometrics. North Holland:Amsterdam 1966.
SINT,P.P.: Klassifikation und Information. Mitteilungsblatt der österreichischen Gesellschaft für Statistik und Informatik 1 (1971),4,1-9.

STOCHASTIC IDENTIFICATION OF DIGITAL FEEDBACK LOOPS USING A TABLE APPROACH

N. A. Lindberger, University of Alaska
Fairbanks, Alaska 99701, USA

This paper builds on earlier published methods for the estimation of the parameters of discrete models of DDC computer-regulated linear plants, influenced by multiple noise sources of rational spectral densities. The basic method implies the fitting of the autocovariance function of an autoregressive-moving average (ARMA) model to the sample covariance of a record of operating data, using maximum likelihood (ML) estimation. This can be done either by the variation of the parameters of the ARMA model or by using the primary parameters of the assumed plant and noise Z-transforms. This paper discusses i) a table approach for the identification of a set of primary parameters out of a multitude of possible structures, and ii) the risk of obtaining ambiguous results.

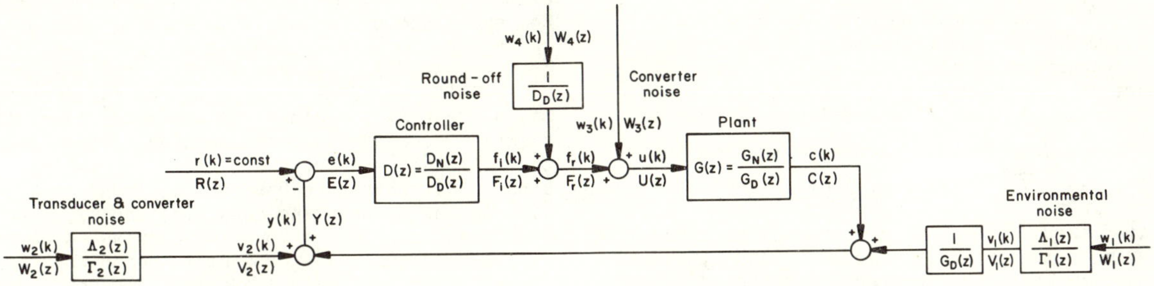

Fig. 1. DISCRETE - TIME MODEL OF COMPUTER - REGULATED PLANT

1. INTRODUCTION

The problem under consideration is the identification by discrete-time models of a linear, computer-controlled process and its linear environment by methods of statistical estimation using natural noise data recorded on-line. The modelling of the noise sources is based on the assumption that they represent autoregressive, moving-average processes, i.e. have rational spectral densities. It was shown in [4] that, if the reference input $r(k)$ is kept constant, the deviation from its mean of the feedback sequence, $\{y(k)\}$, of a single-loop DDC-regulated linear plant satisfies the Z-transform equation:

$$\Gamma_1\Gamma_2(D_NG_N+D_DG_D)Y(z) = \Lambda_1\Gamma_2D_DW_1(z) + \Gamma_1\Lambda_2D_DG_DW_2(z) + \Gamma_1\Gamma_2G_ND_DW_3(z) + \Gamma_1\Gamma_2G_NW_4(z), \tag{1}$$

the quantities of which are defined with reference to Fig. 1. This model, which is the basis for the development of the present paper, presumes four independent white noise sources having transforms $W_1(z),\ldots,W_4(z)$. When translated into the time domain, (1) yields the difference eqn.

$$\sum_{i=0}^{p}\alpha_i y(k-i) = \sum_{\nu=1}^{4}\sum_{j=0}^{r_\nu}\eta_j^{(\nu)}w_\nu(k-j) , \tag{2}$$

which (for any number of noise sources) can be reduced to the form

$$\sum_{i=0}^{p}\alpha_i y(k-i) = \sum_{j=0}^{q}\eta_j w(k-j) \quad ; \quad \alpha_0 = 1 , \tag{3}$$

the autoregressive, moving-average (ARMA) model [5] in the feedback deviation sequence $\{y(k)\}$ (an autoregression) and a unity variance white noise sequence $\{w(k)\}$ (a moving average). The models above have a certain resemblance to the Åström-Bohlin (1965) models referred to in [1],[2], but are more specialized by the fact that a control law connecting $u(k)$ to $y(k)$ is introduced by the controller transform $D(z)$ (see Fig. 1). In another aspect, this model is more general than theirs due to multiple noise sources being allowed.

The identification of the dimension (p,q) of the ARMA model (3) will be presumed in the present context as a basis for the identification of the primary parameters implied in (1),(2), using essentially the same methods of estimation as for the ARMA parameter vectors $\underline{\alpha}$ and $\underline{\eta}$ but maximizing the likelihood function with respect to the primary plant and source parameter vectors $\underline{\kappa}$ and $\underline{\omega}$;

$$\underline{\kappa} = (n_0,\ldots,n_{K_1},d_1,\ldots,d_{K_2},\gamma_{1,1},\ldots,\gamma_{1,p_1},\gamma_{2,1},\ldots,\gamma_{2,p_2}) \tag{4}$$

$$\underline{\omega} = (\sigma_1,\sigma_2,\lambda_{1,1},\ldots,\lambda_{1,q_1},\lambda_{2,1},\ldots,\lambda_{2,q_2}), \tag{5}$$

where the elements are the primary parameters, related to the Z-transforms of Fig. 1 by:

$$G(z) = G_N(z)/G_D(z) = z^{-\tau} \sum_{i=0}^{K_1} n_i z^{-i} \Big/ \sum_{j=0}^{K_2} d_j z^{-j} \quad ; \; d_0=1$$

$$V(z) = \frac{\Lambda_1(z)}{\Gamma_1(z)} W_1(z), \text{ where } \Lambda_1(z) = \sum_{\mu=0}^{q_1} \lambda_{1,\mu} z^{-\mu} ,$$

$$\lambda_{1,0} = 1 \quad ; \quad \Gamma_1(z) = \sum_{\nu=0}^{p_1} \gamma_{1,\nu} z^{-\nu} \quad , \quad \gamma_{1,0} = 1 .$$

There are corresponding expressions for the passing filter Λ_2/Γ_2 of the colored noise source $V_2(z)$ (Λ_1 and Λ_2 may be multiplied by arbitrary integer powers z^{-k}.) The remaining noise sources have been assumed white and of known variance. The unknown variances σ_1^2 and σ_2^2 of $\{w_1(k)\}$ and $\{w_2(k)\}$ are part of $\underline{\omega}$. Assuming a set of integers for τ, $K_1, K_2, p_1, p_2, q_1, q_2$ (the structural parameters) and after having determined a sample covariance function $C_r = \frac{1}{N-r} \sum_{k=1}^{N-r} y(k)y(k+r)$ of the noise-generated sequence $\{y(k)\}$, one can now go ahead and estimate the parameter vectors $\underline{\kappa}$ and $\underline{\omega}$, and subsequently test the structural hypothesis by the Q tests described in [6]. The estimation procedure asymptotically reduces to a minimization of Q, the exponent of the likelihood function, which can be shown to have a chi-square distribution. In the estimation procedure $\underline{\kappa}$ and $\underline{\omega}$ are referred to as the upper and lower parameter vectors, with respect to the regions of the covariance argument r. The upper region is where $r>q$, the lower region is where $r<q$. In the iterative solution of the ML equations, $\underline{\kappa}$ is varied for minimization of the Q pertaining to the upper region, where $\underline{\omega}$ is kept constant, and vice versa. If n_U and n_L are the number of elements of $\underline{\kappa}$ and $\underline{\omega}$, the conditions $n_U \le p$, $n_L \le q$, will allow for estimation and hypothesis testing of the *identifiable* cases. The case of $n_L = q+1$ will be called the *marginal* case; although it does not allow hypothesis testing, it can be made identifiable by various devices. Other cases will be called *non-identifiable*.

2. THE TABLE APPROACH

After having identified the pair (p,q), one wants to find all structures of plant and noise sources that could have given rise to an ARMA model of that dimension. This can be done systematically by means of tables, the generation of which is to be described. The dimensions (p,q) of (3) are related to the structural parameters by:

$$p = p_1 + p_2 + J + max\,(K_2, \tau + K_1) \;, \qquad (6)$$

$$q = J + max(p_2 + q_1, K_2 + p_1 + q_2, K_1 + p_1 + p_2) \;, \qquad (7)$$

where J is the order of the numerator and denominator of the controller transform $D(z)$. The three terms to be maximized in (7) correspond to the noise sources w_1, w_2 and w_3; if any of these is assumed absent, as part of the structural hypothesis, the corresponding term should be deleted. The fourth noise source, w_4, does not enter this expression unless w_3 is absent; however, this possibility has been disregarded due to the insignificant nature of the round-off noise w_4. Under the assumptions: $J \ge 1$, $\tau \ge 1$, $K_1 \ge 0$, $K_2 \ge 0$, it is seen that $p_{min} = 2$, $q_{min} = 1$. This combination will be used for an example in the sequel. To explore what structural hypothesis could have given rise to the $(2,1)$, or any other combination, and knowing for what value of J the operating record was obtained, one proceeds as follows. Starting with the minimum values (1,0,0) for τ, K_1, K_2 one tries to find p_1 and p_2 so as to satisfy (6). For an example, if $J=1$, $p=2$, there is a solution $p_1=p_2=0$. For every one of the possible combinations of p_1 and p_2 one looks for combinations of q_1 and q_2 that satisfy (7), bearing in mind that one or several of the terms subject to the maximization may be deleted. In the $(2,1)$ case one finds the solution $q_1=q_2=0$, i.e. the passing filters for the white noise sources are unity. A set of structural parameters having been found as a solution to the diophantine eqns. (6) and (7), the next step is to write down the corresponding primary parameters. In the example being considered these are n_0, σ_1 and σ_2; one in the upper and two in the lower region. As $n_L = 2 = q+1$, this case will be termed marginal and entered as such in Table 1. By making the assumptions $\sigma_1=0$ or $\sigma_2=0$, n_L can be reduced to 1. These cases are identifiable and marked Yes in Table 1. (The case $\sigma_1=\sigma_2=0$ will be disregarded.)

TABLE 1. Structures yielding ARMA models of dimension (2,1).

Case	J	τ	K_1	K_2	p_1	p_2	q_1	q_2	Cond.	Ident. Class.	Parameters
1	1	1	0	0	0	0	0	0		Marg.	n_0 σ_1 σ_2
2	1	1	0	0	0	0	0	0	$\sigma_1=0$	Yes	n_0 σ_2
3	1	1	0	0 1	0	0	0	0	$\sigma_2=0$	Yes	n_0 (d_1) σ_1

TABLE 2. Structures yielding ARMA models of dimension (2,2).

Case	J	τ	K_1	K_2	p_1	p_2	q_1	q_2	Cond.	Ident. Class.	Parameters
1	1	1	0	0	0	0	0	1		Marg.	n_0 σ_1 σ_2 $\lambda_{2,1}$
2	1	1	0	0	0	0	0	1	$\sigma_1=0$	Yes	n_0 σ_2 $\lambda_{2,1}$
3	1	1	0	0	0	0	1	0		Marg.	n_0 σ_1 σ_2 $\lambda_{1,1}$
4	1	1	0	0	0	0	1	0	$\sigma_2=0$	Yes	n_0 σ_1 $\lambda_{1,1}$
5	1	1	0	0	0	0	1	1		No	n_0 σ_1 σ_2 $\lambda_{1,1}$ $\lambda_{2,1}$
6	1	1	0	1	0	0	0	0		Yes	n_0 d_1 σ_1 σ_2
7	1	1	0	1	0	0	1	0		Marg.	n_0 d_1 σ_1 σ_2 $\lambda_{1,1}$
8	1	1	0	1	0	0	1	0	$\sigma_2=0$	Yes	n_0 d_1 σ_1 $\lambda_{1,1}$

In this manner, one explores all (q_1,q_2) combinations in (7) for every (p_1,p_2) combination found as a solution to (6). The diophantine eqns. (6) and (7) having been solved for one combination of τ, K_1, and K_2, the next step is to increase K_2 by one unit and search for a new solution to (6). In the example being considered, there is again the unique solution $p_1=p_2=0$. One has to search

(7) again for combinations of q_1 and q_2, the value of K_2 being different this time. This gives the remaining case of Table 1 which, having $q_1=q_2=0$, is entered in common with the third case and distinguished by the upper primary parameter (d_1). When trying to raise K_2 to 2, one finds that (6) admits no further solutions. For the example considered the same is true for increases of K_1 and τ. In the general case one increases K_1 by 1 and goes through the whole procedure again, finding all admissible values of K_2, p_1, p_2, q_1 and q_2. This goes on until a limit is found for K_1. Finally, τ is increased by unity and the whole procedure is repeated until a limit is found for τ and the search is complete for one value of J. This algorithm for obtaining the complete solution to (6) and (7) could obviously be computerized. The case $(2,2)$ is presented in Table 2. In Table 3 only the identifiable cases of $(3,3)$ having $\sigma_1, \sigma_2 \neq 0$ have been entered.

TABLE 3. Identifiable structures yielding ARMA models of dimension (3,3).

Case	J	τ	K_1	K_2	p_1	p_2	q_1	q_2	Parameters
1	1	1	0	0	1	0	0	1	$n_0\ \gamma_{1,1}\ \sigma_1\ \sigma_2\ \lambda_{2,1}$
2	1	1	0	1	0	1	0	1	$n_0\ d_1\ \gamma_{2,1}\ \sigma_1\ \sigma_2\ \lambda_{2,1}$
3	1	1	0	0 1	0	1	1	0	$n_0\ (d_1)\ \gamma_{2,1}\ \sigma_1\ \sigma_2\ \lambda_{1,1}$
4	1	1	0	1	1	0	0 1	0	$n_0\ d_1\ \gamma_{1,1}\ \sigma_1\ \sigma_2\ (\lambda_{1,1})$
5	1	1	0	2	0	0	0 1	0	$n_0\ d_1\ d_2\ \sigma_1\ \sigma_2\ (\lambda_{1,1})$
6	1	2 1	0 1	1	0	0	0	1	$n_0\ (n_1)\ d_1\ \sigma_1\ \sigma_2\ \lambda_{2,1}$
7	1	2	0	2	0	0	0 1	0	$n_0\ d_1\ d_2\ \sigma_1\ \sigma_2\ (\lambda_{1,1})$

Cases with $\sigma_1=0$ or $\sigma_2=0$ not included.

The next point of inquiry concerns the number of tables that have to be generated this way, particularly for higher values of J. It is noted that an increase of J by unity will increase p and q by unity, leaving the values of the structural parameters unchanged. Using the notation $J(p,q)$ to classify an ARMA model derived from a process with a J:th order controller, it is noted that the case $(J+1)(p+1,q+1)$ will yield identically the same structural parameter solution as will $J(p,q)$. Therefore, having obtained the table solution for $1(2,1)$, there is no need to analyze $2(3,2)$, $3(4,3)$ etc. However, one must solve $1(2,2)$, $1(2,3)$,...,$1(3,1)$,$1(3,2)$,$1(3,3)$,$1(3,4)$,...,$1(4,1)$, $1(4,2)$,...,etc. The array of tables for $p=P$ will then have the appearance:

$$
\begin{array}{llllll}
1(P,1) & 1(P,2) & 1(P,3)..1(P,P-1) & 1(P,P) & 1(P,P+1)... \\
 & 2(P,2) & 2(P,3)..2(P,P-1) & 2(P,P) & 2(P,P+1)... \\
 & & 3(P,3)..3(P,P-1) & 3(P,P) & 3(P,P+1)... \\
 & & .... \\
 & & (P-1)(P,P-1).................
\end{array}
$$

The first row of these arrays must be generated for each value of $P>2$, $J=1$. For the minimum value $P=2$, the first two entries of the first row are Tables 1 and 2. All the remaining rows, having increasing values of J, can be obtained from the previous array of order $P-1$. For an example, the $(P-1)$st row above is obtained from the $(P-2)$nd row of the array $P-1$ which is obtained from the $(P-3)$rd row of the array $P-2$, etc. The $(P-1)$st row above also is the last one, as $J=(P-1)$ induces the minimum values $p=P$, $q=P-1$.

3. THE AMBIGUITY PROBLEM

The use of the tables in the identification of an entirely unknown plant, subject to unknown disturbances from the environment can be sketched as follows. Using a first order control law (e.g. the digital version of a three-mode controller) the loop under study is tuned to stability from the computer console, and a normal operating record of the feedback variable is obtained while keeping the reference input constant. The record is processed off-line and a sample covariance function is obtained that will serve as an input to the estimation program, which under the hypothesis of an ARMA dimension (p,q) determines the vectors $\underline{\hat{\alpha}}$ and $\underline{\hat{n}}$ and also yields the quadratic forms which are the basis of hypothesis testing. If the natural noise sources influencing the system have rational spectral densities, it will be possible to identify an ARMA model of dimension (p,q) (Chow (1972)[3],Lindberger (1970,1973)[4],[6]). Going to the table $1(p,q)$, one will recognize a number of identifiable cases that could have yielded the ARMA model of dimension (p,q). These are tried by estimation programs using the same sample covariance input as above but maximizing the likelihood function with respect to the parameters listed in the table. Supposing that the hypothesis tests are all negative for the identifiable cases, one could try the marginal cases by one of the following expedients: i) presuming one of the primary parameters as known, e.g. σ_2, which may be interpreted as pertaining to the A/D converter noise variance, ii) artificially increasing the lower region by one unit, thereby gaining the necessary numbers of freedom for a hypothesis test. Alternatively, one can defer these cases to the next stage. This consists in increasing the order of the control law to $J=2$, and obtaining a new sample, thereby increasing p and q by one. The condition $q-n_L \geq 0$ brings about the increase in degrees of freedom sought for in the chi-square tests, making the marginal cases identifiable and some of the non-identifiable cases marginal. If the hypothesis tests are still negative, the increase of J is repeated until all the non-identifiable cases have been brought under the identifiable label and tried.

The problems encountered in this procedure are more likely to be those of obtaining too many model fits rather than none at all. This is the problem of ambiguity in the identification of plant and noise structure [2]. It can be described as follows, with reference to Fig. 2 below.

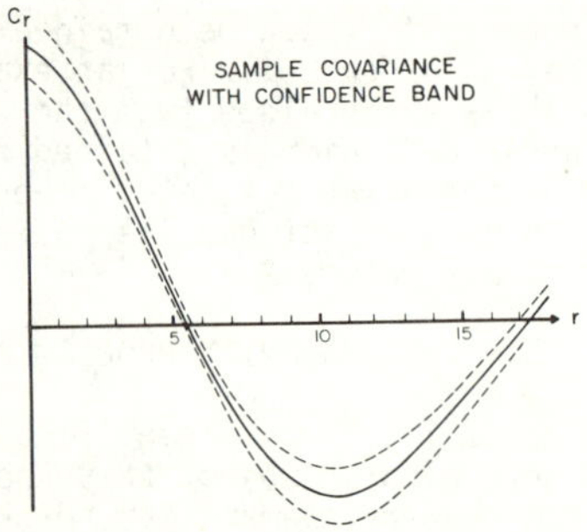

Fig. 2

The sample covariance curve can be visualized as being surrounded by a confidence band, inversely proportional to the square root of the number of observations, within which with a certain probability the true model autocovariance curve will fall. Different models which have their autocovariance curves within this band cannot be expected to be separable, except by increasing the number of observations and narrowing the band. Such models will be called stochastically equivalent. In the presence of many parameters, there is an intrinsic risk of achieving an artificial fit of one model's autocovariance curve within another models confidence band. This may be illustrated by means of the following example, in which one structural model is used for simulating a time series sample which is then identified, deliberately using another configuration of plant and noise sources than the one generating the sample.

Example: Model 1 specifications: $\tau=1$, $K_1=0$, $K_2=1$, $p_1=1$, $q_1=0$, $p_2=q_2=0$, $n_0=.5$, $d_1=-1$, $\gamma_{1,1}=-.8$, $\sigma_1=\sigma_2=.1$, $\sigma_3=\sigma_4=0$. This is a 1st order plant influenced by a first order environmental noise and a white A/D converter noise, the remaining noise sources being zero. Using the digital version of a PI controller $(J=1)$ this structure generates a $(3,3)$ ARMA model.

Model 2 specifications: $\tau=1$, $K_1=0$, $K_2=2$, $p_1=p_2=q_1=q_2=0$. This configuration has five primary parameters n_0' d_1' d_2' σ_1' and σ_2' which were used for ML fitting to a sample covariance curve generated by Model 1, with results as shown in the second row of Table 4 below.

TABLE 4. Computed and Estimated Primary Parameters, Model 2

	n_0'	d_1'	d_2'	σ_1'	σ_2'
Computed	.030	-1.62	.68	.104	.109
Estimated	.031	-1.615	.672	.0994	.1055
Deviations	±.127	± .197	±.069	±.0026	±.0017

Although the chi-square Q-tests give reasonable values, the structural hypothesis of Model 2 must be rejected due to the large deviation of $\hat{n}_0'$. Nevertheless, a covariance curve calculated from the estimates will fall within the dotted lines of Fig. 2. This false fitting can be explained as follows. In the upper region the three parameters $\alpha_1,\alpha_2,\alpha_3$ are uniquely determined by n_0,d_1, and $\gamma_{1,1}$, but also by n_0', d_1', and d_2'. Consequently, the latter three parameters can be solved for and entered into the first row of Table 4. In the lower region, there are three (originally four) conditions to be satisfied by the variation of the two parameters σ_1' and σ_2', but an analysis shows that once the zeroth and first order covariances of the moving average in the RHS of (3) are determined, the 2nd and 3rd order covariance expressions are very nearly satisfied in this case.

It is therefore possible, by a simple calculation, to arrive at an entirely false fit. In the ML estimation procedure, the same false model hypothesis passes the chi-square tests but yields exorbitant variances of the parameter estimates. For a comparison, Table 5 shows the results of an estimation using the right hypothesis.

TABLE 5. True and Estimated Primary Parameters, Model 1.

	n_0	d_1	$\gamma_{1,1}$	σ_1	σ_2
True	.5	-1	-.8	.1	.1
Estimated	.485	-1.019	-.745	.0981	.1007
Deviations	±.039	± .017	±.107	±.0026	±.0016

In summary, no general rule can be given for the separation of stochastically equivalent models, as the negative indication obtained in the present example cannot be guaranteed always to occur. It is recalled that the present objective is the identification of an entirely unknown structure from a record of its feedback variable under the influence of natural noise. With the table approach, there will be ambiguous cases, in the total absence of information of the structure being identified. If such information is available in part, however, the problem can be redefined in less generality and may be amenable to unambiguous solution.

REFERENCES

[1] Åström, K.J., and P. Eykhoff. *System identification - a survey*. Automatica, Vol. 7 (1971), pp. 123-162.

[2] Bohlin, T. *On the problem of ambiguities in maximum likelihood identification*. Automatica, Vol. 7 (1971), pp. 199-210.

[3] Chow, J.C. *On estimating the orders of an autoregressive moving-average process with uncertain observations*. IEEE Trans. Automat. Contr. Vol. AC-17, No. 5 (1972), pp. 707-709.

[4] Lindberger, N.A. *Stochastic estimation and identification of computer-regulated linear plants in a noisy environment*. Diss. Abstr. Int. B, 31, (1970), p. 1967.

[5] ----- *Stochastic modelling of computer regulated linear plants in noisy environments*. Int. J. of Control, (1972), Vol. 16, No. 6, pp. 1009-1019.

[6] ----- *Stochastic identification of computer regulated linear plants in noisy environments*. Int. J. of Control (1973), Vol. 17, No. 1, pp. 65-80.

AN AUTOMATIC MODEL ADJUSTMENT TECHNIQUE

Laurence C. W. Dixon

Numerical Optimisation Centre
The Hatfield Polytechnic - Hatfield
Hertfordshire - The United Kingdom

The problem of fitting a functional form $y=F(z, x)$ to numeric data sets (y_i, z_i) can be separated into two parts. In the first part the functional relationship F must be determined, after this the best values of the parameters x can be determined by optimisation on some measure M of the discrepancy between the model and data. In this paper a strategy for achieving the first objective is outlined. The aim is to select the optimal functional form from a prespecified tree of developable possibilities. The data set (y_i, z_i) is divided into two sets S1 and S2. For any particular functional form F the optimal parameters xo are obtained by optimising a measure M(S1(x)) and the set S2 then provides an independent performance criteria M(S2(xo)) which can be used to help guide the iteration through the tree of possible functional forms. The method is illustrated on the two case studies given by Dr. -Ing. P. Isermann.

1. INTRODUCTION

The identification problem was formulated by Zadeh [9] as:
"Identification is the determination, on the basis of input and output, of a system within a specified class of systems, to which the system under test is equivalent".

Given a set of input vectors z_i and their corresponding outputs y_i (assumed scalar), the choice of a system involves choosing a functional form F and a set of parameters x such that

$$y = F(z, x) \tag{1.1}$$

is a good fit to the data. If we let $y_{mi}^{(k)}(x)$ be the value predicted by the model $F^{(k)}$ for input z_i and parameter values x, then the simplest "Least Squares" measure of fit over a set i S would be given by

$$M^{(k)}(S(x)) = \sum_{i \in S} (y_i - y_{mi}^{(k)}(x))^2. \tag{1.2}$$

Unless otherwise stated this measure of fit will be assumed, though most of the subsequent discussion would apply equally to any alternative definition.

To perform an identification process using Zadeh's definition, we must first define a class of systems. In our approach this will take the form of a tree, it will be a precondition that the tree should be finite and that alternative branches from any node should be defined in advance. In general the earlier models in the tree will be the least complex and tests will be required at each node to determine which if any of the more complex branches is an improvement and also to check whether the existing complexity is necessary. The acceptability tests will be defined in section 2 and the examples of the definition of trees in section 3. The problem as defined above is of course not confined to the sphere of control and a preliminary simpler example will be drawn from another field. The full trees obtained on the case studies proposed by Isermann [7] will then be discussed.

2. ACCEPTABILITY TESTS

The simplest identification problem that can be posed is probably that of fitting a polynomial to a set of data. In this problem the model output is given

$$y_{mi}^{(k)}(x) = \sum_{j=1}^{k} x_j \, z_i^{j-i} \tag{2.1}$$

The problem of fitting polynomials to data has been extensively studied, and standard texts on the subject now exist (e.g. Cheney [3]). In particular the Runge phenomenon is usually quoted which implies that the interpolating polynomial through a set of equidistant data points is often a very bad fit to the function at intermediate points.

Bjork [2] illustrated this point neatly with the example

$$y = 1/(1+25z^2). \tag{2.2}$$

He used 20 equidistant points on the interval (-1, 1) to form his data set S for minimising (1.2) and then introduced an independent measure over 100 equidistant points over the same interval to indicate the accuracy at intermediate points. Whilst the sums of squares of residuals over the minimising set decreased steadily as the order of the polynomial fitted (2.1) increased, the independent measure showed a unique minimum when k=13.

The above phenomenon does not appear to be uniquely associated with the model (2.1) and the basic proposition of this paper is that in any identification problem the data should be divided into two subsets S1 and S2. The optimisation process to estimate the parameters x should then be undertaken on a measure similar to (1.2) over the set S1, whilst the value of a similar measure over the set S2 would be used as a test of acceptability of the modified model.

In constructing a tree of possible models F we will impose a restriction that each new model should be obtainable from a previous model by the addition or deletion of a parameter x_i, in such a way that $x_i=0$ corresponds to the previous model. With this proviso we may then state four conditions for the more complex model to be preferable to the simpler model, namely:

(1) The optimum value of the additional parameter is significantly different from zero. Here we may note that if

$$M = \sum_{i=1}^{m} s_i^2, \qquad (2.3)$$

most optimisation routines will give an estimate of the matrix

$$H = (J^T J)^{-1} \qquad (2.4)$$

at the optimum, where

$$J_{ij} = \frac{\partial s_i}{\partial x_j}. \qquad (2.5)$$

The estimate of the variance of x_i is then given by

$$\sigma_i^2 = H_{ii} M/(m-1) \qquad (2.6)$$

and if

$$x_i < 2(\sigma_i) \qquad (2.7)$$

we are probably justified in concluding that x_i is not significant.

(2) The condition number of the matrix H (2.4) must be less than some bound.

If the condition number of H becomes unbounded then the model cannot theoretically be used to estimate x with any degree of confidence. This will usually prevent the optimisation process from being completed and would hence be applied before test (1) above.

(3) The reduction in the measure M1 should be statistically significant. This is discussed in detail in many references (e.g. Astrom & Eykoff [1] p. 139).

(4) The independent measure M2 should not increase.

The introduction of the measure M2 raises some interesting, as yet unanswered, questions. We have to decide how to divide the available data points between sets S1 and S2. Bjork's [2] allocation of only one fifth of the points to S1 would seem rather extreme for general purposes and in the tests reported here the points have been divided equally. Again the best distribution of the points between the two sets would probably vary with the purpose of the identification, as might the form of the measure M. If the identification is to be used to forecast future events from a time series the distribution and weight given to the points in M2 could be biased to reject models that gave bad estimates to some data points at the most recent end of the series.

3. TREES

Let us now consider the canonical form of a linear time invariant, discrete time model with one input and one output. This is given by Astrom & Eykoff [1] p. 136 as

$$\begin{aligned} & y_{m,i} + a_1 y_{m,i-1} + \dots + a_n y_{m,i-n} \\ & = b_0 u_i + b_1 u_{i-1} + \dots + b_n u_{i-n}. \end{aligned} \qquad (3.1)$$

With this problem the simplest model used as the base contains only a_1 and b_0

$$\text{i.e. } y_{m,i} + a_1 y_{m,i-1} = b_0 u_1. \qquad (3.2)$$

This model was chosen because the optimisation algorithm used, a modified Gauss-Newton approach described by Dixon & Biggs [4], can only be used when n, the dimension of x, is at least 2. The additional parameters are then brought in one at a time and the tests of acceptability described above were applied. In addition when at any test an earlier parameter failed to satisfy the significance test, this was deleted and the modified model run.

Two examples of trees built up in this way will now be described. These correspond to the two original case studies proposed by Isermann [7]. The modification proposed in [8] has not been tested. Both these case studies are of the form (3.1) and the identification study was to isolate those elements of a_i, b_i that were non-zero. Both the studies were undertaken using an input signal 700s in length and the 350 data points were split equally between S1 and S2, the first 175 being allocated to S1, the second to S2. The identification studies were run with a noise/signal ratio of 0.1 and the input signal shown in Fig. 1. As the example II was the more satisfactory and straight forward this will be described first.

Case Study II

The measures M used were the direct sums of squares of the residual errors (1.2) over both sets. The data, i.e. the parameters a, b obtained and the corresponding values of M1 and M2, are given in Table 1. An underlined parameter value indicates that it does not satisfy the significance test 1.

Model 2 (which includes a_2), is an improvement over Model 1.
Model 3 (which includes b_1), is an improvement over Model 2.
Model 4 (which includes a_3), fails on Measure M2.
Model 5 (which includes b_2 but not a_3) fails 3 times in test 1 and the first such parameter is dropped.
Model 6 (which includes b_2 but not b_0) is an improvement over models 3 and 5.
Model 7 (which includes a_3) fails on test 1 compared to Model 6.
Model 8 (which includes b_3) fails on test 1 compared to Model 6.

Table 1

k	a_1	a_2	a_3	b_o	b_1	b_2	b_3	M1	M2
0	-	-	-	-	-	-	-	19.0	23.88
1	-0.93	-	-	.089	-	-	-	5.24	4.135
2	-1.72	0.75	-	.027	-	-	-	3.44	3.255
3	-1.55	0.60	-	.115	.166	-	-	1.88	2.025
4	-2.18	1.70	.49	-.06	.102	-	-	1.55	2.590
5	-1.37	0.45	-	-.03	-.04	.01	-	1.68	1.833
6	-1.35	0.42	-	-	-.12	.20	-	1.71	1.730
7	-1.40	0.52	.04	-	-.11	.11	-	1.71	1.494
8	-1.31	0.39	-	-	-.11	.18	.01	1.71	1.611

The logic has successfully identified Model 6, the correct form for this problem. The tree built by the iterative process is shown as figure 2, where the rejection of b_o in model 5 is shown to be crucial to the success of the search.

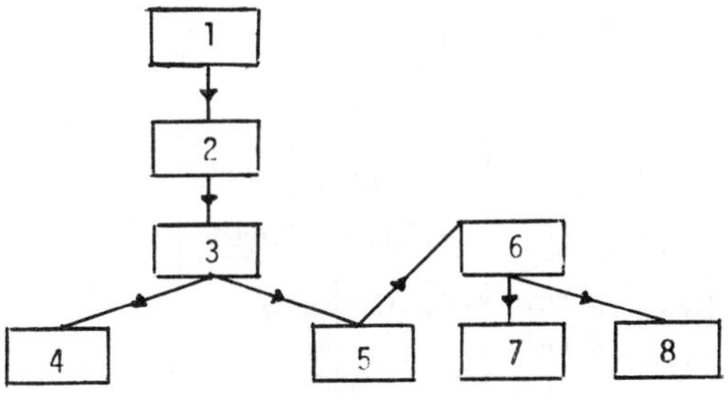

Fig. 2: Isermann Example II

Having identified the form of the model the parameter estimation problem must now be solved more accurately. As the measure (1.1) being used is non-quadratic and the dimension of the problem could be reduced from 4 to 3 by using the fact that the output of this system due to a unit step at t=0 is given by

$$y(t) = K\left(1 + \frac{2T_1}{(T_2-T_1)} e^{-\frac{1}{T_1}t} - \frac{(T_1+T_2)}{(T_2-T_1)} e^{-\frac{1}{T_2}t}\right)$$

the parameter estimations were performed in terms of $x^T = (K, T_1, T_2)$. These are reported in more detail in Dixon [5].

Tests were run with signals of length 175, 750, 3400 and the short signal (175 units) was replicated 9 times, in each test the appropriate value of the required parameters $V=(a_1,a_2,b_1,b_2)$ was calculated and also their estimated variance. These were obtained by the standard statistical approach on the signal replicated 9 times and by using the formula

$$\sigma_V^2 = \frac{M1}{m-1} (\nabla V^T H \nabla V)$$

on the single sets. These results are summarised on the following table where the error and standard error in each parameter are given.

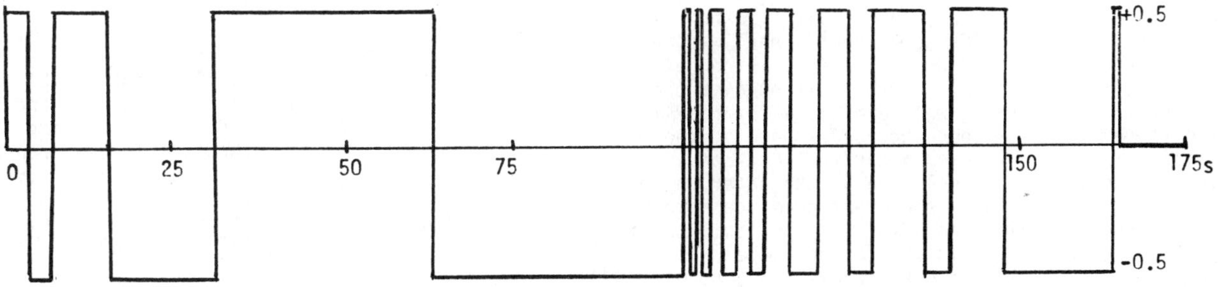

Fig. 1: The Input Signal.

Table 2

	m=175	m=750	m=3400	Replication
Δa_1	0.0043	0.0146	0.0063	0.005
σa_1	±0.06	±0.03	±0.014	±0.02
Δa_2	-0.0024	-0.0112	-0.0035	-0.003
σa_2	±0.05	±0.026	±0.010	±0.019
Δb_1	+0.0011	-0.0044	-0.007	-0.0002
σb_1	±0.006	±0.004	±0.001	±0.014
Δb_2	+0.0016	-0.0032	-0.0012	+0.001
σb_2	±0.009	±0.004	±0.002	±0.016

It is also of interest to note that in order to perform the run with m=3400 data inputs a variable metric algorithm was used instead of the generalised least squares algorithm for storage reasons. That used was that discussed by Dixon [6] and the relative performance is shown below. In this Table IT denotes number of iterations, EFE=No. of function evaluations, SEC=SEC. System 10 CPU time.

Table 3

M	VARIABLE METRIC			GENERALISED LEAST SQUARES		
	IT	EFE	SEC	IT	EFE	SEC
40	12	75	8.3	8	39	5.1
175	11	75	22.9	8	41	16.8
750	10	66	85.3	8	41	74.5
3400	11	72	351.2	OUT OF CORE		

It is interesting to note that the specialised least squares algorithm is not significantly faster than the general purpose optimiser on this problem.

Case Study I

The first case study proposed by Isermann was tackled in an identical manner, except that the Quadratic Measure, Astrom & Eykoff [1] p. 136, was used instead of (1.2) to simplify the numerical optimisation. The same input signal was initially used as before, but is in fact inadequate to identify the system. It was felt however that an unsuccessful tree might be of more interest to the reader than a second successful one, so the initial results have been included. Lack of space precludes the inclusion of the later set of results. The results are presented in the same way as before in Table 4 and Fig. 3. The last row of Table 4 corresponding to the correct model was not found and would, as were models 6-10, have been rejected compared to the simple model 5.

Table 4

k	a_1	a_2	a_3	b_0	b_1	b_2	M1	M2
0	-	-	-	-	-	-	522.	989.
1	0.99	-	-	-0.13	-	-	19.	23.
2	1.89	-0.93	-	-0.10	-	-	2.62	2.27
3	1.89	-0.94	-	-0.04	0.04	-	2.58	2.25
4	2.69	-2.55	0.85	+.005	-	-	0.691	0.704
5	2.69	-2.54	0.85	-	-	-	0.692	0.706
6	2.69	-2.54	0.85	-	.009	-	0.688	0.707
7	2.69	-2.54	0.85	-	-	.004	0.691	0.704
				b_3	b_4	b_5		
8	2.69	-2.54	0.85	+.001	-	-	0.692	0.705
9	2.69	-2.54	0.85	-	-.006	-	0.691	0.710
10	2.70	-2.56	0.86	-	-	-.017	0.680	0.722
*	2.69	-2.55	0.85	+.012	+.006	-.028	0.673	0.708

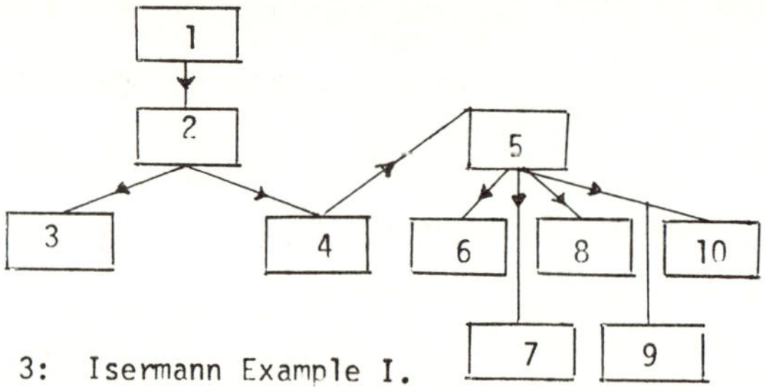

Fig. 3: Isermann Example I.

4. REFERENCES

1. K. J. Astrom and P. Eykoff: System Identification - A Survey. Automatica Vol. 7 pp. 123-168 (1971).
2. H. Bjork: Contribution to the problem of least squares approximation. Royal Institute of Technology, Stockholm Report NA 71.37(1971).
3. E. W. Cheney: Introduction to Approximation Theory. McGraw Hill (1966).
4. L. C. W. Dixon & M. C. Biggs: The advantages of adjoint-control transformation when determining optimal trajectories by Pontryagin's Maximum Principle. The Aeronautical Journal (1972) pp. 169-173.
5. L. C. W. Dixon: Parameter Estimation: The Hague Test Case A II. The Numerical Optimisation Centre, The Hatfield Polytechnic TN.21.
6. L. C. W. Dixon: In F. Lootsma Editor "Numerical Methods for Nonlinear Optimisation" (1972).
7. R. Isermann: Test Case A for evaluation and comparison of different identification and parameter estimation methods.(this conference).
8. R. Isermann: Supplement to Test Case A.(this conference).
9. L. A. Zadeh: From Circuit Theory to System Theory. Proc. IRE 50, 856-865 (1962).

IDENTIFICATION OF COMPLEX SYSTEMS IN TERMS OF REDUCED MODELS

F. Donati - E. Canuto

Gruppo Automatica e Informatica - I.E.N.G.F. - Politecnico di Torino - Italy

This work is concerned with the approximation of systems by means of models within a specified class on the basis of input and output. Discrete linear time-invariant systems are considered. Output data are assumed corrupted by a noise, unknown except for a bound on its energy. A guaranteed cost criterion of approximation is used.

1. INTRODUCTION

This paper is concerned with the determination, on the basis of input and output, of a model within a specified class of models, by which the system under test is approximated in the best way according to a given guaranteed cost criterion.
An algorithm is derived under the following conditions.

1. The system under test is discrete linear time-invariant, with finite memory of known length. Single-variable systems are considered.
2. The input data are noiseless. The output data are corrupted by a noise, unknown except for a bound on its energy.
3. The class of models is taken as the class of discrete linear time-invariant systems, whose weighting sequences belong to a linear subspace spanned by a given basis.
4. The approximation criterion is expressed as a functional of the error between the system and model weighting sequences:

$$J = \sum_{i=0}^{m-1} (h(i) - g(i))^2, \qquad (1)$$

where: $h(i)$ is the system weighting sequence; $g(i)$ is the model weighting sequence; m is the length of the system and model memory.

Since $h(i)$ is determined by system data with uncertainty, for a given model $g(i)$ the functional J can only be evaluated with uncertainty. Then a guaranteed cost criterion is introduced and the "best" approximant model is defined such that it minimizes the maximum admissible value of J within the uncertainty limits.

Remarks - The functional J has been chosen in order to get an approximation in norm [1, 2, 3, 4]. Indeed, if $\hat{J}$ is the guaranteed value of J which is determined by the identification procedure, the error $e(i)$, between the system zero-state response and the identified model one, satisfies the following inequalities, whatever the input $u(i)$ is belonging to l^1 or l^2

$$\sum e^2(i) \le \hat{J} (\sum |u(i)|)^2, \ \forall u \in l^1 \qquad (2)$$

$$(\max_i |e(i)|)^2 \le \hat{J} \sum u^2(i), \forall u \in l^2 \qquad (3)$$

This work has been developed along the following line. An algorithm is introduced to process system input-output data in order to identify the weighting sequence set, whose members model systems which agree with the data and the "a priori" knowledge of the system under test (Section 2). This set is derived by the same computational algorithm which is followed in the least squares identification of an impulse response.

In Section 3 the model which minimizes the guaranteed value of the functional J is determined. A "one-shot" computational algorithm is given, which follows two different lines according as the functional J defined by (1) has a saddle point or only a min-max point, when $h(i)$ is varying within the uncertainty set and $g(i)$ within the class of models.

This method can be extended to multivariable systems. Here single-variable systems are considered for the sake of simplicity.

2. IDENTIFICATION OF THE WEIGHTING SEQUENCE SET

The system input is denoted by $u(i)$ and the resulting output by $y_0(i)$. It is assumed that $u(i)$ is known accurately, but that $y_0(i)$ is obscured by additive noise $d(i)$, which is unknown except for a bound on its energy. Therefore there is available for analysis at the output only

$$y(i) = d(i) + y_o(i) \qquad (4)$$

The following assumptions and notations

are introduced

1. The unknown system weighting sequence $h(i)$ is equal to zero for $i > m$, where m is given. The m-dimensional vector h is used to denote both $h(i)$ and whatever estimation it may have.
2. $u(i)$ is observed for $-m < i < n$, where $n > m$. It is rearranged in the matrix

$$U = \begin{bmatrix} u(o) & u(-1) & \dots & u(-m+1) \\ u(1) & u(o) & \dots & u(-m+2) \\ \vdots & \vdots & & \vdots \\ u(n-1) & u(n-2) & \dots & u(n-m) \end{bmatrix} = [u_1 \; u_2 \dots u_m] \qquad (5)$$

The vectors $u_1, u_2, \dots, u_m$ are linearly independent.

3. $y(i)$ is observed for $o < i < n$. The n-dimensional vector y is used to denote it.
4. The unknown noise $d(i)$ satisfies the bound

$$\sum_{i=1}^{n-1} d^2(i) \leq \rho^2 \qquad (6)$$

5. The vector y_o is used to denote both the unknown output $y_o(i)$, for $o < i < n$, and whatever estimation it may have.

With the above assumptions there results:

$$y_o = Uh \qquad (7)$$

$$(y - y_o)' (y - y_o) \leq \rho^2 \qquad (8)$$

Any y_o and h which satisfy (7),(8) are estimations of the unknown output and weighting sequence respectively.

Let $\hat{y}$ be the orthogonal projection of y on the subspace spanned by the basis $u_1, u_2, \dots, u_m$.

$$\hat{y} = U (U'U)^{-1} U'y \qquad (9)$$

From the projection theorem and from (7) there results

$$y_o' (y - \hat{y}) = 0 \qquad (10)$$

and from (8)

$$(\hat{y} - y_o)' (\hat{y} - y_o) \leq \rho^2 - (y - \hat{y})' (y - \hat{y}) \qquad (11)$$

Let us, now, denote

$$h_o = (U'U)^{-1} U'y = (U'U)^{-1} U' \hat{y} \qquad (12)$$

$$\delta^2 = \rho^2 - (y - \hat{y})' (y - \hat{y}) \qquad (13)$$

$$\Sigma = U'U \qquad (14)$$

Substitution of (7) into (11) yields

$$(h - h_o)' \Sigma (h - h_o) \leq \delta^2 \qquad (15)$$

This inequality defines the set H of the estimations h of the unknown system weighting sequence. H is a bounded convex set contained in the m-dimensional real vector space R^m.

<u>Remark 1</u> - If the vectors $u_o, u_1, \dots, u_m$ were not linearly independent, the matrix Σ would be singular and the set H not bounded.

<u>Remark 2</u> - If there results

$$(y - \hat{y})' (y - \hat{y}) > \rho^2 \qquad (16)$$

it means that the assumptions are not correct.

2. MINIMUM GUARANTEED COST MODEL

The class of models is taken as the class of discrete linear time-invariant finite-memory systems, whose weighting sequences g belong to a linear subspace $M \subset R^m$ spanned by a given basis

$$B = [b_1 \; b_2 \dots b_q], \qquad (17)$$

where B is a $m \times q$-dimensional matrix with $q < m$. There results

$$M = \left\{ g; \; g = Bx, \; x \in R^q \right\}, \qquad (18)$$

where x is the q-dimensional vector of the unknown parameters of the model.

The purpose of this section is to determine two vectors, $\hat{g} \in M$ and $\hat{h} \in H$, such that

$$(\hat{g} - \hat{h})' (\hat{g} - \hat{h}) = \min_{g \in M} \max_{h \in H} (g - h)'(g - h) \qquad (19)$$

where $(g - h)'(g - h)$ is the functional J previously defined by (1); M is the class of models, defined by (18); H is the set, defined by (15), of estimations h of the unknown system weighting sequence.

Since $\hat{g} = B\hat{x}$, we can directly determine $\hat{x}$, i.e. the best model parameters, by substitution of this expression into (19).

It is proved that the solution exists and is unique. Expressions for $\hat{x}$ and $\hat{h}$ are given.

In order to obtain the final result let us claim and prove the following theorems.

<u>THEOREM I</u> - For any fixed $g \in R^m$ a maximum value $\hat{J}$ of the functional J exists, where

$$\hat{J} = \max_{h \in H} (g-h)'(g-h) \tag{20}$$

A vector $\hat{h} = \hat{h}(g)$ which maximizes the functional J satisfies the following necessary and sufficient conditions

$$(\hat{h}-h_o)' \Sigma (\hat{h}-h_o) = \delta^2 \tag{21}$$

$$g = \hat{h} + \lambda \Sigma (\hat{h}-h_o) \tag{22}$$

$$\lambda \leq -d_m^{-1} \tag{23}$$

where d_m is the smallest eigenvalue of Σ and h_o is defined by (12). $\hat{h}$ is not necessarily unique.

Proof - First we prove (21) and (22) are necessary conditions and the same are sufficient conditions when (23) holds.

Then existence is proved. The conditions are derived, under which $\hat{h}$ is not unique, but two different maximums $\hat{h}_1$ and $\hat{h}_2$ exist.

Necessity - Eq. (21) states $\hat{h}$ must belong to the boundary of H. Its necessity is clear and the relative proof is omitted.

To prove the necessity of (22) it is sufficient to remark that

$$\operatorname{Grad}_h (h-h_o)' \Sigma (h-h_o) \Big|_{h=\hat{h}} = 2\Sigma(\hat{h}-h_o) \tag{24}$$

Therefore (22) states $(g-\hat{h})$ must have the same direction as the gradient in $\hat{h}$.

Sufficiency - To prove (21) and (22) are sufficient conditions when (23) is verified, we shall show that there results

$$(h-g)'(h-g) \leq (\hat{h}-g)'(\hat{h}-g), \forall h \in H, \tag{25}$$

if and only if (23) holds.

From (22), the above inequality can be written as

$$(h-\hat{h})'(h-\hat{h}) + 2(\hat{h}-h)'\lambda\Sigma(\hat{h}-h_o) \leq 0. \tag{26}$$

Let us now rewrite (15)

$$(h-\hat{h})'\Sigma(h-\hat{h}) - 2(\hat{h}-h)'\Sigma(\hat{h}-h_o) + + (\hat{h}-h_o)'\Sigma(\hat{h}-h_o) \leq \delta^2. \tag{27}$$

From (21) and (27), it results

$$2(\hat{h}-h)'\Sigma(\hat{h}-h_o) \geq (h-\hat{h})'\Sigma(h-\hat{h}). \tag{28}$$

Now, using (26) and (28), inequality (25) can be written as

$$(h-\hat{h})'(I+\lambda\Sigma)(h-\hat{h}) \leq 0 \tag{29}$$

The last inequality holds for any $(h-\hat{h})$ if and only if the matrix $(I+\lambda\Sigma)$ is negative semidefinite. Since Σ is a positive definite matrix, it follows that λ must be less or equal to $-d_m^{-1}$, where d_m is the smallest eigenvalue of Σ.
This proves the first part of the theorem.

Existence - First the conditions for the existence of a vector $\hat{h}$ which satisfies (21), (22) with $\lambda < -d_m^{-1}$ (Nonsingular Case) are investigated. It is proved that if a solution exists, it is unique. On the contrary, if a solution does not exist for $\lambda < -d_m^{-1}$, it is proved that two solutions $\hat{h}_1$, $\hat{h}_2$ exist for $\lambda = -d_m^{-1}$ (Singular Case).

Nonsingular Case. The matrix $(I+\lambda\Sigma)$ is not singular and (22) can be rewritten as

$$(\hat{h}-h_o) = (I+\lambda\Sigma)^{-1}(g-h_o) \tag{30}$$

Now the substitution of (30) into (21) yields

$$(g-h_o)'(I+\lambda\Sigma)^{-2}\Sigma(g-h_o) = \delta^2 \tag{31}$$

If this equation has solutions $\lambda_i < -d_m^{-1}$, esistence is proved.

Now we prove that if (31) has a solution for $\lambda < -d_m^{-1}$, it is unique; if (31) has no solutions for $\lambda < -d_m^{-1}$, then (21), (22) are satisfied for $\lambda = -d_m^{-1}$.

The left hand side of (31) is a function of λ, which takes only positive values and is singular for $\lambda = -(d_i)^{-1}$, where d_i are the eigenvalues of Σ. Now for $\lambda < -(d_m)^{-1}$ the function is monotonically crescent; therefore if (31) has a solution, it is unique. Since the m-th eigenvalue $d_m(1+\lambda d_m)^{-2}$ of $\Sigma(I+\lambda\Sigma)^{-2}$ tends to infinity as λ tends to $-d_m^{-1}$, the left hand side of (31) tends to infinity as λ tends to $-d_m^{-1}$, if the component of $(g-h_o)$ along the m-th eigendirection t_m of Σ is not zero. Then, if

$$(g-h_o)'t_m \neq 0 \tag{32}$$

a solution exists and is unique.

If there results

$$(g-h_o)'t_m = 0 \tag{33}$$

(31) can be written as

$$(g-h_o)T_r(I+\lambda D_r)^{-2}D_r T_r'(g-h_o) = \delta^2 \tag{34}$$

where T_r is the m x (m - 1) matrix of the

first m - 1 eigenvectors of Σ

$$T_r = [t_1 \; t_2 \ldots t_{m-1}], \tag{35}$$

and D_r is the diagonal matrix of the first m - 1 eigenvalues of Σ.

Let (33) now hold; a solution exists and is unique if the left hand side of (34) computed for $\lambda = -dm^{-1}$ is greater than δ^2.

Finally let us define the set C of the vectors $g \in R^m$, which satisfy (33) and for which the left hand side of (34) computed for $\lambda = -dm^{-1}$ is smaller than δ^2.

Thus, it has been proved that for any $g \in R^m$, $g \notin C$, a solution exists and is unique.

Singular Case - Let us now consider the vectors $g \in C$.

Taking into account (33) and the fact that the eigenvalue corresponding to the eigen vector t_m of $(I - d_m^{-1}\Sigma)$ is zero, (22) can be written as

$$T_r'(g - h_o) = (I - d_m^{-1} D_r)\, T_r'(\hat{h} - h_o), \tag{36}$$

where the above equation is true, whatever $(h - h_o)'t_m$ may be.

From (36), Eq. (21) becomes

$$(g - h_o)'T_r (I - d_m^{-1} D_r)^{-1} D_r T_r'(g - h_o) = \delta^2 - d_m [(\hat{h} - h_o)'t_m]^2. \tag{37}$$

Since by definition the left hand side of (37) is smaller than δ^2 for any $g \in C$, (37) always has a solution for a suitable value of $[(\hat{h} - h_o)'t_m]^2$.

Since $\hat{h}$ is defined by (36), except for its component $(\hat{h} - h_o)'t_m$, and this component is known from (37) except for its sign, it follows that two solutions $\hat{h}_1, \hat{h}_2$ exist and they are different only because of the sign of the component along the eigen direction t_m.

Let us, now, derive the expression of $\hat{J}$. If $g \notin C$, from (22) it follows

$$\hat{J} = (g - h_o)'(1 + \lambda\Sigma)^{-2} \lambda^2 \Sigma^2 (g - h_o) \tag{38}$$

where λ is given as a solution of (31).

A more general expression of $\hat{J}$, which holds for any $g \in R^m$, can be derived from (31). There results

$$\hat{J} = (g - h_o)'T_r (I + \lambda D_r)^{-2} (\lambda^2 D_r^2 - d_m^{-1} D_r) \cdot T_r'(g - h_o) - \delta^2 d_m^{-1}, \tag{39}$$

where $\lambda = -d_m^{-1}$ if $g \in C$

From (39) it follows that $\hat{J}$ takes the same value for the two solutions $\hat{h}_1$, $\hat{h}_2$, if $g \in C$.

Thus the proof of Theorem I is complete.

COROLLARY I - If Euclidean metric is defined on R^m, $\hat{J}$ and λ are continuous functionals on R^m; moreover, $\hat{J}$, λ and $\hat{h}$ are continuously differentiable functionals on $R^m - C$.

Proof - When the continuity of λ on R^m and its differentiability on $R^m - C$ are proved, the above claimed properties of $\hat{J}$ and $\hat{h}$ derive respectively from (39) and (22).

For $g \in (R^m - C)$ the continuity and differentiability of λ derive from (31); then to prove the corollary completely it is sufficient to prove that when $g \notin C$ tends to $g_o \in C$, then λ tends to $-d_m^{-1}$. The proof is very easy and is not given.

COROLLARY II - If $\hat{g}$ belongs to the class M of models (18), $\hat{J}$ is a continuous functional on the Euclidean space E^q of the unknown parameters x. Moreover, $\hat{J}$ admits at least a minimum.

Proof - The first part of Corollary II is a direct consequence of Corollary I. The second part derives from the consideration that $\hat{J}$ is always positive and tends to infinity when $x'x$ tends to infinity.

THEOREM II - If a point of min-max of J exists for $\hat{g} \in M$, $\hat{g} \notin C$ and $\hat{h} = \hat{h}(\hat{g})$, then the pair $\hat{g}$, $\hat{h}$ satisfies the inequality

$$(\hat{g} - h)'(\hat{g} - h) \leq (\hat{g} - \hat{h})'(\hat{g} - \hat{h}) \leq \leq (g - \hat{h})'(g - \hat{h}); \quad g \in M, \; h \in H, \tag{40}$$

which defines a saddle point.

Proof - If $\hat{g} = B\hat{x}$, $\hat{g} \notin C$, is a minimum of $\hat{J} = (g - \hat{h}(g))'(g - \hat{h}(g))$, from differentiation of J, there results

$$d\,x'B\,(B\hat{x} - h) = d\hat{h}'(B\hat{x} - \hat{h}) \tag{41}$$

where on the basis of Corollary I $d\hat{h}$ exists and is infinitesimal, however infinitesimal dx is.

But from the property that $\hat{h}$ is the maximum of J for $g = \hat{g}$, there follows (Theorem I) that $B\hat{x} - \hat{h}$ is orthogonal to the boundary of H in $\hat{h}$. Then considering that $d\hat{h}$ has to belong to the boundary of H, there results

$$d\hat{h}'(B\hat{x}-\hat{h}) = 0 \qquad (42)$$

and

$$dx'B(B\hat{x}-\hat{h}) = 0\ ; \qquad (43)$$

and, since dx' can have any value, it follows

$$B(B\hat{x}-\hat{h}) = 0 \qquad (44)$$

This equation expresses the orthogonality condition of $(B\hat{x}-\hat{h})$ to the subspace M. Then from the orthogonal projection theorem Equation (40) and Theorem II follow.

THEOREM III - If a saddle point of J exists for $\hat{g}=B\hat{x}$ and $\hat{h}$, this is the unique point of min-max.

Proof - The theorem is proved by contradiction, always denoting by $\hat{h}(g)$ the vector $h \in H$, which maximizes J for a given g. Let $\hat{g}$, $\hat{h}(\hat{g})$ be a saddle point and g_1, $h(g_1)$ a min-max point; for the property of a saddle point there results

$$J(\hat{g}, \hat{h}(\hat{g})) \le J(g_1, \hat{h}(\hat{g})) \le J(g_1, \hat{h}(g_1)) \qquad (45)$$

and for the min-max property

$$J(g_1, \hat{h}(g_1)) \le J(\hat{g}, \hat{h}(\hat{g})). \qquad (46)$$

From (45) and (46) it follows

$$J(\hat{g}, \hat{h}(\hat{g})) = J(g_1, \hat{h}(\hat{g})) = J(g_1, \hat{h}(g_1)) \qquad (47)$$

In order that (47) is true, it is necessary that $g_1-\hat{h}(\hat{g})$ is orthogonal to the boundary of H in $\hat{h}(\hat{g})$, what is not possible, since by assumption $\hat{g}-\hat{h}(\hat{g})$ is orthogonal to the boundary of H in $\hat{h}(\hat{g})$ and $\hat{g} \neq g_1$, with $\hat{g}$ and $g_1 \in M$.

This proves the theorem.

COROLLARY III - If there is a saddle point of J for $\hat{g} \in M$, $\hat{g} \notin C$, then there is no min-max point for $g \in C$, $g \in M$.

COROLLARY IV - If there is no saddle point of J for $g \in M$, $g \notin C$, then there is a min-max point for $g \in M$, $g \in C$.

This Corollary follows from Corollary II.

On the basis of Corollaries III and IV we seek, at first, whether the problem has a solution for any $g \in M$, $g \notin C$. If such a solution does not exist, we seek the solution on $M \cap C$, which in this case cannot be empty.

If a solution exists on $M - M \cap C$, the pair $\hat{g}$, $\hat{h}$ has to satisfy (18), (21), (22), (44) with $\lambda < -d_m^{-1}$.

From (18), (22) and (44) there results

$$\hat{x} = \left[B'\Sigma(1+\lambda\Sigma)^{-1}B\right]^{-1}B'\Sigma(I+\lambda\Sigma)^{-1}h_o =$$

$$= A(\lambda)\, h_o \qquad (48)$$

$$\hat{h} = (I+\lambda\Sigma)^{-1}(A(\lambda)+\lambda\Sigma)h_o = D(\lambda)\, h_o. \qquad (49)$$

Then (21) becomes

$$h_o'(D(\lambda)-I)'\Sigma(D(\lambda)-I)\, h_o = \delta^2 \qquad (50)$$

Thus the problem is reduced to seeking the solution of a nonlinear equation (50) in the unknown parameter $\lambda < -d_m^{-1}$. It is a simple matter from a numerical viewpoint, since it is known that the function is monotonically crescent.

When (50) has no solution in the field $\lambda < -d_m^{-1}$, from Corollary IV, a min-max point exists on $M \cap C$. Then, imposed

$$\lambda = -d_m^{-1} \quad \text{and} \quad g = Bx \qquad (51)$$

the min-max point is determined by minimization of $\hat{J}$, given by (39).

There results

$$\hat{x} = (B'T_r(I-d_m^{-1}D_r)^{-1}D_rT_r')^{-1}\cdot$$

$$\cdot B'T_r(I-d_m^{-1}D_r)^{-1}D_rT_r'h_o \qquad (52)$$

and from (36), (37)

$$T_r'\hat{h} = (I-d_m^{-1}D_r)^{-1}(T_r'B\hat{x}-d_m^{-1}D_rT_r'h_o) \qquad (53)$$

$$t_m'\hat{h} = \pm\Big[d_m^{-1}\delta^2 - (B\hat{x}-h_o)'T_r(I-d_m^{-1}D_r)^{-1}\cdot$$

$$\cdot d_m^{-1}D_rT_r'(B\hat{x}-h_o)\Big] \qquad (54)$$

4. CONCLUSION

Systems described by partial data are approximated by models belonging to a given class, according to a guaranteed cost criterion.

Input and output are processed in order to get a class of systems which agree with the "a priori" knowledge and the input-output data of the system under test. This class is described in terms of Property (15), which states a constraint on system weighting sequences. Let us point out that (15) is expressed by means of the weighting sequence h_o identified by the least-squares method.

A model belonging to the class M defined by (18) is determined such that the guaranteed value $\hat{J}$ of the functional J defined by (1) is minimum. The parameters $\hat{x}$ of the optimal model and the weighting sequence $\hat{h}$, which expresses the worst case within the uncertainty limits, are determined by (48) and (49), when (50) has solution $\lambda < -d_m^{-1}$. On the contrary case $\hat{x}$ and $\hat{h}$ are given respectively by (52) and by (53) and (54).

The value $\hat{J}$ of the functional (1) computed for $g = B\hat{x}$ and $h = \hat{h}$ is such to guarantee the error limits stated in (2) and (3), whatever the input and the observation length may be.

5. ACKNOWLEDGMENT

This work was supported in part by C.N.R. and NATO Grant n. 524.

6. REFERENCES

1 F. Donati: "Approssimazione di sistemi lineari in spazi normati" - XI International Automation and Instrumentation Conference, FAST, Milano 1970.

2 M. Milanese: "Identification of uniformly approximating models of systems", Ricerche di Automatica, July 1971.

3 F. Donati: "A dynamic upper bound of the uncertainty in linear system approximation" , USA - Italian Seminar on "Theory and Applications on Variable - Structure Systems", Sorrento, April 1972.

4 M. Milanese, A. Negro: "Uniform approximation of systems. A Banach space approach", JOTA, April 1973.

MODELING A PARTICULAR CLASS OF MULTIPLE-INPUT/MULTIPLE-OUTPUT BLACK BOXES WITH STOCHASTIC INTEGRAL EQUATIONS AND IDENTIFYING THE REQUIRED PARAMETERS**

Earl D. Eyman, Thomas H. Kerr*, and Nan K. Loh

Department of Electrical Engineering
University of Iowa, Iowa
U.S.A.

A method is given for obtaining a mathematical model of a class of black boxes having multiple inputs and multiple outputs in terms of Ito stochastic integral equations. This method is applicable to the class of black boxes having ergodic correlation functions when there is zero applied input. The point of view adopted in this paper is phenomenological in that it is desired that calculations made using the mathematical model should be "close" to what is actually observed at the output of the black box. How "close" is defined in the problem statement [conditions (a), (b), and (c) being satisfied].

1. INTRODUCTION

Kalman-Bucy filtering algorithms and the techniques of stochastic control all begin with a mathematical model. In this paper, an approach for obtaining such a model will be given.

To achieve the objective of modeling and identification, several statistical experiments must be performed and the data processed. The theoretical basis for these experiments, and what may be concluded from the processed data are given in the theorems and corolarries. The modeling and identification proceeds in two distinct phases. The first phase involves testing the correctness of the structure of the assumed mathematical model. If the results of the experiments indicate that the assumed structure of the mathematical model is correct, the second phase of the modeling and identification is entered. Without any further statistical experimentation, the earlier data is processed to yield all of the unknown parameters of the model. The steps required in implimenting the two phases are stated explicitly.

2. STATEMENT OF THE PROBLEM

Suppose that we have a physical black box, as shown in figure 1:

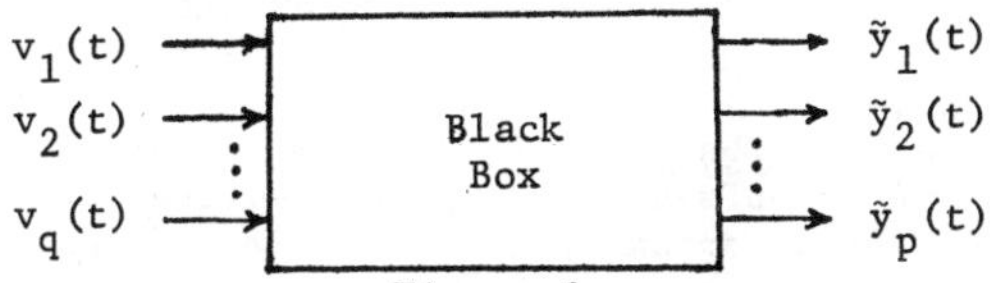

Figure 1

**This work was partially supported by a NASA Grant NGR-001-090.

*Presently at General Electric, Corporate Research & Development Laboratories, Schenectady, New York.

The black box has random process outputs $\tilde{y}_1(t), \tilde{y}_2(t), \ldots, \tilde{y}_p(t)$ and deterministic inputs $v_1(t), v_2(t), \ldots, v_1(t)$ [multiple-input/multiple-output]. (There are noisy disturbances within the black box.)

Suppose that the physical black box satisfies the following restrictions: (R1) for $v(t) = [v_1(t), v_2(t), \ldots, v_q(t)]^T \equiv o$-function, the correlation function matrix is ergodic, (R1') for $v(t) \equiv o$-function, the mean of $\tilde{y}(t)=[\tilde{y}\;(t), \tilde{y}_2(t), \ldots, y_p(t)]^T$ is a constant.

We would like to fit the physical black box to the following mathematical model:

$$dx_t = Fx_t\,dt + Gd\beta_t + Mv(t)dt,\quad x(o)=c, \tag{1}$$

$$y_t = Hx_t + d, \tag{2}$$

where $x(o)$ is a zero mean Gaussian random vector with $E[x(o)x^T(o)] = P$, $\{\beta_t\}_{t\epsilon R}$ is a zero-mean Wiener process with $E[\beta_t\beta_u{}^T]=I\min(t,u)$, and d is a constant.

The stochastic differential equation (1) is symbolism for a mathematically rigorous Ito stochastic integral equation. The above proposed model has six (6) unknown vectors or matrices: $F(nxn)$, $G(nxr)$, $H(pxn)$, $P(nxn)$, $M(nxq)$, and $d(px1)$. (The symbol $\tilde{y}(t)$ represents the actual output of the black box; y_t represents the output of the model.)

The "closeness" criterion is that the actual black box and its mathematical model have:

(a) the same power-spectral density matrices, $S_{\tilde{y}\tilde{y}}(p) = S_{yy}(p)$,
(b) the same constant mean vector $E[y(t)] = E[\tilde{y}(t)]$ for all t,
(c) the same type of sample functions (either continuous or piece-wise continuous).

The closeness criterion comes from a phenomenological motivation. (A drawback is that we require no correspondence between the actual black box and the mathematical model in the higher moments implicit in the ergodic assumption R1).

First it must be determined whether the structure of the mathematical model of equations (1), (2) is "adequate" to describe the behaviour of the actual black box. This question is contorted into a test of hypothesis situation. If the conclusion of the hypothesis test is that the model is adequate, the unknown matrices are then found so as to satisfy (a), (b), and (c).

3. STRUCTURE TEST: PHASE 1

STEP 1: For the actual black box, clamp $v(t) \equiv o$-function.
STEP 2: Record the sample functions of the black box over a long time interval and process to obtain the mean vector and the correlation function matrix. (Recall the restriction R1.)
STEP 3: Obtain the power spectral density matrix of the black box by using the fast fourier transform.
STEP 4: Reexpress or approximate the power spectral density matrix by a matrix having only rational elements [following the methods presented in Laning and Batin (1956, p. 381) or solodovnikov (1960, chapter V)].
STEP 5: Define $d=E[\tilde{y}(t)]$. Condition (b) is satisfied.

Step 6 to Step 10, inclusive, follow an approach given by Bucy and Joseph (1968, p. 25-26, p. 39).

STEP 6: To force condition (a) to be satisfied, set $S_{yy}(p) \triangleq S_{\tilde{y}\tilde{y}}(p)$, where $S_{\tilde{y}\tilde{y}}(p)$ is known from measurement processing, and the form of $S_{yy}(p)$ is known from Theorem 1 below.

STEP 7: Since $H(pI-F)^{-1}GG^T(-pI-F^T)^{-1}H^T=S_{\tilde{y}\tilde{y}}(p)$, perform matrix spectral factorization (Youla, 1961; Davis, 1963; Anderson, 1967; IBM computer program by W.G. Tuel, 1968) to obtain $S_{yy}(p)=W(p)W^T(-p)$, where $W(p)$ is a factor with all of its poles in the left half plane. (We actually factor $S^T_{\tilde{y}\tilde{y}}(p)$ to obtain our result in the desired form.)
STEP 8: Make the association $W(p)=H(pI-F)^{-1}G$, where $W(p)$ is known, but H, F, and G are unknown.
STEP 9: Using the mathematical technique of realization theory, find the required unknown matrices $F(n \times n)$, $G(n \times r)$, $H(p \times n)$ by treating $W(p)$ as if it were a $p \times r$ transfer function matrix. Notice that the pair (F,G) is controllable, (F,H) is observable, and F is stable, i.e., the eigenvalues associated with F have negative real parts.

STEP 10: From Theorem 1, P satisfies the equation

$$FP + PF^T + GG^T = 0 \tag{3}$$

since (F, G) is controllable, a solution of equation (3) is

$$P = \int_0^\infty e^{Fw}\, GG^T\, e^{F^T w}\, dw > 0. \tag{4}$$

STEP 11: If the sample functions of the actual black box are continuous, make $\{\beta_t\}_{t\epsilon R}$ a vector Wiener process. If the sample functions of the actual black box are piece-wise continuous, make $\{\beta_t\}_{t\epsilon R}$ a centered poisson process. ($\{\beta_t\}_{t\epsilon R}$ is still a martingale, and the stochastic integrals have a rigorous definition.) Note that condition (c) is now satisfied. (In all that follows in this paper it will be assumed that the sample functions were continuous; if in reality they are piece-wise continuous, then the necessary modifications in changing from Gaussian to poisson should be made in the remainder of the paper.)

The next step is to determine if the proposed mathematical model equations (1), (2), has an adequate structure for the particular physical black box under consideration. If the structure is adequate, then F, G, H, P, and d obtained from Step 1 to Step 11, are the required matrices which were to be identified.

4. THEORY

THEOREM 1 (Bucy and Joseph): The mathematical model of equations (1)-(2) has

(a) mean $E[y(t)] = d$ for all t;

(b) power spectral density matrix

$$S_{yy}(p) = H(pI-F)^{-1}GG^T(-pI-F^T)^{-1}H^T \tag{5}$$

(c) correlation function matrix

$$R_{yy}(\tau) = \begin{cases} He^{F\tau}PH^T & \text{for } \tau > o, \\ HPe^{-F^T\tau}H^T & \text{for } \tau \leq o; \end{cases} \tag{6}$$

where P satisfies equation (3). ■

LEMMA 1: For $t = s$, y_s is a Gaussian random vector having mean

$$E[y_s] = \int_0^s He^{(s-w)F}Mv(w)\,dw + d \tag{7}$$

and variance

$$\mathrm{Var}[y_s] = He^{Fs}Pe^{F^T s}H^T + \int_0^s He^{(s-w)F}GG^Te^{(s-w)F^T}H^T\,dw \ ■ \tag{8}$$

Remark: Notice that $\mathrm{Var}[y_s]$ is independent of $v(t)$, the input, while $E[y_s]$ depends on the input.

Now the question of whether the mathematical model of equations (1), (2) is adequate for the actual black box under consideration is contorted into a test of hypothesis situation. A criterion (Theorem 2), based on the super position principle that is characteristic of a linear system, is developed and an experiment is designed to determine if equations (1), (2) represent an adequate model (i.e., to determine if this black box satisfies the criterion which was derived from the theoretical mathematical model).

If the criterion is not satisfied, the hypothesis is rejected; if the criterion is satisfied, the hypothesis is accepted. It is possible to calculate, after concluding Step 11, the number of times (trials), N*, the experiment must be performed to have α confidence in the outcome of the hypothesis test (Theorem 3, 4, Corollaries 1, 2).

Partition the gain matrix of the deterministic input, M, of equation (1) into column vectors:

$$M = [m_1 \quad m_2 \quad \dots \quad m_q].$$

Define: $[v(t)]^j = [\delta_{1j}, \delta_{2j}, \dots, \delta_{qj}]^T u_j(t)$, where $u_j(t)$ is a scalar function of time, $u_j(t) \in L^2[o, t_k]$, δ_{ij} is the kronecker delta, $\delta_{ij} = o$ for $i \neq j$, $\delta_{ij} = 1$ for $i=j$, and t_k is some finite time.

Define: $[y(t)]^j$ is the vector response to input $[v(t)]^j$.

THEOREM 2 (Kerr): For input $[v(t)]^j$ (for fixed j), for the mathematical model of equations (1) (2), with Var[y(t)] and E[y(t)] as in Lemma 1, at time t=s, y(s) is a Gaussian random vector with unknown mean of the form

$$E[y(s)]^j = [\int_o^s He^{(s-w)F} u_j(w)dw]m_j + d, \qquad (9)$$

and known variance, Var[y(s)].

Proof: In the expression for Var[y(s)], equation (8), everything is known so Var[y(s)] is known.

$$E[y(s)] = \int_o^s He^{(s-w)F} Mv^j(w)dw + d =$$

$$\int_o^s He^{(s-w)F} M[\delta_{1j}, \delta_{2j}, \dots, \delta_{qj}]^T u_j(w)dw + d$$

$$= [\int_o^s He^{(s-w)F} u_j(w)dw]m_j + d,$$

everything in equ. (9) is known except m_j which renders E[y(s)] unknown. All other conclusions of the theorem follow from Lemma 1. ■

Let $\mu(s)$ represent the true population mean of the random vector y(s). The sample mean given by

$$[\bar{y}(s)]^j = \frac{1}{N} \sum_{n=1}^{N} [y(s)]_n^j. \qquad (10)$$

is a maximum likelihood, sufficient, unbiased, efficient, and consistent statistic for $\mu(s)$. We now provide the physical motivation for the following two theorems. For a deterministic system model, equation (1), (2), with P=0 and G=0, the superposition principle associated with linearity applies; that is, for two different inputs $[v^A(t)]^j$ and $[v^B(t)]^j$ having the responses $[y^A(t)]^j$ and $[y^B(t)]^j$, respectively, for $o \leq t \leq t_k$, the response $[y^{A+B}(t)]^j$ to input $[v^A(t)+v(t)]$ is such that $[y^{A+B}(t)]^j = [y^A(t)]^j + [y^B(t)]^j$ for $o \leq t \leq t_k$. For the stochastic mathematical model of equations (1), (2), with $P \neq 0$, $Q \neq 0$, $G \neq 0$, $[\bar{y}^{A+B}(t)]^j$ should be close to $[\bar{y}^A(t)]^j + [\bar{y}^B(t)]^j$ to a degree described in the following two Theorems.

Define: R(t) = Var[y(t)].

THEOREM 3 (Kerr): For $t=t_i$, $P[\frac{1}{N}||\bar{y}^A(t_i)]^j - [\mu^A(t_i)]^j||^2_{R^{-1}(t_i)} \leq \varepsilon^2] = P[||[\bar{z}^A]^j||^2_I \leq N\varepsilon^2]$

Proof: For fixed ε (in normalized units), for any $t=t_i$,

$$P[\frac{1}{N}||[\bar{y}^A(t_i)]^j - [\mu^A(t_i)]^j||^2_{R^{-1}(t_i)} \leq \varepsilon^2] =$$

$$P[||[\bar{y}^A(t_i)]^j - [\mu^A(t_i)]^j||^2_{R^{-1}(t_i)} \leq N\varepsilon^2].$$

Now make the substitution

$$z^T = [[\bar{y}^A(t_i)]^j - [\mu^A(t_i)]^j]^T R^{-1/2}(t_i), \text{ then}$$

$$z^T z = ||[\bar{y}^A(t_i)]^j - [\mu^A(t_i)]^j||^2_{R^{-1}(t_i)} \text{ and}$$

$$E[zz^T] = R^{-1/2}(t_i)E\{[[\bar{y}^A(t_i)]^j - [\mu^A(t_i)]^j - [\mu^A(t_i)^j]^T\}R^{-1/2}(t_i)$$

$$= R^{-1/2}(t_i)R(t_i)R^{-1/2}(t_i) = I, \text{ and}$$

$$E[z] = E[[\bar{y}^A(t_i)]^j - [\mu^A(t_i)]^j]^T R^{-1/2}(t_i) = 0. ■$$

Corollary 1 (Kerr): For fixed α ($o<\alpha<1$), fixed $\varepsilon>o$, and known $Var[y(t_i)]$ (as in Lemma 1), N* can be specified such that for $N \geq N^*$,

$$\alpha \leq P[\frac{1}{N}||[\bar{y}^A(t_i)]^j - [\mu^A(t_i)]^j||^2_{R^{-1}(t_i)} \leq \varepsilon^2].$$

Proof: By utilizing Theorem 3, it follows that

$$\alpha = P[\frac{1}{N}||[\bar{y}^A(t_i)]^j - [\mu^A(t_i)]^j||^2_{R^{-1}(t_i)} \leq \varepsilon^2] =$$

$$P[||\bar{z}^A||^2_I \leq N\varepsilon^2] = P[||\bar{z}||^2_I \leq N\varepsilon^2].$$

Now $||\bar{z}||^2_I = \bar{z}_1^2 + \bar{z}_2^2 + \dots + \bar{z}_p^2$, where each $\bar{z}_i$ is a Gaussian random variable. Since the associated covariance matrix is I, the $\bar{z}_i$ are uncorrelated; since they are Gaussian they are also independent (having identical unit variance). The sums of the squares of p independent Gaussian unit variance random variables is a chi-square random variable, having p degrees of freedom.

Hence, $||\bar{z}||^2_I = \chi^2_p$, or $\alpha = P[\frac{1}{N}||[\bar{y}^A(t_i)]^j - [\mu^A(t_i)]^j||^2_{R^{-1}(t_i)} \leq \varepsilon^2] = P[\chi^2_p \leq N\varepsilon^2] =$ $\int_o^{N\varepsilon^2} p_{\chi_p^2}(w)dw$. Now α, p, ε^2 are known, and the N* that is required may be read from the commulative distribution function for the chi-square distribution with p degrees of freedom. ■

Remark: Notice that N* is really independent of j, of the input function $u_j(\cdot)$, and of time t_i.

THEOREM 4 (Kerr): For the conditions of Theorem 3 and Corollary 1, $||[\bar{y}^A(t_i)]^j + [\bar{y}^B(t_i)]^j - [\bar{y}^{A+B}(t_i)]^j||_{R^{-1}(t_i)} \leq 3\varepsilon\sqrt{N^*}$.

Proof: $||[\bar{y}^A(t_i)]^j + [\bar{y}^B(t_i)]^j - [\bar{y}^{A+B}(t_i)]^j||_{R^{-1}(t_i)} = ||([\bar{y}^A(t_i)]^j - [\mu^A(t_i)]^j) + ([\bar{y}^B(t_i)]^j - [\mu^B(t_i)]^j) - ([\bar{y}^{A+B}(t_i)]^j - [\mu^A(t_i)]^j - [\mu^B(t_i)]^j)||_{R^{-1}(t_i)} \leq ||[\bar{y}^A(t_i)]^j - [\mu^A(t_i)]^j||_{R^{-1}} + ||[\bar{y}^B(t_i)]^j - [\mu^B(t_i)]^j||_{R^{-1}} + ||[\bar{y}^{A+B}(t_i)]^j - [\mu^A(t_i)]^j - [\mu^B(t_i)]^j||_{R^{-1}} \leq 3\varepsilon\sqrt{N^*}$ ■

Define: $\gamma = \sum_{i=1}^{h} ||[\bar{y}^A(t_i)]^j + [\bar{y}^B(t_i)]^j - [\bar{y}^{A+B}(t_i)]^j||^2_{R^{-1}(t_i)}$.

Corollary 2 (Kerr: For the conditions of Theorem 4, $\gamma \leq 9k\varepsilon^2 N^*$. Proof: From Theorem 4, upon squaring we obtain $||[\bar{y}^A(t_i)]^j + [\bar{y}^B(t_i)]^j - [\bar{y}^{A+B}(t_i)]^j||^2_{R^{-1}(t_i)} \leq 9\varepsilon^2 N^*$. Summing both sides over i from o to k yields

$$\gamma = \sum_{i=1}^{k} 9\varepsilon^2 N^* = 9k\varepsilon^2 N^*. \quad ■$$

Define: An estimator of the columns, m_j, of M is

$$\hat{m}_j = \frac{1}{k \cdot N} \sum_{m=1}^{k} \{[\int_0^{t_m} He^{F(t_m-w)} u_j(w)\,dw]^{\dagger} \cdot (\sum_{n=1}^{N} [\tilde{y}(t_m)]_n - d)\},$$

where N is the number of trials (sample size); $t_1, t_2, \ldots, t_k$, are the observation instants and † is the pseudoinverse.

Remark: Notice that the estimate is linear in the measurements, and makes use of all of the measurements. Note also that the estimator is nonlinear in the deterministic function $u_j(\cdot)$, hence it appears that there is some benefit in choosing u_j optimally (i.e., use $u_j^*(\cdot)$ as a probing function to attempt to get the most information out of the system about m_j in the time allowed, t_k).

THEOREM 5 (Kerr): The above estimator is unbiased.

Proof: $E[\hat{m}_j] = \frac{1}{k} \sum_{m=1}^{k} \{[\int_0^{t_m} He^{F(t_m-w)} u_j(w)\,dw]^{\dagger} (\frac{1}{N} \sum_{n=1}^{N} E[y(t_m)]-d)\} = \frac{1}{k} \sum_{m=1}^{k} \{[\int_0^{t_m} He^{F(t_m-w)} u_j(w)\,dw]^{\dagger} (\frac{1}{N} \sum_{n=1}^{N} [\int_0^{t_m} He^{F(t_m-w)} u_j(w)\,dw) m_j\} = m_j$. ■

5. STRUCTURE TEST: PHASE 2

STEP 12: Calculate N* as in Corollary 1, for the mathematical model of equations (1), (2), using a reasonable fixed choice of ε and α.

STEP 13: Apply input $[v^A(t)]^j$ to the black box, record the corresponding output $[\tilde{y}^A(t_i)]$ at the time points $t=t_i$, i=1, 2, ..., k. Perform this experiment of observation N* times, with a waiting period between trials of 10 times the reciprical of the smallest eigenvalue of the F matrix, before starting the next trial. (This allows the system enough time to decay back into a state unperturbed by external deterministic inputs since F is stable.)

STEP 14: Repeat Step 13 with input $[v^B(t)]^j$.

STEP 15: Repeat Step 14 with input $[v^A(t) + v^B(t)]^j$.

STEP 16: Calculate $[\bar{y}^A(t_i)]^j$, $[\bar{y}^B(t_i)]^j$, and $[\bar{y}^{A+B}(t_i)]^j$ by the above mentioned definition.

$$[\bar{y}^{\alpha}(t_i)]^j = \frac{1}{N^*} \sum_{n=1}^{N^*} [\tilde{y}^{\alpha}(t_i)]^j_n, \quad (i=1, 2, \ldots, k).$$

STEP 17: Calculate γ, as defined following Theorem 4.

STEP 18: If $\gamma > \varepsilon^2 N$, reject the mathematical model of equations (1)-(2) as being representative of the black box; any further modeling of this black box is not persued in this paper. However, if $\gamma < 9k\varepsilon^2 N^*$, accept the hypothesis with α-confidence that the mathematical model of equations (1)-(2) does represent the actual black box, and that five (5) of the required matrix parameters have already been determined. It remains to determine m_j [Step 13-18 must be done for each j (j=1, 2, ..., q).]

STEP 19: Using the input $[v^A(t)]^j$, determine m_j using the estimate defined just before Theorem 5 (j=1, 2, ..., 2).

The modeling and identification of the particular black box is now complete.

REFERENCES

Anderson, B.D.O. (1967), "An Algebraic Solution to the Spectral Factorization Problem", IEEE Trans. Antomat. Contr., AC-12, p. 410.

Tuel, W.G. (1968), "Computer Algorithm for Spectral Factorization of Rational Matrices", IBM J. p.163

Bucy, R.S., and Joseph, P.D. (1968), Filtering for Stochastic Processes with Applications to Guidance. New York: Interscience.

Davis, M.C. (1963), "Factoring the Spectral matrix", IEEE Trans. Automat. Contr., AC-12, p. 296.

Solodovnikov, V.V. (1960), Introduction to the Statistical Dynamics of Automatic Control Systems. New York: Dover.

Kerr, T.H. (1971), "Applying Stochastic Integral Equations to Solve a Particular Stochastic Modeling Problem", Ph.D. Thesis, University of Iowa.

Laning, J.H., and Battin, R.H. (1956), Random Processes in Automatic Control. New York: McGraw-Hill.

Youla, D.C. (1961), "On the Factorization of Rational Matrices", IRE Trans. Inform. Theory, IT-7, p. 172.

ESTIMATION OF THE ORDER OF NONLINEAR SYSTEMS

C.M. Woodside
Division of Systems Engineering
Carleton University
Ottawa K1S 5B6, Canada

INTRODUCTION

A part of the modelling of systems is the choice of a suitable model structure, which may have variable parameters imbedded in it. For a linear dynamic system, methods for estimating directly the structure (the order of the difference or differential equation model) have been given by Lee [1] for the noise-free case and by the author [2] for the noisy case. The essence of both approaches is to find the dimension of a linear subspace containing data points with some heuristic approaches to compensating for measurement noise included in [2].

The system order is estimated as the intrinsic dimensionality of a collection of data points made up from measurements of output. The construction of the data vectors ensures that they lie on a nonlinear manifold of the same dimension as the system order. The estimation problem has been solved by various approaches in recent papers by Trunk [3], Bennett [4] and Fukunaga [5], and at the end of the paper an example using Trunk's algorithm is described. The estimation problem derives from multi-dimensional scaling techniques (see surveys in [3] and [6]) which map data points into linear subspaces of decreasing dimension. Trunk's contribution was to give a statistical (maximum likelihood) estimator for the nonlinear dimensionality of the original, untransformed data points.

The paper mainly considers estimation of the order of nonlinear dynamic systems which (1) are in discrete time and (2) have a special test input, or no input. The generalization to (3) multiple outputs; (4) continuous time; (5) noisy outputs and disturbances; (6) general random operating inputs is trivial in theory but space limitations preclude a more complete treatment. A conceptual framework within which order estimation is feasible is established and conditions of H-observability which the system must satisfy to make the order estimation possible.

Illustration

A discrete time first-order system illustrates the geometric basis of the argument. The system equations are

$$x(k+1) = f[x(k), u(k)]$$
$$y(k) = h[x(k)] \qquad (1)$$

Many transient responses of just two steps are recorded with various independent initial conditions x(0) and a repeated "test program" input $u^*(0)$, $u^*(1)$. In each transient only y(1) and y(2) are needed, and they are functions of x(0) since the input program is always the same.

$$y(1) = F^{(1)}[x(0)], \quad y(2) = F^{(2)}[x(0)]$$

The functions F depend implicitly on f and h in (1), and on $u^*(0)$, $u^*(1)$.

If y(2) is plotted against y(1) and if the combination of the system, the input program u^* and the set of initial conditions are H-observable as defined in the next section, the data points will be on a curve Z in the y(1) - y(2) plane. For the linear system

$$x(k+1) = .9\ x(k) + u(k)$$
$$y(k) = .25\ x(k) \qquad (2)$$

with $u^*(0)=.5$, $u^*(1)=.1$ and initial condition x(0) located in (0,1) the curve Z is the segment Z_L in Figure 1, lying on the line

$$y(2) = .9\ y(1) + .025.$$

With a nonlinear first-order system, the curve is usually nonlinear as with the equations:

$$x(k+1) = .9 \tan^{-1} [x(k)+u(k)]$$
$$y(k) = .25 \exp[x(k)] \qquad (3)$$

With the same inputs and initial conditions, it is the nonlinear curve Z_N in Fig. 1.

To detect that the system (2) or (3) is first-order an estimation procedure must detect that the data points lie on a manifold of dimension unity, a curve in the two-dimensional space. If a three-point input program $u^*(0)$, $u^*(1)$, $u^*(2)$ were used instead, and output data were recorded in threes to give data points in a space with axes y(1), y(2), y(3), the points would still lie on a one-dimensional curved line in the three-dimensional data space. For linear systems, this is the geometrical meaning of the result of Lee [1].

Except in special cases the curve Z is nonlinear for a nonlinear system, as it is a mapping by the system of the space of initial conditions (or states) into the data space. The H-observability condition ensures that this mapping preserves the dimension of the state space. Then the order estimation problem becomes the problem of estimating the dimensionality of a data collection in exactly the terms investigated by Trunk [3].

DIMENSIONALITY AND H-OBSERVABILITY

A system with n states $x_i(k)$, m inputs $u_j(k)$ and one output y(k) can be described by Equation (1) if x and u are interpreted as vectors, and f is a vector of functions. A test period of duration K>n then gives data vectors u and z as follows:

$$u = [u_1(0),\ u_2(0),\ \dots\ u_m(K)]^T$$

$$z = [y(1),\ y(2),\ \dots\ y(K)]^T$$

$$K > n$$

The three vectors x(0), u, z belong to Euclidean spaces of dimension n, mK and K respectively:

$$x(0) \in X \subset E_n$$

$$u \in U \subset E_{mK}$$

$$z \in Z \subset E_K$$

For our purposes, the set X must have dimension n, written as

$$\dim X = n$$

This means x(0) is not confined to any nonlinear manifold in E_n by a relationship among its components. This will be assured if x(0) is randomly chosen from any open subset of E_n, such as $0 \leq |x_i| < \alpha_i$ for all i.

The system equations define a transformation T which generates a data vector z for each pair [x(0),u]

$$z = T[x(0),u],$$

and under a test input u^*, z is determined only by x(0),

$$z = T^*[x(0)] = T[x(0),u^*]$$
$$Z = T^*(X) \qquad (4)$$

For a linear system, T is given in terms of the observability matrix A [1]:

$$z = A\,x(0) + b(u^*)$$

If the system is observable (rank A = n), then dim Z = dim X. For a nonlinear system, the corresponding observability condition is that T be a homeomorphism.

Homeomorphism (Definition)

A transformation is a homeomorphism if it is one-to-one and onto, and it and its inverse are continuous. The argument set and its image are called homeomorphic [7].

The main results of this paper are the following definition and theorem.

H-Observability (Definition)

The system (1) and its input u^* are homeomorphically observable or H-observable, if the transformation T^* defined by (4) is a homeomorphism of X in Z.

Theorem: If
a) X and Z are the sets defined above for the system (1)
b) X is homeomorphic to the unit open, n-dimensional disc U_n, defined by:

$$U_n = \{x: |x|<1,\ x \in E_n\}$$

c) the system (1) is H-observable and of order n,

then dim Z = n.(proof omitted here)

H-observability is an assumption that ensures that the dimensionality of the data collection of z vectors is the same as the order of the system. In practice two weaker assumptions are useful:

1. Under local H-observability, every open neighbourhood in X is homeomorphic to an open neighbourhood in Z under T^*, and dim Z = dim X. This condition may be satisfied by systems which are not H-observable such as the following first-order example:

   $$y(k+1) = 10 \sin[x(k) + u(k)]$$

   $$y(k) = x(k)$$

   X and Z are not one-to-one; adding 2π to x(0) does not change z.
2. In a system which is almost everywhere locally H-observable, there is a small (in some sense) set in X for which local H-observability fails. The condition dim Z = dim X is satisfied provided the test data do not all fall in this set, (e.g. if x(0) is random with probability zero of being in the set).
   A practical test ensuring local H-observability for a certain $[x(0),u^*]$ is the existence and non-singularity of the Jacobian of T^* when z is defined with just n components (K=n).

ESTIMATION OF DIMENSIONALITY

Given a set of points {Z'} known to lie in a subspace Z, an estimate N for the dimensionality of Z can be found by several methods as quoted above. Trunk points out [3] that a curve of any dimensionality can be passed through any set of points; it is the most likely dimensionality that must be found, so he establishes a likelihood function L(N) from the assumptions that over a spherical neighbourhood of any point, Z resembles a linear N-manifold, and the measured points are drawn from a uniform distribution on the interior of an N-sphere. Around each measured point the likelihood of the K' nearest points is calculated. L(N) is obtained under several approximations, and is then shown empirically to give satisfactory performance, at least for $N \leq 3$ [3].

EXAMPLES OF APPLICATIONS

The estimation technique was tested on a second-order simulated system representing a furnace which heats and accumulates an incoming stream of cold material [8]. The states are temperature x_1 and the thermal capacity of furnace and contents, x_2.

$$x_1(k+1) = x_1(k) + [u_1(k)-2\ x_1(t)]/x_2(k)$$

$$x_2(k+1) = x_2(k) + u_2(k)$$

$$y(k) = x_1(k)$$

u_1 = energy rate

u_2 = materials rate

The test input was u_1=1.5, u_2=.5, held constant. With randomly chosen initial conditions 50 simulated transients were generated, and the data vector z was set up with 10 components.

$$z = [y(1),\ \ldots\ y(10)]$$

Trunk's algorithm found log likelihoods shown in Figure 2, with a clear maximum for the correct order 2.

The effect of independent Gaussian measurement noise is shown in Figure 2, the standard deviation being labelled on the likelihood curves for L(N). With 1% or more noise, the order was overestimated as 3.

REFERENCES

[1] Lee, R.C.K., "Optimal Estimation, Identification and Control", M.I.T. Press, 1964.

[2] Woodside, C.M., "Estimation of the Order of Linear Systems", Automatica, Vol. 7, pp. 727-733, 1971. Also in the Preprints, IFAC Symposium on Identification and Process Parameter Estimation.

[3] Trunk, G.V., "Statistical Estimation of the Intrinsic Dimensionality of Data Collections", Information and Control, Vol. 12, pp. 508-525, 1968.

[4] Bennett, R.S., "The Intrinsic Dimensionality of Signal Collections", IEEE Trans. on Information Theory, Vol. IT-15, No. 5, pp. 517-525, Sept. 1969.

[5] Fukunaga, K., Olsen, D.R., "An Algorithm for Finding the Intrinsic Dimensionality of Data", IEEE Trans. on Computers, Vol. C-20, pp. 176-183, 1971.

[6] Guttman, L., "A General Non-Metric Technique for Finding the Smallest Coordinate Space for a Configuration of Points", Psychometrika, Vol. 33, No. 4, pp. 469-506, Dec. 1968.

[7] Kuratowski, K., "Topology, V. I", Academic Press, 1966.

[8] Woodside, C.M., "A Dynamic Model for Electric Melting of Continuous-Fed Pre-Reduced Iron Pellets", pp. 191-197, Proc. AIME, Electric Furnace Conference, 1970.

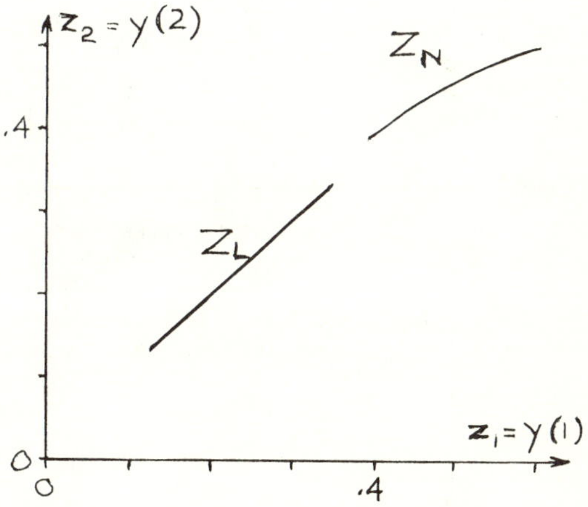

Figure 1 Sets of points Z for the first-order illustration.
- Z_L for the linear example
- Z_N for nonlinear example

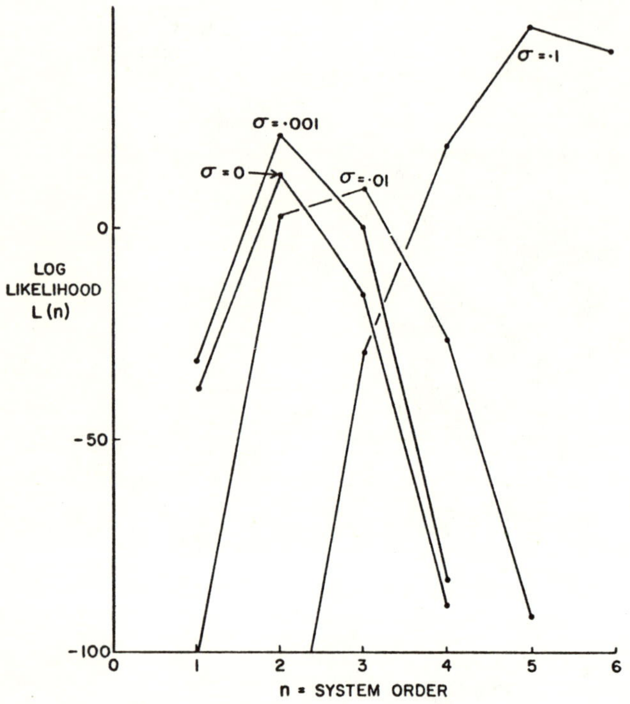

Figure 2 Log Likelihoods of system order n with output measurement errors of variance σ^2, 2nd order example

Errors in Orthogonal Expansion and Relevance to Modelling.

J. Hiller

School of Electrical Engineering
University of New South Wales
Kensington, Australia. 2033.

This paper looks at some of the errors that can arise with orthogonal functions and at some of their advantages. Although orthogonalising is merely one method of matrix inversion, it possesses advantages when an additive structure is used in the realisation. Should a multiplicative process be invoked (recurrence relation), the errors have the familiar problems due to 'marching'; it is shown that this difficulty may be quite formidable even at low orders. The paper concludes with a question as to the relevance of function sets that exhibit high correlations.

1. Introduction.

In their review of system identification Astrom and Eykhoff [1] refer to identification as being one step in the design of a control system. Mention is also made of a separation hypothesis in which design is divided into the two steps of identification and control. Just as the structure envisaged for the control system and the form of performance criterion influence choice of weighting function and norm used in the approximation, so too are these considerations relevant to an assessment of errors in the mechanisation of that approximation. This paper is not concerned with such profundities but refers rather to other difficulties in realisation such as the effect of component drift. The principal question relates to the advantages of proceeding via orthogonalisation of the basis functions.

Orthogonal expansions are often advocated for two reasons:

1) calculations are performed directly without the need for time consuming matrix inversions,
2) sensitivity problems are obviated.

Consider the *curve fitting* problem that arises when an impulse response is known. It may be required to obtain a minimum error in prediction of response at some future time or to minimise the error in approximation of the response over the interval in which it is known. In both instances a strong case may be made for use of a *specified* set of basis functions chosen with regard being given to the statistics of likely impulse responses. This class of problem should be distinguished from the compensator design of Roberts [2] and papers which have recently appeared in the literature on adaptive echo cancellers (e.g. Gersho [3]). When the excitation is non-impulsive there can be a considerable difference between the 'richness' of the representation required for the excitation and response and that of the system which relates these; here there is no such difference. Finally since the aim is to assess orthogonal techniques, the error measure chosen will be the L_2 norm.

2. Some Basic Results.

Let the curve fitting problem be the representation of $f(t)$ in terms of the specified set $\{\phi_i(t)\}$.

$$f(t) \simeq \sum_i a_i \, \phi_i(t) \tag{1}$$

Then for a weighting function $w(t)>0$ this amounts to solution of the linear equations

$$\psi \, . \, a = \gamma \tag{2}$$

where the coefficient matrix will be positive definite. The a_i may be found by a matrix inversion or alternatively an orthonormal basis $\{\theta_i(t)\}$ may be found from the $\{\phi_i(t)\}$

$$\theta_i(t) = \sum_{j=i}^{i} \xi_{ij} \, . \, \phi_j(t) \tag{3}$$

It has been shown by Wilkinson [4] that pivoting is not required in solution of equations having positive definite matrices whence the factorisation

$$\psi = L \, L^T \tag{4}$$

where L is a lower triangular matrix is equivalent* to the operation performed in direct solution of (2). Further, Hiller [5] has shown that ξ and L are related by

$$\xi = L^{-1} \tag{5}$$

Therefore if the aim of the fitting exercise is to determine the a_i the difference between direct inversion and proceeding via the orthogonal set, amounts to the choice between a one-stage and a two-stage process. Alternatively if the approximate representation of $f(t)$ is

---

* Direct inversion involves multiplication by one triangular matrix during sweep out and by another triangular matrix during back substitution. These differ from L^T and L^{-1} merely by diagonal matricies.

is sought, both methods can be considered as having similar complexity. The difference between these two aims is very important in assessing error behaviour.

3. Generation of Approximations to $\delta(t)$.

One of the much quoted advantages of orthogonal functions is that the coefficients of expansion are usually well behaved whereas the a_i may become very large. (See, for example, the discussion in Rice [6] relating expansion in terms of Tchebycheff polynomials to expansion in terms of powers of the independent variable.) If the $\theta_i(t)$ are members of an orthonormal set and

$$b_i = \xi_{i.} \cdot \gamma \tag{6}$$

then the Bessel inequality (Courant & Hilbert [7]) gives

$$\int w(t).\delta^2(t)\,dt - \sum_i b_i^2 \geq 0 \tag{7}$$

whence if the first term on the $l.h.s.$ is finite the b_i satisfy

$$|b_i| \leq \sqrt{\int w(t).\delta^2(t)\,dt} = F \quad . \tag{8}$$

No such inequality applies to the a_i. All that may be written then is that

$$||a|| \leq ||\xi|| \, . \, F \tag{9}$$

from (5), (7), (8), Table 1 gives $||\psi^{-1}||$ for several standard sets of real exponentials. It should be noted from (4), (5) and the rule relating $||\xi||_2$ to the spectral radius of $\xi^T\xi$ that

$$||\xi||_2 = \sqrt{||\psi^{-1}||_2} \tag{10}$$

ORDER	2	3	6	8	10
$\phi_i(t)=e^{-it}$	52.6	1546	$4.56*10^7$	$4.64*10^{10}$	$4.83*10^{13}$
$\phi_i(t)=e^{-2^{i-1}t}$	52.6	551.2	$4.17*10^4$	$2.67*10^5$	$1.26*10^6$
$\phi_i(t)=e^{-i^2t}$	26.0	331.8	$6.14*10^5$	$8.71*10^7$	$1.20*10^{10}$

Table 1 : Variations of $||\psi^{-1}||_2$ with order for three exponential function sets.

It is apparent that even with moderate orders sets such as the 'harmonic exponential' set may lead to large $|a_i|$. The reason for this and the desirability of such sets will be discussed later.

The next point to consider is the way in which the $\theta_i(t)$ are generated. Lubbock and Barker [8] use a network of operational amplifiers and resistors to implement the matrix multiplication by ξ in (3). For many practical function sets where the condition number of ψ is poor this structure can lead to severe overloading problems. It is possible to alleviate this by strategems that amount to pre- and post- multiplication of ξ by diagonal matrices but such moves do not remove the need for impractical precision in components. Note that this is not necessarily a 'high order' problem. As an illustration for the 5th order ψ for the harmonic exponential set

$$\xi = k_1 \cdot \begin{bmatrix} 1.05 & 0 & 0 & 0 & 0 \\ -2.97 & 4.45 & 0 & 0 & 0 \\ 5.46 & -21.82 & 18.18 & 0 & 0 \\ -8.40 & 62.99 & -125.99 & 73.49 & 0 \\ 11.74 & -140.87 & 493.04 & -657.39 & 295.82 \end{bmatrix}$$

(from [4] p.119), the above figures being given to the second decimal place though calculated with 8 decimal multiplication with accumulation. An indication of the sensitivity in such cases is given by $||\xi||$.

The component problems associated with realising direct multiplication by ψ^{-1} are, however, far more severe. As an illustration the upper triangular part of ψ^{-1} for the 5th order harmonic exponential set is

$$k_2 \begin{bmatrix} 248.02 & -2314.81 & 6944.44 & -8333.33 & 3472.22 \\ & 24305.6 & -77777.8 & 97222.2 & -41666.7 \\ & & 259259. & 333333. & 145833. \\ & & & 437500. & -194444. \\ & & & & 87500.0 \end{bmatrix}$$

(from [4] p.119).

Thus when $||\psi^{-1}||$ is large, orthogonalisation of the basis leads to a reduction in the component sensitivity problem in the realisation. It can, however, be that neither approach is practical.

One way out of this difficulty is to realise the $\theta_i(t)$ by means of a recurrence relation to $\theta_{i-1}(t)$ and $\theta_{i-2}(t)$, Trunk & Higgins [9]. This is fine provided such a recurrence relation exists. What happens, however, if the $\phi_i(t)$ are not the impulse responses of lumped systems or $w(t)$ is not constant because of some predective requirement ? Further, marching processes are notorious sources of error propagation — is this relevant here ?

4. Use of Recurrence Relations.

To consider this last point, a series of simulations were carried out for a function set having a simple recurrence relation. When

$$w(t) = u(t) \quad \text{and} \quad \phi_i(t) = e^{-p_i t} \tag{11}$$

then the Laplace transforms of $\theta_i(t)$ and $\theta_{i-1}(t)$ and related by

$$\mathcal{L}\{\theta_i(t)\} = \mathcal{L}\{\theta_{i-1}(t)\} \cdot \sqrt{\frac{p_i}{p_{i-1}}} \cdot \frac{s-p_{i-1}}{s+p_i} \qquad (12)$$

For simplicity the p_i were considered real; this is not a restrictive assumption since an equivalent relation may be established for the complex pole and multiple pole cases, Trunk & Huggins [9]. To show the effect of errors (12) was modified to

$$\frac{\mathcal{L}\{\sigma_i(t)\}}{\mathcal{L}\{\sigma_{i-1}(t)\}} = \sqrt{\frac{p_{i_d}}{p_{i-1_n}} \cdot \frac{s-p_{i-1_n}}{s+p_{i_d}}} \qquad (13)$$

where

$$p_{i_d} \quad \varepsilon \quad p_i\,[1-\tau,\ 1+\tau]$$

$$p_{i_n} \quad \varepsilon \quad p_i\,[1-\tau,\ 1+\tau] \qquad (14)$$

τ is a fractional tolerance variable and the p_{i_d}, p_{i_n} are generated by means of random number generators having a uniform distribution in $[1-\tau,\ 1+\tau]$. Since $\sigma_1(t)$ was found in the same manner as $\theta_1(t)$,

$$\sigma_i(t) \to \theta_i(t) \quad \text{as} \quad \tau \to 0 \qquad (15)$$

If the $\sigma_i(t)$ are used as though they are orthonormal then

$$\sum_i b_i\,\theta_i(t) \quad \text{will become}$$

$$\sum_i c_i\,\sigma_i(t) \quad \Big| \quad c_i = \sum_i b_i \int \sigma_i(t).\theta_j(t)dt \qquad (16)$$

The resulting error may be shown to be

$$b^T \left[I - 2\,\mathcal{C}^T_{\sigma\theta} \cdot \mathcal{C}^T_{\sigma\theta} + \mathcal{C}^T_{\sigma\theta} \cdot \mathcal{C}_{\sigma\sigma} \cdot \mathcal{C}_{\sigma\theta} \right] b, \qquad (17)$$

The matrices $\mathcal{C}_{xy}$ give the cross correlations of the elements of the sets $\{x(t)\}$, $\{y(t)\}$. The norm of the matrix of (17) may be used to indicate the error.

Provided that τ is sufficiently small the change in ξ and $\phi(t)$ will be of order τ i.e.

$$\xi \to \xi + \tau\,\Xi \quad , \quad \phi_i(t) \to \phi_i(t) + \tau\,\Phi_i(t) \qquad (18)$$

also

$$\sigma(t) = (\xi + \tau\,\Xi) \quad (\phi(t) + \tau\,\phi(t)\,) \qquad (19)$$

Then

$$\mathcal{C}_{\sigma\theta} = I + \tau \left[\Xi\,\psi\,\xi^T + \xi\,\mathcal{C}_{\phi\Phi}\,\xi^T \right] + O(\tau^2) \qquad (20)$$

and $\mathcal{C}_{\sigma\sigma}$ may be similarly expanded. Substitution shows that

$$\left\| \left[I - 2\,\mathcal{C}_{\theta\sigma} \cdot \mathcal{C}_{\sigma\theta} + \mathcal{C}_{\sigma\theta} \cdot \mathcal{C}_{\sigma\sigma} \cdot \mathcal{C}_{\sigma\theta} \right] \right\|_2 \quad \text{is } O(\tau^2) \qquad (21)$$

provided that τ is sufficiently small for (18) to hold.

Table 2 gives an indication of the simulation results for the three function sets referred to in Table 1. Distributions were obtained for the norm of the matrix of (17); the figures provided show the likely values by giving the difference between the 0.25 and 0.75 points on the cumulative p.d.f.'s. It is apparent that (21) applies for values of τ such as 10^{-6} although it may not obtain with practical hardware tolerances such as $\tau = 10^{-3}$. The values given in the table were obtained from simulations involving 100 sets of parameters. With such a restricted test one can only ascribe significance to the exponents and not to the mantissae of the elements in the table.

τ	ORDER	2	3	6
10^{-3}	$p_i=i$	$3.2*10^{-5}$	$1.2*10^{-3}$	$1.1*10^{+3}$
10^{-5}		$3.3*10^{-9}$	$1.1*10^{-7}$	$2.8*10^{-3}$
10^{-3}	$p_i=2^{i-1}$	$3.3*10^{-5}$	$1.7*10^{-4}$	$1.4*10^{-3}$
10^{-5}		$3.4*10^{-6}$	$1.7*10^{-8}$	$1.4*10^{-7}$
10^{-3}	$p_i=i^2$	$4.7*10^{-6}$	$2.1*10^{-5}$	$2.9*10^{-2}$
10^{-5}		$4.7*10^{-10}$	$2.1*10^{-9}$	$2.6*10^{-6}$

Table 2 : Dependence on τ and order of the difference between the 0.25 and 0.75 points on a cumulative p.d.f. of

$$\left\| \left[I - 2\,\mathcal{C}_{\theta\sigma}\,\mathcal{C}_{\sigma\theta} + \mathcal{C}_{\theta\sigma}\,\mathcal{C}_{\sigma\theta}\,\mathcal{C}_{\sigma\theta} \right] \right\|_2$$

5. Completeness & Correlation of Basis Functions.

Direct solution of (2) is difficult when ψ is ill-conditioned. Since the norm being used throughout this discussion is L_2 this means that certain of the basis functions are highly correlated. It may be remarked that of the three sets of basis considered in tables 1 & 2 that which leads to the most difficulty is the only set that is complete viz. the harmonic exponential set $p_i = i$. It has been remarked previously, [5], that completeness is a concept that has questionable relevance to practical curve fitting and identification. How significant is it to know that an error measure may be brought to zero with an infinite number of terms if practical approximation is restricted to say 10 terms ? There is no simple answer to such a query. There are situations in which curve fitting is necessary to obtain a model that will be used within a feedback system. There, error measures of the order of 5-10% may be acceptable. Alternatively there are situations where very accurate fitting over an interval is being sought and the error measures being looked to may be less than 1%. (Note that a 1% error in the integral error squared results from a 10% error in approximation throughout the range.) In the first case the widest range of function space (consistent with the expected form of $f(t)$) should be spanned by the basis. This will tend to reduce correlation and lead to well-conditioned ψ . In the second instance small errors can only be achieved by a clustering of the basis and thus an approach towards linear dependence with consequent poor conditioning. It can not be asserted that a high condition number of ψ is always an indication of poor selection of the members of the function set.

6. Conclusion.

It has been the role of this paper to look at some realisation problems associated with use of orthogonal functions. Though it is true that orthogonal expansion possesses advantages such as

directness of expansion (i.e. no inversions),
good sensitivity behaviour in terms of changes to $f(t)$,

These count for little if there are realisation problems that more than offsett them. The results reported here should be considered along with those of Miller [10]. This reference considers the effect of rounding errors on the inversion of ψ . The advantages of an analytical inverse are pointed out and such a rule is given for a restricted class of functions. The computing situation differs in some important aspects from our problem here. There are, for example, cases where significant errors in the elements of ψ do not cause large departures of $\psi^{-1}.\psi$ from the unit matrix ([4] p.119,126). In the case of component inaccuracies it is not practical to consider that the errors in ψ^{-1} will be mainly associated with the small eigenvalues of ψ .

Section 3 looked at some matrix results and showed that the component sensitivity problem is somewhat reduced by orthogonalising. Note however, that this technique does not obviate the problem in the case of very ill conditioned ψ and, that for well conditioned correlation matrices, it is unnecessary. Significant advantages thus only result in restricted cases.

Use of recurrence relations (when these are simple) is also not recommended in the case of highly correlated bases. This should hardly be surprising for, ideally, the same realisation should be obtained with both the multiplicative and additive structures. Since the errors are $O(\tau^2)$ there is a major difference between errors in hardware and software realisations. It is also apparent from the simulation that the errors are dependent on a high power of $||\psi^{-1}||$.

The final part of the paper has looked but briefly at the important question 'are sets leading to ill-conditioned ψ relevant ?' . In cases where such sets are needed, orthogonalisation will alleviate rather than remove accuracy troubles. The basic weakness of modelling which amounts to this type of curve fitting is that it does not involve the error being fed back. The accuracy requirement in this open loop case is much more significant than that in feedback systems such as [3] .

References.

[1] K.J. Astrom & P. Eykhoff, System Identification — A Survey, *Automatica*, 7, 123-162, (1971).

[2] P.D. Roberts, The Application of Orthogonal Functions to Self-Optimising Control System Containing a Digital Compensator, Proc. 3rd I.F.A.C. Congress, London, paper 36C (1966).

[3] A. Gersho, Linear Adaption, Symposium on Computer Processing in Communications, P.I.B., April 1969, pp 653-664.

[4] J.H. Wilkinson, *Rounding Errors in Algebraic Processes*, Prentice Hall, Englewood Cliffs, Chapter 3 (1963).

[5] J. Hiller, Is Orthogonal Expansion Desirable ?, *Int. J. Control*, 5, 11-22 (1967).

[6] J.R. Rice, *The Approximation of Functions, Vol. 1 : Linear Theory*, Addison-Wesley, p. 145 (1964).

[7] R. Courant & D. Hilbert, *Methods of Mathematical Physics, Vol. 1*, Interscience, New York, p. 51 (1953).

[8] J.K. Lubbock & H.A. Barker, A Solution of the Identification Problem, *Trans* A.I.E.E., 83, 166-173 (1964).

[9] G.V. Trunk & W.H. Huggins, Orthogonal Expansion of a Real Rational Function, Having Hurwitz Denominator, I.E.E.E. *Trans*, CT-15, 144-5 (1968).

[10] G. Miller, Least Squares Approximation of Functions by Exponentials, Technical Report Johns Hopkins University, June 1969.

ESTIMATION OF STATE VARIABLE MODELS USING INPUT-OUTPUT DATA

by
S. P. Bingulac and M. Djorović
Institute "Boris Kidrič", Beograd, POB 522, Yugoslavia

S u m m a r y

Starting from comparison, analysis and review of some existing methods for identification of minimal realization of linear time-inveriant multivariable dynamic systems, a modification of a previously published method is suggested which permits considerable saving in computation. The modification is based on an important property of linear systems which allows that the order of the minimal realization be determined in advance by direct processing of available input-output data $u(t), y(t)$. Included is a computational example simulated on a computer, illustrating the applicability of the procedure.

1. INTRODUCTION

Each method for identification of dynamic systems which attempts to determine its mathematical model starting from normal noisy input-output operating records may be thought of as a data processing procedure represented by the following diagram:

$$\{u,y\} \Rightarrow \left\{\begin{matrix}\text{data processing}\\ \text{transformation of}\\ \text{raw data}\end{matrix}\right\} \Rightarrow \left\{\begin{matrix}\text{mathematical}\\ \text{model}\end{matrix}\right\}$$

In the case of identification of multivariable dynamic systems the data processing procedure uses either a set of $m + p$ time functions

$$\begin{aligned} u(t)^T &= [u_1(t) \;\; u_2(t) \; \ldots \; u_m(t)] \\ y(t)^T &= [y_1(t) \;\; y_2(t) \; \ldots \; y_p(t)] \end{aligned} \quad ; \text{ for } t \in [t_o, T] \qquad (1)$$

or discrete values $\{u(t), y(t)\}$; $t = 1, 2, \ldots, N$ (1')
representing normal noisy operating records of a system to be identified, where $u(t) \in R^m$, $y(t) \in R^p$ are control and output vectors of the system, respectively. For time invariant dynamic systems in the continuous domain the mathematical model to be determined by the data processing procedure is usually in the form of a state space representation:

$$S_{pm} : \quad \begin{aligned} \dot{x}(t) &= F \cdot x(t) + G \cdot u(t); \quad x(t_o) = x_o \\ y(t) &= H \cdot x(t) \end{aligned} \qquad (2)$$

where the state vector $x(t) \in R^n$, $n \geq m$; $n \geq p$, while the realization

$$R_{min} = \{F, G, H\} \qquad (3)$$

should be minimal. Similarly, in the discrete domain the model of the considered system may be represented by

$$S_{pm} : \quad \begin{aligned} x(t+1) &= F \cdot x(t) + G \cdot u(t) \\ y(t) &= H \cdot x(t) \end{aligned} \quad ; \text{ for } t = 0, 1, 2, \ldots \qquad (4)$$

Therefore, the general identification problem of a multivariable time invariant dynamic system which starts from input-output data (1) or (1') consists of:

1. Determination of the order n of the minimal realization R_{min},
2. Determination of parameters in the minimal realization, i.e. elements of matrices F, G, and H in (3).
3. Determination of initial condition vector x_o.

The input-output charcteristics of the obtained model (2) or (4) should correspond to the available signals.

So far a number of methods and algorithms /B. Gopinath, (1969); M. A. Budin, (1971); J. E. Ackermann, (1971); M. Djorović, (1972)/ for solving the formulated identification problem has been suggested. Also some other procedures /Ackermann, (1971); Ho, (1966); Mayne, (1968); Rosenbrock, (1970)/ originally suggested for solving some related problems may be partially incorporated in the solution of the identification problem. An interesting identification approach applicable both to linear and a class of nonlinear dynamic systems has been suggested by Mataušek, (1973). All existing identification methods starting from normal input-output operating records differ solely in the form of the applied data processing procedures by which they yield the desired minimal realization. However, some procedures are designed to give state variable model in discrete domain (4), while some other consider the system as continuous (2), although, these differences, as it has been demonstrated /Ho, (1966)/ are not significant. The main distinction, however, is in the form of the applied identification procedure. The advantages and disadvantages of all these procedures may be judged according to the specific criteria such as:

1. The form of obtained minimal realization, i.e. the number of parameters to be determined,
2. Computational complexity of the entire data processing procedure, which might introduce inevitable numerical errors.
3. Possibilities of supressing the existing measurement noises.

4. Theoretical and conceptual clarity and simplicity of the applied data processing procedure.

Having in mind these criteria all existing identification methods may be roughly divided into the following three groups:

1. Selector matrix approach
2. Application of the Bucy canonical form
3. Elimination of uncontrollable modes.

Selector matrix approach

The first group comprises the methods based on the selector matrix approach /Gopinath, (1969); Budin, (1971)/. Gopinath's procedure requires discrete input-output data (1') where N should be sufficiently large and must satisfy

$$N > n(m+2)$$

The unknown order n of the minimal realization, as well as the appropriate selector matrix S are determined by a rather complex computation which is of the trial-and-error nature. Also, the procedure determines $n(n+m)$ parameters, i.e. all elements of matrices F and G. For matrix H it is assumed that rank $H = p$ and the procedure always yields $H = [I_p \mid 0]$ where identity I_p has dimension p. Budin's identification procedure permits the use of the fewer number of data, which depends on the available a priori information about the rank of matrix H. In this case, the required number of data is given by

$$N^* = n(m+2) - q + 1 \leq N$$

where $q = \text{rank } H$.

Also, the order n and the selector matrix S are determined simultaneously by a direct computational procedure. The form of the obtained minimal triple is basically the same as in /Gopinath, (1969)/. The procedure does not require the assumption that rank $H=p$, and in the case of rank $H = q<p$ an appropiate H of the form $H = [H_1 \mid 0]$ is determined where $(p \times q)$ matrix H_1 is obtained as a part of the used selector matrix S.

Bucy canonical form

The second identification approach /Ackermann, (1971)/ uses the Bucy canonical form /Bucy, (1968)/ and is based on the procedure /Ackermann, (1971)-2/ originally applicable only in the case of availability of impulse response matrices in the discrete form. The main property of Bucy canonical form is that the unknown system may be considered as a set of $r \leq p$ coupled subsystems. The order n of the minimal realization is given by

$$n = n_1 + \sum_{i=2}^{r} n_i \qquad (5)$$

where n_1 is the order of the first subsystem which is observable from $y_1(t)$. Each n_i, $i \in [2, r]$ represents the order of a corresponding subsystem which becomes additionally observable from a group of the first i output signals considered together $\{y_1, y_2, \dots, y_{i-1}, y_i\}$ instead of a group of the first i-1 signals $\{y_1, y_2, \dots, y_{i-1}\}$.

The number as well as dimensions of subsystems appearing in Bucy canonical form are highly dependent upon the sequencing of elements y_i of the output vector y(t). In /Ackermann, (1971)/ it has been shown that each order n_i may be determined by processing of all elements of the control vector u(t) and the first i elements y_j, $j \in [1, i]$ of the output vector y(t). Integer r in (5) represents a total number of determined n_i which satisfy $n_i > 0$. Orders n_i determines the structure of the matrix F in the minimal realization (3). The parameters in $\{F, G, H\}$ are determined from corresponding difference equations /Ackermann, (1971)/ which are direct consequence of the Bucy canonical form. In this case the number of parameters to be determined is considerably smaller than that of the method based on the selector matrix approach. Also, the procedure is conceptually clearer and computationally simpler. The total number of parameters, however, as well as the sizes and number of subsystems in matrix F depend upon the way of sequencing the elements of y(t). This property is not natural and may be considered as a disadvantage. The mentioned dependence on sequencing results from coupling terms existing in the assumed form of the matrix F.

Elimination of uncontrollable modes

In the third approach /Djorović, (1972)/ the unknown system S_{pm} is considered as a set of p independent single-output, multi-input subsystems corresponding to each individual element $y_i(t)$ of the output vector y(t). The i-th subsystem contains all modes observable from $y_i(t)$, regardless whether some of them may be observable from some other output coordinates as well. Denoting the order of the i-th subsystem by n^i, it is clear that the unknown order n of the minimal realization satisfies

$$n \leq \sum_{i=1}^{p} n^i \qquad (6)$$

Equality sign in (6) holds only if the same modes do not apper in more then one subsystem, i.e. when the system S_{pm} has already been decoupled. In order to determine the unknown n, an elimination procedure is developed which sequentially detects each common mode appearing in more then one subsystem. Simultaneously, with determination of n, the parameters of the minimal realization are also determined. Since the mentioned elimination procedure deals with the subsystems represented in the Jordan canonical form, this identification approach is conceptually clear and computationally simple. As it is known, the Jordan canonical form perinits an easy access to each individual mode exissting in the system.

The total number of parameters in the minimal realization to be determined is given by:

$$\sum_{i=1}^{p} n^i + nm \qquad (7)$$

Since each n^i is characterised solely by the properties of the corresponding coordinate $y_i(t)$, the total number of parameters (7) is independent of the sequencing of $y_i(t)$ and represents the smallest number of parameters by which the corresponding system may be described. In the obtained minimal realization the coupling existing in the system S_{pm} is represented by a special form of the output matrix H, and not by the coupling terms in F as in the Bucy canonical form.

Some other existing methods which determine the minimal realization (3), starting either from the transfer function matrix Z(s), where

$$Z(s) = H\,(Is - F)^{-1}\,G \qquad (8)$$

or from the impulse response matrices may be in principle also applied in solving the formulated identification problem, provided that additional procedures are incorporated which, using available $\{u,y\}$, calculate Z(s) or impulse responses. However, it is not likely that such a procedure whould be computationally advantageous in respect to the methods discussed above.

The purpose of this paper is to suggest an extention and modification of the previously published method /Djorović, (1972)/ which, based on an important property of linear time invariant dynamic systems, permits in some cases considerably saving in computation. In order to facilitate a better understanding of the suggested modification and extention, the identification method /Djorović, (1972)/ based on the use of elimination procedure will be briefly reviewed in the next section.

2. BACKGROUND

The identification method presented in /Djorović, (1972)/ consists of the two steps:

(i) determination of nonminimal-observable realization

$$R = \{F,G,H\} \qquad (9)$$

(ii) construction of a minimal realization

$$R_{min} = \{\hat{F},\hat{G},\hat{H}\} \qquad (10)$$

by eliminating all uncontrollable modes from the realization R.

In order to determine the nonminimal-observable realization R, the system S_{pm} (2) is considered as a set of p independent single-output, multi-input subsystems S^i_{lm}, $i\in[1,p]$, satisfying differential equations

$$S^i_{lm}: \quad \begin{aligned} \dot{x}^i(t) &= F^i x^i(t) + G^i u(t)\,; \quad & u(t_o) &= u_o \\ y_i(t) &= h^i x^i(t) & x^i(t_o) &= x^i_o \end{aligned} \qquad (11)$$

It has been shown that using the vector u(t) and $y_i(t)$, (the i-th element of the vector y(t)), it is possible to determine the order and parameters in the minimal realization

$$R^i_{min} = \{F^i, G^i, h^i\} \qquad (12)$$

corresponding to i-th subsystem. In addition to measurable u(t) and y(t) the determination of all R^i_{min} $i\in[1,p]$ requires also a suffisient number of functions to be obtained by integrating k times vectors u(t), y(t) and Dirak impulse function $\delta(t)$. These functions may be denoted by

$$\begin{aligned} v^i_r(t) &= \int_{t_o}^{t} v^i_{r-1}(\tau)\,d\tau; \quad & v^i_o(t) &= y_i(t) \\ w^j_r(t) &= \int_{t_o}^{t} w^j_{r-1}(\tau)\,d\tau; & w^j_o(t) &= u_j(t) \\ s_r(t) &= \int_{t_o}^{t} s_{r-1}(\tau)\,d\tau; & s_o(t) &= \delta(t-t_o) \end{aligned} \qquad (13)$$

$i\in[1,p]$; $j\in[1,m]$; $r\in[1,k]$.

Assumption: functions $u_j(t)$ should assure linear independence of all k(m+1) functions $w^j_r(t)$ and $s_r(t)$. Cascades of integratiors (13) may be interpreted as simplified sensitivity models /Bingulac, (1973)/ generating a special form of pseudo sensitivity functions. In /Djorović, (1973)/ it has been shown that instead of integrators an arbitrary first order blocks may be used.

It has been shown that each order n^i of the minimal realization R^i_{min} may be determined from

$$n^i = \nu^{ik} - k(m+1) \qquad k \geq n^i \qquad (14)$$

where $\nu^{ik} = \text{rank}\left[z^{ik}(t)^T \cdot z^{ik}(t)\right] \triangleq G\left[z^{ik}(t)\right]$

while $z^{ik}(t) = \left[v^{ik}(t) \,\vdots\, w^k(t) \,\vdots\, s^k(t)\right]$ and

$$v^{ik}(t) = \left[v^i_1(t) \ldots v^i_k(t)\right]$$

$$w^k(t) = \left[w^1_1(t) \ldots w^1_k(t) \;\; w^2_1(t) \ldots\ldots w^m_k(t)\right].$$

$$s^k(t) = \left[s_1(t) \ldots\ldots s_k(t)\right]$$

For $k = n^i$ the matrix $G\left[z^{in^i}(t)\right]$ is of the full rank and its inversion permits determination of parameters in R^i_{min}. $(n^i \times n^i)$ matrix F^i is in vertical companion form

$$F^i = \left[-f^i \;\middle|\; \begin{matrix} I \\ \hline 0 \end{matrix} \right] \qquad (15)$$

while h^i is n^i-dimensional row vector of the form

$$h^i = [1\ o\ \ldots\ldots\ o]$$

In (15) n^i-dimensional column vector f^i contains coefficients of the characteristic polinomial of the i-th subsystem S^i_{lm}.

The unknown parameters in (12) are determined from identity

$$y_i(t) = - v^{in^i}(t) \cdot f^i + w^{n^i}(t) \cdot G^i + s^{n^i}(t) \cdot \alpha \ , \qquad (15')$$

which is a direct consequence of the mathematical model (11) as well as the assumed form of F^i and h^i in (15). Coefficients in n^i-dimensional column vector α take care about the initial conditions of the considered subsystem S^i_{lm}. Therefore, combining all realizations R^i_{min}, a nonminimal-observable realization (9) of the system S_{pm} may be written as

$$F = \begin{bmatrix} F^1 & & & 0 \\ & F^2 & & \\ & & \ddots & \\ 0 & & & F^p \end{bmatrix} (n_* \times n_*) \ ; \ G = \begin{bmatrix} G^1 \\ G^2 \\ \vdots \\ G^p \end{bmatrix} (n_* \times m) \qquad (16)$$

$$H = \begin{bmatrix} h^1 & & & 0 \\ & h^2 & & \\ & & \ddots & \\ 0 & & & h^p \end{bmatrix} (p \times n_*) \ ; \ n_* = \sum_{i=1}^{p} n^i$$

Note that this form of nonminimal realization is most convenient for the subsequent procedure of constructing R_{min}.

Realization R, given in (15), is by a definition observable but the controllability as well as the minimality cannot be guaranteed. Hence, the corresponding minimal realization may be obtained satisfying the required controllability conditions. This may be done eliminating all uncontrollable modes from system (15). In order to facilitate the detection of all uncontrollable modes the realization $R = \{F, G, H\}$ should be by the nonsingular transformation

$$T = \begin{bmatrix} T^1 & & & 0 \\ & T^2 & & \\ & & \ddots & \\ 0 & & & T^p \end{bmatrix} (n \times n) \qquad (17)$$

transformed into the form

$$\hat{R} = \{\hat{F}, \hat{G}, \hat{H}\} \qquad (18)$$

where the matrices $\hat{F}, \hat{G}$ and $\hat{H}$ defined by

$$\hat{F} = T^{-1}.F.T = \begin{bmatrix} \hat{F}^1 & & & 0 \\ & \hat{F}^2 & & \\ & & \ddots & \\ 0 & & & \hat{F}^p \end{bmatrix} ; \ \hat{G} = T^{-1}.G = \begin{bmatrix} \hat{G}^1 \\ \hat{G}^2 \\ \vdots \\ \hat{G}^p \end{bmatrix}$$

$$\hat{H} = H.T = \begin{bmatrix} \hat{h}^1 & & & 0 \\ & \hat{h}^2 & & \\ & & \ddots & \\ 0 & & & \hat{h}^p \end{bmatrix} \qquad (19)$$

satisfy the relations

$$\begin{aligned} \dot{\hat{x}}(t) &= \hat{F}.\hat{x}(t) + \hat{G}.u(t) \\ y(t) &= \hat{H}.\hat{x}(t) \\ Z(s) &= \hat{H}.(sI-\hat{F})^{-1}.\hat{G} \end{aligned} \qquad (20)$$

Since the matrix F in (15) is in block diagonal form the transformation T may also be in block diagonal form. Each block T^i in (17) should be given by the matrix of eigenvectors corresponding to the eigenvalues of F^i. Therefore, the obtained blocks $\hat{F}^i$ are of the Jordan cononical form. Since blocks F^i are in companion form, the process of calculating eigenvalues and eigenvectors may be considerably simplified. The simplified eigenvector calculation procedure applicable to both single and multiple eigenvalues (real and complex) which is used in this paper is described in /Djorović, (1972)/.

Elimination of uncontrollable modes

In order to describe the procedure for elimination of uncontrollable modes assume that the common eigenvalue λ appears in different blocks $\hat{F}^1, \hat{F}^2, \ldots, \hat{F}^{s_\lambda}$ ($s_\alpha \leqslant p$) of the matrix $\hat{F}$. Each row and column in which the eigenvalue λ appears will be denoted by $j = k_1, k_2, \ldots, k_{s_\lambda}$, where k_{s_λ} satisfy the condition

$$\sum_{i=1}^{s_\lambda - 1} n^i < k_{s_\lambda} \leqslant \sum_{i=1}^{s_\lambda} n^i \qquad (21)$$

n^i being dimension of matrix F^i, i.e. dimension of the minimal realization R^i_{min}. Using notation

$$\hat{G}_\lambda = \begin{bmatrix} \hat{g}_{k_1} \\ \hline \hat{g}_{k_2} \\ \hline \vdots \\ \hline \hat{g}_{k_{s_\lambda}} \end{bmatrix} (s_\lambda \times m); \ \hat{H}_\lambda = \left[\hat{h}_{k_1} \,\middle|\, \hat{h}_{k_2} \,\middle|\, \ldots \,\middle|\, \hat{h}_{k_{s_\lambda}} \right] \ (p \times s_\lambda) \qquad (22)$$

where $\hat{g}_j$ and $\hat{h}_j$ represent corresponding rows and columns of matrices $\hat{G}$ and $\hat{H}$, respectively, differential equation defining s_λ - dimensional vector

$\hat{x}_\lambda(t)$ which contains modes with common eigenvalue λ is

$$\dot{\hat{x}}_\lambda(t) = \lambda \cdot \hat{x}_\lambda(t) + \hat{G}_\lambda \cdot u(t) \tag{23}$$

where

$$\hat{x}_\lambda(t) = \left[\hat{x}_{k_1}(t)\ \hat{x}_{k_2}(t)\ \dots\ \hat{x}_{k_{s_\lambda}}(t)\right]^T \tag{24}$$

According to the principle of superposition, the contribution of the vector $\hat{x}_\lambda(t)$ to the entire output vector $y(t)$ may be expressed by

$$y_\lambda(t) = \hat{H}_\lambda \cdot \hat{x}_\lambda(t) \tag{25}$$

The combining of all state coordinates corresponding to a common eigenvalue λ is motivated by the fact that the possible uncontrollability of the system (20) is a direct consequence of the linear dependence of elements $\hat{x}_j(t)$ of the vector $\hat{x}_\lambda(t)$ (24).

Assuming a linear independence of elements of the control vector $u(t)$, the number of linearly independent elements of $\hat{x}_\lambda(t)$ is given by the rank of matrix $\hat{G}_\lambda$, i.e.

$$\nu_\lambda = \text{rank}\ \hat{G}_\lambda \tag{26}$$

Therefore, if the $\nu_\lambda = s_\lambda$, all coordinates of the vector $\hat{x}_\lambda(t)$ are linearly independent and span the basis of the corresponding subspace, and consequently should be retained in the minimal realization. On the other hand, if $\nu_\lambda < s_\lambda$ then $r_\lambda = s_\lambda - \nu_\lambda$ coordinates of $\hat{x}_\lambda(t)$ are linearly dependent and should be eliminated from the realization (18). Thus, the problem of elimination of uncontrollable modes from the system (20) has been reduced to the elimination of r_λ linearly dependent elements of vector $\hat{x}_\lambda(t)$. At this end, the vector $\hat{x}_\lambda(t)$ should be partitioned in two parts as

$$\hat{x}_\lambda(t) = \begin{bmatrix} \hat{x}^1_\lambda(t) \\ \hline \hat{x}^2_\lambda(t) \end{bmatrix} \begin{matrix} (\nu_\lambda \times 1) \\ \hline (r_\lambda \times 1) \end{matrix} \tag{27}$$

where $\hat{x}^1_\lambda(t)$ contains ν_λ linearly independent elements of $\hat{x}_\lambda(t)$, while $\hat{x}^2_\lambda(t)$ is composed of the remaining r_λ elements.

In order to eliminated r_λ linearly dependent coordinates from $\hat{x}_\lambda(t)$, the vector $\hat{x}^2_\lambda(t)$ may be expresed by a linear transformation of $\hat{x}^1_\lambda(t)$, i.e.

$$\hat{x}^2_\lambda(t) = A_\lambda \cdot \hat{x}^1_\lambda(t) \tag{28}$$

where A_λ is a $(r_\lambda \times \nu_\lambda)$ matrix with unknown coefficients dependent upon the elements of $\hat{G}_\lambda$. Inserting (27) and (28) into (23) yields

$$A_\lambda \cdot \hat{G}^1_\lambda = \hat{G}^2_\lambda \tag{29}$$

which finally leads to

$$A_\lambda = \left[\hat{G}^2_\lambda \cdot \hat{G}^{1T}_\lambda\right] \cdot \left[\hat{G}^1_\lambda \cdot \hat{G}^{1T}_\lambda\right]^{-1} \tag{30}$$

where matrices $\hat{G}^1_\lambda$ and $\hat{G}^2_\lambda$ are defined by

$$\hat{G}_\lambda = \begin{bmatrix} \hat{G}^1_\lambda \\ \hline \hat{G}^2_\lambda \end{bmatrix} \begin{matrix} (\nu_\lambda \times m) \\ \hline (r_\lambda \times m) \end{matrix} \tag{31}$$

From this discussion it follows that the vector $\hat{x}^2_\lambda(t)$ may be eliminated and according to (25) its contribution to $y_\lambda(t)$ may be taken over by the vector $\hat{x}^1_\lambda(1)$. This elimination requires reduction of the triple $\hat{R} = \{\hat{F}, \hat{G}, \hat{H}\}$ and consists of the following steps:

1. Elimination of r_λ rows and columns from matrix $\hat{F}$ (corresponding to elements of vector $\hat{x}^2_\lambda(t)$,
2. Elimination of r_λ rows from matrix $\hat{G}$ (corresponding to matrix $\hat{G}^2_\lambda$),
3. Elimination of r_λ columns from matrix $\hat{H}$ (corresponding to elements of vector $\hat{x}^2_\lambda(t)$.
4. Correction of elements in ν_λ columns of matrix $\hat{H}$ (corresponding to the elements of the vector $\hat{x}^1_\lambda(t)$.

The correction of the step 4. takes care of the contribution of $\hat{x}^2_\lambda(t)$ to the output vector $y(t)$. In accordance with (25) and (27) and using notation

$$\hat{H}_\lambda = \left[\hat{H}^1_\lambda \,\vdots\, \hat{H}^2_\lambda\right] \quad (p \times \nu_\lambda) \quad (p \times r_\lambda) \tag{32}$$

the mentioned correction becomes

$$\hat{H}^{12}_\lambda = \hat{H}^1_\lambda + \hat{H}^2_\lambda \cdot A_\lambda \tag{33}$$

where columns of matrix $\hat{H}^{12}_\lambda$ represent the columns of the matrix $\hat{H}$ in the desired minimal realization(10). By the described reduction procedure the relations(23) and (25) are reduced to

$$\dot{\hat{x}}^1_\lambda(t) = \lambda \cdot \hat{x}^1_\lambda(t) + \hat{G}^1_\lambda \cdot u(t) \tag{34}$$

$$y_\lambda(t) = \hat{H}^{12}_\lambda \cdot \hat{x}^1_\lambda(t)$$

Repeating the described reduction procedure for all other common eigenvalues appearing simultaneously in more then one block of matrix $\hat{F}$, all uncontrollable modes can be eliminated from the realization (18), which finally leads to the minimal realization (10). The total number of detected and eliminated uncontrollable modes may be expressed as

$$r = \sum_{\xi=1}^{\xi_\lambda} r_\xi \tag{36}$$

where ξ_λ represents the number of all common eigenvalues $\lambda = \lambda_\xi$; $\xi \in [1, \xi_\lambda]$, appearing in more that one block of $\hat{F}^i$ for which $s_\lambda > 1$. Each r_ξ in (36) represents the corresponding r_λ which by definition is

$$r_\lambda \triangleq s_\lambda - \nu_\lambda \geqslant 0 \tag{37}$$

Therefore, the order n of the minimal realization(10) may be expressed as

$$n = n_* - r \tag{38}$$

where n_* is given by (16).

On the other hand, for all eigenvalues for which $s_\lambda = 1$ it is clear that the corresponding modes must be retained in the minimal realization. Hence, expressing formally: $\nu_\lambda = 1$ and $r_\lambda = 0$, the summation in (36) may be extended to all different eigenvalues appearing in the whole realization (18), i.e.

$$r = \sum_{\xi=1}^{\xi_m} r_\xi ; \quad \xi_m \geq \xi_\lambda \tag{39}$$

In (39) ξ_m represents the total number of different eigenvalues in (18). Using (39) and (37) equation (38) yields:

$$n = n_* - \sum_{\xi=1}^{\xi_m} s + \sum_{\xi=1}^{\xi_m} \nu_\xi \tag{40}$$

From the definition of s_ξ it follows that

$$\sum_{\xi=1}^{\xi_m} s_\xi = \sum_{i=1}^{p} n^i = n_* \tag{41}$$

Hence, using (41) the order of the minimal realization may be finally expressed as

$$n = \sum_{\xi=1}^{\xi_m} \nu_\xi \tag{42}$$

where ν_ξ for $s_\xi > 1$ is given by (26) while obviously for $s_\xi = 1$, $\nu_\xi = 1$. Therefore, the problem of calculating the order n is reduced to the simple determination of ranks ν_ξ.

3. OTHER POSSIBILITIES FOR DETERMINATION OF n

From the previous section it is celar that prior to calculation of the order n of the minimal realization (10) it is necessary to determine the whole nonminimal-observable realization (18) which then permints obtaining of the desired n. However, since realizations R^i_{min} of subsystems S^i_{lm} and consequently the realization R are obtained by direct processing of input-output operating records $\{u, y\}$, it is natural to expect that the unknown order n of the system S_{pm} may be also determined directly by processing the same signals, without the necessity for the data processing of the form:

$$\{u, y\} \Longrightarrow \{R^i_{min}\} \Longrightarrow \{R\} \Longrightarrow \{\hat{R}\} \Longrightarrow n$$

which is required for application of (42).

Direct determination of the order n

The procedure for determination of the order n of the minimal realization R_{min} by direct processing of functions $\{u(t), y(t)\}$ is based on the following theorem.

Theorem

For each linear time-invariant dynamic system S_{pm}, the order n of the minimal realization may be calculated from

$$n = \nu^k - k(m+1) \quad \text{for} \quad k \geq n \tag{43}$$

where $\nu^k \triangleq \operatorname{rank} G^k$; $G^k \triangleq \left[\overline{z^k(t)^T \cdot z^k(t)}\right]$ $(\eta^k \times \eta^k)$. The $\eta^k = k(p+m+1)$ dimensional row vector $z^k(t)$ is given by

$$z^k(t) = v^k(t) \;\; w^k(t) \;\; s^k(t) \qquad (1 \times \eta^k)$$

$$\text{where} \quad v^k(t) = v^{1k}(t) \ldots v^{pk}(t) \qquad (1 \times pk) \tag{44}$$

$$w^k(t) = w^{1k}(t) \ldots w^{mk}(t) \qquad (1 \times mk)$$

According to (13), elements $v^i_r(t)$, $w^j_r(t)$ and $s_r(t)$ $i \in [1, p]$; $j \in [1, m]$; $r \in [1, k]$ of k - dimensional vectors $v^{ik}(t)$, $w^{jk}(t)$ and $s^k(t)$ are obtained by successive integration of signals $y_i(t)$, $u_j(t)$ and Dirak´s impulse function, respectively. The exact proof of the Theorem is given in /Bingulac, (1972)/. For completeness of the presentation in this paper, only the general outline of the proof is given.

If a $(\eta^k \times \eta^k)$ matrix G^k given by (43) has rank ν^k then there exist

$$\rho^k = \eta^k - \nu^k \tag{45}$$

linear relations of the form

$$C^k \cdot z^k(t)^T = 0 \tag{46}$$

where $(\rho^k \times \eta^k)$ matrix C^k should satisfy

$$\operatorname{rank} C^k = \rho^k \tag{47}$$

According to (43) and (44) equation (45) yields

$$\rho^k = k(p+m+1) - n - k(m+1) = kp - n$$

which then using (16), (38) and (39) becomes

$$\rho^k = \sum_{i=1}^{p} (k - n^i) + \sum_{\xi=1}^{\xi_\lambda} r_\xi \tag{48}$$

Matrix C^k from (46) may be partitioned as

$$C^k = \begin{bmatrix} C^{11} & C^{12} & C^{13} \\ C^{21} & C^{22} & C^{23} \end{bmatrix} \begin{matrix} \}\rho^{k1} \\ \}\rho^{k2} \end{matrix} \tag{49}$$

$$\qquad\quad \underbrace{\phantom{C^{11}}}_{kp} \; \underbrace{\phantom{C^{12}}}_{km} \; \underbrace{\phantom{C^{13}}}_{k}$$

where $\rho^{k1} = \sum_{i=1}^{p} (k - n^i)$ and $\rho^{k2} = \sum_{\xi=1}^{\xi_\lambda} r_\xi$

As a direct consequence of identity (15'), used in the identification of the minimal realization R^i_{min} of all subsystems S^i_{lm}; $i \in [1, p]$, the $(\rho^{k1} \times kp)$ matrix C^{11} appear in a block diagonal form, i.e.:

$$C^{11} = \begin{bmatrix} C^{11}_{1.} & & 0 \\ & \ddots & \\ 0 & & C^{11}_{p} \end{bmatrix} \tag{50}$$

where matrices C^{11}_{i} in the main diagonal are of the form

$$C^{11}_{i} = \begin{bmatrix} a^{i}{}_{a}{}^{i} & & 0 \\ & \ddots & \\ 0 & & a^{i} \end{bmatrix} \quad \left[(k-n^{i}) \times k \right]$$

The $(n^{i}+1)$ - dimensional row vector

$$a^{i} = \left[1 \;\; f^{i}_{n} \; \ldots \; f^{i}_{1} \right]$$

contains unity and n^{i} coefficients from the first column of the matrix F^{i}, given by (15).

On the other hand, $(\rho^{k2} \times kp)$ matrix C^{21} is not block diagonal. For convenience, it may be partitioned in ξ_{λ} parts as:

$$C^{21} = \begin{bmatrix} C^{21}_{1} \\ \hline \vdots \\ \hline C^{21}_{\xi} \\ \hline \vdots \\ \hline C^{21}_{\xi_{\lambda}} \end{bmatrix} \tag{51}$$

Each row of the $(r_{\xi} \times kp)$ matrix C^{21}_{ξ} may be further partitioned /Bingulac, (1972)/ into p k-dimensional row vectors, where in each row among all p vectors there are only $(\nu_{\xi}+1)$ ones with non-zero elements. Moreover the total number of vectors in each row of C^{21} with non-zero elements is s_{ξ} . A typical form of the matrix C^{21}_{ξ} is as follows:

$$C^{21}_{\xi} = \Bigg[\underbrace{\begin{matrix} \overbrace{c^{11}}^{k} & \ldots & \overbrace{c^{1\nu_{\xi}}}^{k} \\ \vdots & & \vdots \\ c^{r_{\xi}1} & \ldots & c^{r_{\xi}\nu_{\xi}} \end{matrix}}_{\nu_{\xi}k} \;\Bigg|\; \underbrace{\begin{matrix} \overbrace{b^{1}_{\xi}}^{k} & & 0 \\ & b^{2}_{\xi} \ddots & \\ 0 & & b^{r_{\xi}}_{\xi} \end{matrix}}_{r_{\xi}k} \;\Bigg|\; \underbrace{\begin{matrix} \\ 0 \\ \\ \end{matrix}}_{(p-s_{\xi})k} \Bigg] \; (r_{\xi} \times pk) \tag{52}$$

The k-dimensional vectors b^{ℓ}_{ξ} , $\ell \in [1, r_{\xi}]$ are coefficients of the polinomial

$$b^{\ell}_{\xi}(s) = \prod_{i=1}^{\nu_{\xi}} \frac{a_{i}(s)}{(s+\lambda_{\xi})} \cdot \frac{a_{1}(s)}{(s+\lambda_{\xi})}$$

where $a_{i}(s)$ is the characteristic polinomial of the matrix F^{i} of subsystem S^{i}_{lm}. Each submatrix C^{21}_{ξ} is introduced by r_{ξ} linearly dependent modes resulting from the appearence of the particular common eigenvalue $\lambda = \lambda_{\xi}$ in s_{ξ} bloks of $\hat{F}$.

Finally, inspecting the form of matrices C^{11} and C^{21} given by (50) and (52) as well as their position in the matrix C^{k} defined by (49), it may be concluded that

$$\text{rank } C^{11} = \rho^{k1} \quad ; \quad \text{rank } C^{21} = \rho^{k2} \tag{53}$$

as well as that rank $C^{k} = \rho^{k1} + \rho^{k2}$ which completes the prof of the Theorem indicating that among η^{k} elements of the vector $z^{k}(t)$ there are only ν^{k} linearly independent ones.

4. DETERMINATION OF THE MINIMAL REALIZATION R_{min}

Since the Theorem of Sec.3 permits direct determination of the order n of the minimal realization, the subsequent procedure of identifying the parameters in the minimal realization R_{min} may be computationally simplified. Because the order n may be determined in advance it could be considered as an additional a priori information, which should be appropriately used in order to allow the corresponding computational simplification. By definition, the order n of the minimal realization represents the number of linearly independent (controllable and observable) modes of the whole system S_{pm}, while n^{i} determines the number of these independent modes contained in each particular subsystem S^{i}_{lm}. Thus, the difference between n and n^{i} represents the number of modes which are not observable from the corresponding $y_{i}(t)$. In order to apply the elimination procedure which leads to the minimal realization R_{min} it is not necessary always to determine the whole nonminimal-observable realization R. Rather, it is more advisable to prosess sequentially each particular $y_{i}(t)$ $i \in [1, q]$, $q \leqslant p$, and also to perform partial elimination procedure until the a priori known n is reached.

Therefore, the simplified identification procedure which uses n as an additional a priori information starts with the determination of order n^{1} and the corresponding minimal realization R^{1}_{min} of subsystem S^{1}_{lm}. This is done exactly as explained in sec.2. When S_{pm} is observable from $y_{1}(t)$, the calculated n^{1} will be equal to the previously determined order n and consequently it is not necessary either to calculate the minimal realizations R^{i}_{min} of other subsystems, or to perform the elimination procedure. Then matrices in the minimal realization (10) of the whole system become

$$\hat{\hat{F}} = \hat{F}^{1} \quad ; \quad \hat{\hat{G}} = \hat{G}^{1} \tag{54}$$

On the other hand, if S_{pm} is not observable in respect to $y_{1}(t)$, which occurs more often in practice, the order n^{1} satisfies $n^{1} < n$. Then in order to determine the minimal realization (10) it is necessary to proceed with the determination of both the order and the minimal realization of the next subsystem, i.e. to calculate n^{2} and matrices in R^{2}_{min}.

For

$$n^{1} + n^{2} < n \tag{55}$$

it is clear that all modes of S_{pm} are not contained in these two subsystems, and consequently it is necessary to proced with the determination of minimal realizations of other remaining subsystems R^i_{min}, $i \in [3,q]$, $q \leq p$ until the condition

$$\sum_{i=1}^{q} n^i \geq n \tag{56}$$

is reached.

However, if

$$n^1 + n^2 > n \tag{57}$$

it is then necessary to perform the partial elimination procedure and from R^1_{min} and R^2_{min} to eliminate r^2 linear dependent modes. Eliminating r^2 modes from R^2_{min}, only

$$n^2_a \triangleq n^2 - r^2 \tag{58}$$

modes from R^2_{min} are independent and should be added in order to form the minimal realization (10). Therefore, the total number of independent modes obtained so far is

$$n^{12} = n^1 + n^2_a \tag{59}$$

If the obtained n^{12} is equal to the previously determined n the sequential identification procedure is completed. Hence matrices in (10) become

$$\hat{F}^{12} = \hat{\hat{F}} = \left[\begin{array}{c|c} \hat{F}^1 & 0 \\ \hline 0 & \hat{F}^2_a \end{array}\right]; \quad \hat{G}^{12} = \hat{\hat{G}} = \left[\begin{array}{c} \hat{G}^1 \\ \hline \hat{G}^2_a \end{array}\right] \tag{60}$$

The $(n^2_a \times n^2_a)$ and $(n^2_a \times m)$ matrices $\hat{F}^2_a$ and $\hat{G}^2_a$ are obtained from $\hat{F}^2$ and $\hat{G}^2$ by eliminating corresponding r^2 linear dependent modes, which is done applying the elimination procedure of sec. 2. If $n^{12} < n$, it is necessary, similarly as in the case of sitisfying (55), to continue sequentially with the determination of minimal realizations R^i_{min} of remaining subsystems as well as with the corresponding partial elimination procedure.

Assume that for $i = q$ the condition

$$n^{1q} = n \tag{61}$$

is satisfied. The value n^{1q} is given by the following recurrence relation

$$n^{1q} = n^{1(q-1)} + n^q_a \tag{62}$$

where $n^{1(q-1)}$ represents the number of linearly independent modes contained in the first (q-1) subsystems, while

$$n^q_a \triangleq n^q - r^q \tag{63}$$

is a number of additional linearly independent modes from subsystem S^q_{lm} which are to be retained in the minimal realization (10). In the most general case matrices in the minimal realization become

$$\hat{F}^{1q} = \hat{\hat{F}} = \begin{bmatrix} \hat{F}^1 & & & & 0 \\ & \ddots & & & \\ & & \hat{F}^i_a & & \\ & & & \ddots & \\ 0 & & & & \hat{F}^q_a \end{bmatrix}; \quad \hat{G}^{1q} = \hat{\hat{G}} = \begin{bmatrix} \hat{G}^1 \\ \vdots \\ \hat{G}^i_a \\ \vdots \\ \hat{G}^q_a \end{bmatrix};$$

$$\hat{\hat{H}} = \left[\begin{array}{c} \hat{\hat{H}}^{1q} \\ \hline H^2 \end{array}\right] \begin{array}{l} (q \times n) \\ [(p-q) \times n] \end{array} \tag{64}$$

In (64) blocks $\hat{F}^i_a$ and $\hat{G}^i_a$ are matrices of dimensions $(n^i_a \times n^i_a)$ and $(n^i_a \times m)$, respectively. The $(q \times n)$ matrix $\hat{\hat{H}}^{1q}$, (the upper part of $\hat{\hat{H}}$), is obtained while the partial elimination procedures are being performed. The only unknown matrix from (64) is the $[(p-q) \times n]$ matrix H^2, i.e. the lower part of $\hat{\hat{H}}$. As it is shown in **Appendix**, the matrix H^2 may be determined using some elements of the vector $z^k(t)$ already applied for the calculation of n (44).

Discussion

In this section the sequential method for identification of a minimal realization of a multivariable linear time-invariant dynamic system has been presented. As it has been shown, the desired minimal realization may be obtained by sequential processing of individual output signals $y_i(t)$, which if compared to method of sec. 2, offers considerably saving in computation. Note that the structure of matrix $\hat{\hat{F}}$ in (64) does not depend upon the sequencing of elements of $y_i(t)$, as it is in the case of the method based on the Bucy canonical form. Moreover, in (64) matrix $\hat{\hat{F}}$ remains decupled while whole coupling existing in the considered system is represented by the output matrix $\hat{\hat{H}}$. The form of matrices $\hat{\hat{F}}$, $\hat{\hat{G}}$ and $\hat{\hat{H}}^{1q}$ are exactly the same as in /Djorović, (1972)/. The number of parameters to be determined in these matrices is given by

$$\sum_{i=1}^{q} n^i + n\,m \tag{65}$$

which is similar to (7). Obviously, the difference in number of parameters between (7) and (65) is

$$\sum_{i=1}^{p-q} n^{(q+i)} \tag{66}$$

Applying the method of sec. 2, these parameters (66) would appear in the lower part of the matrix $\hat{\hat{H}}$ of the corresponding minimal realization. Note that the difference (66) is smaller than the total number of elements of H^2 in (64). Application of the sequential method, however, requires the determination of the whole matrix H^2, i.e. all (p–q)n its parameters.

Considering that the matrix $\hat{\hat{F}}$ is in the Jordan canonical form and taking into account the number of particular modes which are contained in the output signals $y_i(t)$, $i \in [(q+1), p]$, it can be concluded that the number of non-zero elements in the calculated matrix H^2 would exactly be equal to (66). Finally, one comes to the point that the total number of parameters in (64) is the same as (7) and represents the smallest number of parameters by which the considered system may be described.

5. EXPERIMENTAL RESULTS

In order to illustrate the effectiveness of the suggested sequential identification method, a dynamic system S_{pm} with p=3 and m = 2, having a transfer function matrix as

$$Z(s) = \begin{bmatrix} \dfrac{2s+3}{(s+1)(s+2)} & \dfrac{2s+3}{(s+1)^2} \\[2ex] \dfrac{-1}{(s+2)(s+3)} & \dfrac{s+5}{(s+1)(s+3)} \\[2ex] \dfrac{-(3s+7)}{(s+1)(s+3)} & \dfrac{-(3s^2+16s+17)}{(s+1)^2(s+3)} \end{bmatrix}$$

has been simulated on the digital computer. For input signals two uncorrelated pseudo-random signals were used. By simulation of Z(s) corresponding signals $y_i(t)$ have been obtained. According to (13) signals u(t), y(t) have been integrated k=6 times. Initial conditions in the system are assumed to be zero.

Applying the theorem of sec. 3 the value n=4 has been obtained as an unknown order of the minimal realization. According to the procedure of sec. 2, by processing functions u(t), y(t) and $w^k(t)$, $v^{1k}(t)$, for k=4 order and minimal realization of the first subsystem become: $n^1 = 3$, and

$$F^1 = \begin{bmatrix} -4 & 1 & 0 \\ -5 & 0 & 1 \\ -2 & 0 & 0 \end{bmatrix} ; \quad G^1 = \begin{bmatrix} 2 & 2 \\ 2 & 7 \\ 3 & 6 \end{bmatrix} ; \quad h^1 = [1 \quad 0 \quad 0]$$

Since $n^1 < n$, repeating the same procedure for the second subsystem the corresponding minimal realization R^2_{min} has been obtained: $n^2 = 3$,

$$F^2 = \begin{bmatrix} -6 & 1 & 0 \\ -11 & 0 & 1 \\ -6 & 0 & 0 \end{bmatrix} ; \quad G^2 = \begin{bmatrix} 0 & 1 \\ -1 & 7 \\ -1 & 10 \end{bmatrix} ; \quad h^2 = [1 \quad 0 \quad 0]$$

Since $n^1 + n^2 = 6 > n = 4$ (sec. 4), it is necessary to perform the first partial elimination procedure. The required Jordan canonical forms for these two subsystems are

$$\hat{F}^1 = \begin{bmatrix} -1 & 1 & 0 \\ 0 & -1 & 0 \\ 0 & 0 & -2 \end{bmatrix} ; \quad \hat{G}^1 = \begin{bmatrix} 1 & 1 \\ 0 & 1 \\ 1 & 0 \end{bmatrix} ; \quad \hat{h}^1 = [1 \quad 1 \quad 1]$$

$$\hat{F}^2 = \begin{bmatrix} -1 & 0 & 0 \\ 0 & -2 & 0 \\ 0 & 0 & -3 \end{bmatrix} ; \quad \hat{G}^2 = \begin{bmatrix} 0 & 2 \\ -1 & 0 \\ 1 & -1 \end{bmatrix} ; \quad \hat{h}^2 = [1 \quad 1 \quad 1]$$

Eliminating the uncontrollable modes from the second subsystem, the reduced triple $\{\hat{F}^{12}, \hat{G}^{12}, \hat{H}^{12}\}$ becomes

$$\hat{F}^{12} = \begin{bmatrix} -1 & 1 & 0 & 0 \\ 0 & -1 & 0 & 0 \\ 0 & 0 & -2 & 0 \\ 0 & 0 & 0 & -3 \end{bmatrix} ; \quad \hat{G}^{12} = \begin{bmatrix} 1 & 1 \\ 0 & 1 \\ 1 & 0 \\ 1 & -1 \end{bmatrix} ;$$

$$\hat{H}^{12} = \begin{bmatrix} 1 & 1 & 1 & 0 \\ 0 & 2 & -1 & 1 \end{bmatrix}$$

Since the order $n^{12} = 4$ of the matrix $\hat{F}^{12}$ is equal to the previously determined order of the system, the sequential procedure is completed. Thus, matrices $\hat{\hat{F}}$ and $\hat{\hat{G}}$ of the desired minimal realization become

$$\hat{\hat{F}} = \hat{F}^{12} \quad \text{and} \quad \hat{\hat{G}} = \hat{G}^{12}$$

Matrix $\hat{\hat{H}}$ from (64) may be written as

$$\hat{\hat{H}} = \begin{bmatrix} \hat{\hat{H}}^{12} \\ \hline h^2 \end{bmatrix}$$

According to the Appendix and using the obtained $\hat{\hat{F}}$, $\hat{\hat{G}}$, $\hat{\hat{H}}^{1q}$ as well as signals {u(t), y(t)} and their integrals, vector

$$h^2 = [-2 \quad -2 \quad 0 \quad -1]$$

is obtained; in what way the whole identification procedure has been completed. The form of vector h^2 indicates that the order of the third subsystem is $n^3 = 3$, as well as that the mode with the eigenvalue equal to −2 is not contained in $y_3(t)$.

Therefore, the total number of parameters by which the simulated system may be represented in the state space domain is equal to

$$\sum_{i=1}^{2} n^i + n\,m + n^3 = 6 + 8 + 3 = 17$$

Note that the method of sec. 2 would lead to the same state space representation having also 17 parameters. Application of the method based on the Bucy canonical form, however, would yield a realization with 19 parameters.

6. REFERENCES

Ackermann, J. E. (1971), "Die minimale Ein-Ausgangs Beschreibung von Mehrgrössensystemen und ihre Bestimmung aus Ein-Ausgangs Messungen", Regelungstechnik, Heft 5, pp. 203-206.

Ackermann J.E. and Bucy, R.S, (1971), "Canonical Minimal Realization of a Matrix of Impulse Response Sequences", Information and Control, Vol. 19, No. 3, pp. 224-231.

Bingulac S. P., Djorović, M., (1972), "On the Determination of order of Multivariable Linear Dynamic Systems", Publication of the Institute "B. Kidrič".

Bingulac, S. P. and Djorović, M., (1973), "On the Generation of Sensitivity ans Pseudosensitivity Functions for Multivariable Systems", III IFAC Simposium on Sensitivity, Adaptivity and Optimality Ischia, Italy.

Bucy, R.S. (1968), "Canonical Forms for Multivariable Systems", IEEE Trans., Vol AC-13, No. 5, pp. 567-569.

Budin M.A., (1971), "Minimal Realization of Discrete Linear Systems from Input-Output Observations", IEEE Trans., Vol. AC-16, No. 5, pp. 395-401.

Cruz, J. B. Jr. (1971), "Feedback Systems", McGraw Hill.

Djorović M. and Bingulac, S. P. (1972), "Order Determination and Parameter Identification of Time--Invariant State Variable Model", V-th IFAC Congress, Paris, paper No. 38.5.

Djorović, M. (1973), "Identification of Single-Input, Single-Output Systems", MS Thesis, University of Belgrade, EE Department, Beograd.

Gopinath, B., (1969), "On the Identification of Linear Time-Invariant Systems from Input-Output Data", The Bell System Tech. Journal, Vol. 48, No. 5, pp. 1101-1113.

Ho. B. L. and Kalman, R. E., (1966), "Effective Construction of Linear State-Variable Models from Input-Output Functions", Regelungstechnik, Heft 12, pp. 545-548.

Mataušek, M. R. and Milovanović M. D. (1973), "Identification by Pseudosensitivity Functions and Quasilinearization", III IFAC Symposium on Identification and System Parameter Estimation", The Hague, The Netherlands.

Mayne, D. Q. (1968), "Computational Procedure for the Minimal Realisation of Transfer Functions Matrix", Proc. IEE, Vol. 115, No. 2. pp. 325-327.

Rosenbrock, H. H. (1970), "State Space and Multivariable Theory", Nelson and Sons Ltd., London.

7. APPENDIX

System S_{pm} having minimal realization (64) may be described by

$$\dot{x}(t) = \hat{\hat{F}} \cdot x(t) + \hat{\hat{G}} \cdot u(t) \; ; \quad x(t_o) = x_o \qquad (a.1)$$

$$y(t) \triangleq \begin{bmatrix} y^1(t) \\ \hline y^2(t) \end{bmatrix} = \begin{bmatrix} \hat{H}^{1q} \\ \hline H^2 \end{bmatrix} \cdot x(t) \qquad (a.2)$$

Since matrices $\hat{\hat{F}}, \hat{\hat{G}}$ and vector x_o are known, by solving the system of differential equation (a.1) the unknown matrix H^2 may be calculated from

$$H^2 = \left[\overline{y^2(t) \cdot x(t)^T} \right] \cdot \left[\overline{x(t) \cdot x(t)^T} \right]^{-1} \qquad (a.3)$$

which is a direct consequence of (a.2).

However, there is a more convenient possibility for determination of H^2 which does not require the solution of any differential equations, but rather uses some of the elements of the vector $z^k(t)$ already applied in the determination of the order n. Applying the Laplace transformation and by appropriate partitioning of the state vector x(t) and matrices $\hat{H}^{1q}$ and H^2, eq. (a.2) may be written as:

$$\begin{aligned} y^1(s) &= \hat{H}_1^{1q} \cdot x^1(t) + \hat{H}_2^{1q} \cdot x^2(t) \\ y^2(s) &= H_1^2 \cdot x^1(t) + H_2^1 \cdot x^2(t) \end{aligned} \qquad (a.4)$$

Since by definition rank $\hat{H}^{1q} = q$, the partitioning of $\hat{H}^{1q}$ and x(t) may be always done in such a way that (q x q) matrix $\hat{H}_1^{1q}$ is regular. Thus, eliminating q-dimensional vector $x^1(t)$, eq. (a.4) may be written as

$$y^2(s) = H_1^2 \left[\hat{H}_1^{1q} \right]^{-1} \cdot y^1(s) + \left\{ H_2^2 - H_1^2 \left[\hat{H}_1^{1q} \right]^{-1} \cdot \hat{H}_2^{1q} \right\} x^2(s) \qquad (a.5)$$

Taking into account the diagonal form of $\hat{\hat{F}}$, the state coordinates become

$$x_i(s) = \frac{g^i \cdot u(s)}{(s + \lambda_i)}$$

g^i being i-th row of $\hat{\hat{G}}$. Multiplying both sides of (a.5) by a polinomial

$$a(s) = \prod_{i=q+1}^{n} (s + \lambda_i)$$

divided by $s^{(n-q)}$ eq. (a.5) leads to a linear expression relating vectors $\{y(s), u(s)\}$ to some elements of the vector $z^k(s)$, defined by (44), from which the unknown matrices H_1^2 and H_2^2 may be easily determined.

The case of single eigenvalues is assumed here. By a simple extention of the above derived formulae the case of multiple eigenvalues may be also treated.

DESIGN AND CHARACTERISATION OF OPTIMAL TEST SIGNALS FOR LINEAR SINGLE INPUT-SINGLE OUTPUT PARAMETER ESTIMATION

G.C. Goodwin and R.L. Payne
Department of Computing and Control
Imperial College, London SW7 2BT.

The choice of test signal in an identification experiment has a significant bearing upon the achievable accuracy. This paper shows that, for large record lengths, the optimal signal for linear single input-single output parameter estimation may be characterised by its spectral properties. This fact is shown to lead to a general test signal design algorithm. Realizations of optimal test signals can be obtained with either amplitude or energy constraints. The design method is applied to several examples and it is shown to lead to improved estimation accuracy.

1. INTRODUCTION

Until recently most interest in identification was directed towards the data analysis problem. However, there now exists a growing recognition of the importance of experiment design. This paper presents a unified treatment of the test signal design problem for linear discrete time single input single output parameter estimation.

Several previous authors have discussed the problem of choice of test signal for restricted classes of systems and cost functions. Levin (1960) showed that a signal having an impulsive autocorrelation function is the best test signal for the estimation of the discrete pulse response of a linear system in the presence of white output noise when the input in energy or amplitude constrained. Cumming (1970) experimentally demonstrated that for a system described by a simple transfer function the input band width should be approximately equal to the system bandwidth. The book by Box and Jenkins (1970) contains an interesting discussion of the problem and some preliminary analytic results. Other authors; Levadi (1966), Litman and Huggins (1963), Nahi and Wallis (1969), Goodwin et al (1973), Reid (1972), Keviczky (1972), Mehra (1970) and Aoki and Staley (1970), have proposed design algorithms based on the selection of an input sequence.

In this paper it is shown that the optimal test signal is characterised by its spectral properties and this is shown to lead to a general design method.

2. MODEL DESCRIPTION

A linear single input-single output noisy dynamic system may be described by a transfer function model of the form

$$(1-z^{-1})^p y_t = \frac{B(z^{-1})}{A(z^{-1})} z^{-r} (1-z^{-1})^q u_t + \frac{D(z^{-1})}{C(z^{-1})} w_t \qquad (2.1)$$

The parameters appearing in the polynomial operators $A(z^{-1})$, $B(z^{-1})$, $C(z^{-1})$, $D(z^{-1})$ may be estimated by analysis of system data. A measure of the accuracy of estimation is the parameter covariance matrix. An expression for the Cramer Rao Lower Bound on the Covariance matrix is developed in the next section.

3. CRAMER RAO LOWER BOUND

From Eq. (2.1) the total set of parameters to be estimated is

$$\beta^T = \{a_1 \ldots a_n, \; b_o \ldots b_m, \; c_1 \ldots c_k, \; d_1 \ldots d_l\} \qquad (3.1)$$

Assuming that w_t is normally distributed and that N observations of the input/output record will be used to estimate β, it can be shown that the log likelihood function is given by:

$$L = -\frac{N}{2}\log(2\pi) - N\log\sigma - \frac{1}{2\sigma^2}\sum_{t=1}^{N} \varepsilon_t^2 \qquad (3.2)$$

where $\varepsilon_1 \ldots \varepsilon_N$ is the residual sequence defined by:

$$\varepsilon_t = \frac{C(z^{-1})}{D(z^{-1})}\{(1-z^{-1})^p y_t - \frac{B(z^{-1})}{A(z^{-1})} z^{-r} (1-z^{-1})^q u_t\} \qquad (3.3)$$

The Cramer Rao inequality states that the covariance matrix of any unbiased estimator $\hat{\beta}$ of β satisfies the inequality:

$$\mathrm{cov}(\hat{\beta}) \geq I_\beta^{-1} \qquad (3.4)$$

where equality is achieved for an efficient estimator and I_β is the Fisher information matrix given by:

$$I_\beta = E_\beta[(\frac{\partial L}{\partial \beta})(\frac{\partial L}{\partial \beta})^T] \qquad (3.5)$$

An expression for $(\frac{\partial L}{\partial \beta})$ can be obtained by differentiating (3.2) with respect to the relevant parameters:

$$\frac{\partial L}{\partial \beta} = -1/\sigma^2 \sum_{t=1}^{N} \varepsilon_t \frac{\partial \varepsilon_t}{\partial \beta} \qquad (3.6)$$

The derivatives of ε_t with respect to the parameters can be obtained by differentiation of Eq. (3.3):

$$\frac{\partial \varepsilon_t}{\partial a_i} = \frac{C(z^{-1})B(z^{-1})z^{-(i+r)}(1-z^{-1})^q}{D(z^{-1})A^2(z^{-1})} u_t \qquad (3.7)$$

$$\frac{\partial \varepsilon_t}{\partial b_i} = \frac{C(z^{-1})z^{-(i+r)}(1-z^{-1})^q}{D(z^{-1})A(z^{-1})}u_t \tag{3.8}$$

$$\frac{\partial \varepsilon_t}{\partial c_i} = \frac{z^{-i}}{C(z^{-1})}\varepsilon_t \tag{3.9}$$

$$\frac{\partial \varepsilon_t}{\partial d_i} = \frac{-z^{-i}}{D(z^{-1})}\varepsilon_t \tag{3.10}$$

It is observed that Eqs. (3.7) and (3.8) are independent of $\{\varepsilon_t\}$. Substituting (3.7) to (3.10) into (3.5) and taking expectations with respect to the distribution of $\{\varepsilon_t\}$ leads to the deterministic result:

$$\hat{I}_\beta = 1/\sigma^2 \sum_{t=1}^{N} \left[\begin{array}{cc|c} (\frac{\partial \varepsilon_t}{\partial a})(\frac{\partial \varepsilon_t}{\partial a})^T & (\frac{\partial \varepsilon_t}{\partial a})(\frac{\partial \varepsilon_t}{\partial b})^T & \\ (\frac{\partial \varepsilon_t}{\partial b})(\frac{\partial \varepsilon_t}{\partial a})^T & (\frac{\partial \varepsilon_t}{\partial b})(\frac{\partial \varepsilon_t}{\partial b})^T & 0 \\ \hline \multicolumn{2}{c|}{0} & R \end{array}\right] \tag{3.11}$$

where 0 is a null matrix and R refers to the parameters in C and D, and is independent of $\{u_t\}$.

Eqs. (3.7) and (3.8) can be expressed as follows:

$$S_t = z^{-1}B(z^{-1})x_t \tag{3.12}$$

$$T_t = -A(z^{-1})x_t \tag{3.13}$$

$$x_t = z^{-r}\frac{B'(z^{-1})}{A'(z^{-1})}u_t \tag{3.14}$$

where

$$S_t = \frac{\partial \varepsilon_t}{\partial a_1}; \quad \frac{\partial \varepsilon_t}{\partial a_i} = S_{t-i+1} \tag{3.15}$$

$$T_t = \frac{\partial \varepsilon_t}{\partial b_o}; \quad \frac{\partial \varepsilon_t}{\partial b_i} = T_{t-i} \tag{3.16}$$

and

$$B'(z^{-1}) = C(z^{-1})(1-z^{-1})^q \tag{3.17}$$

$$A'(z^{-1}) = D(z^{-1})A^2(z^{-1}) \tag{3.18}$$

Substituting (3.15) and (3.16) into (3.11) yields

$$\hat{I}_\beta = \frac{1}{\sigma^2}\sum_{t=1}^{N} \left[\begin{array}{c|c|c} \begin{matrix} S_t^2 & \cdots & S_t S_{t-n+1} \\ \vdots & & \vdots \\ S_{t-n+1}S_t & \cdots & S_{t-n+1}^2 \end{matrix} & \begin{matrix} S_t T_t & \cdots & S_t T_{t-m} \\ \vdots & & \vdots \\ S_{t-n+1}T_t & \cdots & S_{t-n+1}T_{t-m} \end{matrix} & 0 \\ \hline \begin{matrix} T_t S_t & \cdots & T_t S_{t-n+1} \\ \vdots & & \vdots \\ T_{t-m}S_t & \cdots & T_{t-m}S_{t-n+1} \end{matrix} & \begin{matrix} T_t^2 & \cdots & T_t T_{t-m} \\ \vdots & & \vdots \\ T_{t-m}T_t & \cdots & T_{t-m}^2 \end{matrix} & 0 \\ \hline 0 & 0 & R \end{array}\right] \tag{3.19}$$

It is observed that the right hand side of Eq. (3.19) depends upon the sample autocorrelation and crosscorrelation of $\{S_t\}$ and $\{T_t\}$. For large N and assuming stationarity, Eq. (3.19) can be written in terms of the second moment matrices of $\{S_t\}$ and $\{T_t\}$ as follows:

$$\hat{I}_\beta = \frac{N}{\sigma^2}\left[\begin{array}{cc|c} \Gamma_{SS} & \Gamma_{ST} & 0 \\ \Gamma_{TS} & \Gamma_{TT} & \\ \hline \multicolumn{2}{c|}{0} & R \end{array}\right] \tag{3.20}$$

where Γ_{SS}, Γ_{ST}, Γ_{TS} and Γ_{TT} are Toeplitz matrices e.g.:

$$\Gamma_{ST} = \begin{bmatrix} \gamma_{ST}(0) & \cdots & \gamma_{ST}(m) \\ \vdots & & \vdots \\ \gamma_{ST}(1-n) & \cdots & \gamma_{ST}(0) \end{bmatrix} \tag{3.21}$$

where from Eqs. (3.12) to (3.14)

$$\gamma_{SS}(\tau) = B(z^{-1})B(z)\gamma_{xx}(\tau) \tag{3.22}$$

$$\gamma_{ST}(\tau) = \gamma_{TS}(-\tau) = -z^{-1}B(z^{-1})A(z)\gamma_{xx}(\tau) \tag{3.23}$$

$$\gamma_{TT}(\tau) = A(z^{-1})A(z)\gamma_{xx}(\tau) \tag{3.24}$$

and

$$\gamma_{xx}(\tau) = \frac{B'(z^{-1})B'(z)}{A'(z^{-1})A'(z)}\gamma_{uu}(\tau) \tag{3.25}$$

The significance of Eqs. (3.20) to (3.25) is, firstly, that the test signal only affects the system model parameters, a and b, and secondly, that the Cramer Rao Lower Bound for large N, depends upon only the second moments (auto-correlation or power spectrum) of the test signal.

4. DESIGN CRITERIA

As is the case in all optimisation problems, the relative importance of various conflicting requirements must be ultimately expressed as a scalar function. For the case of parameter estimation, the scalar function should reflect the final use to which the model will be put. For example the scalar might be the mean square error in model step response, the performance of a minimum variance controller, or an optimal control cost function. These functions obviously depend upon $\hat{\beta}$.

The scalar function may be expanded about the "true" parameter values, β^o as follows:

$$S(\hat{\beta}) = s(\beta^o) + [\frac{\partial s}{\partial \beta}]\Delta\beta + \tfrac{1}{2}\Delta\beta^T[\frac{\partial^2 s}{\partial \beta^2}]\Delta\beta + \ldots \tag{4.1}$$

where $s(\hat{\beta})$ is the value of S obtained using the estimated parameters $\hat{\beta}$ and where $\hat{\beta}$ is expressed as $\beta^o + \Delta\beta$.

Assuming an unbiased estimator is used and neglecting terms in $\Delta\beta^3$ and higher, the expected value of $s(\hat{\beta})$ over the distribution of $\hat{\beta}$ can readily be shown to be:

$$E_{\hat{\beta}}\{s(\hat{\beta})\} = s(\beta^o) + \tfrac{1}{2}\mathrm{Trace}\{[\frac{\partial^2 s}{\partial \beta^2}]\mathrm{cov}_\beta(\hat{\beta})\}$$

The expected value of the ultimate cost function s can therefore be optimised by designing the test signal in the parameter estimation experiment so as to minimise the trace of the weighted parameter covariance matrix. The weighting matrix, $\frac{\partial^2 s}{\partial \beta^2}$, which will normally be positive definite for sensibly chosen functions s, is readily calculated using model sensitivity equations.

In the sequel the trace of the weighted covariance matrix will be used for optimal experiment design. It is trivial matter to include other cost functions such as determinant of covariance matrix if desired. However a cautionary note regarding the use of the trace of the Information Matrix as a cost function is in order. The trace of the Information Matrix, in principle, avoids the difficulties inherent in the matrix inversion in Eq. (3.4) and leads to analytic results. However a simple example will serve to show that this cost function is inappropriate. For the case of a weighting sequence model and white output noise the Information Matrix (Eq. (3.20)) reduces to the second moment matrix of the input. Hence maximation of the trace of the Information Matrix is equivalent to maximation to the test signal energy, independent of the spectral properties. This conclusion is false as it is well known (Levin 1960) that the optimal input for this example has an impulsive autocorrelation function. The correct result is achieved using the trace or determinant of the covariance matrix as a cost function.

5. THE CONDITIONS OF OPTIMALITY

The optimal test signal is characterised by the autocorrelation function which minimises:

$$J = \mathrm{Trace}(W\hat{I}_\beta^{-1}) \tag{5.1}$$

where W is a weighting matrix, $\hat{I}_\beta^{-1}$ is the Cramer Rao Lower Bound on the parameter covariance matrix and $\hat{I}_\beta$ is the Information Matrix given by Eqs. (3.20) to (3.24).

In the determination of the optimal input autocorrelation function it is necessary to ensure that the second moment matrix remains positive definite and that the input energy remain finite. These two objectives can be achieved by factorisation of the second moment matrix and by constraining the energy in the factor, i.e.

$$\Gamma_{uu} = U^T U \tag{5.2}$$

where U is a lower triangular semi infinite toeplitz matrix of the form:

$$U = \begin{bmatrix} u_1 & 0 & 0 & \cdot & \cdot & \cdot & 0 & \cdot & \cdot & \cdot & \cdot \\ u_2 & u_1 & 0 & & & & \cdot & & & & \\ \cdot & u_2 & u_1 & \cdot & & & \cdot & & & & \\ \cdot & \cdot & u_2 & \cdot & \cdot & \cdot & \cdot & & & & \\ u_N & \cdot & \cdot & & \cdot & \cdot & \cdot & & & & \\ 0 & u_N & \cdot & & & \cdot & 0 & & & & \\ \cdot & 0 & u_N & & & & u_1 & \cdot & & & \\ \cdot & \cdot & \cdot & & & & & \cdot & \cdot & & \\ \cdot & \cdot & \cdot & & & & & \cdot & \cdot & & \end{bmatrix} \tag{5.3}$$

where N is chosen large enough that truncation errors are insignificant and where the factor energy is constrained, i.e.

$$\sum_{i=1}^{N} u_i^2 = \gamma_{uu}(0) = 1.0 \text{ (say)} \tag{5.4}$$

Given the sequence $\{u_i\}$ Eqs. (5.3), (5.2) and (3.20) to (3.24) can be used to construct the Information Matrix. This is formally equivalent to using $\{u_i\}$ in Eqs. (3.12), (3.13), (3.14) and (3.19) to construct $\hat{I}_\beta$.

The energy constraint on $\{u_i\}$ can be achieved by the use of generalised polar coordinates; $(u_1, u_2 \ldots u_N)$ is constrained to lie on an N dimension hypersphere.

$$\begin{aligned} u_N &= \sin\Theta_{N-1} \\ u_{N-1} &= \sin\Theta_{N-2}\cos\Theta_{N-1} \\ &\vdots \\ u_2 &= \sin\Theta_1\cos\Theta_2 \ldots \cos\Theta_{N-1} \\ u_1 &= \cos\Theta_1\cos\Theta_2 \ldots \cos\Theta_{N-1} \end{aligned} \tag{5.5}$$

The Jacobian matrix $(\frac{\partial u_t}{\partial \Theta i})$ follows from Eq. (5.5). The characterisation of the optimal test signal is now seen to be a standard constrained optimisation problem; viz. minimisation of (5.1) subject to (5.5), (3.12) to (3.14) and (3.19) The necessary conditions for optimality are readily shown to be:

$$\mathrm{grad}_\Theta(i) = 0, \; i = 1, \ldots N-1 \tag{5.6}$$

where

$$\mathrm{grad}_\Theta(i) = \sum_{t=1}^{N} \mathrm{grad}_u(t)\frac{\partial u_t}{\partial \Theta_i} \tag{5.7}$$

$$\mathrm{grad}_u(t) = B'(z)\lambda_t \tag{5.8}$$

$$\lambda_t = \frac{1}{A'(z)}\{B(z)\gamma_t - A(z)\rho_t\} \tag{5.9}$$

$$\gamma_t = 2\sum_{i=1}^{n}\sum_{j=1}^{n+m+1} F_{ij}\upsilon_j(t+i) \tag{5.10}$$

$$\rho_t = 2 \sum_{i=n+1}^{n+m+1} \sum_{j=1}^{n+m+1} F_{ij} \upsilon_j(t+i-1) \quad (5.11)$$

$$\upsilon_j(t) = j^{th} \text{ element of } [S_t..S_{t-n+1}, T_t..T_{t-m}] \quad (5.12)$$

$$F = -(\hat{I}_\beta)^{-2} W \quad (5.13)$$

$$S_t = z^{-1}B(z^{-1})x_t \quad (5.14)$$

$$T_t = -A(z^{-1})x_t \quad (5.15)$$

$$x_t = \frac{B'(z^{-1})}{A'(z^{-1})} u_t \quad (5.16)$$

Together with the boundary conditions

$$x_t = 0; \ t \leq 0 \quad (5.17)$$

$$\upsilon_j(t) = 0; \quad t > N; \quad j = 1,...n+m+1 \quad (5.18)$$

$$\lambda_t = 0; \ t > N \quad (5.19)$$

6. THE DESIGN ALGORITHM

The design algorithm uses an efficient gradient method and is described in the following steps:

i. Set $\Theta_i = 0$; $i = 1,..., N-1$

This gives

$u_1 = 1$

$u_t = 0$; $t = 2...N$

ii. Solve the <u>scalar</u> difference equation (5.16) in forward time using (5.17) as boundary conditions

iii. Substitute x_t into (5.14) and (5.15) to evaluate S_t and T_t, $t = 1...N$

iv. From $\hat{I}_\beta$ using (3.19) (Note R is not required)

v. Evaluate the cost function, (5.1)

vi. Determine F using (5.13)

vii. Evaluate γ_y and ρ_t, $t = 1,...N$ using (5.10), (5.11) and (5.18)

viii. Solve the <u>scalar</u> difference equation (5.9) in reverse time from the boundary conditions given in (5.19)

ix. Evaluate $grad_u(t)$, $t = 1,...N$ using (5.8)

x. Evaluate $grad_\Theta(i)$, $i = 1,...N$ using (5.7)

xi. Adjust Θ_i in the negative gradient direction using a linear search to minimise J in the search direction (using steps (ii) to (v))

xii. Go to (ii) and iterate until (5.6) is satisfied

xiii. Evaluate optimal autocorrelation using (5.2).

7. REALIZATION OF OPTIMAL TEST SIGNALS

Two types of constraint will be considered, namely, total signal energy and instantaneous signal amplitude. For the case of constrained energy, the algorithm described in sections (5) and (6) leads to a possible realization $u_1,..,u_N$. For the case of amplitude constraints the algorithm may be suitably modified to lead to an amplitude constrained test signal. The modification simply entails deletion of the polar coordinate transformation (5.5) and replacement of Eqs. (5.6) and (5.7) by an orthogonality condition relating $grad_u(t)$ and the constraint.

For large sequence lengths and for the usual practical case of constrained input amplitude, it is more appropriate, however, to realize the test signal via the optimal autocorrelation. It is first observed from Eqs. (3.20) to (3.25) that $\gamma_{uu}(0)$ simply acts as a scaling factor on the part of $\hat{I}_\beta$ relating to the system parameters. Hence, if the signal amplitude is symmetrically constrained, it follows that the cost function (5.1) will be minimised if $\gamma_{uu}(0)$ is maximised i.e. the test signal is saturated at the constraints. It only remains to realize a binary signal having the given autocorrelation. A possible approximate method for doing this is to use the Arcsine Law (Papoulis 1965) and to represent the signal as the saturated output of a linear system driven by white noise.

8. EXAMPLES

Three examples will be presented to show the improvement in estimation accuracy possible with an optimal test signal.

Fig. (1) shows the optimal autocorrelation for a simple first order model $\frac{1}{1-0.95z^{-1}}$ with white output noise. The predicted improvement in the trace of the parameter covariance matrix compared with the use of a test signal having an impulsive autocorrelation function is 11 to 1. It is observed from Fig. (1) that the input energy is primarily low frequency which is consistent with the slow response of the given system.

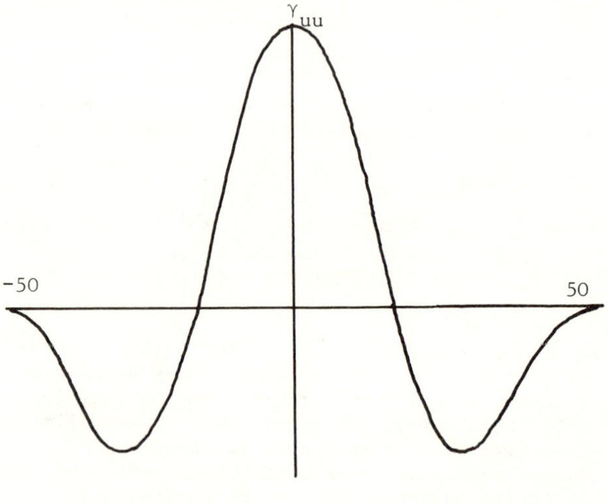

<u>Fig. (1)</u>

Fig. (2) shows the optimal autocorrelation function for a system

$\frac{1}{1-05z^{-1}}$ with white output noise. The predicted improvement in the trace of the parameter covariance matrix compared with the use of a test signal having an impulsive autocorrelation function is 1.7 to 1. It is observed from Fig. (2) that the optimal autocorrelation has high frequency components which is consistent with the fast response of the given system.

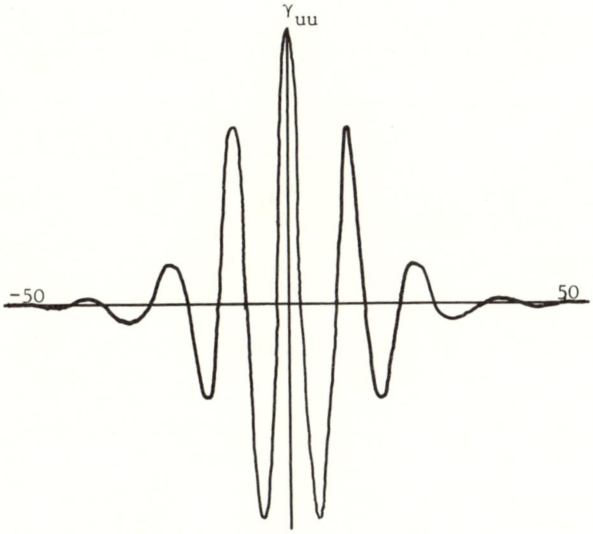

Fig. (2)

To show the effect of the noise model on the optimal test signal, a weighting function system model was chosen with noise model $\frac{1}{1-0.95z^{-1}}$. The optimal test signal autocorrelation is shown in Fig. (3). It is to be noted that the optimal test signal is significantly different from the well known optimal impulsive autocorrelation result for the weighting function model with white output noise (Levin 1960).

The predicted improvement in estimation accuracy has been born out in simulated identification experiments with known models.

9. CONCLUSIONS

This paper has shown that the optimal test signal for linear single input single output parameter estimation is characterised by its autocorrelation function. This fact has been shown to lead to an efficient algorithms for test signal design. Furthermore a procedure has been described for arriving at the most appropriate design criteria for any given ultimate use of the model.

The test signal design and characterisation algorithm has been applied to several examples and has been shown to lead to significant improvement in estimation accuracy.

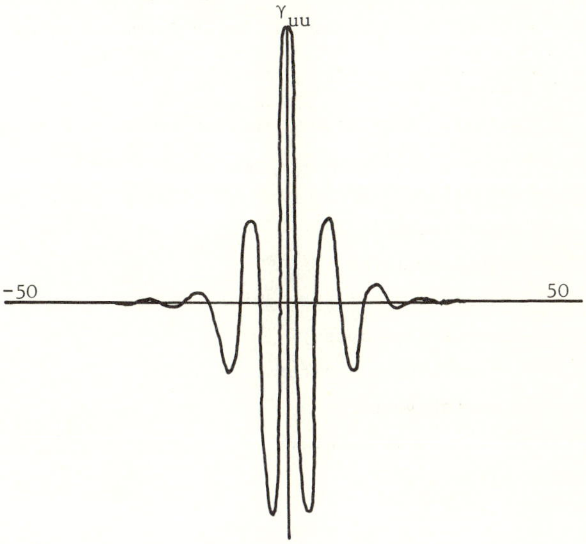

Fig. (3)

10. ACKNOWLEDGEMENT

The authors wish to thank Professor D.Q. Mayne for advice and helpful suggestions.

11. LIST OF PRINCIPLE SYMBOLS

$A(z^{-1})$ = polynomial in z^{-1} = $1+a_1z^{-1}+\ldots+a_nz^{-n}$

$A'(z^{-1})$ = polynomial in z^{-1} of order $2n+l$

a = vector of coefficients of A

$B(z^{-1})$ = polynomial in z^{-1} = $b_o+b_1z^{-1}+\ldots+b_mz^{-m}$

$B'(z^{-1})$ = polynomial in z^{-1} of order $k + q$

b = vector of coefficients of B

$C(z^{-1})$ = polynomial in z^{-1} = $1+cz^{-1}+\ldots c_kz^{-k}$

c = vector of coefficients of C

$D(z^{-1})$ = polynomial in z^{-1} = $1+d_1z^{-1}+\ldots+d_lz^{-l}$

d = vector of coefficients of D

E = expected value operator

F = $(n+m+1,\ n+m+1)$ matrix = $-\hat{I}_\beta^{-2}W$

J = scalar cost function = trace$(I_\beta^{-1}W)$

k = order of C

L = log-likelihood function

l = order of D

m = order of B

N = number of data points

n = order of A

p = sum of integrations in system and noise models

q = number of integrations in noise model

R^{-1} = σ^2/N times Cramer Rao Lower Bound on c, d parameters

r = time delay ≥ 1

S_t = partial derivative of t-th residual w.r.t. a_1

s = scalar function of β

T_t = partial derivative of t-th residual w.r.t. b_o

u_t = input sequence

W = weighting matrix

x_t = output of B/A filter driven by u_t

y_t = noisy system output

z = forward shift operator

β = parameter vector

$\Gamma_{..}$ = second moment matrix

$\gamma_{..}$ = element of $\Gamma_{..}$

γ_t = adjoint variable

ε_t = t-th residual

Θ_i = polar coordinate

λ_t = adjoint variable

υ_t = vector of a and b partial derivatives of ε_t

σ^2 = variance of residuals

w_t = normally distributed white noise source

12. REFERENCES

Aoki, M. and Staley, R.M., 1970, *On Input Signal Synthesis in Parameter Identification,* Automatica, 6, 431.

Box, G.E.P. and Jenkins, G.M., 1970, *Time Series Analysis Forecasting and Control,* (San-Francisco, Holden-Day), pp. 416-420.

Cumming, I.G., 1970, *Frequency of Input Signal in Identification,* IFAC Symposium, June.

Goodwin, G.C., Murdoch, J.C., and Payne, R.L., 1973, *Optimal Test Signal Design for Linear SISO System Identification,* Int. J. Control, 17, 45.

Keviczby, L., 1972, *On Some Questions of Input Signal Synthesis,* Report 7226(B), Lund Institute of Technology, Division of Automatic Control.

Levadi, V.S., 1966, *Design of Input Signals for Parameter Estimation,* IEEE Trans., 11, 205.

Levin, M.J., 1960, *Optimum Estimation of Impulse Response in the Presnece of Noise,* IRE Trans., Circuit Theory, 7, 50.

Litman, S., and Huggins, N.H., 1963, *Growing Exponentials as a Probing Signal for System Identification,* Proc. Inst. Elect. Engrs., 51, 917.

Mehra, R.K., 1970, *Optimal Inputs for Linear System Identification,* Systems Control Inc., Technical Memorandum.

Nahi, N.E. and Wallis, D.E., 1969, *Optimal Inputs for Parameter Estimation in Dynamic Systems with White Observation Noise,* Proceedings of the Joint Automation Control Conference, University of Colorado, U.S.A., paper IV-A5.

Papoulis, A., 1965, *Probability Random Variables and Stochastic Processes,* McGraw Hill, p. 483.

ON INPUT SIGNAL SYNTHESIS FOR LINEAR DISCRETE-TIME SYSTEMS

L.Keviczky

Technical University, Budapest
Automatic Control Department
Budapest
Hungary

Cs.Bányász

Computer and Automation Institute
Hungarian Academy of Sciences
Budapest
Hungary

In our paper it is investigated how to generate optimal input signal series for the identification of linear discrete-time system in order to improve the accuracy of estimate. The determinant of the covariance matrix or the information matrix's inverse are considered as a measure of the error in the parameter estimate. We suggest very simple methods for the minimization of these criteria in case of amplitude constrained input signal.

INTRODUCTION

In the practical parameter estimation problems the question of suitability or optimality of input signal for identification purposes occurs very often. The input signal synthesis has been investigated by many authors and they pointed out that besides the persistently exciting condition other properties of input can also be demanded in order to fulfil some optimality criteria regarding to the goodness of parameter estimation. The relation between the input synthesis and the sensitivity analysis was examined by Rault /1968/ and Inoue /1970/. Many authors have dealt with this problem approaching it from statistical aspects. Levadi /1966/ investigated continuous system : approximating the output by linear series-expansion he set the minimization of covariance matrix's trace as an aim. Nahi /1969/ dealt with the maximization of information matrix's trace likewise for continuous systems. These works led to difficult computation methods /Fredholm equation,variation technique, and so on/ which can not be performed easily. Aoki /1970/ called the attention to the fact that the linear discrete-time systems can be rewritten easily into the form having linearity in parameters and it is more advisable to use the optimum input synthesis there. He elaborated a method for the maximization of information matrix's trace when the equation error is white noise. This procedure minimizes asymptotically the Cramer-Rao lower bound, but its realization is getting more and more difficult concerning the computation technique for many samplings. On the basis of detailed analysis of experimental design's methods we have pointed out the methods of optimum input's synthesis are different in case of static and dynamic identification and suggested the determinant of information matrix as a criterion;c.f. Keviczky /1972/. In this paper procedures elaborated on the basis of mentioned principle are presented.

In our investigations linear discrete-time system is used which can be described by stochastic difference equation

$$A/z^{-1}/\ y/t/=B/z^{-1}/\ u/t/+C/z^{-1}/\lambda e/t/\ , \qquad /1/$$

where $\{y/t/,u/t/,e/t/,\ t=1,2,\ldots,N\}$ are the measured output signal; the applied input signal and disturbance sequence /with normal distribution, zero mean, variance one and it is independent of the input/, respectively. Here t means the discrete-time of the system /natural number/ taking sampling time unit. Furthermore in equation /1/

$$A/z^{-1}/ = 1+a_1 z^{-1}+\ldots+a_n z^{-n}$$
$$B/z^{-1}/ = b_o+b_1 z^{-1}+\ldots+b_m z^{-m} \quad ; \quad m \leq n \qquad /2/$$
$$C/z^{-1}/ = 1+c_1 z^{-1}+\ldots+c_k z^{-k} \quad ; \quad k \leq n$$

where z^{-1} is the backward shift operator;c.f. Åström /1970/. Using /2/ the system equation /1/ can be written in the following form, too :

$$y/t/ = \underline{f}^T/u,y,t/\underline{p} + \lambda\, C/z^{-1}/\, e/t/ \,, \qquad /3/$$

where T means the transposition and

$$\underline{f}/u,y,t/=[u/t/,u/t-1/,\ldots,u/t-m/,-y/t-1/,\ldots$$
$$\ldots,-y/t-n/]^T = [\,u/t/,\ \underline{g}^T/t-1/]^T \qquad /4/$$

and

$$\underline{p} = [\,b_o,b_1,\ldots,b_m,a_1,\ldots,a_n]^T \;. \qquad /5/$$

Assuming e/t/ has normal distribution it can be deduced the information matrix regarding to the estimation of parameter vector $\underline{p}$ is, for N samplings;c.f. Keviczky /1972/ :

$$\underline{\underline{J}}_N = \frac{1}{\lambda^2} \sum_{t=1}^{N} \underline{f}/u^F,x^F,t/\ \underline{f}^T/u^F,x^F,t/ \qquad /6/$$

where

$$\underline{f}/u^F,x^F,t/=[u^F/t/,u^F/t-1/,\ldots,u^F/t-m/,$$
$$-x^F/t-1/,\ldots,-x^F/t-n/]^T \;. \qquad /7/$$

Here

$$C/z^{-1}/\ u^F/t/=u/t/ \quad ; \quad C/z^{-1}/\ x^F/t/=x/t/ \qquad /8/$$

and

$$A/z^{-1}/\ x/t/ = B/z^{-1}/\ u/t/ \qquad /9/$$

/i.e. x/t/ is the output without noise/.

Let the covariance metrix of the parameter estimate $\underline{p}_N$, obtained from N samplings, be :

$$\underline{\underline{K}}_N = E\{\ /\hat{\underline{p}}_N - \underline{p}/\ /\hat{\underline{p}}_N - \underline{p}/^T\} \qquad /10/$$

where E{...} and ^ mean the expected and estimated value, respectively. The Cramer-Rao lower bound gives a limit for $\underline{\underline{K}}_N$ according to which $\underline{\underline{K}}_N \geq \underline{\underline{J}}_N^{-1}$ where the inequality sign indicates the difference matrix is non-negative definite. It is abvious that the strength of this equality is characterized in the same way by the determinant as the trace of these two matrices.

Moreover, since the maximization of $\underline{\underline{J}}_N$'s trace is also used for the minimization of $\underline{\underline{J}}_N^{-1}$'s trace which is only asymptotically efficient, it is more worth maximizing the $\underline{\underline{J}}_N$'s determinant since this directly minimizes the determinant of $\underline{\underline{J}}_N^{-1}$, too. In the following the minimization of the determinant of the covariance matrix, for white noise equation error, and of the information matrix's inverse, for general noise, is presented. The input signal u/t/ is assumed an amplitude constrained signal :

$$-U \leq u/t/ \leq +U \qquad /11/$$

CASE OF LEAST-SQUARES STRUCTURE

First, let us consider the case of least-squares /LS/ structure when the equation error - the term $\lambda C/z^{-1}/\ e/t/$ in /3/ is white noise, i.e. $C/z^{-1}/\equiv 1$. In this case the well-known LS estimation coincides the maximum likelihood /ML/ one and gives an unbiased estimate. The well--known recursive version of the method is

$$\hat{\underline{p}}_{N+1}=\hat{\underline{p}}_N + \underline{\underline{R}}_{N+1}[\,y/N+1/-\underline{f}^T/u,y,N+1/\hat{\underline{p}}_N]\,\underline{f}/u,y,N+1/ \qquad /12/$$

where

$$\underline{\underline{R}}_{N+1}=\underline{\underline{R}}_N - \frac{\underline{\underline{R}}_N\underline{f}/u,y,N+1/\ \underline{f}^T/u,y,N+1/\ \underline{\underline{R}}_N}{1+\underline{f}^T/u,y,N+1/\ \underline{\underline{R}}_N\ \underline{f}/u,y,N+1/} =$$
$$= \frac{1}{\lambda^2}\ \underline{\underline{K}}_{N+1} \qquad /13/$$

and the covariance matrix

$$\underline{\underline{K}}_{N+1} = \left[\sum_{t=1}^{N+1} \underline{f}/u,y,t/\ \underline{f}^T/u,y,t/\right]^{-1} \lambda^2 \qquad /14/$$

It can be deduced that the increasing rate of $\underline{\underline{K}}_N^{-1}$'s determinant is

$$\frac{|\underline{\underline{K}}_{N+1}^{-1}|}{|\underline{\underline{K}}_N^{-1}|} = 1+\underline{f}^T/u,y,N+1/\ \underline{\underline{R}}_N\ \underline{f}/u,y,N+1/ \qquad /15/$$

if the sampling number N is changed to N+1;c.f. Keviczky /1972/. Here |...| means the determinant of a matrix. A locally optimum strategy can be formed for the maximization of $|\underline{\underline{K}}^{-1}|$ /which is equivalent to the minimization of $|\underline{\underline{K}}|$/ if the quadratic form on the right side of /15/

is maximized by u/N+1/ in every step. /This is a so-called locally optimum strategy because only the next step is optimized in every time./ Partitionate $\underline{\underline{R}}_N$ according to /4/ :

$$\underline{\underline{R}}_N = \begin{bmatrix} r_N & \underline{d}_N^T \\ \underline{d}_N & \underline{\underline{Q}}_N \end{bmatrix} = \frac{1}{\lambda^2} \underline{\underline{K}}_N \qquad /16/$$

By this designations we get for /15/ :

$$\frac{|\underline{\underline{K}}_{N+1}^{-1}|}{|\underline{\underline{K}}^{-1}|} = u^2/N+1/\ r_N + 2u/N+1/\ \underline{g}^T/N/\underline{d}_N + \underline{g}^T/N/\ \underline{\underline{Q}}_N\ \underline{g}/N/ + 1 \qquad /17/$$

i.e. this means a parabola as a function of u/N+1/ having its vertex down /since r_N is absolutely positive/. It can be seen easily the following expression gives the optimum value of u^o/N+1/ - on a constrained region given by /11/ - ensuring the global maximum of /17/ :

$$u^o/N+1/ = \begin{cases} +U\ ,\ \text{if}\ \ \underline{g}^T/N/\underline{d}_N > 0 \\ -U\ ,\ \text{if}\ \ \underline{g}^T/N/\underline{d}_N \leqq 0 \end{cases} \qquad /18/$$

In this way the optimal input signal can be generated by the on-line connection with the process for identification purpose. u^o/t/ depends on only y/t-1/,...,y/t-n/, consequently, on values e/t-1/,...,e/t-n/ and the independence of input signal and the measurement noise is valid in this case, too. The algorithm /18/ can be realized easily because only the data applied so far are needed to generate the new u^o/t/. Instead of the local minimization of the determinant of the covariance matrix the local minimization of its trace can also be chosen since

$$tr/\underline{\underline{R}}_{N+1}/ = tr/\underline{\underline{R}}_N/ - \frac{\underline{f}^T/u,y,N+1/\underline{\underline{R}}_N\underline{\underline{R}}_N\underline{f}/u,y,N+1/}{1+\underline{f}^T/u,y,N+1/\underline{\underline{R}}_N\underline{f}/u,y,N+1/} \qquad /19/$$

and the second term of /19/'s right side is a second order rational fractional function of u^o/N+1/./Here tr/.../ means the trace of a matrix./ Thus, to generate optimum input /i.e. to minimize tr/$\underline{\underline{R}}$/ or tr/$\underline{\underline{K}}$/, namely, tr/$\underline{\underline{K}}$/ is proportional to tr/$\underline{\underline{R}}$/ / the global maximum of a much more difficult function than /17/ has to be determined therefore it is reasonable to use /18/.

It can be established from the comparison of equation /6/ and /14/ that the same algorithm can be used for the minimization of $\underline{\underline{J}}^{-1}$'s determinant /i.e. for the maximization of $|\underline{\underline{J}}|$ / as what was used in /18/ for the minimization of $|\underline{\underline{K}}|$. Since $C/z^{-1}/\equiv 1$, $u^F/t/=u/t/$ and $x^F/t/=x/t/$. So $\underline{f}$/u,y,t/ must be replaced by $\underline{f}$/u,x,t/. Here x/t/ is unknown and it can be produced by prediction :

$$\hat{x}/t/ = \sum_{i=o}^{m} \hat{b}_i\ u/t-i/ - \sum_{i=1}^{n} \hat{a}_i\ \hat{x}/t-i/ \qquad /20/$$

Obviously, now $\underline{g}/u,\hat{x},N/$ is in /18/ instead of $\underline{g}/N/=\underline{g}/u,y,N/$. The local maximization of information matrix's determinant does not need the knowledge of output but needs $\hat{x}$/t/,i.e. the parameter estimates $\hat{a}_i,\hat{b}_i$. Thus, this strategy can be performed by on-line way /at the same time with the identification/ but it can be done by off-line way, too, in the knowledge of apriori parameters' estimates. This means the optimum input sequences /so-called D-optimum/ can be generated in advance to the identification. Unfortunately, we have to know much enough to the synthesis /the parameter estimates themselves/ and this strategy can be realized only by the succesive application of the off-line identification methods.

THE CASE OF MAXIMUM LIKELIHOOD STRUCTURE

Such an algorithm which minimizes locally the determinant of the covariance matrix for the general form of system equation /3/ can not be constructed similarly to /18/ but the algorithm suggested for the local maximization of information matrix's determinant in LS structure can be generalized for this case. Comparing /6/ and

/14/ it can be seen $\underline{f}/u^F,x^F,t/$ corresponds to $\underline{f}/u,y,t/$, formally. Since x/t/ is unknown $\underline{f}/u^F,\hat{x}^F,t/$ can be determined from the predicted value $\hat{x}/t/$. By the filtering equations /8/ :

$$\underline{f}/u^F,\hat{x}^F,N+1/=[u/N+1/-q \ , \ \underline{g}^T/u^F,x^F,N/]^T \qquad /21/$$

where

$$q = \sum_{i=1}^{N} \hat{c}_i \, u^F/t-i/ \qquad /22/$$

Now the optimum $u^o/N+1/$ is computed according to

$$u^o/N+1/ = \begin{cases} +U \ , \text{ if } \ \underline{g}^T/N/\underline{d}_N > q \, r_N \\ \\ -U \ , \text{ if } \ \underline{g}^T/N/\underline{d}_N \leqq q \, r_N \end{cases} \qquad /23/$$

/It is to be mentioned that now r_N and $\underline{d}_N$ issue from the partitionation of $\underline{\underline{R}}_N=/\lambda^2\underline{\underline{J}}_N/^{-1}$./ Since there is no good on-line method to estimate the coefficients of $A/z^{-1}/,B/z^{-1}/$ and $C/z^{-1}/$ the off-line input signal synthesis - suggested for the LS structure, too - should be applied.

In a special case, when $C/z^{-1}/\cong\frac{1}{H/z^{-1}/}$, Hastings' /1969/ on-line method can be used to estimate the parameters of

$$H/z^{-1}/ = 1+h_1z^{-1}+\ldots+h_sz^{-s} \ . \qquad /24/$$

The formula /23/ is also valid taking into account that

$$q =-\sum_{i=1}^{s} \hat{h}_i \, u/t-i/ \qquad . \qquad /25/$$

SUMMARY AND CONCLUSIONS

In this paper a locally optimum algorithm is suggested to solve the problem of input signal synthesis by means of which in case of white noise equation error the minimization of covariance matrix's determinant, in other cases the maximization of the information matrix's determinant can be performed easily. The local optimality is an analogous concept with the "one-stage" stochastic control. It would be a considerably more difficult problem concerning the computation technique to ensure the global optimality corresponding to the "N-stage" control. Our feeling is that the suggested algorithms are very useful in the off-line input signal synthesis to improve the identification results step by step. The efficiency of the elaborated methods is supported by several simulation examples, unfortunately, they need more space to present them;c.f. Keviczky /1972/.

REFERENCES

Aoki,M. and R.M.Staley /1968/. On Approximate Input Signal Synthesis in Plant Parameter Identification,First Hawaii Int.Conf. on Syst.Sc., University of Hawaii,363.

Aoki,M. and R.M.Staley /1970/. On Input Signal Synthesis in Parameter Identification,Automatica,6,431.

Åström,K.J. and P.Eykhoff /1970/. System Identification-A Survey,IFAC,Prague.

Hastings,R.J. and M.W.Sage /1969/. Recursive Generalised Least-Squares Procedure for On-line Identification of Process Parameter,Proc.IEE, 116,2057.

Inoue,K.,K.Ogino and Y.Sawaragi /1970/. Sensitivity Synthesis of Optimal Input for Parameter Identification,IFAC,Prague.

Keviczky,L. and Cs.Bányász /1972/. Optimal Identification by Simulation of the Information Obtained from the Processes,Summer Computer Simulation Conference,San Diego.

Keviczky,L. /1972/. On Some Questions of Input Signal Synthesis,Report 7226/B/,Lund Institute of Technology,Division of Automatic Control.

Levadi,V.S. /1966/. Design of Input Signals for Parameter Estimation,IEEE Trans.Aut.Control, AE-11,205.

Nahi,N.E. and D.E.Wallis /1969/. Optimal Inputs for Parameter Estimation in Dynamic Systems with White Observation Noise,Proc.JACC,506.

Rault,A.,Pouliquen,R. and Richalet,J. /1968/. Sensitivity and Identification,IFAC,Dubrovnik.

SELECTION OF PERIODIC TEST SIGNALS FOR ESTIMATION OF LINEAR SYSTEM DYNAMICS

A. van den Bos
Dept. of Applied Physics
Delft University of Technology
Delft, Netherlands.

This paper discusses the influence of the dynamic properties of test signals on the error of system parameter estimates computed from noisy observations of the responses. The parameters are the coefficients of the differential equation and the time delay. The measure of estimation error is the trace of the minimum variance bound (MVB) for the estimation of the parameters. The computation of the MVB is discussed and a numerical procedure for synthesis of optimal test signals which minimize the trace of the MVB is described. In a numerical example it is found that the accuracy with a maximum length binary sequence is comparable to the optimum only for a rather limited range of fundamental frequency. Estimation of time delay influences the accuracy of the remaining parameters unfavourably. In the example it is also found that the power of the optimal spectra is concentrated in a very small number of harmonics.

INTRODUCTION

In practice the choice of a particular test signal for measurement of system dynamics is usually not determined by accuracy considerations. For example, maximum length binary sequences are frequently used for crosscorrelation measurement of the impulse response since they have a suitable autocorrelation function and are easy to generate, to synchronize and to introduce into the system. The choice of the bandwidth and the length of the sequence is determined by the required time resolution of the measurement and the expected duration of the impulse response respectively. However, this particular choice of bandwidth and length of the sequence does not imply that the sequence so obtained is a suitable one for accurate estimation of one or more parameters (time constants, coefficients of differential equation, etc.) of the system.

Generally the error in system parameter estimates computed from noise corrupted responses to test signals depends on the dynamic properties of the test signal. This offers the opportunity to manipulate the estimation error by selection of the test signal. The problem considered in this paper is the synthesis of periodic test signals which minimize a measure of the error in estimation of the coefficients of the differential equation and the time delay of continuous linear systems. The aim is to compare the minimum error with the error of estimates computed from noisy responses to the usual test signals like maximum length binary sequences and binary multifrequency signals. A further aim is to investigate how a priori knowledge about the system and the noise can be used for synthesis of appropriate test signals.

A suitable measure of estimation error of parameter estimates is the trace of their covariance matrix. Generally the elements of the covariance matrix are functions of the dynamic properties of the test signal. Furthermore these elements depend on the particular estimator used. So, the test signal which minimizes the trace is specific for this estimator. In order to avoid this difficulty in this research the measure of error in estimation is taken as the trace of the minimum variance bound (MVB) for the estimation of the parameters. The particular test signal which minimizes the trace of the MVB is optimal.

Åström (1967) discusses the computation of the MVB for the estimation of linear system and noise parameters from noise corrupted responses to known inputs. The system parameters are the coefficients of the differential equation describing the system. The noise parameters describe the noise generating process. The results apply to a wide class of inputs. The procedure mainly consists of the computation of the information matrix. The MVB is the inverse of this matrix.

Using Åström's procedure it is shown in Section 1 that those elements of the MVB which represent the covariances of the estimates of the noise parameters are independent of the test signal. On the other hand the elements which represent the MVB for the estimation of the system parameters do not depend on whether or not the noise parameters are known. Hence for the computation of the optimal test signal only the MVB for the estimation of the system parameters needs be considered.

Section 2 applies the results of Section 1 to the particular case that the test signal is periodic. The parameters are the coefficients of the differential equation and the time delay of the system. It is shown that in this case the elements of the information matrix are relatively simple functions of the parameters, the power spectrum of the noise and the power spectrum of the test signal. In practice input or output power is always restricted. The power spectrum and the corresponding test signals which minimize the trace of the MVB under the power constraint are optimal. The trace of the MVB is nonlinear in the power spectrum of the test signal and even in simple cases closed form solutions for the optimal spectrum are difficult to obtain. It is shown in Section 3, however, that the numerical computation of the gradient is relatively simple. This offers the opportunity to apply powerful numerical optimization techniques.

Now the question arises whether for the particular problem at hand there exist estimators which at least asymptotically achieve the MVB. A previous paper by the author describes an explicit least squares estimator for the coefficients of the differential equation of the system from noisy responses to periodic test signals. Time delay is estimated by repeating the procedure for a number of values of the time delay and selecting according to goodness of fit. See Van den Bos (1970). This procedure is summarized in Section 4. The accuracy of this procedure can be improved by adding a second computational step. This second step is a weighted least square procedure with weights computed from the estimated residuals of the first step. See Van den Bos (1973).
In Section 4 the numerical results are given of two computer experiments with the two-step procedure. These results show that at least in the cases considered the MVB is actually achieved.

In Section 5 optimal input spectra are computed for a second-order system with white Gaussian noise at the output. The computations are carried out for a number of fundamental frequencies of the input. For comparison purposes the case of known and unknown time delay is considered separately. The traces of the MVB for a maximum length binary sequence and a binary multifrequency signal under the same conditions are next computed.

The synthesis of test signals minimizing some measure of error in estimation of system parameters has been discussed by several authors. Levadi (1966) considers the case that the system weighting function is either linear in the parameters or may be linearized with respect to the deviations of the parameter estimates from a reference value. For this case power constrained test signals are computed minimizing the trace of the covariance matrix of the least squares estimator. The weighting function is a complicated nonlinear expression in the coefficients of the differential equation, at least for orders higher than one. As a result the procedure is hard to generalize and computationally intractable for systems having an arbitrary number of poles and zero's. Aoki and Staley (1970) discuss the synthesis of optimal time constrained and energy constrained pulses for estimation of the coefficients of the difference equation of the system from the white noise corrupted transient response. The measure of estimation error is the reciprocal value of the trace of the information matrix. Cumming (1970) uses discrete interval binary noise as test signal and investigates the influence of the switching parameter on the MVB for estimation of the coefficients of the system difference equation.

1. MINIMUM VARIANCE BOUND

Let the system be described by

$$a_N y^{(N)}(t)+a_{N-1}y^{(N-1)}(t)+\ldots+a_o y(t) = \\ =u^{(M)}(t-\tau)+b_{M-1}u^{(M-1)}(t-\tau)+\ldots+ \\ + b_o u(t-\tau) \qquad (1.1)$$

where $u(t)$ is the input, $y(t)$ is the response to $u(t)$ and τ denotes time delay. In what follows (1.1) will be denoted by

$$y(t) = H_S(j\omega)u(t) \qquad (1.2)$$

where

$$H_S(j\omega) = \frac{B(j\omega)}{A(j\omega)}\exp(-j\omega\tau) \qquad (1.3)$$

with

$$A(j\omega) = a_N(j\omega)^N+a_{N-1}(j\omega)^{N-1}+\ldots+a_o \qquad (1.4)$$

and

$$B(j\omega) = (j\omega)^M+b_{M-1}(j\omega)^{M-1}+\ldots+b_o \qquad (1.5)$$

where ω is frequency in rads^{-1}. Furthermore let

$$z(t) = y(t) + h(t) \qquad (1.6)$$

be observed at the output. The disturbance $h(t)$ is assumed to be a stationary Gaussian stochastic process described by

$$h(t) = H_D(j\omega)e(t) \qquad (1.7)$$

where $e(t)$ is a zero mean stationary Gaussian process with autocovariance function $E\, e(t)e(t') = \delta(t-t')$ and

$$H_D(j\omega) = \frac{D(j\omega)}{C(j\omega)}$$

with

$$D(j\omega) = (j\omega)^K+d_{K-1}(j\omega)^{K-1}+\ldots+d_o$$

and

$$C(j\omega)=c_L(j\omega)^L+c_{L-1}(j\omega)^{L-1}+\ldots+c_o.$$

Furthermore assume that u(t) and e(t) are independent and that D(s) has all zero's in the left half-plane.

Now let u(t) and z(t) be observed for $0 < t < T$ and let the vector θ be defined by

$$\theta = (\theta_S' \ \ \theta_D')'$$

where

$$\theta_S=(a_o\ldots a_N \ b_o\ldots b_{M-1}\tau)' \qquad (1.8)$$

denotes the vector of system parameters and

$$\theta_D=(c_o\ldots c_L \ d_o\ldots d_{K-1})'$$

denotes the vector of parameters of the noise. Then using a procedure due to Åström (1967) the MVB for the estimation of θ is computed as follows. First it is observed that the logarithm of the likelihood function of e(t) for $0 < t < T$ may be written

$$\ln L=-\tfrac{1}{2}\int_0^T e^2(t)dt + \text{constant} \qquad (1.9)$$

Generally the relation between the logarithm of the likelihood function and the MVB of an unbiased estimator $\hat{\theta}$ of the parameters θ of the likelihood function is

$$E(\hat{\theta}-\theta)(\hat{\theta}-\theta)' \geq P^{-1}$$

where $P = [p_{ij}]$ is the information matrix defined by

$$p_{ij} = E\,\frac{\partial \ln L}{\partial\theta_i}\,\frac{\partial \ln L}{\partial\theta_j}$$

If the second-order derivatives of ln L with respect to θ exist this may be written

$$p_{ij}=-E\,\frac{\partial^2 \ln L}{\partial\theta_i\,\partial\theta_j} \qquad (1.10)$$

See Kendall and Stuart (1967). Now it follows from (1.9) and (1.10) that

$$p_{ij} = E\int_0^T \frac{\partial e(t)}{\partial\theta_i}\,\frac{\partial e(t)}{\partial\theta_j}\,dt +$$

$$+ E\int_0^T e(t)\,\frac{\partial^2 e(t)}{\partial\theta_i\,\partial\theta_j}\,dt \qquad (1.11)$$

Combining (1.2), (1.6) and (1.17) and rearranging

$$e(t)=\frac{1}{H_D(j\omega)}\{z(t)-H_S(j\omega)u(t)\}$$

Hence

$$\frac{\partial e(t)}{\partial\theta_S}= -\frac{1}{H_D(j\omega)}\,\frac{\partial H_S(j\omega)}{\partial\theta_S}\,u(t) \qquad (1.12)$$

and

$$\frac{\partial e(t)}{\partial\theta_D} = -\frac{1}{H_D(j\omega)}\,\frac{\partial H_D(j\omega)}{\partial\theta_D}\,e(t) \qquad (1.13)$$

Furthermore

$$\frac{\partial^2 e(t)}{\partial\theta_S\,\partial\theta_D} = \frac{H_S(j\omega)}{H_D^2(j\omega)}\,\frac{H_D(j\omega)}{\partial\theta_D}\,\frac{H_S(j\omega)}{\partial\theta_S}\,u(t) \qquad (1.14)$$

Since by assumption u(t) and e(t) are independent it follows from (1.10)-(1.14) that

$$E\,\frac{\partial^2 \ln L}{\partial\theta_S\,\partial\theta_D} = 0.$$

Hence P may be written

$$P=-\begin{pmatrix} E\,\dfrac{\partial^2 \ln L}{\partial\theta_S^2} & 0 \\ 0 & E\,\dfrac{\partial^2 \ln L}{\partial\theta_D^2} \end{pmatrix}$$

consequently the MVB is

$$P^{-1}=-\begin{pmatrix} \left(E\,\dfrac{\partial^2 \ln L}{\partial\theta_S^2}\right)^{-1} & 0 \\ 0 & \left(E\,\dfrac{\partial^2 \ln L}{\partial\theta^2}\right)^{-1} \end{pmatrix}^{-1} \qquad (1.15)$$

From this expression it follows that, for the model of system and noise assumed here, those elements of P^{-1} which represent the MVB of the system parameter estimates do not depend on whether or not the noise parameters are known. Furthermore it follows from (1.11) and (1.13) that those elements of P^{-1} which represent the MVB of the noise parameter estimates do not depend on u(t). So the only elements of P^{-1} which can be manipulated by selection of u(t) are those of the matrix

$$Q^{-1} = -\left(E\,\frac{\partial^2 \ln L}{\partial\theta_S^2}\right)^{-1} \qquad (1.16)$$

Consequently the particular u(t) which minimizes the trace of Q^{-1} also minimizes the trace of P^{-1}. Therefore in what follows only the minimization of the trace of Q^{-1} is considered.

In view of (1.10) and (1.11) the matrix Q is described by

$$Q = E\int_0^T \frac{\partial e(t)}{\partial\theta_S}\left(\frac{\partial e(t)}{\partial\theta_S}\right)'\,dt + E\int_0^T e(t)\,\frac{\partial^2 e(t)}{\partial\theta_S^2}\,dt \qquad (1.17)$$

Since by assumption e(t) and u(t) are independent it follows from (1.12) that the second term of (1.17) is zero and hence

$$Q = E \int_0^T \frac{\partial e(t)}{\partial \theta_S} \left(\frac{\partial e(t)}{\partial \theta_S}\right)' dt \qquad (1.18)$$

The elements of the vector $\frac{\partial e(t)}{\partial \theta_S}$ in this expression are computed from (1.3) and (1.12). These elements are

$$\frac{\partial e(t)}{\partial a_i} = (j\omega)^i \frac{H_S(j\omega)}{H_D(j\omega)A(j\omega)} u(t) \qquad (1.19)$$

$$\frac{\partial e(t)}{\partial b_i} = -(j\omega)^i \frac{H_S(j\omega)}{H_D(j\omega)B(j\omega)} u(t) \qquad (1.20)$$

$$\frac{\partial e(t)}{\partial \tau} = j\omega \frac{H_S(j\omega)}{H_D(j\omega)} u(t) \qquad (1.21)$$

2. PERIODIC TEST SIGNALS

It is observed that (1.19), (1.20) and (1.21) are of the form $H(j\omega)u(t)$ where $H(j\omega)$ is a transfer function. Now assume that $u(t)$ is periodic with period T_o. Furthermore let T be an integral multiple of T_o. Then for two periodic signals $H_1(j\omega)u(t)$ and $H_2(j\omega)u(t)$

$$\int_0^T H_1(j\omega)u(t) \; H_2(j\omega)u(t)dt =$$

$$= T \sum_{k=-\infty}^{\infty} \mathrm{Re}\{H_1(jk\omega_o)H_2^{\star}(jk\omega_o)\}S(k\omega_o) \qquad (2.1)$$

where $S(k\omega_o)$ is the power of the k-th harmonic of $u(t)$ and $\omega_o = 2\pi/T_o$. Now using (2.1) the elements of the information matrix Q can easily be computed from (1.18)-(1.21). For example, the diagonal elements of Q are

$$q_{ii} = T \sum_{k=-\infty}^{\infty} (k\omega_o)^{2i} \left|\frac{H_S(jk\omega_o)}{A(jk\omega_o)}\right|^2 \frac{S(k\omega_o)}{S_D(k\omega_o)} \qquad i=0,\dots,N$$

$$q_{ii} = T \sum_{k=-\infty}^{\infty} (k\omega_o)^{2(i-N-1)} \left|\frac{H_S(jk\omega_o)}{B(jk\omega_o)}\right|^2 \frac{S(k\omega_o)}{S_D(k\omega_o)} \qquad i=N+1,\dots,N+M$$

and

$$q_{ii} = T \sum_{k=-\infty}^{\infty} (k\omega_o)^2 \left|H_S(jk\omega_o)\right|^2 \frac{S(k\omega_o)}{S_D(k\omega_o)} \qquad i=N+M+1$$

where $S_D(\omega) = |H_D(j\omega)|^2$ denotes the power spectrum of the noise. The expressions for the off-diagonal elements of Q are similar. Note that all q_{ij} are independent of the magnitude of τ since τ causes merely a displacement of the signals (1.19), (1.20) and (1.21). Furthermore observe that the elements of Q can be made arbitrarily large by increasing the power of $u(t)$. Finally, for what follows it is important to note that in view of (2.1) all elements of Q are linear in the $S(k\omega_o)$.

3. NUMERICAL MINIMIZATION OF THE MINIMUM VARIANCE BOUND

Given the system and noise characteristics the problem is to minimize $\mathrm{tr}\, Q^{-1}$ with respect to the power spectrum of $u(t)$ described by

$$\dots, S(-k\omega_o), \dots, S(-\omega_o), S(0), S(\omega_o), \dots, S(k\omega_o), \dots$$

subject to the power constraint

$$\sum_{k=-\infty}^{\infty} \lambda_k \, S(k\omega_o) = 1 \qquad (3.1)$$

and the requirement that

$$S(k\omega_o) \geq 0 \quad k=0,\pm 1, \pm 2,\dots \qquad (3.2)$$

where the λ_k are positive weighting factors. The constraint (3.1) is an output power constraint if

$$\lambda_k = |H_S(jk\omega_o)|^2$$

Alternatively (3.1) is an input power constraint if $\lambda_k = 1$ for all k. The latter constraint is chosen in this research. Now the constraints (3.1) and (3.2) are both removed by substituting

$$S(k\omega_o) = x_k^2 \Big/ \sum_{l=-\infty}^{\infty} x_l^2 \qquad (3.3)$$

and minimizing $\mathrm{tr}\, Q^{-1}$ with respect to $x = (\dots, x_{-k}, \dots, x_{-1}, x_o, x_1, \dots, x_k, \dots)'$.
The gradient of $\mathrm{tr}\, Q^{-1}$ with respect to x is computed as follows. First it is observed that in view of (3.3)

$$\frac{\partial \mathrm{tr} Q^{-1}}{\partial x_k} = \sum_l \frac{\partial \mathrm{tr} Q^{-1}}{\partial S(l\omega_o)} \frac{\partial S(l\omega_o)}{\partial x_k}$$

$$= \frac{2x_k}{\sum_l x_l^2} \left\{\frac{\partial \mathrm{tr} Q^{-1}}{\partial S(k\omega_o)} - \sum_l \frac{\partial \mathrm{tr} Q^{-1}}{\partial S(l\omega_o)} S(l\omega_o)\right\} \qquad (3.4)$$

Furthermore

$$\mathrm{tr} Q^{-1} = \frac{1}{\det Q} \sum_{i=1}^{N+M+1} \det Q_{ii}$$

where Q_{ij} is the matrix obtained by eliminating the i-th row and j-th column from Q. It follows that

$$\frac{\partial \mathrm{tr} Q^{-1}}{\partial S(k\omega_o)} = -\frac{1}{\det Q} \frac{\partial \det Q}{\partial S(k\omega_o)} \mathrm{tr}\, Q^{-1} + \frac{1}{\det Q} \sum_i \frac{\partial \det Q_{ii}}{\partial S(k\omega_o)} \qquad (3.5)$$

The derivatives in this expression may be written

$$\frac{\partial \det Q}{\partial S(k\omega_o)} = \sum_i \sum_j (-1)^{i+j} \det Q_{ij} \frac{\partial q_{ij}}{\partial S(k\omega_o)} \qquad (3.6)$$

and

$$\frac{\partial \det Q_{ii}}{\partial S(k\omega_o)} = \sum_m \sum_n (-1)^{m+n} \det Q_{ii,mn} \frac{\partial q_{ii,mn}}{\partial S(k\omega_o)} \qquad (3.7)$$

where $Q_{ij,mn}$ is the matrix obtained by eliminating the m-th row and n-th column from Q_{ij} and $q_{ij,mn}$ is the (m,n)-th element of Q_{ij}. Note that the elements $q_{ij,mn}$ are a subset of the elements q_{ij}.

Now in order to obtain the gradient of $\mathrm{tr}Q^{-1}$ with respect to x at some point x^o, first the values of $S(k\omega_o)$ are computed for x^o with the aid of (3.3). The elements of Q are calculated next as described in Section 2. Subsequent inversion with the Gauss elimination method yields $\mathrm{tr}Q^{-1}$, detQ and the cofactors $(-1)^{i+j}\det Q_{ij}$. The derivatives $\frac{\partial q_{ij}}{\partial S(k\omega_o)}$ follow from (1.18)-(1.21) and (2.1).

Since the q_{ij} are linear in the $S(k\omega_o)$, the derivatives $\frac{\partial q_{ij}}{\partial S(k\omega_o)}$ are independent of $S(k\omega_o)$ and need be computed only once. Now using (3.6) the first term of (3.5) can be calculated. The second term of (3.5) is evaluated using (3.7). Inversion of Q_{ii} yields the cofactors $(-1)^{m+n}\det Q_{ii,mn}$. The derivatives $\frac{\partial q_{ii,mn}}{\partial S(k\omega_o)}$ are known since they are a subset of the $\frac{\partial q_{ii}}{\partial S(k\omega_o)}$. So $\frac{\partial \det Q_{ii}}{\partial S(k\omega_o)}$, and hence the second term of (3.5), can be computed. This completes the computation of the gradient of $\mathrm{tr}Q^{-1}$ with respect to $S(k\omega_o)$. Finally, using (3.4) the gradient of $\mathrm{tr}Q^{-1}$ with respect to x is obtained.

Now the application of gradient techniques like the steepest descent method or the conjugate gradient method is straightforward.

4. ESTIMATION USING PERIODIC TEST SIGNALS

Let the system under test be described by (1.1). Let u(t) be a periodic test signal with period T_o and let y(t) denote the steady state response to u(t). Define the Fourier coefficient of the k-th harmonic of u(t) by

$$\gamma_{ku} = \alpha_{ku} - j\beta_{ku} = \frac{1}{T_o}\int_0^{T_o} u(t)\exp(-jk\omega_o t)dt$$

where $\omega_o = 2\pi/T_o$ and define γ_{ky} correspondingly. For the moment let τ be known and assume $\tau = 0$. Then using (1.4) and (1.5)

$$A(jk\omega_o)\gamma_{ky} = B(jk\omega_o)\gamma_{ku} \qquad (4.1)$$

From (1.4), (1.5) and (4.1) it follows that each harmonic gives rise to two real inhomogeneous linear equations in the coefficients of the differential equation. Now, suppose that u(t) and y(t) are exactly known. Then the γ_{ku} and γ_{ky} can be computed for a sufficiently large number of harmonics and the resulting equations (4.1) can be solved for $a_o,\ldots,a_N,b_o,\ldots,b_{M-1}$.

Next suppose that the observations of y(t) are corrupted by noise and are described by (1.6). Furthermore let the harmonics taken into consideration have the harmonic numbers $k_1,\ldots,k_L$. Then the Fourier coefficients $\gamma_{k_i y}$ may be estimated by

$$\hat{\gamma}_{k_i y} = \frac{1}{JT_o}\int_0^{JT_o} z(t)\exp(-jk\omega_o t)dt \qquad i=1,\ldots,L$$

where JT_o is the observation time. Under very mild conditions $\hat{\gamma}_{ky}$ is a consistent, unbiased least squares estimator of γ_{ky}. See Levin (1959). Now $\gamma_{k_i u}$ and $\hat{\gamma}_{k_i y}$, $i = 1,\ldots,L$ do not satisfy (4.1). However, the computation of the least squares solution $\hat{a}_o,\ldots,\hat{b}_{M-1}$ for $a_o,\ldots,b_{M-1}$ in (4.1) for all harmonics considered is straightforward. This least squares solution is a closed form solution since (4.1) is linear in the unknowns. Moreover, the estimator is consistent if $\hat{\gamma}_{ky}$ is consistent. See Van den Bos (1970).

The variance of $\hat{a}_o,\ldots,\hat{b}_{M-1}$ can be reduced in an additional computational step. Define the residuals ε_k and η_k by

$$\begin{matrix}\mathrm{Re}\\ \mathrm{Im}\end{matrix}\; A(jk\omega_o)\hat{\gamma}_{ky} - B(jk\omega_o)\gamma_{ku} = \begin{matrix}\varepsilon_k\\ \eta_k\end{matrix} \qquad (4.2)$$

If the covariance matrix of the ε_{k_i} and η_{k_i} is known, the least squares estimator $\hat{a}_o,\ldots,\hat{b}_{M-1}$ can be replaced by a more accurate weighted least squares estimator. It can be shown that asymptotically the covariance matrix of the ε_{k_i} and η_{k_i} is diagonal with diagonal elements $\mathrm{var}\,\varepsilon_{k_i} = \mathrm{var}\,\eta_{k_i}$. See Van den Bos (1973). These variances, however, are usually not known and must therefore be estimated. Let $\hat{\varepsilon}_{k_i}$ and $\hat{\eta}_{k_i}$ denote the values of ε_{k_i} and η_{k_i} obtained by replacing $a_o,\ldots,b_{M-1}$ in (4.2) by the estimates $\hat{a}_o,\ldots,\hat{b}_{M-1}$. Now both var ε_{k_i} and var η_{k_i} are estimated by

$$s^2_{k_i} = \frac{1}{2k_w+1}\sum_{l=-k_w}^{k_w} \tfrac{1}{2}(\hat{\varepsilon}^2_{k_i+l} + \hat{\eta}^2_{k_i+l}) \qquad (4.3)$$

This is an average over frequency of the squares of the measured residuals. Note that the calculation of (4.3) requires the additional evaluation of $\hat{\gamma}_{k_i+j}$ for i=1,...,L and $j = \pm 1,\ldots,\pm k_w$. Now the weighted least squares estimates $\hat{\hat{a}}_o,\ldots,\hat{\hat{b}}_{M-1}$ can be computed using the reciprocal values of the $s^2_{k_i}$ as weights. See Van den Bos (1973).

Numerical example
In order to investigate the efficiency of the two-step estimator $\hat{\hat{a}}_o,\ldots,\hat{\hat{b}}_{M-1}$ the following computer experiment has been carried out. The system is described by $a_2y^{(2)}(t)+a_1y^{(1)}(t)+a_oy(t) = u(t-\tau)$, with $a_2=0.25$, $a_1 = 1.25$, $a_o = 1.0$ and $\tau = 0.7854$s. The test signal is a multifrequency binary signal (MFBS), i.e., a periodic two-level signal which has the major part of its power concentrated in a relatively small number of dominant harmonics. For a discussion of MFBS see Van den Bos (1967 and 1970). The power of the MFBS used here is one. The dominant harmonics have the harmonic numbers 1, 15 and 31. The power of these harmonics is $S(\omega_o) = 0.248$, $S(15\omega_o) = 0.267$ and $S(31\omega_o) = 0.236$, i.e., 75% of the total power. The fundamental frequency is 0.125 rads^{-1}. The observations of the response are corrupted by an additive Gaussian process having a power spectrum $S_D(\omega) = 0.033$.

Fourty independent records of two periods each were generated and the estimates of a_2, a_1, a_o and τ were computed using the two-step least squares procedure. Time delay was estimated by repeating the procedure for a number of values of time delay and selecting according to goodness of fit. In the first step of the procedure only the three dominant harmonics of the test signal have been taken into consideration. The results of the experiment are shown in Table 1. The first column shows the average and standard error of the estimates. The second column shows in parentheses the standard deviation corresponding to the MVB for the case considered. Comparison of the standard error and the minimum standard deviation shows that these quantities agree very well. So it is concluded that at least in the case considered the MVB is achieved by the two-step estimator.

Table 1

$a_o=1.0$	$a_1=1.25$	$a_2=0.25$	$\tau=0.7854$
$\hat{\hat{a}}_o$	1.00±0.04	(0.04)	
$\hat{\hat{a}}_1$	1.20±0.10	(0.09)	
$\hat{\hat{a}}_2$	0.23±0.12	(0.13)	
$\hat{\hat{\tau}}$	0.81±0.10	(0.10)	

5. NUMERICAL EXAMPLES

The system considered is described by

$$a_2y^{(2)}(t)+a_1y^{(1)}(t)+a_oy(t) = u(t-\tau)$$

where the values of the parameters are $a_2=0.25$, $a_1=1.25$, $a_o=1$ and τ is arbitrary. The observed output is

$$z(t) = y(t) + h(t)$$

where $y(t)$ is the steady state response to the periodic input $u(t)$ and $h(t)$ is a stationary white Gaussian process with power spectral density

$$S_{hh}(\omega) = S_o$$

where S_o is a constant. In what follows it is assumed that the observation time is an integral number of periods. The power of all inputs is equal to one.

Since $u(t)$ is periodic time delay can only be estimated modulo the period of the fundamental. Therefore it is assumed that the time delay consists of the sum of a known integral multiple of the period of the fundamental and an unknown fraction of this period. Ambiguity in the interpretation of the estimates of this fraction may still arise if their standard deviation is comparable to the period of the fundamental. Therefore it is assumed that the standard deviation of the time delay estimates is small compared with the period of the fundamental. Asymptotically this condition is always met.

In order to investigate the effect of estimation of time delay upon the variance of the coefficient estimates all computations were carried out both for unknown and for known time delay. First the minimum values of the trace of the MVB and the corresponding input spectra are determined for a number of fundamental frequencies of the input. For the same set of frequencies the trace of the MVB is computed for a maximum length binary sequence (MLBS) and the multifrequency binary signal (MFBS) described in Section 4. The computation is based on the dominant harmonics only. The MLBS considered here has 63 steps in a period and only its first 63 harmonics, representing nearly 90% of its total power, are taken into consideration.

Figure 1 shows the results for joint estimation of coefficients and time delay. The trace of the MVB, normalized with respect to observation time and intensity of the noise, is plotted as a function of the fundamental frequency. In what follows fundamental frequency will be denoted by ω_o. In the same figure the optimal spectrum is shown for $\omega_o = 0.5$ rads^{-1}. In order to give an idea of the bandwidth of the system its corner frequencies are given by ω_{c_1} and ω_{c_2}.

Note that the bandwidth of the system and that of the optimal spectrum are approximately equal. Also note that the optimal test signal has three harmonics only. These are the fundamental, the second harmonic and the eighth harmonic.

With respect to the MLBS and MFBS it follows that these signals give results comparable to the optimum only for fundamental frequencies of about 0.1 rads^{-1}. As an illustration the standard deviations of the MVB estimates obtained with the optimal input, the MFBS and the MLBS are given in Table 2 for $\omega_o = 0.125$ rads^{-1} and for $\omega_o = 0.5$ rads^{-1}. Note that for $\omega_o=0.5$ rads^{-1} the deviations with the MFBS and the MLBS from the optimum are quite serious.

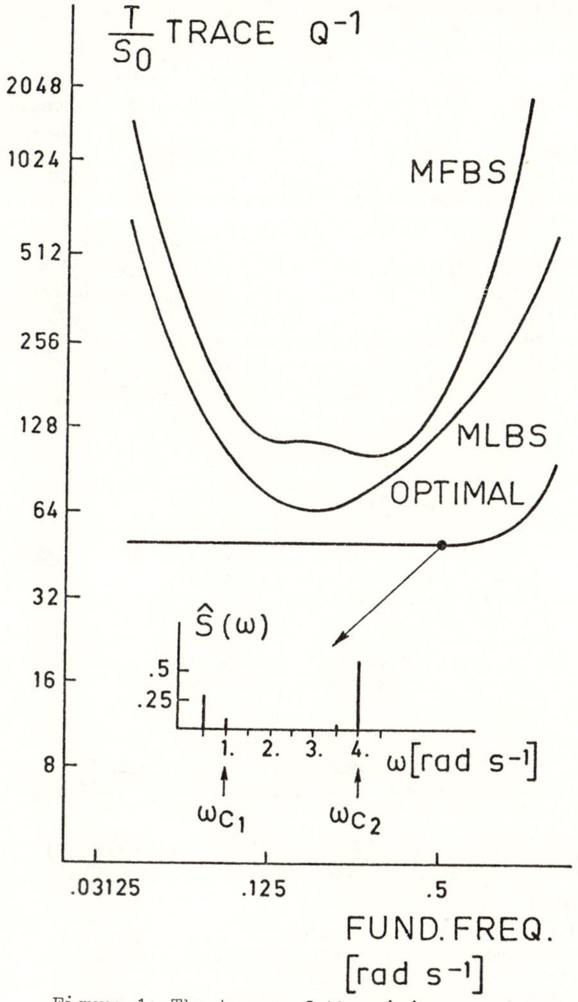

Figure 1: The trace of the minimum variance bound as a function of fundamental frequency for the case of unknown time delay.

Table 2

$\sqrt{\frac{T}{S_o}}\,\sigma$

$\omega_o = 0.125$ rads^{-1}

parameter	optimal	MLBS	MFBS
a_o	2.50	2.75	2.04
a_1	3.88	3.71	5.12
a_2	4.50	5.71	7.44
τ	2.93	4.14	5.60

$\omega_o = 0.5$ rads^{-1}

parameter	optimal	MLBS	MFBS
a_o	2.43	6.45	3.35
a_1	3.96	6.25	6.20
a_2	4.51	5.64	10.08
τ	2.93	3.44	4.17

Figure 2 and Table 3 representing the case of known time delay show corresponding results for joint estimation of the coefficients. Once more it is observed that the MLBS and MFBS yield results comparable to the optimum only in a limited frequency range. Again the optimal spectrum is shown for $\omega_o = 0.5$ rads^{-1}. Note that the bandwidth of the optimal spectrum is less than the system bandwidth.

Table 3

$\sqrt{\frac{T}{S_o}}\,\sigma$

$\omega_o = 0.0625$ rads^{-1}

parameter	optimal	MLBS	MFBS
a_o	1.89	2.04	1.92
a_1	2.77	2.84	3.40
a_2	2.26	2.71	2.82

$\omega_o = 0.5$ rads^{-1}

parameter	optimal	MLBS	MFBS
a_o	2.06	6.45	3.35
a_1	2.75	6.25	5.08
a_2	2.43	3.64	8.83

Comparing the results of Table 2 and Table 3 it is clear that the estimation of time delay causes an increase of the variance of some coefficient estimates. In particular this applies to the estimate of a_2. This is equivalent with the observation that estimates of a_2 and estimates of τ are strongly covariant. For example, for $\omega_o = 0.125$ rads^{-1} the correlation coefficient of these estimates is as high as -0.81, -0.94 and -0.91 for the optimal signal, the MFBS and the MLBS respectively. This means that overestimation of time delay usually yields underestimation of a_2. Although this may not be surprising it may easily give rise to an incorrect interpretation of the measurement results.

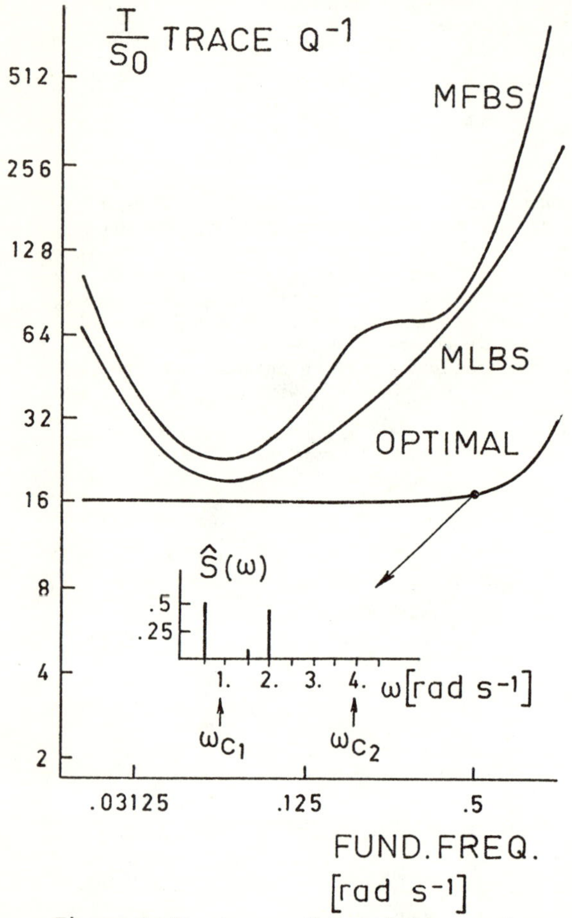

Figure 2: The trace of the minimum variance bound as a function of fundamental frequency for the case of known time delay.

6. CONCLUSIONS

In this paper the influence of the spectral properties of the test signal on the minimum variance bound for the estimation of system parameters has been investigated. The parameters were the coefficients of the differential equation and the time delay.

Optimum signals have been synthesized which minimized the trace of the minimum variance bound of the parameter estimates. In the cases considered the variance of estimates with a particular maximum length binary sequence or binary multifrequency signal is comparable to the optimum only for a rather limited range of fundamental frequency. Outside this range the variance increases rapidly. The numerical examples also show that estimation of time delay may increase the variance of the remaining parameters substantially. This is equivalent with the observation that in the cases considered time delay estimates were strongly covariant with the estimates of some of the remaining parameters.

A practical conclusion to be drawn from this research is that in order to find an appropriate fundamental frequency it seems worthwile to carry out a priori computations of the minimum variance bound for a number of different fundamental frequencies of the input. These computations can be based on a priori knowledge about system and noise. In addition to information about the appropriate frequency such computations also provide information about possibly strong covariances between the estimates.

REFERENCES

Aoki, M. and R.M. Staley (1970). On input signal synthesis in parameter identification. Automatica, 6, 431-440.

Åström, K.J. (1967). On the achievable accuracy in identification problems. Proc. IFAC Symp. Identification in autom. control systems, Prague, paper 1.8.

Cumming, I.G. (1970). Frequency of input signal in identification. Proc. IFAC Symp. Identification and process parameter estimation, Prague, paper 7.8.

Kendall M.G. and A. Stuart (1967). The advanced theory of statistics vol. II. London, Griffin Comp.

Levadi, V.S. (1966). Design of input signals for parameter estimation. IEEE Trans. autom. control, AC-11, 205-211.

Levin, M.J. (1959). Estimation of the characteristics of linear systems in the presence of noise. Report, Dept. of electrical engineering, Columbia University, New York.

Van den Bos, A. (1967). Construction of binary multifrequency test signals. Proc. IFAC Symp. Identification in autom. control systems, Prague, paper 4.6.

Van den Bos, A. (1970). Estimation of linear system coefficients from noisy responses to binary multifrequency test signals. Proc. IFAC Symp. Identification and process parameter estimation, Prague, paper 7.2.

Van den Bos, A. (1973). Estimation of parameters of linear systems using periodic test signals. Report, Coop. Centre for Meas. and Control, Delft University of Technology, Delft, Netherlands.

HARMONIC ANALYSIS VIA BINARY TEST SIGNAL WITH QUASI SINUSOIDAL AUTOCORRELATION FUNCTION

M. NOUGARET, Maître-Assistant H.BACK, J.C.CAUJOLLE Graduate students

Laboratoire d'Automatique I.N.P.G.
Campus Universitaire
- 38 - ST MARTIN D'HERES (France)

It is well-known that the input output correlation function (cross-correlation function) of a linear stationnary dynamic system is equal to the response of the tested system to the auto-correlation function of the input signal. This property have been used, in the past, for identification purposes, with pseudo-random binary sequences (PRBS). The autocorrelation function of the PRBS being approximately an impulse, the input output correlation function gives thus approximately the impulse response of the tested linear system. The purpose of this paper is to show how to obtain periodic binary sequences which autocorrelation functions are approximately sinusoïdal. Thus when these particular binary sequences are injected at the input of a linear system, the cross-correlation between the binary test signal and the system's output gives approximately the harmonic response of the plant.
A method is thus given which permits harmonic analysis via the use of low level binary test signal. Many examples have been worked out in numerical simulation and some experimental results of this identification technique applied to a pilot distillation column are given.

1. INTRODUCTION

Let h(t) be the impulse response of a linear stationnary system, with h(t) = 0 for t < 0. For any input u(t), the output S(t) is given by the familiar convolution integral

$$S(t) = \int_0^\infty h(x) . u(t-x)\, dx \qquad (1.1)$$

Defining respectively the input auto-correlation and the input-output cross-correlation functions as:

$$C_{uu}(\tau) = \lim_{T\to\infty} \frac{1}{T}\int_0^T u(t) . u(t-\tau)\, dt$$

$$C_{Su}(\tau) = \lim_{T\to\infty} \frac{1}{T}\int_0^T u(t-\tau) . S(t)\, dt$$

It is then easy to show, using equation (1.1) and the above definitions, that the following relation holds :

$$C_{Su}(\tau) = \int_0^\infty h(x) . C_{uu}(\tau - x)\, dx \qquad (1.2)$$

Thus it appears that the input-output cross-correlation process produces a function which is the response of the system to the forcing input $C_{uu}(\tau)$. When the output is corrupted by an additive noise, b(t), the noise-input cross-correlation function $C_{bu}(\tau)$ must be added to the right member of equation (1.2). Using equation (1,2) for identification purposes, a traditionnal way is to apply an input signal u(t) with an impulse-like auto-correlation function, the cross-correlation process producing thus approximately the impulse response of the linear system. A pseudo-random binary sequence, input signal, exhibiting an impulse-like auto-correlation function, is usually used as a testing signal for identification purposes. In this paper, a binary test signal, with an approximately sinusoïdal auto-correlation function is constructed. This gives a mean to achieve harmonic analysis of a linear process, using a binary (and consequently easy to multiply and shift) low-level signal, via the cross-correlation process.

2. BINARY SEQUENCES WITH A QUASI SINUSOÏDAL AUTOCORRELATION FUNCTION

Let T be the duration of a timing clock pulse, a binary sequence of length N is made up of N clock pulses having either the values +1 or -1. Thus, for a given length N, we have 2^N possible sequences. The problem is to find, among all these sequences, the one which has an autocorrelation function which is, at least in some respect, approximately sinusoïdal. Obviously, an approach to the problem is to compute the autocorrelation function for each of the 2^N sequences, and to choose the best one accordingly. Unfortunately, for N = 10, the computation of all the autocorrelation functions requires 45 secondes using an I.B.M. computer ; as the interesting values of N are greater than 100, the computation time for the 2^{100} possible sequences is prohibitive. Thus, for large values of N, a direct search is not possible and an other way must be found.
To tackle the problem we have used the technique of DELTA modulation of a signal. The basic idea underlying the technique can be explained as follows :
Consider the modulation scheme shown in figure 1

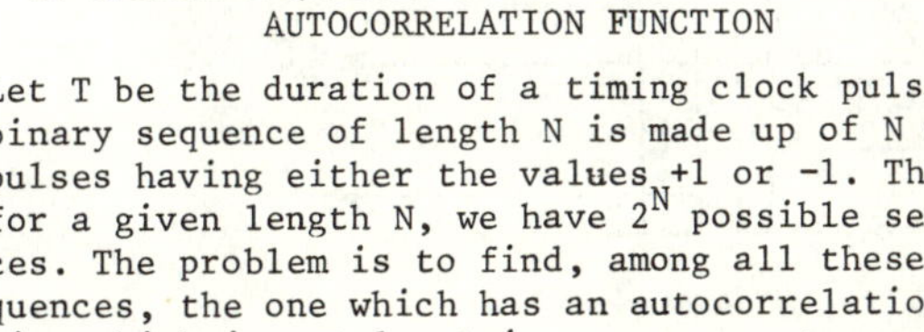

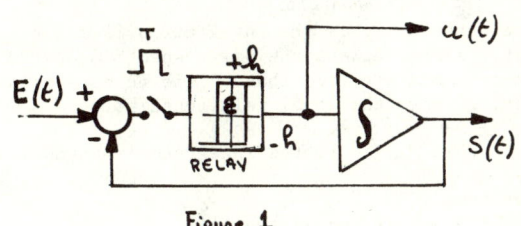

Figure 1

2.1 The basic idea.

In the scheme of figure 1, the error signal e(t) is sampled every T seconds, according to the polarity of e(t), a pulse having an amplitude of + h or − h is produced. The signal u(t) is thus a binary sequence of clock pulses with either the value + h or − h. The signal u(t) is integrated by the output integrator, yielding thus a signal S(t) which is made up of successive ramps with slopes + h or − h.
If we realize now that the system shown in figure 1 represents a closed loop servo system, it is then clear that the output signal is tracking the input signal E(t). When the sampling period is small enough, the output signal S(t) is fairly close to the input signal E(t). The binary signal u(t) is called the DELTA modulation representation of E(t); note that to "demodulate" the signal u(t), i.e to recover the signal E(t) from its DELTA modulation representation, it is sufficient to pass it through an integrator which yields S(t). To illustrate the DELTA modulation concept, we have shown, on figure 2, all the interesting signals.

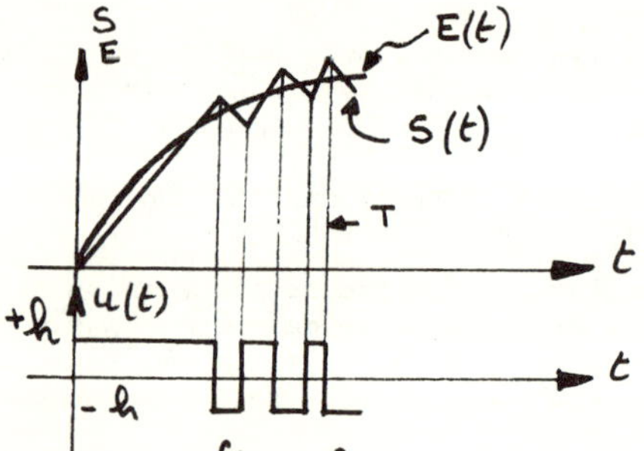

Figure 2.

We shall now proceed to explain why, if E(t) is a sinusoïdal time function, the autocorrelation function $C_{uu}(\tau)$ of the DELTA modulation representation of E(t) : (i.e the binary sequence u(t)) must be close to a sinusoïdal function. Let To be the period of E(t), such that:

$$E(t) = \sin \omega_0 t \qquad \text{and} \qquad \omega_0 T_0 = 2\pi.$$

Provided that the sampling period T is small enough, as compared with To, due to the tracking property of the modulation scheme, we get the approximation equation :

$$S(t) \simeq E(t)$$

Tacking then the power spectrum of these two signals, we have

$$|\mathbb{S}(\omega)|^2 \simeq |\mathbb{E}(\omega)|^2$$

On the other hand, as the integration is a linear system, between the power spectrum of the input sequence u(t) and that of the integrator's output, we have the usual relationship:

$$|\mathbb{S}(\omega)|^2 = \frac{1}{\omega^2} |\mathbb{U}(\omega)|^2$$

(Recall that $[j\omega]^{-1}$ is the integrator's harmonic gain).
We get then :

$$|\mathbb{U}(\omega)|^2 \simeq \omega^2 |\mathbb{E}(\omega)|^2 \qquad (2.1)$$

x note $|\mathbb{X}(\omega)|^2$ denotes the power spectrum of X(t) and so on

As E(t) is a sinusoïdal function, its spectrum is a Dirac function centered at the angular frequency ω_0. According to the equation (2.1), the power spectrum of u(t) must be also close to a Dirac function centered on the angular frequency ω_0. So the binary sequence u(t) which has a Dirac power spectrum, must also have, at least approximately an autocorrelation function which is close to a sinusoïdal function.

2.2 Some results.

The search of the binary sequence u(t) is made by programming an algorithm which is the digital simulation of the basic delta modulation scheme shown on figure 1. For a prescribed value of N, which is there equal to the ratio T/To the search is conducted by an optimization on the pulse amplitude, h, and on the hysteresis, ε, of the relay. A quadratic criterium is computed as

$$C = \int_0^{T_0} [E(t) - S(t)]^2 dt \qquad \text{with } E(t) = \sin \omega_0 t$$

When this criterium is minimized, via a suitable choice of h and ε, the binary sequence u(t) is kept and its autocorrelaiton function is computed. As predicted, the autocorrelation function of u(t) is then quasi sinusoïdal. For values of N greater than 100, the autocorrelation functions are roughly the sum of three basic components :
- a strong cosine function with period To.
- a small high frequency term.
- a small impulse component.

To precise the harmonic content of the autocorrelation function the Fourier transform of the autocorrelation function have been computed. The spectrum shows three kinds of terms.
- a strong harmonic component with period To.
- a small flat spectrum (associated with the impulse component of the autocorrelation function)
- a high frequency component

Let's note that the high frequency component is centered at the clock's frequency, and even if its amplitude is not small, it can be neglected because it is filtered by the plant's transfer function when the cross-correlation process is performed.

To illustrate the results, we have shown on figure 3, the binary sequences u(t) obtained for N = 2000 and N = 180, together with their autocorrelation functions. The autocorrelation functions are shown only on a quarter of period To. The amplitude spectrum of these autocorrelation functions are also displayed.

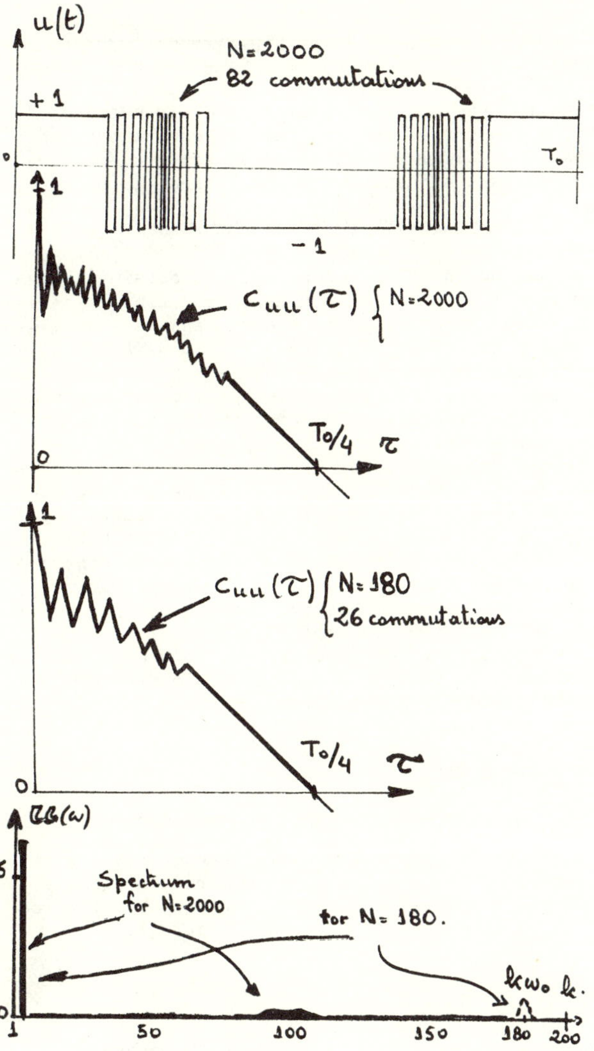

figure 3.

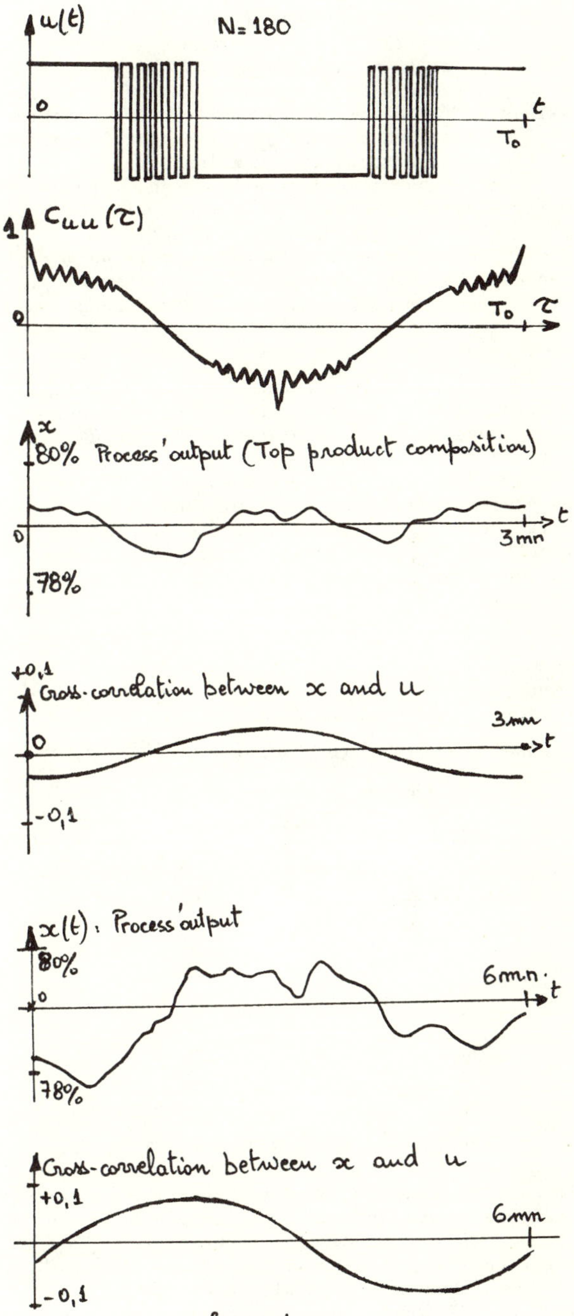

figure 4.

2.3 Remarks.

The search of the binary sequence u(t) has been also conducted, using dynamic programming, in order to choose optimally, at each clock's period, the appropriate value + h or - h. The resulting sequence u(t) is much more oscillatory than the sequence obtained via the delta modulation' scheme, without a significant improvement in the autocorrelation function. Thus the Delta modulation scheme gives a binary sequence u(t) which presents a small number of commutations while yielding an autocorrelation function which is sufficiently close to a sinusoïdal function. (When used for identification purpose via the cross-correlation process specified by equation 1.2).

3. HARMONIC ANALYSIS VIA BINARY SEQUENCES.

As stated in the introduction, the harmonic analysis of a process can be achieved using the general structure shown in figure 5. The binary test signal u(t) may be either generated on line using the delta modulation scheme or stored in binary form into the computer's memory. Different values of the auto-correlation function's period, To, are obtained by modifying the clock's period T. The cross-correlation function can be computed on-line because of the simplicity and ease of the multiplicative and shift operations via the use of the binary sequence u(t).
The influence of a broad band noise is very small because the input signal u(t) has almost all is power spectrum packed on a single frequency, so the noise-test input cross-correlation function $C_{bu}(\;)$ is generally very small. The influence of drift have been studied and to take drift into account, the most convenient way have been found to compute it by integration of the output signal S(t) on a whole period To. The digital simulation of the identification scheme shown on figure 5 have been used for the harmonic analysis of a first order plant with delay. The simulation results have shown that the precision of the harmonic analysis was better than 1%.

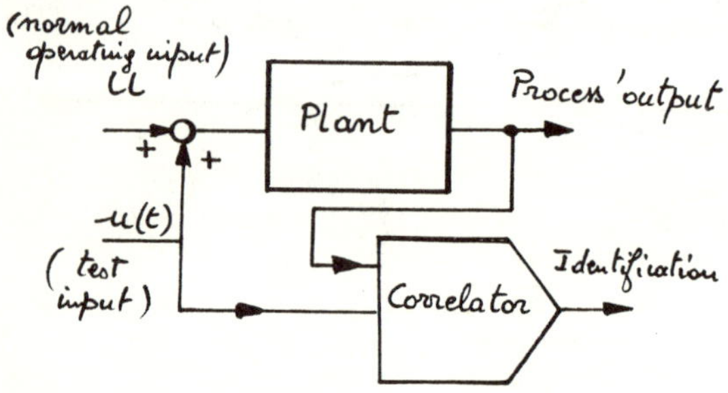

figure 5

4. EXPERIMENTAL RESULTS ON A PILOT PLANT.

The method have been used to perform the harmonic analysis of a pilot distillation plant at the "Laboratoire d'Automatique de Grenoble". This plant consists in a small packed distillation column linked to a digital process computer (MAT 01, Télémécanique). The binary sequence u(t) have been added on the reflux flow set point control. The amplitude of the input test signal have been adjusted until the output' signal was almost totally concealed into the normal output's noise of the process. Nevertheless, due to the high filtering effectiveness of the cross-correlation process, it has been possible to get a very good harmonic response. Figure 4 shows, for two different frequencies, the output signal (the top product composition) and the outputs of the cross-correlator program. It can be noticed that the normal operation noise amplitude on the concentration output is about 1% peak to peak.

5. CONCLUSIONS.

We have shown in this paper how to obtain a binary signal which autocorrelation function is close to a sine function. This binary sequence can thus be used as a low-level test signal, particurlaly well suited for digital treatment because of its binary nature. The two basic operations involved in cross-correlation ie : multiplication and delay are easily performed using this binary test signal. The method may find numerous applications due to the high effectiveness of the cross-correlation process when extracting a sine signal from a noise corrupted output. An application have been made on a pilot distillation plant, others applications may involve on line static optimization via extremum seeking technique, or automatic self-adaptive control system. In that latter case the binary low-level signal is used to excite continuously the control system in order to appreciate its on-line operation.

REFERENCES.

P.A.N. BRIGGS, K.R.GODFREY, P.H.HAMMOND

Estimation of process dynamic characteristics by correlation methods using pseudo-random signals. I.F.A.C. (Prague) Symposium on the identification in Automatic Control Systems, June 1967.

W.D.T.DAVIES, J.L.DOUCE.

On-line system identification in the presence of drift.
I.F.A.C. (Prague) Symposium on the identification in Automatic Control Systems, June 1967.

J.CHERUY, A.MENENDEZ

A comparison on three dynamic identification methods
J.A.C.C. (Boulder, Colorado) August 1969.

Patrick L. Smith
The Aerospace Corporation
El Segundo, California 90009, U. S. A.

Average statistical divergence is proposed as a figure of merit for ranking test inputs used in identifying the unknown parameters of a system. Average divergence is a concept taken from communication theory; it can be computed a priori from a recursion relation derived in this paper. Two closed-form analytic examples are presented.

The development in this paper is for linear multistage processes and is applicable to on-line nonstationary adaptive filtering problems. The average divergence of a general multistage process that has unknown parameters can be calculated recursively.

1. TEST INPUT EVALUATION

The modern approach to control system design requires the use of a parametric system model. As frequency analysis is the cornerstone of classical control system design, so has parametric system identification become equally important in modern control system design (Anstrom, 1970). Particularly useful is the technique referred to as "adaptive filtering" (Magill, 1963, 1965; Hilborn, 1969; Lee, 1964), which combines on-line system parameter identification and state estimation.

This paper is concerned with situations where it is possible to apply a test input to the unknown system to increase the convergence rate of the corresponding adaptive filter.[1] A test input figure of merit is defined; it ranks test inputs by their effect on the convergence rate of the adaptive filter.

Test input evaluation is emphasized in this paper. In practice, test inputs are selected from a class of test functions (e.g., sine waves, up-down steps). Usually a few parameters, such as amplitude and frequency, remain to be determined. A scalar figure of merit is derived in this paper that can be used to optimize the test input parameters, while satisfying various constraints on the state of the system.

Test input evaluation is part of a larger problem: test input synthesis. For classical frequency analysis, test inputs are usually sine waves. Theoretically, only the input frequency, not the amplitude of the sine wave input, is a factor in determining the transfer function of a noise-free linear system. In practice, however, the sine wave amplitude is adjusted to achieve a compromise between a large signal-to-noise output ratio and the limitation of output excursions to linear regions.

Limiting output excursions caused by a test input in on-line adaptive filtering is probably the most critical constraint on the selection of a test input. The conflict is like that of determining if a person is really asleep. One should nudge him softly enough to avoid waking him if he is asleep, but hard enough to get his attention if he is awake.

[1] This problem is the dynamic analog to experimental design in statistics (Kendall, 1961).

The development in this paper is for linear multistage processes; multistage process models are well suited to digital computation.

A linear multistage process is described by

$$\underline{x}(i+1) = \Phi(i)\underline{x}(i) + \Gamma(i)\underline{u}(i) + \underline{r}(i) \quad (1)$$

$$\underline{z}(i) = H(i)\underline{x}(i) + \underline{w}(i) \quad (2)$$

where

$\underline{x}(i)$	$= n \times 1$	vector of state variables (at time index i)
$\underline{u}(i)$	$= p \times 1$	vector of inputs
$\underline{z}(i)$	$= m \times 1$	vector of outputs
$\underline{r}(i)$	$= n \times 1$	vector of random inputs
$\underline{w}(i)$	$= m \times 1$	vector of output noises
$\Phi(i)$	$= n \times n$	state transition matrix
$\Gamma(i)$	$= n \times p$	input matrix
$H(i)$	$= m \times n$	output matrix

and also where

$\{\underline{r}(i) : i = 1, 2, \cdots\}$ are independent multivariate normal random variables distributed $N_n(\underline{0}, R(i))$, respectively.[2]

$\{\underline{w}(i) : i = 1, 2, \cdots\}$ are independent multivariate normal random variables distributed $N_m(\underline{0}, R(i))$, respectively.

$\underline{x}(1)$, the initial state, is a multivariate normal random variable distributed $N_n(\overline{\underline{x}}, M(1))$.

[2] The notation $N_n(\underline{0}, R(i))$ means that $\underline{x}(i)$ is an n-dimensional multivariate normal random variable with mean $\underline{0}$ and variance $R(i)$.

$\underline{r}(i)$, $\underline{w}(j)$, $\underline{x}(1)$ are independent for all i and j.

The corresponding Kalman filter for the multistage process specified by Eqs. (1) and (2) is

$$\underline{v}(i) = \underline{z}(i) - H(i)\hat{\underline{x}}'(i)$$

$$\hat{\underline{x}}'(i) = \Phi(i)\hat{\underline{x}}'(i) + \Phi(i)K(i)\underline{v}(i) + \Gamma(i)\underline{u}(i) \quad (3)$$

$$B(i) = H(i)M(i)H^T(i) + R(i)$$

$$K(i) = M(i)H^T(i)B^{-1}(i)$$

$$M(i+1) = \Phi(i)\big(I - K(i)H(i)\big)M(i)\Phi^T(i) + Q(i)$$

where $\hat{\underline{x}}'(i) = n \times 1$ vector of one-step-ahead extrapolated state variable estimates

$\underline{v}(i) = m \times 1$ vector measurement residual or innovations term

$K(i) = n \times m$ Kalman filter gain matrix

and also where (a) the initial condition for Eq. (3) is $\hat{\underline{x}}'(1) = \overline{\underline{x}}$; (b) the estimation error $\underline{e}(i) = \hat{\underline{x}}'(i) - \underline{x}(i)$ is a multivariate normal random variable distributed $N_n\big(\underline{0}, M(i)\big)$; and (c) the measurement residuals or innovations process $\{v(i) : i = 1, 2, \cdots\}$ are independent multivariate normal random variables distributed $N_m\big(\underline{0}, B(i)\big)$, respectively (the "innovations property" of the Kalman filter).

In many problems, the system matrices $\Phi(i)$, $\Gamma(i)$, $H(i)$, $R(i)$, and/or $Q(i)$ contain unknown parameters. By convention, these unknown parameters are denoted by the $r \times 1$ vector $\underline{\theta}$.

The test input may be a small perturbation applied to an on-line regulating control system or an input profile applied to a system off-line. The on-line problems are of particular interest, since a test signal causes the system to act counter to the purpose of the regulator. Hence, one must select a test input that excites the system sufficiently for adequate identification without unduly upsetting the regulating mechanism. Identification of biological systems is one obvious area where test inputs must be carefully chosen. Chemical and thermal on-line process control problems also require cautious test signal evaluation.

The problem of selecting input signals to enhance the system identification process has been examined by several authors (Aoki, 1970; Smith, 1972; Cumming, 1970; Levadi, 1966; Gagliardi, 1967; Mehra, 1972; Arimoto, 1971; Mosca, 1972). Aoki (1970) and Mehra (1972) have obtained some general results for off-line identification of a large class of systems where the input minimizing the trace of the Fisher information matrix is said to be optimal. In a previous paper, the author (1972) employed the same performance index, but used the Kalman filter representation, which allowed a significant reduction in the amount of computation required. The result from each approach is essentially the same: for equal input energy, the optimal input is the eigenvector corresponding to the largest eigenvalue of a large sensitivity matrix.

Unfortunately, the Fisher information matrix is generally a function of the unknown parameters. Therefore, a problem arises: to solve for the optimal input, the true values of the unknown parameters must be known. Furthermore, for maximum likelihood parameter estimation, validity of the Fisher information matrix as a figure of merit depends ultimately on the Central Limit Theorem.

In adaptive filtering, however, the transient response is the main concern. An asymptotic, steady-state figure of merit for input evaluation is not valid in this case; a different approach to input evaluation is required for adaptive filtering.

To obtain a practical figure of merit for test inputs, the author modified an approach first suggested by Gagliardi (1967). The important features of Gagliardi's formulation are as follows: (a) $\underline{\theta}$ is assumed to be a member of a finite set $\{\underline{\theta}_j : j = 1, \cdots, \ell\}$; (b) the input is selected from a finite set of possible inputs; and (c) the figure of merit for each input is the probability of correctly determining true parameter value from a multiple hypothesis test.

The practical difficulty in Gagliardi's formulation is that the figure of merit is a difficult probability to compute for realistic systems (Hancock, 1966). In this study, the author retains items a and b in Gagliardi's approach, and defines a new figure of merit with almost as much intuitive appeal as item c, but easier to compute.

Sections 2 and 3 of this paper review optimal adaptive filtering theory and develop a figure of merit for test input evaluation, using a concept from communication theory. Some analytical examples are provided in Section 4 to illustrate the theory. Finally, the results for linear multistage processes are presented in Section 5.

2. OPTIMAL ADAPTIVE FILTERING THEORY

This section presents an adaptive approach to estimating the state of a linear multistage process described by an unknown parameter vector $\underline{\theta}$. Knowledge of $\underline{\theta}$ completely specifies the dynamic model and statistics of the process. Consequently, an adaptive filter must estimate the value of $\underline{\theta}$. Adaptive filtering is very similar to system identification. However, an adaptive filter computes the parameter estimates and the state vector estimates on-line, whereas system identification is usually performed off-line; parallel computation is generally used in adaptive filtering, whereas iterative computation is generally used in identification. The adaptive filter can be used off-line for system identification, but the iterative identification techniques cannot, in general, be used for on-line filtering in high performance systems. The primary application of adaptive filtering is in stochastic control. Pattern recognition, failure mode analysis, and signal detection are other related areas in which similar adaptive techniques are useful (Hancock, 1966; Ho, 1968; Prabhu, 1970; Newbold, 1967).

To limit the amount of parallel computation required, one must assume that $\underline{\theta}$ comes from a finite set of known parameter vectors $\{\underline{\theta}_j : j = 1, \cdots, \ell\}$. Most systems of interest can be adequately approximated by this model. If there is prior information about $\underline{\theta}$, it is incorporated into a discrete density function, $P(\underline{\theta} = \underline{\theta}_j)$. $P(\underline{\theta} = \underline{\theta}_j)$ is assigned a uniform discrete density function if there is no prior information.

The following notation is required:

$\hat{\underline{x}}_j'(i)$ = one-step-ahead predicted estimate of the state vector $\underline{x}(i)$ computed from a Kalman filter assuming $\underline{\theta} = \underline{\theta}_j$

$\underline{v}_j(i)$ = measurement residual corresponding to $\hat{\underline{x}}_j'(i)$

$B_j(i)$ = covariance of $\underline{v}_j(i)$ assuming $\underline{\theta} = \underline{\theta}_j$

$p_j(k)$ = $p(\underline{v}_j(k), \underline{v}_j(k-1), \cdots, \underline{v}_j(1) | \underline{\theta})$, the likelihood of the residuals $\{\underline{v}_j(i) : i = 1, \cdots, k\}$ assuming $\underline{\theta} = \underline{\theta}_j$

If $\underline{\theta} = \underline{\theta}_j$, then, from the innovations property of the Kalman filter, it follows that the likelihood is given by

$$p_j(k) = (2\pi)^{-km/2} \prod_{i=1}^{k} |B_j(i)|^{-1/2}$$

$$\times \exp\left\{-\frac{1}{2} \sum_{i=1}^{k} \underline{v}_j^T(i) B_j^{-1}(i) \underline{v}_j(i)\right\}$$

Two optimal adaptive filters are defined: the Bayes adaptive filter (Magill, 1963) and the maximum likelihood adaptive filter. The Bayes adaptive filter is appropriate when the confidence in the prior discrete density $P(\underline{\theta} = \underline{\theta}_j)$ is high. Otherwise, the maximum likelihood adaptive filter is recommended.

The Bayes estimate of $\underline{x}(i)$ is given by

$$\widehat{\underline{x}}'(k) = \sum_{j=1}^{\ell} \alpha_j(k) \hat{\underline{x}}_j'(k)$$

$$\text{where } \alpha_j(k) = p_j(k) P(\underline{\theta} = \underline{\theta}_j) / \sum_{j=1}^{\ell} p_j(k) P(\underline{\theta} = \underline{\theta}_j)$$

The weighting coefficients $\alpha_j(k)$ can be computed in a recursive manner (Sengbush, 1969) as follows:

$$\alpha_j(0) = \alpha_j'(0) = P(\underline{\theta} = \underline{\theta}_j)$$

$$\alpha_j'(k) = \alpha_j'(k-1) |B_j(k)|^{-1/2}$$

$$\times \exp\left\{-\frac{1}{2} \underline{v}_j^T(k) B_j^{-1}(k) \underline{v}_j(k)\right\}$$

$$\alpha_j(k) = \alpha_j'(k) / \sum_{j=1}^{\ell} \alpha_j'(k)$$

If, as more measurements are made, the $\underline{\theta}$ estimate converges to $\underline{\theta}$, the Bayes state estimate will converge to the optimal state estimate. This implies that the $\alpha_j(k)$ corresponding to the true value of $\underline{\theta}$ should converge to 1 and the other coefficients should converge to 0 as $k \to \infty$. This is in fact the case for stationary processes (Magill, 1963). If the process is nonstationary, the weighting coefficients may not converge; convergence may be simply precluded by the time-variable nature of the problem. It is shown in Magill (1963) and Aoki (1967) that $\widehat{\underline{x}}'(k)$ is the minimum variance estimate of $\underline{x}(k)$ (under the assumption that $P(\underline{\theta} = \underline{\theta}_j)$ is the correct prior distribution).

The maximum likelihood estimate of $\underline{x}(k)$ is given by $\widetilde{\underline{x}}'(k) = \hat{\underline{x}}_j'(k)$ where $p_j(k) > p_i(k)$ for all $i \neq j$ (Kendall, 1961).

Both types of adaptive filters are composed of ℓ filters of the Kalman form in parallel, which "try out" all the possible values of $\underline{\theta}$. In the Bayes adaptive filter, the output is a weighted average of the outputs of the individual filters; in the maximum likelihood adaptive filter, the output is the state estimate of the filter whose output appears most likely.

During the first few measurements, the maximum likelihood adaptive estimate may exhibit some "jitter," i.e., relatively large changes in the state estimate over a short period of time. The Bayes adaptive estimate, by comparison, has a relatively smooth response. The absence of jitter is an important consideration in many process control problems. For a nonuniform prior discrete density, the initial jitter of the maximum likelihood adaptive estimate may be even more pronounced. A comparison of the results for the maximum likelihood and Bayes adaptive filters, with respect to convergence and jitter, suggests the use of a hybrid adaptive filter. If initial jitter is a problem, but the reliability of the prior discrete density is low, the Bayes adaptive estimate can be used initially until the maximum likelihood estimate has settled. In many cases, this approach allows one to trade off transient jitter against long-term convergence.

The maximum likelihood adaptive filter is basically a parallel implementation of a multiple hypothesis test on the measurements. The aim in this study is to choose the input sequence that increases the probability of selecting the correct alternative. In this regard, selecting a favorable input seemed similar to signal design in a communication theory; this analogy suggested to the author that certain techniques in communication might be applicable to the input synthesis problem. Section 3 is a discussion of pertinent concepts from communication theory.

3. INFORMATION AND DIVERGENCE

Information theory, a branch of probability and statistics, is very important in modern communication theory, which treats signals as random processes. This section develops the theory of information as required for input evaluation. The definitions, for the most part, are taken directly from Kullback (1959).

The average information contained in $\{\underline{z}(i) : i = 1, \cdots, N\}$ for discriminating $\underline{\theta}_j$ against $\underline{\theta}_k$ is defined as the following expectation conditioned on $\underline{\theta}_j$:

$$I(j;k) = E_j\left[\log \frac{p_j(N)}{p_k(N)}\right]$$

A scalar criterion for pair-wise discrimination among ℓ alternatives is the average divergence, which is defined as

$$D = \sum_{j=1}^{\ell}\sum_{k=1}^{\ell} P(\underline{\theta} = \underline{\theta}_j)P(\underline{\theta} = \underline{\theta}_k)\big(I(j;k) + I(k;j)\big)$$

D, then, is the average amount of information contained in the measurements. D is independent of the true value of $\underline{\theta}$ and can be computed a priori.

D is selected as the figure of merit for a test input. For linear multistage processes, a simplified expression for D can be obtained. Define the following notation for the normalized average covariance of a sequence of measurement residuals:

$$C(j;k) = E_k\left[\sum_{i=1}^{N} \underline{v}_j^T(i)B_j^{-1}(i)\underline{v}_j(i)\right]$$

$$= \sum_{i=1}^{N} \mathrm{Tr}\left[B_j^{-1}(i)E_k\left[\underline{v}_j(i)\underline{v}_j^T(i)\right]\right] \tag{4}$$

It follows immediately that $C(j;j) = mN$. The relationship between D and $\{C(j;k) ; j, k = 1, \cdots, \ell\}$ is

$$D = \frac{1}{2}\sum_{j=1}^{\ell}\sum_{k=1}^{\ell} P(\underline{\theta} = \underline{\theta}_j)P(\underline{\theta} = \underline{\theta}_k)\big(C(j;k) + C(k;j)\big) \tag{5}$$

This expression is straightforward to derive and is left to the reader. The significance is that C(j;k) and C(k;j) are the average normalized covariances of the residuals resulting from assuming the wrong hypothesis in the pair-wise discrimination between $\underline{\theta}_k$ and $\underline{\theta}_j$. The larger C(k;j) and C(j;k), the easier it is to discriminate between $\underline{\theta}_k$ and $\underline{\theta}_j$.

The divergence is related to the concept of entropy. See Kullback (1959) for more details.

4. ANALYTIC EXAMPLES

Before presenting the general results, it is insightful to examine two simple processes for which closed-form expressions for the divergence can be obtained. In both examples it is assumed that

$$r = n = 1 \qquad \ell = 2$$

$$P(\theta = \theta_1) = P(\theta = \theta_2) = 1/2$$

EXAMPLE 1

Consider the following scalar multistage process:

$$x(i + 1) = \theta u(i) + r(i)$$

$$z(i) = x(i) + w(i)$$

where the noise covariances Q(i) and R(i) are assumed to be constant. In the steady state, one finds that M(i) = Q and B(i) = Q + R. Assume M(1) = Q so that the innovations sequence is stationary. The Kalman filters corresponding to each alternative are

$$\hat{x}_j'(i + 1) = \theta_j u(i) \qquad \text{and} \qquad j = 1, 2$$

If $\theta = \theta_1$, then one finds that

$$v_2(1) = z(1) - \hat{x}_2'(1) = x(1) + w(1)$$

$$E_1[v_2^2(1)] = Q + R$$

$$v_2(i) = z(i) - \hat{x}_2'(i)$$

$$= (\theta_2 - \theta_1)u(i-1) + r(i-1) + w(i)$$

$$E_1[v_2^2(i)] = (\theta_2 - \theta_1)^2 u^2(i-1) + R + Q$$

for $i = 2, \cdot \cdot \cdot, N$. The same expressions hold for $E_2[v_1^2(i)]$. From Eqs. (4) and (5), one finds that

$$C(2;1) = C(1;2) = N + \frac{(\theta_2 - \theta_1)^2}{Q + R} \sum_{j=2}^{N} u^2(j-1)$$

$$D = \frac{(\theta_1 - \theta_2)^2}{4} \frac{\sum_{j=2}^{N} u^2(j-1)}{Q + R}$$

Either of the following effects will increase D:

Increase the difference between θ_2 and θ_1

Increase $\sum_{i=2}^{N} u^2(i-1)/(R+Q)$, the input signal-to-noise ratio.

Clearly, these effects will increase the divergence in more complex processes as well.

EXAMPLE 2

The equations for this example are summarized in Table 1.

If $\theta = \theta_1$, then one finds that

$$e_2(1) = \hat{x}_1'(1) - x(1) = -x(1)$$

$$E_1[e_2^2(1)] = R$$

$$e_2^2(i) = \hat{x}_2'(2) - x(2) = u(1) - x(1) - u(1) = -x(1)$$

$$E_1[e_2^2(2)] = R$$

$$e_2(i) = \hat{x}_2'(i) - x(i) = -x(1) - \sum_{j=1}^{i-2} u(j)$$

$$E_1[e_2^2(i)] = R + \left(\sum_{j=1}^{i-2} u(j)\right)^2$$

for $i = 3, \cdot \cdot \cdot, N$. From the fact that

$$v_2(i) = x(i) + w(i) - \hat{x}_2'(i) = -e_2(i) + w(i)$$

Table 1. Equations for Example 2

Parameter value	$\theta_1 = 1$		$\theta_2 = 0$	
Process model	$x(i+1) = x(i) + u(i)$		$x(i+1) = u(i)$	
	$\overline{x} = 0$		$\overline{x} = 0$	
	$z(i) = x(i) + w(i)$		$z(i) = x(i) + w(i)$	
	$M_1(1) = R$		$M_2(1) = 0$	
Kalman filter equations	$\hat{x}_1'(i+1) = \hat{x}_1'(i) + K_1(i)v_1(i) + u(i)$		$\hat{x}_2'(i+1) = u(i)$	
	$K_1(i) = \big(M_1(i)\big)/\big(B_1(i)\big)$			
	$M_1(i) = R/i$		$M_2(i) = 0$	
	$B_1(i) = \big((i+1)/i\big) R$	(6)	$B_2(i) = R$	(7)

it follows that

$$E_1[v_2^2(1)] = E_1[v_2^2(2)] = 2R$$

$$E_1[v_2^2(i)] = \left(\sum_{j=1}^{i-2} u(i)\right)^2 + 2R$$

If $\theta = \theta_2 = 0$, then it is easy to show that $e_1(i+1)$ satisfies the following difference equation:

$$e_1(1) = \hat{x}_1'(1) - x(1) = 0$$

$$e_1(2) = (1/2)\, w(1)$$

$$e_1(i+1) = \frac{1}{i+1} e_1(i) + u(i-1) + \frac{1}{i+1} w(i)$$

for $i = 2, \cdots, N-1$. The closed form of this last equation is

$$e_1(i) = \frac{1}{i} \sum_{j=1}^{i-2} (j+2)u(j) + \frac{1}{i} \sum_{j=1}^{i-1} w(j)$$

for $i = 3, \cdots, N$. As before, it follows that

$$E_2[v_1^2(1)] = R$$

$$E_2[v_1^2(2)] = \frac{1}{4} R + R = \frac{5}{4} R$$

$$E_2[v_1^2(i)] = \left(\frac{1}{i} \sum_{j=1}^{i-2} (j+2)u(j)\right)^2 + \frac{i^2 + i - 1}{i^2} R$$

for $i = 3, 4, \cdots, N$. From Eqs. (4), (5), (6), and (7) it follows that

$$D = \sum_{i=1}^{N} \frac{i^2 + i - 1}{4i\,(i+1)}$$

$$+ \frac{1}{4R} \sum_{i=3}^{N} \left[\frac{1}{i(i+1)} \left(\sum_{j=1}^{i-2} (j+2)u(j)\right)^2 + \left(\sum_{j=1}^{i-2} u(j)\right)^2 \right]$$

Due to the transient response of the system, the earlier inputs increase the divergence proportionately more than the later inputs. Furthermore, the divergence is nonzero even for zero inputs. This arises from the fundamental structural differences between the two plants corresponding to θ_1 and θ_2.

5. DIVERGENCE FOR LINEAR PROCESSES

A recursive equation for $E_k[\underline{v}_j(i)\underline{v}_j^T(i)]$ is derived in this section. D follows immediately from Eqs. (4) and (5). By definition, it follows that

$$\underline{v}_j(i) = -H_k(i)\underline{e}_j(i) + \underline{w}(i)$$

$$E_k\left[\underline{v}_j(i)\underline{v}_j^T(i)\right] = H_k(i)E_k\left[e_j(i)e_j^T(i)\right]H_k^T(i) + R_k(i)$$

The notations $\Phi_k(i)$, $\Gamma_k(i)$, etc., mean the particular values of $\Phi(i)$, $\Gamma(i)$, etc., corresponding to $\underline{\theta}_k$. It remains to derive an expression for $E_k[\underline{e}_j(i)\underline{e}_j^T(i)]$.

Assume $\underline{\theta} = \underline{\theta}_k$ and define the following $\ell n \times 1$ composite vector:

$$\underline{y}_k^T(i) = \left[\underline{x}^T(i)\ \underline{e}_1^T(i) \cdots \underline{e}_{k-1}^T(i)\ \underline{e}_{k+1}^T(i) \cdots \underline{e}_\ell^T(i)\right]$$

A recursive relation for $\underline{y}_k(i)$ is

$$\underline{y}_k(i+1) = A_k(i)\underline{y}_k(i) + B_k(i)\underline{u}(i) + C(i)\underline{r}(i) + F_k(i)\underline{w}(i)$$

$$\text{where } A_k(i) = \begin{bmatrix} A_k^{11}(i) & \cdots & A_k^{\ell 1}(i) \\ \vdots & & \vdots \\ A_k^{1\ell}(i) & \cdots & A_k^{\ell\ell}(i) \end{bmatrix}$$

and where the nonzero submatrices of $A_k(i)$ are

$$A_k^{11}(i) = \Phi_k(i)$$

$$A_k^{jj}(i) = \begin{cases} \Phi_{j-1}(i)\left(I - K_{j-1}(i)H_{j-1}(i)\right) & j = 2, \ldots, k \\ \Phi_j(i)\left(I - K_j(i)H_j(i)\right) & j = k+1, \ldots, \ell \end{cases}$$

$$A_k^{1j}(i) = \begin{cases} \Phi_{j-1}(i) - \Phi_k(i) - \Phi_{j-1}(i)K_{j-1}(i) \times \left(H_{j-1}(i) - H_k(i)\right); & j = 2, \cdots, k \\ \Phi_j(i) - \Phi_k(i) - \Phi_j(i)K_j(i)\left(H_j(i) - H_k(i)\right) & j = k+1, \cdots, \ell \end{cases}$$

$$B_k(i) = \begin{bmatrix} \Gamma_j(i) - \Gamma_k(i) \\ ----- \\ 0 \\ ----- \\ \vdots \\ ----- \\ 0 \end{bmatrix}$$

$$C_k(i) = \begin{bmatrix} I \\ ----- \\ 0 \\ ----- \\ \vdots \\ ----- \\ 0 \end{bmatrix}$$

$$F_k(i) = \begin{bmatrix} 0 \\ ----- \\ \Phi_1(i)K_1(i) \\ ----- \\ \vdots \\ ----- \\ \Phi_\ell(i)K_\ell(i) \end{bmatrix}$$

It follows that $E_k[\underline{x}(i)\underline{x}^T(i)]$ and $\{E_k[\underline{e}_j(i)\underline{e}_j^T(i)] : j = 1, \cdots, k-1, k+1, \cdots, \ell\}$ are the diagonal submatrices of $E_k[\underline{y}_k^T(i)\underline{y}_k^T(i)]$

where

$$E_k\left[\underline{y}_k(i)\underline{y}_k^T(i)\right] = \mathrm{cov}_k\left[\underline{y}_k(i)\right] + \overline{\underline{y}}_k(i)\overline{\underline{y}}_k^T(i)$$

and where

$$\overline{y}_k(i) = E_k[\underline{y}_k(i)]$$

$$\overline{\underline{y}}_k(i+1) = A_k(i)\overline{\underline{y}}_k(i) + B_k(i)\underline{u}(i) \qquad (8)$$

$$\mathrm{cov}_k\left[\underline{y}_k(i+1)\right] = A_k(i)\mathrm{cov}_k\left[\underline{y}_k(i)\right]A_k^T(i) + C(i)Q_k(i)C^T(i) + F_k(i)R_k(i)F_k^T(i) \qquad (9)$$

Computing D for linear multistage processes is straightforward, if laborious. For example, evaluating a particular input sequence for $n = 2$, $\ell = 3$ requires evaluations of Eqs. (8) and (9) for the mean and covariance of $\{\underline{y}_j(i) : j = 1, 2, 3\}$ where $\underline{y}_j(i)$ is a 6×1 vector.

6. SUMMARY

The use of the divergence as a figure of merit for test input evaluation has the following advantages: (a) it can be computed a priori; (b) it is straightforward to compute for linear multistage processes; (c) it is applicable to non-stationary processes.

For large-dimensional systems containing several unknown parameters, extensive computation is required. In the author's experience, however, higher order processes can often be approximated by lower order processes, at least for the purpose of test input evaluation. This approach is analogous to typical approximations in classical frequency analysis.

REFERENCES

Anstrom, K. J., and P. Eykhoff (1970), "System Identification--A Survey, " 2nd IFAC Symposium on Identification and Process Parameter Estimation, held 15-20 June 1970, Preprints [Academia (Czechoslovakia), Prague], Paper No. 0.1.

Aoki, M. (1967), Optimization of Stochastic Systems (Academic Press, New York).

Aoki, M., and T. M. Staley (1970), "On Input Signal Synthesis in Parameter Estimation, " Automatica, 6, 431.

Arimoto, S., and H. Kimura (1971), "Optimum Input Test Signals for System Identification--An Information Theoretical Approach, " Int. J. Syst. Sci., 1, 279.

Cumming, I. G. (1970), "Frequency of Input Signal in Identification, " 2nd IFAC Symposium on Identification and Process Parameter Estimation, held 15-20 June 1970, Preprints [Academia (Czechoslovakia), Prague], Paper No. 7.8.

Gagliardi, R. M. (1967), "Input Selection for Parameter Identification in Discrete Systems, " IEEE Trans. Auto. Control, AC-12, 597.

Hancock, J. C., and P. A. Wintz (1966), Signal Detection Theory (McGraw-Hill Book Co., Inc. New York).

Hilborn, C. G., and D. G. Lainiotis (1969), "Optimal Estimation in the Presence of Unknown Parameters, " IEEE Trans. Syst. Sci. & Cybern., SSC-5, 38.

Ho, Y. C., and A. K. Agrawala (1968), "On Pattern Classification Algorithms--Introduction and Survey, " IEEE Trans. Auto. Control, AC-13, 676.

Kendall, M. G., and A. Stuart (1961), The Advanced Theory of Statistics (Hafner Publishing Co., New York), 2 and 3.

Kullback, S. (1959), Information Theory and Statistics (John Wiley and Sons, Inc., New York).

Lee, R. C. K. (1964), Optimal Estimation, Identification and Control (MIT Press, Cambridge, Massachusetts).

Levadi, V. S. (1966), "Design of Input Signals for Parameter Estimation," IEEE Trans. Auto. Control, AC-11, 205.

Magill, D. T. (1963), Optimal Adaptive Estimation of Sampled Stochastic Processes, Systems Theory Laboratory, Stanford Electronics Laboratories, Stanford, California, Technical Report No. 6302-3.

Magill, D. T. (1965), "Optimal Adaptive Estimation of Sampled Stochastic Processes," IEEE Trans. Auto. Control, AC-10, 434.

Mehra, R. K. (1972), "Optimal Inputs for Linear System Identification," 13th Joint Automatic Control Conference, Stanford, California, held 16-18 August 1972, Preprints of Technical Papers (American Institute of Aeronautics and Astronautics, Inc., New York), 811.

Mosca, E. (1972), "Probing Signal Design for Linear Channel Identification," IEEE Trans. Inform. Theory, IT-18, 481.

Newbold, P. M., and Y. C. Ho (1967), Detection of Changes in Characteristics of a Gauss-Markov Process, Div. Engr. and Appl. Phys., Harvard, Cambridge, Massachusetts, Technical Report No. 531.

Prabhu, K. P. S. (1970), "On the Detection of a Sudden Change in System Parameters," IEEE Trans. Inform. Theory, IT-16, 497.

Sengbush, R. L., and D. G. Lainiotis (1969), "Simplified Parameter Quantization Procedure for Adaptive Estimation," IEEE Trans. Auto. Control, AC-14, 424.

Smith, P. L. (1972), Synthesis of an Optimal Input to Enhance the Identification of a Linear Multistage System, Aerospace Corporation Report No. TR-0172(2441-02)-5.

GAME THEORETIC DESIGN OF INPUT SIGNALS FOR PARAMETER IDENTIFICATION

Yoshikazu SAWARAGI & Katsuya OGINO

Dept. of Applied Mathematics & Physics
Faculty of Engineering, Kyoto University
Kyoto, JAPAN

A new concept of designing an optimal input for accurate parameter identification is developed from the standpoint of game theory, where the input is constrained to the semi-closed form with positive feedback. Firstly, the property of parameter-inidentifiability is defined and discussed to investigate whether the design problem makes sense. Two game theoretic approaches to the problem of input design for parameter identification are, then, presented. Secondly, the lower bound of the error criterion preserved by the optimal inputs is derived, and the structure of the equation error is examined by the use of ρ-sensitivity. As shown by the numerical examples, the present game theoretic designs are effective under the worst case of equation error.

INTRODUCTION

It is well recognized that the choice of an input signal for parameter identification is one of the essential factors which affect the identification accuracy. The design problem of an optimal input for accurate parameter identification, thus, arises and is the problem of choosing an input signal which optimizes some measure of the identification accuracy against computational round off error and observation noise under various constraints on the input and output.

There have been various attempts to attack this problem. While Levin (1960), Levadi (1966), Aoki et al. (1969) and Nahi (1969) adopted the statistical approaches, Rault et al. (1969) analyzed the sensitivity aspects appeared in the identification problem from the deterministic viewpoint. In our previous works (Inoue et al. (1970), Sawaragi et al. (1972)), the more and most sensitive inputs were introduced.

In order to increase the accuracy of parameter identification, it is desirable to adopt an input such that it maximizes the difference in the output between the actual system and its model system, even under the worst case of system uncertainty. Accounting by contraries the superiority properties of feedback configuration in system designs, a new concept of game theoretic designs of input signals for the system with large uncertainties are developed, where the inputs is constrined to the form of semi-closed with positive feedback. In order to investigate whether the design problem of the input makes sense, the property of parameter-inidentifiability is first defined and examined by the use of parameter invariance. The design problem of an optimal input is, then, precisely formulated as a game, where the input and the equation error are assumed to be the antagonists. Standing on the practical situation, another game theoretic design of an optimal input are presented under constraints on feedback gains of the input. Secondly, the lower bound of the error criterion preserved by the present game theoretic designs is derived, and the relation between error criterion and equation error is analyzed by the use of ρ-sensitivity. Numerical examples are presented to show the effectiveness of the present game theoretic approaches even under the worst case. Although the parameter identification by the model method is presupposed in this paper, the concrete method is not necessarily specified.

FUNDAMENTAL NOTIONS

Let the model system of the actual system to be identified be represented by

$$\dot{x} = f(t, x, u, q), \quad x(0) = x_0 \tag{1}$$

where $x = col.(x_1, \cdots, x_n)$, $u = col.(u_1, \cdots, u_r)$ and $q = col.(q_1, \cdots, q_p)$ are respectively the state vector, the input vector and the nominal constant parameter vector. Suppose that the actual system has the same structure and initial condition as those of the model system and is described as

$$\dot{x} = f(t, x, u, q) + w(t, x, u, \Delta q) \quad x(0) = x_0 \tag{2}$$

where nx1-vector $w(t, x, u, \Delta q)$ represents the equation error caused by the uncertain parameter difference Δq .

The observation system is given by

$$z = h(t, x), \quad t \in [0, T] \tag{3}$$

where $z = col.(z_1, \cdots, z_m)$ is the output vector to be observed for the purpose of the parameter identification.

As shown by the solid lines in Fig. 1, the fundamental procedure of the parameter identification by the model method is, given a certain input to the two systems, to determine the parameter difference Δq by processing the observed output difference Δz .

When it is possible to choose an input such that it makes the output difference Δz as large as possible even under the worst case of equation error $w(t, x, u, \Delta q)$, the input possibly increases the identification accuracy against computational round off error and observation noise. The fundamental idea of making the output difference large in this paper is to adopt an input signal with positive feedback, accounting by contraries the superior properties of negative feedback in the design of optimal control systems (for example; less sensitivity, distortion reduction, systems stabilization and so on.)

As shown by the broken and solid lines in Fig. 1. the input is, thus in the following, constrained to the semi-closed form as

$$u(t, z) = u_0(t) + G(t) z \tag{4}$$

When the semi-closed input $u(t, z)$ is chosen in such a way that it maximizes the error criterion

$$J_{er}(u, w) = V_0(\Delta z) \tag{5}$$

the system possibly becomes more sensitive, and the output difference Δz will be expected to be made as large as possible if the property of the unknown equation error $w(t, x, u, \Delta q)$ is not taken into consideration. As the equation error $w(t, x, u, \Delta q)$ is unknown, it seems reasonable to assume at the stage of input design that the equation error takes on values which minimizes the error criterion J_{er} against the demand of designer especially when the equation error is large. The game theoretic approach to the problem of input design, thus, appears to be a very natural one, and the designer should take the advantage of its effectiveness even under the worst case of equation error.

In the special case when the equation error does not affect the error criterion J_{er} , that is, when the error criterion is independent of the equation error, the design problem of an optimal input for parameter identification does not make sense, and the parameter identification cannot be accomplished in practice. This case is called parameter-inidentifiability which is to be examined at the stage of design.

[Definition 1] Parameter-Inidentifiability
When the error criterion J_{er} is invariant with respect to equation error $w(t, x, u, \Delta q)$, that is

$$J_{er}(u, w) = J_{er}(u, 0) \text{ for given } u \tag{6}$$

then, the parameter is called unidentifiable.

In what follows, discussions are focused on a linear system with linear observation for the brevity of presentation.

LINEAR SYSTEM WITH LINEAR OBSERVATION

Consider a linear model system with linear observation

$$\dot{x} = Ax + Bu, \quad x(0) = x_0 - \Delta x_0 \tag{7}$$

$$z = Cx \quad , t \in [0, T] \tag{8}$$

where A, B and C are respectively nxn-, nxr- and mxn-matrices continuous in t.

Suppose that the actual system to be identified is expressed as

$$\dot{x} = (A + \Delta A) x + (B + \Delta B) u, \quad x(0) = x_0 \tag{9}$$

where ΔA and ΔB represent the parameter difference between the two systems. It is to be noted that the initial condition of the model system is, as shown in (7), artificially perturbed to avoid the trivial case in the following design.

The error criterion (5) is given here in the form of

$$J_{er}(u, w) = \int_0^T \Delta z' Q \Delta z \, dt \tag{10}$$

where Q is mxm-symmetric positive definite matrix.

When the input constrained to the semi-closed form

$$u(t, z) = u_0(t) + G(t) z \tag{11}$$

is applied to the two systems, the output difference between the two systems is given by

$$\Delta z = C \Delta x \tag{12}$$

where Δx is the difference in the state between the two systems and is dominated by the state dispersion equation

$$\Delta \dot{x} = A \Delta x + B v + I w, \quad \Delta x(0) = \Delta x_0 \tag{13}_1$$

$$v \triangleq GC\Delta x \qquad (13)_2$$

$$w \triangleq \Delta B u_0 + E(x+\Delta x)\ ;\ Z \triangleq \Delta A + \Delta BGC, \qquad (13)_3$$

where (13) is obtained by subtracting (7) from (9), and x and I are respectively the state of the model system and nxn-unit matrix. As shown in Fig. 2, v and w in (13) can respectively be regarded as the input and the unknown equation error of the state dispersion system, and the identification is to be performed on the dispersion system (13) with observed output Δz in (12). It is to be noted that the state dispersion system (13) plays an important role in the following sections.

By applying the theorem of parameter invariance (McClamroch et al. (1967)) to the system (13) with output (12) and the error criterion(10), we obtain the following necessary and sufficient conditions for the parameter-inidentifiability of linear systems:

$$E = P_0^{-1}(t)\, S_1(t) \qquad (14)_1$$

$$\Delta B u_0 + E x = 0 \qquad (14)_2$$

$$\dot{P}_0 + (A+BGC)'P_0 + P_0(A+BGC) + C'QC = 0\ ,\quad P_0(T)=0 \qquad (14)_3$$

where x and S are respectively the state of the model system and a skew symmetric matrix. The above conditions can be used to gain some insights into the equation error by which the parameter becomes unidentifiable.

GAME THEORETIC DESIGNS OF OPTIMAL INPUTS

When the input of the semi-closed form in (11) is applied to the actual system and its model system, $u_0(t)$ appears only in the unknown equation error w as shown in (13). Therefore, in the case of linear systems, we cannot adopt $u_0(t)$ for the purpose of accurate parameter identification. Although by letting $u_0(t)\equiv 0$ in (11), the design problem is henceforth forcussed on the design of the optimal feedback gain $G^*(t)$, we can adopt $u_0(t)$ for other purposes, for example, for the state regulation.

As mentioned before, the fundamental idea of input design here is to choose an input such that it makes output difference Δz as large as possible. When the input $u(t,z)$ is applied to the actual system and its model system, the output difference Δz is, as shown in Fig. 2, dominated by the state dispersion equation (13) which is driven by the input v and equation error w . Therefore, in order to make the output difference Δz large even under the worst case of unknown large equation error, it is natural to regard the equation error w as an control by which the system uncertainty antagonizes to the designer and minimizes the error criterion.

As a game, the problem of optimal input design for accurate parameter identification can thus be formulated as follows:

Game Theoretic Design I (GTD-I)
For the system and the output

$$\Delta\dot{x} = A\Delta x + Bv + Iw,\quad \Delta x(0)=\Delta x_0 \qquad (15)$$

$$\Delta z = C\Delta x \qquad (16)$$

with payoff functional

$$J(v,w) = J_{er}(v,w) - \int_0^T (v'R_1 v - w'R_2 w)\,dt \qquad (17)$$

find the control v^* and w^* which are optimal in the sense that, for any other controls v , w , there holds

$$J(v,w^*) \le J(v^*,w^*) \le J(v^*,w) \qquad (18)$$

i.e., find a saddle point of $J(v,w)$ if it exists.

R_1 and R_2 in (17) are respectively suitably chosen rxr- and nxn-symmetric positive definite matrices. The equation error w in $(13)_3$ appears as the forcing term in the state dispersion equation $(13)_1$. Therefore, accounting the practical situation such as the limitation of the capacity of observation implement, the consuming energy term of w is added to the payoff functional (17).

In a quite similar form to the one sided output regulator problem, the optimal controls are given in the state feedback form of

$$v^* = R_1^{-1}B'K\Delta x \qquad (19)$$

$$w^* = -R_2^{-1}K\Delta x \qquad (20)$$

and the value of the optimal payoff is given by

$$J(v^*,w^*) = \Delta x_0'\, K(0)\, \Delta x_0 \qquad (21)$$

where $K(t)$ is the solution of the matrix Ricatti equation

$$\dot{K} = -A'K - KA + K(R_2^{-1} - BR_1^{-1}B')K - C'QC,\quad K(T)=0 \qquad (22)$$

It is to be noted that a necessary and sufficient condition for the saddle point (v^*,w^*) to exist and to be unique is that the matrix $K(t)$ solving (22) exist for all $t \in [0,T]$. Unlike the one-sided regulator problem,the existence of the solution of (22) is not always assured. This problem is examined by Ho et al. (1965), Rhodes et al. (1969) and Kimura (1970).

From $(13)_2$ and (19), the optimal input $u^*(t,Z)$ in (11) for parameter identification is given by

$$u^*(t,Z) = G^*(t)Z, \quad G^*(t) = R_1^{-1}B'KC^{-1} \tag{23}$$

when the matrix C is nonsingular. In the case when C is not nonsingular, however, the optimal input $u^*(t,Z)$ is implemented by the state feedback as

$$u^*(t,Z) = R_1^{-1}B'KZ \tag{24}$$

and its realization needs the state estimation. As it is a very difficult task to estimate the state of the actual system here, it is desirable to develop another realizable design method of an optimal input in this case.

To avoid this difficulty, let's impose the constraints of constant on the feedback gain $G(t)$. Then, the design problem of an optimal input can be formulated as a game under the constraints of constant feedback gains (Natrajan et al. (1971)) as follows:

Game Theoretic Design II (GTD-II)
For the system (15), output (16) and the payoff functional (17), find a saddle point of $J(v,w)$ under the constraints

$$v = GC\Delta x, \quad \dot{G} = 0 \tag{25}$$

$$w = E\Delta x, \quad \dot{E} = 0 \tag{26}$$

on v and w.

In a quite similar form to Natrajan et al. (1971), the optimal gains $G \triangleq (g_1, g_2, \cdots, g_m)$ and $E \triangleq (e_1, e_2 \cdots \cdots e_n)$ are obtained as follows by applying the variational procedure.

$$\begin{pmatrix} g_1^* \\ g_2^* \\ \vdots \\ g_m^* \end{pmatrix} = N_1^{-1}(0,T)\int_0^T \begin{pmatrix} \Delta z_1 I_r \\ \Delta z_2 I_r \\ \vdots \\ \Delta z_m I_r \end{pmatrix} B'P\Delta x\, dt \tag*{$(27)_1$}$$

$$N_1(0,T) \triangleq \int_0^T [\Delta z_1 I_r \cdots \Delta z_m I_r]' R_1 [\Delta z_1 I_r \cdots \Delta z_m I_r]\, dt \tag*{$(27)_2$}$$

$$\begin{pmatrix} e_1^* \\ e_2^* \\ \vdots \\ e_n^* \end{pmatrix} = -N_2^{-1}(0,T)\int_0^T \begin{pmatrix} \Delta x_1 I_n \\ \Delta x_2 I_n \\ \vdots \\ \Delta x_n I_n \end{pmatrix} P\Delta x\, dt \tag*{$(28)_1$}$$

$$N_2(0,T) \triangleq \int_0^T [\Delta x_1 I_n \cdots \Delta x_n I_n]' R_2 [\Delta x_1 I_n \cdots \Delta x_n I_n]\, dt \tag*{$(28)_2$}$$

where I_r and I_n are respectively rxr- and nxn-unit matrices, and $N_1^{-1}(0,T)$ and $N_2^{-1}(0,T)$ are assumed to exist. nxn-matrix $P(t)$ in the above equations is the solution of matrix differential equation

$$\begin{aligned} \dot{P} = & -P(A+BGC+E) - (A+BGC+E)'P \\ & -(C'QC - C'G'R_1GC + E'R_2E), \\ & P(T) = 0 \end{aligned} \tag{29}$$

The value of the optimal payoff is given by

$$J(v^*, w^*) = \Delta x_0' P(0) \Delta x_0 \tag{30}$$

In this case the optimal input $u^*(t,z)$ in (11) for parameter identification is given by

$$u^*(t,z) = G^* z(t) \tag{31}$$

In the case when B , R_1 and R_2 are time invariant, the conditions (27) and (28) are simplified to

$$G^* = R_1^{-1}B'\left[\int_0^T P\Delta x \cdot \Delta z' dt\right]\cdot\left[\int_0^T \Delta z \Delta z' dt\right]^{-1} \tag{32}$$

$$E^* = -R_2^{-1}\left[\int_0^T P\Delta x \cdot \Delta x' dt\right]\cdot\left[\int_0^T \Delta x \cdot \Delta x' dt\right]^{-1} \tag{33}$$

EVALUATION OF ERROR CRITERION AND EQUATION ERROR

The error criterion is evaluated, in this section, when the optimal inputs obtained in the preceding section are adopted for parameter identification, and a relation between the equation error and the error criterion is analyzed by the use of ρ-sensitivity (McClamroch et al. (1969)).

Let the optimal controls obtained in the game theoretic designs in the preceding section rewrite as

$$v^* = V^* \Delta x \tag{34}$$

$$w^* = W^* \Delta x \tag{35}$$

where $V^* = R_1^{-1}B'K$ and $W^* = -R_2^{-1}K$ in the case of GTD-I, and $V^* = G^*C$ and $W^* = E^*$ in the case of GTD-II. Then, at the saddle point, that is, when the optimal controls (34) and (35) are applied to the system (15), the error criterion takes value of

$$J_{er}(v^*, w^*) = \Delta x_0' L_1(0) \Delta x_0 \tag{36}$$

where nxn-symmetric matrix $L_1(t)$ is the solution of matrix differential equation

$$L_1' = -F_1'L_1 - L_1F_1 - C'QC, \quad L_1(T) = 0 \qquad (37)_1$$

$$F_1 \triangleq A + BV^* + W^* \qquad (37)_2$$

In order to evaluate the error criterion when the optimal inputs $u^*(t, z)$ in (23), (24) and (31) are adopted for the parameter identification, construct the following combined system which consists of the state dispersion equation (13) and the model equation (7):

$$\begin{cases} \Delta\dot{x} = A\Delta x + BV^*\Delta x + (\Delta A + \Delta BV^*)(x + \Delta x), \quad \Delta x(0) = \Delta x_0 & (38)_1 \\ \dot{x} = Ax + BV^*x, \quad x(0) = x_0 & (38)_2 \end{cases}$$

Then, taking account of the second inequality of the saddle point condition in (18) for the combined system (38), we obtain the lower bound for error criterion, when the optimal input $u^*(t, z)$ is adopted to the identification, as follows:

$$J_{er}(u^*, w) \geqslant I(0) \triangleq (\Delta x_0', x_0')L_2(0)\begin{pmatrix}\Delta x_0 \\ x_0\end{pmatrix} \qquad (39)$$

2nx2n-matrix $L_2(t)$ is the solution of the matrix differential equation

$$L_2' = -\begin{pmatrix} F_2 & E \\ 0 & F_1 \end{pmatrix}' L_2 - L_2\begin{pmatrix} F_2 & E \\ 0 & F_1 \end{pmatrix} + \begin{pmatrix} X(E - W^*) & XE \\ 0 & 0 \end{pmatrix}' + \begin{pmatrix} X(E - W^*) & XE \\ 0 & 0 \end{pmatrix} + \begin{pmatrix} E'R_2E - C'QC - W^*R_2W^* & E'R_2E \\ E'R_2E & E'R_2E \end{pmatrix}$$

$$L_2(T) = 0 \qquad (40)_1$$

$$F \triangleq A + BV^*, \quad E \triangleq \Delta A + \Delta BV^* \qquad (40)_2$$

$$F_2 \triangleq F_1 + E \qquad (40)_3$$

where in the case of GTD-I, $X = K(t)$ in (22), and in the case of GTD-II, $X = P(t)$ in (29).

In order to gain deep insight into the equation error, ρ-sensitivity is newly defined here as follows:

[Definition 2] ρ-Sensitivity
For some real number $\rho\ (\geqslant 1)$, some optimal input $u^*(t, z)$ and some equation error E, the equation error E is said to be ρ-sensitive, if the lower bound of error criterion $I(0)$ in (39) satisfies the inequality

$$I(0) \geqslant \rho J_{er}(v^*, w^*) \qquad (41)$$

The basis of the ρ-sensitivity here is to investigate the structure of equation error by which the lower bound of the error criterion $I(0)$ in (39) is increased by more than a factor of ρ in comparison with the known error criterion $J_{er}(v^*, w^*)$ in (36).

By the same way as McClamroch et al. (1969) we obtain the following conditions for ρ-sensitivity.

[Theorem 1] The necessary and sufficient conditions for $I(0) \geqslant \rho J_{er}(v^*, w^*)$ is given by

$$L_3(0) - (\rho - 1)\begin{pmatrix} L_1(0) & 0 \\ 0 & 0 \end{pmatrix} \geqslant 0 \qquad (42)$$

where 2nx2n-matrix $L_3(t)$ is the solution of the matrix differential equation

$$L_3' = -\begin{pmatrix} F_2 & E \\ 0 & F \end{pmatrix}' L_3 - L_3\begin{pmatrix} F_2 & E \\ 0 & F \end{pmatrix} + \begin{pmatrix} (X - L_1)(E - W^*) & (X - L_1)E \\ 0 & 0 \end{pmatrix}' + \begin{pmatrix} (X - L_1)(E - W^*) & (X - L_1)E \\ 0 & 0 \end{pmatrix} + \begin{pmatrix} E'R_2E - W^*R_2W^* & E'R_2E \\ E'R_2E & E'R_2E \end{pmatrix}$$

$$L_3(T) = 0 \qquad (43)$$

[Theorem 2] The sufficient conditions for $I(0) \geqslant \rho J_{er}(v^*, w^*)$ is given by

$$\begin{pmatrix} (X - \rho L_1)(E - W^*) & (X - \rho L_1)E \\ 0 & 0 \end{pmatrix} + \begin{pmatrix} (X - \rho L_1)(E - W^*) & (X - \rho L_1)E \\ 0 & 0 \end{pmatrix}' + \begin{pmatrix} E'R_2E + (\rho - 1)C'QC - W^*R_2W^* & E'R_2E \\ E'R_2E & E'R_2E \end{pmatrix} \leq 0$$

$$\text{for all } t \in [0, T] \qquad (44)$$

From the above conditions, especially from the more practical sufficient conditions (44), useful informations are obtained at the stage of input design and performing identification.

NUMERICAL EXAMPLES

For illustrative example, consider the following first order model, observation system and error criterion:

$$\dot{x} = ax + bu, \quad x(0) = 1 \qquad (45)$$

$$z = x \qquad (46)$$

$$J_{er} = \int_0^1 2z^2\, dt \qquad (47)$$

Numerical results are shown in Figs. 3-5 for the case when $a = b = 0.5$, $\Delta Z_0 = 0.2$, and $R_1 = R_2 = 1$. Fig. 3 shows the locus of contours of constant J near the saddle point corresponding to GTD-II, where the saddle point is obtained by iteratively solving (15) with (25) and (26), (29), (32) and (33). The values of error criterion, corresponding to the optimal feedback input $u^*(z)$ and a open-loop input $u_0(t)$ which consumes the same energy as that of optimal feedback input of model system, versus two kinds of parameter variations are shown in Figs. 4 and 5 respectively for GTD-I and GTD-II. The superiority of the optimal input $u^*(z)$ obtained by the present game theoretic designs over open-loop input $u_0(t)$ is clearly understood from Figs. 4 and 5.

CONCLUSIONS

From the deterministic point of view, game theoretic approaches to the design problem of optimal inputs for accurate parameter identification are presented, where the equation error caused by parameter error and the positive feedback input are assumed to be antagonists.

The parameter-inidentifiability is first defined and analyzed to investigate whether the design problem makes sense. Secondly, then, the design problem of an optimal input is formulated in the framework of game theory. Furthermore, the error criterion preserved by the present game theoretic designs are evaluated, and the structure of equation error are analyzed by the use of $\int$-sensitivity. By giving simple examples, the effectiveness of the present game theoretic approaches even under the worst case are shown. Although the stochastic approach is to be adopted when the observation noise is not negligible, the separation theorem in the stochastic differential games (Behn et al. (1968)) ensures the effectiveness of the present game theoretic approaches in the case of Gaussian additive noise.

All numerical computations were performed by HITAC 5020 at Data Processing Center in Kyoto University.

ACKNOWLEDGEMENT

The authors wish to express their gratitude to Prof. T. Ono for his discussions of the subjects of this paper.

REFERENCES

AOKI,M. & STALY,M.R. (1969); On input signal synthesis in parameter identification, Proc. of the 4th IFAC Congress, Paper 26.2.

BEHN,R.D. & YU-CHI HO (1968); On a class of linear stochastic differential games, Trans. IEEE, AC-13, 227.

HO,Y.C. , BRYSON,A.E. & BARON,S. (1965); Differential games and optimal pursuit-evasion strategies, Trans. IEEE, AC-10,385.

INOUE,K., OGINO,K. & SAWARAGI,Y. (1970); Sensitivity synthesis of optimal input for parameter identification, Proc. of the 2nd Prague IFAC Symposium on Identification and process Parameter Estimation, Paper 9.7.

KIMURA,H. (1970); A game theoretic approach to control and stabilization of systems with disturbances (in Japanese), Trans. SICE, 6, 366.

LEVADI,V.S. (1966); Design of input signals for parameter estimation, Trans. IEEE, AC-11, 205.

LEVIN,M.J. (1960); Optimum estimation of impulse response in the presence of noise, Trans. IRE, CT-7, 50.

MCCLAMROCH,N.H.,AGGARWAL,J.K. & CLARK,L.G. (1967) On parameter invariance in linear control systems, Int. J. Contr., 5, 361: (1969) Sensitivity of linear control systems to large parameter variations, Automatica, 5, 257.

NAHI,N.E. (1969); Optimal inputs for parameter estimation in dynamic systems with white observation noise, Proc. of the 10th JACC, Paper IV-A5.

NATRAJAN,T. & PIERRE,D.A. (1971); Differential games with constraints on feedback gains, Proc. of the 12th JACC, Paper 6-C1.

RAULT,A., POULIQUEN,R. & RICHALET,J. (1969);Sensitivizing inputs and identification, Proc. of the 4th IFAC Congress, Paper 5.2.

ROHDES,I.B. & LUENBERGER,D.G. (1969); Differential games with imperfect state information, Trans. IEEE, AC-14,29.

SAWARAGI,Y. & OGINO,K. (1972); Sensitivity approach to optimal input synthesis for parameter identification of bilinear system, Proc. of the 5th IFAC Congress, Paper 31.2.

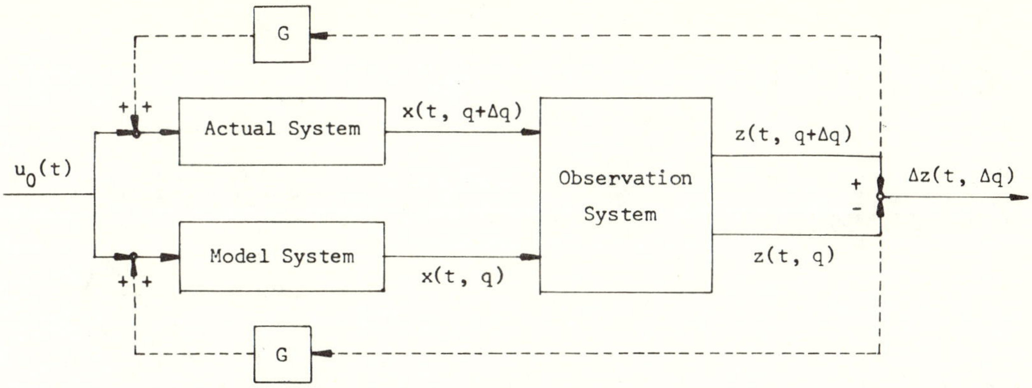

Fig. 1 Parameter identification by model method

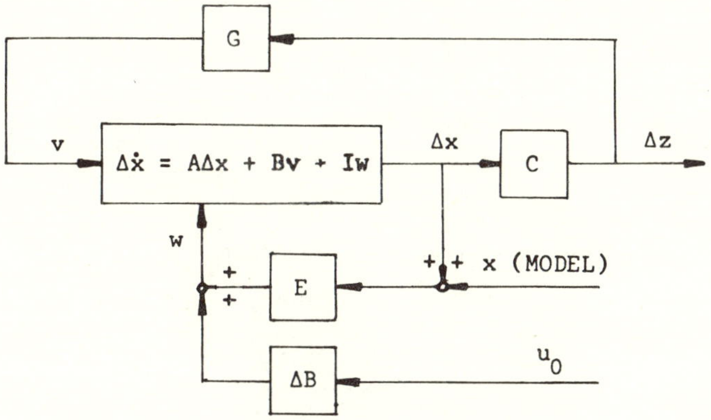

Fig. 2 State dispersion system with v & w

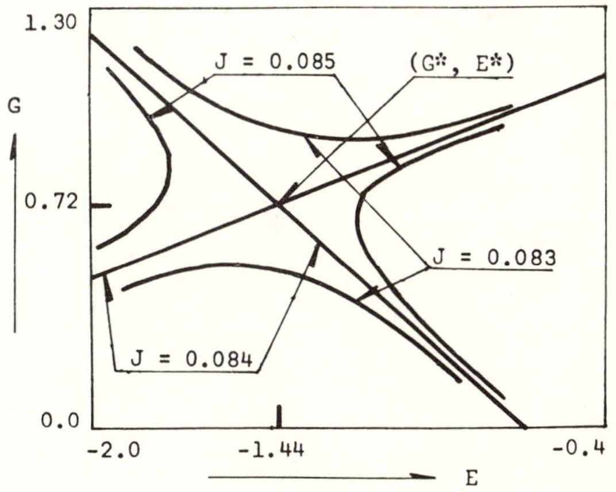

Fig. 3 Locus of contours of constant J near saddle point

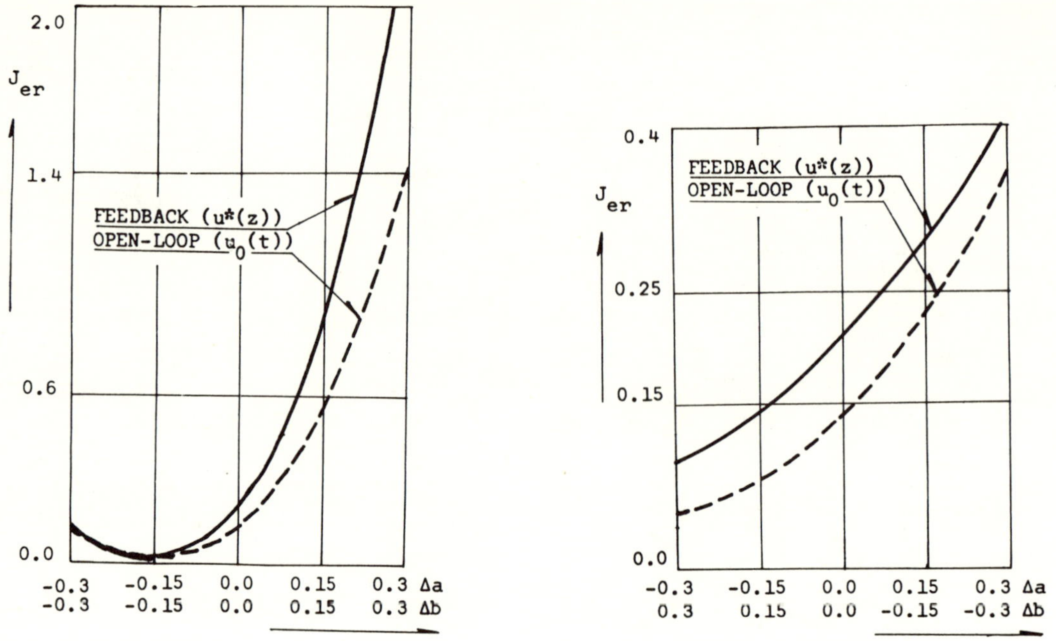

Fig. 4 Comparison of error criterion corresponding to GTD-I

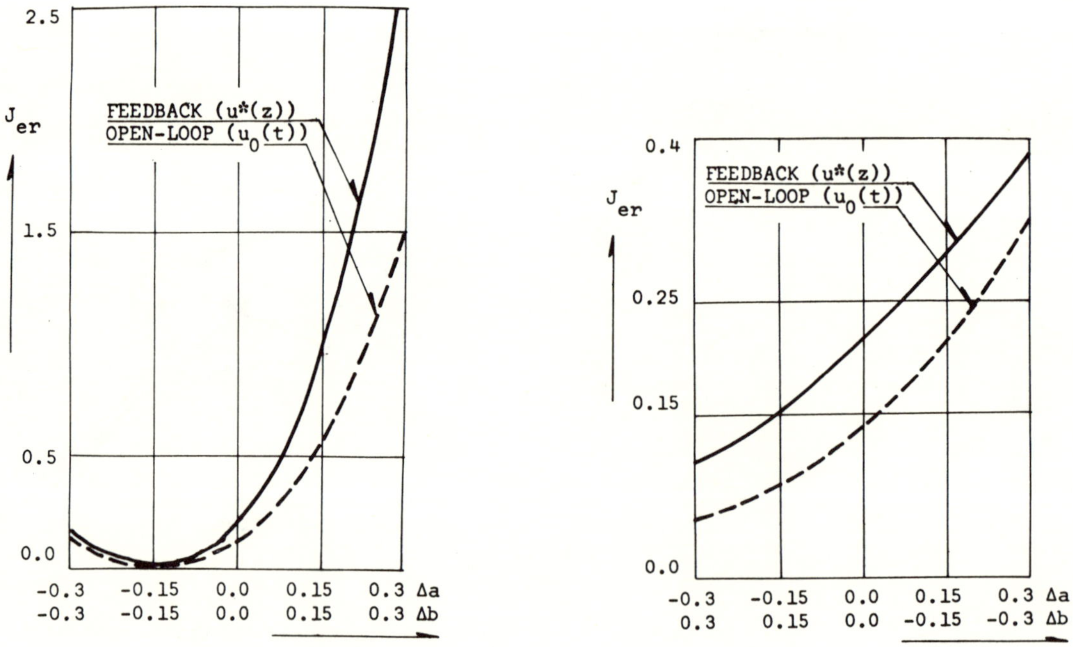

Fig. 5 Comparison of error criterion corresponding to GTD-II

A METHOD FOR MODELLING THE PROCESS DYNAMICS DIRECTLY IN THE TIME DOMAIN

Prof.Dr.A.Frigyes

Department for Process Control
Technical University Budapest, Hungary

The time function $\alpha/t/$-termed "rise function"- is introduced. It has in many respects similar properties as the phase-frequency function. With the help of $\alpha/t/$ it is possible to designate the "decisive time interval"/DTI/ which is the portion of the step response having the major influence on the dynamic behaviour of the system. Methods are presented for matching the DTI of the step response of the system to that of a suitably selected model structure, and the goodness of the modelling is tested.

Let W/s/ be the transfer function of a linear system,v/t/and w/t/its unit step response and unit impulse response,resp. Introduce a function in the time domain:

$$\alpha/t/=\frac{t.w/t/}{v/t/} \qquad /1/$$

This function - termed "rise function"- has the following main properties:

1° $\alpha/t/$ is independent of the gain.

2° $\frac{\pi}{2}\lim_{t\to o}\alpha/t/=-\lim_{\omega\to\infty}\arg W/j\omega/ \qquad /2a/$

$\frac{\pi}{2}\lim_{t\to\infty}\alpha/t/=-\lim_{\omega\to o}\arg W/j\;/ \qquad /2b/$

where arg $W/j\omega/$ is the argument /phase angle/ of the frequency characteristic.

3° If $W/s/=K/s^n$ n=0,1,2....then $\alpha/t/=n=$const $0<t<\infty$

Remember, that in this case arg $W/j\omega/=-n\frac{\pi}{2}$

4° Using the new variables $\tau=\ln t$ and $u=\ln v$,

$$\alpha/\tau/=du/d\tau \qquad /3/$$

/Remember that according to the theorem formulated by Bode

$$\varphi/\omega_x/=\frac{\pi}{2}\left|\frac{da}{d\Omega}\right|_{\Omega=o}+\frac{1}{\pi}\int_{-\infty}^{+\infty}\left\{\left|\frac{da}{d\Omega}\right|-\left|\frac{da}{d\Omega}\right|_{\Omega=o}\right\}\ln\coth\left|\frac{\Omega}{2}\right|d\Omega \qquad /4/$$

a is the natural logarithm of the amplitude of the frequency function, φ is its phase and $\Omega=\ln\omega/\omega_x$ .It is **interesting** that Eq./3/ corresponds to the first term of the Bode theorem/.

In general there is a similarity between the phase function of the Bode diagram and the rise function. The similarity can be visualized if in the plot ot the latter the $\tau=\ln t$ variable is used, and if the axes of one of the plots are directed opposit. Fig.1 and 2 demonstrate, as illustrative examples, the rise functions and the phase functions resp. of some typical structures.
The similarity is based upon the following facts: The value of $\alpha=2$ has the same significance as the value $\varphi=-\pi$ of the phase function. For instance if the rise function of the open control loop is $\alpha<2$ or $\alpha>2$ in the whole frequency range, then the closed loop is structurally /with every value of the loop gain/ stable or unstable, respectively. If the rise function intersects the $\alpha=2$ value once, then the closed loop may be stable or unstable depending on the value of the gain. If the rise function intersects the $\alpha=2$ value twice,then the closed loop is conditionally stable.The nearer the time constants of the system are lying,the steeper the rise function is decreasing.

With the help of the rise function the structure of the system can be estimated. This is one of the adventages of the rise function, but because of the limited size of the present paper it is not possible to go into details.

On the other hand we may try to choose a simple structure for the model, the frequency characteristic of which can be matched to the characteristic of the system at least in its middle frequency region. This means, that the slope of the amplitude characteristic of the model has to vary at least between -20 dB/decade and -60 dB/decade, and the phase characteristic between $-\pi/2$ and $-3\pi/2$. The simplest structure which fulfills this requirement is

$$W_m/s/=\frac{k}{a_3s^3+a_2s^2+a_1s}=\frac{K}{Ts/T^2s^2+2\zeta Ts+1/} \qquad /5/$$

Using this structure as a model, the identification problem is to determine the constants K,T and ζ . This will not be done in the frequency domain, but in the time domain.

The step response of the system is v/t/ and its rise function is $\alpha/t/$. Tests on systems with widely various structures have shown, that there exists a time interval, - termed "decisive time interval"/DTI/ - which has the same significance, as the middle-frequency range of the frequency domain. The latter is characterized as the frequency interval, where the value of the phase function is about $-\pi$.

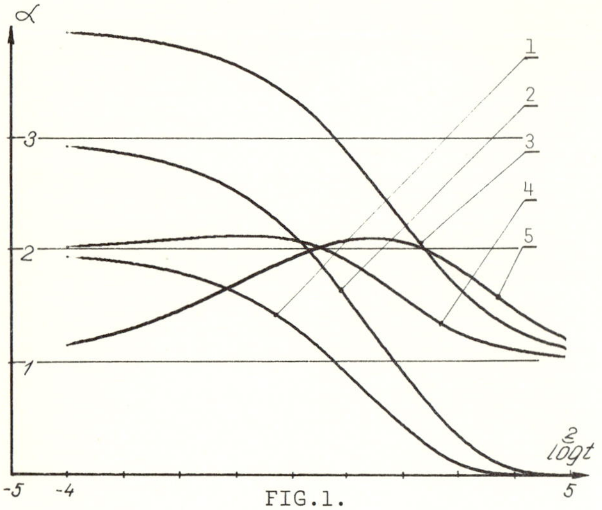

FIG.1.

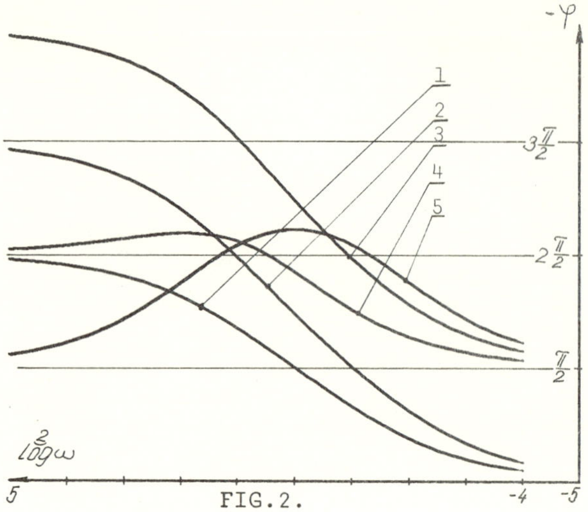

FIG.2.

The transfer functions related to the curves in Figs.1.and 2.are:

1. $\frac{1}{/1+2s//1+0,5s/}$ 2. $\frac{1}{/1+0,4s//1+0,8s//1+3,17s/}$ 3. $\frac{1}{s/1+0,63s/^2./1+2,52s/}$

4. $\frac{1+0,25s}{s/1+s/^2}$ 5. $\frac{/1+0,32s/^2}{s/1+3,2s/^2}$

According to the tests on various structures the DTI can be designated by their rise functions, and this will be the interval, where 2 α/t/ 1,6. If the constants of the model are determined so,that its step response $v_m/t/$ or rise function $\alpha_m/t/$ matches the step response v/t/ or rise function α/t/ of the system exactly in the DTI,then the approximation of the system by the model will be appropriate. The rise functions of structures according to Eq./5/ are plotted in FIG.3.

Two questions can be raised:

1° Which methods are recommended to fit the time function of the model to that of the system in the DTI?

2° How to estimate the goodness of the modelling?

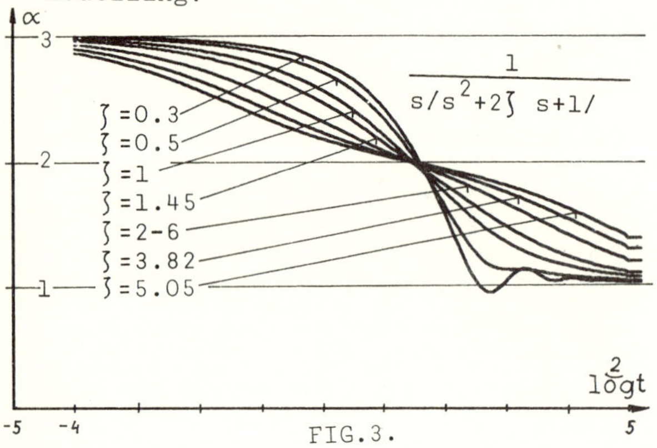

FIG.3.

As to the first question,two methods have been examined. To follow the steps of the first method, see Fig.4. The rise functions α/τ/ and $\alpha_m/\tau/$ of the system, and of the model resp. are plotted,the latter with some arbitrarily selected initial parameter T and ζ./The rise function does not depend on the third parameter,K/.

Due to the logarithmic scale of time,varying the value of T, only the location of the curve $\alpha_m/\tau/$ changes, but not its shape. On the other hand varying the damping factor ζ,the slope of the curve changes. One way to fit these curves is to minimize the mean square error measured along one of the coordinate axes. This two-variable optimization procedure converges fast because of the above mentioned properties of the rise function curve of the model.

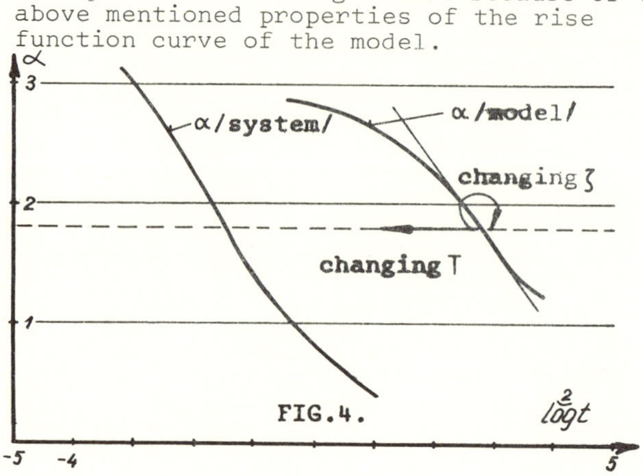

FIG.4.

Another way to fit the curves is based upon the fact, that the slope of the rise function of the model is approximately constant in the DTI, and depends only on the damping factor/see.Fig.3/.Using a single precalculated function between the average slope of the rise function of the system and the damping factor,one can immediately get the damping factor of the model from the slope of the rise function of the system and then,through a simple shifting procedure,its time constant T. To determine the gain K of the model,the comparison of the step responses is needed.

According to the second method the step responses of the system and those of the model are fitted directly.In this case the rise function is used only to designate the DTI in which the step response of the model has to approximate that of the system. The step response of the model fulfills the following differential equation

$$a_3\frac{d^3v_m}{dt^3}+a_2\frac{d^2v_m}{dt^2}+a_1\frac{dv_m}{dt}=1 \qquad /6/$$

Let us integrate Eq./6/ twice from -0 to t,and consider,that the initial conditions at t=-0 are all equal to zero:

$$a_3\frac{dv_m}{dt}+a_2v_m+a_1\int_{-0}^{t}v_m dt=\frac{t^2}{2} \qquad /7/$$

We denote

$$\frac{dv_m}{dt}=z_3/t/;v_m=z_2/t/;\quad \int_{-0}^{t}v_m dt=z_1/t/$$

Replacing them in Eq./7/ we get

$$a_3z_3/t/+a_2z_2/t/+a_1z_1/t/=\frac{t^2}{2} \qquad /8/$$

Suppose,that we know the step response of a system having exactly the same structure as the model.In this case Eq./8/holds for any value of the variable t. Let us choose three values t_1,t_2,t_3,and denote

$$z_k/t_i/=z_{ik}$$

Introducing the following matrices

$$\underline{\underline{Z}}=\begin{pmatrix}z_{11}&z_{12}&z_{13}\\z_{21}&z_{22}&z_{23}\\z_{31}&z_{32}&z_{33}\end{pmatrix},\underline{\underline{a}}=\begin{pmatrix}a_1\\a_2\\a_3\end{pmatrix},\underline{\underline{b}}=\frac{1}{2}\cdot\begin{pmatrix}t_1^2\\t_2^2\\t_3^2\end{pmatrix},$$

we my write the matrix equation, and its solution as

$$\underline{\underline{Z}}\cdot\underline{\underline{a}}=\underline{\underline{b}},\quad \underline{\underline{a}}=\underline{\underline{Z}}^{-1}\cdot\underline{\underline{b}} \qquad /9/$$

/In practical cases $\underline{\underline{Z}}$ is nonsingular/. Suppose,that for some values a_1,a_2,a_3 of the coefficients we solve the differential equation/7/. The matrix equation/9/ stands for the three linear equations,which hold simultaneously.These equations contain the coefficients of the differential equation, as well as the z_{ik}-s, which are derived from the solution of the differential equation.Thus the solution of the matrix equation/9/will give back the initially choosen coefficients a_1,a_2 and a_3.
Now we are puting in Eq./9/the z_{ik}-s derived from the step response of the system:

$$z_{i1}=\int_{-0}^{t_i}v\,dt,z_{i2}=v/t_i/,z_{i3}=\left.\frac{dv}{dt}\right|_{t=t_i}$$

where the t_i-s have to be in the DTI of the system. According to investigations carried out on systems of various structures, the best selection for the t_i-s seem to be those for which $\alpha/t_1/=2$, $\alpha/t_2/=1{,}8$ and $\alpha/t_3/=1{,}6$.The solution will be the set of the parameters of the model related to the transfer function expressed by Eq./5/.

And now the question about the goodness of the modelling.The following test has been performed.The gain of the system - considered to be an open loop - was set to its critical value,i.e.the value at which the closed loop is on the limit of stability.
By the fitting procedure we get the step response $v_m/t/$ for the model which contains of course, the gain,too. Then forming from this model a closed control loop,one must multiply the gain of the model by the value c,to bring the closed loop on the limit of stability. c is a possible estimate of the goodness of the modelling.The general structure of the systems modelled with widely varying parameters was

$$W/s/=\frac{1+sT_d}{s^i/1+sT_1//1+sT_2//1+sT_3//1+sT_4//1+sT_5/} \qquad /10/$$

where i equals either 0 or 1./If there exists in the system a pure transportation lag,then it is possible to recognize it directly from the step response and one has to make the model of the remainder part only./According to the few illustrative examples in Tabl.1 the first method gives a good approximation only if the parameters of the system are such,that the shape of the middle-frequency portion of its Bode diagram does not differ very much from that of the model. This occurs mainly,if the system contains an integrating element /i=1/,or there is one dominant pole. On the other hand, if the system is proportional /i=o/ and has more poles all equal or lying close to each other,and they are not compensated by zeros of appropriate values,then the first method does not give a good approximation with a model having the structure of Eq./5/. The second method gives a better approximation,and is acceptable even in the latter case. As illustrative example,Fig.5.demonstrates how the Bode diagrams of the model obtained by the second method approximate those of the system.The structure of one system is close to its model, the other is quite different from it.

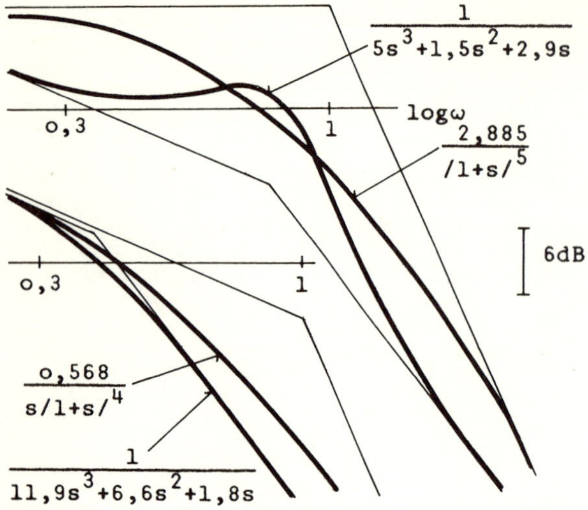

FIG.5.

It must be pointed out,that even in the latter case one can obtain acceptable results by the first method too,if other structures are used for the model.The decision of which structures are to be used for the model can be made by the examination of the rise function.

The second method has the disadvantage, that it needs the derivative of the step response. To overcome this problem Eq./6/ has to be integrated three times. In this case $\int_0^t \int_0^t v/t/dt\, dt$ must be evaluated instead of dv/dt. Nevertheless,the fitting procedure is similar to that as demonstrated above. The need of the derivative of the step response can cause difficulties at the determination of the rise function itself. In this case one my define the "second order rise function" as

$$\gamma /t/=\frac{t.v/t/}{\int_0^t v/t/dt} \qquad /11/$$

which has properties very similar to the rise function defined by Eq./1/. The DTI can be designated with the help of γ /t/ too. This is the time interval in which the value of γ/t/ is about 3.

Conclusion

By means of the time functions according to Eqs/1/ and /11/it is possible to designate the "decisive time interval" which is the portion of the step response having the major influence on the dynamic behaviour of the system. There are several methods - all carried out directly in the time domain - to fit the time functions of various models of simple structures to the time function of the system exactly in this interval. Such a model can be the starting point of any procedure of synthesis,parameter setting,etc.of a control system.The adventages of the methods are:

- At establishing the dynamic properties of a system there is no need of the whole step response function,only of its beginning part up to that moment t at which α/t/ reaches the value 1,6.
- Using a transient test on the system the input step can be switched off at this moment. Therefore the input step - with respect of the saturation - can be greater, and the time of the test is as short as possible. So it is more probable that no disturbances will occur during the test.
- This method,due to its simplicity can be used in computer control systems as an identification and parameter setting algorithm for operator control and interaction executed directly from the process control keyboard.

i	T_1	T_2	T_3	T_4	T_5	T_d	c_1	c_2
o	1	1	16	16	-	-	1,o22	o,997
o	1	1	2	16	-	-	o,9o6	o,984
o	1	16	16	16	-	-	1,o79	1,o36
1	1	1	1	16	-	-	o,952	o,962
1	1	1	16	16	-	-	o,951	o,977
1	1	2	16	16	-	-	o,938	o,974
o	1	1	4	8	64	-	o,923	o,985
1	1	1	1	1	16	-	o,941	o,958
1	1	4	4	8	64	-	o,943	o,962
o	1	1	16	-	-	-	1,o42	1,o14
o	1	4	64	-	-	-	1,o19	1,oo3
1	1	1	16	-	-	-	o,96o	o,973
1	1	4	64	-	-	-	o,973	o,984
o	1	1	1	1	-	o,25	1,o46	1,o1o
o	1	1	1	16	-	-	o,885	o,98o
o	1	1	2	2	-	-	o,766	o,96o
1	1	1	1	1	-	-	o,845	1,oo7
1	1	16	16	16	-	-	o,889	o,98o
o	1	4	4	8	16	-	o,842	o,991
1	1	1	1	1	1	-	o,817	1,o48
o	1	1	1	-	-	-	1,179	1,o82
1	4	4	4	-	-	-	o,897	o,977
o	1	1	1	1	-	o,125	o,893	o,967
o	1	1	1	1	-	o,75	1,184	1,o84
o	1	1	1	1	-	-	o,699	o,94o
o	1	1	1	2	-	-	o,75o	o,954
o	1	1	1	1	1	-	o,461	o,89o
o	1	1	1	1	2	-	o,447	o,916
o	1	1	1	1	16	-	o,719	1,o11
o	1	1	1	1	-	4	1,251	1,112

c_1 corresponds to the modelling by the first method,c_2 to that by the second one

TAB.1.

Acknowledgment.

The author would like to thank Mr.P BAKONYI, Mr. L.LANGER and Miss K.TIHANYI for their help and assistance.

Reference. A.Frigyes:Examination of the closed control loops on the basis of some characteristics in the step response of the open control loops.
Periodica Polytechnica El.Eng.Vol.14.No.3.

=

THE ESTIMATION AND CONTROL IN A FUZZY ENVIRONMENT

a round table discussion*)

Organizer and chairman: M.M Gupta

Written contributions:

Decision and control in a fuzzy environment: A rationale.
M.M. Gupta, P.N. Nikiforuk and K. Kanai

On the fuzzy-mathematical programming.
K. Asai and H. Tanaka

The kth optimal policy algorithm for decision making in fuzzy environments.
Lai-wo Fung and K.S. Fu

Self-organization theory as the basis of direct complex modelling after the experimental data.
A.G. Ivakhnenko

An approach to the identification of human characteristics by applying the fuzzy integral.
M. Sugeno and T. Terano

Adaptive optimal controller and control responses of man.
H. Tamura

Some behaviors of composite fuzzy automata in random environment.
K. Tanaka

A linguistic approach to decision-making.
L.A. Zadeh

---

*) It has been recognized that round table discussions can be means, by which many symposium participants can actively be involved and by which each participant gets a fair chance of bringing in his knowledge, experience and opinions.
The Local Organizing Committee of this symposium has gratefully accepted the initiative of Dr. M.M. Gupta to organize such a round table discussion and to invite introductory statements by a number of engineers/scientists.

The Editor

Decision and Control in a Fuzzy Environment: A Rationale

M.M. Gupta, P.N. Nikiforuk, and K. Kanai

Systems and Adaptive Control Research Group

Department of Mechanical Engineering

University of Saskatchewan

Saskatoon, Canada

Abstract

One of the most inovative aspects of modern control engineering is undoubtedly the prevalence of rigorous mathematical theory. The design of deterministic or stochastic optimal control policies using optimal control theory is based upon the assumption that an exact mathematical model of the process to be controlled is available to the designer either in a deterministic or stochastic sense.. However, as is well known, such a model cannot be obtained for most processes. Situations of this type exist in many decision making areas such as forecasting, economic planning, management, medical diagnostic and pattern recognition. This may, in fact, be the reason why the relation between theory and practice has recently become most tenuous and why control theory which made space mission highly successful is not directly applicable to many humanistic processes.

In general, Mechanistic type systems are amenable to exact quantitative mathematical analysis, while humanisitc type processes are too fuzzy to be amendable to such exact mathematical techniques. It is imperative, therefore, for complex, fuzzily defined humanistic processes not to make use of exact models or conventional mathematical analysis as is being done to-day. The desirability of introducing mathematical concepts which reflect characterisitcs inherent to fuzzy humanistic processes has become apparent since the introduction of Fuzzy Sets in 1965. Researchers at many of the world's leading institutions are now working on the creation of the foundations of a new and exciting field viz. Fuzzy Automata. This new field has evolved from the pioneer work of Lofti.A.Zadeh.

These newly developed concepts of fuzzy automata seem to be extremely useful for humanistic type processes, and are finding applications in many field [1,4,7,29,30,32, 43]. For example, the concepts of fuzzy sets can be used to define a complex process exactly rather than giving an inexact description using deterministic or probabilistic approach. Also the Fuzzy Automata can be used to generate decision policies for such fuzzy processes.

In terms of the future development, the most dramatic results very likely will emerge from the implementation of fuzzy algorithms in the control of industrial processes (such as chemical processes and power systems), pattern classification, designing decision processes for social sciences, economic processes and management, and in the medical diagnostic area. It is envisaged that the fuzzy automata which underline much of human thinking promises have a revolutionary impact on the future of systems engineering.

* This work is supported by the National Research Council of Canada, Grants A-5625 and A-1080, and the Defence Research Board of Canada, Grants 4003 -62 and 9781-04.

Bibliography

1. Asai, K., and Kitajima, S., "A Method for Optimizing Control of Multimodal Systems Using Automata", Information Science, 1972.

2. Asai, K., and Kitajima, S., "Optimizing Control Using Fuzzy Automata", Automatica, Vol. 8, pp. 101-104, 1972.

3. Bellman, R.E., Kalaba, R., and Zadeh, L.A., "Abstraction and Pattern Classification", J. Math. Anal. Appl., Vol. 13, pp. 1-7, Jan. 1966.

4. Bellman R.E., and Zadeh, L.A., "Decision Making in a Fuzzy Environment", Management Science, Vol. 17, No.4, pp. B.141 --B.164 Dec. 1970.

5. Brown, J.G., "Fuzzy Sets on Booleau Lattices", Ballistic Res. Lab. Aberdee, Md. Rep. 1957, Jan. 1969.

6. Chang, S.K., "On the Execution of Fuzzy Programs Using Finite-State Machines", IEEE Trans. Comput., Vol. C-21, pp. 241-253, March 1972.

7. Chang, S.S.L., "Fuzzy Dynamic Programming and Approximate Optimization of Partially Known Systems". Proceedings of the Second Hawaii International Conference on System Science, January 1969.

8. Chang, S.S.L., " Fuzzy Dynamic Programming and the Decision Making Process", Proc., 3rd. Princeton Conf. Information Sciences and Systems, pp. 200-203, 1969.

9. Chang, S.S.L., and Zadeh, L.A., "On Fuzzy Mapping and Control", IEEE Trans. on Systems, Man and Cybernetics, Vol. 2, No.1, pp. 30-34, Jan.1972.

10. DeLuca, A. and Termini, S. "A Definition of a Non-Probabilistic Entropy in the Setting of Fuzzy Sets Theory" Inform. Contr. Vol. 20, No.4, pp. 301-312, May, 1972.

11.Fu, K.S., and Li, T.J., "On the Behavior of Learning Automata and its Applications", Purdue Univ., Lafayette, Ind., Tech. Rep. TR-EE 68-20, Aug. 1968.

12. Groguen, J.A., "L-Fuzzy Sets", J. Math. Anal. Appl., Vol. 18, pp. 145-175, Apr, 1967.

13. Groguen, J.A., "The Logic of Inexact Concepts", Syn., Vol. 19, pp. 325-373, 1969.

14. Lakoff, G. "Hedges: A Study in Meaning Criteria and the Logic of Fuzzy Concepts", in Proc. 8th. Reg. Meet. Chicago Linguist, Soc., 1972.

15. Lee, E.T., and Zadeh, L.A. "Note on Fuzzy Languages", Electronic Res. Lab. Univ. California, Berkley, E.R.L. Rep. 69-7, Nov. 1969.

16. Lee, E.T., and Zadeh, L.A., "Notes on Fuzzy Languages", Inform. Science, Vol. 1, pp. 421-434, Oct. 1969.

17. Lee, R.C.T., "Fuzzy Logic and the Resolution Principle", J. Ass. Comput. Mach., Vol. 19, pp. 109-119, 1972.

18. Leientz, B.P., "On Time Dependent Fuzzy Sets", Inf. Science, Vol. 6, No. 4, pp. 367-76, Oct. 1972.

19. Mizumoto, M., Toyoda, J., Tanaka, K., "Some Considerations on Fuzzy Automata" J. Comut. Syst. Sciences , Vol. 3, No. 4, pp. 409-422, Nov. 1969.

20. Mizumoto, M., Toyoda, J., and Tanaka, K. "Fuzzy Languages" Trans. Inst. Elec. Communication Eng. Japan (C) (in Japanese), Vol. 53-C, pp. 333-340, May 1970.

21. Mizumoto, M., Toyoda, J. and Tanaka, K., "General Formulation of Formal Grammar", Information Sciences, Vol. 4, pp. 87-100. 1972

22. Ruspini, E.H., "Optimization in Sample Description: Data Reduction and Pattern Recognition Using Fuzzy Clustering" 1971 Workshop on Pattern Recognition , Anaheim, California, Oct. 27,1971.

23. Santos, E., "Fuzzy Algorithms", Inform. Contr. Vol. 17, pp. 326-339, 1970.

24. Sugeno, M., "Fuzzy Integrals and its Application", Symposium of Fuzzy System Theory, Tokyo, Dec. 1971.

25. Sugeno, M., "Fuzzy Measure and Fuzzy Integral", Transactions of the Society of Instruments and Control Engineers (In Japan) Vol. 8, pp. 218-226. 1972.

26. Tamura, S., and Tanaka, K. "Pattern Classification Based on Fuzzy Relations" IEEE Tran. on Systems, Man and Cybernetics, SMC-1, 61, 1971.

27. Tanaka, K., et. al. "Fuzzy Automata Theory and its Applications to Control Systems", Control Engineering of Japan Association of Automatic Control Engineers, Vol. 14, No. 9, 1970.

28. Tsuji, H., Mizumoto, M., Toyoda, J. and Tanaka, K. "Interaction Between Random Environment and Fuzzy Automata with a Variable Structure", Trans. IECE of Japan, Vol. 55-D, No. 2, February, 1972.

29. Wee, W.G., "On Generalization of Adaptive Algorithms and Application of the Fuzzy Set Concept to Pattern Classification", Purdue Univ., Lafayette, Ind. Tech. Rep. TR-EE-67-7, July 1967.

30. Wee, W.G. and Fu, K.S., "A Formation of Fuzzy Automata and Its Application as a Model of Learning Systems" IEEE Trans. Systems Science and Cybernetics, Vol. SMC-5, No. 3, pp. 215-223, July 1969.

31. Zadeh, L.A., "Fuzzy Sets", Inform. Control, Vol. 8, pp. 338-353, June 1965.

32. Zadeh, L.A. "Fuzzy Algorithms," Information Control, Vol. 12, pp. 94-102,Feb. 1968.

33. Zadeh, L.A., "Probability Measures of Fuzzy Events", J. Math. Anal. Appl., Vol. 10, pp. 421-427, Aug. 1968.

34. Zadeh. L.A., "Toward a Theory of Fuzzy Systems", Electronics Res.Lab. Univ. California, Berkley, E.R.L. Rep. No. 69-2, June 1969.

35. Zadeh, L.A., "Fuzzy Languages and Their Relation to Human and Machine Intelligence", Proc. Conf. Man and Computer,1970; also electron Res. Lab., Univ. of California, Berkley, Memo. M-320, 1971.

36. Zadeh, L.A."Similarity Relations and Fuzzy Orderings", Inform. Sci., Vol. 3, pp. 177-200, 1971.

ON THE FUZZY - MATHEMATICAL PROGRAMMING

Kiyoji Asai and Hideo Tanaka
College of Engineering
University of Osaka Prefecture
Mozu-Umemachi, Sakai, Osaka, Japan.

Abstract
Many decision problems are of the form that "Decide the optimal strategy in a sense in order to attain some goals". In general a set of strategies and a set of goals are fuzzy. Such a problem have been already researched by R.E. Bellman and L.A. Zadeh[1]. The authors commenced a study of this subject using the theory of fuzzy sets[2] and applied this problem as the mathematical programming based on the idea of level set.

1. Definition and Formulation

Let $X = \{x\}$ denote a set of alternatives. A Fuzzy Constraint in X is denoted by C, which is characterized by a membership function $\mu_C(x)$. Similarly, a Fuzzy Goal in X is denoted by G, which is characterized by a membership function $\mu_G(x)$.

For the example of a Fuzzy Constraint, even if a corporation has x_1 dollars as a reserve fund, the possible amount of invested capital of the corporation may be substantially less than x_1 dollars, or more than x_1 dollars. Since it depends on the judgement of the corporation, the class of the possible amount may be a fuzzy set. Similarly, a set of goals of the corporation may be not characterized by a familiar characteristic function.

A Fuzzy Decision D is defined as a fuzzy set resulting from intersection of C and G, which is characterized by a membership function $\mu_D(x) = \mu_C(x) \wedge \mu_G(x)$, where $a \wedge b$ denotes min [a,b]. And the optimal decision is defined as any alternative in x which maximizes $\mu_D(x)$.

In what follows, we will reformulate this problem using the following level set.

[Definition 1]

For α in [0,1], an α level set of a constraint C is denoted by C_α and is a non-Fuzzy Set in X defined by

$$C_\alpha = \{x \mid \mu_C(x) \geq \alpha\}$$

where x is assumed to be an element in R^n.

We may obtain the following reformulation to [1].

[Proposition 1]

$$\sup_x \mu_D(x) = \sup_\alpha [\alpha \wedge \max_{C_\alpha} \mu_G(x)]$$

where $\max\limits_{C_\alpha}$ means $\max\limits_{x \in C_\alpha}$ for brevity.

Given n constraints $C^1, \cdots, C^n$, the following proposition may be significant.

[Proposition 2]

$$\sup_{\alpha_1 \cdots \alpha_n} \alpha_1 \wedge \cdots \cdots \cdots \wedge \alpha_n \wedge \max_{C^1_{\alpha 1} 0 \cdots 0 C^n_{\alpha_n}} \mu_G(x)$$

$$= \sup_\alpha \alpha \wedge \max_{C^1_\alpha 0 \ldots 0 C^n_\alpha} \mu_G(x),$$

where $C^1_{\alpha_1} 0 \ldots 0 C^n_{\alpha_n} = C^1_{\alpha_1} \cap \cdots \cdots \cap C^n_{\alpha_n}$

Even if n Fuzzy Constraints $C^1, \ldots, C^n$ are given, α level set may be defined by $C^1{}_\alpha \cap \ldots \ldots \cap C^n_\alpha$. Therefore let us consider only one constraint in the sequel.

2. Fuzzy - Mathematical Programming

In the usual mathematical programming, the set which satisfies a given constraint is characterized by a familiar characteristic function of the set. As described in §1., the constraint set may be "fuzzy" in many practical problems. Also, Fuzzy Goal may be the generalized form of a given performance function and then the membership function of a Fuzzy Goal may be derived from a given performance function by a normalization which leaves the linear ordering unaltered[1].

Let the membership function of a Fuzzy Goal $\mu_G(x)$ be regarded as f(x) which is the normalized form of a given performance function $\hat{f}(x)$ on $\overline{s(c)}$, $\overline{s(c)}$ = closure of $s(c) = \{x \mid \mu_C(x) > 0\}$. Therefore $f(x) \in [0,1]$ and $\max\limits_{\overline{s(c)}} f(x) = 1$. It is assumed that $\mu_C(x)$ and f(x) are continuous and that there is an x such that $\max\limits_{s(c)} \mu_C(x) = 1$ and that $\overline{s(c)}$ is a bounded set.

From the proposition 1, the Fuzzy - Mathematical Programming may be the problem that "Determine the optimal pair (α^*, x^*) such that

$$\alpha^* \wedge f(x^*) = \sup_\alpha [\alpha \wedge \max_{C_\alpha} f(x)]".$$

If the optimal α^* can be determined, this problem may be reduced to the usual mathematical programming, since it remains only to determine the optimal x* such that maximizes f(x) in the set C_{α^*}. Also, this formulation is analogous to the Fuzzy Integration [3].

In order to investigate the properties of this problem, let us suppose first the following assumptions.

(i) $\max_{C_\alpha} f(x)$ is α - continuous, i.e. if

$$\lim_{\alpha_n \to \infty} C_{\alpha_n} = C\alpha, \quad \lim_{\alpha_n \to \infty} \left[\max_{C\alpha_n} f(x) \right] = \max_{C\alpha} f(x).$$

(ii) The optimal α^* is unique.

The following lemmas 1,2 and theorems 1,2 can be shown under these assumptions.

[Lemma 1] α^* is optimal if and only if

$$\alpha^* = \max_{C_{\alpha^*}} f(x).$$

[Theorem 1]

$$\sup_\alpha [\alpha \wedge \max_{C_\alpha} f(x)] = \max_T f(x)$$

where $T = \{x | f(x) - \mu_c(x) = 0\}$.

From the theorem 1, the Fuzzy - Mathematical Programming may be reduced to the usual one in this case.

Let the dual formulation of this problem be defined by

$$\inf_\alpha (\alpha \vee \max_{C_\alpha} f(x))$$

where $a \vee b = \max [a,b]$.

[Lemma 2] α^* is optimal for the dual form if and only if

$$\alpha^* = \max_{C_{\alpha^*}} f(x).$$

[Theorem 2]

$$\inf_\alpha \left[\alpha \vee \max_{C_\alpha} f(x)\right] = \sup_\alpha \left[\alpha \wedge \max_{C_\alpha} f(x)\right].$$

It is previously assumed that (i) and (ii) are satisfied. In what follows let us derive the sufficient conditions for α^* - uniquness and α- continuity.

[Definition 2] $f(x)$ is strongly fuzzy convex if and only if for every pair of point x_1, x_2 in $\overline{s(c)}$ $f(x)$ satisfies the inequality

$$f(\lambda x_1 + (1 - \lambda) x_2) > f(x_1) \wedge f(x_2) \text{ for } 0<\lambda<1.$$

[Theorem 3] If $\max_{C_\alpha} f(x)$ is α - continuous and $f(x)$ is strongly fuzzy convex, α^* is unique.

[Theorem 4] If $\mu_c(x)$ is strongly fuzzy convex, $\max_{C_\alpha} f(x)$ is α- continuous.

[Comment] If there exists $\{C\alpha_n\} \ni C_\alpha$ such that C_α can be approximated by $C\alpha_n \in C\alpha_n$ as much exactly as one like, $\max_{C_\alpha} f(x)$ is α- continuous.

References

[1] R.E. Bellman, and L.A. Zadeh; Decision Making in a Fuzzy Environment, Management Science, Vol. 17, No. 4, B - 141/B - 164(1970).

[2] L.A. Zadeh; Fuzzy Sets, Inf. and cont. Vol. 8 (1965).

[3] M. Sugeno; Fuzzy Integration and it's Application, Symposium of fuzzy system in Tokyo, Dec. (1971).

THE kth OPTIMAL POLICY ALGORITHM FOR DECISION MAKING IN FUZZY ENVIRONMENTS†

Lai-wo Fung
School of Electrical Engineering
Purdue University
West Lafayette, Indiana 47907
U.S.A.

K. S. Fu
School of Electrical Engineering
Purdue University
West Lafayette, Indiana 47907
U.S.A.

ABSTRACT

The kth optimal policy algorithm is a natural generalization of the optimal routing problem. We consider a finite-state automaton characterized by a transition-cost matrix and definite initial and final states. The kth optimal policy is the kth best valued sequence of transitions from the intiial state to the final state, assuming that the total cost is the sum of transition costs between intermediate states in the policy.

This problem has been studied by several authors, and at this stage the Bellman-Kalaba algorithm appears superior to all the others for larger networks. However, the total amount of computation is still excessive for practical problems due to its two-stage process, namely, a direct extension of the dynamic programming approach and the backward subtraction routine. In this paper we propose a new algorithm which incorporates the Dijkstra labeling method with our optimal policy elimination routine. In essence, the algorithm is recursive: we start with the 1st optimal policy, then we construct an automaton with the 1st optimal policy eliminated, so that the next 1st optimal policy of this new automaton will be the 2nd optimal policy of the original automaton. Other sub-optimal policies can be generated in the same way. Advantages and drawbacks of the recursive scheme will be discussed. On the other hand, we shall show that the algorithm can be extended to fuzzy automata.

Practical application of the kth optimal policy algorithm will be studied. In the first step, we consider as many uncertainties in the model as we can using the fuzzy set approach or the risk function approach. We ask the decision maker to provide a range of allowable deviations (from the nominally optimal value) in which he would choose a policy of his preference or his best judgment. The kth optimal policy algorithm is employed to enumerate all possible optimal policies inside the prescribed range. In this manner, human factors in decision-making processes can best be considered.

In the fuzzy set approach to decision making, the problem of assigning meaningful membership functions to the variables in the model is of utmost significance. We shall discuss the possibility of using the sensitivity analysis, first studied by Bellman and Kalaba on the basis of the kth optimal policy algorithm, for the identification of membership functions in a fuzzy decision-making model.

†This work was supported by the National Science Foundation Grant GK-36721.

I. FUZZY GOALS AND CONSTRAINTS

Recently the problem of decision-making in an environment in which the goals, the constraints and consequences of possible actions are not precisely known has aroused considerable interest. Fung [1] reviews several existing approaches to optimal decision and control under this kind of non-stochastic uncertainties. The innovation of fuzzy set theory [2] has opened a broad field of decision and control problems which are studied in a framework quite different from the traditional probabilistic models. Many promising results in control [3], decision-algorithm [4,5] and semantics [5] solved with the tool of fuzzy set appeared in the past, and we expect more to be achieved after the several fundamental problems on fuzzy sets studied in this paper have been solved.

The nature of imprecision in real-world decision problems was examined by Bellman and Zadeh [6]. They pointed out that, in dealing quantitatively with imprecision, we are tacitly accepting the premise that imprecision--whatever its nature--can be equated with randomness, and consequently we usually employ the concepts and techniques of probability theory and, particularly, the tools provided by statistical decision theory, stochastic control theory and information theory. Noting that the main distinction between human intelligence and machine intelligence lies in the ability of humans to manipulate fuzzy concepts and respond to fuzzy instructions, Bellman and Zadeh [6] contend that fuzziness is a major source of imprecision in many decision processes. Fuzziness is differentiated from randomness in that the former is associated with fuzzy sets; that is, classes in which there is no sharp transition from membership to membership. By a fuzzy goal they mean an objective which can be characterized as a fuzzy set in an appropriate space, and a fuzzy constraint is simply a fuzzy set of possible actions. We shall base our discussions on the contents of their paper.

We assume that the decision process is a multi-stage process [6-8] which can be represented by a sequence of transitions in a time-invariant finite-state deterministic system. The state x_t, at time $t = 0,1,2,\ldots$, ranges over a finite set $X = \{\sigma_1,\sigma_2,\ldots,\sigma_n\}$, and the input u_t ranges over $U = \{\alpha_1,\alpha_2,\ldots,\alpha_m\}$. The temporal evolution of the system is described by the state equation

$$x_{t+1} = f(x_t,u_t), \qquad t = 0,1,2,\ldots \tag{1}$$

in which f is a function (mapping) from X x U to X. Thus $f(x_t,u_t)$ represents the successor state of x_t for input u_t. We assume that the initial state x_0 is given and the goal of the decision process is a fuzzy set G^N in T, where N is the termination time of the process and $T \subset X$ is the set of final (terminal) states. Suppose that the system makes a sequence of transitions from the initial state x_0 to a final state $x_N \in T$ under the sequence of inputs $u_0,u_1,\ldots,u_{N-1}$. Then the most general performance function, i.e., criterion of optimality, can be expressed as

$$J\left[\{u_t\}_{t=0}^{N-1}\right] = J\left[\varphi(x_N),\{L(x_t,u_t)\}_{t=0}^{N-1}\right] \tag{2}$$

where $L(x_t,u_t)$ is the cost at state x_t due to the input u_t.

We assume that at time t the input u_t is subjected to a fuzzy constraint C^t which is a fuzzy set in space U, characterized by a membership function $\mu_{C^t}(u_t)$. Bellman, et.al. [6] used the following criterion of optimality:

$$\max_{\{u_t\}_{t=0}^{N-1}} J\left[\{u_t\}_0^{N-1}\right]$$

where

$$J\left[\{u_t\}_0^{N-1}\right] = \mu_{G^N}(x_N) \wedge \{\mu_{C^0}(u_0)\wedge\cdots\wedge\mu_{C^{N-1}}(x_{N-1})\} \tag{3}$$

where x_N is expressed as a function of x_0 and $u_0,u_1,\ldots u_{N-1}$ through the iteration of (1). The criterion is based on the definition of decision which is the confluence of goals and constraints. In particular, here "confluence" acquires the meaning of intersection when the concept of fuzzy set theoretic "and" is interpreted as

$$\mu_{A\cap B} = \mu_A \wedge \mu_B. \tag{4}$$

The weakness of this representation is that we do not allow any tradeoff between $\mu_A(x)$ and $\mu_B(x)$ so long as $\mu_A(x) > \mu_B(x)$ or vice versa. Therefore, in the optimality criterion (3), if any one step of transition in a sequence of transitions has a low membership in the constraint set, the entire decision will be considered to be unsatisfactory no matter how high the memberships of the other transitions. The lack of any tradeoff at all on the memberships of the intermediate transitions in the decision process results in a highly insensitive optimality criterion which virtually depends on the worst stage of the whole process.

To partially remove this kind of bias in the hard sense "and" and "or" functions, algebraic product and sum are proposed; in particular, a fuzzy decision D might be expressed as a convex combination of the goals and constraints, with the weighting coefficients reflecting the relative importance of the constituent terms. In this paper we shall assume that every transition stage in a decision process is of equal importance, and the relative importance of the goal and that of the entire process are weighted by a coefficient α. Then the optimality criterion takes the familiar form of

$$J\left[\{u_t\}_0^{N-1}\right] = \alpha\,\mu_{G^N}(x_N)+(1-\alpha)\frac{1}{N}\sum_{t=0}^{N-1}\mu_{C^t}(u_t). \tag{5}$$

In both cases of (3) and (5), it is easy to show that

$$0 \le J\left[\{u_t\}_0^{N-1}\right] \le 1 \tag{6}$$

for all input sequences $\{u_t\}_0^{N-1}$. Let D_N be the fuzzy decision set on the space U x U x...x U (N times), which contains all sequences $\{u_t\}_0^{N-1}$ to bring the system from initial state x_0 through (1) to some final state $x_N \in T$. We do not assume that the termination stage N is fixed and we use $D \equiv \bigcup_{\text{all } N} D_N$ to denote the general fuzzy decision set. Since (6) is always satisfied, we may regard the performance function J as a membership function of the fuzzy decision set D; that is,

$$\mu_D(\{u_t\}_0^{N-1}) \equiv J\left[\{u_t\}_{t=0}^{N-1}\right] \tag{7}$$

provided $\{u_t\}_0^{N-1} \in D_N$, $N = 1,2,\ldots$. In this way we have unified the concept of performance function (or optimality criterion) and the fuzzy decision set membership function, and we can treat the problem of optimal control in a fuzzy environment in exactly the same manner as we deal with fuzzy decision-making problems. On the other hand, from (7) it appears to be more natural to employ optimality criterion (5) instead of (3) which was previously used in [6] and [7].

The optimal decision $\tilde{d} \in D$, $\tilde{d} = \{\tilde{u}_t\}_{t=0}^{N-1}$ for some N, can be defined as

$$J\left[\{u_t\}_{t=0}^{N-1}\right] \ge J\left[\{u_t\}_{t=0}^{M-1} \in D_M,\ M \ge 1\right] \tag{8}$$

It will be understood that all possible policies in D are loopless sequences of transitions; that is, no state of the system is visited more than once in any sequence. Furthermore, there is no transition between any two states of the final state set T. Let $T = \{\sigma_{\ell+1},\ldots,\sigma_n\} \subset X = \{\sigma_1,\ldots\sigma_\ell,\sigma_{\ell+1},\ldots,\sigma_n\}$. Then the decision set D contains sequences of lengths not greater than ℓ; i.e., $N < (\ell + 1)$. For convenience of description, consider the equivalent finite-state model in which the state set X and final state set T are defined above and transition from state σ_i, $i \leq \ell$, to state $\sigma_j \in X$ is associated with an input constraint membership μ_{ij}. Assuming that the constraint set C and goal set G have time-invariant membership functions, i.e.,

$$\mu_{G^t} = \mu_G \text{ and } \mu_{C^t} = \mu_C,$$

and that there is a unique input u which drives the system from state σ_j, it is obvious that

$$\mu_{ij} = \mu_C(u) \quad \text{if } f(\sigma_i,u) = \sigma_j.$$

If there is no transition from state σ_i to σ_j, $\mu_{ij} = \infty$ for $j = 1,2,\ldots,n$ and $\mu_{ij} = \infty$ for $(\ell + 1) \leq i,\ j \leq n$.

Since the unspecified termination stage N appears in the optimality criterion (5), ordinary dynamic programming procedures must be slightly modified to obtain the optimal decision d. Denote the initial state of the system by σ_1, say, and the maximum value of J for some N-stage decision sequence starting from σ_i and ending in any final state in T by R_i^N, $i \leq \ell$. The flowchart in Figure 1 describes the algorithm for finding the optimal decision. The amount of computations required by this algorithm, in the worst case, is ℓ times the amount for solving the ordinary type of optimal decision problem in a system of identical size where the termination stage N does not appear explicitly in J as in (5). Other algorithms for solving the optimal decision problem of (5) do exist; we mention this particular one because it can be used to solve other more complicated problems later in this paper.

II. MEASUREMENT OF MEMBERSHIP FUNCTIONS AND COLLECTIVE DECISION MAKING

At the present stage, we still have many important fundamental problems in the application of the fuzzy set theory to decision making; some interesting ones are listed in the following:

(1) How can we establish a measurement scale for the membership functions of fuzzy sets?

(2) Even if we have such measurements, the best policy obtained in previous works [6,7] is very subjective since the assignment of membership functions depends on the individual who evaluates the situation in his best judgment. Can we reduce the amount of subjectivity by considering the judgments of more individuals for the same problem and obtaining the best compromise from the aggregate of opinions?

(3) How can we solve the decision-making problem in a mixed environment (i.e., both fuzziness and quantitative criteria are involved)?

In this section we shall very briefly examine how we can establish a measurement scale for the membership functions of fuzzy sets and then proceed to propose a simple method of reducing subjectivity which often occurs in the assignment of memberships to elements of fuzzy sets. The discussions will be restricted to decision-making problems in this paper.

The difficulty in estimating memberships lies on two characteristics of the membership function:

(1) The membership of an element of a fuzzy set is not a chance phenomenon; it only indicates the "degree of belonging" of the element to the set. It often depends exclusively on the idiosyncracy of the problem designer, and at this stage we hardly know any satisfactory method with which we can construct a measurement scale for these quantities. Furthermore, it is not likely that these quantities possess statistical properties which we can estimate by repeatedly asking the problem-designer to provide his estimates at various times. That is, multi-observation by a single problem-designer does not reduce excessive subjectivity in the estimates.

(2) The formulation of subjective probability in establishing a measurement scale for memberships is not applicable because here the memberships of all the elements of the fuzzy set do not normalize to unity and the measure of memberships of disjoint subsets of the fuzzy set is not additive.

How to define the membership function of subsets of a given fuzzy set has not yet been formulated. We contend that it must depend on the applications of the function and is not merely a matter of definition. Suppose G is a fuzzy set in X with membership function μ_G, and let 2^X be the set of all subsets of X. Then we say that a fuzzy set $\tilde{G}$ in 2^X with membership $\mu_{\tilde{G}}$ which maps from 2^X to the interval [0,1]. It is still uncertain what form this function $\mu_{\tilde{G}}$ should take; however, the two forms proposed below should be quite useful.

$$\text{(a)}\quad \mu_{\tilde{G}}\left(\{x_1,x_2,\ldots,x_n\} \in 2^X\right) \equiv \frac{1}{n}\sum_{i=1}^{n} \mu_G(x_i) \tag{9}$$

This form has the intuitive meaning of "average membership" for a subset $\{x_1,\ldots,x_n\}$ in 2^X.

(b) Define $\mu_{\tilde{G}}(\{x_i\}) \equiv \mu_G(x_i)$, $x_i \in X$

$$\mu_{\tilde{G}}(\{x_i,x_j\}) \equiv \mu_G(x_i)+\mu_G(x_j)-\mu_G(x_i)\mu_G(x_j),$$

$x_i,x_j \in X$

and, in general,

$$\mu_G(\{x_i\} \cup S) \equiv \mu_G(S)+\mu_G(x_i)-\mu_G(S)\mu_G(x_i) \quad (10)$$

for $x_i \in X$, $S \in 2^X$ and $x_i \notin S$. It is easy to show that this definition is well formulated because the operation in (10) is associative. That is, we have

$$\mu_G(\{x_i,x_j,x_k\}) = \mu_G(\{x_i\} \cup \{x_j,x_k\})$$

$$= \mu_G(\{x_i,x_j\} \cup \{x_k\})$$

for any $x_i,x_j,x_k \in X$ which are all distinct. By induction it follows that $\mu_G(S)$ is unique under any permutation of the elements in S in the recursive construction (10). This function also has some meaning of "average membership", and it arises from the "soft-sense" concept of union of fuzzy sets in the following form [6]:

$$\mu_{A \oplus B}(x) = \mu_A(x) + \mu_B(x) - \mu_A(x)\mu_B(x), \quad x \in X$$

where A and B are both fuzzy sets in X.

When the above concept of associated fuzzy sets is applied to the input set U, some kind of measurement scale might be established. Let C be the fuzzy constraint set in U and let $\tilde{C}$ be the fuzzy set induced by C in 2^U. Suppose optimality criterion (5) and definition (9) are adopted. Then $\mu_{\tilde{C}}(\{u_1,u_2,\ldots,u_k\} \in 2^U)$ has the intuitive meaning of average membership of the set of inputs $u_1,u_2,\ldots,u_k$, or the average membership of the ensemble of transitions corresponding to these inputs, in the associated constraint set $\tilde{C}$. Suppose there is a preferential ordering between any two elements of 2^U, i.e., let $\gtrsim$ be a binary relation defined on $2^U \times 2^U$. We wish to establish sufficient and necessary conditions such that there exists a function $\mu_{\tilde{C}}(.)$: $2^U \to [0,1]$ and for all S_1, $S_2 \in 2^U$,

$$\text{(i) } S_1 \gtrsim S_2 \text{ iff } \mu_{\tilde{C}}(S_1) \geq \mu_{\tilde{C}}(S_2) \quad (11)$$

$$\text{(ii) } \mu_{\tilde{C}}(S_1 \cup S_2) = \gamma_{12}\mu_{\tilde{C}}(S_1)+\gamma_{21}\mu_{\tilde{C}}(S_2) \quad (12)$$

where

$$\gamma_{12} = \frac{|S_1|}{|S_1| + |S_2|}, \quad \gamma_{21} = \frac{|S_2|}{|S_1| + |S_2|}$$

with $|S|$ being the cardinality of set S. A special case of condition (ii) is the definition of average membership (9):

$$\mu_{\tilde{C}}(\{u_1,\ldots,u_k\}) = \frac{1}{k} \sum_{i=1}^{k} \mu_C(u_i) \quad (13)$$

Unfortunately, we still do not know what kind of necessary and sufficient conditions on the preferential ordering $\gtrsim$ should be satisfied in order to construct such a membership function $\mu_{\tilde{C}}$ (or μ_C). We believe that it should be a very challenging problem for future studies in the fundamental properties of fuzzy sets. (Interested readers should consult Krantz, et.al., [9] for a detailed treatise of the theory of measurements.)

Once these conditions are found, experiments can be set up to construct a measurement scale for the membership function μ_C. One simple way to achieve this is to acquire information about the preferential orderings on some selected elements of 2^U from a rational subject (one whose preferential orderings always satisfy the prescribed necessary and sufficient conditions), and to set up inequalities in view of (11). Solving these linear inequalities and normalizing the quantities, one may obtain values of the membership function $\mu_C(u)$ for specific elements $u \in U$. Note that since $u \in U$ and their corresponding transitions in the decision problem need not be any real quantities, i.e., they can represent a cause of transition, a particular action, a subjective preference, the quality of the transition, or anything appropriate in the problem, the elements $u \in U$ should more generally be called transitional events. Furthermore, we think that the implications of "preferential orderings" are wider than those of "fuzzy constraints."

Let us now present a practical simple solution to the estimation problem without actually constructing a measurement scale. As we have remarked earlier, repeated assessments of the value of the membership function for a given element does not reduce excessive subjectivity in the estimates. Instead, sufficient expertise of several individuals may help to obtain an objectively more acceptable optimal decision. More specifically, we propose the concept of multi-observer for solving decision-making problems in a fuzzy environment. The method of solution is analogous to the collective decision model we have studied in [10]. Here several individuals are asked to provide their estimates of the membership of transitional events in the problem, a restricted decision-region in which a few of the best policies are enumerated is obtained, and the problem designer makes his ultimate choice from the restricted decision-region. The only difference between the two models is that the quantities (such as distances, costs and times of shipment) in the first model [10] which concerns management decision problems can often be accurately determined while fuzzy membership functions in the present model are more prone to subjective uncertainties, which makes the problem more subtle to tackle. On the other hand, due to the appearance of termination stage N in the performance function (5), the kth shortest route algorithm [11] is used to solve the present problem. Details of the new k^{th} best policy algorithm are given in [10]. We shall only briefly outline the principle.

The main body of the algorithm is the same as the general decision-making algorithm presented in [10]. The new k^{th} best policy algorithm works in the following manner. After the first best policy is found, augment the decision problem (which in our case is a finite-state system) so that the sequence of transitions corresponding to the first policy is deleted from the original problem; this augmentation procedure is repeated every time the kth best policy is obtained, and the best policy for the problem after the kth augmentation will be the (k+1)st best policy of the original problem. In every stage of seeking the best policy from the augmented problem, the algorithm given in Section I is used instead of other conventional methods such as dynamic programming or Dijkstra labelling method. Our present augmentation algorithm is specially designed for problems with optimal criterion (5) (or generalized (5)). It is feasible only for small systems and small values of k; thus it is suitable for the collective decision-making problems discussed in this section because only a few of the best policies with respect to both criteria need be enumerated before the problem-designer can reach a decision.

A simple collective decision problem involving two individuals is given in [10]. Two sets of memberships are assigned to the transitional events of the problem, each corresponding to one of the individuals. It is shown that there is no difficulty in obtaining a compromise from the aggregate. Furthermore, there is no difficulty in generalizing the problem to the case involving more than two individuals, each contributing to the ultimate decision with unequal significances. Since the method of the general algorithm can be used to solve the collective decision problem of this section as well as the mixed decision problem of Section III, we shall illustrate the method with more details in the following section.

III. DECISION-MAKING IN A MIXED ENVIRONMENT

The third question we pose in Section II concerns the most general form of application of the kth best policy algorithm: how can we solve the decision-making problem in a mixed environment (i.e., both fuzziness and quantitative criteria are involved)? We can distinguish two types of problems of this nature.

First Type

We consider ordinary decision-making problems where there is only one quantitative criterion to be optimized. Furthermore, there is also a qualitative criterion which must be optimized at the same time. The latter has a prominent feature it is assessed globally; i.e., the quality of the entire decision sequence is considered instead of the qualities of the transitional events as we have studied in the previous section.

Assume that we want to plan a trip for vacation. We have a starting point and a destination, and we are supposed to find a route with the least mileage. However, we also desire that the trip should have the lowest accident rate and the most beautiful scenery. Since it is not always possible to find such a trip which satisfies all criteria, we have to allow a certain amount of tradeoff in making our decision. For instance, suppose the shortest route is 100 miles long; we may be content with routes within 15 miles of this range. Then we can proceed to enumerate all the shortest routes no longer than 115 miles and select the route which is qualitatively most appealing. Such a route is "optimal" in the sense of a tradeoff between mileage and subjective preference. It might be noted that a global qualitative criterion has the following advantages:

(1) The quality of policies in some problems must be considered in the entirety of the policies instead of stagewise assessments. The average membership criterion suggested in (5) may not be appropriate in some problems.

(2) No quantitative scale of the quality criterion is needed--here we only require that the enumerated routes in the allowed range be ranked, and all other possible routes outside this prescribed range can be discarded from consideration.

Furthermore, this method can be extended to problems with several quantitative criteria and one global qualitative criterion. In this case, the allowed range is just the restricted decision-region studied in [10], and it is only necessary to rank the policies within the restricted decision-region, which are enumerated by the kth best policy algorithm.

Second Type

In this category the qualitative criterion (or criteria) is not globally assessed. Every transitional event of the problem is associated with some quantitative and fuzzy constraints. The overall quality of the decision sequences may be the average membership of the transitions in the sequences as we have studied in Sections I and II, while the overall quantitative criterion may be the total cost, total distance, or total time of the transitions. Since quantitatively good policies are usually not qualitatively good policies, this type of problems is the kind of decision-making problem with several conflicting criteria studied in Section 3 of [10]. It can be solved with the general algorithm (finding the restricted decision-region from which the ultimate decision is made) of [10] with the following features: the new k^{th} best policy algorithm for fuzzy criterion which we discuss in Section II is used to enumerate the best policies with respect to the qualitative criterion (or criteria), whereas ordinary dynamic programming or Dijkstra labelling method is used to enumerate the best policies with respect to the quantitative criteria.

Figure 2 shows the transition diagram of a decision system in which the quantitative criterion is γ_1, and the qualitative criterion γ_2 is the membership function of the transitions. The average membership criterion (5) is used for entire decision sequences, whereas total cost criterion is used with respect to γ_1. We assume that both criteria γ_1 and γ_2 are equally significant for making a decision. Figure 3 depicts the policies and Figure 4 shows the manner in which these policies are enumerated. Note that we are required to make a compromise between minimizing total cost (γ_1) and maximizing membership (γ_2) of the policies.

The restricted decision-region for this example is arbitrarily chosen to be [0,5.0] with respect to γ_1 and [0.65,1.00] with respect to γ_2. In this region there are seven policies which are supposed to be the best policies, one of which will be selected as the ultimate decision. From Figure 4 we see that these are the first few best policies common to both criteria. It is easy to show that some of these policies, namely, 15, 16 and 3 can be discarded from the ultimate selection. For instance, policy 15 is absolutely worse than 2 because 15 is worse than 2 with respect to both criteria (15 has average membership of 0.77 compared to 0.78 of 2, and 15 has total cost of 4.4 compared to 4.3 of 2). Similarly, 16 and 3 can be rejected when compared to 4. On the other hand, policies 1 and 2 are said to be <u>incomparable</u> since 1 is better than 2 with respect to γ_1 but 1 is worse than 2 with respect to γ_2. Unless some specific multi-criterion comparison scheme is used (such as [12]), the ultimate selection between 2 and 1 depends on many subjective factors and is not easy to determine. Similarly, it is easy to see that 2 and 4, 4 and 5, 1 and 4, 1 and 5, and 2 and 5 are all incomparable pairs; furthermore, 1, 2, 4, 5 form an <u>incomparable set</u> because any pair of policies in the set is an incomparable pair. The ultimate optimal decision will be chosen from the incomparable set. There is a simple algorithm to find the incomparable set from any given set of policies. Figure 5 is a flowchart of the algorithm. Finding the incomparable set from the policies in the restricted decision-region is a further sep of reducing the decision region, and it makes the ultimate decision-making process much easier. Particularly, when the pairwise comparison algorithm of [12] is employed to rank policies in a multi-criterion space, our k^{th} best policy algorithm and the incomparable set algorithm can be very efficiently coupled with the pairwise comparison algorithm to achieve decision-making in a mixed environment involving both quantitative and fuzzy criteria (constraints).

IV. CONCLUSIONS AND REMARKS

Throughout this paper we have emphasized that the problem-designer has no difficulty to choose the best decision out of the small subset of best policies enumerated by the kth best policy algorithm. The ultimate choice can be made arbitrarily. A well-defined man-machine procedure has been proposed by Stankard, et.al., [12] which provides <u>pairwise comparisons</u> of policies in the restricted decision-region, and thus selects the optimal policy from the comparisons. We might view the problem in this way: since it is not practical to consider pairwise comparisons of <u>all</u> possible policies in the problem, we need to have an algorithm which yields the relatively small subset of best policies from which the ultimate decision can be selected through pairwise comparisons. Thus the method of [12] can be applied efficiently with the reduction of decision space studied in this paper.

REFERENCES

1. Fung, L. W., "On the Approaches of Optimal Control Systems Under Non-Stochastic Uncertainties," Technical Report TR-EE 72-26, School of Electrical Engineering, Purdue University, Lafayette, Indiana, August 1972.

2. Zadeh, L. A., "Fuzzy Set," <u>Inform. and Control</u>, Vol. 8, June 1965.

3. Chang, S. S. L., and Zadeh, L. A., "On Fuzzy Mapping and Control," <u>IEEE Trans. on Systems, Man and Cybernetics</u>, Vol. SMC-2, January 1972.

4. Zadeh, L. A., "On Fuzzy Algorithms," Electron. Res. Lab., Univ. of California, Berkeley, Memo M-325, 1971.

5. Chang, S. K., "On the Execution of Fuzzy Programs Using Finite-State Machines," <u>IEEE Trans. on Computers</u>, Vol. C-12, pp. 94-102, 1968.

6. Bellman, R., and Zadeh, L. A., "Decision Making in a Fuzzy Environment," <u>Management Science</u>, 1970.

7. Chang, S. S. L., "Computer Aid to Decision Making, Its Organization and Algorithm," <u>IFIP Congress</u>, Ljubljand, Yugoslavia, August 1971.

8. Elmaghraby, S., <u>Some Network Models in Management Science</u>, Springer-Verlag, Berlin, 1970.

9. Krantz, D. H., Luce, R. D., Suppes, P., and Tversky, A., <u>Foundations of Measurements</u>, Academic Press, 1972.

10. Fung, L. W., and Fu, K. S., "Decision-Making in a Fuzzy Environment," Technical Report TR-EE 73-22, School of Electrical Engineering, Purdue University, Lafayette, Indiana, May 1973.

11. Yen, J. Y., "Finding the k Shortest Loopless Paths in a Network," <u>Management Science</u>, Vol. 17, No. 11, 1971.

12. Stankard, M. E., Maier-Rothe, C., and Gupta, S. K., "Choosing Between Multiple Objective Alternatives: A Programming Approach," <u>Management Science Center</u>, University of Pennsylvania, December 1969.

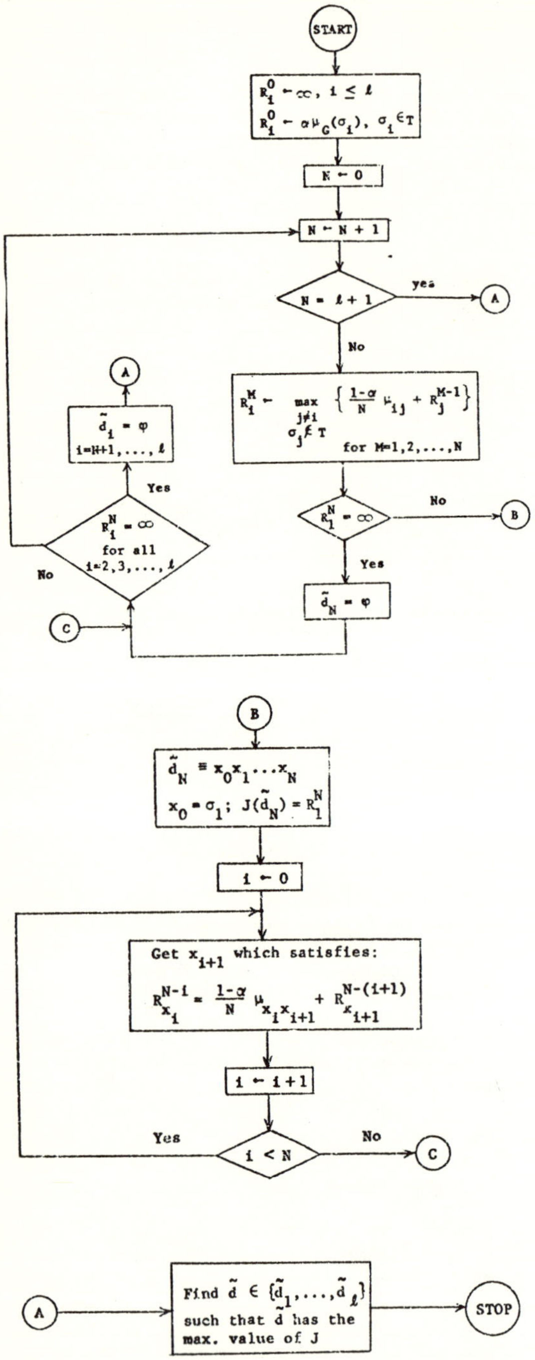

An Algorithm Which Maximizes Average Membership [5]

Figure 1

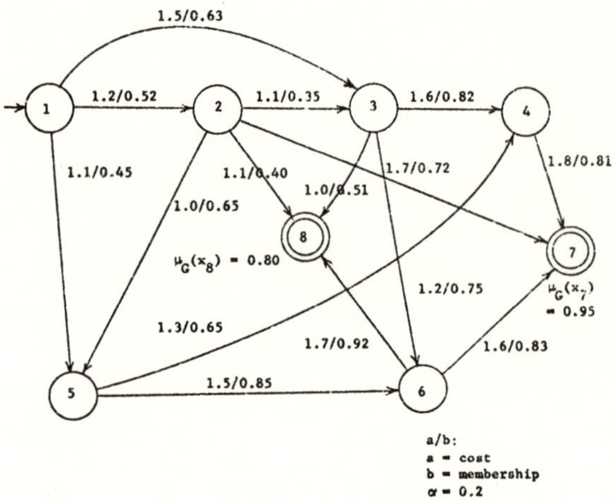

Figure 2

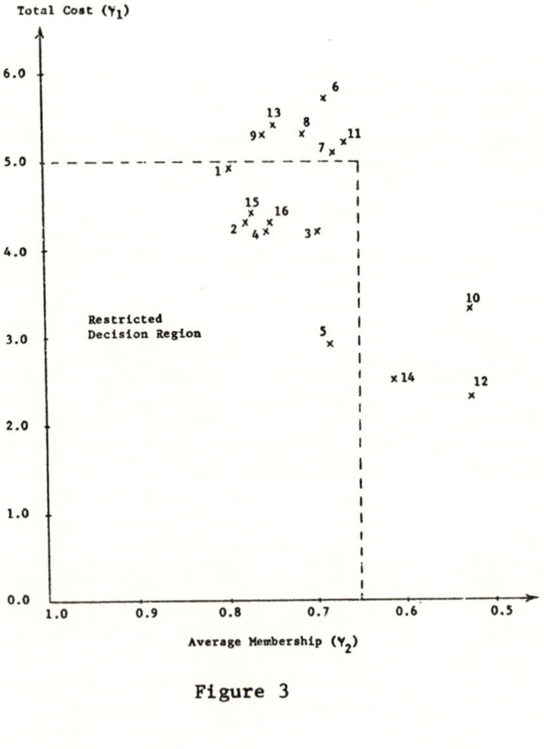

Figure 3

Rank	Policy Number	& Average Membership	Policy Number	Total Cost
1	1	0.79	12	2.3
2	2	0.78	14	2.5
3	15	0.77	5	2.9
4	9	0.76	10	3.3
5	4	0.758	3, 4	4.2
6	16	0.752		
7	13	0.748	2, 16	4.3
8	8	0.716		
9	3	0.699	15	4.4
10	6	0.69	1	4.9
11	5	0.686	7	5.1
⋮				

Policy No.	Sequence of Transitions
1	1347
2	1367
3	1547
4	1567
5	127
6	12347
7	12367
8	12547
9	12567
10	1238
11	12368
12	128
13	12568
14	138
15	1368
16	1568

Figure 4

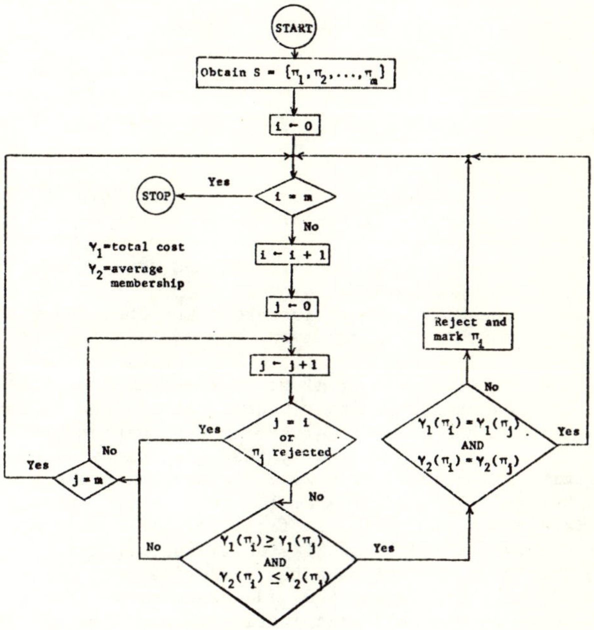

Figure 5

SELF-ORGANIZATION THEORY AS THE BASIS OF DIRECT COMPLEX SYSTEM MODELLING AFTER THE EXPERIMENTAL DATA

A.G. Ivakhnenko

Professor, Institute of Cybernetics

Prospect Nauky 109, Kiev, USSR

Self-organization theory is the basis of the software for the new trend in the engineering cybernetics which is called as the direct modelling of complex system after the experimental data.
The principal differences and advantages of the theory in comparison with the deterministic approach are explained.
The main information about the obtained results in the self-organization theory and direct modelling is conveyed.

The direct modelling of complex system from the limited number of experimental data is the modern "break through" in engineering cybernetics. If the sixties are marked by the development of optimization ideas (Pontryagin's maximum principle, Belman's dynamic programming etc.) which have put aside the problems of the classical control theory (the problems of stabilizing, program and servo system) then the seventies and the eighties are facing the wide development of the ideas of the self-organization theory, the use of which is the basis of the direct modelling mentioned above.
The theoretical significance of the self-organization theory corresponds to the level of the deterministic optimizing methods whereas its practical significance exceeds them considerably. The advantages of the self-organization theory for complex systems are predetermined by the fact that this theory enables us to solve the problems which can not be solved by means of other deterministic methods due to their large dimension. For example, this theory enables us to solve the problems having up to 200 input arguments (variables) which is impossible for deterministic methods as well as it is impossible for them to predict the course of complex random processes for 10 years ahead.
But the main advantage is that the self-organization approach allows us to take decisions without demanding from a man to introduce not very reliable apriori information into the statement of the problem. With the deterministic approach, for example, a man has to invent a system of equations which as if they describe all the elements of complex system. No one element can be left without mathematical description. With such "forced" invention man as a rule makes the more mistakes the more the object is complex and non-understandable to him. Many mistakes are caused by reduction of number characteristics of variables. The natural researchers' tendency to the simple linear relationships with one or two variables brings them to the area of the low accuracy of the model.
For example some people try to predict the earthquakes after one argument (the change of current value of the intensity of Earth's magnetic field), but they won't obtain the accuracy in this case. The exact models should have some optimum number of argument, should be non-linear (optimum complexity) and should take into account the past values of arguments i.e. they should be described by the equations with the delayed arguments. It is clear that the man can not invent so complex equations with the help of deterministic appraoch. The equations of optimum complexity can be obtained only as a result of self-organization method using a computer.
The solution of modelling problem by the self-organization method has already shown that the complex systems in the fields of economics, ecology and industrial production should be described by non-linear and differential (more exact-difference) equations with delayed arguments. Only in such case we can achieve the model accuracy needed for prediction or control.
At the same time we should bear in mind that the self-organization approach is complete open for introducing any reliable apriori information. In the very "self-organizing" algorithms it is enough to show to the computer only the means of the problem solution (the list of all possible variables with great excess) experimental data for all these variables (15 - 20 points) as well as goals (criterium) of solution of the problem. The goal can be the obtaining of the accurate prediction or the obtaining the physical model of the object for control of the latter. If the data are not too noise strongly mixed with, the computer itself shows the unique model, which is the best one for prediction or another also unique model, exibiting the cause and effect relationships, i.e. the physical model of the object. In such case it will discover the block structure of the object and will find the nonlinear equations of optimum complexity (with delayed arguments) for every block. According to our desire the computer can find also more simple equations (for example, linear equations or equations without delayed arguments) but in such case it will show how catastrophically the accuracy reduces due to the simplifications.

If we have the reliable information about the block structure of the object or about the equations of separate links then we of course tell it to the computer. It will simplify and accelerate the calculations and will increase the level of permisible noises.
If the noise is small the computer does not require such prompt and it finds the structure and the equations by itself without man's help. Practically such information reduces to the fact that the separate list of arguments (instead of the general list as in the previous case) is indicated to synthesise the equations of every variable in the self-organization program.
Hence we can speak about the advantages (certainly only under the condition of the presense of noise of nonaccurate measurement of variables etc.) of so-called combined method when a man introduces into the computer program all his reliable knowledge (but no more!) and the rest will be found by the computer using the self-organization method after the small number of experimental data.
From the foregoing it follows that the self-organization approach calls a man for modesty as in Galilei and Kopernikus times; a man is assigned a rather passive part in solving the problem (though he is a main customer) the computer tends in the best way to approach a goal shown by a man. The self-organization methods revive the belief into the creative abilities of the computer to discover the new knowledge or "to think" by its own way.
The computer under conditions of free decision making becomes a man's teacher and friend. To justify and ground such view the self-organization theory had to look through and modify many at the first sight fundamental positions in applied mathematics, which were considered to be absolutely stable up to our days.
So beginning from the Shanon's works (for continuous systems) and McCullock and Pitts' (for sample-data systems) the so-called "principle of the multiplicity of mathematical models" was proclaimed in applied mathematics. It was said that from any set of data it was possible to find the infinite agregate of equally accurate models.
The physical model is unique, but it was claimed that we cannot find it among many of its reflections in principle. Absurdity of such statement is evidently.
The analogous misunderstanding in applied mathematics is the statement about the impossibility of unique value of the great number of coefficients in mathematical model from a small number of interpolation points (i.e. from a small number of experimental data). The self-organization theory had to begin its way by showing the limits of the validity regions of the above-mentioned and some other common statements for they were obtained only with criteria or one "external addition" (in terms of St. BEER) that is minimum of the mean-square error calculated for all given points. Already with two criteria it is possible to find only a unique model as well as to determine the coefficient estimates of the model of any complexity, having only several points at our disposal. Three criteria will be needed to solve any of three interrelated problems of the model optimization etc. The obtained optimal model corresponds to the second (heuristic) criterium. If accurate prediction is required we shall get a model specific for prediction. It is often "meaningless" i.e. it can not be explained from our human point of view. If synthesis of the physical model is required (which in spite of the general-accepted opinion is not always the best for prediction) we shall get the model which reflects the whole picture of the cause-effect relationships. The computer will discover the natural law which objectively exists in the observed object.
In such a manner the Newton's second law has been "discovered" by the computer anew (on the basis of five points of the falling apples trajectory). Then some practical problems of self-organization were solved; that of the model of photosynthesis of organic substances in plants, the model of mixing two flows, the model of the Britain national economy, the model of organization of a branch of cellulose-paper industry and some other examples.
The self-organization principle has got an important support in a recently published work of the Nobel prize winner for 1971 the British scientist Dennis Gabor [1] . According to this work the self-organization algorithms can be determined as those in which the "freedom of choice of subsequent decisions" is provided for. The principle of nonfinal decisions being realized in perceptrons (where not a unique optimal solution goes from one layer to another but a number of solutions close to it), and also the selecter's work in developing new species of plants and animals can be the examples of such algorithms. "The freedom of choice" is provided, by the fact, that a number of decisions of foregoing choice is produced to the next choice but not the unique decision.
A number of algorithms based on the self-organization principle has been developed under the guidance of the author [2,3] .
First it should be mentioned here the Group Method of Data Handling where the input arguments and intermediate variables combines in pairs and complexity of combinations increases in each layer of information processing (like mass selection) until the unique model of optimum complexity will be obtained.
Secondly this is the method for determination harmonical trends of optimum complexity for predictions. No arguments pairs are present but the "freedom of choice of subsequent decisions" is provided. In the first layer of selection not the unique harmonic (as in deterministic approach) is determined but several harmonics (equal to $F > 1$) with aliquant periods are determined. In the second layer F harmonic of this layer fit to all selected harmonics of the first layer and all the F best sums are selected from all results and etc. From layer to layer only the $F > 1$ the most accurate combinations are transmitted but it often turns out that the best unique decision of the first layer is lost.
The result of accuracy increase is striking because the sum of harmonics giving minimum of

mean-square error never consists of the harmonic where each correspond to minimum of error.
Have we increased the degree of freedom $F > 1$ the accuracy begins growing quickly. This, in fact, is the sentence of death for all deterministic theories based on the deep belief in the existing of an unique choice, the optimum of which could be estimated simultaneously with this choice.
According to the self-organization theory the optimum of each decision can be estimated only the subsequent decision, theoretically on the infinity or at the very end of system operating if the system under consideration is of the momentary action.
The third example of the algorithm based on the principle of self-organization is the algorithm of optimum control synthesis, based on the sorting out the apices of the multidimensional control cube. Here like in previous case at each moment not a unique control is chosen but $F > 1$ of them which are close to the optimum control and each of them combines with F possible controls at the next moment of time etc. The estimate of controls sequence is given only on the basis of prognosis of future which is sufficiently for apart.
The forth example can be the algorithm of searching for more effective group of characteristic for the problems of pattern recognitions and automatic classification according to the "criterion of a number of settled argument". Here again the use of the necessity calculation of "freedom of choice of subsequent decisions" F allows to obtain the extremely good results, which are not accessible in case of deterministic choice with $F = 1$. Thus you should not connect the realization of self-organization principle only with the Group Method Data Handling (GMDH) algorithms, the mass selection exists for a long time as well as perceptrons and besides many other algorithms based on the self-organization principles have been developed (or are being developed).
In the self-organization algorithms mentioned above it turns out that accuracy (effectiveness) of decisions changes monotonously with the degree growth of freedom of choice F.
First the accuracy is increasing, reaches to maximum and then begins to decrease slowly. The excess of freedom is also bad for the accuracy as well as the "rigid decision" with $F = 1$. There exists optimum of freedom of choice of subsequent decisions.
Only with the help of self-organization theory it is possible to explain mathematically strictly the disadvantages of the so-called "rigid planning", With "rigid planning" at some initial instant the unique decision and the unique way for subsequent moment of time is chosen. The variety of ways depending on unexpected obstacles is not considered. Mathematically it means that the degree of freedom of subsequent decisions F is taken equal to one, which does not correspond to optimum value of this important criterion. For this reason the self-organization theory denies the idea of "unique decision" and "unique right way" if these conceptions are connected with the estimate of their optimum value at the moment of decision making.

The only right decision and the only right way could be found out from the point of view of the future i.e. by means of longterm prediction.
As it was mentioned above the theoretical optimum can be estimated either in the end of way (if the end exists) or in the infinity (i.e. with $T_Y = \infty$). In this way we could come to agosticism i.e. to the acception of impossibility to estimate the optimum neither at the moment of decision making nor at some other moment after it, for the prediction for infinite time is impossible. However asymptotic law of prediction optimization saves the position: with constant increasing of lead time of the prediction the estimates of optimum decision change less and less. Practically we can think that with sufficient prediction time we obtain constant and exact estimates of the decisions taken at the given moment (from the point of view of the future) and we can choose the best of them.
The prediction is like a searchlight lighting up the way before the locomotive. If we know the brake section there is no need of lighting the whole way up to the horisont.
In connection with these ideas of the self-organization theory the optimum control with prediction optimization becomes more important. With such control the estimate of control sequences (way) is given by the prediction for some number of time intervals ahead (for example, for economics problems for five years ahead). The chosen optimum control according to the asymtotic law does not change with subsequent increase of prediction lead.
The research has shown that the stability of a closed control system with the optimalized prediction increases monotonously with the increase of prediction time as well. In control systems with the prediction optimization the requirements to accuracy of the medium and longterm prediction increase and in this case self-organization approach provides the most accurate prediction. The regularization theory and the theory of differential prediction are the main achievements in the field of above-mentioned prediction.
The regularization theory discovers a number of necessary "external additions" and the ways of their construction.
Practically it solves the important problem about the optimum division of the available points into the checking and training samples of data. There are many useful findings in the regularization theory. As an example we can mention the construction of the so-called the "third" and the "seventh" methods of regularization.
The third method comes to the minimization models complexity and is advised for synthesis of "remainders".
The seventh method comes to the minimization of trends error and is advised for synthesis of trends. Random process is represented as a sum of trend and remainder. Thus the combined regularization (the third method- for the remainder and the seventh one- for the trend) is the optimum one. The theory of system differential (difference) regression equations obtained by the

GMDH method or in other words on the regression · equations for time derivatives of all variables, taken part in the model. These equations like all regression equations in some sense extract the deterministic part of random processes but as distinct from algebraic and integral regression equations they do not contain any information about integration constance i.e. they are less connected to the object from which they were obtained. Differential regression equations obtained on one object are valued for all other simular objects. Multiple integration of differential equations year by year, step by step gives us the long-term prediction. If the equations were accurate we could forecast random processes for thousand years ahead in the same way as we make prediction solar and lunar eclipse. But the errors usually are summarized and prediction goals either to infinity or drops to the time trend of the variable. It seems that nothing useful will be obtained from the differential regression equations since the calculation procedure is not stable. But the self-organization principle saves the situation in such case too: algorithms of selection choose such complexity of mathematical description (i.e. the number of selection and the degree of complexity of partial descriptions) when multiple differential prediction turns out not only stable but even the more accurate on the given interval of prognosis prediction. It turns out that multiple differential prediction posseses the smallest "smoothing action" essensual disadvantage of known algorithms of prediction.
The high accuracy of medium and longterm system multiple differential prediction with extraction of time trends of optimal complexity shows once more the limited possibilities of deterministic approach. When investigating the dynamic equations of complex system elements, the man "forgets" to take into account not only the delayed arguments and, of course, does not use the criterion of "freedom of choice of subsequent decisions". It is the mirror of human deterministic approach to modelling and is the best apparatus but only to describing the simple enough systems.
Let's note the following dialectics: the self-organization approach stresses the complexity and randomness of laws of surrounding world and brings to the exact prediction equations, i.e. it widens the sphere of validity of exact deterministic computations. It turns out that the random processes which are not suitable for being forecasted from our point of view at all and actually can be forecasted exactly enough.
It turns out that Laplace was right: everything can be calculated.
The future automatical management control systems (AMCS) will use the model self-organization algorithms as the main software for multipurpose computers. Now it is the time for design and application of special purpose computers designed for convénient use of self-organization algorithms. A sample of such computer has been already tested in the Institute of Cybernetics of Ukrainian Academy of Sciences [4].
Special-purpose computing complexes are designed abroad too. ADAPTRONIX company has built the Hypercomp-80 computer for direct modelling of complex system after a small number of experimental data. The computer is used for modelling of the multililinked systems with a great number of variables in pattern recognition and classification, economic modelling problems, process control systems, ėnvironment pollution calculation, research into natural resources, ecological models of city planning and so on. This list certainly does not exhaust all possible applications of the self-organization algorithms which are used by this computer.
The self-organization methods make it possible to solve the problem of indirect measurement of variables i.e. to construct filters which, using a number of experimental sample data for a set of variables, enables us to compute the current values of another set of variables (i.e. those which are without the reach for direct measurements). Importance of such filters and programs for AMCS is very great.

REFERENCES

[1] "Freedom of Solution Choice" in the Subsequent Moments of Time as an Important Criterion of Optimality of Control for Complex Random Processes.

[2] A.G. Ivakhnenko, Systems of Heuristic Self-Organization in the Engineering Cybernetics. "Tekhnika", Kiev, 1971.

[3] A.G. Ivakhnenko at al., Articles and programs in the Ukrainian journal "Avtomatika" for 1971-1972.

[4] Khrushceva N.V., Kravets T.D., Reutskaya N.I., Specialized Computer for Direct Simulation of Complex Systems by Experimental Data.

AN APPROACH TO THE IDENTIFICATION OF HUMAN CHARACTERISTICS BY APPLYING THE FUZZY INTEGRAL

Michio Sugeno and Toshiro Terano
Department of Control Engineering
Faculty of Engineering,
Tokyo Institute of Technology, Japan

1. Introduction

Lately, it has become of general interest to control the complicated systems such as social, biological and economical systems. These systems have so many uncertain elements that it is difficult to make their mathematical models. Of course, it is possible to make the stochastic models by using the probability measure, if a plenty of data on such systems and the environment surrounding them are available.
However, it is doubtful that the stochastic models are suitable for the mechanism of a human decision or evaluation process including many unclear factors, because these factors mainly depend on man's subjectivity. The authors have proposed the concept of fuzzy measure and fuzzy integral as a means of measuring the fuzziness in such fuzzy systems as described above.

2. The fuzzy measure and fuzzy integral

Let x be a certain set and $\mathcal{F}$ be a family of subsets of x such that 1) ϕ, $x \in \mathcal{F}$ 2) if $F_i \in \mathcal{F}$ and $F_i \subset F_j$ $(i \leq j)$ or $F_i \supset F_j$, then $\lim_{n\to\infty} F_n \in \mathcal{F}$

Definition 1 A set function g having the following properties is called a fuzzy measure.
1) $0 \leq g(F) \leq 1$, $F \in \mathcal{F}$, especially $g(\phi) = 0$, $g(X) = 1$
2) if $F, F' \in \mathcal{F}$ and $F \subset F'$, then $g(F) \leq g(F')$
3) if $F_n \in \mathcal{F}$, then $\lim_{n\to\infty} g(F_n) = g(\lim_{n\to\infty} F_n)$

$(X, \mathcal{F}, g)$ is called the fuzzy measure space. Practically, σ-algebra β is adopted instead of $\mathcal{F}$. The fuzzy measure has all properties of the probability measure except additivity. Therefore, the latter is included in a class of fuzzy measures.

Definition 2 A fuzzy integral of h on $(X, \mathcal{F}, g)$ is defined as follows.
Let $h : X \to [0,1]$

$$⨍_X h(x) \circ g(\cdot) = \sup_{\alpha\in[0,1]} [\alpha \wedge g(F_\alpha)], \text{ where } F_\alpha = \{x \mid h(x) \geq \alpha\}$$

If X is a finite set K and $0 \leq h(s_1) \leq h(s_2) \leq \cdots \leq h(s_n) \leq 1$, $s_i \in K$, then the fuzzy integral on $(K, 2^K, g)$ has the expression

$$⨍_K h(s) \circ g(\cdot) = \max_{1\leq i\leq n} [h(s_i) \wedge g(F_i)], \quad F_i = \{s_i, s_{i+1}, \ldots, s_n\}$$

Some Properties of the fuzzy integral

1) $⨍(h_1 \vee h_2) \circ g(\cdot) \geq ⨍ h_1 \circ g(\cdot) \vee ⨍ h_2 \circ g(\cdot)$
2) $⨍(h_1 \wedge h_2) \circ g(\cdot) \leq ⨍ h_1 \circ g(\cdot) \wedge ⨍ h_2 \circ g(\cdot)$
3) $\lim_{n\to\infty} ⨍ h_n \circ g(\cdot) = ⨍ \lim_{n\to\infty} h_n \circ g(\cdot)$
4) $\left| \int h(x)\, dp - ⨍ h(x) \circ p(\cdot) \right| \leq 1/4$, where p is a probabability measure.

Definition 3 Fuzzy measure g satisfying 2') instead of 2) in definition 1 is written as g_λ.
2') if $F, F' \in \mathcal{F}$ and $F \cap F' = \phi$, then
$g(F \cup F') = g(F) + g(F') + \lambda g(F) g(F')$, $-1 \leq \lambda \leq \infty$

In the case X is a finite set K,

$$g_\lambda(K') = (1/\lambda) \left[\prod_{s_i \in K'} (1 + \lambda g^i) - 1 \right], \quad K' \subset K$$

where $g' = H(s_1)$, $g^i = \dfrac{H(s_{i+1}) - H(s_i)}{1 + \lambda H(s_i)}$,

and $0 \leq H(s_1) \leq H(s_2) \leq \cdots \leq H(s_n) = 1$
Here, H(s) is called the fuzzy distribution function.

3. The identification of human characteristics in grading the similarity of patterns

Let us consider, as an example of fuzzy systems, the mechanism of the human decision making process where several patterns are distinguished.
Let P be a set of one dimensional fuzzy patterns such as shown in Fig. 1, where A and B are the standard patterns. Let $h_c : K \to [0,1]$ be a characteristic function of pattern C. We now define

$$\psi(C) = ⨍_K h_C(s) \circ g_\lambda(\cdot), \quad C \in P$$

$$w(C) = \frac{\psi(C) - \psi(B)}{\psi(A) - \psi(B)}$$

Here, $w : P \to [0,1]$ is a membership function of a fuzzy set of patterns such as similiar to A and not similiar to B. Let $d : P \to [0,1]$, $d(A) = 1$, $d(B) = 0$ be one which human determines, and let

$$J = \sqrt{1/N \sum_{i=1}^{N} (w(C_i) - d(c_i))^2}, \quad N = \text{number of patterns.}$$

Now, man's subjective measure g_λ is identified so that the value of J becomes small. A simple experiment was performed. The results are shown in Fig. 2 where $\min_{g_\lambda} J = 0.10$ and $\lambda = -0.53$. The fuzzy distribution function H of optimal g_λ is also shown in Fig. 3.

4. Conclusions

The experimental results show that the similarity of patterns obtained from the model approximately agrees with one evaluated by several persons. It is expected that the concept of fuzzy measure and fuzzy integral will be a powerful means when the decision or control problems are treated in the fuzzy environments.

References

1) L.A. Zadeh : Fuzzy Sets, Information and Control, 8, 338/353, 1965.

2) M. Sugeno : Fuzzy Measure and Fuzzy Integral, Transactions of the Society of Instrument and Control Engineers (in Japan), 8, 218/226, 1972.

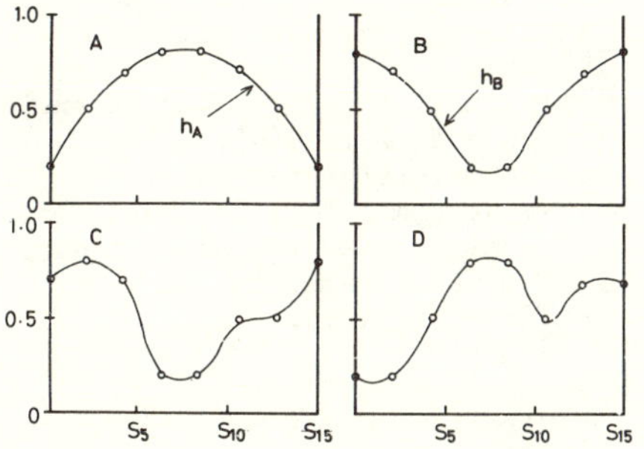

Fig. 1 Some test patterns

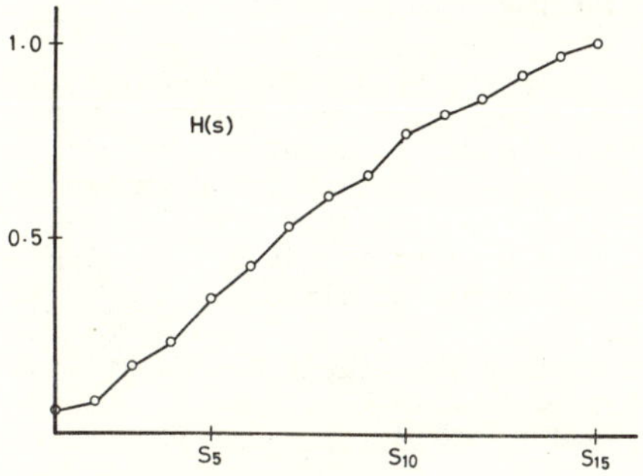

Fig. 3 Fuzzy distribution function of optimal g_λ

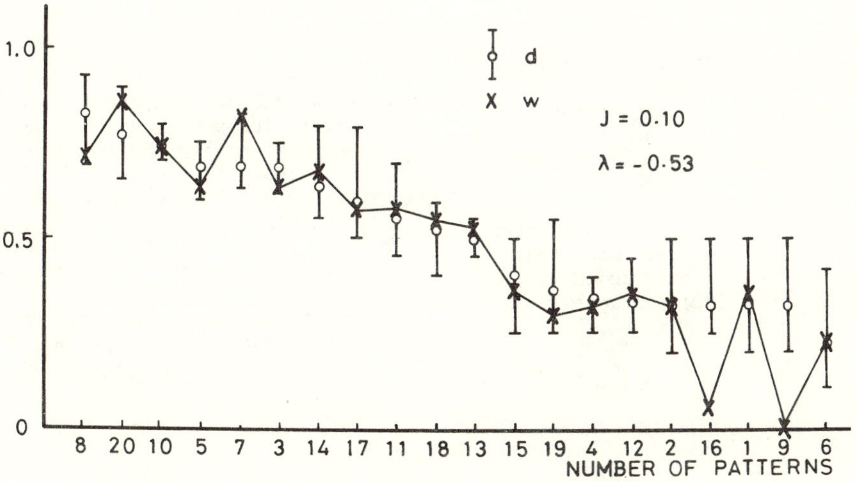

Fig. 2 Comparison of experiment and theory

ADAPTIVE OPTIMAL CONTROLLER AND CONTROL RESPONSES OF MAN

Hiroshi Tamura

Osaka University

Faculty of Engineering Science

Toyonaka, Osaka, Japan

The control movement of man in manual control system can be classified in several modes of control depending on the controller, controlled plant and training. The differences and the changes of mode from one to another are considered to be made by subjects to choose an appropriate signal for the response identifications.

The manual control system under consideration is furnished with CRT display, a hand-driven controller and a linear controlled plant of various order $G(s) = 1/(s+1)s^n$. The task to be implemented by the subject is compensatory tracking to step input, which is applied to the system with sufficiently long interval of time.

The control action to a step input can be divided into three processes, the initial process, the control process, and the stabilizing process.

The control mode of a man in the control and stabilizing process changes with the order and type of the plant to be controlled and with the controller.

In the control process, two types of control modes are observed. The one, named the linear mode (L), is observed for the plants of the lower order (1st and 2nd). For the higher order plants (2nd and 3rd), the bang-bang mode (BB) is dominant.

In the stabilizing process, four modes are observed. Firstly, the damped bang-bang mode (DB), secondly the steady state bang-bang mode (SB), thirdly, the bang-zero mode (BO) and lastly, the spike mode (SP).

The L mode is observed with untrained subjects, while the DB mode with well trained ones. The SB mode is observed when the subject is about to loose the system stability. The SP mode is very effective in controlling the higher order plants, but this mode is only seen when the controller is furnished with the sensor for the correct position of zero-output.

The limit of control was determined by the number n and δ in G(s), at which the subject failed to maintain stability. The limit of control thus determined is dependent mainly on the mode of control that the subject adopted. So far as the mode is same, the limit of control does not differ very much.

As the conclusion it has to be emphasized that control output of man should not be formulated as the response to the input supplied from the display. This formulation implies that the subjet is passively controlled by the display signal. Conversely, we have to take notice to the fact, that he is actively controlling the unknown (fuzzy) environment

The control behavior has to be understood to have two roles. The one is the command signal to govern the plant, and the second is the test signal to identify the response characteristics. With increase in the order of the plant, the latter aspect becomes essential. Thus the plant is controllable only when the subject adopts the mode of control which is more powerful in identifying the response to his command.

The above features are worth considering in the design of adaptive optimal controller.

SOME BEHAVIORS OF COMPOSITE FUZZY AUTOMATA IN RANDOM ENVIRONMENT

K. TANAKA
Dep't of Information & Computer Sciences
Faculty of Engineering Science
University of Osaka
Toyonaka, Osaka 560, Japan

1. Fuzzy automata can be employed as a model of learning systems operating in an unknown environment on account of its simplicity in design and computation [1]. It has already been known that the learning algorithm in automatic control of a time-discrete stochastic system employing either a pessimistic or an optimistic automaton does not always converge.

A composite fuzzy automaton which combines both characteristics of a pessimistic and an optimistic automaton may be applicable as a learning scheme for guarantee of convergence. The learning behavior of composite fuzzy automata is described by nonstationary fuzzy transition matrices with convergent property. It will be shown that, by the use of a linear reinforcement algorithm, the expedient performance of composite fuzzy automata can also be obtained [2].

2. Moreover, composite fuzzy automata with a variable structure in nonstationary environment as well as in stationary one have the optimal behavior even if the linear reinforcement techniques are applied to modify the state grades of fuzzy automata [3] [4].

On the contrary, the probabilistic automata shows an optimal behavior only when the nonlinear reinforcement techniques are applicable to modify the state transition probability [5].

[1] W.G. Wee, K.S. Fu: 'A Formulation of Fuzzy Automata and Its Application as a Model of Learning Systems', IEEE Trans. on SSC, Vol. SSC-5, No. 3, pp. 215-223 (July, 1969)

[2] K. Tanaka, et al: 'Fuzzy Automata Theory and its Application to Control Systems', Control Engineering of Japan Association of Automatic Control Engineers, Vol. 14, No. 9(1970)

[3] H. Tsuji, M. Mizumoto, J. Toyoda & K. Tanaka: 'Interaction between Random Environment and Fuzzy Automata with a Variable Structure', Trans. IECE of Japan, vol. 55-D, No.2 (Feb. 1972)

[4] M. Mizumoto, J. Toyoda & K. Tanaka: 'An Automaton in the Nonstationary Random Environment' Infor. Sciences (In Press)

[5] B. Chandrasekaran, D.W. Shen: 'Adaption of Stochastic Automata in Nonstationary Environment', Proc. 1967 Nat'l Electronics conference, p. 39(1967)

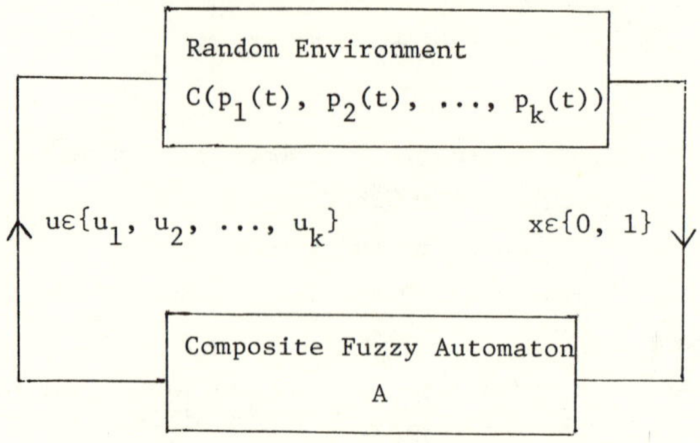

Interaction between Random Environment & a Fuzzy Automaton

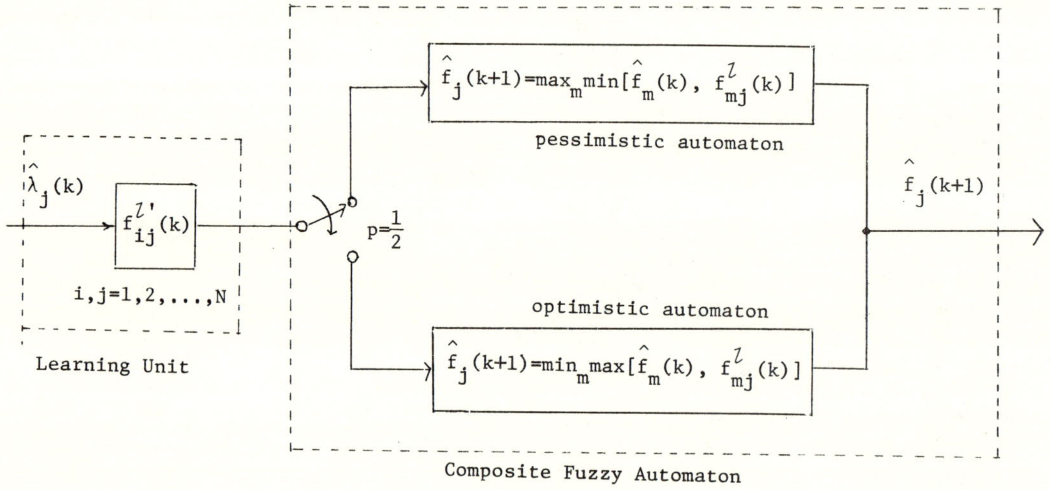

Learning Scheme

A LINGUISTIC APPROACH TO DECISION-MAKING

L.A. Zadeh

Department of Electrical Engineering

Computer Sciences, and Electronics

University of California

Berkley, California 94720, U.S.A.

Abstract

In the linguistic approach to decision-making, some of the variables entering into the decision process are allowed to be linguistic, that is, are allowed to take on values which are not numbers but sentances in a natural or artifical language. For example, a linguistic variable x may have as its values: small, large, not small, not very small, extremely small, more or less small, etc. These values are, in effect, labels of fuzzy sets whose membership functions can be computed from the knowledge of the membership functions of the primary terms small and large, and the definition of the hedges such as very, extremely, more or less, etc.

The use of linguistic variables is motivated by the fact that in many real-world decision processes the goals, the constraints, the interrelation between variables and the underlying probabilities are not known with sufficient precision to be susceptible of characterization in numerical terms. For example, we may know that an event is very likely without being able to specify whether its probability is 0.7 or 0.8 or 0.9. In this case, the probability may be assumed to be a linguistic variable with two possible values very likely and not very likely.

In effect, the concept of a linguistic variable serves to provide a basis for an approximate analysis of decision processes which are too complex or too ill-defined to be amenable to treatment by conventional numerically-orientated techniques. The present paper is a preliminary account of some of the main aspects of the linguistic approach. In particular, it is shown how ill-defined goals, constraints and probabilities may be described in linguistic terms and how approximate computations with linguistic variables may be performed. In general, the results of computations is the membership function of a fuzzy set. This membership function is approximated to by a linguistic label whose meaning can be computed by a procedure described in the paper. However, a general approach to this approximation problem has not yet been developed.

This work was supported in part by the Navy Electronic Systems Command under contract N00039-71-c-0255 and the Army Research office, Durham N.C., under Grant DA-ARD-D-31-124-71G174.

References

1. L. A. Zadeh, "Outline of a New Approach to the Analysis of Complex Systems and Decision Processes," IEEE Trans. on Systems, Man and Cybernetics, Vol. SMC - 3, No. 1, January 1973, pp. 28-44.

THE APPLICABILITY OF IDENTIFICATION

a round table discussion*)

Organizer and chairman: K.R. Godfrey

Secretary: G.C. Goodwin

Written contributions:

The applicability of identification.
K.R. Godfrey

Data problems in practical identification studies.
G.C. Goodwin

Using identification.
T. Bohlin

Applying identification techniques.
D.W. Clarke

Summary of the discussions:

Industry university confrontation on process identification
K.R. Godfrey and G.C. Goodwin

*) The Local Organizing Committee of this symposium is very grateful to Dr. K.R. Godfrey for accepting the responsibility to organize this round table discussion. It is a real service to the scientific and engineering community that he and Dr. Goodwin wrote the summary about the discussions that is included in these Proceedings.

THE APPLICABILITY OF IDENTIFICATION

K.R. Godfrey
Department of Engineering
University of Warwick
Coventry CV 4 7AL, U.K.

The aim of this Symposium is to present, discuss and summarise recent advances in modelling methods, quantitative evaluation of parameters and practical results of such methods. Although parameter estimation is only one stage in implementing a control scheme, its importance is very fundamental. Only by having a good idea of the dynamics of a process can the full value of a sophisticated control scheme be realised. Feedback-type control can be better tuned while feedforward (predictive) control can be designed much more effectively. The World effort in the identification field can perhaps be gauged from the very large number of summaries (250) submitted to this Symposium following the first call for papers.

As in so many areas of control, identification theory has outstripped practice over the past few years, but it is an encouraging sign that there is a relatively large number of practical applications being reported at the Symposium. What is perhaps baffling to the potential user is the very large number of different techniques with which he is faced, all of them apparently sound theoretically and capable of producing answers on simulated data, but with many untried on real industrial data.

The user is faced with a large number of problems even before he gets to the selection of the parameter estimation technique. This roundtable discussion session will be concerned with these problems, as well as the identification methods themselves. The opening speakers have all had considerable experience with applying (or attempting to apply) identification techniques in practice and each of them will be giving a short talk to open the discussion. Much of the session will be open to the floor, however, and comments from those members of the audience who have used identification techniques will be very welcome. Please don't be afraid to contribute.

The difficulties of applying identification in practice are most clearly brought out in an application with which I have been concerned - that of of a steelworks blast furnace. Basically, a blast furnace is a device into which solids in the form of cokes, ores and fluxes are tipped semi-continuously. These gradually descend through the various regions of the furnace until they form a pool of molten iron (and slag) at the bottom of the furnace. A continuous blast of hot air is fed upwards through the solids. The iron is emptied out of the furnace only at discrete intervals of time (at casts) and the ladles of iron are then taken to the steel making furnace.

Measurements made of the solids are usually confined to the ore/coke ratio and the very important coke moisture, while the blast flow, temperature and humidity, and injection rates of oxygen and oil are measured at the blast inlet and percentages of carbon monoxide, carbon dioxide and hydrogen are measured at the blast outlet (i.e. in the top gas). Iron quality is only measurable after a cast, and from the steelmaking point of view, Silicon and Sulphur contents of the iron are the all-important variables. Shortterm control for off-specification iron is achieved by altering the blast temperature and/or humidity, (occasionally oil injection rate also) and long-term regulation is achieved by altering the ore/coke ratio.

The difficulties encountered in trying to identify a blast furnace are numerous, and the main ones are: -

- Very long dynamics: up to six hours time-constant for the blast inputs, and about eight hours dead-time and up to a day time-constant for the solids inputs.
- Frequently more than one blast input is altered simultaneously, making identification from normal operating records very difficult.
- Manual control is applied form the output to the blast variables. In theory, this could be tolerated if what is required is a control system to operate in conjunction with the present manual control. In practice, the manual control is not consistent, the quality varying widely from one shift to another.
- Casting is not regular, and cast intervals depend on the availability of the steelmaking furnace to accept the iron produced. This means that records have to be interpolated.
- Unmeasured variables: there is insufficient data on the quality of the solids feed and there is no information between casts of iron quality (no in-furnace measurements are possible).
- Breaks in records: the very long time-constants involved means that records have to be collected over periods of weeks. Very often over such a period, one or more of the measurements will be unavailable for a short time. These breaks prove inconvenient in the analysis of normal operating records, and can prove disastrous if perturbation signal experiments are being carried out.

These difficulties mean that the task of getting the data into a form suitable for parameter estimation far exceeds that of actually analysing the data. The blast furnace is of course a bad example, but the difficulties outlined above are present to a greater or lesser degree in many applications. The message in identification ismuch the same as in other branches of control technology - our measurements so often are inadequate for the sophistication of our theoretical methods. It seems to me that quite enough linear identification methods are now available. What has been lacking in the past are methods of dealing with common occurrences, such as drift and feedback, but this is a situation which is rapidly being overcome. In the field of non-linear system identification, there is a need for techniques to cope with specific classes of non-linear identification, such as those based on Volterra integral expansions for example, while very interesting theoretically, appear to be too cumbersome to achieve wide application in practice.

DATA PROBLEMS IN PRACTICAL IDENTIFICATION STUDIES

G.C. Goodwin

Department of Computing and Control
Imperial College
London, England

The objective of these introductory comments will be to highlight some of the data problems that can arise in practical identification experiments. Two representative examples will be used to illustrate some of the problems. The first example concerns the modelling of the frequency power characteristics of a 120 MW boiler unit. The second example relates to the identification of a manufacturing process.

The two examples clearly show that real process data is often far from ideal for identification purposes. Some of the problems arise from drift, manual adjustments, spurious data points and hidden feedback. If these problems remain undetected then serious errors can be introduced in the estimation procedure.

It will be shown that some of the data errors can be eliminated using an interactive data manipulation program. For the case of the manufacturing process, this will be shown to lead to an acceptable solution. However for the boiler unit, the data problems are of a more fundamental nature and cannot be simply resolved. For example, the boiler data was collected under closed loop conditions and this can lead to ambiguity difficulties. This fact should be considered if the estimation results are to be interpreted correctly.

Another important consideration in data collection is the choice of test signal. It is highly desirable, if possible, to inject a known input perturbation. However, due to operating constraints this is often impossible and one has no alternative but to rely upon normal operating data. If this course is taken, then one needs to ensure that the input is suitable for parameter estimation. The minimal requirements are that the input should be persistently exciting and independent of the output and disturbances but it is also important that the input should adequately excite the plant dynamics. To illustrate this point a comparison will be made between the estimation accuracy achieved with the given data for the two examples and the theoretical lower bound on identification accuracy which could be achieved with an optimally designed input signal. The difference is not so marked for the other example where an externally generated PRBN signal was injected.

To conclude, some brief comments will be made on the ultimate use to which the models fitted to the two example data sets were put. The model for the manufacturing process was used to design a control system which was found to perform according to the specification when implemented. The boiler model has not, as yet, been used pending further investigation of the system.

USING IDENTIFICATION

Torsten Bohlin

Royal Institute of Technology

Stockholm, Sweden

Identifying physical processes differs basically from designing an identification method. Design means going from mathematical assumptions to a mathematical solution in the form of an algorithm. The constraints the designer has to face are mainly computational.

Using identification methods means going from physical reality to assumptions, i.e. choose a method, and also to design of an experiment. The constraints are varying and generally ill-defined. There are

. Fundamental limitations of identification. Basically, all one can do is fit a model to data which is not always sufficient for the purpose of the modelling.
. Technical and economic constraints. Often it is impossible or expensive to measure the right variables and to perturb the process in an efficient way.
. Constraints on the analysis. Only a few identification problems have indeed been solved, and still fewer solutions are available in the form of computer programs.

Under well defined conditions ideal experiments can be designed, assumptions tested, and methods compared. However, in practice conditions are undefined and one is often forced to use data from experiments that are so far from the ideal that one is in doubt whether identification by any method will make sense.

It is my experience that a good method, correctly used, can cope with surprisingly ugly-looking data, but also that correct use is difficult and, today, requires much insight into the fundamentals of identification and much experience.

This state of the art is unfortunate but not inevitable. It is possible to analyse the problems of application more systematically and also to present the results of this analysis as rules and guidelines for applying identification. In particular, it is possible to describe the consequences of a number of well-known phenomena in physical data, caused by technical/economic constraints: e.g. input errors, uncontrolled input, indirectly measured output, feedback, short experiment, effects of external changes. Each one of these cases has its particular, more or less serious consequences for the feasibility to identify.

To advance the use of identification it is essential that more be done along this line. Generally, we need more:

. Statements of what effects the basic limitations have, i.e. which of the assumptions a method is based on that can be put to test, and which can not.
. Means to carry out such tests.
. Means to decide what kind of information that can be extracted from a given experiment and, in particular, to test identifiability.
. Assessments of particular methods with respect to the extent they introduce extra constraints that are not inherent in the data, and use assumptions that are not verifiable.

In conclusion, the success or failure of identification in practice is more a question of how it is used than one of choosing a method. There is much that theoreticians can do to aid in this respect and that would advance the applicability of existing methods substantially.

REFERENCE

Bohlin, T. "Using statistical identification methods". IBM Nordic Laboratory, Technical paper 18.206, Lidingö, Sweden.

APPLYING IDENTIFICATION TECHNIQUES

D.W. Clarke

Department of Engineering Science,

Oxford University

England

The choice of an identification technique should depend principally upon the economic benefits likely to be attained by its success, as a function of the likely costs of the experiments and the analysis. The original hope that a general algorithm would be developed which would automatically identify a wide variety of systems, and hence eliminate the need for a sound technical understanding of specific systems, has not been realised. However, many systems, such as those involving chemically reacting components, are difficult to model a priori, are highly nonlinear, and are crucial to the economic operation of a process. These are probably best analysed using as much physical information as available, with their unknown parameters obtained by as specific an experiment as possible, and the final model tuned as a whole using input/output data. Methods such as Extended Kalman Filtering may be appropriate in the final stages of analysis. Economic Systems show a similar complexity, and detailed analysis is best done with an acceptable economic theory in mind. If a process is continuously operating about a relatively fixed setpoint, and the product is such that it is beneficial to minimise instantaneous fluctuations in some output variable, then a general linear dynamic model is a suitable candidate for identification, as the model would then be available for controller design. These models have the advantage that only minimal a priori knowledge may be necessary in the identification, and hence success is possible without using specialised expertise. The comments below are specific to identifying linear models, though some may be extended to the more general situation.

It is tempting to use normal operating records alone as data, as this minimises the interference with the plant or its operators. Unfortunately, these records rarely have the correct characteristics of input signal for adequate identification and checking of system models, but the disturbances present are probably more typical than those obtained using designed experiments. Experimental design and executing depends a lot upon the particular process and its operators. Their cooperation and knowledge can be helpful in determining approximate system behavior, and this preliminary work is usually amply rewarded by later rapid progress. Moreover, frequently systems are found which have been badly designed from a control point of view, where input controls have little effect in comparison with disturbances, and the real improvements are to be made by installing new actuators or transducers. The greatest benefit from using identification comes with systems which are heavily disturbed or which drift. Steady-state operation may be difficult to achieve, and feedback control during the experiment may be necessary; to sort out the resulting data, it is important to have a consistent control law, which is impossible if the feedback is manual.

The use of PRBS as a test-signal and cross-correlation as an algorithm seems appealing computationally, but the weighting-sequence models produced have a high d.c. gain variance, drifting data may invalidate the analysis, and an appreciable amount of data has to be discarded because of the assumptions of the methods. A PRBS with a relatively long bit interval is ofter more useful, or more specifically "optimal" test-signals may be employed. It is practically convenient however, to retain the two-level basis of the PRBS, though with other signals than PRBS a weighting-sequence model must generally be obtained by noise-magnifying deconvolution methods, and so the data is best analysed using other methods.

After collecting the data the first thing to do is obviously to look at it and see if the input has any effect upont the output: identification programs obey the 'rubbish in - rubbish out' rule. Most algorithms do not work well with the drifting data which is frequently obtained in masstransfer systems such as distillation columns. If the data drifts, check whether the bandwidth of the drift is much less than the system bandwidth. If so, the drift may be easily eliminated by simply high-pass filtering the output data, but otherwise differencing of the data may be necessary, where the resulting sequences tend to lose their original visual structure, making it more difficult to follow the input/output relationship. The next stage is to discover possible ranges for system order and time delay; the weighting sequence model gives a good guide of these parameters. If, however, the input signal is not 'white', the data should be 'prewhitened' so that cross-correlation can be directly employed. One nice way is to consider the input data to be white noise passing through a linear transfer-function, to estimate its parameters, and then to filter input and output by its inverse. It is also useful to cross-correlate backwards to check whether some feedback exists. In all this preliminary data handling a visual display and an interactive program are most helpful.

There are several choices of parameter estimation programs: Maximumlikelihood, Generalised-Least-Squares, Instrumental Variables, etc., each of which has its own combination of speed and efficiency. If your data is heavily disturbed and expensive to obtain, computing costs may be relatively small, and a statistically optimal, slow method is best. If you have much data with little noise, a computationally quick method may be superior.

The estimated model must be checked for adequacy, and better identification algorithms produce a range of statistics such as parameter covariance matrices and 'residuals' which help considerably Each parameter should in principle be checked individually to see whether its elimination seriously degrades the model, and the model as a whole should be checked against other proposed models of the same or lower order. The residuals should be tested both for 'whiteness' and for independence from the input. In practise, however, life is too short to try out all possible tests, and data is often such that an excessively high order a model is chosen using them. The most effective 'engineering' tests are derived by using the model, driven by the input alone, as a deterministed predictor and by looking to see whether the prediction errors appear to be unrelated to the input. Further insight is obtained by splitting the data in two, checking the two models obtained, and using each model to predict the output over the other half of data. A final verification can be obtained by a further experiment (taken on a day with different weather conditions) and using the previous model as a predictor here.
If you are going to use the model in controller design, remember that it is not exact, so do not choose a highly sensitive controller algorithm, and instead allow it to be flexible enough for some degree of on-line tuning.

INDUSTRY-UNIVERSITY CONFRONTATION ON PROCESS IDENTIFICATION

K.R. Godfrey
Inter-University Institute of Engineering Control, University of Warwick, Coventry CV4 7AL, U.K.

G.C. Goodwin
Department of Computing and Control, Imperial College of Science and Technology, London, SW7 2BT, U.K.

The Third IFAC Symposium on Identification and Process Parameter Estimation in Delft (June 1973) saw a feature quite novel to IFAC - a session in which industrial participants could give their views on identification and parameter estimation in general, could describe their experiences with using the techniques and could question a panel of experts on any aspect of the techniques. The session proved an outstanding success with a lively discussion over a wide range of topics, lasting well over three hours instead of the scheduled hour. The need for this type of forum at IFAC Symposia was clearly established.

PANEL CONTRIBUTIONS

The session opened with some short prepared contributions from a panel of experts with the objective of putting forward ideas for subsequent discussion. The Chairman, K.R. GODFREY (Inter-University Institute of Engineering Control, University of Warwick) made no apology for the fact that he had drawn the panel from Universities;they all had considerable experience in applying identification methods in practice.

GODFREY illustrated the problems facing the user by describing briefly an application of experimental identification to a steelworks blast furnace. This work was complementary to physico-chemical modelling of the furnace being carried out by a British Steel Corporation team. The difficulties here included very long dynamics (up to a day time-constant for the solids inputs), manual feedback control from the output (iron quality) to the blast inputs, simultaneous alteration of more than one blast input (making identification from normal operating records almost impossible) and unmeasured variables - insufficient data on the quality of solids feed and no information of iron quality between casts. Casting could be irregular, meaning that records had to be interpolated. These difficulties meant that the task of getting the data into a form suitable for parameter estimation far exceeded that of the estimation itself.

T. BOHLIN (Royal Institute of Technology,Stockholm, Sweden) took these points further. He felt that very great care was needed in using normal operating records for identification. However, he added that although practical data was usually far from ideal, there were a number of situations where existing theory could prove helpful. Feedback and drift were not dangerous as they could be detected by inspection. Inferring unmeasurable output variables from other measurements was dangerous as the model structure cannot be verified. Long dynamics were dangerous because other process conditions could change while the measurements were in progress. Other external disturbances were not dangerous as these could be modelled as part of the noise process. BOHLIN felt that the choice of model was not too important - a good method well used can cope with these difficulties. He considered that experience was necessary to use any of the methods effectively, and identification was not just a matter of collecting data and applying algorithms.

D.W. CLARKE (Department of Engineering Science, Oxford University, U.K.) contrasted the test case using simulated data which had been presented earlier in the week with the situation existing in real life. In the test case, the process was assumed to be stationary, the disturbance was well-defined and there was a good deal of data. In real life, the process is sometimes variable, the disturbances (drift etc.) are never well-defined and often only a small amount of data is available. (CLARKE quoted a case on which his team had been working in which each response measurement cost over £2, so that one could not think in terms of thousands of data points). In practical situations, the type of test signal that can be applied is often restricted, while the form of the process model is not known - different structures must be tried and compared against each other. Error measure statistics must be included in the program package, which should ideally include a variety of different methods. The methods should include the estimation of a noise model.

The importance of model structure was stressed by W. LEONHARD (Technische Universität, Braunschweig, GFR) who began by describing the well-known least squares method of parameter estimation in the presence of uncorrelated (white) noise. Several different methods of dealing with the practical case where the noise is not white, were then discussed, but LEONHARD considered the usual procedure of identifying polynomials B/A for the plant and D/C for the noise to be quite artificial, because noise normally occurs anywhere in the plant and not only at the output. He advocated the use of a least squares procedure in which the model order is gradually increased and with the model now comprising both the plant and the noise generator. Experimental work showed that this method usually gives sufficiently accurate results.

The panel contributions concluded with G.C. GOODWIN (Department of Computing and Control, Imperial College, London U.K.) describing briefly two applications of identification, one on a 120 MW boiler unit and the other on a manufacturing unit in the glass industry. Both applications had highlighted the problems associated with drift, hidden feedback and spurious data points. The need to examine the process data prior to identification by an interactive data manipulation program was stressed. Even this did not enable the fundamental difficulties of the power unit data to be resolved. The records from this unit were collected under closed loop conditions and this led to ambiguity difficulties. GOODWIN also stressed the need to choose the test signal with care. The minimal requirements are that it should be persistently exciting and independent of the output and disturbances, but it should also adequately excite the plant dynamics. All too often, normal operating records do not possess these requirements. The model obtained for the glass process had been used to design a control system which performed according to specification when implemented.

APPLICATIONS IN THE GLASS AND CHEMICAL INDUSTRIES

The session continued with two short contributions by speakers from industry outlining the state of the art (as the speakers saw it) in their own industries.

B.V. RAJA RAO (Pilkington Bros., St. Helens, U.K.) knew of six identification applications in the glass industry, three of which had been successful and three unsuccessful. Turning to an application to a glass furnace, RAJA RAO said that, as for the steelworks blast furnace, one of the main difficulties was long dynamics (typically 6-8 hours dead-time and 20-30 hours time-constant). The model obtained was used for training and he would like to extend it to aid instrumentation positioning and production information and the design of future plants. Simple models (gain, dead-time and one or two time-constants) had proved adequate and the experiments had been successful in the face of resistance on the part of plant management. RAJA RAO considered that 70% of the effort went into designing and collecting data, 10% in getting the model, 10% in designing a control system and 10% in tuning the system on-line.

By contrast, T.W. OERLEMANNS (Royal Dutch Shell Laboratories, Amsterdam, Netherlands) was relatively pessimistic when discussing applications in the chemical and petroleum industries. Why did so few industries have any activities in the identification field? One reason was certainly the cost; software writing is expensive. Another was that the models were required to operate over a large range; most of those discussed at the Symposium were obtained for small perturbations. The benefits of modelling were considerable, he continued. With a good process simulation, control schemes could be tested and failure of part of a multivariable control scheme could be examined without damaging the process. OERLEMANNS considered that the tuning of controllers in the chemical and petroleum industries left much to be desired, even where a physico-chemical model of the process is available, The idea of self-tuning regulation [1] discussed at this Symposium, seemed very interesting.

GENERAL DISCUSSION

The session was now thrown open to the floor for more general discussion, and it was pleasant to note that OERLEMANNS' gloomy view of the state of the art in the chemical and petroleum industries did nothing to dampen the liveliness of the debate.

Black box versus physico-chemical modelling.
Opening the discussion E. PAVLIK (Siemens AG, Karlsruhe,GFR) said that he was concerned with instrument development and process modelling, and that he felt that the panel had given too much emphasis to black-box modelling. Surely the more usual case is to have a model of the process based on the physics and chemistry and to use the experiments to estimate the unknown parameters, by hill climbing. D.W. CLARKE disagreed with this. Although in a few large industries, process models are available, these are still the exception rather than the rule. Indeed, there is often little a priori knowledge of the process;- examples of this are in the brewing industry and in sewage treatment. It would be nice, he continued, to have a "library" of process models available; we are working towards that end and as such, process identifiers are a selfexterminating breed. K.R. GODFREY agreed with CLARKE's view and stressed the need to get away from the term "black-box". There is almost always some knowledge available and this must be used. T. BOHLIN noted that operator experience is useful as a basis for plant modelling, and this had been used with success in his own work on paper plant modelling and control.

Adaptive control.
E. PAVLIK continued by noting that models had to be able to follow variations in parameters, and he was concerned about the stability of adaptive control. L.G.M. van AARLE (Royal Dutch Shell Laboratories, Amsterdam, Netherlands) continuing this theme, said that in distillation column measurements adaptive control was needed to cope with changes in feed flow and feed composition, but otherwise there was not very much need for adaptive control in the petroleum industry. G.DUYFJES (Dutch State Mines, Geleen, Netherlands) contrasted the chemical industry in which he worked with the petroleum industry. In the chemical industry, there is a wide variety of processes, there are many distillation columns and feedforward control is widely used. Some adaptation is essential. He had had some success with using a simple static model of a process and updating the parameters dynamically. (A paper on this is due to appear in the September 1973 issue of Automatica). He cited an example of a process in which it had taken a whole work-shift (eight-hours) to change from one level to another manually, but this is now completed within one hour using a computer.
A. JOHNSON (Royal Dutch Shell, Netherlands) sug-

gested that one approach to adaptive control would be to use one form of controller for regulation and a different form for large changes. T.BOHLIN said that this type of approach had in fact been used in the paper industry, the second form of controller design being used for grade changes. J. MATHER (Pilkington Bros.,St.Helens, U.K.) said that in glass furnaces control problems usually arose only when changing from one level to another. P.A.N. BRIGGS (Warren Spring Laboratory, Stevenage U.K.) went so far as to say that in the process industries considerable control effort was only justified when the process variables are changing. M. HEALEY (University College, Cardiff, U.K.) contrasted this situation with that of aircraft control, where different control laws are designed for different conditions, dependent on the environment. Frequently, gains are switched to achieve the adaptive control. Changing parameters are the rule, rather than the exception in aerospace applications.

J. RICHALET (GERBIOS, Velizy-Villacoublay, France) stressed the need not to forget those processes whose throughput is dependent on consumer demand, and quoted the case of a steam generator in the electricity supply industry, where the load could quickly be reduced by a half, so that the controller dynamics must be changed. The Chairman noted that much the same situation occurred in the gas industry, although in an application with which he had been concerned, quite large load changes could be dealt with using a fixed-parameter feedforward controller (designed using step-response identification), but using feedback trim to cope with the variable dynamics.

Cost benefit.

The Chairman asked the participants to consider further the cost benefit of identification and improved control, a topic which had surfaced once or twice during the preceding discussion. It became clear that how to justify spending money on an identification and control exercise was something that troubled many of the industrial participants. The Chairman wondered whether control personnel were really in that much worse a position than other company staff who used carefully thought-out arguments for spending company money, but who often could only guess at the returns. P. KELLY (Unilever Research Laboratory, Port Sunlight, U.K.) had a word of caution for the participants. He gave an example, where following an identification and optimisation exercise the process operating staff had been reduced to a minimum, and the product quality had deteriorated as a result. Further, improved instrumentation - also costly - was often needed for identification and control.

E. PAVLIK quoted an example which his Company (Siemens) had undertaken on the control of a steam superheater. They had made a 250-amplifier model of such a superheater, and were able to show to customers what profit would result from improved control. J. RICHALET considered this to be an isolated case. Models are very much dependent on the situation. Quoting an example of an extractive distillation column for two-product separation (and recognising that even this is a somewhat simpler situation than usual), he had found a static model based on physical equations to be adequate for set-point control. For this well-behaved system, a more complex dynamic model scarcely seemed justified. PAVLIK conceded that the superheater situation is still an exceptional case, but by no means unique. He contrasted it with a steelworks blast furnace where it is much more difficult to demonstrate conclusively how much profit would result from improved control.

D.W. CLARKE summarised the situation from the point of view of the Symposium by noting that physico-chemical models of a process are expensive in expertise, while experimental models are less expensive. The software effort in programming the different identification techniques could be reduced by making use of computer packages already available at Universities in the U.K., Sweden, and elsewhere.

Distributed parameter system identification.

Turning to the next topic of discussion, M.M. HAMZA (University of Calgary, Canada) wondered why had more not been heard of distributed parameter system identification at this session? G.C. GOODWIN provided the key to the solution to Hamza's question by quoting the work on control system design for a cold rolling mill. A lumped model of the mill was used at first, but little progress had been made until a distributed parameter model based on the physics and mechanics of rolling, had been used. A hybrid computer simulation of the mill had been set up and a somewhat simplified model derived from the simulation used for the control system design. C. BRAVINGTON (British Steel Corporation, Corporate Laboratories, London U.K.) agreed with Goodwin but added that distributed models for rolling processes were rather easier to construct than for other processes.

J. RICHALET quoted the case of an iron ingot, in which the temperature is far from uniform. Control must, however, be implemented in a lumped mode because of the difficulties of temperature measurement within the ingot. A.JOHNSON (Royal Dutch Shell, Netherlands) wondered whether a distributed parameter model can help in deciding the placement of sensors. It could indeed, remarked the Chairman, if you had any choice;- all too often, you have not. It appeared, he said, that it was highly desirable to build physico-chemical models of processes with distributed parameters, but experimental identification and subsequent control often had to be implemented in a lumped mode because of the placement of transducers.

Choice of identification method.

R.F.G. ROOMAN (Agfa-Gevaert, Belgium) said that he was involved in the control of a number of relatively small processes. He was a little confused about the multiplicity of identification methods described at the Symposium and wondered if the panel could give guidance as to which method to use. Is a correlation-based method good enough?

T. BOHLIN said that a correlation method should produce the right answer if used correctly. The choice between methods does not matter in the ideal case presented in one of the test cases earlier in the week, but it does matter when dealing with real data. For data without too much noise, any method would probably give much the same answer. For more difficult cases, a rugged method was essential, and he recommended maximum-likelihood estimation for single input-single output processes, while for multivariable processes generalised least squares estimation scores on the grounds of reduced computing requirements. K.R. GODFREY noted that it was as well to put all these methods into context, and one could quite often get satisfactory experimental models by curve-fitting to step response functions. It was difficult to estimate how often this was so, though.

R.F.G. ROOMAN expressed concern at the effect of feedback on identification. T. BOHLIN agreed that this was a matter for concern. With automatic closed loop control, one can still identify the process provided that the control is not too tight[2] but one can be in trouble with manual feedback, which can be quite variable[3].

CLOSING REMARKS

With the session now well into the lunch period, it was with some reluctance that the Chairman drew the discussion to a close. The debate had proved wide ranging. It was to be hoped that the controversy over black-box versus physico-chemical modelling had been resolved to some extent, as had the role of distributed parameter models in identification and control. The contrasting importance of adaptive control in the process industries and in the aerospace industries was interesting, and perhaps explained the relative lack of work on adaptive control this side of the Atlantic. The cost benefit discussion had been a particularly lively one and the Chairman hoped that industrial workers would make use of the computer packages now available in Universities. He also hoped that the guidelines on which technique to use in practice were helpful.

The session had proved highly satisfactory, he continued, with the participants from industry playing such an active role. Their views had been much appreciated and it was clear that such a forum should be considered for future IFAC Symposia.

REFERENCES

[1] Cegrell, T and Hedqvist, T. Successful Adaptive Control of Paper Machines, Paper PV-2.

[2] Bohlin, T. On the Problem of Ambiguities in Maximum-Likelihood Parameter Estimation. Automatica, 1971, 1, pp. 199-210.

[3] Bohlin, T. Using Statistical Identification Techniques. IBM Nordic Laboratory, Lindingö, Sweden, Technical Paper 18.206

COMPARISON AND EVALUATION OF SIX ON-LINE IDENTIFICATION AND PARAMETER ESTIMATION METHODS WITH THREE SIMULATED PROCESSES

R. Isermann; U. Baur; W. Bamberger; P. Kneppo; H. Siebert;
University of Stuttgart
7000 Stuttgart, W.-Germany
Keplerstr. 17.

1. INTRODUCTION

In past years many different identification and parameter estimation methods for dynamic processes have been described in the literature. The relations between many of these methods are relatively well known as far as theoretical background is concerned. Missing, however, are general comparisons of the properties of the different methods. For these reasons this paper deals with comparisons of the performance and of the computational expense of each of six different identification methods. On-line methods only are considered because they are especially needed for the efficient utilization of process computers in the control, supervision and optimization of technical processes and for the analysis of biological processes.

Identification methods can be classified, according to the type of model which results, in the following ways:

- Nonparametric identification methods, which primarily lead to nonparametric models, e.g. Fourier analysis and correlation analysis.
- Parametric identification methods, leading primarily to parametric models, such as parameter estimation methods.

Within each class many different identification methods are known, but only a few authors have made comparisons of the properties of these methods.

Following methods have been compared for simulated processes:

-CLARKE (1967): Generalized least squares with maximum likelihood.

-HASTINGS-JAMES and SAGE (1969): Recursive generalized least squares, maximum likelihood.

-VAN DEN BOS (1970): Fourieranalysis and least squares parameter estimation, maximum likelihood.

-MEHRA (1971): Maximum likelihood, correlation techniques.

-SARIDIS (1972): Stochastic approximation, maximum likelihood, extended Kalman-filter.

-GENTIL (1972): Least squares, generalized least squares, instrumental variables, maximum likelihood.

GUSTAVSSON (1970) compared methods of least squares, maximum likelihood, prior knowledge fitting, correlation and spectral analysis for several real and simulated processes. Fourier analysis for nonperiodic test signals with correlation analysis for stochastic test signals were compared theoretically by ISERMANN (1971,a,c). For the identification of the process gain a comparison of efficient parameter estimation, measurement with one step function, Fourier-analysis with rectangular pulses and correlation analysis is given by ISERMANN (1972 b).

So different methods have been tested and compared by using different processes, different numbers of data sequences, different noise-to-signal ratios and different error measures. Application purposes of the identified models were not regarded. Therefore it is obvious that hardly more general results can be derived from all these interesting papers. Mostly processes of relatively low order with not more than four unknown parameters were used. However, identification and especially control of low order processes in general is no big problem. If we restrict the considerations to the linear case, processes which cause problems in identification and control are of higher order or belong to the class of distributed parameters, perhaps with time delays.

A comparison of different identification methods by theoretical analysis of the variances of the results is rather complicated and very extensive to do.

For these reasons the comparisons in this paper are done by simulation on digital computers. Six different on-line versions of both classes of identification methods are compared for three typical linear, single-input/single-output processes. Simulated processes are used because the models and disturbances then are known exactly and, additionally, the test signals as well as the noise-to-signal ratio can be varied easily. The identification methods, which all result finally in a parametric model, are:

1. Least squares
2. Generalized least squares (Hastings-James, Sage, 1969)
3. Instrumental variables (Wong-Polak, 1966; Young, 1970)
4. Stochastic approximation (Saridis-Stein, 1968; Isermann-Siebert, 1972)
5. Correlation analysis with least squares parameter estimation (Isermann-Baur, 1972)
6. Fourier analysis using a model with three unknown parameters (Isermann-Bamberger, 1971)

Methods 1, 2, 3 and 4 are typical representatives of the parameter estimation methods which lead to parametric models in a direct way. According to most practical cases only such methods are regarded which do not need a priori information on noise statistics.

Method 5 is a combination of correlation analysis and least squares parameter estimation, which first results in a nonparametric model (impulse response) and then in a parametric model.

Method 6 uses Fourier analysis for first identifying values of a nonparametric model (frequency response) and then uses simple relationships for calculating three unknown parameters.

As ultimate goal of the identification the design of a discrete control algorithm is regarded.

2. TEST CASES FOR EVALUATION AND COMPARISON OF DIFFERENT IDENTIFICATION METHODS

2.1. Simulated processes

As test cases for the evaluation and comparison of different identification methods three simulated linear timeinvariant processes, Fig.2.1, are used:

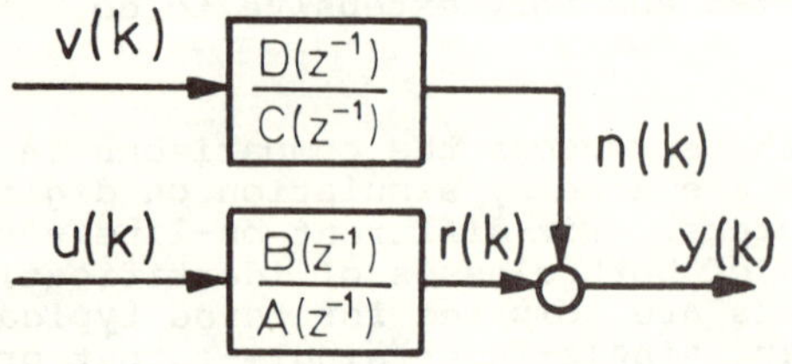

Fig. 2.1. Block diagram of the simulated processes.

I) Second-order oscillating process:

$$G_1(z^{-1}) = \frac{B(z^{-1})}{A(z^{-1})} = \frac{b_1 z^{-1} + b_2 z^{-2}}{1 + a_1 z^{-1} + a_2 z^{-2}} \quad (2.1)$$

$a_1 = -1.5; \quad a_2 = 0.7;$
$b_1 = 1.0; \quad b_2 = 0.5$
sampling interval: $T_0 = 2s$.

II) Second-order nonminimum phase process:

$$G_2(z^{-1}) = \frac{B(z^{-1})}{A(z^{-1})} = \frac{b_1 z^{-1} + b_2 z^{-2}}{1 + a_1 z^{-1} + a_2 z^{-2}} \quad (2.2)$$

$a_1 = -1.425; \quad a_2 = 0.496;$
$b_1 = -0.102; \quad b_2 = 0.173$
sampling interval: $T_0 = 2s$.

III) Third-order low pass delay process:

$$G_3(z^{-1}) = \frac{B(z^{-1})}{A(z^{-1})} = \frac{b_1 z^{-1} + b_2 z^{-2} + b_3 z^{-3}}{1 + a_1 z^{-1} + a_2 z^{-2} + a_3 z^{-3}} z^{-d} \quad (2.3)$$

$a_1 = -1.500; \quad a_2 = 0.705; \quad a_3 = -0.100;$
$b_1 = 0.065; \quad b_2 = 0.048; \quad b_3 = -0.008$
$d = 1$
sampling interval: $T_0 = 4s$.

Processes II and III are the two processes proposed earlier, ISERMANN (1972 a), to be used as test cases. They have been derived from continuous processes with a zero-order hold. Process I was used by ÅSTRÖM, BOHLIN (1966), HASTINGS-JAMES, SAGE (1969), GUSTAVSSON (1970) and GENTIL (1972). No corresponding simple continuous process with zero-order hold seems to exist for this process.

The output r(k) of the process to be identified is contaminated by an autocorrelated discrete noise n(k) which is generated by a noise filter $G_v(z^{-1})$ driven by a discrete white noise process v(k) with normal distribution, zero mean and variance $\sigma_v^2=1$. The noise filter is different from the processes G_1, G_2 and G_3, because this is the general case. The continuous transfer function $G_v(s)$ is the same for all three processes and the discrete transfer function is:

$$G_v(z^{-1}) = \frac{D(z^{-1})}{C(z^{-1})} = \frac{d_1 z^{-1}}{1 + c_1 z^{-1} + c_2 z^{-2}} \quad (2.4)$$

$c_1=-1.027;\ c_2=0.264$, $d_1=0.0114\gamma$ } for processes I and II.

$c_1=-0.527;\ c_2=0.0695$, $d_1=0.0117\gamma$ } for process III

γ depends on the signal-to-noise ratio

$$\eta = \frac{\sqrt{\overline{n^2(k)}}}{K\,u_o} \qquad (2.5)$$

(see section 8.1). K, gain.

2.2. Identification task

The input/output behavior of the processes G_1, G_2 and G_3 are to be identified. As a priori knowledge it is assumed

- the structures and the orders of the processes G_1, G_2, G_3 are known
- $n(k)$ is a correlated zero-mean random variable.

The input signal is an arbitrary artificial signal and is limited by $-u_o/2 \leq u(k) \leq u_o/2$ with $u_o = 1$.

For two noise-to-signal ratios $\eta = 0.1$ and 0.2 and the measuring time periods T_M = 350; 1400; 7000s the errors of the estimated parameters and of the resulting impulse responses are to be shown by taking following error definitions:

a) Squared relative parameter errors

$$\delta_{\Sigma 1} = \left\| \frac{\Delta\Theta_i}{\Theta_i} \right\| = \left[\sum_{i=1}^{p} \left[\frac{\Delta\Theta_i}{\Theta_i} \right]^2 \right]^{\frac{1}{2}} \qquad (2.6)$$

b) Mean squared parameter errors related to mean squared true parameters

$$\delta_{\Sigma 2} = \frac{\|\Delta\Theta_i\|}{\|\Theta_i\|} = \left[\frac{\sum_{i=1}^{p} \Delta\Theta_i^2}{\sum_{i=1}^{p} \Theta_i^2} \right]^{\frac{1}{2}} \qquad (2.7)$$

c) Mean squared impulse response errors related to mean squared true impulse function values.

$$\delta_g = \left[\frac{\overline{\Delta g^2(k)}}{\overline{g^2(k)}} \right]^{\frac{1}{2}} = \left[\frac{\sum_{k=0}^{\infty} \Delta g^2(k)}{\sum_{k=0}^{\infty} g^2(k)} \right]^{\frac{1}{2}} \qquad (2.8)$$

Hereby:

Θ_i: exact parameter values, $i=1,2,\ldots,p$. $(p=2m)$ $(\Theta_1,\Theta_2,\ldots,\Theta_p = a_1,a_2,\ldots a_m, b_1,b_2,\ldots b_m)$

$\hat{\Theta}_i$ = estimated parameters

$\Delta\Theta_i = \hat{\Theta}_i - \Theta_i$; $\Delta g(k) = \hat{g}(k) - g(k)$.

For getting better comparisons of the different identification methods 5 usable runs with different noise data sets are to be made. Then the standard deviations $\sigma_{\delta\Sigma 1}$, $\sigma_{\delta\Sigma 2}$ and $\sigma_{\delta g}$ of the errors are to be calculated.

As initial value of the parameter vector $\underline{\Theta}(0) = 0$ has to be taken.

2.3. Design of a control algorithm

As the result of an identification procedure only can be judged by applying it according to the ultimate goal of the identification, the design of a control algorithm is regarded as one possible example. Therefore the identified models II and III are used to optimize the parameters p_o, p_1 and p_2 of the three mode control algorithm

$$u(k)=u(k-1)+p_o y(k)+p_1 y(k-1)+p_2 y(k-2) \qquad (2.9)$$

taking a step disturbance $n(t)=1(t)$ at the output and minimizing the performance index

$$PI = \sum_{k=0}^{\infty} y^2(k). \qquad (2.10)$$

This leads to optimal parameters $\hat{p}_o, \hat{p}_1$ and $\hat{p}_2$ and to the closed loop controlled variable $\hat{y}(k)$. These parameters $\hat{p}_o$, $\hat{p}_1$ and $\hat{p}_2$ are then used to determine the controlled variable $y(k)$ with the exact model (the real process). Then the error of the controlled variable is

$$\Delta y(k) = \hat{y}(k) - y(k) \qquad (2.11)$$

and the closed loop output error

$$\delta_y = \left[\frac{\sum_{k=0}^{\infty} \Delta y^2(k)}{\sum_{k=0}^{\infty} y_o^2(k)} \right]^{\frac{1}{2}} \qquad (2.12)$$

is to be calculated for each data run. $y_o(k)$ is the controlled variable of the exact model, Fig.2.3, with its optimized parameters p_o^o, p_1^o and p_2^o. Then the mean squared parameter errors of the controller

$$\delta_{\Sigma p} = \frac{\|\Delta p_i\|}{\|p_i^o\|} = \left[\frac{\sum_{i=0}^{2} \Delta p_i^2}{\sum_{i=0}^{2} \left[p_i^o\right]^2} \right]^{\frac{1}{2}} \qquad (2.13)$$

are to be determined and analog to section 2.2 the standard deviations $\sigma_{\delta y}$ and $\sigma_{\delta \Sigma p}$ for 5 runs should be shown. Model I is not used for checking the closed loop behavior, because with this second-order model no absolute optimal controller parameters do exist and any control parameters could be used.

3. THE METHOD OF GENERALIZED LEAST SQUARES (GLS)

The symbols used in this paper first are explained for least squares estimation (LS).

It is assumed that a linear process can be described by the model

$$y(k) + a_1 y(k-1) + \ldots + a_m y(k-m) = b_1 u(k-d-1) + \ldots + b_m u(k-d-m) \quad (3.1)$$

respectively by

$$\underline{y}_k \underline{a} = \underline{u}_{k-d} \underline{b} \quad (3.2)$$

with

$$\left.\begin{aligned} \underline{y}_k &= [y(k)\ \ y(k-1)\ \ldots\ y(k-m)] \\ \underline{u}_{k-d} &= [u(k-d-1)\ \ u(k-d-2) \ldots u(k-d-m)] \\ \underline{a}^T &= [1\ \ a_1\ \ a_2\ \ldots\ a_m] \\ \underline{b}^T &= [b_1\ \ b_2\ \ldots\ b_m] \end{aligned}\right\} \quad (3.3)$$

or by the pulse transfer function

$$G_M(z^{-1}) = \frac{Y(z)}{U(z)} = \frac{B_M(z^{-1})}{A_M(z^{-1})} = \frac{b_1 z^{-1} + \ldots + b_m z^{-m}}{1 + a_1 z^{-1} + \ldots + a_m z^{-m}} z^{-d}. \quad (3.4)$$

Taking the measured input u(k) and measured output y(k) of the real process the generalized error used for parameter estimation is defined as

$$\begin{aligned} e(k) &= y(k) + a_1 y(k-1) + \ldots + a_m y(k-m) \\ &\quad - b_1 u(k-d-1) - \ldots - b_m u(k-d-m) \\ &= y(k) - y_M(k) \end{aligned} \quad (3.5)$$

where

$$y_M(k) = \underline{\psi}(k)\underline{\Theta} \quad (3.6)$$

is the prediction of y(k) of the model based on the observations y(k-m),..., y(k-1), with

$$\begin{aligned} \underline{\psi}(k) &= [-y(k-1) \ldots -y(k-m) \mid u(k-d-1) \ldots u(k-d-m)] \\ \underline{\Theta}^T &= [a_1\ a_2\ \ldots\ a_m \mid b_1\ \ldots\ b_m]. \end{aligned} \quad (3.7)$$

Minimizing the loss function

$$V = \sum_{k=m+d}^{N+m+d} e^2(k) \quad (3.8)$$

and using the notations

$$\underline{y}^T = [y(m+d)\ \ y(m+d+1)\ \ldots\ y(m+d+N)] \quad (3.9)$$

$\underline{\Psi}$ = [1]

the least squares estimate follows as

$$\hat{\underline{\Theta}} = [\underline{\Psi}^T \underline{\Psi}]^{-1} \underline{\Psi}^T \underline{y} = \underline{P}\, \underline{\Psi}^T \underline{y} \quad (3.11)$$

$$\text{with} \quad \underline{P} = [\underline{\Psi}^T \underline{\Psi}]^{-1} \quad (3.12)$$

The recursive least squares estimate then is obtained writing eq.(3.11) in the partitioned form and introducing the matrix inversion lemma, FRIEDMANN(1954), LEE(1964)

$$\begin{aligned} \hat{\underline{\Theta}}(k+1) &= \underline{\Theta}(k) + [\underline{\psi}(k+1)\underline{P}(k)\underline{\psi}^T(k+1)+1]^{-1} \cdot \\ &\quad \cdot \underline{P}(k)\underline{\psi}^T(k+1)[y(k+1) - \underline{\psi}(k+1)\hat{\underline{\Theta}}(k)] \end{aligned} \quad (3.13)$$

with

$$\begin{aligned} \underline{P}(k+1) &= \underline{P}(k)\Big[\underline{I} - \underline{\psi}^T(k+1)\underline{\psi}(k+1)\underline{P}(k) \cdot \\ &\quad \cdot [\underline{\psi}(k+1)\underline{P}(k)\underline{\psi}^T(k+1)+1]^{-1}\Big]. \end{aligned} \quad (3.14)$$

As known the parameter estimates are biased if the residuals e(k) are correlated. The method of generalized least squares tries to overcome this problem by introducing filters $F(z^{-1})$, Fig.3.1, for getting uncorrelated residuals, see CLARKE (1967).

1) The first step consists in applying LS estimation on the model

$$\underline{y}_k\ \underline{a} = \underline{u}_{k-d}\ \underline{b} + w(k) \quad (3.15)$$

where w(k) are correlated random variables. The results are biased estimates $\hat{\underline{\Theta}}_1$.

---

[1]

$$\underline{\Psi} = \begin{bmatrix} -y(m+d-1) & -y(m+d-2) & \ldots & -y(d) & u(m-1) & u(m-2) & \ldots & u(0) \\ -y(m+d-2) & -y(m+d-1) & \ldots & -y(d+1) & u(m) & u(m-1) & \ldots & u(1) \\ \vdots & & & & \vdots & & & \\ -y(m+d+N-1) & \ldots & & -y(d+N) & u(m+N-1) & \ldots & & u(N) \end{bmatrix} \quad (3.10)$$

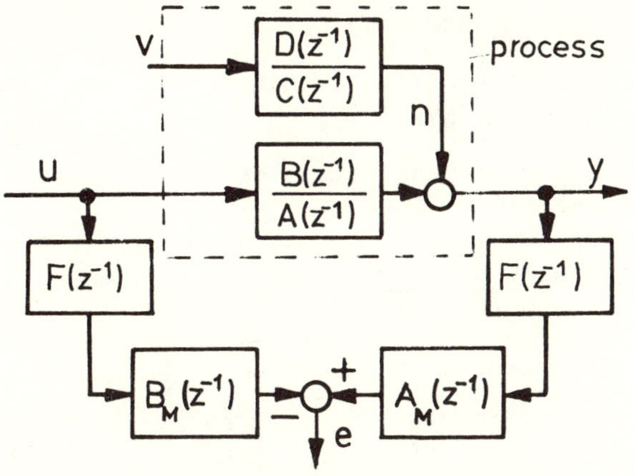

Fig. 3.1. Model used for generalized least squares estimation.

2) Then the residuals are calculated and analyzed by autoregression, assuming a model

$$w(k)=-f_1 w(k-1)-f_2 w(k-2)-\dots-f_\nu w(k-\nu)+e(k) \quad (3.16)$$

respectively

$$w(k) = \underline{\xi}(k)\underline{f} + e(k) \quad (3.17)$$

where e(k) are uncorrelated random variables and the order ν has to be chosen appropriately. If v(k), the input of the noise filter, as assumed is uncorrelated and equal to e(k), the filter $F(z^{-1})$ should have the structure

$$F(z^{-1}) = C(z^{-1})D^{-1}(z^{-1})A^{-1}(z^{-1}). \quad (3.18)$$

Therefore the assumed filter equation (3.16) has to be an approximation of eq.(3.18).

LS estimation of the filter parameters, using the notations [1] leads to

$$\hat{\underline{f}} = \left[\underline{\Xi}^T\underline{\Xi}\right]^{-1}\underline{\Xi}^T\underline{w} = \underline{Q}\,\underline{\Xi}^T\underline{w} \quad (3.20)$$

3) The inputs and outputs are filtered

$$\tilde{u}(k-d) = \underline{u}_{k-d}\hat{\underline{f}} + u(k-d)$$
$$\tilde{y}(k) = \underline{y}_k\hat{\underline{f}} + y(k) \quad (3.21)$$

4) A new LS fit is made with the filtered in-and outputs and with new matrices $\underline{\Psi}$.

5) Repeat from 2).

Algorithms for recursive generalized least squares estimation have been developed by HASTINGS-JAMES and SAGE (1969) using the same equations as for eq.(3.13) and (3.14).

$$\hat{\underline{\theta}}(k+1) = \hat{\underline{\theta}}(k) + \left[\tilde{\underline{\psi}}(k+1)\tilde{\underline{P}}(k)\tilde{\underline{\psi}}^T(k+1)+1\right]^{-1}\cdot$$
$$\cdot\tilde{\underline{P}}(k)\tilde{\underline{\psi}}^T(k+1)\left[\tilde{y}(k+1)-\tilde{\underline{\psi}}(k+1)\hat{\underline{\theta}}(k)\right] \quad (3.22)$$

$$\tilde{\underline{P}}(k+1) = \tilde{\underline{P}}(k)\left[\underline{I} - \tilde{\underline{\psi}}^T(k+1)\tilde{\underline{\psi}}(k+1)\tilde{\underline{P}}(k)\cdot\left[\tilde{\underline{\psi}}(k+1)\tilde{\underline{P}}(k)\tilde{\underline{\psi}}^T(k+1)+1\right]^{-1}\right] \quad (3.23)$$

$$\hat{\underline{f}}(k+1) = \hat{\underline{f}}(k) + \left[\underline{\xi}(k+1)\underline{Q}(k)\underline{\xi}^T(k+1)+1\right]^{-1}\cdot$$
$$\cdot\underline{Q}(k)\underline{\xi}^T(k+1)\left[w(k+1)-\underline{\xi}(k+1)\hat{\underline{f}}(k)\right] \quad (3.24)$$

$$\underline{Q}(k+1) = \underline{Q}(k)\left[\underline{I} - \underline{\xi}^T(k+1)\underline{\xi}(k+1)\underline{Q}(k)\cdot\left[\underline{\xi}(k+1)\underline{Q}(k)\underline{\xi}^T(k+1)+1\right]^{-1}\right]. \quad (3.25)$$

The initial matrices $\underline{P}(0)$ and $\underline{Q}(0)$ can be chosen as diagonal matrices with elements as large as no instability occurs. We noticed that the diagonal elements can be the larger the smaller the noise. The initial values of the parameter vector $\underline{\theta}(0)$ can be zero.

An exponential weighting of past data using a weighting factor ρ, HASTINGS-JAMES and SAGE (1969), in the terms

$\left[\tilde{\underline{\psi}}(k+1)\tilde{\underline{P}}(k)\tilde{\underline{\psi}}^T(k+1)+\rho\right]$ of eq.(3.22),(3.23)

$\tilde{\underline{P}}(k+1) = \frac{1}{\rho}\tilde{\underline{P}}(k)\left[\underline{I}-\dots\right]$, eq.(3.23)

and in the analog terms of eq.(3.24) and eq.(3.25) turned out to improve the estimates if it is used for the first 100 or 200 samples with ρ=0.99. This prevents the first estimates to become too bad and thus improves the convergence.

---

[1]
$$\begin{bmatrix} w(m+d) \\ w(m+d+1) \\ \vdots \\ w(m+d+N) \end{bmatrix} = \begin{bmatrix} -w(m+d-1) & \cdots & -w(m+d-\nu) \\ -w(m+d) & & -w(m+d-\nu+1) \\ \vdots & & \vdots \\ -w(m+d+N-1) & & -w(m+d-\nu+N) \end{bmatrix} \begin{bmatrix} f_1 \\ f_2 \\ \vdots \\ f_\nu \end{bmatrix} + \begin{bmatrix} e(m+d) \\ e(m+d+1) \\ \\ e(m+d+N) \end{bmatrix}$$

$$\underline{w} = \underline{\Xi}\cdot\underline{f} + \underline{e} \quad (3.19)$$

4. THE METHOD OF INSTRUMENTAL VARIABLES (IVA)

Using eq.(3.1) as a model and taking the generalized error for LS estimation

$$\underline{e} = \underline{y} - \underline{\Psi}\ \underline{\Theta} \tag{4.1}$$

with

$$\underline{e}^T = \left[e(m+d)\ \ e(m+d+1)\ldots e(m+d+N)\right] \tag{4.2}$$

and with eq.(3.7), (3.9), (3.10), unbiased estimates can be obtained by premultiplying eq.(4.1) with $\underline{W}^T$

$$\underline{W}^T\underline{e} = \underline{W}^T\underline{y} - \underline{W}^T\underline{\Psi}\ \underline{\Theta} \tag{4.3}$$

where $\underline{W}$ is called instrumental matrix which satisfies

$$E\{\underline{W}^T\underline{e}\} = 0 \tag{4.4}$$

$$E\{\underline{W}^T\underline{\Psi}\} \quad \text{nonsingular.} \tag{4.5}$$

The elements of $\underline{W}$ therefore are chosen to be uncorrelated with the residuals $\underline{e}$, e.g. REIERSOL (1941), DURBIN (1954), KENDALL and STUART (1961). Then from eq.(4.3)

$$\hat{\underline{\Theta}} = \left[\underline{W}^T\underline{\Psi}\right]^{-1}\underline{W}^T\underline{y}\ . \tag{4.6}$$

The input variables u(k) probably were first used as instrumental variables, JOSEPH, LEWIS and TOU (1961). WONG and POLAK (1967) and YOUNG (1969) showed that there exist optimal instrumental variables and they used the calculated, undisturbed output signal as instrumental variables, taking the parameter estimates as the parameters of an auxiliary model. If h is the output of the auxiliary model, the instrumental matrix then becomes

$$\underline{W} = \left[\begin{array}{ccc|ccc} -h(m+d-1) & \ldots & -h(d) & u(m-1) & \ldots & u(0) \\ -h(m+d-2) & \ldots & -h(d+1) & u(m) & \ldots & u(1) \\ \vdots & & \vdots & \vdots & & \vdots \\ -h(m+d+N-1) & \ldots & -h(d+N) & u(m+N-1) & \ldots & u(N) \end{array}\right] \tag{4.7}$$

A recursive algorithm of eq.(4.6) is obtained analog to eq.(3.13) and (3.14)

$$\hat{\underline{\Theta}}(k+1) = \hat{\underline{\Theta}}(k) + \left[\underline{\psi}(k+1)\underline{P}(k)\underline{w}^T(k+1)+1\right]^{-1}\cdot \underline{P}(k)\underline{w}^T(k+1)\left[y(k+1)-\underline{\psi}(k+1)\hat{\underline{\Theta}}(k)\right] \tag{4.8}$$

with

$$\underline{P}(k+1) = \underline{P}(k)\left[\underline{I} - \underline{w}^T(k+1)\underline{w}(k+1)\underline{P}(k)\cdot \left[\underline{\psi}(k+1)\underline{P}(k)\underline{w}^T(k+1)+1\right]^{-1}\right] \tag{4.9}$$

$$\underline{P}(k) = \left[\underline{W}^T(k)\underline{\Psi}(k)\right]^{-1} \tag{4.10}$$

$$\underline{w}(k) = \left[-h(k-1)\ldots -h(k-m) \mid u(k-d-1)\ldots u(k-d-m)\right] \tag{4.11}$$

A block diagram of the method is shown in Fig.4.1.

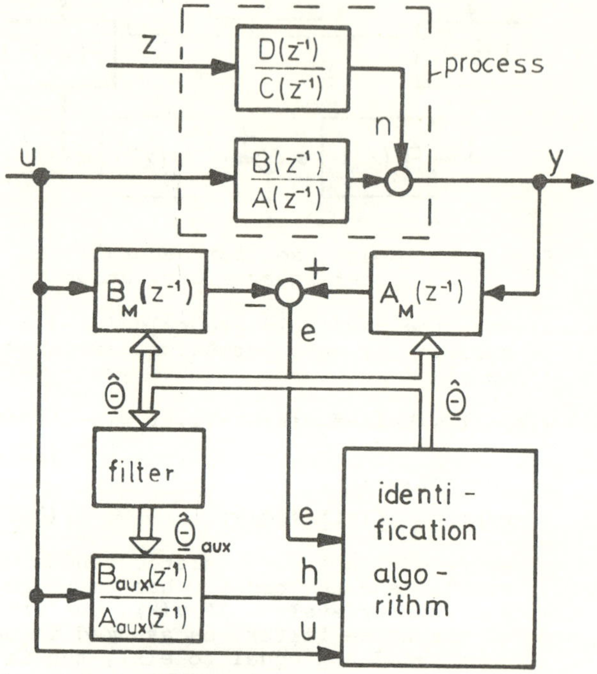

Fig. 4.1. Block diagram of on-line instrumental variable identification.

YOUNG (1972) introduced a time delay and a low pass filter before updating the auxiliary model, to ensure that the auxiliary model parameters are not correlated with e at the same instant and to smooth the estimates. We used the low pass filter

$$\hat{\underline{\Theta}}_{aux}(k) = (1-\gamma)\hat{\underline{\Theta}}_{aux}(k-1) + \gamma\hat{\underline{\Theta}}(k) \tag{4.12}$$

$$\gamma = 0.03 \ldots 0.05.$$

γ has to be chosen in order to prevent instability in the estimation. The initial matrix $\underline{P}(0)$ can be chosen as a diagonal matrix with elements as large as possible. The initial values of the parameter vector $\hat{\underline{\Theta}}(0)$ of the model and of $\hat{\underline{\Theta}}_{aux}(0)$ of the auxiliary model can be zero.

5. THE METHOD OF STOCHASTIC APPROXIMATION (STA)

Surveys on stochastic approximation can be found in SAKRISON (1966), ALBERT and GARDNER (1967) and WASAN (1969). Most of stochastic approximation algorithms used for recursive identification of dynamic systems are based on the ROBBINS-MONROE- or KIEFER-WOLFOWITZ-Algorithms. They are computationally very simple. The recursive least squares algorithm can also be regarded as a special case of stochastic approximation.

Without any correction these basic stochastic approximation algorithms lead to biased estimates, e.g. SARIDIS and STEIN (1968 a), if the generalized error, eq. (3.5), is used for minimization. SARIDIS and STEIN (1968 b) showed that unbiased estimates g(ν) of the impulse response can be obtained by minimizing the output error and they proposed the algorithm

$$\underline{\hat{G}}(k+\ell) = \underline{\hat{G}}(k-1) + \gamma(k)\, \underline{U}(k+\ell) \cdot \left[y(k+\ell) - \underline{U}^T(k+\ell)\, \underline{\hat{G}}(k-1)\right] \quad (5.1)$$

with

$$\underline{\hat{G}}^T = \left[g_1, g_2, \ldots, g_\ell\right] \quad (5.2)$$

$$\underline{U}^T(k) = \left[u(k-1)\ u(k-2)\ \ldots u(k-\ell)\right] \quad (5.3)$$

$$\gamma(k) = \gamma(\zeta) = \frac{1}{\zeta(k)} ;$$

$$\zeta(k) = \frac{k-1}{\ell+1} ; \quad k = 1,\ \ell+2,\ 2\ell+3, \ldots \quad (5.4)$$

If there are 2m unknown parameters of the parameter vector $\underline{\theta}$, ℓ=2m values of the impulse response $g_{\underline{i}}$ are estimated by eq. (5.1) and the parameters a_i and b_i of $\underline{\theta}$ are calculated by using fundamental relationships. The algorithm (5.1) converges in the mean-square sense to the true parameter values if following conditions are fulfilled:

1) The input u(k) is an independent random variable with

   $E\{u(k)\} = 0.$

2) The noise n(k) is an independent random variable with

   $E\{n(k)\} = 0.$

As only 2m values of the impulse response are used, not all information on the process dynamics is included in estimating the parameters a_i and b_i. Therefore we extended the algorithm (5.1) to estimate ℓ>2m impulse response values and took approximately $\ell = T_S/T_O$ with T_S, settling time and T_O, sampling interval.

Additionally we did not take the straight forward calculation of the parameters a_i and b_i from the g_i, because this can lead to large variances. We used a least squares parameter estimation for finding the a_i and b_i from the noisy g_i. The method is described in the following section as "method 2".

It turned out to give better convergence if the correction factor γ of eq.(5.1) for the first estimations is not used according to eq.(5.4) but to Fig. 5.1.

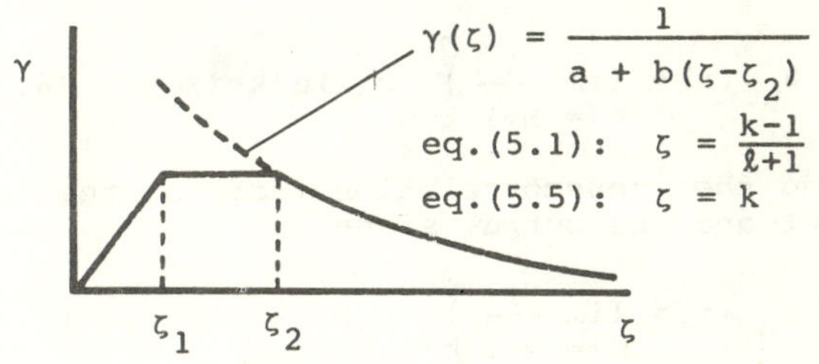

Fig. 5.1. Correction factor γ(k) of the stochastic approximation algorithm.

Then the residuals at the beginning of the estimation, which can be very large, are not weighted so much and the first corrections for $\zeta<\zeta_1$ are damped.

We also updated algorithm (5.1) after each sample interval so leading to an especially simple form

$$\underline{\hat{G}}(k+1) = \underline{\hat{G}}(k) + \gamma(k+1)\, \underline{U}(k+1) \cdot \left[y(k+1) - \underline{U}^T(k+1)\, \underline{\hat{G}}(k)\right] \quad (5.5)$$

with $\gamma(k) = 1/k$; k= 1,2,3,... (here ζ=k).

This algorithm gives biased estimates, but the bias are relatively small also for large noise-to-signal ratio. The initial values of the parameter vector $\underline{\hat{G}}(0)$ can be zero.

6. CORRELATION ANALYSIS WITH LEAST SQUARES PARAMETER ESTIMATION (COR)

Before applying parameter estimation methods as least squares, instrumental variables or maximum likelihood on sets of input and output data, one has to know a priori the structure, the order and the time delay of a parametric process model or one has to assume and to test them by iterative procedures. For each iteration one has to handle the whole data set and the computational effort therefore can build up tremendously. A way to reduce this computations is to first identify a nonparametric model and then to apply parameter estimation on this model, which already contains a data reduction.

In this section an identification method for linear processes is described which first determines an impulse response by correlation analysis and then estimates the parameters of the wanted parametric model with the method of least squares.

If the input signal is a stationary random variable, then its autocorrelation function (acf) is

$$\Phi_{uu}(\tau) = \lim_{N\to\infty} \frac{1}{N+1} \sum_{k=0}^{N} u(k)u(k-\tau) \qquad (6.1)$$

and the crosscorrelation (ccf) of the input and the output signal

$$\Phi_{uy}(\tau) = \lim_{N\to\infty} \frac{1}{N+1} \sum_{k=0}^{N} u(k-\tau)\ y(k). \qquad (6.2)$$

The convolution equation

$$\Phi_{uy}(\tau) = \sum_{\nu=0}^{\infty} g(\nu)\ \Phi_{uu}(\tau-\nu), \qquad (6.3)$$

relates both together with $g(\nu)$ as the impulse response of the considered process.

From eq.(6.3) we get $g(\nu)$ as follows:

$$\begin{bmatrix} \Phi_{uy}(-M) \\ \vdots \\ \Phi_{uy}(-1) \\ \Phi_{uy}(0) \\ \Phi_{uy}(1) \\ \vdots \\ \Phi_{uy}(M) \end{bmatrix} = \begin{bmatrix} \Phi_{uu}(-M) & \dots & \Phi_{uu}(-M-\ell) \\ \vdots & & \vdots \\ \Phi_{uu}(-1) & \dots & \Phi_{uu}(-1-\ell) \\ \Phi_{uu}(0) & \dots & \Phi_{uu}(-\ell) \\ \Phi_{uu}(1) & \dots & \Phi_{uu}(1-\ell) \\ \vdots & & \vdots \\ \Phi_{uu}(M) & \dots & \Phi_{uu}(M-\ell) \end{bmatrix} \begin{bmatrix} g(0) \\ \vdots \\ \vdots \\ g(\ell) \end{bmatrix}$$

$$\underline{\hat{\Phi}}_{uy} = \underline{\Phi}_{uu} \cdot \underline{\hat{g}} \qquad (6.4)$$

$$\underline{\hat{g}} = \left[\underline{\hat{\Phi}}^T_{uu}\underline{\hat{\Phi}}_{uu}\right]^{-1} \underline{\hat{\Phi}}^T_{uu}\ \underline{\hat{\Phi}}_{uy}\ . \qquad (6.5)$$

It is assumed now that $\Phi_{uu}(\tau)$ is known exactly. If the input is discrete white noise, a PRBS with spectral density $S_{uu}(\omega=0)=u_0^2\lambda/4$, eq.(6.5) simplifies to

$$\hat{g}(\tau) = \frac{1}{\Phi_{uu}(0)}\ \Phi_{uy}(\tau) = \frac{\lambda}{S_{uu}(0)}\ \Phi_{uy}(\tau). \qquad (6.6)$$

Hence white noise is preferred as input signal if small computational expense is wanted to determine the impulse response.

The ccf can be obtained recursively by

$$\Phi_{uy}(\tau,k) = \hat{\Phi}_{uy}(\tau,k-1) + \frac{1}{k+1}\cdot \left[u(k-\tau)y(k)-\hat{\Phi}_{uy}(\tau,k-1)\right] \qquad (6.7)$$

which follows from eq. (6.2).

The parameters a_i and b_i of the parametric model eq.(3.1)

$$y(k) = -a_1y(k-1) - a_2y(k-2) - \dots - a_my(k-m) + b_1u(k-d-1) + b_2u(k-d-2) + \dots + b_mu(k-d-m) \qquad (6.8)$$

can be estimated from the impulse response values by two methods.

<u>Method 1.</u>

First the time delay d is separated. Then eq.(6.8) furnishes an impulse response if $u(0)=1$ and $u(k)=0$ for $k=1,2,\dots$ [1])

The equation of the model then is

$$\underline{g}_M = \underline{Q}\ \underline{\theta} \qquad (6.10)$$

and the residuals are defined as

$$\underline{e} = \underline{\hat{g}} - \underline{g}_M = \underline{\hat{g}} - \underline{Q}\ \underline{\theta}\ . \qquad (6.11)$$

$\underline{\hat{g}}$ is the impulse response vector which follows from eq.(6.5) or (6.6). Minimization of the error function

---

[1])

$$\begin{bmatrix} g(1) \\ g(2) \\ g(3) \\ \vdots \\ g(\ell) \end{bmatrix} = \left[\begin{array}{ccccc|cccc} 0 & 0 & 0 & \dots & 0 & u(0) & 0 & 0 \dots & 0 \\ -g(1) & 0 & 0 & \dots & 0 & 0 & u(0) & 0 \dots & 0 \\ -g(2) & -g(1) & 0 & \dots & 0 & 0 & 0 & u(0) \dots & 0 \\ \vdots & \vdots & \vdots & & \vdots & & & & \\ -g(\ell-1) & -g(\ell-2) & -g(\ell-3) & \dots & -g(\ell-m) & 0 & 0 & 0 \dots & 0 \end{array}\right] \begin{bmatrix} a_1 \\ a_2 \\ \vdots \\ a_m \\ b_1 \\ b_2 \\ \vdots \\ b_m \end{bmatrix}$$

$$\underline{g} = \underline{Q}\ \underline{\theta} \qquad (6.9)$$

$$V = \sum_{k=1}^{\ell} e^2(k) \tag{6.12}$$

leads to the LS estimate

$$\hat{\underline{\theta}} = \left[\underline{Q}^T \underline{Q}\right]^{-1} \underline{Q}^T \hat{\underline{g}}. \tag{6.13}$$

This estimate is consistent since with the assumptions made

$$E\{\hat{g}(\tau) - g_o(\tau)\}$$

$$= \frac{1}{\Phi_{uu}(0)} \cdot \left[E\{\hat{\Phi}_{uy}(\tau)\} - \Phi^o_{uy}(\tau)\right] = 0 \tag{6.14}$$

$$E\{\hat{\underline{\theta}} - \underline{\theta}_o\} = 0 \tag{6.15}$$

and

$$\lim_{N\to\infty} E\{\left[\hat{g}(\tau) - g_o(\tau)\right]^2\} = 0 \tag{6.16}$$

$$\lim_{N\to\infty} E\{(\hat{\underline{\theta}} - \underline{\theta}_o)^2\} = 0. \tag{6.17}$$

The index $_o$ notes the true values.

Method 2.

With method 1 the parameters b_i are only estimated from the few values $\hat{g}(1),\ldots \hat{g}(m)$ and can therefore have larger variances. This is avoided by making the least squares fit for the response function of the model

$$\hat{y}(q) = \sum_{\nu=0}^{\infty} \hat{g}(q-\nu)\, u(\nu) \tag{6.18}$$

to any input which is exciting the process also for $\nu>0$. Then eq.(6.8) leads to

[1])

and the estimates are

$$\hat{\underline{\theta}} = \left[\underline{R}^T \underline{R}\right]^{-1} \underline{R}^T \hat{\underline{y}} \tag{6.20}$$

Eq. (6.18) and (6.20) become especially simple if a step input (for the model) with u(0), u(1), u(2),...=1 is used.

An additional advantage of this 2nd method is, that the noise present in the values $\hat{g}(\tau)$ is smoothed, because of eq.(6.18). Several tests with simulated data verified the advantages of method 2.

This identification method can be used off-line or on-line. In the on-line version the ccf is updated with eq.(6.7) after each sample and eq.(6.13) or (6.18) and (6.20) are calculated as often as necessary. Thus this method is relatively flexible in the distribution of the computations over time, what is especially advantageous for process computers. Some other advantages are:

1. Only little a priori knowledge about the structure of the process model is necessary to identify $g(\tau)$.
2. Short computation time and small computer storage.
3. The impulse response as an intermediate nonparametric result leads to:
   a) The performance of the result can be evaluated immediately.
   b) Preselection of structure and order of the parametric model.
   c) Short computation time for testing the order and the time delay.
   d) The impulse response is correct also for linear processes with distributed parameters.
4. Initial values of matrices and parameters are not necessary.
5. Instability not possible.
6. The accuracy of the result can be controlled by changing the number ℓ of impulse response samples ($\ell \geq 2m$). Thus the computational expense can be reduced by accepting a loss of accuracy of the model.

---

[1])

$$\begin{bmatrix} y(1) \\ y(2) \\ y(3) \\ \vdots \\ y(\ell) \end{bmatrix} = \left[\begin{array}{cccc|cccc} 0 & 0 & 0 \ldots & 0 & u(0) & 0 & 0 & \ldots 0 \\ -y(1) & 0 & 0 \ldots & 0 & u(1) & u(0) & 0 & \ldots 0 \\ -y(2) & -y(1) & 0 \ldots & 0 & u(2) & u(1) & u(0) & \ldots 0 \\ \vdots & \vdots & \vdots & \vdots & \vdots & \vdots & \vdots & \vdots \\ -y(\ell) & -y(\ell-1) & & -y(\ell-m) & u(\ell) & u(\ell-1) & & u(\ell-m) \end{array}\right] \begin{bmatrix} a_1 \\ a_2 \\ \vdots \\ a_m \\ b_1 \\ b_2 \\ \vdots \\ b_m \end{bmatrix}$$

$$\hat{\underline{y}} = \underline{R} \cdot \underline{\theta} \tag{6.19}$$

7. THREE PARAMETER IDENTIFICATION USING FOURIER ANALYSIS (3PI)

If one replaces some not too different time constants by equal time constants one can reduce the variances of the parameter estimates for higher order processes taking into account a loss of information about detailed knowledge of process dynamics, ISERMANN (1973 a). Hence it can be advantageous to replace the exact model of a low pass process by the model

$$G(s) = \frac{K}{(1+Ts)^n} e^{-T_D s} \qquad (7.1)$$

where the small time constants are replaced by a time delay. n is assumed to be known. K, T and T_D have to be estimated. A sequence of m rectangular pulses is chosen as input signal, Fig.7.1.

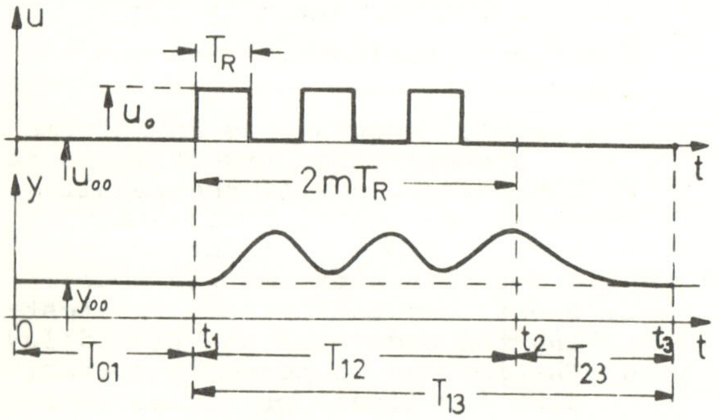

Fig. 7.1. Input and output signal for method 3PI.

Taking appropriate pulse width $T_R=\pi/\omega_\nu$ so that $0.6 \geq |G(i\omega_\nu)| \geq 0.2$, this sequence gives strong excitation in the middle frequency region, for which the identified model has to be most accurate if the model is used for calculating the closed loop behavior, ISERMANN (1971 d, 1971 a). Then also an effective estimation of the gain K is possible.

The frequency response is determined by

$$G(i\omega) = \frac{Y(i\omega)}{U(i\omega)} = \frac{\int_{t_1}^{t_3} [y(t)-y_{oo}] \exp(-i\omega t) dt}{\int_{t_1}^{t_3} [u(t)-u_{oo}] \exp(-i\omega t) dt} = |G(i\omega_\nu)| e^{-i\Phi(\omega_\nu)} \qquad (7.2)$$

for $\omega_\nu=\pi/T_R$ and $\omega=0$. The reference value u_{oo} is always known, but the reference value y_{oo} in general has to be identified by

$$y_{oo} = \frac{1}{T_{01}} \int_0^{t_1} y(t)\, dt. \qquad (7.3)$$

The gain is easily obtained from eq.(7.2) with $\omega=0$

$$K = G(0) = \int_{t_1}^{t_3} [y(t)-y_{oo}] dt \Big/ \int_{t_1}^{t_3} [u(t)-u_{oo}] dt. \qquad (7.4)$$

The time constant is obtained by

$$T = \frac{1}{\omega_\nu} \left[\left[\frac{K}{|G(i\omega_\nu)|} \right]^{\frac{2}{n}} - 1 \right]^{\frac{1}{2}} \qquad (7.5)$$

and the time delay in a direct way (not iterative as with other methods)

$$T_D = \frac{1}{\omega_\nu} \left[\phi(\omega_\nu) - n \text{ arc tg } \omega_\nu T \right] \qquad (7.6)$$

which both follow eq.(7.1). Further details are given in ISERMANN (1973 a). To obtain very simple algorithms not all frequencies are taken for evaluation and therefore it cannot be expected that the parameter estimates are efficient. The parameters are available after each sequence of pulses. Eq.(7.1) is only one example. The method can be used also for other continuous processes with three unknown parameters.

8. COMPARISON OF SIX ON-LINE IDENTIFICATION METHODS [1)]

Following abbreviations are used for signification of the identification methods:

LS	-	least squares
GLS	-	generalized least squares
IVA	-	instrumental variables
STA	-	stochastic approximation
COR	-	correlation analysis with least squares parameter estimation
3PI	-	three parameter identification with Fourier analysis.

Each of the methods LS, GLS, IVA, STA and COR have been applied on each of the processes I, II, III, using PRBS test signals,

- taking 5 data sets
- each with 3 identification time periods
- with two noise-to-signal ratios.

1) Specifications of the identification methods, Appendix A, and tables of all results, Appendix B, can be obtained from the authors. The results of more than 180 identification runs and 300 controller parameter optimizations are used.

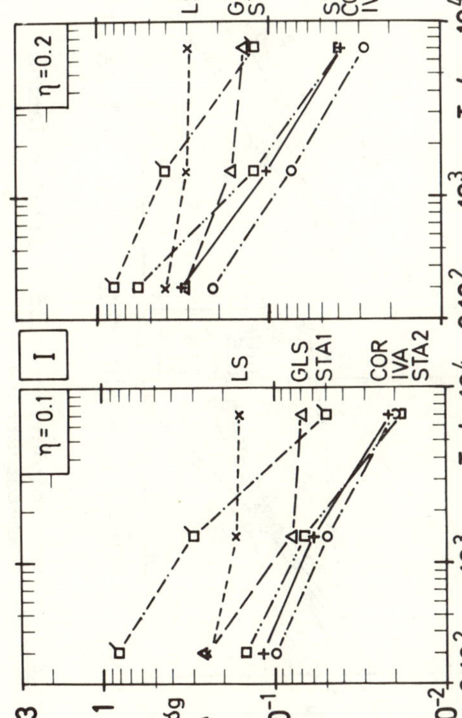

Fig.8.1. Standard deviation of impulse response error.

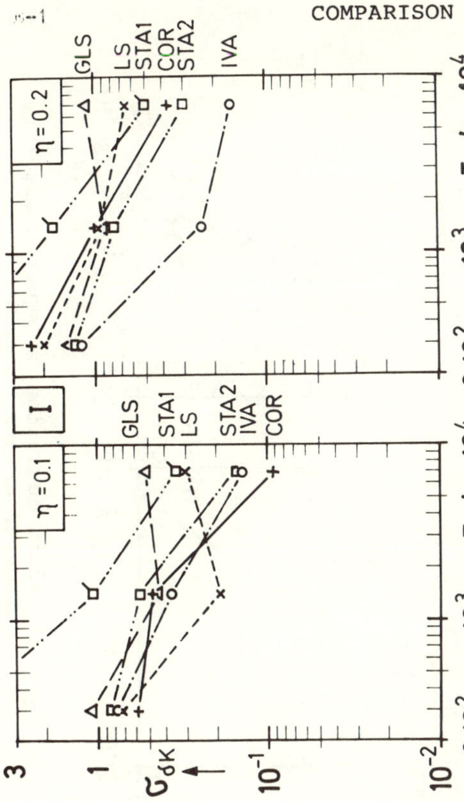

Fig.8.2. Standard deviation of gain error.

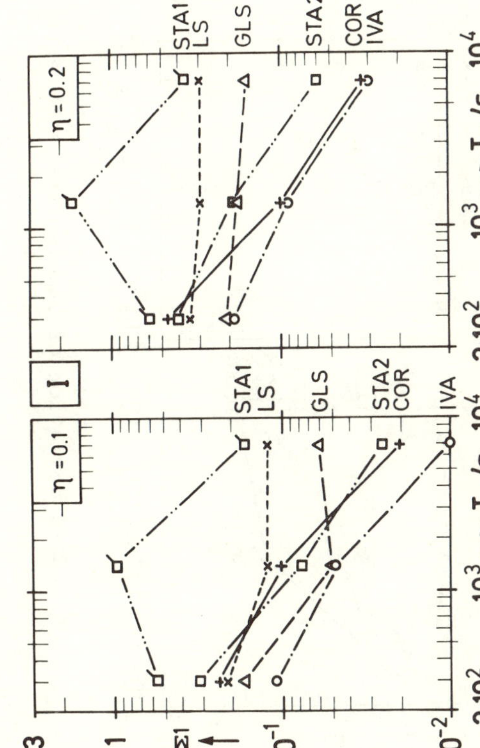

Fig.8.3. Standard deviation of squared relative parameter errors.

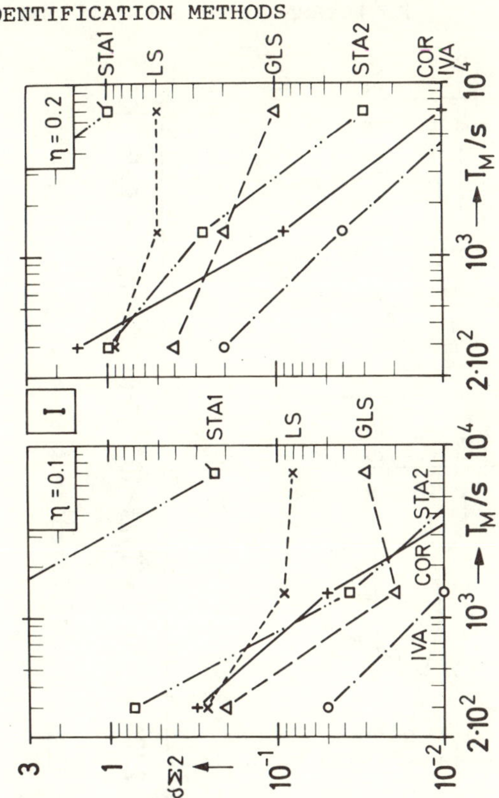

Fig.8.4. Standard deviation of squared absolute parameter errors.

PROCESS I

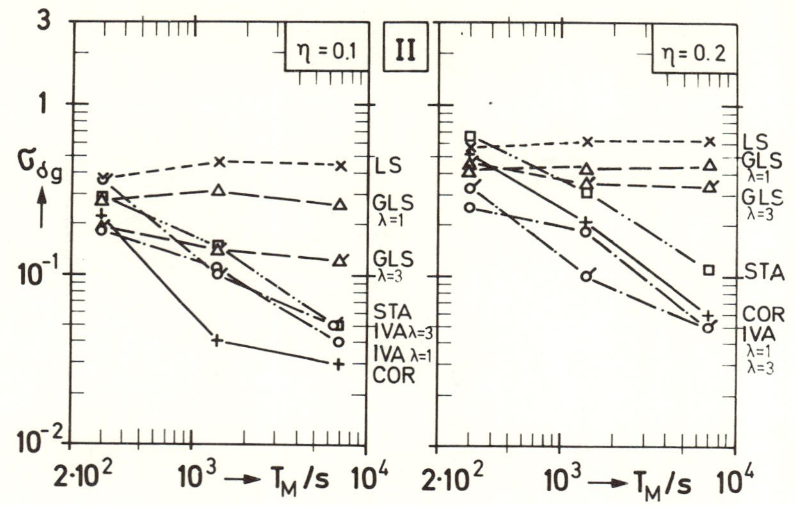

Fig.8.5. Standard deviation of impulse response error.

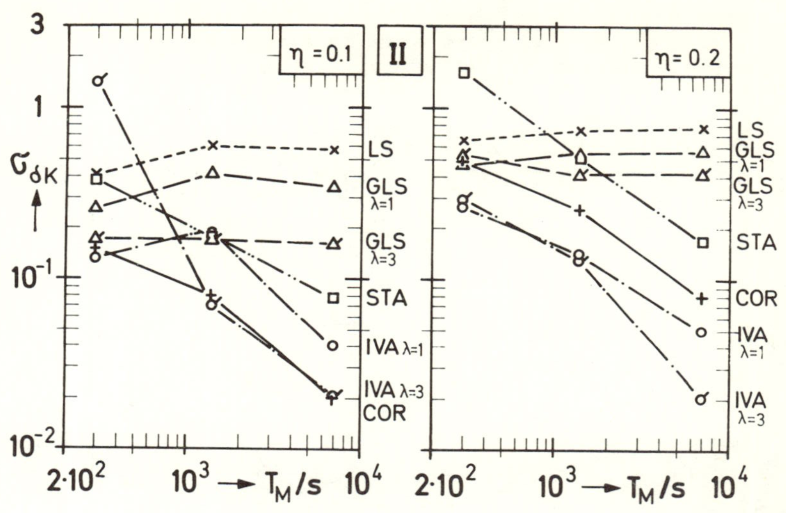

Fig.8.6. Standard deviation of gain error.

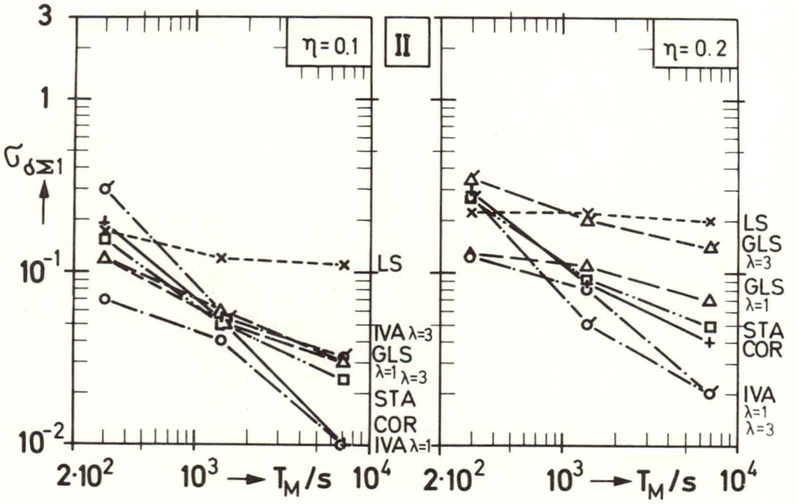

Fig.8.7. Standard deviation of squared relative parameter errors.-

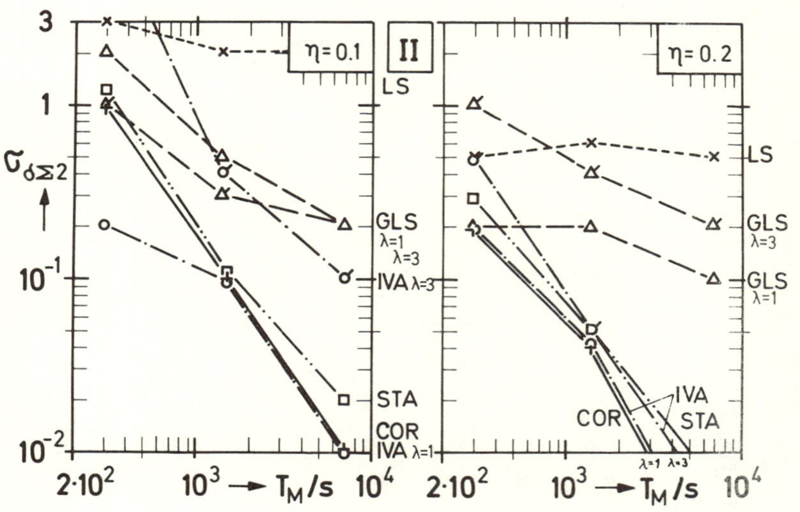

Fig.8.8. Standard deviation of squared absolute parameter errors.

PROCESS II

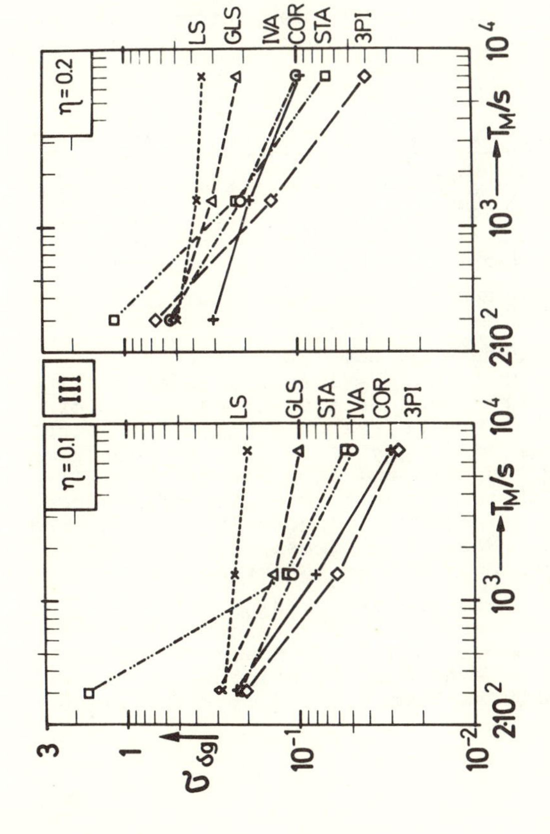

Fig.8.9. Standard deviation of impulse response error.

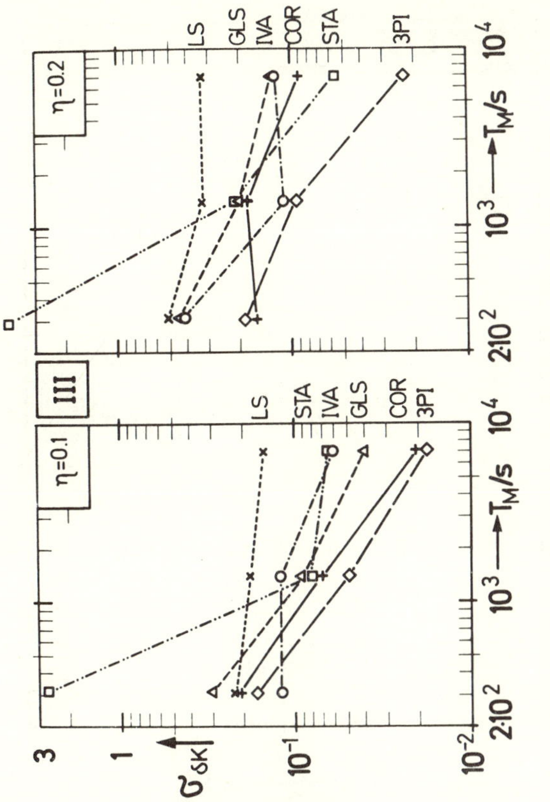

Fig.8.10. Standard deviation of gain error.

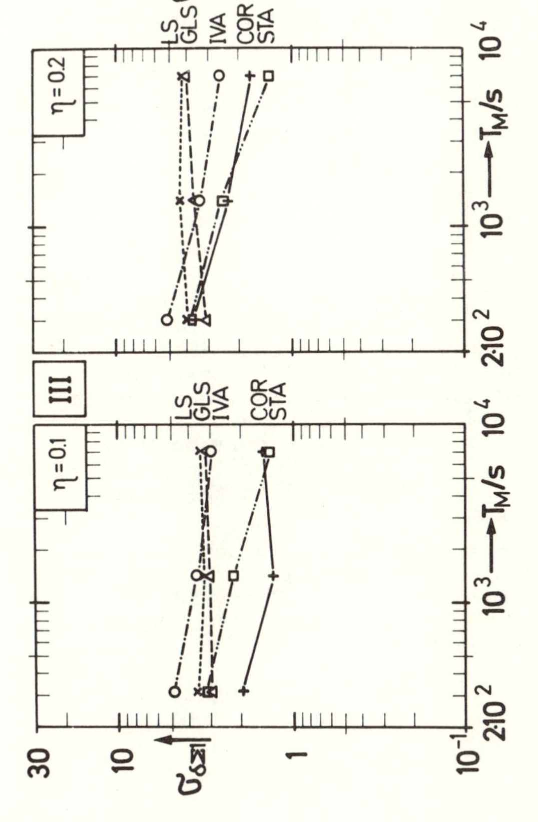

Fig.8.11. Standard deviation of squared relative parameter errors.

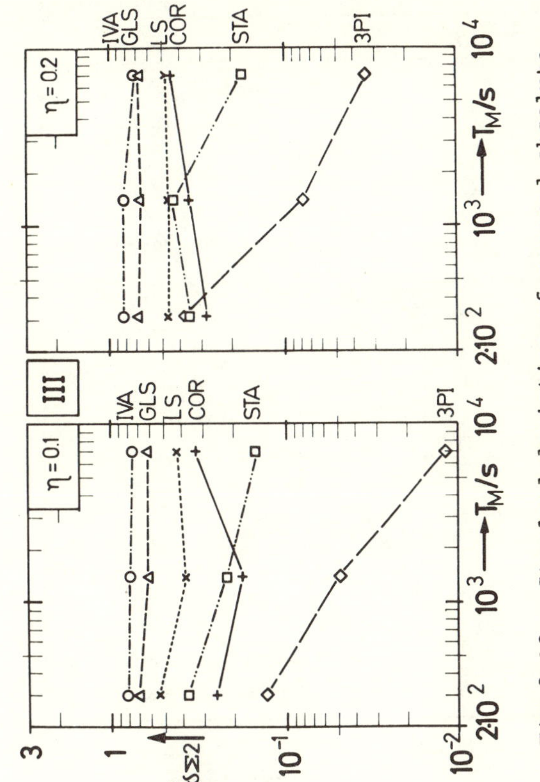

Fig.8.12. Standard deviation of squared absolute parameter errors.

PROCESS III

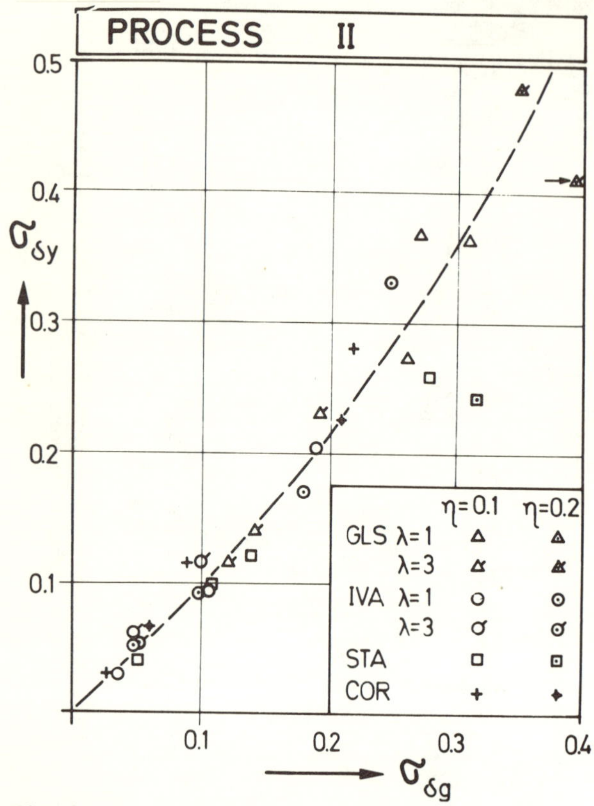

Fig.8.13. Standard deviation of closed loop output error for process II.

8.1. Results for processes I,II and III

Summarized results for process I

Fig.8.1 to 8.4 show:

- IVA and COR are most accurate.
- STA 2, according to algorithm (5.5), which leads to small bias, gives much better results than STA 1, algorithm (5.1). This results from correcting the parameters after each interval k and not after each $(\ell+1)$ intervals, as for STA 1, eq.(5.1). The results of STA 2 are almost as good as those of IVA and COR. Therefore only results of STA 2 are shown for processes II and III. STA 2 in the following is noted as STA.
- The estimates of GLS do not converge essentially or diverge for identification time T_M>1400s. This indicates that GLS tends to give bias, which, however, are smaller than those of LS.

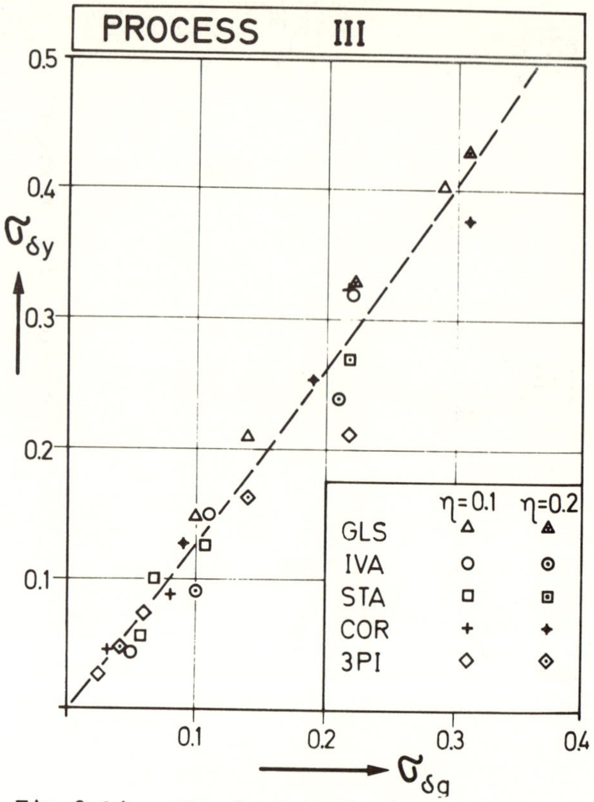

Fig.8.14. Standard deviation of closed loop output error for process III.

Summarized results for process II

For GLS a clock interval λ/T_o=3 was also used, because with λ/T_o=1 the errors of the impulse response were very large. Fig. 8.5 to 8.8 and 8.13 demonstrate:

- COR and IVA are most accurate, followed by STA.
- Increase of the clock interval of the PRBS from λ/T_o=1 to 3 yielded a slight improvement in the GLS estimates of the impulse response g(k), the gain K, $\sigma_{\delta\Sigma 1}$ and $\sigma_{\delta\Sigma 2}$ for η=0.1, and deteriorated $\sigma_{\delta\Sigma 1}$ and $\sigma_{\delta\Sigma 2}$ for η=0.2. All GLS runs tend to give no convergence for T_M>1400s, especially for g(k) and K. However, GLS estimates in general are better than LS estimates.
- The diagram $\sigma_{\delta y}$ =f($\sigma_{\delta g}$), Fig.8.13, expresses a definite relationship between $\sigma_{\delta y}$ and $\sigma_{\delta g}$. However, there exists no corresponding clear relationship between $\sigma_{\delta y}$ and $\sigma_{\delta\Sigma 1}$ or $\sigma_{\delta\Sigma 2}$.

Summarized results for process III

The sampling interval for process III originally proposed in the test case task, ISERMANN (1972a), was $T_o=2s$. However, we did not obtain reasonable results with GLS. Therefore we increased the sampling interval to $T_o=4s$. The following tables show for IVA :

COR and STA, larger errors for IVA, GLS and LS.

•• Slow decrease of $\sigma_{\delta\Sigma 2}$ only for STA and almost no decrease for all other methods. Smallest errors for STA and COR, larger errors for LS, GLS and IVA. Errors of LS are smaller than those of GLS and IVA.

Table 8.1. Results with IVA using different sampling intervals T_o and clock intervals λ.

IVA		$T_o = 4s;\ \lambda = T_o$				$T_o = 4s;\ \lambda = 2T_o$				$T_o = 2s;\ \lambda = T_o$			
η	T_M [s]	$\sigma_{\delta g}$	$\sigma_{\delta\Sigma 1}$	$\sigma_{\delta\Sigma 2}$	$\sigma_{\delta K}$	$\sigma_{\delta g}$	$\sigma_{\delta\Sigma 1}$	$\sigma_{\delta\Sigma 2}$	$\sigma_{\delta K}$	$\sigma_{\delta g}$	$\sigma_{\delta\Sigma 1}$	$\sigma_{\delta\Sigma 2}$	$\sigma_{\delta K}$
0.1	312	0.22	4,81	0.80	0.12	0.25	7.73	0.99	0.07	0.48	5.00	1.01	0.62
	1412	0.11	3.46	0.78	0.12	0.10	4.06	0.86	0.05	0.26	3.35	0.93	0.14
	7012	0.05	2.90	0.75	0.06	0.04	3.43	0.82	0.01	0.07	2.09	0.85	0.04

(a) Increase of $T_o=2s$ to 4s for $\lambda/T_o=1$ gives smaller impulse response errors $\sigma_{\delta g}$ and gain errors $\sigma_{\delta K}$, and approximately the same parameter errors $\sigma_{\delta\Sigma 1}$ and $\sigma_{\delta\Sigma 2}$.

(b) Increase of the clock interval λ from T_o to $2T_o$ gives smaller gain errors $\sigma_{\delta K}$, approximately the same impulse response errors $\sigma_{\delta g}$, and larger parameter errors.

The following results belong to the case $T_o=4s$, Fig.8.9 to 8.12 and 8.14.

- 3 PI gives most accurate results with respect to all error measures.
- With all methods the impulse response and the gain factor are estimated with about the same variances and the same convergence as for the other two processes. However, the single process parameter estimates show either very slow or even no convergence and have large errors.
- According to the estimates of the impulse response and the gain it follows from $\sigma_{\delta g}$ and $\sigma_{\delta K}$:
  - •• With respect to a steady convergence COR is next best after 3PI, followed by IVA.
  - •• STA shows relatively large errors for short measuring time, but small errors for the largest measuring time.
  - •• GLS tends to indicate biased estimates, however, with smaller variances than the estimates of LS.
- Based on the estimates of the single process parameters it follows from $\sigma_{\delta\Sigma 1}$ and $\sigma_{\delta\Sigma 2}$:
  - •• Slow decrease of $\sigma_{\delta\Sigma 1}$ for COR, STA and IVA. Almost no decrease for GLS and LS. Smallest errors for

- Diagram $\sigma_{\delta y}=f(\sigma_{\delta g})$, Fig.8.14, expresses again a definite relationship, also for 3PI. There exists no clear relationship between $\sigma_{\delta y}$ and $\sigma_{\delta\Sigma 1}$ or $\sigma_{\delta\Sigma 2}$.

8.2. Discussion and evaluation of the results

The results of the different identification methods are now discussed and evaluated for all three processes together.

8.2.1. Identification

Based on the four different errors used previously it follows:

Impulse function error $\sigma_{\delta g}$

The impulse response errors depend linearly on the process parameter errors, eq.(6.9).

COR and IVA resulted in impulse responses with the smallest variances, except for process III, where 3PI led to somewhat better results. For the largest identification time, STA gave very good estimates also with about the same variances as COR and IVA. However, for the shortest identification time, the variances of STA were about as large as those for GLS and LS, or with process III, even larger.

The standard deviation $\sigma_{\delta g}$ decreases steadily in $1/\sqrt{T_M}$ or even more for COR, IVA and STA which is not the case for GLS. For all processes GLS showed either slow or no convergence, indicating biased parameter estimates which also lead to biased impulse response values. The reason for this behavior is assumed to be found in the use of an autogressive process to describe the residuals, eq. (3.16). This autogressive process does not describe the residuals accurately enough, as shown by eq.(3.18).

LS leads to biased impulse response values as is seen clearly in the results. The standard deviation $\sigma_{\delta g}$ essentially does not decrease for T_M>300s. The standard deviation for T_M=300s, however, is about the same for LS, GLS, IVA, and COR in the case of process III. Hence the bias-free identification methods are only advantageous for longer measurement time periods.

Process gain error $\sigma_{\delta K}$

The process gain error depends also on all process parameter errors, eq.(8.7). Further, the gain is given by

$$K = \sum_{k=0}^{\infty} g(k).$$

Hence, it is not surprising, that the diagrams for $\sigma_{\delta K}$ lead to about the same conclusions as those for $\sigma_{\delta g}$, except for process I, where LS gain estimates are in part better than GLS estimates. If the clock interval λ of the PRBS is increased, more accurate estimates of K can be obtained, see tables for process III and ISERMANN (1972b).

Parameter errors $\sigma_{\delta\Sigma 1}$ and $\sigma_{\delta\Sigma 2}$

Both parameter errors, eq.(2.6) and (2.7), summarize the deviations of the true and the estimated parameters. In $\sigma_{\delta\Sigma 1}$, the relative errors of all a_i parameters and b_i parameters are weighted equally. And in $\sigma_{\delta\Sigma 2}$, the absolute errors of all a_i and b_i parameters are weighted equally. $\sigma_{\delta\Sigma 2}$ is therefore more relevant for the input-output behavior than $\sigma_{\delta\Sigma 1}$, since the error of the output variable is linearly dependent on the absolute parameter errors. However, if the absolute errors of the (often small) b_i parameters are small in comparison to the errors of the a_i parameters, essentially only the last ones influence $\sigma_{\delta\Sigma 2}$.

In the case of process I $\sigma_{\delta\Sigma 1}$ and $\sigma_{\delta\Sigma 2}$ show about the same results, since a_i and b_i parameters have the same magnitude. This is not true for process II, where some slight differences exist. Best results are given by IVA followed by COR and STA 2. The slope of convergence is sometimes greater than $1/\sqrt{T_M}$. GLS exhibits slow or even no convergence , indicating bias that are, however, smaller than for LS. The relative positions of these methods agree only in part with those resulting from $\sigma_{\delta g}$.

$\sigma_{\delta\Sigma 1}$ for process III shows very slow convergence for COR, IVA and STA and no convergence for GLS and LS. The variances are very large. This occurs also with $\sigma_{\delta\Sigma 2}$, where particularly the a_i parameters exert influence. Contrary to all other methods which use error minimization procedures, 3PI shows very good convergence. Notice, however, that the model used for this method and hence the parameters are different from the true model.

Explanations for the fact, that for process III the input-output behavior, indicated by $\sigma_{\delta g}$, is identified accurately, but not the single process parameters, indicated by $\sigma_{\delta\Sigma 1}$, are found to be:

- Since this behavior was not observed with processes I and II, which have 4 unknown parameters, this fact must result from the model for process III, which has 6 unknown parameters.
- IVA and GLS minimize the equation error, COR and STA the impulse response error, or respectively, the step response error. Indeed, the estimates of each parameter obtained for T_M= 7000s with IVA and GLS on the one hand and with COR and STA on the other hand approximately concentrate themselves around two different values. Hence, for finite identification time periods different error measures lead to different parameter sets which all result in small impulse response errors $\sigma_{\delta g}$.
- Table 8.5 shows the results of "marathon runs" for an identification time period of T_M=520 000s (N=130 000 or 7500 times of the settling time) with IVA, COR and STA, all with the same noise sequence. The parameters of STA show very good agreement with the true values. However, the parameter errors $\delta_{\Sigma 1}$ of COR and IVA are much larger, though in all cases the errors δ_g of the impulse response and the error of the gain K are extremely small. Hence, also for very small impulse response errors, very different estimates of parameters resulted.
- In Fig. 8.15 the convergence of IVA, COR and STA is represented for the "marathon runs". In all cases the terms

$$1 + \sum_{i=1}^{3} a_i \quad \text{and} \quad \sum_{i=1}^{3} b_i$$

have relative errors of average magnitude and converge very slowly, sometimes not directly, to the true values. However, the single parameters have large relative errors, except b_1 which is estimated relative exactly. Influenced by the noise they converge or diverge simultaneously during short time periods, so as to compensate each other, resulting in approximately the same input-output behavior, and converge very slowly only for long time periods. This indicates that very flat slopes exist in the multidimensional parameter surface. Notice, that the noise sequence n(k) is the same for all three runs shown in Fig. 8.15.

8.2.2. Identification and control

The results for the standard deviation $\sigma_{\delta y}$ of the closed loop output error appear to be approximately linearly dependent on the standard deviation $\sigma_{\delta g}$ of the impulse response error for $0 \leq \sigma_{\delta g} \leq 0.2$. For processes II and III $\sigma_{\delta y}/\sigma_{\delta g} \approx 1.05$ and

		PROCESS MODEL PARAMETERS									
Process III		a_1	a_2	a_3	b_1	b_2	b_3				K
true values		-1.500	0.705	-0.100	0.065	0.048	-0.008				1.000
	N	$\hat{a}_1$	$\hat{a}_2$	$\hat{a}_3$	$\hat{b}_1$	$\hat{b}_2$	$\hat{b}_3$	$\delta_{\Sigma 1}$	$\delta_{\Sigma 2}$	δ_g	$\hat{K}$
IVA	$1{,}3 \cdot 10^5$	-0.941	-0.038	0.155	0.064	0.084	0.026	1.938	0.337	0.011	0.989
STA	$1{,}3 \cdot 10^5$	-1.490	0.689	-0.093	0.065	0.048	-0.007	0.065	0.0001	0.005	0.999
COR	$1{,}3 \cdot 10^5$	-1.205	0.322	0.026	0.065	0.068	0.010	0.967	0.091	0.005	1.004

$\eta = 0.1$; $T_0 = 4s$; $\lambda/T_0 = 1$. IVA: $\underline{P}(0) = 1 \cdot \underline{I}$. COR: $\ell = 18$. STA: $\ell = 15$.
(same noise sequence for all 3 methods)

Table 8.2.
Estimated parameters for N = 130 000.

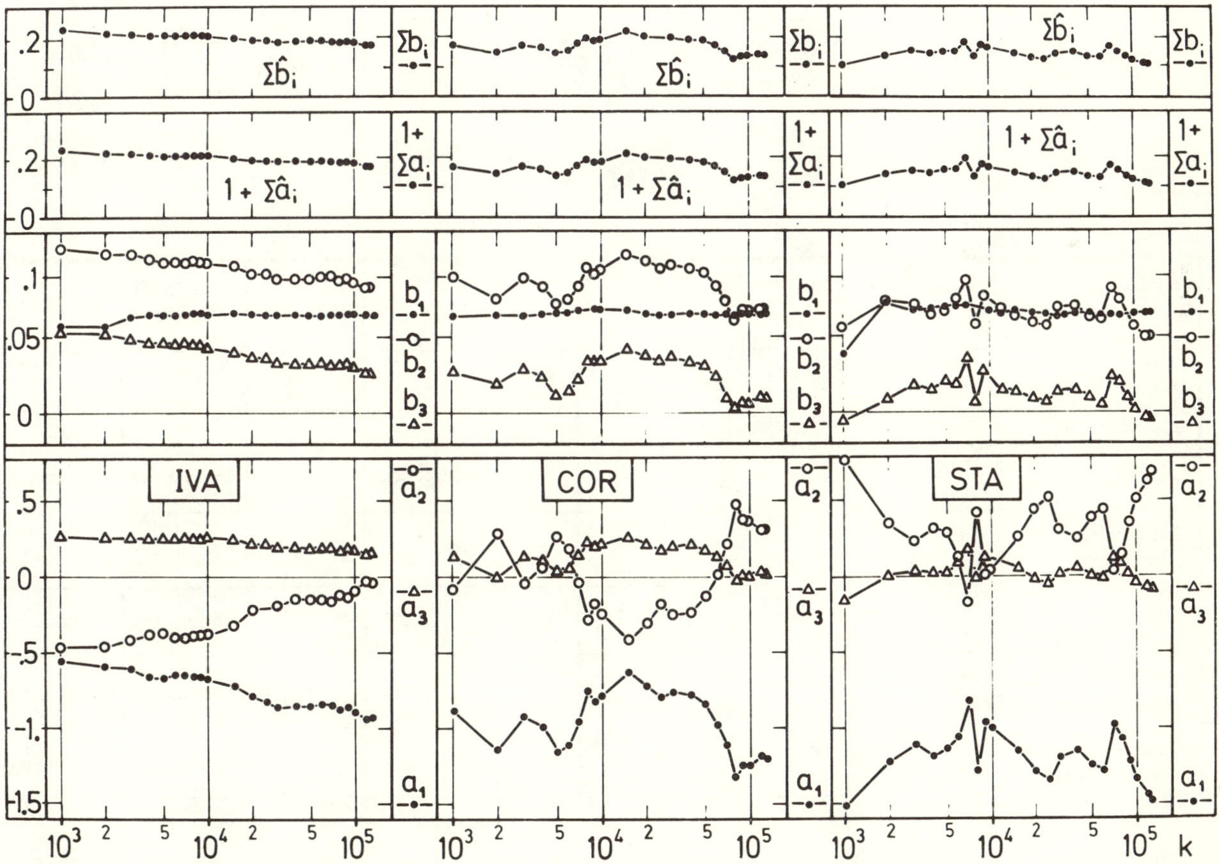

Fig.8.15. Estimated parameters for large identification time periods up to N = 130 000.

1.3 are valid within this range. However, the results do not show correspondingly clear relationships between $\sigma_{\delta y}$ and the standard deviations $\sigma_{\delta\Sigma 1}$ and $\sigma_{\delta\Sigma 2}$ of the parameter errors. Hence, the impulse response error δ_g of the identified model is relevant for the behavior of the closed loop containing a three-mode controller with optimized parameters. The closed loop behavior with this controller depends on the input-output behavior of the process, which of course is represented by the impulse response. With respect to our final goal of the identification, the identification methods therefore have to be judged in terms of the impulse response error.

No clear connection between the standard deviation $\sigma_{\delta p\Sigma}$ of the controller parameters and $\sigma_{\delta\Sigma 1}$ or $\sigma_{\delta\Sigma 2}$ seems to exist. However, $\sigma_{\delta p\Sigma} = f(\sigma_{\delta g})$ shows an approximately linear relationship for each single identification method.

8.3. Conclusions

In table 8.3 the crucial properties of all six identification methods considered here are listed. The types of processes to which the methods can be applied and the types of possible input signals are given. The performance of the identification methods is expressed by the standard deviation $\bar{\sigma}_{\delta g}$ of the impulse response, averaged for processes I,II and III for the measuring time period T_M=1400s, and by the resulting standard deviation $\bar{\sigma}_{\delta y}$ of the closed loop output error, averaged for processes II and III. For 3PI, of course,only values for process III are used. To compare the computation time for one identification run, we used the time needed for 1000 input-output data pairs with a CDC 6600 computer, and set the time for COR to 100%, i.e. 1.7s for processes I and II and 1.95s for process III. Using the preselected a priori factors, given at the bottom of the tables of the individual identification methods, some of the methods did not give reasonable results, because of a failure of convergence or even instability. For a total amount of about 30 runs for each method the percentage of successful runs is given as overall reliability of the method.

Table 8.3. Comparison of six on-line identification methods.

	PROCESSES	INPUT SIGNAL	PERFORMANCE				COMPUTATION TIME ONE RUN		OVERALL RELIABILITY OF METHOD		A PRIORI FACTORS	SEARCH FOR ORDER	DELAY
			$\bar{\sigma}_{\delta g}$		$\bar{\sigma}_{\delta y}$		PROCESS I + II	III	%				
η			0.1	0.2	0.1	0.2	-	-	0.1	0.2			
COR	linear	white noise PRBS	0.08	0.17	0.11	0.25	100 %	100 %	100	100	ℓ	iterative	iterative
IVA	linear in parameters	any noise PRBS	0.09	0.16	0.13	0.21	153	236	90	80	$\underline{P}(0);\gamma$	iterative	iterative
STA	linear	white noise PRBS	0.10	0.22	0.13	0.25	106	108	90	85	$\zeta_1;\zeta_2;a;b;\ell$	iterative	iterative
GLS	linear in parameters	any noise PRBS	0.18	0.31	0.31	0.72	194	287	95	90	$\underline{P}(0);\underline{Q}(0);\nu$	iterative	iterative
LS	linear in parameters	any noise PRBS	0.29	0.44	-	-	145	230	100	100	$\underline{P}(0)$	iterative	iterative
3 PI	linear	sequence of rect.pulses	0.06	0.14	0.07	0.16	-	≈5	100	100	ω_ν	iterative	no, direct

The a priori factors needed before starting a method and the kind of procedure necessary for the search of the order of the model or the time delay are also listed.

For process III 3PI led to the best performance, extraordinarily short computation time, 100% overall reliability and direct detection of the time delay. This method, however, is restricted to linear processes with three unknown parameters and requires a special input signal to obtain these good results.

For general linear processes, COR shows most advantages in comparison to IVA, STA, GLS and LS. Very good performance, shortest computation time, 100% overall reliability (no problems with poor convergence or instability), only one a priori factor ℓ (no initial matrices). Since the impulse response (or other input response y(k)) is an intermediate nonparametric model, the computational expense for search of order and time delay is very small, so that the computation time, in addition to the values given in table 8.6, decreases considerably in comparison to IVA and GLS. COR is not restricted to white noise input, however, with this input the computational expense is the smallest.

IVA resulted in very good performance, approximately as COR. The computation time is longer than for COR and the overall reliability was worst of all methods. An initial matrix and a filter factor have to be known a priori. For nonlinear processes which are linear in the parameters IVA in general is probably better than GLS.

The results of STA also show very good performance for large identification time periods and the computation time is almost as short as for COR. However, the overall reliability is not very good and 5 factors have to be known a priori. Since the impulse response is an intermediate result, STA has the same advantages as COR, concerning the search of the process order and the time delay. For short identification time periods STA furnished only results with large variances.

GLS led to relatively bad performance, which, however, was better than that of LS. The computation time is the longest, the overall reliability was better than for IVA and STA, two initial matrices and the filter order have to be known a priori.

The worst performance was obtained with LS. However, for small identification time period T_M=300s the results of LS have approximately the same variances as for all other methods. The computation time is approximately the same as for IVA. Overall reliability was very good. One initial matrix has to be known a priori.

Depending on the number of unknown parameters in the three processes used as test cases, the effective identification methods COR, IVA and STA have shown different behavior in estimating the single process parameters. For the second-order processes I and II, both of which have 4 unknown parameters, the input-output behavior in form of the impulse responses and all parameters have been estimated accurately, each with good convergence. However, for the third-order process III, which has six unknown parameters, only the impulse response has been identified accurately, with a good steady convergence. The estimates of the parameters converged very slowly and the reason was suggested to be the very flat slopes of the multidimensional parameter surface. This also indicates that for higher order processes many different parameter sets are possible, if small errors in the input-output behavior are permitted as was shown for continuous processes by ISERMANN (1971b).

It is interesting to note, that all online methods which use the same a priori knowledge of the process and theoretically result in consistent estimates, COR, IVA, STA and 3PI, lead to about the same performance ($\sigma_{\delta g}$; $\sigma_{\delta\Sigma 1}$; $\sigma_{\delta\Sigma 2}$; $\sigma_{\delta y}$), though the methods are very different. Hence, it is suggested that identification methods using the same a priori knowledge of the process model result in about the same performance, if the laws for good identification are applied. The main differences among the identification methods then are to be seen in the kind of input signal, in the computational expense, in the overall reliability and in the assumption of specific a priori known factors.

A shortened version of a general test case for identification methods which can be used by different authors has been prepared [1]). Based on the results of this paper it seems to be sufficient, to use only one noise-to-signal ratio.

9. LITERATURE

ÅSTRÖM,K.J. and BOHLIN,T. (1966).Numerical identification of linear dynamic systems from normal operating records. IFAC-Symposium of Selfadaptive Control Systems, Teddington.

[1]) can be obtained from the authors.

ÅSTRÖM,K.J. and EYKHOFF,P. (1971). System Identification - A Survey. IFAC-Symposium on Identification, Prague 1970. Automatica pp. 123-162.

ALBERT,A.E. and GARDNER, L.A. (1967). Stochastic Approximation and nonlinear regression. MIT-Press, Cambridge, Mass.

VAN DEN BOS,A. (197o). Estimation of linear system coefficients from noisy responses to binary multifrequency test signals. IFAC-Symposium on Identification, Prague, paper 7.2.

CLARKE,D.W. (1967). Generalized-least-squares estimation of the parameters of a dynamic model. Preprints IFAC-Symposium on Identification, Prague, paper 3.17.

DURBIN,J.(1954). Errors in variables.Rev. Int.Statist.Inst., Vol22, pp. 23-32.

FRIEDMANN,B.(1954). Principles and techniques of applied mathematics, Wiley, New York.

GENTIL,S.(1972). Etude comparative de diverses méthodes statistiques d'identification de systèmes dynamiques. Thesis Université Scientifique de Grenoble.

GUSTAVSSON, I.(1970). Comparison of different methods for identification of linear models for industrial processes. Preprints IFAC-Symposium on Identification, Prague, paper 11.4.

HASTINGS-JAMES,R. and SAGE,M.W.(1969). Recursive generalized-least-squares procedure for on-line identification of process parameters. Proc. IEE, Vol. 116, pp. 2057-2062.

ISERMANN,R. (1971a). Experimentelle Analyse der Dynamik von Regelsystemen -Identifikation I. University textbook Nr. 515, Bibliographisches Institut Mannheim.

ISERMANN,R. (1971b). Theoretische Analyse der Dynamik industrieller Prozesse-Identifikation II. University textbook Nr. 764, Bibliographisches Institut, Mannheim.

ISERMANN,R. (1971c). Vergleich der Genauigkeiten und Mindestmesszeiten einiger Identifikationsverfahren (Comparison of some identification methods). Regelungstechnik und Prozessdatenverarbeitung, Vol.19, pp. 339-344.

ISERMANN,R.(1971d). Required accuracy of mathematical models of linear time-invariant controlled elements. IFAC-Automatica, Vol.7, pp. 333-341.

ISERMANN,R.(1972a). Test cases for evaluation and comparison of different identification and parameter-estimation methods. Multi solution test cases distributed by the Organizing Committee of the 3rd IFAC-Symposium on Identification, The Hague, 1973.

ISERMANN,R.(1972b). Identification of the static behavior of very noisy dynamic processes. Preprints IEEE-Conference on Cybernetics and Society, Washington D.C. and Regelungstechnik und Prozessdatenverarbeitung (1973).

ISERMANN,R. (1973a). Identification of very noisy dynamic processes using models with few parameters. IFAC-Symposium on Identification, The Hague.

IZAWA,K. and FURUTA, K.(1967). Measurement of Plant Dynamics. Bulletin of the JSME, Vol.10, No.37, pp. 68-76

JOSEPH,P., LEWIS,J. and TOU,J. (1961). Plant identification in the presence of disturbances and application to digital adaptive systems. Trans.AIEE (Applications and Industry), Vol.80, pp. 18-24.

KENDALL,M.G. and STUART,A. (1961). The advanced theory of statistics. Vol.2, Griffin, London.

KRONMAC,R. (1964). Evaluation of a pseudorandom normal number generator. Journal of the Association for Computing Machinery, Vol.11, No.3,pp.357-363.

LEE,R.C.K. (1964). Optimal estimation, identification and control. MIT-press, Cambridge, Mass.

MEHRA,R.K. (1971). On-line identification of linear dynamic systems with applications to Kalman filtering. IEEE Trans. Aut.Contr., AD-16, pp. 12-21.

REIERSOL, O.(1941). Confluence analysis by means of lag moments and other methods of confluence analysis. Econometrica, pp. 1-23.

SAKRISON,D.J. (1966). Stochastic Approximation: A recursive method for solving regression problems. Advan.Commun. Syst.2, pp. 51-106.

SARIDIS,G.N. and STEIN,G. (1968 a,b). Stochastic approximation algorithms for linear system identification. IEEE Trans.Aut.Contr.Vol.AC-13, pp.515-523 and 592-594.

SARIDIS,G.N. (1972). Comparison of five popular identification algorithms. Proc. of IEEE Conference on Decision and Control and 11th Symposium on Adaptive Processes, New Orleans.

WONG,K.Y. and POLAK,E. (1967). Identification of linear discrete time systems using an instrumental variable method. IEEE Tr.Aut.Contr. AC-12, pp.707-719.

WASAN,M.T.(1969). Stochastic Approximation. Cambridge at the University Press, Great Britain.

YOUNG,P.C. (1969). An instrumental variable method for real-time identification of a noisy process. IFAC-Congress Warsaw 1969 and IFAC-Automatica, Vol.6, pp. 271-287 (1970).

YOUNG,P.C. (1972). Lectures on parameter estimation. Summer school "Theory and Practice of Systems Modeling and Identification" in Toulouse.

Jan 20, 1973

TEST-CASES FOR EVALUATION AND COMPARISON OF DIFFERENT IDENTIFICATION AND PARAMETER-ESTIMATION METHODS USING SIMULATED PROCESSES.

For the evaluation and comparison of identification and parameter-estimation methods, the following test-cases are proposed, the chief consideration being that if several authors take the same examples, a direct comparison of different identification methods will be much easier.

For an assignment of an identification procedure the ultimate use of the estimates is very important. Therefore as one example of a final goal, the synthesis of a digital control algorithm was chosen. The test-cases are equally well suited for direct parameter-estimation or for identification methods using correlation-analysis or Fourier-analysis. [1)]

1. The Processes.

The identification task comprises the identification of the dynamic behavior of three linear, timeinvariant discrete simulated processes $G_1(z^{-1})$, $G_2(z^{-1})$ and $G_3(z^{-1})$, Fig.1. $G_1(z^{-1})$ is a process proposed by ÅSTRÖM and BOHLIN (1966), $G_2(z^{-1})$ and $G_3(z^{-1})$ and the noise filter $G_v(z^{-1})$ are derived from continuous processes.

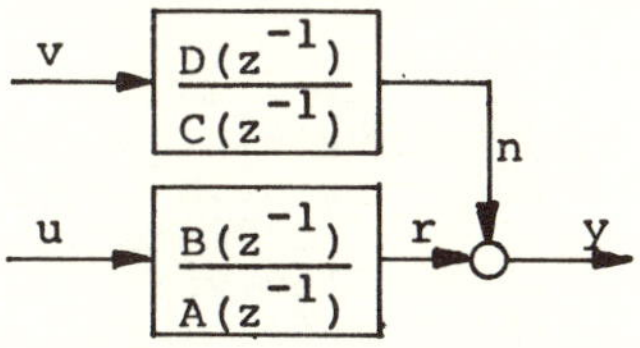

Fig. 1. Block diagram of simulated processes.

Process I: Second-order oscillating process

$$G_1(z^{-1}) = \frac{B(z^{-1})}{A(z^{-1})} = \frac{b_1 z^{-1} + b_2 z^{-2}}{1+a_1 z^{-1}+a_2 z^{-2}} \quad (1)$$

$a_1 = -1.5$; $a_2 = 0.7$;
$b_1 = 1.0$; $b_2 = 0.5$
sampling interval: T_o=2s.

Process II: Second-order nonminimum phase process

$$G_2(z^{-1}) = \frac{B(z^{-1})}{A(z^{-1})} = \frac{b_1 z^{-1}+b_2 z^{-2}}{1+a_1 z^{-1}+a_2 z^{-2}} \quad (2)$$

$a_1 = -1.425$; $a_2 = 0.496$;
$b_1 = -0.102$; $b_2 = 0.173$
sampling interval: T_o=2s.

Process III: Third-order low pass delay process

$$G_3(z^{-1}) = \frac{B(z^{-1})}{A(z^{-1})} = \frac{b_1 z^{-1}+b_2 z^{-2}+b_3 z^{-3}}{1+a_1 z^{-1}+a_2 z^{-2}+a_3 z^{-3}} z^{-d} \quad (3)$$

a) sampling interval T_o=2s:

a_1= -2.084; a_2= 1.422; a_3= -0.316
b_1= 0.018; b_2= 0.010; b_3= -0.006
d = 2

b) sampling interval T_o=4s:

a_1= -1.500; a_2= 0.705; a_3= -0.100
b_1= 0.065; b_2= 0.048; b_3= -0.008
d = 1

For process III use either T_o=2s or T_o=4s or both.

- Noise Filter: Second-order low pass process

The noise signal n(k) is generated by the noise filter

$$G_v(z^{-1}) = \frac{D(z^{-1})}{C(z^{-1})} = \frac{d_1 z^{-1}}{1+c_1 z^{-1}+c_2 z^{-2}} \quad (4)$$

a) sampling interval T_o=2s:

c_1= -1.027; c_2= 0.264
d_1= 0.0114γ

b) sampling interval T_o=4s:

c_1= -0.527; c_2= 0.0695
d_1= 0.0117γ

(γ depends on the noise-to-signal ratio η)

which is driven by a discrete white-noise process v(k) with normal distribution, zero mean and variance σ_v^2=1. For identification methods which require $C(z^{-1}) = A(z^{-1})$ one may use

$$G_v(z^{-1}) = \frac{D'(z^{-1})}{A(z^{-1})} \approx \frac{D(z^{-1})}{C(z^{-1})} \quad (5)$$

2. Identification task.

2.1. Identify the dynamic behavior of the given three processes with the following a priori knowledge:

A) • The structure, the order and the time delay of the models are known. The parameters $a_1, a_2, \ldots, b_1, b_2, \ldots$ are unknown.
• n(k) is a correlated zero-mean random variable.

[1)] Even if identification results for only one or two of the proposed processes are obtained, without synthesis of control algorithms, this will help to compare different methods more easily.

B) • It is only known that the process can be described by a linear difference equation. The order and the time delay of the process model is not known.
• $n(k)$ is a correlated zero-mean random variable.

The input signal is an arbitrary artificial signal, generated by a test signal generator or process computer and is limited by $-u_o/2 \leq u(k) \leq u_o/2$ with $u_o=1$. The test signal should be chosen as to minimize the measuring time.

2.2. Show the dependence of the errors of the estimated parameters and of the errors of the resulting impulse response on the measuring time periods T_M= 350; 1400 and 7000s for the noise-to signal ratio (with gain K)

$$\eta = \frac{n_{eff}}{u_o} = \frac{\left[\overline{n^2(k)}\right]^{\frac{1}{2}}}{u_o K} = 0.1 \qquad (6)$$

by taking the following error definitions

• error of a single parameter Θ_i:

$$\delta_{\Theta i} = \frac{\Delta\Theta_i}{\Theta_i} = \frac{\hat{\Theta}_i - \Theta_i}{\Theta_i} \qquad (7)$$

• squared relative parameter errors:

$$\delta_\Sigma = \left\|\frac{\Delta\Theta_i}{\Theta_i}\right\| = \left[\sum_{i=1}^{p}\left[\frac{\Delta\Theta_i}{\Theta_i}\right]^2\right]^{\frac{1}{2}} \qquad (8)$$

with Θ_i: exact parameter values, $i = 1,2,\ldots,p$.
$\hat{\Theta}_i$: parameter estimate
p: number of all parameters to be identified.

• r.m.s. error of the impulse response $g(k)$

$$\delta_g = \left[\overline{\Delta g^2(k)}/\overline{g^2(k)}\right]^{\frac{1}{2}}$$

$$= \sum_{k=0}^{M} \Delta g^2(k) / \sum_{k=0}^{M} g^2(k) \qquad (9)$$

with: $\Delta g(k) = \hat{g}(k) - g(k)$
$g(k)$: exact impulse response
$\hat{g}(k)$: impulse response, resulting from the identification.

In order to obtain better comparisons of the different identification methods, at least 5 usable runs with different noise data sets should be made. Then the standard deviations $\sigma_{\delta\Sigma}$ and $\sigma_{\delta g}$ of the errors are to be calculated.

Example: $\sigma_{\delta g} = \left[\frac{1}{5}\sum_{\alpha=1}^{5}(\delta_g^2)_\alpha\right]^{\frac{1}{2}}$

For presentation of the results, tables and diagrams of the same form as used in our paper should be utilized where possible.

3. Synthesis of a Control Algorithm

As example of the application of the identified models, design control algorithms for processes II and III with the same sampling interval as the processes and show the error of the closed loop controlled variable $y(k)$ caused by the errors of the identified process model.

3.1. Three-mode-controller

In order to get comparable results for the evaluation of the identified models a definite control algorithm is first proposed

$$u(k) = u(k-1)+p_o y(k)+p_1 y(k-1)+p_2 y(k-2) \qquad (10)$$

Then
a) Take the identified models for T_M = 350; 1400; 7000s.
b) Take a step disturbance $n(t)=1(t)$
c) Determine the optimal controller parameters $\hat{p}_o, \hat{p}_1$ and $\hat{p}_2$ for the identified model, minimizing

$$PI = \sum_{k=0}^{\infty} \hat{y}^2(k) \qquad (11)$$

d) Calculate the related r.m.s. error of the controlled variable

$$\delta_y = \left[\frac{\sum_{k=0}^{\infty} \Delta y^2(k)}{\sum_{k=0}^{\infty} y_o^2(k)}\right]^{\frac{1}{2}} \qquad (12)$$

with
$\Delta y(k) = \hat{y}(k) - y(k)$
$\hat{y}(k)$: controlled variable with the identified model and $\hat{p}_o, \hat{p}_1, \hat{p}_2$.
$y(k)$: controlled variable with the exact model (real process) and $\hat{p}_o, \hat{p}_1, \hat{p}_2$.
$y_o(k)$: controlled variable with the exact model and its optimal controller parameters p_o^o, p_1^o, p_2^o.

e) For each identification run and each identification time period one δ_y is obtained. Determine the standard deviation of δ_y for at least 5 runs

$$\sigma_{\delta y} = \left[\frac{1}{5}\sum_{\alpha=1}^{5}(\delta_y)_\alpha^2\right]^{\frac{1}{2}}$$

3.2. High performance controller

Develop a high performance controller taking into account
• stochastic disturbances $v(k)$ as in the identification task (or any other disturbance signals)
• manipulated variable $u(k)$ limited to $-2u_o \leq u(k) \leq 2u_o$.

Determine δ_y and $\sigma_{\delta y}$.

R. Isermann, University of Stuttgart
Keplerstr. 17, 7000 Stuttgart, W.-Germany

RESULTS FOR TEST-CASE A: COMPARISON OF DIFFERENT IDENTIFICATION AND PARAMETER-ESTIMATION METHODS USING SIMULATED PROCESSES

R. Isermann and U. Baur
University of Stuttgart
7000 Stuttgart, W.-Germany

1. TEST CASE A

The test case task [1] was sent out by the Local Organizing Committee of the 3rd IFAC Symposium on Identification and System Parameter Estimation on July 31, 1972 to those who intended to participate in the multi solution test case. The original version of the test case task consisted in two processes, a lowpass delay process and a nonminimum phase process. In a second version [2] also an oscillating process was included. The following results refer to both versions and the processes therefore are indicated by number I, II, III:

- Process I: Second-order oscillating process
- Process II: Second-order nonminimum phase process
- Process III: Third-order lowpass delay process

2. CONTRIBUTIONS TO THE TEST-CASE

Table 1 shows the authors, who submitted their results before and during the Symposium. The methods they have used as well as references and some details are also given. The symbols used are explained in [1], [2] or [3].

Following abbreviations are used:

- LS - Least squares
- GLS - Generalized least squares
- IVA - Instrumental variables
- STA - Stochastic approximation
- COR - Correlation analysis with least squares parameter estimation
- ML - Maximum Likelihood

Indices indicate the numbers of the authors given in table 1.

Table 1. Contributions to the test-case.

AUTHORS	IDENTIFIC. METHODS	REF.	OFF-LINE	ON-LINE	PROCESS I	PROCESS II	PROCESS III	SAMPL. TIME FOR PROC. III	NUMBER OF IDENTI. RUNS	ORDER KNOWN ?	REMARKS
1.DIXON	MODEL ADJUSTMENT	[4]	X	-	-	X	X	2s	1	no yes	
2.FURHT	ML,quasilinearis	[5]	X	-	-	X	X	2s	5;1	yes	
3.GUSTAVSSON	ML,off line	[6]	X	-	X	X	-	-	3	yes	clock time $\lambda=3T_o$
	ML,on-line	[7]	-	X	X	X	-	-	3	yes	" "
4.ISERMANN et al.	LS,GLS,IVA, STA,COR	[3]	-	X	X	X	X	4s	5	yes	
5.KASTLAN	GLS	[8]	X	-	-	X	X	2s	2	yes no	ord.of $G_v(z^{-1})$ P II: $\nu = 5$; PIII: $\nu = 4$
6.WITTENMARK	SELFTUNING CONTROLLER	[9]	-	X	X	-	-	-	5	yes	closed loop !

As input signal the authors 2, 4, 6 have used PRBS with clocktime $\lambda = T_o$, author 5 has used $\lambda = 2T_o$ and author 3: $\lambda = 3T_o$ (T_o, sampling time). Author 1 has used a special binary sequence.

3. RESULTS

Figures 1,2,3 and 4 show the results. Please notice the remarks in table 1 before comparing the results. Since not all authors have calculated the parameter errors $\delta_{\Sigma 1}$ respectively their standard deviations $\sigma_{\delta\Sigma 1}$, or could not calculate them because of using models with higher order than the simulated processes (GUSTAVSSON, WITTENMARK) only the impulse response errors δ_g respectively their standard deviations $\sigma_{\delta g}$ are compared.

GUSTAVSSON calculated the best possible variances of the impulse response errors $\sigma_{\delta g}$ for all 3 processes according to the Cramér-Rao lower bound, ÅSTRÖM [10]. However, it must be emphasized, that the Cramér-Rao lower bounds do not fit exactly to the identified processes, because they have been calculated

a) for white noise input u(k) (not PRBS)
b) for the model

$$A'Y(z) = B'U(z) + D'E(z)$$

with $A'=AC$; $B'=BC$; $D'=AD$

and not for the model

$$Y(z) = \frac{B}{A}U(z) + \frac{D}{C}E(z)$$

Therefore the shown Cramér-Rao lower bounds are an approximation. GUSTAVSSON first calculated the theoretical Cramér-Rao covariance matrix $\mathrm{cov}(\hat{\Theta})$ for the parameters $\hat{\Theta}$. Then he generated random variables of the parameter errors with the same covariance matrix $\mathrm{cov}(\hat{\Theta})$ and determined the standard deviation $(\sigma_{\delta g})_{C-R}$ of the impulse response error using Monte Carlo simulations of 750 different impulse responses.

WITTENMARK used another error for the closed loop behavior than proposed in [1] or [2]:

$$\delta'_y = \left[\Sigma y^2(k) - \Sigma y_o^2(k)\right] / \Sigma y_o^2(k).$$

Using the controller [9]

$$\frac{U(z)}{Y(z)} = \frac{\alpha_1 + \alpha_2 z^{-1} + \ldots + \alpha_m z^{-m+1}}{1 + \beta_1 z^{-1} + \ldots + \beta_m z^{-\ell}}$$

he obtained for process I; $\eta=0.1$; 5 identification runs; known order; $\sigma_{\delta y}=0.052$; 0.041; 0.008 for $T_M=200;700;3500$s. Since these results are valid for the output variable of the closed loop they cannot directly be compared with the results of the other authors.

DIXON [4] only calculated $\delta_{\Sigma 1}$ for process II. For one run he obtained with $T_M=175;750;3400$s:

$$\delta_{\Sigma 1} = 0.015;\ 0.017;\ 0.013 \quad \text{for } \eta = 0.1$$
$$0.017;\ 0.096;\ 0.095 \quad \text{for } \eta = 0.2.$$

4. SUMMARY OF THE COMPARISON

To obtain a realistic picture only those results should be compared for which several identification runs, at least 3, have been made and for which therefore mean squared values of the impulse response errors could be calculated. That means, for process II the results of GLS_5 and for process III of ML_2 and GLS_5 have not the same weight as the other results. Referring to the standard deviation $\sigma_{\delta g}$ of the impulse response error following results can be obtained:

4.1. Comparison of on-line identification methods

Process I and II: 4 unknown parameters

- Long identification time ($T_M=7000$s; $N=3500$ for $T_0=2$s)
  Best results: IVA_4, COR_4, STA_4, ML_3 (→close to minimum variance bound)
  Worst results: GLS_4; LS_4 (→bias)
- Short identification time ($T_M=350$s; $N=175$)
  Best results: IVA_4, ML_3, COR_4
  Worst results: STA_4, GLS_4, LS_4 (→differences of all methods are smaller than for long identif. time)

Process III: 6 unknown parameters

- Long identification time
  Best results: COR_4, IVA_4, STA_4 (→close to minimum variance bound)
  Worst results: GLS_4, LS_4.
- Short identification time
  Best results: COR_4, IVA_4, GLS_4, LS_4
  Worst results: STA_4 (→differences of all methods are smaller than for long identif. time)

LS and GLS do not converge for longer identification time, thus indicating biased estimates. The reason for the bias with the GLS method is, that in the test case task no special assumptions about the disturbance filter $G_v(z^{-1})$ have been made. By long division it follows for the filter of Process I

$$F(z^{-1}) = 1+0.47z^{-1}+0.27z^{-2}+0.08z^{-3}-0.07z^{-4} - 0.16z^{-5}-0.20z^{-6}-0.18z^{-7}-0.13z^{-8}\ldots$$

This expression obviously cannot be described by an autoregressive process of lower order.

4.2. Comparison of off-line and on-line methods

Process I:

ML_3: off-line better than on-line (→especially for short identif.time)

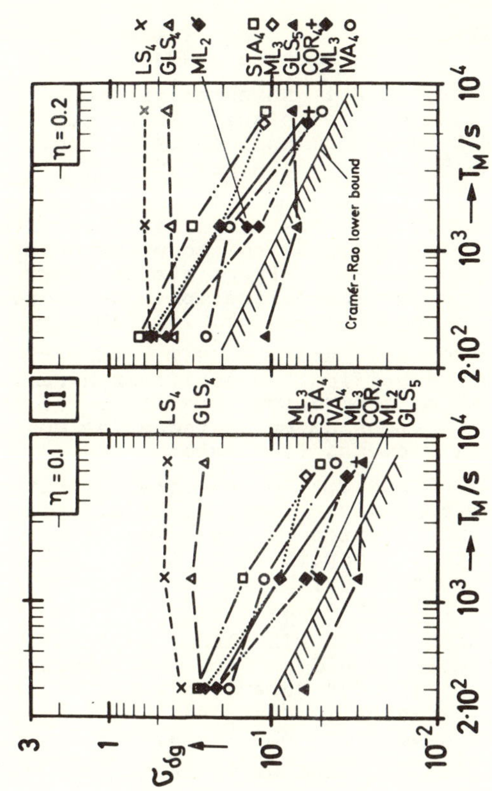

Fig.2. Impulse reponse errors for process II.

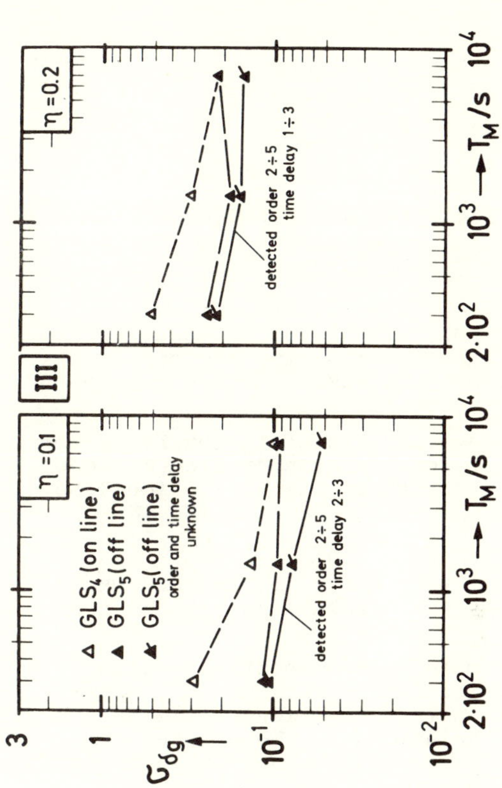

Fig.4. Impulse response errors for process III. Comparisons GLS off-line and on-line.

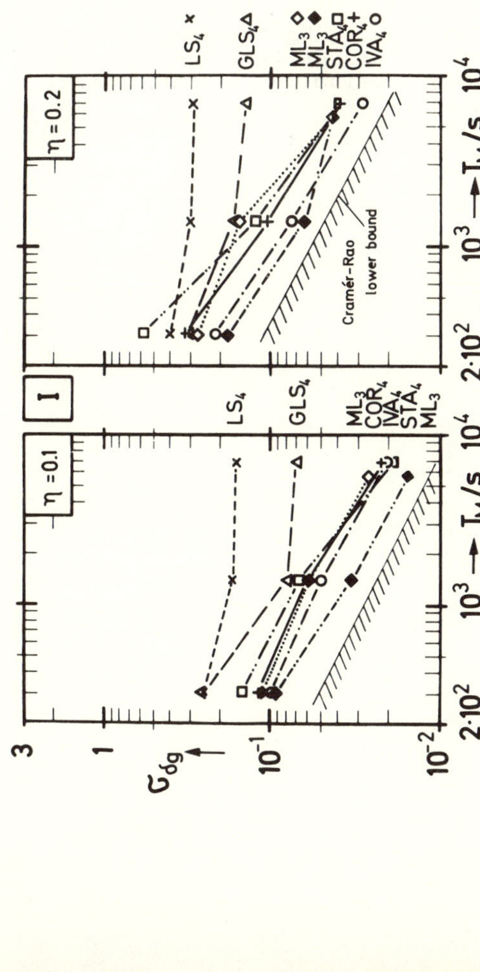

Fig.1. Impulse response errors for process I.

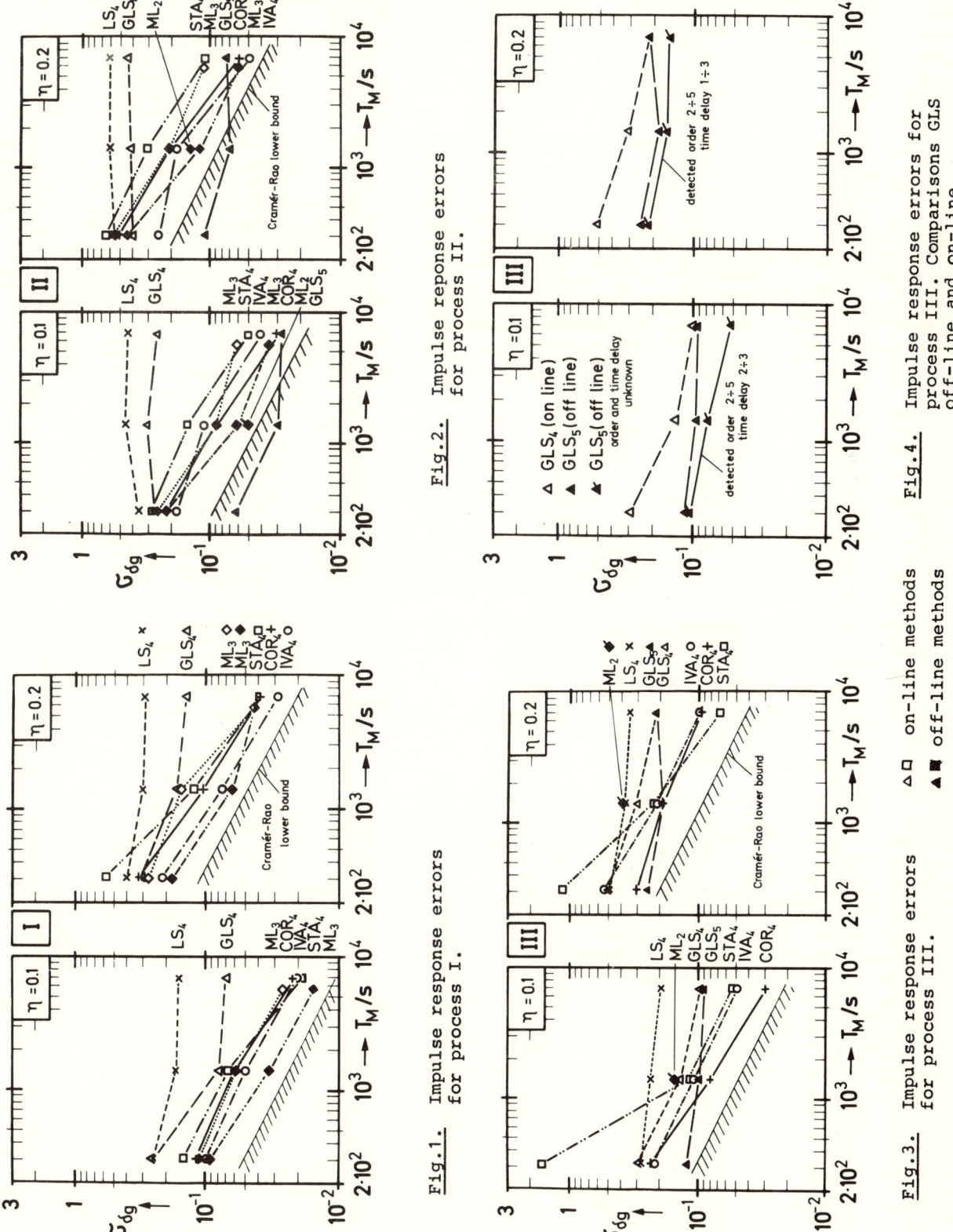

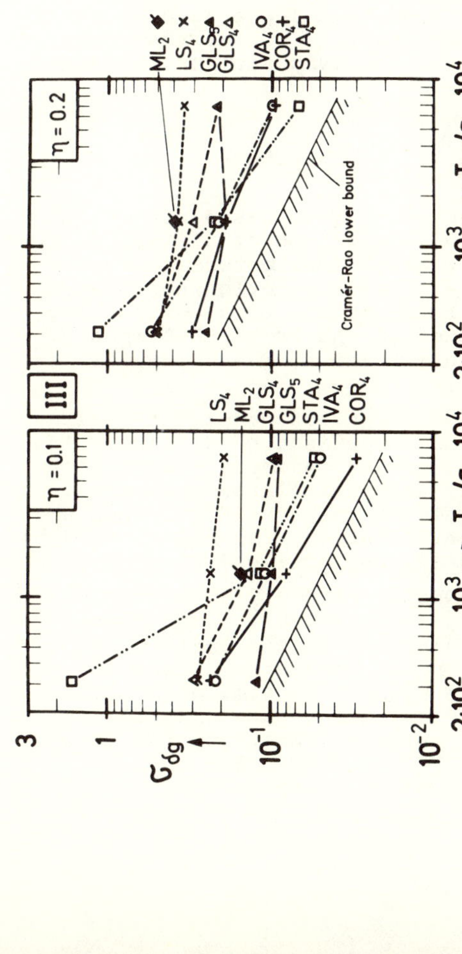

Fig.3. Impulse response errors for process III.

△□ on-line methods
▲■ off-line methods

<u>Process II:</u>

ML_3 (and ML_2): off-line better than on-line

<u>Process III:</u>

GLS_4: off-line better than on-line (2 runs only!)
(→ especially for short identific.time)

4.3. <u>Overall comparison</u>

All consistent identification methods (COR, IVA, ML, STA) lead, at least for larger identification time periods to approximately the same accuracy of the input-output model.

Off-line methods furnish in general somewhat better results than on-line methods, especially for short identification time periods.

<u>Remarks:</u>

We would like to thank all authors who contributed their solutions to this test case.

This test case will be continued also after the 3rd IFAC Symposium on Identification. Authors who would like to submit their solutions to the general comparison should send us their results for a further publication.

A second test case task will be prepared and will be send to all who have shown their interest in filling out the form distributed at the Symposium or who contact us.

5. LITERATURE

[1] ISERMANN,R.: Test cases for evaluation and comparison of different identification and parameter-estimation methods. Multi solution test cases distributed by the Local Organizing Committee of the 3rd IFAC-Symposium on Identification, The Hague, 1973, (1972).

[2] ISERMANN,R.: Test cases for evaluation and comparison of different identification and parameter-estimation methods using simulated processes. Preprints of the 3rd IFAC-Symposium on Identification, The Hague, 1973, paper E-2, pp. 1081 - 1082.

[3] ISERMANN,R.; BAUR,U.; BAMBERGER,W.; KNEPPO,P; SIEBERT,H.: Comparison and evaluation of six on-line identification and parameter estimation methods with three simulated processes. Preprints of the 3rd IFAC-Symposium on Identification, The Hague, 1973, paper E-1., pp. 1061- 1080. and IFAC-Automatica, 1974.

[4] DIXON,L.C.W.: An automatic model adjustment technique. Preprints IFAC-Symposium on Identification, The Hague, 1973, paper TS-9,pp. 973-976.

[5] FURTH,B.P.: Maximum likelihood identification of Åström model by quasilinearization. Preprints IFAC-Symposium on Identification, The Hague, 1973, paper TM-6, pp. 737-740.

[6] ÅSTRÖM,K.J. and BOHLIN,T.: Numerical identification of linear dynamic systems from normal operating records. IFAC-Symposium on Selfadaptive Control Systems, Teddington, 1966.

[7] SÖDERSTRÖM,T.: An on-line algorithm for approximate maximum likelihood identification of linear dynamic systems. Report 7308, Lund Institute of Technology, Division of Automatic Control, 1973.

[8] KOHLAS,J. and FIECHTER,A.: Identifikation von dynamischen Systemen: Schätzung der Parameter von linearen Differenzengleichungen. BBC Forschungsbericht KLR-72-24, 1972.

[9] ÅSTRÖM,K.J. and WITTENMARK, B.: On self-tuning regulators. IFAC-Automatica, Vol.9, (1973), pp. 185-199.

[10] ÅSTRÖM,K.J.: On the achievable accuracy in identification problems. IFAC-Symposium on Identification, Prague, 1967.

APPLICATIONS OF IDENTIFICATION METHODS IN POWER GENERATION AND DISTRIBUTION

R. BAEYENS and B. JACQUET
LABORELEC
Belgium

SUMMARY

The first part of this survey paper specifies the nature of processes, the purpose of identification, and the main difficulties encountered in this field.

The second part is a review of the main applications, classified according to identification methods.

Some general concluding remarks are formulated.

1. INTRODUCTION

The systematic interconnection of networks, the increase in size of base load power plants, the use of fast starting units, the more frequent start-ups of medium power plants have conferred greater importance, new dimensions, and at the same time increased complexity on control problems.

This situation requires a good knowledge of processes - generally unrealizable by one mathematical analysis only- and of their environment.

If nowadays identification presents and imperative necessity, this need is very recent and appeared very progressively. That can partially explain that few applications have furnished really conclusive results or have been used to control purposes. Another reason is that the applications of identification methods in power plants and networks often set many problems.

- processes are generally multivariable, complex and very interactive.
- they are industrial processes and it is necessary to carry out the experiments during normal operation.
- For reasons of security and efficiency, the processes will be perturbed as little as possible, and during the minimum time, the regulators often operating.
- the theorical knowledge is often limited or models are very complex and usable with difficulty.

The nature of processes, their a priori knowledge, the purpose of identification, are very important in the realization of identification. They will form the subject of the first part of the survey paper.

The second part will be a review of the main applications we know in real processes.

Some general concluding remarks will be formulated.

2. PURPOSE OF IDENTIFICATION

When identification is performed, one must clearly bear in mind the purposes it serve. In the field of power generation and transport the final goal is often to design control systems but the model can also be used in forecast problems or in improving knowlegde of processes.

2.1 A first application in control concerns the design of faster and more accurate controls to keep controlled variable transient deviations as small as possible, in order to improve operating efficiency. If, for example, one can reduce superheated steam temperature deviations, temperature set point can be increased with consequent appreciable increase in unit efficiency. If this is the case, it might be sufficient to have a linear input-output model, representative of small deviations, completed by a statistical knowledge of noise.

2.2 Designing a control system for optimal transition from one state to another - start-up of an unit, transition from one load to another one - requires a model suitable in a very large area - the model must be generally non-linear.

2.3 The purpose of identification can be a better knowlegde of a process or the checking of a mathematical model obtained with some hypothesis and simplifications. Such a model must be representative of physical phenomena.

It will be used in systems design.

2.4 A model is also necessary for prediction : determining temperature evolution in processes where a dangerous excursion is to be feared.
Overload of transformers, lines, cables - requires a generally non-linear model which must be elementary, owing to its real-time utilization.

3. NATURE OF PROCESSES EXPERIMENTAL PROBLEMS.

3.1 Thermal power plants.

During the last years, many analytical studies have concerned drum-boilers and once-through boilers.
Models representative of small deviations around a given operating point have been obtained. They are complex and show the interactive and multivariable character of plants specially of once-through boilers.
Generally, they permit a good comprehension of the behaviour, but quantitative estimation of parameters is still often unsatisfactory.

Making experiments sets some problems.

- reasons of security and efficiency can justify the realization of tests during normal operation, regulators operating.
- large changes in inputs are prohibitive.
- dynamics are slow - one or several minutes - and requires long tests - some problems due to non-stationary phenomena can exist.
- number of transducers is often insufficient specially in older plants.
- interactive natures of processes pratically excludes using natural deviations.

The most applications have concerned super-heaters, regulators generally operating (4, 5, 14, 17, 34, 48, 50, 51, 52, 89, 96, 99, 106, 107).
In a few applications the whole boiler has been considered. The required model is generally linear and representative of small deviations (1, 33, 35, 45, 55, 62, 77, 81, 105).
In the reports (1, 33 , 81).
a physical model, valid for important deviation has been obtained.
One finds an utilization of identification to control purposes in (69 , 47).

3.2 Nuclear power reactors.

Nuclear Power Reactors were intensively studied before they were actually installed. Analog and digital computer models were developed to a great extent and were described in various books and articles.

The highly interactive nature of the process requires very intensive experimental studies, but experimentation possibilities are quite limited because many transducers are connected to security devices.
Non-stationarities make also experimentation difficult.

However some experimental results have been obtained with natural noise measurements and even with injected noise. (6, 8, 10, 12, 13, 15, 16, 36, 42, 44, 48, 58, 65, 68, 71, 72, 78, 79, 80, 82, 83, 84, 86, 88, 92, 93, 95, 108).

3.3 Networks.

The operation of networks leads to several identification problems, the distinction between them being the values of time constants involved in the process.
Short term dynamics - 100 ms to a few seconds - includes all aspects of network stability.
Medium term dynamics - a few seconds to one minute - include all aspects related to the rotating energy of the network when dealing with power - frequency control.
Long term dynamics concern the thermal behaviour of transformers, cables, overhead lines, the time constants varying between a few minutes and several hours.

Mathematical models of networks, including turbine dynamics and power-frequency control have been obtained, they show the multivariable aspect of the process.
Owing to the quantity of parameters involved and the high dimensionality of the process, these models are not directly utilizable and can only give guidelines to the experimental studies.

Making experiments presents some difficulties

- desirable duration of experiments is uncompatible with the non-stationnarity of processes.
- existence of non-linearities.
- practical impossibility of exciting the process.
- natural perturbations used for identification present a narrow frequency band.
- some important variables are unmeasurable: load variations specially.

Many applications have been realized in this field - turbines (46, 60, 73).
- turboalternators (31, 56, 64, 90, 90)
- and networks (11, 23, 39, 57, 59, 74, 75, 94, 98, 102).

4. APPLICATIONS.

Applications have been classified following identification methods.

a. Deterministic methods.

b. Methods using correlation functions or spectral analysis.
These methods a and b used off-line furnish linear and non parametric models.

c. Parametric methods.

4.1 Deterministic methods.

These methods require process excitation by means of a predetermined signal : step, impulse or periodical perturbation.

4.11 Step Response Methods.

This method is the most widely used. Quite simple to implement, it requires no special equipment and can be easily interpreted. The step response may be graphically analysed. Graphs allow a frequency transfer function of the $\frac{e^{-p\tau}}{(1 + pT)^n}$ type to be fitted to the step response.
(Baeyens 5, Broida 18, Debelle et al. 26, de Larminat et al. 27, de Reuck 30, EDF 32, Quentin 76).

These methods only apply to damped processes but permit to tune directly regulator parameters.
(Chaussard 20, de Reuck 30).

More elaborated methods than graphical methods were equally used.

They compute the coefficients of a series expansion of the transfer function near the origin of the Laplace plane. Using then a Padé approximation of the series expansion, a rational form of the transfer function is obtained.
(Aubrun 3, Baeyens 4, Bidoul 14, Debelle 24, de Reuck 29, Le Couturier et al. 63, Sagaspe 85, Simoyu 89, Unbehauen et al. 96, Waha 99).
These methods require a computer.

Altough more complex, they fit well to processes with overshoot , non-minimum phase behaviour and may be adapted if the input signal is not a step.

Unhappily these methods are very sensitive to noise and have bad precision in the high frequencies due to the limited frequency band of the input signal . Worth mentioning is also the sensitivity to nonlinearities.

In order to avoid these inconvenients steps tests have to be repeated with different amplitudes. This procedure lengthens considerably experimentation time.

Other methods Fourier transform the input and output signal and then fit a rational transfer function to the frequency response.
(Bidoul 14, Iserman 50, 51, Unbehauen et al. 96, Welfonder 106, Welfonder et al.107)

These methods have been applied to superheaters (Chaussard 19, Debelle 24, 25, Debelle et al. 26, de Reuck 30, Isermann 52, 53, Jacquet 54, Simoyu 89, Waha 99) as to drum boilers (Chaussard 19, Debelle 24, Waha 99, EDF 32, Jacquet 54, Waha 104).. They were equally used to identify the hydraulic regulation loops of a turbine (Van de Meulebroeke 97).

However, the identification of the power frequency transfer function of a power network by step response gave unsatisfactory results.
(Cuenod et al. 23, Preminger et al. 74).

Thermal dynamic behaviour of power lines has also been studied by step responses. (Delcomminette 28).

Step responses have been used also in nuclear power plants (Barzacchi et al.10, Bliselius et al. 15) and gave good results for the power reactivity and coolant temperature reactivity transfer functions.

The multi-step method (Menahem 66).

This method uses a sequence of steps with different amplitudes applied at different instants. The major advantage is a better use of testing time.

The identification uses a least-squares method. The choice of the input signal sequence is important since it can reduce computing time. It can also be chosen in order to shape the frequency characteristic of the input signal.

4.12 Impulse methods.

These methods have been very rarely used.

Certain papers (Bliselius et al. 15, Isermann 53, Unbehauen et al. 96) relate the application of rectangular pulses of a fixed duration, as these pulses are repeated periodically, this method will be considered as a frequency method.

4.13 Frequency methods.

The application of a sinusoïdal input at different frequencies allows a determination of the frequency response.

As a boiler is a very slow response process, this method would be too long to be practical. Also, the amplitude of the sinusoïd must be important unless correlation techniques are used.
Nevertheless certain authors (Isermann 50, 52) identify the parameters of a superheater model by measuring the frequency response of the process at two different frequencies with good precision.

Sometimes periodically rectangular pulses are used as it was the case of a superheater (Isermann 53, Unbebauen 96, Welfonder 106) and a nuclear power reactor (Bliselius 15).

Also a periodical signal containing a certain number of harmonics of the fundamental have been used (Bliselius 15) this method can in certain cases compensate non linear effects.

In the case of a drum-boiler Caseau et al. used a deterministic signal with a specially studied spectral shape; this method gave good results. (112)

Resuming, it can be said that these methods have been widely used in the power generation industry to identify the dynamic behaviour of superheaters and nuclear reactors.These methods apply to linear monovariable processes with little interaction. The regulation loops must be unoperating and a good signal to noise ratio is needed. This is of course a serious drawback.On the other side these methods are very simple and easy to apply.

4.2 Correlation techniques and spectral analysis

4.21 Identification methods using correlation functions or spectral analysis provide a linear input-output model. The results are given in the form of weighting responses at different times or transfer functions at different frequencies.
These methods have been largely used in power plants and networks for the following reasons :

- a linear black-box model is adequate for many applications,
- they don't require an important a priori knowledge of processes,
- many systems are continually perturbed by random noise utilizable for identification of open loop systems,
- identification during normal operation with regulators operating, is possible, provided that noise is injected at the inputs,
- they provide statistical estimation of disturbances on regulated variables - important for control problems,
- they give an estimation of coherency between inputs and outputs.

On the other hand, they are direct methods and not interative, the results are not given in a mathematical form, and therefore not easily applicable to further uses, the experience time is generally long for power plants and networks if one doesn't use special noise.

4.22 Natural disturbances.

When using natural disturbances one identifies global process, including regulating loops and internal feedbacks. To identify the physical process, only the regulators must be unoperating. Disturbances must be random and have a sufficiently large spectrum.

They have been specially used in networks where load variations are approximatively random.

The main applications have concerned the determination of the power-frequency transfer function (Basaldella 11, Farmer and al 39, Koszelnick and al 59, Quazza 75, Trybula and al 94).
The basic difficulty here is the impossibility of measuring load so as to obtain the transfer function as a ratio of spectra.
Several methods have been proposed to circumvent this difficulty, and specially an indirect method using relations between spectra and interspectra of frequency and power exchanges of interconnected zones.
Some shortcomings still prevent obtaining quite satisfactory results - the contrast between the need for long records and the non-stationary character of the power systems, the existence of many nonlinearities,the small amplitude of higher

frequency components in the load variations (Quazza 75).

Correlation functions have been used to determine active and reactive power variations with voltage in different substations and at different times (Waha 102) primary regulating energy (Waha) load dispatch factor (Farmer 39) frequency bias(Vitek 98).

Stanton (90, 91) has used natural disturbances for identification of turbo-alternators as a black-box model with two inputs and two outputs.
In (91), he has considered a machine operating in parallel with an interconnected system.
The machine was not subject to load-frequency control during tests, the voltage regulator was practically inoperative.
During one test the governor was blocked - identification of turbo alternator in open loop - during two tests, it was operating, identification concerned the whole system including governor.
These data were also used by Jenkins (56) (Spectral analysis) Lindahl (64) and Desai (31).

Power plants are very interactive processes and cannot be identified by using normal operating data.
On the other hand, many applications have concerned nuclear reactors - they have specially furnished the transfer function at zero power, power to reactivity transfer function at power and the transfer function of the steam generator.
(Ball and al. 8, Bastl 12, Bliselius and al 15, Boardman 16, Eurola 36, Godfrey 42, 44, Kolb and al. 58, Miida and al. 68, Pluta 72, Rajagopal and al. 78, Randall and al. 80, Ricker and al. 82, Schultz 86, Seifritz and al. 88, Thie 93, Uhrig 95, Wilkie 108).

Resuming, some interesting results have been obtained when using natural noise but many conditions must be satisfied to obtain very good results :

- large spectrum of noise,
- possibility of long experiments to obtain suitable estimation of transfer functions, good stationnarity of processes,
- inoperating regulators if the identification concerns only the physical process.

4.23 Forced perturbation.

When using forced perturbations, it is theoretically possible to identify a process, regulators operating.Choice of their amplitude must result of different considerations :

- the coherency between input and output must be sufficiently high in the whole spectrum considered,
- but amplitude must stay moderate for reasons of exploitation and also for validity of linear models.

Ideal spectrum of deviations depends in fact on noise affecting outputs variables, on feedback loops but generally, injected noise presents an uniform spectrum in the concerned area and is a binary, ternary or multi-level noise. Multi-level noise reproduces approximatively normal sollicitations and seems to be more adequate to obtain a linear model representative of small deviations.
Nevertheless, binary noise is more used, owing to facilities of mathematical analysis . As for natural noise, injected random noise requires a long duration experiment in power plants where dynamics are rather slow.

Random noise has been used by Laurans and al (62) for identifying a drum-boiler.
The regulators were operating,uncorrelated, noise was injected at different inputs at the same time, the experiment time was five hours.
Two state random disturbances and spectral analysis have been used by Boardman (17) to determine the dynamic characteristics of superheaters. He made different experiments, the duration of each experiment was between 12 and 48 hours depending upon the dominant time constant of the response being studied and the effectiveness of the disturbance. Dynamics of once-through boiler has been determined by using two-level noise, regulators operating (Jacquet 55), experiment time was six hours.

Use of pseudo-random-binary sequences (P.R.B.S) reduces the experiment time. P.R.B.S. is a deterministic, periodic signal characterized by a correlation function simular to the white noise of a limited bandwith. Its period is choosen a little greater than the settling time of the process.With P.R.B.S. one forsakes the statistical aspect in favour of the deterministic aspect.

During the last years, P.R.B.S. has been largely used for identifying nuclear reactors.
Bliselius and al. have obtained satisfactory results on a heavy water moderated reactor. The transfer functions confirmed the theoritical analysis (15).
Kolb and al. applied this method to a PWR reactor and obtained by model fitting the reactivity coefficients but with low accuracy (58).
Rajagopal obtained also the reactivity coefficients by the same method (78, 79).

4.24 General considerations on correlation and spectral analysis.

4.241 Field of applications.

Correlation and spectral analysis can be theoretically used to identify multivariable systems even with the regulators operating.

Owing to the limited number of applications on complex and multivariable processes, it is not possible to specify nowadays their exact domains of application. Some very promising results have been obtained in various fields.

4.242 Accuracy.

Some expressions have been formulated to calculate confidence intervals of transfer functions in function of coherency and number of degrees of freedom.
Boardman gives in (17) confidence intervals for transfer functions of a superheater.
Laurans and al. estimate in (62) that error estimations generally furnish rather pessimist results and use a continuity criterium of transfer function phase in function of frequency.

4.243 Correlation or spectral analysis.

Methods using correlation functions or spectra have been both largely used. They don't set the same problems and don't necessary provide the same results, but it is not possible, when considering the applications, to compare the results.

The utilization of either seems to be dependent of computation facilities and also of the nature of used disturbances.

In spectral analysis, the Fast Fourier Transform reduced considerably the computation time, it is generally applied when natural disturbances are used.

When using binary signal, correlation functions are largely used owing to the simplifications introduced in the convolution equations.
With P.R.B.S., one can envisage an easier determination of the impulse weighting in the parametric form by minimization of a mean square error criterion on its period. (Labarrière 61).

4.244 Use of an a priori knowledge.

These methods necessitate determination of a great number of parameters. It is possible to use an a priori knowledge of the process to reduce this number and obtain a better estimation.
In (106, 107) Welfonder has applied two periodical test signals to identify by correlation methods the parameters of a superheater.
The a priori knowledge can be used for a better determination of the injected noise or/and of the transfer function.

4.3 Parametric methods.

Identification by parametric methods determine unknown parameter of the transfer function or of the differential equations system describing the process.

This assumes the model of the process known at start by analysis of the principles governing the process or by experimentation. These models need not to be linear.

These methods are mostly used to gain insight into the process or to make sufficiently precise models available in order to predict the reactions of the process in certain circumstances or also to determine an optimal control strategy.

It must be recognized that these methods are, up to now, not frequently used in the power generation domain.

4.31 Least square error methods.

This relatively simple method can be performed on-line by using and analog computer as it was the case with Mizutani and al. (69) and allowed to identify certain parameters of the boiler and an adoptive control.

In this case, they identified the time constant between the feedwater flow and the drum level and also the time constant relating the coal supply to the drum pressure.

This case seemed to give good results in spite of the fact the derivatives of certain signals had to be calculated.

A least square error method by Bereznaï and al. was also applied to a nuclear reactor (13).

Eklund (33, 34, 35) determined the parameters of a simple non linear model of a drum boiler with good success.

In the study of thermal contraints of a turbine in starting condition, coefficients of a model of the turbine walls were determined by such a method (46) (60) (73). This work was performed by Gorelik, Duel, Pokhoriler and al.

Borget and al. (111) used a very elegant method to determine impulse responses of a drum boiler. They used an orthogonal polynomial expansion of the impulse response. (Laguerre polynomials).

The weight of every polynomial was determined by a least square error method.

This method leans itself to on line processing and is likely to be applied also to a power network. It is simple to apply but convergence problems arise when the process possesses complex poles.

4.32 Identification using modern control theory.

These methods have been tremendously studied these last thirteen years, but their applications are rather limited in the electricity production.

There are many reasons for this ; the principal seems to be the complexity of a production unit needing very much computing time.

Furthermore, few algorithms have been studied on real processes limiting automatically their potentiel use.

Classical and spectral methods could seem then much more attractive. However these methods cannot always furnish the information sought.

If a quantitative physical insight into the process is sought or if the non-linear behaviour of the process has to be studied modern methods can provide us this information. Several methods have anyhow been used.

4.321 Maximum likelihood method.

This method applied only to linear models Eklund (33) has used this method to identify the transfer functions of a 160 MW monobloc (drum boiler and turbine).

This method is interesting because it is possible to determine the "optimal" order of the system and provide in the same time a description of the nature of the noise perturbing the process.

Identification has been achieved by injection of pseudo random binary noise, regulation loops in manual.

Multivariable variable system identification has been tried too but computing times were excessive.

Gustavson (48) and Olsson (71) applied it to a nuclear reactor.

Lindahl and Ljung (64) identified the active power/frequency transfer function of a turbo-generator set by this method, but the "order test" didn't seem to give satisfaction.

Irving - Logeay and Roquefort (49) also used the maximum likelihood method to identify a turbo alternator set with his voltage control loop by introducing pseudo random binary noise on the voltage set point.

4.322 Identification by solving a two boundary value problem.

Desai (31) identifies the transfer function of a 50 MW turboset connected to a power network by measuring the natural random fluctuations with this method.

A first order model for the open loop system gave good results. A higher order model for the closed loop system gave lesser good results.

This seems to indicate that this method can only be applied to low order systems.

4.33 "Model" method.

A minimisation of structural distance between the process to be identified and its model is used in a method described by Richalet and al. (81).

This method whose advantages are :
- its iterative nature,
- the direct useful form of the results,

- performing computing algorithms,
- non-linear models can be used too,
- on-line implementation of the method is possible and can be used for multivariable processes,

has been used to obtain impulse responses of a boiler (Richalet 81, Quentin 77) and a non-linear model of a turbojet (81).

4.34 Quasilinearisation.

Quasilinearisation has been used to identify parameters of a non-linear model of a power transformer by Germay and al. (40).

This model describes the thermal behaviour of the cupper massa.

This model had to predict the cupper temperature in order to keep it in due limits by unloading the transformer in time.

Quasilinearisation provided results which could predict the cupper temperature with a better than 5°C error. However simple linear discrete models behaved practically as good as the non-linear model.

4.35 Minimum variance estimation of parameters.

This method, an extension of linear Kalman filtering, has been used to identify reactivity coefficients of nuclear power reactors by Habegger and al (6).

This method was equally used in state estimation of a power network by Wolters (110) and many other investigators.

5. COMPARISON OF DIFFERENT METHODS FOR DETERMINATION OF LINEAR MODELS.

Some authors have used different methods to identify a linear process.

5.1 Welfonder (107) has determined the transfer function of a superheater by five methods. The regulator loops were opened during the experiments.

1) Identification by step responses - 5 steps of 10 minutes each and determination of transfer function as ratio of the Laplace transforms of the step responses and the excitation.

2) Identification by impulse responses - pulse duration ≃ 0,4 - 0,7 times the rise time of the step response. 5 impulses - experiment time : 73 min.

3) Identification by rectangular oscillations and direct frequency response measurement. 13 frequency response points over 3-6 periods experiment time : 6 h.

4) Identification by correlation method and two periodical test signals. This method uses the a priori knowledge of the process, the parameters of the model are determinated from the two results - experiment time : 4 h 1/4.

5) Identification by correlation method and binary random test sequences experiment time : 4 h 1/4.

The different methods have provided good results specially the method using the a priori knowledge of the process.

To identify such a linear process - regulators loops being open the deterministic methods using step or impulse functions are valid, they are easy to use, the amplitude of the sollicitations can be rather important to eliminate noise effects.

Correlation methods using P.R.B.S. don't produce better results in such circumstances.

5.2 Gustavsson (48) has identified a superheater by two methods :
- prior knowledge fitting,
- maximum likelihood.

The input was of pulse type.
The prior knowledge fitting method has produced models of higher order than the maximum likelihood method.

5.3 He has also measured the dynamic characteristics of a nuclear reactor. The process was disturbed by P.R.B.S. signals on the reactivity and the nuclear power was measured.
Three methods have been used :
- Least squares method,
- Maximum likelihood method,
- Spectral analysis.

The maximum likelihood has produced reasonable low order models but is rather time consuming.
The least squares method produced models of higher order.
The spectral analysis has given rather bad estimates for noisy data.

5.4 Irving and al have used for voltage regulator identification, maximum likelihood method, and spectral analysis (49).

The principal conclusions of this comparison are :

- Spectral analysis identification method has an easier realisation and a relative low computing time but supplies less precise results because of the large number of parameters to be identified.

- Maximum likelihood method is more complicated and requires larger computing times but gives directly model parameters.

They estimate that these two methods are complementary : spectral analysis method may rough out the data in a preliminary treatment time. In a second stage, use of the maximum likelihood method with identified system order approximatively deduced from the first model.

5.5 Bliselius and al. have used three methods for transfer function measurements of a nuclear reactor:step function, trapeze waves and pseudo-random binary sequences (P.R.B.S) (15).

- A frequency analysis of the step responses have not give satisfactory results for frequencies above 0,01 Hz. Non-linearities and reactor feedbacks with long time constant may have a negative effect on the step measurements.

- Trapeze waves have provided good accuracy in the frequency analysis but the measurements are time consuming and there might be some change in reactor condition between measurements at different frequencies.

- The cross correlation method using P.R.B.S. has been an accurate and practical method for obtaining transfer functions over the frequency range of main interest. Small perturbations can be applied which reduces the influence of non-linearities.

5.6 There is no general response to the answer :
Which method of identification is to use ?
The choice of the method depends on many factors :
- purpose of the model,
- nature of the required model - parametric or non-parametric, linear or non-linear.
- nature of the process,
- possibility of opening the regulators loops,
- possibility of sollicitating the process,
- a priori knowledge of the process.

6. Some general concluding remarks.

6.1 Concerning the results of identification.

The number of applications to power plants and networks is not yet very high. Obtained results are generally encouraging but conclusive results are still scarce. Publications relating their utilization to concrete realizations,specially the control,are few.

6.2 Concerning the necessity of identification

Many reasons justify the necessity of identification. Basic requirements of quality of service and economy of electric power supply impose a better control of networks and power plants. To this end, often the user will be much more interested in global simple models obtained with minimum effort, and easy to use rather than complicated, precide models obtained with great efforts.

In this field identification will be useful in two domains :
- Physical investigation of several still obscure parts of the processes involved for design purposes.
- Obtainment of simple global models of the multivariable process. These models have to be simple if they are to be useful for control purposes.

Nowadays, hierarchical control of the boiler and turbine to meet the requirements of the power network is an important objective to many people.

For the network control real time identification and estimation methods will be applied in order to avoid overload of power devices and to ameliorate the quality of service.

6.3 Concerning the processes.

Identification methods are difficult to apply on account of the highly multivariable nature of many processes and their non-stationnarity.
They are industrial processes and generally the experiments must be realized during normal operation, with regulators in service.
Production units don't always possess adequate instrumentation for the sake of identification (response times of many

transducers are slow).

6.4 Concerning the methods.

The classical methods providing linear black-box -models have been the most widely used because they are easier to apply, and linear global models were sufficient. At present there is a great need to use more elaborate methods which can provide parametric and non-linear models. But they have not yet been widely used because of high computional requirements and difficulties of application.

There is no universal method to advise for all the applications. The choice of the method depends on the purpose and the process.

Comparison between different methods are rare it would be highly desirable to have comparisons available.

6.5 Concerning future developments.

Multivariable methods have to be developed and real time applications of identification methods must be investigated.

Much more start-ups and big load variations are demanded to intermediate power units, this necessitates the study of non-linear models. Modern control theory methods will find here an important domain of applications.

There will be also a need for knowledge of the requirements to make the different identification methods work.

The practical important methods must be investigated in order to determine their domain of application.

Another important item is the development of more precise and reliable transducers in order to meet identification requirements.

The instrumentation of the power units must be developed.

6.6 Concerning this survey.

Bibliography is certainly incomplete.
We have tried to gather as much information as we could but as in many industries many works in this field are not always published.

Several methods deserved maybe more place, some were even ignored.

Anyhow we have tried to describe some of the most used identification methods in the electric power systems.

The help of Mr. Waha is gratefully aknowledged and we thank also many institutions which provided material for this survey.

REFERENCES

Signification of the symbols in parentheses

1. Conventional power plant boilers.
2. Nuclear reactors.
3. Power networks.

A. Identification methods in general.
B. Step, impulse and frequency responses.
C. Correlation and spectral analysis-natural noise.
D. Correlation and spectral analysis-injected noise.
E. Parametric identification.
F. Direct synthesis of the regulation loops.
G. Comparison of identification methods.

-Åström K.J.
1. Modelling and Identification of Power System Components (1, 2, 3, E).
Symposium Brown-Boveri.
Real-time Control of Electric Power Systems (Editor E. Handschin) 1972.

-Åström K.J. - Olsson.
2. Final report for project process control (1, 2, E).
Report 6919 - Lund-Institute of technology.

-Aubrun M. - Himbert C.
3. Méthode temporelle d'identification des systèmes linéaires (1, B).
C.R. Acad. Sc. Paris t. 268 (Avril 69).

-Baeyens R.
4. Méthode des surfaces (1, B)
Laborelec 2/4814.
5. Méthodes graphiques pour la détermination des fonctions de transfert d'un processus-Abaques du deuxième ordre (1, B).
Laborelec 2/401/26 - 2/4992.

-Bailey R.E., Habegger L.J.
6. Minimum variance estimation of parameters and states in nuclear Power systems (2, E)
IFAC - Warszawa 1969.

-Balakrishnan A.V. - Peterka V.
7. Identification in automatic Control systems (A).
IFAC - Warszawa 1969.

-Ball R.M., Batch M.L.
8. Measurement of noise in three Pressurized-water reactors (2, C).
Noise analysis in nuclear systems symposium Florida 1963.

-Barraud A. - de Larminat Ph.
9. Process identification using its weighting sequences matrix (1, E).
IFAC Düsseldorf 1971.

-Barzacchi - Giordano - Santese.
10. Synthesis of a non-interacting Control System for a Nuclear Power Plant.
IFAC Düsseldorf 1968 (2, B).

-Basaldella F., Di Caprio U.
11. Use of correlation techniques for the estimation of dynamical parameters of an electrical power system in the normal operation (3, C).
IFAC The Hague 1973.

-Bastl W.
12. Korrelationsverfahren in der Kernreaktor-messtechnik (2, C, D).
Interkama 1965.

-Bereznaï - Sinha.
13. Adaptive Control and identification of nuclear reactor (2, E).
IFAC Prague 1970.

-Bidoul J.
14. Méthodes numériques pour l'identification des processus et la synthèse de leurs commandes automatiques (1, B).
ULB 1970.

-Bliselius P.A., Vollmer H., Akerhielm F.
15. Experimental and theoritical dynamic study of the Agesta Nuclear power station (2, B, D).
AE - 376, 1969.

-Boardman F.D.
16. Noise measurements in the Dounreay Fast Reactor (2,C).
Noise analysis in nuclear systems symposium Florida 1963.

-Boardman K.D.
17. The measurement of the dynamic behaviour of a superheater by random signal testing (1, D).
IBRA Bruxelles 1966.

-Broïda V.
18. L'extrapolation des réponses indicielles apériodiques (B).
Automatisme n° 3, mars 1969.

-Chaussard R.
19. Utilisation des méthodes d'analyse indicielle et du critère de Naslin dans le cas des centrales thermiques (1, 3, F).
Automatisme sept. 64.

-Chaussard - Grauvogel - Davoust.
20. Rational adjustement of the controls of a thermal power station (1, 8, F).
IFAC London 1966.

-Cuenod M. - Sage A.
21. Comparison of some methods used for process identification (A).
IFAC Prague 1967 (survey paper).

-Cuenod - Durling - Valisalo.
22. Analysis of random process of hybrid computers (3, C).
5th International Analogue Computation meeting - Lausanne 67.

-Cuenod M. - Quazza G.
23. Dynamic statistical analysis of electric power systems control (3, C).
Automatisme Tome XIV n° 3 Mars 69.

-Debelle J.
24. Détermination de la transmittance d'un système linéaire à partir de sa réponse indicielle (1, B).
Revue A, IX, (1967).

25. Réglage automatique des générateurs de vapeur (1, B, D).
Chance 1968.

-Debelle J. - Waha J.P.
26. L'analyse des processus et son application à la détermination des fonctions de transfert des chaudières (1, B-D).
AIM Liège 1970.

-de Larminat - Mezencev.
27. Méthodes d'identification des systèmes apériodiques pour l'étude des réponses impulsionnelles (B).
C.R. Acad. Sc. 6 fév.67.

-Delcomminette P.
28. Mise en valeur de la capacité effective de transport des lignes, câbles souterrains et transformateurs de puissance (4, B).
Conférence SRBE 1972.

-de Reuck J.P.
29. La détermination des paramètres échantillonés d'un processus non oscillant par identification à partir des surfaces d'erreur indicielles (B).
ULB - Juin 1970.

30. Controller up-dating and process parameter estimation by straight-forward graphic step response analysis (1, B, F).
IFAC Prague 1970.

-Desai.
31. Estimation of turbo-alternator transfer functions by variational theory using normal operating data non sequentially (3, E).
Internat. J. Control, 14 n° 3 Sept. 71.

-E.D.F.
32. Notes sur la détermination expérimentale de transmittances dans les générateurs de vapeur + Abaques (1, B, F).
EDF 1964 - 1966.

-Eklund K.
33. Linear drum boiler-turbine models (1,3,E).
Lund Inst. Aut. Cont. 7117, 1971.

34. A comparison of a drum boiler-turbine model to measurements and models obtained by identification (1, E).
IFAC Paris 1972.

35. Identification of drum-boiler dynamics (1, E).
IFAC The Hague 1973.

-Eurola T.
36. Reactor-noise experiments on Halden Boiling water reactor (2, C).
Noise analysis in nuclear systems symposium Florida 1963.

-Eykhoff P. - van der Grinten P.M.E.M., Kwakernaak H., Veltman B.
37. Systems Modelling and Identification (A).
IFAC London 1966.

-Eykhoff P.
38. Process parameter and state estimation (A).
IFAC - Prague 1967 - Survey paper.

-Farmer - Myerscough - Ashmole.
39. The determination of power system response to frequency and load fluctuations (1,3,C).
PSCC Grenoble 72.

-Germay N.
40. Echauffement des transformateurs (3, E).
Descartes IE/SM/1/70 Janvier 70.

-Giras, I.C. - Mutafelija, B.A.
41. Power plant parameter identification (1, 2, E).
J.A.C.C. 1969.

-Godfrey K.R.
42. The theory of the correlation method of dynamic analysis and its application to industrial processes and nuclear power plant (2, D).
Measurement and Control vol.2 May 69.

43. Some difficulties encountered in the pratical application of pseudo-random sequences for dynamic analysis (2, D).
Conference on Industrial applications of dynamic Modelling, Durham 1969.

44. The application of pseudo-random sequences to industrial processes and nuclear power plant (2, D).
IFAC, Prague 1970.

-Godin P. - Davoust G.
45. Identification par entrées-sorties de systèmes linéaires multivariables (1, D).
IFAC - Düsseldorf oct. 1968.

-Gorelik - Duel.
46. Approximate equations for turbine warm up (1, E).
Thermal Engineering 1968 n° 2.

-Grumbach - Robinson - Sato - Versluis.
47. Adaptive control of the neutron flux distribution based on on-line process identification (2, E, F).
Enlarged Halden Programme group meetings on computer control.
Loen Norway 1972.

-Gustavson I.
48. Comparison of different methods for identification of linear models for industrial processes (1, G).
IFAC Prague 1970.

-Irving E., Logeay Y., Roquefort Y.
49. Power Grid Network Identification (3, D, E)
PSCC Grenoble 1972.

-Isermann R.
50. Optimal series of determined test-signals to measure the dynamics of plants under small disturbances (B).
IFAC Prague Juin 1967.

51. Experimentelle analyse der dynamik von regelsystemen (A).
Bibliographisches Institut - Zurich 1971.

52. Mathematische modellen für das dynamische Verhalten dampfbeheizter Wärmerubertruger.
Regelungs technik Heft 1 (1970) (1, B).

53. Messung der Frequenz gange eines Dampfuberhitzers durch Eingäbe von rechteckschwingungen (1, B).
MSR 9 (1966).

-Jacquet B.
54. Etude de la dépression au foyer de la chaudière de Rodenhuize 2 (1, D).
Laborelec 2/4730 (1968).

55. Essais d'identification d'une chaudière à circulation forcée (1, B, D).
Laborelec (à paraître).

-Jenkins G.M. and Watts D.G.
56. Spectral Analysis and its applications (C, D).
Holden Day San Francisco (1968).

-Kohlas.
57. Parameter estimation applied to Power-lines.
IFAC The Hague 1973.

-Kolb M.,Raschti M.A., Robinson E., Weber A.
58. Noise analysis and pseudostochastic perturbation techniques for reactor parameter estimation.
IAEA/SL - 168.

-Koszelnik M., Malkiewicz J., Trybula St.
59. A method to determine the transfer function of power systems (3, C).
IFAC Warszawa 1969.

-Kozlov.
60. A dynamic model of the heating of the casing wall in a steam turbine (1, E).
Thermal Engineering n° 9, 1969.

-Labarrère - Roche - Gimonet - Krief.
61. Identification method for multivariable systems derived from correlation methods (1, D).
IFAC Prague 1970.

-Laurans - Labarrère - Menahem.
62. Sur l'identification des systèmes industriels par la méthode de corrélation. Application à l'identification d'une chaudière de centrale thermique (1, D).
IFAC-IFIP Toronto 1968.

-Le Couturier - Pouliquen - Quentin.
63. Aperçus sur une méthode d'identification numérique des systèmes industriels à partir de leur réponse indicielle (B).
Automatisme, Tome XVI Mai 71.

-Lindahl - Ljung.
64. Estimation of Power Generator dynamics from normal operating data (3, D, E).
IFAC The Hague 1973.

-Lipinski - Vacroux.
65. Parameter identification in a non linear reactor system (2, E).
IEEE Trans. N.S. Fev. 70.

-Menahem M.
66. Process dynamic identification by the multistep method (1, B).
IFAC Warszawa 1969.

67. Méthodes expérimentales, anciennes et nouvelles, d'analyse dynamique des systèmes industriels (A).
Mesucora Paris 1967.

-Miida J., Sumita K., Kuroda Y.
68. Noise analysis of nuclear reactor systems in Japan (2, C, D).
Noise analysis in nuclear systems symposium Florida 1963.

-Mizutani K. - Kitami T.
69. On-Line real-time identification of parameters of thermal power plant (1, E).
IFAC Prague 1970.

-Nieman R.E., Fisher D.G., Seborg D.E.
70. A review of process identification and parameter estimation techniques (A).
Int. J. Control 1971 vol.13, n° 2.

-Olsson G.
71. Modelling and identification of nuclear power reactor dynamics from multivariable experiments (1, E).
IFAC The Hague 1973.

-Pluta P.R.
72. Preliminary Results of Vallecitos Boiling Water Reactor Noise Analysis (2, C). Noise analysis in nuclear systems symposium Florida 1963.

-Pokhoriler - Katsnel'son - Vikulov.
73. Using a model to determine the heat transfer coefficient and thermal stresses in turbine elements (E). Thermal Engineering n° 9, 1969.

-Preminger J. - Park G.L.
74. Analysis of dynamic stability of a power system under deterministic load changes (3, B). IFAC Warszawa 1969.

-Quazza G.
75. Control problems in electric power systems (3). IFAC Warszawa 1969.

-Quentin J.F.
76. Quelques perfectionnements apportés à la méthode d'identification de V.Strejc (B). Automatisme mai 67.

-Quentin J.F., Testud J.L., Richalet J.
77. Identification of thermal power plants (1, E). IFAC The Hague 1973.

-Rajagopal V. - Gallagher J.M.
78. Some applications of dynamic measurements in pressurized water reactor nuclear power plants (2, C, D). IEEE Trans. N.S. - 14 n° 2 April 67.

-Rajagopal V.
79. Reactor-noise measurements on Saxton reactor (2, C, D). Noise analysis in nuclear systems symposium Florida 1963.

-Randall R.L. and Griffin C.W.
80. Application of power spectra to reactor-system analysis (2, C). Noise analysis in nuclear systems symposium Florida 1963.

-Richalet J., Rault A., Pouliguen R.
81. Identification des processus par la méthode du modèle (E). Gordon and Breach 1971 (book).

-Ricker C.W., Ery D.N.,Mawn E.R.,Hanauer S.H.
82. Investigation of negative reactivity measurement by neutron fluctuation analysis (2, C). Noise analysis in nuclear systems symposium Florida 1963.

-Roggenbauer H.
83. Real-time nuclear power plant parameter identification with a process computer (2). IFAC The Hague 1973.

-Roman, Hsu, Habegger.
84. Parameter identification in a non linear reactor system (2, E). IEEE Trans. N.S., Feb. 71.

-Sagaspe J.P.
85. Principe d'une méthode d'identification d'un système linéaire à partir de sa réponse impulsionnelle (B). C.R. Acad. Sc. Paris t.272 (juin 1971).

-Schultz, M.A.
86. Shutdown reactivity measurements using noise techniques (2, D). Noise analysis in nuclear systems symposium Florida 1963.

-Schweppe F.C.
87. Role of system identification in electric power systems (2, 3, E). PSCC Grenoble 1972.

-Seifritz W., Stegemann D.
88. Reactor noise analysis (2). Atomic En, Revue vol. 9N1 1971.

-Simoyu.
89. Détermination of transfer functions coefficients of linearised units of control systems (1, B). Automatika y Telemecanika, Tome XVIII n° 6, 1957.

-Stanton K.N.
90. Measurement of turbo-alternator transfer functions using normal operating date (3, C). Proc. IEE 110 (11) (1963).

91. Estimation of turbo-alternator transfer functions using normal operating date (3, C). Proc. IEE 112 (9) 1965.

-Summer A.
92. Optimal identification of some parameters of a nuclear reactor dynamical system (2, E). IFAC London 1966.

-Thie J.A.
93. Noise sources in Power Reactors (2, C). Noise analysis in nuclear systems symposium Florida 1963.

-Trybula S., Malkiewicz J., Koszelnik M.
94. Parameters estimation of controlled power systems (3, C). IFAC The Hague 1973.

-Uhrig R.E.
95. Random noise techniques in nuclear reactor systems (2) (book).
The Ronald Press Company, New York 1970.

-Unbehauen - Schlegel.
96. Estimation of the accuracy in the identification of control systems using deterministic test signals (1, B).
IFAC Prague 1967.

-Van de Meulebroeke F.
97. Etude du modèle dynamique des régulations de vitesse des turbo-groupes (3, B).
Rapport Laborelec 1972 - 2/402/26.

-Vitek V. - Josefus J.
98. Statistical methods in the automation of the operation control of power systems (3, C).
IFAC London 1966.

-Waha J.P.
99. Analyse de la dynamique de processus industriels par des essais transitoires (1, B, C).
Laborelec 1965.

100. Analyse statistique de la dynamique des installations de production d'énergie électrique (1, D).
IBRA, Bruxelles 1966.

101. Méthodologie d'analyse expérimentale de processus industriels (1, A).
IBRA, Bruxelles 1968.

102. Analyse des coefficients de sensibilité des puissances active et réactive à la tension (3, B, E).
Laborelec 2/4559 et 4604 (1967) 2/4705 (1968).

103. Multivariable technical control systems - Survey of applications in power plants and power distribution systems (1, A).
IFAC Düsseldorf 1971.

104. Etude expérimentale de la dynamique d'une chaudière en démarrage (1, B, D).
Laborelec 2/4932.

-Wolker, R.W.
105. Modelling of a 500 MW boiler turbine unit from dynamic response measurements (1, E).
Conference on Industrial Application of Dynamic modelling Durham sept. 69.

-Welfonder E.
106. Correlation method for identification of disturbed control systems by periodical test signal (1, B, D).
IFAC Prague 1967.

-Welfonder E. - Hasenkopf O.
107. Comparison of deterministic and statistical methods for system identification by test-measurements at a disturbed industrial control system (1, G).
IFAC Prague 1970.

-Wilkie J.D.F.
108. Off-line statistical identification of the kinetics of a nuclear reactor (2, E).
IFAC Prague 1967.

-Wismer D.A., Mehra R.
109. Modeling and identification of industrial process (1, E).
JACC 1971.

-Wolters E.J.
110. A tracking state-estimator (3, E).
Electrical Utilities - International Group Meeting at The Hague, Oct. 71.

-Borget G - Faure P.
111. Identification des systèmes.
Une méthode d'identification déterministe de systèmes dynamiques linéaires stationnaires par approximation fonctionnelle.
E.D.F. Série C n° 2 - 1971.

-Caseau P. - Godin P. - Malhouitre G.
112. Simulation numérique d'un générateur de vapeur.
AIM - Liège 1971.

PARAMETER ESTIMATION IN BIOLOGICAL SYSTEMS: A SURVEY

George A. Bekey

University of Southern California
Los Angeles, California, U.S.A.

This paper presents a survey of the applications of parameter identification methods to physiological systems, including compartment models, cardiovascular, respiratory, nervous endocrine, sensory and other systems. Some of the reasons for the limited acceptance of system identification methods in biology is presented. References are included by the area of application and cross-referenced by identification method.

1. INTRODUCTION

During the past twenty years, mathematical description of biological systems has become increasingly fashionable. Mathematical models are being used to represent a large variety of physiological processes. Several recent books attest to the increasing acceptance of mathematical descriptions of biological phenomena [1-8]. However, systematic techniques of parameter estimation from input-output data have been applied in only a limited number of cases, and the results obtained from their application are the subject of some controversy in the field. Only in compartmental analysis have the notions of parameter estimation been widely accepted and used, as described in Section 3 below. Among the reasons for lack of acceptance for system identification techniques in biology are the following:

a. A tradition of suspicion for values not directly measured. Biology in general and physiology in particular have developed primarily as descriptive experimental sciences. Perhaps for this reason, values of biological parameters obtained by a computer from an analysis of input-output data, rather than measured directly by the experimenter are considered suspect until they are verified in the biological laboratory.

b. Biological systems exhibit great variability. As every experimenter knows, in every biological experiment there is variation from trial to trial in a given animal, and certainly from animal to animal. The variability of the data is such that parameters obtained from applying identification techniques to ensemble averaged data may have little or no biological significance and therefore be suspect by biologists. Some identification techniques require that initial estimates of the parameters be available within say 20% of the correct values in order that the algorithms converge. Such is the case with some of the Newton-Raphson techniques and their variations. In many biological systems, knowledge of a parameter with an accuracy of $\pm$ 20% is more than the experimenter hopes to achieve. If a parameter is known with that accuracy, no other identification is required.

c. Inputs and outputs of biological systems are difficult to isolate. Identification techniques designed for systems where input and output signals can be clearly isolated may not work in the biological situation at all. The very nature of the living organism, particularly in higher animals, leads to such a complex of interconnections of subsystems that isolation of a portion of the system may destroy its natural function and thus lead to identification of a purely hypothetical process. On the other hand, when input and output can be isolated, often by exceedingly clever surgical procedures, and the system in question is approximately linear, then frequency response methods have been applied with considerable success.

d. Observability problems. In many cases, in those biological systems where parameters cannot be measured either directly or indirectly, so that one would have to resort to parameter estimation procedures, the state variables required for such estimates are also inaccessible.

e. Time variation. Even in most cases where one may be fortunate enough to be able to reach the appropriate state variables, the measurement time which is required to obtain consistent parameter estimates may be so long that stationarity can no longer be assumed. Biological systems at all times display an interaction of dynamic processes in different time scales with time constants ranging from milliseconds to years. The problem of isolation for measurement purposes requires not only physical isolation but time isolation, so that effects more rapid than those under investigation can be considered instantaneous while those slower are considered constant.

f. Non-linearity. Parameter estimation procedures have had their greatest success in linear systems. In that range of biological processes where linear, time-invariant models are appropriate, frequency response techniques have been used with some success. However, the vast majority of biological processes display a wide range of intrinsic non-linearities, in the presence of which parameter estimation procedures may not behave as expected. In addition to "normal" non-linearities such as threshold and saturation effects (which are present in all biological systems) there are a multitude of other non-linearities involving frequency dependent behavior, adaptation, inhibitory and excitatory phenomena dependent in non-linear ways on feedback signals, rectification phenomena, parametric feedback of nonlinearities, and state-dependent transport delays.

g. Experimental difficulties. Experimental difficulties associated with obtaining good quality input-output data from animal preparations in the laboratory are often such that a researcher may find himself with a system with twelve unknown parameters and five data points, a situation hardly conducive to the use of sophisticated techniques.

h. Lack of uniqueness. Parameter estimates in nonlinear systems are not unique and this lack of uniqueness is an embarassment to the investigator.

i. Cost. An additional important phenomenon is simply that of computation cost. In contrast to the aerospace and process industries, biological research is consistently hampered by the lack of funds and good identification algorithms are often expensive in computer time.

In summary then, while mathematical modeling is prevalent in biological research at the present time, systematic parameter identification is not. Perhaps the most fundamental reason for this phenomenon is that biologists feel much greater confidence in directly observed or directly measured parameters; they distrust the results of indirect measurements somewhat, and they distrust the output of complex and mysterious computer algorithms even further. Perhaps one additional general remark can be made: research in biology is aimed at a furthering of knowledge concerning the behavior of the organism, rather than at control. Hence, input-output models which may match behavior fairly well over some range of stimuli are of little interest to a researcher who is really concerned with mechanisms and their interpretation.

2. SCOPE

The major portion of this paper is concerned with a review of applications of identification techniques to physiological systems. Furthermore, we restrict ourselves to parameter estimation in systems rather than the analysis of signal properties. For this reason such interesting problems as signal analysis in electromyography, electroencephalography and electrocardiography are not included in the paper. Furthermore, statistical methods used in research on automated diagnosis, such as Bayes techniques or linear discriminant analysis are also not included.

A large proportion of the mathematical modeling in biology is done by applying the techniques of regression analysis to the fitting of data. This too is considered to be a straightforward application of statistical methods and is not included in this survey.

It should also be noted that by restricting the scope to physiological systems, we specifically omit discussion of ecological problems, population dynamics and human operator models. Parameter estimation techniques have been applied in the study of human operators so extensively and the literature is so vast, that it was not feasible to include it in the present survey.

The sections which follow are organized by the areas of application, rather than by identification method. The concluding section of the paper provides a set of cross-references by identification technique so that the interested reader may find entries to the literature either from techniques, or from the system he is interested in, i.e., cardiovascular, muscular, etc.

Since publications in this field are scattered throughout the biological and engineering literature, this survey has undoubtedly omitted a number of important papers. The author hereby apologizes for these omissions; he would appreciate receiving references to additional papers describing applications of system identification to biological processes.

3. MULTICOMPARTMENTAL MODELS OF BIOLOGICAL SYSTEMS

One of the most useful and widely used representations of biological systems is based on the notion of a compartment. A compartment generally represents a quantity of a substance within the organism which has uniform and distinguishing kinetics of transformation or transport. Such compartments may represent volumes which are clearly identifiable and localized, (for example, the lungs) or they may be non-contiguous and highly distributed, such as the capillary bed, the tissue or the red blood cells. One can speak, for example, of a red cell compartment and discuss the movement of water or chemical species into and out of it. The most widely used models of biological systems consist of compartments represented by systems of linear first-order differential equations. Thus, an n-compartment system could be represented by

$$\dot{x} = Ax + Bu \qquad (1)$$

where x is an n-vector representing the concentration of a particular material in each of the n compartments, u is an r-vector representing inputs into the various compartments, A is a matrix of rate constants representing the transfer from the ith to the jth compartment and B is an n by r vector. It is typical in problems of this type that the concentrations of a particular substance are measured in several of the compartments but that the transfer coefficients, i.e., the elements of the matrix A are unknown. Extensive work has gone on in the determination of techniques for identifying the elements of the matrix A.

3.1 Exponential Fitting. Provided all the eigenvalues of A are distinct, the solutions of the system of equation (1), i.e. the values of concentration in any compartment, will be represented by a sum of exponentials, e.g.,

$$x_j(t) = \sum_{i=1}^{n} a_o + a_i e^{-k_i t} \quad , j=1,2,\ldots n \quad (2)$$

where a_o and a_i are real and k_i is real and positive. In many cases, the number of exponentials (n) may also be unknown. A number of graphical and computational techniques have been developed for identifying the exponents k_i and the coefficients a_i [A1-A5]. Since such techniques are useful in the study of systems involving radioactive tracers, it is not surprising to find extensive discussion of exponential fitting in the tracer literature (for example A6). A common technique is known as the "peeling" method. It assumes that the system eigenvalues are sufficiently separated so that one can "peel off" the exponential with the longest time-constant by fitting data for large values of time only, subtracting this exponential from the data, fitting the tail of the residual with another exponential, and so forth. In many cases, this technique is made considerably more efficient by working with data plotted on semi-log paper in order to obtain straight line fits.

Recently, formal parameter estimation techniques have been applied to this problem. Mancini and Pilo [A4] have applied a Gauss-Newton iterative procedure alternated with steepest descent to find the values of the unknown coefficients in the exponential fit in the presence of experimental noise. Lemaitre and Melenge [A5] have presented a variation of this technique by using classical linear regression to find the a_i and a direct search algorithm to find the exponents k_i.

Perhaps the most sophisticated work in fitting models of this type has been done by Mones Berman and his associates [A8, A9]. The method consists of using least-squares techniques to identify the unknown parameters in equation (2). The algorithm is widely utilized in the United States, as a component of a large system analysis program available from the National Institutes of Health under the name of SAAM [A9]. Many other references to the method are given in the literature [A10, A11]. A particularly interesting least-squares method, applied to chemical reaction data where there is wide variation between reaction time constants was published by Chandler et al. [A12, See also A13].

3.2 Other Techniques. Many other techniques have been utilized for identification of compartment systems, and related problems such as those concerned with pharmacokinetics, tracer kinetics, and chemical reactions. Among these are a variety of gradient procedures including steepest descent [A14], Gauss-Newton methods [A15], and Newton-Raphson methods [A16]. Quasilinearization [A17] has been applied to the fitting of nonlinear models of drug metabolism to experimental kinetic data. It is known that certain drugs undergo reactions mediated by enzymes, and such processes are described by means of nonlinear differential equations. Maximum likelihood estimation of parameters in multiexponential fits has also been tried [A18].

Of particular interest is the study of Davis and Ottaway [A19] which applied a number of optimization procedures to the fitting of solutions of linear compartment models. Specifically, the methods used were those of Fletcher and Powell, Fletcher and Reeves and the simplex method of Nadler and Mead.

4. CARDIOVASCULAR SYSTEM

The cardiovascular system has been of interest to mathematically oriented physiologists for many years and a vast literature concerns mathematical models and computer simulations of the entire system, and of its component parts including the heart, the pulmonary circulation, the systemic circulation, portions of individual blood vessels, and the microcirculation. In general, such models are constructed by using the basic equations of hydrodynamics suitably modified when necessary for the presence of elastic vessels, non Newtonian flow in small diameter tubes and so forth. Parameter values are generally obtained from experiment or by trial and error adjustment until simulated flows and pressures agree with measurements, in magnitude if not wave shape. However, many other techniques have been used in recent years. There are numerous studies aimed at the determination of transfer functions of portions of the circulatory system. Typical of these are those which use dye concentration measurements in and out of a vascular bed, and use their Fourier transforms for transfer function estimates [C1]. An alternate approach has been based on approximating blood pressure transients across the carotid pressure sensors by analytic functions of time, and then evaluating Laplace transforms [C2].

Gradient methods have been increasingly applied in recent years. A continuous gradient method has been used to identify parameters in a model of a single blood vessel from pressure and flow data [C3]. An iterative gradient method has been used to estimate cardiac output non-invasively from respiratory data [C11, C12]. An iterative algorithm combining steepest descent and a Gauss-Newton procedure was applied to the estimation of model parameters in the microcirculation [C4]. Newton-Raphson methods have been applied to the estimation of circulation parameters such as resistances and compliances, working directly with the differential equation of the system [C5] and to Z-transform models using sampled data [C6]. The essential parameters of the aorta, namely the geometric taper, area,hoop elasticity and effective loss factor have been computed using experimental amplitude ratio and phase shift data from both dogs and chickens [C7] by means of a learning model technique. One of the most ambitious parameter estimation projects used Marquardt's method to obtain the values of ten compliances and three resistances in a mathematical model of the arterial system [C8]. This large parameter estimation procedure was mechanized on a hybrid computer. The criterion function was based on matching a number of pressures and flow signals measured from dogs.

As a final illustration in the cardiovascular area, correlation techniques have been used to obtain large transfer functions relating mean arterial pressure and suprarenal blood flow in dogs. Deconvolution from the correlation function led to weighting functions of the system, which were fitted by transfer functions of various types [C9].

In some cases, a combination of excessive computer time, slow convergence of certain algorithms, high sensitivity to initial parameter estimates, and general frustration with estimation algorithms have led investigators to abandon parametric methods entirely, to return to sinusoidal forcing and manual fitting of transfer functions to Bode diagrams. In at least one such example, some six different identification algorithms were tried and abandoned [C10].

5. RESPIRATORY SYSTEM

As with the circulatory system, the respiratory system has been the subject of intensive research and numerous mathematical models have been developed in recent years. However, with this system too relatively few systematic approaches to parameter estimation have been published.

One of the earliest attempts at applying identification procedures to the respiratory system was published by Clynes in 1960. Subjects were asked to rapidly inhale or rapidly exhale, thus generating approximate step function inputs and the resulting changes in heart rate were recorded. Transfer functions were fitted separately to the inhalatory and exhalatory data [R1]. The effect of respiration on heart rate also has been studied using spectral analysis [R2] and digital filtering [R3].

Common respiratory system parameters, such as airway resistance and compliance, have been computed using least-squares techniques applied to measured flow and pressure data [R4]. Least-squares methods in the frequency domain have been used to obtain parameters in assumed transfer functions representing the relation between alveolar CO_2 concentration and tidal volume [R5, R6]. Newton-Raphson algorithms were used to minimize a frequency domain error function between model and data. The same authors have also applied an adaptive step-size random search algorithm to the same problem [R7].

A particularly interesting application of parameter identification techniques to the respiratory system is described by Feinberg et al. [R8]. The Fletcher Powell method was used to estimate such parameters as lung tissue compliance, airway resistance (both in the lower airway and in the mid airways) and a complex parameter known as the index of bronchial compliance. Plots of the resulting parameter values on appropriate parameter planes made it possible to diagnose obstructive lung diseases such as chronic bronchitis, asthma and emphysema using only airflow at the mouth and alveolar pressure data. This paper is particularly interesting since it illustrates the potential value of parameter estimation as a diagnostic tool.

6. THE NERVOUS SYSTEM

Difficulties in obtaining experimental data and the statistical nature of nerve signals have probably contributed to the relative paucity of systematic identification techniques in this field. The firing patterns of sensory neurons as a function of stimulus intensity have been modeled, and their parameters estimated using steepest descent and stochastic approximation methods [N1]. Neuron dynamics have been studied using random search techniques [N2]. The study is of particular interest because of the fact that the state variables, (i.e., the transmembrane potential and the threshold potential of the cell)are defined statistically only and are incorporated in a model by means of random variations in the model parameters. Theoretical and experimental interspike interval distribution functions were compared at each iteration where the theoretical interspike interval distribution function depends on the particular parameters.

A neuronal system of the retina of a catfish has also been studied by using the

Wiener functional approach [N3]. This powerful technique was first applied to nonlinear biological systems by Stark [N4]. [See also N5].

7. ENDOCRINE AND METABOLIC SYSTEM

Pioneering work in the application of frequency domain techniques to the development of models of the adrenocortical system have been done by Yates, Urquhart, and Lee [e.g., E1, E2]. Five parameters in a model of blood glucose regulation were evaluated using a neighborhood search program [E3]. Identification of the glucose regulation system has also been studied by generalized recursive least-squares methods [E4]. This study is of particular interest because it implements a feedback control system to provide adaptive control of blood glucose in a human subject using a PDP-8 computer on-line. The adaptive algorithm implemented in this study combines the functions of prefiltering, identification and control. Perhaps the most ambitious parameter identification study in connection with blood glucose regulation was done by Charette et al. who applied quasilinearization techniques in a highly nonlinear model of this system [E5].

8. VISUAL SYSTEM

Only a few references are included to illustrate some of the advanced identification procedures being applied in this field. Orthogonal Wiener functionals, mentioned above in Section 6 in connection with nervous system research were originally applied by Stark to the study of pupillary control system nonlinearities [V1]. In spite of the power of this technique, it has been seldom used in biological system identification because an orthogonal expansion lacks any close correspondence with the physiological properties being studied, even though it may provide excellent input-output fits to the data. Kalman filter equations have been used to identify parameters in a model of the lens accommodation system in the eye of the cat [V2]. The approximate characteristics of the linear filter assumed to generate the electroretinogram in the frog have been derived by fitting experimental frequency response characteristics [V3].

9. MUSCLE

Three publications have been located in the area of muscle physiology. Transfer functions of load-moving muscle systems, of the human ocular motor system, and of several types of mechanoreceptors have been obtained using swept frequency signals and correlators. The Fourier transform of the input autocorrelation function and the input-output crosscorrelation were used to obtain the transfer function [M1]. A differential equation with three unknown parameters relating muscle force to the mean rectified value of the electromyogram has been obtained by using steepest descent methods [M2]. Parameters of a striated muscle including elastic components and viscoelastic properties, as well as nonlinearities were obtained by using a least-squares method for nonlinear discrete system identification [M3].

10. OTHER SYSTEMS

There are of course numerous other papers in the literature describing techniques of various degrees of sophistication applied to a variety of biological systems. For example, linear and nonlinear programming techniques have been applied with considerable success to such diverse problems as parameter identification in the temperature regulation system of the dolphin and in the intraocular pressure control system of the rabbit [O1]. Certain parameters of the kidney including properties of the renal arteries and glomerular pressure were identified by steepest descent techniques [O2]. Pattern search procedures were used to determine optimum parameter values in a model of the patient-artificial kidney system [O3].

11. CROSS REFERENCE BY TECHNIQUE

In this section, the various applications described above are listed by the identification technique employed. In some cases, a reference may appear under two techniques indicating that both methods were used.

a. Non-parametric methods. Correlation and spectral analysis: E9, R2, M1. Fitting of frequency response data: C1, C2, E1, E2, V3.

b. Least-squares method. Time domain: A7, A12, C6, C10, R4, E4, M3. Frequency domain: R5, R6.

c. Search methods. Pattern Search, Neighborhood search: A5, E3, O3. Random search: C10, R7, N2.

d. Gradient method. Steepest descent: A4, A14, C3, C4, C10, C11, N1, M2, O2. Gauss-Newton: A4, A15, C4. Marquardt's: C8. Newton-Raphson: A16, C5. Quasilinearization: A17, E5. Fletcher-Powell and related methods: A19, R8.

e. Learning model technique.: C7.

f. Estimation based method. Maximum likelihood: A18, Kalman filtering: V2. Stochastic approach N1.

g. Linear and non-linear programming. :O1.

h. Wiener functional. : N3, N5, V1.

12. CONCLUSION

System identification has a fertile field of application in the biological sciences. However, it is evident from a survey of the dates in the references given in this survey that most of the sophisticated techniques have been applied relatively recently, and that most of the applications appearing in the literature have come about as a result of physiological research, where parameter estimates have been obtained as a reluctant substitute for the more desirable direct measurement of parameters. However, there is some indication that parameter estimation techniques may have a significant and useful future as guides to diagnosis and as an aid in the noninvasive estimation of meaningful clinical parameters.

REFERENCES

A1 Perl, W.A.: A method for curve fitting by exponential functions. Int. J. Appl. Radiat. Isotop. 8, (1960) 211-222.

A2 Cornell, R.G.: A method for fitting linear combinations of exponentials. Biometrics 18, (1962) 104-113.

A3 Parson, D.H.: Biological problems involving sums of exponential functions of time: a mathematical analysis that reduces experimental time. Math. Biosci. 2, (1968) 123-128.

A4 Mancini, P. and Pilo, A.: A computer program for multiexponential fitting by the peeling method. Comp. & Biomed. Res. 3, (1970) 1-14.

A5 Lemaitre, A. and Melenge, J.: An efficient method for multiexponential fitting with a computer. Comp. & Biomed. Res. 4, (1971) 555-560.

A6 Sheppard, C.W.: Basic Principles of the Tracer Method, Wiley and Sons, Inc., 1962.

A7 Berman, M., Weiss, M.F., and Shahn, E.: Some formal approaches to the analysis of kinetic data in terms of linear compartment systems. Biophys. J. 2, (1962) 289.

A8 Berman, M., Shahn, E. and Weiss, M.F.: The routine fitting of kinetic data to models: a mathematical formalism for digital computers. J. Biophys. & Biochem. 2, (1962) 275.

A9 Berman, M. and Weiss, M.: Users Manual for SAAM, National Institute of Arthritis and Metabolic Diseases, National Institutes of Health, Bethesda, Maryland (1967).

A10 Waxman, A.D., Leins, P.A., Siemsen, J.K.: On vivo dynamic studies of hepatocyte function: a computer method for the interpretation of Rose Bengal kinetics. Comp. & Biomed. Res. 5 (1972) 1-13.

A11 Llaurado, J.G. and Smith, G.A.: Some observations on the effects of time curtailing on the parameters of decaying exponentials fitted to biomedical data. Internat'l. J. of Biomed. Comp. 2 (1971) 265-275.

A12 Chandler, J.P., Hill, D.E., and Spring, H.O.: A program for efficient integration of rate equations and least-squares fitting of chemical reaction data. Comp. & Biomed. Res. 5, (1972) 515-534.

A13 Cannon, J.R. & Filmer, D.L.: A numerical experiment on the determination of unknown parameters in an analytic system of ordinary differential equations. Math. Biosci. 3, (1968) 267-274.

A14 Benetazzo, L., Buonardo and Clemente, G.: A functional model of early thyroxine distribution in man. Med. & Biol. Engineering 10, (1972) 337-346.

A15 Glass, N.R.A.: A technique for fitting nonlinear models to biological data. Ecology 48, (1967) 1010.

A16 Buell, J.R., Kalaba, R., Yakush, A., Ruspini, E.H.: A program for identification of linear systems. Comp. Programs in Biomed. 2, (1971) 8-15.

A17 Buell, J. and Kalaba, R.: Quasilinearization and the fitting of nonlinear models of drug metabolism to experimental kinetic data. Math. Biosci. 5, (1969) 121-132.

A18 Sandor, T. and Wilson, G.D.: Maximum likelihood estimation of parameters in multiexponential fits when data follow a Poisson distribution. Comp. Programs in Biomed. 2 (1972) 111-117.

A19 Davis, R.H. and Ottaway, J.H.: Application of optimization procedures to tracer kinetic data. Math. Biosci. 13, (1972) 265-282.

C1 Coulam, C.M., Warner, H.R., Marshall, H.W. and Bassingthwaighte, J.B.: A steady-state transfer function analysis of portions of the circulatory system using indicator dilution techniques. Comp. & Biomed. Res 1. (1967) 124-138.

C2 Suga, H., Oshima, M.: Measurement of the transfer function of the carotid sinus pressure control system with its feedback loop physiologically closed. Med. & Biol. Engineering 9, (1971) 147-150.

C3 Bekey, G.A., Merritt, M.J., Assali, N.S.: Synthesis and optimization of a model of cardiovascular dynamics. Proc. 4th Internat'l Congress for Analog Computation, Presses Academiques Europeenes, Brussels, (1965) 383-388.

C4 Rothe, C.F. and Williams, B.F.: Dynamic characteristics of the renal arterial microvasculature of the dog obtained by simulation, IEEE Trans. on Biomed. Engineering, 19, (1972) 213-220.

C5 Dennison, J.C. and Dick, D.E.: Cardiovascular parameter estimation. Rocky Mountain Bioengineering Symposium, (1972).

C6 Burrus, C., Park, T.W. and Watt, T.B.: A digital parameter-identification technique applied to biological signals. IEEE Trans. on Biomed. Engineering 18 (1971).

C7 Strano, J.J., Welkowitz, W. and Fich, S.: Measurement and utilization of in vivo blood-pressure transfer functions of dog and chicken aortas. IEEE Trans. on Biomed. Engineering 19, (1972) 261-171.

C8 Sims, J.B.: Estimation of arterial system parameters from dynamic records. Comp. & Biomed. Res. 5, (1972) 131-147.

C9 Szűcs, B., Monos, E. and Csaki, F.: On a method for identification of the cardiovascular system. 2nd Prague IFAC Symposium (1970) 11.5.

C10 Yamashiro, S.M., Sato, T., Grodins, F.S.: Comparison of parameter identification methods applied to a study of peripheral vascular function. Fed. Proc. 30, (1971) 698.

C11 Maloney, J.C.: Estimation of cardiac output using respiratory measurements and real-time computation. A dissertation presented to the University of Southern California (1971).

C12 Bekey, G.A. and Maloney, J.C.: On-line estimation of cardiac output using respiratory models and measurements. Fed. Proc. 29 (1970)946ABS.

R1 Clynes, M.: Respiratory control of heart rate: laws derived from analog computer simulation. IEEE Trans. on Med. Electronics, 7, (1960) 2-14.

R2 Womack, B.F.: The analysis of respiratory sinus arrhythmia using spectral analysis and digital filtering. IEEE Trans. on Biomed. Engineering 18, (1971) 399-409.

R3 Womack, B.F.: Identification of models of respiratory sinus arrhythmia in humans: Three approaches. Joint Automatic Control Conference (1972).

R4 Wald, A., Jason, D., Murphy, T.W. and Mazzia, V.D.B.: A computer system for respiratory parameters. Comp. in Biomed. Res. 2 (1969) 411-429.

R5 Stoll, P.J. and Meditch, J.S.: Frequency response studies of human and avian respiratory regulation. Joint Automatic Control Conference (1972).

R6 Stoll, P.J. and Meditch, J.S.: Least-squares estimation of respiratory system parameters. Math. Biosci. 8, (1970) 307-321.

R7 Jelinek, C.O., Meditch, J.S. and Stoll, P.J.: Respiratory system parameter estimation using a random search method. Proc. 5th Hawaii Internat'l Conf. on System Sciences (1972) 247-279.

R8 Feinberg, B.N., Chester, E.H., and Schoeffler, J.D.: Parameter estimation: a diagnostic aid for lung disease. Instrumentation Technology (1970) 40-46.

N1 Sclabassi, R. and Moore, G.P.: Parameter estimation methods for sensory neuron models, 8th ICMBE (1969).

N2 Sclabassi, R.J.: The statistical investigation of hypotheses in neurophysiology. A dissertation presented to the University of Southern California (1971).

N3 Marmarelis, P.A.: Nonlinear identification of bioneuronal systems through white-noise stimulation. Joint Automatic Control Conference (1970) 117-126.

N4 Stark, L.: Neurological Control Systems: Studies in Bioengineering. Plenum Press, New York (1968).

N5 Marmarelis, P., Naka, K.I.: White-noise analysis of a neuron chain: an application of the Wiener theory. Science 175, (1972) 1276-1278.

E1 Urquhart, J. and Li, C.C.: Dynamic testing and modeling of the adreno-cortical secretory response to corticotropin. Annals N.Y. Acad. Sci. 156, (1969) 756-778.

E2 Yates, F.E., Brennan, R.D., Urquhart, J.: Adrenal glucocorticoid control system. Fed. Proc. 28 (1969) 7-83.

E3 Gatewood, L.C., Ackerman, E., Roseavear, J.W., Molnar, G.D. and Burns, T.W.: Tests of a mathematical model of the blood-glucose regulatory system. Comp. & Biomed. Res. 2 (1968) 1-14.

E4 Pagurek, B., Riordon, J.S. and Mahmoud, S.: Adaptive control of the human glucose-regulatory system. Med. & Biol. Engineering 10, (1972) 752-761.

E5 Charette, W.P., Kadish, A.H. and Sridhar, R.: Modeling and control aspects of glucose homeostasis.

Math. Biosci. Supplement I: Hormonal Control Systems (1969). 115-137.

V1 Stark, L.: The pupillary control system: its non-linear adaptive and stochastic engineering design characteristics. Automatica 5, (1969) 655-676.

V2 O'Neill, William, D., Sanathanan, C.K. an and Brodkey, J.S.: A minimum variance time optimal, control system model of human lens accommodation. IEEE Trans. on Systems Sci. & Cybernetics. SSC-5 (1969) 290-299.

V3 Levett, J.: The approximate characteristics of the linear filters in the electroretinogram of the frog. Internat'l J. of Biomed. Comp. 2, (1971) 189-199.

M1 Williams, W.J., Gesink, J.W. and Stern, M.M.: Biological system transfer function extraction using swept-frequency and correlation techniques. Med. & Biol. Engineering 10 (1972) 609-620.

M2 Coggshall, J.C. & Bekey, G.A.: EMG-force dynamics in human skeletal muscle. Med. & Biol. Engineering 8 (1970) 265-270.

M3 Inbar, G.F. and Baskin, R.J.: Parameter identification analysis of muscle dynamics. Math. Biosci. 7, (1970) 61-79.

O1 Reid, M.H., MacKay, R.S.: Applications of a new technique for system identification to dolphin temperature regulation and rabbit intraocular pressure control. Med. & Biol. Engineering 6, (1968) 269-290.

O2 Rothe, C.F., Nash, F.D.: Renal arterial compliance and conductance measurement using on-line self-adaptive analog computation of model parameters. Med. & Biol. Engineering 6 (1968) 53-69.

O3 Abbrecht, P.H. and Prodany, N.W.: A model of the patient-artificial kidney system. IEEE Trans. on Biomed. Engineering 18, (1971) 257-264.

GENERAL REFERENCES.

1. Brown, J., Jacobs, J.E., Stark, L.: Biomedical Engineering, F.A. Davis Co., Philadelphia (1971).

2. Clynes, M. and Milsum, J.H.: Biomedical Engineering Systems, McGraw-Hill Book Co., New York, (1970).

3. Grodins, F.S.: Control Theory and Biological Systems, Columbia University Press, (1963).

4. Milhorn, H.T.: The Application of Control Theory to Physiological Systems, W.B. Saunders Co., Philadelphia (1966).

5. Milsum, J.H.: Biological Control Systems Analysis, Electronics Science Series. McGraw-Hill Book Co., New York (1966).

6. Plonsey, R.: Bioelectric Phenomena, McGraw-Hill Book Co., New York, (1969).

7. Stark, L.: Neurological Control Systems: Studies in Bioengineering, Plenum Press, New York, (1968).

8. Schwan, H.P. (ed.): Biological Engineering, McGraw-Hill Book Co., New York, (1969).

AIRCRAFT PERFORMANCE MEASUREMENTS IN NONSTEADY FLIGHTS

J.A. Mulder

Aeronautical Department

Delft University of Technology

Delft

The Netherlands

This paper discusses some aspects of the determination of aircraft performance from measurements in nonsteady flight conditions. It is shown that additional variables must be measured compared to the conventional technique of measuring performance during steady flight. The gain is a remarkable reduction in required flight test time. Performance has been derived by means of maximum likelihood estimation. The Cramér-Rao lower bound theorem has been applied to calculate the maximally achievable accuracies of performance derived from nonsteady flight data.

1. INTRODUCTION

Aircraft flight dynamics are usually studied by considering the motions about the centre of gravity and those of the centre of gravity itself separately.
Stability and control characteristics are related primarily to the motions of the aircraft about the centre of gravity. These motions are often described by a linearized dynamic model, valid only for ''small'' deviations from a condition of steady rectilinear flight.
The aerodynamic parameters in the model, the so-called stability and control derivatives, can be estimated from measurements of the response of the aircraft following small control surface deflections in steady rectilinear flight.
Much work has been done by now in this field e.g. Wolowicz (1966), Taylor et al. (1969) and Grove et al. (1972).
In symmetrical flight conditions the motion of the centre of gravity characterises the aircraft's performance. Performance is defined here as the rate of climb in steady rectilinear flight conditions in an atmosphere at rest relative to the earth, see e.g. AGARD, Flight Test Manual, Vol. I.
Performance used to be and still usually is measured in steady straight flight conditions. This, However, is very time consuming due to the many parameters involved, like aircraft wheight and configuration, engine power, airspeed, atmospheric air pressure (altitude) and air temperature.
In Gerlach (1964) a method was described in which the stability and control derivatives and performance could be obtained simultaneously from measurements during one nonsteady manoeuvre. This method could significantly reduce the amount of time usually required for performance measurements. This is important because present day flight test programs are very costly.
Very few authors deal with performance measurements in nonsteady flight, e.g. Klopfenstein (1965).
In view of the amount of work done recently in the field of obtaining stability and control derivatives this paper is devoted mainly to the measurement of aircraft performance in nonsteady flight.
An important aspect of the flight test technique is the so-called flight path reconstruction. In the flight path reconstruction the time histories of variables which cannot well be directly measured in flight are calculated. In addition the zero-shifts of the inertial transducers are derived. These zero-shifts of the inertial transducers would significantly reduce the accuracy of performance derived from nonsteady flight measurements if they were not corrected for, Jonkers et al. (1972).
Hosman (1971) solves the flight path reconstruction problem by applying regression analysis. In this paper maximum likelihood estimation is used to this end.
In the paper referance is made to a series of flight tests performed with the laboratory aircraft of the Delft University of Technology a De Havilland DHC-2 ''Beaver'', Fig. 1.
The results of these flight tests are presented in Gerlach (1971).

The paper contains the following Sections.
In Section 3 performance measurements in nonsteady flights are compared to the conventional technique of measuring performance in steady straight flight conditions. Some aspects of the instrumentation system are described.
The maximally achievable accuracies of the flight path reconstruction and of performance are calculated in Section 4 by applying the Cramér-Rao lower bound theorem to a numerical simulation experiment. It follows that errors of performance measurements are mainly due to the vertical movement of the atmosphere relative to the earth during the flight tests.
In Section 5 the flight path reconstruction problem has been solved by means of maximum likelihood estimation. Results of actual flight tests are presented in this Section.

2. NOTATION

A specific force, quantity sensed by an accelerometer

C	rate of climb
E	mathematical expectation operator
f	vector function in (3-2)
g	acceleration due to gravity
Δh	change of altitude relative to some reference altitude
H	matrix in (4-12)
J	Fisher information matrix
L	logarithm of the likelihood function
m	aircraft mass
n	number of measurements in a nonsteady manoeuvre
p	likelihood function
P	engine power
q	rate of pitch
r	correlation function
R	measurement error covariance matrix
t	time
u	input vector to the system (4-3)
U	horizontal component of V
V	airspeed
w_w	steady vertical wind component
x	state vector
x_a	augmented state vector
X	aerodynamic force along X-axis
y	observation vector
z	measurement vector
Z	aerodynamic force along Z-axis
$(\alpha_t)\alpha$	(total) angle of attack
β	parameter vector
δ_{ij}	kronecker delta
γ	flight path angle
τ	argument of r
θ	pitch angle
$\hat{a}$	estimate of a
δa	"small" error of a

index e relative to earth

1 nonsteady flight conditions

2 steady flight conditions

m measured quantity

The system of axis used is a body-fixed, rectangular, right-handed system X Y Z, X- and Z-axes in the plane of symmetry of the aircraft, Fig. 3, the X-axis chosen parallel to the aircraft wing chord, the origin is in the aircraft's centre of gravity.

3. THE MEASUREMENT OF AIRCRAFT PERFORMANCE

In Section 1 of the present paper aircraft performance has been defined as the rate of climb in steady rectilinear flight conditions in an atmosphere at rest relative to the earth. Performance thus defined is a function of several parameters like aircraft wheight and configuration (landing gear up or down, degree of extension of wing leading edge and trailing edge flaps, etc.), engine power, airspeed, atmospheric air pressure (altitude) and air temperature.
As an example may serve Fig. 2, presenting the performance at climbing power of the laboratory aircraft of the Delft University of Technology, a De Havilland DHC-2 "Beaver" as a function of airspeed at one representative aircraft configuration (flaps up), wheight and altitude.

3.0 Performance measurements in steady flight conditions

From the definition of performance follows that, in principle at least, performance could be measured directly in flight by measuring the rate of climb in steady rectilinear flight.
Atmospheric air pressure decreases with increasing altitude. Rate of climb could follow from the direct measurement of the rate of change of atmospheric air pressure.
Due to various limitations, however, the resulting data are of limited accuracy and therefore not suited for accurate performance measurements. More accurate rate of climb data follow from the change of altitude Δh during some time interval Δt in steady straight flight. Δh can be derived from the change in atmospheric air pressure during the time interval Δt. The rate of climb follows easily from

$$C_e = \frac{\Delta h}{\Delta t} \tag{3-1}$$

In order to arrive at sufficiently accurate data time intervals of 30 to 60 seconds are used in practice depending on the nominal value of performance. In actual flight tests however, the pilot may need several minutes to establish a steady flight condition. This makes performance measurements in steady flight very time consuming, due to the many parameters, the influence of which has to be determined.
In addition the accuracy of the results is limited since exactly steady rectilinear flight conditions are difficult to establish and maintain during the time interval required for sufficient accuracy.
In Section 3.1 and alternative to performance measurements in steady straight flights is discussed. By measuring additional variables in non steady flight manoeuvres a remarkable reduction in flight test time results. One of the variables which must be obtained is the angle of

attack. Section 3.2 deals with some aspects of the measurement of the angle of attack in flight. Finally Section 3.3 describes the instrumentation system used for the flight test series mentioned in Section 1.

3.1 Performance measurements in quasi-steady flight conditions

The necessity to establish and maintain a steady straight flight condition during the measurements may be eliminated by measuring additional variables. Which variables must be measured follows from the equations of motion, Etkin (1959), Fig. 3,

$$\Delta\dot{h} = C_e \quad (3\text{-}2)$$

$$\dot{U} = A_x \cos\Theta + A_z \sin\Theta \quad (3\text{-}3)$$

$$\dot{C}_e = A_x \sin\Theta - A_z \cos\Theta - g \quad (3\text{-}4)$$

$$\dot{\Theta} = q \quad (3\text{-}5)$$

The equations are valid for symmetrical flight conditions, relative to a flat earth. U denotes the horizontal speed with respect to undisturbed air, Δh the change of altitude relative to the altitude at time $t = t_0$ and C_e the rate of climb relative to the earth.
A_x and A_z are the quantities sensed by accelerometers having their sensitive axes along the body fixed X- resp. Z-axis, Fig. 3.
From the following relations:

$$A_x = \frac{X}{m}$$

$$A_z = \frac{Z}{m}$$

in which m denotes aircraft mass and X and Z the components of the total aerodynamic force acting on the aircraft along the X- resp. Z-axis, follows that A_x and A_z may be denominated as ''specific forces'', i.e. external nonfield forces per unit of mass, Gerlach (1964).
In steady flight $\dot{U} = 0$, then the pitch angle Θ follows from (3-3):

$$\Theta = -\arctan\frac{A_x}{A_z} \quad (3\text{-}6)$$

In Fig. 3, which is valid for steady as well as nonsteady symmetrical flight conditions, the total angle of attack α_t has been defined as the angle between the body X-axis and the air velocity V, the velocity of the centre of gravity of the aircraft relative to the undisturbed air ''far away'' from the aircraft.
V can be derived with sufficient accuracy from total and static air pressure measurements as described in Section 3.3.
Assuming for the time being α_t can somehow be measured directly in flight, the rate of climb C relative to the atmosphere may be derived from

$$C = V \sin(\Theta - \alpha_t) \quad (3\text{-}7)$$

Performance has been defined as usual in an atmosphere at rest relative to the earth. During actual flight tests however, it turns out that the air mass surrounding the aircraft may move with a low, approximately constant velocity relative to the earth even in an apparently quiet atmosphere. Performance defined relative to the surrounding air mass remains invariant. It is, therefore, the latter performance which should, ideally, be determined.
From Fig. 3 follows the relation between the vertical speed relative to earth C_e, the vertical speed relative to the surrounding air mass C and the vertical wind component relative to earth w_w:

$$C_e = C + w_w \quad (3\text{-}8)$$

In the presence of a vertical wind component w_w it follows that equation (3-7) would yield rate of climb C relative to the surrounding air mass. In practice, however, the accuracy of C derived from equation (3-7) is limited due to the difficulties related to the measurement of α_t in flight, Section 3.2.
Equation (3-1) results in C_e. Thus rate of climb relative to earth instead of rate of climb relative to the surrounding air mass follows from the conventional measuring technique in a series of steady straight flights, which reduces the accuracy of the measured performance.
From (3-6) and (3-7) follows that C could be derived from A_x, A_z, α_t and V at any single instant in time during steady straight flight, provided the latter variables just indicated are known at that instant.
This would result in a reduction of flight test time compared to the conventional measuring technique based on (3-1). However, a further reduction in flight test time proves to be possible.
The variables in the equations of motion during nonsteady flight are now denoted by the index 1. In steady flight conditions the index 2 is used. It will be clear from the text whether variables without index 1 or 2 refer to steady or nonsteady flight conditions.
Per definition:

$$\dot{U}_2 = \dot{C}_{e_2} = \dot{\Theta}_2 = 0 \quad (3\text{-}9)$$

whereas (3-5) and (3-7) transform into:

$$\Theta_2 = -\arctan\frac{A_{x_2}}{A_{z_2}} \quad (3\text{-}10)$$

$$C_2 = V_2 \sin(\Theta_2 - \alpha_{t_2}) \quad (3\text{-}11)$$

Nonsteady flight conditions are very often compared with conditions in steady straight flight at equal values of the parameters - such as aircraft configuration, engine power, etc. -

affecting performance under the additional constraint:

$$\alpha_{t_2} = \alpha_{t_1} \qquad (3\text{-}12)$$

The notion of quasi-steady flight is now introduced and defined as follows.
Quasi-steady flight is a nonsteady flight condition such that the specific forces A_{x_1} and A_{z_1} and airspeed V_1 differ only by ''small'' amounts from A_{x_2}, A_{z_2} and V_2 of the corresponding steady flight condition.
Quasi-steady flight conditions permit the use of (3-11) and (3-12) to derive first estimates of pitch angle θ_2 and rate of climb C2 related to the corresponding steady straight flight condition by substitution of:

$$A_{x_2} = A_{x_1}$$

$$A_{z_2} = A_{z_1}$$

$$V_2 = V_1$$

resulting in:

$$\theta_2 \cong - \arctan \frac{A_{x_1}}{A_{z_1}} \qquad (3\text{-}13)$$

$$C_2 \cong V_1 \sin (\theta_2 - \alpha_{t_1}) \qquad (3\text{-}14)$$

(3-13) and (3-14) yield the possibility to measure performance (characterizing steady flight conditions) in quasi-steady flight conditions. This remarkably reduces the flying time required for performance measurements. As an example may serve the following manoeuvre developed during the flight test series with the ''Beaver'' laboratory aircraft.
At the start of the manoeuvre the aircraft is kept in a steady flight condition at some aircraft configuration and engine power. By a smooth and gradual nose-down rotation about the Y-axis, the aircraft is accelerated forward through the whole range of airspeeds of interest in this particular case, from about 35 m/sec up to 65 m/sec, in approximately one minute.
This results in a forward acceleration of about 0.5 m/sec^2, small enough as to permit the use of (3-13) and (3-14).
At four time instants, equally spaced, with respect to air speed, short nonsteady pull up - push down manoeuvres are performed, each taking 3 to 4 seconds.
The time histories of some variables during the manoeuvre are presented in Fig. 4.
The measurements during the pull up - push down manouevres are not directly used for the calculation of performance. They serve for the identification of the parameters in a nonlinear, algebraic aerodynamic model from which the stability and control derivatives mentioned in Section 1 are derived. A discussion of the model and the identification method is given in Gerlach (1964).
In the present flight test technique the aerodynamic model was used to correct for the small differences between A_{x_1}, A_{z_1} and V_1 on one hand and A_{x_2}, A_{z_2} and V_2 on the other.
Moreover knowledge of the parameters in the aerodynamic model permits the performance measurements to be corrected for small differences between actual values of engine power, altitude and temperature during the flight tests and some nearby set of nominal ''standard conditions'' for which performance has to be known.
A discussion of the aerodynamic model is thought to be somewhat beyond the scope of the present paper. The correction technique itself has been described in Mulder (1973).
In the following the conventional measuring technique is compared with the technique of measuring in quasi-steady flight conditions.
Conventionally one just measures - in principle - the altitude increment during some time interval of say one minute in steady flight additionally to airspeed.
The measurement of performance in quasi-steady flight conditions requires the measurement of the specific forces and the total angle of attack α_t in addition to V. However, the time interval required to obtain merely one point in the performance versus airspeed curve in the case of the conventional technique, results in the complete curve if measurements are made in quasi-steady flight. This may result in an order of magnitude reduction in required flight test time.

3.2 The measurement of the angle of attack

From Section 3.1 follows that for performance measurements in quasi-steady flight conditions the total angle of attack α_t must be known in addition to the specific forces A_x and A_z and the airspeed V.
In principle α_t can be measured by means of a vane installed on a beam mounted on the nose of the fuselage or on the wing some distance before the leading edge of the wing. However, α_t has been defined as the angle between the body-fixed X-axis and the air velocity vector V, which is the direction of the undisturbed airflow ''far away '' from the aircraft. The presence of the aircraft and of the beam itself may cause small but significant deviations of the direction of the airflow at the location of the vane. In steady flight these deviations are a function of the nominal value of the total angle of attack α_t. The vane must be calibrated in flight to allow correction for these deviations.

Usually the calibration of the vane is performed in a series of steady horizontal flights at different airspeeds and consequently at different values of the total angle of attack α_t.
Assume first the atmosphere to be at rest relative to the earth. From Fig. 3 the following relation may be deduced:

$$\alpha = \theta - \gamma \qquad (3\text{-}15)$$

α is called the angle of attack, γ the flight path angle. (3-15) holds for all symmetrical, steady as well as nonsteady flight conditions. In horizontal flight the flight path angle equals zero thus:

$$\alpha = \theta \qquad (3\text{-}16)$$

In steady flight θ can easily be measured, e.g. by means of a pendulum.
As long as the atmosphere is assumed to be at rest it follows from Fig. 3 that:

$$\alpha_t = \alpha \qquad (3\text{-}17)$$

Thus the calibration of the vane in a series of steady horizontal flights is based upon the assumption the atmosphere is at rest relative to the earth. A little reflection shows that (3-17) still holds true in the presence of a horizontal wind component.
In the presence of a vertical wind component, w_w, however, (3-17) is not longer valid, a fact which follows immediately from Fig. 3. Since there are no means to measure w_w, the assumption has to be made that $w_w = 0$.
This reduces the accuracy of the calibration in a series of steady horizontal flights.
Moreover it is at least questionable whether a calibration in a series of steady flights holds equally well in quasi-steady or nonsteady manoeuvring flight conditions.
The real problem in applying the technique of performance measurements in quasi-steady flight lies in the determination of the total angle of attack α_t.
In the present flight test technique it was decided to circumvent the difficulties associated with the direct measurement of the total angle of attack α_t by means of a vane.
Instead the angle of attack α is obtained from (3-15), valid in steady as well as nonsteady flight conditions and also in the presence of a vertical wind component. By substituting α_1, instead of α_{t_1} in (3-14) this relation transforms into, see Fig. 3:

$$C_{e_2} \cong V_1 \sin(\theta_2 - \alpha_1) \qquad (3\text{-}18)$$

Here lies the crux of the problem. From (3- 8) it follows:

$$C_2 = C_{e_2} - w_w \qquad (3\text{-}19)$$

The vertical wind component cannot be measured. Hence it is assumed that $w_w = 0$. From (3-19) follows this introduces an error w_w in the performance C_2.
The angle of attack α is obtained as the difference between angle of pitch θ and flight path angle γ. Neither θ nor γ are measured directly for maximum accuracy, due to limitations in present day instrumentation techniques, but rather follow from what has been called the flight path reconstruction. The flight path reconstruction is based on the equations of motion (3-2)..(3-5) and constitutes in fact the key feature of the present flight test method.
We can look upon (3-2)..(3-5) as representing a system with state vector x defined by:

$$x = \text{col}\left[\Delta h \quad U \quad C_e \quad \theta\right] \qquad (3\text{-}20)$$

In the instrumentation system the rate of pitch q is measured additionally to A_x, A_z, Δh and V. Assuming the state vector $x(t_o)$ at the start of the nonsteady manoeuvre to be known sufficiently well, the time histories of the components of the state vector follow by numerical integration of (3-2)..(3-5).
The flight path angle γ is then calculated, see Fig. 3, from:

$$\gamma = \arctan \frac{C_e}{U} \qquad (3\text{-}21)$$

and α from (3-15).
In Section 4 the flight path reconstruction method will be described in greater detail. It follows that the flight path reconstruction can be formulated as a statistical estimation problem. The problem is to estimate the initial state vector $x(t_o)$ and constant bias errors in the time histories of A_x, A_z and q from noisy measurements of Δh and V. To see this some aspects of the instrumentation system have to be discussed first.

3.3 Some aspects of the instrumentation system

The instrumentation system employed in the flight test series under discussion comprises several high accuracy inertial and air pressure transducers. Analog signals from the transducers are conditioned to range from zero to 10.000 mV dc and then filtered by identical fourth order low-pass filters. The dc outputs of the filters are sampled and digitized by an analog to digital convertor (digital voltmeter) which has a resolution of 0.01°/o of full scale i.e. 1 mV. The digital outputs of the voltmeter are stored on magnetic tape.
Filtering the analog signals from the transducers prior to the sampling is thought to be essential in present flight test technique. The frequency contents of the transducer outputs due to aircraft

motions generated by the pilot range usually from zero up to about 3 rad/sec.
These frequencies are separated by the low-pass filters from the higher frequencies due to e.g. structural vibrations, electromagnetic interference from other onboard systems and transducer noise.
The fourth order linear filters each consist of two identical second order filters connected in series. The second order filters have an undamped natural frequency of 19 rad/sec and damping ratio of 0.691 chosen such as to obtain a constant gain and a reasonably linear phase characteristic in the range of the pilot induced frequency contents of the signal.
This effectively results in a constant and for all transducers equal time shift of the filtered signals relative to the transducer output signals.
The inertial transducers i.e. the accelerometers and pitch rate gyro are sampled at a rate of 10 times per second.
Other transducers such as the air pressure transducers measuring the total air pressure and the change of static pressure (from which V and Δh are derived) are sampled at the lower rate of 5 times per second.
A typical property of the inertial transducers is the zero-shift between calibration curves from different calibrations. Fig. 4 illustrates this for two types of accelerometers. The Donner type 4310 was used in the present series of tests. The magnitude of the zero-shifts of these accelerometers is typically in the order of 5 - 10 mV corresponding to 0.01 - 0.02 m/sec^2.
The pitch rate gyro behaves similarly. For this transducer zero-shifts in the order of 5 mV (0.02 °/sec) are common.
Apart from the zero-shifts, the remaining random errors of A_x, A_z and q after applying the proper calibration results probably are in the order of 1 mV.
In contrast with the inertial transducers, the zero-shifts of the pressure transducers are easily measured in flight before and after every measuring run and can subsequently be corrected for. The remaining errors of these transducers are assumed to be adequately represented by ''white noise'' processes.
They amount to several mV's corresponding to 0.20 m and 0.15 m/sec in Δh and V respectively.
The noisy transducer outputs are filtered and thus the recorded noise from the transducers can - in principle - no longer be considered as ''white''.
The sampling interval of 0.2 sec, however, is such that the correlation between successive measurement errors is small. From the filter characteristics mentioned follows:

$$r(0.2) \cong 0.2\, r(0)$$

$$r(0.4) \cong 0$$

4. MAXIMALLY ACHIEVABLE ACCURACIES OF THE FLIGHT PATH RECONSTRUCTION

From Section 3 follows that the main purpose of the flight path reconstruction is the estimation of the time histories of the angle of attack. This variable is derived from components of the state vector x that cannot well be measured in flight directly, i.e. C_e and Θ. (3-21) and (3-15) yield the flight path angle γ and finally the angle of attack α.
The mathematical model of the flight path reconstruction is now completed.
The zero-shifts of the inertial transducers ΔA_{x_o}, ΔA_{z_o} and Δq_o are defined by:

$$\begin{aligned} A_x &= A_{x_m} + \Delta A_{x_o} \\ A_z &= Z_{z_m} + \Delta A_{z_o} \\ q &= q_m + \Delta q_o \end{aligned} \qquad (4\text{-}1)$$

Relations (4-1) are substituted in (3-3), (3-4) and (3-5). We define the input vector of the system (3-2)..(3-5) by:

$$u = \text{col}\,[A_{x_m} \quad A_{z_m} \quad q_m] \qquad (4\text{-}2)$$

The zero-shifts ΔA_{x_o}, ΔA_{z_o} and Δq_o are unknown as are the components of the initial state vector $x(t_o)$. Thus a parameter vector β may be defined:

$$\beta = \text{col}\,[\Delta A_{x_o} \quad \Delta A_{z_o} \quad \Delta q_o \quad \Delta h(t_o) \quad U(t_o) \quad C_e(t_o) \quad \Theta(t_o)]$$

(3-2)..(3-5) may now be written as:

$$\dot{x} = f(x, u, \beta) \qquad (4\text{-}3)$$

in which f denotes a vector function of appropriate dimension. The observation vector y is defined:

$$y = \text{col}\,[\Delta h \;\; U] = Hx \qquad (4\text{-}4)$$

From Fig. 3 follows:

$$U = V \cos \gamma \qquad (4\text{-}5)$$

In order to keep the discussion in this Section as simple as possible and because γ is small we assume initially:

$$U_m = V_m$$

Δh and U are measured at descrete times at a rate of 5 per second. The measurement errors are assumed to be adequately represented by an additive independent gaussian random sequence with zero mean, thus we have at time t_i during the manoeuvre:

$$z(t_i) = y(t_i) + \varepsilon(t_i) \qquad (4\text{-}6)$$

where z represents the measurement vector and:

$$E\ \varepsilon(t_i) = 0$$

$$E\ \varepsilon(t_i)\ \varepsilon^T(t_j) = \delta_{ij}R$$

where R is the measurement error covariance matrix. Due to the high quality of the inertial transducers, the random errors of the input vector u are very small compared to the random measurement errors of the observation vector and are therefore assumed to be zero.
The flight path reconstruction can now be formulated as a statistical estimation problem: estimate the parameter vector β of the system (3-2)..(3-5) from the noisy measurements $z(t_i)$.
The assumption that the input vector u is measured with infinite accuracy results in the case of additive independent gaussian measurement noise in $z(t_i)$ into a very simple expression for the likelihood function of the measurements i.e. the probability density function of the measurements conditioned on the parameter vector β and the measurement error covariance matrix R, e.g. Grove et al. (1972):

$$p(z(t_o), \ldots, z(t_n) ; \beta, R) =$$

$$(2\pi.\det R)^{-\frac{n}{2}} \exp \left\{-\tfrac{1}{2} \sum_{i=0}^{n} [z(t_i) - y(t_i)]^T R^{-1} [z(t_i - y(t_i)]\right\} \qquad (4\text{-}7)$$

In Section 5 the present estimation problem has been solved by maximizing L, the logarithm of the likelihood function (4-7) with respect to the parameter vector β. The resulting estimate β is called the maximum likelihood estimate of the parameter vector β. Once β has been derived, the time histories of the components of the state vector follow directly from numerical integration of (3-2)..(3-5), whence γ and finally α follow from (3-21) and (3-15).

As important question in any estimation problem is what accuracies might reasonably be expected. A lower bound on the covariance matrix of the estimation error of β is given by the Cramér-Rao inequality, Aström(1967),

$$E\ [\hat{\beta} - \beta]\ [\hat{\beta} - \beta]^T \geqq J^{-1} \qquad (4\text{-}8)$$

J represents the Fisher information matrix defined as:

$$J = E \left[\frac{\partial L}{\partial \beta}\right] \left[\frac{\partial L}{\partial \beta}\right]^T \qquad (4\text{-}9)$$

Under certain general conditions maximum likelihood estimates are consistent and asymptotically efficient, Cramér (1946).
This means that the accuracy of the estimated parameter β tends to the maximally achievable accuracy as the number of measurements tends to infinity:

$$E\ [\hat{\beta} - \beta]\ [\hat{\beta} - \beta]^T = J^{-1} \qquad (4\text{-}10)$$

$$n \longrightarrow \infty$$

Because of (4-10) it was thought that some insight into the estimation problem could be gained by first evaluating J^{-1} by means of a numerical simulation experiment, before actually solving the estimation problem.
The information matrix can be written as, Grove et al. (1972):

$$J = \sum_{i=0}^{n} \left[\frac{\partial y(t_i)}{\partial\ \beta}\right] R^{-1} \left[\frac{\partial y(t_i)}{\partial\ \beta}\right]^T \qquad (4\text{-}11)$$

From (4-4) it follows:

$$\frac{\partial y}{\partial \beta_j} = H \frac{\partial x}{\partial \beta_j}, \quad j = 1,\ldots,7 \qquad (4\text{-}12)$$

The partial derivatives $\frac{\partial x}{\partial \beta_j}$ follow from the sensitivity equations:

$$\frac{d}{dt}\left(\frac{\partial x_i}{\partial \beta_j}\right) = \sum_{k=1}^{4} \frac{\partial f_i}{\partial x_k} \cdot \frac{\partial x_k}{\partial \beta_j} + \frac{\partial f_i}{\partial \beta_j}$$

$$\frac{\partial x_i}{\partial \beta_j}(t_o) = 0 \qquad (4\text{-}13)$$

$$i = 1,\ldots,4$$
$$j = 1,2,3$$

$$\frac{d}{dt}\left(\frac{\partial x_i}{\partial \beta_j}\right) = \sum_{k=1}^{4} \frac{\partial f_i}{\partial x_k} \cdot \frac{\partial x_k}{\partial \beta_j}$$

$$\frac{\partial x_i}{\partial \beta_j}(t_o) = \delta_{i(j-3)} \qquad (4\text{-}14)$$

$$i = 1,\ldots,4$$
$$j = 4,\ldots,7$$

In the simulation experiment $x(t_o)$ is known in addition to the time histories of the components of u.
Thus J^{-1} can now be calculated by numerically integrating (3-2)...(3-5), (4-13) and (4-14).
The results obtained for a representative measurement error covariance matrix:

$$R = \begin{bmatrix} (0.20)^2 & 0 \\ 0 & (0.15)^2 \end{bmatrix}$$

are presented in Table 1. It appears that the correlation coefficient of estimation errors of ΔA_{x_o} and ΔA_{z_o} almost equals one. This also follows clearly from the one-sigma concentration ellipse of the marginal two dimensional distribution of the estimation errors of ΔA_{x_o} and ΔA_{z_o}, Fig. 6.

From:

$$\delta x = \left[\frac{\partial x}{\partial \beta}\right]^T \delta\beta \qquad (4\text{-}15)$$

follows that the covariance matrix of the state vector as a function of time may be deduced from the covariance matrix of the parameter vector β according to

$$E\ \delta\hat{x}\delta\hat{x}^T = \left[\frac{\partial x}{\partial \beta}\right]^T E\ \delta\hat{\beta}\delta\hat{\beta}^T \left[\frac{\partial x}{\partial \beta}\right] \qquad (4\text{-}16)$$

The maximally achievable accuracies follow by substitution of:

$$E\ \delta\hat{\beta}\delta\hat{\beta}^T = J^{-1} \qquad (4\text{-}17)$$

The results are presented in Fig. 7 and Fig. 8. The accuracies of other variables of interest also follow from the covariance matrix of the parameter vector β. This has been shown in the appendix for the flight path angle γ, the angle of attack α and C_{e_2}, the rate of climb in steady flight relative to earth.
By substitution of (4-17) follow the maximally achievable accuracies of γ and α, see Fig. 8.

From the maximally achievable accuracy of C_{e_2} the maximally achievable accuracy of performance i.e. rate of climb in steady flight relative to the surrounding atmosphere can be derived. Errors of C_{e_2} and C_2 satisfy (3-19):

$$\delta C_2 = \delta C_{e_2} - w_w \qquad (4\text{-}18)$$

In (4-18) the vertical wind component w_w now is interpreted as a random variable which has different values during different flight test manoeuvres but is constant during any particular manoeuvre. It is assumed that $E(w_w) = 0$. Because of $E(\delta\hat{C}_{e_2} w_w) = 0$ it follows that:

$$E(\delta\hat{C}_2)^2 = E(\delta\hat{C}_{e_2})^2 + E(w_w)^2 \qquad (4\text{-}19)$$

Thus the maximally achievable accuracy of performance (C_2) is a function of the maximally achievable accuracy of C_{e_2} and of the standard deviation of the vertical wind component w_w.
This has been illustrated in Fig. 9.
Vertical wind components of 0.1 to 0.2 m/sec are not at all uncommon during flight tests. Thus it follows from Fig. 9 that errors in performance derived from measurements in nonsteady flight must be ascribed primarily to the presence of unknown vertical wind components.
This conclusion holds only true however for the flight test series discussed in this paper, i.e. a low performance aircraft employing a high accuracy instrumentation system.

5. RESULTS OF FLIGHT TESTS

The flight path reconstruction problem has been solved by maximum likelihood estimation. L, the logarithm of the likelihood function (4-7) is maximized with respect to the parameter vector β subject to the constraints (3-2)..(3-5).
Once β is known, x, γ and α follow directly from the equations of motion (3-2)..(3-5), (3-21) and (3-15).
L has been maximized employing the well known iteration scheme called ''quasi-linearization'' or ''modified Newton-Raphson'', e.g. Taylor et al. (1969).
In the algorithm starting values of the components of β must be specified. In most cases convergence was achieved within 10 iterations by substituting:

$$\Delta A_{x_o} = \Delta A_{z_o} = \Delta q_o = \Delta h(t_o) = 0$$

$$U(t_o) = V_m(t_o)$$

The nonsteady manoeuvres flown in the flight tests under consideration start from a very nearly steady flight condition, thus a first rough estimate of the remaining elements of β, viz. the pitch angle and rate of climb at the start of the manoeuvre, $\Theta(t_o)$ and $C_e(t_o)$, could be obtained from (3-6) and (3-1). Already after the first iteration the flight path angle γ follows accurately enough from (3-21) as to permit replacing $U_m = V_m$ by the more accurate relation:

$$U_m = V_m \cos\gamma$$

After convergence of the iteration has been achieved, the covariance matrix of the measurement errors can be estimated from

$$\hat{R} = \frac{1}{n}\sum_{i=0}^{n} \begin{bmatrix} \Delta h_m(t_i) - \Delta\hat{h}(t_i) \\ U_m(t_i) - \hat{U}(t_i) \end{bmatrix} \cdot \begin{bmatrix} \Delta h_m(t_i) - \Delta\hat{h}(t_i) \\ U_m(t_i) - \hat{U}(t_i) \end{bmatrix}^T$$

The covariance matrix of the parameters then follows:

$$E\,[\hat{\beta} - \beta]\,[\hat{\beta} - \beta]^T =$$

$$= \left[\sum_{i=0}^{n} \left[\frac{\partial \hat{y}(t_i)}{\partial \beta}\right] \hat{R}^{-1} \left[\frac{\partial \hat{y}(t_i)}{\partial \beta}\right]^T\right]^{-1}$$

and finally by substitution in (4-16) the covariance matrix of δx is obtained as a function of time. The resulting covariance matrices from actual flight test data closely resembled the

corresponding matrices derived in Section 4 by applying the Cramér-Rao lower bound to a simulation experiment.
The correlation functions of the random sequences $\Delta h_m(t_i) - \Delta\hat{h}(t_i)$ and $U_m(t_i) - \hat{U}(t_i)$ were calculated in order to check the assumption made in Section 4 concerning independence. For one particular manoeuvre, representative in this respect, however, for many others, the results are presented in Fig. 10. The cause of the correlation between measurement errors of U has not yet been established.
From 7 nonsteady manoeuvres performance was derived. At six values of V the standard deviations of C_{e_2} were plotted in Fig. 9.

The magnitude of the standard deviations compared to the curve $\sigma_{w_w} = 0$ demonstrates the influence of the vertical wind on the accuracy of performance. It also clearly establishes w_w as the primary source of the scatter in the experimental results.
A similar scatter was obtained from the conventional measuring technique in steady flights Hosman (1971).

6. CONCLUSIONS

In this paper a method has been described to measure aircraft performance, characterizing steady flight conditions of the aircraft from measurements in nonsteady or quasi-steady flight conditions. This yields a remarkable reduction in flight test time required because from one measuring run the complete performance curve can be derived compared to the conventional measuring technique which would yield just one point in the curve in the same time.
A flight path reconstruction is essential in the analysis of nonsteady flight measurements. From the flight path reconstruction follow the time histories of those variables which cannot well be measured in flight directly by means of present day instrumentation techniques, in addition to the zero shifts of the inertial transducers which have to be corrected for.
The flight path reconstruction problem has been formulated as a statistical estimation problem. The Cramér-Rao lower bound theorem has been applied to a simulation experiment yielding the maximally achievable accuracies of the zero-shifts, the time histories and of performance. Thus considerable insight into the estimation problem could be gained before actually solving the flight path reconstruction problem itself.
The flight path estimation problem has been solved by means of maximum likelihood estimation. It is concluded that with the instrumentation system used for the flight tests discussed in this paper, the unknown vertical wind component is the major error source in the performance data.
From seven nonsteady flights followed that the standard deviation of the vertical atmospheric wind component amounted to about 0.2 m/sec when the measurements were made.

REFERENCES

AGARD, flighttestmanual, volume I.

Aström, K.J. (1967),
''On the achievable accuracy in identification problems''
IFAC Symposium on identification in automatic control systems, Prague.

Cramér, H. (1946),
''Mathematical methods of statistics''
Princeton University Press, Princeton.

Etkin, B. (1959),
''Dynamics of flight''
John Wiley and Sons, New York.

Gerlach, O.H. (1964),
''Analyse van een mogelijke methode voor het meten van prestatie en stabiliteits- en besturingseigenschappen van een vliegtuig in niet-stationaire symmetrische vluchten''
Doctoral Thesis (In Dutch with summary in English'.
Rep. VTH-117, Delft University of Technology.

Gerlach, O.H. (1971),
''The determination of stability derivatives and performance characteristics from dynamic manoeuvres''
Rep. VTH-163, Delft University of Technology.

Grove, R.D., R.L. Bowles and S.C. Mayhew (1972),
''A procedure for estimating stability and control parameters from flight test data by using maximum likelihood methods employing a real-time digital system''
NASA TN D-6735.

Hosman, R.A.J.W. (1971),
''A method to derive angle of pitch, flight path angle and angle of attack from measurements in nonsteady flight''
Rep. VTH-156, Delft University of Technology.

Jonkers, H.L., J.A. Mulder and K. van Woerkom (1972),
''Measurements in nonsteady flight: instrumentation and analysis''
7th International Aerospace Instrumentation Symposium, Cranfield, England.

Klopfenstein, H.B. (1965),
"Obtaining airplane drag data from nonsteady flight"
AIAA paper no. 65-211.

Mulder, J.A. (1973),
"Maximum likelihood estimation of performance from nonsteady flight data"
Rep. VTH 174, Delft University of Technology.
To be published.

Tailor, Jr., L.W., K.W. Ilif and B.G. Powers (1969),
"A comparison of Newton-Raphson and other methods for determining stability derivatives from flight data"
AIAA paper no. 69-315.

Wolowicz, C.H. (1966),
"Considerations in the determination of stability and control derivatives and dynamic characteristics from flight data"
AGARD Report 549 - Part I.

APPENDIX

The covariance matrix of estimation errors of the flight path angle γ_1, the angle of attack α_1 and the rate of climb C_{e_2} in the "corresponding" steady flight condition can be obtained from the covariance matrix of estimation errors of the parameter vector β at some time t_i during the nonsteady manoeuvre.
The augmented state vector is defined as:

$$x_a = \text{col}\,[\Delta h \; U \; C_e \; \theta \; \Delta A_{x_o} \; \Delta A_{z_o} \; \Delta q_o]$$

From (4-15) follows:

$$\delta x_a = \left[\frac{\partial x_a}{\partial \beta}\right]^T \delta\beta \qquad \text{(a-1)}$$

$$\frac{\partial x_{a_i}}{\partial \beta_j} = \frac{\partial x_i}{\partial \beta_j} \qquad \begin{array}{l} i = 1,\ldots,4 \\ j = 1,\ldots,7 \end{array}$$

$$\frac{\partial x_{a_i}}{\partial \beta_j} = \delta_{ij} \qquad \begin{array}{l} i = 5, 6, 7 \\ j = 1,\ldots,7 \end{array}$$

Thus:

$$E\, \delta\hat{x}_a \delta\hat{x}_a^T = \left[\frac{\partial x_a}{\partial \beta}\right]^T E\, \delta\hat{\beta}\delta\hat{\beta}^T \left[\frac{\partial x_a}{\partial \beta}\right] \qquad \text{(a-2)}$$

The flight path angle γ_1 and the angle of attack α_1 during the manoeuvre are obtained from (3-21) and (3-15)

$$\gamma_1 = \arctan \frac{C_{e_1}}{U_1} \qquad \text{(a-3)}$$

and

$$\alpha_1 = \theta_1 - \gamma_1 \qquad \text{(a-4)}$$

The rate of climb in steady flight relative to earth follows from (3-18):

$$C_{e_2} \cong V_1 \sin(\theta_2 - \alpha_1) = U_1 \tan(\theta_2 - \alpha_1) \qquad \text{(a-5)}$$

In (3-18) the pitch angle in steady flight follows from

$$\theta_2 \cong -\,\text{acrtan} \frac{A_{x_1}}{A_{z_1}} \qquad \text{(3-13)}$$

after substitution of

$$\begin{aligned} A_{x_1} &= A_{x_{m_1}} + \Delta\hat{A}_{x_o} \\ A_{z_1} &= A_{z_{m_1}} + \Delta\hat{A}_{z_o} \end{aligned} \qquad \text{(a-6)}$$

The relations (a-3), (a-4), (a-5), (3-13) and (a-6) are linearized for "small" errors in γ_1, α_1, C_{e_2} and U_1, C_{e_2}, θ_1, ΔA_{x_o}, ΔA_{z_o} and subsequently written as:

$$\begin{bmatrix} \delta\gamma_1 \\ \delta\alpha_1 \\ \delta C_{e_2} \end{bmatrix} = S^T \delta x_a \qquad \text{(a-7)}$$

in which S denotes a matrix of appropriate dimension. The covariance matrix of γ_1, α_1 and C_{e_2} can now be derived from (a-2) and (a-7):

$$E \begin{bmatrix} \delta\hat{\gamma}_1 \\ \delta\hat{\alpha}_1 \\ \delta\hat{C}_{e_2} \end{bmatrix} \begin{bmatrix} \delta\hat{\gamma}_1 \\ \delta\hat{\alpha}_1 \\ \delta\hat{C}_{e_2} \end{bmatrix}^T =$$

$$= S^T \left[\frac{\partial x_a}{\partial \beta}\right]^T E\, \delta\hat{\beta}\delta\hat{\beta}^T \left[\frac{\partial x_a}{\partial \beta}\right] S \qquad \text{(a-8)}$$

ΔA_{x_o}	ΔA_{z_o}	Δq_o	$\Delta h(t_o)$	$U(t_o)$	$C_e(t_o)$	$\theta(t_o)$	
0.0028 m/sec^2	0.98	0.25	0.65	-0.16	-0.90	0.78	ΔA_{x_o}
	0.0005 m/sec^2	0.26	0.53	-0.13	-0.81	0.76	ΔA_{z_o}
		0.0004^o/sec	0.04	-0.75	-0.11	-0.40	Δq_o
			0.045 m	-0.03	-0.87	0.57	$\Delta h(t_o)$
				0.026 m/sec	0.85	0.42	$U(t_o)$
					0.006 m/sec	-0.76	$C_e(t_o)$
						0.018^o	$\theta(t_o)$

Table 1. Standard deviations and simple correlation coefficients of the parameters in a flight path reconstruction problem derived from the Cramér-Rao lower bound.

Fig. 1. The De Havilland DHC-2 ''Beaver'' laboratory aircraft of the Delft University of Technology.

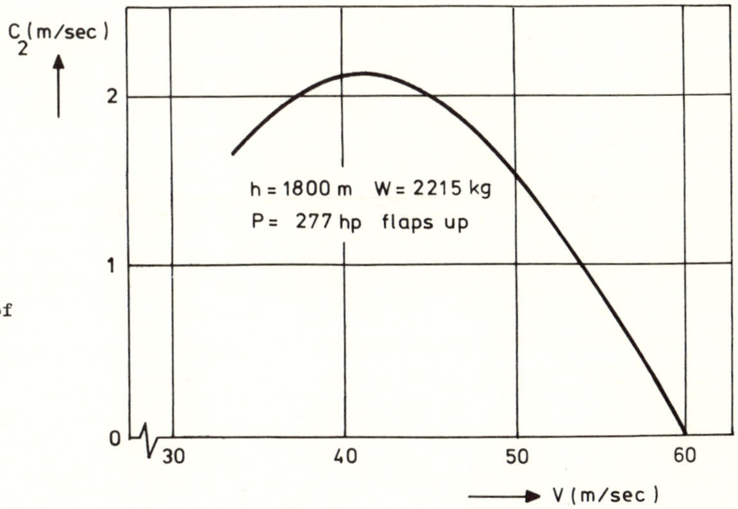

Tig. 2 Performance as a function of airspeed of the ''Beaver'' laboratory aircraft.

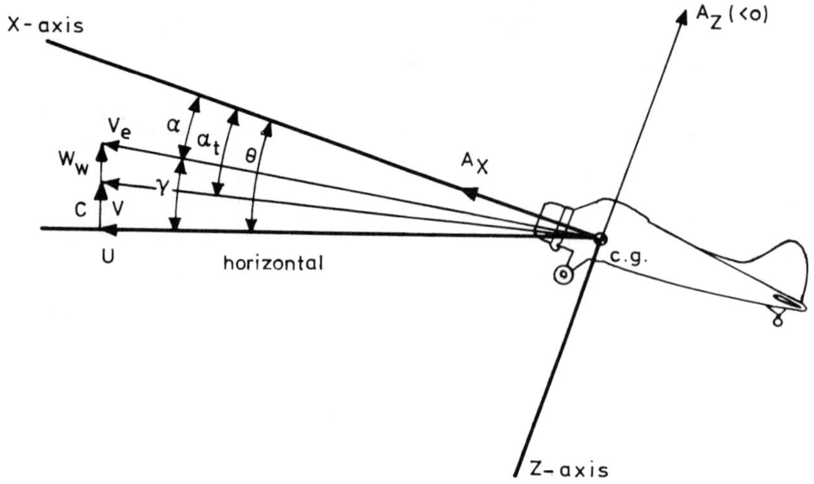

Fig. 3 Definition of several variables in the body-fixed axes system.

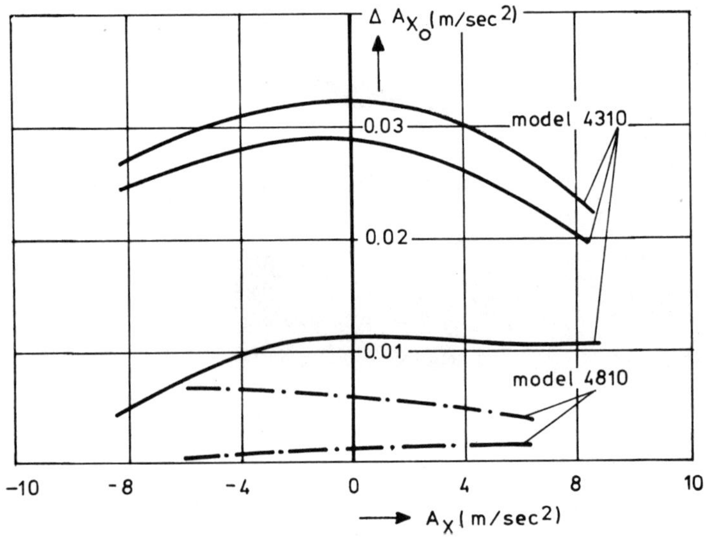

Fig. 4
Mutual deviations between successive calibration curves of two different types of longitudinal accelerometers (Donner).

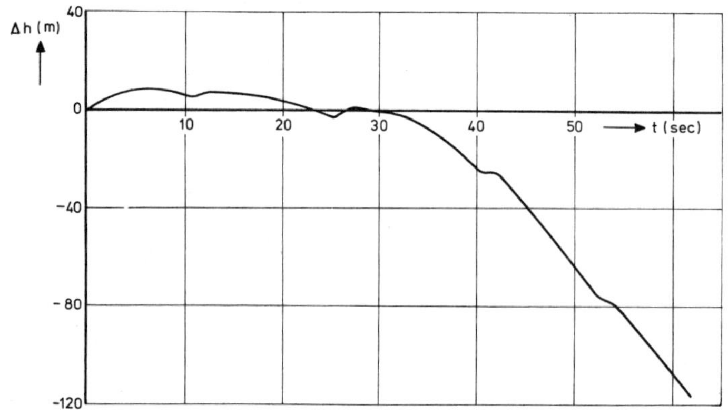

Fig. 5 Time histories of some variables during a non-steady manoeuvre.

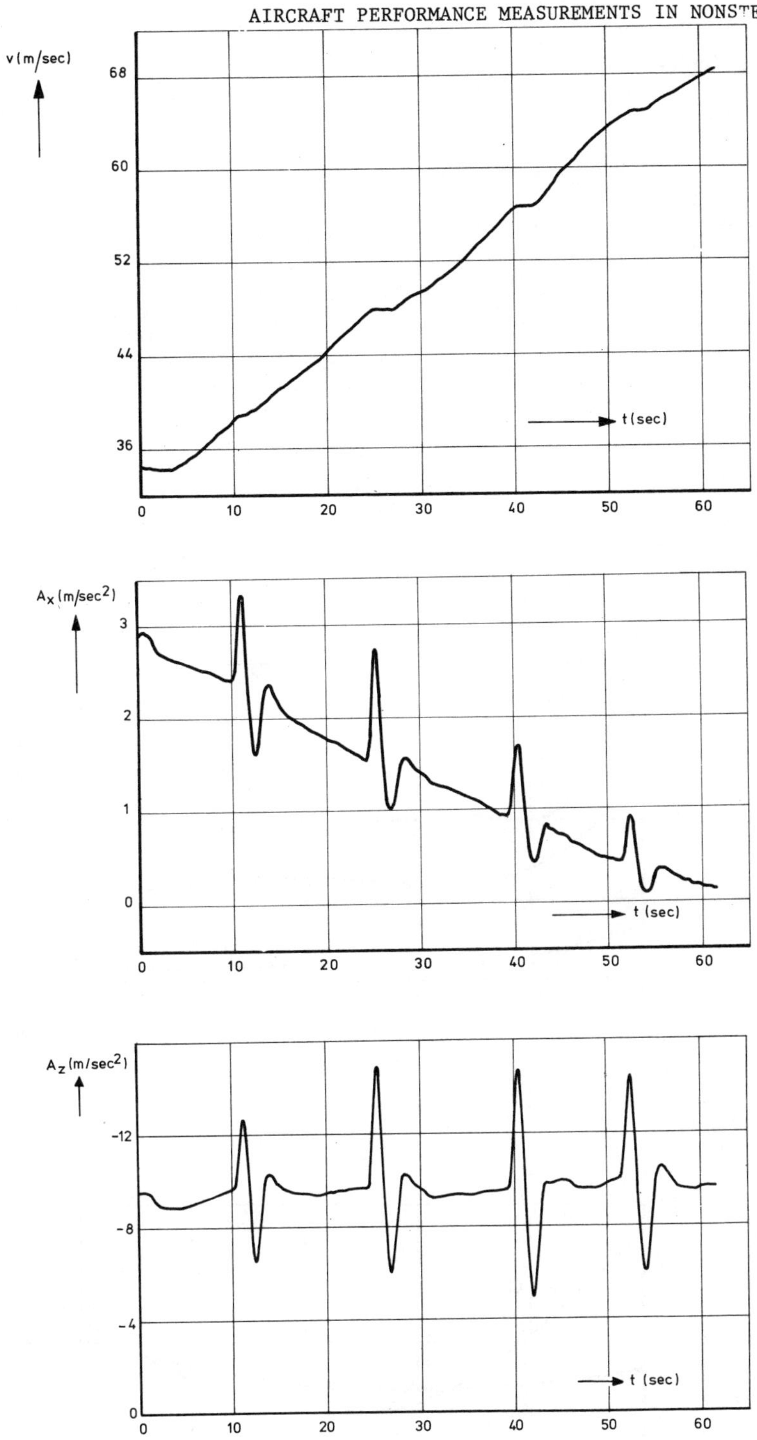

Fig. 5 Continued.

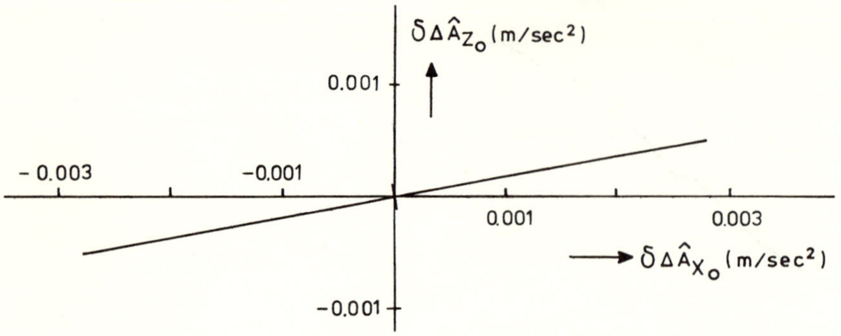

Fig. 6
The two axis of the one-sigma concentration ellipse of the marginal two dimensional probability density function of estimation errors of the zero-shifts of the vertical and longitudinal accelerometers.

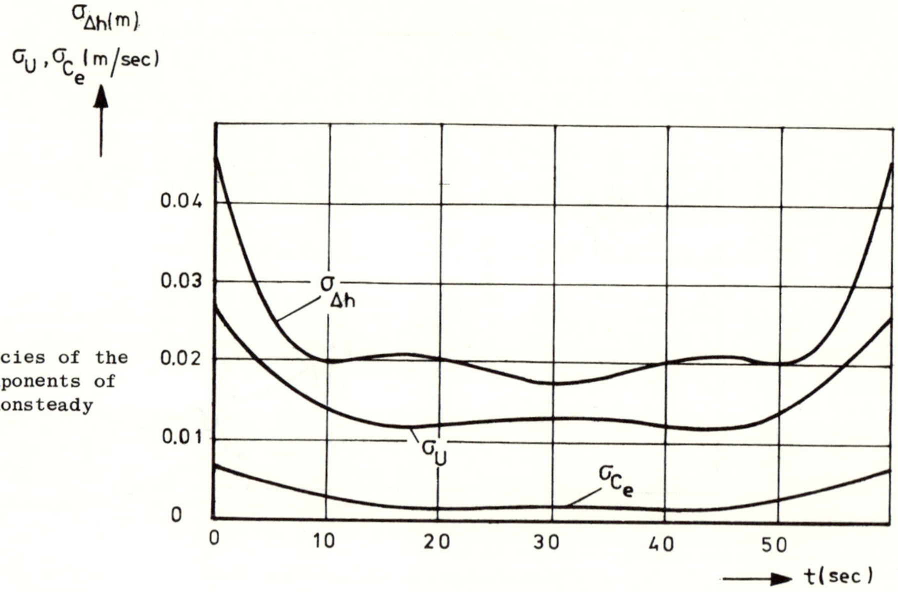

Fig. 7
Maximally achievable accuracies of the time histories of three components of the state vector during a nonsteady manoeuvre.

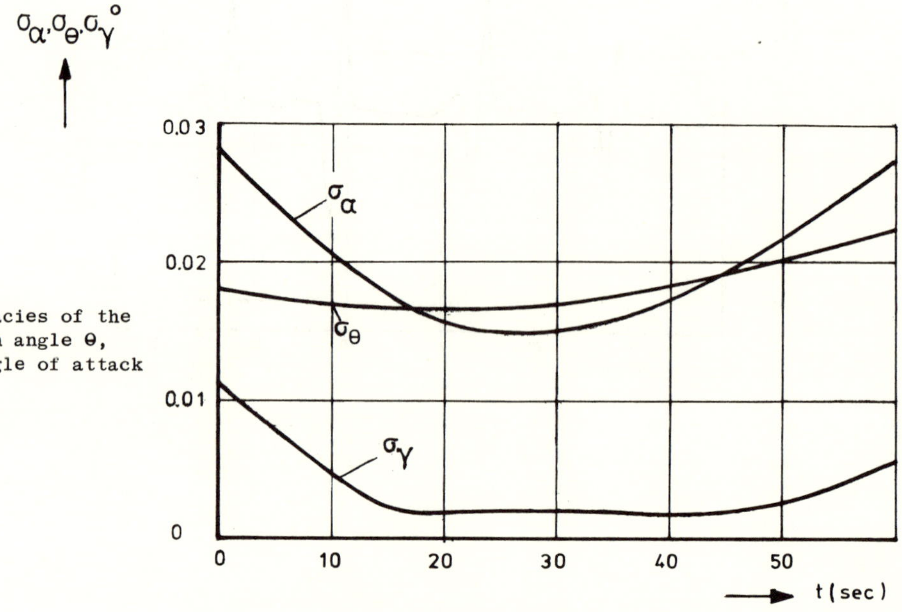

Fig. 8
Maximally achievable accuracies of the time histories of the pitch angle θ, flight path angle γ and angle of attack α.

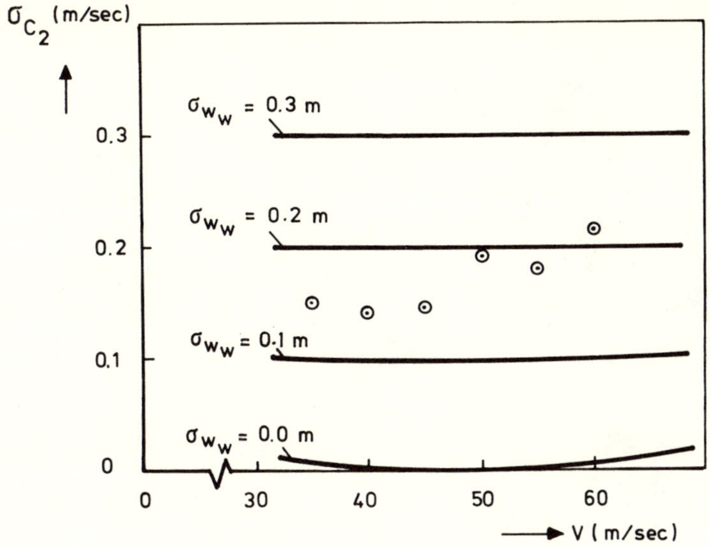

Fig. 9 Influence of a vertical wind component on the accuracy of performance. Encircled points denote the standard deviations of performance as actually derived from seven nonsteady manoeuvres.

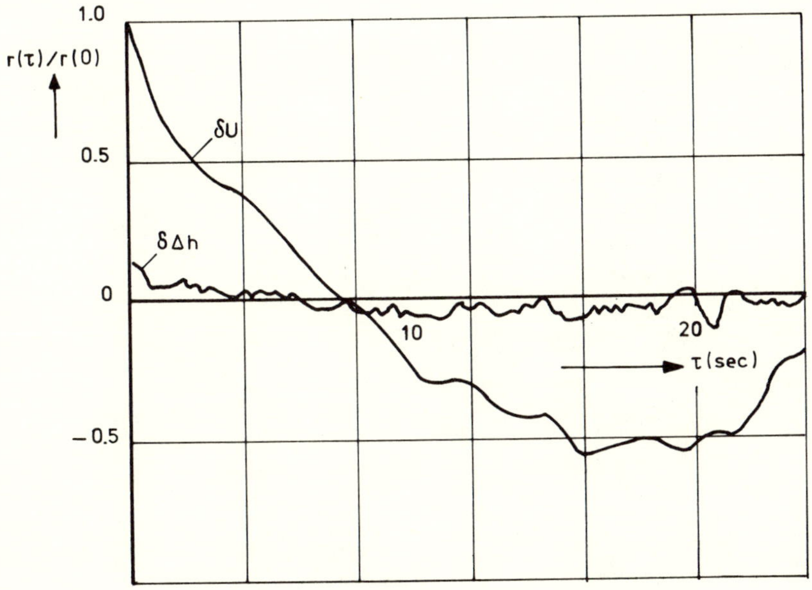

Fig. 10 Correlation functions of the measurement errors determined from an actual manoeuvre.

IDENTIFICATION OF PHYSICAL PARAMETERS OF A METALLURGICAL REHEATING FURNACE

D. VOLPERT - M. GAUVRIT - D. RAMBACH

CERT/D.E.R.A. - 2, av. E. Belin - 31, TOULOUSE (FRANCE)

INTRODUCTION

In a metallurgical plant, the reheating furnace is up-stream above the rolling mill. It must yield, at a given rate, slabs which can be rolled in good conditions. The conditions can be summed up in the following rule : the temperature of the slab at the out put of the furnace must be homogeneous and near a given value. This consideration leads the conduct of the furnace. To be able to predict the comportement of the slab in the furnace, one must define a model of the furnace and identify some parameter of this model.

I - MODELING OF THE FURNACE

Inside of the slab, the mean of heat transfer is conduction and the temperature follows the Fourier law :

$$\frac{\partial \Theta}{\partial t} = \frac{1}{C\rho}\left(\frac{\partial \lambda \frac{\partial \Theta}{\partial x}}{\partial x} + \frac{\partial \lambda \frac{\partial \Theta}{\partial y}}{\partial y} + \frac{\partial \lambda \frac{\partial \Theta}{\partial z}}{\partial z}\right) \qquad (1)$$

- C is the specific heat
- λ is the thermal conductibility
- ρ, the specific mass, is supposed constant.

Heat transfer between the furnace and the product is mainly done by radiation : a model of radiation is given by the Stefan Boltzman law :

$$\lambda \frac{\partial \Theta}{\partial x}\bigg| = \varepsilon \times K \times (T^4 - \Theta^4) \qquad (2)$$

- K is the radiation constant of a black body
- ε express that it is not a perfect black body
- T is the temperature in Kelvin degree of the vault of the furnace
- Θ is the temperature of the upper surface of the slab.

One can add to the flux computed on equation (2) a term of convection

$$\phi_{\text{convection}} = \alpha\,(T - \Theta) \qquad (3)$$

where α is an ill known coefficient.

If the slab are not joinning there is a lateral exchange which can be expressed by the following formula :

$$\lambda \frac{\partial \Theta}{\partial y} \Bigg| = \varepsilon \times K \times \left(1 - \frac{\sin \phi_x + \sin \psi_x}{2}\right)\left(T^4 - \Theta^4\right) \quad (4)$$

at the surface

point x

with :

$tg\ \phi_x = x / d$

$tg\ \psi_x = (e - x) / d$

d is the space between two slabs

e is the thickness of the slabs.
cf. Figure 1

The set of equation (1) , (2) and (4) make up the model. To be able to solve numerically the system, we have substituated to it a finite difference equation systeme. The numerical integration had to be consistant, convergent and stable. For that, we have chosen the alternating direction method of Peaceman and Rachford.

II - IDENTIFICATION

The identification was based on measurements done on an experimental furnace. Temperature has been continuously measured on eight different points of a slab, inside and on the surface.

The parameters to identify are mainly $\lambda(\Theta)$ and $C(\Theta)$. We have caracterized them through straight segments.(Figure 2)

To fix these parameters, one uses the measured temperature on the surface of the slab as boundary condition and integrate equation (1) to have the temperature inside. The comparison then from the temperature measured at a given point and the temperature computed brings out a relatively good criterion of the accuracy of the identification. The identification can be expressed in terms of minimization of a fonction of some parameters. To do that, we have used a Fibonnacci algorithm parameter being taken into account three by three. This constitute the identification of the parameters of the slab.

Since we have the temperature distribution inside the slab, we can now use the Stefan-Boltzman law (2) "up side down" to compute the temperature of the furnace. Compared to the assigned temperature, this allow us to identify the heat transfer in the furnace.

Figure 3 gives the evolution of some temperatures :

1 measured temperature of the upper surface

2 measured temperature of the bottom surface

3 measured temperature of an inside point

4 identified temperature of the same point

5 computed temperature of the furnace

6 assigned temperature of the furnace

The results of the identification have showned that the flux on lateral faces could be considered as a constant ratio of the flux of the upper surface.

In an industrial plant, one can consider that the flux on the bottom face is equal to zero. This hypothesis must be review in each particular case and the flux can be identified.

III - CONCLUSION

The following test has shown that the characterization is good enough.

- First, we have asumed λ and C constant and we have identified the value of these constants
- Then, we have assume that C had a pick and we have identified it
- Then, we have assume that λ was made of two straight lines
- And finally, we have identified the slope of C before the pick.

Each step brought an improvement in the identification until the complexity of the model was large enough. This work is an important step in the probleme of the direct digital control of the slab reheating furnace. Now, it is possible to test and fix the parameters of the simple model from which the process computer will predict the temperature.

BIBLIOGRAPHY

J. RICHALET, A. RAULT, R. POULIQUEN (1971) "*Théorie du modèle*". Gordon and Breach. Paris.

A.G. BUTKOVSKYI (1969) "*Distributed control systems*". American Elsevier Publishing Company. New York.

R. KISSEL "Chauffage dans les fours à longerons, combustion, oxydation" *Bulletin Heurtey n° 35*.

M. GAUVRIT, J.F. LE MAITRE, R. KISSEL, G. GOSSE Juin 1971 "Control of a slab reheating furnace" *Proceeding of the third IFAC/IFIP Conference*. Helsinki.

M. GAUVRIT, D. VOLPERT, J.F. LE MAITRE, R. KISSEL, G. GOSSE "Idenfification et optimisation des fours". *Rapport de fin d'étude*. Contrat D.G.R.S.T.

figure 1

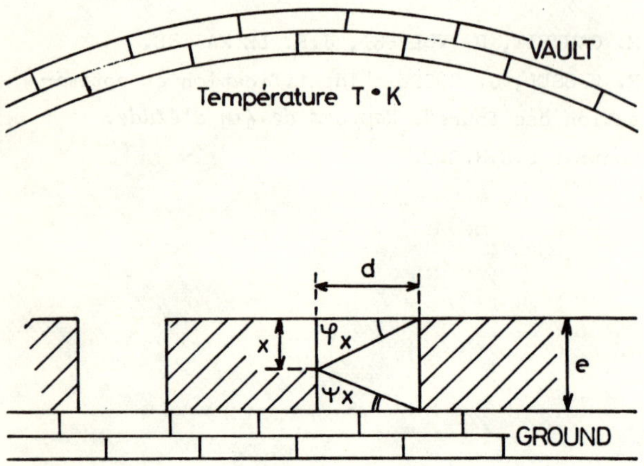

figure 3

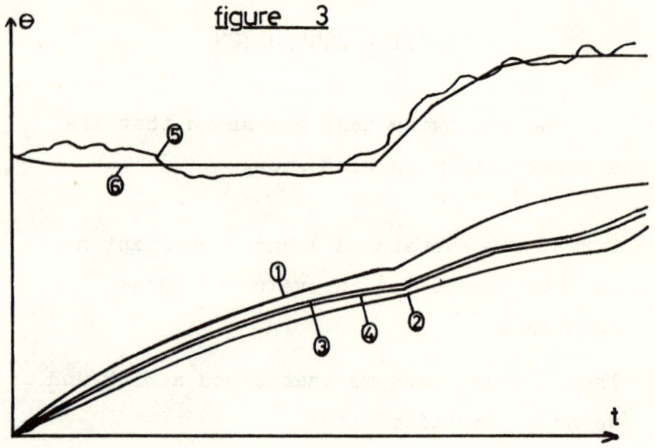

figure 2

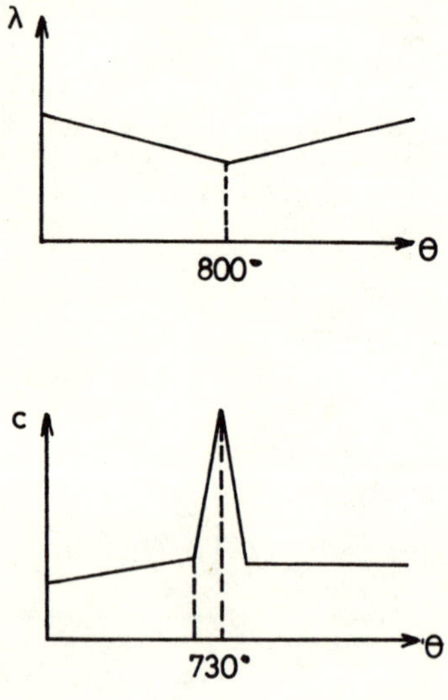

IDENTIFICATION OF PROCESS DYNAMICS IN THE GLASS INDUSTRY

B. V. Raja Rao, J. Mather and B. D. Lyons

Pilkington Brothers Limited
Watson Street Laboratories,
St. Helens, Lancashire, England. WA10 3TT

Two case studies on the application of modern process identification and parameter estimation techniques in the Glass Industry. In the first example the dynamics of a glass forming machine are identified using generalised least squares method and in the second case the dynamics of a glass melting process are identified by using analog computer (experimental) technique. The results are compared with those produced by maximum likelihood method. The process models obtained are used to compute the control laws which are then used to set up the controllers.

INTRODUCTION

The purpose of this short paper is to present two case studies on the development of dynamic models in the glass industry using modern parameter estimation techniques. Though the technical progress in the glass industry has been considerable in recent years with the development of the float process for making flat glass [1], the application of modern control engineering techniques are still in the early stages. Some of the developments in instrumentation and process control are discussed in Tooley [2], Mouly [3, 21 Ref.] and Raja Rao [4, over 300 Ref.] .

In the first example, the dynamic model of a glass slab forming process (for eventual automatic control of slab thickness) is discussed, while the second example is concerned with the study of the dynamics of a glass tank for optimum scheduling of raw materials. A brief description of the glass manufacturing process is given in Section 2 and the two case studies are presented in Sections 3 and 4. The results are summarised in the last section.

GLASS MANUFACTURING PROCESS

A block diagram of a typical glass manufacturing line is shown in Fig. 1. The basic glass production process, starting from batch, melting and forming followed by lehr procedures has not altered greatly although there have been modifying development in each, viz: float process, Fig. 2. Raw materials (sand, soda, limestone) and cullet (rejected glass) are mixed to required proportions and fed through hoppers into the glass melting tank. The molten glass then passes through a forming machine where a particular type of product is formed. There are different forming machines that produce a variety of glass products. The glass then passes through a zone of controlled cooling and heating (lehr) before it is inspected and stored in a warehouse. Different colouring components (e.g. iron oxide, cobalt oxide, etc.) will be added to the batch in small proportions to produce a required colour in the final product.

CASE STUDIES

Two practical examples of the applications of modern system identification and parameter estimation techniques in the glass industry are presented. No attempt is made to survey or present material on general identification methods as there are excellent surveys available already [5, 6]. These methods have not been applied so far in the glass industry, although a glass production line is similar to a paper or cement plant [7]. The following procedure was adopted to obtain the process model and to design the control system.

Design of experiment
Data collection
Data analysis
Parameter estimation
Conversion from discrete to continuous model
Design of controller
Implementation and evaluation of controller

In the first example the dynamics of a glass ribbon forming machine are obtained using the generalised least squares method [8] and in the second case an analog computer was used to obtain the transfer function of a glass melting tank.

IDENTIFICATION OF FORMING MACHINE DYNAMICS FROM PRBS TESTS DURING NORMAL OPERATION

This case study is concerned with a glass slab forming machine. It is required to maintain the glass ribbon thickness at a constant value, while the input glass flow rate decreases as a direct result of decreasing glass level in the tank. After a brief study of the process, the input and output variables were identified as drawing speed and thickness respectively. From knowledge of the process, a test signal (a form of PRBS) was designed and injected into the motor speed control signal. The main consideration here was that the nature and the amplitude of the extra injected signal was such that neither production rate nor the glass quality suffered noticeably. The input and output process data was recorded using an analog tape recorder and is shown in Fig. 3. The deviations from the mean levels in the input and output are shown in Figs. 4(a) and 4(b) respectively.

Once the appropriate data has been collected the process identification and parameter estimation problems are fairly simple in that they need only off-line computation. A knowledge of the process behaviour also indicated that a first or second order model with a time delay would be adequate for the purpose of achieving the required control.

The process parameters of both the models (first order with delay and second order with delay) have been estimated using the generalised least squares method [8]. As a first step both these models were selected. When the controllers were designed and constructed it was found that the first order model of the form

$$\frac{Y(s)}{U(s)} = \frac{K \exp(-s\tau_d)}{(1 + sT_1)}$$

where $K = 3$, $T_1 = 16$, $\tau_d = 2$; Y = thickness, U = motor speed

would give a similar performance to the second order model. Hence a first order model was finally selected and used for the design and construction of the controller. The controller designed using the process model has been functioning very well and has contributed to producing better quality product. This is mainly due to the fact that the product thickness could be predicted from the model and can be controlled to a tighter specification.

IDENTIFICATION OF GLASS TANK DYNAMICS FROM INPUT - OUTPUT DATA

One of the problems of continuously operated glass furnaces is the efficient change of glass composition without draining and refilling the furnace. These changeovers may be made by simultaneous changes in a number of batch components, but the changes can often be considered to be reasonably independent. Two methods have been used successfully to identify the furnace response to batch changes.

In the first method, smoothed step-response curves were calculated from radioactive tracer trials and have been analysed by graphical methods [9, 10]. An analog computer program based on the Strejc method [10] was given in Madden and Mason [11]. Approximate transfer functions were obtained relating glass properties to batch composition. The structure of these transfer functions was compatible with the known mixing properties of the furnace.

In other cases, analog computer simulations of mixing models have been used to fit transfer functions to input and output data from actual changes in composition [12]. Fig. 5 shows a typical case in which a batch change in excess of that required was fed before changing to the new target batch. The noise in the output data is largely due to sampling limitations and measurement errors, but did not significantly affect the parameter estimation. Models of varying degrees of complexity have been examined, but simple second- and third-order models with time delays were found to be adequate for determining optimum batch schedules.

These transfer functions were of the form

$$\frac{K \exp(-\tau_d s)}{(1 + Ts)^n}$$

which represents the response of n mixers in series, each with a time constant

$$T = \text{constant} \times \frac{\text{furnace capacity}}{\text{furnace throughput}}$$

For a glass processing line with a tank capacity of 1000 tons operating at a load of 2000 tons/week the typical parameters of the transfer function are

$K = 1$, $T = 35$ hours, $\tau_d = 6$ hours, $n = 2$

Practical considerations including the time-scale preclude the possibility of continuous control action, but the simulation of these models have been valuable to Production personnel in estimating the response of the furnace to changes in batch composition and predicting when the new glass will be within specification, and therefore saleable.

CONCLUDING REMARKS

Two case studies on the application of modern process identification and parameter estimation techniques in the glass industry are presented in this paper. In the first example, the dynamics of a glass forming machine are identified using generalised least squares method and in the second case, the dynamics of a glass melting tank are obtained using analog computer techniques. The control actions implemented using the process models have functioned successfully and contributed to producing better quality glass compared to the previous practice.

ACKNOWLEDGEMENTS

The authors wish to thank the Directors of Pilkington Brothers Ltd. and Dr. D. S. Oliver, Director of Group Research and Development for permission to publish this paper. They would also like to thank Dr. G. C. Goodwin of Imperial College of Science & Technology, London, for his helpful advice and assistance with the parameter estimation programs

REFERENCES

1. L. A. B. Pilkington, "The Float Glass Process - Review Lecture", Proc. Royal Society, A314, pp. 1-25, 1969.

2. F. V. Tooley, "Handbook of Glass Manufacture", Vol.1 & 2, Ogden, New York, 1953/61.

3. R. J. Mouly, "Systems Engineering in Glass Industry", Survey Paper, 8th Int. Congress on Glass, 1-6 July, 1968, London.

4. B. V. Raja Rao, "A Survey of Automatic Control in Glass Industry", to be presented at IFAC Conference on Automatic Control in Glass, September, 1973, Lafayette, U.S.A.

5. K. J. Astrom and P. Eykhoff, "System Identification - A Survey", Automatica, Vol.7, pp. 123-162, March 1971.

6. I. Gustavsson, "Comparison of Different Methods for Identification of Industrial Processes", Automatica, Vol.8, pp. 127-142, 1972.

7. L. A. Madonna, "The Glass Packaging Line - Chemical Systems Approach", The Glass Industry, Vol.48, pp. 543-547, October, 1967.

8. D. W. Clarke, "Generalised Least Squares Estimation of the Parameters of a Dynamic Model", 1st IFAC Symposium on Identification Techniques, Prague, 1967.

9. D. M. H. Platt, "The Response of a Glass Tank to Variations in the Batch Composition", The Glass Industry, Vol.49, pp. 428-432, August, 1968.

10. V. Strejc, "Approximate Determination of the Control Characteristics of an Aperiodic Response Process", Automatisme, Vol.5, pp. 109-111, No.3, 1960.

11. B. G. Madden and D. A. Mason, "A Program for On-line Process Identification", Proc. AICA, Vol.12, pp. 118-121, No.3, 1970.

12. B. J. Hoetink, "Process Dynamics of a Glass Furnace following a step change of one of the Batch Components", Glass Tech., Vol.10, pp. 84-89, June, 1969.

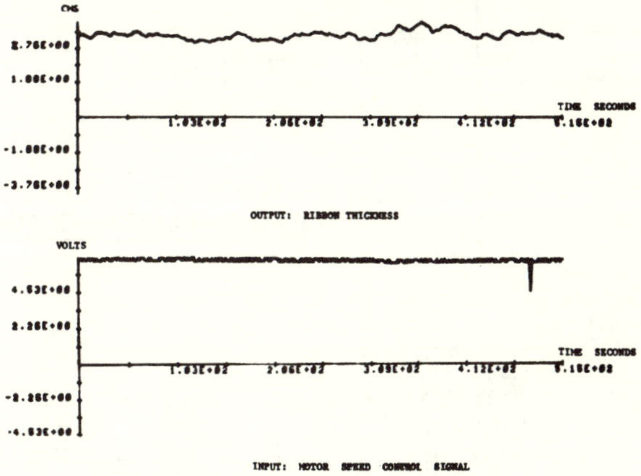

FIGURE 3 - PROCESS RECORDS

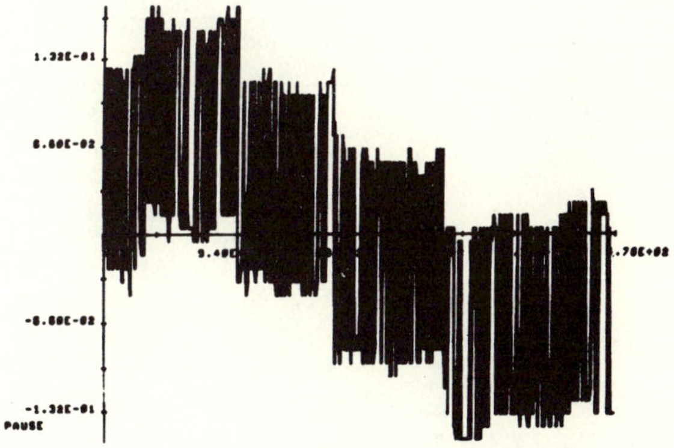

FIGURE 4 (a) - INPUT (MEAN LEVEL REMOVED)

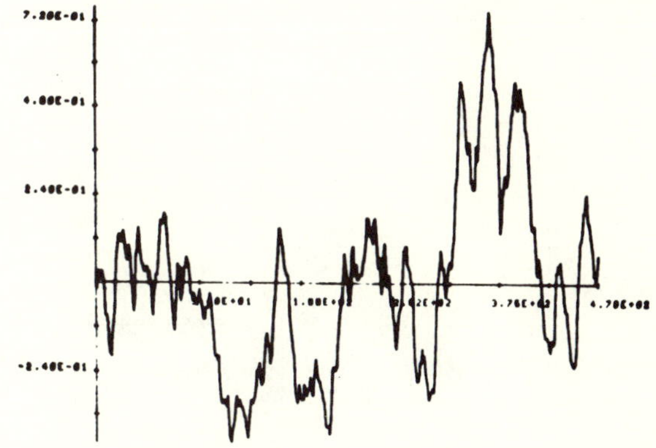

FIGURE 4 (b) - OUTPUT (MEAN LEVEL REMOVED)

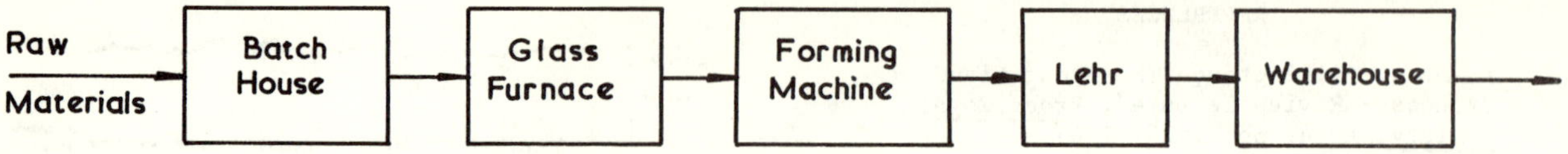

FIG I. BLOCK DIAGRAM OF A GLASS PLANT

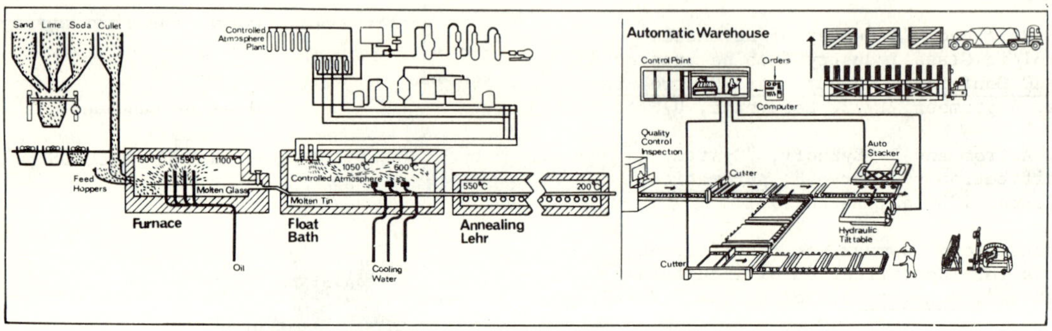

FIG. 2 *A float glass line, showing batch preparation, feeding, melting furnace, float bath with lehr, automatic cutting, storage, packing and dispatch.*

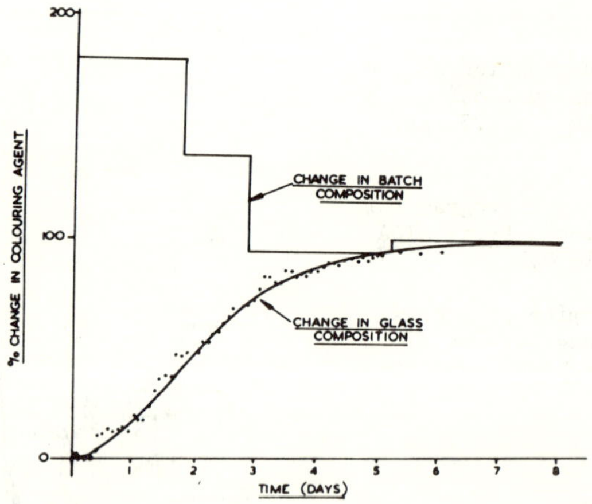

FIGURE 5 - GLASS TANK INPUT-OUTPUT DATA

ADAPTIVE ESTIMATION AND CONTROL OF LINEAR SYSTEMS

Otto J.M. Smith

University of California
Berkeley, California

Differential equation coefficients are most easily estimated as discrete-time equation coefficients. The inverse z transform thus yields the s-plane poles and their residues. Discrete-time signal matrices do not require integrations, and therefore save both computer storage and computer time. The mean of the noise is estimated to remove any coefficient-estimate-bias due to non-zero-mean signals. The many equations due to many samples are averaged to the few equations needed for the few coefficients by using a few-by-many noise-free matrix derived from the signals in a noise-free model. After averaging, the noise component in the equations has an expected value of zero, so the expected values of the estimated coefficients equal the true values. The variance of these unbiased estimates can be minimized on the last iteration by computing the estimated noise as the system output minus the model output, and then filtering the measurements with a noise rejection filter.

Pure dead time is easily estimated as n zeros at the origin of the inverse-z-plane, and these transform directly to e^{-nsT}. The effect of noise on the estimate is minimized by also estimating the noise, i.e., by selecting the number of unknown poles equal to the sum of the number of system poles plus the number of noise spectrum poles. The "energy" associated with each pole is a function of the s-plane residue times the time constant. When the noise level is low, the poles with the lowest "energies" can be sorted out as the noise poles. When the noise level is very high, the noise poles can be sorted out from a knowledge of the noise spectrum. The estimated system poles and their associated residues are very accurate and essentially unaffected by large errors in the noise estimates.

The transfer function for an unknown system can be computed from the input and output signals even when the output is contaminated with noise.

Figure 1 shows the block diagram of the unknown system and a model used to generate noise-free state variables. x(t) and w(t) are the measured input and output signals.

1. SYSTEM EQUATIONS

For K successive measurements uniformly spaced at T seconds apart, define vectors

$$X = [x(1),\dots,x(K)]^T$$
$$Y = [y(1),\dots,y(K)]^T$$
$$W = [w(1),\dots,w(K)]^T$$
$$E = [e(1),\dots,e(K)]^T \qquad (1)$$

Let delayed values of X and W be "state variables." Define

$$X_\tau = [x(2-\tau),\dots,x(K-\tau+1)]^T$$
$$Y_\tau = [y(2-\tau),\dots,y(K-\tau+1)]^T$$
$$W_\tau = [w(2-\tau),\dots,w(K-\tau+1)]^T \qquad (2)$$

Let the numerator transfer function have NA zeros and NA+1 constant coefficients A_j and deliver the signal

$$v(t) = A_1\, x(t) + A_2\, x(t-1) +,\dots,+ A_{NA}\, x(t-NA+1) + A_{NA+1}\, x(t-NA) \qquad (3)$$

Let the denominator transfer function be represented by its inverse with constant coefficients B_j.

$$v(t) + C(m) = B_1\, y(t) + B_2\, y(t-1)+,\dots,+ B_{NB}\, y(t-NB+1) + B_{NB+1}\, y(t-NB) \qquad (4)$$

The discrete-time differential equation for Fig. 1 for all K measurements is

$$\begin{bmatrix} X_1\ X_2,\dots,X_{NA+1}\ -Y_1\ -Y_2,\dots,-Y_{NB+1} & \begin{matrix}1.0\\ \dots\\ 1.0\end{matrix}\end{bmatrix}$$

$$\begin{bmatrix} A_1 \\ A_2 \\ \cdots \\ A_{NA+1} \\ B_1 \\ B_2 \\ \cdots \\ B_{NB+1} \\ C(m) \end{bmatrix} = 0 \qquad (5)$$

It is convenient to set $B_1 = 1.0$ and define the coefficient vector

$$C = [A_1,\ldots,A_{NA+1} \quad B_2,\ldots,B_{NB+1} \quad C(m)]^T \qquad (6)$$

Define the noise-free signal matrix which is K by (NA+NB+2):

$$\hat{S}^T = \begin{bmatrix} X_1 \; X_2,\ldots, \; X_{NA+1} & -Y_2,\ldots,-Y_{NB+1} & \begin{matrix}1.0\\ \cdots \\ 1.0\end{matrix} \end{bmatrix} \qquad (7)$$

The differential equation is

$$\hat{S}^T C = Y_1 \qquad (8)$$

The measurements w(t) and vectors W_T are noisy. The noisy signal matrix using W instead of Y is

$$S^T = \begin{bmatrix} X_1,\ldots,X_{NA+1} & -W_2,\ldots,-W_{NB+1} & \begin{matrix}1.0\\ \cdots \\ 1.0\end{matrix} \end{bmatrix} \qquad (9)$$

The differential equation in terms of dirty signals is

$$S^T C = W_1 + R \qquad (10)$$

where R is the vector of equivalent noise at the location of v(t) in the system.

2. ESTIMATION

The first estimate for C is given by the least squares solution:

$$SS^T C = SW_1 + SR \qquad (11)$$

SR must be discarded because it is unknown and unmeasurable.

$$\tilde{C} = (SS^T)^{-1} SW_1 \qquad (12)$$

Insert the estimated coefficients $\tilde{C}$ into a model of the system and generate noise-free estimates of the output y(t) and Y_T, and the vector of equation error E.

$$E = S^T \tilde{C} - W_1 \qquad (13)$$

$\tilde{y}(t)$ satisfies the simulation or implicit or regression equation

$$\tilde{Y}_1 = \tilde{S}^T \tilde{C}.$$

Assemble a noise-free signal matrix using $\tilde{Y}$ instead of Y, and use it for averaging equation (10). Define the noise-free signal matrix

$$\tilde{S}^T = \begin{bmatrix} X_1,\ldots,X_{NA+1} & -\tilde{Y}_2,\ldots, & -\tilde{Y}_{NB+1} & \begin{matrix}1.0\\ \cdots \\ 1.0\end{matrix} \end{bmatrix} \qquad (15)$$

Eliminate W_1 between equations (10) and (13)

$$S^T(C-\tilde{C}) = -E + R \qquad (16)$$

Let the correction in the coefficient vector be

$$\Delta C = (C - \tilde{C}) \qquad (17)$$

The equation to be solved for this correction is

$$S^T(\Delta C) = -E + R \qquad (18)$$

Average using $\tilde{S}$

$$\tilde{S}S^T(\Delta C) = -\tilde{S}E + \tilde{S}R \qquad (19)$$

Discard the unmeasurable and uncomputable $\tilde{S}R$.

$$(\Delta C) = -(\tilde{S}S^T)^{-1} \tilde{S}E \qquad (20)$$

This is the MINIMUM-ERROR-PROJECTION-ESTIMATE, abbreviated MEPEST.

3. PROOF

To derive the properties of this estimator, assume that a "best" estimate $\hat{C}$ exists. Then from equation (10) a "best" estimate of computed equation error and transformed noise is the K-vector

$$R = S^T\hat{C} - W_1 \qquad (21)$$

Define a reference "direction" D as a noise-free K-vector of dynamic error only, using again the hypothetical "best" estimate $\hat{C}$ and the noise-free signals y(t). From equation (8),

$$D = \hat{S}^T\hat{C} - Y_1 \qquad (22)$$

Now minimize the projection of the error R in the direction D. This picks out of R only the dynamic error and discriminates against the noise, which does not appear in D.

The scalar error to be minimized is the inner product of R and D, i.e., $\langle R,D \rangle$. This is the cross correlation between the dynamic error R and a noise-free signal. This is the projection of R on D times the norm of D.

$$\varepsilon = R^T D = D^T R = <R,D> = ||R||\cdot||D|| \cos \angle R,D \quad (23)$$

To minimize ε,

$$\frac{\partial \varepsilon}{\partial \hat{c}} = \frac{\partial D}{\partial \hat{c}}^T R + \frac{\partial R}{\partial \hat{c}}^T D = 0 \quad (24)$$

From equation (21)

$$\frac{\partial R}{\partial \hat{c}} = S^T \quad \text{(noisy or "dirty" state variables)} \quad (25)$$

From equation (22)

$$\frac{\partial D}{\partial \hat{c}} = \hat{S}^T \quad \text{(noise-free state varibles)} \quad (26)$$

Therefore

$$\hat{S}R + SD = 0 \quad (27)$$

To interpret this equation, assume that D is zero, and note that the correlation between R, i.e., the "best" E, and the noise-free state variables is zero. The noise free state variables $\hat{S}$ are not available; the best estimate for them is the noise-free state variables $\tilde{S}$ derived from the "model" driven by the input.

Substitute for R from equations (21) and (13), where $\Delta C = \hat{C} - \tilde{C}$, the change;

$$R = (S^T\hat{C} - W_1) = (S^T\hat{C} - S^T\tilde{C} + E) = S^T(\Delta C) + E \quad (28)$$

Then

$$\tilde{S}(S^T(\Delta C) + E) = -SD \quad (29)$$

If the order of ΔC is large enough, a solution exists for ΔC (good fit) such that $D \cong 0$.

$$\therefore (\tilde{S}S^T)(\Delta C) = -\tilde{S}E \quad (30)$$

and

$$(\Delta C) = -(\tilde{S}S^T)^{-1}\tilde{S}E \quad (31)$$

This is the MEPEST (minimum-error-projection-estimate) given in equation (20). It is also called the error-decomposition method or the instrumental matrix method.

4. ALTERNATE DERIVATION

From equation (28)

$$S^T(\Delta C) = R - E \quad (32)$$

Equation (32) is exactly true, whereas (27) is only useful to find the "best" ΔC. To solve (32), minimize the effect of R by multiplying by Φ, the inverse covariance matrix of R, where

$$\Phi^{-1} = \text{Exp}(RR^T) = \text{Covariance Matrix of } R \quad (33)$$

$$\Phi S^T(\Delta C) = -\Phi E + \Phi R \quad (34)$$

Further, to minimize the effect of ΦR, cross-correlate with the noise-free signal matrix $\tilde{S}$.

$$\tilde{S}\Phi S^T(\Delta C) = -\tilde{S}\Phi E + \tilde{S}\Phi R \quad (35)$$

Since $\text{Exp}(\tilde{S}\Phi R)$ is now zero, remove this uncomputable component to find the "best" $\Delta\hat{C}$.

$$\tilde{S}\Phi S^T(\Delta\hat{C}) = -\tilde{S}\Phi E \quad (36)$$

$$(\Delta\hat{C}) = -(\tilde{S}\Phi S^T)^{-1}(\tilde{S}\Phi E) \quad (37)$$

In this case, the <u>Averaging Instrumental Matrix</u> is $(\tilde{S}\Phi)$.

5. INTERPRETATION

The A and B coefficients in equations (3) and (4) form sampled-data or discrete-time polynomials. The roots of the B polynomial are the resonant poles of the system and the roots of the A polynomial are the zeros. The transfer function is

$$G(z^{-1}) = \frac{A_1 + A_2 z^{-1} +,\ldots,+ A_{NA+1} z^{-NA}}{B_1 + B_2 z^{-1} +,\ldots,+ B_{NB+1} z^{-NB}} \quad (38)$$

The denominator can be factored into roots r_j and the transference can be expanded in partial fractions to form

$$G(z^{-1}) = \sum_{j=1}^{NB} \frac{(Res)_j}{z^{-1} - r_j} \quad (39)$$

where z^{-1} is the time shift operator defined by $f(t-1) = z^{-1}f(t)$. The LaPlace s-plane poles are at

$$s_j T = -\ln_e r_j \quad (40)$$

Using the inverse z transform, each residue in the s-plane is equal to the corresponding inverse z residue divided by the pole location r_j.

$$R_{j(s\text{-plane})} = (Res)_j / r_j \quad (41)$$

The transference is

$$G(s) = \sum_{j=1}^{NB} \frac{R_j}{(sT) - (sT)_j} \quad (42)$$

Clearing of fractions will yield the transference numerator and denominator polynomials if desired.

The estimator is basically a pole and residue calculator. If the sampling rate is too low, high frequency poles will be distorted or lost, but the low frequency poles and residues will be correct, even though the polynomial coefficients will not be recognizable.

If the sample length is too short, the low frequency poles may be in the wrong locations or missing, but the high frequency resonances and residues will not be significantly in error. To obtain the transfer function of a "stiff" system with a ratio of pole frequencies of more than R = 10,000, two sets of data may be required. For a maximum system pole frequency of f_{max}, a minimum sampling period of

$$T_{min} = 1/(3f_{max})$$

should be used, and more than $1.5\sqrt{R}$ data points should be measured. For the second run, the total length of data should be at least $2/f_{min}$, where f_{min} is the minimum system pole frequency. If this also has $1.5\sqrt{R}$ data points, then the sampling period is $1.25/(f_{min}\sqrt{R})$.

Each set of data is processed separately to obtain transferences as in equation (42). The complete transfer function is the sum of the terms from both sets, after removing the duplicates of the midfrequency poles.

6. OFFSET

For the expected value of $\tilde{S}\Phi R$ in equation (35) to be zero, R should be zero mean, because the signals can have non-zero components. Since at the convergent condition R is the transformed noise, the noise should be zero mean. For this reason, the constant C(m) must be specified and solved for as an unknown. Improved performance can be obtained by also removing the constant output offset from W and $\tilde{Y}$ before assembling the signal matrices.

7. OVERESTIMATION

If the assumed number of poles NB and number of zeros NA are larger than the correct values for the unknown system, there will be pairs of coincident poles and zeros introduced into the numerator and denominator polynomials (3) and (4). These would seem to make the vector C non-unique, but instead these extra poles and zeros perform the function of slightly fitting the noise, and they permit the remaining poles and their residues to model the unknown system quite accurately. A noise-free minimum-phase system with NB poles can be estimated with only 2 + 2(NB) coefficients in the vector C. When the data is noisy, it is better to use 4 + 2(NB) coefficients in C. After C is estimated and G(s) in equation (42) is obtained, the two poles with the smallest energies should be removed from the summation and "discarded" as noise poles. These two poles are often "unstable," since there is no constraint of stability imposed upon them.

Another advantage of removing the noise poles is that the model $\tilde{A}$ and $\tilde{B}$ in Figure 1 is stable and has the expected order of the unknown system.

8. NOISE FILTER

Overestimation provides extra poles in the estimate which have been identified as noise poles. A noise rejection filter can be constructed from this knowledge. Let the noise poles be at m_1 and m_2. The corresponding zeros are

$$(z^{-1} - m_1)(z^{-1} - m_2) \tag{43}$$

A no-phase-shift filter will also have non-minimum-phase zeros which are complex conjugates about the real frequency axis of the minimum-phase zeros. The transference of the rejection filter is

$$F(z^{-1}) = (z-m_1)(z-m_2)(z^{-1}-m_1)(z^{-1}-m_2) \tag{44}$$

$$F(z^{-1}) = pz^2 - (1 + p)sz + (1 + s^2 + p^2) - (1 + p)sz^{-1} + pz^{-2} \tag{45}$$

where s is the sum and p is the product of the roots.

$$s = m_1 + m_2 \tag{46}$$

$$p = m_1 m_2 \tag{47}$$

Normalize by using the coefficients

$$F_1 = -(1+p)s/(1 + s^2 + p^2) \tag{48}$$

$$F_2 = p/(1 + s^2 + p^2) \tag{49}$$

The noise rejection filter is

$$F(z^{-1}) = F_2 z^2 + F_1 z + 1 + F_1 z^{-1} + F_2 z^{-2} \tag{50}$$

This is an "unrealizable" filter because of the z operators with positive exponents. But with data stored in a vector, z^{-2} selects data from 2 units earlier in time, and z^2 selects data from 2 units later in time.

This filter can be used for the matrix Φ in equations (34) - (37).

$$\Phi = \begin{bmatrix} 1.0 & F_1 & F_2 & & & \\ F_1 & 1.0 & F_1 & & & \\ F_2 & F_1 & 1.0 & F_2 & & \\ & & \cdots & & & \\ & & & F_2 & F_1 & 1.0 & F_1 \\ & & & & F_2 & F_1 & 1.0 \end{bmatrix} \tag{51}$$

This filter minimizes the spectral components of the vector E that occur near the frequencies of the "estimated noise poles." The purpose of this filter is to minimize the variance of ($S\Phi R$) in equation (35), so that the removal of this term from the equation produces the least error in the estimate.

The noise filter is combined with the noise-free signal matrix from the model to produce the "averaging" matrix M.

$$M = \tilde{S}\Phi \tag{52}$$

The new estimate from equation (37) is

$$(\Delta C) = -(MS^T)^{-1}(ME) \tag{53}$$

There is another way to obtain the noise rejection filter. After several iterations, a "steady state" is reached in which ΔC is nearly zero. From equation 18, this implies that the vector E has converged to a good estimate of the vector R. The Φ in equation 33 can therefore be calculated from the inverse of the covariance matrix of E. This is useful on the last iteration to improve the estimate.

9. CLOSED - LOOP ADAPTIVE CONTROL

Fig. 3 shows a closed-loop system with unity feedback. The output w is subtracted from the input to produce an error signal v which is processed by the adjustable controller transfer function (A/B) to yield the actuating signal x which enters the unknown, slowly time-varying process with transfer function (N/D). Consider the case when ρ is zero mean and

$$\rho_o = 0 \tag{54}$$

The desired forward-path transfer function from v to w is (G/M). The desired closed-loop transfer function from the input to w is (G/H).

$$\frac{G}{H} = \frac{G}{G+M} \qquad \frac{G}{M} = \frac{G}{H-G} \tag{55}$$

We wish to select an error function that is linear in both A and B. For this purpose, consider a pseudo-input to the system to be u(t), and a pseudo output which contains the information about the unknown process is

$$P = \left(\frac{AN}{BD}\right)u. \tag{56}$$

This is compared with the desired pseudo-output (G/M)u. When this difference is expressed in terms of the measurables w and x, it is linear in A and B.

$$e = p - \frac{G}{M}u \tag{57}$$

$$e = Aw - \frac{G}{M}(Bx) = Aw - B\left(\frac{G}{M}x\right) \tag{58}$$

The above neglects the initial conditions in B, G, and M. Let s(t) be the signal delivered by x(t) passing through (G/M).

$$s(t) = \frac{G}{M}x(t) \tag{59}$$

The diminished signal matrix S_B analogous to a part of eq.(9) is

$$S_B = -\begin{bmatrix} s_2(1) & s_2(2) & \cdots & s_2(k) \\ s_3(1) & s_3(2) & \cdots & s_3(k) \\ \cdots & & & \\ s_n(1) & s_n(2) & & s_n(k) \end{bmatrix} \tag{60}$$

Let

$$S_A = \begin{bmatrix} {}_1(1) & \cdots & u_1(k) \\ \cdots & & \\ u_n(1) & \cdots & u_n(k) \end{bmatrix} \tag{61}$$

and the noisy signal matrix as in eq.(9) is

$$S^T = \begin{bmatrix} S_A^T & S_B^T & \begin{matrix} 1 \\ \vdots \\ 1 \end{matrix} \end{bmatrix} \tag{62}$$

The vector associated with the coefficient B(1) is

$$S_1^T = \begin{bmatrix} s_1(1) & s_1(2) & \cdots & s_1(k) \end{bmatrix} \tag{63}$$

The coefficient vector C is the same as equation (6).

From eq.(58) for k measurements,

$$E = S^T C - S_1 B_1 \tag{64}$$

A better error $\hat{E}$ can be produced by an optimum $\hat{C}$ which must be solved for.

$$\hat{E} = S^T\hat{C} - S_1 B_1 \tag{65}$$

The difference between these error vectors yields

$$\hat{E} = S^T(\Delta C) + E \tag{66}$$

This better error $\hat{E}$ is to be minimized.

Note that all of the state variables and all of the error vectors have noise in them. To obtain a noise-free matrix, one must start with the input which is outside of the closed loop, and which does not have any signal which correlates with ρ. The best estimate of the noise-free output of the system is

$$\tilde{y}(t) = \left(\frac{G}{H}\right)(\text{Input}) \quad (67)$$

The best estimate of the noise-free s(t) is

$$\tilde{s}(t) = \frac{G}{M}\tilde{x} = \frac{G}{M}\left(\frac{A}{B}\tilde{v}\right) = \frac{A}{B}\left(\frac{G}{M}\tilde{v}\right) = \frac{A}{B}\tilde{y} = \left(\frac{A}{B}\right)\left(\frac{G}{H}\right)_{\text{Input}} \quad (68)$$

The noise-free instrumental matrix is

$$\tilde{S} = \begin{bmatrix} \tilde{y}_1(1) & \tilde{y}_1(2) & \cdots & \tilde{y}_1(k) \\ \tilde{y}_2(1) & \tilde{y}_2(2) & \cdots & \tilde{y}_2(k) \\ & \cdots & & \\ \tilde{y}_n(1) & y_n(2) & & \tilde{y}_n(k) \\ -\tilde{s}_2(1) & -\tilde{s}_2(2) & & -\tilde{s}_2(k) \\ & \cdots & & \\ -\tilde{s}_n(1) & -\tilde{s}_n(2) & & -\tilde{s}_n(k) \\ 1 & 1 & & 1 \end{bmatrix} \quad (69)$$

Eq.(66) is solved using the averaging matrix $\tilde{S}$ from (69) as in equation (19):

$$\tilde{S}\hat{E} = 0 = (\tilde{S}\, S^T)(\Delta C) + \tilde{S}E \quad (70)$$

The changes in the controller are given by all but the last term of ΔC.

$$\Delta C = -(\tilde{S}\, S^T)^{-1}\, \tilde{S}\, E \quad (71)$$

(G/H) cannot be specified as 1.0, because then (G/M) would have to be infinite. Realistic constraints and limitations must be considered for both gain and bandwidth or step-response rise time for both (G/H) and (G/M) and they must be consistent with eq.(55).

ΔC are the changes to be made in the adaptive controller to compensate for the drifting changes in (N/D). Because initial conditions have not been rigorously treated, the changes ΔC should be introduced slowly, and the parametrically-driven transient caused by these changes should be allowed to die out before additional data is accumulated.

10. ADAPTATION WITHOUT E EXPLICIT

First compute from C(m) the value of ρ_0, the non-zero-mean of ρ. Subtract this from w(t) in Fig. 3 before assembling the matrix S_A in eq.(61) and S in eq.(62). Then solve eq.(65) using $\tilde{S}$ as the instrumental matrix.

$$\hat{E} = S^T\hat{C} - S_1B_1 \quad (65)$$

$$\tilde{S}\hat{E} = 0 = \tilde{S}\, S^T\hat{C} - \tilde{S}\, S_1B_1 \quad (72)$$

$$\hat{C} = (\tilde{S}\, S^T)^{-1}\, \tilde{S}\, S_1\, B_1 \quad (73)$$

This yields the new optimum value of $\hat{C}$, not the change ΔC. The advantage is that E is not computed, and the only signals used from the process are the input, the output w, and x. One model is required for (G/M) to generate s(t), one model for (G/H) to generate $\tilde{y}(t)$, and one model for (A/B) to generate $\tilde{s}(t)$. The unknown process parameters are never estimated.

11. SUMMARY

A new method has been presented for deriving an estimator which uses a model to obtain an error vector containing both noise and dynamic mistake and minimizes the projection of this error vector in a direction containing only dynamic mistake, to yield unbiased estimates for the model. The model is a ratio of polynomials in z, but it has its highest accuracy when interpreted as poles and residues in the z-plane, and the corresponding poles and residues in the s-plane. A key to success is that every constraint should appear as an unknown to be estimated.

A slowly time-varying process can be held to model performance by a slowly time-varying compensator which does not require estimation of the unknown process parameters.

12. PRIMARY REFERENCES

Kreidieh (Ward), Mrs. Rabab Abdul Kader, (July 12, 1972) "Estimation of Economic Systems," Ph.D. dissertation, University of California, Berkeley.

Montjean, Dominique, (June 2, 1972) "Estimation of Sampled-Data Systems with Minimum Storage and Noise Filtering," M.S. University of California.

Reiersol, O., (1945), "Confluence Analysis by Means of Instrumental Sets of Variables," Arkiv for Mathematik, Astronomi och fysik, 32.

Sargan, J., (July 1958), "The Estimation of Economic Relationships Using Instrumental Variables," Econ. 26, No. 3, 393.

Smith, Otto J.M., (Sept. 1, 1970), "Method, Apparatus and Systems for the Identification of the Relationship Between Two Signals," U.S. Patent No. 3,526,761.

Smith, Otto J.M., (1965), "Identifier for an Unknown Process," Proc. of IFAC Tokyo Symposium on Systems Engineering for Control System Design, 255.

Wonnacott, Ronald J. and Thomas H. Wonnacott, (1970), Econometrics, John Wiley and Sons, Inc. pp. 149-170, 337-400.

13. SECONDARY REFERENCES

Bass, Ronald M., (1969), "Testing Function Techniques in System Identification," Ph.D. Dissertation, Univ. of California.

Bass, Ronald M., (Oct. 1969), "Linear System Identification Using a New Canonical Form," Technical Memorandum TM-120, National Resource Analysis Center Systems Evaluation Div., Executive Office of the President, Office of Emergency Preparedness.

Bhumiratana, Amarit, (Sept. 4, 1970), "System Parameters and Initial State Estimation," M.S. University of California, Berkeley.

Boel, Rene K., (May 30, 1972), "Estimation of Continuous Time Systems," M.S., Univ. of California, Berkeley.

Boel, Rene K., and Smith, Otto J.M., (June 1972), "Estimation of Continuous Time Systems," Research Report, Dept. of Elec. Engr. & Comp. Sci., Univ. of California, Berkeley.

Bonivento, C., and Guidorzi, R. (Aug. 11-13, 1971), "Parametric Identification of Linear Multivariable Systems", Twelfth Joint Automatic Control Conf. of the American Automobile Control Council: Reprints of Technical Papers. Paper No. 5-C6, 342.

Brundy, James M., and Jorgenson, Dale W. (Aug. 1971), "Efficient Estimation of Simultaneous Equations by Instrumental Variables." The Review of Econ. & Stat., LIII, No. 3, 207.

Coutinho, J.A.M., (June 1963), "Identification Machine for a High Order System," M.S. thesis, Univ. of California, Berkeley.

Coutinho, J.A.M., (Dec. 1964), "Identification Machine Convergence Properties," Ph.D. Dissertation, Univ. of California, Berkeley.

daSilva Forte, Jorge, (June 1970), "Computer Model of the U.S. Balance of Payments," M.B.A., Univ. of California, Berkeley.

Darovskikh, L.N., (Sept. 1959), "Experimental Determination of Automatic Control Systems Links Transfer Function by Means of Standard Electronic Models," Automation and Remote Control (Eng. Translation).

Durbin, J., (1954), "Errors in Variables," Review of the Inst. of Int. Stat., 22, 23.

Ganapathy, S., and Krishna, G., (Nov. 1962), "A Method of Determining the Transfer Function of a Linear System in Terms of Its Poles and Zeros from the Frequency Response," AIEE Appli. and Indust.

Goldberger, A.S., (1964), Econometric Theory, Wiley, 284.

Goodman, T.P., and Reswick, J.B., (Feb. 1956), "Determination of System Characteristics from Normal Operating Records," Trans. of the ASME, 78, 259.

Goodman, T.P. and Hillsley, R.H., (Nov. 1958), "Continuous Measurement of Characteristics of Systems with Random Inputs: A Step Toward Self Optimizing Control," ASME, 80, 1839.

Hauge, James A., (June 1971), "Adaptive Control" Ph.D. Dissertation, Univ. of California, Berkeley.

Joseph, P., Lewis, J., and Tou, J., (March 1961) "Plant Identification in the Presence of Disturbances and Application to Digital Adaptive Systems," AIEE Trans. Appli. and Indust., 80, 18

Kitamori, T., (1960), "Application of Orthogonal Functions to the Determination of Process Dynamic Characteristics and to the Construction of Self-Optimizing Control Systems," Intl. Fed. of Auto. Control, 610.

Lacoss, Richard T., (July 1967), "Identification of Systems Using the Method of Instrumental Variables," Ph.D. Dissertation, Univ. of Calif., Berkeley.

Leovov, Y.P. and Lipatov, L.N., (1959), "Application of Statistical Method of Determining System Transfer Function," Automation and Remote Control (Eng. Translation), 20, No. 9.

Levin, M.J., (March 1960), "Optimum Estimation of Impulse Response in the Presence of Noise," IRE Trans. on Circuit Theory, 7, 1, 50.

Levin, M.J., (July 1964), "Estimation of a System Pulse Transfer Function in the Presence of Noise," IEEE Trans. on Auto. Control, AC-9, No. 3, 229.

Liviatan, N., (Jan. 1963), "Consistent Estimation of Distributed Lags," Intl. Econ. Review, 4.

Malinvaud, E., (1970), "Statistical Methods of Econometrics." Studies in Math. & Managerial Econ., 6.

Matter, Gabriel, (June 8, 1971), "Instrumental Variable Identification of Dynamic Tchebytcheff Nonlinearity," M.S., Univ. of California, Berkeley.

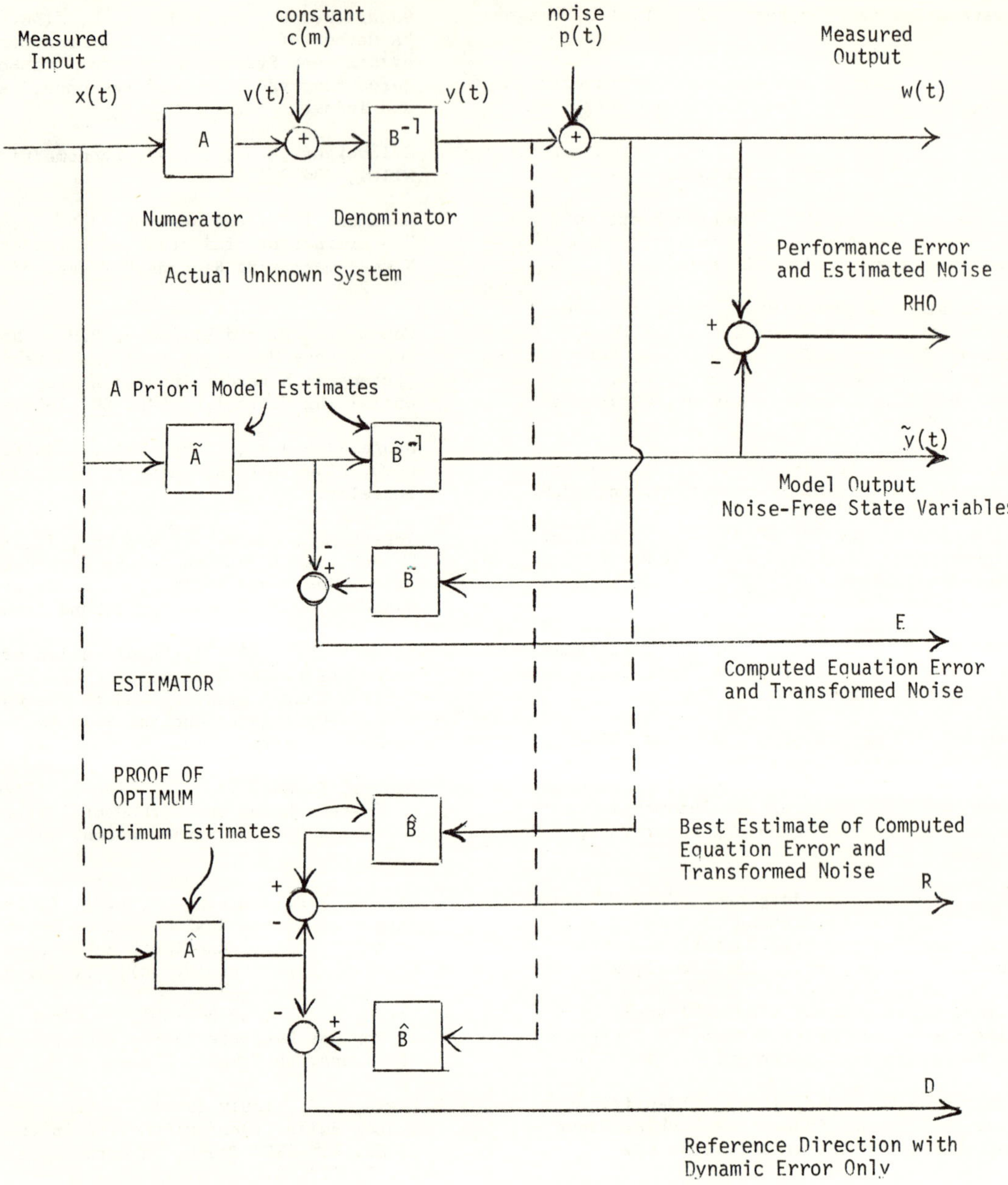

MINIMUM ERROR PROJECTION
ESTIMATOR SIGNALS

Figure 1

Plackett, R.L., (1950), "Some Theorems in Least Squares," Biometrika, 37, 149.

Reiersol, O., (1941), "Confluence Analysis by Means of Lag Moments and Other Methods of Confluence Analysis," Econometrica, 9, No. 1, 1.

Rucker, Richard, (June 1963), "Identifier and Automatic Coefficient Calculator of a System Differential Equation with Additive Noise," M.S. thesis, Univ. of California, Berkeley.

Rucker, R., (Aug. 1963), "Real Time System Identification in the Presence of Noise," Paper No. 23, IRE Wescon Convention Record.

Sakrison, D.J., (Oct. 1967), "The Use of Stochastic Approximation to Solve the System Identifi-

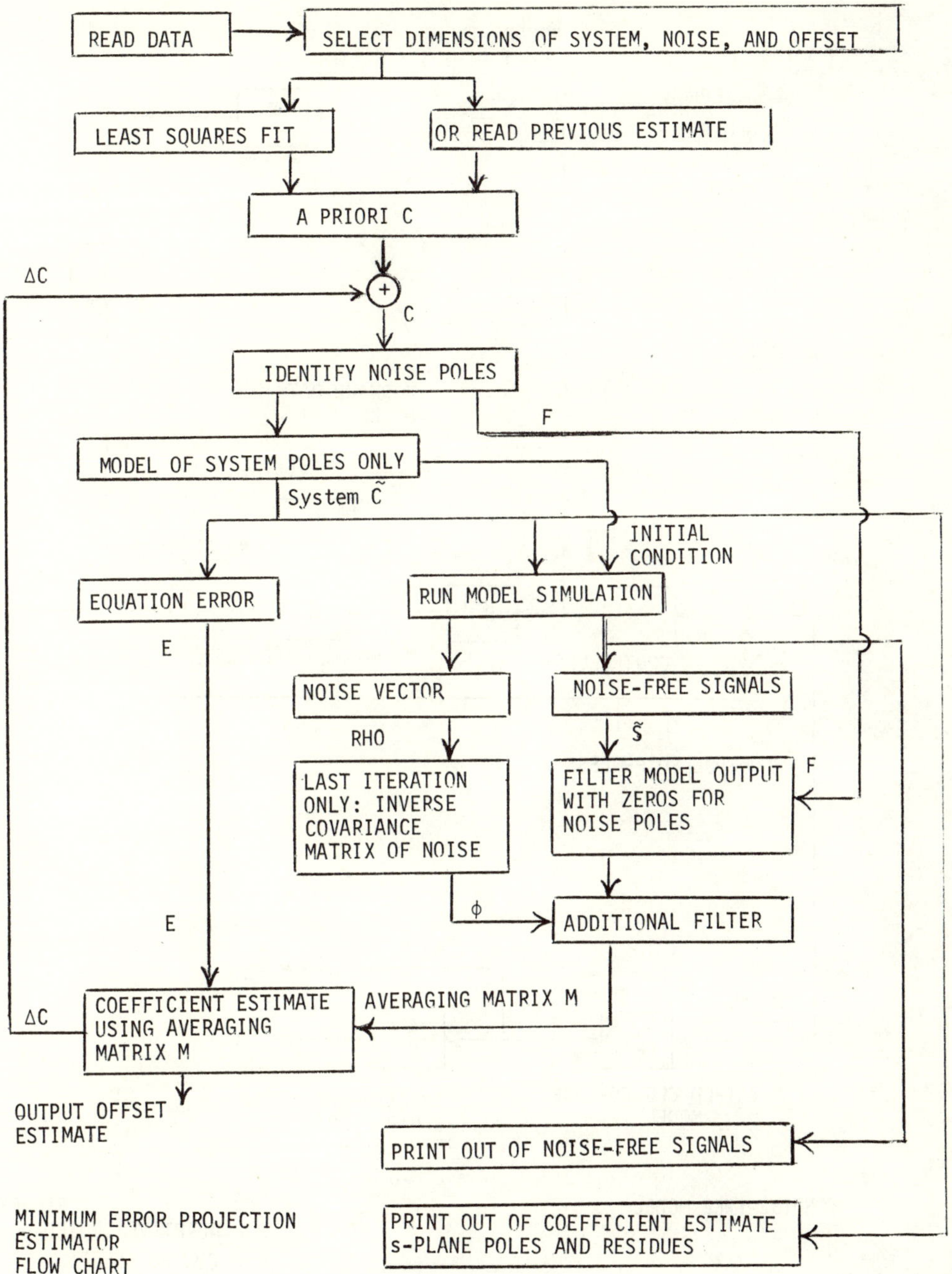

MINIMUM ERROR PROJECTION ESTIMATOR FLOW CHART

Figure 2

cation Problem", IEEE Trans. on Auto. Control, AC-12, No. 5.

Schoenemann, P.T., (June 1965), "An Identification Machine for Dynamic Biological Data," Ph.D. Dissertation, Univ. of California, Berkeley.

Wong, Kwan Yui, (June 1966), "Estimation of Parameters of Linear Systems Using Instrumental Variable Method," Ph.D. Dissertation, Univ. of California, Berkeley.

K.Y. Wong and E. Polak, (Dec. 1967), "Identification of Linear Discrete Time Systems Using the

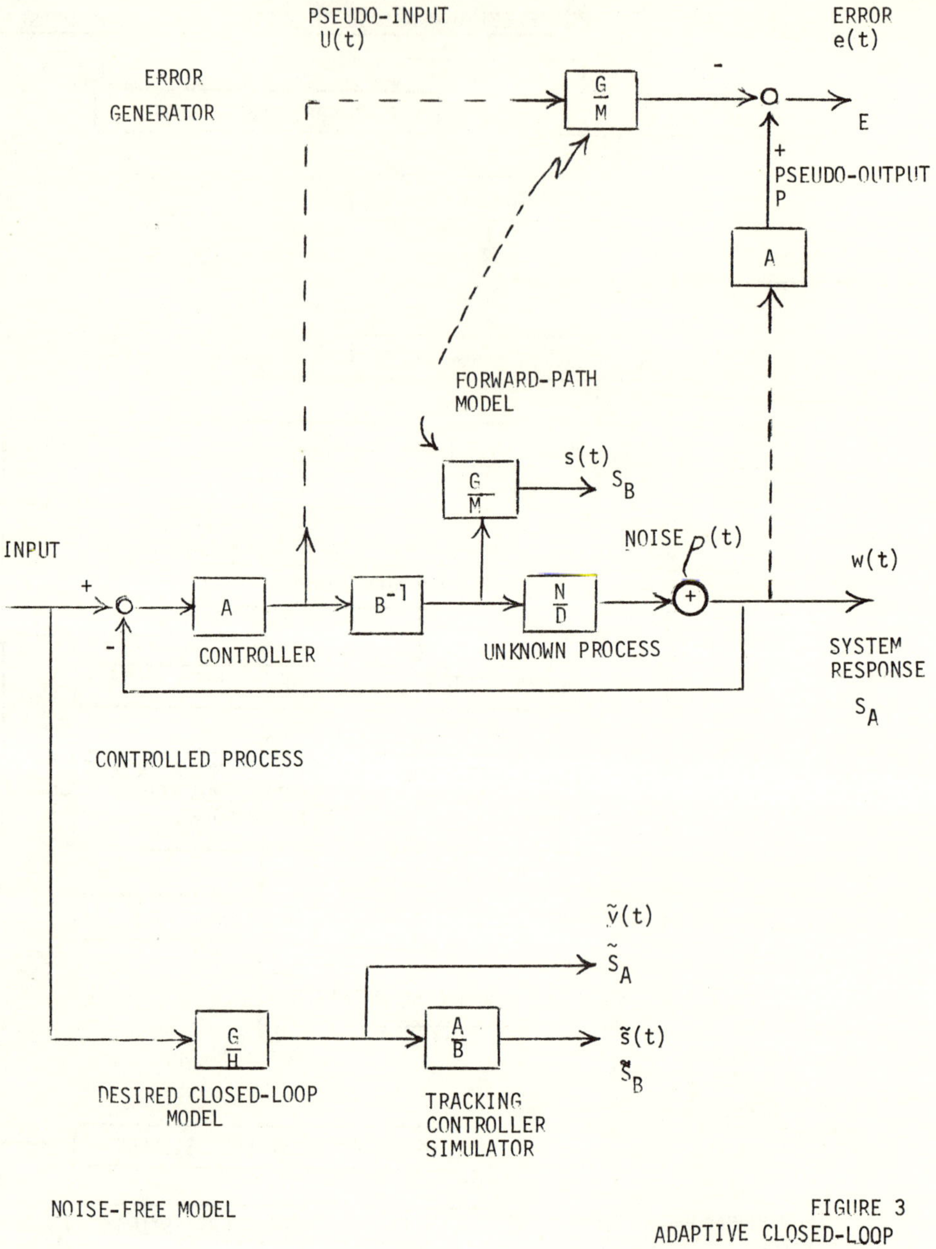

FIGURE 3
ADAPTIVE CLOSED-LOOP
CONTROL SYSTEM

Instrumental Variable Method", IEEE Trans. on Auto. Control, AC-12, No. 6, 707.

Woodside, C.M. (1970), "Estimation of the Order of Linear Systems.", Reprints of the Second Prague IFAC Symposium on Identification and Process Parameter Estimation. Paper 1.7. Prague.

IDENTIFICATION OF PARAMETRIC MODELS USING CORRELATION ANALYSIS

D. C. Williams and D. G. Delaney

School of Control Engineering
University of Bradford, G.B.

A correlation matching method is introduced for the identification of parametric models in Laplace Transfer Function and Difference Equation forms. The basic procedure of the proposed method is the matching of correlation response functions - calculated from a mathematical model - to cross correlation functions evaluated from the system input/output data records. Implementation using interactive computer graphics allows an operator/analyst to match the model's dynamic characteristics to those of the system and at the same time gain deeper understanding of the system's behaviour. Practical results are presented for the identification of a boiler-turbine plant together with those obtained from an economic system.

1. INTRODUCTION

This paper describes an approach to multi-input/multi-output system identification using the correlation matching technique. The basic procedure in the proposed method is the matching of correlation response functions - calculated from a mathematical model - to cross correlation functions evaluated from experimental input/output data. The correlation matching is implemented using interactive computer graphics with which an operator/analyst adjusts model structure and parameters in order to match the model's dynamics to those of the system. Once the operator is satisfied with the structure of the model the final parameter adjustment may be carried out automatically by computational minimisation of a suitable error criterion.

The interactive graphic implementation allows the operator to gain a deeper understanding of the structural dynamics of the system than would be possible using, say, a more sophisticated 'black-box' approach. A further advantage of the correlation matching method is its independence of the model's mathematical formulation, which may therefore be determined by the subsequent use of the model - for control system design[1] or alternatively in forecasting future behaviour[2].

The following sections introduce the correlation matching procedure for the identification of single and multivariable system models formulated parsimoniously either in continuous or discrete form. Practical results are then presented for a boiler-turbine system identified in Laplace Transfer Function form, and secondly a Difference Equation model developed from economic data.

2. CORRELATION MATCHING

For the single input/single output system shown in Fig. 1a, the sampled output y(i) may be related to the input u(i) and disturbance noise e(i) by the discrete form of the convolution equation :

$$y(i) = \sum_{j=0}^{M} w(j).u(i-j) + e(i) \qquad (1a)$$

where w(j) is the system weighting function and w(j) 0 as j M (settling time)

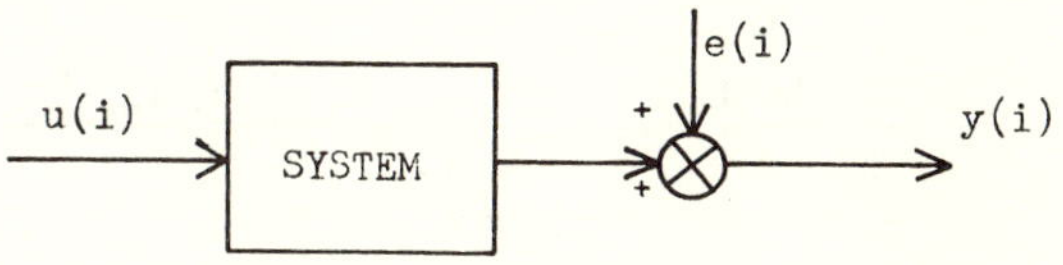

Fig. 1a. Single input/single output system.

Alternatively the relationship may be specified in difference equation form as :

$$y(i) = \frac{B(z^{-1})}{1 + A(z^{-1})}.u(i) + e(i) \qquad (1b)$$

where $1 + A(z^{-1}) = 1 + a_1z^{-1} + a_2z^{-2} + .. + a_nz^{-n}$

$$B(z^{-1}) = b_1z^{-1} + b_2z^{-2} + .. + b_nz^{-n}$$

and z^{-1} represents the discrete time shift operator :

$$z^{-d}.u(i) \rightarrow u(i-d)$$

The discrete correlation functions of u(i) and y(i) may be defined as :

$$\text{Autocorrelation} \quad Ruu(k) = \frac{1}{N-k}\sum_{i=1}^{N-k} u(i).u(i+k)$$

$$\text{Crosscorrelation} \quad Ruy(k) = \frac{1}{N-k}\sum_{i=1}^{N-k} u(i).y(i+k) \qquad (2)$$

and are related by the Wiener-Hopf equation[3]:

$$Ruy(k) = \sum_{j=0}^{M} w(j).Ruu(k-j) + Rue(k) \qquad (3)$$

where Rue(k) represents the unmeasurable statistical correlation between the input u(i) and the disturbance noise e(i). This equation includes the weighting function and is comparable with the discrete convolution equation (1a). A similar relationship may be developed for the difference equation formulation as follows. Firstly the vector of difference equations

$$\underline{y}(i) = \frac{B(z^{-1})}{1 + A(z^{-1})} . \underline{u}(i) + \underline{e}(i) \quad (4)$$

is premultiplied on both sides by $\underline{u}^T(i-k)$ - thus

$$LHS = \underline{u}^T(i-k).\underline{y}(i) = N.Ruy(k) \text{ , and}$$

$$RHS = \frac{B(z^{-1})}{1 + A(z^{-1})} . \underline{u}^T(i-k).\underline{u}(i) + \underline{u}^T(i-k).\underline{e}(i)$$

$$= \frac{B(z^{-1})}{1 + A(z^{-1})} .N.Ruu(k) + N.Rue(k) \text{ , giving}$$

$$Ruy(k) = \frac{B(z^{-1})}{1 + A(z^{-1})} .Ruu(k) + Rue(k) \quad (5)$$

The Wiener-Hopf relationship can therefore be represented in difference equation form. Both correlation equations ((3) and (5)) may be described diagrammatically, as in Fig. 1b. in which

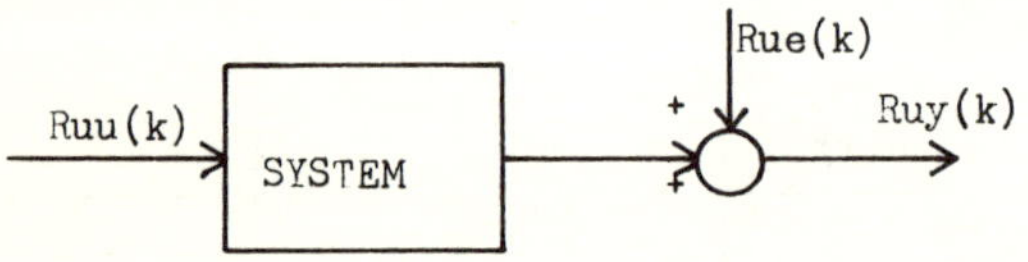

Fig. 1b. Correlation function relationship.

the cross correlation Ruy(k) is the result of passing the autocorrelation Ruu(k) through the system This analogy is now harnessed by the correlation matching identification technique, shown in Fig. 1c. in which a model is used to generate a correlation response Ruy (k) as a result of the autocorrelation at the model's input.

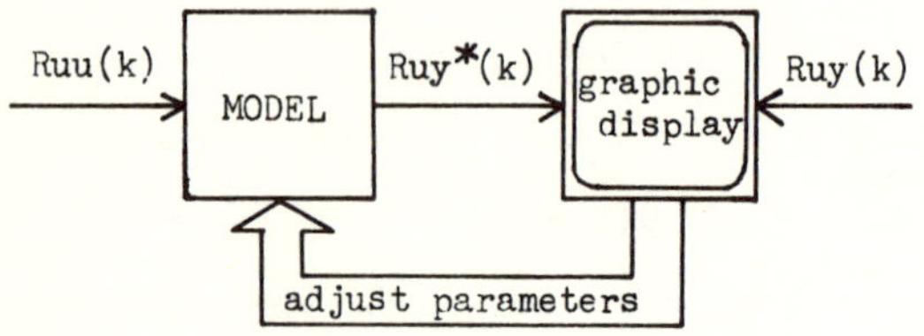

Fig. 1c. Correlation matching procedure.

Thus for the weighting function relationship :

$$Ruy^*(k) = \sum_{j=0}^{M} m(j).Ruu(k) \quad (6a)$$

where the model weighting function coefficients m(j) may be evaluated from the discrete inverse of a suitable Laplace Transfer Function formulation m(s)

$$m(j) = m(\tau) \text{ at } \tau = j\Delta T, \text{ where } m(\tau) = \mathcal{L}^{-1}(m(s))$$

and ΔT is the sampling interval.

In this case the parameters of the Laplace Transfer Function (including gain, time constants, dead time, natural frequency and damping) are adjusted in order to match the correlation response Ruy (k) to the cross correlation Ruy(k).

The difference equation model may also be used to generate a correlation response function :

$$Ruy^*(k) = \frac{\hat{B}(z^{-1})}{1 + \hat{A}(z^{-1})} .Ruu(k) \quad (6a)$$

where $\hat{A}(z^{-1})$ and $\hat{B}(z^{-1})$ represent the model.

In both cases the correlation response Ruy*(k) is compared with the cross correlation Ruy(k) and the model parameters adjusted in such a way as to minimise the difference between the two functions.

Thus as $Ruy^*(k) \rightarrow Ruy(k)$

then a) $m(j) \rightarrow w(j)$ (7)

and b) $\frac{\hat{B}(z^{-1})}{1 + \hat{A}(z^{-1})} \rightarrow \frac{B(z^{-1})}{1 + A(z^{-1})}$

The correlation matching may be implemented using interactive computer graphics[4] in which a human operator adjusts the model on the basis of the displayed correlation functions. Having obtained an acceptable visual match between the correlation response Ruy*(k) and the cross correlation Ruy(k) it may be desirable to evaluate the 'goodness of fit' quantitatively. A suitable error criterion for this is the I.S.E. (integral of squared error) between the functions

$$I.S.E. = \sum_{k=-M}^{M} (Ruy(k) - Ruy(k))^2 \quad (8)$$

The lower limit of k=-M is chosen because both Ruy(k) and Ruu(k) - hence Ruy*(k) - will often have significant non-zero values for $k < 0$, especially when normal operating records are used.

Once the operator is satisfied with the structure of the model the final parameter adjustment may be carried out automatically by computational minimisation of this error criterion.

3. MULTIVARIABLE SYSTEMS

The multivariable system with r inputs and m outputs may be described by the weighting function relationship as :

$$y_p(i) = \sum_{\ell=1}^{r} \sum_{j=0}^{M} w_{p\ell}(j).u_\ell(i-j) \text{ , } p = 1,2,..,m \quad (9)$$

together with the correlation equation :

$$Ru_q y_p(k) = \sum_{\ell=1}^{r} \sum_{j=0}^{M} w_{p\ell}(j).Ru_q u_\ell(k-j), \; q=1,..,r \quad (10)$$

The multivariable difference equation may be defined using a dummy output variable $\underline{v}$ as follows :

$$y_p(i) = \sum_{\ell=1}^{r} v_{p\ell}(i) \text{ , where} \tag{11}$$

$$v_{p\ell}(i) = \frac{B_{p\ell}(z^{-1})}{1 + A_{p\ell}(z^{-1})} \cdot u_\ell(i)$$

and the corresponding correlation equation :

$$Ru_q y_p(k) = \sum_{\ell=1}^{r} Ru_q v_{p\ell}(k) \text{ , where} \tag{12}$$

$$Ru_q v_{p\ell}(k) = \frac{B_{p\ell}(z^{-1})}{1 + A_{p\ell}(z^{-1})} \cdot Ru_q u_\ell(k)$$

This particular difference equation formulation includes the multivariate least squares simplification that

$$A_p(z^{-1}) = A_{p1}(z^{-1}) = \ldots = A_{pr}(z^{-1})$$

The correlation matching technique is now given for the identification of a multivariable weighting function/Laplace Transfer Function model and its difference equation equivalent.

A weighting function model $m_{pq}(j)$, derived from a suitable transfer function $m_{pq}(s)$, is used to calculate the correlation response functions

$$Ru_q y_p^*(k) = \sum_{\ell=1}^{r} \sum_{j=0}^{M} m_{p\ell}(j) \cdot Ru_q u_\ell(k-j) \text{ , } q=1,..,r \tag{13}$$

The transfer function parameters are now adjusted so that each correlation response matches its corresponding cross correlation.

The difference equation solution involves the multivariable model $\hat{A}_{pq}(z^{-1})$, $\hat{B}_{pq}(z^{-1})$ from which the correlation response functions are obtained :

$$Ru_q y_p^*(k) = \sum_{\ell=1}^{r} Ru_q v_{p\ell}^*(k) \text{ , where} \tag{14}$$

$$Ru_q v_{p\ell}^*(k) = \frac{\hat{B}_{p\ell}(z^{-1})}{1 + \hat{A}_{p\ell}(z^{-1})} \cdot Ru_q u_\ell(k) \text{ , } q = 1,\ldots,r$$

Again, as for the single variable solution, the correlation response functions are compared with the cross correlation functions and the model parameters adjusted to minimise the error.

$$\text{Thus as } Ru_q y_p^*(k) \rightarrow Ru_q y_p(k)$$

$$\text{then a) } m_{pq}(j) \rightarrow w_{pq}(j) \tag{15}$$

$$\text{and b) } \frac{\hat{B}_{pq}(z^{-1})}{1 + \hat{A}_{pq}(z^{-1})} \rightarrow \frac{B_{pq}(z^{-1})}{1 + A_{pq}(z^{-1})}$$

A quantitative evaluation of the goodness of fit may be calculated from the I.S.E. criterion :

$$ISE(p) = \sum_{q=1}^{r} \sum_{k=-M}^{M} (Ru_q y_p(k) - Ru_q y_p^*(k))^2 \tag{16}$$

The correlation matching may be carried out separately for each output as a series of multi-input single output problems.[5] As the method accounts for each input simultaneously any requirement for uncorrelated input signals is obviated. A further advantage is that the experimental data collection period becomes independent of the number of inputs - a necessity when using pseudo random chaincodes for multivariable system identification.[6]

4. PRACTICAL RESULTS

The correlation matching technique was applied to data supplied by I.F.A.C. as Test Case B - normal operating records obtained from a pilot scale boiler-turbine unit. A computer program for examining and correlating these data records over suitably chosen time periods was used to obtain correlations between selected process variables. Fig. 2a shows the crosscorrelation function relating steam pressure to turbine governor valve position. The correlation response of a first order Laplace Transfer Function is also shown in Fig. 2a.

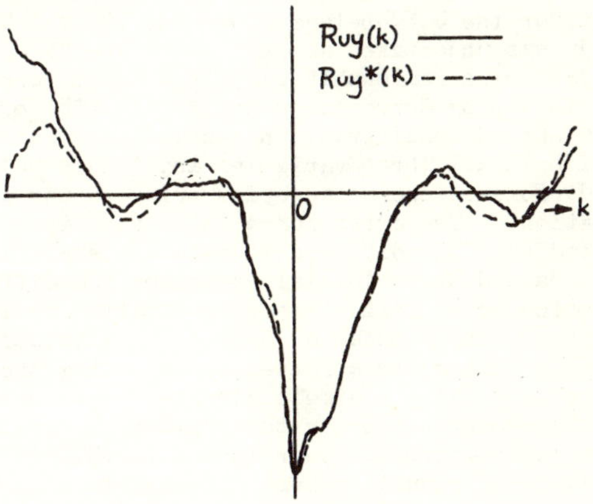

Fig. 2a Governor valve/Steam pressure

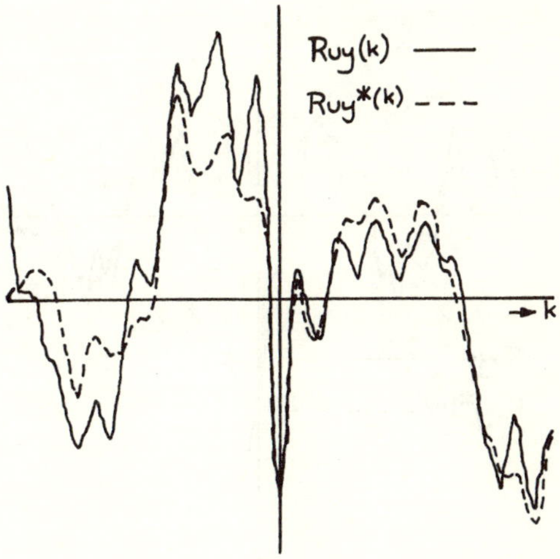

Fig. 2b Spray water/Steam pressure

Initial matching attempts for this model suggested that more than one input was required to achieve an acceptable fit. This was verified by the actual plant data which indicated an influence on the attemporator spray due to changes in governor valve setting. A 2-input/single output model was therefore chosen to relate steam pressure to both governor valve and spray water variables. The cross correlation and correlation response of the second path - spray water input - are shown in Fig. 2b. The model parameters of the resulting best fit transfer functions are given in Table 1.

Table 1 - Transfer Function Models

Input	Transfer function
Gov. valve	$\frac{-[illegible].6}{1 + 70s}$
Spray water	$\frac{+0.02}{s^2 - [illegible]}$ [illegible]

Data records for the multivariable difference equation identification example were obtained from quarterly statistics of Employment (input 1), Gross Domestic Product GDP (input 2) and Unemployment for the U.K. between 1955 and 1972. The study was undertaken as part of an IBM UK Scientific Centre Fellowship project - Short Term Forecasting using Correlation Analysis. The original data showed considerable non-stationary characteristics in all three variables and it was necessary to difference the data series prior to the identification. The cross correlation functions calculated from the differenced data are shown in Fig. 3a and 3b. Initial values of the difference equation model parameters were obtained by least squares using a model order n = 3. The correlation matching was affected by adjusting the model parameters using a simple direct-search optimisation procedure in order to minimise the cross correlation/correlation response I.S.E. criterion (17). The resulting best fit correlation response functions are shown in Fig. 3a and 3b together with the corresponding cross correlations.

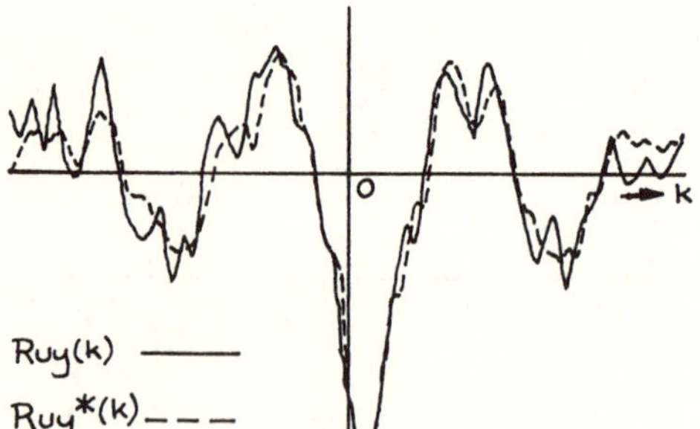

Fig. 3a Employment/Unemployment

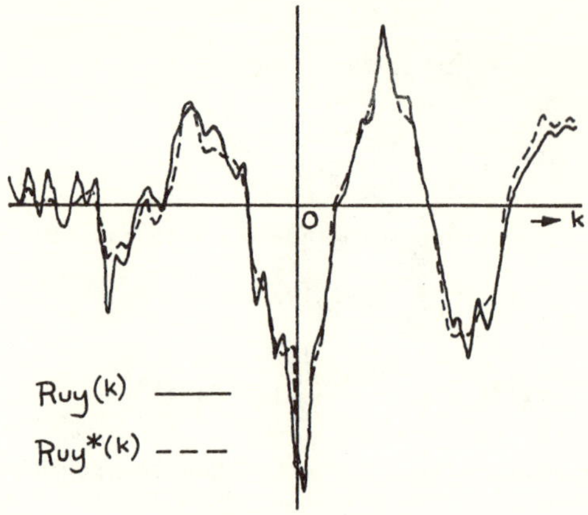

Fig. 3b GDP/Unemployment

These results indicate the feasibility of minimising the I.S.E. criterion automatically for high order systems when manual parameter adjustment would be rather laborious. The parameters of the identified model are given in Table 2.

Table 2 Difference Equation Models

Unempl. / Empl.	$\frac{-1.4(z^{-1}+1.5z^{-2}+0.5z^{-3})}{1-0.1z^{-1}-0.23z^{-2}+0.1z^{-3}}$
Unempl. / GDP	$\frac{-0.1(z^{-1}+0.72z^{-2}+0.4z^{-3})}{1-0.1z^{-1}-0.23z^{-2}+0.1z^{-3}}$

5. CONCLUSIONS

The results indicate the effectiveness of the correlation matching method for the identification of multivariable systems from normal operating data. Implementation by interactive computer graphics offers a flexible approach to identification by which a complex model may be constructed in progressive stages. Initially, suitable data periods during which relevant variables exhibit minimal non-stationary characteristics, may be convenientiy isolated. A preliminary evaluation of the model structure may be realised from an examination of the resulting correlation functions. Inadequacies in the proposed model may necessitate alteration in structure or point to the inclusion of other input variables. The computer communicates the results of each stage in graphic form to the operator. Using this information together with his experience the operator can decide whether to continue with the analysis or to repeat a previous stage. The operator is therefore able to extract the maximum information from the data and hence increase the effectiveness of the identification.

REFERENCES

1. Delaney D.G and Williams D.C. in I.Chem.E. Sym. series No. 35, pp 6:14, 1972.
2. Bray J., J.R.Stat.Soc., Vol 134(2), 1971.
3. Astrom K-J and Eykhoff P., Automatica 7, 1971.
4. Williams D.C. Electronics Letters, 6,No 15,1970.
5. Williams D.C. PhD Thesis, Bradford Univ., 1972.
6. Briggs P.A.N. and Godfrey K.R., Proc.I.E.E. 113(7), 1259 - 1267, 1966.

Reduction of Multivariable Systems

M. LABARRERE - J.P. KRIEF - B. GIMONET

CERT/D.E.R.A. - 2, av. E. Belin - 31, TOULOUSE (FRANCE)

ABSTRACT

Many identification methods of multivariable systems conduct to the impulse responses which are difficult to use in regulation and optimization theory. The problem is to find an equivalent transfer or state matrix as simple as possible which represents this system. This reduction appears also interesting at the step of the realization of a controller. The authors propose a simple solution which as given good results in constructing model of a given order.

This problem has received many solutions in monovariable case. However, the algorithm of Ho-Kalman (1966) solve the multivariable case, but this theory is complex and without any criterium on the results, and its application to real problem is delicate. The authors present an algorithm, using least square technics, which delivers as good results with an application easier.

1. INTRODUCTION

Classical theories of identification and optimization use a least mean square criterium, so it may be interesting to use the same criterium to simplify the identification model or the expression of the controller. The algorithm presented is one application to the multivariable case of the procedure developped by Steiglitz and Mc Bride (1965) in order to identify linear systems. It presents also some analogies with the Ho-Kalman (1966) theory.

To find a simplify form of a transfert matrix, we may use the developpement of the original matrix in the Z^{-1} or P^{-1} form. The use of the Z domain give better results and presents the advantage of being a direct representation of the impulse response.

2. PRESENTATION OF THE METHOD

Let us consider a transfer matrix H(Z), relating the input and output samples of a system :

$$\vec{S}(Z) = H(Z).\vec{E}(Z)$$

H(Z) can be developped in Z^{-k} terms :

$$H(Z) = H_1 Z^{-1} + H_2 Z^{-2} + \dots + H_p Z^{-p} + \dots$$

If we note "r" the rank of the approximation desired of this system, the problem is to find a transfer matrix $\frac{N(Z)}{d(Z)}$ so that :

$$\frac{N(Z)}{d(Z)} = \sum_{i}^{r} \frac{N_i^*}{Z^{-1} - Z_i} \quad \text{with rank } N_i = 1$$

If we want that degree of d(Z) equal r, we must note that the condition degree of d(Z) equal r is not suffisant, not necessary.

Thus, if we set a polynomial d(Z) which degree is r, the problem consists in finding the coefficients of this polynomial $dj(j\in 1,r)$ and the constant matrices $Ni(i\in 1,r)$ of rank 1 which minimize the criterium :

$$C = \left\| H(Z) - \frac{N(Z)}{d(Z)} \right\|^2$$

Thus, we consider an expansion limited to p terms or impulse responses of p points :

$$H(Z) = \sum_{k=1}^{p} H_k Z^{-k}$$

and

$$\frac{N(Z)}{d(Z)} = \sum_{k=1}^{p} H'_k Z^{-k}$$

The criterium becomes :

$$C = \sum_{k=1}^{p} \sum_{i}^{n_s} \sum_{j}^{n_e} (h_{ijk} - h'_{ijk})^2$$

where h_{ijk} is a term of the matrix H_k

h'_{ijk} is a term of the matrix H'_k

Let us forget for a while the constraint of the rank, and solve the problem as in the method of Steiglitz by an iterative method in order to linearize the equation ; if we consider a step i+1 :

$$C_i = \left\| d_{i+1}(Z) . \frac{H(Z)}{d_i(Z)} - \frac{N_{i+1}(Z)}{d_i(Z)} \right\|^2$$

where $d_{i+1}(Z)$ and $N_{i+1}(Z)$ are unknown.

The solution to this minimization is given by the least square. If $d_i(Z)$ converges, we thus obtain the minimum of the criterium chosen C.

A good initialization of this procedure is given by the polynomial $d_o(Z)$ such as $d_o(Z).H(Z)$ is a polynomial matrix of degree (r-1), that's to say such as all the others matrices coefficients are equal to zero as it can be seen on the following matrices. If we note :

$$d(Z) = \sum_{l=1}^{r+1} d_l \, Z^{l-1}$$

$$N(Z) = \sum_{k=0}^{r-1} N_{k+1} \, Z^k$$

We can draw the figure on next column.

The least square solution of Part II delivers the initial polynomial $d_o(Z)$. As in the monovariable case, this very good initialization leads to a quick convergence and a solution of the minimization of the criterium without constraints of rank.

If we transpose the Ho-Kalman method in the transfer form, we could see that this initial polynomial $d_o(Z)$ is almost the same as the one of Ho-Kalman, which uses different weight.

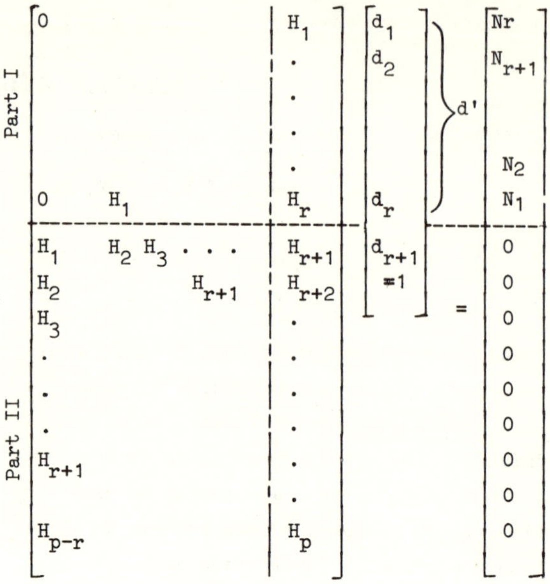

3. SOLUTION WITH THE ORDER CONSTRAINT

In order to resolve the order constraint, we expand the transfert matrix $\frac{N(Z)}{d(Z)}$ in elementary terms and we introduce in the resolution system the constraint equations :

$$\frac{N(Z)}{d_i(Z)} = \frac{N_1^*}{d_{1+Z^{-1}}} + \frac{N_2^*}{d_{2+Z^{-1}}} + \ldots + \frac{N_r^*}{d_{r+Z^{-1}}}$$

where $d_1, \ldots d_r$ are given by the preceding step. The constraint of rank 1 of the matrices N_j^* is written by the proportionality of the lines.

The constraint problem which is non linear can also be solved iteratively by linearization.

At a step, we compute the best coefficients of proportionality of differents lines then, at the following step, we introduce in the resolution system the condition that the lines are proportional with these coefficients. The resolution system by least square give a new denominators d_i, new numerators N_i^* and so new coefficients of proportionality.

In fact, some attention must be paid to the choice of the weights to give to supplementary lines.

4. RESULTS AND CONCLUSION

The procedure has been tested in simplifying high order system with two inputs and two outputs. The results are obviously better than the ones of Ho-Kalman, which do not use the same criterium. When we obtain good criteriums, the improvements are not significant, in every way, the interest of this method resides in its realization.

REFERENCES

HO B.L., KALMAN R.E. (1966). "Effective construction of linear state variable models from input-output functions". *Regelungstechnik*, 14

STEIGLITZ K., Mc BRIDE L.E. (1965). "A technique for identification of linear systems". *IEEE, Trans. on Auto. Cont.*, Vol. AC-10 pp 461-464

GODIN P., IRVING E. (1970). "Identification et régulation multivariable par le filtre de Kalman" Journées AFCET.

AUTHORS INDEX

For an explanation of the paper-number code see the Editorial, page iii

CORRECTIONS AND ADDITIONS

p. 1 S-1

Author's complete address:
Institute of Control Sciences,
Profsoyuznaya 81, Moskva GSP-312, USSR.

p. 117 C-3

Address:
R.K. Mehra, Pierce Hall,
Harvard University
Cambridge, Mass. 02138, USA.

p. 145 C-5

Top of page: C-3 → C-5

p. 167 C-6

Address authors:
Laboratoire d'Automatique de Grenoble, B.P. 15,
Av. Félix Viallet, 38040 Grenoble, France.

p. 221-230 PB-6

Page 225, Table 4.1: $C_1 = 2\ s^{-1}$, $C_2 = 180\ s^{-1}$.
Pages 226 and 227 should be interchanged.
In Table 6.1, p. 226, first column, third row, the wall volume, V_W, has to be 35 ml.
The example given in Fig. 7.3, p. 229, is generated by the computer with one of the patch cords broken. Therefore, Section 7.3.2. should be read as follows:

7.3.1. Dynamic sensitivity of P_{LV}

After initialization of the sensitivity model at the chosen operating point, the sensitivity of P_{LV} as a function of time to a certain parameter can be determined. Some results are presented in Fig. 7.3. In the same way sensitivity coefficients of other behavior functions can be determined and, besides, the sensitivity in different operating points can be studied. It may be concluded from the results in Fig. 7.3 that the sensitivity of the pressure to variations in the parameters changes drastically at the moment of transition from contraction to relaxation. Therefore, the end of the first timewindow should be coupled with the change of sign of the velocity. The parameters η, C_1 and T_1 may be estimated during the first timewindow, the parameters T_2 and T_3 in the second one.

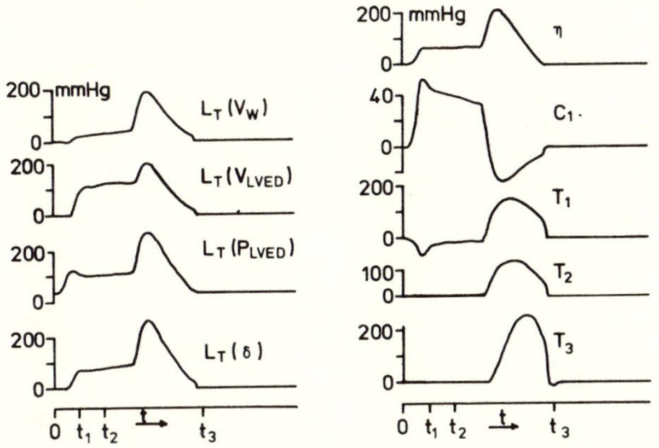

Fig. 7.3 Sensitivity coefficient $\phi_r = \delta P_{LV}/\delta \ln p$ for various parameters. Left panel: the parameter equals L_T. Changes in L_T may be caused by either V_W, V_{LVED}, P_{LVED} or δ. The time marks t_1, t_2, t_3 correspond to those in Fig. 3.3.
Right panel: sensitivity of P_{LV} to changes in the Type C parameters (indicated in the figure). Note the marked change in the sensitivity coefficient at the moment of transition from contraction to relaxation.

p. 303 PC-8

Last line:
2. Without filter o should read:
2. Without filter/12.5 o

p. 317 PE-1

N. Verheyden is now associated with the Royal Belgian Institute of Natural Sciences, Vautierstraat 31, B-1040 Brussels, Belgium.

p. 339 PE-4

Author's address:
J.T. Wagner, 2711 Calle Aventura
Miraleste, Calif. 90732, USA.

p. 435 PT-3

Top of page: PT-2 → PT-3

p. 593 TC-1

Equation (11) should read:

$$\sigma^2_{rel} = K_1 K_2^{-1} K_3.$$

p. 639-648 TD-1

page 640 column 2, line 3 from top:
a silver plate → silver plates
page 643 column 1, line 17 from top:

$$\frac{\partial E(K-k)}{\partial k} \rightarrow \frac{E(K-k)}{\partial \kappa}$$

page 643 column 2, line 10 from bottom:
k → N
page 644 column 1, line 12 from top:
k → N

p. 687 TD-6

Top of page: head missing TD-6

p. 721 TM-4

Top of page: TM-3 → TM-4

p. 744 TM-7

Figure 2 $\alpha_o \rightarrow -\alpha_o$
$\alpha_1 \rightarrow -\alpha_1$

p. 759-762 TM-11

Eq. (28): $-\partial\phi^+(k-1)\partial a_1'$ should be $-\partial\phi^+(k-1)/\partial a_1'$

Eq. (29): on the 4th line from above:
index (k-i)-(-M+r) should be (k-1)-(-M+r)

p. 771 TM-14
Top of page: TM-13 → TM-14

p. 980-981 TS-10
Equation (37) should be written as

$$(g-h_o)'T_r(I-d_m^{-1}D_r)^{-2}D_r T'_r (g-h_o) = \\ = \delta^2 - d_m\left[(\hat{h}-h_o)' t_m\right]^2$$

Equation (39) should be written as

$$\hat{J} = (g-h_o)'T_r(I+\lambda D_r)^{-2}(\lambda^2 D_r{}^2 - d_m\lambda^2 D_r)\cdot \\ \cdot T_r'(g-h_o) + \delta^2 d_m \lambda^2$$

Equation (48) should be written as

$$\hat{x} = \left[B'\lambda\Sigma(I+\lambda\Sigma)^{-1}B\right]^{-1} B'\lambda\Sigma(I+\lambda\Sigma)^{-1} h_o = \\ = A(\lambda)\ h_o$$

Equation (49) should be written as

$$\hat{h} = (I+\lambda\Sigma)^{-1}(BA(\lambda)+\lambda\Sigma)\ h_o = D(\lambda)\ h_o$$

Equation (52) should be written as

$$\hat{x} = (B'T_r(I-d_m{}^{-1}D_r)^{-1}D_r d_m{}^{-1}T_r')^{-1}\cdot \\ \cdot B'T_r(I-d_m{}^{-1}D_r)^{-1}D_r d_m{}^{-1}T_r' h_o$$

Equation (54) should be written as

$$t_m'(\hat{h}-h_o) = \pm\left[d_m{}^{-1}\delta^2 - (B\hat{x}-h_o)'\cdot \right. \\ \left. \cdot T_r(I-d_m{}^{-1}D_r)^{-2} d_m{}^{-1} D_r T_r'(B\hat{x}-h_o)\right]^{1/2}$$

p. 1031 TT-5
Pagenumber 3031 → 1031

p. 1107 S-4
Authors' address:
Laborélec, Rhode Saint Genese, Belgium.

ACKNOWLEDGEMENTS

INTERNATIONAL FEDERATION OF AUTOMATIC CONTROL

President
J.C. Lozier - USA

1st Vice-President
U.A. Luoto - Finland

2nd Vice-President
Y. Sawaragi- Japan

Past President
V. Broïda - France

Treasurer
M. Cuenod - Switzerland

Members:
G. Ferrate Pascual - Spain
M.N. Özdas - Turkey

Academician A.M. Letov - USSR
T. Vamos - Hungary
M. Nałecz - Poland
J.H. Westcott - UK

Honorary Secretary
M.A. Kaaz - G.F.R.

Deputy Secretary
Mrs. L. Schröder - G.F.R.

Honorary Editor
J.F. Coales

Honorary Editor
P. Eykhoff - Netherlands

INTERNATIONAL PROGRAM COMMITTEE (I.P.C.)

K.J. Åström - Sweden
A.V. Balakrishnan - USA
J.G. Balchen - Norway
S.P. Bingulac - Yugoslavia
P.H. Hammond - UK
M. Installe - Belgium
V. Peterka - Czechoslovakia
N.S. Rajbman - USSR
K. Reinisch - GDR
J.E.N. Richalet - France
G. Schweizer - FRG
V. Strejc - Czechoslovakia
Ya. Z. Tsypkin - USSR
J.P. Waha - Belgium
T.J. Williams - USA
+ Members of the L.O.C.

LOCAL ORGANIZING COMMITTEE (L.O.C.)

P. Eykhoff - Chairman
A.J.W. van den Boom - Secretary
A. van den Bos
P.M.E.M. van der Grinten
J.M. van der Kamp - Congr. Office
J. van Kruiselbergen - Congr. Office
J. Ligthart
C.H. Loos
T.W. Oerlemans
J.E. Rijnsdorp
L. de Schamphelaere - Belgium
B.P. Veltman

Members of the I.P.C. and L.O.C., from left to right Messrs:

Rijnsdorp, Veltman, Oerlemans, Installe, Bingulac, Boiten,*) Van den Boom, Eykhoff, Åström, Strejc, Peterka, Van den Bos, Loos

*) Chairman Dutch Nat. Member Org. of IFAC.